SCHAUM'S SOLVED PROBLEMS SERIES

2000 SOLVED PROBLEMS IN

PHYSICAL CHEMISTRY

by

Clyde R. Metz, Ph.D.
College of Charleston

McGRAW-HILL PUBLISHING COMPANY
New York St. Louis San Francisco Auckland Bogotá Caracas Hamburg Lisbon London Madrid Mexico Milan Montreal New Delhi Oklahoma City Paris San Juan São Paulo Singapore Sydney Tokyo Toronto

Clyde R. Metz, Ph.D., *Professor of Chemistry, College of Charleston*

Clyde Metz received his Ph.D. from Indiana University. He is a coauthor of a general chemistry textbook and related laboratory materials, solutions manuals, and study guides and has published several student research papers. Dr. Metz is active in the Division of Chemical Education of the American Chemical Society; is a member of the Electrochemical Society, Alpha Chi Sigma, Sigma Xi, Tau Beta Pi, Phi Lambda Upsilon, and the South Carolina Academy of Science; and is a Fellow of the Indiana Academy of Science.

Project supervision was done by The Total Book.

Library of Congress Cataloging-in-Publication Data

Metz, Clyde R.
2000 solved problems in physical chemistry / by Clyde Metz.
p. cm.—(Schaum's solved problems series)
ISBN 0-07-041717-2(HC)
ISBN 0-07-041716-4(PBK)
1. Chemistry, Physical and theoretical—Problems, exercises, etc. I. Title.
II. Title: Two thousand solved problems in physical chemistry. III. Series.
QD456.M39 1990
541.3'076—dc20 89-2616
CIP

1 2 3 4 5 6 7 8 9 0 SHP/SHP 8 9 4 3 2 1 0 9

ISBN 0-07-041717-2{HC}
ISBN 0-07-041716-4{PBK}

CONTENTS

To the Student

OK, so now you have your first homework assignment due in physical chemistry. Where do you begin? How do you solve the problems? What should the answers that you submit look like? These questions, as well as several others, are likely to be on your mind. The purpose of this book is to help you to prepare your assignments and to review for examinations.

The first step in doing a homework assignment is to understand the material before trying to solve any problems. Reading the chapter in your text several times, looking at any example problems in the text, and rewriting your lecture notes will help in understanding the material. Now you are ready to take the second step, to solve the problems.

Unless a problem is particularly difficult, usually the data can be used with an equation or two to give the desired answer. The general format used in this book—identification of the proper equation or relationship, substitution of the data (with proper units), and the calculation of the desired answer (to the proper number of significant figures)—is the format that you should use in solving your problems. Although the method for solving a problem should be recognizable within a few minutes, the actual solving may take longer.

This book will not serve as an answer book to your text, because all the problems are original. However, it will serve as a "how to" manual, because the problems are similar in content and style to those found in your physical chemistry textbook. The organization of this book is exactly parallel to that found in *Schaum's Outline of Theory and Problems of Physical Chemistry*, 2d ed., McGraw-Hill (1989), and the two books make a complete set of supplementary materials for examination reviews and for learning to solve problems in physical chemistry.

I would like to acknowledge the tremendous contribution made by my wife, Jennie, in the preparation of the manuscript from my handwritten scrawlings.

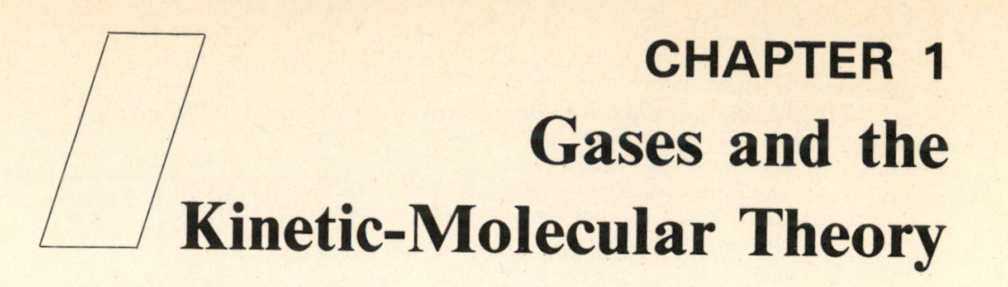

CHAPTER 1
Gases and the Kinetic-Molecular Theory

1.1 TEMPERATURE AND PRESSURE

1.1 The normal boiling points of liquid nitrogen, oxygen, and helium are −195.8 °C, −297.332 °F, and 4.25 K, respectively. Which element has the lowest boiling point? Which element will remain a liquid after samples of the other two elements have evaporated?

▮ The temperature (T) in kelvins at which N_2 boils is

$$T/(\mathrm{K}) = T/(^\circ\mathrm{C}) + 273.15 = -195.8 + 273.15 = 77.4 \tag{1.1}$$

The boiling point of O_2 in units of Celsius degrees is

$$T/(^\circ\mathrm{C}) = \tfrac{5}{9}[T/(^\circ\mathrm{F}) - 32] = \tfrac{5}{9}(-297.332 - 32) = -182.962 \tag{1.2}$$

which upon substitution into *(1.1)* gives

$$T/(\mathrm{K}) = -182.962 + 273.15 = 90.188$$

Helium has the lowest point at 4.25 K. Nitrogen will boil at 77.4 K, leaving the oxygen sample in the liquid state. The oxygen will not boil until it reaches a temperature of 90.188 K.

1.2 At what temperature will both the Celsius and Fahrenheit scales read the same value?

▮ Substituting $T/(^\circ\mathrm{C}) = T/(^\circ\mathrm{F}) = T$ into *(1.2)* and solving gives

$$T = \tfrac{5}{9}(T - 32) = -40$$

Both temperature scales are equivalent at $-40\ ^\circ\mathrm{C} = -40\ ^\circ\mathrm{F}$.

1.3 Combine *(1.1)* and *(1.2)* to define the Rankine temperature scale (°R), the absolute Fahrenheit temperature scale.

▮ The general form of the defining equation is

$$T/(^\circ\mathrm{R}) = T/(^\circ\mathrm{F}) + k$$

Solving *(1.2)* for $T/(^\circ\mathrm{F})$ in terms of $T/(^\circ\mathrm{C})$ and substituting gives

$$T/(^\circ\mathrm{R}) = \tfrac{9}{5}[T/(^\circ\mathrm{C})] + 32 + k$$

Solving *(1.1)* for $T/(^\circ\mathrm{C})$ in terms of $T/(\mathrm{K})$ and substituting gives

$$T/(^\circ\mathrm{R}) = \tfrac{9}{5}[T/(\mathrm{K}) - 273.15] + 32 + k$$

Recognizing that $T/(^\circ\mathrm{R}) = T/(\mathrm{K}) = 0$ at absolute zero,

$$0 = \tfrac{9}{5}(0 - 273.15) + 32 + k$$

$$k = 459.67$$

Thus the Rankine temperature scale is defined as

$$T/(^\circ\mathrm{R}) = T/(^\circ\mathrm{F}) + 459.67$$

1.4 The vapor pressure of water is 4.58 torr at 0 °C and 760.00 torr at 100 °C. Assume that a water temperature scale (°W) is based on this thermodynamic property with $0\ ^\circ\mathrm{C} = 0\ ^\circ\mathrm{W}$ and $100\ ^\circ\mathrm{C} = 100\ ^\circ\mathrm{W}$. Find the relation between this hypothetical scale and the Celsius scale. Express 50 °C in units of °W. The vapor pressure of water at 50 °C is 92.51 torr.

▮ If y_2 and y_1 represent the values of a thermodynamic property at fixed temperatures T_2 and T_1, respectively, then the temperature T that corresponds to the value y for the thermodynamic property is defined as

$$T = T_1 + \frac{y - y_1}{y_2 - y_1}(T_2 - T_1) \tag{1.3}$$

In this case, the fixed temperatures are 0 °W and 100 °W, giving

$$T = 0\,°\text{W} + \frac{P - 4.58\ \text{torr}}{760.00\ \text{torr} - 4.58\ \text{torr}}(100\,°\text{W} - 0\,°\text{W})$$

which can be written as

$$T/(°\text{W}) = (0.132\,38)[P/(\text{torr}) - 4.58]$$

On this scale, 50 °C is equivalent to

$$T/(°\text{W}) = (0.132\,38)(92.51 - 4.58) = 11.64$$

Note that this hypothetical temperature scale is nonlinear.

1.5 The temperature dependence of the length (l) of a platinum rod is given by

$$l = l_0\{1 + \alpha[T/(°\text{C})] + \beta[T/(°\text{C})]^2\}$$

where $\alpha = 8.68 \times 10^{-6}$, $\beta = 1.3 \times 10^{-9}$, and l_0 is the length at 0 °C. A nonlinear temperature scale (°Pt) is defined in terms of this thermodynamic property such that 0 °Pt = 0 °C and 100 °Pt = 100 °C. Find the relation between this hypothetical scale and the Celsius scale. Express 50 °C and 500 °C in units of °Pt.

▌ The fixed points in *(1.3)* are 0 °Pt, where $y(0\,°\text{Pt}) = l_0$, and 100 °Pt, where $y(100\,°\text{Pt}) = l(100\,°\text{C})$ and $y = l$. Thus,

$$T = 0\,°\text{Pt} + \frac{l - l_0}{l(100\,°\text{C}) - l_0}(100\,°\text{Pt} - 0\,°\text{Pt})$$

$$T/(°\text{Pt}) = \frac{l_0\{1 + \alpha[T/(°\text{C})] + \beta[T/(°\text{C})]^2\} - l_0}{l_0[1 + \alpha(100) + \beta(100)^2] - l_0}(100) = \frac{\alpha[T/(°\text{C})] + \beta[T/(°\text{C})]^2}{\alpha + 100\beta}$$

Expressing 50 °C and 500 °C on this temperature scale shows that the scale is nonlinear:

$$T/(°\text{Pt}) = \frac{(8.68 \times 10^{-6})(50) + (1.3 \times 10^{-9})(50)^2}{8.68 \times 10^{-6} + (100)(1.3 \times 10^{-9})} = 49.6$$

$$T/(°\text{Pt}) = \frac{(8.68 \times 10^{-6})(500) + (1.3 \times 10^{-9})(500)^2}{8.81 \times 10^{-6}} = 534$$

1.6 A hypothetical absolute temperature scale (°A) is defined by assigning 0 °C = 100 °A. What will be the normal boiling point of water on this scale?

▌ The values of the thermodynamic property in *(1.3)* can be represented by $y_1 = y(0\,°\text{A}) = y(0\ \text{K})$ and $y_2 = y(100\,°\text{A}) = y(0\,°\text{C}) = y(273.15\ \text{K})$. For each temperature scale,

$$T = 0\ \text{K} + \frac{y - y_1}{y_2 - y_1}(273.15\ \text{K} - 0\ \text{K})$$

$$T = 0\,°\text{A} + \frac{y - y_1}{y_2 - y_1}(100\,°\text{A} - 0\,°\text{A})$$

Solving the first equation for $(y - y_1)/(y_2 - y_1)$ and substituting into the second equation gives

$$T/(°\text{A}) = \{[T/(\text{K})]/(273.15)\}(100) = (0.366\,10)[T/(\text{K})]$$

On this hypothetical scale, 100.00 °C = 373.15 K is

$$T/(°\text{A}) = (0.366\,10)(373.15) = 136.61$$

1.7 Assume that a hypothetical temperature scale (°B) is based on the melting point (0 °B = 5.5 °C) and the normal boiling point (100 °B = 80.1 °C) of benzene. What is the freezing point of water on this scale?

▌ The value of the thermodynamic property in *(1.3)* can be represented by $y_1 = y(5.5\,°\text{C}) = y(0\,°\text{B})$ and $y_2 = y(80.1\,°\text{C}) = y(100\,°\text{B})$. For each temperature scale,

$$T = 5.5\,°\text{C} + \frac{y - y_1}{y_2 - y_1}(80.1\,°\text{C} - 5.5\,°\text{C}) = 5.5\,°\text{C} + \frac{y - y_1}{y_2 - y_1}(74.6\,°\text{C})$$

$$T = 0\,°\text{B} + \frac{y - y_1}{y_2 - y_1}(100\,°\text{B} - 0\,°\text{B}) = \frac{y - y_1}{y_2 - y_1}(100\,°\text{B})$$

Solving the first equation for $(y - y_1)/(y_2 - y_1)$ and substituting into the second equation gives

$$T/(°\mathrm{B}) = \frac{T/(°\mathrm{C}) - 5.5}{74.6}(100)$$

On this hypothetical scale, 0 °C is

$$T/(°\mathrm{B}) = \frac{0 - 5.5}{74.6}(100) = -7.4$$

1.8 Express each of the following values of pressure in units of pascals: (*a*) 32 psi, (*b*) 745 torr, (*c*) 14.6 mbar, (*d*) 12.7 atm, (*e*) 850 kPa, and (*f*) $8 \times 10^4\ \mathrm{N \cdot m^{-2}}$. Which pressure is the greatest?

▌ Using the conversion factors given in Table 1-1 gives

(*a*) $(32\ \mathrm{psi})[(6.89 \times 10^3\ \mathrm{Pa})/(1\ \mathrm{psi})] = 2.2 \times 10^5\ \mathrm{Pa}$
(*b*) $(745\ \mathrm{torr})[(1\ \mathrm{atm})/(760\ \mathrm{torr})][(101\,325\ \mathrm{Pa})/(1\ \mathrm{atm})] = 9.93 \times 10^4\ \mathrm{Pa}$
(*c*) $(14.6\ \mathrm{mbar})[(10^{-3}\ \mathrm{bar})/(1\ \mathrm{mbar})][(10^5\ \mathrm{Pa})/(1\ \mathrm{bar})] = 1.46 \times 10^3\ \mathrm{Pa}$
(*d*) $(12.7\ \mathrm{atm})[(101\,325\ \mathrm{Pa})/(1\ \mathrm{atm})] = 1.29 \times 10^6\ \mathrm{Pa}$
(*e*) $(850\ \mathrm{kPa})[(10^3\ \mathrm{Pa})/(1\ \mathrm{kPa})] = 8.5 \times 10^5\ \mathrm{Pa}$
(*f*) $(8 \times 10^4\ \mathrm{N \cdot m^{-2}})[(1\ \mathrm{Pa})/(1\ \mathrm{N \cdot m^{-2}})] = 8 \times 10^4\ \mathrm{Pa}$

The greatest pressure is 12.7 atm.

Table 1-1

$1\ \mathrm{Pa} = 1\ \mathrm{N \cdot m^{-2}} = 1\ \mathrm{kg \cdot m^{-1} \cdot s^{-2}}$
$1\ \mathrm{bar} = 10^5\ \mathrm{Pa}$
$1\ \mathrm{atm} = 101\,325\ \mathrm{Pa}$
$1\ \mathrm{atm} = 760\ \mathrm{torr}$
$1\ \mathrm{psi} = 6894.7572\ \mathrm{Pa}$

1.9 Calculate the pressure exerted by a 760-mm column of mercury.

▌ The pressure (P) is given by

$$P = f/A = mg/A = mgh/V = \rho gh \tag{1.4}$$

where f is force, A is area, m is mass, g is the gravitational acceleration, h is the height of the liquid column, V is the volume, and ρ is the density of the liquid. Substituting numerical values gives

$$P = (13.5955\ \mathrm{g \cdot cm^{-3}})\frac{10^{-3}\ \mathrm{kg}}{1\ \mathrm{g}}\left(\frac{10^2\ \mathrm{cm}}{1\ \mathrm{m}}\right)^3(9.806\,65\ \mathrm{m \cdot s^{-2}})(760\ \mathrm{mm})\frac{10^{-3}\ \mathrm{m}}{1\ \mathrm{mm}}\left(\frac{1\ \mathrm{N}}{1\ \mathrm{kg \cdot m \cdot s^{-2}}}\right)$$
$$= 101\,328\ \mathrm{N \cdot m^{-2}} = 101\,328\ \mathrm{Pa}$$

1.10 Dibutyl phthalate is commonly used as a liquid in manometers. What pressure is equivalent to a centimeter of this liquid?

▌ The pressure is given by *(1.4)* as

$$P = (1.0465\ \mathrm{g \cdot cm^{-3}})\frac{10^{-3}\ \mathrm{kg}}{1\ \mathrm{g}}\left(\frac{10^2\ \mathrm{cm}}{1\ \mathrm{m}}\right)^3(9.806\,65\ \mathrm{m \cdot s^{-2}})(1\ \mathrm{cm})\frac{10^{-2}\ \mathrm{m}}{1\ \mathrm{cm}}\left(\frac{1\ \mathrm{N}}{1\ \mathrm{kg \cdot m \cdot s^{-2}}}\right)$$
$$= 102.63\ \mathrm{N \cdot m^{-2}}$$

1.11 What is the approximate mass of the atmosphere of the earth? Assume the radius of the earth to be 6370 km.

▌ The surface area of the earth is given by

$$A = 4\pi r^2 = 4\pi(6370\ \mathrm{km})^2\left(\frac{10^3\ \mathrm{m}}{1\ \mathrm{km}}\right)^2 = 5.10 \times 10^{14}\ \mathrm{m^2}$$

The mass of air that produces a pressure of 101 325 Pa is given by *(1.4)* as

$$m = \frac{PA}{g}$$
$$= \frac{(101\,325\ \mathrm{Pa})[(1\ \mathrm{kg \cdot m^{-1} \cdot s^{-2}})/(1\ \mathrm{Pa})](5.10 \times 10^{14}\ \mathrm{m^2})}{9.81\ \mathrm{m \cdot s^{-2}}} = 5.27 \times 10^{18}\ \mathrm{kg}$$

1.12 At what altitude will the atmospheric pressure decrease from 1.00 bar to 0.50 bar? Assume that the average molar mass of air is $29\ \text{g}\cdot\text{mol}^{-1}$.

▮ The atmospheric pressure at a height h above sea level is related to the atmospheric pressure at sea level (P_0) by the *barometric formula*

$$P = P_0\, e^{-gMh/RT} \tag{1.5}$$

where M is the molar mass and R is the gas constant. Solving *(1.5)* for h and substituting values gives

$$h = \frac{-RT\ln(P/P_0)}{gM}$$

$$= \frac{-(8.314\ \text{J}\cdot\text{K}^{-1}\cdot\text{mol}^{-1})[(1\ \text{kg}\cdot\text{m}^2\cdot\text{s}^{-2})/(1\ \text{J})](298\ \text{K})\ln(0.50/1.00)}{(9.81\ \text{m}\cdot\text{s}^{-2})(29\times10^{-3}\ \text{kg}\cdot\text{mol}^{-1})} = 6.0\times10^3\ \text{m}$$

1.13 At sea level (Charleston, S.C.) the pressure of nitrogen in the atmosphere is 0.80 bar and the pressure of oxygen is 0.20 bar at 25 °C. What are the respective pressures at an altitude of 1600 m (Denver, Col.) at 25 °C? Which city has the greater oxygen-to-nitrogen ratio in the air?

▮ For each gas, *(1.5)* gives

$$P(N_2) = (0.80\ \text{bar})\exp\left(-\frac{(9.81\ \text{m}\cdot\text{s}^{-2})(28\times10^{-3}\ \text{kg}\cdot\text{mol}^{-1})(1600\ \text{m})}{(8.314\ \text{J}\cdot\text{K}^{-1}\cdot\text{mol}^{-1})(298\ \text{K})[(1\ \text{kg}\cdot\text{m}^2\cdot\text{s}^{-2})/(1\ \text{J})]}\right) = 0.67\ \text{bar}$$

$$P(O_2) = (0.20)\exp\left(-\frac{(9.81)(32\times10^{-3})(1600)}{(8.314)(298)}\right) = 0.16\ \text{bar}$$

At sea level,

$$P(O_2)/P(N_2) = (0.20\ \text{bar})/(0.80\ \text{bar}) = 0.25$$

and at 1600 m,

$$P(O_2)/P(N_2) = 0.16/0.67 = 0.24$$

The air in Charleston is slightly richer in oxygen than the Denver air.

1.14 Compare the atmospheric pressure values predicted by *(1.5)* for dry air with the following actual values:

h/(km)	0	10	20	30	40	50	60	70	80	90	100
P/(Pa)	101 325	27 998	5600	1267	320	100	28	7.2	1.3	0.25	0.056

Assume $M = 29.0\times10^{-3}\ \text{kg}\cdot\text{mol}^{-1}$ and $T = 237\ \text{K}$.

▮ A plot of $\ln[P/(\text{Pa})]$ against $h/(\text{km})$ is shown in Fig. 1-1. The symbols represent the actual values of pressure, and the line represents the values calculated using *(1.5)*. The agreement is excellent, especially considering the fact that the values of T range from 190 K to 290 K.

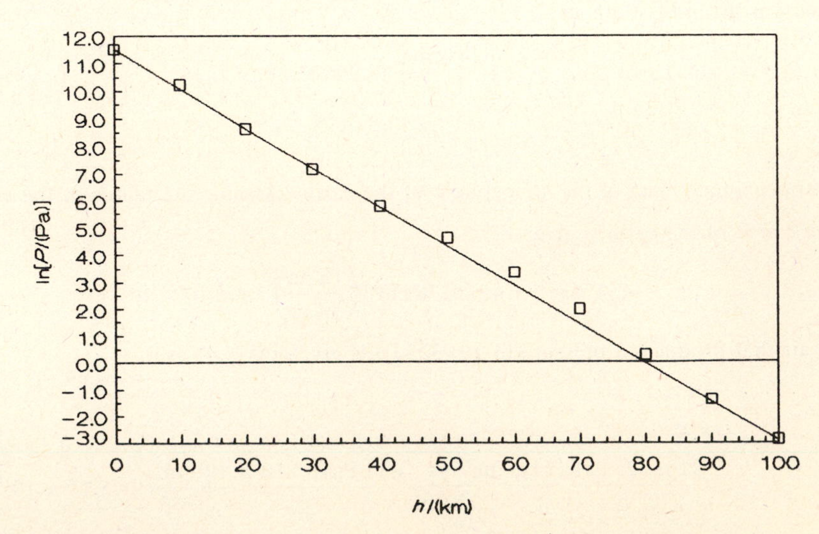

Fig. 1-1

1.15 How is the pressure predicted by *(1.5)* affected by the small decreases in g and in T that occur as the altitude increases?

▌ As T decreases with increasing h, the exponential term in *(1.5)* becomes slightly smaller, and so P decreases somewhat faster than if T were constant. As g decreases with increasing h, the exponential term in *(1.5)* becomes slightly larger, and so P decreases somewhat slower than if g were constant. (The two decreases offset each other to a certain extent.)

1.2 LAWS FOR IDEAL GASES

1.16 Identify which of the following plots for a fixed amount of an ideal gas at constant temperature will be linear: **(a)** P against V, **(b)** P against $(1/V)$, **(c)** PV against P, and **(d)** $(1/P)$ against V. Determine the slope of the straight line in each linear plot.

▌ The pressure–volume behavior for a fixed amount of an ideal gas under isothermal conditions is given by *Boyle's law* as

$$PV = k \qquad (1.6)$$

where k is a constant. The linear forms of *(1.6)* are **(b)** $P = k(1/V)$ with the slope equal to k, **(c)** $PV = k$ with the slope equal to 0, and **(d)** $1/P = (1/k)V$ with the slope equal to $1/k$.

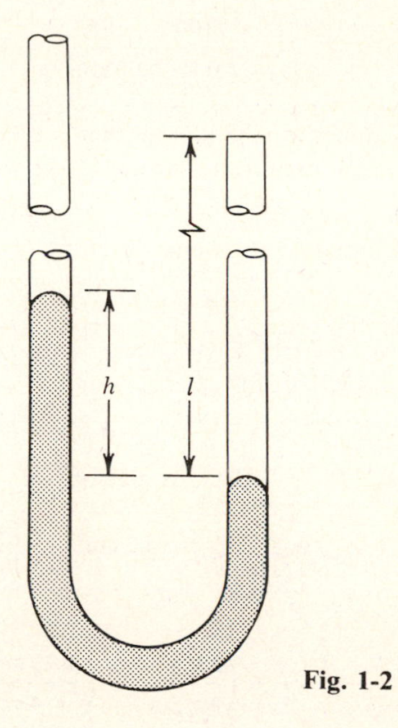

Fig. 1-2

1.17 The apparatus used by Boyle is shown schematically in Fig. 1-2. The length of the trapped air in the right arm is given by l (measured in arbitrary units), and the pressure difference in the arms is determined from h, the difference in the heights of the mercury columns in the arms. If the atmospheric pressure on the left arm is $29\frac{2}{16}$ inches of mercury, show that the following data agree with *(1.6)*:

l	44	38	32	26	20
h/(inches Hg)	$2\frac{13}{16}$	$7\frac{14}{16}$	$15\frac{1}{16}$	$25\frac{3}{16}$	$41\frac{9}{16}$

▌ The pressure of the trapped air is given by $P = 29\frac{2}{16} + h$, and the volume is directly proportional to l. If *(1.6)* is valid, then the product Pl should be constant. For the first data point,

$$Pl = (29\tfrac{2}{16} + 2\tfrac{13}{16})(44) = 1410$$

Likewise, the other products of Pl are all equal to 1410. Under these conditions, *(1.6)* is valid.

1.18 What pressure is required to compress an ideal gas from 10.0 ft^3 and 1.00 atm to 2.00 ft^3 under isothermal conditions?

▌ In terms of two different pressures and volumes, *(1.6)* can be written as

$$P_1 V_1 = P_2 V_2 \tag{1.7}$$

Solving *(1.7)* for P_2 and substituting the data gives

$$P_2 = (1.00\ \text{atm})\frac{10.0\ \text{ft}^3}{2.00\ \text{ft}^3} = 5.00\ \text{atm}$$

The pressure on the gas sample must be increased to 5.00 atm (= 5.07 bar).

1.19 A commercial gas cylinder contains 75 L of helium at 15 bar (gauge pressure). Assuming ideal gas behavior for the isothermal expansion, how many 3.0-L balloons at a pressure of 1.1 bar can be filled by the gas in this cylinder?

▌ Assuming that the atmospheric pressure is 1 bar, the initial and final pressures of the helium are

$$P_1 = 15\ \text{bar} + 1\ \text{bar} = 16\ \text{bar} \qquad \text{and} \qquad P_2 = 1.1\ \text{bar}$$

The final volume of the helium is given by *(1.7)* as

$$V_2 = (75\ \text{L})\frac{16\ \text{bar}}{1.1\ \text{bar}} = 1100\ \text{L}$$

Subtracting the 75 L that will remain in the cylinder, the number of balloons is

$$\frac{1100\ \text{L} - 75\ \text{L}}{3.0\ \text{L}\cdot\text{balloon}^{-1}} = 340\ \text{balloons}$$

1.20 A diver at a depth of 45 m exhales a bubble of air that is 1.0 cm in radius. Assuming ideal gas behavior, what will be the radius of this bubble as it breaks the surface of the water?

▌ The volume of the bubble at 45 m is $V_1 = \frac{4}{3}\pi r^3 = \frac{4}{3}\pi(1.0\ \text{cm})^3 = 4.2\ \text{cm}^3$. The original pressure on the bubble is the sum of the pressure exerted by the water, given by *(1.4)*, plus the atmospheric pressure:

$$\begin{aligned} P_1 &= \rho g h + 1.0\ \text{bar} \\ &= (1.00\ \text{g}\cdot\text{cm}^{-3})\frac{10^{-3}\ \text{kg}}{1\ \text{g}}\left(\frac{10^2\ \text{cm}}{1\ \text{m}}\right)^3(9.81\ \text{m}\cdot\text{s}^{-2})(45\ \text{m})\frac{1\ \text{N}}{1\ \text{kg}\cdot\text{m}\cdot\text{s}^{-2}}\left(\frac{1\ \text{bar}}{101\,325\ \text{N}\cdot\text{m}^{-2}}\right) + 1.0\ \text{bar} \\ &= 5.4\ \text{bar} \end{aligned}$$

The volume of the bubble at $P_2 = 1.0\ \text{bar}$ is given by *(1.7)* as

$$V_2 = (4.2\ \text{cm}^3)\frac{5.4\ \text{bar}}{1.0\ \text{bar}} = 23\ \text{cm}^3$$

which corresponds to

$$r = \left(\frac{3V}{4\pi}\right)^{1/3} = \left(\frac{3(23\ \text{cm}^3)}{4\pi}\right)^{1/3} = 1.8\ \text{cm}$$

1.21 A 1.103-L flask containing nitrogen at a pressure of 710.6 torr is connected to an evacuated flask of unknown volume. The nitrogen, which acts ideally, is allowed to expand into the combined system of both flasks isothermally. If the final pressure of the nitrogen is 583.1 torr, determine the volume of the evacuated flask.

▌ The initial and final volumes of the nitrogen are

$$V_1 = 1.103\ \text{L} \qquad \text{and} \qquad V_2 = 1.103\ \text{L} + V$$

where V is the volume of the evacuated flask. Substituting the data into *(1.7)* and solving for V gives

$$(710.6\ \text{torr})(1.103\ \text{L}) = (583.1\ \text{torr})(1.103\ \text{L} + V)$$
$$V = 0.241\ \text{L}$$

1.22 A McLeod gauge (a device for measuring low pressures in vacuum systems) is shown in Fig. 1-3. As the mercury is drawn into the gauge, a fixed volume of gas (V_{bulb}) is trapped in the bulb and compressed into a capillary tube at the top of the bulb. The two distances h_1 and h_2 are measured, and the pressure is given by *(1.7)* as

$$P/(\text{torr}) = \frac{[(h_1 + h_2)/(\text{mmHg})]\,V_{\text{gas}}}{V_{\text{bulb}}}$$

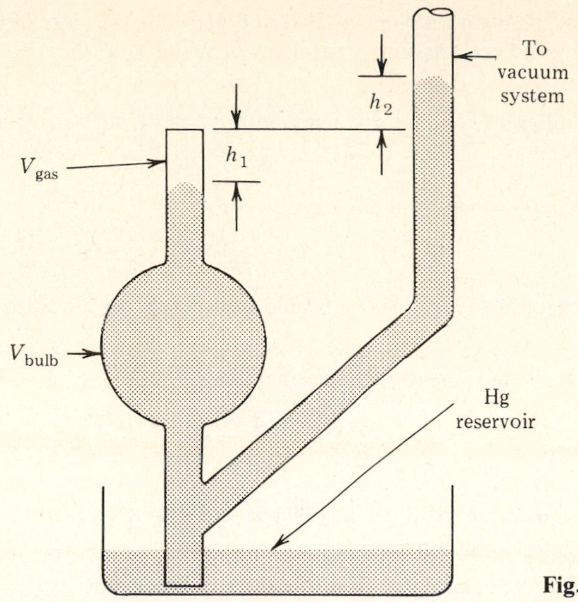

Fig. 1-3

Determine the pressure in a vacuum system given that $h_1 = 35$ mm, $h_2 = 82$ mm, $V_{bulb} = 100.3$ cm^3, and the radius of the capillary tube is 0.25 mm.

▮ The volume of the compressed gas is

$$V_{gas} = 4\pi r^2 h_1 = 4\pi(0.25\text{ mm})^2(35\text{ mm})\left(\frac{10^{-1}\text{ cm}}{1\text{ mm}}\right)^3 = 2.7 \times 10^{-2}\text{ cm}^3$$

The pressure is

$$P/(\text{torr}) = \frac{(35 + 82)(2.7 \times 10^{-2}\text{ cm}^3)}{100.3\text{ cm}^3} = 3.1 \times 10^{-2}$$

which is equivalent to

$$P = (3.1 \times 10^{-2}\text{ torr})\frac{1\text{ atm}}{760\text{ torr}}\left(\frac{101\,325\text{ Pa}}{1\text{ atm}}\right) = 4.1\text{ Pa}$$

1.23 Derive an expression for $(\partial V/\partial P)_{T,n}$ for an ideal gas. Evaluate the *isothermal compressibility*

$$\kappa = \left(-\frac{1}{V}\right)\left(\frac{\partial V}{\partial P}\right)_{T,n} \tag{1.8}$$

for an ideal gas at 1.00 bar.

▮ Solving *(1.6)* for V and taking the derivative with respect to P gives

$$\left(\frac{\partial V}{\partial P}\right)_{T,n} = \left(\frac{\partial(k/P)}{\partial P}\right)_{T,n} = -\frac{k}{P^2} = -\frac{PV}{P^2} = -\frac{V}{P}$$

The isothermal compressibility is given by *(1.8)* as

$$\kappa = \left(-\frac{1}{V}\right)\left(-\frac{V}{P}\right) = \frac{1}{P} = \frac{1}{1.00\text{ bar}} = 1.00\text{ bar}^{-1}$$

1.24 Identify which of the following plots for a fixed amount of an ideal gas at constant pressure will be linear: **(a)** V vs. T, **(b)** V vs. $1/T$, **(c)** V/T vs. T, and **(d)** VT vs. V. Determine the slope of the straight line in each linear plot.

▮ The volume-temperature behavior for a fixed amount of an ideal gas under isobaric conditions is given by *Charles's law* as

$$V = kT \tag{1.9}$$

where k is a constant. The linear forms of *(1.9)* are **(a)** $V = kT$ with the slope equal to k and **(c)** $V/T = k$ with the slope equal to 0.

1.25 A rubber balloon was filled with helium at 25 °C and placed in a beaker of liquid nitrogen at −196 °C. The volume of the cold helium was 15 cm^3. Assuming ideal gas behavior and isobaric conditions, what was the volume of the helium at 25 °C?

▮ In terms of two different volumes and temperatures, *(1.9)* can be written as

$$V_2/T_2 = V_1/T_1 \tag{1.10}$$

Solving *(1.10)* for V_2 and substituting the data gives

$$V_2 = (15\ \text{cm}^3)[(298\ \text{K})/(77\ \text{K})] = 58\ \text{cm}^3$$

1.26 To what temperature must an ideal gas be heated under isobaric conditions to increase the volume from 100 cm^3 at 25 °C to 1.00 dm^3?

▮ Solving *(1.10)* for T_2 and substituting the data gives

$$T_2 = (298\ \text{K})\frac{(1.00\ \text{dm}^3)[(10\ \text{cm})/(1\ \text{dm})]^3}{100\ \text{cm}^3} = 2980\ \text{K}$$

1.27 The volume of 1.00 mol of an ideal gas at 1.00 bar and 0 °C decreased to 21.1 L when placed in a methanol–ice bath. What was the temperature of the bath?

▮ For 1.00 mol of an ideal gas at 1.00 bar, *(1.9)* can be written as

$$V = V_0\{1 + \alpha[T/(°\text{C})]\}$$

where $V_0 = 22.7$ L and $\alpha = 3.66 \times 10^{-3}$. Solving for temperature and substituting the data gives

$$T/(°\text{C}) = \frac{V/V_0 - 1}{\alpha} = \frac{(21.1\ \text{L})/(22.7\ \text{L}) - 1}{3.66 \times 10^{-3}} = -19$$

The temperature of the bath was −19 °C.

1.28 Derive an expression for $(\partial V/\partial T)_{P,n}$ for an ideal gas. Evaluate the *cubic expansion coefficient*

$$\alpha = \frac{1}{V}\left(\frac{\partial V}{\partial T}\right)_P \tag{1.11}$$

for an ideal gas at 0 °C.

▮ Solving *(1.9)* for V and taking the derivative with respect to T gives

$$\left(\frac{\partial V}{\partial T}\right)_{P,n} = \left(\frac{\partial(kT)}{\partial T}\right)_{P,n} = k = \frac{V}{T}$$

The cubic expansion coefficient is

$$\alpha = \frac{1}{V}\left(\frac{V}{T}\right) = \frac{1}{T} = \frac{1}{273\ \text{K}} = 3.66 \times 10^{-3}\ \text{K}^{-1}$$

1.29 For a fixed amount of an ideal gas under constant-volume conditions, the pressure–temperature behavior is given by *Amonton's law* as

$$P = kT \tag{1.12}$$

where k is a constant. Evaluate k in terms of the isothermal compressibility, see *(1.8)*, and the cubic expansion coefficient, see *(1.11)*.

▮ Taking the derivative of P with respect to T gives

$$\left(\frac{\partial P}{\partial T}\right)_V = k$$

Expressing the same derivative in terms of $(\partial V/\partial T)_P$ and $(\partial V/\partial P)_T$ (see Table 6-3) gives

$$k = \left(\frac{\partial P}{\partial T}\right)_V = -\frac{(\partial V/\partial T)_P}{(\partial V/\partial P)_T} = \frac{(1/V)(\partial V/\partial T)_P}{-(1/V)(\partial V/\partial P)_T} = \frac{\alpha}{\kappa}$$

1.30 The pressure in an ideal gas thermometer increased from 589 torr to 785 torr when the thermometer bulb was transferred from an ice-water bath at 0 °C to a beaker of boiling water. Assuming that the volume of the thermometer bulb did not change, what was the temperature of the boiling water?

▮ In terms of two different pressures and temperatures, *(1.12)* can be written as

$$P_2/T_2 = P_1/T_1 \tag{1.13}$$

Solving *(1.13)* for T_2 and substituting the data gives

$$T_2 = (273\ \text{K})\frac{785\ \text{torr}}{589\ \text{torr}} = 364\ \text{K} = 91\ ^\circ\text{C}$$

1.31 The pressure of air in an automobile tire at 15 °C was 32 psig. After the tire warmed to 35 °C from driving, what was the air pressure? Did it exceed the recommended maximum pressure of 34 psig?

▮ Assuming an atmospheric pressure of 15 psi, the original pressure was $P_1 = 32\ \text{psi} + 15\ \text{psi} = 47\ \text{psi}$. Solving *(1.13)* for P_2 and substituting the data gives

$$P_2 = (47\ \text{psi})\frac{308\ \text{K}}{288\ \text{K}} = 50\ \text{psi} = 35\ \text{psig}$$

which slightly exceeds the recommended maximum pressure.

1.32 According to *Gay-Lussac's law of combining volumes*, 2 L of hydrogen gas will react with 1 L of oxygen gas (measured at the same temperature and pressure) to produce steam. What volume of steam will be formed?

▮ The relationship between the volume and the amount of substance (commonly known as the "number of moles," n) for an ideal gas under constant pressure and temperature conditions is given by *Avogadro's hypothesis*

$$V = kn \tag{1.14}$$

where k is constant. From the balanced chemical equation describing the reaction,

$$2H_2(g) + O_2(g) \rightarrow 2H_2O(g)$$

we can easily see that $k = 1$ in this case, so 2 L or 2 mol of hydrogen gas will react with 1 L or 1 mol of oxygen gas to produce 2 L or 2 mol of steam.

1.33 A toy balloon originally held 1.00 g of helium gas and had a radius of 10.0 cm. During the night, 0.25 g of the gas effused from the balloon. Assuming ideal gas behavior under these constant pressure and temperature conditions, what was the radius of the balloon the next morning?

▮ In terms of two different volumes and amounts of substance, *(1.14)* can be written as

$$V_1/n_1 = V_2/n_2 \tag{1.15}$$

Solving *(1.15)* for V_2 and substituting the data gives

$$V_2 = \tfrac{4}{3}\pi(10.0\ \text{cm})^3\frac{(0.75\ \text{g})/(4.00\ \text{g}\cdot\text{mol}^{-1})}{(1.00\ \text{g})/(4.00\ \text{g}\cdot\text{mol}^{-1})} = 3100\ \text{cm}^3$$

which corresponds to a radius of

$$r = (3V/4\pi)^{1/3} = [3(3100\ \text{cm}^3)/4\pi]^{1/3} = 9.0\ \text{cm}$$

1.34 Derive an expression for $(\partial V/\partial n)_{T,P}$ for an ideal gas. What will be the volume change corresponding to an increase of 0.01 mol at 25 °C and 1.00 bar? The molar volume of an ideal gas under these conditions is 24.8 L · mol^{-1}.

▮ Taking the derivative of *(1.14)* with respect to n gives

$$\left(\frac{\partial V}{\partial n}\right)_{T,P} = \left(\frac{\partial kn}{\partial n}\right)_{T,P} = k = \frac{V}{n}$$

The value of k under these conditions is

$$k = \frac{V}{n} = \frac{24.8\ \text{L}}{1.00\ \text{mol}} = 24.8\ \text{L}\cdot\text{mol}^{-1}$$

The volume change is found by integrating the expression for $(\partial V/\partial n)_{T,P}$:

$$\int_{V_1}^{V_2} dV = \int_{n_1}^{n_2} k\,dn$$

$$\Delta V = k\,\Delta n = (24.8\ \text{L}\cdot\text{mol}^{-1})(0.01\ \text{mol}) = 0.2\ \text{L}$$

1.35 The pressure of a fixed amount of an ideal gas is 0.750 bar. What will be the pressure after the volume of the gas is tripled and the absolute temperature is doubled?

▮ Combining *(1.6)* and *(1.9)* gives the *combined gas law*

$$PV/T = k \tag{1.16}$$

where k is a constant. For two different sets of conditions, *(1.16)* can be written as

$$P_1V_1/T_1 = P_2V_2/T_2 \tag{1.17}$$

Solving *(1.17)* for P_2 and substituting the data gives

$$P_2 = (0.750 \text{ bar})\frac{V_1}{3V_1}\left(\frac{2T_1}{T_1}\right) = 0.500 \text{ bar}$$

1.36 An ideal gas sample had a volume of 1.00 L at sea-level conditions of 1.01 bar and 18 °C. As the gas enters the thermosphere at 500 km, the volume increases to 3.5×10^{11} L at a pressure of 1.5×10^{-11} bar. What is the temperature of the gas at this altitude?

▮ The final temperature of the gas is given by *(1.17)* as

$$T_2 = (291 \text{ K})\frac{1.5 \times 10^{-11} \text{ bar}}{1.01 \text{ bar}}\left(\frac{3.5 \times 10^{11} \text{ L}}{1.00 \text{ L}}\right) = 1500 \text{ K}$$

1.37 The volume of one mole of an ideal gas at STP (273.15 K and exactly 1 atm) is 22.4138 L. The "standard ambient temperature and pressure" (SATP) is defined as 298.15 K and exactly 1 bar. What is the molar volume under these new standard conditions?

▮ Solving *(1.17)* for V_2 and substituting the data gives

$$V_2 = (22.4138 \text{ L})\frac{(1 \text{ atm})[(1.013\,25 \text{ bar})/(1 \text{ atm})]}{1 \text{ bar}}\left(\frac{298.15 \text{ K}}{273.15 \text{ K}}\right) = 24.7894 \text{ L}$$

1.38 What is the mass of air in a 250-mL Erlenmeyer flask (actual volume is 267 mL) at a typical laboratory pressure of 715 torr and a temperature of 21 °C? Assume air to be an ideal gas having an average molar mass of 29.0 g · mol^{-1}.

▮ The pressure–volume–temperature–amount of substance relationship for an ideal gas is given by the *ideal gas law*

$$PV = nRT \tag{1.18}$$

where R is given in Table 1-2. The amount of air is

$$n = \frac{PV}{RT} = \frac{(715 \text{ torr})[(1 \text{ atm})/(760 \text{ torr})](267 \text{ mL})[(10^{-3} \text{ L})/(1 \text{ mL})]}{(0.0821 \text{ L} \cdot \text{atm} \cdot \text{K}^{-1} \cdot \text{mol}^{-1})(294 \text{ K})} = 0.0104 \text{ mol}$$

which corresponds to

$$m = nM = (0.0104 \text{ mol})(29.0 \text{ g} \cdot \text{mol}^{-1}) = 0.302 \text{ g}$$

Table 1-2

R
8.314 41 J · K^{-1} · mol^{-1}
8.314 41 m^3 · Pa · K^{-1} · mol^{-1}
0.083 144 1 L · bar · K^{-1} · mol^{-1}
0.082 056 8 L · atm · K^{-1} · mol^{-1}

1.39 What volume of hydrogen will be produced at 0.963 bar and 28 °C from the reaction of 6.0 g of zinc with excess 6 M hydrochloric acid? Assume that hydrogen is an ideal gas under these conditions.

▌ The amount of zinc reacting is

$$n = \frac{m}{M} = \frac{6.0\text{ g Zn}}{65.39\text{ g}\cdot\text{mol}^{-1}} = 0.092\text{ mol Zn}$$

The chemical equation $Zn(s) + 2HCl(aq) \rightarrow ZnCl_2(aq) + H_2(g)$ shows that the amount of hydrogen produced is

$$(0.092\text{ mol Zn})\frac{1\text{ mol H}_2}{1\text{ mol Zn}} = 0.092\text{ mol H}_2$$

Solving *(1.18)* for V and substituting the data gives

$$V = \frac{(0.092\text{ mol})(0.083\,14\text{ L}\cdot\text{bar}\cdot\text{K}^{-1}\cdot\text{mol}^{-1})(301\text{ K})}{0.963\text{ bar}} = 2.4\text{ L}$$

1.40 The relative humidity in a room is 65%. Assuming ideal gas behavior, what mass of water vapor is in the air at 25 °C if the room measures 4 m × 4 m × 3 m? The vapor pressure of liquid water is 23.8 torr at 25 °C.

▌ The *relative humidity* is the actual vapor pressure of water vapor in the air divided by the equilibrium vapor pressure of water at that temperature and is usually expressed as a percentage. The pressure of the water vapor in this case is

$$P = (0.65)(23.8\text{ torr}) = 15.5\text{ torr}$$

which upon substitution into *(1.18)* gives the amount of water as

$$n = \frac{(15.5\text{ torr})[(1\text{ atm})/(760\text{ torr})][(101\,325\text{ Pa})/(1\text{ atm})][(4\text{ m})(4\text{ m})(3\text{ m})]}{(8.314\text{ m}^3\cdot\text{Pa}\cdot\text{K}^{-1}\cdot\text{mol}^{-1})(298\text{ K})}$$

$$= 40\text{ mol}$$

which corresponds to

$$m = nM = (40\text{ mol})(18\text{ g}\cdot\text{mol}^{-1}) = 700\text{ g}$$

1.41 The vapor pressure of liquid water is 23.8 torr at 25 °C. By what factor does the molar volume of water increase as it vaporizes to form an ideal gas under these conditions? The density of liquid water is $0.997\text{ g}\cdot\text{cm}^{-3}$ at 25 °C.

▌ Substituting *(1.18)* for the volume of the vapor into the expression for the ratio of the volumes gives

$$\frac{V(\text{gas})}{V(\text{liquid})} = \frac{RT/P}{M/\rho} = \frac{RT\rho}{MP}$$

$$= \frac{(0.083\,14\text{ L}\cdot\text{bar}\cdot\text{K}^{-1}\cdot\text{mol}^{-1})(298\text{ K})(0.997\text{ g}\cdot\text{cm}^{-3})[(10^3\text{ cm}^3)/(1\text{ L})]}{(18.0\text{ g}\cdot\text{mol}^{-1})(23.8\text{ torr})[(1\text{ atm})/(760\text{ torr})][(1.013\,25\text{ bar})/(1\text{ atm})]}$$

$$= 4.32 \times 10^4$$

1.42 What is the value of the gas constant expressed in units of $\text{L}\cdot\text{torr}\cdot\text{K}^{-1}\cdot\text{mol}^{-1}$?

▌ Using appropriate conversion factors gives

$$R = (8.314\,41\text{ m}^3\cdot\text{Pa}\cdot\text{K}^{-1}\cdot\text{mol}^{-1})\frac{10^3\text{ L}}{1\text{ m}^3}\left(\frac{1\text{ atm}}{101\,325\text{ Pa}}\right)\left(\frac{760\text{ torr}}{1\text{ atm}}\right)$$

$$= 62.3632\text{ L}\cdot\text{torr}\cdot\text{K}^{-1}\cdot\text{mol}^{-1}$$

1.43 Prepare a plot of PV/T against P to determine R from the following data for CO_2 at 0 °C:

P/(atm)	0.500	0.100	0.010
V/(L · mol^{-1})	44.672	223.980	2241.204

▌ The respective values of $(PV/T)/(\text{L}\cdot\text{atm}\cdot\text{K}^{-1}\cdot\text{mol}^{-1})$ are 0.081 771, 0.081 998, and 0.082 050. The intercept of the plot shown in Fig. 1-4 gives $R = 0.082\,055\text{ L}\cdot\text{atm}\cdot\text{K}^{-1}\cdot\text{mol}^{-1}$.

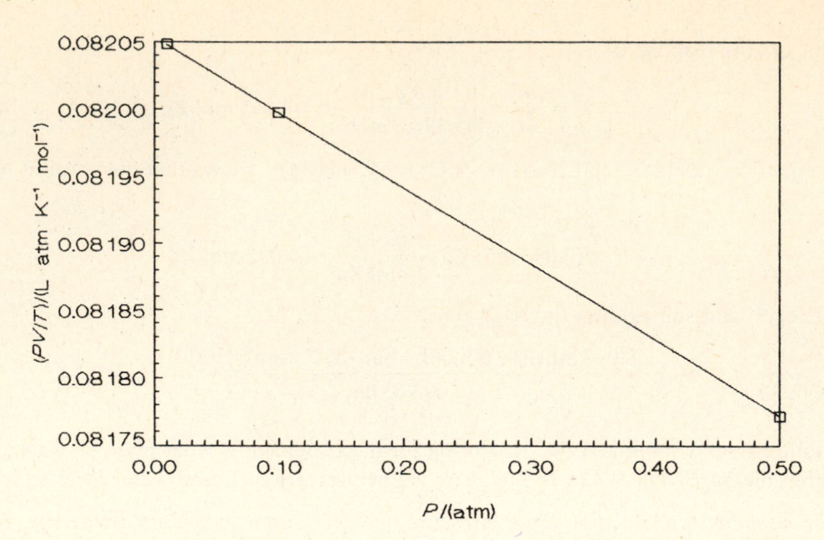

Fig. 1-4

1.44 A 5.00-L flask containing N_2 at 1.00 bar and 25 °C is connected to a 4.00-L flask containing N_2 at 2.00 bar and 0 °C. After the gases are allowed to mix, keeping both flasks at their original temperatures, what will be the final pressure and the amount of N_2 in the 5.00-L flask? Assume ideal gas behavior.

▮ The total amount of N_2 in the combined system is found by using *(1.18)* for each flask:

$$n = n(\text{hot}) + n(\text{cold}) = \frac{P(\text{hot})V(\text{hot})}{RT(\text{hot})} + \frac{P(\text{cold})V(\text{cold})}{RT(\text{cold})}$$

$$= \frac{(1.00\ \text{bar})(5.00\ \text{L})}{(0.083\,14\ \text{L}\cdot\text{bar}\cdot\text{K}^{-1}\cdot\text{mol}^{-1})(298\ \text{K})} + \frac{(2.00\ \text{bar})(4.00\ \text{L})}{(0.083\,14\ \text{L}\cdot\text{bar}\cdot\text{K}^{-1}\cdot\text{mol}^{-1})(273\ \text{K})}$$

$$= 0.554\ \text{mol}$$

Recognizing that $P(\text{hot}) = P(\text{cold})$ after mixing, *(1.18)* gives

$$\frac{n(\text{hot})RT(\text{hot})}{V(\text{hot})} = \frac{n(\text{cold})RT(\text{cold})}{V(\text{cold})}$$

$$\frac{n(\text{hot})(298\ \text{K})}{5.00\ \text{L}} = \frac{[0.554\ \text{mol} - n(\text{hot})](273\ \text{K})}{4.00\ \text{L}}$$

Solving for $n(\text{hot})$ gives $n(\text{hot}) = 0.296\ \text{mol}$, which corresponds to

$$P(\text{hot}) = \frac{(0.296\ \text{mol})(0.083\,14\ \text{L}\cdot\text{bar}\cdot\text{K}^{-1}\cdot\text{mol}^{-1})(298\ \text{K})}{5.00\ \text{L}} = 1.47\ \text{bar}$$

1.45 The valve on a commercial cylinder of N_2 gas was left slightly open so that a small amount of gas leaked into the laboratory. The leak rate, dP/dt, is proportional to the difference between the pressure in the tank and the atmospheric pressure. The initial leak rate was 1.0 g · s^{-1}, and the initial pressure in the 7.28-m^3 tank was 17 180 kPa. What was the pressure of N_2 in the tank after 1.00 h? Assume ideal gas behavior and laboratory conditions of 25 °C and 101 kPa.

▮ The equation for the leak rate is

$$\frac{dP}{dt} = -k(P - 101\ \text{kPa})$$

where k can be evaluated from the initial leak rate data.

$$\frac{dP}{dt} = \frac{d(nRT/V)}{dt} = \frac{RT}{V}\frac{dn}{dt} = \frac{RT}{MV}\frac{dm}{dt} = -k(P - 101\ \text{kPa})$$

$$k = \frac{-RT(dm/dt)}{MV(P - 101\ \text{kPa})}$$

$$= \frac{-(8.314\ \text{m}^3\cdot\text{Pa}\cdot\text{K}^{-1}\cdot\text{mol}^{-1})(298\ \text{K})(-1.0\ \text{g}\cdot\text{s}^{-1})[(10^{-3}\ \text{kPa})/(1\ \text{Pa})]}{(28\ \text{g}\cdot\text{mol}^{-1})(7.28\ \text{m}^3)(17\,180\ \text{kPa} - 101\ \text{kPa})}$$

$$= 7.1\times10^{-7}\ \text{s}^{-1}$$

Integrating the equation for the leak rate and substituting the data gives

$$\ln \frac{P - 101\ \text{kPa}}{17\,180\ \text{kPa} - 101\ \text{kPa}} = -kt = -(7.1 \times 10^{-7}\ \text{s}^{-1})(3600\ \text{s})$$

Solving gives $P = 17\,140$ kPa.

1.46 What will be the volume of helium needed in a balloon to lift 1.00 kg of mass? Assume that the temperature is 25 °C, the pressure is 1.00 bar, air has an average molar mass of 29 g · mol^{-1}, and both gases act ideally.

▌ The *lifting capacity* of a balloon, Δm, is determined by subtracting the mass of helium in the balloon from the mass of the air that is displaced by the balloon.

$$\begin{aligned} \Delta m &= m(\text{air}) - m(\text{He}) = n(\text{air})M(\text{air}) - n(\text{He})M(\text{He}) \\ &= n[M(\text{air}) - M(\text{He})] \\ &= (PV/RT)[M(\text{air}) - M(\text{He})] \end{aligned} \qquad (1.19a)$$

Solving *(1.19a)* for V and substituting the data gives

$$V = \frac{(8.314\ \text{m}^3 \cdot \text{Pa} \cdot \text{K}^{-1} \cdot \text{mol}^{-1})(298\ \text{K})(1.00\ \text{kg})[(10^3\ \text{g})/(1\ \text{kg})]}{(1.00\ \text{bar})[(10^5\ \text{Pa})/(1\ \text{bar})](29\ \text{g} \cdot \text{mol}^{-1} - 4\ \text{g} \cdot \text{mol}^{-1})} = 0.99\ \text{m}^3$$

1.47 What will be the temperature difference needed in a hot-air balloon to lift 1.00 kg of mass? Assume that the volume of the balloon is 100 m^3, the temperature of the ambient air is 25 °C, the pressure is 1.00 bar, and air is an ideal gas with an average molar mass of 29 g · mol^{-1}.

▌ For a hot-air balloon, the lifting capacity, Δm, is determined by subtracting the mass of the hot air from the mass of the cold air that is displaced by the balloon.

$$\begin{aligned} \Delta m &= m(\text{cold}) - m(\text{hot}) = [n(\text{cold}) - n(\text{hot})]M \\ &= \left(\frac{PV}{RT(\text{cold})} - \frac{PV}{RT(\text{hot})}\right)M \\ &= \frac{PVM}{R}\left(\frac{1}{T(\text{cold})} - \frac{1}{T(\text{hot})}\right) \end{aligned} \qquad (1.19b)$$

Solving *(1.19b)* for $1/T(\text{hot})$ and substituting the data gives

$$\begin{aligned} \frac{1}{T(\text{hot})} &= \frac{1}{298\ \text{K}} - \frac{(8.314\ \text{m}^3 \cdot \text{Pa} \cdot \text{K}^{-1} \cdot \text{mol}^{-1})(1.00\ \text{kg})[(10^3\ \text{g})/(1\ \text{kg})]}{(1.00\ \text{bar})[(10^5\ \text{Pa})/(1\ \text{bar})](100\ \text{m}^3)(29\ \text{g} \cdot \text{mol}^{-1})} \\ &= 3.33 \times 10^{-3}\ \text{K}^{-1} \end{aligned}$$

giving $T(\text{hot}) = 301\ \text{K} = 28\ °\text{C}$. The temperature difference needed is 3 °C.

1.48 Assuming ideal gas behavior, what is the density of carbon dioxide gas at the sublimation temperature of −78.5 °C? The pressure is 1.01 bar.

▌ The density of an ideal gas is given by

$$\rho = \frac{m}{V} = \frac{nM}{nRT/P} = \frac{PM}{RT} \qquad (1.20)$$

Substituting the data gives

$$\rho = \frac{(1.01\ \text{bar})(44.01\ \text{g} \cdot \text{mol}^{-1})}{(0.083\,14\ \text{L} \cdot \text{bar} \cdot \text{K}^{-1} \cdot \text{mol}^{-1})(194.7\ \text{K})} = 2.75\ \text{g} \cdot \text{L}^{-1}$$

1.49 How many molecules of gas are in a 0.535-L vacuum system at a pressure of 3.5 × 10^{-7} torr and 25 °C? Assume ideal gas behavior.

▌ Using *(1.18)* gives

$$n = \frac{(3.5 \times 10^{-7}\ \text{torr})[(1\ \text{atm})/(760\ \text{torr})](0.535\ \text{L})}{(0.0821\ \text{L} \cdot \text{atm} \cdot \text{K}^{-1} \cdot \text{mol}^{-1})(298\ \text{K})} = 1.0 \times 10^{-11}\ \text{mol}$$

which corresponds to

$$N = nL = (1.0 \times 10^{-11}\ \text{mol})(6.022 \times 10^{23}\ \text{mol}^{-1}) = 6.0 \times 10^{12}$$

where L is Avogadro's constant.

1.50 Assuming ideal gas behavior, how many atoms of Ar are contained in a typical human breath of 0.5 L at 1.0 bar and 37 °C? Air consists of 1% Ar atoms. Assuming that the argon atoms from the last breath of Plato have been distributed randomly throughout the atmosphere (5×10^{18} m^3), how long would it take to breathe one of these atoms? A typical adult breath rate is 10 min^{-1}.

▮ The number of Ar atoms in one breath using *(1.18)* is

$$n = \frac{(1.0\ \text{bar})(0.5\ \text{L}\cdot\text{breath}^{-1})}{(0.083\,14\ \text{L}\cdot\text{bar}\cdot\text{K}^{-1}\cdot\text{mol}^{-1})(310\ \text{K})} = 0.02\ \text{mol air}\cdot\text{breath}^{-1}$$

$$N = (0.01)(0.02\ \text{mol}\cdot\text{breath}^{-1})(6.022 \times 10^{23}\ \text{mol}^{-1}) = 1 \times 10^{20}\ \text{breath}^{-1}$$

The concentration of the atoms from the last breath of Plato is

$$\frac{(1\ \text{breath})(1 \times 10^{20}\ \text{breath}^{-1})}{5 \times 10^{18}\ \text{m}^3} = 20\ \text{m}^{-3}$$

giving the number of breaths needed as

$$\frac{1}{(20\ \text{m}^{-3})(0.5\ \text{L}\cdot\text{breath}^{-1})[(10^{-3}\ \text{m}^3)/(1\ \text{L})]} = 100\ \text{breaths}$$

which is equivalent to

$$\frac{100\ \text{breaths}}{10\ \text{breaths}\cdot\text{min}^{-1}} = 10\ \text{min}$$

1.51 Derive the relationship between the *molecular density* of an ideal gas (the number of molecules per unit volume), N^*, and the temperature and pressure of the gas. Evaluate N^* for a sample of N_2 at 1.00 atm and 25 °C.

▮ Substituting *(1.18)* into the expression for N^* in terms of N and V gives

$$N^* = \frac{N}{V} = \frac{nL}{V} = \frac{(PV/RT)L}{V} = \frac{PL}{RT} \tag{1.21}$$

For this sample of N_2,

$$N^* = \frac{(1.00\ \text{atm})[(101\,325\ \text{Pa})/(1\ \text{atm})](6.022 \times 10^{23}\ \text{mol}^{-1})}{(8.314\ \text{m}^3\cdot\text{Pa}\cdot\text{K}^{-1}\cdot\text{mol}^{-1})(298\ \text{K})} = 2.46 \times 10^{25}\ \text{m}^3$$

1.52 The density of dry air at 25 °C and 76.00 cmHg is 0.001 185 g · mL^{-1}. Assuming ideal gas behavior, calculate the average molar mass of air under these conditions.

▮ Solving *(1.20)* for M and substituting the data gives

$$M = \frac{(0.001\,185\ \text{g}\cdot\text{mL}^{-1})[(10^3\ \text{mL})/(1\ \text{L})](0.082\,06\ \text{L}\cdot\text{atm}\cdot\text{K}^{-1}\cdot\text{mol}^{-1})(298.2\ \text{K})}{(76.00\ \text{cmHg})[(10\ \text{mm})/(1\ \text{cm})][(1\ \text{torr})/(1\ \text{mmHg})][(1\ \text{atm})/(760\ \text{torr})]}$$

$$= 29.00\ \text{g}\cdot\text{mol}^{-1}$$

1.53 A Dumas bulb used for determining the molar mass of a gas had a mass of 65.2649 g empty, 65.3324 g when filled with natural gas at a temperature of 23 °C and a pressure of 741 torr, and 170.15 g when filled with water. What is the molar mass of natural gas? The density of water is 0.998 g · mL^{-1} at 23 °C.

▮ The volume of the bulb is

$$V = \frac{m}{\rho} = \frac{170.15\ \text{g} - 65.2649\ \text{g}}{0.998\ \text{g}\cdot\text{mL}^{-1}} = 105.1\ \text{mL}$$

and the density of the natural gas is

$$\rho = \frac{m}{V} = \frac{65.3324\ \text{g} - 65.2649\ \text{g}}{105.1\ \text{mL}} = 6.42 \times 10^{-4}\ \text{g}\cdot\text{mL}^{-1}$$

The molar mass, using *(1.20)*, is

$$M = \frac{(6.42 \times 10^{-4}\ \text{g}\cdot\text{mL}^{-1})[(10^3\ \text{mL})/(1\ \text{L})](0.0821\ \text{L}\cdot\text{atm}\cdot\text{K}^{-1}\cdot\text{mol}^{-1})(298\ \text{K})}{(741\ \text{torr})[(1\ \text{atm})/(760\ \text{torr})]}$$

$$= 16.0\ \text{g}\cdot\text{mol}^{-1}$$

1.54 Suppose the conditions of 1.000 bar, 300.0 K, and 25.00 dm^3 were chosen to define the molar volume of an ideal gas. Calculate the "revised" gas constant, R', and the "revised" molar mass of O_2, M'.

▌ Solving *(1.18)* for R' and substituting the data gives

$$R' = \frac{(1.00\ \text{bar})(25.00\ \text{dm}^3)[(1\ \text{L})/(1\ \text{dm}^3)]}{(1\ \text{mol})(300.0\ \text{K})} = 0.083\,33\ \text{L}\cdot\text{bar}\cdot\text{K}^{-1}\cdot\text{mol}^{-1}$$

The revised molar mass will be given by

$$M' = M\frac{R'}{R} = (31.9988\ \text{g}\cdot\text{mol}^{-1})\frac{0.083\,33\ \text{L}\cdot\text{bar}\cdot\text{K}^{-1}\cdot\text{mol}^{-1}}{0.083\,14\ \text{L}\cdot\text{bar}\cdot\text{K}^{-1}\cdot\text{mol}^{-1}} = 32.07\ \text{g}\cdot\text{mol}^{-1}$$

1.3 MIXTURES OF IDEAL GASES

1.55 A 1.00-mol sample of O_2 and a 3.00-mol sample of H_2 are mixed isothermally in a 125.3-L container at 125 °C. Assuming ideal gas behavior, calculate the partial pressure of each gas and the total pressure of the gaseous mixture.

▌ For each gas, *(1.18)* gives

$$P(O_2) = \frac{(1.00\ \text{mol})(0.083\,14\ \text{L}\cdot\text{bar}\cdot\text{K}^{-1}\cdot\text{mol}^{-1})(398\ \text{K})}{125.3\ \text{L}} = 0.264\ \text{bar}$$

$$P(H_2) = \frac{(3.00)(0.083\,14)(398)}{125.3} = 0.792\ \text{bar}$$

The total pressure of the gaeous mixture, P_t, in terms of the partial pressures, P_i, is given by *Dalton's law*

$$P_t = \sum P_i = 0.264\ \text{bar} + 0.792\ \text{bar} = 1.056\ \text{bar} \qquad (1.22)$$

1.56 After the gaseous mixture described in Problem 1.55 undergoes reaction to form water, what will be the partial pressure of each gas and the total pressure of the resulting mixture? Assume isothermal conditions and ideal gas behavior.

▌ From the chemical equation describing the reaction, $2H_2(g) + O_2(g) \rightarrow 2H_2O(g)$, we can see that the limiting reactant is the oxygen, so the resulting gaseous mixture will contain

$$n(O_2) = 1.00\ \text{mol} - 1.00\ \text{mol} = 0.00\ \text{mol}$$

$$n(H_2) = 3.00\ \text{mol} - (1.00\ \text{mol}\ O_2)[(2\ \text{mol}\ H_2)/(1\ \text{mol}\ O_2)] = 1.00\ \text{mol}$$

$$n(H_2O) = (1.00\ \text{mol}\ O_2)[(2\ \text{mol}\ H_2O)/(1\ \text{mol}\ O_2)] = 2.00\ \text{mol}$$

The respective partial pressures are given by *(1.18)* as

$$P(H_2) = \frac{(1.00\ \text{mol})(0.083\,14\ \text{L}\cdot\text{bar}\cdot\text{K}^{-1}\cdot\text{mol}^{-1})(398\ \text{K})}{125.3\ \text{L}} = 0.264\ \text{bar}$$

$$P(H_2O) = \frac{(2.00)(0.083\,14)(398)}{125.3} = 0.528\ \text{bar}$$

and the total pressure is given by *(1.22)* as

$$P_t = 0.264\ \text{bar} + 0.528\ \text{bar} = 0.792\ \text{bar}$$

1.57 After the gaseous mixture described in Problem 1.56 is allowed to cool to 25 °C, what will be the partial pressure of each gas and the total pressure of the gaseous mixture? The vapor of liquid water is 23.8 torr at 25 °C.

▌ Assuming that the condensed water occupies a negligible volume in the container and that the H_2 acts ideally, the partial pressure of the H_2 is given by *(1.18)* as

$$P(H_2) = \frac{(1.00\ \text{mol})(0.083\,14\ \text{L}\cdot\text{bar}\cdot\text{K}^{-1}\cdot\text{mol}^{-1})(298\ \text{K})}{125.3\ \text{L}} = 0.198\ \text{bar}$$

The partial pressure of the water vapor is

$$P(H_2O) = (23.8\ \text{torr})\frac{1\ \text{atm}}{760\ \text{torr}}\left(\frac{1.01\ \text{bar}}{1\ \text{atm}}\right) = 0.0316\ \text{bar}$$

The total pressure is given by *(1.22)* as

$$P_t = 0.198\ \text{bar} + 0.0316\ \text{bar} = 0.230\ \text{bar}$$

1.58 The pressure of a gaseous mixture of NH_3 and N_2 decreased from 1.50 kPa to 1.00 kPa after the NH_3 was absorbed from the mixture. Assuming ideal gas behavior, find the composition of the original mixture.

▮ The partial pressure of N_2 is 1.00 kPa. Using *(1.22)* gives

$$P(NH_3) = P_t - P(N_2) = 1.50\text{ kPa} - 1.00\text{ kPa} = 0.50\text{ kPa}$$

The mole fraction, x_i, of a gas in a gaseous mixture is given by Dalton's law as

$$x_i = P_i/P_t \tag{1.23}$$

Substituting the data into *(1.23)* gives

$$x(N_2) = (1.00\text{ kPa})/(1.50\text{ kPa}) = 0.67$$

$$x(NH_3) = 1 - x(N_2) = 1 - 0.67 = 0.33$$

1.59 Assume that the hydrogen gas described in Problem 1.39 is collected over water at 25 °C. What would be the volume of the "wet" H_2 under these conditions? The vapor pressure of water is 0.0376 bar at 25 °C.

▮ The total pressure of the H_2 and water vapor will be 0.963 bar. The partial pressure of H_2 is given by *(1.22)* as

$$P(H_2) = P_t - P(H_2O) = 0.963\text{ bar} - 0.0376\text{ bar} = 0.925\text{ bar}$$

Assuming ideal gas behavior, the volume of the collected H_2 gas is given by *(1.18)* as

$$V = \frac{(0.092\text{ mol})(0.083\,14\text{ L}\cdot\text{bar}\cdot\text{K}^{-1}\cdot\text{mol}^{-1})(301\text{ K})}{0.925\text{ bar}} = 2.5\text{ L}$$

1.60 Assume that the 4.00-L flask in the system described in Problem 1.44 contained O_2 instead of N_2. What is the composition and what is the partial pressure of the N_2 in the final mixture?

▮ Using *(1.18)* for each gas gives

$$n(N_2) = \frac{(1.00\text{ bar})(5.00\text{ L})}{(0.083\,14\text{ L}\cdot\text{bar}\cdot\text{K}^{-1}\cdot\text{mol}^{-1})(298\text{ K})} = 0.202\text{ mol}$$

$$n(O_2) = \frac{(2.00)(4.00)}{(0.083\,14)(298)} = 0.352\text{ mol}$$

The mole fraction of N_2 is

$$x(N_2) = \frac{n(N_2)}{n(N_2) + n(O_2)} = \frac{0.202\text{ mol}}{0.202\text{ mol} + 0.352\text{ mol}} = 0.365$$

The final pressure of the system does not depend on whether only N_2 was present in both flasks or both N_2 and O_2 were present. Thus *(1.23)* gives

$$P(N_2) = x(N_2)P_t = (0.365)(1.47\text{ bar}) = 0.537\text{ bar}$$

1.61 A 255-mL sample of Ne at 735 torr is mixed with a 395-mL sample of Ar at 705 torr, and the resulting mixture is compressed to a final volume of 315 mL. Assuming ideal gas behavior and isothermal conditions, calculate the partial pressure of each gas, the total pressure, and the composition of the final mixture.

▮ Using *(1.7)* for each gas gives

$$P(\text{Ne}) = (735\text{ torr})\frac{255\text{ mL}}{315\text{ mL}} = 595\text{ torr} \quad \text{and} \quad P(\text{Ar}) = 705\frac{395}{315} = 884\text{ torr}$$

The total pressure of the mixture is given by *(1.22)* as

$$P_t = 595\text{ torr} + 884\text{ torr} = 1479\text{ torr}$$

The mole fractions, using *(1.23)*, are

$$x(\text{Ne}) = \frac{595\text{ torr}}{1479\text{ torr}} = 0.402 \quad \text{and} \quad x(\text{Ar}) = 1 - x(\text{Ne}) = 1 - 0.402 = 0.598$$

1.62 A 1.00-g sample of liquid water is added to 1.00 L of N_2 gas at 1.01 bar and 25 °C. To what volume must the system be adjusted in order for all the water to evaporate? Assuming ideal gas behavior, what will be the total pressure of the system? The vapor pressure of water is 0.0316 bar at 25 °C.

▮ The volume of the system will have to be increased so that the pressure of the water vapor is 0.0316 bar. Using *(1.18)* gives

$$V = \frac{[(1.00\ \text{g})/(18.0\ \text{g}\cdot\text{mol}^{-1})](0.083\ 14\ \text{L}\cdot\text{bar}\cdot\text{K}^{-1}\cdot\text{mol}^{-1})(298\ \text{K})}{0.0316\ \text{bar}} = 43.6\ \text{L}$$

The pressure of the N_2 under these conditions will be given by *(1.7)* as

$$P(N_2) = (1.01\ \text{bar})\frac{1.00\ \text{L}}{43.6\ \text{L}} = 0.0232\ \text{bar}$$

Using *(1.22)* gives

$$P_t = 0.0316\ \text{bar} + 0.0232\ \text{bar} = 0.0548\ \text{bar}$$

1.63 A mixture of ethane (C_2H_6) and ethene (C_2H_4) occupied 35.5 L at 1.000 bar and 405 K. This mixture reacted completely with 110.3 g of O_2 to produce CO_2 and H_2O. What was the composition of the original mixture? Assume ideal gas behavior.

▮ The total amount of the original mixture is given by *(1.18)* as

$$n_t = \frac{(1.000\ \text{bar})(35.5\ \text{L})}{(0.083\ 14\ \text{L}\cdot\text{bar}\cdot\text{K}^{-1}\cdot\text{mol}^{-1})(405\ \text{K})} = 1.054\ \text{mol}$$

The amount of O_2 that underwent reaction was

$$n(O_2) = m/M = (110.3\ \text{g})/(32.00\ \text{g}\cdot\text{mol}^{-1}) = 3.447\ \text{mol}$$

From the chemical equations

$$C_2H_4(g) + 3O_2(g) \rightarrow 2CO_2(g) + 2H_2O(g) \qquad \text{and} \qquad C_2H_6(g) + \tfrac{7}{2}O_2(g) \rightarrow 2CO_2(g) + 3H_2O(g)$$

we can see that

$$n(C_2H_4)\frac{3\ \text{mol}\ O_2}{1\ \text{mol}\ C_2H_4} + n(C_2H_6)\frac{\tfrac{7}{2}\ \text{mol}\ O_2}{1\ \text{mol}\ C_2H_6} = 3.447\ \text{mol}$$

Solving the above equation with

$$n_t = n(C_2H_4) + n(C_2H_6) = 1.054\ \text{mol}$$

gives $n(C_2H_4) = 0.484$ mol and $n(C_2H_6) = 0.570$ mol. The mole fractions, using *(1.23)*, are

$$x(C_2H_4) = (0.484\ \text{mol})/(1.054\ \text{mol}) = 0.459$$

$$x(C_2H_6) = 1 - x(C_2H_4) = 1 - 0.459 = 0.541$$

1.64 The total pressure of a mixture of H_2 and O_2 is 1.00 bar. The mixture is allowed to react to form water, which is completely removed to leave only pure H_2 at a pressure of 0.35 bar. Assuming ideal gas behavior and that all pressure measurements were made under the same temperature and volume conditions, calculate the composition of the original mixture.

▮ For the original mixture, *(1.18)* gives

$$n_t = \frac{(1.00\ \text{bar})V}{RT} = 1.00(V/RT)\ \text{mol}$$

After the reaction, the amount of excess H_2 is given by *(1.18)* as

$$n(H_2) = \frac{(0.35\ \text{bar})V}{RT} = 0.35(V/RT)\ \text{mol}$$

Thus the amount of O_2 and N_2 in the original mixture that underwent reaction is

$$n(O_2) + n(H_2) = 1.00(V/RT)\ \text{mol} - 0.35(V/RT)\ \text{mol} = 0.65(V/RT)\ \text{mol}$$

From the chemical equation $2H_2(g) + O_2(g) \rightarrow 2H_2O(l)$ we can see that

$$n(H_2) = n(O_2)[(2\ \text{mol}\ H_2)/(1\ \text{mol}\ O_2)]$$

so substituting into the equation for $n(O_2) = n(H_2)$ gives

$$n(O_2) + 2n(O_2) = 0.65(V/RT)\ \text{mol}$$

Solving gives $n(O_2) = 0.22(V/RT)$ mol and $n(H_2) = 0.43(V/RT)$ mol. The original mixture contained

$$n(H_2) = 0.43(V/RT)\text{ mol} + 0.35(V/RT)\text{ mol} = 0.78(V/RT)\text{ mol}$$
$$n(O_2) = 0.22(V/RT)\text{ mol}$$

which corresponds to

$$x(H_2) = \frac{n(H_2)}{n_t} = \frac{0.78(V/RT)}{1.00(V/RT)} = 0.78$$
$$x(O_2) = 1 - x(H_2) = 1 - 0.78 = 0.22$$

1.65 Find the relationship between the partial pressure of an ideal gas in a gaseous mixture and the concentration of the gas expressed in terms of molarity, C. What is the concentration of water vapor in a sample of air at 45 °C such that $P(H_2) = 35$ torr?

▮ Molarity is defined as the amount of substance present in one liter of solution. Solving *(1.18)* for n_i/V gives

$$C = \frac{n_i}{V} = \frac{P_i}{RT} = \frac{(35\text{ torr})[(1\text{ atm})/(760\text{ torr})]}{(0.0821\text{ L}\cdot\text{atm}\cdot\text{K}^{-1}\cdot\text{mol}^{-1})(318\text{ K})}$$
$$= 1.8 \times 10^{-3}\text{ mol}\cdot\text{L}^{-1} = 1.8 \times 10^{-3}\text{ M}$$

1.66 A gaseous mixture at 25 °C and 1.01 bar contains 75.0 g of He and 75.0 g of Ne. Assuming ideal behavior, calculate the partial volume of each gas and the total volume of the gaseous mixture.

▮ The amount of each gas in the mixture is

$$n(\text{He}) = \frac{m(\text{He})}{M(\text{He})} = \frac{75.0\text{ g}}{4.00\text{ g}\cdot\text{mol}^{-1}} = 18.8\text{ mol}$$
$$n(\text{Ne}) = 75.0/20.2 = 3.71\text{ mol}$$

which upon substitution into *(1.18)* gives each partial volume as

$$V(\text{He}) = \frac{(18.8\text{ mol})(0.083\,14\text{ L}\cdot\text{bar}\cdot\text{K}^{-1}\cdot\text{mol}^{-1})(298\text{ K})}{1.01\text{ bar}} = 461\text{ L}$$
$$V(\text{Ne}) = \frac{(3.71)(0.083\,14)(298)}{1.01} = 91.0\text{ L}$$

The total volume of the gaseous solution is given by *Amagat's law of partial volumes* as

$$V_t = \sum V_i = 461\text{ L} + 91.0\text{ L} = 552\text{ L} \qquad (1.24)$$

1.67 What is the mole fraction of He in the mixture described in Problem 1.66?

▮ The mole fraction of a gas can be determined from the partial volume of the gas and the total volume of the mixture using Amagat's law as

$$x_i = V_i/V_t \qquad (1.25)$$
$$x(\text{He}) = (461\text{ L})/(552\text{ L}) = 0.835$$

1.68 A mixture of He and CO_2 has a volume of 63.5 mL at 1.00 bar and 28 °C. The system containing the mixture is cooled in liquid nitrogen, and the remaining gas is evacuated. The system is restored to 1.00 bar and 28 °C, and the volume is 40.5 mL. What was the composition of the original mixture?

▮ When the system was cooled in liquid nitrogen, the CO_2 condensed and the He is the gas that was evacuated. Thus $V(CO_2) = 40.5$ mL. Using *(1.25)* gives

$$x(CO_2) = (40.5\text{ mL})/(63.5\text{ mL}) = 0.638$$
$$x(\text{He}) = 1 - x(CO_2) = 1 - 0.638 = 0.362$$

1.69 Air contains 18.18 ppm by volume of Ne and 0.934% by volume of Ar. What is the ratio of the number of Ar atoms to the number of Ne atoms?

▮ The number of atoms for a given gas is proportional to the amount of that gas, which in terms, is proportional to the mole fraction of that gas. Recognizing that *(1.25)* shows a direct proportion between x_i and V_i, the ratio of the atoms is

$$\frac{N(\text{Ar})}{N(\text{Ne})} = \frac{V(\text{Ar})}{V(\text{Ne})} = \frac{9.34 \times 10^{-3}}{18.18 \times 10^{-6}} = 514$$

1.70 The composition of air inhaled by a human is 21% by volume O_2 and 0.03% CO_2, and that of exhaled air is 16% O_2 and 4.4% CO_2. Assuming a typical volume of 7200 L · day^{-1}, what mass of O_2 is used by the body and what mass of CO_2 is generated by the body each day? Assume ideal gas behavior for the air at 37 °C and 1.00 bar.

▮ The amounts of gases are given by *(1.18)* as

$$n(\text{O}_2, \text{in}) = \frac{(1.00\ \text{bar})(0.21)(7200\ \text{L})}{(0.083\ 14\ \text{L} \cdot \text{bar} \cdot \text{K}^{-1} \cdot \text{mol}^{-1})(310\ \text{K})} = 59\ \text{mol}$$

$$n(\text{O}_2, \text{out}) = \frac{(1.00)(0.16)(7200)}{(0.083\ 14)(310)} = 45\ \text{mol}$$

$$n(\text{CO}_2, \text{in}) = \frac{(1.00)(0.0003)(7200)}{(0.083\ 14)(310)} = 0.08\ \text{mol}$$

$$n(\text{CO}_2, \text{out}) = \frac{(1.00)(0.044)(7200)}{(0.083\ 14)(310)} = 12\ \text{mol}$$

The mass of O_2 used is

$$m(\text{O}_2) = [n(\text{O}_2, \text{in}) - n(\text{O}_2, \text{out})]M(\text{O}_2)$$
$$= (59\ \text{mol} - 45\ \text{mol})(32.0\ \text{g} \cdot \text{mol}^{-1}) = 450\ \text{g}$$

and the mass of CO_2 generated is

$$m(\text{CO}_2) = [n(\text{CO}_2, \text{out}) - n(\text{CO}_2, \text{in})]M(\text{CO}_2)$$
$$= (12\ \text{mol} - 0.08\ \text{mol})(44.0\ \text{g} \cdot \text{mol}^{-1}) = 530\ \text{g}$$

1.71 A 0.0200-mol sample of N_2O_4 is placed in a 0.499-L container at 25 °C. The pressure of the sample is 1.192 bar. Find the fraction of N_2O_4 that undergoes dissociation to form NO_2.

▮ The total amount of gas present after the equilibrium has been established is given by *(1.18)* as

$$n_t = \frac{(1.192\ \text{bar})(0.499\ \text{L})}{(0.083\ 14\ \text{L} \cdot \text{bar} \cdot \text{K}^{-1} \cdot \text{mol}^{-1})(298\ \text{K})} = 0.0240\ \text{mol}$$

Letting x represent the amount of N_2O_4 that dissociates, we can see that the equilibrium amounts are

equation	$N_2O_4(g) \rightleftarrows$	$2NO_2(g)$
$n_{\text{initial}}/(\text{mol})$	0.0200	
$n_{\text{change}}/(\text{mol})$	$-x$	$+2x$
$n_{\text{equilibrium}}/(\text{mol})$	$0.0200 - x$	$2x$

giving

$$n_t = (0.0200 - x)\ \text{mol} + 2x\ \text{mol} = (0.0200 + x)\ \text{mol} = 0.0240\ \text{mol}$$

Solving for x gives $x = 0.0040$ mol. The fraction of N_2O_4 that has dissociated is

$$\alpha = (0.0040\ \text{mol})/(0.0200\ \text{mol}) = 0.20$$

1.72 The density of air is 3.8×10^{-12} kg · m^{-3} at an altitude of 300 km above sea level. The temperature is 960 K, and the pressure is 1.6×10^{-8} mbar. Assuming ideal gas behavior, calculate the average molar mass of air at this altitude. If the air consists of N_2 molecules and O atoms, find $x(N_2)$.

▮ Solving *(1.20)* for the average molar mass, $\bar{M}$, and substituting the data gives

$$\bar{M} = \frac{(3.8 \times 10^{-12}\ \mathrm{kg \cdot m^{-3}})(8.314\ \mathrm{m^3 \cdot Pa \cdot K^{-1} \cdot mol^{-1}})(960\ \mathrm{K})}{(1.6 \times 10^{-8}\ \mathrm{mbar})[(10^{-3}\ \mathrm{bar})/(1\ \mathrm{mbar})][(10^5\ \mathrm{Pa})/(1\ \mathrm{bar})]}$$
$$= 0.019\ \mathrm{kg \cdot mol^{-1}} = 19\ \mathrm{g \cdot mol^{-1}}$$

The average molar mass of a gaseous mixture is given by

$$\bar{M} = \sum x_i M_i \tag{1.26}$$

Applying *(1.26)* gives

$$19\ \mathrm{g \cdot mol^{-1}} = [x(\mathrm{N_2})](28.0\ \mathrm{g \cdot mol^{-1}}) + [x(\mathrm{O})](16.0\ \mathrm{g \cdot mol^{-1}})$$

and recognizing that $x(\mathrm{N_2}) + x(\mathrm{O}) = 1$ gives $x(\mathrm{N_2}) = 0.25$.

1.73 Assume the composition of dry air to be $x(\mathrm{N_2}) = 0.7808$, $x(\mathrm{O_2}) = 0.2095$, $x(\mathrm{Ar}) = 0.0093$, and $x(\mathrm{CO_2}) = 0.0003$. Calculate the average molar mass and density of dry air at 25.0 °C and 1.000 bar.

▮ For dry air, *(1.26)* gives

$$\bar{M} = (0.7808)(28.01\ \mathrm{g \cdot mol^{-1}}) + (0.2095)(32.00\ \mathrm{g \cdot mol^{-1}})$$
$$+ (0.0093)(39.95\ \mathrm{g \cdot mol^{-1}}) + (0.0003)(44.01\ \mathrm{g \cdot mol^{-1}})$$
$$= 28.96\ \mathrm{g \cdot mol^{-1}}$$

Using *(1.20)* gives

$$\rho = \frac{(1.000\ \mathrm{bar})(28.96\ \mathrm{g \cdot mol^{-1}})}{(0.083\,14\ \mathrm{L \cdot bar \cdot K^{-1} \cdot mol^{-1}})(298.2\ \mathrm{K})} = 1.168\ \mathrm{g \cdot L^{-1}}$$

1.74 The vapor pressure of water is 0.0316 bar at 25 °C. Calculate the average molar mass and density of water-saturated air at 25.0 °C and 1.000 bar. See Problem 1.73 for additional information.

▮ The respective mole fractions of the gases in wet air are given by *(1.22)* as

$$x(\mathrm{N_2}) = \frac{P(\mathrm{N_2})}{P_t} = \frac{x(\mathrm{N_2, dry})P_t(\mathrm{dry})}{P_t}$$
$$= (0.7808)(1.000\ \mathrm{bar} - 0.0316\ \mathrm{bar})/(1.000\ \mathrm{bar}) = 0.7561$$
$$x(\mathrm{O_2}) = (0.2095)(1.000 - 0.0316)/1.000 = 0.2029$$
$$x(\mathrm{Ar}) = (0.0093)(1.000 - 0.0316)/1.000 = 0.0090$$
$$x(\mathrm{CO_2}) = (0.0003)(1.000 - 0.0316)/1.000 = 0.0003$$
$$x(\mathrm{H_2O}) = 0.0316/1.000 = 0.0316$$

Applying *(1.26)* gives

$$\bar{M} = (0.7561)(28.01\ \mathrm{g \cdot mol^{-1}}) + (0.2029)(32.00\ \mathrm{g \cdot mol^{-1}}) + (0.0090)(39.95\ \mathrm{g \cdot mol^{-1}})$$
$$+ (0.0003)(44.01\ \mathrm{g \cdot mol^{-1}}) + (0.0317)(18.01\ \mathrm{g \cdot mol^{-1}})$$
$$= 28.61\ \mathrm{g \cdot mol^{-1}}$$

Using *(1.20)* gives

$$\rho = \frac{(1.000\ \mathrm{bar})(28.61\ \mathrm{g \cdot mol^{-1}})}{(0.083\,14\ \mathrm{L \cdot bar \cdot K^{-1} \cdot mol^{-1}})(298.2\ \mathrm{K})} = 1.154\ \mathrm{g \cdot L^{-1}}$$

1.75 The average molar mass of a mixture of C_2H_4 and C_2H_6 is 28.31 g · mol^{-1}. What is the composition of this mixture?

▮ In terms of the unknown mole fractions, *(1.26)* gives

$$28.31\ \mathrm{g \cdot mol^{-1}} = [x(\mathrm{C_2H_4})](28.05\ \mathrm{g \cdot mol^{-1}}) + [x(\mathrm{C_2H_6})](30.07\ \mathrm{g \cdot mol^{-1}})$$

Recognizing that $x(\mathrm{C_2H_4}) + x(\mathrm{C_2H_6}) = 1$ gives $x(\mathrm{C_2H_4}) = 0.871$ and $x(\mathrm{C_2H_6}) = 0.129$.

1.76 The average molar mass of the vapor above solid NH_4Cl is nearly 26.5 g · mol^{-1}. What is the composition of this vapor?

❚ The molar mass of $NH_4Cl(s)$ is nearly twice that of the vapor. If we assume that the reaction $NH_4Cl(s) \rightarrow NH_3(g) + HCl(g)$ takes place, then $x(NH_3) = x(HCl) = 0.500$. Substitution into *(1.26)* gives

$$\bar{M} = (0.500)(17.0\ \text{g}\cdot\text{mol}^{-1}) + (0.500)(36.5\ \text{g}\cdot\text{mol}^{-1}) = 28.8\ \text{g}\cdot\text{mol}^{-1}$$

Thus the vapor must consist of HCl(g) and $NH_3(g)$.

1.77 The average molar mass of the vapor above liquid acetic acid, CH_3COOH, is $101\ \text{g}\cdot\text{mol}^{-1}$ at 1.01 bar and 118.5 °C. Assuming that a dimerization reaction occurs, determine the partial pressure of the monomer and of the dimer in the vapor, and calculate the equilibrium constant defined as

$$K_P = \frac{[P(\text{monomer})/(\text{bar})]^2}{P(\text{dimer})/(\text{bar})}$$

❚ For the vapor, *(1.26)* gives

$$101\ \text{g}\cdot\text{mol}^{-1} = [x(\text{monomer})](60.05\ \text{g}\cdot\text{mol}^{-1}) + [x(\text{dimer})](120.10\ \text{g}\cdot\text{mol}^{-1})$$

Recognizing that $x(\text{monomer}) + x(\text{dimer}) = 1$ gives $x(\text{monomer}) = 0.32$ and $x(\text{dimer}) = 0.68$. The respective partial pressures are given by *(1.23)* as

$$P(\text{dimer}) = (0.68)(1.01\ \text{bar}) = 0.69\ \text{bar}$$
$$P(\text{monomer}) = (0.32)(1.01\ \text{bar}) = 0.32\ \text{bar}$$

giving

$$K_P = (0.32)^2/0.69 = 0.15$$

1.4 REAL GASES

1.78 Estimate the critical temperature and volume for O_2 given the following data:

$T/(°C)$	−154.5	−140.2	−129.9	−123.3	−120.4
$\rho(l)/(g\cdot cm^{-3})$	0.9758	0.8742	0.7781	0.6779	0.6032
$\rho(g)/(g\cdot cm^{-3})$	0.0385	0.0805	0.1320	0.2022	0.2701

❚ The *rectilinear diameter law of Cailletet and Mathias* states that plots of the orthobaric densities of the liquid and gas against temperature will intersect at the critical temperature. To help in locating this point graphically, a plot of the average of the densities is also included as predicted by the mathematical statement of the rectilinear law

$$\frac{\rho(l) + \rho(g)}{2} = A + B[T/(°C)] + C[T/(°C)]^2 + \cdots \qquad (1.27)$$

where A, B, and C are constants. The three plots are shown in Fig. 1-5. The point of intersection is $T_c =$

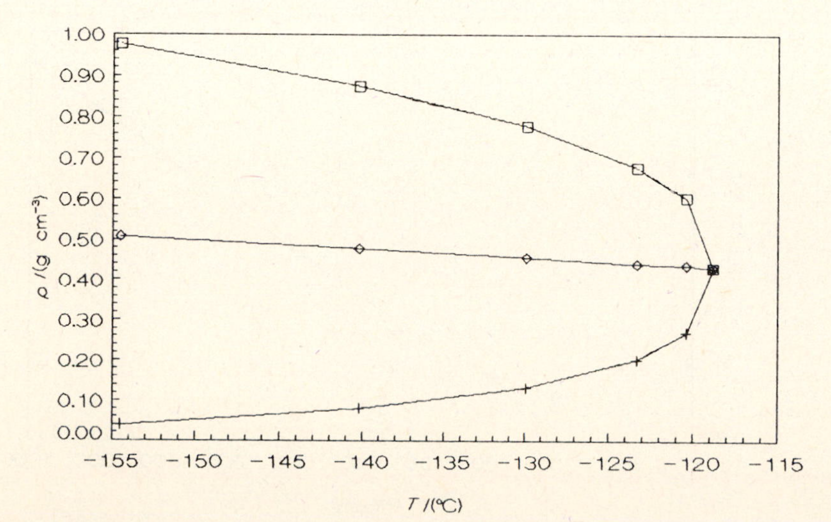

Fig. 1-5

$-118.8\ °C$ at $\rho_c = 0.430\ g \cdot cm^{-3}$. The critical volume is

$$V_c = \frac{M}{\rho_c} = \frac{32.00\ g \cdot mol^{-1}}{0.430\ g \cdot cm^{-3}} = 74.4\ cm^3 \cdot mol^{-1}$$

1.79 The orthobaric densities for liquid and gaseous N_2 can be given by

$$\rho(l)/(g \cdot cm^{-3}) = -88.17 - (1.5989)[T(°C)] - (9.631 \times 10^{-3})[T/(°C)]^2 - (1.941 \times 10^{-5})[T/(°C)]^3$$

$$\rho(g)/(g \cdot cm^{-3}) = 86.78 + (1.5666)[T(°C)] + (9.447 \times 10^{-3})[T/(°C)]^2 + (1.902 \times 10^{-5})[T(°C)]^3$$

Find the critical temperature.

▌ Solving the two equations for the temperature at which $\rho(l) = \rho(g)$ gives $T_c = -143.5\ °C$.

1.80 The *compressibility factor*, Z, for a real gas is given by

$$Z = PV/nRT \tag{1.28}$$

and is a measure of the ideal behavior of the gas. Use the following data reported by Amagat to prepare a plot of Z versus P for O_2 at 0 °C:

P/(atm)	1	100	200	300	400	500
$V_m/(L \cdot mol^{-1})$	22.4138	0.2077	0.1024	0.0719	0.0589	0.0518

P/(atm)	600	700	800	900	1000
$V_m/(L \cdot mol^{-1})$	0.0474	0.0444	0.0421	0.0403	0.0389

where V_m is the molar volume.

▌ As a sample calculation, consider the data given for $P = 300$ atm:

$$Z = \frac{PV}{nRT} = \frac{(300\ atm)(0.0719\ L)}{(1.00\ mol)(0.0821\ L \cdot atm \cdot K^{-1} \cdot mol^{-1})(273\ K)} = 0.962$$

The plot appears in Fig. 1-6.

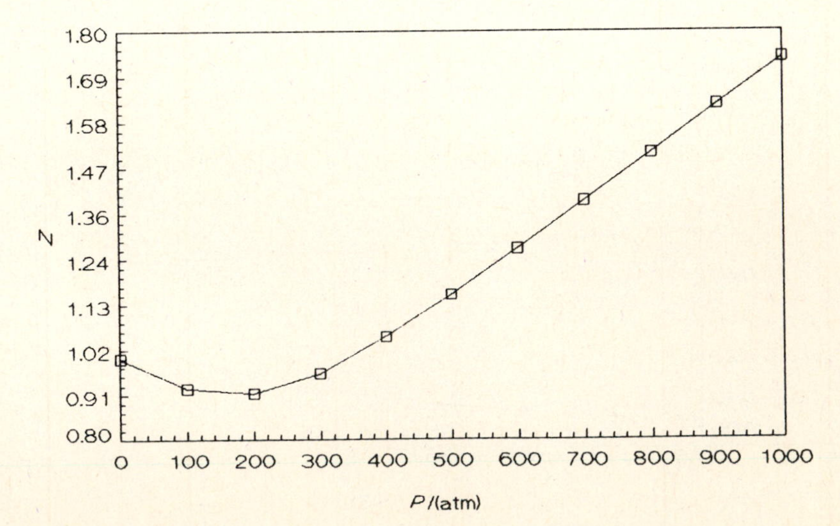

Fig. 1-6

1.81 The density of steam at 100.0 °C and 1.013 25 bar is 0.5974 kg · m^{-3}. Calculate the compressibility factor from these data.

▮ Substituting $m = nM$ and $\rho = m/V$ into *(1.28)* gives

$$Z = \frac{PV}{nRT} = \frac{PV}{(m/M)/RT} = \frac{PM}{(m/V)RT} = \frac{PM}{\rho RT} \qquad (1.29)$$

The value of Z is

$$Z = \frac{(1.013\,25\ \text{bar})(18.0152\ \text{g}\cdot\text{mol}^{-1})[(10^{-3}\ \text{kg})/(1\ \text{g})][(10^5\ \text{Pa})/(1\ \text{bar})]}{(0.5975\ \text{kg}\cdot\text{m}^{-3})(8.314\ \text{m}^3\cdot\text{Pa}\cdot\text{K}^{-1}\cdot\text{mol}^{-1})(373.2\ \text{K})}$$

$$= 0.9848$$

1.82 A commercial cylinder contains 6.91 m^3 of O_2 at 15.18 MPa and 21 °C. The critical constants for O_2 are $T_c = -118.4$ °C and $P_c = 50.1$ atm. Determine the reduced temperature and pressure for O_2 under these conditions.

▮ The *reduced temperature* is defined as

$$T_r = T/T_c \qquad (1.30)$$

and the *reduced pressure* is defined as

$$P_r = P/P_c \qquad (1.31)$$

Under the given conditions, the reduced temperature and pressure are

$$T_r = \frac{(21 + 273)\ \text{K}}{(-118.4 + 273)\ \text{K}} = 1.90$$

$$P_r = \frac{(15.18\ \text{MPa})[(10^6\ \text{Pa})/(1\ \text{MPa})]}{(50.1\ \text{atm})[(101\,325\ \text{Pa})/(1\ \text{atm})]} = 2.99$$

1.83 The *law of corresponding states* implies that all real gases act similarly under the same reduced conditions. Thus a plot such as that in Fig. 1-7 can be used to determine the compressibility factor for any gas. Using the results of Problem 1.82, determine Z for O_2 under the given conditions and calculate the amount of O_2 in the gas cylinder.

▮ Using $T_r = 1.90$ and $P_r = 2.99$, the value of Z read from the plot is 0.94. Solving *(1.28)* for n and substituting the data give

$$n = \frac{(15.18\ \text{MPa})[(10^6\ \text{Pa})/(1\ \text{MPa})](6.91\ \text{m}^3)}{(0.94)(8.314\ \text{m}^3\cdot\text{Pa}\cdot\text{K}^{-1}\cdot\text{mol}^{-1})(294\ \text{K})} = 4.57 \times 10^4\ \text{mol}$$

1.84 Use *(1.28)* to determine the pressure of a molar sample of CO_2 in a 0.0949-L container at 90 °C. The critical constants for CO_2 are $T_c = 31$ °C and $P_c = 72.9$ atm.

▮ To obtain a value of Z from Fig. 1-7, the value of P_r must be known. In this case, it is the pressure that is unknown, so P_r cannot be calculated from the data. An iterative process will be used to obtain the value of P. Starting with $Z = 1.00$, *(1.28)* gives

$$P = \frac{(1.00)(1.00\ \text{mol})(0.0821\ \text{L}\cdot\text{atm}\cdot\text{K}^{-1}\cdot\text{mol}^{-1})(363\ \text{K})}{0.0949\ \text{L}} = 314\ \text{atm}$$

Using this value of pressure, *(1.30)* and *(1.31)* give

$$T_r = \frac{363\ \text{K}}{304\ \text{K}} = 1.19 \qquad P_r = \frac{314\ \text{atm}}{72.9\ \text{atm}} = 4.31$$

which corresponds to a value of $Z = 0.64$. Repeating the process using this value of Z gives

$$P = \frac{(0.64)(1.00)(0.0821)(363)}{0.0949} = 201\ \text{atm} \qquad P_r = \frac{201}{72.9} = 2.76$$

The corresponding value of Z is 0.56. Using $Z = 0.56$ gives $P = 176$ atm, $P_r = 2.41$, and $Z = 0.56$. Because Z has not changed in this last step, the answer is $P = 176$ atm. (The observed pressure is 175 atm.)

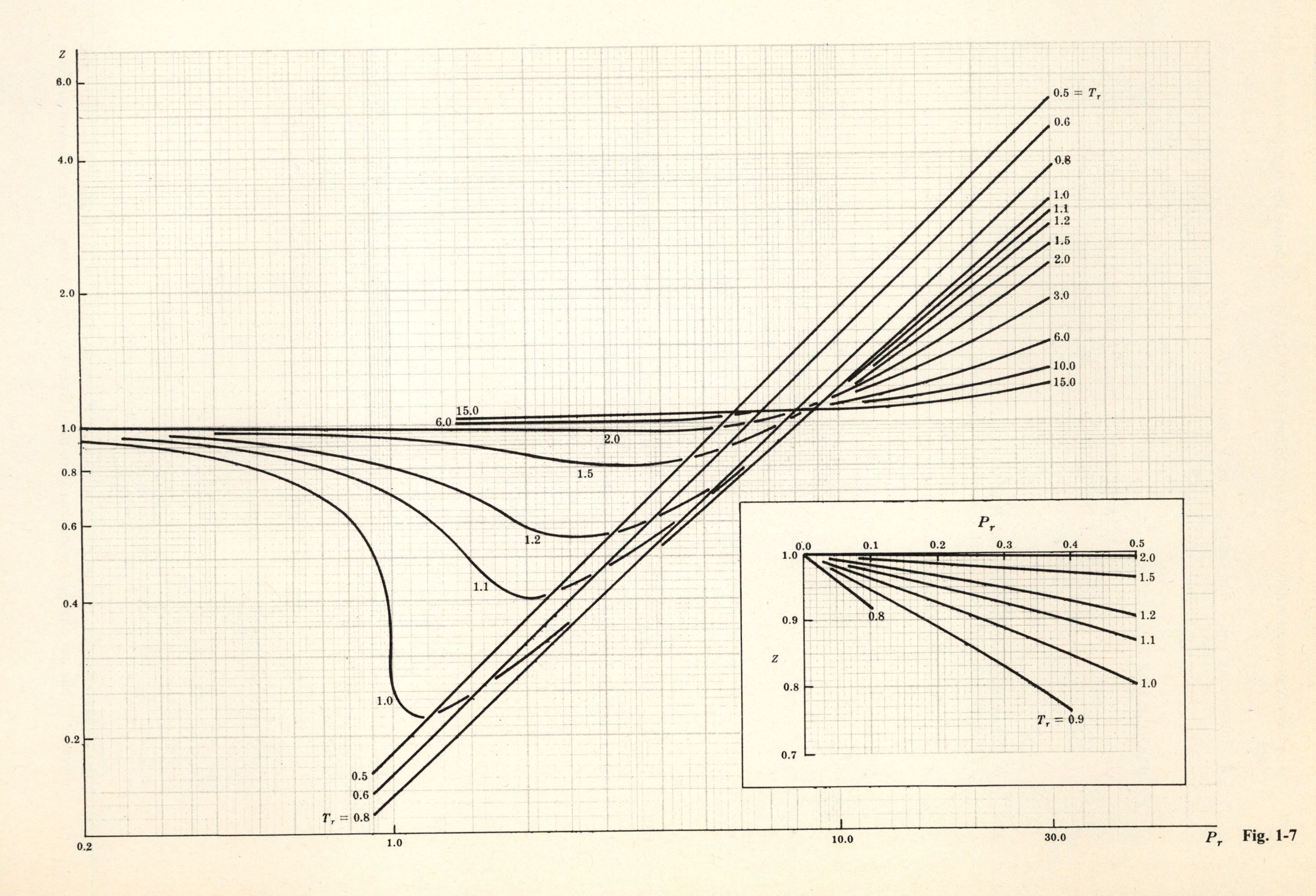
Z
6.0
4.0
2.0
1.0
0.8
0.6
0.4
0.2
0.2
1.0
10.0
30.0
P_r
0.5 = T_r
0.6
0.8
1.0
1.1
1.2
1.5
2.0
3.0
6.0
10.0
15.0
15.0
6.0
2.0
1.5
1.2
1.1
1.0
0.5
0.6
T_r = 0.8
P_r
0.0
0.1
0.2
0.3
0.4
0.5
1.0
0.9
Z
0.8
0.7
2.0
1.5
1.2
1.1
1.0
0.8
T_r = 0.9

Fig. 1-7

1.85 Using the molar P-V data given in Problem 1.80, determine B_p and C_p in the *pressure virial equation* for O_2 at 0 °C

$$PV_m = RT + B_pP + C_pP^2 + D_pP^3 + \cdots \tag{1.32}$$

▌ The values of B_p and C_p can be determined graphically. If *(1.32)* is written as

$$\frac{PV_m - RT}{P} = B_p + C_pP + \cdots$$

we can see that a plot of $(PV_m - RT)/P$ against P will give B_p as the intercept and C_p as the limiting slope of the linear portion. An analysis of the plot shown in Fig. 1-8 gives $B_p = -2.331 \times 10^{-2}\ \mathrm{L \cdot mol^{-1}}$ and $C_p = 6.836 \times 10^{-5}\ \mathrm{L \cdot atm^{-1} \cdot mol^{-1}}$. (A least squares analysis of the data yields the values given in Table 1-3, which are more accurate.)

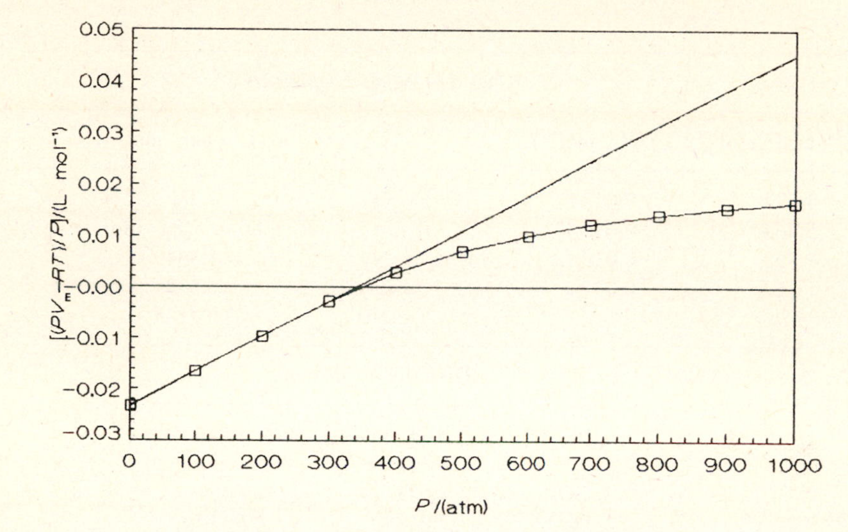

Fig. 1-8

1.86 Using the molar P-V data given in Problem 1.80, determine B_v and C_v in the *volume virial equation* for O_2 at 0 °C

$$PV_m = RT + B_v\frac{1}{V_m} + C_v\left(\frac{1}{V_m}\right)^2 + D_v\left(\frac{1}{V_m}\right)^3 + \cdots \tag{1.33}$$

▌ The values of the virial coefficients can be determined graphically. If *(1.33)* is written as

$$V_m(PV_m - RT) = B_v + C_v\frac{1}{V_m} + \cdots$$

we can see that a plot of $V_m(PV_m - RT)$ against $1/V_m$ will give B_v as the intercept and C_v as the limiting slope of the linear portion. An analysis of the plot shown in Fig. 1-9 gives $B_v = -0.4937\ \mathrm{L^2 \cdot atm \cdot mol^{-2}}$ and

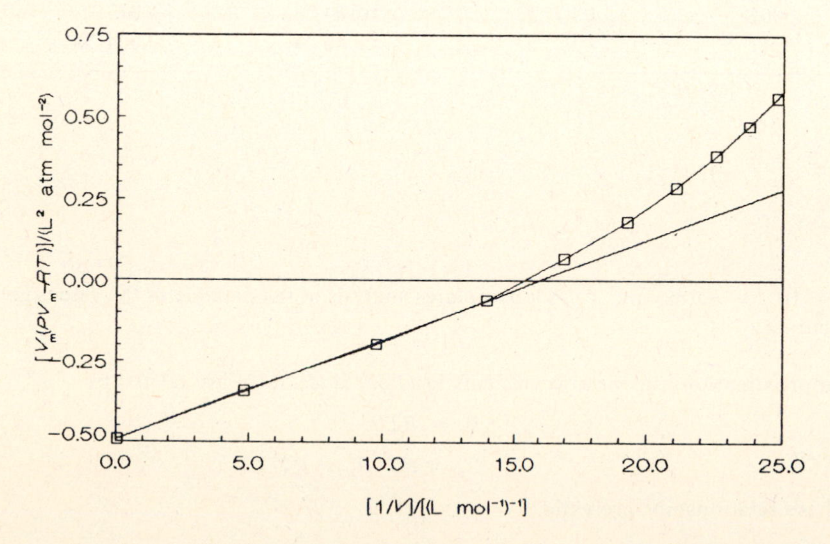

Fig. 1-9

Table 1-3

Virial coefficients

	$B_p/(L \cdot mol^{-1})$	$C_p/(L \cdot atm^{-1} \cdot mol^{-1})$	$D_p/(L \cdot atm^{-2} \cdot mol^{-1})$	$E_p/(L \cdot atm^{-3} \cdot mol^{-1})$
O_2(0 °C)	-2.777×10^{-2}	1.1073×10^{-4}	-9.906×10^{-8}	3.257×10^{-11}
CO_2(90 °C)	-7.57×10^{-2}	-1.1×10^{-5}		

	$B_v/(L^2 \cdot atm \cdot mol^{-2})$	$C_v/(L^3 \cdot atm \cdot mol^{-3})$	$D_v/(L^4 \cdot atm \cdot mol^{-4})$	$E_v/(L^5 \cdot atm \cdot mol^{-5})$
O_2(0 °C)	−0.5168	4.166×10^{-2}	-1.533×10^{-3}	6.467×10^{-5}
CO_2(90 °C)	−2.26	0.161		

van der Waals coefficients

	$a/(L^2 \cdot atm \cdot mol^{-2})$	$b/(L \cdot mol^{-1})$
CO_2	3.592	0.042 67
H_2O	5.464	0.030 49
N_2	1.390	0.039 13
O_2	1.362	0.031 9

	$a/(L^2 \cdot bar \cdot mol^{-1})$	$b/(L \cdot mol^{-1})$
O_2	1.380	0.031 9
	($0.1380\ m^3 \cdot Pa \cdot mol^{-2}$)	($3.19 \times 10^{-5}\ m^3 \cdot mol^{-1}$)
N_2	1.364	0.039 13
Cl_2	6.579	0.056 22
Ne	0.2135	0.017 09

Dieterici coefficients

	$a/(L^2 \cdot atm \cdot mol^{-2})$	$b/(L \cdot mol^{-1})$
CO_2	4.621	0.0463
H_2O	7.011	0.0330
O_2	1.748	0.0345

Redlich–Kwong coefficients

	$a/(L^2 \cdot K^{1/2} \cdot atm \cdot mol^{-2})$	$b/(L \cdot mol^{-1})$
O_2	17.16	0.0221

Berthelot coefficients

	$a/(L^2 \cdot K \cdot atm \cdot mol^{-2})$	$b/(L \cdot mol^{-1})$
O_2	211	3.17×10^{-2}

Beattie–Bridgeman coefficients

	$A_0/(L^2 \cdot atm \cdot mol^{-2})$	$a/(L \cdot mol^{-1})$	$B_0/(L \cdot mol^{-1})$	$b/(L \cdot mol^{-1})$	$c/(L \cdot K^3 \cdot mol^{-1})$
CO_2	5.0065	0.071 32	0.104 76	0.072 35	66.00×10^4
O_2	1.4911	0.025 62	0.046 24	0.004 208	4.80×10^4

$C_v = 3.093 \times 10^{-2}\ L^3 \cdot atm \cdot mol^{-3}$. (A least squares analysis of the data yields the values given in Table 1-3, which are more accurate.)

1.87 To a good approximation, the virial coefficients in *(1.32)* and *(1.33)* are related by

$$B_v = RTB_p \tag{1.34}$$

$$C_v = RT(B_p^2 + RTC_p) \tag{1.35}$$

Show that these relationships are valid.

▮ Substituting the values of B_p and C_p found in Table 1-3 gives

$$B_v = (0.082\,06\ \text{L}\cdot\text{atm}\cdot\text{K}^{-1}\cdot\text{mol}^{-1})(273.15\ \text{K})(-2.777\times10^{-2}\ \text{L}\cdot\text{mol}^{-1})$$
$$= -0.6213\ \text{L}^2\cdot\text{atm}\cdot\text{mol}^{-2}$$
$$C_v = (0.082\,06\ \text{L}\cdot\text{atm}\cdot\text{K}^{-1}\cdot\text{mol}^{-1})(273.15\ \text{K})[(-2.777\times10^{-2}\ \text{L}\cdot\text{mol}^{-1})^2$$
$$+ (0.082\,06\ \text{L}\cdot\text{atm}\cdot\text{K}^{-1}\cdot\text{mol}^{-1})(273.15\ \text{K})(1.1073\times10^{-4}\ \text{L}\cdot\text{atm}^{-1}\cdot\text{mol}^{-1})]$$
$$= 7.291\times10^{-2}\ \text{L}^3\cdot\text{atm}\cdot\text{mol}^{-3}$$

The calculated values of B_v and C_v are close to those given in Table 1-3.

1.88 Calculate the pressure of a molar sample of CO_2 confined to 0.0949 L at 90 °C.

▮ Solving *(1.33)* for P and substituting the data from Table 1-3 gives

$$P = RT/V_m + B_v/V_m^2 + C_v/V_m^3$$
$$= \frac{(0.0821\ \text{L}\cdot\text{atm}\cdot\text{K}^{-1}\cdot\text{mol}^{-1})(363\ \text{K})}{0.0949\ \text{L}\cdot\text{mol}^{-1}} + \frac{-2.26\ \text{L}^2\cdot\text{atm}\cdot\text{mol}^{-2}}{(0.0949\ \text{L}\cdot\text{mol}^{-1})^2} + \frac{0.161\ \text{L}^3\cdot\text{atm}\cdot\text{mol}^{-3}}{(0.0949\ \text{L}\cdot\text{mol}^{-1})^3}$$
$$= 251\ \text{atm}$$

This value is 43% greater than the observed value of 175 atm.

1.89 Calculate the volume of a molar sample of CO_2 at 175 atm and 90 °C.

▮ Solving *(1.32)* for V_m and substituting the data from Table 1-3 gives

$$V_m = RT/P + B_p + C_pP$$
$$= \frac{(0.0821\ \text{L}\cdot\text{atm}\cdot\text{K}^{-1}\cdot\text{mol}^{-1})(363\ \text{K})}{175\ \text{atm}} + (-7.57\times10^{-2}\ \text{L}\cdot\text{mol}^{-1}) + (-1.1\times10^{-5}\ \text{L}\cdot\text{atm}^{-1}\cdot\text{mol}^{-1})(175\ \text{atm})$$
$$= 0.092\ \text{L}\cdot\text{mol}^{-1}$$

This value agrees well with the observed value of $0.0949\ \text{L}\cdot\text{mol}^{-1}$.

1.90 Determine the compressibility factor for O_2 at 300 atm and 0 °C. Compare this value to that determined from Fig. 1-7. The critical constants for O_2 are $T_c = -118.4$ °C and $P_c = 50.1$ atm.

▮ Combining *(1.28)* and *(1.32)* gives

$$Z = 1 + \frac{B_p}{RT}P + \frac{C_p}{RT}P^2 + \cdots \qquad (1.36)$$
$$= 1 + \frac{(-2.777\times10^{-2}\ \text{L}\cdot\text{mol}^{-1})(300\ \text{atm})}{(0.0821\ \text{L}\cdot\text{atm}\cdot\text{K}^{-1}\cdot\text{mol}^{-1})(273\ \text{K})}$$
$$+ \frac{(1.1073\times10^{-4}\ \text{L}\cdot\text{atm}^{-1}\cdot\text{mol}^{-1})(300\ \text{atm})^2}{(0.0821\ \text{L}\cdot\text{atm}\cdot\text{K}^{-1}\cdot\text{mol}^{-1})(273\ \text{K})}$$
$$+ \frac{(-9.906\times10^{-8}\ \text{L}\cdot\text{atm}^{-2}\cdot\text{mol}^{-1})(300\ \text{atm})^3}{(0.0821\ \text{L}\cdot\text{atm}\cdot\text{K}^{-1}\cdot\text{mol}^{-1})(273\ \text{K})}$$
$$= 0.96$$

The reduced variables are found using *(1.30)* and *(1.31)*:

$$T_r = \frac{273\ \text{K}}{(-118.4 + 273.2)\ \text{K}} = 1.76 \qquad P_r = \frac{300\ \text{atm}}{50.1\ \text{atm}} = 5.99$$

giving $Z = 0.95$. Both values agree well with $Z = 0.962$ determined in Problem 1.80.

1.91 The compressibility factor for O_2 at low pressures and 0 °C is a linear function as shown by the following data:

P/(atm)	1.00	2.00	3.00	4.00	5.00
Z	1.0000	0.9990	0.9980	0.9971	0.9961

Determine B_p for O_2 under these conditions.

▮ The linear form of *(1.36)* at low pressures is

$$Z = 1 + (B_p/RT)P$$

Thus a plot of Z against P will have a slope equal to B_p/RT. From the plot shown in Fig. 1-10, the slope is $-9.7 \times 10^{-4}\ \text{atm}^{-1}$, giving

$$B_p = (0.0821\ \text{L} \cdot \text{atm} \cdot \text{K}^{-1} \cdot \text{mol}^{-1})(273\ \text{K})(-9.7 \times 10^{-4}\ \text{atm}^{-1})$$
$$= -2.17 \times 10^{-2}\ \text{L} \cdot \text{mol}^{-1}$$

This value agrees well with that given in Table 1-3.

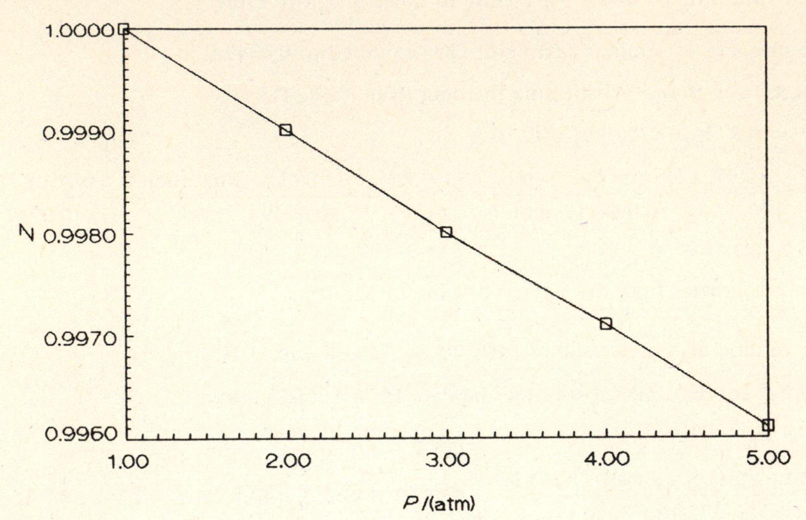

Fig. 1-10

1.92 At the *Boyle temperature* for a real gas, the plot of Z against P is tangent to the ideal gas line at $P = 0$ and only gradually increases above $Z = 1$. Derive an expression for the Boyle temperature using *(1.36)*, and determine the value for O_2 using the following data:

$T/(\text{K})$	350	400	450	500	550
$(B_v/RT)/(\text{cm}^3 \cdot \text{mol}^{-1})$	−7.27	−1.85	1.87	4.65	7.23

▮ Neglecting the higher pressure terms in *(1.36)* and differentiating with respect to P gives

$$\frac{\partial Z}{\partial P} = \frac{B_p}{RT} = \frac{B_v/RT}{RT}$$

Setting $\partial Z/\partial P = 0$ gives $B_v/RT = 0$. Thus a plot of B_v/RT versus T will give the Boyle temperature where $B_v/RT = 0$. From Fig. 1-11, $T_{\text{Boyle}} = 424\ \text{K}$.

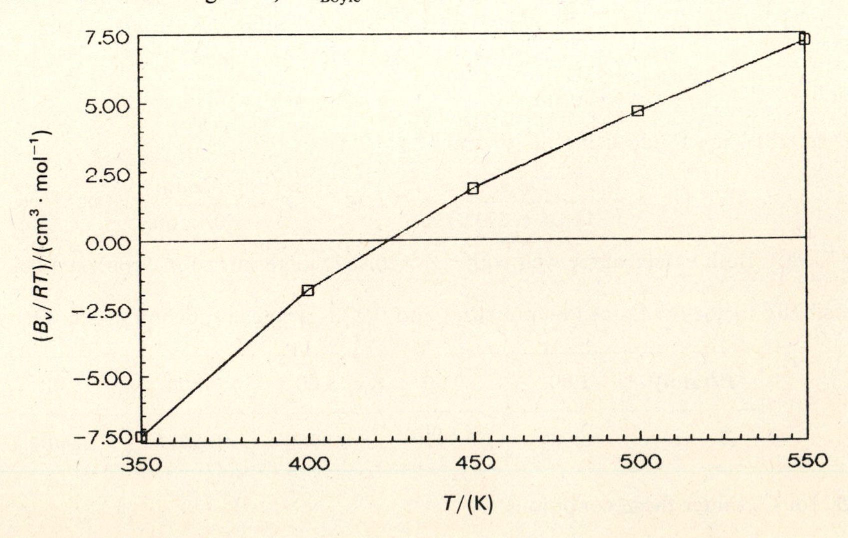

Fig. 1-11

1.93 Assume that the equation

$$PV_m = RT(1 + aP) \tag{1.37}$$

can be used to describe a real gas at low pressures where a is a constant. Derive the expression for the cubic expansion coefficient. Compare the result to that for an ideal gas.

▮ Solving *(1.37)* for V_m and differentiating with respect to T gives

$$\left(\frac{\partial V_m}{\partial T}\right)_P = \left[\frac{\partial}{\partial T}\left(\frac{RT(1 + aP)}{P}\right)\right]_P = \frac{R(1 + aP)}{P} = \frac{V_m}{T}$$

Substituting into *(1.11)* gives

$$\alpha = \frac{1}{V_m}\left(\frac{\partial V_m}{\partial T}\right)_P = \frac{1}{V_m}\left(\frac{V_m}{T}\right) = \frac{1}{T}$$

This is the same result as obtained for an ideal gas (see Problem 1.28).

1.94 Using *(1.37)*, derive the expression for the isothermal compressibility. Compare the result to that for an ideal gas.

▮ Solving *(1.37)* for V_m and differentiating with respect to P gives

$$\left(\frac{\partial V_m}{\partial P}\right)_T = \left(\frac{\partial(RT)(1/P + a)}{\partial P}\right)_T = \frac{-RT}{P^2} = \frac{-PV_m}{1 + aP}\left(\frac{1}{P^2}\right) = \frac{-V_m}{P(1 + aP)}$$

Substituting into *(1.8)* gives

$$\kappa = -\frac{1}{V_m}\left(\frac{\partial V_m}{\partial P}\right)_T = -\frac{1}{V_m}\left(\frac{-V_m}{P(1 + aP)}\right) = \frac{1}{P(1 + aP)}$$

The denominator contains a $(1 + aP)$ term that does not appear in the result for an ideal gas (see Problem 1.23).

1.95 The second virial coefficient is related to the strength of the intermolecular forces within a substance by

$$B_v = 2\pi LRT \int_0^\infty (1 - e^{-U/kT})r^2\,dr \tag{1.38}$$

where k is Boltzmann's constant, r is the distance between the molecules, and U is the potential energy function describing the strength of the forces. Assuming a *hard-sphere model* given by

$$U = \begin{cases} \infty & \text{for } r \le \sigma \\ 0 & \text{for } r > \sigma \end{cases}$$

where σ is the distance of closest approach between the molecules, derive the expression for B_v. Determine B_v for O_2 at 0 °C assuming $\sigma = 0.35$ nm.

▮ The integral in *(1.38)* can be split into two parts:

$$B_v = 2\pi LRT\left(\int_0^\sigma (1 - e^{-U/kT})r^2\,dr + \int_\sigma^\infty (1 - e^{-U/kT})r^2\,dr\right)$$

$$= 2\pi LRT\left(\int_0^\sigma r^2\,dr + 0\right) = 2\pi LRT\frac{\sigma^3}{3}$$

This result appears in many calculations and is given the symbol b_0.

$$b_0 = 2\pi LRT(\sigma^3/3) \tag{1.39}$$

For O_2,

$$B_v = b_0 = 2\pi(6.022 \times 10^{23}\ \text{mol}^{-1})(0.0821\ \text{L}\cdot\text{atm}\cdot\text{K}^{-1}\cdot\text{mol}^{-1})(273\ \text{K})$$

$$\times\left[\frac{(0.35\ \text{nm})^3[(10^{-9}\ \text{m})/(1\ \text{nm})]^3[(10^3\ \text{L})/(1\ \text{m}^3)]}{3}\right]$$

$$= 1.21\ \text{L}^2\cdot\text{atm}\cdot\text{mol}^{-2}$$

This rather poor model gives a value of B_v that is considerably in error compared to the value given in Table 1-3.

1.96 Calculations using *(1.38)* are improved considerably if a *square-well model* given by

$$U = \begin{cases} \infty & \text{for } r < \sigma \\ -\varepsilon & \text{for } \sigma \le r \le R\sigma \\ 0 & \text{for } r > R\sigma \end{cases}$$

where R and ε are constants is used to describe the molecular attractions. For this model, the expression for B_v becomes

$$B_v = b_0[1 - (R^3 - 1)(e^{\varepsilon/kT} - 1)]$$

Assuming $R = 1.9$, $\sigma = 0.35$ nm, and $\varepsilon = 8 \times 10^{-22}$ J, determine B_v for O_2 at 0 °C.

▮ Substituting the values gives

$$B_v = (1.21\ \text{L}^2 \cdot \text{atm} \cdot \text{mol}^{-2})\left\{1 - [(1.9)^3 - 1]\left[\exp\left(\frac{8 \times 10^{-22}\ \text{J}}{(1.38 \times 10^{-23}\ \text{J} \cdot \text{K}^{-1})(273\ \text{K})}\right) - 1\right]\right\}$$

$$= -0.47\ \text{L}^2 \cdot \text{atm} \cdot \text{mol}^{-2}$$

which agrees fairly well with the value given in Table 1-3.

1.97 A common potential energy function used to describe the intermolecular forces in *(1.38)* is the *Lennard-Jones "6-12" model* given by

$$U = 4\varepsilon[(\sigma/r)^{12} - (\sigma/r)^6] \tag{1.40}$$

where σ and ε are constants. Using $\varepsilon = 1.629 \times 10^{-21}$ J and $\sigma = 0.346$ nm for O_2, prepare a plot of this function over the range $0.300\ \text{nm} \le r \le 1.000\ \text{nm}$.

▮ As a sample calculation, consider $r = 0.400$ nm:

$$U = 4(1.629 \times 10^{-21}\ \text{J})\left[\left(\frac{0.346\ \text{nm}}{0.400\ \text{nm}}\right)^{12} - \left(\frac{0.346\ \text{nm}}{0.400\ \text{nm}}\right)^{6}\right] = -1.586 \times 10^{-21}\ \text{J}$$

The plot is shown in Fig. 1-12.

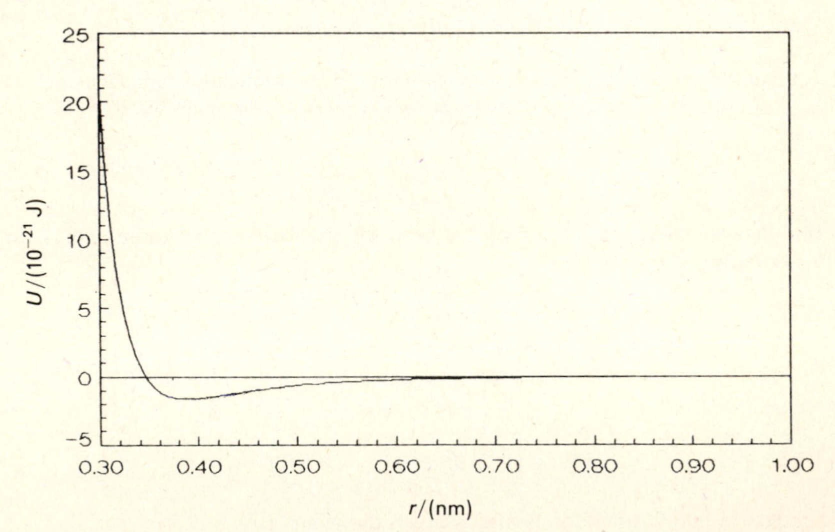

Fig. 1-12

1.98 At what value of r will the Lennard-Jones potential be a minimum? What is the value of U at this distance?

▮ The minimum value of U will occur when $\partial U/\partial r = 0$. Differentiating *(1.40)* with respect to r gives

$$\frac{\partial U}{\partial r} = 4\varepsilon\left(\frac{-12\sigma^{12}}{r^{13}} - \frac{-6\sigma^6}{r^7}\right)$$

Setting the derivative equal to zero and solving for r gives $r(\text{min}) = 2^{1/6}\sigma$. The value of $U(\text{min})$ is

$$U(\text{min}) = 4\varepsilon\left[\left(\frac{\sigma}{2^{1/6}\sigma}\right)^{12} - \left(\frac{\sigma}{2^{1/6}\sigma}\right)^{6}\right] = 4\varepsilon(\tfrac{1}{4} - \tfrac{1}{2}) = -\varepsilon$$

1.99 The value of the second virial coefficient can be determined by using *(1.38)* and *(1.40)*. Application of the law of corresponding states allows the evaluation of B_v graphically (see Fig. 1-13) as a function of $\ln T^*$ for $0.70 \le T^* \le 100.00$, where

$$T^* = \frac{T}{\varepsilon/k} \tag{1.41}$$

$$B_v = b_0 B^* \tag{1.42}$$

Using $\varepsilon/k = 117.5$ K, determine B_v for O_2 at 0 °C.

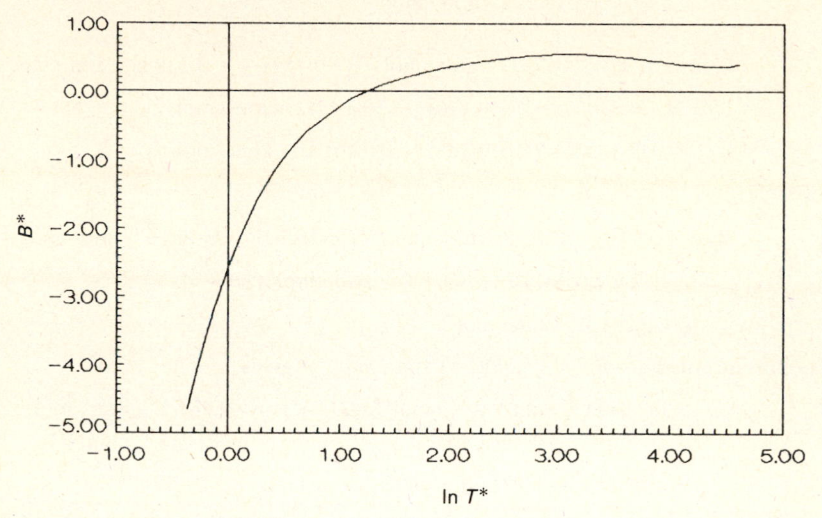

Fig. 1-13

▮ At this temperature, $T^* = (273\text{ K})/(117.5\text{ K}) = 2.32$, which corresponds to $B^* = -0.381$. Using $b_0 = 1.21\text{ L}^2\cdot\text{atm}\cdot\text{mol}^{-2}$ (see Problem 1.95) gives

$$B_v = (1.21\text{ L}^2\cdot\text{atm}\cdot\text{mol}^{-2})(-0.381) = -0.461\text{ L}^2\cdot\text{atm}\cdot\text{mol}^{-2}$$

which agrees fairly well with the value given in Table 1-3.

1.100 The Boyle temperature and the critical temperature are related to the parameter ε/k used with the Lennard-Jones potential by

$$T_{\text{Boyle}} = 3.42(\varepsilon/k) \qquad \text{and} \qquad T_c = 1.30(\varepsilon/k)$$

Using $\varepsilon/k = 117.5$ K for O_2, determine T_{Boyle} and T_c.

▮ Substituting the value of ε/k gives

$$T_{\text{Boyle}} = (3.42)(117.5\text{ K}) = 402\text{ K}$$

which agrees with the value of 424 K determined in Problem 1.92, and

$$T_c = (1.30)(117.5\text{ K}) = 153\text{ K} = -120\text{ °C}$$

which agrees well with the value of -118.8 °C determined in Problem 1.78.

1.101 For a mixture of gases at relatively low pressures, *(1.33)* becomes

$$PV_m = RT + B_v(1/V_m)$$

where

$$B_v = x_1^2 B_1 + x_2^2 B_2 + 2x_1 x_2 B_{12} \tag{1.43}$$

The value of B_{12} can be obtained from Fig. 1-13 using

$$\varepsilon_{12}/k = [(\varepsilon_1/k)(\varepsilon_2/k)]^{1/2} \qquad b_{0,12} = (b_{0,1}^{1/3} + b_{0,2}^{1/3})^3/8$$

Calculate the pressure of a 1.00-mol sample of air confined to 0.218 L at 0 °C. Assume $x(N_2) = 0.80$ and $x(O_2) = 0.20$. Compare the result to that predicted by the ideal gas law and to the observed value of 100.0 atm. For O_2, $\varepsilon/k = 117.5$ K and $b_0 = 1.21\text{ L}^2\cdot\text{atm}\cdot\text{mol}^{-2}$; and for N_2, $\varepsilon/k = 95.0$ K and $b_0 = 1.43\text{ L}^2\cdot\text{atm}\cdot\text{mol}^{-2}$.

▌ For the mixture, $\varepsilon(\text{mixture})/k = [(95.0\ \text{K})(117.5\ \text{K})]^{1/2} = 106\ \text{K}$, giving

$$T^*(N_2) = \frac{273\ \text{K}}{95.0\ \text{K}} = 2.87 \qquad T^*(O_2) = \frac{273\ \text{K}}{117.5\ \text{K}} = 2.32 \qquad T^*(\text{mixture}) = 273/106 = 2.58$$

The respective values of B^* from Fig. 1-13 are $B^*(N_2) = -0.146$, $B^*(O_2) = -0.381$, and $B^*(\text{mixture}) = -0.257$. For the mixture,

$$b_0(\text{mixture}) = \frac{[(1.43\ \text{L}^2\cdot\text{atm}\cdot\text{mol}^{-2})^{1/3} + (1.21\ \text{L}^2\cdot\text{atm}\cdot\text{mol}^{-2})^{1/3}]^3}{8}$$
$$= 1.32\ \text{L}^2\cdot\text{atm}\cdot\text{mol}^{-2}$$

giving

$$B_v(\text{mixture}) = (1.32\ \text{L}^2\cdot\text{atm}\cdot\text{mol}^{-2})(-0.257) = -0.339\ \text{L}^2\cdot\text{atm}\cdot\text{mol}^{-2}$$
$$B_v(N_2) = (1.43)(-0.146) = -0.209\ \text{L}^2\cdot\text{atm}\cdot\text{mol}^{-2}$$
$$B_v(O_2) = (1.21)(-0.381) = -0.461\ \text{L}^2\cdot\text{atm}\cdot\text{mol}^{-2}$$

Using *(1.43)* gives

$$B_v = (0.80)^2(-0.209\ \text{L}^2\cdot\text{atm}\cdot\text{mol}^{-2}) + (0.20)^2(-0.461\ \text{L}^2\cdot\text{atm}\cdot\text{mol}^{-2})$$
$$+ (2)(0.80)(0.20)(-0.339\ \text{L}^2\cdot\text{atm}\cdot\text{mol}^{-2})$$
$$= -0.261\ \text{L}^2\cdot\text{atm}\cdot\text{mol}^{-2}$$

Solving for the pressure using $B_v = -0.261\ \text{L}^2\cdot\text{atm}\cdot\text{mol}^{-2}$ gives

$$P = \frac{(0.0821\ \text{L}\cdot\text{atm}\cdot\text{K}^{-1}\cdot\text{mol}^{-1})(273\ \text{K})}{0.218\ \text{L}\cdot\text{mol}^{-1}} + \frac{-0.261\ \text{L}^2\cdot\text{atm}\cdot\text{mol}^{-2}}{(0.218\ \text{L}\cdot\text{mol}^{-1})^2}$$
$$= 103\ \text{atm} - 5.49\ \text{atm} = 98\ \text{atm}$$

This is 2% lower than the observed value. The ideal gas law value, 103 atm, is 3% greater than the observed value.

1.102 The B_{12} term in *(1.43)* is often approximated using Amagat's law by

$$B_{12} = (B_1 + B_2)/2$$

Rework Problem 1.101 using this approximation for B_{12}.

▌ Substituting the values for $B(N_2)$ and $B(O_2)$ determined in Problem 1.101 gives

$$B_v(\text{mixture}) = [(-0.2092\ \text{L}^2\cdot\text{atm}\cdot\text{mol}^{-2}) + (-0.461\ \text{L}^2\cdot\text{atm}\cdot\text{mol}^{-2})]/2$$
$$= -0.335\ \text{L}^2\cdot\text{atm}\cdot\text{mol}^{-2}$$

giving

$$B_v = (0.80)^2(-0.209\ \text{L}^2\cdot\text{atm}\cdot\text{mol}^{-2}) + (0.20)^2(-0.461\ \text{L}^2\cdot\text{atm}\cdot\text{mol}^{-2})$$
$$+ (2)(0.80)(0.20)(-0.335\ \text{L}^2\cdot\text{atm}\cdot\text{mol}^{-2})$$
$$= -0.259\ \text{L}^2\cdot\text{atm}\cdot\text{mol}^{-2}$$

The predicted pressure is

$$P = \frac{(0.0821\ \text{L}\cdot\text{atm}\cdot\text{K}^{-1}\cdot\text{mol}^{-1})(273\ \text{K})}{0.218\ \text{L}\cdot\text{mol}^{-1}} + \frac{-0.259\ \text{L}^2\cdot\text{atm}\cdot\text{mol}^{-2}}{(0.218\ \text{L}\cdot\text{mol}^{-1})^2} = 97\ \text{atm}$$

which is 3% lower than the observed value.

1.103 A rougher approximation for the B_{12} term in *(1.43)* according to Dalton's law is $B_{12} = 0$. Rework Problem 1.101 using the approximation for B_{12}.

▌ For the mixture,

$$B_v = (0.80)^2(-0.209\ \text{L}^2\cdot\text{atm}\cdot\text{mol}^{-2}) + (0.20)^2(-0.461\ \text{L}^2\cdot\text{atm}\cdot\text{mol}^{-2})$$
$$= -0.152\ \text{L}^2\cdot\text{atm}\cdot\text{mol}^{-2}$$

giving

$$P = \frac{(0.0821\ \text{L}\cdot\text{atm}\cdot\text{K}^{-1}\cdot\text{mol}^{-1})(273\ \text{K})}{0.218\ \text{L}\cdot\text{mol}^{-1}} + \frac{-0.152\ \text{L}^2\cdot\text{atm}\cdot\text{mol}^{-2}}{(0.218\ \text{L}\cdot\text{mol}^{-1})^2} = 99.6\ \text{atm}$$

which is 0.4% lower than the observed value.

1.104 Assuming the van der Waals equation of state to describe CO_2, calculate the pressure of 1.00 mol confined to 0.0949 L at 90 °C.

▮ The *van der Waals equation of state* is

$$(P + an^2/V^2)(V - nb) = nRT \tag{1.44a}$$

Solving *(1.44a)* for pressure and substituting values of a and b from Table 1-3 give

$$P = \frac{(1.00\ \text{mol})(0.0821\ \text{L}\cdot\text{atm}\cdot\text{K}^{-1}\cdot\text{mol}^{-1})(363\ \text{K})}{0.0949\ \text{L} - (1.00\ \text{mol})(0.042\,67\ \text{L}\cdot\text{mol}^{-1})} - \frac{(3.592\ \text{L}^2\cdot\text{atm}\cdot\text{mol}^{-2})(1.00\ \text{mol})^2}{(0.0949\ \text{L})^2}$$

$$= 172\ \text{atm}$$

This value is 2% lower than the observed value of 175 atm.

1.105 Assuming the van der Waals equation to be valid, determine the molar volume of H_2O at 1.013 25 bar and 100.0 °C. Calculate the density of steam under these conditions.

▮ The polynomial form of *(1.44a)* is

$$PV^3 - n(bP + RT)V^2 + n^2aV - n^3ab = 0 \tag{1.44b}$$

Substituting the data for H_2O from Table 1-3 gives

$$\begin{aligned}(1.00\ \text{atm})V^3 &- (1.00\ \text{mol})[(0.030\,49\ \text{L}\cdot\text{mol}^{-1})(1.00\ \text{atm})\\ &+ (0.0821\ \text{L}\cdot\text{atm}\cdot\text{K}^{-1}\cdot\text{mol}^{-1})(373.2\ \text{K})]V^2\\ &+ (1.00\ \text{mol})^2(5.464\ \text{L}^2\cdot\text{atm}\cdot\text{mol}^{-2})V\\ &- (1.00\ \text{mol})^3(5.464\ \text{L}^2\cdot\text{atm}\cdot\text{mol}^{-2})(0.030\,49\ \text{L}\cdot\text{mol}^{-1}) = 0\end{aligned}$$

$$V^3 - (30.67)V^2 + 5.46V - 0.1666 = 0$$

Solving this cubic equation gives $V = 30.49\ \text{L}$. The density is

$$\rho = \frac{m}{V} = \frac{(18.015\ \text{g}\cdot\text{mol}^{-1})[(10^{-3}\ \text{kg})/(1\ \text{g})]}{(30.49\ \text{L}\cdot\text{mol}^{-1})[(10^{-3}\ \text{m}^3)/(1\ \text{L})]} = 0.5908\ \text{kg}\cdot\text{m}^{-3}$$

which is 1% lower than the observed value of $0.5974\ \text{kg}\cdot\text{m}^{-3}$.

1.106 Repeat the calculations of Problem 1.83 assuming O_2 to be a van der Waals gas.

▮ Substituting the data given in Table 1-3 into *(1.44b)* and solving the polynomial for n gives

$$\begin{aligned}(15.18\times10^6\ \text{Pa})(6.91\ \text{m}^3) &- n[(3.19\times10^{-5}\ \text{m}^3\cdot\text{mol}^{-1})(15.18\times10^6\ \text{Pa})\\ &+ (8.314\ \text{m}^3\cdot\text{Pa}\cdot\text{K}^{-1}\cdot\text{mol}^{-1})(294\ \text{K})](6.91\ \text{m}^3)^2\\ &+ n^2(0.1380\ \text{m}^3\cdot\text{Pa}\cdot\text{mol}^{-2})(6.91\ \text{m}^3)\\ &- n^3(0.1380\ \text{m}^3\cdot\text{Pa}\cdot\text{mol}^{-2})(3.19\times10^{-5}\ \text{m}^3\cdot\text{mol}^{-1}) = 0\end{aligned}$$

$$5.01\times10^9 - (1.398\times10^5)n + 0.954n^2 - (4.40\times10^{-6})n^3 = 0$$

$$n = 4.81\times10^4\ \text{mol}$$

which is 5% greater than the answer to Problem 1.83.

1.107 The van der Waals constant a is a correction factor to the ideal gas law for the intermolecular attractions within a substance. Match the following values of $a/(\text{L}^2\cdot\text{atm}\cdot\text{mol}^{-2})$—0.2107, 5.464, 18.00, and 24.06—with the gases C_6H_6, $C_6H_5CH_3$, Ne, and H_2O.

▮ The gas with the weakest intermolecular forces will be Ne, which has only London forces resulting from 10 electrons. Thus $a(\text{Ne}) = 0.2107\ \text{L}^2\cdot\text{atm}\cdot\text{mol}^{-2}$. Water will have stronger intermolecular forces than Ne, even though both molecules have 10 electrons, because of the polar nature of the molecule and because of the hydrogen bonding present. Thus $a(H_2O) = 5.464\ \text{L}^2\cdot\text{atm}\cdot\text{mol}^{-2}$. Both C_6H_6 and $C_6H_5CH_3$ have very strong London forces because of the large number of electrons in the molecules and because of the delocalized π bonding present in the ring. Because $C_6H_5CH_3$ also has a dipole moment, it should have the largest value of a. Thus $a(C_6H_5CH_3) = 24.06\ \text{L}^2\cdot\text{atm}\cdot\text{mol}^{-2}$ and $a(C_6H_6) = 18.00\ \text{L}^2\cdot\text{atm}\cdot\text{mol}^{-2}$.

1.108 The van der Waals constant b is a correction factor to the ideal gas law for the intrinsic volume of the molecule. Match the following values of $b/(\text{L}\cdot\text{mol}^{-1})$—0.017 09, 0.030 49, 0.1154, and 0.1463—with the gases C_6H_6, $C_6H_5CH_3$, H_2O, and Ne.

▌ The smallest molecule is atomic Ne, so $b(\text{Ne}) = 0.017\,09\ \text{L}\cdot\text{mol}^{-1}$. Both of the hydrocarbons are much larger than H_2O, so $b(H_2O) = 0.030\,49\ \text{L}\cdot\text{mol}^{-1}$. The addition of the CH_3 group to the benzene ring will make it larger, so $b(C_6H_5CH_3) = 0.1436\ \text{L}\cdot\text{mol}^{-1}$ and $b(C_6H_6) = 0.1154\ \text{L}\cdot\text{mol}^{-1}$.

1.109 Calculate the molecular diameter of O_2 from $b = 0.0319\ \text{L}\cdot\text{mol}^{-1}$.

▌ The van der Waals constant b is related to the molecular volume of the substance by

$$b = 4LV \tag{1.45}$$

The volume of one molecule is given by *(1.45)* as

$$V = \frac{(0.0319\ \text{L}\cdot\text{mol}^{-1})[(10^{-3}\ \text{m}^3)/(1\ \text{L})]}{4(6.022\times10^{-23}\ \text{mol}^{-1})} = 1.32\times10^{-29}\ \text{m}^3$$

Recognizing that $V = \frac{4}{3}\pi(\sigma/2)^3$, where σ is the molecular diameter, gives

$$\sigma = 2[3(1.32\times10^{-29}\ \text{m}^3)/4\pi]^{1/3} = 2.93\times10^{-10}\ \text{m} = 0.293\ \text{nm}$$

1.110 Using the van der Waals equation, find the minimum in the PV versus P curve for 1.00 mol of O_2 at 0 °C.

▌ The minimum in the curve will occur at $\partial(PV)/\partial P = 0$. Solving *(1.44a)* for P gives for one mole of gas

$$P = RT/(V_m - b) - a/V_m^2 \tag{1.44c}$$

and taking the derivative with respect to P gives

$$\frac{\partial(PV_m)}{\partial P} = V_m\frac{\partial P}{\partial P} + P\frac{\partial V_m}{\partial P} = V_m + \frac{P}{\partial P/\partial V_m}$$

$$= V_m + \frac{RT/(V_m - b) - a/V_m^2}{-RT/(V_m - b)^2 + 2a/V_m^3}$$

Setting the derivative equal to zero and rearranging the equation into polynomial form give

$$(bRT - a)V_m^2 + 2abV_m - ab^2 = 0$$

Substituting the values of a and b from Table 1-3 gives

$$[(0.0319\ \text{L}\cdot\text{mol}^{-1})(0.0821\ \text{L}\cdot\text{atm}\cdot\text{K}^{-1}\cdot\text{mol}^{-1})(273\ \text{K}) - 1.362\ \text{L}^2\cdot\text{atm}\cdot\text{mol}^{-2}]V_m^2$$
$$+ (2)(1.362\ \text{L}^2\cdot\text{atm}\cdot\text{mol}^{-2})(0.0319\ \text{L}\cdot\text{mol}^{-1})V_m - (1.362\ \text{L}^2\cdot\text{atm}\cdot\text{mol}^{-2})(0.0319\ \text{L}\cdot\text{mol}^{-1})^2 = 0$$

$$-0.647V_m^2 + 0.0869V_m + (-0.001\,386) = 0$$

Solving gives $V_m = 0.0185\ \text{L}\cdot\text{mol}^{-1}$ and $0.1158\ \text{L}\cdot\text{mol}^{-1}$. The first solution yields a negative pressure on substitution into *(1.44c)*, but using $V_m = 0.1158\ \text{L}\cdot\text{mol}^{-1}$ gives

$$P = \frac{(0.0821\ \text{L}\cdot\text{atm}\cdot\text{K}^{-1}\cdot\text{mol}^{-1})(273\ \text{K})}{0.1158\ \text{L}\cdot\text{mol}^{-1} - 0.0319\ \text{L}\cdot\text{mol}^{-1}} - \frac{1.362\ \text{L}^2\cdot\text{atm}\cdot\text{mol}^{-2}}{(0.1158\ \text{L}\cdot\text{mol}^{-1})^2} = 166\ \text{atm}$$

This value agrees with the minimum shown in Fig. 1-6.

1.111 Derive an equation relating a and b to Z. Calculate Z for 1.00 mol of CO_2 at 175 atm, 0.0949 L, and 90 °C.

▌ The desired result is obtained by multiplying *(1.44c)* by V_m/RT, giving

$$Z = \frac{PV_m}{RT} = \frac{V_m}{RT}\left(\frac{RT}{V_m - b} - \frac{a}{V_m^2}\right) = \frac{V_m}{V_m - b} - \frac{a}{V_mRT}$$

Substituting the data for CO_2 gives

$$Z = \frac{0.0949\ \text{L}\cdot\text{mol}^{-1}}{0.0949\ \text{L}\cdot\text{mol}^{-1} - 0.042\,67\ \text{L}\cdot\text{mol}^{-1}} - \frac{3.592\ \text{L}^2\cdot\text{atm}\cdot\text{mol}^{-2}}{(0.0949\ \text{L}\cdot\text{mol}^{-1})(0.0821\ \text{L}\cdot\text{atm}\cdot\text{K}^{-1}\cdot\text{mol}^{-1})(363\ \text{K})}$$
$$= 0.55$$

This value agrees well with the one determined in Problem 1.84.

1.112 Evaluate the van der Waals constants for O_2 using $T_c = -118.4\ °C$ and $P_c = 50.1$ atm.

▌ An analysis of the van der Waals equation at the critical point gives

$$a = 27R^2T_c^2/64P_c \tag{1.46}$$

$$b = RT_c/8P_c \tag{1.47}$$

Substituting into *(1.46)* and *(1.47)* gives

$$a = \frac{27(0.0821\ \text{L}\cdot\text{atm}\cdot\text{K}^{-1}\cdot\text{mol}^{-1})^2(154.8\ \text{K})^2}{64(50.1\ \text{atm})}$$

$$= 1.360\ \text{L}^2\cdot\text{atm}\cdot\text{mol}^{-2} = 1.378\ \text{L}^2\cdot\text{bar}\cdot\text{mol}^{-2}$$

$$b = \frac{(0.0821\ \text{L}\cdot\text{atm}\cdot\text{K}^{-1}\cdot\text{mol}^{-1})^2(154.8\ \text{K})}{8(50.1\ \text{atm})} = 0.0317\ \text{L}\cdot\text{mol}^{-1}$$

These values agree well with those given in Table 1-3.

1.113 Estimate P_c, V_c, and T_c for H_2O.

▌ The analysis of the van der Waals equation at the critical point described in Problem 1.112 also gives

$$V_{m,c} = 3b = 3(0.030\,49\ \text{L}\cdot\text{mol}^{-1}) = 0.091\,47\ \text{L}\cdot\text{mol}^{-1} \tag{1.48}$$

$$T_c = 8a/27Rb = \frac{8(5.464\ \text{L}^2\cdot\text{atm}\cdot\text{mol}^{-2})}{27(0.0821\ \text{L}\cdot\text{atm}\cdot\text{K}^{-1}\cdot\text{mol}^{-1})(0.030\,49\ \text{L}\cdot\text{mol}^{-1})} = 647\ \text{K} \tag{1.49}$$

$$P_c = a/27b^2 = \frac{5.464\ \text{L}^2\cdot\text{atm}\cdot\text{mol}^{-2}}{27(0.030\,49\ \text{L}\cdot\text{mol}^{-1})^2} = 217.7\ \text{atm} \tag{1.50}$$

The values of T_c and P_c agree well with the observed values of 647 K and 218.3 atm.

1.114 The van der Waals equation can be expressed in a virial equation form as

$$PV_m = RT + (bRT - a)\frac{1}{V_m} + b^2RT\left(\frac{1}{V_m}\right)^2 + \cdots \tag{1.51}$$

Compare *(1.51)* to *(1.33)* and evaluate the second and third virial coefficients for O_2 at 0 °C.

▌ Upon comparison of *(1.51)* with *(1.33)* we see that

$$B_v = bRT - a$$

$$= (0.0319\ \text{L}\cdot\text{mol}^{-1})(0.0821\ \text{L}\cdot\text{atm}\cdot\text{K}^{-1}\cdot\text{mol}^{-1})(273\ \text{K}) - 1.362\ \text{L}^2\cdot\text{atm}\cdot\text{mol}^{-2}$$

$$= -0.647\ \text{L}^2\cdot\text{atm}\cdot\text{mol}^{-2}$$

$$C_v = b^2RT = (0.0319\ \text{L}\cdot\text{mol}^{-1})^2(0.0821\ \text{L}\cdot\text{atm}\cdot\text{K}^{-1}\cdot\text{mol}^{-1})(273\ \text{K})$$

$$= 2.28 \times 10^{-2}\ \text{L}^3\cdot\text{atm}\cdot\text{mol}^{-3}$$

Both calculated values agree fairly well with the respective values of $-0.5168\ \text{L}^2\cdot\text{atm}\cdot\text{mol}^{-2}$ and $4.166 \times 10^{-2}\ \text{L}^3\cdot\text{atm}\cdot\text{mol}^{-3}$ determined in Problem 1.86.

1.115 Calculate the Boyle temperature using *(1.51)* for O_2.

▌ Recalling the results of Problem 1.92, the Boyle temperature occurs at $B_v = 0$. Setting the expression for B_v from *(1.51)* equal to zero and solving for T_{Boyle} gives

$$T_{\text{Boyle}} = a/bR \tag{1.52}$$

For O_2,

$$T_{\text{Boyle}} = \frac{1.362\ \text{L}^2\cdot\text{atm}\cdot\text{mol}^{-2}}{(0.0319\ \text{L}\cdot\text{mol}^{-1})(0.0821\ \text{L}\cdot\text{atm}\cdot\text{K}^{-1}\cdot\text{mol}^{-1})} = 520\ \text{K} = 247\ °\text{C}$$

This value is high compared to the value of 424 K determined in Problem 1.92.

1.116 Determine the equation for the cubic expansion coefficient for a van der Waals gas. Evaluate α for O_2 at 0 °C and 100.0 atm, where $V_m = 0.2077\ \text{L}\cdot\text{mol}^{-1}$. Compare this value to that predicted for an ideal gas.

▮ In order to evaluate the $(\partial V_m/\partial T)_P$ term in *(1.11)*, we use the equality

$$\left(\frac{\partial V_m}{\partial T}\right)_P = -\frac{(\partial P/\partial T)_{V_m}}{(\partial P/\partial V_m)_T}$$

Finding the two respective derivatives of *(1.44c)* and substituting give

$$\left(\frac{\partial V_m}{\partial T}\right)_P = -\frac{R/(V_m - b)}{-RT/(V_m - b)^2 + 2a/V_m^3}$$

Substituting into *(1.11)* and simplifying give

$$\alpha = \frac{R(V_m - b)V_m^2}{RTV_m^3 - 2aV_m^2 + 4abV_m - 2ab^2}$$

$$= \frac{(0.0821\ \mathrm{L\cdot atm\cdot K^{-1}\cdot mol^{-1}})(0.2077\ \mathrm{L\cdot mol^{-1}} - 0.0319\ \mathrm{L\cdot mol^{-1}})(0.2077\ \mathrm{L\cdot mol^{-1}})^2}{\left[\begin{array}{l}(0.0821\ \mathrm{L\cdot atm\cdot K^{-1}\cdot mol^{-1}})(273\ \mathrm{K})(0.2077\ \mathrm{L\cdot mol^{-1}})^3 \\ \quad -2(1.362\ \mathrm{L^2\cdot atm\cdot mol^{-2}})(0.2077\ \mathrm{L\cdot mol^{-1}})^2 \\ \quad\quad +4(1.362\ \mathrm{L^2\cdot atm\cdot mol^{-2}})(0.0319\ \mathrm{L\cdot mol^{-1}})(0.2077\ \mathrm{L\cdot mol^{-1}}) \\ \quad\quad\quad -2(1.362\ \mathrm{L^2\cdot atm\cdot mol^{-2}})(0.0319\ \mathrm{L\cdot mol^{-1}})^2\end{array}\right]}$$

$$= 5.34 \times 10^{-3}\ \mathrm{K^{-1}}$$

This value is considerably greater than the $3.66 \times 10^{-3}\ \mathrm{K^{-1}}$ predicted for an ideal gas (see Problem 1.28).

1.117 The van der Waals equation can be used to describe a gaseous mixture where

$$a = [a_1^{1/2}x_1 + a_2^{1/2}x_2]^2 \qquad b = b_1x_1^2 + b_2x_2^2 + 2b_{12}x_1x_2$$

$$b_{12} = [b_1^{1/3} + b_2^{1/3}]^3/8$$

Calculate the pressure of 1.00 mol of air confined to a volume of 0.218 L at 0 °C. Assume $x(N_2) = 0.80$ and $x(O_2) = 0.20$.

▮ The value of b_{12} is

$$b(\text{mixture}) = [(0.0391\ \mathrm{L\cdot mol^{-1}})^{1/3} + (0.0319\ \mathrm{L\cdot mol^{-1}})^{1/3}]^3/8$$
$$= 0.0354\ \mathrm{L\cdot mol^{-1}}$$

giving

$$b = (0.0391\ \mathrm{L\cdot mol^{-1}})(0.80)^2 + (0.0319\ \mathrm{L\cdot mol^{-1}})(0.20)^2 + 2(0.0354\ \mathrm{L\cdot mol^{-1}})(0.80)(0.20)$$
$$= 0.0376\ \mathrm{L\cdot mol^{-1}}$$

Likewise,

$$a = [(1.390\ \mathrm{L^2\cdot atm\cdot mol^{-2}})^{1/2}(0.80) + (1.362\ \mathrm{L^2\cdot atm\cdot mol^{-2}})^{1/2}(0.20)]^2$$
$$= 1.384\ \mathrm{L^2\cdot atm\cdot mol^{-2}}$$

Substituting into *(1.44c)* gives

$$P = \frac{(0.0821\ \mathrm{L\cdot atm\cdot K^{-1}\cdot mol^{-1}})(273\ \mathrm{K})}{0.218\ \mathrm{L\cdot mol^{-1}} - 0.0376\ \mathrm{L\cdot mol^{-1}}} - \frac{(1.384\ \mathrm{L^2\cdot atm\cdot mol^{-2}})}{(0.218\ \mathrm{L\cdot mol^{-1}})^2} = 95.1\ \mathrm{atm}$$

which is 5% lower than the observed value of 100.0 atm.

1.118 Use *(1.44a)* to calculate the pressure of 0.80 mol of N_2 confined to 0.218 L and 0 °C. Repeat the calculation for 0.20 mol of O_2 under the same conditions. Use these partial pressures to predict the pressure of 1.00 mol of air under these conditions.

▮ For each gas, *(1.44c)* gives

$$P(N_2) = \frac{(0.80\ \mathrm{mol})(0.0821\ \mathrm{L\cdot atm\cdot K^{-1}\cdot mol^{-1}})(273\ \mathrm{K})}{0.218\ \mathrm{L} - (0.80\ \mathrm{mol})(0.0391\ \mathrm{L\cdot mol^{-1}})} - \frac{(1.390\ \mathrm{L^2\cdot atm\cdot mol^{-2}})(0.80\ \mathrm{mol})^2}{(0.218\ \mathrm{L})^2}$$
$$= 77.3\ \mathrm{atm}$$

$$P(O_2) = \frac{(0.20)(0.0821)(273)}{0.218 - (0.20)(0.0319)} - \frac{(1.362)(0.20)^2}{(0.218)^2} = 20.0 \text{ atm}$$

Using *(1.22)* gives

$$P_t = 77.3 \text{ atm} + 20.0 \text{ atm} = 97.3 \text{ atm}$$

which is 3% lower than the observed value of 100.0 atm.

1.119 The P-V-T behavior of real gases may be described by the Dieterici equation of state. Calculate the pressure of 1.00 mol of O_2 confined to 0.208 L at 0 °C.

▌ The *Dieterici equation of state* is

$$P = \frac{RT}{V_m - b} e^{-(a/V_m RT)} \tag{1.53}$$

The pressure given by *(1.53)* is

$$\begin{aligned} P &= \frac{(0.0821 \text{ L}\cdot\text{atm}\cdot\text{K}^{-1}\cdot\text{mol}^{-1})(273 \text{ K})}{0.208 \text{ L}\cdot\text{mol}^{-1} - 0.0345 \text{ L}\cdot\text{mol}^{-1}} \\ &\quad \times \exp\left(-\frac{1.748 \text{ L}^2\cdot\text{atm}\cdot\text{mol}^{-2}}{(0.208 \text{ L}\cdot\text{mol}^{-1})(0.0821 \text{ L}\cdot\text{atm}\cdot\text{K}^{-1}\cdot\text{mol}^{-1})(273 \text{ K})}\right) \\ &= 88.8 \text{ atm} \end{aligned}$$

This value is 11% lower than the observed value of 100.0 atm.

1.120 Assuming that CO_2 obeys the Dieterici equation, calculate the molar volume at 90 °C and 175 atm.

▌ Under these conditions *(1.53)* becomes

$$\begin{aligned} 175 \text{ atm} &= \frac{(0.0821 \text{ L}\cdot\text{atm}\cdot\text{K}^{-1}\cdot\text{mol}^{-1})(363 \text{ K})}{V_m - 0.0463 \text{ L}\cdot\text{mol}^{-1}} \\ &\quad \times \exp\left(-\frac{4.621 \text{ L}^2\cdot\text{atm}\cdot\text{mol}^{-2}}{V_m(0.0821 \text{ L}\cdot\text{atm}\cdot\text{K}^{-1}\cdot\text{mol}^{-1})(363 \text{ K})}\right) \end{aligned}$$

Solving gives $V_m = 0.0581 \text{ L}\cdot\text{mol}^{-1}$. This value is 61% greater than the observed value of $0.0949 \text{ L}\cdot\text{mol}^{-1}$.

1.121 Using $T_c = -118.4$ °C and $P_c = 50.1$ atm, evaluate the Dieterici constants for O_2.

▌ An analysis of *(1.53)* at the critical point gives

$$a = 4R^2T_c^2/e^2P_c \qquad b = RT_c/e^2P_c$$

Substituting the critical point data gives

$$a = \frac{4(0.0821 \text{ L}\cdot\text{atm}\cdot\text{K}^{-1}\cdot\text{mol}^{-1})^2(154.8 \text{ K})^2}{e^2(50.1 \text{ atm})} = 1.745 \text{ L}^2\cdot\text{atm}\cdot\text{mol}^{-2}$$

$$b = \frac{(0.0821 \text{ L}\cdot\text{atm}\cdot\text{K}^{-1}\cdot\text{mol}^{-1})(154.8 \text{ K})}{e^2(50.1 \text{ atm})} = 0.0343 \text{ L}\cdot\text{mol}^{-1}$$

These values are in good agreement with those given in Table 1-3.

1.122 Using the Dieterici equation, estimate P_c, V_c, and T_c for H_2O.

▌ The analysis of the Dieterici equation at the critical point described in Problem 1.121 also gives

$$V_c = 2b = 2(0.0330 \text{ L}\cdot\text{mol}^{-1}) = 0.0660 \text{ L}\cdot\text{mol}^{-1}$$

$$T_c = \frac{a}{4bR} = \frac{7.011 \text{ L}^2\cdot\text{atm}\cdot\text{mol}^{-2}}{4(0.0330 \text{ L}\cdot\text{mol}^{-1})(0.0821 \text{ L}\cdot\text{atm}\cdot\text{K}^{-1}\cdot\text{mol}^{-1})} = 647 \text{ K}$$

$$P_c = \frac{a}{4e^2b^2} = \frac{7.011 \text{ L}^2\cdot\text{atm}\cdot\text{mol}^{-2}}{4e^2(0.0330 \text{ L}\cdot\text{mol}^{-1})^2} = 218 \text{ atm}$$

The values of T_c and P_c agree well with the observed values of 647 K and 218.3 atm.

1.123 The Dieterici equation can be expressed in a virial equation form as

$$PV_m = RT + (bRT - a)\left(\frac{1}{V_m}\right) + \left(b^2RT - ab + \frac{a^2}{2RT}\right)\left(\frac{1}{V_m}\right)^2 \tag{1.54}$$

Compare *(1.54)* to *(1.33)*, and evaluate the second and third virial coefficients for O_2 at 0 °C.

▮ Upon comparison of *(1.54)* with *(1.33)*, we see that

$$\begin{aligned} B_v &= bRT - a \\ &= (0.0345\ \text{L}\cdot\text{mol}^{-1})(0.0821\ \text{L}\cdot\text{atm}\cdot\text{K}^{-1}\cdot\text{mol}^{-1})(273\ \text{K}) - 1.748\ \text{L}^2\cdot\text{atm}\cdot\text{mol}^{-2} \\ &= -0.975\ \text{L}^2\cdot\text{atm}\cdot\text{mol}^{-2} \\ C_v &= b^2RT - ab + a^2/2RT \\ &= (0.0345\ \text{L}\cdot\text{mol}^{-1})^2(0.0821\ \text{L}\cdot\text{atm}\cdot\text{K}^{-1}\cdot\text{mol}^{-1})(273\ \text{K}) \\ &\quad - (1.748\ \text{L}^2\cdot\text{atm}\cdot\text{mol}^{-2})(0.0345\ \text{L}\cdot\text{mol}^{-1}) + \frac{(1.748\ \text{L}^2\cdot\text{atm}\cdot\text{mol}^{-2})^2}{2(0.0821\ \text{L}\cdot\text{atm}\cdot\text{K}^{-1}\cdot\text{mol}^{-1})(273\ \text{K})} \\ &= 3.45\times10^{-2}\ \text{L}^3\cdot\text{atm}\cdot\text{mol}^{-3} \end{aligned}$$

Both values agree fairly well with the values given in Table 1-3.

1.124 Using *(1.54)*, calculate the Boyle temperature for O_2.

▮ Setting the expression for B_v in *(1.54)* equal to zero and solving for T_{Boyle} gives *(1.52)*:

$$T_{Boyle} = \frac{a}{bR} = \frac{1.748\ \text{L}^2\cdot\text{atm}\cdot\text{mol}^{-2}}{(0.0345\ \text{L}\cdot\text{mol}^{-1})(0.0821\ \text{L}\cdot\text{atm}\cdot\text{K}^{-1}\cdot\text{mol}^{-1})} = 617\ \text{K}$$

This value is very high compared with the value of 424 K determined in Problem 1.92.

1.125 Using the Redlich-Kwong equation of state, calculate the pressure of 1.00 mol of O_2 confined to 0.208 L at 0 °C.

▮ The *Redlich-Kwong equation of state* is

$$[P + a/V_m(V_m + b)T^{1/2}](V_m - b) = RT \tag{1.55}$$

Solving *(1.55)* for P and substituting the data gives

$$\begin{aligned} P &= RT/(V_m - b) - a/[V_m(V_m + b)T^{1/2}] \\ &= \frac{(0.0821\ \text{L}\cdot\text{atm}\cdot\text{K}^{-1}\cdot\text{mol}^{-1})(273\ \text{K})}{0.208\ \text{L}\cdot\text{mol}^{-1} - 0.0221\ \text{L}\cdot\text{mol}^{-1}} \\ &\quad - \frac{17.16\ \text{L}^2\cdot\text{K}^{1/2}\cdot\text{atm}\cdot\text{mol}^{-2}}{(0.208\ \text{L}\cdot\text{mol}^{-1})(0.208\ \text{L}\cdot\text{mol}^{-1} - 0.0221\ \text{L}\cdot\text{mol}^{-1})(273\ \text{K})^{1/2}} \\ &= 93.7\ \text{atm} \end{aligned}$$

This value is 6% lower than the observed value of 100.0 atm.

1.126 Using $T_c = -118.4$ °C and $P_c = 50.1$ atm, evaluate the Redlich-Kwong constants for O_2.

▮ An analysis of *(1.55)* at the critical point gives

$$a = \frac{R^2T_c^{5/2}}{9(2^{1/3} - 1)P_c} \qquad b = \frac{(2^{1/3} - 1)RT_c}{3P_c}$$

Substituting the critical point data gives

$$\begin{aligned} a &= \frac{(0.0821\ \text{L}\cdot\text{atm}\cdot\text{K}^{-1}\cdot\text{mol}^{-1})^2(154.8\ \text{K})^{5/2}}{9(2^{1/3} - 1)(50.1\ \text{atm})} \\ &= 17.15\ \text{L}^2\cdot\text{K}^{1/2}\cdot\text{atm}\cdot\text{mol}^{-2} \\ b &= \frac{(2^{1/3} - 1)(0.0821\ \text{L}\cdot\text{atm}\cdot\text{K}^{-1}\cdot\text{mol}^{-1})(154.8\ \text{K})}{3(50.1\ \text{atm})} \\ &= 2.20\times10^{-2}\ \text{L}\cdot\text{mol}^{-1} \end{aligned}$$

These values are in good agreement with those given in Table 1-3.

1.127 The *Berthelot equation of state* is given by

$$(P + a/TV_m^2)(V_m - b) = RT \tag{1.56}$$

where

$$a = \frac{27R^2T_c^3}{64P_c} \qquad b = \frac{RT_c}{8P_c}$$

Using $T_c = -118.4\,°C$ and $P_c = 50.1$ atm, calculate the pressure of 1.00 mol of O_2 confined to 0.208 L at 0 °C.

▌ The values of the Berthelot constants are

$$a = \frac{27(0.0821\ \text{L}\cdot\text{atm}\cdot\text{K}^{-1}\cdot\text{mol}^{-1})^2(154.8\ \text{K})^3}{(64)(50.1\ \text{atm})} = 211\ \text{L}^2\cdot\text{atm}\cdot\text{K}\cdot\text{mol}^{-2}$$

$$b = \frac{(0.0821\ \text{L}\cdot\text{atm}\cdot\text{K}^{-1}\cdot\text{mol}^{-1})(154.8\ \text{K})}{8(50.1\ \text{atm})} = 3.17\times10^{-2}\ \text{L}\cdot\text{mol}^{-1}$$

Solving *(1.56)* for P and substituting the data gives

$$\begin{aligned} P &= RT/(V_m - b) - a/TV_m^2 \\ &= \frac{(0.0821\ \text{L}\cdot\text{atm}\cdot\text{K}^{-1}\cdot\text{mol}^{-1})(273\ \text{K})}{0.208\ \text{L}\cdot\text{mol}^{-1} - 0.0317\ \text{L}\cdot\text{mol}^{-1}} - \frac{211\ \text{L}^2\cdot\text{atm}\cdot\text{K}\cdot\text{mol}^{-2}}{(273\ \text{K})(0.208\ \text{L}\cdot\text{mol}^{-1})^2} \\ &= 109\ \text{atm} \end{aligned}$$

This value is 9% greater than the observed value of 100.0 atm.

1.128 Determine the value of the compressibility factor at the critical point, Z_c, using *(1.56)*.

▌ An analysis of *(1.56)* at the critical point gives

$$V_{m,c} = 3b \qquad T_c = \left(\frac{8a}{27bR}\right)^{1/2} \qquad P_c = \left(\frac{Ra}{216b^3}\right)^{1/2}$$

For 1.00 mol of gas, *(1.28)* gives

$$Z_c = \frac{P_cV_{m,c}}{RT_c} = \frac{(Ra/216b^3)^{1/2}(3b)}{R(8a/27bR)^{1/2}} = \frac{3}{8}$$

1.129 Determine the Boyle temperature for O_2 using *(1.56)*.

▌ The Boyle temperature for a gas obeying *(1.56)* is given by

$$\begin{aligned} T_{\text{Boyle}} &= (a/Rb)^{1/2} \\ &= \left(\frac{211\ \text{L}^2\cdot\text{atm}\cdot\text{K}\cdot\text{mol}^{-2}}{(0.0821\ \text{L}\cdot\text{atm}\cdot\text{K}^{-1}\cdot\text{mol}^{-1})(3.17\times10^{-2}\ \text{L}\cdot\text{mol}^{-1})}\right)^{1/2} = 284\ \text{K} \end{aligned}$$

which is very low compared to the value of 424 K determined in Problem 1.92.

1.130 The P-V-T behavior of real gases can be described by the virial form of the *Beattie–Bridgeman equation of state*

$$PV_m = RT + \beta\frac{1}{V_m} + \gamma\left(\frac{1}{V_m}\right)^2 + \delta\left(\frac{1}{V_m}\right)^3 + \cdots \tag{1.57}$$

where

$$\beta = RT\left(B_0 + \frac{A_0}{RT} - \frac{c}{T^3}\right) \qquad \gamma = RT\left(-B_0b + \frac{A_0}{RT} - \frac{B_0c}{T^3}\right) \qquad \delta = RT\left(\frac{B_0bc}{T^3}\right)$$

Using the data given in Table 1-3, calculate the pressure of 1.00 mol of O_2 confined to 0.208 L at 0 °C.

▌ The values of the three coefficients in *(1.57)* are

$$\begin{aligned} \beta &= (0.0821\ \text{L}\cdot\text{atm}\cdot\text{K}^{-1}\cdot\text{mol}^{-1})(273\ \text{K}) \\ &\quad\times\left(0.046\,24\ \text{L}\cdot\text{mol}^{-1} - \frac{1.4911\ \text{L}^2\cdot\text{atm}\cdot\text{mol}^{-2}}{(0.0821\ \text{L}\cdot\text{atm}\cdot\text{K}^{-1}\cdot\text{mol}^{-1})(273\ \text{K})} - \frac{4.80\times10^4\ \text{L}\cdot\text{K}^3\cdot\text{mol}^{-1}}{(273\ \text{K})^3}\right) \\ &= -0.508\ \text{L}^2\cdot\text{atm}\cdot\text{mol}^{-2} \end{aligned}$$

$$\gamma = (0.0821\ \text{L}\cdot\text{atm}\cdot\text{K}^{-1}\cdot\text{mol}^{-1})(273\ \text{K})$$
$$\times\Bigg(-(0.046\,24\ \text{L}\cdot\text{mol}^{-1})(4.208\times 10^{-3}\ \text{L}\cdot\text{mol}^{-1})$$
$$+\frac{(1.4911\ \text{L}^2\cdot\text{atm}\cdot\text{mol}^{-2})(0.025\,62\ \text{L}\cdot\text{mol}^{-1})}{(0.0821\ \text{L}\cdot\text{atm}\cdot\text{K}^{-1}\cdot\text{mol}^{-1})(273\ \text{K})}$$
$$-\frac{(0.046\,24\ \text{L}\cdot\text{mol}^{-1})(4.80\times 10^{4}\ \text{L}\cdot\text{K}^3\cdot\text{mol}^{-1})}{(273\ \text{K})^3}\Bigg)$$
$$= 3.14\times 10^{-2}\ \text{L}^3\cdot\text{atm}\cdot\text{mol}^{-3}$$
$$\delta = (0.0821\ \text{L}\cdot\text{atm}\cdot\text{K}^{-1}\cdot\text{mol}^{-1})(273\ \text{K})$$
$$\times\left(\frac{(0.046\,24\ \text{L}\cdot\text{mol}^{-1})(4.208\times 10^{-3}\ \text{L}\cdot\text{mol}^{-1})(4.80\times 10^{4}\ \text{L}\cdot\text{K}^3\cdot\text{mol}^{-1})}{(273\ \text{K})^3}\right)$$
$$= 1.03\times 10^{-5}\ \text{L}^4\cdot\text{atm}\cdot\text{mol}^{-4}$$

These values agree fairly well with the values of B_v, C_v, and D_v given in Table 1-3. Solving *(1.57)* for P and substituting the data give

$$P = \frac{RT}{V_m} + \beta\left(\frac{1}{V_m}\right)^2 + \gamma\left(\frac{1}{V_m}\right)^3 + \delta\left(\frac{1}{V_m}\right)^4$$
$$= \frac{(0.0821\ \text{L}\cdot\text{atm}\cdot\text{K}^{-1}\cdot\text{mol}^{-1})(273\ \text{K})}{0.208\ \text{L}\cdot\text{mol}^{-1}} + \frac{-0.508\ \text{L}^2\cdot\text{atm}\cdot\text{mol}^{-2}}{(0.208\ \text{L}\cdot\text{mol}^{-1})^2}$$
$$+\frac{3.14\times 10^{-2}\ \text{L}^3\cdot\text{atm}\cdot\text{mol}^{-3}}{(0.208\ \text{L}\cdot\text{mol}^{-1})^3} + \frac{1.03\times 10^{-5}\ \text{L}^4\cdot\text{atm}\cdot\text{mol}^{-4}}{(0.208\ \text{L}\cdot\text{mol}^{-1})^4}$$
$$= 99.5\ \text{atm}$$

This value agrees very well with the observed value of 100.0 atm.

1.131 The pressure form of *(1.57)* is

$$V_m = \frac{RT}{P} + \frac{\beta}{RT} + \gamma' P + \delta' P^2 + \cdots$$

where

$$\gamma' = \frac{1}{RT}\left[\frac{\gamma}{RT} - \left(\frac{\beta}{RT}\right)^2\right]$$
$$\delta' = \frac{1}{(RT)^2}\left[\frac{\delta}{RT} - \frac{3\beta\gamma}{(RT)^2} + 2\left(\frac{\beta}{RT}\right)^3\right]$$

Calculate the molar volume of CO_2 at 90 °C and 175 atm.

I The values of the five coefficients are

$$\beta = (0.0821\ \text{L}\cdot\text{atm}\cdot\text{K}^{-1}\cdot\text{mol}^{-1})(363\ \text{K})$$
$$\times\left(0.104\,76\ \text{L}\cdot\text{mol}^{-1} - \frac{5.0065\ \text{L}^2\cdot\text{atm}\cdot\text{mol}^{-2}}{(0.0821\ \text{L}\cdot\text{atm}\cdot\text{K}^{-1}\cdot\text{mol}^{-1})(363\ \text{K})} - \frac{66.00\times 10^4\ \text{L}\cdot\text{K}^3\cdot\text{mol}^{-1}}{(363\ \text{K})^3}\right)$$
$$= -2.296\ \text{L}^2\cdot\text{atm}\cdot\text{mol}^{-2}$$
$$\gamma = (0.0821\ \text{L}\cdot\text{atm}\cdot\text{K}^{-1}\cdot\text{mol}^{-1})(363\ \text{K})$$
$$\times\Bigg(-(0.104\,76\ \text{L}\cdot\text{mol}^{-1})(0.072\,35\ \text{L}\cdot\text{mol}^{-1})$$
$$+\frac{(5.0065\ \text{L}^2\cdot\text{atm}\cdot\text{mol}^{-2})(0.071\,32\ \text{L}\cdot\text{mol}^{-1})}{(0.0821\ \text{L}\cdot\text{atm}\cdot\text{K}^{-1}\cdot\text{mol}^{-1})(363\ \text{K})}$$
$$-\frac{(0.104\,76\ \text{L}\cdot\text{mol}^{-1})(66.00\times 10^{4}\ \text{L}\cdot\text{K}^3\cdot\text{mol}^{-1})}{(363\ \text{K})^3}\Bigg)$$
$$= 8.81\times 10^{-2}\ \text{L}^3\cdot\text{atm}\cdot\text{mol}^{-3}$$

$$\delta = (0.0821\ \mathrm{L\cdot atm\cdot K^{-1}\cdot mol^{-1}})(363\ \mathrm{K})$$
$$\times\left(\frac{(0.104\,76\ \mathrm{L\cdot mol^{-1}})(0.072\,35\ \mathrm{L\cdot mol^{-1}})(66.00\times10^{4}\ \mathrm{L\cdot K^{3}\cdot mol^{-1}})}{(363\ \mathrm{K})^{3}}\right)$$
$$= 3.12\times10^{-3}\ \mathrm{L^{4}\cdot atm\cdot mol^{-4}}$$

$$\gamma' = \left(\frac{1}{(0.0821\ \mathrm{L\cdot atm\cdot K^{-1}\cdot mol^{-1}})(363\ \mathrm{K})}\right)$$
$$\times\left[\frac{8.81\times10^{-2}\ \mathrm{L^{3}\cdot atm\cdot mol^{-3}}}{(0.0821\ \mathrm{L\cdot atm\cdot K^{-1}\cdot mol^{-1}})(363\ \mathrm{K})} - \left(\frac{-2.296\ \mathrm{L^{2}\cdot atm\cdot mol^{-2}}}{(0.0821\ \mathrm{L\cdot atm\cdot K^{-1}\cdot mol^{-1}})(363\ \mathrm{K})}\right)^{2}\right]$$
$$= -1.000\times10^{-4}\ \mathrm{L\cdot atm^{-1}\cdot mol^{-1}}$$

$$\delta' = \left(\frac{1}{(0.0821\ \mathrm{L\cdot atm\cdot K^{-1}\cdot mol^{-1}})(363\ \mathrm{K})}\right)^{2}$$
$$\times\left[\frac{3.12\times10^{-3}\ \mathrm{L^{4}\cdot atm\cdot mol^{-4}}}{(0.0821\ \mathrm{L\cdot atm\cdot K^{-1}\cdot mol^{-1}})(363\ \mathrm{K})} - \frac{3(-2.296\ \mathrm{L^{2}\cdot atm\cdot mol^{-2}})(8.81\times10^{-2}\ \mathrm{L^{3}\cdot atm\cdot mol^{-3}})}{(0.0821\ \mathrm{L\cdot atm\cdot K^{-1}\cdot mol^{-1}})^{2}(363\ \mathrm{K})^{2}}\right.$$
$$\left.+2\left(\frac{-2.296\ \mathrm{L^{2}\cdot atm\cdot mol^{-2}}}{(0.0821\ \mathrm{L\cdot atm\cdot K^{-1}\cdot mol^{-1}})(363\ \mathrm{K})}\right)^{3}\right]$$
$$= -1.42\times10^{-7}\ \mathrm{L^{2}\cdot atm\cdot mol^{-1}}$$

The molar volume is

$$V_{\mathrm{m}} = \frac{(0.0821\ \mathrm{L\cdot atm\cdot K^{-1}\cdot mol^{-1}})(363\ \mathrm{K})}{175\ \mathrm{atm}}$$
$$+\frac{-2.296\ \mathrm{L^{2}\cdot atm\cdot mol^{-2}}}{(0.0821\ \mathrm{L\cdot atm\cdot K^{-1}\cdot mol^{-1}})(363\ \mathrm{K})}$$
$$+(-1.000\times10^{-4}\ \mathrm{L\cdot atm^{-1}\cdot mol^{-1}})(175\ \mathrm{atm})$$
$$+(-1.42\times10^{-7}\ \mathrm{L\cdot atm^{-2}\cdot mol^{-1}})(175\ \mathrm{atm})^{2}$$
$$= 0.0714\ \mathrm{L\cdot mol^{-1}}$$

which is 25% lower than the observed value of $0.0949\ \mathrm{L\cdot mol^{-1}}$.

1.132 Using *(1.57)*, determine the Boyle temperature for O_2.

▮ Upon comparing *(1.57)* with *(1.33)*, we see that $B_v = \beta$. Setting $\beta = 0$ at the Boyle temperature gives

$$B_0 - A_0/RT_{\mathrm{Boyle}} - c/T^{3}_{\mathrm{Boyle}} = 0$$
$$B_0T^{3}_{\mathrm{Boyle}} - (A_0/R)T^{2}_{\mathrm{Boyle}} - c = 0$$

Substituting the Beattie–Bridgeman constants from Table 1-3 gives

$$(0.046\,24\ \mathrm{L\cdot mol^{-1}})T^{3}_{\mathrm{Boyle}} - \left(\frac{1.4911\ \mathrm{L^{2}\cdot atm\cdot mol^{-2}}}{0.0821\ \mathrm{L\cdot atm\cdot K^{-1}\cdot mol^{-1}}}\right)T^{2}_{\mathrm{Boyle}} - (4.80\times10^{4}\ \mathrm{L\cdot K^{3}\cdot mol^{-1}}) = 0$$

Solving gives $T_{\mathrm{Boyle}} = 399\ \mathrm{K}$, which is 6% lower than the value of 474 K determined in Problem 1.92.

1.133 Use the following density–pressure data at 105.1 °C to determine the molar mass of acetic acid.

$P/(\mathrm{atm})$	0.04	0.06	0.08	0.10	0.12
$\rho/(\mathrm{g\cdot L^{-1}})$	0.0939	0.1449	0.1980	0.2532	0.3101
$P/(\mathrm{atm})$	0.14	0.16	0.18	0.20	
$\rho/(\mathrm{g\cdot L^{-1}})$	0.3684	0.4290	0.4891	0.5225	

Why is the value different from $60.05\ \mathrm{g\cdot mol^{-1}}$?

I At low pressures *(1.32)* can be written as

$$PV = n(RT + B_pP) = (m/M)(RT + B_pP)$$

giving

$$\rho/P = M/RT - [MB_p/(RT)^2]P \tag{1.58}$$

Thus a plot of ρ/P versus P will be linear, and the intercept will be equal to M/RT.

Values of ρ/P were calculated from the above data and plotted in Fig. 1-14. The intercept is 2.221 $g \cdot L^{-1} \cdot atm^{-1}$, giving

$$\begin{aligned} M &= RT(\text{intercept}) \\ &= (0.0821\ L \cdot atm \cdot K^{-1} \cdot mol^{-1})(378.3\ K)(2.221\ g \cdot L^{-1} \cdot atm^{-1}) \\ &= 68.98\ g \cdot mol^{-1} \end{aligned}$$

The discrepancy is the result of dimerization of acetic acid molecules.

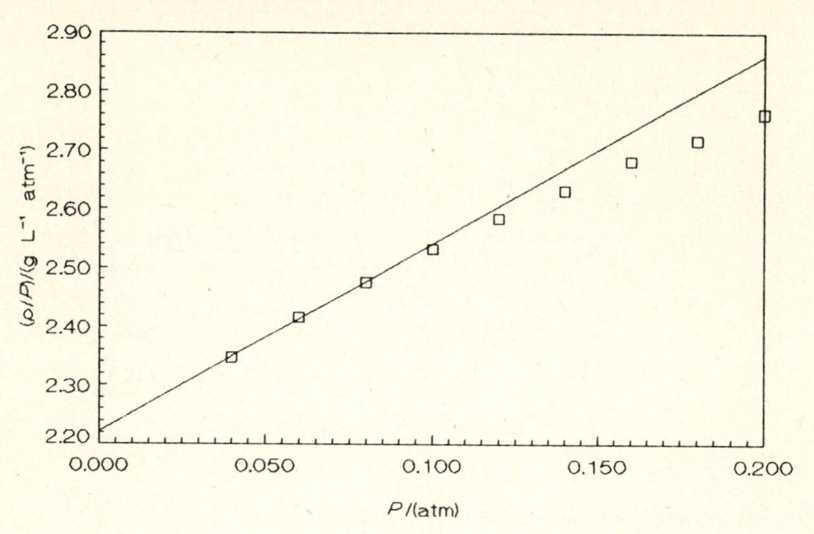

Fig. 1-14

1.134 Use the following density-pressure data at 0 °C to determine the molar mass of SO_2.

P/(atm)	1.0000	0.5000	0.1000	0.0100	0.0010	0.0001
$(\rho/P)/(g \cdot L^{-1} \cdot atm^{-1})$	2.926 682	2.892 407	2.864 974	2.858 800	2.858 183	2.858 121

Using the atomic mass of O as 15.9994 $g \cdot mol^{-1}$, calculate the atomic mass of S.

I As suggested by *(1.58)*, Fig. 1-15 is a plot of ρ/P versus P. The intercept is 2.858 114 $L^{-1} \cdot atm^{-1}$, giving

$$\begin{aligned} M(SO_2) &= (0.082\,056\,8\ L \cdot atm \cdot K^{-1} \cdot mol^{-1})(273.15\ K)(2.858\,114\ g \cdot L^{-1} \cdot atm^{-1}) \\ &= 64.0612\ g \cdot mol^{-1} \end{aligned}$$

The atomic mass of S is

$$M(S) = 64.0612\ g \cdot mol^{-1} - 2(15.9994\ g \cdot mol^{-1}) = 32.0624\ g \cdot mol^{-1}$$

1.135 Using the data in Problem 1.134, determine the second virial coefficient for SO_2.

I According to *(1.58)*, the slope of the plot of ρ/P against P will be equal to $-MB_p/(RT)^2$, giving

$$\begin{aligned} B_p &= -(RT)^2(\text{slope})/M \\ &= -\frac{[(0.082\,056\,8\ L \cdot atm \cdot K^{-1} \cdot mol^{-1})(273.15\ K)]^2(0.068\,585\ g \cdot L^{-1} \cdot atm^{-2})}{64.0612\ g \cdot mol^{-1}} \\ &= -0.537\,86\ L \cdot mol^{-1} \end{aligned}$$

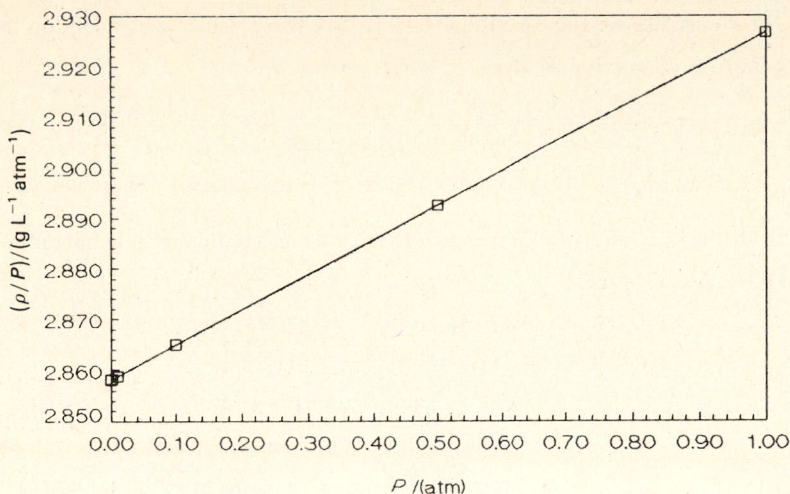

Fig. 1-15

1.5. KINETIC-MOLECULAR THEORY

1.136 Calculate the average translational kinetic energy ($\bar{\varepsilon}_{trans}$) for an H_2 molecule and for 1.00 mol of H_2 ($\bar{E}_{trans}$) at 25.00 °C.

▮ The average translational kinetic energy of a gaseous molecule is given by

$$\bar{\varepsilon}_{trans} = \tfrac{1}{2}m\overline{v^2} = \tfrac{3}{2}kT \tag{1.59}$$

Substituting into *(1.59)* gives

$$\bar{\varepsilon}_{trans} = \tfrac{3}{2}(1.381 \times 10^{-23}\ \text{J} \cdot \text{K}^{-1})(298.15\ \text{K}) = 6.175 \times 10^{-21}\ \text{J}$$

On a molar basis, multiplying *(1.59)* by Avogadro's constant gives

$$\bar{E}_{trans} = L\bar{\varepsilon}_{trans} = \tfrac{3}{2}LkT = \tfrac{3}{2}RT \tag{1.60}$$

which upon substitution gives

$$\bar{E}_{trans} = \tfrac{3}{2}(8.314\ \text{J} \cdot \text{K}^{-1} \cdot \text{mol}^{-1})(298.15\ \text{K}) = 3718\ \text{J} \cdot \text{mol}^{-1}$$

1.137 Calculate the root-mean-square speed (v_{rms}) for H_2 at 25.00 °C.

▮ The *root-mean-square speed* for a gas is defined as

$$v_{rms} = (\overline{v^2})^{1/2} \tag{1.61}$$

Substituting *(1.59)* and *(1.60)* into *(1.61)* gives

$$v_{rms} = \left(\frac{2\bar{\varepsilon}_{trans}}{m}\right)^{1/2} = \left(\frac{3kT}{m}\right)^{1/2} = \left(\frac{2\bar{E}_{trans}}{M}\right)^{1/2} = \left(\frac{3RT}{M}\right)^{1/2} \tag{1.62}$$

For H_2, substituting $\bar{\varepsilon}_{trans} = 6.175 \times 10^{-21}$ J (see Problem 1.136) gives

$$v_{rms} = \left(\frac{2(6.175 \times 10^{-21}\ \text{J})[(1\ \text{kg} \cdot \text{m}^2 \cdot \text{s}^{-2})/(1\ \text{J})]}{(2.0158 \times 10^{-3}\ \text{kg} \cdot \text{mol}^{-1})/(6.022 \times 10^{23}\ \text{mol}^{-1})}\right)^{1/2} = 1921\ \text{m} \cdot \text{s}^{-1}$$

1.138 Calculate $\bar{E}_{trans}$ and v_{rms} for gaseous He at 25.00 °C. To what temperature must H_2 be cooled so that it has the same v_{rms} as He at 25.00 °C?

▮ For ideal gases, *(1.60)* shows that $\bar{E}_{trans}$ is a function of temperature only, so $\bar{E}_{trans} = 3718\ \text{J} \cdot \text{mol}^{-1}$ for He just as it was for H_2 (see Problem 1.136). The v_{rms} for He is given by *(1.62)* as

$$v_{rms} = \left(\frac{2(3718\ \text{J} \cdot \text{mol}^{-1})[(1\ \text{kg} \cdot \text{m}^2 \cdot \text{s}^{-2})/(1\ \text{J})]}{4.0026 \times 10^{-3}\ \text{kg} \cdot \text{mol}^{-1}}\right)^{1/2} = 1363\ \text{m} \cdot \text{s}^{-1}$$

Solving *(1.62)* for $T(H_2)$ in terms of $v_{rms}(\text{He})$ gives

$$T(H_2) = M(H_2)[v_{rms}(\text{He})]^2/3R$$

$$= \frac{(2.0158 \times 10^{-3}\ \text{kg} \cdot \text{mol}^{-1})(1363\ \text{m} \cdot \text{s}^{-1})^2}{3(8.314\ \text{J} \cdot \text{K}^{-1} \cdot \text{mol}^{-1})[(1\ \text{kg} \cdot \text{m}^2 \cdot \text{s}^{-2})/(1\ \text{J})]} = 150.1\ \text{K}$$

1.139 Calculate v_{rms} for Ne at 10.0 K, 100.0 K, and 1000.0 K. Are these values dependent on the pressure of the gas?

▮ Substituting into *(1.62)* for 10.0 K gives

$$v_{rms} = \left(\frac{3(8.314\ \text{J}\cdot\text{K}^{-1}\cdot\text{mol}^{-1})[(1\ \text{kg}\cdot\text{m}^2\cdot\text{s}^{-2})/(1\ \text{J})](10.0\ \text{K})}{20.179\times10^{-3}\ \text{kg}\cdot\text{mol}^{-1}}\right)^{1/2} = 111\ \text{m}\cdot\text{s}^{-1}$$

Likewise, $v_{rms} = 352\ \text{m}\cdot\text{s}^{-1}$ at 100.0 K and $1112\ \text{m}\cdot\text{s}^{-1}$ at 1000.0 K. There is no pressure dependence.

1.140 What is the ratio of the v_{rms} of H_2 to the v_{rms} of He at 25.00 °C? Is this ratio temperature-dependent?

▮ For the two gases, *(1.62)* gives

$$\frac{v_{rms}(H_2)}{v_{rms}(He)} = \frac{[3RT/M(H_2)]^{1/2}}{[3RT/M(He)]^{1/2}} = \left[\frac{M(He)}{M(H_2)}\right]^{1/2}$$

$$= [(4.0026\ \text{g}\cdot\text{mol}^{-1})/(2.0158\ \text{g}\cdot\text{mol}^{-1})]^{1/2} = 1.4091$$

This value is the same as calculated from the results of Problems 1.137 and 1.138.

$$v_{rms}(H_2)/v_{rms}(He) = (1921\ \text{m}\cdot\text{s}^{-1})/(1363\ \text{m}\cdot\text{s}^{-1}) = 1.409$$

There is no temperature dependence for this ratio.

1.141 The v_{rms} of a certain gas at 25 °C is $411\ \text{m}\cdot\text{s}^{-1}$. Determine the molar mass of the gas.

▮ Solving *(1.62)* for M and substituting gives

$$M = \frac{3RT}{v_{rms}^2} = \frac{3(8.314\ \text{J}\cdot\text{K}^{-1}\cdot\text{mol}^{-1})[(1\ \text{kg}\cdot\text{m}^2\cdot\text{s}^{-2})/(1\ \text{J})](298\ \text{K})}{(411\ \text{m}\cdot\text{s}^{-1})^2}$$

$$= 4.40\times10^{-2}\ \text{kg}\cdot\text{mol}^{-1}$$

1.142 The v_{rms} of a sample of O_2 is $575\ \text{m}\cdot\text{s}^{-1}$. Determine the temperature of the gas.

▮ Solving *(1.62)* for T and substituting gives

$$T = \frac{Mv_{rms}^2}{3R} = \frac{(32.00\times10^{-3}\ \text{kg}\cdot\text{mol}^{-1})(575\ \text{m}\cdot\text{s}^{-1})^2}{3(8.314\ \text{J}\cdot\text{K}^{-1}\cdot\text{mol}^{-1})[(1\ \text{kg}\cdot\text{m}^2\cdot\text{s}^{-2})/(1\ \text{J})]} = 424\ \text{K}$$

1.143 Using *(1.62)*, derive equations relating the increase in the v_{rms} of a gas for a relatively small increase and for a relatively large increase in temperature. Use both equations to calculate the v_{rms} increase for a sample of Ne as the temperature is increased from 100.0 K to 100.1 K.

▮ For the small temperature change, where the change in v_{rms} is small, *(1.62)* gives

$$\frac{dv_{rms}}{dT} = \frac{1}{2}\left(\frac{3RT}{M}\right)^{-1/2}\frac{3R}{M} = \frac{1}{2}\left(\frac{M}{3RT}\right)^{1/2}\frac{3R}{M}$$

$$= \tfrac{1}{2}(3R/M)(1/v_{rms})$$

Solving for the small change dv_{rms} and substituting $v_{rms} = 352\ \text{m}\cdot\text{s}^{-1}$ (see Problem 1.139) gives

$$dv_{rms} = (3R/2Mv_{rms})\,dT$$

$$= \frac{3(8.314\ \text{J}\cdot\text{K}^{-1}\cdot\text{mol}^{-1})[(1\ \text{kg}\cdot\text{m}^2\cdot\text{s}^{-2})/(1\ \text{J})](0.1\ \text{K})}{2(20.179\times10^{-3}\ \text{kg}\cdot\text{mol}^{-1})(352\ \text{m}\cdot\text{s}^{-1})} = 0.2\ \text{m}\cdot\text{s}^{-1}$$

For the large temperature change, *(1.62)* gives

$$\Delta v_{rms} = (3R/M)^{1/2}(T_2^{1/2} - T_1^{1/2})$$

which gives

$$\Delta v_{rms} = \left(\frac{3(8.314\ \text{J}\cdot\text{K}^{-1}\cdot\text{mol}^{-1})[(1\ \text{kg}\cdot\text{m}^2\cdot\text{s}^{-2})/(1\ \text{J})]}{20.179\times10^{-3}\ \text{g}\cdot\text{mol}^{-1}}\right)^{1/2}$$

$$\times [(100.1\ \text{K})^{1/2} - (100.0\ \text{K})^{1/2}]$$

$$= 0.2\ \text{m}\cdot\text{s}^{-1}$$

1.144 Using *(1.59)*, derive equations relating the increase in $\bar{\varepsilon}_{trans}$ of a gas for a relatively small increase and for a relatively large increase in temperature. Use both equations to calculate the $\bar{\varepsilon}_{trans}$ increase for a sample of Ne as the temperature is increased from 100.0 K to 100.1 K.

▮ For Ne at 100.0 K, *(1.59)* gives

$$\bar{\varepsilon}_{\text{trans}} = \tfrac{3}{2}kT = \tfrac{3}{2}(1.381 \times 10^{-23}\ \text{J} \cdot \text{K}^{-1})(100.0\ \text{K}) = 2.071 \times 10^{-21}\ \text{J}$$

For the small temperature change, where the change in $\bar{\varepsilon}_{\text{trans}}$ is small, *(1.59)* gives

$$\frac{d\bar{\varepsilon}_{\text{trans}}}{dT} = \tfrac{3}{2}k = \frac{\bar{\varepsilon}_{\text{trans}}}{T}$$

Solving for the small change $d\bar{\varepsilon}_{\text{trans}}$ and substituting gives

$$d\bar{\varepsilon}_{\text{trans}} = \frac{\bar{\varepsilon}_{\text{trans}}}{T}\,dT = \frac{(2.071 \times 10^{-21}\ \text{J})(0.1\ \text{K})}{100.0\ \text{K}} = 2 \times 10^{-24}\ \text{J}$$

For the large temperature change, *(1.59)* gives

$$\Delta\bar{\varepsilon}_{\text{trans}} = \tfrac{3}{2}k\,\Delta T = \tfrac{3}{2}(1.381 \times 10^{-23}\ \text{J} \cdot \text{K}^{-1})(0.1\ \text{K}) = 2 \times 10^{-24}\ \text{J}$$

1.145 How much energy is needed to raise the temperature of Ar from 0 °C to 25 °C? By what ratio is the v_{rms} increased? Is this the same ratio by which the average translational kinetic energy is increased?

▮ All of the energy involved in this process will be translational energy. From *(1.60)*,

$$\Delta\bar{E}_{\text{trans}} = \tfrac{3}{2}R\,\Delta T = \tfrac{3}{2}(8.314\ \text{J} \cdot \text{K}^{-1} \cdot \text{mol}^{-1})(25\ \text{K}) = 312\ \text{J} \cdot \text{mol}^{-1}$$

The ratio of the v_{rms} values is given by *(1.62)* as

$$v_{\text{rms}}(25\ ^\circ\text{C})/v_{\text{rms}}(0\ ^\circ\text{C}) = [(298\ \text{K})/(273\ \text{K})]^{1/2} = 1.04$$

The ratio of the values of $\bar{E}_{\text{trans}}$ is given by *(1.60)* as

$$\bar{E}_{\text{trans}}(25\ ^\circ\text{C})/\bar{E}_{\text{trans}}(0\ ^\circ\text{C}) = (298\ \text{K})/(273\ \text{K}) = 1.09$$

The ratio of the v_{rms} values is the square root of the ratio of the values of $\bar{E}_{\text{trans}}$.

1.146 The *escape velocity*, the velocity required by an object to escape from the gravitational field of a body, is given by

$$v_e = (2gr)^{1/2} \qquad (1.63)$$

where $r = 6.37 \times 10^6$ m for Earth. At what temperature will the v_{rms} of an H_2 molecule and of an N_2 molecule attain the escape velocity?

▮ Substituting the expression for v_e into *(1.62)* and solving for T gives for H_2

$$T = \frac{2Mgr}{3R} = \frac{2(2.02 \times 10^{-3}\ \text{kg} \cdot \text{mol}^{-1})(9.81\ \text{m} \cdot \text{s}^{-2})(6.37 \times 10^6\ \text{m})}{3(8.314\ \text{J} \cdot \text{K}^{-1} \cdot \text{mol}^{-1})[(1\ \text{kg} \cdot \text{m}^2 \cdot \text{s}^{-2})/(1\ \text{J})]}$$
$$= 1.02 \times 10^4\ \text{K}$$

Likewise, $T(N_2) = 1.41 \times 10^5$ K.

1.147 A He atom at 25 °C is released from the surface of the Earth to travel upwards. Assuming that it undergoes no collisions with other molecules, how high will it travel before coming to rest?

▮ The translational kinetic energy of the atom was determined in Problem 1.136 to be 6.175×10^{-21} J. The altitude at which the kinetic energy is converted to potential energy is

$$h = \frac{\bar{\varepsilon}_{\text{trans}}}{mg}$$
$$= \frac{(6.175 \times 10^{-21}\ \text{J})[(1\ \text{kg} \cdot \text{m}^2 \cdot \text{s}^{-2})/(1\ \text{J})]}{[(4.0026 \times 10^{-3}\ \text{kg} \cdot \text{mol}^{-1})/(6.022 \times 10^{23}\ \text{mol}^{-1})](9.81\ \text{m} \cdot \text{s}^{-2})}$$
$$= 9.47 \times 10^4\ \text{m}$$

1.148 A wall of a container is struck at 90 °C by 1.0×10^{23} N_2 molecules each second. What is the total force that the molecules exert on the wall if the speed of the molecules is 450 m · s^{-1}? Calculate the pressure if the area of the wall is 10.0 cm^2.

▮ The force exerted by one molecule is related to the rate of change of momentum. In 1 s, the force exerted by N molecules is

$$f = N\frac{d(mv)}{dt} = N\frac{2mv}{1\ \text{s}}$$

$$= \frac{(1.0 \times 10^{23})(2)[(28.0 \times 10^{-3}\ \text{kg} \cdot \text{mol}^{-1})/(6.022 \times 10^{23}\ \text{mol}^{-1})](450\ \text{m} \cdot \text{s}^{-1})}{1\ \text{s}}$$

$$= 4.2\ \text{kg} \cdot \text{m} \cdot \text{s}^{-2} = 4.2\ \text{N}$$

The corresponding pressure is

$$P = \frac{f}{A} = \frac{4.2\ \text{N}}{(10.0\ \text{cm}^2)(10^{-2}\ \text{m}/1\ \text{cm})^2} = 4200\ \text{N} \cdot \text{m}^{-2} = 4200\ \text{Pa}$$

1.149 Calculate the pressure needed to confine 1.0×10^{25} gas molecules, each with a mass of 1.0×10^{-25} kg and a v_{rms} of $1.0 \times 10^3\ \text{m} \cdot \text{s}^{-1}$ in a 1.0-m^3 container. Calculate the temperature of the gas.

▮ For an ideal gas,

$$PV = Nm\overline{v^2}/3 = nM\overline{v^2}/3 \qquad (1.64)$$

Solving *(1.64)* for P and substituting gives

$$P = \frac{(1.0 \times 10^{25})(1.0 \times 10^{-25}\ \text{kg})(1.0 \times 10^3\ \text{m} \cdot \text{s}^{-1})^2[(1\ \text{Pa})/(1\ \text{kg} \cdot \text{m}^{-1} \cdot \text{s}^{-2})]}{3(1.0\ \text{m}^3)}$$

$$= 330\,000\ \text{Pa} = 3.3\ \text{bar}$$

The temperature can be determined using *(1.18)* as

$$T = \frac{PV}{nR} = \frac{(330\,000\ \text{Pa})(1.0\ \text{m}^3)}{[(1.0 \times 10^{25})/(6.022 \times 10^{23}\ \text{mol}^{-1})](8.314\ \text{m}^3 \cdot \text{Pa} \cdot \text{K}^{-1} \cdot \text{mol}^{-1})}$$

$$= 2400\ \text{K}$$

1.150 The pressure needed to confine 1.00 mol of N_2 molecules to 24.8 L is 1.00 bar. Calculate the v_{rms} and the temperature for this sample of gas assuming ideal gas behavior.

▮ Solving *(1.64)* for $\overline{v^2}$ and combining with *(1.61)* gives

$$v_{\text{rms}} = \left(\frac{3PV}{nM}\right)^{1/2}$$

$$= \left(\frac{3(1.00\ \text{bar})\left(\frac{10^5\ \text{Pa}}{1\ \text{bar}}\right)\left(\frac{1\ \text{kg} \cdot \text{m}^{-1} \cdot \text{s}^{-2}}{1\ \text{Pa}}\right)(24.8\ \text{L})\left(\frac{10^{-3}\ \text{m}^3}{1\ \text{L}}\right)}{(1.00\ \text{mol})(28.0 \times 10^{-3}\ \text{kg} \cdot \text{mol}^{-1})}\right)^{1/2}$$

$$= 515\ \text{m} \cdot \text{s}^{-1}$$

The temperature is given by *(1.62)* as

$$T = \frac{Mv_{\text{rms}}^2}{3R} = \frac{(28.0 \times 10^{-3}\ \text{kg} \cdot \text{mol}^{-1})(515\ \text{m} \cdot \text{s}^{-1})^2}{3(8.314\ \text{J} \cdot \text{K}^{-1} \cdot \text{mol}^{-1})[(1\ \text{kg} \cdot \text{m}^2 \cdot \text{s}^{-2})/(1\ \text{J})]} = 298\ \text{K}$$

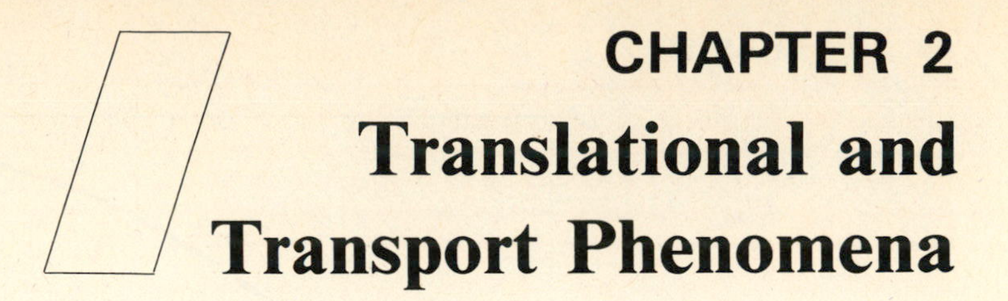

CHAPTER 2 Translational and Transport Phenomena

2.1 VELOCITY AND ENERGY DISTRIBUTIONS OF GASES

2.1 The fraction of molecules in 1 mol (dN/L) having a velocity in the x direction between v_x and $v_x + dv_x$ is given by the *Maxwell relation* as

$$\frac{dN/L}{dv_x} = \left(\frac{m}{2\pi kT}\right)^{1/2} e^{-mv_x^2/2kT} \tag{2.1}$$

Show that this distribution function is normalized, i.e., that the sum of the fractions is unity.

▌ Solving *(2.1)* for dN/L and integrating from $v_x = -\infty$ to $v_x = \infty$ gives

$$\int \frac{dN}{L} = \int_{-\infty}^{\infty} \left(\frac{m}{2\pi kT}\right)^{1/2} e^{-mv_x^2/2kT}\, dv_x$$

Transforming variables by letting $x = (m/2kT)^{1/2} v_x$ gives

$$\int \frac{dN}{L} = \left(\frac{m}{2\pi kT}\right)^{1/2} \left(\frac{2kT}{m}\right)^{1/2} \int_{-\infty}^{\infty} e^{-x^2}\, dx$$

The integral is equal to $2(\pi^{1/2}/2)$ (see Table 2-1), giving

$$\int \frac{dN}{L} = \left(\frac{m}{2\pi kT}\right)^{1/2} \left(\frac{2kT}{m}\right)^{1/2} \pi^{1/2} = 1$$

Table 2-1

$$\int_0^{\infty} e^{-x^2}\, dx = \frac{\pi^{1/2}}{2}$$

$$\int_0^{\infty} x^{1/2} e^{-ax}\, dx = \frac{(\pi/a)^{1/2}}{2a}$$

$$\int_0^{\infty} x^n e^{-ax^2}\, dx = \begin{cases} \dfrac{(1)(3)(5)\cdots(n-1)}{2(2a)^{n/2}} \left(\dfrac{\pi}{a}\right)^{1/2} & \text{for even } n \ge 2 \\ \dfrac{[(n-1)/2]!}{2a^{(n+1)/2}} & \text{for odd } n \ge 1 \end{cases}$$

$$\int_{-\infty}^{\infty} x^n e^{-ax^2}\, dx = \begin{cases} 2\displaystyle\int_0^{\infty} x^n e^{-ax^2}\, dx & \text{for even } n \\ 0 & \text{for odd } n \end{cases}$$

$$\frac{2}{\pi^{1/2}} \int_0^{z} e^{-x^2}\, dx = \operatorname{erf}(z) \qquad \text{(see Fig. 2-1)}$$

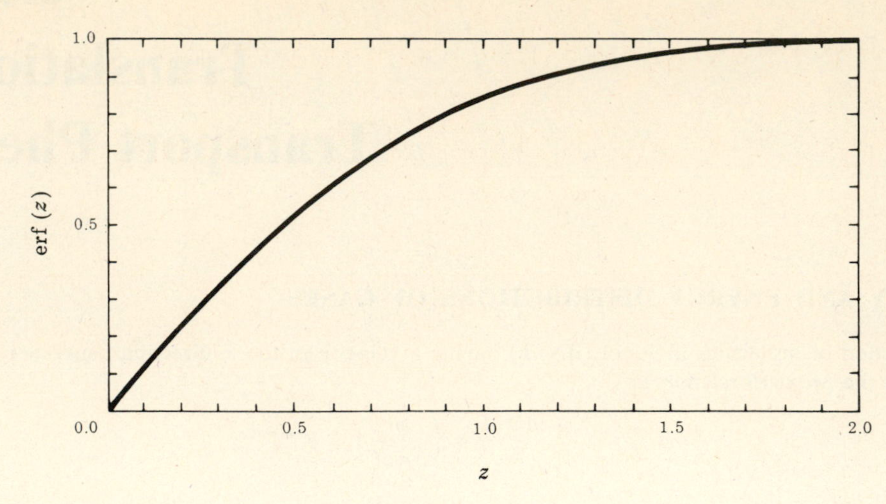

Fig. 2-1

2.2 Prepare a plot of $(dN/L)/dv_x$ versus v_x for N_2 molecules at 298 K over the velocity range $-2000 \leq v_x/(\text{m}\cdot\text{s}^{-1}) \leq 2000$.

▌ A sample calculation using *(2.1)* appears below for $500\ \text{m}\cdot\text{s}^{-1}$.

$$m = \frac{(28.0\ \text{g}\cdot\text{mol}^{-1})[(10^{-3}\ \text{kg})/(1\ \text{g})]}{6.022\times 10^{23}\ \text{mol}^{-1}} = 4.65\times 10^{-26}\ \text{kg}$$

$$\frac{dN/L}{dv_x} = \left(\frac{(4.65\times 10^{-26}\ \text{kg})[(1\ \text{J})/(1\ \text{kg}\cdot\text{m}^2\cdot\text{s}^{-2})]}{2\pi(1.381\times 10^{-23}\ \text{J}\cdot\text{K}^{-1})(298\ \text{K})}\right)^{1/2}$$

$$\times \exp\left(-\frac{(4.65\times 10^{-26}\ \text{kg})(500\ \text{m}\cdot\text{s}^{-1})^2[(1\ \text{J})/(1\ \text{kg}\cdot\text{m}^2\cdot\text{s}^{-2})]}{2(1.381\times 10^{-23}\ \text{J}\cdot\text{K}^{-1})(298\ \text{K})}\right)$$

$$= 3.27\times 10^{-4}\ \text{s}\cdot\text{m}^{-1}$$

The graph is shown in Fig. 2-2.

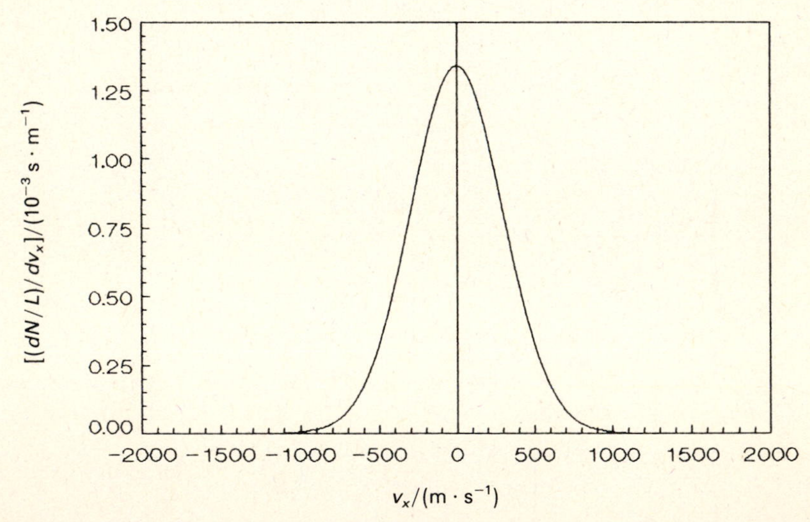

Fig. 2-2

2.3 Evaluate $\overline{v_x}$ and $\overline{p_x}$.

▌ The average value of any function $h(u)$ is defined as

$$\bar{h} = \int h(u)\frac{dN}{L} \tag{2.2}$$

For v_x, substituting $h(u) = v_x$ and *(2.1)* into *(2.2)* and following the same procedure as in Problem 2.1 gives

$$\overline{v_x} = \int_{-\infty}^{\infty} v_x\left(\frac{m}{2\pi kT}\right)^{1/2} e^{-mv_x/2kT}\,dv_x = \left(\frac{m}{2\pi kT}\right)^{1/2}\frac{2kT}{m}\int_{-\infty}^{\infty} x\,e^{-x^2}\,dx$$

The integral is equal to zero (see Table 2-1). Thus $\overline{v_x} = 0$, which agrees with the plot shown in Fig. 2-2.

Recognizing that momentum is given by the product of mass and velocity,

$$\overline{p_x} = \overline{mv_x} = m\overline{v_x} = m(0) = 0$$

2.4 Evaluate $\overline{p_x^2}$. Evaluate the root-mean-square momentum in one dimension for N_2 molecules at 298 K.

▮ Letting $h(u) = p_x^2 = (mv_x)^2$ and following the same procedure as in Problem 2.3 gives

$$\overline{p_x^2} = \int_{-\infty}^{\infty} (mv_x)^2 \left(\frac{m}{2\pi kT}\right)^{1/2} e^{-mv_x/2kT}\, dv_x$$

$$= m^2\left(\frac{m}{2\pi kT}\right)^{1/2}\left(\frac{2kT}{m}\right)^{3/2}\int_{-\infty}^{\infty} x^2\, e^{-x^2}\, dx$$

$$= m^2\left(\frac{m}{2\pi kT}\right)^{1/2}\left(\frac{2kT}{m}\right)^{3/2}\left(\frac{\pi^{1/2}}{2}\right) = mkT$$

For N_2 at 298 K, with $m = 4.65 \times 10^{-26}$ kg (see Problem 2.2)

$$(\overline{px^2})^{1/2} = (mkT)^{1/2}$$

$$= \left[(4.65 \times 10^{-26}\ \text{kg})(1.381 \times 10^{-23}\ \text{J} \cdot \text{K}^{-1})(298\ \text{K})\left(\frac{1\ \text{kg} \cdot \text{m}^2 \cdot \text{s}^{-2}}{1\ \text{J}}\right)\right]^{1/2}$$

$$= 1.38 \times 10^{-23}\ \text{kg} \cdot \text{m} \cdot \text{s}^{-1}$$

2.5 The deviation of a variable u from the average value is given by $u - \bar{u}$. The square of the *standard deviation*, σ_u, can be determined by

$$\sigma_u^2 = \overline{(u - \bar{u})^2} \qquad (2.3a)$$

Show that

$$\sigma_u^2 = \overline{u^2} - \bar{u}^2 \qquad (2.3b)$$

and determine $\sigma(v_x)$.

▮ Expanding *(2.3a)* gives

$$\sigma_u^2 = \overline{(u - \bar{u})^2} = \overline{u^2 - 2u\bar{u} + \bar{u}^2} = \overline{u^2} - 2\overline{u}\bar{u} + \bar{u}^2 = \overline{u^2} - \bar{u}^2$$

Substituting the expressions for $\overline{v_x}$ and $\overline{v_x^2} = \overline{p_x^2}/m^2$ determined in Problems 2.3 and 2.4 gives

$$\sigma(v_x)^2 = \overline{v_x^2} - \bar{v}_x^2 = \frac{mkT}{m^2} - 0 = \frac{kT}{m}$$

Thus $\sigma(v_x) = (kT/m)^{1/2}$.

2.6 Consider a hypothetical set of 100 molecules having the velocities shown in the following table:

N	5	20	35	22	10	5	3
$v/(\text{m} \cdot \text{s}^{-1})$	100	200	300	400	500	600	700

Prepare a plot of N against v, and determine the most probable velocity (v_{mp}), the average velocity ($\bar{v}$), and the root-mean-square speed (v_{rms}) for this sample of gas.

▮ The plot is shown in Fig. 2-3. The most probable velocity corresponds to the maximum of the distribution curve giving $v_{mp} = 300\ \text{m} \cdot \text{s}^{-1}$. For this sample, the average velocity is given by

$$\bar{v} = (\textstyle\sum v)/N \qquad (2.4)$$

$$= \frac{[5(100) + 20(200) + 35(300) + 22(400) + 10(500) + 5(600) + 3(700)]\ \text{m} \cdot \text{s}^{-1}}{100}$$

$$= 339\ \text{m} \cdot \text{s}^{-1}$$

and the root-mean-square speed is given by

$$v_{rms} = (\overline{v^2})^{1/2} = [(\textstyle\sum v^2)/N]^{1/2} \qquad (2.5)$$

$$= \frac{\{[5(100)^2 + 20(200)^2 + 35(300)^2 + 22(400)^2 + 10(500)^2 + 5(600)^2 + 3(700)^2]\ \text{m}^2 \cdot \text{s}^{-2}\}^{1/2}}{100}$$

$$= 365\ \text{m} \cdot \text{s}^{-1}$$

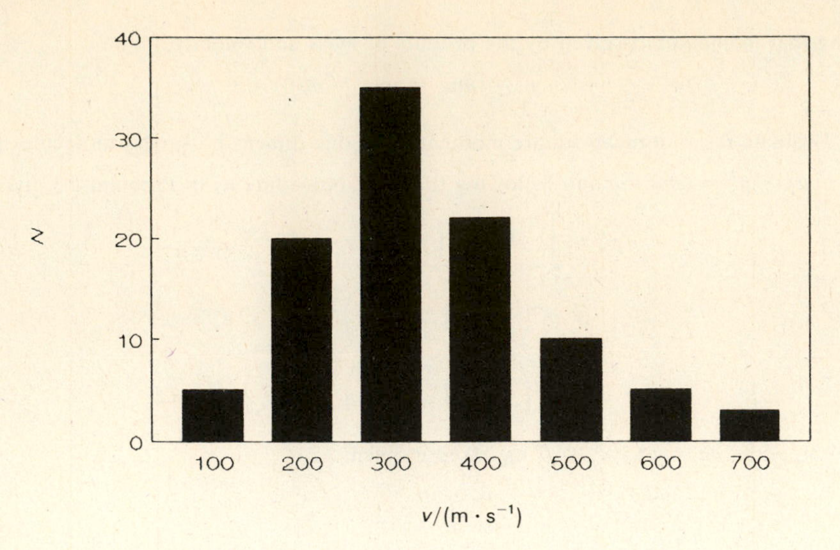

Fig. 2-3

2.7 The fraction of molecules in 1 mol having a three-dimensional velocity v between v and $v + dv$ is given by the *Maxwell-Boltzmann relation* as

$$\frac{dN/L}{dv} = 4\pi v^2\left(\frac{m}{2\pi kT}\right)^{3/2} e^{-mv^2/2kT} \tag{2.6}$$

Prepare a plot of this distribution function against v for N_2 molecules at 100 K and at 1000 K over the velocity range $0 \le v/(\text{m}\cdot\text{s}^{-1}) \le 2500$. What is the general trend of the distribution function as the temperature increases?

▌ Substituting $m = 4.65 \times 10^{-26}$ kg (see Problem 2.2), 100 K, and 500 m · s^{-1} into *(2.6)* gives

$$\frac{dN/L}{dv} = 4\pi(500\ \text{m}\cdot\text{s}^{-1})^2\left(\frac{(4.65\times10^{-26}\ \text{kg})[(1\ \text{J})/(1\ \text{kg}\cdot\text{m}^2\cdot\text{s}^{-2})]}{2\pi(1.381\times10^{-23}\ \text{J}\cdot\text{K}^{-1})(100\ \text{K})}\right)^{3/2}$$
$$\times \exp\left(-\frac{(4.65\times10^{-26}\ \text{kg})(500\ \text{m}\cdot\text{s}^{-1})^2[(1\ \text{J})/(1\ \text{kg}\cdot\text{m}^2\cdot\text{s}^{-2})]}{2(1.381\times10^{-23}\ \text{J}\cdot\text{K}^{-1})(100\ \text{K})}\right)$$
$$= 5.79 \times 10^{-4}\ \text{s}\cdot\text{m}^{-1}$$

The graph is shown in Fig. 2-4. As the temperature increases, the distribution becomes more widely spread over the range of velocities.

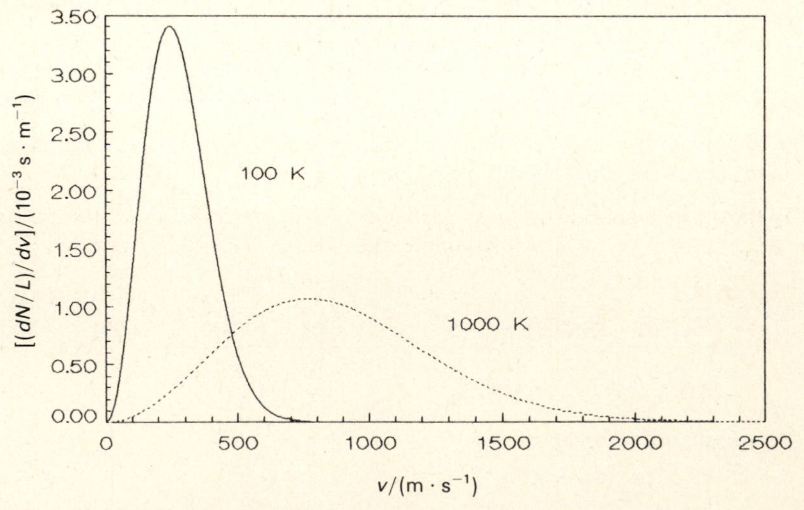

Fig. 2-4

2.8 Prepare a plot of $(dN/L)/dv$ at 298 K for He, N_2, and CO_2 over the velocity range $0 \le v/(\text{m}\cdot\text{s}^{-1}) \le 2500$. What is the general trend of distribution function as the molar mass increases?

▌ The respective masses are

$$m(\mathrm{He}) = \frac{(4.00\ \mathrm{g \cdot mol^{-1}})[(10^{-3}\ \mathrm{kg})/(1\ \mathrm{g})]}{6.022 \times 10^{23}\ \mathrm{mol^{-1}}} = 6.64 \times 10^{-27}\ \mathrm{kg}$$

$$m(\mathrm{CO_2}) = \frac{(44.01)(10^{-3})}{6.022 \times 10^{23}\ \mathrm{mol^{-1}}} = 7.31 \times 10^{-26}\ \mathrm{kg}$$

and $m(\mathrm{N_2}) = 4.65 \times 10^{-26}$ kg (see Problem 2.2). For a sample calculation of $(dN/L)/dv$, see Problem 2.7. The three plots are shown in Fig. 2-5. As the molar mass increases, the distribution becomes less widely spread over the range of velocities.

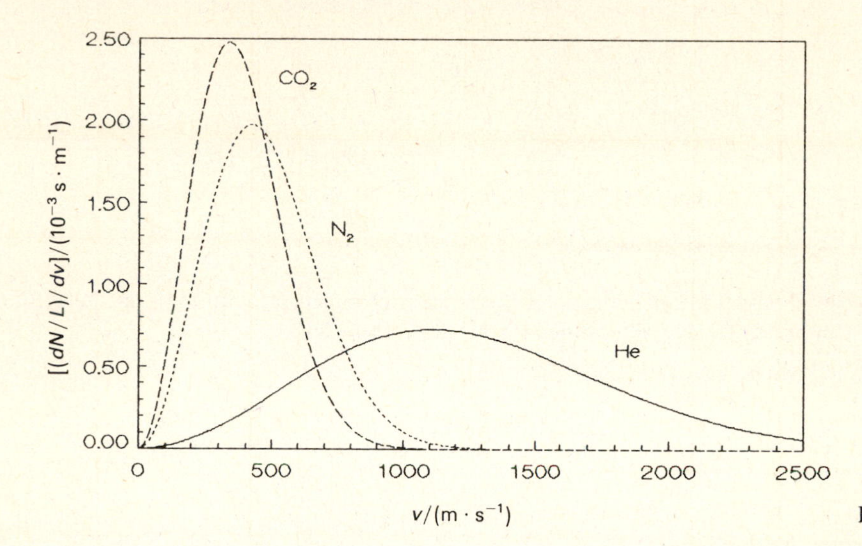

Fig. 2-5

2.9 A flask contains 1.00 mol of N_2 molecules at 100 K. How many molecules have a velocity in the range of 500.0–500.1 $\mathrm{m \cdot s^{-1}}$? Repeat the calculation for the range of 500.00-500.01 $\mathrm{m \cdot s^{-1}}$. What is the trend in the number of molecules in the velocity range as the range decreases? How many molecules have a velocity exactly equal to 500 $\mathrm{m \cdot s^{-1}}$?

▌ The distribution function at 100 K for N_2 (see Problem 2.7) is $(dN/L)/dv = 5.79 \times 10^{-4}\ \mathrm{s \cdot m^{-1}}$. Solving for dN and substituting $dv = 0.1\ \mathrm{m \cdot s^{-1}}$ gives

$$dN = (5.79 \times 10^{-4}\ \mathrm{s \cdot m^{-1}})(6.022 \times 10^{23}\ \text{molecules})(0.1\ \mathrm{m \cdot s^{-1}}) = 3 \times 10^{19}\ \text{molecules}$$

Likewise for $dv = 0.01\ \mathrm{m \cdot s^{-1}}$,

$$dN = (5.79 \times 10^{-4})(6.022 \times 10^{23})(0.01) = 3 \times 10^{18}\ \text{molecules}$$

As dv decreases, the value of dN decreases proportionally. If v is exactly 500 $\mathrm{m \cdot s^{-1}}$, then $dv = 0$, giving $dN = 0$.

2.10 A flask contains 1.00 mol of N_2 molecules at 100 K. How many molecules have a velocity in the range of 300.0–300.1 $\mathrm{m \cdot s^{-1}}$? Calculate the ratio of the number of molecules within the 0.1 $\mathrm{m \cdot s^{-1}}$ range of velocity at 300.0 $\mathrm{m \cdot s^{-1}}$ to that at 500.0 $\mathrm{m \cdot s^{-1}}$ (see Problem 2.9).

▌ At 300.0 $\mathrm{m \cdot s^{-1}}$, *(2.6)* gives

$$dN = (6.022 \times 10^{23}\ \text{molecules})(4\pi)(300.0\ \mathrm{m \cdot s^{-1}})^2\left(\frac{(4.65 \times 10^{-26}\ \mathrm{kg})[(1\ \mathrm{J})/(1\ \mathrm{kg \cdot m^2 \cdot s^{-2}})]}{2\pi(1.381 \times 10^{-23}\ \mathrm{J \cdot K^{-1}})(100\ \mathrm{K})}\right)^{3/2}$$

$$\times \left[\exp\left(-\frac{(4.65 \times 10^{-26}\ \mathrm{kg})(300.0\ \mathrm{m \cdot s^{-1}})^2[(1\ \mathrm{J})/(1\ \mathrm{kg \cdot m^2 \cdot s^{-2}})]}{2(1.381 \times 10^{-23}\ \mathrm{J \cdot K^{-1}})(100\ \mathrm{K})}\right)\right](0.1\ \mathrm{m \cdot s^{-1}})$$

$$= 2 \times 10^{20}\ \text{molecules}$$

The ratio is

$$\frac{dN(300.0\ \mathrm{m \cdot s^{-1}})}{dN(500.0\ \mathrm{m \cdot s^{-1}})} = \frac{2 \times 10^{20}\ \text{molecules}}{3 \times 10^{19}\ \text{molecules}} = 7$$

2.11 Calculate the value of v at which the two curves cross in Fig. 2-4.

▌ At the crossing point, *(2.6)* gives for the two temperatures

$$\frac{dN/L}{dv} = 4\pi v\left(\frac{m}{2\pi kT_1}\right)^{3/2} e^{-mv^2/2kT_1} = 4\pi v\left(\frac{m}{2\pi kT_2}\right)^{3/2} e^{-mv^2/2kT_2}$$

which simplifies to

$$(1/T_1)^{3/2}\, e^{-mv^2/2kT_1} = (1/T_2)^{3/2}\, e^{-mv^2/2kT_2}$$

The value of v can be determined by combining terms, taking logarithms, and solving for v^2:

$$(T_2/T_1)^{3/2} = \exp\left[-(mv^2/2k)(T_2^{-1} - T_1^{-1})\right]$$

$$\tfrac{3}{2}\ln(T_2/T_1) = -(mv^2/2k)(T_2^{-1} - T_1^{-1})$$

$$v^2 = \frac{-3k\ln(T_2/T_1)}{m(T_2^{-1} - T_1^{-1})}$$

$$= \frac{-3(1.381\times10^{-23}\ \mathrm{J\cdot K^{-1}})[(1\ \mathrm{kg\cdot m^2\cdot s^{-2}})/(1\ \mathrm{J})]\ln[(1000\ \mathrm{K})/(100\ \mathrm{K})]}{(4.65\times10^{-26}\ \mathrm{kg})[(1000\ \mathrm{K})^{-1} - (100\ \mathrm{K})^{-1}]}$$

$$= 2.28\times10^5\ \mathrm{m^2\cdot s^{-2}}$$

$$v = 477\ \mathrm{m\cdot s^{-1}}$$

2.12 Derive a general equation for the ratio of the number of molecules with the same velocity range at two different velocities. Confirm the value determined for N_2 in Problem 2.10.

▌ Applying *(2.6)* for two velocities, v_2 and v_1, with $dv_1 = dv_2$ gives

$$\frac{dN(v_2)}{dN(v_1)} = \frac{4\pi L v_2^2(m/2\pi kT)^{3/2}\, e^{-mv_2^2/2kT}\, dv_2}{4\pi L v_1^2(m/2\pi kT)^{3/2}\, e^{-mv_1^2/2kT}\, dv_1}$$

$$= \left(\frac{v_2}{v_1}\right)^2 e^{-m(v_2^2-v_1^2)/2kT} \qquad (2.7)$$

For N_2 at 100.0 K, *(2.7)* gives

$$\frac{dN(300.0\ \mathrm{m\cdot s^{-1}})}{dN(500.0\ \mathrm{m\cdot s^{-1}})} = \left(\frac{300.0\ \mathrm{m\cdot s^{-1}}}{500.0\ \mathrm{m\cdot s^{-1}}}\right)^2$$

$$\times\exp\left(-\frac{(4.65\times10^{-26}\ \mathrm{kg})[(300.0\ \mathrm{m\cdot s^{-1}})^2 - (500.0\ \mathrm{m\cdot s^{-1}})^2][(1\ \mathrm{J})/(1\ \mathrm{kg\cdot m^2\cdot s^{-2}})]}{2(1.381\times10^{-23}\ \mathrm{J\cdot K^{-1}})(100.0\ \mathrm{K})}\right)$$

$$= 5.33$$

Within the rules of significant figures, the values of these ratios are the same.

2.13 At a given temperature there are two velocities that have the same value of $(dN/L)/dv$, one on each side of the maximum of the distribution function (see Fig. 2-4). Derive a general equation relating these velocities. Determine the velocity that has the same value of $(dN/L)/dv$ as does 1 mol of N_2 molecules at 100 K having a velocity of $100\ \mathrm{m\cdot s^{-1}}$.

▌ For the two velocities, v_1 and v_2, *(2.6)* gives

$$\frac{dN/L}{dv} = 4\pi v_1^2\left(\frac{m}{2\pi kT}\right)^{3/2} e^{-mv_1^2/2kT} = 4\pi v_2^2\left(\frac{m}{2\pi kT}\right)^{3/2} e^{-mv_2^2/2kT}$$

which simplifies to

$$v_1^2\, e^{-mv_1^2/2kT} = v_2^2\, e^{-mv_2^2/2kT} \qquad (2.8)$$

Substituting the data gives

$$(100\ \mathrm{m\cdot s^{-1}})^2\exp\left(-\frac{(4.65\times10^{-26}\ \mathrm{kg})(100.0\ \mathrm{m\cdot s^{-1}})^2[(1\ \mathrm{J})/(1\ \mathrm{kg\cdot m^2\cdot s^{-2}})]}{2(1.381\times10^{-23}\ \mathrm{J\cdot K^{-1}})(100.0\ \mathrm{K})}\right)$$

$$= v_2^2\exp\left(-\frac{(4.65\times10^{-26}\ \mathrm{kg})v_2^2[(1\ \mathrm{J})/(1\ \mathrm{kg\cdot m^2\cdot s^{-2}})]}{2(1.381\times10^{-23}\ \mathrm{J\cdot K^{-1}})(100.0\ \mathrm{K})}\right)$$

This equation is not easily solved for v_2. (Several approximate methods are available to solve for v^2, such as successive approximation techniques, graphical methods, iterative techniques similar to the Newton–Raphson method, and inexpensive computer software programs.) The solution to this equation is $v_2 = 427\ \mathrm{m\cdot s^{-1}}$. Note that this value agrees with the plot at 100 K in Fig. 2-4 where $(dN/L)/dv = 1.32\times10^{-3}\ \mathrm{s\cdot m^{-1}}$ for both velocities.

2.14 At what temperature will the value of $(dN/L)/dv$ be the same for $400\ \mathrm{m \cdot s^{-1}}$ and $1200\ \mathrm{m \cdot s^{-1}}$ for 1 mol of N_2 molecules?

▌ Solving for T in *(2.8)* by collecting terms and taking logarithms gives

$$T = \frac{-m(v_2^2 - v_1^2)}{2k \ln (v_1^2/v_2^2)}$$

$$= \frac{-(4.65 \times 10^{-26}\ \mathrm{kg})[(1200\ \mathrm{m \cdot s^{-1}})^2 - (400\ \mathrm{m \cdot s^{-1}})^2][(1\ \mathrm{J})/(1\ \mathrm{kg \cdot m^2 \cdot s^{-2}})]}{2(1.381 \times 10^{-23}\ \mathrm{J \cdot K^{-1}}) \ln [(400\ \mathrm{m \cdot s^{-1}})^2/(1200\ \mathrm{m \cdot s^{-1}})^2]}$$

$$= 981\ \mathrm{K}$$

Note that this temperature is quite close to the plot at 1000 K in Fig. 2-4, where $(dN/L)/dv$ is approximately $6 \times 10^{-4}\ \mathrm{s \cdot m^{-1}}$ at both of these velocities.

2.15 What is the limiting slope of $(dN/L)/dv$ as v approaches 0 and as it approaches ∞?

▌ Differentiating *(2.6)* with respect to v gives the slope as

$$\frac{d}{dv}\left(\frac{dN/L}{dv}\right) = 4\pi\left(\frac{m}{2\pi kT}\right)^{3/2} e^{-mv^2/2kT}\left(2v - \frac{mv^3}{kT}\right)$$

Taking the limit as $v \to 0$ gives

$$\lim_{v\to 0}\ (\text{slope}) = 4\pi\left(\frac{m}{2\pi kT}\right)^{3/2} \lim_{v\to 0}\left(\frac{2v - mv^3/kT}{e^{mv^2/2kT}}\right) = 0$$

To find the limit as $v \to \infty$, it is necessary to apply l'Hôpital's rule

$$\lim_{v\to 0}\ (\text{slope}) = 4\pi\left(\frac{m}{2\pi kT}\right)^{3/2} \lim_{v\to 0}\left[\frac{2v - mv^3/kT}{e^{mv^2/2kT}}\right]$$

$$= 4\pi\left(\frac{m}{2\pi kT}\right)^{3/2} \lim_{v\to 0}\left(\frac{2 - 3mv^2/kT}{(mv/kT)\, e^{mv^2/2kT}}\right)$$

$$= 4\pi\left(\frac{m}{2\pi kT}\right)^{3/2} \lim_{v\to 0}\left(\frac{-6mv/kT}{(m/kT)(1 + mv^2/kT)\, e^{mv^2/2kT}}\right)$$

$$= 4\pi\left(\frac{m}{2\pi kT}\right)^{3/2} \lim_{v\to 0}\left(\frac{-6}{(mv/kT)(3 + mv^2/kT)\, e^{mv^2/2kT}}\right) = 0$$

The curves in Fig. 2-4 show that the limiting slopes are both zero.

2.16 Derive an expression for $\overline{v^3}$, the "skewness" of $(dN/L)/dv$.

▌ Letting $h(u) = v^3$ and following the same procedure as in Problem 2.1 gives

$$\overline{v^3} = \int_0^\infty v^3 4\pi v^2\left(\frac{m}{2\pi kT}\right)^{3/2} e^{-mv^2/2kT}\, dv$$

$$= 4\pi\left(\frac{m}{2\pi kT}\right)^{3/2}\left(\frac{2kT}{m}\right)^3 \int_0^\infty x^5\, e^{-x^2}\, dx$$

$$= 4\pi\left(\frac{m}{2\pi kT}\right)^{3/2}\left(\frac{2kT}{m}\right)^3 (1) = 4\pi\left(\frac{2kT}{\pi m}\right)^{3/2}$$

2.17 Derive an equation for $\bar{v}$. Calculate $\bar{v}$ for N_2 at 298 K.

▌ Substituting $h(u) = v$ into *(2.2)* and following the same procedure as in Problem 2.1 gives

$$\bar{v} = \int_0^\infty v 4\pi v^2\left(\frac{m}{2\pi kT}\right)^{3/2} e^{-mv^2/2kT}\, dv$$

$$= 4\pi\left(\frac{m}{2\pi kT}\right)^{3/2}\left(\frac{2kT}{m}\right)^2 \int_0^\infty x^3\, e^{-x^2}\, dx$$

$$= 4\pi\left(\frac{m}{2\pi kT}\right)^{3/2}\left(\frac{2kT}{m}\right)^2\left(\frac{1}{2}\right) = \left(\frac{8kT}{\pi m}\right)^{1/2} = \left(\frac{8RT}{\pi M}\right)^{1/2} \qquad (2.9)$$

For N_2 at 298 K,

$$\bar{v} = \left(\frac{8(8.314\ \mathrm{J\cdot K^{-1}\cdot mol^{-1}})[(1\ \mathrm{kg\cdot m^2\cdot s^{-2}})/(1\ \mathrm{J})](298\ \mathrm{K})}{\pi(28.0\times 10^{-3}\ \mathrm{kg\cdot mol^{-1}})}\right)^{1/2} = 475\ \mathrm{m\cdot s^{-1}}$$

2.18 Derive an equation for the root-mean-square speed. Calculate v_{rms} for N_2 at 298 K.

▌ The root-mean-square speed is defined by *(1.61)* as $v_{rms} = (\overline{v^2})^{1/2}$. Substituting $h(u) = v^2$ into *(2.2)* and following the same procedure as in Problem 2.1 give

$$\overline{v^2} = \int_0^\infty v^2 4\pi v^2 \left(\frac{m}{2\pi kT}\right)^{3/2} e^{-mv^2/2kT}\, dv$$

$$= 4\pi\left(\frac{m}{2\pi kT}\right)^{3/2}\left(\frac{2kT}{m}\right)^{5/2}\int_0^\infty x^4 e^{-x^2}\, dx$$

$$= 4\pi\left(\frac{m}{2\pi kT}\right)^{3/2}\left(\frac{2kT}{m}\right)^{5/2}\left(\frac{3\pi^{1/2}}{8}\right) = \frac{3kT}{m} \qquad (2.10)$$

Thus,

$$v_{rms} = (3kT/m)^{1/2} = (3RT/M)^{1/2} \qquad (1.62)$$

For N_2 at 298 K,

$$v_{rms} = \left[\frac{3(8.314\ \mathrm{J\cdot K^{-1}\cdot mol^{-1}})[(1\ \mathrm{kg\cdot m^2\cdot s^{-2}})/(1\ \mathrm{J})](298\ \mathrm{K})}{28.0\times 10^{-3}\ \mathrm{kg\cdot mol^{-1}}}\right] = 515\ \mathrm{m\cdot s^{-1}}$$

2.19 Show that $\overline{v^2} \neq \bar{v}^2$. How is the difference between these terms related to the standard deviation?

▌ The difference between these terms is shown by *(2.3b)* to be σ_v^2. Substituting *(2.9)* and *(2.10)* gives

$$\sigma_v^2 = \overline{v^2} - \bar{v}^2 = \frac{3kT}{m} - \frac{8kT}{\pi m} = \left(3 - \frac{8}{\pi}\right)\left(\frac{kT}{m}\right)$$

which is nonzero.

2.20 Derive an equation for the most probable velocity. Calculate v_{mp} for N_2 at 298 K.

▌ Setting the derivative of $(dN/L)/dv$ with respect to v found in Problem 2.15 equal to zero and solving for v_{mp} gives

$$4\pi\left(\frac{m}{2\pi kT}\right)^{3/2} e^{-mv_{mp}^2/2kT}\left(2v_{mp} - \frac{mv_{mp}^3}{kT}\right) = 0$$

$$v_{mp} = (2kT/m)^{1/2} = (2RT/M)^{1/2} \qquad (2.11)$$

For N_2 at 298 K,

$$v_{mp} = \left(\frac{2(8.314\ \mathrm{J\cdot K^{-1}\cdot mol^{-1}})[(1\ \mathrm{kg\cdot m^2\cdot s^{-2}})/(1\ \mathrm{J})](298\ \mathrm{K})}{28.0\times 10^{-3}\ \mathrm{kg\cdot mol^{-1}}}\right)^{1/2} = 421\ \mathrm{m\cdot s^{-1}}$$

This value of v_{mp} agrees with the maximum shown for the N_2 plot in Fig. 2-5.

2.21 Find the ratio $v_{rms}:\bar{v}:v_{mp}$. Confirm this ratio for N_2 molecules at 298 K.

▌ Substituting *(1.62)*, *(2.9)*, and *(2.11)* gives

$$v_{rms}:\bar{v}:v_{mp} = (3RT/M)^{1/2}:(8RT/\pi M)^{1/2}:(2RT/M)^{1/2}$$

$$= 3^{1/2}:(8/\pi)^{1/2}:2^{1/2} = 1.000:0.921:0.817$$

Substituting the values determined in Problems 2.18, 2.17, and 2.20 gives the ratio for N_2 as

$$v_{rms}:\bar{v}:v_{mp}:(515\ \mathrm{m\cdot s^{-1}}):(475\ \mathrm{m\cdot s^{-1}}):(421\ \mathrm{m\cdot s^{-1}}) = 1.000:0.922:0.817$$

2.22 Determine the ratio of the average velocity of He to N_2. Is this ratio independent of temperature?

▌ Using *(2.9)* for each gas gives the ratio as

$$\frac{v(\mathrm{He})}{v(\mathrm{N_2})} = \frac{[8RT/\pi M(\mathrm{He})]^{1/2}}{[8RT/\pi M(\mathrm{N_2})]^{1/2}} = \left[\frac{M(\mathrm{N_2})}{M(\mathrm{He})}\right]^{1/2} = \left(\frac{28.0\ \mathrm{g\cdot mol^{-1}}}{4.00\ \mathrm{g\cdot mol^{-1}}}\right)^{1/2} = 2.65$$

The ratio is independent of temperature.

2.23 At what temperature will the average velocity of N_2 be equal to that of He at 298 K?

▌ Equating $\bar{v}(\text{He}) = \bar{v}(N_2)$ as given by *(2.9)* gives

$$[8RT(\text{He})/\pi M(\text{He})]^{1/2} = [8RT(N_2)/\pi M(N_2)]^{1/2}$$

which upon simplification and substituting of data gives

$$T(N_2) = T(\text{He})\frac{M(N_2)}{M(\text{He})} = (298\ \text{K})\frac{28.0\ \text{g}\cdot\text{mol}^{-1}}{4.00\ \text{g}\cdot\text{mol}^{-1}} = 2090\ \text{K}$$

2.24 At what temperature will the average velocity of N_2 molecules become equal to the root-mean-square speed at 25 °C?

▌ Equating the expressions for $\bar{v}$ and v_{rms} given by *(2.9)* and *(1.62)* gives

$$[8RT(\bar{v})/\pi M]^{1/2} = [3RT(v_{\text{rms}})/M]^{1/2}$$

Solving for $T(\bar{v})$ and substituting the data gives

$$T(\bar{v}) = 3\pi T(v_{\text{rms}})/8 = 3\pi(298\ \text{K})/8 = 351\ \text{K}$$

2.25 What is the molar mass of a gas having molecules that have a value of $v_{\text{mp}} = 346\ \text{m}\cdot\text{s}^{-1}$ (the speed of sound in dry air) at 25 °C?

▌ Solving *(2.11)* for M and substituting the data gives

$$\begin{aligned} M &= 2RT/v_{\text{mp}}^2 \\ &= \frac{2(8.314\ \text{J}\cdot\text{K}^{-1}\cdot\text{mol}^{-1})[(1\ \text{kg}\cdot\text{m}^2\cdot\text{s}^{-2})/(1\ \text{J})](298\ \text{K})[(10^3\ \text{g})/(1\ \text{kg})]}{(346\ \text{m}\cdot\text{s}^{-1})^2} \\ &= 41.1\ \text{g}\cdot\text{mol}^{-1} \end{aligned}$$

2.26 The fraction of molecules in 1 mol of gas having a velocity $v \geq c$ is given by

$$\frac{N_{v\geq c}}{L} = \frac{2}{\pi^{1/2}}\left(\frac{c}{v_{\text{mp}}}\right)e^{-(c/v_{\text{mp}})^2} + 1 - \text{erf}\left(\frac{c}{v_{\text{mp}}}\right) \tag{2.12}$$

where the definition of the *error function* erf (z) is given in Table 2-1. Show that the median velocity v_{med} is given by

$$v_{\text{med}} = 1.5382(kT/m)^{1/2} = 1.5382(RT/M)^{1/2} \tag{2.13}$$

▌ The value of the argument c/v_{mp} in *(2.12)* is

$$c/v_{\text{mp}} = 1.5382(RT/M)^{1/2}/(2RT/M)^{1/2} = 1.088$$

giving

$$\frac{N_{v\geq v_{\text{med}}}}{L} = \frac{2}{\pi^{1/2}}(1.088)\,e^{-(1.088)^2} + 1 - \text{erf}\,(1.088) = 0.50$$

where erf (1.088) was evaluated using Fig. 2-1. This is the expected result because the median of any property represents the value at which half the distribution has greater values and half has lesser values.

2.27 What fraction of molecules have velocities within the range $v_{\text{mp}} \leq v \leq v_{\text{rms}}$?

▌ Following the procedure shown in Problem 2.26 for the fraction of molecules with $v \geq v_{\text{mp}}$ gives $c/v_{\text{mp}} = v_{\text{mp}}/v_{\text{mp}} = 1.000$.

$$\frac{N_{v\geq v_{\text{mp}}}}{L} = \frac{2}{\pi^{1/2}}(1.000)\,e^{-(1.000)^2} + 1 - \text{erf}\,(1.000) = 0.58$$

and for the fraction with $v \geq v_{\text{rms}}$ it gives

$$\frac{c}{v_{\text{mp}}} = \frac{v_{\text{rms}}}{v_{\text{mp}}} = \frac{(3RT/M)^{1/2}}{(2RT/M)^{1/2}} = 1.225$$

$$\frac{N_{v\geq v_{\text{rms}}}}{L} = \frac{2}{\pi^{1/2}}(1.225)\,e^{-(1.225)^2} + 1 - \text{erf}\,(1.225) = 0.39$$

Subtracting the values gives

$$\frac{N_{v_{\text{mp}}\leq v\leq v_{\text{rms}}}}{L} = 0.58 - 0.39 = 0.19$$

2.28 Determine the escape velocity of gaseous molecules from the earth. Calculate the fraction of N_2 molecules and the fraction of H_2 molecules that have $v \geq v_e$ at 298 K.

▮ The escape velocity from the earth is given by *(1.63)* as

$$v_e = (2gr)^{1/2} = [2(9.81\ \text{m}\cdot\text{s}^{-2})(6.37 \times 10^6\ \text{m})]^{1/2} = 1.12 \times 10^4\ \text{m}\cdot\text{s}^{-1}$$

At 298 K, $v_{mp} = 421\ \text{m}\cdot\text{s}^{-1}$ for N_2 (see Problem 2.20), and for H_2, *(2.11)* gives

$$v_{mp}(H_2) = \left(\frac{2(8.314\ \text{J}\cdot\text{K}^{-1}\cdot\text{mol}^{-1})[(1\ \text{kg}\cdot\text{m}^2\cdot\text{s}^{-2})/(1\ \text{J})](298\ \text{K})}{2.02 \times 10^{-3}\ \text{kg}\cdot\text{mol}^{-1}}\right)^{1/2}$$

$$= 1570\ \text{m}\cdot\text{s}^{-1}$$

For $c \gg v_{mp}$, *(2.12)* becomes

$$\frac{N_{v\geq c}}{L} = \left(\frac{2}{\pi^{1/2}}\right) e^{-(c/v_{mp})^2}\left(\frac{c}{v_{mp}} + \frac{v_{mp}}{c}\right) \tag{2.14}$$

Substituting the data gives

$$\frac{N_{v\geq v_e(N_2)}}{L} = \frac{2}{\pi^{1/2}}\left\{\exp\left[-\left(\frac{1.12 \times 10^4\ \text{m}\cdot\text{s}^{-1}}{421\ \text{m}\cdot\text{s}^{-1}}\right)^2\right]\right\}\left(\frac{1.12 \times 10^4\ \text{m}\cdot\text{s}^{-1}}{421\ \text{m}\cdot\text{s}^{-1}} + \frac{421\ \text{m}\cdot\text{s}^{-1}}{1.12 \times 10^4\ \text{m}\cdot\text{s}^{-1}}\right)$$

$$= 1.3 \times 10^{-306}$$

$$\frac{N_{v\geq v_e(H_2)}}{L} = \frac{2}{\pi^{1/2}}\left\{\exp\left[-\left(\frac{1.12 \times 10^4}{1570}\right)^2\right]\right\}\left(\frac{1.12 \times 10^4}{1570} + \frac{1570}{1.12 \times 10^4}\right)$$

$$= 6.5 \times 10^{-22}$$

This means that very few N_2 molecules can escape but many H_2 molecules can escape.

2.29 A simple sketch of a molecular velocity selector is shown in Fig. 2-6. The relation between the rotation frequency ν and the velocity of the molecules that successfully pass through the slits is given by

$$v = (360°/\theta)l\nu \tag{2.15}$$

where l is the distance between the slits and θ is the slit displacement angle. What is the rotation frequency needed to pass N_2 molecules at 298 K with $v_{mp} = 421\ \text{m}\cdot\text{s}^{-1}$ through a selector with $l = 10.0$ cm and $\theta = 2.00°$?

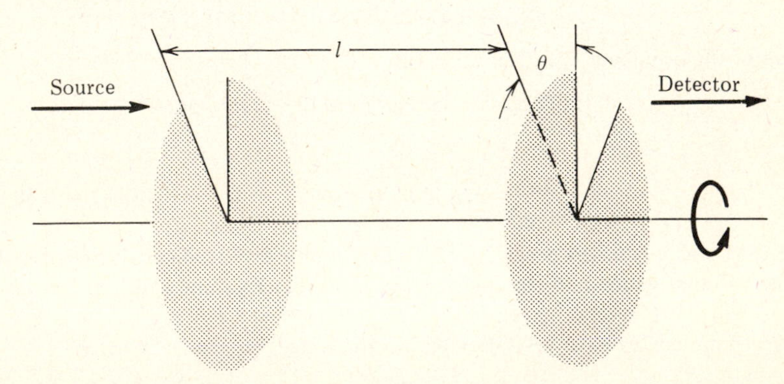

Fig. 2-6

▮ Solving *(2.15)* for ν and substituting the data gives

$$\nu = \frac{v_{mp}\theta}{(360°)l} = \frac{(421\ \text{m}\cdot\text{s}^{-1})(2.00°)[(10^2\ \text{cm})/(1\ \text{m})]}{(360°)(10.0\ \text{cm})} = 23.4\ \text{s}^{-1}$$

2.30 The velocity selector described in Problem 2.29 was used to analyze the distribution of velocities in a sample of CO_2 at 298 K. Use the following rotation frequency–detector relative intensity (I/I_0) data to prepare a plot of the distribution of velocities.

$\nu/(\text{s}^{-1})$	5	10	15	20	25	30	35	40	45	50	60
I/I_0	18.4	59.2	93.1	100.0	81.6	53.5	28.5	12.7	4.7	1.5	0.4

▮ Substituting $\nu = 5\ \text{s}^{-1}$ into *(2.15)* gives

$$v = \frac{360°}{2°}(10.0\ \text{cm})\frac{10^{-2}\ \text{m}}{1\ \text{cm}}(5\ \text{s}^{-1}) = 90\ \text{m}\cdot\text{s}^{-1}$$

The plot is shown in Fig. 2-7. [Compare it to the plot for CO_2 shown in Fig. 2-5.]

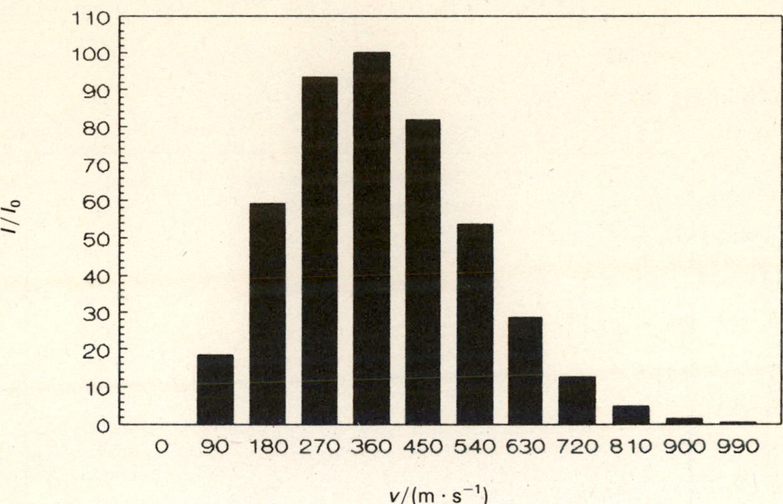

Fig. 2-7

2.31 The fraction of gaseous molecules in 1 mol with translational kinetic energy along the x direction between $\varepsilon_{\text{trans},x}$ and $\varepsilon_{\text{trans},x} + d\varepsilon_{\text{trans},x}$ is given by

$$\frac{dN/L}{d\varepsilon_{\text{trans},x}} = \frac{1}{(\pi kT\varepsilon_{\text{trans},x})^{1/2}}\,e^{-\varepsilon_{\text{trans},x}/kT} \tag{2.16}$$

Derive an equation for the average translational kinetic energy in one dimension, and calculate the value for N_2 at 298 K.

▮ Substituting $h(u) = \varepsilon_{\text{trans},x}$ into *(2.2)* and following the same procedure as in Problem 2.1 gives

$$\begin{aligned}\bar{\varepsilon}_{\text{trans},x} &= \int_0^\infty \varepsilon_{\text{trans},x}\frac{1}{(\pi kT\varepsilon_{\text{trans},x})^{1/2}}\,e^{-\varepsilon_{\text{trans},x}/kT}\,d\varepsilon_{\text{trans},x} \\ &= \frac{1}{(\pi kT)^{1/2}}\int_0^\infty \varepsilon_{\text{trans},x}^{1/2}\,e^{-\varepsilon_{\text{trans},x}/kT}\,d\varepsilon_{\text{trans},x} \\ &= \frac{1}{(\pi kT)^{1/2}}(kT)^{3/2}\int_0^\infty x^{1/2}\,e^{-x}\,dx \\ &= \frac{1}{(\pi kT)^{1/2}}(kT)^{3/2}\frac{\pi^{1/2}}{2} = \frac{kT}{2}\end{aligned} \tag{2.17}$$

For any gas at 298 K,

$$\varepsilon_{\text{trans},x} = (1.381\times 10^{-23}\ \text{J}\cdot\text{K}^{-1})(298\ \text{K})/2 = 2.06\times 10^{-21}\ \text{J}$$

or $1239\ \text{J}\cdot\text{mol}^{-1}$.

2.32 Show that the maximum of a plot of $\varepsilon_{\text{trans},x}\,(dN/L)/d\varepsilon_{\text{trans},x}$ against $\varepsilon_{\text{trans},x}$ is equal to $\bar{\varepsilon}_{\text{trans},x}$. Prepare a plot of $\varepsilon_{\text{trans},x}\,(dN/L)/d\varepsilon_{\text{trans},x}$ for $0 \le \varepsilon_{\text{trans},x}/(10^{-21}\ \text{J}) \le 20$ at 298 K.

▮ Differentiating *(2.16)* multiplied by $\varepsilon_{\text{trans},x}$ with respect to x gives

$$\begin{aligned}\frac{d}{dx}\left(\frac{\varepsilon_{\text{trans},x}(dN/L)}{d\varepsilon_{\text{trans},x}}\right) &= \frac{d}{dx}\left(\frac{\varepsilon_{\text{trans},x}^{1/2}}{(\pi kT)^{1/2}}\,e^{-\varepsilon_{\text{trans},x}/kT}\right) \\ &= \frac{1}{(\pi kT)^{1/2}}\left(\frac{1}{2\varepsilon_{\text{trans},x}^{1/2}} - \frac{\varepsilon_{\text{trans},x}^{1/2}}{kT}\right)e^{-\varepsilon_{\text{trans},x}/kT}\end{aligned}$$

Setting the derivative equal to zero at $\varepsilon_{\text{trans},x} = \varepsilon_{\max}$ gives

$$\frac{1}{2\varepsilon_{\max}^{1/2}} - \frac{\varepsilon_{\max}^{1/2}}{kT} = 0 \qquad \varepsilon_{\max} = \frac{kT}{2} = \bar{\varepsilon}_{\text{trans},x}$$

As a sample calculation, consider $\varepsilon_{\text{trans},x} = 2.00 \times 10^{-21}$ J:

$$\frac{\varepsilon_{\text{trans},x}(dN/L)}{d\varepsilon_{\text{trans},x}} = \frac{(2.00 \times 10^{-21}\ \text{J})^{1/2}}{[\pi(1.381 \times 10^{-23}\ \text{J}\cdot\text{K}^{-1})(298\ \text{K})]^{1/2}} \exp\left(-\frac{2.00 \times 10^{-21}\ \text{J}}{(1.381 \times 10^{-23}\ \text{J}\cdot\text{K}^{-1})(298\ \text{K})}\right)$$
$$= 0.242$$

The graph is shown in Fig. 2-8.

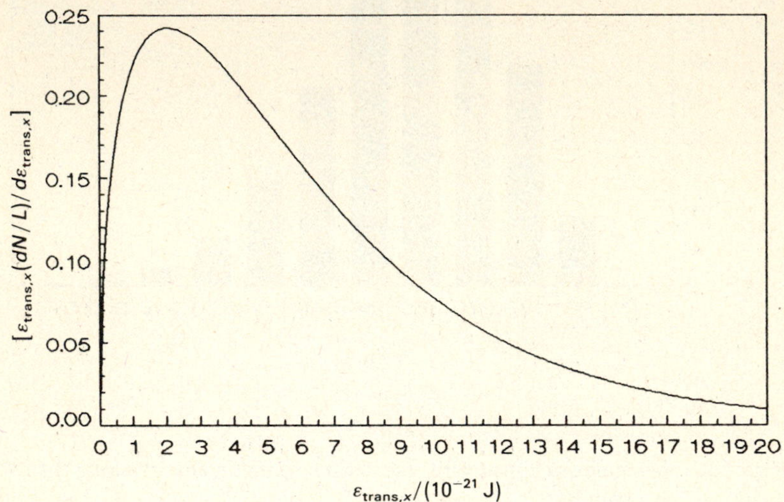

Fig. 2-8

2.33 The fraction of molecules in 1 mol having a translational kinetic energy between $\varepsilon_{\text{trans}}$ and $\varepsilon_{\text{trans}} + d\varepsilon_{\text{trans}}$ is given by

$$\frac{dN/L}{d\varepsilon_{\text{trans}}} = \frac{2\varepsilon_{\text{trans}}^{1/2}}{\pi^{1/2}(kT)^{3/2}} e^{-\varepsilon_{\text{trans}}/kT} \tag{2.18}$$

Prepare plots of $(dN/L)/d\varepsilon_{\text{trans}}$ against $\varepsilon_{\text{trans}}$ for the range $0 \le \varepsilon_{\text{trans}}/(10^{-20}\ \text{J}) \le 5.00$ for N_2 at 100 K, 298 K, and 1000 K. How will the plots differ for O_2?

▌ As a sample calculation, consider $\varepsilon_{\text{trans}} = 1.00 \times 10^{-20}$ J at 298 K:

$$\frac{dN/L}{d\varepsilon_{\text{trans}}} = \frac{2(1.00 \times 10^{-20}\ \text{J})^{1/2}}{\pi^{1/2}[(1.381 \times 10^{-23}\ \text{J}\cdot\text{K}^{-1})(298\ \text{K})]^{3/2}} \exp\left(-\frac{1.00 \times 10^{-20}\ \text{J}}{(1.381 \times 10^{-23}\ \text{J}\cdot\text{K}^{-1})(298\ \text{K})}\right)$$
$$= 3.76 \times 10^{19}\ \text{J}^{-1}$$

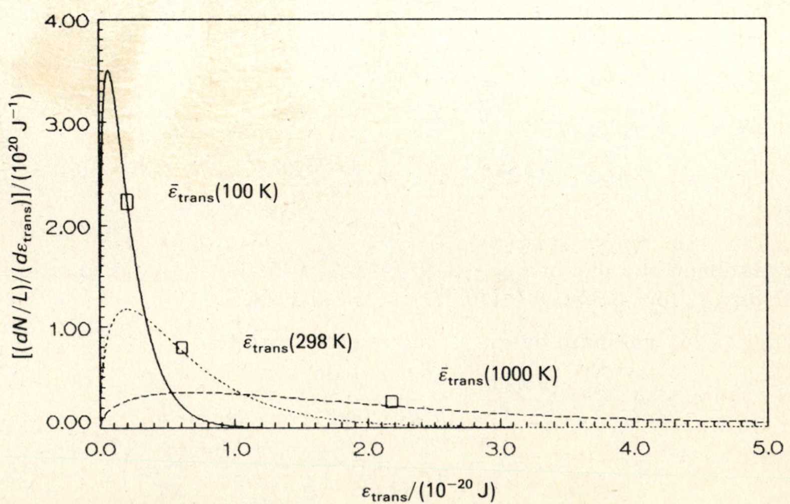

Fig. 2-9

The graph is shown in Fig. 2-9. The same curves are valid for O_2 because *(2.18)* is a function of temperature only and is not a function the identity of the gas.

2.34 Show that the expression for $(dN/L)/d\varepsilon_{trans}$ is normalized.

▮ Solving *(2.18)* for dN/L and integrating over the range $0 \le \varepsilon_{trans} \le \infty$ gives

$$\int \frac{dN}{L} = \int_0^\infty \frac{2\varepsilon_{trans}^{1/2}}{\pi^{1/2}(kT)^{3/2}} e^{-\varepsilon_{trans}/kT}\, d\varepsilon_{trans}$$

$$= \frac{2}{\pi^{1/2}(kT)^{3/2}} \int_0^\infty \varepsilon_{trans}^{1/2} e^{-\varepsilon_{trans}/kT}\, d\varepsilon_{trans}$$

$$= \frac{2}{\pi^{1/2}(kT)^{3/2}} (kT)^{3/2} \int_0^\infty x^{1/2} e^{-x}\, dx$$

$$= [2/\pi^{1/2}(kT)^{3/2}](kT)^{3/2}(\pi^{1/2}/2) = 1$$

2.35 What is the limiting slope of $(dN/L)/d\varepsilon_{trans}$ as ε_{trans} approaches 0 and as it approaches ∞?

▮ The slope of *(2.18)* is found by differentiating with respect to $d\varepsilon_{trans}$, giving

$$\frac{d}{d\varepsilon_{trans}}\left(\frac{dN/L}{d\varepsilon_{trans}}\right) = \frac{d}{d\varepsilon_{trans}}\left(\frac{2\varepsilon_{trans}^{1/2}}{\pi^{1/2}(kT)^{3/2}} e^{-\varepsilon_{trans}/kT}\right)$$

$$= \frac{2}{\pi^{1/2}(kT)^{3/2}}\left(\frac{1}{2\varepsilon_{trans}^{1/2}} - \frac{\varepsilon_{trans}^{1/2}}{kT}\right) e^{-\varepsilon_{trans}/kT}$$

As $\varepsilon_{trans} \to 0$,

$$\lim_{\varepsilon_{trans}\to 0} (\text{slope}) = \frac{2}{\pi^{1/2}(kT)^{3/2}} \lim_{\varepsilon_{trans}\to 0} \left[\frac{1/2\varepsilon_{trans}^{1/2} - \varepsilon_{trans}^{1/2}/kT}{e^{\varepsilon_{trans}/kT}}\right] = \infty$$

To find the limit as $\varepsilon_{trans} \to \infty$ it is necessary to apply l'Hôpital's rule,

$$\lim_{\varepsilon_{trans}\to\infty} (\text{slope}) = \frac{2}{\pi^{1/2}(kT)^{3/2}} \lim_{\varepsilon_{trans}\to\infty} \left[\frac{1/2\varepsilon_{trans}^{1/2} - \varepsilon_{trans}^{1/2}/kT}{e^{\varepsilon_{trans}/kT}}\right]$$

$$= \frac{2}{\pi^{1/2}(kT)^{3/2}} \lim_{\varepsilon_{trans}\to\infty} \left[\frac{-1/4\varepsilon_{trans}^{3/2} + 1/2kT\varepsilon_{trans}^{1/2}}{(1/kT)\, e^{\varepsilon_{trans}/kT}}\right] = 0$$

2.36 What fraction of molecules have energies in the range of 6.17×10^{-21} J and 6.18×10^{-21} J at 298 K? How does the fraction depend on $d\varepsilon_{trans}$?

▮ Solving *(2.18)* for dN/L and substituting $d\varepsilon_{trans} = 1 \times 10^{-23}$ J gives

$$\frac{dN}{L} = \frac{2(6.17 \times 10^{-21}\ \mathrm{J})^{1/2}}{\pi^{1/2}[(1.381 \times 10^{-23}\ \mathrm{J \cdot K^{-1}})(298\ \mathrm{K})]^{3/2}} \exp\left(-\frac{6.17 \times 10^{-21}\ \mathrm{J}}{(1.381 \times 10^{-23}\ \mathrm{J \cdot K^{-1}})(298\ \mathrm{K})(1 \times 10^{-23}\ \mathrm{J})}\right)$$

$$= 7 \times 10^{-4}$$

The fraction is directly proportional to $d\varepsilon_{trans}$.

2.37 Derive an expression for $\bar{\varepsilon}_{trans}$. Also derive an expression for $\bar{E}_{trans}$, the average molar translational kinetic energy.

▮ Substituting $h(u) = \varepsilon_{trans}$ into *(2.2)* and following the same procedure as in Problem 2.1 gives

$$\bar{\varepsilon}_{trans} = \int_0^\infty \varepsilon_{trans} \frac{2\varepsilon_{trans}^{1/2}}{\pi^{1/2}(kT)^{3/2}} e^{-\varepsilon_{trans}/kT}\, d\varepsilon_{trans}$$

$$= \frac{2}{\pi^{1/2}(kT)^{3/2}} \int_0^\infty \varepsilon_{trans}^{3/2} e^{-\varepsilon_{trans}/kT}\, d\varepsilon_{trans}$$

$$= \frac{2}{\pi^{1/2}(kT)^{3/2}} (2)(kT)^{5/2} \int_0^\infty x^4 e^{-x^2}\, dx$$

$$= \frac{2}{\pi^{1/2}(kT)^{3/2}} (2)(kT)^{5/2} \frac{3\pi^{1/2}}{8} = \tfrac{3}{2}kT \qquad (2.19a)$$

For 1 mol of gas,

$$\bar{E}_{trans} = L(\tfrac{3}{2}kT) = \tfrac{3}{2}RT \qquad (2.19b)$$

2.38 Calculate ε_{trans} at 100 K, 298 K, and 1000 K. Indicate these values in Fig. 2-9.

▌ At 100 K, *(2.19a)* gives

$$\bar{\varepsilon}_{trans}(100\text{ K}) = \tfrac{3}{2}(1.381 \times 10^{-23}\text{ J}\cdot\text{K}^{-1})(100\text{ K}) = 2.07 \times 10^{-21}\text{ J}$$

Likewise,

$$\bar{\varepsilon}_{trans}(298\text{ K}) = 6.17 \times 10^{-21}\text{ J} \qquad \text{and} \qquad \bar{\varepsilon}_{trans}(1000\text{ K}) = 2.07 \times 10^{-20}\text{ J}$$

See Fig. 2-9.

2.39 Consider a system such that at some initial time all of the molecules in 1 mol of Ne have the same velocity, $275\text{ m}\cdot\text{s}^{-1}$. As time passes, a Maxwell distribution is established. Determine the equilibrium temperature of the system.

▌ The initial translational kinetic energy of the system is given by

$$E_{trans} = L(\tfrac{1}{2}mv^2) = \tfrac{1}{2}Mv^2$$

$$= \tfrac{1}{2}(20.18 \times 10^{-3}\text{ kg}\cdot\text{mol}^{-1})(275\text{ m}\cdot\text{s}^{-1})^2 \frac{1\text{ J}}{1\text{ kg}\cdot\text{m}^2\cdot\text{s}^{-2}}$$

$$= 763\text{ J}\cdot\text{mol}^{-1}$$

Because the energy is conserved, once equilibrium is reached $\bar{E}_{trans}$ will also be equal to 763 mol^{-1}. Thus *(2.19b)* gives

$$T = \frac{2\bar{E}_{trans}}{3R} = \frac{2E_{trans}}{3R} = \frac{2(763\text{ J}\cdot\text{mol}^{-1})}{3(8.314\text{ J}\cdot\text{K}^{-1})} = 61.2\text{ K}$$

2.40 The average translational kinetic energy is $3717\text{ J}\cdot\text{mol}^{-1}$ for both He and N_2 at 298 K. Calculate the root-mean-square speed for each of these gases.

▌ For 1 mol of He, *(1.59)* gives

$$\overline{v^2(\text{He})} = \frac{2\bar{E}_{trans}}{M(\text{He})} = \frac{2(3717\text{ J}\cdot\text{mol}^{-1})[(1\text{ kg}\cdot\text{m}^2\cdot\text{s}^{-2})/(1\text{ J})]}{4.003 \times 10^{-3}\text{ kg}\cdot\text{mol}^{-1}}$$

$$= 1.857 \times 10^6\text{ m}^2\cdot\text{s}^{-2}$$

or $v_{rms} = 1363\text{ m}\cdot\text{s}^{-1}$. Likewise, for N_2, $\overline{v^2(N_2)} = 2.654 \times 10^5\text{ m}\cdot\text{s}^{-2}$ and $v_{rms} = 515\text{ m}\cdot\text{s}^{-1}$.

2.41 Find the ratio of $\bar{\varepsilon}_{trans}$ to $\varepsilon_{trans,mp}$, the maximum in the plot of $(dN/L)/d\varepsilon_{trans}$ against ε_{trans}.

▌ Setting the derivative of $(dN/L)/d\varepsilon_{trans}$ found in Problem 2.35 equal to zero at $\varepsilon_{trans} = \varepsilon_{trans,mp}$ gives

$$1/2\varepsilon_{trans,mp}^{1/2} - \varepsilon_{trans,mp}^{1/2}/kT = 0$$

$$\varepsilon_{trans,mp} = kT/2 \tag{2.20}$$

The desired ratio is found using *(2.19a)* and *(2.20)* as

$$\bar{\varepsilon}_{trans}/\varepsilon_{trans,mp} = (3kT/2)/(kT/2) = 3$$

2.42 Determine the standard deviation of the translational kinetic energy. What is the ratio of the standard deviation to $\bar{\varepsilon}_{trans}$?

▌ The square of the standard deviation is given by *(2.3b)* as $\sigma(\varepsilon_{trans})^2 = \overline{\varepsilon_{trans}^2} - \bar{\varepsilon}_{trans}^2$. Substituting $h(u) = \varepsilon_{trans}^2$ into *(2.2)* and following the same procedures as in Problem 2.1 gives

$$\overline{\varepsilon_{trans}^2} = \int_0^\infty \varepsilon_{trans}^2 \frac{2\varepsilon_{trans}^{1/2}}{\pi^{1/2}(kT)^{3/2}} e^{-\varepsilon_{trans}/kT}\, d\varepsilon_{trans}$$

$$= \frac{2}{\pi^{1/2}(kT)^{3/2}} \int_0^\infty \varepsilon_{trans}^{5/2} e^{-\varepsilon_{trans}/kT}\, d\varepsilon_{trans}$$

$$= \frac{2}{\pi^{1/2}(kT)^{3/2}} (2)(kT)^{7/2} \int_0^\infty x^6 e^{-x^2}\, dx$$

$$= \frac{2}{\pi^{1/2}(kT)^{3/2}} (2)(kT)^{7/2} \frac{15\pi^{1/2}}{16} = \frac{15(kT)^2}{4}$$

Subtracting *(2.19a)* gives $\sigma(\varepsilon_{\text{trans}})^2 = 15(kT)^2/4 - (\frac{3}{2}kT)^2 = \frac{3}{2}(kT)^2$, giving $\sigma(\varepsilon_{\text{trans}}) = (\frac{3}{2})^{1/2}kT$. The ratio is

$$\frac{\sigma(\varepsilon_{\text{trans}})}{\bar{\varepsilon}_{\text{trans}}} = \frac{(\frac{3}{2})^{1/2}kT}{\frac{3}{2}kT} = \left(\frac{2}{3}\right)^{1/2} = 0.816$$

2.43 The fraction of molecules in 1 mol of gas having a translational kinetic energy $\varepsilon_{\text{trans}} \geq \varepsilon'$ is

$$\frac{N_{\varepsilon_{\text{trans}} \geq \varepsilon'}}{L} = 2\left(\frac{\varepsilon'}{\pi kT}\right)^{1/2} e^{-\varepsilon'/kT} + 1 - \operatorname{erf}\left[\left(\frac{\varepsilon'}{kT}\right)^{1/2}\right] \tag{2.21}$$

Determine the fraction of molecules that have $\varepsilon_{\text{trans}} \geq \varepsilon_{\text{trans,mp}}$. Repeat the calculation for $\varepsilon_{\text{trans}} \geq \bar{\varepsilon}_{\text{trans}}$.

▮ Substituting *(2.20)* into *(2.21)* gives

$$\frac{N_{\varepsilon_{\text{trans}} \geq \varepsilon_{\text{trans,mp}}}}{L} = 2\left(\frac{kT/2}{\pi kT}\right)^{1/2} e^{-(kT/2)/kT} + 1 - \operatorname{erf}\left[\left(\frac{kT/2}{kT}\right)^{1/2}\right] = 0.79$$

Likewise, using *(2.19a)* gives $(N_{\varepsilon_{\text{trans}} \geq \bar{\varepsilon}_{\text{trans}}})/L = 0.39$.

2.44 Determine the number of molecules in 1 mol of gas having a translational energy between $0.9kT$ and $1.1kT$.

▮ The fraction of molecules that have $\varepsilon' \geq 0.9kT$ is given by *(2.21)* as

$$\frac{N_{\varepsilon_{\text{trans}} \geq \varepsilon'}}{L} = 2\left(\frac{0.9kT}{\pi kT}\right)^{1/2} e^{-0.9kT/kT} + 1 - \operatorname{erf}\left[\left(\frac{0.9kT}{kT}\right)^{1/2}\right] = 0.63$$

Likewise, for $\varepsilon'' \geq 1.1kT$, the fraction is $N_{\varepsilon_{\text{trans}} \geq \varepsilon''}/L = 0.53$. Subtracting the values gives

$$\frac{N_{\varepsilon' \leq \varepsilon_{\text{trans}} \leq \varepsilon''}}{L} = 0.63 - 0.53 = 0.10$$

$$N_{\varepsilon' \leq \varepsilon_{\text{trans}} \leq \varepsilon''} = (6.022 \times 10^{23}\ \text{molecules})(0.10) = 6.0 \times 10^{22}\ \text{molecules}$$

2.45 In the limiting case where $\varepsilon' > kT$, *(2.21)* becomes a special case of the *Boltzmann distribution law*:

$$N_i/N_j = (g_i/g_j)\, e^{-(\varepsilon_i - \varepsilon_j)/kT} \tag{2.22}$$

where g represents the *statistical weight* or *degeneracy* of the energy level. Compare the number of molecules with $\varepsilon_{\text{trans}} = 4.00kT$ to $\varepsilon_{\text{trans}} = 3.00kT$.

▮ For translational kinetic energy, $g = 1$. Substituting the two values of energy into *(2.22)* gives

$$\frac{N(4.00kT)}{N(3.00kT)} = \frac{1}{1} e^{-(4.00kT - 3.00kT)/kT} = 0.368$$

2.46 Calculate the ratio of the numbers of molecules separated by a translational kinetic energy of 1×10^{-21} J at 0 K, at 100 K, at 298 K, and at 1000 K. What does the trend in these answers indicate?

▮ Using *(2.22)* at 0 K gives

$$\frac{N_j}{N_i} = \frac{1}{1} \exp\left(-\frac{1 \times 10^{-21}\ \text{J}}{(1.381 \times 10^{-23}\ \text{J})(0\ \text{K})}\right) = 0$$

Likewise $N_j/N_i = 0.485$ at 100 K, 0.784 at 298 K, and 0.930 at 1000 K. As the temperature increases, the population of the higher-energy state approaches that of the lower-energy state.

2.47 Repeat the calculations of Problem 2.46 for an energy separation of 1×10^{-23} J and for an energy separation of 1×10^{-19} J. What does the trend in these answers indicate?

▮ For $\Delta\varepsilon_{\text{trans}} = 1 \times 10^{-23}$ J, $N_j/N_i = 0$ at 0 K, 0.993 at 100 K, 0.998 at 298 K, and 0.999 at 1000 K. (See Problem 2.46 for a sample calculation.) For $\Delta\varepsilon_{\text{trans}} = 1 \times 10^{-19}$ J, $N_j/N_i = 0$ at 0 K, 3.5×10^{-32} at 100 K, 2.8×10^{-11} at 298 K, and 7.2×10^{-4} at 1000 K. The same general trend as described in Problem 2.46 with increasing temperature is observed for both of these sets of results. At a given temperature, the population of the higher energy state approaches that of the lower energy state as the difference in energy decreases.

2.48 At what temperature will the ratio of the number of molecules separated by a translational kinetic energy of $1.000\ \text{kJ} \cdot \text{mol}^{-1}$ be equal to 0.25?

▮ The difference in energy is

$$\varepsilon_j - \varepsilon_i = (1.000 \text{ kJ} \cdot \text{mol}^{-1}) \frac{10^3 \text{ J}}{1 \text{ kJ}} \left(\frac{1 \text{ mol}}{6.022 \times 10^{23}} \right)$$

$$= 1.661 \times 10^{-21} \text{ J}$$

Taking logarithms of both sides of *(2.22)*, solving for T, and substituting the data gives

$$T = \frac{-(\varepsilon_j - \varepsilon_i)}{k \ln (N_j/N_i)} = \frac{-1.661 \times 10^{-21} \text{ J}}{(1.381 \times 10^{-23} \text{ J} \cdot \text{K}^{-1}) \ln 0.25} = 87 \text{ K}$$

2.2 COLLISION PARAMETERS

2.49 Calculate the molecular density of each gas in a mixture of H_2 and N_2 at 1.00 bar and 25 °C such that $x(N_2) = 0.25$.

▮ For each gas *(1.21)* gives

$$N^*(N_2) = \frac{LP(N_2)}{RT} = \frac{(6.022 \times 10^{23} \text{ mol}^{-1})(0.25 \text{ bar})[(10^5 \text{ Pa})/(1 \text{ bar})]}{(8.314 \text{ m}^3 \cdot \text{Pa} \cdot \text{K}^{-1} \cdot \text{mol}^{-1})(298 \text{ K})}$$

$$= 6.1 \times 10^{24} \text{ m}^{-3}$$

$$N^*(H_2) = \frac{(6.022 \times 10^{23})(0.75)(10^5)}{8.314(298)} = 1.82 \times 10^{25} \text{ m}^{-3}$$

2.50 Calculate the reduced mass for a system consisting of a mixture of H_2 and N_2 molecules.

▮ The reduced mass (μ) is defined as

$$\mu = m_i m_j/(m_i + m_j) = M_i M_j/(M_i + M_j)L \tag{2.23}$$

Substituting the respective molar masses gives

$$\mu = \frac{(2.02 \times 10^{-3} \text{ kg} \cdot \text{mol}^{-1})(28.01 \times 10^{-3} \text{ kg} \cdot \text{mol}^{-1})}{(2.02 \times 10^{-3} \text{ kg} \cdot \text{mol}^{-1} + 28.01 \times 10^{-3} \text{ kg} \cdot \text{mol}^{-1})(6.022 \times 10^{23} \text{ mol}^{-1})}$$

$$= 3.13 \times 10^{-27} \text{ kg}$$

2.51 Calculate the average collision diameter for a system consisting of a mixture of H_2 and N_2 molecules given the individual collision diameter is 0.271 nm for H_2 and 0.373 nm for N_2.

▮ The average collision diameter is defined as

$$\sigma_{ij} = (\sigma_i + \sigma_j)/2 \tag{2.24}$$

Substituting the individual values gives

$$\sigma(N_2, H_2) = (0.373 \text{ nm} + 0.271 \text{ nm})/2 = 0.322 \text{ nm}$$

2.52 How many collisions occur between the H_2 molecules in the mixture described in Problem 2.49?

▮ The total number of collisions per unit time per unit volume that occur between like molecules in a binary mixture is given by

$$Z_{ii} = (N_i^{*2} \pi \sigma_i^2 / 2^{1/2})(8kT/\pi m)^{1/2} \tag{2.25}$$

Substituting $N^*(H_2) = 1.82 \times 10^{25} \text{ m}^{-3}$ (see Problem 2.49) and $\sigma(H_2) = 0.271$ nm (see Problem 2.51) gives

$$Z(H_2, H_2) = [(1.82 \times 10^{25} \text{ m}^{-3})^2 (\pi)(2.71 \times 10^{-10} \text{ m})^2 / 2^{1/2}]$$

$$\times \left(\frac{8(1.381 \times 10^{-23} \text{ J} \cdot \text{K}^{-1})(298 \text{ K})[(1 \text{ kg} \cdot \text{m}^2 \cdot \text{s}^{-2})/(1 \text{ J})]}{\pi(2.02 \times 10^{-3} \text{ kg} \cdot \text{mol}^{-1})/(6.022 \times 10^{23} \text{ mol}^{-1})} \right)^{1/2}$$

$$= 9.55 \times 10^{34} \text{ m}^{-3} \cdot \text{s}^{-1}$$

2.53 Repeat the calculations of Problem 2.52 for the N_2 molecules in the mixture. Why is the collision number so much smaller than $Z(H_2, H_2)$?

▌ Substituting $N^*(N_2) = 6.1 \times 10^{24}\ m^{-3}$ (see Problem 2.49) and $\sigma(N_2) = 0.373$ nm (see Problem 2.51) into *(2.25)* gives

$$Z(N_2, N_2) = [(6.1 \times 10^{24})^2(\pi)(3.73 \times 10^{-10})^2/2^{1/2}]\left(\frac{8(1.381 \times 10^{-23})(298)}{\pi(28.01 \times 10^{-3})/(6.022 \times 10^{23})}\right)^{1/2}$$
$$= 5.5 \times 10^{33}\ m^{-3} \cdot s^{-1}$$

The value of $Z(N_2, N_2)$ reflects the square of the factor of 3 between $N^*(N_2)$ and $N^*(H_2)$.

2.54 How many collisions occur between the H_2 molecules with the N_2 molecules in the mixture described in Problem 2.49? What is the total number of collisions taking place in the mixture?

▌ The total number of collisions per unit time per unit volume that occur between unlike molecules in a binary mixture is given by

$$Z_{ij} = N_i^* N_j^* \pi\sigma_{ij}^2(8kT/\pi\mu)^{1/2} \tag{2.26}$$

Substituting the results of Problems 2.49, 2.51, and 2.50 gives

$$Z(H_2, N_2) = (1.82 \times 10^{25}\ m^{-3})(6.1 \times 10^{24}\ m^{-3})(\pi)(3.22 \times 10^{-10}\ m)^2$$
$$\times \left(\frac{8(1.381 \times 10^{-23}\ J \cdot K^{-1})(298\ K)[(1\ kg \cdot m^2 \cdot s^{-2})/(1\ J)]}{\pi(3.13 \times 10^{-27}\ kg)}\right)^{1/2}$$
$$= 6.62 \times 10^{34}\ m^{-3} \cdot s^{-1}$$

The total number of collisions will be given by

$$Z(\text{total}) = Z_{ii} + Z_{jj} + Z_{ij} \tag{2.27}$$

Substituting the results of Problems 2.52 and 2.53 gives

$$Z(\text{total}) = Z(H_2, H_2) + Z(N_2, N_2) + Z(H_2, N_2)$$
$$= 9.55 \times 10^{34}\ m^{-3} \cdot s^{-1} + 5.5 \times 10^{33}\ m^{-3} \cdot s^{-1} + 6.62 \times 10^{34}\ m^{-3} \cdot s^{-1}$$
$$= 1.672 \times 10^{35}\ m^{-3} \cdot s^{-1}$$

2.55 Consider a mixture of N_2 and H_2. At what relative concentrations will the number of collisions between unlike molecules be equal to the number of collisions between like molecules?

▌ Equating $Z(H_2, N_2)$ given by *(2.26)* to the sum of $Z(H_2, H_2)$ and $Z(N_2, N_2)$ given by *(2.25)* gives

$$N^*(N_2)N^*(He)(\pi)[\sigma(N_2, He)]^2\left(\frac{8kT}{\pi\mu}\right)^{1/2} = \frac{[N^*(N_2)]^2(\pi)[\sigma(N_2)]^2}{2^{1/2}}\left(\frac{8kT}{\pi m(N_2)}\right)^{1/2}$$
$$+ \frac{[N^*(H_2)]^2(\pi)[\sigma(H_2)]^2}{2^{1/2}}\left(\frac{8kT}{\pi m(H_2)}\right)^{1/2}$$

Cancelling common terms and substituting $\mu = 3.13 \times 10^{-27}$ kg (see Problem 2.50) and $\sigma(N_2, H_2) = 0.322$ nm, $\sigma(N_2) = 0.373$ nm, and $\sigma(H_2) = 0.271$ nm (see Problem 2.51) gives

$$\frac{N^*(N_2)N^*(H_2)(0.322\ nm)^2}{(3.13 \times 10^{-27}\ kg)^{1/2}} = \frac{[N^*(N_2)]^2(0.373\ nm)^2}{2^{1/2}[(28.01 \times 10^{-3}\ kg \cdot mol^{-1})/(6.022 \times 10^{23}\ mol^{-1})]^{1/2}}$$
$$+ \frac{[N^*(H_2)]^2(0.271\ nm)^2}{2^{1/2}[(4.00 \times 10^{-3}\ kg \cdot mol^{-1})/(6.022 \times 10^{23}\ mol^{-1})]^{1/2}}$$
$$(1.85 \times 10^{12})N^*(N_2)N^*(H_2) = (4.56 \times 10^{11})[N^*(N_2)]^2 + (6.37 \times 10^{11})[N^*(H_2)]^2$$
$$[N^*(N_2)]^2 - 4.06N^*(N_2)N^*(He) + 1.40[N^*(He)]^2 = 0$$

This equation can be factored as

$$[N^*(N_2) - 0.38N^*(H_2)][N^*(N_2) - 3.68N^*(H_2)] = 0$$

giving two sets of relative concentration conditions,

$$N^*(N_2) = 0.38N^*(H_2) \quad \text{and} \quad N^*(N_2) = 3.68N^*(He)$$

2.56 What is the number of collisions that occur in a sample of He at 1.00 bar and 25 °C? The collision diameter of He is 0.218 nm.

▮ Using *(1.21)* gives

$$N^*(\mathrm{He}) = \frac{(6.022\times 10^{23}\ \mathrm{mol^{-1}})(1.00\ \mathrm{bar})[(10^5\ \mathrm{Pa})/(1\ \mathrm{bar})]}{(8.314\ \mathrm{m^3\cdot Pa\cdot K^{-1}\cdot mol^{-1}})(298\ \mathrm{K})} = 2.43\times 10^{25}\ \mathrm{m^{-3}}$$

For a pure gas, the number of collisions is given by *(2.25)* as

$$Z(\mathrm{He}) = [(2.43\times 10^{25}\ \mathrm{m^{-3}})^2(\pi)(2.18\times 10^{-10}\ \mathrm{m})^2/2^{1/2}]$$
$$\times\left(\frac{8(1.381\times 10^{-23}\ \mathrm{J\cdot K^{-1}})(298\ \mathrm{K})[(1\ \mathrm{kg\cdot m^2\cdot s^{-2}})/(1\ \mathrm{J})]}{\pi(4.00\times 10^{-3}\ \mathrm{kg\cdot mol^{-1}})/(6.022\times 10^{23}\ \mathrm{mol^{-1}})}\right)^{1/2}$$
$$= 7.83\times 10^{34}\ \mathrm{m^{-3}\cdot s^{-1}}$$

2.57 Repeat the calculations of Problem 2.56 for He at 10.00 bar and 25 °C and also at 1.00 bar and 2980 K. Which has the greater effect on the collision number, temperature or pressure?

▮ At 10.00 bar and 25 °C,

$$N^*(\mathrm{He}) = \frac{(6.022\times 10^{23})(10.00)(10^5)}{(8.314)(298)} = 2.43\times 10^{26}\ \mathrm{m^{-3}}$$

$$Z(\mathrm{He}) = \frac{(2.43\times 10^{26})^2(\pi)(2.18\times 10^{-10})^2}{2^{1/2}}\left(\frac{8(1.381\times 10^{-23})(298)}{\pi(4.00\times 10^{-3})/(6.022\times 10^{23})}\right)^{1/2}$$
$$= 7.83\times 10^{36}\ \mathrm{m^{-3}\cdot s^{-1}}$$

and at 1.00 bar and 2980 K, $N^* = 2.43\times 10^{24}\ \mathrm{m^{-3}}$ and $Z(\mathrm{He}) = 2.48\times 10^{33}\ \mathrm{m^{-3}\cdot s^{-1}}$. The 10-fold pressure increase caused a 100-fold change in $Z(\mathrm{He})$, while the 10-fold increase in temperature caused a $10^{3/2}$-fold change.

2.58 At what temperature will the collison frequency in N_2 be equal to that in He at 25 °C if both gases are at a pressure of 1.00 bar? The collision diameter is 0.373 nm for N_2 and 0.218 nm for He.

▮ Equating *(2.25)* for the two gases gives

$$\frac{[LP/RT(\mathrm{He})]^2(\pi)[\sigma(\mathrm{He})]^2}{2^{1/2}}\left(\frac{8kT(\mathrm{He})}{\pi m(\mathrm{He})}\right)^{1/2} = \frac{[LP/RT(\mathrm{N_2})]^2(\pi)[\sigma(\mathrm{N_2})]^2}{2^{1/2}}\left(\frac{8kT(\mathrm{N_2})}{\pi m(\mathrm{N_2})}\right)^{1/2}$$

and solving for $T(\mathrm{N_2})$ gives

$$T(\mathrm{N_2}) = T(\mathrm{He})\left(\frac{\sigma(\mathrm{N_2})}{\sigma(\mathrm{He})}\right)^{4/3}\left(\frac{M(\mathrm{He})}{M(\mathrm{N_2})}\right)^{1/3}$$
$$= (298\ \mathrm{K})\left(\frac{0.373\ \mathrm{nm}}{0.218\ \mathrm{nm}}\right)^{4/3}\left(\frac{4.00\ \mathrm{g\cdot mol^{-1}}}{28.01\ \mathrm{g\cdot mol^{-1}}}\right)^{1/3} = 319\ \mathrm{K}$$

2.59 How many collisions will a single H_2 molecule undergo with other H_2 molecules in the mixture described in Problem 2.49?

▮ The number of collisions per unit time between a given molecule and like molecules in a binary mixture is given by

$$z_{ii} = N_i^*\pi\sigma_i^2(2^{1/2})\left(\frac{8kT}{\pi m_i}\right)^{1/2} \tag{2.28}$$

Substituting $N^*(\mathrm{H_2}) = 1.82\times 10^{25}\ \mathrm{m^{-3}}$ (see Problem 2.49) and $\sigma(\mathrm{H_2}) = 0.271\ \mathrm{nm}$ (see Problem 2.51) gives

$$z(\mathrm{H_2},\mathrm{H_2}) = (1.82\times 10^{25}\ \mathrm{m^{-3}})(\pi)(2.71\times 10^{-10}\ \mathrm{m})^2(2^{1/2})$$
$$\times\left(\frac{8(1.381\times 10^{-23}\ \mathrm{J\cdot K^{-1}})(298\ \mathrm{K})[(1\ \mathrm{kg\cdot m^2\cdot s^{-2}})/(1\ \mathrm{J})]}{\pi(2.02\times 10^{-3}\ \mathrm{kg\cdot mol^{-1}})/(6.022\times 10^{23}\ \mathrm{mol^{-1}})}\right)^{1/2}$$
$$= 1.05\times 10^{10}\ \mathrm{s^{-1}}$$

2.60 Repeat the calculations of Problem 2.59 for the N_2 molecules in the mixture. Why is this collision number so much smaller than $z(\mathrm{H_2},\mathrm{H_2})$?

▮ Substituting $N^*(\mathrm{N_2}) = 6.1\times 10^{24}\ \mathrm{m^{-3}}$ (see Problem 2.49) and $\sigma(\mathrm{N_2}) = 0.373\ \mathrm{nm}$ (see Problem 2.51) into *(2.28)* gives

$$z(\mathrm{N_2},\mathrm{N_2}) = (6.1\times 10^{24})(\pi)(3.73\times 10^{-10})^2(2^{1/2})\left(\frac{8(1.381\times 10^{-23})(298)}{\pi(28.01\times 10^{-3})/(6.022\times 10^{23})}\right)^{1/2}$$
$$= 1.8\times 10^{9}\ \mathrm{s^{-1}}$$

Much of the difference in the z_{ii} values is the result of the factor of 3 between $N^*(N_2)$ and $N^*(H_2)$. The remainder of the difference is the factor of 14 between the molar masses (i.e., the N_2 molecules are moving more slowly).

2.61 What is the number of collisions that occur between a single H_2 molecule with the N_2 molecules in the mixture described in Problem 2.49? What is the total number of collisions that a single H_2 molecule undergoes in the system?

▮ The number of collisions per unit time between a given molecule and unlike molecules in a binary mixture is given by

$$z_{ij} = N_j^* \pi \sigma_{ij}^2 (8kT/\pi\mu)^{1/2} \tag{2.29}$$

Substituting the results of Problems 2.49–2.51 gives

$$z(H_2, N_2) = (6.1 \times 10^{24}\ \mathrm{m^{-3}})(\pi)(3.22 \times 10^{-10}\ \mathrm{m})^2 \times \left(\frac{8(1.381 \times 10^{-23}\ \mathrm{J \cdot K^{-1}})(298\ \mathrm{K})[(1\ \mathrm{kg \cdot m^2 \cdot s^{-2}})/(1\ \mathrm{J})]}{\pi(3.13 \times 10^{-27}\ \mathrm{kg})}\right)^{1/2}$$
$$= 3.6 \times 10^9\ \mathrm{s^{-1}}$$

The total number of collisions will be given by

$$z_i(\text{total}) = z_{ii} + z_{ij} \tag{2.30}$$

Substituting the results of Problem 2.58 gives

$$z(H_2, \text{total}) = 1.05 \times 10^{10}\ \mathrm{s^{-1}} + 3.6 \times 10^9\ \mathrm{s^{-1}} = 1.41 \times 10^{10}\ \mathrm{s^{-1}}$$

2.62 Repeat the calculations of Problem 2.62 for a single N_2 molecule.

▮ The number of collisions for a N_2 molecule with H_2 molecules is given by *(2.29)* as

$$z(N_2, H_2) = (1.82 \times 10^{25})(\pi)(3.22 \times 10^{-10})^2 \left(\frac{8(1.381 \times 10^{-23})(298)}{\pi(3.13 \times 10^{-27})}\right)^{1/2}$$
$$= 1.08 \times 10^{10}\ \mathrm{s^{-1}}$$

giving

$$z(N_2, \text{total}) = 1.8 \times 10^9\ \mathrm{s^{-1}} + 1.08 \times 10^{10}\ \mathrm{s^{-1}} = 1.26 \times 10^{10}\ \mathrm{s^{-1}}$$

2.63 How many collisions does a single He atom have with other He atoms in the system described in Problem 2.56?

▮ For a pure gas, the number of collisions is given by *(2.28)*. Substituting the results of Problem 2.56 gives

$$z(\mathrm{He}) = (2.43 \times 10^{25}\ \mathrm{m^{-3}})(\pi)(2.18 \times 10^{-10})^2(2^{1/2}) \times \left(\frac{8(1.381 \times 10^{-23}\ \mathrm{J \cdot K^{-1}})(298\ \mathrm{K})[(1\ \mathrm{kg \cdot m^2 \cdot s^{-2}})/(1\ \mathrm{J})]}{\pi(4.00 \times 10^{-3}\ \mathrm{kg \cdot mol^{-1}})/(6.022 \times 10^{23}\ \mathrm{mol^{-1}})}\right)^{1/2}$$
$$= 6.44 \times 10^9\ \mathrm{s^{-1}}$$

2.64 The number of molecular collisions per unit time between a gas and a surface of area A is given by

$$Z_{A,i} = N_i^*(kT/2\pi m)^{1/2} A \tag{2.31}$$

Calculate the rate at which the He atoms in the system described in Problem 2.56 collide with a unit area of the surface of their container.

▮ The rate is given by *(2.31)* as

$$\frac{Z_{A,i}}{A} = N_i^* \left(\frac{kT}{2\pi m}\right)^{1/2}$$
$$= (2.43 \times 10^{25}\ \mathrm{m^{-3}})\left[\frac{(1.381 \times 10^{-23}\ \mathrm{J \cdot K^{-1}})[(1\ \mathrm{kg \cdot m^2 \cdot s^{-2}})/(1\ \mathrm{J})]}{2\pi(4.00 \times 10^{-3}\ \mathrm{kg \cdot mol^{-1}})/(6.022 \times 10^{23}\ \mathrm{mol^{-1}})}\right]^{1/2}$$
$$= 7.63 \times 10^{27}\ \mathrm{m^{-2} \cdot s^{-1}}$$

2.65 What is the mass of CO_2 molecules in air that strike a leaf having a total surface area of 100 cm^2 during 1 day's time? Assume that the atmospheric pressure is 1.00 bar, the temperature is 25 °C, and $x(CO_2) = 3 \times 10^{-4}$.

▮ Using *(1.21)* gives

$$N^*(CO_2) = \frac{(6.022 \times 10^{23}\ \text{mol}^{-1})(1.00\ \text{bar})(3 \times 10^{-4})[(10^5\ \text{Pa})/(1\ \text{bar})]}{(8.314\ \text{m}^3 \cdot \text{Pa} \cdot \text{K}^{-1} \cdot \text{mol}^{-1})(298\ \text{K})}$$

$$= 7 \times 10^{21}\ \text{m}^{-3}$$

and using *(2.31)* gives

$$Z_A(CO_2) = (7 \times 10^{21}\ \text{m}^{-3})\left(\frac{(1.381 \times 10^{-23}\ \text{J} \cdot \text{K}^{-1})(298\ \text{K})[(1\ \text{kg} \cdot \text{m}^2 \cdot \text{s}^{-2})/(1\ \text{J})]}{2\pi(44.01 \times 10^{-3}\ \text{kg} \cdot \text{mol}^{-1})/(6.022 \times 10^{23}\ \text{mol}^{-1})}\right)^{1/2}$$

$$\times (100\ \text{cm}^2)[(10^{-2}\ \text{m})/(1\ \text{cm})]^2$$

$$= 7 \times 10^{21}\ \text{s}^{-1}$$

For 24 h,

$$m = (7 \times 10^{21}\ \text{s}^{-1})\frac{60\ \text{s}}{1\ \text{min}}\left(\frac{60\ \text{min}}{1\ \text{h}}\right)\left(\frac{24\ \text{h}}{1\ \text{day}}\right)\left(\frac{44.01 \times 10^{-3}\ \text{kg} \cdot \text{mol}^{-1}}{6.022 \times 10^{23}\ \text{mol}^{-1}}\right)$$

$$= 40\ \text{kg}$$

2.66 A beam of K atoms is generated by heating K metal in an oven at 425 °C. The dimensions of the rectangular exit slit of the oven are 1.0 cm × 15 μm. How many K atoms are in this "atomic current"? The vapor pressure of K in the oven is 7.05 torr.

▮ Using *(1.21)* gives

$$N^*(K) = \frac{(6.022 \times 10^{23}\ \text{mol}^{-1})(7.05\ \text{torr})\left(\frac{1\ \text{atm}}{760\ \text{torr}}\right)\left(\frac{101\,325\ \text{Pa}}{1\ \text{atm}}\right)}{(8.314\ \text{m}^3 \cdot \text{Pa} \cdot \text{K}^{-1} \cdot \text{mol}^{-1})(698\ \text{K})}$$

$$= 9.75 \times 10^{22}\ \text{m}^{-3}$$

Substituting into *(2.31)* gives

$$Z_A(K) = (9.75 \times 10^{22}\ \text{m}^{-3})\left(\frac{(1.381 \times 10^{-23}\ \text{J} \cdot \text{K}^{-1})(698\ \text{K})[(1\ \text{kg} \cdot \text{m}^2 \cdot \text{s}^{-2})/(1\ \text{J})]}{2\pi(39.1 \times 10^{-3}\ \text{kg} \cdot \text{mol}^{-1})/(6.022 \times 10^{23}\ \text{mol}^{-1})}\right)^{1/2}$$

$$\times (1\ \text{cm})(15\ \mu\text{m})\left(\frac{10^{-2}\ \text{m}}{1\ \text{cm}}\right)\left(\frac{10^{-6}\ \text{m}}{1\ \mu\text{m}}\right)$$

$$= 2.2 \times 10^{18}\ \text{s}^{-1}$$

2.67 Consider a typical 1.0-m³ laboratory vacuum system that has a "pinhole" leak in it. During the period of 1 h, 0.1 Pa of air leaked into the system. What is the radius of the hole? Assume that $M = 29.0\ \text{g} \cdot \text{mol}^{-1}$ for air and that the laboratory pressure and temperature are 1.00 bar and 25 °C, respectively.

▮ Using *(1.18)* to determine the amount of air that leaked into the system gives

$$n = \frac{PV}{RT} = \frac{(0.1\ \text{Pa})(1.0\ \text{m}^3)}{(8.314\ \text{m}^3 \cdot \text{Pa} \cdot \text{K}^{-1} \cdot \text{mol}^{-1})(298\ \text{K})} = 4 \times 10^{-5}\ \text{mol}$$

This corresponds to

$$Z_A(\text{air}) = \frac{nL}{t} = \frac{(4 \times 10^{-5}\ \text{mol})(6.022 \times 10^{23}\ \text{mol}^{-1})}{(1\ \text{h})[(3600\ \text{s})/(1\ \text{h})]} = 7 \times 10^{15}\ \text{s}^{-1}$$

For the air in the laboratory, using *(1.21)* gives

$$N^*(\text{air}) = \frac{(6.022 \times 10^{23}\ \text{mol}^{-1})(1.00\ \text{bar})[(10^5\ \text{Pa})/(1\ \text{bar})]}{(8.314\ \text{m}^3 \cdot \text{Pa} \cdot \text{K}^{-1} \cdot \text{mol}^{-1})(298\ \text{K})} = 2.43 \times 10^{25}\ \text{m}^{-3}$$

Solving *(2.31)* for A and substituting the above results gives

$$A = \frac{7 \times 10^{15}\ \text{s}^{-1}}{2.43 \times 10^{23}\ \text{m}^{-3}}\left(\frac{2\pi(29.0 \times 10^{-3}\ \text{kg} \cdot \text{mol}^{-1})/(6.022 \times 10^{23}\ \text{mol}^{-1})}{(1.381 \times 10^{-23}\ \text{J} \cdot \text{K}^{-1})(298\ \text{K})[(1\ \text{kg} \cdot \text{m}^2 \cdot \text{s}^{-2})/(1\ \text{J})]}\right)^{1/2}$$

$$= 2 \times 10^{-12}\ \text{m}^2$$

which corresponds to a radius of

$$r = (A/\pi)^{1/2} = [(2 \times 10^{-12}\ \text{m}^2)/\pi]^{1/2} = 8 \times 10^{-7}\ \text{m}$$

2.68 How long will it take for O_2 molecules at a pressure of 1×10^3 Pa and a temperature of 25 °C to cover 1% of a 1-m² surface of a metal that adsorbs each molecule that strikes the surface? Assume that the area of an O_2 molecule is 0.098 nm².

▮ The number of molecules that are needed to cover 1% of the surface is

$$N = (0.01)(1\ \text{m}^2)/(0.098 \times 10^{-18}\ \text{m}^2) = 1 \times 10^{17}$$

giving

$$Z_A(\text{O}_2) = (1 \times 10^{17})/t$$

For the gaseous sample, using *(1.21)* gives

$$N^*(\text{O}_2) = \frac{(6.022 \times 10^{23}\ \text{mol}^{-1})(1 \times 10^3\ \text{Pa})}{(8.314\ \text{m}^3 \cdot \text{Pa} \cdot \text{K}^{-1} \cdot \text{mol}^{-1})(298\ \text{K})} = 2 \times 10^{23}\ \text{m}^{-3}$$

Substituting into *(2.31)* and solving for t gives

$$Z_A(\text{O}_2) = (1 \times 10^{17})/t$$

$$= (2 \times 10^{24}\ \text{m}^{-3})\left(\frac{(1.381 \times 10^{-23}\ \text{J} \cdot \text{K}^{-1})(298\ \text{K})[(1\ \text{kg} \cdot \text{m}^2 \cdot \text{s}^{-2})/(1\ \text{J})]}{2\pi(32.0 \times 10^{-3}\ \text{kg} \cdot \text{mol}^{-1})/(6.022 \times 10^{23}\ \text{mol}^{-1})}\right)^{1/2}$$

$$t = 5 \times 10^{-10}\ \text{s}$$

2.69 Consider a system of He atoms at 25 °C. At what pressure will the number of molecular collisions per unit time on a 1.0-m² surface be equal to the number of collisions that occur within a 1.0-m³ volume? The collision diameter of He is 0.218 nm.

▮ Equating *(2.25)* and *(2.31)*, canceling common terms, solving for N_i^*, and substituting the data gives

$$\frac{N_i^{*2}\pi\sigma_i^2}{2^{1/2}}\left(\frac{8kT}{\pi m}\right)^{1/2} = N_i^*\left(\frac{kT}{2\pi m}\right)^{1/2} A$$

$$N^*(\text{He}) = \frac{A}{\pi\sigma_i^2 8^{1/2}} = \frac{1.0\ \text{m}^2}{\pi(2.18 \times 10^{-10}\ \text{m})^2 8^{1/2}} = 2.4 \times 10^{18}$$

Using *(1.21)* gives

$$P = \frac{(8.314\ \text{m}^3 \cdot \text{Pa} \cdot \text{K}^{-1} \cdot \text{mol}^{-1})(298\ \text{K})(2.4 \times 10^{18})}{6.022 \times 10^{23}\ \text{mol}^{-1}} = 9.9 \times 10^{-3}\ \text{Pa}$$

2.70 The vacuum system described in Problem 2.67 is connected to the vacuum pump by a tube of radius = 1.0 cm. How long will it take for the pump to decrease the pressure of the air in the system from 1.00 bar to 0.50 bar? Assume $M = 29\ \text{g} \cdot \text{mol}^{-1}$ for air and $T = 25\ °\text{C}$.

▮ The rate of change in pressure will be given by *(1.18)* as

$$\frac{dP}{dt} = \frac{d}{dt}\left(\frac{nRT}{V}\right) = \frac{RT}{V}\left(\frac{dn}{dt}\right) = \frac{RT}{V}\left(\frac{d(N/L)}{dt}\right)$$

$$= \frac{RT}{LV}\frac{dN}{dt} = \frac{RT}{LV}(-Z_{A,i}) = -\frac{RT}{LV}N_i^*\left(\frac{kT}{2\pi m}\right)^{1/2} A$$

Substituting *(1.21)* for N_i^* gives

$$\frac{dP}{dt} = -\frac{RT}{LV}\left(\frac{LP}{RT}\right)\left(\frac{kT}{2\pi m}\right)^{1/2} A = -P\left(\frac{kT}{2\pi m}\right)^{1/2}\left(\frac{A}{V}\right)$$

Integrating both sides gives

$$\ln\frac{P_2}{P_1} = -\left(\frac{kT}{2\pi m}\right)^{1/2}\left(\frac{A}{V}\right)t \tag{2.32}$$

Solving *(2.32)* for t and substituting the data gives

$$t = \frac{-1.0\ \text{m}^3}{\pi(1.0 \times 10^{-2}\ \text{m})^2}\ln\frac{0.50\ \text{bar}}{1.00\ \text{bar}}\left(\frac{2\pi(29 \times 10^{-3}\ \text{kg} \cdot \text{mol}^{-1})/(6.022 \times 10^{23}\ \text{mol}^{-1})}{(1.381 \times 10^{-23}\ \text{J} \cdot \text{K}^{-1})(298\ \text{K})[(1\ \text{kg} \cdot \text{m}^2 \cdot \text{s}^{-2})/(1\ \text{J})]}\right)^{1/2}$$

$$= 19\ \text{s}$$

2.71 Repeat Problem 2.70 for a similar system in which the pressure was decreased from 0.50 bar to 0.25 bar. What would be the effect on the time if the 1.0-cm tube were replaced by a 1.0-mm tube?

▮ As can be seen in Problem 2.70, the time required to decrease the pressure by a factor of 2 is 19 s. If $r = 1.0$ mm, then t increases to 1900 s (=32 min).

2.72 The pressure of air in the cabin of a space vehicle is 1.05 bar at 15 °C. A hole, $r = 5$ mm, is formed by a meteorite, and the cabin pressure begins to decrease. What will be the pressure in the cabin at the end of 3.5 min, the time required for the pilot to attach and pressurize a helmet? Assume the cabin volume to be 4 m^3 and $M = 29\ \text{g}\cdot\text{mol}^{-1}$ for air.

▮ The cabin pressure at the end of 3.5 min is given by *(2.32)* as

$$\ln\frac{P_2}{1.05\ \text{bar}} = -\left(\frac{(1.381\times10^{-23}\ \text{J}\cdot\text{K}^{-1})(288\ \text{K})[(1\ \text{kg}\cdot\text{m}^2\cdot\text{s}^{-2})/(1\ \text{J})]}{2\pi(29\times10^{-3}\ \text{kg}\cdot\text{mol}^{-1})/(6.022\times10^{23}\ \text{mol}^{-1})}\right)^{1/2}$$

$$\times\frac{\pi(5\times10^{-3}\ \text{m})^2}{4\ \text{m}^3}(3.5\ \text{min})\frac{60\ \text{s}}{1\ \text{min}} = -0.5$$

$$P_2 = 0.6\ \text{bar}$$

2.73 A sample of tungsten in a *Knudsen cell* at 3000 K lost 0.95 mg of mass in 15 h. The radius of the hole in the cell is 0.75 mm. What is the vapor pressure of W under these conditions?

▮ The vapor pressure as determined by this method is given by

$$P = \frac{\Delta m/\Delta t}{A}\left(\frac{2\pi RT}{M}\right)^{1/2} \tag{2.33}$$

Substituting the data gives

$$P = \frac{\{(0.9\ \text{mg})[(10^{-6}\ \text{kg})/(1\ \text{mg})]\}/\{(15\ \text{h})[(3600\ \text{s})/(1\ \text{h})]\}}{\pi(0.75\times10^{-3}\ \text{m})^2}$$

$$\times\left(\frac{2\pi(8.314\ \text{J}\cdot\text{K}^{-1})(3000\ \text{K})[(1\ \text{kg}\cdot\text{m}^2\cdot\text{s}^{-2})/(1\ \text{J})]}{183.85\times10^{-3}\ \text{kg}\cdot\text{mol}^{-1}}\right)^{1/2}$$

$$= 9.19\times10^{-3}\ \text{kg}\cdot\text{m}^{-1}\cdot\text{s}^{-2} = 9.19\times10^{-3}\ \text{Pa}$$

2.74 The nominal diameter of a hole in a Knudsen cell is 0.1 mm. The actual value was determined by observing the mass change of water in the cell at 50.00 °C. Given that the vapor pressure of water at this temperature is 92.51 torr, determine the actual diameter of the hole if 73.8 mg of water escaped during a period of 12.53 min.

▮ Solving *(2.33)* for A and substituting the data gives

$$A = \frac{\{(73.8\ \text{mg})[(10^{-6}\ \text{kg})/(1\ \text{mg})]\}/\{(12.53\ \text{min})[(60\ \text{s})/(1\ \text{min})]\}}{(92.51\ \text{torr})[(1\ \text{atm})/(760\ \text{torr})][(101\,325\ \text{Pa})/(1\ \text{atm})]}$$

$$\times\left(\frac{2\pi(8.314\ \text{J}\cdot\text{K}^{-1})(323.15\ \text{K})[(1\ \text{kg}\cdot\text{m}^2\cdot\text{s}^{-2})/(1\ \text{J})]}{18.015\times10^{-3}\ \text{kg}\cdot\text{mol}^{-1}}\right)^{1/2}$$

$$= 7.70\times10^{-9}\ \text{m}^2$$

$$r = (A/\pi)^{1/2} = [(7.70\times10^{-9}\ \text{m}^2)/\pi]^{1/2} = 4.95\times10^{-5}\ \text{m} = 0.0495\ \text{mm}$$

giving $d = 0.0990$ mm.

2.75 Use *(2.33)* to determine the number of molecules that escape from a 1.0 m^2 surface of liquid H_2O at 25 °C each second. The vapor pressure of H_2O at this temperature is 3173 Pa. Use *(2.31)* to determine the number of vapor molecules that strike the surface of the liquid under the same conditions. What conclusion can be made?

▮ Solving *(2.33)* for Δm and substituting the data gives

$$\Delta m = (3173\ \text{Pa})\frac{1\ \text{kg}\cdot\text{m}^{-1}\cdot\text{s}^{-2}}{1\ \text{Pa}}(1.0\ \text{m}^2)(1\ \text{s})\left(\frac{18.015\times10^{-3}\ \text{kg}\cdot\text{mol}^{-1}}{2\pi(8.314\ \text{J}\cdot\text{K}^{-1}\cdot\text{mol}^{-1})(298\ \text{K})[(1\ \text{kg}\cdot\text{m}^2\cdot\text{s}^{-2})/(1\ \text{J})]}\right)^{1/2}$$

$$= 3.4\ \text{kg}$$

which is equivalent to

$$(3.4\ \text{kg})\frac{10^3\ \text{g}}{1\ \text{kg}}\left(\frac{1\ \text{mol}}{18.015\ \text{g}}\right)\left(\frac{6.022\times10^{23}}{1\ \text{mol}}\right) = 1.1\times10^{26}$$

For the water vapor, *(1.21)* gives

$$N^*(H_2O) = \frac{(6.022 \times 10^{23}\ \text{mol}^{-1})(3173\ \text{Pa})}{(8.314\ \text{m}^{-3} \cdot \text{Pa} \cdot \text{K}^{-1} \cdot \text{mol}^{-1})(298\ \text{K})} = 7.712 \times 10^{23}\ \text{m}^{-3}$$

Using *(2.31)* gives

$$Z_A(H_2O) = (7.712 \times 10^{23}\ \text{m}^{-3}) \left(\frac{(1.381 \times 10^{-23}\ \text{J} \cdot \text{K}^{-1} \cdot \text{mol}^{-1})(298\ \text{K})[(1\ \text{kg} \cdot \text{m}^2 \cdot \text{s}^{-2})/(1\ \text{J})]}{2\pi(18.015 \times 10^{-3}\ \text{kg} \cdot \text{mol}^{-1})/(6.022 \times 10^{23}\ \text{mol}^{-1})} \right)^{1/2}$$

$$= 1.1 \times 10^{26}\ \text{s}^{-1}$$

Under equivalent conditions, the rate of evaporation is equal to the rate of condensation.

2.76 Use *(2.31)* to derive *Graham's law of effusion.* Consider a toy balloon that is filled with an equimolar mixture of He and N_2. What are the initial relative rates of effusion of the gases?

▌ The rate of effusion for a given gas is directly related to $Z_{A,i}$, giving

$$\frac{\text{Rate}_i}{\text{Rate}_j} = \frac{Z_{A,i}}{Z_{A,j}} = \frac{N_i^*(kT/2\pi m_i)^{1/2}A}{N_j^*(kT/2\pi m_j)^{1/2}A} = \frac{(LP_i/RT)/m_i^{1/2}}{(LP_j/RT)/m_j^{1/2}} = \frac{P_i/M_i^{1/2}}{P_j/M_j^{1/2}} \tag{2.34}$$

For the equimolar mixture, $P(\text{He}) = P(N_2) = P$, giving

$$\frac{\text{Rate (He)}}{\text{Rate }(N_2)} = \frac{P/(4.00\ \text{g} \cdot \text{mol}^{-1})^{1/2}}{P/(28\ 01\ \text{g} \cdot \text{mol}^{-1})^{1/2}} = 2.65$$

2.77 At what composition will the rates of effusion of the gases from the toy balloon described in Problem 2.76 become equal?

▌ The ratio of the partial pressures will be given by *(2.34)* as

$$\frac{P(\text{He})}{P(N_2)} = \frac{\text{Rate (He)}}{\text{Rate }(N_2)} \left(\frac{[M(\text{He})]^{1/2}}{[M(N_2)]^{1/2}} \right) = (1) \frac{(4.00\ \text{g} \cdot \text{mol}^{-1})^{1/2}}{(28.01\ \text{g} \cdot \text{mol}^{-1})^{1/2}} = 0.378$$

2.78 Assume that the air in the space vehicle cabin described in Problem 2.72 consists of $x(O_2) = 0.20$. What is the composition of the air that comes out the hole?

▌ The ratio of the rates of effusion is given by *(2.34)* as

$$\frac{\text{Rate }(N_2)}{\text{Rate }(O_2)} = \frac{(0.80)P/(28.01\ \text{g} \cdot \text{mol}^{-1})^{1/2}}{(0.20)P/(32.00\ \text{g} \cdot \text{mol}^{-1})^{1/2}} = 4.3$$

The mole fraction of O_2 in the exiting gas will be

$$x(O_2) = 1/(1 + 4.3) = 0.19$$

2.79 A given amount of O_2 effused from a container in 32.5 min. It took 23.0 min for the same amount of another gas to effuse from the same container under the same conditions. What is the molar mass of the second gas?

▌ Solving *(2.34)* for the molar mass of the unknown gas and substituting $P_j = P(O_2)$ gives

$$M_j = \left[\frac{\text{Rate }(O_2)}{\text{Rate}_j} \left(\frac{P_j}{P(O_2)} \right) \right]^2 M(O_2) = \left(\frac{\text{Rate }(O_2)}{\text{Rate}_j} \right)^2 M(O_2)$$

Recognizing that the time required for effusion is inversely proportional to the rate,

$$M_j = \left(\frac{t_j}{t(O_2)} \right)^2 M(O_2) = \left(\frac{23.0\ \text{min}}{32.5\ \text{min}} \right)^2 (32.00\ \text{g} \cdot \text{mol}^{-1}) = 16.0\ \text{g} \cdot \text{mol}^{-1}$$

2.80 Consider a tube that has a fritted disk sealed in its center so that gases can effuse through it in both directions. If the left-hand side contains H_2 at a pressure of P and a temperature T and the right-hand side contains He at a pressure of $2P$ and a temperature T, what is the initial ratio of the rates of effusion?

▌ Using *(2.34)* gives

$$\frac{\text{Rate }(H_2)}{\text{Rate (He)}} = \frac{P/(2.01\ \text{g} \cdot \text{mol}^{-1})^{1/2}}{2P/(4.00\ \text{g} \cdot \text{mol}^{-1})^{1/2}} = 0.705$$

2.81 Consider the glass tube system described in Problem 2.80 except that both sides contain He at the same pressure, the left-hand side at 100 °C and the right-hand side at 0 °C. What is the initial ratio of the rates of effusion? Describe the equilibrium condition of the system.

▌ The rate of effusion for each gas is directly related to $Z_{A,i}$. Taking a ratio and substituting *(2.31)* gives

$$\frac{\text{Rate}(T_2)}{\text{Rate}(T_1)} = \frac{Z_{A,i}(T_2)}{Z_{A,i}(T_1)} = \frac{N^*(T_2)(kT_2/2\pi m)^{1/2}A}{N^*(T_1)(kT_1/2\pi m)^{1/2}A} = \frac{(LP/RT_2)T_2^{1/2}}{(LP/RT_1)T_1^{1/2}}$$

$$= (T_1/T_2)^{1/2} = [(273\ \text{K})/(373\ \text{K})]^{1/2} = 0.856$$

(Note that the net flow occurs from the cooler side of the disk to the warmer side.) When equilibrium is reached, the ratio of the rates will be unity as a result of a pressure difference built up on the warmer side given by

$$\frac{\text{Rate}(T_2)}{\text{Rate}(T_1)} = \frac{(LP_2/RT_2)T_2^{1/2}}{(LP_1/RT_1)T_1^{1/2}} = \frac{P_2}{P_1}\left(\frac{T_1}{T_2}\right)^{1/2} = 1$$

$$P_2/P_1 = (T_2/T_1)^{1/2} = [(373\ \text{K})/(273\ \text{K})] = 1.17$$

2.82 The *mean free path* (λ_i) of a molecule in a binary gaseous mixture is given by

$$\lambda_i = \bar{v}_i/(z_{ii} + z_{ij}) \qquad (2.35)$$

Calculate the mean free path of a hydrogen molecule in the mixture described in Problem 2.49.

▌ Using *(2.9)* gives

$$\bar{v}(\text{H}_2) = \left[\frac{8(8.314\ \text{J}\cdot\text{K}^{-1}\cdot\text{mol}^{-1})(298\ \text{K})[(1\ \text{kg}\cdot\text{m}^2\cdot\text{s}^{-2})/(1\ \text{J})]}{\pi(2.01\times10^{-3}\ \text{kg}\cdot\text{mol}^{-1})}\right]^{1/2} = 1770\ \text{m}\cdot\text{s}^{-1}$$

Substituting the results of Problems 2.59 and 2.61 into *(2.35)* gives

$$\lambda(\text{H}_2) = \frac{1770\ \text{m}\cdot\text{s}^{-1}}{1.05\times10^{10}\ \text{s}^{-1} + 3.6\times10^{9}\ \text{s}^{-1}} = 1.26\times10^{-7}\ \text{m}$$

2.83 Repeat the calculations of Problem 2.82 for a N_2 molecule. Find the ratio $\lambda(\text{H}_2)/\lambda(\text{N}_2)$.

▌ As in Problem 2.82,

$$\bar{v}(\text{N}_2) = [8(8.314)(298)/\pi(28.01\times10^{-3})]^{1/2} = 475\ \text{m}\cdot\text{s}^{-1}$$

$$\lambda(\text{N}_2) = 475/(1.8\times10^{9} + 1.08\times10^{10}) = 3.77\times10^{-8}\ \text{m}$$

where the values of $z(\text{N}_2,\text{N}_2)$ and $z(\text{N}_2,\text{H}_2)$ were determined in Problems 2.60 and 2.62. The desired ratio is

$$\lambda(\text{H}_2)/\lambda(\text{N}_2) = (1.26\times10^{-7}\ \text{m})/(3.77\times10^{-8}\ \text{m}) = 3.34$$

2.84 What is the mean free path of a He atom in the system described in Problem 2.56?

▌ For a pure gas, *(2.35)* becomes

$$\lambda_i = \bar{v}_i/z_{ii} = 1/\pi 2^{1/2}\sigma_i^2 N_i^* \qquad (2.36)$$

Substituting the results of Problem 2.56 gives

$$\lambda(\text{He}) = \frac{1}{\pi(2^{1/2})(2.18\times10^{-10}\ \text{m})^2(2.43\times10^{25}\ \text{m}^{-3})} = 1.95\times10^{-7}\ \text{m}$$

2.85 What is the average time between collisions for a He atom in the upper atmosphere where $T = 1000$ K and $P = 1\times10^{-12}$ bar? Assume $\sigma(\text{He}) = 0.218$ nm.

▌ Using *(1.21)* and *(2.36)* gives

$$N^*(\text{He}) = \frac{(6.022\times10^{23}\ \text{mol}^{-1})(1\times10^{-12}\ \text{bar})[(10^5\ \text{Pa})/(1\ \text{bar})]}{(8.314\ \text{m}^3\cdot\text{Pa}\cdot\text{K}^{-1}\cdot\text{mol}^{-1})(1000\ \text{K})} = 7\times10^{12}\ \text{m}^{-3}$$

$$\lambda(\text{He}) = \frac{1}{\pi(2^{1/2})(2.18\times10^{-10}\ \text{m})^2(7\times10^{12}\ \text{m}^{-3})} = 7\times10^{5}\ \text{m}$$

The average velocity is given by *(2.9)* as

$$\bar{v}(\text{He}) = \left[\frac{8(8.314\ \text{J}\cdot\text{K}^{-1}\cdot\text{mol}^{-1})(1000\ \text{K})[(1\ \text{kg}\cdot\text{m}^2\cdot\text{s}^{-2})/(1\ \text{J})]}{\pi(4.00\times10^{-3}\ \text{kg}\cdot\text{mol}^{-1})}\right]^{1/2} = 2\times10^{3}\ \text{m}\cdot\text{s}^{-1}$$

The average time between collisions is

$$\bar{t} = \lambda/\bar{v} = (7 \times 10^5\ \text{m})/(2 \times 10^3\ \text{m} \cdot \text{s}^{-1}) = 400\ \text{s}$$

2.86 At what pressure will the mean free path be equal to the collision diameter for He at 25 °C? Assume $\sigma(\text{He}) = 0.218\ \text{nm}$.

▮ Solving *(2.36)* for N_i^* and substituting the data gives

$$N_i^* = \frac{1}{\pi(2^{1/2})(2.18 \times 10^{-10}\ \text{m})^3} = 2.17 \times 10^{28}\ \text{m}^{-3}$$

where $\lambda_i = \sigma_i$. Solving *(1.21)* for P and substituting the data gives

$$P = \frac{(2.17 \times 10^{28}\ \text{m}^{-3})(8.314\ \text{m}^3 \cdot \text{Pa} \cdot \text{K}^{-1} \cdot \text{mol}^{-1})(298\ \text{K})}{6.022 \times 10^{23}\ \text{mol}^{-1}}$$

$$= 8.93 \times 10^7\ \text{Pa} = 893\ \text{bar}$$

2.87 An evacuated space will be a good thermal insulator when the mean free path of any gas present is greater than the thickness between the walls of the container. To what pressure should N_2 at 25 °C be evacuated to produce a good insulator between walls 1 cm apart? Assume $\sigma(N_2) = 0.373\ \text{nm}$.

▮ Solving *(2.36)* for N_i^* and substituting the data gives

$$N^*(N_2) = 1/\pi(2^{1/2})(3.73 \times 10^{-10}\ \text{m})^2(1 \times 10^{-2}\ \text{m}) = 2 \times 10^{20}\ \text{m}^{-3}$$

Solving *(1.21)* for P and substituting the data gives

$$P = \frac{(2 \times 10^{20}\ \text{m}^{-3})(8.314\ \text{m}^3 \cdot \text{Pa} \cdot \text{K}^{-1} \cdot \text{mol}^{-1})(298\ \text{K})}{6.022 \times 10^{23}\ \text{mol}^{-1}} = 0.8\ \text{Pa}$$

2.88 What are the odds that a H_2 molecule travels 1 mm without undergoing a collision in the system described in Problem 2.49? Repeat the calculations for a distance of 1 nm.

▮ The number of molecules that a given molecule will collide with during the time interval dt is given by $dN = N(z_{ii} + z_{ij})\,dt$, which upon integration and substituting of *(2.35)* and $t = x/\bar{v}_i$ gives the odds that a collision will occur:

$$N/N_0 = e^{(z_{ii}+z_{ij})t} = e^{(\bar{v}_i/\lambda_i)(x/\bar{v}_i)} = e^{x/\lambda_i}$$

The odds of not undergoing a collision are $N_0/N = e^{-x/\lambda_i}$. Substituting $\lambda_i = 1.26 \times 10^{-7}\ \text{m}$ (see Problem 2.82) gives for 1 mm

$$N_0/N = e^{-(1\times10^{-3}\ \text{m})/(1.26\times10^{-7}\ \text{m})} = 10^{-3446}$$

and for 1 nm

$$N_0/N = e^{-(1\times10^{-9}\ \text{m})/(1.26\times10^{-7}\ \text{m})} = 0.992$$

2.3 TRANSPORT PROPERTIES

2.89 The rate of transport of heat $(\partial q/\partial t)$ between two surfaces of area A separated by a distance z is given by *Fourier's law* as

$$\frac{\partial q}{\partial t} = -k_T A \frac{\partial T}{\partial z} \tag{2.37}$$

where q is heat, t is time, and k_T is the coefficient of thermal conductivity. What will be the heat transport between two surfaces separated by 0.50 cm of air such that the temperature difference is 2.5 °C? Assume $k_T = 0.0242\ \text{J} \cdot \text{m}^{-1} \cdot \text{s}^{-1} \cdot \text{K}^{-1}$ for air.

▮ Assuming $\partial T/\partial z$ to be constant,

$$\frac{\partial T}{\partial z} = \frac{\Delta T}{\Delta z} = \frac{-2.5\ \text{K}}{(0.50\ \text{cm})[(10^{-2}\ \text{m})/(1\ \text{cm})]} = -5.0 \times 10^2\ \text{K} \cdot \text{m}^{-1}$$

The negative sign is introduced because the heat flow occurs from the warmer surface to the cooler surface while z is defined as the distance from the cooler surface to the warmer surface. Because the area of the surface is not given, *(2.37)* is rearranged to solve for the *flux* or *flow* defined as $(1/A)(\partial q/\partial t)$, giving

$$\frac{1}{A}\frac{\partial q}{\partial t} = -k_T\frac{\partial T}{\partial z} = -(0.0242\ \text{J}\cdot\text{m}^{-1}\cdot\text{s}^{-1}\cdot\text{K}^{-1})(-5.0\times 10^2\ \text{K}\cdot\text{m}^{-1})$$

$$= 12\ \text{J}\cdot\text{s}^{-1}\cdot\text{m}^{-2} = 12\ \text{W}\cdot\text{m}^{-2}$$

2.90 Repeat the calculations of Problem 2.89 assuming the surfaces to be separated by solid silver with $k_T = 427\ \text{J}\cdot\text{m}^{-1}\cdot\text{s}^{-1}\cdot\text{K}^{-1}$. Compare the results.

▮ Substituting the value of k_T for silver into the expression for the flux gives

$$\frac{1}{A}\frac{\partial q}{\partial t} = -(427\ \text{J}\cdot\text{m}^{-1}\cdot\text{s}^{-1}\cdot\text{K}^{-1})(-5.0\times 10^2\ \text{K}\cdot\text{m}^{-1}) = 2.1\times 10^5\ \text{J}\cdot\text{s}^{-1}\cdot\text{m}^{-2}$$

an increase of 1.8×10^4.

2.91 Consider a Dewar flask sitting in a laboratory at 298 K containing liquid N_2 at 77 K. Neglecting radiation, what is the heat flow into the container from the room? Assume that the height of the flask is 35 cm, the outside diameter is 10.0 cm, the wall separation is 1.0 cm, and the space between the walls contains air ($k_T = 0.0242\ \text{J}\cdot\text{m}^{-1}\cdot\text{s}^{-1}\cdot\text{K}^{-1}$) at 10^{-6} bar.

▮ The heat flow between concentric cylinders of radius r_i and length l is given by

$$\frac{\partial q}{\partial t} = \frac{-2\pi k_T l\,\Delta T}{\ln(r_2/r_1)} = \frac{-2\pi(0.0242\ \text{J}\cdot\text{m}^{-1}\cdot\text{s}^{-1}\cdot\text{K}^{-1})(35\ \text{cm})[(10^{-2}\ \text{m})/(1\ \text{cm})](77\ \text{K} - 298\ \text{K})}{\ln[(5.0\ \text{cm})/(4.0\ \text{cm})]}$$

$$= 53\ \text{J}\cdot\text{s}^{-1}$$

2.92 If a molecule of a gas can be considered to be a rigid, hard sphere,

$$k_T = \frac{25}{32}\left(\frac{kT}{\pi m}\right)^{1/2}\frac{C_{V,\text{m}}}{L\sigma^2} = \frac{25\pi}{64}\bar{v}\frac{N^*}{L}C_{V,\text{m}}\lambda \tag{2.38}$$

where $C_{V,\text{m}}$ is the molar heat capacity of the gas under constant volume conditions. Predict k_T for water vapor at 25 °C assuming $\sigma = 0.50$ nm and $C_{V,\text{m}} = 25.26\ \text{J}\cdot\text{K}^{-1}\cdot\text{mol}^{-1}$ using this model.

▮ Substituting the data into *(2.38)* gives

$$k_T = \frac{25}{32}\left(\frac{(1.381\times 10^{-23}\ \text{J}\cdot\text{K}^{-1})(298\ \text{K})[(1\ \text{kg}\cdot\text{m}^2\cdot\text{s}^{-2})/(1\ \text{J})]}{\pi(18.02\times 10^{-3}\ \text{kg}\cdot\text{mol}^{-1})/(6.022\times 10^{23}\ \text{mol}^{-1})}\right)^{1/2}$$

$$\times\frac{25.26\ \text{J}\cdot\text{K}^{-1}\cdot\text{mol}^{-1}}{(6.022\times 10^{23}\ \text{mol}^{-1})(5.0\times 10^{-10}\ \text{m})^2}$$

$$= 0.027\ \text{J}\cdot\text{m}^{-1}\cdot\text{s}^{-1}\cdot\text{K}^{-1}$$

2.93 For hard, polyatomic molecules, *(2.38)* becomes

$$k_T = \frac{5}{16}\left(\frac{kT}{\pi m}\right)^{1/2}\frac{C_{V,\text{m}} + \frac{9}{4}R}{L\sigma^2} \tag{2.39}$$

Repeat the calculations of Problem 2.92 for water and compare the results.

▮ Using *(2.39)* gives

$$k_T = \frac{5}{16}\left(\frac{(1.381\times 10^{-23}\ \text{J}\cdot\text{K}^{-1})(298\ \text{K})[(1\ \text{kg}\cdot\text{m}^2\cdot\text{s}^{-2})/(1\ \text{J})]}{\pi(18.02\times 10^{-3}\ \text{kg}\cdot\text{mol}^{-1})/(6.022\times 10^{23}\ \text{mol}^{-1})}\right)^{1/2}$$

$$\times\frac{25.26\ \text{J}\cdot\text{K}^{-1}\cdot\text{mol}^{-1} + \frac{9}{4}(8.314\ \text{J}\cdot\text{K}^{-1}\cdot\text{mol}^{-1})}{(6.022\times 10^{23}\ \text{mol}^{-1})(5.0\times 10^{-10}\ \text{m})^2}$$

$$= 0.019\ \text{J}\cdot\text{m}^{-1}\cdot\text{s}^{-1}\cdot\text{K}^{-1}$$

Although both values are in the correct order of magnitude, the value predicted by *(2.39)* agrees much better with the experimental value of $0.0181\ \text{J}\cdot\text{m}^{-1}\cdot\text{s}^{-1}\cdot\text{K}^{-1}$.

2.94 For N_2, $k_T/(\text{J}\cdot\text{m}^{-1}\cdot\text{s}^{-1}\cdot\text{K}^{-1}) = 0.0386$ at 500.0 K and 0.0631 at 1000.0 K. Assuming $C_{V,\text{m}}/(\text{J}\cdot\text{K}^{-1}\cdot\text{mol}^{-1}) = 21.266$ and 24.383, respectively, at these temperatures, calculate σ for the N_2 molecule. Comment on the two values obtained.

▮ Rearranging *(2.38)* for σ^2 and substituting the data for 500.0 K gives

$$\sigma^2 = \frac{25}{32}\left(\frac{(1.381\times10^{-23}\,\mathrm{J\cdot K^{-1}})(500.0\,\mathrm{K})[(1\,\mathrm{kg\cdot m^2\cdot s^{-2}})/(1\,\mathrm{J})]}{\pi(28.01\times10^{-3}\,\mathrm{kg\cdot mol^{-1}})/(6.022\times10^{23}\,\mathrm{mol^{-1}})}\right)^{1/2}$$

$$\times\frac{21.266\,\mathrm{J\cdot K^{-1}\cdot mol^{-1}}}{(6.022\times10^{23}\,\mathrm{mol^{-1}})(0.0386\,\mathrm{J\cdot m^{-1}\cdot s^{-1}\cdot K^{-1}})}$$

$$= 1.55\times10^{-19}\,\mathrm{m^2}$$

This is equivalent to $\sigma = 0.394\,\mathrm{nm}$. Likewise, at 1000.0 K,

$$\sigma^2 = \frac{25}{32}\left(\frac{(1.381\times10^{-23})(1000.0)}{\pi(28.01\times10^{-3})/(6.022\times10^{23})}\right)^{1/2}\left(\frac{24.383}{(6.022\times10^{23})(0.0631)}\right)$$

$$= 1.54\times10^{-19}\,\mathrm{m^2}$$

giving $\sigma = 0.393\,\mathrm{nm}$. The molecular diameter is essentially constant.

2.95 At 25 °C, $k_T = 8.7\,\mathrm{mJ\cdot m^{-1}\cdot s^{-1}\cdot K^{-1}}$ for Kr. Assuming $\sigma = 0.42\,\mathrm{nm}$, determine $C_{V,\mathrm{m}}$ for this gas.

▮ Solving *(2.38)* for $C_{V,\mathrm{m}}$ and substituting the data gives

$$C_{V,\mathrm{m}} = \frac{32}{25}\left(\frac{\pi(83.80\times10^{-3}\,\mathrm{kg\cdot mol^{-1}})/(6.022\times10^{23}\,\mathrm{mol^{-1}})}{(1.381\times10^{-23}\,\mathrm{J\cdot K^{-1}})(298\,\mathrm{K})[(1\,\mathrm{kg\cdot m^2\cdot s^{-2}})/(1\,\mathrm{J})]}\right)^{1/2}$$

$$\times(8.7\times10^{-3}\,\mathrm{J\cdot m^{-1}\cdot s^{-1}\cdot K^{-1}})(6.022\times10^{23}\,\mathrm{mol^{-1}})(4.2\times10^{-10}\,\mathrm{m})^2$$

$$= 12.2\,\mathrm{J\cdot K^{-1}\cdot mol^{-1}}$$

The actual value is $12.47\,\mathrm{J\cdot K^{-1}\cdot mol^{-1}}$.

2.96 The coefficient of viscosity of gaseous Ar at 25 °C is $2.10\times10^{-5}\,\mathrm{Pa\cdot s}$. Express this value in units of $\mathrm{N\cdot s\cdot m^{-2}}$, $\mathrm{kg\cdot m^{-1}\cdot s^{-1}}$, and poise ($1\,\mathrm{P} = 1\,\mathrm{g\cdot cm^{-1}\cdot s^{-1}}$).

▮ Making the necessary conversions gives

$$(2.10\times10^{-5}\,\mathrm{Pa\cdot s})\frac{1\,\mathrm{N\cdot m^{-2}}}{1\,\mathrm{Pa}} = 2.10\times10^{-5}\,\mathrm{N\cdot s\cdot m^{-2}}$$

$$(2.10\times10^{-5}\,\mathrm{N\cdot s\cdot m^{-2}})\frac{1\,\mathrm{kg\cdot m\cdot s^{-2}}}{1\,\mathrm{N}} = 2.10\times10^{-5}\,\mathrm{kg\cdot m^{-1}\cdot s^{-1}}$$

$$(2.10\times10^{-5}\,\mathrm{kg\cdot m^{-1}\cdot s^{-1}})\frac{10^3\,\mathrm{g}}{1\,\mathrm{kg}}\left(\frac{1\,\mathrm{m}}{10^2\,\mathrm{cm}}\right)\left(\frac{1\,\mathrm{P}}{1\,\mathrm{g\cdot cm^{-1}\cdot s^{-1}}}\right) = 2.10\times10^{-4}\,\mathrm{P}$$

2.97 For gaseous molecules that can be considered to be rigid, hard spheres, the coefficient of viscosity (η) is given by

$$\eta = \frac{5\bar{v}m}{32(2^{1/2})\sigma^2} = \frac{5\pi}{32}N^*\bar{v}m\lambda \qquad (2.40)$$

Using $\sigma = 0.353\,\mathrm{nm}$ for O_2 at 1.00 bar and 25 °C, calculate η. Repeat the calculations for a pressure of 10.00 bar.

▮ For O_2 at 298 K, *(2.9)* gives

$$\bar{v} = \left(\frac{8(8.314\,\mathrm{J\cdot K^{-1}\cdot mol^{-1}})(298\,\mathrm{K})[(1\,\mathrm{kg\cdot m^2\cdot s^{-2}})/(1\,\mathrm{J})]}{\pi(32.00\times10^{-3}\,\mathrm{kg\cdot mol^{-1}})}\right)^{1/2} = 444\,\mathrm{m\cdot s^{-1}}$$

Substituting into *(2.40)* gives

$$\eta = \frac{5(444\,\mathrm{m\cdot s^{-1}})(32.00\times10^{-3}\,\mathrm{kg\cdot mol^{-1}})/(6.022\times10^{-23}\,\mathrm{mol^{-1}})}{32(2^{1/2})(3.53\times10^{-10}\,\mathrm{m})^2}$$

$$= 2.09\times10^{-5}\,\mathrm{kg\cdot m^{-1}\cdot s^{-1}} = 2.09\times10^{-5}\,\mathrm{Pa\cdot s}$$

None of the above calculations involve pressure, so $\eta = 2.09\times10^{-5}\,\mathrm{Pa\cdot s}$ at both 1.00 bar and 10.00 bar.

2.98 Repeat the calculations of Problem 2.97 for O_2 at 500.0 K. In general, what is the temperature dependence of η for a gas?

▮ For the new temperature, *(2.9)* and *(2.40)* give

$$\bar{v} = \left(\frac{8(8.314\ \text{J}\cdot\text{K}^{-1}\cdot\text{mol}^{-1})(500.0\ \text{K})[(1\ \text{kg}\cdot\text{m}^2\cdot\text{s}^{-2})/(1\ \text{J})]}{\pi(32.00\times10^{-3}\ \text{kg}\cdot\text{mol}^{-1})}\right)^{1/2} = 575\ \text{m}\cdot\text{s}^{-1}$$

$$\eta = \frac{5(575\ \text{m}\cdot\text{s}^{-1})(32.00\times10^{-3}\ \text{kg}\cdot\text{mol}^{-1})/(6.022\times10^{-23}\ \text{mol}^{-1})}{32(2^{1/2})(3.53\times10^{-10}\ \text{m})^2}$$

$$= 2.71\times10^{-5}\ \text{kg}\cdot\text{m}^{-1}\cdot\text{s}^{-1} = 2.71\times10^{-5}\ \text{Pa}\cdot\text{s}$$

Inspection of these two equations shows that $\eta \propto T^{1/2}$.

2.99 If intermolecular attractions are considered, the temperature dependence of the coefficient of viscosity is given by the *Sutherland equation*

$$\eta = k_s T^{1/2}/(1 + S/T) \qquad (2.41)$$

where k_s and S are constants for a given gas. Determine the Sutherland constants for methane from the following data.

$T/(°\text{C})$	0	100.0	200.5	284
$\eta/(10^{-5}\ \text{Pa}\cdot\text{s})$	1.026	1.331	1.605	1.813

Use *(2.41)* to predict η at 20 °C for CH_4.

▮ Equation *(2.41)* can be transformed into

$$\frac{T^{1/2}}{\eta} = \frac{1}{k_s} + \frac{S}{k_s}\left(\frac{1}{T}\right)$$

which means that a plot of $T^{1/2}/\eta$ against $1/T$ will be linear with an intercept of $1/k_s$ and a slope of S/k_s (see Fig. 2-10). The intercept is $1.007\times10^6\ \text{K}^{1/2}\cdot\text{Pa}^{-1}\cdot\text{s}^{-1}$, giving

$$k_s = \frac{1}{\text{intercept}} = \frac{1}{1.007\times10^6\ \text{K}^{1/2}\cdot\text{Pa}^{-1}\cdot\text{s}^{-1}} = 9.93\times10^{-7}\ \text{Pa}\cdot\text{s}\cdot\text{K}^{-1/2}$$

and the slope is $1.65\times10^8\ \text{K}^{3/2}\cdot\text{Pa}^{-1}\cdot\text{s}^{-1}$, giving

$$S = k_s(\text{slope}) = (9.93\times10^{-7}\ \text{Pa}\cdot\text{s}\cdot\text{K}^{-1/2})(1.65\times10^8\ \text{K}^{3/2}\cdot\text{Pa}^{-1}\cdot\text{s}^{-1}) = 164\ \text{K}$$

At 20 °C, *(2.41)* gives

$$\eta = \frac{(9.93\times10^{-7}\ \text{Pa}\cdot\text{s}\cdot\text{K}^{-1/2})(293\ \text{K})^{1/2}}{1 + (164\ \text{K})/(293\ \text{K})} = 1.090\times10^{-5}\ \text{Pa}\cdot\text{s}$$

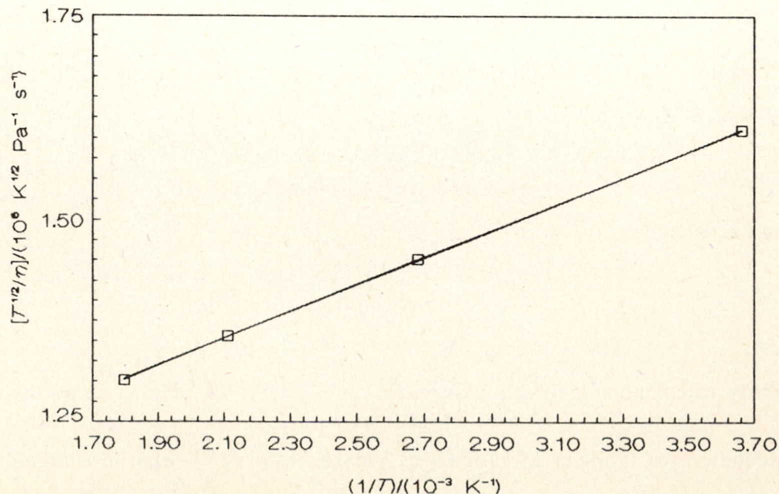

Fig. 2-10

2.100 Determine the collision diameter for gaseous H_2O given $\eta = 8.61 \times 10^{-6}$ Pa · s at 0 °C.

▮ For H_2O at 273 K, using *(2.9)* gives

$$\bar{v} = \left(\frac{8(8.314\ \text{J}\cdot\text{K}^{-1}\cdot\text{mol}^{-1})(273\ \text{K})[(1\ \text{kg}\cdot\text{m}^2\cdot\text{s}^{-2})/(1\ \text{J})]}{\pi(18.02\times10^{-3}\ \text{kg}\cdot\text{mol}^{-1})}\right)^{1/2} = 566\ \text{m}\cdot\text{s}^{-1}$$

Solving *(2.40)* for σ^2 and substituting the data gives

$$\sigma^2 = \frac{5(566\ \text{m}\cdot\text{s}^{-1})(18.02\times10^{-3}\ \text{kg}\cdot\text{mol}^{-1})/(6.022\times10^{23}\ \text{mol}^{-1})}{32(2^{1/2})(8.61\times10^{-6}\ \text{Pa}\cdot\text{s})[(1\ \text{kg}\cdot\text{m}^{-1}\cdot\text{s}^{-1})/(1\ \text{Pa}\cdot\text{s})]}$$

$$= 2.17\times10^{-19}\ \text{m}^2$$

giving $\sigma = 0.466$ nm.

2.101 The coefficient of viscosity of H_2 at 0 °C is 8.53×10^{-6} Pa · s. What is the value of η for deuterium under these conditions? For D_2, $M = 4.0280\ \text{g}\cdot\text{mol}^{-1}$.

▮ A reasonable assumption to make is $\sigma(H_2) = \sigma(D_2)$. Thus inspection of *(2.9)* and *(2.40)* shows that $\eta(D_2)/\eta(H_2) = M(D_2)^{1/2}/M(H_2)^{1/2}$. Solving for $\eta(D_2)$ and substituting the data gives

$$\eta(D_2) = (8.53\times10^{-6}\ \text{Pa}\cdot\text{s})\left(\frac{4.0280\ \text{g}\cdot\text{mol}^{-1}}{2.0158\ \text{g}\cdot\text{mol}^{-1}}\right)^{1/2} = 1.206\times10^{-5}\ \text{Pa}\cdot\text{s}$$

which agrees well with the experimental value of 1.18×10^{-5} Pa · s.

2.102 The temperature dependence of the coefficient of viscosity of a liquid is given by

$$\log \eta = A/T + B \tag{2.42}$$

where A and B are constants for the liquid. Use the following data to determine A and B for water.

T/(°C)	10	20	30	40
$\eta/(10^{-3}$ Pa · s)	1.307	1.002	0.7975	0.6529

Predict η at 25 °C for H_2O.

▮ According to *(2.41)*, a plot of $\log \eta$ against $1/T$ will be linear with a slope of A and an intercept of B. Such a plot is shown in Fig. 2-11. From the plot, $A = 889.9$ K and $B = -6.032$. At 25 °C,

$$\log[\eta/(\text{Pa}\cdot\text{s})] = \frac{889.9\ \text{K}}{298\ \text{K}} + (-6.032) = -3.046$$

giving $\eta = 0.900 \times 10^{-3}$ Pa · s.

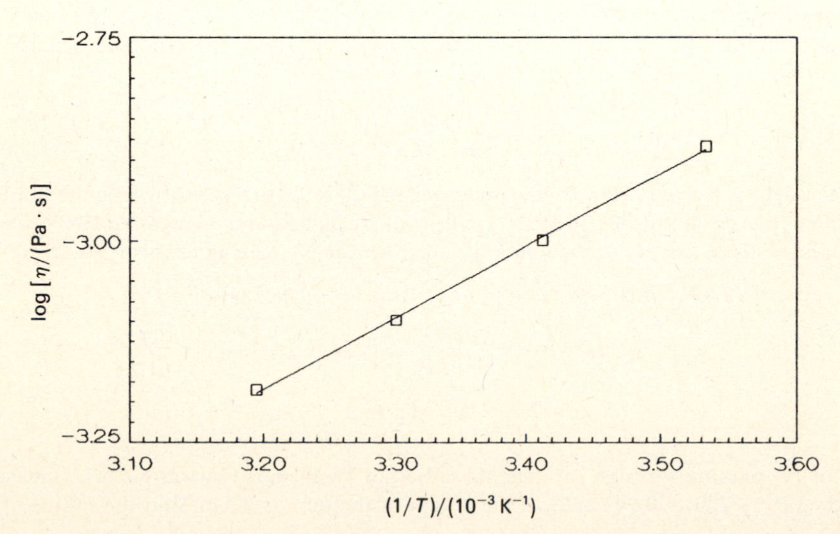

Fig. 2-11

2.103 Equation *(2.42)* is often written in the form

$$\eta = C e^{\Delta E(\text{viscosity})/RT} \tag{2.43}$$

where ΔE(viscosity) and C are constants for a given liquid. Given $A = 312.5$ K and $B = -4.900$ for liquid NH_3, determine ΔE(viscosity).

▮ Taking the natural logarithm of both sides of *(2.43)* gives

$$\ln \eta = \ln C + \frac{\Delta E(\text{viscosity})}{RT}$$

Converting to base 10 logarithms and equating to *(2.42)* gives

$$\ln \eta = \ln C + \frac{\Delta E(\text{viscosity})}{2.303RT} = \frac{A}{T} + B$$

Upon a term-by-term comparison,

$$\Delta E(\text{viscosity}) = A(2.303)R = (312.5\ \text{K})(2.303)(8.314\ \text{J}\cdot\text{K}^{-1}\cdot\text{mol}^{-1})$$
$$= 5984\ \text{J}\cdot\text{mol}^{-1}$$

2.104 Find the coefficient of viscosity for liquid NH_3 at -50.0 °C given $\Delta E(\text{viscosity}) = 5984\ \text{J}\cdot\text{mol}^{-1}$ and $\eta = 2.76 \times 10^{-4}\ \text{Pa}\cdot\text{s}$ at -40.0 °C.

▮ An alternative form of *(2.43)* is

$$\ln \frac{\eta_2}{\eta_1} = \frac{\Delta E(\text{viscosity})}{R}\left(\frac{1}{T_2} - \frac{1}{T_1}\right)$$

Substituting the data gives

$$\ln \frac{\eta_2}{2.76 \times 10^{-4}\ \text{Pa}\cdot\text{s}} = \frac{5984\ \text{J}\cdot\text{mol}^{-1}}{8.314\ \text{J}\cdot\text{K}^{-1}\cdot\text{mol}^{-1}}\left(\frac{1}{223.2\ \text{K}} - \frac{1}{233.2\ \text{K}}\right) = 0.138$$

Taking antilogarithms and solving for η_2 gives

$$\eta_2 = (1.15)(2.76 \times 10^{-4}\ \text{Pa}\cdot\text{s}) = 3.17 \times 10^{-4}\ \text{Pa}\cdot\text{s}$$

2.105 For many liquids, the value of ΔE(viscosity) is related to the internal energy change for vaporization ($\Delta_{\text{vap}}U$) by $\Delta E(\text{viscosity}) = 0.3\,\Delta_{\text{vap}}U$. Using the results of Problem 2.103, estimate $\Delta_{\text{vap}}U$ for NH_3.

▮ Solving the above relationship for $\Delta_{\text{vap}}U$ and substituting the results of Problem 2.103 gives

$$\Delta_{\text{vap}}U = \frac{\Delta E(\text{viscosity})}{0.3} = \frac{5984\ \text{J}\cdot\text{mol}^{-1}}{0.3} = 2 \times 10^4\ \text{J}\cdot\text{mol}^{-1}$$

This estimate agrees very well with the experimental value of $21.35\ \text{kJ}\cdot\text{mol}^{-1}$.

2.106 The coefficient of viscosity for a gas can be determined by measuring how long it takes for a given volume of gas to flow through a capillary tube of radius R and length l under a pressure difference (*Poiseuille's law* for compressible fluids)

$$\eta = \frac{\pi R^4 t}{16Vl}\left(\frac{P_i^2 - P_f^2}{P_0}\right) \tag{2.44}$$

where P_i is the inlet pressure, P_f is the outlet pressure, and P_0 is the pressure at which the fluid volume is measured. Under the same pressure conditions, identical volumes of Ne and N_2 at 0 °C required 109.5 s and 61.2 s, respectively, to flow through the tube. Given $\eta = 1.66 \times 10^{-5}\ \text{Pa}\cdot\text{s}$ for N_2, calculate η for Ne.

▮ Taking a ratio of *(2.44)* for the two gases and solving for η(Ne) gives

$$\eta(\text{Ne}) = \eta(\text{N}_2)\frac{t(\text{Ne})}{t(\text{N}_2)} = (1.66 \times 10^{-5}\ \text{Pa}\cdot\text{s})\frac{109.5\ \text{s}}{61.2\ \text{s}}$$
$$= 2.97 \times 10^{-5}\ \text{Pa}\cdot\text{s}$$

2.107 The volume of N_2 passing through the tube described in Problem 2.106 was 5.0 mL (measured at 1.00 bar) in a cell such that $(P_i^2 - P_f^2) = 0.90\ \text{bar}^2$. If the length of the tube is 2.5 m, find the radius of the capillary tubing.

▌ Solving *(2.44)* for R^4 and substituting the data gives

$$R^4 = \frac{16\left[(5.0\ \text{mL})\dfrac{1\ \text{cm}^3}{1\ \text{mL}}\left(\dfrac{10^{-2}\ \text{m}}{1\ \text{cm}}\right)^3\right](2.5\ \text{m})(1.66 \times 10^{-5}\ \text{Pa}\cdot\text{s})(1.00\ \text{bar})}{\pi(0.90\ \text{bar}^2)[(10^5\ \text{Pa})/(1\ \text{bar})](61.2\ \text{s})}$$

$$= 1.9 \times 10^{-16}\ \text{m}^4$$

Solving for R gives $R = 1.2 \times 10^{-4}$ m.

2.108 The coefficient of viscosity of a liquid can be measured by determining the length of time for a given volume of the liquid to drain through the capillary portion of an *Ostwald viscometer*, where

$$\eta = \pi R^4\,\Delta P\,t/8Vl \tag{2.45}$$

Under identical conditions, the time required for samples of ethanol and water to drain were 64.5 and 42.6 s, respectively. Given $\rho/(\text{g}\cdot\text{mL}^{-1}) = 0.789\,45$ and 0.9982, respectively, and $\eta = 1.002 \times 10^{-3}$ Pa · s for H_2O, determine η for CH_3CH_2OH.

▌ If a comparison method is used, *(2.45)* becomes

$$\eta/\eta_0 = \rho t/\rho_0 t_0 \tag{2.46}$$

where the subscript refers to the reference substance. Solving *(2.46)* for $\eta(CH_3CH_2OH)$ and substituting the data gives

$$\eta(\text{CH}_3\text{CH}_2\text{OH}) = (1.002 \times 10^{-3}\ \text{Pa}\cdot\text{s})\left(\frac{0.789\,45\ \text{g}\cdot\text{mL}^{-1}}{0.9982\ \text{g}\cdot\text{mL}^{-1}}\right)\left(\frac{64.5\ \text{s}}{42.6\ \text{s}}\right)$$

$$= 1.200 \times 10^{-3}\ \text{Pa}\cdot\text{s}$$

2.109 How long will it take for a sample of mercury to flow through the viscometer described in Problem 2.108? For Hg at 0 °C, $\eta = 1.554 \times 10^{-3}$ Pa · s and $\rho = 13.5462$ g · mL^{-1}.

▌ Solving *(2.46)* for t(Hg) and substituting the data gives

$$t(\text{Hg}) = (42.6\ \text{s})\frac{0.9982\ \text{g}\cdot\text{mL}^{-1}}{13.5462\ \text{g}\cdot\text{mL}^{-1}}\left(\frac{1.554 \times 10^{-3}\ \text{Pa}\cdot\text{s}}{1.002 \times 10^{-3}\ \text{Pa}\cdot\text{s}}\right) = 4.87\ \text{s}$$

2.110 The *falling-sphere technique* for measuring the coefficient of viscosity of a liquid is based on a balance between viscous drag and the force of gravity:

$$\eta = 2r_b(\rho_b - \rho)g/9v \tag{2.47}$$

where v is the velocity of the falling sphere, g is the gravitational constant, and the subscript b refers to the sphere or bead. A small steel bead required 0.27 s to sink through a distance of 1.00 m in water at 20 °C. Given $\rho = 0.9982$ g · mL^{-1}, $\rho_b = 7.83$ g · mL^{-1}, $r_b = 0.50$ mm, find η for water.

▌ The velocity is

$$v = (1.00\ \text{m})/(0.27\ \text{s}) = 3.7\ \text{m}\cdot\text{s}^{-1}$$

which gives

$$\eta = \frac{2(0.50\ \text{mm})^2\left(\dfrac{10^{-3}\ \text{m}}{1\ \text{mm}}\right)^2(7.83\ \text{g}\cdot\text{mL}^{-1} - 0.9982\ \text{g}\cdot\text{mL}^{-1})\dfrac{10^3\ \text{kg}\cdot\text{m}^{-3}}{1\ \text{g}\cdot\text{mL}^{-1}}(9.81\ \text{m}\cdot\text{s}^{-2})}{9(3.7\ \text{m}\cdot\text{s}^{-1})}$$

$$= 1.01 \times 10^{-3}\ \text{kg}\cdot\text{m}^{-1}\cdot\text{s}^{-1} = 1.01 \times 10^{-3}\ \text{Pa}\cdot\text{s}$$

2.111 Often comparison methods are used to determine η for a liquid using the falling-sphere technique (see Problem 2.110). Using water as the reference, a glass ball required 38.3 s to sink through the marked distance on the viscometer at 20 °C. In ethanol, the same ball required 39.3 s. Given $\eta = 1.002 \times 10^{-3}$ Pa · s for H_2O and $\rho/(\text{g}\cdot\text{mL}^{-1}) = 0.9982, 0.789\,45$, and 2.24 for H_2O, ethanol, and glass, respectively, find η for CH_3CH_2OH.

▌ For the comparison method, *(2.47)* becomes

$$\eta/\eta_0 = (\rho_b - \rho)t/(\rho_b - \rho_0)t_0 \tag{2.48}$$

where the subscript zero refers to the reference liquid. Solving for $\eta(CH_3CH_2OH)$ and substituting the data gives

$$\eta(\text{CH}_3\text{CH}_2\text{OH}) = (1.002 \times 10^{-3}\ \text{Pa}\cdot\text{s})\frac{(2.24\ \text{g}\cdot\text{mL}^{-1} - 0.789\,45\ \text{g}\cdot\text{mL}^{-1})(39.3\ \text{s})}{(2.24\ \text{g}\cdot\text{mL}^{-1} - 0.9982\ \text{g}\cdot\text{mL}^{-1})(38.3\ \text{s})}$$

$$= 1.20 \times 10^{-3}\ \text{Pa}\cdot\text{s}$$

2.112 Two spheres of identical diameter were released at the same time and allowed to sink the same distance in water ($\rho = 1.00\ \mathrm{g \cdot mL^{-1}}$). One ball was steel ($\rho = 7.83\ \mathrm{g \cdot mL^{-1}}$), and the other was glass ($\rho = 2.24\ \mathrm{g \cdot mL^{-1}}$). What is the ratio of the time required for the balls to sink? Would this ratio change if the liquid was changed to ethanol ($\rho = 0.79\ \mathrm{g \cdot mL^{-1}}$)?

▮ The ratio of the times required by the two spheres will be given by *(2.47)* as

$$t_{b2}/t_{b1} = (\rho_{b1} - \rho)/(\rho_{b2} - \rho)$$

Substituting the data gives for water

$$\frac{t(\text{glass})}{t(\text{steel})} = \frac{7.83\ \mathrm{g \cdot mL^{-1}} - 1.00\ \mathrm{g \cdot mL^{-1}}}{2.24\ \mathrm{g \cdot mL^{-1}} - 1.00\ \mathrm{g \cdot mL^{-1}}} = 5.51$$

In ethanol, the ratio is slightly different:

$$\frac{t(\text{glass})}{t(\text{steel})} = \frac{7.83 - 0.79}{2.24 - 0.79} = 4.86$$

2.113 Determine the terminal velocity of a 1.00-mm-diameter bead as it falls through a sample of ethanol at 20 °C. The density of ethanol is $0.789\,45\ \mathrm{g \cdot mL^{-1}}$, the density of the glass bead is $2.24\ \mathrm{g \cdot mL^{-1}}$, and the coefficient of viscosity of ethanol is $1.200 \times 10^{-3}\ \mathrm{Pa \cdot s}$.

▮ Solving *(2.47)* for v and substituting the data gives

$$v = \frac{2(0.50\ \mathrm{mm})^2\left(\dfrac{10^{-3}\ \mathrm{m}}{1\ \mathrm{mm}}\right)^2(2.24\ \mathrm{g \cdot mL^{-1}} - 0.789\,45\ \mathrm{g \cdot mL^{-1}})\dfrac{10^3\ \mathrm{kg \cdot m^{-3}}}{1\ \mathrm{g \cdot mL^{-1}}}(9.81\ \mathrm{m \cdot s^{-2}})}{9(1.200 \times 10^{-3}\ \mathrm{Pa \cdot s})[(1\ \mathrm{kg \cdot m^{-1} \cdot s^{-1}})/(1\ \mathrm{Pa \cdot s})]}$$

$$= 0.66\ \mathrm{m \cdot s^{-1}}$$

2.114 The coefficient of viscosity for a substance can be measured using *Newton's law*:

$$F = \frac{\partial p_y}{\partial t} = -\eta A \frac{\partial v_y}{\partial z} \tag{2.49}$$

where F is force and p_y is momentum. An apparatus based on *(2.49)* consists of two concentric cylinders with the space between them filled by the substance being considered. The outer cylinder of radius r_o is revolved slowly, and the torque produced on the inner cylinder of radius r_i is measured. Calculate η for air at 0 °C and 1.00 bar using the following data: $r_o = 10.0$ cm, $r_i = 9.9$ cm, torque $= 1.11 \times 10^{-6}$ J, cylinder length = 10.0 cm, and outer cylinder speed = 1 rev/min.

▮ The velocity of the outer cylinder is

$$v = 2\pi(0.100\ \mathrm{m})/60\ \mathrm{s} = 1.05 \times 10^{-2}\ \mathrm{m \cdot s^{-1}}$$

giving

$$\frac{\partial p_y}{\partial z} = \frac{-v}{r_o - r_i} = \frac{-1.05 \times 10^{-2}\ \mathrm{m \cdot s^{-1}}}{0.100\ \mathrm{m} - 0.099\ \mathrm{m}} = -10.5\ \mathrm{s^{-1}}$$

The force corresponding to the torque on the inner cylinder is

$$F = \frac{(1.11 \times 10^{-6}\ \mathrm{J})[(1\ \mathrm{kg \cdot m^2 \cdot s^{-2}})/(1\ \mathrm{J})]}{0.099\ \mathrm{m}} = 1.12 \times 10^{-5}\ \mathrm{kg \cdot m \cdot s^{-2}}$$

Solving *(2.49)* for η and substituting the data gives

$$\eta = \frac{-(1.12 \times 10^{-5}\ \mathrm{kg \cdot m \cdot s^{-2}})}{[2\pi(0.099\ \mathrm{m})(0.100\ \mathrm{m})](-10.5\ \mathrm{s^{-1}})}$$

$$= 1.71 \times 10^{-5}\ \mathrm{kg \cdot m^{-1} \cdot s^{-1}} = 1.71 \times 10^{-5}\ \mathrm{Pa \cdot s}$$

2.115 Use *(2.38)* and *(2.40)* to predict k_T/η for molecules that can be described as rigid, hard spheres. Calculate this ratio for Ar using $C_{V,\mathrm{m}} = 12.47\ \mathrm{J \cdot K^{-1} \cdot mol^{-1}}$. Compare this value to the actual value using $\eta = 2.10 \times 10^{-5}\ \mathrm{Pa \cdot s}$ and $k_T = 1.74 \times 10^{-2}\ \mathrm{J \cdot m^{-1} \cdot s^{-1} \cdot K^{-1}}$ for Ar at 0 °C and 1.00 bar. Repeat the calculations for CO_2 given $\eta = 1.366 \times 10^{-5}\ \mathrm{Pa \cdot s}$, $k_T = 1.46 \times 10^{-2}\ \mathrm{J \cdot m^{-1} \cdot s^{-1} \cdot K^{-1}}$, and $C_{V,\mathrm{m}} = 28.7\ \mathrm{J \cdot K^{-1} \cdot mol^{-1}}$ under the same conditions. Comment.

▌ Dividing (2.38) by (2.40) gives the desired ratio as

$$\frac{k_T}{\eta} = \frac{(25\pi/64)\bar{v}(N^*/L)C_{V,m}\lambda}{(5\pi/32)N^*\bar{v}m\lambda} = \frac{2.5C_{V,m}}{Lm} = \frac{2.5C_{V,m}}{M}$$

The predicted value for Ar is

$$\frac{k_T}{\eta} = \frac{(2.5)(12.47\ \mathrm{J\cdot K^{-1}\cdot mol^{-1}})[(1\ \mathrm{kg\cdot m^2\cdot s^{-2}})/(1\ \mathrm{J})]}{39.948\times 10^{-3}\ \mathrm{kg\cdot mol^{-1}}}$$
$$= 7.80\times 10^2\ \mathrm{m^2\cdot K^{-1}\cdot s^{-2}}$$

and the actual value for Ar is

$$\frac{k_T}{\eta} = \frac{(1.74\times 10^{-2}\ \mathrm{J\cdot m^{-1}\cdot s^{-1}\cdot K^{-1}})[(1\ \mathrm{kg\cdot m^2\cdot s^{-2}})/(1\ \mathrm{J})]}{(2.10\times 10^{-5}\ \mathrm{Pa\cdot s})[(1\ \mathrm{kg\cdot m^{-1}\cdot s^{-1}})/(1\ \mathrm{Pa\cdot s})]}$$
$$= 8.29\times 10^2\ \mathrm{m^2\cdot K^{-1}\cdot s^{-2}}$$

The predicted value is 6% low. Likewise for CO_2, the respective values are

$$\frac{k_T}{\eta} = \frac{(2.5)(28.7)}{44.01\times 10^{-3}} = 1.63\times 10^3\ \mathrm{m^2\cdot K^{-1}\cdot s^{-2}}$$
$$\frac{k_T}{\eta} = \frac{1.46\times 10^{-2}}{1.366\times 10^{-5}} = 1.07\times 10\ \mathrm{m^2\cdot K^{-1}\cdot s^{-2}}$$

The predicted value is 52% high. The rigid, hard sphere model is quite good for Ar, but not for CO_2.

2.116 A sample of Ar at 2.00 bar and 25 °C passed through a capillary tube into a laboratory at 1.00 bar. The length of the capillary was 91 cm, and the diameter was 0.50 mm. Given that $\eta = 2.22\times 10^{-5}\ \mathrm{Pa\cdot s}$, find the flow rate of the amount of Ar, $n(\mathrm{Ar})/t$.

▌ For an ideal gas, the P_0 term in (2.44) can be replaced by (1.18), giving

$$v = \frac{\pi R^4}{16\eta l}\left(\frac{P_i^2 - P_f^2}{nRT/V}\right)t$$

which upon rearrangment gives

$$\frac{n(\mathrm{Ar})}{t} = \frac{\pi(2.5\times 10^{-4}\ \mathrm{m})^4[(2.00\ \mathrm{bar})^2 - (1.00\ \mathrm{bar})^2][(10^5\ \mathrm{Pa})/(1\ \mathrm{bar})]^2}{16(2.22\times 10^{-5}\ \mathrm{Pa\cdot s})(0.91\ \mathrm{m})(8.314\ \mathrm{m^3\cdot Pa\cdot K^{-1}\cdot mol^{-1}})(298\ \mathrm{K})}$$
$$= 4.6\times 10^{-4}\ \mathrm{mol\cdot s^{-1}}$$

2.117 Fluid flow in a tube is often described in terms of the *Reynolds number* given by

$$\mathrm{Re} = 2R\bar{v}\rho/\eta \qquad (2.50)$$

If $\mathrm{Re} < 2000$, the flow is called *laminar.* What is the maximum average velocity of water in a 2.00-cm diameter tube that will still be laminar? Assume $\rho = 1.00\ \mathrm{g\cdot mL^{-1}}$ and $\eta = 8.9\times 10^{-4}\ \mathrm{Pa\cdot s}$ for H_2O.

▌ Solving (2.50) for $\bar{v}$ and substituting the data gives

$$\bar{v} = \frac{(<2000)(8.9\times 10^{-4}\ \mathrm{Pa\cdot s})[(1\ \mathrm{kg\cdot m^{-1}\cdot s^{-1}})/(1\ \mathrm{Pa\cdot s})]}{2(1.00\times 10^{-2}\ \mathrm{m})(1.00\ \mathrm{g\cdot mL^{-1}})[(10^3\ \mathrm{kg\cdot m^{-3}})/(1\ \mathrm{g\cdot mL^{-1}})]} < 0.089\ \mathrm{m\cdot s^{-1}}$$

2.118 During laminar flow (see Problem 2.117), the velocity of the fluid at a distance from the center of the tube (r) is given by

$$v = (\Delta P/4\eta l)(R^2 - r^2) \qquad (2.51)$$

At what locations in the tube will the fluid have **(a)** the greatest velocity, **(b)** the least velocity, and **(c)** a velocity equal to $\bar{v}$?

▌ Inspection of (2.51) shows that v will be **(a)** a maximum at $r = 0$, the center of the tube, and **(b)** a minimum ($v = 0$) at $r = R$, the wall of the tube. The average velocity will be given by $\bar{v} = V/At$, which upon substitution of (2.45) and setting equal to (2.51) gives

$$\bar{v} = \frac{V}{At} = \frac{\pi R^4 \Delta P t/8\eta l}{At} = \frac{\Delta P}{4\eta l}(R^2 - r^2)$$

Canceling common terms and substituting $A = \pi R^2$ gives

$$r^2 = R^2 - \frac{\pi R^4}{2A} = R^2 - \frac{\pi R^4}{2\pi R^2} = \frac{R^2}{2}$$

(c) At $r = R/2^{1/2}$, v will be equal to $\bar{v}$.

2.119 Human blood ($\rho = 1.2\ \text{g} \cdot \text{mL}^{-1}$, $\eta = 4 \times 10^{-3}\ \text{Pa} \cdot \text{s}$) is passing through a vein that has a diameter of 5.0 mm. If the flow rate is $50\ \text{mL} \cdot \text{min}^{-1}$, find the pressure drop per unit length along the vein and determine whether or not the flow is laminar.

▮ Solving *(2.45)* for $\Delta P/l$ and substituting the data gives

$$\frac{\Delta P}{l} = \frac{\left(\dfrac{50\ \text{mL}}{1\ \text{min}}\right)\left(\dfrac{10^{-6}\ \text{m}^3}{1\ \text{mL}}\right)\left(\dfrac{1\ \text{min}}{60\ \text{s}}\right)(8)(4 \times 10^{-3}\ \text{Pa} \cdot \text{s})\left(\dfrac{1\ \text{kg} \cdot \text{m}^{-1} \cdot \text{s}^{-1}}{1\ \text{Pa} \cdot \text{s}}\right)}{\pi(2.5 \times 10^{-3}\ \text{m})^4} = 200\ \text{Pa} \cdot \text{m}^{-1}$$

The average velocity is

$$\bar{v} = \frac{V/t}{A} = \frac{\left(\dfrac{50\ \text{mL}}{1\ \text{min}}\right)\left(\dfrac{10^{-6}\ \text{m}^3}{1\ \text{mL}}\right)\left(\dfrac{1\ \text{min}}{60\ \text{s}}\right)}{\pi(2.5 \times 10^{-3}\ \text{m})^2} = 4 \times 10^{-2}\ \text{m} \cdot \text{s}^{-1}$$

which upon substitution into *(2.50)* gives

$$\text{Re} = \frac{2(2.25 \times 10^{-3}\ \text{m})(4 \times 10^{-2}\ \text{m} \cdot \text{s}^{-1})(1.2\ \text{g} \cdot \text{mL}^{-1})\left(\dfrac{10^3\ \text{kg} \cdot \text{m}^{-3}}{1\ \text{g} \cdot \text{mL}^{-1}}\right)}{(4 \times 10^{-3}\ \text{Pa} \cdot \text{s})[(1\ \text{kg} \cdot \text{m}^{-1} \cdot \text{s}^{-1})/(1\ \text{Pa} \cdot \text{s})]} = 60$$

The flow is laminar.

2.120 A small hole developed in the bottom of a 20-L plastic carboy filled with water [$h(H_2O) = 35$ cm]. During a 10.0-s period, a 50.0-mL sample of water was collected from the leak. Given $\rho = 1.00\ \text{g} \cdot \text{mL}^{-1}$ and $\eta = 1.002 \times 10^{-3}\ \text{Pa} \cdot \text{s}$ for H_2O, find the radius of the hole. The thickness of the plastic is 5 mm.

▮ The pressure difference is given by *(1.4)* as

$$\begin{aligned}\Delta P &= \rho g h \\ &= (1.00\ \text{g} \cdot \text{mL}^{-1})\frac{10^3\ \text{kg} \cdot \text{m}^{-3}}{1\ \text{g} \cdot \text{mL}^{-1}}(9.81\ \text{m} \cdot \text{s}^{-2})(35 \times 10^{-2}\ \text{m})\frac{1\ \text{Pa}}{1\ \text{kg} \cdot \text{m}^{-1} \cdot \text{s}^{-2}} \\ &= 3400\ \text{Pa}\end{aligned}$$

Solving *(2.45)* for R^4 and substituting the data gives

$$\begin{aligned}R^4 &= \frac{(50.0\ \text{mL})[(10^{-6}\ \text{m}^3)/(1\ \text{mL})](8)(1.002 \times 10^{-3}\ \text{Pa} \cdot \text{s})(5 \times 10^{-3}\ \text{m})}{\pi(3400\ \text{Pa})(10.0\ \text{s})} \\ &= 2 \times 10^{-14}\ \text{m}^4\end{aligned}$$

The radius of the hole is $4 \times 10^{-4}\ \text{m} = 0.4\ \text{mm}$.

2.121 The rate of diffusion in the z direction is given by *Fick's first law* as

$$\frac{1}{A}\frac{\partial N^*}{\partial t} = -D\frac{\partial c}{\partial z} \qquad (2.52)$$

where D is the diffusion coefficient and c is the concentration. Given $D = 1.6 \times 10^{-5}\ \text{m}^2 \cdot \text{s}^{-1}$ for Ar at 0 °C and 1.00 atm, find the flow as a result of diffusion through a pressure gradient of $0.01\ \text{atm} \cdot \text{cm}^{-1}$.

▮ The concentration gradient in *(2.52)* is

$$\begin{aligned}\frac{\partial c}{\partial z} &= \frac{L}{RT}\frac{\partial P}{\partial z} \\ &= \frac{(6.022 \times 10^{23}\ \text{mol}^{-1})(-0.10\ \text{atm} \cdot \text{cm}^{-1})[(101\,325\ \text{Pa})/(1\ \text{atm})][(10^2\ \text{cm})/(1\ \text{m})]}{(8.314\ \text{m}^3 \cdot \text{Pa} \cdot \text{K}^{-1} \cdot \text{mol}^{-1})(273\ \text{K})} \\ &= -3 \times 10^{25}\ \text{m}^{-4}\end{aligned}$$

The flow is given by *(2.52)* as

$$\frac{1}{A}\frac{\partial N^*}{\partial t} = -(1.6 \times 10^{-5}\ \mathrm{m^2 \cdot s^{-1}})(-3 \times 10^{25}\ \mathrm{m^{-4}}) = 5 \times 10^{20}\ \mathrm{m^{-2} \cdot s^{-1}}$$

or $8 \times 10^{-4}\ \mathrm{mol \cdot m^{-2} \cdot s^{-1}}$.

2.122 A hollow iron cube, 1.0 m on a side, containing hydrogen gas at a concentration c_0 is placed in an oven at 500 °C. What is the concentration in the cube after 1 day? The wall thickness of the cube is 0.10 cm, and the diffusion coefficient of H_2 in Fe is $1.5 \times 10^{-9}\ \mathrm{m^2 \cdot s^{-1}}$ at 500 °C.

▌ Assuming the concentration of H_2 outside the cube is maintained at zero, the concentration gradient is $\partial c/\partial z = c/z$, where z is the wall thickness. The flow in terms of concentration is $\partial N^*/\partial t = \partial cV/\partial t = V(\partial c/\partial t)$. Substituting these results into *(2.52)* gives

$$\frac{V}{A}\frac{\partial c}{\partial t} = -D\frac{c}{z}$$

which upon integration gives

$$\ln (c/c_0) = (-DA/Vz)t \tag{2.53}$$

Taking the antilogarithm of each side, solving for c, and substituting the data gives

$$c = c_0\, e^{-(DA/Vz)t} = c_0 \exp\left(-\frac{(1.5 \times 10^{-9}\ \mathrm{m^2 \cdot s^{-1}})(6.0\ \mathrm{m^2})}{(1.0\ \mathrm{m^3})(1.0 \times 10^{-3}\ \mathrm{m})}(3600\ \mathrm{s})\right) = 0.968 c_0$$

The concentration has decreased by 3.2%.

2.123 How long will it take for the concentration of H_2 in the cube described in Problem 2.122 to decrease by one-half?

▌ Solving *(2.53)* for t and substituting $c = c_0/2$ gives

$$t = \frac{-Vz}{DA}\ln\frac{c}{c_0} = \frac{-(1.0\ \mathrm{m^3})(1.0 \times 10^{-3}\ \mathrm{m})}{(1.5 \times 10^{-9}\ \mathrm{m^2 \cdot s^{-1}})(6.0\ \mathrm{m^2})}\ln\frac{c_0/2}{c_0} = 7.7 \times 10^4\ \mathrm{s}$$

or roughly 21 days.

2.124 The rate of diffusion of Na^+ through a cell membrane is $1.8\ \mathrm{pmol \cdot cm^{-2} \cdot s^{-1}}$ when the concentrations of Na^+ on the two sides of the membrane are 145 and 12 mM, respectively. Determine the *permeability coefficient* defined as

$$P = D/z \tag{2.54}$$

▌ The concentration gradient and the flow for this system are

$$\frac{\partial c}{\partial z} = \frac{(12\ \mathrm{mM} - 145\ \mathrm{mM})\dfrac{10^{-3}\ \mathrm{mol \cdot L^{-1}}}{1\ \mathrm{mM}}\left(\dfrac{10^3\ \mathrm{L}}{1\ \mathrm{m^3}}\right)(6.022 \times 10^{23}\ \mathrm{mol^{-1}})}{z}$$

$$= -(8.01 \times 10^{25})/z\ \mathrm{m^{-3}}$$

$$\frac{1}{A}\frac{\partial N^*}{\partial t} = (1.8\ \mathrm{pmol \cdot cm^{-2} \cdot s^{-1}})\frac{10^{-12}\ \mathrm{mol}}{1\ \mathrm{pmol}}(6.022 \times 10^{23}\ \mathrm{mol^{-1}})\left(\frac{10^2\ \mathrm{cm}}{1\ \mathrm{m}}\right)^2$$

$$= 1.1 \times 10^{16}\ \mathrm{m^{-2} \cdot s^{-1}}$$

Substituting the above results and *(2.52)* into *(2.54)* gives

$$P = \frac{-(1/A)(\partial N^*/\partial t)/(\partial c/\partial z)}{z}$$

$$= \frac{-(1.1 \times 10^{16}\ \mathrm{m^{-2} \cdot s^{-1}})/[-(8.01 \times 10^{25})/z]\mathrm{m^{-3}}}{z}$$

$$= 1.4 \times 10^{-10}\ \mathrm{m \cdot s^{-1}}$$

2.125 A layer of ethanol is in contact with a layer of water. The diffusion coefficient of CH_3CH_2OH into H_2O is $2.4 \times 10^{-5}\ \mathrm{cm^2 \cdot s^{-1}}$ at 25 °C. What will be the relative concentration of CH_3CH_2OH at a distance of 1.0 cm into the H_2O layer after 1.0 h?

▮ For this type of diffusion process, the integrated form of *Fick's second law*

$$\frac{\partial c}{\partial t} = D\frac{\partial^2 c}{\partial z^2} \tag{2.55}$$

gives the relative concentration (c/c_0) at a distance x from the boundary at a given time as

$$\frac{c}{c_0} = 0.5\left[1 - \operatorname{erf}\left(\frac{x}{(4Dt)^{1/2}}\right)\right] \tag{2.56}$$

Substituting the data gives

$$\frac{c}{c_0} = 0.5\left[1 - \operatorname{erf}\left(\frac{(1.0\ \text{cm})[(10^{-2}\ \text{m})/(1\ \text{cm})]}{\left[4(2.4\times10^{-5}\ \text{cm}^2\cdot\text{s}^{-1})\left(\frac{10^{-2}\ \text{m}}{1\ \text{cm}}\right)^2(1.0\ \text{h})\left(\frac{3600\ \text{s}}{1\ \text{h}}\right)\right]^{1/2}}\right)\right]$$

$$= 0.5[1 - \operatorname{erf}(1.70)] = 0.5(1 - 0.98) = 0.01$$

where the error function was evaluated using Fig. 2-1.

2.126 An equimolar mixture of H_2 and D_2 diffused through a porous disk. What is the composition of the gas that passes through the disk? For D_2, $M = 4.0280\ \text{g}\cdot\text{mol}^{-1}$ and for H_2, $M = 2.015\,65\ \text{g}\cdot\text{mol}^{-1}$.

▮ For this process, the relative rates of diffusion are given by *Graham's law of diffusion*,

$$\frac{\text{Rate}_i}{\text{Rate}_j} = \left(\frac{\rho_j}{\rho_i}\right)^{1/2} = \left(\frac{M_j}{M_i}\right)^{1/2} \tag{2.57}$$

Substitution of the data gives

$$\frac{\text{Rate}\,(H_2)}{\text{Rate}\,(D_2)} = \left(\frac{4.0280\ \text{g}\cdot\text{mol}^{-1}}{2.015\,65\ \text{g}\cdot\text{mol}^{-1}}\right)^{1/2} = 1.4136$$

The mole fraction of H_2 in the exiting gas is

$$x(H_2) = \frac{1.4136}{1.4136 + 1} = 0.5857$$

2.127 Natural uranium can be enriched in the concentration of ^{235}U by diffusion of gaseous UF_6. The number of enriching stages (N_s) needed to change the concentration from $[x(^{235}U)]_0/[x(^{238}U)]_0$ to an enriched value $x(^{235}U)/x(^{238}U)$ is given by

$$N_s = \frac{\log\left(\frac{x(^{235}U)/x(^{238}U)}{[x(^{235}U)]_0/[x(^{238}U)]_0}\right)}{\log\alpha}$$

where α is a separation factor based on the system. Assuming $\alpha = 1.0030$, find the number of stages needed to enrich natural uranium $[x(^{235}U) = 0.0072]$ to $x(^{235}U) = 0.90$.

▮ Substituting the data gives

$$N_s = \frac{\log[(0.90/0.10)/(0.0072/0.9928)]}{\log 1.0030} = 2400$$

2.128 For a mixture of rigid, hard-sphere gases, the diffusion coefficient of gas i into gas j is given by

$$D_{ij} = \frac{3(2^{1/2})\pi}{64}\lambda_i\bar{v}_i = \frac{3}{8\pi^{1/2}}\left(\frac{RT}{M_i}\right)^{1/2}\left(\frac{1}{\sigma_{ij}^2 N^*}\right) \tag{2.58}$$

Predict $D(O_2, N_2)$ for an equimolar mixture at 1.00 atm and 0 °C using $\sigma(O_2) = 0.353$ nm and $\sigma(N_2) = 0.373$ nm.

▮ For the mixture, *(1.21)* and *(2.24)* give

$$N^* = \frac{(101\,325\ \text{Pa})(6.022\times10^{23}\ \text{mol}^{-1})}{(8.314\ \text{m}^3\cdot\text{Pa}\cdot\text{K}^{-1}\cdot\text{mol}^{-1})(273\ \text{K})} = 2.69\times10^{25}\ \text{m}^{-3}$$

$$\sigma(O_2, N_2) = (0.353\ \text{nm} + 0.373\ \text{nm})/2 = 0.363\ \text{nm}$$

which upon substituting into *(2.58)* gives

$$D(O_2, N_2) = \frac{3}{8\pi^{1/2}}\left(\frac{(8.314\ J\cdot K^{-1}\cdot mol^{-1})(273\ K)[(1\ kg\cdot m^2\cdot s^{-2})/(1\ J)]}{32.00\times 10^{-3}\ kg\cdot mol^{-1}}\right)^{1/2}$$

$$\times\frac{1}{(3.63\times 10^{-10}\ m)^2(2.69\times 10^{25}\ m^{-3})}$$

$$= 1.59\times 10^{-5}\ m^2\cdot s^{-1}$$

2.129 The "self-diffusion" coefficient for O_2 at 1.00 atm and 0 °C is 0.187 $cm^2\cdot s^{-1}$. Calculate the collision diameter of O_2.

▮ For O_2 at 1.00 atm and 0 °C, *(1.21)* gives

$$N^*(O_2) = \frac{(101\ 325\ Pa)(6.022\times 10^{23}\ mol^{-1})}{(8.314\ m^3\cdot Pa\cdot K^{-1}\cdot mol^{-1})(273\ K)} = 2.69\times 10^{25}\ m^{-3}$$

Rearranging *(2.58)* for σ^2 and substituting the data gives

$$\sigma^2 = \frac{3}{8\pi^{1/2}}\left(\frac{(8.314\ J\cdot K^{-1}\cdot mol^{-1})(273\ K)[(1\ kg\cdot m^2\cdot s^{-2})/(1\ J)]}{32.00\times 10^{-3}\ kg\cdot mol^{-1}}\right)^{1/2}$$

$$\times\frac{1}{(0.187\ cm^2\cdot s^{-1})[(10^{-2}\ m)/(1\ cm)]^2(2.69\times 10^{25}\ m^{-3})}$$

$$= 1.12\times 10^{-19}\ m^2$$

The collision diameter for O_2 determined from diffusion measurements is 0.335 nm. Note that this value differs slightly from that determined from thermal conductivity measurements (0.406 nm) and from viscosity measurements (0.353 nm).

2.130 What is the value of D_{ii} for O_2 at 10.00 atm and 0 °C? See Problem 2.129 for additional data.

▮ For this higher-pressure system, N^* will increase by a factor of 10 to $2.69\times 10^{26}\ m^{-3}$. By inspection of *(2.58)*, $D_{ii}\propto 1/N^*$, so $D = 0.0187\ cm^2\cdot s^{-1}$.

2.131 For very dilute solutions of solute i in liquid solvent j, the *Stokes–Einstein equation* gives the diffusion coefficient as

$$D_{ij}^{\infty} = 2kT/6\pi\eta_j\sigma_i \qquad (2.59)$$

for spherical solute molecules with $r_i > r_j$. [For spherical solute molecules with $r_i \approx r_j$, the 6 in *(2.59)* should be replaced with a 4.] Using $D^{\infty}(C_{12}H_{22}O_{11}, H_2O) = 5.2\times 10^{-6}\ cm^2\cdot s^{-1}$ and $\eta(H_2O) = 0.8904$ cP at 25 °C, determine the spherical collision diameter of a sucrose molecule in water.

▮ Solving *(2.59)* for $\sigma(C_{12}H_{22}O_{11})$ and substituting the data gives

$$\sigma(C_{12}H_{22}O_{11}) = \frac{2(1.381\times 10^{-23}\ J\cdot K^{-1})(298\ K)[(1\ m^3\cdot Pa)/(1\ J)]}{6\pi(0.8904\ cP)\left(\frac{10^{-2}\ P}{1\ cP}\right)\left(\frac{0.1\ Pa\cdot s}{1\ P}\right)(5.2\times 10^{-6}\ cm^2\cdot s^{-1})\left(\frac{10^{-2}\ m}{1\ cm}\right)^2}$$

$$= 9.4\times 10^{-10}\ m = 0.94\ nm$$

2.132 Using the data and results of Problem 2.131, predict the value of the diffusion coefficient of sucrose in water at 20 °C. At this temperature, $\eta(H_2O) = 1.002$ cP.

▮ Substituting into *(2.59)* gives

$$D^{\infty}(C_{12}H_{22}O_{11}, H_2O) = \frac{2(1.381\times 10^{-23}\ J\cdot K^{-1})(293\ K)[(1\ m^3\cdot Pa)/(1\ J)]}{6\pi(1.002\ cP)\left(\frac{10^{-2}\ P}{1\ cP}\right)\left(\frac{0.1\ Pa\cdot s}{1\ P}\right)(9.4\times 10^{-10}\ m)}$$

$$= 4.6\times 10^{-10}\ m^2\cdot s^{-1}$$

2.133 Use *(2.59)* to determine $D^{\infty}(N_2, H_2O)$ at 1.00 atm and 25 °C. For water, $\eta = 8.904\times 10^{-4}\ Pa\cdot s$, and for N_2, $\sigma = 0.373$ nm.

▮ Because the nitrogen molecule is approximately the same size as the water molecule, the factor of 4 should be used in the denominator of *(2.59)*.

Substituting the data gives

$$D^\infty(N_2, H_2O) = \frac{2(1.381\times10^{-23}\ \text{J}\cdot\text{K}^{-1})(298\ \text{K})[(1\ \text{m}^3\cdot\text{Pa})/(1\ \text{J})]}{4\pi(8.904\times10^{-4}\ \text{Pa}\cdot\text{s})(3.73\times10^{-10}\ \text{m})}$$

$$= 1.97\times10^{-9}\ \text{m}^2\cdot\text{s}^{-1}$$

2.134 For the special case of self diffusion in a liquid, *(2.59)* becomes

$$D_{ii}^\infty = \frac{kT}{2\pi\eta}\left(\frac{L}{V_m}\right)^{1/3} \tag{2.60}$$

Determine D_{ii}^∞ for ethanol at 20.0 °C given $\eta = 1.200$ cP and $\rho = 0.789\,45\ \text{g}\cdot\text{mL}^{-1}$.

▌ The molar volume is

$$V_m = \frac{M}{\rho} = \frac{46.07\ \text{g}\cdot\text{mol}^{-1}}{(0.789\,45\ \text{g}\cdot\text{mL}^{-1})[(10^6\ \text{mL})/(1\ \text{m}^3)]} = 5.832\times10^{-5}\ \text{m}^3\cdot\text{mol}^{-1}$$

which upon substituting into *(2.60)* gives

$$D^\infty(CH_3CH_2OH, CH_3CH_2OH) = \frac{(1.381\times10^{-23}\ \text{J}\cdot\text{K}^{-1})(293.2\ \text{K})[(1\ \text{m}^3\cdot\text{Pa})/(1\ \text{J})]}{2\pi(1.200\ \text{cP})[(10^{-3}\ \text{Pa}\cdot\text{s})/(1\ \text{cP})]}$$

$$\times\left(\frac{6.022\times10^{23}\ \text{mol}^{-1}}{5.832\times10^{-5}\ \text{m}^3\cdot\text{mol}^{-1}}\right)^{1/3}$$

$$= 1.17\times10^{-9}\ \text{m}^2\cdot\text{s}^{-1}$$

2.135 The diffusion coefficient of carbon in iron is $3\times10^{-8}\ \text{cm}^2\cdot\text{s}^{-1}$ at 500 °C. What is the *root-mean-square distance* $(\overline{z^2})^{1/2}$ that carbon will move in iron after 1 yr (1 yr = 3.16×10^7 s)? How long would it take carbon to move this same distance at 27 °C, where $D = 4\times10^{-21}\ \text{m}^2\cdot\text{s}^{-1}$?

▌ The root-mean-square distance is given by the *Einstein–Smoluchowski equation* as

$$(\overline{z^2})^{1/2} = (2Dt)^{1/2} \tag{2.61}$$

Substituting the data gives

$$(\overline{z^2})^{1/2} = \left[2(3\times10^{-8}\ \text{cm}^2\cdot\text{s}^{-1})\left(\frac{10^{-2}\ \text{m}}{1\ \text{cm}}\right)^2(3.16\times10^7\ \text{s})\right]^{1/2}$$

$$= 1\times10^{-2}\ \text{m} = 1\ \text{cm}$$

Solving *(2.61)* for t and substituting the data gives

$$t = \frac{[(\overline{z^2})^{1/2}]^2}{2D} = \frac{(1\times10^{-2}\ \text{m})^2}{2(4\times10^{-21}\ \text{m}^2\cdot\text{s}^{-1})} = 1\times10^{16}\ \text{s} = 3\times10^8\ \text{yr}$$

CHAPTER 3

First Law of Thermodynamics

3.1 INTERNAL ENERGY AND ENTHALPY

3.1 Place the following values of energy in order of increasing size: **(*a*)** 6.3 kcal, **(*b*)** 97 erg, **(*c*)** $5000\ \text{cm}^{-1}$, **(*d*)** $0.3\ \text{Pa}\cdot\text{m}^3$, **(*e*)** 3.7 L · atm, **(*f*)** 0.01 u, **(*g*)** 14 eV, and **(*h*)** 246 N · m.

▮ The conversion factors between the various energy units are given in Table 3-1.

Table 3-1

energy unit	symbol	conversion factor
Joule	J	$1\ \text{N}\cdot\text{m}^{a} = 1\ \text{Pa}\cdot\text{m}^{3a} = 1\ \text{kg}\cdot\text{m}^2\cdot\text{s}^{-2a}$
Erg	erg	$10^{-7}\ \text{J}^{a}$
Calorie	cal	$4.184\ \text{J}^{a}$
Electronvolt	eV	$1.602\,189\,2 \times 10^{-19}\ \text{J}$
Liter-atmosphere	L · atm	$101.325\ \text{J}^{a}$
Wavenumber	cm^{-1}	$1.986\,477 \times 10^{-23}\ \text{J}$
Atomic mass unit	u (or amu)	$1.492\,442 \times 10^{-10}\ \text{J}$

[a] Exact conversion factor

(*a*) $$(6.3\ \text{kcal})\frac{10^3\ \text{cal}}{1\ \text{kcal}}\left(\frac{4.184\ \text{J}}{1\ \text{cal}}\right) = 2.6 \times 10^4\ \text{J}$$

(*b*) $$(97\ \text{erg})\frac{10^{-7}\ \text{J}}{1\ \text{erg}} = 9.7 \times 10^{-6}\ \text{J}$$

(*c*) $$(5000\ \text{cm}^{-1})\frac{1.986 \times 10^{-23}\ \text{J}}{1\ \text{cm}^{-1}} = 1 \times 10^{-19}\ \text{J}$$

(*d*) $$(0.3\ \text{Pa}\cdot\text{m}^3)\frac{1\ \text{J}}{1\ \text{Pa}\cdot\text{m}^3} = 0.3\ \text{J}$$

(*e*) $$(3.7\ \text{L}\cdot\text{atm})\frac{101.325\ \text{J}}{1\ \text{L}\cdot\text{atm}} = 370\ \text{J}$$

(*f*) $$(0.01\ \text{u})\frac{1.492 \times 10^{-10}\ \text{J}}{1\ \text{u}} = 1 \times 10^{-12}\ \text{J}$$

(*g*) $$(14\ \text{eV})\frac{1.602 \times 10^{-19}\ \text{J}}{1\ \text{eV}} = 2.2 \times 10^{-18}\ \text{J}$$

(*h*) $$(246\ \text{N}\cdot\text{m})\frac{1\ \text{J}}{1\ \text{N}\cdot\text{m}} = 246\ \text{J}$$

The energies in order of increasing size are $c < g < f < b < d < h < e < a$.

3.2 A heating element was rated as "150 W." What amount of energy will the heating element provide each minute?

▮ Multiplying the power by time gives energy

$$(150\ \text{W})\frac{1\ \text{J}\cdot\text{s}^{-1}}{1\ \text{W}}(60\ \text{s}) = 9.0 \times 10^3\ \text{J}$$

3.3 List the contributions to the internal energy (U) of a system at a temperature $T > 0\ \text{K}$. Which of these are present at $T = 0\ \text{K}$?

▮ At $T > 0$ K, the possible contribution will include **(a)** translational kinetic energy, **(b)** rotational kinetic energy, **(c)** vibrational kinetic energy, **(d)** electronic kinetic energy, **(e)** nuclear kinetic energy, **(f)** various forms of potential energy as a result of the position of the system, and **(g)** energy as the result of having mass. At $T = 0$ K, the contribution will include ***c–g***.

3.4 The *enthalpy* (H) and *enthalpy change* (ΔH) are defined as

$$H = U + PV \tag{3.1a}$$

$$\Delta H = \Delta U + \Delta(PV) \tag{3.1b}$$

Calculate $\Delta H - \Delta U$ for heating 1.00 mol of Zn(s) from 25 °C to 98 °C at 1.00 bar. The Zn undergoes a volume change from $9.16\ \text{mL}\cdot\text{mol}^{-1}$ to $9.22\ \text{mL}\cdot\text{mol}^{-1}$ during this temperature increase.

▮ For a process carried out under isobaric conditions, *(3.1b)* gives

$$\Delta H - \Delta U = \Delta(PV) = P\,\Delta V \tag{3.2}$$

$$= \left[(1.00\ \text{bar})\frac{10^5\ \text{Pa}}{1\ \text{bar}}\right]$$

$$\times\left[(1.00\ \text{mol})(9.22\ \text{mL}\cdot\text{mol}^{-1} - 9.16\ \text{mL}\cdot\text{mol}^{-1})\frac{1\ \text{cm}^3}{1\ \text{mL}}\left(\frac{10^{-6}\ \text{m}^3}{1\ \text{cm}^3}\right)\right]\left(\frac{1\ \text{J}}{1\ \text{m}^3\cdot\text{Pa}}\right)$$

$$= 6 \times 10^{-3}\ \text{J}$$

3.5 Repeat the calculations of Problem 3.4 for 1.00 mol of an ideal gas undergoing the same temperature increase.

▮ For an ideal gas, substituting *(1.18)* into *(3.1b)* gives

$$\Delta H - \Delta U = \Delta(PV) = \Delta(nRT) \tag{3.3}$$

$$= (1.00\ \text{mol})(8.314\ \text{J}\cdot\text{K}^{-1}\cdot\text{mol}^{-1})(371\ \text{K} - 298\ \text{K}) = 607\ \text{J}$$

3.6 Repeat the calculations of Problem 3.5, assuming that the gas acts as a real gas. Assume $Z = 0.95$ at 25 °C and $Z = 0.97$ at 98 °C. How much of an error is introduced by assuming ideal gas behavior in this type of calculation?

▮ Substituting *(1.28)* into *(3.1b)* gives

$$\Delta H - \Delta U = \Delta(PV) = \Delta(ZnRT) \tag{3.4}$$

$$\Delta H - \Delta U = (1.00\ \text{mol})(8.314\ \text{J}\cdot\text{K}^{-1}\cdot\text{mol}^{-1})[(371\ \text{K})(0.97) - (298\ \text{K})(0.95)]$$

$$= 638\ \text{J}$$

Although there is a 5% difference in these results, the 31-J difference is only 1–2% of the typical values of ΔH and ΔU involved in these processes.

3.7 For the chemical equation $C(s) + CO_2(g) \rightarrow 2CO(g)$, $\Delta H = 172.459$ kJ at 1.00 bar and 25.0 °C. Find ΔU for this reaction. Assume ideal gas behavior for CO_2 and CO. The density of graphite is $2.25\ \text{g}\cdot\text{cm}^{-3}$.

▮ The ΔV term in *(3.2)* will be given by

$$\Delta V = V(\text{CO}) - [V(\text{C}) + V(\text{CO}_2)] = \frac{n(\text{CO})RT}{P} - \left(n(\text{C}) + \frac{n(\text{CO}_2)RT}{P}\right)$$

$$= \frac{(2\ \text{mol})(8.314\ \text{m}^3\cdot\text{Pa}\cdot\text{K}^{-1}\cdot\text{mol}^{-1})(298.2\ \text{K})}{(1.00\ \text{bar})[(10^5\ \text{Pa})/(1\ \text{bar})]}$$

$$-\left(\frac{(1\ \text{mol})(12.01\ \text{g}\cdot\text{mol}^{-1})}{(2.25\ \text{g}\cdot\text{cm}^{-3})[(10^6\ \text{cm}^3)/(1\ \text{m}^3)]} + \frac{(1\ \text{mol})(8.314\ \text{m}^3\cdot\text{Pa}\cdot\text{K}^{-1}\cdot\text{mol}^{-1})(298.2\ \text{K})}{(1.00\ \text{bar})[(10^5\ \text{Pa})/(1\ \text{bar})]}\right)$$

$$= 4.96 \times 10^{-2}\ \text{m}^3 - (5.34 \times 10^{-6}\ \text{m}^3 + 2.479 \times 10^{-2}\ \text{m}^3)$$

$$= 2.478 \times 10^{-2}\ \text{m}^3$$

Solving *(3.2)* for ΔU and substituting gives

$$\Delta U = \Delta H - P\,\Delta V$$

$$= 172.459\ \text{kJ} - (1.00\ \text{bar})(2.478 \times 10^{-2}\ \text{m}^3)\frac{10^5\ \text{Pa}}{1\ \text{bar}}\left(\frac{1\ \text{J}}{1\ \text{m}^3\cdot\text{Pa}}\right)\left(\frac{1\ \text{kJ}}{10^3\ \text{J}}\right)$$

$$= 172.459\ \text{kJ} - 2.478\ \text{kJ} = 169.981\ \text{kJ}$$

3.8 A chemical equation describing a reaction can be written as

$$\nu_A A + \nu_B B + \cdots \rightarrow \nu_Z Z + \nu_Y Y + \cdots$$

where ν_i represents the stoichiometric coefficients. If the contributions to the $\Delta(PV)$ term are neglected for all substances except gases, *(3.2)* becomes

$$\Delta H - \Delta U = RT\,\Delta n_g = RT\,\Delta \nu_{i,g} \qquad (3.5)$$

where $\nu_{i,g}$ represents the stoichiometric coefficients of only the gaseous reactants and products (positive for products and negative for reactants). Repeat the calculations of Problem 3.7 using *(3.5)*.

▌ For the chemical reaction being considered,

$$\Delta n_g = \nu(CO) + \nu(CO_2) = 2\ \text{mol} + (-1\ \text{mol}) = 1\ \text{mol}$$

Solving *(3.5)* for ΔU and substituting the data gives

$$\Delta U = \Delta H - RT\,\Delta n_g$$
$$= 172.459\ \text{kJ} - (8.314 \times 10^{-3}\ \text{kJ} \cdot \text{K}^{-1} \cdot \text{mol}^{-1})(298.2\ \text{K})(1\ \text{mol})$$
$$= 169.980\ \text{kJ}$$

3.9 How are the degrees of freedom distributed among the translational, rotational, and vibrational motions of a monatomic molecule such as He(g), I(g), or Na(g)?

▌ A gaseous molecule will have a total of 3Λ degrees of freedom, where Λ is the number of atoms in the molecule. Three degrees of freedom are required to describe the translational motion of the center of mass of the molecule in the x, y, and z directions. Thus, there will be no other degrees of freedom, since $3(1) - 3 = 0$.

3.10 Repeat Problem 3.9 for a diatomic molecule such as N_2(g), CO(g), or NaCl(g) or a linear polyatomic molecule such as CO_2(g).

▌ In addition to the three translational degrees of freedom, linear molecules require two degrees of freedom to describe the rotational motion of the molecule about its center of mass. For the diatomic molecule, the remaining degree of freedom $[3(2) - 3 - 2 = 1]$ is used to describe the single intramolecular vibration about the center of mass. For the linear polyatomic molecule, the number of intramolecular vibrational degrees of freedom is $3\Lambda - 5$. (Note that some of these vibrational degrees of freedom can be degenerate.)

3.11 Repeat Problem 3.9 for a nonlinear polyatomic molecule such as H_2O(g) or CH_3CH_3(g).

▌ In addition to the three translational degrees of freedom, nonlinear molecules require three degrees of freedom to describe the rotational motion of the molecule about its center of mass. This leaves $3\Lambda - 6$ internal degrees of freedom for the molecule. For a simple molecule such as H_2O, these internal degrees will be associated with intramolecular vibrational motion about its center of mass $[3(3) - 6 = 3$ vibrational degrees of freedom$]$. For more complicated molecules such as CH_3CH_3, these internal degrees will be associated with intramolecular rotation of atoms along a chemical bond as well as intramolecular vibrational motion about its center of mass (1 intramolecular rotation degree of freedom and 17 vibrational degrees of freedom). (Note that some of these vibrational degrees of freedom can be degenerate.)

3.12 The *thermal energy* of an ideal gas represents the difference between the internal energy of the gas at some temperature T and $T = 0\ \text{K}$ (see Problem 3.3):

$$E(\text{thermal}) = E_T - E_0 \qquad (3.6)$$

The three major contributions to E(thermal) below 1000 K result from the possible translational, rotational, and vibrational motions of the molecules (see Table 3-2). Determine E(thermal) for 1 mol for Ne at 25 °C and 1.00 bar. Repeat the calculations for Ar.

▌ An ideal monatomic gas has only three translational degrees of freedom (see Problem 3.9). Thus,

$$E(\text{thermal}) = 3(\tfrac{1}{2}RT) = \tfrac{3}{2}RT \qquad (3.7)$$
$$= \tfrac{3}{2}(8.314\ \text{J} \cdot \text{K}^{-1} \cdot \text{mol}^{-1})(298\ \text{K}) = 3718\ \text{J} \cdot \text{mol}^{-1}$$

Because *(3.7)* does not depend on the identity of the substance, $E(\text{thermal, Ne}) = E(\text{thermal, Ar}) = 3718\ \text{J} \cdot \text{mol}^{-1}$ at 25 °C, neglecting any electronic contribution.

Table 3-2

$$E(\text{thermal, trans}) = \sum \tfrac{1}{2}RT$$

$$E(\text{thermal, rot}) = \sum \tfrac{1}{2}RT$$

$$E(\text{thermal, vib}) = \sum \frac{RTx_j}{e^{x_j} - 1}$$

$$\text{where } x_j = \frac{h\nu_j}{kT} = \frac{(1.4388)[\bar{\nu}_j/(\text{cm}^{-1})]}{T/(\text{K})} = \frac{\Theta_{v,j}}{T}$$

$$E(\text{thermal, elec or nuc}) = RT\frac{\sum g_j(\varepsilon_j/kT)\, e^{-\varepsilon_j/kT}}{\sum g_j\, e^{-\varepsilon_j/kT}}$$

where g_j is the degeneracy of the energy level

3.13 For atomic Cl(g), calculate the electronic contribution to the thermal energy at 25 °C given that the degeneracy of the ground level ($\varepsilon_0 = 0$) is 4 and the degeneracy of the first excited level ($\varepsilon_1 = 882.36\ \text{cm}^{-1}$) is 2. (The other excited levels are too high in energy to be of importance.)

▮ The energy of the first excited level is

$$(882.36\ \text{cm}^{-1})\frac{1.986\,477 \times 10^{-23}\ \text{J}}{1\ \text{cm}^{-1}} = 1.752\,79 \times 10^{-20}\ \text{J}$$

From Table 3-2,

$$E(\text{thermal, elec}) = RT\frac{g_0(\varepsilon_0/kT)\, e^{-\varepsilon_0/kT} + g_1(\varepsilon_1/kT)\, e^{-\varepsilon_1/kT}}{g_0\, e^{-\varepsilon_0/kT} + g_1\, e^{-\varepsilon_1/kT}}$$

$$= (8.314\ \text{J}\cdot\text{K}^{-1}\cdot\text{mol}^{-1})(298\ \text{K})$$

$$\times \frac{\left[4\frac{0}{kT}e^{-(0)/kT} + 2\left(\frac{1.752\,79 \times 10^{-20}\ \text{J}}{(1.381 \times 10^{-23}\ \text{J}\cdot\text{K}^{-1}\cdot\text{mol}^{-1})(298\ \text{K})}\right) \times \exp[-(1.752\,79 \times 10^{-20}\ \text{J})/(1.381 \times 10^{-23}\ \text{J}\cdot\text{K}^{-1})(298\ \text{K})]\right]}{\left[4\, e^{-(0)/kT} + 2\exp\left(-\frac{1.752\,79 \times 10^{-20}\ \text{J}}{(1.381 \times 10^{-23}\ \text{J}\cdot\text{K}^{-1})(298\ \text{K})}\right)\right]}$$

$$= 74.16\ \text{J}\cdot\text{mol}^{-1}$$

3.14 Repeat the calculations of Problem 3.13 for atomic I(g) given the following electronic level data: $g_0 = 4$, $\varepsilon_0 = 0$; $g_1 = 2$, $\varepsilon_1 = 1.51 \times 10^{-19}$ J. How does an increase in ε_1 change the value of E(thermal, elec)? Repeat the calculations for I(g) at 1000.0 K. How does E(thermal, elec) vary with an increase in temperature?

▮ At 298 K,

$$E(\text{thermal, elec}) = (8.314\ \text{J}\cdot\text{K}^{-1}\cdot\text{mol}^{-1})(298\ \text{K})$$

$$\times \frac{\left[4\frac{0}{kT}e^{-(0)/kT} + 2\left(\frac{1.51 \times 10^{-19}\ \text{J}}{(1.381 \times 10^{-23}\ \text{J}\cdot\text{K}^{-1}\cdot\text{mol}^{-1})(298\ \text{K})}\right) \times \exp[-(1.51 \times 10^{-19}\ \text{J})/(1.381 \times 10^{-23}\ \text{J}\cdot\text{K}^{-1})(298\ \text{K})]\right]}{\left[4\, e^{-(0)/kT} + 2\exp\left(-\frac{1.51 \times 10^{-19}\ \text{J}}{(1.381 \times 10^{-23}\ \text{J}\cdot\text{K}^{-1})(298\ \text{K})}\right)\right]}$$

$$= 5.33 \times 10^{-12}\ \text{J}\cdot\text{mol}^{-1}$$

For an increase in ε_1, the electronic thermal energy decreases. Likewise, at 1000.0 K,

$$E(\text{thermal, elec}) = (8.314)(1000.0)$$

$$\times \frac{\left[4\frac{0}{kT}e^{-(0)/kT} + 2\left(\frac{1.51 \times 10^{-19}}{(1.381 \times 10^{-23})(1000.0)}\right) \times \exp[-(1.51 \times 10^{-19})/(1.381 \times 10^{-23})(1000.0)]\right]}{4\, e^{-(0)/kT} + 2\exp[-(1.51 \times 10^{-19})/(1.381 \times 10^{-23})(1000.0)]}$$

$$= 0.809\ \text{J}\cdot\text{mol}^{-1}$$

As the temperature increases, the electronic thermal energy increases.

3.15 Determine E(thermal) for 1 mol of ICl(g) at 25 °C and 1.00 bar neglecting electronic contributions. For this gas, $\bar{\nu} = 382.18\ \text{cm}^{-1}$.

▌ An ideal diatomic gas has three translational, two rotational, and one vibrational degrees of freedom (see Problem 3.10). Thus,

$$E(\text{thermal}) = 3(\tfrac{1}{2}RT) + 2(\tfrac{1}{2}RT) + \frac{RTx}{e^x - 1} = \tfrac{5}{2}RT + \frac{RTx}{e^x - 1} \tag{3.8}$$

For this gas,

$$x = (1.4388)(382.18)/298 = 1.85$$

giving

$$E(\text{thermal}) = \tfrac{5}{2}(8.314\ \text{J}\cdot\text{K}^{-1}\cdot\text{mol}^{-1})(298\ \text{K}) + \frac{(8.314\ \text{J}\cdot\text{K}^{-1}\cdot\text{mol}^{-1})(298\ \text{K})(1.85)}{e^{1.85} - 1}$$
$$= 7053\ \text{J}\cdot\text{mol}^{-1}$$

3.16 Repeat the calculations of Problem 3.15 for CO(g). For CO, $\bar{\nu} = 2169.52\ \text{cm}^{-1}$. How does the value of E(thermal, vib) change with increasing $\bar{\nu}$?

▌ For CO(g),

$$x = (1.4388)(2169.52)/298 = 10.5$$

giving

$$E(\text{thermal}) = \tfrac{5}{2}(8.314\ \text{J}\cdot\text{K}^{-1}\cdot\text{mol}^{-1})(298\ \text{K}) + \frac{(8.314\ \text{J}\cdot\text{K}^{-1}\cdot\text{mol}^{-1})(298\ \text{K})(10.5)}{e^{10.5} - 1}$$
$$= 6198\ \text{J}\cdot\text{mol}^{-1}$$

As $\bar{\nu}$ increases, the value of E(thermal, vib) decreases at a given temperature.

3.17 Determine E(thermal) for 1 mol of CS_2(g), a linear molecule, at 25 °C and 1.00 bar. For this gas, $\bar{\nu}_1 = 657.98\ \text{cm}^{-1}$, $\bar{\nu}_2 = 395.93\ \text{cm}^{-1}$ (doubly degenerate), and $\bar{\nu}_3 = 1535.35\ \text{cm}^{-1}$.

▌ An ideal linear polyatomic gas has three translational, two rotational, and $3\Lambda - 5$ vibrational degrees of freedom (see Problem 3.10). Thus,

$$E(\text{thermal}) = 3(\tfrac{1}{2}RT) + 2(\tfrac{1}{2}RT) + \sum_{i=1}^{3\Lambda-5} \frac{RTx_i}{e^{x_i} - 1} = \tfrac{5}{2}RT + \sum_{i=1}^{3\Lambda-5} \frac{RTx_i}{e^{x_i} - 1} \tag{3.9}$$

For this gas,

$$x_1 = (1.4388)(657.98)/298 = 3.175$$

Likewise, $x_2 = 1.911$ and $x_3 = 7.409$, giving

$$E(\text{thermal}) = \tfrac{5}{2}(8.314\ \text{J}\cdot\text{K}^{-1}\cdot\text{mol}^{-1})(298\ \text{K})$$
$$+(8.314\ \text{J}\cdot\text{K}^{-1}\cdot\text{mol}^{-1})(298\ \text{K})\left(\frac{3.175}{e^{3.175} - 1} + \frac{(2)(1.911)}{e^{1.911} - 1} + \frac{7.409}{e^{7.409} - 1}\right)$$
$$= 8197\ \text{J}\cdot\text{mol}^{-1}$$

3.18 Determine E(thermal) for 1 mol of O_3(g), a nonlinear molecule, at 25 °C and 1.00 bar. For this gas, $\Theta_{v,1} = 1583\ \text{K}$, $\Theta_{v,2} = 1014\ \text{K}$, and $\Theta_{v,3} = 1501\ \text{K}$.

▌ A simple ideal nonlinear polyatomic gas has three translational, three rotational, and $3\Lambda - 6$ vibrational degrees of freedom (see Problem 3.11). Thus,

$$E(\text{thermal}) = 3(\tfrac{1}{2}RT) + 3(\tfrac{1}{2}RT) + \sum_{i=1}^{3\Lambda-6} \frac{RTx_i}{e^{x_i} - 1} = 3RT + \sum_{i=1}^{3\Lambda-6} \frac{RTx_i}{e^{x_i} - 1} \tag{3.10}$$

For this gas,

$$x_1 = (1583\ \text{K})/(298\ \text{K}) = 5.312$$

Likewise, $x_2 = 3.402$ and $x_3 = 5.037$, giving

$$E(\text{thermal}) = 3(8.314\ \text{J}\cdot\text{K}^{-1}\cdot\text{mol}^{-1})(298\ \text{K})$$
$$+(8.314\ \text{J}\cdot\text{K}^{-1}\cdot\text{mol}^{-1})(298\ \text{K})\left(\frac{5.312}{e^{5.312} - 1} + \frac{3.403}{e^{3.403} - 1} + \frac{5.037}{e^{5.037} - 1}\right)$$
$$= 7874\ \text{J}\cdot\text{mol}^{-1}$$

3.19 Calculate ΔU for heating 1 mol of I(g) from 25 °C to 1000 K.

▮ For this heating process,

$$\Delta U = \Delta E(\text{thermal}) \tag{3.11}$$

From Problems 3.12 and 3.14,

$$E(\text{thermal}, 298\text{ K}) = 3718\text{ J}\cdot\text{mol}^{-1} + 5.33\times10^{-12}\text{ J}\cdot\text{mol}^{-1} = 3718\text{ J}\cdot\text{mol}^{-1}$$

At 1000 K, using *(3.7)* and the result of Problem 3.14 gives

$$E(\text{thermal}, 1000\text{ K}) = \tfrac{3}{2}(8.314\text{ J}\cdot\text{K}^{-1}\cdot\text{mol}^{-1})(1000\text{ K}) + 0.809\text{ J}\cdot\text{mol}^{-1}$$
$$= 12\,472\text{ J}\cdot\text{mol}^{-1}$$

Thus,

$$\Delta U = 12\,472\text{ J}\cdot\text{mol}^{-1} - 3718\text{ J}\cdot\text{mol}^{-1} = 8754\text{ J}\cdot\text{mol}^{-1}$$

3.20 Repeat the calculation of Problem 3.19 for 1 mol of ICl(g). Why is this internal energy change larger?

▮ From Problem 3.15, $E(\text{thermal}, 298\text{ K}) = 7053\text{ J}\cdot\text{mol}^{-1}$. At 1000 K,

$$x = (1.4388)(382.18)/1000 = 0.550$$

giving

$$E(\text{thermal}) = \tfrac{5}{2}(8.314\text{ J}\cdot\text{K}^{-1}\cdot\text{mol}^{-1})(1000\text{ K}) + \frac{(8.314\text{ J}\cdot\text{K}^{-1}\cdot\text{mol}^{-1})(1000\text{ K})(0.550)}{e^{0.550}-1}$$
$$= 27\,027\text{ J}\cdot\text{mol}^{-1}$$

Thus, $\Delta U = 27\,027\text{ J}\cdot\text{mol}^{-1} - 7053\text{ J}\cdot\text{mol}^{-1} = 19\,974\text{ J}\cdot\text{mol}^{-1}$. The additional energy ($11\,220\text{ J}\cdot\text{mol}^{-1}$) is absorbed by the molecules in the form of rotational and vibrational kinetic energy.

3.21 What is the classical limit of $E(\text{thermal, vib})$ as $T \to \infty$?

▮ For each degree of vibrational energy,

$$\lim_{T\to\infty}[E(\text{thermal, vib})] = \lim_{T\to\infty}\frac{RTx}{e^x-1} = \lim_{T\to\infty}\frac{RT\Theta_v/T}{e^{\Theta_v/T}-1} = \lim_{T\to\infty}\frac{R\Theta_v}{e^{\Theta_v/T}-1}$$

As $T\to\infty$, $e^{\Theta_v/T}\to 0$. Using a series expansion for the denominator ($e^x = 1 + x + x^2/2 + \cdots$) gives

$$\lim_{T\to\infty}[E(\text{thermal, vib})] = \frac{R\Theta_v}{1+\Theta_v/T+\cdots-1} = RT$$

(Classically, vibrational energy contains two terms—one kinetic and one potential, each equal to $\frac{1}{2}RT$.)

3.22 The *thermal enthalpy* of an ideal gas is defined as

$$H(\text{thermal}) = H_T - E_0 = E(\text{thermal}) + RT \tag{3.12}$$

Determine $H(\text{thermal})$ for 1 mol of I(g) at 298 K.

▮ Substituting $E(\text{thermal}) = 3718\text{ J}\cdot\text{mol}^{-1}$ (see Problem 3.19) into *(3.12)* gives

$$E(\text{thermal}) = 3718\text{ J}\cdot\text{mol}^{-1} + (8.314\text{ J}\cdot\text{K}^{-1}\cdot\text{mol}^{-1})(298\text{ K}) = 6197\text{ J}\cdot\text{mol}^{-1}$$

3.23 Calculate ΔH for heating 1 mol of I(g) from 25 °C to 1000 K.

▮ For this heating process,

$$\Delta H = \Delta H(\text{thermal}) \tag{3.13}$$

At 1000 K, substituting the results of Problem 3.19 into *(3.12)* gives

$$H(\text{thermal}, 1000\text{ K}) = 12\,472\text{ J}\cdot\text{mol}^{-1} + (8.314\text{ J}\cdot\text{K}^{-1}\cdot\text{mol}^{-1})(1000\text{ K}) = 20\,786\text{ J}\cdot\text{mol}^{-1}$$

Substituting $H(\text{thermal}) = 6197\text{ J}\cdot\text{mol}^{-1}$ (see Problem 3.22) into *(3.13)* gives

$$\Delta H = 20\,786\text{ J}\cdot\text{mol}^{-1} - 6197\text{ J}\cdot\text{mol}^{-1} = 14\,589\text{ J}\cdot\text{mol}^{-1}$$

3.24 An alternate definition for $H(\text{thermal})$ is $H(\text{thermal}) = H_T - H_0$. What is the relation between this definition and the one given by *(3.12)*?

▮ For 1 mol of an ideal gas at absolute zero, *(3.1a)* gives

$$H_0 = E_0 + PV = E_0 + RT = E_0$$

Thus, $H(\text{thermal}) = H_T - H_0 = H_T - E_0$. The definitions are the same.

3.25 A thermodynamic table lists $(H^\circ_T - H^\circ_{298})/(\text{kJ}\cdot\text{mol}^{-1}) = 14.592$ for I(g) at 1000 K and -6.197 at 0 K (where ° refers to a pressure of 1.00 bar). What is $H^\circ(\text{thermal})$ for I(g) at 1000 K?

▮ Substituting the result of Problem 3.24 into *(3.12)* gives

$$H^\circ(\text{thermal}) = H^\circ_T - E^\circ_0 = H^\circ_T - H^\circ_0 = H^\circ_T - H^\circ_0 + H^\circ_{298} - H^\circ_{298}$$
$$= (H^\circ_T - H^\circ_{298}) - (H^\circ_0 - H^\circ_{298}) \tag{3.14}$$

Substituting the data gives

$$H^\circ(\text{thermal}) = 14.592\ \text{kJ}\cdot\text{mol}^{-1} - (-6.197\ \text{kJ}\cdot\text{mol}^{-1}) = 20.789\ \text{kJ}\cdot\text{mol}^{-1}$$

3.26 A thermodynamic table lists $(H^\circ_T - H^\circ_{298})/(\text{kJ}\cdot\text{mol}^{-1}) = 63.582$ at 1500 K for NH_3(g) and 32.637 at 1000 K. What is ΔH° for cooling 1 mol of NH_3 from 1500 K to 1000 K?

▮ Combining *(3.13)* and *(3.14)* gives

$$\Delta H^\circ = \Delta H^\circ(\text{thermal}) = \Delta[(H^\circ_T - H^\circ_{298}) - (H^\circ_0 - H^\circ_{298})]$$
$$= (H^\circ_{T_2} - H^\circ_{298}) - (H^\circ_{T_1} - H^\circ_{298}) \tag{3.15}$$

Substituting the data gives

$$\Delta H^\circ = 32.637\ \text{kJ}\cdot\text{mol}^{-1} - 63.582\ \text{kJ}\cdot\text{mol}^{-1} = -30.945\ \text{kJ}\cdot\text{mol}$$

The negative sign on the value means that the heat must be removed from the system.

3.27 Find $E^\circ(\text{thermal})$ for 1 mol of ICl(g) at 1.00 bar at 1000 K given $(H^\circ_T - H^\circ_{298})/(\text{kJ}\cdot\text{mol}^{-1}) = 26.006$ at 1000 K and -9.555 at 0 K.

▮ Combining *(3.12)* and *(3.14)* gives

$$E^\circ(\text{thermal}) = H^\circ(\text{thermal}) - RT = (H^\circ_T - H^\circ_{298}) - (H^\circ_0 - H^\circ_{298}) - RT \tag{3.16}$$

Substituting the data gives

$$E^\circ(\text{thermal}) = 26.006\ \text{kJ}\cdot\text{mol}^{-1} - (-9.555\ \text{kJ}\cdot\text{mol}^{-1}) - (8.314\times10^{-3}\ \text{kJ}\cdot\text{K}^{-1}\cdot\text{mol}^{-1})(1000\ \text{K})$$
$$= 27.247\ \text{kJ}\cdot\text{mol}^{-1}$$

which is in good agreement with the results of Problem 3.20.

3.2 HEAT CAPACITY

3.28 The *heat capacity at constant volume* (C_V) is defined as

$$C_V = \left(\frac{\partial U}{\partial T}\right)_V \tag{3.17}$$

From the plot of the molar thermal energy at 1.00 bar against T for Cl_2(g) shown in Fig. 3-1, determine the standard state molar heat capacity at constant volume at 1000 K and 1.00 bar.

▮ The slope of the tangent to the plot at 1000 K will be equal to $C^\circ_{V,\text{m}}$. From the graph, $C^\circ_{V,\text{m}} = 29.04\ \text{J}\cdot\text{K}^{-1}\cdot\text{mol}^{-1}$.

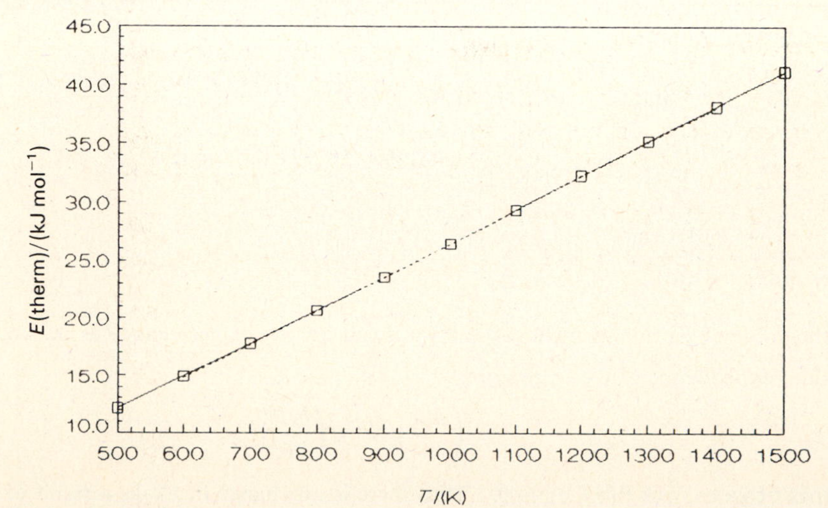

Fig. 3-1

3.29 Calculate ΔU for heating 1 mol of Cl_2 from 25 °C to 125 °C under constant-volume conditions. Assume $C_{V,\mathrm{m}}$ is constant over this temperature range.

▌ Rearranging *(3.17)* and integrating for a constant-volume process gives

$$\Delta U = \int_{T_1}^{T_2} C_V \, dT \tag{3.18}$$

Substituting the value of $C_{V,\mathrm{m}}$ from Table 3-3 gives

$$\Delta U = nC_{V,\mathrm{m}}\,\Delta T = (1.00\ \mathrm{mol})(25.635\ \mathrm{J\cdot K^{-1}\cdot mol^{-1}})(398\ \mathrm{K} - 298\ \mathrm{K}) = 2560\ \mathrm{J}$$

Table 3-3

		$C^\circ_{\mathrm{m},298}$ ($\mathrm{J\cdot K^{-1}\cdot mol^{-1}}$)	Equation *(3.19)* constants			
			$a/(\mathrm{J\cdot K^{-1}\cdot mol^{-1}})$	$b/(\mathrm{J\cdot K^{-2}\cdot mol^{-1}})$	$c/(\mathrm{J\cdot K^{-3}\cdot mol^{-1}})$	$c'/(\mathrm{J\cdot K\cdot mol^{-1}})$
$Cl_2(g)$	$C^\circ_{V,\mathrm{m}}$	25.635	22.970			
	$C^\circ_{P,\mathrm{m}}$	33.949	31.284	10.144×10^{-3}	-40.38×10^{-7}	
$Pb(s)$	$C^\circ_{P,\mathrm{m}}$		22.13	11.72×10^{-3}		0.96×10^5
$H_2O(l)$	$C^\circ_{V,\mathrm{m}}$	74.7				
	$C^\circ_{P,\mathrm{m}}$	75.291				
$Hg(l)$	$C^\circ_{P,\mathrm{m}}$	27.983				
$O_2(g)$	$C^\circ_{V,\mathrm{m}}$	21.041				
	$C^\circ_{P,\mathrm{m}}$	29.355	25.849	12.98×10^{-3}	-38.62×10^{-7}	
$FeCl_3(s)$	$C^\circ_{P,\mathrm{m}}$	96.65				
$N_2(g)$	$C^\circ_{P,\mathrm{m}}$	29.125	27.565	5.230×10^{-3}	-0.04×10^{-7}	
$ICl(g)$	$C^\circ_{P,\mathrm{m}}$	35.548				
$CH_3OH(g)$	$C^\circ_{P,\mathrm{m}}$	43.89				
$Cu(s)$	$C^\circ_{P,\mathrm{m}}$	24.435				
$Ne(g)$	$C^\circ_{P,\mathrm{m}}$	12.472				

3.30 Heat capacity data are often given in various polynomial forms such as

$$C = a + bT + cT^2 + \cdots \tag{3.19a}$$

$$C = a + bT + c'T^{-2} \tag{3.19b}$$

Repeat the calculations of Problem 3.29 and compare the results.

▌ Substituting *(3.19a)* into *(3.18)* and integrating gives

$$\Delta U = a(T_2 - T_1) + \frac{b}{2}(T_2^2 - T_1^2) + \frac{c}{3}(T_2^3 - T_1^3)$$

Substituting the data from Table 3-3 gives

$$\Delta U = (22.970\ \mathrm{J\cdot K^{-1}\cdot mol^{-1}})(398\ \mathrm{K} - 298\ \mathrm{K}) + \frac{10.144\times 10^{-3}\ \mathrm{J\cdot K^{-1}\cdot mol^{-1}}}{2}$$

$$\times[(398\ \mathrm{K})^2 - (298\ \mathrm{K})^2] + \frac{-40.38\times 10^{-7}\ \mathrm{J\cdot K^{-3}\cdot mol^{-1}}}{2}[(398\ \mathrm{K})^3 - (298\ \mathrm{K})^3]$$

$$= 2601\ \mathrm{J\cdot mol^{-1}}$$

These values differ by 1.6%.

3.31 How does the value of C_V for an ideal gas and for a van der Waals gas change as the volume changes?

▌ The volume dependence of C_V is given by

$$\left(\frac{\partial C_V}{\partial V}\right)_T = T\left(\frac{\partial^2 P}{\partial T^2}\right)_V \tag{3.20}$$

For both types of gases, $(\partial^2 P/\partial T^2)_V = 0$. Thus there is no change in C_V as a result of a volume change.

3.32 The *heat capacity at constant pressure* (C_P) is defined as

$$C_P = \left(\frac{\partial H}{\partial T}\right)_P \tag{3.21}$$

Show that

$$C_P^\circ = \left(\frac{\partial H^\circ(\text{thermal})}{\partial T}\right)_P = \left(\frac{\partial(H_T^\circ - H_{298}^\circ)}{\partial T}\right)_P \tag{3.22}$$

▌ The derivative of H°(thermal) with respect to T can be found from *(3.12)* as

$$\left(\frac{\partial H^\circ(\text{thermal})}{\partial T}\right)_P = \left(\frac{\partial(H_T^\circ - E_0^\circ)}{\partial T}\right)_P = \left(\frac{\partial H_T^\circ}{\partial T}\right)_P - \left(\frac{\partial E_0^\circ}{\partial T}\right)_P = \left(\frac{\partial H_T^\circ}{\partial T}\right)_P = C_P^\circ$$

Likewise,

$$\left(\frac{\partial(H_T^\circ - H_{298}^\circ)}{\partial T}\right)_P = \left(\frac{\partial H_T^\circ}{\partial T}\right)_P - \left(\frac{\partial H_{298}^\circ}{\partial T}\right)_P = \left(\frac{\partial H_T^\circ}{\partial T}\right)_P = C_P^\circ$$

3.33 Using the plot of $C_{P,\text{m}}$ against T for $Cl_2(g)$ at 1.00 bar shown in Fig. 3-2, determine ΔH° for heating 1 mol from 500 K to 1500 K under constant-pressure conditions.

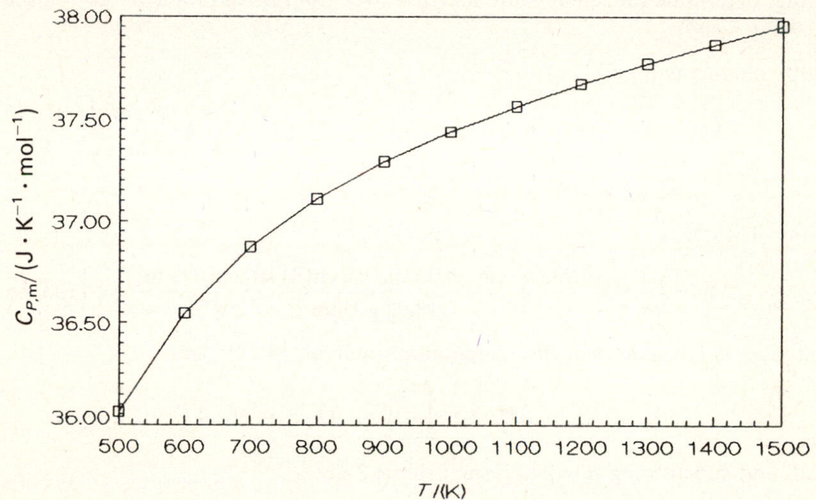

Fig. 3-2

▌ Rearranging *(3.21)* and integrating for a constant-pressure process gives

$$\Delta H = \int_{T_1}^{T_2} C_P \, dT \tag{3.23}$$

The value of the integral will be equal to the area under the curve given in Fig. 3-2. The value of ΔH_m° is 37.318 kJ · mol^{-1}.

3.34 Calculate the molar heat capacity at constant pressure at 400 K and 1.00 bar.

▌ Substituting the data from Table 3-3 into *(3.19a)* gives

$$\begin{aligned} C_{P,\text{m}}^\circ &= 22.13\ \text{J}\cdot\text{K}^{-1}\cdot\text{mol}^{-1} + (11.72\times10^{-3}\ \text{J}\cdot\text{K}^{-2}\cdot\text{mol}^{-1})(500\ \text{K}) + (0.96\times10^{5}\ \text{J}\cdot\text{K}\cdot\text{mol}^{-1})(500\ \text{K})^{-2} \\ &= 28.37\ \text{J}\cdot\text{K}^{-1}\cdot\text{mol}^{-1} \end{aligned}$$

3.35 A 150-W heater was placed in 1.00 kg of methanol and turned on for exactly 1 min. The temperature increased by 3.54 °C. Assuming that all the heat is absorbed by the methanol, calculate the molar heat capacity for $CH_3OH(l)$.

▌ Assuming $C_{P,\text{m}}$ to be constant over this small temperature interval, *(3.23)* gives

$$\Delta H = n\int_{T_1}^{T_2} C_{P,\text{m}}\, dT = nC_{P,\text{m}}\,\Delta T$$

Solving for $C_{P,\mathrm{m}}$ and substituting the data gives

$$C_{P,\mathrm{m}} = \frac{(150\ \mathrm{W})[(1\ \mathrm{J\cdot s^{-1}})/(1\ \mathrm{W})](60.0\ \mathrm{s})}{\{(1.00\ \mathrm{kg})[(10^3\ \mathrm{g})/(1\ \mathrm{kg})][(1\ \mathrm{mol})/(32.04\ \mathrm{g})]\}(3.54\ \mathrm{K})}$$
$$= 81.5\ \mathrm{J\cdot K^{-1}\cdot mol^{-1}}$$

3.36 For $Li_2SO_4(s)$, $(H^\circ - H^\circ_{298})/(\mathrm{kJ\cdot mol^{-1}}) = 43.975$ at 600 K and 80.437 at 800 K. Estimate the molar heat capacity of Li_2SO_4 over this temperature range.

▮ Assuming that $C^\circ_{P,\mathrm{m}}$ is constant over this temperature interval, *(3.23)* gives

$$\Delta H^\circ = n\int_{T_1}^{T_2} C^\circ_{P,\mathrm{m}}\,dT = nC^\circ_{P,\mathrm{m}}\,\Delta T$$

Solving for $C^\circ_{P,\mathrm{m}}$ and using *(3.15)* gives

$$C^\circ_{P,\mathrm{m}} = \frac{(80.437\ \mathrm{J\cdot mol^{-1}} - 43.975\ \mathrm{J\cdot mol^{-1}})[(10^3\ \mathrm{J})/(1\ \mathrm{kJ})]}{800\ \mathrm{K} - 600\ \mathrm{K}}$$
$$= 182\ \mathrm{J\cdot K^{-1}\cdot mol^{-1}}$$

3.37 The solar radiation energy falling on a water puddle is $1.0\ \mathrm{cal\cdot cm^{-2}\cdot min^{-1}}$. Assuming that the radiation simply heats the puddle, determine the temperature increase over 1.0-h period for a 1.0-$\mathrm{m^2}$ puddle that is 2 cm deep. For H_2O, $\rho = 0.995\ \mathrm{g\cdot cm^{-3}}$.

▮ The enthalpy change will be

$$\Delta H = (1.0\ \mathrm{cal\cdot cm^{-2}\cdot min^{-1}})(1.0\ \mathrm{m^2})(1.0\ \mathrm{h})\frac{4.184\ \mathrm{J}}{1\ \mathrm{cal}}\left(\frac{10^2\ \mathrm{cm}}{1\ \mathrm{m}}\right)^2\left(\frac{60\ \mathrm{min}}{1\ \mathrm{h}}\right)$$
$$= 2.5\times 10^6\ \mathrm{J}$$

and the amount of water is

$$n = \frac{\rho V}{M} = \frac{(0.995\ \mathrm{g\cdot cm^{-3}})(1.0\ \mathrm{m^2})(2\ \mathrm{cm})[(10^2\ \mathrm{cm})/(1\ \mathrm{m})]}{18.02\ \mathrm{g\cdot mol^{-1}}} = 1100\ \mathrm{mol}$$

Assuming that $C_{P,\mathrm{m}}$ is constant over this temperature interval, *(3.23)* gives

$$\Delta H = n\int_{T_1}^{T_2} C_{P,\mathrm{m}}\,dT = nC_{P,\mathrm{m}}\,\Delta T$$

Solving for ΔT and substituting the data from Table 3-3 gives

$$\Delta T = \frac{2.5\times 10^6\ \mathrm{J}}{(1100\ \mathrm{mol})(75.291\ \mathrm{J\cdot K^{-1}\cdot mol^{-1}})} = 30\ \mathrm{K}$$

3.38 Exactly 10 kJ of energy was absorbed by 1 mol of $Cl_2(g)$ at 298 K. What is the final temperature of the $Cl_2(g)$?

▮ As in Problem 3.30,

$$\Delta H = a(T_2 - T_1) + \frac{b}{2}(T_2^2 - T_1^2) + \frac{c}{3}(T_2^3 - T_1^3)$$

Substituting the data from Table 3-3 gives

$$\begin{aligned} 10\,000\ \mathrm{J\cdot mol^{-1}} &= (31.284\ \mathrm{J\cdot K^{-1}\cdot mol^{-1}})(T - 298\ \mathrm{K}) \\ &\quad + \tfrac{1}{2}(10.144\times 10^{-3}\ \mathrm{J\cdot K^{-2}\cdot mol^{-1}})[T^2 - (298\ \mathrm{K})^2] \\ &\quad + \tfrac{1}{3}(-40.38\times 10^{-7}\ \mathrm{J\cdot K^{-3}\cdot mol^{-1}})[T^3 - (298\ \mathrm{K})^3] \end{aligned}$$

Solving this cubic equation for T gives $T = 584\ \mathrm{K}$.

3.39 The pressure dependence of C_P is given by

$$\left(\frac{\partial C_P}{\partial P}\right)_T = T\left(\frac{\partial^2 V}{\partial T^2}\right)_P \tag{3.24}$$

How does the value of C_P for an ideal gas change as the pressure changes?

▌ For 1 mol of an ideal gas,

$$\left(\frac{\partial^2 V}{\partial T^2}\right)_P = T\left[\frac{\partial}{\partial T}\left(\frac{\partial(RT/P)}{\partial T}\right)_P\right]_P = \left(\frac{\partial(R/P)}{\partial T}\right)_P = 0$$

Thus, $C_{P,\mathrm{m}}$ for an ideal gas is independent of pressure.

3.40 Over the temperature range $273 \le T/(\mathrm{K}) \le 373$, the molar volume of liquid mercury can be given by $V_\mathrm{m} = \alpha + \beta T + \gamma T^2$, where $\alpha = 1.4031 \times 10^{-5}\ \mathrm{m^3 \cdot mol^{-1}}$, $\beta = 2.6193 \times 10^{-9}\ \mathrm{m^3 \cdot K^{-1} \cdot mol^{-1}}$, and $\gamma = 1.15 \times 10^{-13}\ \mathrm{m^3 \cdot K^{-2} \cdot mol^{-1}}$. Determine $C_{P,\mathrm{m}}$ for Hg(l) at 25 °C and 1.0×10^5 bar.

▌ The change in the heat capacity with respect to pressure is given by *(3.24)* as

$$\left(\frac{\partial C_{P,\mathrm{m}}}{\partial T}\right)_T = -T\left(\frac{\partial^2 V_\mathrm{m}}{\partial T^2}\right)_P = -T\left[\frac{\partial}{\partial T}\left(\frac{\partial(\alpha + \beta T + \gamma T^2)}{\partial T}\right)_P\right]_P$$

$$= -T(\partial(\beta + 2\gamma T)/\partial T)_P = -T(2\gamma)$$

Integrating gives $\Delta C_{P,\mathrm{m}} = -T(2\gamma)\,\Delta P$, and substitution from Table 3-3 gives

$$C_{P,\mathrm{m}} = 27.983\ \mathrm{J \cdot K^{-1} \cdot mol^{-1}} - (298\ \mathrm{K})(2)(1.15 \times 10^{-13}\ \mathrm{m^3 \cdot K^{-2} \cdot mol^{-1}})$$

$$\times (1.0 \times 10^5\ \mathrm{bar} - 1\ \mathrm{bar})\frac{10^5\ \mathrm{Pa}}{1\ \mathrm{bar}}\left(\frac{1\ \mathrm{J}}{1\ \mathrm{Pa \cdot m^3}}\right)$$

$$= 27.30\ \mathrm{J \cdot K^{-1} \cdot mol^{-1}}$$

3.41 For real gases, the integrated form of *(3.24)* is often estimated by

$$C_{P,\mathrm{m}}(P) = C^\circ_{P,\mathrm{m}} - \frac{81R}{32}\left(\frac{P_\mathrm{r}}{T_\mathrm{r}^3}\right)$$

where P_r is the reduced pressure given by *(1.31)* and T_r is the reduced temperature given by *(1.30)*. Given $T_\mathrm{c} = -118.4$ °C and $P_\mathrm{c} = 50.8$ bar, find the molar heat capacity at 10.5 bar and 25 °C.

▌ The reduced temperature and pressure are

$$T_\mathrm{r} = \frac{298\ \mathrm{K}}{(-118.4 + 273)\ \mathrm{K}} = 1.93 \qquad P_\mathrm{r} = \frac{10.5\ \mathrm{bar}}{50.8\ \mathrm{bar}} = 0.207$$

Substituting the data from Table 3-3 gives

$$C_{P,\mathrm{m}} = 29.355\ \mathrm{J \cdot K^{-1} \cdot mol^{-1}} + \frac{81(8.314\ \mathrm{J \cdot K^{-1} \cdot mol^{-1}})}{32}\left(\frac{0.207}{(1.93)^3}\right)$$

$$= 29.961\ \mathrm{J \cdot K^{-1} \cdot mol^{-1}}$$

3.42 The specific heat (c) is the heat capacity expressed on a mass basis

$$c = C_\mathrm{m}/M \qquad (3.25)$$

Find the specific heat of $FeCl_3$(s).

▌ Substituting the data from Table 3-3 and $M = 162.21\ \mathrm{g \cdot mol^{-1}}$ gives

$$c = \frac{96.65\ \mathrm{J \cdot K^{-1} \cdot mol^{-1}}}{162.21\ \mathrm{g \cdot mol^{-1}}} = 0.5958\ \mathrm{J \cdot K^{-1} \cdot g^{-1}}$$

3.43 Originally the calorie energy unit referred to the energy needed to increase the temperature of exactly 1 g of liquid water by 1 C°. However, careful measurements showed that 1.008 cal was needed at 0 °C, 0.998 cal at 35 °C, and 1.007 cal at 99 °C to produce this temperature change. Thus, the calorie has undergone several redefinitions. Determine the specific heat and the molar heat capacity of liquid water at 25 °C and 1 atm given that exactly 100 cal produces a temperature increase of 1.00172 K in exactly 100 g of water.

▌ Assuming the heat capacity to be constant, *(3.23)* gives

$$\Delta H = m\int_{T_1}^{T_2} c\,dT = mc\,\Delta T$$

Solving for c and substituting the data gives

$$c = \frac{100.000\ \mathrm{cal}}{(100.000\ \mathrm{g})(1.001\,72\ \mathrm{K})} = 0.998\,28\ \mathrm{cal \cdot K^{-1} \cdot g^{-1}} = 4.1768\ \mathrm{J \cdot K^{-1} \cdot g^{-1}}$$

The molar heat capacity is given by *(3.25)* as

$$C_m = (4.1768\ \mathrm{J\cdot K^{-1}\cdot g^{-1}})(18.0154\ \mathrm{g\cdot mol^{-1}}) = 75.247\ \mathrm{J\cdot K^{-1}\cdot mol^{-1}}$$

3.44 Assuming the composition of air to be $x(N_2) = 0.800$ and $x(O_2) = 0.200$, determine the molar heat capacity expression for air.

▌ Letting the weighted average value of $C_{P,m}$ represent the heat capacity gives

$$\begin{aligned}\overline{C_{P,m}} &= x(O_2)C_{P,m}(O_2) + x(N_2)C_{P,m}(N_2)\\ &= (0.200)[25.849\ \mathrm{J\cdot K^{-1}\cdot mol^{-1}} + (12.98\times10^{-3}\ \mathrm{J\cdot K^{-2}\cdot mol^{-1}})T - (38.62\times10^{-7}\ \mathrm{J\cdot K^{-3}\cdot mol^{-1}})T^2]\\ &\quad + (0.800)[27.565\ \mathrm{J\cdot K^{-1}\cdot mol^{-1}} + (5.230\times10^{-3}\ \mathrm{J\cdot K^{-2}\cdot mol^{-1}})T - (0.04\times10^{-7}\ \mathrm{J\cdot K^{-3}\cdot mol^{-1}})T^2]\\ &= 27.222\ \mathrm{J\cdot K^{-1}\cdot mol^{-1}} + (6.780\times10^{-3}\ \mathrm{J\cdot K^{-2}\cdot mol^{-1}})T - (7.76\times10^{-7}\ \mathrm{J\cdot K^{-3}\cdot mol^{-1}})T^2\end{aligned}$$

3.45 The heat capacity of a calorimeter (commonly called the *calorimeter constant*) was determined by heating the calorimeter and its contents using an electrical heater. If $\Delta T = 1.222\ \mathrm{K}$ as 1.25 A of electricity at 3.26 V was passed through the heater immersed in 137.5 g of water in the calorimeter for 175 s, determine the calorimeter constant.

▌ Using *(3.23)* where $C_P = C_P(\text{calorimeter}) + C_P(\text{contents})$ and the values from Table 3-3 gives

$$C_P(\text{calorimeter}) = \Delta H/\Delta T - C_P(\text{contents}) \tag{3.26}$$

$$= \frac{(1.25\ \mathrm{A})(175\ \mathrm{s})(3.26\ \mathrm{V})[(1\ \mathrm{J\cdot s^{-1}\cdot A^{-1}})/(1\ \mathrm{V})]}{1.222\ \mathrm{K}} - \frac{137.5\ \mathrm{g}}{18.02\ \mathrm{g\cdot mol^{-1}}}(75.291\ \mathrm{J\cdot K^{-1}\cdot mol^{-1}})$$

$$= 9\ \mathrm{J\cdot K^{-1}}$$

3.46 Often a calorimeter is designed so that the $C(\text{contents})$ term in *(3.26)* is either neglected $[C(\text{contents}) \ll C(\text{calorimeter})]$ or incorporated into the $C(\text{calorimeter})$ term $[C(\text{contents}) \approx \text{constant}]$. An adiabatic bomb calorimeter (used to measure heats of combustion) is of this type. Given $C_V(\text{calorimeter}) = 5345\ \mathrm{J\cdot K^{-1}}$, determine the internal energy change corresponding to a temperature increase of 4.26 K in this calorimeter.

▌ Using *(3.18)* gives

$$\Delta U = \int_{T_1}^{T_2} C_V\,dT = C_V(\text{calorimeter})\,\Delta T = (5345\ \mathrm{J\cdot K^{-1}})(4.26\ \mathrm{K})\frac{10^{-3}\ \mathrm{kJ}}{1\ \mathrm{J}}$$

$$= 22.8\ \mathrm{kJ}$$

3.47 Given $C_{P,m} = 33.949\ \mathrm{J\cdot K^{-1}\cdot mol^{-1}}$ for $Cl_2(g)$ at 25 °C. Determine $C_{V,m}$ for this gas.

▌ The *heat capacity difference* $(C_P - C_V)$ is given by

$$C_P - C_V = \left[P + \left(\frac{\partial U}{\partial V}\right)_T\right]\left(\frac{\partial V}{\partial T}\right)_P \tag{3.27}$$

For 1 mol of an ideal gas, *(3.27)* becomes

$$C_{P,m} - C_{V,m} = R \tag{3.28}$$

Thus,

$$C_{V,m} = C_{P,m} - R = 33.949\ \mathrm{J\cdot K^{-1}\cdot mol^{-1}} - 8.314\ \mathrm{J\cdot K^{-1}\cdot mol^{-1}} = 25.635\ \mathrm{J\cdot K^{-1}\cdot mol^{-1}}$$

3.48 Determine $C_{P,m}$ as a function of temperature for $Cl_2(g)$, assuming ideal gas behavior.

▌ Rearranging *(3.28)* and substituting the data given in Table 3-3 gives

$$\begin{aligned}C_{P,m}/(\mathrm{J\cdot K^{-1}\cdot mol^{-1}}) &= C_{V,m}/(\mathrm{J\cdot K^{-1}\cdot mol^{-1}}) + R/(\mathrm{J\cdot K^{-1}\cdot mol^{-1}})\\ &= 22.970 + (10.144\times10^{-3})[T/(\mathrm{K})] - (40.38\times10^{-7})[T/(\mathrm{K})]^2 + 8.314\\ &= 31.284 + (10.144\times10^{-3})[T/(\mathrm{K})] - (40.38\times10^{-7})[T/(\mathrm{K})]^2\end{aligned}$$

3.49 For 1 mol of a real gas, solid, or liquid, *(3.27)* becomes

$$C_{P,m} - C_{V,m} = \alpha^2 VT/\kappa \tag{3.29}$$

where α is the cubic expansion coefficient, see *(1.11)*, and κ is the isothermal compressibility, see *(1.8)*. Given $\alpha = 1.81\times10^{-4}\ \mathrm{K^{-1}}$, $\kappa = 3.4\times10^{-11}\ \mathrm{m^2\cdot N^{-1}}$, and $\rho = 13.59\ \mathrm{g\cdot cm^{-3}}$, find $C^\circ_{V,m}$ under these conditions for Hg(l).

▮ Solving *(3.29)* for $C^\circ_{V,\mathrm{m}}$ and substituting the data from Table 3-3 gives

$$C^\circ_{V,\mathrm{m}} = 27.983\ \mathrm{J \cdot K^{-1} \cdot mol^{-1}} - \frac{(1.81 \times 10^{-4}\ \mathrm{K^{-1}})^2[(200.59\ \mathrm{g \cdot mol^{-1}})/(13.59\ \mathrm{g \cdot cm^{-3}})][(10^{-2}\ \mathrm{m})/(1\ \mathrm{cm})]^3(298\ \mathrm{K})}{(3.4 \times 10^{-11}\ \mathrm{m^2 \cdot N^{-1}})[(1\ \mathrm{N \cdot m})/(1\ \mathrm{J})]}$$
$$= 23.7\ \mathrm{J \cdot K^{-1} \cdot mol^{-1}}$$

3.50 If a real gas can be considered to be accurately described by the van der Waals equation of state, *(1.44a)*, then to a good approximation *(3.27)* can be written as

$$C_{P,\mathrm{m}} - C_{V,\mathrm{m}} = R\left(1 + \frac{2a}{R^2T^2}P\right) \tag{3.30}$$

Find $C_{V,\mathrm{m}}$ for $Cl_2(g)$ at 10.00 bar and 25 °C. Compare this result to that of Problem 3.47.

▮ Rearranging *(3.30)* and substituting the data from Tables 1-3 and 3-3 gives

$$C_{V,\mathrm{m}} = 33.949\ \mathrm{J \cdot K^{-1} \cdot mol^{-1}} - (8.314\ \mathrm{J \cdot K^{-1} \cdot mol^{-1}}) \times \left(1 + \frac{2(6.579\ \mathrm{bar \cdot L^2 \cdot mol^{-2}})[(10^5\ \mathrm{Pa})/(1\ \mathrm{bar})]^2[(10^{-3}\ \mathrm{m^3})/(1\ \mathrm{L})]^2}{(8.314\ \mathrm{m^3 \cdot Pa \cdot K^{-1} \cdot mol^{-1}})^2(298\ \mathrm{K})^2}(10.00\ \mathrm{bar})\right)$$
$$= 23.85\ \mathrm{J \cdot K^{-1} \cdot mol^{-1}}$$

There is a −7% change as a result of including the term for real gas behavior.

3.51 Another method to find the heat capacity difference for a real gas is

$$C_{P,\mathrm{m}} - C_{V,\mathrm{m}} = R\left[1 + \frac{27}{32}\left(\frac{P_\mathrm{r}}{T_\mathrm{r}^3}\right)\right] \tag{3.31}$$

Repeat the calculations of Problem 3.50 for $Cl_2(g)$ given $P_\mathrm{c} = 77.1$ bar and $T_\mathrm{c} = 144$ °C. Compare the results of the two calculations.

▮ Using *(1.30)* and *(1.31)* gives

$$T_\mathrm{r} = \frac{298\ \mathrm{K}}{(144 + 273)\ \mathrm{K}} = 0.715 \qquad P_\mathrm{r} = \frac{10.00\ \mathrm{bar}}{77.1\ \mathrm{bar}} = 0.130$$

which upon substitution into *(3.31)* gives

$$C_{V,\mathrm{m}} = 33.949\ \mathrm{J \cdot K^{-1} \cdot mol^{-1}} - (8.314\ \mathrm{J \cdot K^{-1} \cdot mol^{-1}})\left[1 + \frac{27}{32}\left(\frac{0.130}{(0.715)^3}\right)\right]$$
$$= 23.85\ \mathrm{J \cdot K^{-1} \cdot mol^{-1}}$$

The two values are the same.

3.52 The *heat capacity ratio*

$$\gamma = C_P/C_V \tag{3.32}$$

was determined for cyanogen, C_2N_2, as 1.177. What is $C_{P,\mathrm{m}}$ for this gas?

▮ Assuming ideal gas behavior, substitution of *(3.28)* into *(3.32)* gives

$$\gamma = C_{P,\mathrm{m}}/(C_{P,\mathrm{m}} - R) = 1.177$$

Solving gives $C_{P,\mathrm{m}} = 55.3\ \mathrm{J \cdot K^{-1} \cdot mol^{-1}}$. (The actual value is $56.746\ \mathrm{J \cdot K^{-1} \cdot mol^{-1}}$.)

3.53 The speed of sound (v_s) through a substance is given by

$$v_\mathrm{s} = (\gamma/\rho\kappa)^{1/2} \tag{3.33}$$

where γ is the heat capacity ratio and κ is the isothermal compressibility. Given $\kappa = 4.57 \times 10^{-10}\ \mathrm{m^2 \cdot N^{-1}}$ and $\rho = 997.0479\ \mathrm{kg \cdot m^{-3}}$, determine $C_{V,\mathrm{m}}$ for $H_2O(l)$. The speed of sound in liquid water at 25 °C is $1496.7\ \mathrm{m \cdot s^{-1}}$.

▮ Solving *(3.33)* for γ and substituting the data gives

$$\gamma = v_\mathrm{s}^2\rho\kappa$$
$$= (1496.7\ \mathrm{m \cdot s^{-1}})^2(997.0479\ \mathrm{kg \cdot m^{-3}})(4.57 \times 10^{-10}\ \mathrm{m^2 \cdot N^{-1}})\frac{1\ \mathrm{N}}{1\ \mathrm{kg \cdot m \cdot s^{-2}}}$$
$$= 1.021$$

Solving *(3.32)* for $C_{V,m}$ and substituting the data from Table 3-3 gives

$$C_{V,m} = \frac{75.291\ \text{J}\cdot\text{K}^{-1}\cdot\text{mol}^{-1}}{1.021} = 73.7\ \text{J}\cdot\text{K}^{-1}\cdot\text{mol}^{-1}$$

3.54 For an ideal gas, *(3.33)* becomes

$$v_s = (RT\gamma/M)^{1/2} \tag{3.34}$$

Predict the speed of sound in dry air at 25 °C.

▌ The heat capacity of air can be calculated using the results of Problem 3.44 and *(3.28)* as

$$\overline{C_{P,m}} = (27.222\ \text{J}\cdot\text{K}^{-1}\cdot\text{mol}^{-1}) + (6.780\times10^{-3}\ \text{J}\cdot\text{K}^{-2}\cdot\text{mol}^{-1})(298\ \text{K}) - (7.76\times10^{-3}\ \text{J}\cdot\text{K}^{-3}\cdot\text{mol}^{-1})(298\ \text{K})^2$$
$$= 29.175\ \text{J}\cdot\text{K}^{-1}\cdot\text{mol}^{-1}$$

$$\overline{C_{V,m}} = 29.175\ \text{J}\cdot\text{K}^{-1}\cdot\text{mol}^{-1} - 8.314\ \text{J}\cdot\text{K}^{-1}\cdot\text{mol}^{-1} = 20.861\ \text{J}\cdot\text{K}^{-1}\cdot\text{mol}^{-1}$$

The heat capacity ratio is given by *(3.32)* by

$$\gamma = \frac{29.175\ \text{J}\cdot\text{K}^{-1}\cdot\text{mol}^{-1}}{20.861\ \text{J}\cdot\text{K}^{-1}\cdot\text{mol}^{-1}} = 1.3985$$

Using *(3.34)* gives

$$v_s = \left(\frac{(8.314\ \text{J}\cdot\text{K}^{-1}\cdot\text{mol}^{-1})[(1\ \text{kg}\cdot\text{m}^2\cdot\text{s}^{-2})/(1\ \text{J})](298\ \text{K})(1.3985)}{29.00\times10^{-3}\ \text{kg}\cdot\text{mol}^{-1}}\right)^{1/2}$$
$$= 346\ \text{m}\cdot\text{s}^{-1}$$

The actual value is 346.29 $\text{m}\cdot\text{s}^{-1}$.

3.55 According to the classical *equipartition of energy theory*, the contributions to the molar heat capacity at constant volume are $R/2$ for each degree of translational freedom, $R/2$ for each degree of rotational freedom, and R for each degree of translational freedom. Predict $C_{P,m}$ for O(g), O_2(g), and O_3(g) using this theory.

▌ The value of $C_{P,m}$ for each gas will be given by adding R to the predicted value of $C_{V,m}$ using the above contributions, see *(3.28)*. The monatomic gas will have only three translational degrees of freedom (see Problem 3.9), giving

$$C_{P,m}(\text{O}) = 3(R/2) + R = \tfrac{5}{2}R$$

The diatomic molecule will have three translational, two rotational, and one vibrational degree of freedom (see Problem 3.10), giving

$$C_{P,m}(\text{O}_2) = 3(R/2) + 2(R/2) + 1(R) + R = \tfrac{9}{2}R$$

Ozone is a nonlinear molecule having three translational, three rotational, and three vibrational degrees of freedom (see Problem 3.11), giving

$$C_{P,m}(\text{O}_3) = 3(R/2) + 3(R/2) + 3(R) + R = 7R$$

Note: The values for O_2 and O_3 are too high at ordinary temperatures (see Problem 3.56).

3.56 Predict γ for air using the equipartition of energy theory. Repeat the calculation including only translational and rotational contributions. What conclusion can be drawn?

▌ Assuming air to be an ideal diatomic gas, the contributions described in Problem 3.55 would predict

$$\overline{C_{V,m}} = 3(R/2) + 2(R/2) + 1(R) = \tfrac{7}{2}R$$

Substituting *(3.28)* into *(3.32)* gives

$$\gamma = \frac{\frac{7}{2}R + R}{\frac{7}{2}R} = \frac{9}{7} = 1.286$$

which is too low compared to 1.3985 (see Problem 3.54). Repeating the calculations, neglecting the vibrational contributions, gives

$$\overline{C_{V,m}} = 3(R/2) + 2(R/2) = \tfrac{5}{2}R$$

$$\gamma = \frac{\frac{5}{2}R + R}{\frac{5}{2}R} = \frac{7}{5} = 1.4$$

which is very close to the actual value. Thus, at ordinary temperatures the vibrational contributions are not nearly as great as predicted by the classical theory. Inclusion of a *small* amount of vibrational energy decreases the values of γ by only 1-5%.

3.57 The heat capacity ratio for $SO_2(g)$ was measured at 25 °C to be 1.279. Determine the shape of the SO_2 molecule.

▮ If SO_2 were a rigid linear polyatomic molecule, γ would be 1.4 (see Problem 3.56). This observed heat capacity ratio is considerably less than that value. For a rigid nonlinear polyatomic molecule, adding up the contributions described in Problem 3.55 and using *(3.28)* and *(3.32)* gives

$$C_{V,\mathrm{m}} = 3(R/2) + 3(R/2) = 3R$$

$$\gamma = \frac{3R + R}{3R} = 1.33$$

which is close to the experimental value. Thus SO_2 is nonlinear.

3.58 Predict $C^{\circ}_{P,\mathrm{m}}$ for Ne at 25 °C. Repeat the calculation for 1000 K. The actual value is 20.786 J · K^{-1} · mol^{-1} at both temperatures. What can be concluded about any electronic contribution to $C^{\circ}_{P,\mathrm{m}}$?

Table 3-4

$$C_V(\text{thermal, trans}) = \sum R/2$$

$$C_V(\text{thermal, rot}) = \sum R/2$$

$$C_V(\text{thermal, vib}) = \sum \frac{Rx_j^2\, e^{x_j}}{(e^{x_j} - 1)^2}$$

$$\text{where } x_j = \frac{h\nu_j}{kT} = \frac{1.4388\,[\bar{\nu}_j/(\mathrm{cm}^{-1})]}{T/(\mathrm{K})} = \frac{\Theta_{v,j}}{T}$$

$$C_V(\text{thermal, elec or nuc}) = R\left[\frac{\sum g_j\left(\dfrac{\varepsilon_j}{kT}\right)^2 e^{-\varepsilon_j/kT}}{\sum g_j\, e^{-\varepsilon_j/kT}} - \left(\frac{\sum g_j \dfrac{\varepsilon_j}{kT}\, e^{-\varepsilon_j/kT}}{\sum g_j\, e^{-\varepsilon_j/kT}}\right)^2\right]$$

▮ The contributions to the heat capacity from the various molecular degrees of freedom are given in Table 3-4. For an ideal monatomic gas, there are only three translational degrees of freedom (see Problem 3.9), giving

$$C^{\circ}_{V,\mathrm{m}} = 3(\tfrac{1}{2}R) = 3(\tfrac{1}{2})(8.314\ \mathrm{J\cdot K^{-1}\cdot mol^{-1}}) = 12.472\ \mathrm{J\cdot K^{-1}\cdot mol^{-1}}$$

Substituting this value into *(3.28)* gives

$$C^{\circ}_{P,\mathrm{m}} = 12.472\ \mathrm{J\cdot K^{-1}\cdot mol^{-1}} + 8.314\ \mathrm{J\cdot K^{-1}\cdot mol^{-1}} = 20.786\ \mathrm{J\cdot K^{-1}\cdot mol^{-1}}$$

The above calculations are temperature-independent, so the value of $C^{\circ}_{P,\mathrm{m}}$ would be predicted to be the same at both temperatures (as observed). It appears that any electronic contribution is negligible at these temperatures because the predicted value is the same as the actual value.

3.59 Prepare a plot of $C_{V,\mathrm{m}}(\text{thermal, elec})$ for $I(g)$ over the range of 500 K to 3500 K. The ground electronic energy level ($\varepsilon_0 = 0$) has a degeneracy of 4, and the first excited level ($\varepsilon_1 = 1.51 \times 10^{-19}$ J) has a degeneracy of 2.

▮ As a sample calculation, consider $T = 2000$ K

$$C_{V,\mathrm{m}}(\text{thermal, elec}) = (8.314\ \mathrm{J\cdot K^{-1}\cdot mol^{-1}})$$

$$\times\left\{\left[\frac{\left[4[0/kT]^2\, e^{-(0)/kT} + 2\left[\dfrac{1.51\times10^{-19}\ \mathrm{J}}{(1.381\times10^{-23}\ \mathrm{J\cdot K^{-1}})(2000\ \mathrm{K})}\right]^2 \times \exp[-(1.51\times10^{-19}\ \mathrm{J})/(1.381\times10^{-23}\ \mathrm{J\cdot K^{-1}})(2000\ \mathrm{K})]\right]}{4\, e^{-(0)/kT} + 2\exp[-(1.51\times10^{-19}\ \mathrm{J})/(1.381\times10^{-23}\ \mathrm{J\cdot K^{-1}})(2000\ \mathrm{K})]}\right]\right.$$

$$\left. + \left[\frac{\left[4[0/kT]^2\, e^{-(0)/kT} + 2\left[\dfrac{1.51\times10^{-19}\ \mathrm{J}}{(1.381\times10^{-23}\ \mathrm{J\cdot K^{-1}})(2000\ \mathrm{K})}\right] \times \exp[-(1.51\times10^{-19}\ \mathrm{J})/(1.381\times10^{-23}\ \mathrm{J\cdot K^{-1}})(2000\ \mathrm{K})]\right]}{4\, e^{-(0)/kT} + 2\exp[-(1.51\times10^{-19}\ \mathrm{J})/(1.381\times10^{-23}\ \mathrm{J\cdot K^{-1}})(2000\ \mathrm{K})]}\right]^2\right\}$$

$$= 0.524\ \mathrm{J\cdot K^{-1}\cdot mol^{-1}}$$

The plot is shown in Fig. 3-3.

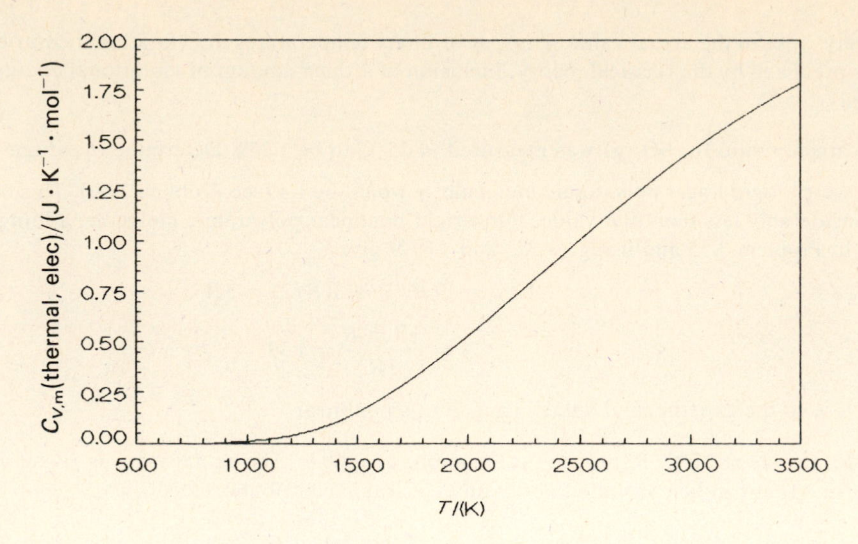

Fig. 3-3

3.60 Confirm the value of γ given in Problem 3.52 for cyanogen. The vibrational data for this linear molecule are 2328.5 cm^{-1}, 850.6 cm^{-1}, 2149 cm^{-1}, 507.2 cm^{-1} (doubly degenerate), and 240 cm^{-1} (doubly degenerate).

I The values of x_i for the vibrational motions of the molecule are

$$x_1 = \frac{(1.4388)(2328.5)}{298\text{ K}} = 11.237$$

Likewise, $x_2 = 4.105$, $x_3 = 10.371$, $x_4 = 2.448$, and $x_5 = 1.158$. The sum of the contributions listed in Table 3-4 is

$$C_{V,\text{m}}(\text{thermal}) = 3(\tfrac{1}{2}R) + 2(\tfrac{1}{2}R)$$
$$+R\left[\frac{(11.237)^2\,e^{11.237}}{(e^{11.237}-1)^2} + \frac{(4.105)^2\,e^{4.105}}{(e^{4.105}-1)^2} + \frac{(10.371)^2\,e^{10.371}}{(e^{10.371}-1)^2} + \frac{2(2.448)^2\,e^{2.448}}{(e^{2.448}-1)^2} + \frac{2(1.158)^2\,e^{1.158}}{(e^{1.158}-1)^2}\right]$$
$$= (5.8248)R$$

giving $C_{V,\text{m}}(\text{thermal}) = (5.8248)(8.314\text{ J}\cdot\text{K}^{-1}\cdot\text{mol}^{-1}) = 48.430\text{ J}\cdot\text{K}^{-1}\cdot\text{mol}^{-1}$. Using *(3.28)* and *(3.32)* gives

$$\gamma = \frac{48.430\text{ J}\cdot\text{K}^{-1}\cdot\text{mol}^{-1} + 8.314\text{ J}\cdot\text{K}^{-1}\cdot\text{mol}^{-1}}{48.430\text{ J}\cdot\text{K}^{-1}\cdot\text{mol}^{-1}} = 1.172$$

which differs only by -0.4%.

3.61 Determine the vibrational characteristic temperature for iodine monochloride.

I The molar heat capacity at constant volume is given by *(3.28)* as

$$C_{V,\text{m}} = 35.548\text{ J}\cdot\text{K}^{-1}\cdot\text{mol}^{-1} - 8.314\text{ J}\cdot\text{K}^{-1}\cdot\text{mol}^{-1} = 27.234\text{ J}\cdot\text{K}^{-1}\cdot\text{mol}^{-1}$$

Subtracting the translational and rotational contributions gives the vibrational contribution as

$$C_{V,\text{m}}(\text{thermal, vib}) = 27.234\text{ J}\cdot\text{K}^{-1}\cdot\text{mol}^{-1} - 3(\tfrac{1}{2})(8.314\text{ J}\cdot\text{K}^{-1}\cdot\text{mol}^{-1}) - 2(\tfrac{1}{2})(8.314\text{ J}\cdot\text{K}^{-1}\cdot\text{mol}^{-1})$$
$$= 6.448\text{ J}\cdot\text{K}^{-1}\cdot\text{mol}^{-1}$$

Setting the expression for the vibration contribution equal to $6.448\text{ J}\cdot\text{K}^{-1}\cdot\text{mol}^{-1}$ and solving for x gives

$$C_{V,\text{m}}(\text{thermal, vib}) = \frac{(8.314\text{ J}\cdot\text{K}^{-1}\cdot\text{mol}^{-1})x^2\,e^x}{(e^x-1)^2} = 6.448\text{ J}\cdot\text{K}^{-1}\cdot\text{mol}^{-1}$$
$$x = 1.769$$

This corresponds to

$$\Theta_v = xT = (1.769)(298\text{ K}) = 527\text{ K}$$

The accepted value is 549.88 K.

3.62 What is the value of $\bar{\nu}$ for a degree of vibrational motion for which $C_{V,\text{m}}(\text{thermal, vib}) = R/2$ at 25 °C?

▌ Setting the expression for the vibrational contribution equal to $R/2$ and solving for x gives

$$C_{V,\mathrm{m}}(\text{thermal, vib}) = \frac{Rx^2 e^x}{(e^x - 1)^2} = \frac{R}{2}$$

$$x = 2.983$$

This corresponds to

$$\bar{\nu}/(\mathrm{cm}^{-1}) = \frac{x[T/(\mathrm{K})]}{1.4388} = \frac{(2.983)(298)}{1.4388} = 618$$

3.63 Use the equipartition of energy theory (see Problem 3.55) to predict the heat capacity of metallic silver.

▌ The silver atoms will have three degrees of vibrational freedom. Thus, according to this classical theory, $C_{V,\mathrm{m}} = 3R = 24.943\ \mathrm{J \cdot K^{-1} \cdot mol^{-1}}$. Although it can be seen from Fig. 3-4 that $C_{V,\mathrm{m}}$ approaches $3R$ at high temperatures, the value of $C_{V,\mathrm{m}}$ is generally less than $3R$, is a function of temperature, and approaches zero as the temperature decreases.

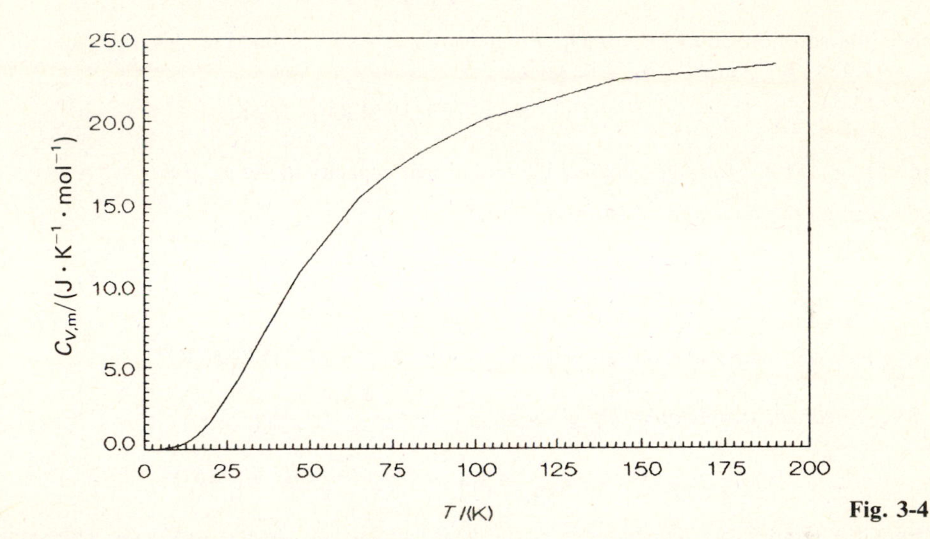

Fig. 3-4

3.64 Determine the number of atoms in the empirical formula of aluminum fluoride given $C_{P,\mathrm{m}} = 115.311\ \mathrm{J \cdot K^{-1} \cdot mol^{-1}}$ at 2500 K (just below the melting point).

▌ An extension of the application of the equipartition of energy theory to more complicated solids (see Problem 3.63) is to multiply the predicted value by Λ, the number of atoms in the empirical formula of the substance. Assuming $C_{V,\mathrm{m}} \approx C_{P,\mathrm{m}}$, the number of atoms is

$$\Lambda = \frac{C_{P,\mathrm{m}}}{3R} = \frac{115.311\ \mathrm{J \cdot K^{-1} \cdot mol^{-1}}}{3(8.314\ \mathrm{J \cdot K^{-1} \cdot mol^{-1}})} = 4.62$$

The value of Λ from this sample calculation contains two errors: the assumption that $C_{V,\mathrm{m}} \approx C_{P,\mathrm{m}}$ and the assumption that only vibrational contributions are present at this relatively high temperature. In actuality, $C_{V,\mathrm{m}} < C_{P,\mathrm{m}}$, which will decrease the numerator in the equation given above, as would subtracting the electronic contribution, thus yielding a value of Λ much closer to 4, the actual value in $AlF_3(s)$.

3.65 Estimate the specific heat of Fe(s) at 25 °C and 1.00 bar.

▌ Combining *(3.25)* with the results of Problem 3.63 gives the *law of Dulong and Petit* for metal elements:

$$cM \approx 3R \qquad (3.35)$$

Solving *(3.35)* for C and substituting the data gives

$$C = \frac{3(8.314\ \mathrm{J \cdot K^{-1} \cdot mol^{-1}})}{55.85\ \mathrm{g \cdot mol^{-1}}} = 0.4466\ \mathrm{J \cdot K^{-1} \cdot g^{-1}}$$

The actual value is $0.444\ \mathrm{J \cdot K^{-1} \cdot g^{-1}}$, which is in good agreement considering that $C_{V,\mathrm{m}} < C_{P,\mathrm{m}}$ and that possible electronic contributions were neglected.

3.66 A rather soft, silvery metal was observed to have a specific heat of $0.215\ \text{cal} \cdot \text{K}^{-1} \cdot \text{g}^{-1}$. Identify the metal.

▌ Solving *(3.35)* for M and substituting the data gives

$$M = \frac{3(8.314\ \text{J} \cdot \text{K}^{-1} \cdot \text{mol}^{-1})}{(0.215\ \text{cal} \cdot \text{K}^{-1} \cdot \text{g}^{-1})[(4.184\ \text{J})/(1\ \text{cal})]} = 27.7\ \text{g} \cdot \text{mol}^{-1}$$

The only pure metallic elements having molar masses in this range are Na, Mg, and Al. (The data are for Al.)

3.67 One of the more successful attempts to describe the heat capacity of metals was the *Einstein heat capacity relationship* given by

$$C_{V,\text{m}} = 3R\left(\frac{\Theta_\text{E}}{T}\right)^2 \frac{e^{\Theta_\text{E}/T}}{(e^{\Theta_\text{E}/T} - 1)^2} \tag{3.36}$$

where Θ_E is known as the *Einstein characteristic temperature.* Using the plot for Ag given in Fig. 3-4, determine Θ_E.

▌ The value of Θ_E is not a constant over large temperature ranges. The usual approach to finding a reasonable value of Θ_E is to find the temperature at which $C_{V,\text{m}} = \frac{3}{2}R$, one-half the classical value, and calculate Θ_E from

$$\Theta_\text{E}/T = 2.983 \tag{3.37}$$

which is the solution of *(3.36)* for Θ_E/T when $C_{V,\text{m}} = \frac{3}{2}R$. From Fig. 3-4, the value of T corresponding to $C_{V,\text{m}} = 12.47\ \text{J} \cdot \text{K}^{-1} \cdot \text{mol}^{-1}$ is 54 K, giving

$$\Theta_\text{E} = (2.983)(54\ \text{K}) = 161\ \text{K}$$

3.68 Using $\Theta_\text{E} = 161\ \text{K}$ for Ag, calculate the molar heat capacity of Ag at 75 K.

▌ Substituting the data into *(3.36)* gives

$$C_{V,\text{m}} = 3(8.314\ \text{J} \cdot \text{K}^{-1} \cdot \text{mol}^{-1})\left(\frac{161\ \text{K}}{75\ \text{K}}\right)^2 \frac{e^{(161\ \text{K})/(75\ \text{K})}}{(e^{(161\ \text{K})/(75\ \text{K})} - 1)^2}$$
$$= 17.22\ \text{J} \cdot \text{K}^{-1} \cdot \text{mol}^{-1}$$

which agrees well with the reported value of $16.90\ \text{J} \cdot \text{K}^{-1} \cdot \text{mol}^{-1}$ at 74.56 K.

3.69 The *Debye heat capacity relationship* given by

$$C_{V,\text{m}} = \frac{9R}{(\Theta_\text{D}/T)^3} \int_0^{\Theta_\text{D}/T} \frac{x^4\, e^x}{(e^x - 1)^2}\, dx = 3R\vartheta(\Theta_\text{D}/T) \tag{3.38}$$

describes the temperature dependence of the heat capacity of a metal very well. The symbol Θ_D is the *Debye characteristic temperature*, and values of the *Debye function*, $\vartheta(\Theta_\text{D}/T)$, are given in Fig. 3-5. Determine Θ_D for Ag.

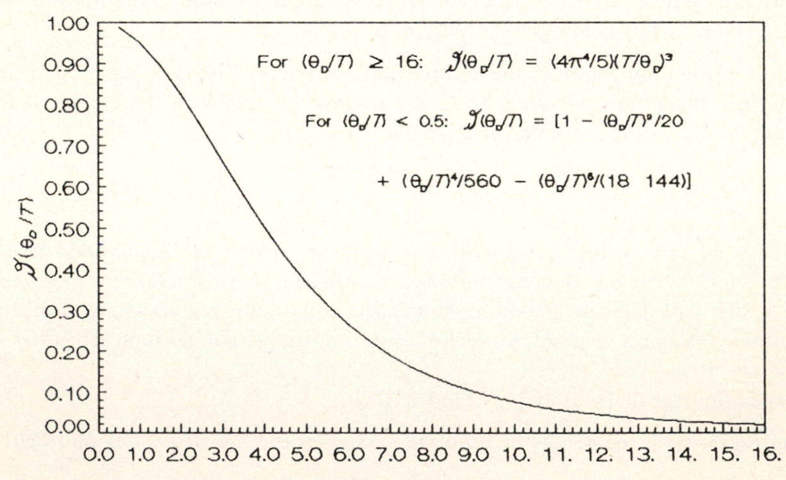

Fig. 3-5

▌ The relation between Θ_E and Θ_D is given by

$$\Theta_\text{E} = (\tfrac{3}{5})^{1/2}\Theta_\text{D} \tag{3.39}$$

Solving for Θ_D and substituting $\Theta_\text{E} = 161\ \text{K}$ (see Problem 3.67) gives

$$\Theta_\text{D} = (\tfrac{5}{3})^{1/2}(161\ \text{K}) = 208\ \text{K}$$

3.70 Evaluate $C_{V,m}/3R$ using the Debye theory for Ag at 104 K.

▌ The value of $C_{V,m}/3R$ is simply $\vartheta(\Theta_D/T)$, see *(3.38)*. At 104 K,

$$\Theta_D/T = (208\ \mathrm{K})/(104\ \mathrm{K}) = 2.00$$

From Fig. 3-5, $\vartheta(\Theta_D/T) = 0.83$. This point, as well as others, are shown in Fig. 3-6 along with experimental values and values determined using *(3.36)* for comparison.

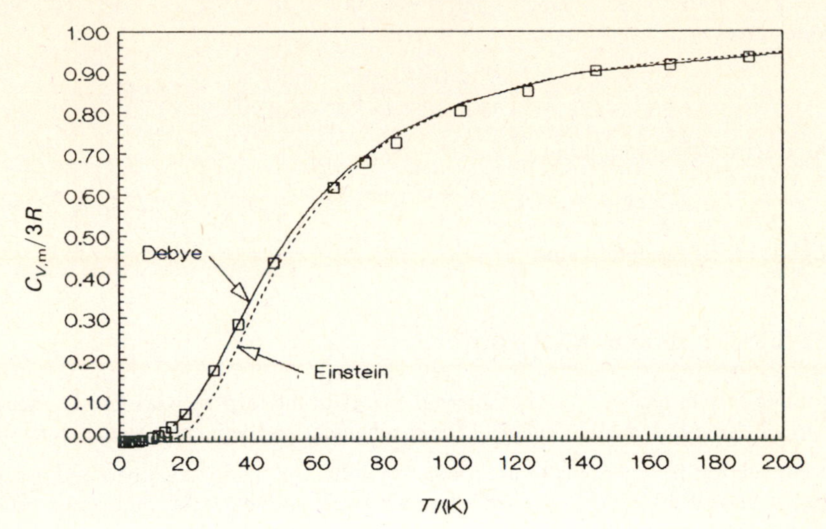

Fig. 3-6

3.71 What is the high temperature limit of *(3.38)*?

▌ As $T \to \infty$, $\Theta_D/T \to 0$. The insert of Fig. 3-5 indicates that as $\Theta_D/T \to 0$, $\vartheta(\Theta_D/T) \to 1$. Thus *(3.38)* shows that $C_{V,m} \to 3R$.

3.72 At extremely low temperatures, there is an electronic contribution to $C_{V,m}$ for metals given by

$$C_{V,m} = \alpha T^3 - \gamma T \tag{3.40}$$

Use the following data to determine γ for Ag(s) at 1.00 bar:

$T/(\mathrm{K})$	1.35	2	3	4	5	6	7
$C_{V,m}/(\mathrm{J \cdot K^{-1} \cdot mol^{-1}})$	0.001 06	0.002 62	0.006 57	0.0127	0.0213	0.0373	0.0632

▌ A linear form of *(3.40)* is $C_{V,m}/T = \alpha T^2 - \gamma$. Thus a plot of $C_{V,m}/T$ against T^2 will be linear, with a slope equal to α and an intercept equal to $-\gamma$ (see Fig. 3-7). From the intercept, $\gamma = -8.9 \times 10^{-4}\ \mathrm{J \cdot K^{-2} \cdot mol^{-1}}$.

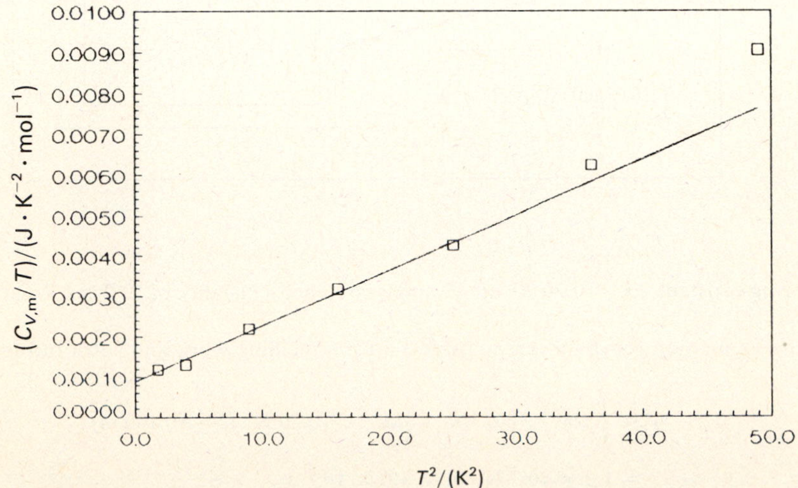

Fig. 3-7

3.73 At what temperature will the electronic contribution in *(3.40)* become less than the vibrational contribution? What is the numerical value of the electronic contribution at 25 °C for Ag(s)?

▌ Setting the two contributions equal to each other and solving for T gives

$$\alpha T^3 = -\gamma T \qquad T = (-\gamma/\alpha)^{1/2}$$

For Ag, $\alpha = 1.37 \times 10^{-4}\ \mathrm{J \cdot K^{-4} \cdot mol^{-1}}$ and $\gamma = -8.9 \times 10^{-4}\ \mathrm{J \cdot K^{-2} \cdot mol^{-1}}$ (see Problem 3.72). Substituting these values gives

$$T = \left(\frac{8.9 \times 10^{-4}\ \mathrm{J \cdot K^{-2} \cdot mol^{-1}}}{1.37 \times 10^{-4}\ \mathrm{J \cdot K^{-4} \cdot mol^{-1}}}\right)^{1/2} = 2.5\ \mathrm{K}$$

At 298 K, the electronic contribution is

$$C_{V,\mathrm{m}}(\mathrm{elec}) = -(-8.9 \times 10^{-4}\ \mathrm{J \cdot K^{-2} \cdot mol^{-1}})(298\ \mathrm{K}) = 0.27\ \mathrm{J \cdot K^{-1} \cdot mol^{-1}}$$

which is roughly 1% of the total value.

3.3 INTERNAL ENERGY, WORK AND HEAT

3.74 A thermodynamic system received 725 J of internal energy in the form of work from a mechanical reservoir. What are $\Delta U(\mathrm{mech})$, the internal energy change of the mechanical reservoir, and w, the *work* for the system?

▌ The sign convention for energy is to use a positive value for energy absorbed and a negative value for energy released. Thus, $\Delta U(\mathrm{mech}) = -725\ \mathrm{J}$. The relation between w and $\Delta U(\mathrm{mech})$ is shown in Fig. 3-8. For the system,

$$w = -\Delta U(\mathrm{mech}) = -(-725\ \mathrm{J}) = 725\ \mathrm{J}$$

The sign convention for w for a system is the same as for any form of energy: $w > 0$ for work done on the system by the mechanical reservoir in the surroundings and $w < 0$ for work done by the system on the mechanical reservoir in the surroundings.

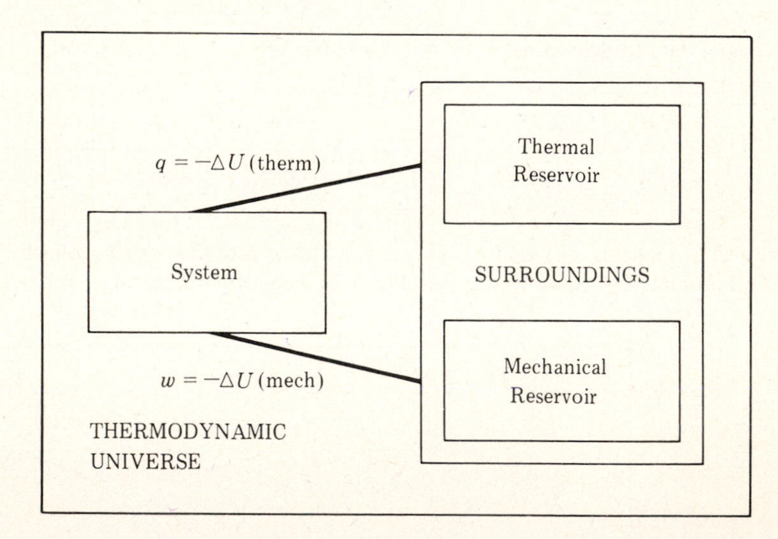

Fig. 3-8

3.75 A spring has a spring constant $k = 0.20\ \mathrm{N \cdot m^{-1}}$ and is stretched a distance of 10.0 cm. What is the work involved?

▌ Integrating the expression for $đw$ given in Table 3-5 for stretching work and substituting the data gives

$$w = \int đw = \int_0^l kl\,dl = \tfrac{1}{2}kl^2 = \tfrac{1}{2}(0.20\ \mathrm{N \cdot m^{-1}})(0.100\ \mathrm{m})^2$$

$$= 1.0 \times 10^{-3}\ \mathrm{N \cdot m} = 1.0 \times 10^{-3}\ \mathrm{J}$$

Table 3-5

Mechanical work	$đw = F_{ext}\,dl$	F_{ext} = external force
		l = displacement
Stretching work	$đw = kl\,dl$	kl = tension
		l = displacement
Gravitational work	$đw = mg\,dl$	m = mass
		g = gravitational constant
		l = displacement
Expansion work	$đw = -P_{ext}\,dV$	P_{ext} = external pressure
		V = volume
Surface work	$đw = \gamma\,dA$	γ = surface tension
		A = area
Electrochemical cell work	$đw = \Delta V\,dQ$	ΔV = electric potential difference
	$= I\,\Delta V\,dt$	Q = quantity of electricity
		T = electric current
		t = time

3.76 An ideal gas underwent an expansion from 1.0 m^3, 5.00 bar, and 500 K to 10.0 m^3, 0.1069 bar, and 108 K. During this expansion, the following $P-V$ data were obtained:

$V/(\mathrm{m}^3)$	1.0	1.5	2.0	2.5	3.0	3.5	4.0	4.5	5.0	5.5
$P_{ext}/(\mathrm{bar})$	5.0000	2.5404	1.5713	1.0825	0.7983	0.6171	0.4938	0.4056	0.3402	0.2901

$V/(\mathrm{m}^3)$	6.0	6.5	7.0	7.5	8.0	8.5	9.0	9.5	10.0
$P_{ext}/(\mathrm{bar})$	0.2509	0.2195	0.1939	0.1728	0.1552	0.1402	0.1275	0.1165	0.1069

Calculate the work done by the system.

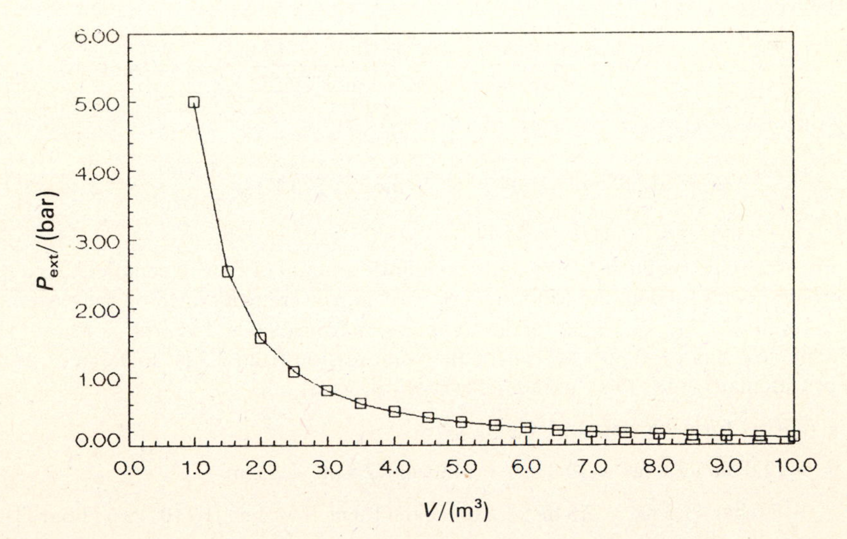

Fig. 3-9

▌ A plot of the data appears in Fig. 3-9. From Table 3-5, for expansion work

$$w = -\int_{V_1}^{V_2} P_{ext}\,dV = -(\text{area})$$

Using Simpson's rule, the area under curve is 5.885 bar · m^3, giving

$$w = -(5.885\ \mathrm{bar \cdot m^3})\frac{10^5\ \mathrm{Pa}}{1\ \mathrm{bar}}\left(\frac{1\ \mathrm{J}}{1\ \mathrm{Pa \cdot m^3}}\right) = -5.885 \times 10^5\ \mathrm{J}$$

3.77 A sample of an ideal gas underwent an expansion against a constant external pressure of 1.0 bar from 1.0 m^3, 10.0 bar, and 273 K to 10.0 m^3, 1.0 bar, and 273 K. What was the work done by the system on the surroundings?

▮ For an isothermal expansion in which P_{ext} is constant,

$$w = -\int_{V_1}^{V_2} P_{ext}\, dV = -P_{ext}(V_2 - V_1) \tag{3.41a}$$

Substituting the data gives

$$w = -(1.0\text{ bar})(10.0\text{ m}^3 - 1.0\text{ m}^3)\frac{10^5\text{ Pa}}{1\text{ bar}}\left(\frac{1\text{ J}}{1\text{ m}^3\cdot\text{Pa}}\right) = -9.0\times 10^5\text{ J}$$

3.78 The external pressure on the system described in Problem 3.77 was increased to 10.0 bar, and the gas sample was isothermally compressed from 10.0 m^3 and 1.0 bar to 1.0 m^3 and 10.0 bar. Calculate the work done on the system by the surroundings. Is w a *path* or a *state* (point) *function*?

▮ For the compression process, *(3.41a)* gives

$$w = -(10.0\text{ bar})(1.0\text{ m}^3 - 10.0\text{ m}^3)\frac{10^5\text{ Pa}}{1\text{ bar}}\left(\frac{1\text{ J}}{1\text{ m}^3\cdot\text{Pa}}\right) = 9.0\times 10^6\text{ J}$$

The process described in Problem 3.77 and the process described in this problem constitute a cyclic process for the system. Because

$$w_{total} = -9.0\times 10^5\text{ J} + 9.0\times 10^6\text{ J} = 8.9\times 10^6\text{ J} \neq 0$$

w is a path function.

3.79 What is the least amount of work needed to restore the system described in Problem 3.77 to the original P-V conditions?

▮ The most efficient process will be one that is carried out reversibly, i.e., $P_{ext} = P_{int} \pm dP$. Substituting *(1.18)* into the expression for expansion work under isothermal conditions gives

$$w = -\int_{V_1}^{V_2} P_{ext}\, dV = -\int_{V_1}^{V_2}\frac{nRT}{V}\, dV = -nRT\ln\frac{V_2}{V_1} \tag{3.42}$$

For this system *(1.18)* gives

$$n = \frac{(10.0\text{ bar})(1.0\text{ m}^3)[(10^5\text{ Pa})/(1\text{ bar})]}{(8.314\text{ m}^3\cdot\text{Pa}\cdot\text{K}^{-1}\cdot\text{mol}^{-1})(273\text{ K})} = 440\text{ mol}$$

Substituting into *(3.42)* gives

$$w = -(440\text{ mol})(8.314\text{ J}\cdot\text{K}^{-1}\cdot\text{mol}^{-1})(273\text{ K})\ln\frac{1.0\text{ m}^3}{10.0\text{ m}^3} = 2.3\times 10^6\text{ J}$$

3.80 Reversible processes (see Problem 3.79) require an infinite amount of time to complete. As a reasonable approximation to a reversible compression, the following series of steps were performed on the system described in Problem 3.77 (all isothermally): $P_{ext} = 2.0$ bar for the compression from 10.0 m^3 to 5.0 m^3, $P_{ext} = 4.0$ bar for the compression from 5.0 m^3 to 2.5 m^3, $P_{ext} = 8.0$ bar for the compression from 2.5 m^3 to 1.3 m^3, and $P_{ext} = 10.0$ bar for the final step. Calculate w for this "pseudoreversible" process.

▮ Applying *(3.41a)* for each step and adding gives

$$\begin{aligned} w_{total} &= -[(2.0\text{ bar})(5.0\text{ m}^3 - 10.0\text{ m}^3) + (4.0\text{ bar})(2.5\text{ m}^3 - 5.0\text{ m}^3) \\ &\quad + (8.0\text{ bar})(1.3\text{ m}^3 - 2.5\text{ m}^3) + (10.0\text{ bar})(1.0\text{ m}^3 - 1.3\text{ m}^3)][(10^5\text{ Pa})/(1\text{ bar})][(1\text{ J})/(1\text{ m}^3\cdot\text{Pa})] \\ &= 3.3\times 10^6\text{ J} \end{aligned}$$

3.81 As 1 mol of liquid water is heated from 15 °C to 25 °C, it expands slightly against the atmosphere. Calculate the work for this process given $\alpha = 2.0\times 10^{-4}\text{ K}^{-1}$, $\rho = 1.00\text{ g}\cdot\text{cm}^{-3}$, and $P_{ext} = 1.00$ bar.

▮ For this expansion process,

$$dV = \left(\frac{\partial V}{\partial T}\right)_P dT = n\alpha V_m\, dT$$

which upon substitution into the expression for work gives

$$w = -\int_{V_1}^{V_2} P_{ext}\,dV = -P_{ext}\int_{V_1}^{V_2} dV = -P_{ext}\int_{T_1}^{T_2} n\alpha V_m\,dT \approx -P_{ext}\,n\alpha V_m(T_2 - T_1)$$

Substituting the data gives

$$w = -(1.00\text{ bar})(1.00\text{ mol})(2.0\times10^{-4}\text{ K}^{-1})\left(\frac{18\text{ g}\cdot\text{mol}^{-1}}{1\text{ g}\cdot\text{cm}^{-3}}\right)\left(\frac{1\text{ m}^3}{10^6\text{ cm}^3}\right)(298\text{ K} - 288\text{ K})\left(\frac{10^5\text{ Pa}}{1\text{ bar}}\right)\left(\frac{1\text{ J}}{1\text{ m}^3\cdot\text{Pa}}\right)$$
$$= 3.6\times10^{-3}\text{ J}$$

3.82 The density of ice at 0 °C is 0.915 $\text{g}\cdot\text{cm}^{-3}$ and that of liquid water at 0 °C is 0.999 87 $\text{g}\cdot\text{cm}^{-3}$. Calculate the work for melting 1 mol of ice at 1.00 bar.

▮ For this isobaric, isothermal process,

$$w = -\int_{V_1}^{V_2} P_{ext}\,dV = -P_{ext}(V_2 - V_1) = -P_{ext}n\left(\frac{M}{\rho_{liq}} - \frac{M}{\rho_{ice}}\right)$$
$$= -(1.00\text{ bar})(1.00\text{ mol})(18.02\text{ g}\cdot\text{mol}^{-1})$$
$$\times\left(\frac{1}{0.999\,87\text{ g}\cdot\text{cm}^{-3}} - \frac{1}{0.915\text{ g}\cdot\text{cm}^{-3}}\right)\left(\frac{10^{-6}\text{ m}^3}{1\text{ cm}^3}\right)\left(\frac{10^5\text{ Pa}}{1\text{ bar}}\right)\left(\frac{1\text{ J}}{1\text{ m}^3\cdot\text{Pa}}\right)$$
$$= 0.17\text{ J}$$

Note that the surroundings are doing work on the system during the melting process for water—this is not the case for most substances.

3.83 A system consisting of 1 mol of an ideal gas undergoes an isothermal expansion at 298 K from 1.00 bar to a lower pressure while generating 100 J of work. What is the final pressure of the system if the external pressure is constant at 0.10 bar?

▮ For this isothermal, isobaric process, *(3.41a)* can be put into a more useful form by substituting *(1.18)* to give

$$w = -P_{ext}nRT(1/P_2 - 1/P_1) \qquad (3.41b)$$

Rearranging and substituting the data gives

$$\frac{1}{P_2} = \frac{-(-100\text{ J})[(1\text{ m}^3\cdot\text{Pa})/(1\text{ J})]}{(0.10\text{ bar})(1.00\text{ mol})(8.314\text{ m}^3\cdot\text{Pa}\cdot\text{K}^{-1}\cdot\text{mol}^{-1})(298\text{ K})} + \frac{1}{1.00\text{ bar}}$$
$$= 1.40\text{ bar}^{-1}$$

giving $P_2 = 0.71$ bar.

3.84 Consider the reversible compression of 1 mol of an ideal gas from 22.4 L to 10.0 L at 0 °C. Calculate the work needed for this process. Repeat the calculations for a second compression from 10.0 L to 1.00 L at 0 °C.

▮ For the compressions, *(3.42)* gives

$$w = -(1.00\text{ mol})(8.314\text{ J}\cdot\text{K}^{-1}\cdot\text{mol}^{-1})(273\text{ K})\ln\frac{10.0\text{ L}}{22.4\text{ L}} = 1831\text{ J}$$

$$w = -(1.00\text{ mol})(8.314\text{ J}\cdot\text{K}^{-1}\cdot\text{mol}^{-1})(273\text{ K})\ln\frac{1.00\text{ L}}{10.0\text{ L}} = 5226\text{ J}$$

3.85 Repeat the calculations of Problem 3.84, for O_2, assuming it to be a van der Waals gas, see *(1.44a)*. Compare the results.

▮ The expression for reversible, isothermal work for 1 mol of a van der Waals gas is

$$w = -RT\ln\frac{V_{m,2} - b}{V_{m,1} - b} - a\left(\frac{1}{V_{m,2}} - \frac{1}{V_{m,1}}\right) \qquad (3.43)$$

Substituting the data for both compressions and values for a and b from Table 1-3 into *(3.43)* gives

$$w = -(8.314\text{ J}\cdot\text{K}^{-1}\cdot\text{mol}^{-1})(273\text{ K})\ln\frac{10.0\text{ L}\cdot\text{mol}^{-1} - 0.0319\text{ L}\cdot\text{mol}^{-1}}{22.4\text{ L}\cdot\text{mol}^{-1} - 0.0319\text{ L}\cdot\text{mol}^{-1}}$$
$$-(1.362\text{ L}^2\cdot\text{atm}\cdot\text{mol}^{-2})\left(\frac{101.325\text{ J}}{1\text{ L}\cdot\text{atm}}\right)\left(\frac{1}{10.0\text{ L}\cdot\text{mol}^{-1}} - \frac{1}{22.4\text{ L}\cdot\text{mol}^{-1}}\right)$$
$$= 1826\text{ J}\cdot\text{mol}^{-1}$$

$$w = -(8.314)(273)\ln\frac{1.00 - 0.0319}{10.0 - 0.0319} - (1.362)(101.325)\left(\frac{1}{1.00} - \frac{1}{10.0}\right)$$

$$= 5169\ \text{J}\cdot\text{mol}^{-1}$$

For the first compression there is very little difference between the values calculated using the ideal gas law and the van der Waals equation because the gas is acting rather ideal. But as the gas becomes highly compressed during the second compression, the difference between the values calculated using the two equations of state becomes significant.

3.86 Repeat the calculations of Problem 3.85 for O_2 assuming that the gas obeys the Redlich-Kwong equation of state, see *(1.55)*. Compare the results.

▌ The expression for the reversible, isothermal work for 1 mol of a Redlich-Kwong gas is

$$w = -RT\ln\frac{V_{m,2} - b}{V_{m,1} - b} + \frac{a}{bT^{1/2}}\ln\frac{V_{m,2}(V_{m,1} + b)}{V_{m,1}(V_{m,2} + b)}$$

Substituting the data for both compressions and the data from Table 1-3 gives

$$w = -(8.314\ \text{J}\cdot\text{K}^{-1}\cdot\text{mol}^{-1})(273\ \text{K})\ln\frac{10.0\ \text{L}\cdot\text{mol}^{-1} - 0.0221\ \text{L}\cdot\text{mol}^{-1}}{22.4\ \text{L}\cdot\text{mol}^{-1} - 0.0221\ \text{L}\cdot\text{mol}^{-1}}$$

$$+\frac{(17.16\ \text{L}^2\cdot\text{K}^{1/2}\cdot\text{atm}\cdot\text{mol}^{-2})[(101.325\ \text{J})/(1\ \text{L}\cdot\text{atm})]}{(0.0221\ \text{L}\cdot\text{mol}^{-1})(273\ \text{K})^{1/2}}$$

$$\times\ln\frac{(10.0\ \text{L}\cdot\text{mol}^{-1})(22.4\ \text{L}\cdot\text{mol}^{-1} + 0.0221\ \text{L}\cdot\text{mol}^{-1})}{(22.4\ \text{L}\cdot\text{mol}^{-1})(10.0\ \text{L}\cdot\text{mol}^{-1} + 0.0221\ \text{L}\cdot\text{mol}^{-1})}$$

$$= 1827\ \text{J}\cdot\text{mol}^{-1}$$

$$w = -(8.314)(273)\ln\frac{1.00 - 0.0221}{10.0 - 0.0221} + \frac{(17.16)(101.325)}{(0.0221)(273)^{1/2}}\ln\frac{1.00(10.0 + 0.0221)}{10.00(1.00 + 0.0221)}$$

$$= 5179\ \text{J}\cdot\text{mol}^{-1}$$

Both values of work calculated using the Redlich-Kwong equation of state agree to within 0.2% with those calculated using the van der Waals equation.

3.87 Repeat the calculations of Problem 3.85 for O_2 assuming the gas to obey the virial equation of state, see *(1.33)*. Compare the results to those of Problems 3.85 and 3.86.

▌ The expression for the reversible, isothermal work for 1 mol of a gas for which the virial equation is valid is

$$w = -RT\ln\frac{V_{m,2}}{V_{m,1}} + B_v\left(\frac{1}{V_{m,2}} - \frac{1}{V_{m,1}}\right) + \cdots$$

Substituting the data for both compressions and the data from Table 1-3 gives

$$w = -(8.314\ \text{J}\cdot\text{K}^{-1}\cdot\text{mol}^{-1})(273\ \text{K})\ln\frac{10.0\ \text{L}\cdot\text{mol}^{-1}}{22.4\ \text{L}\cdot\text{mol}^{-1}}$$

$$+(-0.5168\ \text{L}^2\cdot\text{atm}\cdot\text{mol}^{-2})\left(\frac{101.325\ \text{J}}{1\ \text{L}\cdot\text{atm}}\right)\left(\frac{1}{10.0\ \text{L}\cdot\text{mol}^{-1}} - \frac{1}{22.4\ \text{L}\cdot\text{mol}^{-1}}\right)$$

$$= 1828\ \text{J}\cdot\text{mol}^{-1}$$

$$w = -(8.314)(273)\ln\frac{1.00}{10.0} + (-0.5168)(101.325)\left(\frac{1}{1.00} - \frac{1}{10.0}\right)$$

$$= 5179\ \text{J}\cdot\text{mol}^{-1}$$

All three equations of state describing O_2 as a real gas give essentially the same answers.

3.88 A thermodynamic system transferred 725 J of internal energy in the form of heat to a thermal reservoir. What is $\Delta U(\text{therm})$, the internal energy change of the thermal reservoir, and q, the *heat* for the system?

▌ Using the usual sign convention for energy for the thermal reservoir gives $\Delta U(\text{therm}) = 725\ \text{J}$. The relation between q and $\Delta U(\text{therm})$ is shown in Fig. 3-8. For the system

$$q = -\Delta U(\text{therm}) = -(725\ \text{J}) = -725\ \text{J}$$

The sign convention for q for a system is the same as for any form of energy: $q > 0$ for heat transferred from the thermal reservoir in the surroundings to the system and $q < 0$ for heat absorbed by the thermal reservoir in the surroundings from the system.

3.89 For the processes described in Problem 3.77 and 3.78, $q = -w$. Is heat a path or a state function?

▮ For the expansion process, $q = -w = -(-9.0 \times 10^5\ \text{J}) = 9.0 \times 10^5\ \text{J}$, and for the compression process, $q = -w = -(9.0 \times 10^6\ \text{J}) = -9.0 \times 10^6\ \text{J}$. The two processes constitute a cyclic process for the system, but because

$$q_{\text{total}} = 9.0 \times 10^5\ \text{J} + (-9.0 \times 10^6\ \text{J}) = -8.9 \times 10^6\ \text{J} \neq 0$$

it cannot be a state function. Like work, heat is a path function.

3.90 What will be the temperature change of 1.00 mol of $CH_3OH(g)$ as exactly 100 J is added to it under constant-volume conditions?

▮ Under constant-volume conditions

$$q_v = \Delta U \tag{3.44}$$

Assuming ideal gas behavior and substituting the data from Table 3-3 into *(3.28)* gives

$$C_{V,\text{m}} = 43.89\ \text{J} \cdot \text{K}^{-1} \cdot \text{mol}^{-1} - 8.314\ \text{J} \cdot \text{K}^{-1} \cdot \text{mol}^{-1} = 35.58\ \text{J} \cdot \text{K}^{-1} \cdot \text{mol}^{-1}$$

Using *(3.18)* and *(3.44)* gives $q_v = \Delta U = nC_{V,\text{m}}\,\Delta T$. Solving for ΔT and substituting the data gives

$$\Delta T = \frac{100.0\ \text{J}}{(1.00\ \text{mol})(35.58\ \text{J} \cdot \text{K}^{-1} \cdot \text{mol}^{-1})} = 2.81\ \text{K}$$

The temperature will increase by 2.81 K.

3.91 How much heat would be required to heat the system described in Problem 3.90 by 2.81 K under constant-pressure conditions?

▮ Under constant-pressure conditions,

$$q_P = \Delta H \tag{3.45}$$

Using *(3.23)* and *(3.45)* gives

$$q_P = \Delta H = nC_{P,\text{m}}\,\Delta T = (1.00\ \text{mol})(43.89\ \text{J} \cdot \text{K}^{-1} \cdot \text{mol}^{-1})(2.81\ \text{K}) = 123\ \text{J}$$

3.92 A 1-mol sample of Cu(s) at 75.0 °C was placed in contact with another 1-mol sample of Cu(s) at 25.0 °C. What is the final temperature of the two blocks of copper? Which sample contains the larger amount of internal energy at the end of the process?

▮ Heat is the energy that is transferred between a system and its surroundings during a change in the state of the system. It is transferred as a result of a difference in temperature between the system and its surroundings. Unless work is done, heat transfer will be directed from the point of higher temperature to the point of lower temperature. For this type of calculation, it does not matter whether one sample of copper is considered the system and the second is considered the thermal reservoir in the surroundings, or one sample is considered to transfer heat to a thermal reservoir that in turn transfers the heat to the second copper sample, because the energy will be conserved,

$$\sum q_i = 0 \tag{3.46}$$

In this case, letting T be the final temperature of both samples, substituting the data of Table 3-3 into *(3.23)* gives

$$(1.00\ \text{mol})(24.435\ \text{J} \cdot \text{K}^{-1} \cdot \text{mol}^{-1})(T - 298.2\ \text{K}) + (1.00\ \text{mol})(24.435\ \text{J} \cdot \text{K}^{-1} \cdot \text{mol}^{-1})(T - 348.2\ \text{K}) = 0$$

$$T = 323.2\ \text{K} = 50.0\ ^\circ\text{C}$$

Both samples contain the same amount of internal energy because they are of the same chemical composition, are at the same temperature, and contain the same amount of substance.

3.93 Repeat the calculations of Problem 3.92 assuming that the warmer sample contains 2.00 mol of Cu.

▮ Using *(3.23)* and *(3.46)* gives

$$(1.00\ \text{mol})(24.435\ \text{J} \cdot \text{K}^{-1} \cdot \text{mol}^{-1})(T - 298.2\ \text{K}) + (2.00\ \text{mol})(24.435\ \text{J} \cdot \text{K}^{-1} \cdot \text{mol}^{-1})(T - 348.2\ \text{K}) = 0$$

$$T = 321.5\ \text{K} = 58.4\ ^\circ\text{C}$$

Even though the larger sample contains twice as much internal energy as the smaller sample, additional energy is not transferred because there is no temperature difference between the samples.

3.94 A 35.41-g sample of Cu(s) at 96.73 °C was added to a calorimeter containing 148.3 g of $H_2O(l)$ at 25.20 °C. The final temperature of the system was 26.66 °C. Assuming no heat loss to the surroundings, what is the calorimeter constant?

▮ Letting ΔH represent the heat absorbed by the calorimeter and its contents and using the data from Table 3-3, *(3.23)*, and *(3.46)* gives

$$\Delta H + (35.41 \text{ g Cu}) \frac{1 \text{ mol}}{63.546 \text{ g Cu}} (24.435 \text{ J} \cdot \text{K}^{-1} \cdot \text{mol}^{-1})(299.81 \text{ K} - 369.88 \text{ K}) = 0$$

$$\Delta H = 954.1 \text{ J}$$

Substituting into *(3.26)* gives

$$C_P(\text{calorimeter}) = \frac{954.1 \text{ J}}{299.81 \text{ K} - 298.35 \text{ K}} - (148.3 \text{ g } H_2O) \frac{1 \text{ mol}}{18.02 \text{ g } H_2O} (75.291 \text{ J} \cdot \text{K}^{-1} \cdot \text{mol}^{-1})$$

$$= 34 \text{ J} \cdot \text{K}^{-1}$$

3.95 A thermodynamics system receives 14.6 kJ of electrical energy in the form of work, produces 2.6 kJ of expansion work, and dissipates 10.9 kJ of heat to the surroundings. What is the internal energy change of the system? In what ways might this internal energy change be compensated for by the system?

▮ The relation among ΔU, q, and w is given by the *first law of thermodynamics* as

$$\Delta U = q + w \qquad (3.47)$$

For the process described, $q = -10.9 \text{ kJ}$ and $w = 14.6 \text{ kJ} + (-2.6 \text{ kJ}) = 12.0 \text{ kJ}$, giving

$$\Delta U = -10.9 \text{ kJ} + 12.0 \text{ kJ} = 1.1 \text{ kJ}$$

This increase in internal energy of the system could be in the form of a temperature increase, a phase transformation, and/or a chemical change.

3.96 Although q and w are path functions, ΔU is a state function. Show that this is true for the cyclic process described in Problems 3.77, 3.78, and 3.89.

▮ The total work for the cyclic process was $w = 8.9 \times 10^6 \text{ J}$, and the total heat was $q = -8.9 \times 10^6 \text{ J}$. Using *(3.47)* gives

$$\Delta U = -8.9 \times 10^6 \text{ J} + 8.9 \times 10^6 \text{ J} = 0$$

In general, $\oint dX = 0$ for a state function.

3.97 The stretched spring described in Problem 3.75 is placed in 100.0 g of water and released. What is the temperature increase of the water?

▮ For this process, $q = 0$ and *(3.47)* gives $\Delta U = w = 1.0 \times 10^{-3} \text{ J}$. Substituting the data of Table 3-3 into *(3.18)* and solving for ΔT gives

$$\Delta T = \frac{\Delta U}{nC_{V,\text{m}}} = \frac{1.0 \times 10^{-3} \text{ J}}{(100.0 \text{ g})[(1 \text{ mol})/(18.02 \text{ g})](74.7 \text{ J} \cdot \text{K}^{-1} \cdot \text{mol}^{-1})} = 2.4 \times 10^{-6} \text{ K}$$

3.98 For the expansion process described in Problem 3.78, $q = 0$. The system was restored to the original conditions by a process in which $q = 4.2 \times 10^5 \text{ J}$. How much work was needed to complete the process?

▮ For a cyclic process, $\Delta U = 0$. Using *(3.47)* gives

$$0 = (0 + 4.2 \times 10^5 \text{ J}) + (-5.885 \times 10^5 \text{ J} + w)$$

$$w = 1.7 \times 10^5 \text{ J}$$

3.99 Water dripping out of a water tank increased in temperature by 0.25 °C as a result of falling to the ground. How high off the ground is the water tank?

▮ For this process, $q = 0$, and w will be given by mgl (see Table 3-5). Using *(3.18)* and *(3.47)* gives

$$nC_{V,\text{m}} \Delta T = 0 + mgl$$

Solving for l and substituting the data for 1 mol of H_2O gives

$$l = \frac{(1.00\text{ mol})(74.7\text{ J}\cdot\text{K}^{-1}\cdot\text{mol}^{-1})(0.25\text{ K})[(1\text{ kg}\cdot\text{m}^2\cdot\text{s}^{-2})/(1\text{ J})]}{(18.02\times10^{-3}\text{ kg})(9.8\text{ m}\cdot\text{s}^{-2})} = 106\text{ m}$$

3.100 A "1/4-hp" electric motor uses 187 W of electrical energy while delivering 35 J of work each second. How much energy must be dissipated in the form of friction (heat)?

▮ For this process, $w = 35\text{ J}$ and

$$\Delta U = (187\text{ W})[(1\text{ J}\cdot\text{s}^{-1})/(1\text{ W})](1\text{ s}) = 187\text{ J}$$

Using *(3.47)* gives

$$q = \Delta U - w = 187\text{ J} - 35\text{ J} = 152\text{ J}$$

3.101 Consider a system consisting of 1 mol of a diatomic gas contained in a piston. What is the temperature change of the gas if $q = 50.0\text{ J}$ and $w = -100.0\text{ J}$?

▮ For this process, *(3.47)* gives

$$\Delta U = 50.0\text{ J} + (-100.0\text{ J}) = -50.0\text{ J}$$

Assuming $C_{V,m} = \frac{5}{2}R$, *(3.18)* gives

$$\Delta T = \frac{\Delta U}{\frac{5}{2}R} = \frac{-50.0\text{ J}}{\frac{5}{2}(8.314\text{ J}\cdot\text{K}^{-1}\cdot\text{mol}^{-1})} = -3.61\text{ K}$$

The gas will cool by 3.61 °C.

3.102 In general, ΔU for any process can be calculated using

$$\Delta U = \int_{T_1}^{T_2} C_v\,dT + \int_{V_1}^{V_2}\left(\frac{\partial U}{\partial V}\right)_T dV \qquad (3.48)$$

Calculate ΔU for the heating and expansion of 1.000 mol of O_2(g) from 25.0 °C, 5.00 L, 4.93 bar to 125.0 °C, 6.75 L, and 4.90 bar. Assume that the O_2 is an ideal gas.

▮ For an ideal gas, $(\partial U/\partial V)_T = 0$. Substituting the data from Table 3-3 into *(3.48)* gives

$$\begin{aligned}\Delta U &= nC_{V,m}\,\Delta T\\ &= (1.000\text{ mol})(21.041\text{ J}\cdot\text{K}^{-1}\cdot\text{mol}^{-1})(398.2\text{ K} - 298.2\text{ K})\\ &= 2104\text{ J}\end{aligned}$$

3.103 For a van der Waals gas, $(\partial U/\partial V)_T = an^2/V^2$. Repeat the calculations of Problem 3.102 assuming that the O_2 obeys the van der Waals equation.

▮ The second integral in *(3.48)* for a van der Waals gas becomes

$$\int_{V_1}^{V_2}\left(\frac{\partial U}{\partial V}\right)_T dV = \int_{V_1}^{V_2}\frac{an^2}{V^2}\,dV = -an^2\left(\frac{1}{V_2} - \frac{1}{V_1}\right)$$

Thus *(3.48)* gives

$$\begin{aligned}\Delta U &= nC_{V,m}\,\Delta T - an^2\left(\frac{1}{V_2} - \frac{1}{V_1}\right)\\ &= 2104\text{ J} - (1.380\text{ L}^2\cdot\text{bar}\cdot\text{mol}^{-2})(1.000\text{ mol})^2\left(\frac{10^2\text{ J}}{1\text{ L}\cdot\text{bar}}\right)\left(\frac{1}{6.75\text{ L}} - \frac{1}{5.00\text{ L}}\right)\\ &= 2104\text{ J} - 7\text{ J} = 2097\text{ J}\end{aligned}$$

3.104 For condensed states of matter (e.g., solids, liquids, and aqueous solutions), $(\partial U/\partial V)_T = (\alpha/\kappa)T - P$, which is often approximated by $(\partial U/\partial V)_T \approx -P$. Find ΔU for heating 1.000 mol of liquid water from 25.0 °C, 1.00 bar, and 18.07 mL to 75.0 °C, 2.00 bar, and 18.48 mL.

▮ The second integral in *(3.48)* for this approximation becomes

$$\int_{V_1}^{V_2}\left(\frac{\partial U}{\partial V}\right)_T dV = \int_{V_1}^{V_2} -P\,dV \approx -\bar{P}\,\Delta V$$

Thus *(3.48)* gives

$$\begin{aligned}\Delta U &= nC_{V,\mathrm{m}}\,\Delta T - n\bar{P}\,\Delta V_{\mathrm{m}}\\ &= (1.000\ \mathrm{mol})(74.7\ \mathrm{J\cdot K^{-1}\cdot mol^{-1}})(348.2\ \mathrm{K} - 298.2\ \mathrm{K})\\ &\quad -(1.000\ \mathrm{mol})(1.50\ \mathrm{bar})(18.48\ \mathrm{mL\cdot mol^{-1}} - 18.07\ \mathrm{mL\cdot mol^{-1}})\frac{10^{-3}\ \mathrm{L}}{1\ \mathrm{mL}}\left(\frac{10^{2}\ \mathrm{J}}{1\ \mathrm{L\cdot bar}}\right)\\ &= 3740\ \mathrm{J} - 0.06\ \mathrm{J} = 3740\ \mathrm{J}\end{aligned}$$

3.105 In general, ΔH for any process can be calculated using

$$\Delta H = \int_{T_1}^{T_2} C_P\,dT + \int_{P_1}^{P_2}\left(\frac{\partial H}{\partial P}\right)_T dP \tag{3.49}$$

Calculate ΔH for the process described in Problem 3.102 assuming that O_2 is an ideal gas.

▌ For an ideal gas, $(\partial H/\partial P)_T = 0$. Substituting the data from Table 3-3 into *(3.49)* gives

$$\begin{aligned}\Delta H &= nC_{V,\mathrm{m}}\,\Delta T = (1.000\ \mathrm{mol})(29.355\ \mathrm{J\cdot K^{-1}\cdot mol^{-1}})(398.2\ \mathrm{K} - 298.2\ \mathrm{K})\\ &= 2936\ \mathrm{J}\end{aligned}$$

3.106 For a van der Waals gas, $(\partial H/\partial P)_T = n(b - 2a/RT)$. Repeat the calculations of Problem 3.105.

▌ The second integral in *(3.49)* for a van der Waals gas becomes

$$\int_{P_1}^{P_2}\left(\frac{\partial H}{\partial P}\right)_T dP = \int_{P_1}^{P_2} n\left(b - \frac{2a}{RT}\right)dP \approx n\left(b - \frac{2a}{RT}\right)\Delta P$$

Thus *(3.49)* gives

$$\begin{aligned}\Delta H &= nC_{P,\mathrm{m}}\,\Delta T + n\left(b - \frac{2a}{RT}\right)\Delta P\\ &= 2936\ \mathrm{J} + (1.000\ \mathrm{mol})\\ &\quad\times\left[0.0319\ \mathrm{L\cdot mol^{-1}} - \frac{2(1.380\ \mathrm{L^2\cdot bar\cdot mol^{-2}})}{(0.083\,14\ \mathrm{L\cdot bar\cdot K^{-1}\cdot mol^{-1}})(348.2\ \mathrm{K})}\right]\\ &\quad\times(4.90\ \mathrm{bar} - 4.93\ \mathrm{bar})[(10^2\ \mathrm{J})/(1\ \mathrm{L\cdot bar})]\\ &= 2936\ \mathrm{J} + 0.19\ \mathrm{J} = 2936\ \mathrm{J}\end{aligned}$$

3.107 Often the second integral in *(3.49)* for a real gas is approximated by $(\partial H/\partial P)_T = -C_P\mu$, where μ is the Joule–Thomson coefficient. Given $\bar{\mu} = 0.25\ \mathrm{K\cdot bar^{-1}}$ for $O_2(g)$, repeat the calculations of Problem 3.105.

▌ The second integral in *(3.49)* becomes

$$\int_{P_1}^{P_2}\left(\frac{\partial H}{\partial P}\right)_T dP = \int_{P_1}^{P_2}(-C_P\mu)\,dP \approx -\overline{C_P\mu}\,\Delta P$$

Thus *(3.49)* gives

$$\begin{aligned}\Delta H &= nC_{P,\mathrm{m}}\,\Delta T - n\overline{C_{P,\mathrm{m}}}\bar{\mu}\,\Delta P\\ &= 2936\ \mathrm{J} - (1.000\ \mathrm{mol})(29.355\ \mathrm{J\cdot K^{-1}\cdot mol^{-1}})(0.25\ \mathrm{K\cdot bar^{-1}})(4.90\ \mathrm{bar} - 4.93\ \mathrm{bar})\\ &= 2936\ \mathrm{J} + 0.22\ \mathrm{J} = 2936\ \mathrm{J}\end{aligned}$$

3.108 For condensed states of matter, $(\partial H/\partial P)_T = V(1 - \alpha T)$, which is often approximated by $(\partial H/\partial P)_T \approx V$. Find ΔH for the process described in Problem 3.104. For $H_2O(l)$, $C_{P,\mathrm{m}} = 75.291\ \mathrm{J\cdot K^{-1}\cdot mol^{-1}}$.

▌ The second integral in *(3.49)* for this approximation becomes

$$\int_{P_1}^{P_2}\left(\frac{\partial H}{\partial P}\right)_T dP = \int_{P_1}^{P_2} V\,dP \approx \bar{V}\,\Delta P$$

Thus *(3.49)* gives

$$\begin{aligned}\Delta H &= nC_{P,\mathrm{m}}\,\Delta T + n\bar{V}_{\mathrm{m}}\,\Delta P\\ &= (1.000\ \mathrm{mol})(75.291\ \mathrm{J\cdot K^{-1}\cdot mol^{-1}})(348.2\ \mathrm{K} - 298.2\ \mathrm{K})\\ &\quad +(1.000\ \mathrm{mol})(18.28\ \mathrm{mL\cdot mol^{-1}})(2.00\ \mathrm{bar} - 1.00\ \mathrm{bar})\frac{10^{-3}\ \mathrm{L}}{1\ \mathrm{mL}}\left(\frac{10^{2}\ \mathrm{J}}{1\ \mathrm{L\cdot bar}}\right)\\ &= 3760\ \mathrm{J} + 1.8\ \mathrm{J} = 3760\ \mathrm{J}\end{aligned}$$

3.109 In older thermodynamic tables, ΔH values were listed for 1 atm and 20 °C, and in newer tables the values are listed for 1 bar and 25 °C. Assuming typical values of $C_{P,\mathrm{m}} = 50\ \mathrm{J \cdot K^{-1} \cdot mol^{-1}}$ and $V_\mathrm{m} = 25\ \mathrm{L \cdot mol^{-1}}$, calculate a typical correction factor that can be applied to the older values to generate newer ones for ideal gases. Which contribution is more important, the pressure change or the temperature change? If a typical value of ΔH in the older table is $200\ \mathrm{kJ \cdot mol^{-1}}$, how important is the correction factor?

▌ Assuming ideal gas behavior, the second integral in *(3.49)* is zero, and therefore the pressure change is not important. The correction factor will be

$$\Delta H = nC_{P,\mathrm{m}}\,\Delta T = (1\ \mathrm{mol})(50\ \mathrm{J \cdot K^{-1} \cdot mol^{-1}})(5\ \mathrm{K}) = 250\ \mathrm{J}$$

This correction factor is only 0.1% of a typical value of ΔH.

3.110 Repeat the calculations for Problem 3.109 for a solid using typical values of $C_{P,\mathrm{m}} = 50\ \mathrm{J \cdot K^{-1} \cdot mol^{-1}}$ and $V_\mathrm{m} = 25\ \mathrm{mL \cdot mol^{-1}}$.

▌ The correction factor (see Problem 3.108) will be

$$\begin{aligned}\Delta H &= nC_{P,\mathrm{m}}\,\Delta T + nV_\mathrm{m}\,\Delta P \\ &= (1\ \mathrm{mol})(50\ \mathrm{J \cdot K^{-1} \cdot mol^{-1}})(5\ \mathrm{K}) + (1\ \mathrm{mol})(25\ \mathrm{mL \cdot mol^{-1}})(1\ \mathrm{bar} - 1.013\,25\ \mathrm{bar})\frac{10^{-3}\ \mathrm{L}}{1\ \mathrm{mL}}\left(\frac{10^2\ \mathrm{J}}{1\ \mathrm{L \cdot bar}}\right) \\ &= 250\ \mathrm{J} + 0.025\ \mathrm{J} = 250\ \mathrm{J}\end{aligned}$$

The temperature change generates the greater portion of the correction factor, and the correction factor is only about 0.1% of the typical value of ΔH.

3.4 SPECIFIC APPLICATIONS OF THE FIRST LAW

3.111 Consider a 2-mol sample of an ideal diatomic gas undergoing a reversible isothermal expansion at 298 K from 1.00 bar and 49.6 L to 0.66 bar and 75.0 L. Determine q, w, ΔU, and ΔH for this process.

▌ For a *reversible isothermal expansion of an ideal gas*, $(\partial U/\partial V)_T = (\partial H/\partial P)_T = 0$. Thus *(3.48)* and *(3.49)* give

$$\Delta U = \Delta H = 0 \qquad (3.50a)$$

From *(3.47)*,

$$q = \Delta U - w = -w \qquad (3.50b)$$

and from *(3.42)*,

$$w = -nRT\ln\frac{V_2}{V_1} = nRT\ln\frac{P_2}{P_1} \qquad (3.50c)$$

$$w = -(2.00\ \mathrm{mol})(8.314\ \mathrm{J \cdot K^{-1} \cdot mol^{-1}})(298\ \mathrm{K})\ln\frac{75.0\ \mathrm{L}}{49.6\ \mathrm{L}} = -2050\ \mathrm{J}$$

3.112 Repeat the calculations of Problem 3.111 for a system at 185 K. How does the temperature influence the results?

▌ Using *(3.50a)* and *(3.50b)* gives

$$\Delta U = \Delta H = 0 \qquad q = -w$$

From *(3.50c)*,

$$w = -(2.00\ \mathrm{mol})(8.314\ \mathrm{J \cdot K^{-1} \cdot mol^{-1}})(185\ \mathrm{K})\ln\frac{75.0\ \mathrm{L}}{49.6\ \mathrm{L}} = -1270\ \mathrm{J}$$

Only the value of the work changes.

3.113 During a reversible isothermal compression of 1.00 mol of an ideal gas originally at 0 °C, 0.525 bar, and 43.2 L, 975 J of heat was removed. What is the final pressure of the gas?

▌ From *(3.50b)*, $w = -q = -(-975\ \mathrm{J}) = 975\ \mathrm{J}$. Substituting into *(3.50c)* and solving for P_2 gives

$$(1.00\ \mathrm{mol})(8.314\ \mathrm{J \cdot K^{-1} \cdot mol^{-1}})(273\ \mathrm{K})\ln\frac{P_2}{0.525\ \mathrm{bar}} = 975\ \mathrm{J}$$

$$P_2 = 0.807\ \mathrm{bar}$$

3.114 Assume that the gas in the system described in Problem 3.111 is O_2 and that it obeys the van der Waals equation of state. Repeat the calculations and compare the results.

▮ For a *reversible isothermal expansion of a van der Waals gas,*

$$\Delta U = -an^2\left(\frac{1}{V_2} - \frac{1}{V_1}\right) \tag{3.51a}$$

$$\Delta U = -(1.380\ \text{L}^2\cdot\text{bar}\cdot\text{mol}^{-2})(2.00\ \text{mol})^2\frac{10^2\ \text{J}}{1\ \text{L}\cdot\text{bar}}\left(\frac{1}{75.02} - \frac{1}{49.6\ \text{L}}\right) = 3.8\ \text{J}$$

$$\Delta H = n(b - 2a/RT)\,\Delta P \tag{3.51b}$$

$$\Delta H = (2.00\ \text{mol})\left(0.0319\ \text{J}\cdot\text{mol}^{-1} - \frac{2(1.380\ \text{L}^2\cdot\text{bar}\cdot\text{mol}^{-2})}{(0.083\,14\ \text{L}\cdot\text{bar}\cdot\text{K}^{-1}\cdot\text{mol}^{-1})(298\ \text{K})}\right)$$
$$\times(0.66\ \text{bar} - 1.00\ \text{bar})[(10^2\ \text{J})/(1\ \text{L}\cdot\text{bar})]$$
$$= 5.4\ \text{J}$$

$$w = -n\left[RT\ln\left(\frac{V_2 - nb}{V_1 - nb}\right) + an\left(\frac{1}{V_2} - \frac{1}{V_1}\right)\right] \tag{3.51c}$$

$$w = -(2.00\ \text{mol})\left[(8.314\ \text{J}\cdot\text{K}^{-1}\cdot\text{mol}^{-1})(298\ \text{K})\ln\frac{75.0\ \text{L} - (2.00\ \text{mol})(0.0319\ \text{L}\cdot\text{mol}^{-1})}{49.6\ \text{L} - (2.00\ \text{mol})(0.0319\ \text{L}\cdot\text{mol}^{-1})}\right.$$
$$\left.+(1.380\ \text{L}^2\cdot\text{bar}\cdot\text{mol}^{-2})(2.00\ \text{mol})\left(\frac{1}{75.0\ \text{L}} - \frac{1}{49.6\ \text{L}}\right)\right]$$
$$= -2050\ \text{J}$$

$$q = nRT\ln\frac{V_2 - nb}{V_1 - nb} \tag{3.51d}$$

$$q = (2.00\ \text{mol})(8.314\ \text{J}\cdot\text{K}^{-1}\cdot\text{mol}^{-1})(298\ \text{K})\ln\frac{75.0\ \text{L} - (2.00\ \text{mol})(0.0319\ \text{L}\cdot\text{mol}^{-1})}{49.6\ \text{L} - (2.00\ \text{mol})(0.0319\ \text{L}\cdot\text{mol}^{-1})}$$
$$= 2050\ \text{J}$$

For a van der Waals gas, the values of ΔU and ΔH are no longer zero, but slightly positive. The changes in q and w in this case are not significant.

3.115 Consider the expansion described in Problem 3.111 to be performed irreversibly against a constant external pressure of 0.50 bar. Determine q, w, ΔU, and ΔH for this process. Compare the results for the two processes.

▮ For *isothermal isobaric expansion of an ideal gas,* $(\partial U/\partial V)_T = (\partial H/\partial P)_T = 0$. Thus *(3.48)* and *(3.49)* give

$$\Delta U = \Delta H = 0 \tag{3.52a}$$

From *(3.47)*,

$$q = \Delta U - w = -w \tag{3.52b}$$

and from *(3.41a)*,

$$w = -P_{\text{ext}}\,\Delta V = -P_{\text{ext}}\,nRT\left(\frac{1}{P_2} - \frac{1}{P_1}\right) \tag{3.52c}$$

$$w = -(0.50\ \text{bar})(75.0\ \text{L} - 49.6\ \text{L})\frac{10^2\ \text{J}}{1\ \text{L}\cdot\text{bar}} = -1300\ \text{J}$$

There is no change in the values of the state functions, ΔU and ΔH, for the system. However, only 62% as much work was generated by the irreversible process.

3.116 In an attempt to decrease the amount of work lost by changing from a reversible process to an irreversible process, the process described in Problem 3.115 was modified by **(*a*)** increasing $C_{V,\text{m}}$ for the gas (by changing gases), **(*b*)** increasing P_{ext}, and **(*c*)** increasing the temperature. Which of these changes will increase the work output?

▮ **(*a*)** None of the calculations given by *(3.52)* depend on $C_{V,\text{m}}$, so this modification made no change. **(*b*)** The value of w is directly proportional to P_{ext}, see *(3.52c)*, and so the value of the work produced will increase. However, the greatest pressure permitted for P_{ext} will be 0.66 bar. **(*c*)** The value of w is directly proportional to T, see *(3.52c)*, and so the value of the work produced will increase.

3.117 During an isothermal isobaric compression of 1.00 mol of an ideal gas originally at 0 °C, 0.525 bar, and 43.2 L, 975 J of heat was removed. What is the final pressure of the gas? Assume that the external pressure on the system is 2.00 bar.

▮ From *(3.52b)*, $w = -q = -(-975\ \text{J}) = 975\ \text{J}$. Substituting into *(3.52c)* and solving for P_2 gives

$$-(2.00\ \text{bar})(1.00\ \text{mol})(8.314\ \text{J}\cdot\text{K}^{-1}\cdot\text{mol}^{-1})(273\ \text{K})\left(\frac{1}{P_2} - \frac{1}{0.525\ \text{bar}}\right) = 975\ \text{J}$$

$$P_2 = 0.592\ \text{bar}$$

3.118 A 10.0-g sample of $Cl_2(g)$ was heated under constant-volume conditions from 25.0 °C to 500.0 °C. Calculate q, w, ΔU, and ΔH for this process. Assume ideal gas behavior.

▮ For an *isochoric heating of an ideal gas*, $(\partial U/\partial V)_T = (\partial H/\partial P)_T = 0$ and $dV = 0$. Thus,

$$w = 0 \qquad (3.53a)$$

$$q = \Delta U = \int_{T_1}^{T_2} C_V\, dT \qquad (3.53b)$$

$$\begin{aligned} q &= (10.0\ \text{g})[(1\ \text{mol})/(70.906\ \text{g})] \\ &\quad \times \int_{298.2\ \text{K}}^{773.2\ \text{K}} [(22.970\ \text{J}\cdot\text{K}^{-1}\cdot\text{mol}^{-1}) + (10.144\times10^{-3}\ \text{J}\cdot\text{K}^{-2}\cdot\text{mol}^{-1})T \\ &\quad -(40.38\times10^{-7}\ \text{J}\cdot\text{K}^{-3}\cdot\text{mol}^{-1})T^2]\, dT \\ &= (10.0\ \text{g})\frac{1\ \text{mol}}{70.906\ \text{g}}\{(22.970\ \text{J}\cdot\text{K}^{-1}\cdot\text{mol}^{-1})(773.2\ \text{K} - 298.2\ \text{K}) \\ &\quad +\tfrac{1}{2}(10.144\times10^{-3}\ \text{J}\cdot\text{K}^{-1}\cdot\text{mol}^{-1})[(773.2\ \text{K})^2 - (298.2\ \text{K})^2] \\ &\quad -\tfrac{1}{3}(40.38\times10^{-7}\ \text{J}\cdot\text{K}^{-3}\cdot\text{mol}^{-1})[(773.2\ \text{K})^3 - (298.2\ \text{K})^3]\} \\ &= 1820\ \text{J} \end{aligned}$$

$$\Delta H = \int_{T_1}^{T_2} C_P\, dT = \Delta U + V\,\Delta P \qquad (3.53c)$$

$$\begin{aligned} \Delta H &= (10.0)\frac{1}{70.906}\int_{298.2\ \text{K}}^{773.2\ \text{K}} [31.284 + (10.144\times10^{-3})T - (40.38\times10^{-7})T^2]\, dT \\ &= (10.0)\frac{1}{70.906} \\ &\quad \times\{(31.284)(773.2 - 298.2) + \tfrac{1}{2}(10.144\times10^{-3})[(773.2)^2 - (298.2)^2] \\ &\quad -\tfrac{1}{3}(40.38\times10^{-7})[(773.2)^3 - (298.2)^3]\} \\ &= 2380\ \text{J} \end{aligned}$$

3.119 If the volume of the system described in Problem 3.118 is 2.00 L, find the change in the pressure of the system.

▮ Solving *(3.53c)* for ΔP and substituting the data gives

$$\Delta P = \frac{\Delta H - \Delta U}{V} = \frac{(2380\ \text{J} - 1820\ \text{J})[(1\ \text{L}\cdot\text{bar})/(10^2\ \text{J})]}{2.00\ \text{L}} = 2.8\ \text{bar}$$

3.120 A 2.00-mol sample of an ideal gas is heated from 1.03 bar, 41.2 L, and 255 K to 1.25 bar, 42.2 L, and 317 K against a constant external pressure of 1.00 bar. What are ΔU, ΔH, q, and w for this process? Assume $C_{P,\text{m}} = 7R/2$.

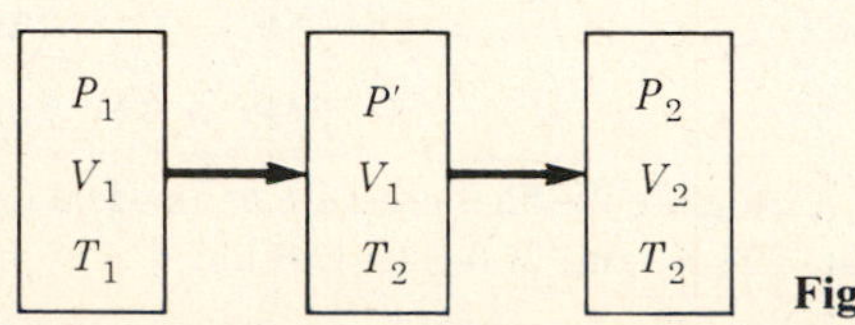

Fig. 3-10

▮ For an *isobaric heating of an ideal gas*, the process is considered to occur in two steps: an isochoric heating and an isobaric isothermal expansion (see Fig. 3-10). Combining *(3.53)* and *(3.52)* gives

$$\Delta U = \int_{T_1}^{T_2} C_V\,dT + 0 = \int_{T_1}^{T_2} C_V\,dT \tag{3.54a}$$

$$\Delta U = (2.00\ \text{mol})(\tfrac{5}{2})(8.314\ \text{J}\cdot\text{K}^{-1}\cdot\text{mol}^{-1})(317\ \text{K} - 255\ \text{K}) = 2580\ \text{J}$$

$$\Delta H = \int_{T_1}^{T_2} C_P\,dT + 0 = \int_{T_1}^{T_2} C_P\,dT \tag{3.54b}$$

$$\Delta H = (2.00)(\tfrac{7}{2})(8.314)(317 - 255) = 3610\ \text{J}$$

$$w = 0 + (-P_{\text{ext}}\,\Delta V) = -P_{\text{ext}}\,\Delta V \tag{3.54c}$$

$$w = -(1.00\ \text{bar})(42.2\ \text{L} - 41.2\ \text{L})\left(\frac{10^2\ \text{J}}{1\ \text{L}\cdot\text{bar}}\right) = -100\ \text{J}$$

$$q = \int_{T_1}^{T_2} C_V\,dT + P_{\text{ext}}\,\Delta V \tag{3.54d}$$

$$q = 2580\ \text{J} + 100\ \text{J} = 2680\ \text{J}$$

3.121 A 1.00-g sample of ice melted at 0 °C to form liquid water. The heat of fusion of ice is 6.0095 kJ · mol^{-1}. Given $\rho = 0.9168\ \text{g}\cdot\text{cm}^{-3}$ for ice and $0.999\,842\,5\ \text{g}\cdot\text{cm}^{-3}$ for liquid water, calculate ΔU, ΔH, q, and w for this process. Assume $P_{\text{ext}} = 1.00$ atm.

For an *isothermal isobaric phase change,*

$$q = \Delta H \tag{3.55a}$$

$$q = (1.00\ \text{g})\frac{1\ \text{mol}}{18.02\ \text{g}}(6.0095\ \text{kJ}\cdot\text{mol}^{-1})\frac{10^3\ \text{J}}{1\ \text{kJ}} = 333\ \text{J}$$

$$w = -P_{\text{ext}}\,\Delta V \tag{3.55b}$$

$$w = -(1.00\ \text{atm})\left(\frac{1.00\ \text{g}}{0.999\,842\,5\ \text{g}\cdot\text{cm}^{-3}} - \frac{1.00\ \text{g}}{0.9168\ \text{g}\cdot\text{cm}^{-3}}\right)\left(\frac{10^{-3}\ \text{L}}{1\ \text{cm}^3}\right)\left(\frac{101.325\ \text{J}}{1\ \text{L}\cdot\text{atm}}\right)$$
$$= 9.18 \times 10^{-3}\ \text{J}$$

$$\Delta U = q + w \tag{3.55c}$$

$$\Delta U = 333\ \text{J} + 0.009\,18\ \text{J} = 333\ \text{J}$$

3.122 Repeat the calculations of Problem 3.121 for the evaporation of 1.00 g H_2O at 100 °C given that the heat of evaporation is 40.656 kJ · mol^{-1} and $V/(\text{ft}^3\cdot\text{lb}^{-1}) = 0.016\,719$ for liquid and 26.799 for steam.

Using *(3.55)* gives

$$q = \Delta H = (1.00\ \text{g})\frac{1\ \text{mol}}{18.02\ \text{g}}(40.656\ \text{kJ}\cdot\text{mol}^{-1}) = 2.26\ \text{kJ}$$

$$w = -(1.00\ \text{atm})(1.00\ \text{g})\frac{1\ \text{lb}}{454\ \text{g}}(26.799\ \text{ft}^3\cdot\text{lb}^{-1} - 0.016\,719\ \text{ft}^3\cdot\text{lb}^{-1})\left(\frac{28.3\ \text{L}}{1\ \text{ft}^3}\right)\left(\frac{101.325\ \text{J}}{1\ \text{L}\cdot\text{atm}}\right)$$
$$= -169\ \text{J}$$

$$\Delta U = 2.26\ \text{kJ} + (-169\ \text{J})\frac{10^{-3}\ \text{kJ}}{1\ \text{J}} = 2.09\ \text{kJ}$$

3.123 For a *reversible adiabatic expansion of an ideal gas,* the P-V-T relationships describing the final conditions with respect to the original conditions are

$$P_1 V_1^{\gamma} = P_2 V_2^{\gamma} \tag{3.56a}$$

$$T_1^{C_V/R} V_1 = T_2^{C_V/R} V_2 \tag{3.56b}$$

$$T_2^{C_P/R}/P_2 = T_1^{C_P/R}/P_1 \tag{3.56c}$$

where γ is given by *(3.32)*. Calculate the final pressure of an ideal diatomic gas that undergoes a reversible adiabatic compression from 1.00 bar, 298 K, and 25.0 L to 1.00 L.

For an ideal diatomic gas, $C_{P,\text{m}} = 7R/2$ and $C_{V,\text{m}} = 5R/2$, neglecting vibrational contributions. Using *(3.32)* gives

$$\gamma = (7R/2)/(5R/2) = 1.4$$

Solving *(3.56a)* for P_2 and substituting the data gives

$$P_2 = (1.00\ \text{bar})\frac{(25.0\ \text{L})^{1.4}}{(1.00\ \text{L})^{1.4}} = 90.6\ \text{bar}$$

3.124 Determine the temperature of the gas after the reversible adiabatic compression described in Problem 3.123.

▌ Substituting the data into *(3.56b)* and solving for T_2 gives

$$(298\ \text{K})^{(5R/2)/R}(25.0\ \text{L}) = T_2^{(5R/2)/R}(1.00\ \text{L})$$

$$T_2^{5/2} = (3.83 \times 10^7\ \text{K})^{5/2} \qquad T_2 = 1080\ \text{K}$$

3.125 A gas originally at 1.10 bar and 298 K underwent a reversible adiabatic expansion to 1.00 bar and 287 K. What is the molar heat capacity of the gas?

▌ The value of $C_{P,\text{m}}$ can be determined using *(3.56c)*:

$$(T_2/T_1)^{C_P/R} = P_2/P_1$$

$$C_P = R\frac{\ln(P_2/P_1)}{\ln(T_2/T_1)} = R\frac{\ln[(1.00\ \text{bar})/(1.10\ \text{bar})]}{\ln[(287\ \text{K})/(298\ \text{K})]} = 2.53R$$

$$= (2.53)(8.314\ \text{J}\cdot\text{K}^{-1}\cdot\text{mol}^{-1}) = 21.0\ \text{J}\cdot\text{K}^{-1}\cdot\text{mol}^{-1}$$

The gas probably consists of diatomic or linear polyatomic molecules.

3.126 Show that a plot of P–V data for a reversible adiabatic expansion of an ideal gas will have a greater value of slope $(\partial P/\partial V)$ than a plot for a reversible isothermal expansion of an ideal gas.

▌ For the reversible adiabatic process, *(3.56a)* gives $PV^\gamma = P_1V_1^\gamma = k$, where k is a constant. Taking the differential gives

$$\left(\frac{\partial P}{\partial V}\right)_q = \left(\frac{\partial(k/V^\gamma)}{\partial V}\right)_q = k(-\gamma)V^{-(\gamma+1)} = PV^\gamma(-\gamma)V^{-(\gamma+1)} = -\frac{\gamma P}{V}$$

Likewise, for the reversible isothermal process, $PV = P_1V_1 = k$, and

$$\left(\frac{\partial P}{\partial V}\right)_T = \left(\frac{\partial(k/V)}{\partial V}\right)_T = k(-1)V^{-2} = PV(-1)V^{-2} = -\frac{P}{V}$$

Taking a ratio of these results gives

$$\frac{(\partial P/\partial V)_q}{(\partial P/\partial V)_T} = \frac{-\gamma P/V}{-P/V} = \gamma$$

Recognizing that $\gamma > 1$ indicates that the slope of the plot of P versus V data for a reversible adiabatic process will "fall faster" than a plot of P versus V for a reversible isothermal process.

3.127 Calculations involving *(3.56)* assume that the heat capacity of the gas does not change as the temperature changes. A student realized that that assumption might lead to an error in determining the value of T_2 from given values of T_1, P_1, and P_2. To check his idea, he performed a calculation for the reversible adiabatic compression of 1.00 mol of $Cl_2(g)$ from 1.00 bar and 298 K to 10.00 bar using *(3.19)*. To simplify the calculations, he used an iterative process involving a $C_{P,\text{m}}$ value determined at the average temperature. Was his idea valid?

▌ Assuming a constant value of $C_{P,\text{m}}$ at 298 K,

$$C_{P,\text{m}}/(\text{J}\cdot\text{K}^{-1}\cdot\text{mol}^{-1}) = 31.284 + (10.144 \times 10^{-3})(298) - (40.38 \times 10^{-7})(298)^2$$

$$= 33.949$$

for the calculation using *(3.56c)* gives

$$T_2^{(33.949\ \text{J}\cdot\text{K}^{-1}\cdot\text{mol}^{-1})/(8.314\ \text{J}\cdot\text{K}^{-1}\cdot\text{mol}^{-1})} = (298\ \text{K})^{(33.949\ \text{J}\cdot\text{K}^{-1}\cdot\text{mol}^{-1})/(8.314\ \text{J}\cdot\text{K}^{-1}\cdot\text{mol}^{-1})}(10.00\ \text{bar})/(1.00\ \text{bar})$$

$$= 1.27 \times 10^{11}$$

$$T_2 = 524\ \text{K}$$

The average temperature is

$$\bar{T} = (198\ \text{K} + 524\ \text{K})/2 = 411\ \text{K}$$

giving

$$C_{P,\text{m}}/(\text{J}\cdot\text{K}^{-1}\cdot\text{mol}^{-1}) = 31.284 + (10.144\times10^{-3})(411) - (40.38\times10^{-7})(411)^2 = 34.771$$

Repeating the calculation using *(3.56c)* gives

$$T_2^{34.771/8.314} = (298)^{34.771/8.314}\frac{10.00}{1.00} = 2.22\times10^{11}$$

$$T_2 = 517\text{ K}$$

Repeating the calculation gives $\bar{T} = 408$ K, $C_{P,\text{m}} = 34.751\text{ J}\cdot\text{K}^{-1}\cdot\text{mol}^{-1}$, and $T_2 = 517$ K. The student was correct in that a small error, about 1%, existed.

3.128 The P-V data given in Problem 3.76 were for the reversible adiabatic expansion of 120 mol of a monatomic gas from 1.0 m^3, 5.00 bar, and 500 K to 10.0 m^3, 0.1069 bar, and 108 K. Calculate q, w, ΔU, and ΔH for this process.

▌ For a *reversible adiabatic expansion of an ideal gas,*

$$q = 0 \tag{3.57a}$$

$$w = \Delta U = \int_{T_1}^{T_2} C_V\,dT \tag{3.57b}$$

$$w = \Delta U = (120\text{ mol})(\tfrac{3}{2})(8.314\text{ J}\cdot\text{K}^{-1}\cdot\text{mol}^{-1})(108\text{ K} - 500\text{ K}) = -5.87\times10^5\text{ J}$$

$$\Delta H = \int_{T_1}^{T_2} C_P\,dT \tag{3.57c}$$

$$\Delta H = (120)(\tfrac{5}{2})(8.314)(108 - 500) = -9.78\times10^5\text{ J}$$

Note that the graphical solution for w determined in Problem 3.76 agrees well with the calculated value.

3.129 Repeat the calculations of Problem 3.128 for Ne(g), assuming that the gas obeys the van der Waals equation of state. Compare the results.

▌ For the *reversible adiabatic expansion of a van der Waals gas,* both *(3.56a-c)* and *(3.57a-c)* are slightly changed. The final temperature will be given by

$$T_2^{C_V/R}(V_2 - nb) = T_1^{C_V/R}(V_1 - nb) \tag{3.58a}$$

Substituting the data from Tables 1-3 and 3-3 gives

$$T_2^{(12.472\text{ J}\cdot\text{K}^{-1}\cdot\text{mol}^{-1})/(8.314\text{ J}\cdot\text{K}^{-1}\cdot\text{mol}^{-1})}\left(10.0\text{ m}^3 - (120\text{ mol})(0.017\,09\text{ J}\cdot\text{mol}^{-1})\frac{10^{-3}\text{ m}^3}{1\text{ L}}\right)$$

$$= (500\text{ K})^{(12.472\text{ J}\cdot\text{K}^{-1}\cdot\text{mol}^{-1})/(8.314\text{ J}\cdot\text{K}^{-1}\cdot\text{mol}^{-1})}\left(1.0\text{ m}^3 - (120\text{ mol})(0.017\,09\text{ J}\cdot\text{mol}^{-1})\frac{10^{-3}\text{ m}^3}{1\text{ L}}\right)$$

$$T_2 = 107\text{ K}$$

which corresponds to a small difference. For this adiabatic process,

$$q = 0 \tag{3.58b}$$

$$\Delta U = w = \int_{T_1}^{T_2} C_V\,dT - an^2\left(\frac{1}{V_2} - \frac{1}{V_1}\right) \tag{3.58c}$$

$$\Delta U = w = (120\text{ mol})(12.472\text{ J}\cdot\text{K}^{-1}\cdot\text{mol}^{-1})(107\text{ K} - 500\text{ K})$$
$$-(0.2135\text{ L}^2\cdot\text{bar}\cdot\text{mol}^{-2})(120\text{ mol})^2\left(\frac{1}{10.0\text{ m}^3} - \frac{1}{1.0\text{ m}^3}\right)\left(\frac{10^{-3}\text{ m}^3}{1\text{ L}}\right)\left(\frac{10^2\text{ J}}{1\text{ L}\cdot\text{bar}}\right)$$
$$= -5.88\times10^5\text{ J}$$

This also is a small change. The enthalpy change will be

$$\Delta H = \Delta U + n\left[\left(\frac{RT_2}{V_2 - nb} - \frac{an}{V_2^2}\right)V_2 - \left(\frac{RT_1}{V_1 - nb} - \frac{an}{V_1^2}\right)V_1\right] \tag{3.58d}$$

$$\Delta H = -5.88 \times 10^5\ \text{J} + (120\ \text{mol})$$
$$\times \left[\left(\frac{(0.083\,14\ \text{L}\cdot\text{bar}\cdot\text{K}^{-1}\cdot\text{mol}^{-1})(107\ \text{K})[(10^{-3}\ \text{m}^3)/(1\ \text{L})]}{10.0\ \text{m}^3 - (120\ \text{mol})(0.017\,09\ \text{L}\cdot\text{mol}^{-1})[(10^{-3}\ \text{m}^3)/(1\ \text{L})]}\right.\right.$$
$$\left. - \frac{(0.2135\ \text{L}^2\cdot\text{bar}\cdot\text{mol}^{-2})(120\ \text{mol})}{(10.0\ \text{m}^3)^2[(10^3\ \text{L})/(1\ \text{m}^3)]^2}\right)(10.0\ \text{m}^3)$$
$$- \left(\frac{(0.083\,14\ \text{L}\cdot\text{bar}\cdot\text{K}^{-1}\cdot\text{mol}^{-1})(500\ \text{K})[(10^{-3}\ \text{m}^3)/(1\ \text{L})]}{1.0\ \text{m}^3 - (120\ \text{mol})(0.017\,09\ \text{L}\cdot\text{mol}^{-1})[(10^{-3}\ \text{m}^3)/(1\ \text{L})]}\right.$$
$$\left.\left. - \frac{(0.2135\ \text{L}^2\cdot\text{bar}\cdot\text{mol}^{-2})(120\ \text{mol})}{(1.0\ \text{m}^3)^2[(10^3\ \text{L})/(1\ \text{m}^3)]^2}\right)(1.0\ \text{m}^3)\right]\left(\frac{10^3\ \text{L}}{1\ \text{m}^3}\right)\left(\frac{10^2\ \text{J}}{1\ \text{L}\cdot\text{bar}}\right)$$
$$= -9.81 \times 10^5\ \text{J}$$

This is a change of about 0.3%.

3.130 A 3.18-mol sample of an ideal gas underwent an adiabatic expansion from 298 K, 15.00 bar, and 5.25 L to 2.50 bar against a constant external pressure of 1.00 bar. What is the final temperature of the system? Assume $C_{V,\text{m}} = 5R/2$ for the gas.

▮ The P–V–T relations for the *isobaric adiabatic expansion of an ideal gas* are

$$T_2 = T_1 - \frac{P_{\text{ext}}\,\Delta V}{nC_{V,\text{m}}} \tag{3.59a}$$

and

$$T_2 = T_1\,\frac{C_{V,\text{m}} + P_{\text{ext}}R/P_1}{C_{V,\text{m}} + P_{\text{ext}}R/P_2} \tag{3.59b}$$

Substituting the data into *(3.59b)* gives

$$T_2 = (298\ \text{K})\,\frac{5R/2 + (1.00\ \text{bar})R/(15.00\ \text{bar})}{5R/2 + (1.00\ \text{bar})R/(2.50\ \text{bar})} = 264\ \text{K}$$

3.131 What was the final volume of the system described in Problem 3.130? How would the value of ΔV change as $C_{V,\text{m}}$ changes, keeping all other values constant?

▮ Solving *(3.59b)* for ΔV and substituting the data gives

$$\Delta V = \frac{-(3.18\ \text{mol})(\tfrac{5}{2})(8.314\ \text{J}\cdot\text{K}^{-1}\cdot\text{mol}^{-1})(264\ \text{K} - 298\ \text{K})}{(1.00\ \text{bar})[(10^2\ \text{J})/(1\ \text{L}\cdot\text{bar})]} = 22.5\ \text{L}$$

The final volume is

$$V_2 = \Delta V + V_1 = 22.5\ \text{L} + 5.25\ \text{L} = 27.8\ \text{L}$$

The value of ΔV is directly proportional to $C_{V,\text{m}}$, so for gases with higher values of $C_{V,\text{m}}$ (e.g., nonlinear polyatomic molecules), ΔV will be higher, and for gases with lower values of $C_{V,\text{m}}$ (e.g., monatomic molecules), ΔV will be less.

3.132 For the process described in Problem 3.130, calculate q, w, ΔU, and ΔH.

▮ For the *isobaric adiabatic expansion of an ideal gas,*

$$q = 0 \tag{3.60a}$$

$$w = -P_{\text{ext}}\,\Delta V \tag{3.60b}$$

$$w = -(1.00\ \text{bar})(22.5\ \text{L})[(10^2\ \text{J})/(1\ \text{L}\cdot\text{bar})] = -2250\ \text{J}$$

$$\Delta U = \int_{T_1}^{T_2} C_V\,dT = w \tag{3.60c}$$

$$\Delta U = (3.18\ \text{mol})(\tfrac{5}{2})(8.314\ \text{J}\cdot\text{K}^{-1}\cdot\text{mol}^{-1})(264\ \text{K} - 298\ \text{K}) = -2250\ \text{J}$$

$$\Delta H = \int_{T_1}^{T_2} C_P\,dT \tag{3.60d}$$

$$\Delta H = (3.18)(\tfrac{7}{2})(8.314)(264 - 298) = -3150\ \text{J}$$

3.133 The Clément–Désormes experiment consists of an open isobaric adiabatic expansion from P_1 to P_2 followed by an isochoric heating from P_2 to P_3. Use the following data to determine $C_{V,\mathrm{m}}$ for the gas studied: $P_1 = 763.1$ torr, $P_2 = 757.0$ torr, and $P_3 = 758.7$ torr.

▮ The $C_{V,\mathrm{m}}$–P relation is

$$C_{V,\mathrm{m}} = R\frac{P_2[1 - P_3/P_1]}{P_3 - P_2} \tag{3.61}$$

$$C_{V,\mathrm{m}} = (8.314\ \mathrm{J \cdot K^{-1} \cdot mol^{-1}})\frac{(757.0\ \mathrm{torr})[1 - (758.7\ \mathrm{torr})/(763.1\ \mathrm{torr})]}{758.7\ \mathrm{torr} - 757.0\ \mathrm{torr}}$$

$$= 21\ \mathrm{J \cdot K^{-1} \cdot mol^{-1}}$$

3.134 A sample of $N_2(g)$ underwent a *Joule–Thomson expansion* from 2.00 bar and 25.367 °C to 1.00 bar and 25.147 °C. Determine μ, the *Joule–Thomson coefficient* for $N_2(g)$ under these conditions.

▮ Assuming $dT = 25.147\ °\mathrm{C} - 25.367\ °\mathrm{C} = -0.220\ \mathrm{K}$ and $dP = 1.00\ \mathrm{bar} - 2.00\ \mathrm{bar} = -1.00\ \mathrm{bar}$, the value of μ is

$$\mu = \left(\frac{\partial T}{\partial P}\right)_H \tag{3.62}$$

$$\mu = \frac{-0.220\ \mathrm{K}}{-1.00\ \mathrm{bar}} = 0.220\ \mathrm{K \cdot bar^{-1}}$$

3.135 Nitrogen is to be cooled from 25 °C to −195 °C and 1.00 bar by a one-step process involving a Joule–Thomson expansion. Assuming a constant value of 0.75 K · bar^{-1}, what must be the initial pressure of the gas?

▮ The integral of *(3.62)* is

$$\Delta T = \int_{P_1}^{P_2} \mu\, dP = \mu(P_2 - P_1)$$

Solving for P_1 and substituting the data gives

$$P_1 = P_2 - \frac{\Delta T}{\mu} = 1.00\ \mathrm{bar} - \frac{78\ \mathrm{K} - 298\ \mathrm{K}}{0.75\ \mathrm{K \cdot bar^{-1}}} = 290\ \mathrm{bar}$$

3.136 For a van der Waals gas,

$$\mu = \frac{2a/RT - b}{C_{P,\mathrm{m}}} \tag{3.63}$$

Calculate μ for $N_2(g)$ at 25 °C. Compare these values to the one determined in Problem 3.134.

▮ Substituting the data from Tables 1-3 and 3-3 into *(3.63)* gives

$$\mu = \frac{\dfrac{2(1.364\ \mathrm{L^2 \cdot bar \cdot mol^{-2}})}{(0.083\,14\ \mathrm{L \cdot bar \cdot K^{-1} \cdot mol^{-1}})(298\ \mathrm{K})} - 0.039\,13\ \mathrm{L \cdot mol^{-1}}}{(29.125\ \mathrm{J \cdot K^{-1} \cdot mol^{-1}})[(1\ \mathrm{L \cdot bar})/(10^2\ \mathrm{J})]} = 0.244\ \mathrm{K \cdot bar^{-1}}$$

The two values agree to within 10%.

3.137 At the Joule–Thomson inversion temperature, $\mu = 0$. What does this mean? Determine the inversion temperature for N_2 assuming that the gas obeys the van der Waals equation of state.

▮ If $\mu = 0$, the gas will neither increase or decrease in temperature as it expands. Solving *(3.63)* for this temperature and substituting the data from Tables 1-3 and 3-3 gives

$$T = \frac{2a}{Rb} = \frac{2(1.364\ \mathrm{L^2 \cdot bar \cdot mol^{-2}})}{(0.083\,14\ \mathrm{L \cdot bar \cdot K^{-1} \cdot mol^{-1}})(0.039\,13\ \mathrm{L \cdot mol^{-1}})} = 839\ \mathrm{K}$$

3.138 Determine the value of the *isothermal Joule–Thomson coefficient*, $\phi = (\partial H/\partial P)_T$, for $N_2(g)$. See Problem 3.136 for additional data.

▮ Using the properties of partial derivations gives

$$\phi = \left(\frac{\partial H}{\partial P}\right)_T = \frac{-(\partial T/\partial P)_H}{(\partial T/\partial H)_P} = \left(\frac{\partial H}{\partial P}\right)_H \left(\frac{\partial H}{\partial T}\right)_P = -\mu C_P$$

$$= -(0.244\ \mathrm{K \cdot bar^{-1}})(29.125\ \mathrm{J \cdot K^{-1} \cdot mol^{-1}})[(1\ \mathrm{L \cdot bar})/(10^2\ \mathrm{J})]$$

$$= -0.071\ \mathrm{L \cdot mol^{-1}}$$

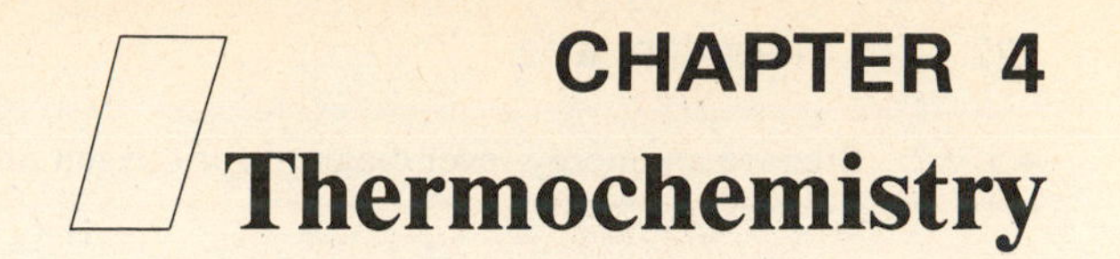

CHAPTER 4
Thermochemistry

4.1 HEAT OF REACTION

4.1 When 4 mol of solid $KClO_3$ is carefully heated, 3 mol of solid $KClO_4$ and 1 mol of solid KCl are formed. There is also a transfer of 144.08 kJ from the system (the reaction mixture) to the thermal reservoir in the surroundings under constant-pressure conditions. Write the thermochemical equation for this chemical reaction. Is this an endothermic or exothermic reaction?

▮ Because heat is transferred from the system to the surroundings (exothermic), $\Delta_r H = -144.08$ kJ. The thermochemical equation is

$$4KClO_3(s) \rightarrow 3KClO_3(s) + KCl(s) \qquad \Delta_r H = -144.08 \text{ kJ}$$

The r subscript refers to "reaction."

4.2 At 25 °C, 282.4 kJ is needed to produce 1 mol of gaseous NaF from solid NaF. Write the thermochemical equation for this process.

▮ Even though this is a physical change, the following thermochemical equation is used to describe the process:

$$NaF(s) \rightarrow NaF(g) \qquad \Delta_r H_{298} = 282.4 \text{ kJ}$$

The temperature is indicated as a subscript and is usually given in K. If no temperature is given, it is 298 K.

4.3 As 1 mol of solid NaCl is dissolved in 100 mol of liquid H_2O at 25 °C and 1 bar to produce a ~0.6 M solution, 4.087 kJ is absorbed from the surroundings. Write the thermochemical equation describing this process.

▮ The thermochemical equation used to describe this dissolution process is

$$NaCl(s) \xrightarrow{100\ H_2O} NaCl(\text{in } 100\ H_2O) \qquad \Delta_r H^\circ_{298} = 4.087 \text{ kJ}$$

The ° superscript refers to standard state conditions, 1 bar.

4.4 Prepare an energy-level diagram for the reaction

$$CH_4(g) + Cl_2(g) \rightarrow CH_3Cl(g) + HCl(g) \qquad \Delta_r H^\circ = -98.33 \text{ kJ}$$

▮ The abscissa (horizontal axis) will be labeled "Progress of reaction," which means that the formulas of the reactants will be given at the left and the formulas of the products at the right. The ordinate (vertical axis) will be labeled "Energy," and the relative energy levels of the reactants and products are shown by straight lines. Between these lines will be an indication of the enthalpy change. The diagram is shown in Fig. 4-1.

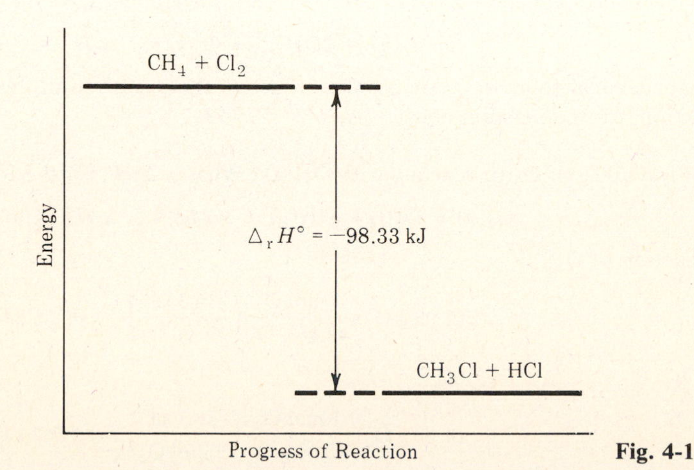

Fig. 4-1

4.5 Prepare an energy-level diagram for oxygen and hydrogen given the following thermochemical equations at 25 °C.

$$\tfrac{1}{2}H_2(g) \rightarrow H(g) \qquad \Delta_r H^\circ = 217.965\ \text{kJ}$$
$$\tfrac{1}{2}O_2(g) \rightarrow O(g) \qquad \Delta_r H^\circ = 249.170\ \text{kJ}$$
$$\tfrac{3}{2}O_2(g) \rightarrow O_3(g) \qquad \Delta_r H^\circ = 142.7\ \text{kJ}$$
$$\tfrac{1}{2}H_2(g) + \tfrac{1}{2}O_2(g) \rightarrow OH(g) \qquad \Delta_r H^\circ = 38.95\ \text{kJ}$$
$$\tfrac{1}{2}H_2(g) + O_2(g) \rightarrow HO_2(g) \qquad \Delta_r H^\circ = 10.5\ \text{kJ}$$
$$H_2(g) + \tfrac{1}{2}O_2(g) \rightarrow H_2O(g) \qquad \Delta_r H^\circ = -241.818\ \text{kJ}$$
$$H_2(g) + O_2(g) \rightarrow H_2O_2(g) \qquad \Delta_r H^\circ = -136.31\ \text{kJ}$$

▮ The diagram will be labeled "Energy" along the vertical axis. The elements in their normal standard states are assigned energy values of 0. The diagram is shown in Fig. 4-2.

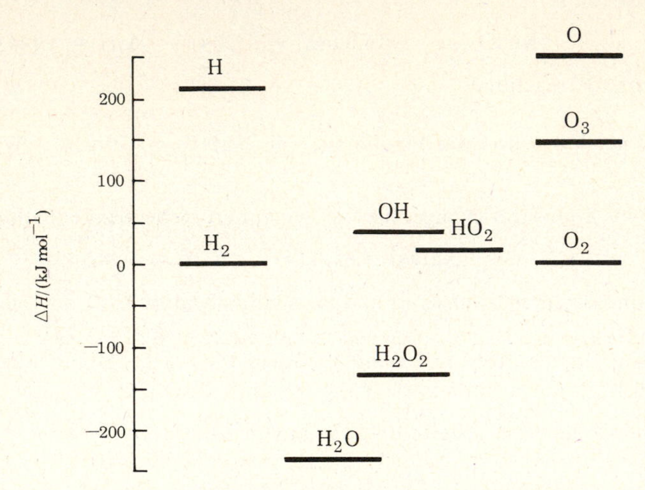

Fig. 4-2

4.6 Use the thermochemical equation

$$2OH(g) + O_2(g) \rightarrow 2HO_2(g) \qquad \Delta_r H^\circ = -56.9\ \text{kJ}$$

to determine $\Delta_r H^\circ$ for

$$OH(g) + \tfrac{1}{2}O_2(g) \rightarrow HO_2(g) \qquad \Delta_r H^\circ = ?$$

▮ The first equation states that 56.9 kJ of energy will be released when 2 mol of OH undergoes reaction with 1 mol of O_2 to produce 2 mol of HO_2. For the reaction of only one-half of the amount of reactants (1 mol of OH with $\frac{1}{2}$ mol of O_2) to produce only half the amount of products (1 mol of HO_2), only half of the energy will be involved, −28.5 kJ.

4.7 Use the thermochemical equation given in Problem 4.6 to determine $\Delta_r H^\circ$ for

$$2HO_2(g) \rightarrow 2OH(g) + O_2(g) \qquad \Delta_r H^\circ = ?$$

▮ This thermochemical equation is simply the reverse of the one given in the previous problem. Because $\Delta_r H^\circ$ is a state function, for the reaction above, $\Delta_r H^\circ = 56.9$ kJ.

4.8 What is $\Delta_r H^\circ$ for 1.00 g of $O_3(g)$ reacting with OH(g)? What is $\Delta_r H^\circ$ for the formation of 1.00 g of $O_2(g)$?

$$O_3(g) + OH(g) \rightarrow H(g) + 2O_2(g) \qquad \Delta_r H^\circ = 36.3\ \text{kJ}$$

▮ For the reaction of O_3,

$$(1.00\ \text{g}\ O_3)\frac{1\ \text{mol}\ O_3}{48.00\ \text{g}\ O_3}\left(\frac{36.3\ \text{kJ}}{1\ \text{mol}\ O_3}\right) = 0.756\ \text{kJ}$$

and for the O_2,

$$(1.00\ \text{g}\ O_2)\frac{1\ \text{mol}\ O_2}{32.00\ \text{g}\ O_2}\left(\frac{36.3\ \text{kJ}}{2\ \text{mol}\ O_2}\right) = 0.567\ \text{kJ}$$

4.9 As 1.00 g of $O_3(g)$ decomposed to form $O_2(g)$ under constant pressure conditions, 2.97 kJ was transferred to the surroundings. Find $\Delta_r H°$ for

$$2O_3(g) \rightarrow 3O_2(g) \qquad \Delta_r H° = ?$$

▮ Converting the data from a mass to a molar basis gives

$$\frac{-2.97\ \text{kJ}}{1.00\ \text{g O}_3}\left(\frac{48.00\ \text{g O}_3}{1\ \text{mol O}_3}\right)(2\ \text{mol O}_3) = -285\ \text{kJ}$$

4.10 Which oxidizing agent, O_2, O_3, or H_2O_2, will generate the greatest amount of energy for 1 mol of $H_2(g)$?

$$2H_2(g) + O_2(g) \rightarrow 2H_2O(g) \qquad \Delta_r H° = -483.636\ \text{kJ}$$

$$3H_2(g) + O_3(g) \rightarrow 3H_2O(g) \qquad \Delta_r H° = -868.2\ \text{kJ}$$

$$H_2(g) + H_2O_2(g) \rightarrow 2H_2O(g) \qquad \Delta_r H° = -347.33\ \text{kJ}$$

▮ For 1 mol of H_2,

$$(1\ \text{mol H}_2)\frac{-483.636\ \text{kJ}}{2\ \text{mol H}_2} = -241.818\ \text{kJ}$$

$$(1\ \text{mol H}_2)\frac{-868.2\ \text{kJ}}{3\ \text{mol H}_2} = -289.4\ \text{kJ}$$

$$(1\ \text{mol H}_2)\frac{-347.33\ \text{kJ}}{1\ \text{mol H}_2} = -347.33\ \text{kJ}$$

The greatest amount of energy will be released by using $H_2O_2(g)$.

4.11 For the reactions given in Problem 4.10, which oxidizing agent will generate the greatest amount of energy for 1.00 g of oxidizing agent?

▮ For 1.00 g of the oxidizing agents,

$$(1.00\ \text{g O}_2)\frac{1\ \text{mol O}_2}{32.00\ \text{g O}_2}\left(\frac{-483.636\ \text{kJ}}{1\ \text{mol O}_2}\right) = -15.11\ \text{kJ}$$

$$(1.00\ \text{g O}_3)\frac{1\ \text{mol O}_3}{48.00\ \text{g O}_3}\left(\frac{-868.2\ \text{kJ}}{1\ \text{mol O}_3}\right) = -18.09\ \text{kJ}$$

$$(1.00\ \text{g H}_2\text{O}_2)\frac{1\ \text{mol H}_2\text{O}_2}{34.01\ \text{g H}_2\text{O}_2}\left(\frac{-347.33\ \text{kJ}}{1\ \text{mol H}_2\text{O}_2}\right) = -10.21\ \text{kJ}$$

The greatest amount of energy will be released by using $O_3(g)$.

4.12 Which of the three reactions given in Problem 4.10 will generate the greatest amount of energy on a total mass basis of reactants? This value indicates which reaction might be used in rocket propulsion.

▮ For each of the reactions, the total mass of reactants is

$$m = (2\ \text{mol H}_2)\frac{2.02\ \text{g H}_2}{1\ \text{mol H}_2} + (1\ \text{mol O}_2)\frac{32.00\ \text{g O}_2}{1\ \text{mol O}_2} = 36.04\ \text{g}$$

$$m = (3\ \text{mol H}_2)\frac{2.02\ \text{g H}_2}{1\ \text{mol H}_2} + (1\ \text{mol O}_3)\frac{48.00\ \text{g O}_3}{1\ \text{mol O}_3} = 54.06\ \text{g}$$

$$m = (1\ \text{mol H}_2)\frac{2.02\ \text{g H}_2}{1\ \text{mol H}_2} + (1\ \text{mol H}_2\text{O}_2)\frac{34.01\ \text{g H}_2\text{O}_2}{1\ \text{mol H}_2\text{O}_2} = 36.03\ \text{g}$$

giving

$$\frac{-483.636\ \text{kJ}}{36.04\ \text{g}} = -13.42\ \text{kJ}\cdot\text{g}^{-1}$$

$$\frac{-868.2\ \text{kJ}}{54.06\ \text{g}} = -16.06\ \text{kJ}\cdot\text{g}^{-1}$$

$$\frac{-347.33\ \text{kJ}}{36.03\ \text{g}} = -9.640\ \text{kJ}\cdot\text{g}^{-1}$$

The H_2–O_3 reaction generates the greatest amount of energy on a total mass basis of reactants.

4.13 Gaseous ozone is bubbled through a water–ice mixture at 0 °C. As the $O_3(g)$ decomposes to form $O_2(g)$, the enthalpy of reaction is absorbed by the resulting ice. Given that the heat of fusion of ice is $6.0095\ \text{kJ} \cdot \text{mol}^{-1}$, determine the mass of ice that melts for each gram of O_3 that decomposes.

$$2O_3(g) \rightarrow 3O_2(g) \qquad \Delta_r H^\circ = -285.4\ \text{kJ}$$

▮ The amount of energy released by the decomposition is

$$(1.00\ \text{g O}_3)\frac{1\ \text{mol O}_3}{48.00\ \text{g O}_3}\left(\frac{-285.4\ \text{kJ}}{2\ \text{mol O}_3}\right) = -2.97\ \text{kJ}$$

The mass of ice needed to absorb this heat is

$$(2.97\ \text{kJ})\frac{1\ \text{mol}}{6.0095\ \text{kJ}}\left(\frac{18.02\ \text{g}}{1\ \text{mol}}\right) = 8.91\ \text{g}$$

4.14 What is the minimum mass of cooling water needed to remove the heat of reaction for each pound of acetylene reacting with $HCl(g)$ to form vinyl chloride? Assume that the inlet temperature of the cooling water is 25 °C, the outlet temperature is 35 °C, and $C_{P,\text{m}} = 75.291\ \text{J} \cdot \text{K}^{-1} \cdot \text{mol}^{-1}$.

$$C_2H_2(g) + HCl(g) \rightarrow CH_2{=}CHCl(g) \qquad \Delta_r H^\circ = -98.8\ \text{kJ}$$

▮ The amount of heat transferred to the water will be

$$(1.00\ \text{lb C}_2\text{H}_2)\frac{454\ \text{g}}{1\ \text{lb}}\left(\frac{1\ \text{mol C}_2\text{H}_2}{26.02\ \text{g C}_2\text{H}_2}\right)\left(\frac{-98.8\ \text{kJ}}{1\ \text{mol C}_2\text{H}_2}\right) = -1720\ \text{kJ}$$

Solving *(3.23)* for n and substituting the data gives

$$n = \frac{(1720\ \text{kJ})[(10^3\ \text{J})/(1\ \text{kJ})]}{(75.291\ \text{J} \cdot \text{K}^{-1} \cdot \text{mol}^{-1})(10\ \text{K})} = 2300\ \text{mol}$$

which is equivalent to 41 kg or 91 lb.

4.15 When molecular beams of $OH(g)$ and $H_2(g)$ at 25 °C undergo reaction, the energy released cannot be dissipated quickly and so the product gases, $H_2O(g)$ and $H(g)$, are warmer than 25 °C. Determine the temperature of the product gases given $C_{P,\text{m}}/(\text{J} \cdot \text{K}^{-1} \cdot \text{mol}^{-1}) = 33.577$ for $H_2O(g)$ and 20.784 for $H(g)$.

$$OH(g) + H_2(g) \rightarrow H_2O(g) + H(g) \qquad \Delta_r H^\circ = -62.80\ \text{kJ}$$

▮ The heat capacity of the product gases is

$$\begin{aligned} C_P &= n(\text{H}_2\text{O})C_{P,\text{m}} + n(\text{H})C_{P,\text{m}}(\text{H}) \\ &= (1\ \text{mol})(33.577\ \text{J} \cdot \text{K}^{-1} \cdot \text{mol}^{-1}) + (1\ \text{mol})(20.784\ \text{J} \cdot \text{K}^{-1} \cdot \text{mol}^{-1}) \\ &= 54.361\ \text{J} \cdot \text{K}^{-1} \end{aligned}$$

Solving *(3.23)* for ΔT and substituting the data gives

$$\Delta T = \frac{(62.80\ \text{kJ})[(10^3\ \text{J})/(1\ \text{kJ})]}{54.361\ \text{J} \cdot \text{K}^{-1}} = 1155\ \text{K}$$

The temperature of the product gases will be $T = 1155\ \text{K} + 298\ \text{K} = 1453\ \text{K}$ (assuming no heat loss to the surroundings and constant heat capacity).

4.16 To a first approximation, a flame can be considered to be an adiabatic, isobaric process in which the heat of reaction is used to heat the product gases to the *flame temperature.* What would be the maximum temperature of a methane flame?

$$CH_4(g) + 2O_2(g) \rightarrow CO_2(g) + 2H_2O(g) \qquad \Delta_r H^\circ_{298} = -802.34\ \text{kJ}$$

Assume

$$C_{P,\text{m}}/(\text{J} \cdot \text{K}^{-1} \cdot \text{mol}^{-1}) = 25.460 + (43.497 \times 10^{-3})[T/(\text{K})] - (148.32 \times 10^{-7})[T/(\text{K})]^2$$

for $CO_2(g)$ and

$$C_{P,\text{m}} = 30.605 + (9.615 \times 10^{-3})[T/(\text{K})] + (11.8 \times 10^{-7})[T/(\text{K})]^2$$

for $H_2O(g)$.

▮ The heat capacity of the product gases is

$$C_P = n(CO_2)C_{P,m} + n(H_2O)C_{P,m}(H_2O)$$
$$= (1\ \text{mol})[(25.460\ \text{J}\cdot\text{K}^{-1}\cdot\text{mol}^{-1}) + (43.497\times10^{-3}\ \text{J}\cdot\text{K}^{-2}\cdot\text{mol}^{-1})T - (148.32\times10^{-7}\ \text{J}\cdot\text{K}^{-3}\cdot\text{mol}^{-1})T^2]$$
$$+ (2\ \text{mol})[(30.605\ \text{J}\cdot\text{K}^{-1}\cdot\text{mol}^{-1}) + (9.615\times10^{-3}\ \text{J}\cdot\text{K}^{-2}\cdot\text{mol}^{-1})T + (11.8\times10^{-7}\ \text{J}\cdot\text{K}^{-3}\cdot\text{mol}^{-1})T^2]$$
$$= 86.670\ \text{J}\cdot\text{K}^{-1} + (6.2727\times10^{-2}\ \text{J}\cdot\text{K}^{-2})T - (1.247\times10^{-5}\ \text{J}\cdot\text{K}^{-3})T^2$$

Assuming that 802.34 kJ is absorbed by the product gases, *(3.23)* gives

$$(802.34\ \text{kJ})\frac{10^3\ \text{J}}{1\ \text{kJ}} = (86.670\ \text{J}\cdot\text{K}^{-1})(T - 298\ \text{K}) + \tfrac{1}{2}(6.2727\times10^{-2}\ \text{J}\cdot\text{K}^{-2})[T^2 - (298\ \text{K})^2]$$
$$- \tfrac{1}{3}(1.247\times10^{-5}\ \text{J}\cdot\text{K}^{-3})[T^3 - (298\ \text{K})^3]$$

Solving for T gives the flame temperature as 3423 K. This value is not realistic because the empirical heat capacity expressions are not valid at these high temperatures, the system is not truly adiabatic, and there is incomplete combustion.

4.17 Assume that a human requires "2500 cal" of energy each day for metabolic activity. What mass of sucrose ($C_{12}H_{22}O_{11}$) or what mass of ethanol (C_2H_5OH) is needed to provide this energy?

$$C_{12}H_{22}O_{11}(s) + 12O_2(g) \rightarrow 12CO_2(g) + 11H_2O(l) \qquad \Delta_rH° = -5.647\ \text{MJ}$$
$$C_2H_5OH(l) + 3O_2(g) \rightarrow 2CO_2(g) + 3H_2O(l) \qquad \Delta_rH° = -1.371\ \text{MJ}$$

▮ The "calorie" used in nutrition is actually a kilocalorie, so the energy is

$$(2500\ \text{kcal})\frac{10^3\ \text{cal}}{1\ \text{kcal}}\left(\frac{4.184\ \text{J}}{1\ \text{cal}}\right) = 1.0\times10^7\ \text{J}$$

This energy can be supplied by

$$(1.0\times10^7\ \text{J})\frac{1\ \text{mol}\ C_{12}H_{22}O_{11}}{5.647\ \text{MJ}}\left(\frac{1\ \text{MJ}}{10^6\ \text{J}}\right)\left(\frac{342\ \text{g}\ C_{12}H_{22}O_{11}}{1\ \text{mol}\ C_{12}H_{22}O_{11}}\right) = 610\ \text{g}\ C_{12}H_{22}O_{11}$$

$$(1.0\times10^7\ \text{J})\frac{1\ \text{mol}\ C_2H_5OH}{1.371\ \text{MJ}}\left(\frac{1\ \text{MJ}}{10^6\ \text{J}}\right)\left(\frac{46.0\ \text{g}\ C_2H_5OH}{1\ \text{mol}\ C_2H_5OH}\right) = 340\ \text{g}\ C_2H_5OH$$

4.18 If the human body is considered to be a closed system, what would be the temperature increase resulting from the energy intake described in Problem 4.17? Assume a mass of 75 kg and a specific heat of $4\ \text{J}\cdot\text{K}^{-1}\cdot\text{g}^{-1}$. If the body temperature is to be maintained at a constant value by the evaporation of water, what mass of water must evaporate? Assume that the heat of vaporization of water is $44\ \text{kJ}\cdot\text{mol}^{-1}$.

▮ Solving *(3.23)* for ΔT and substituting the data gives

$$\Delta T = \frac{1.0\times10^7\ \text{J}\cdot\text{day}^{-1}}{(75\ \text{kg})[(10^3\ \text{g})/(1\ \text{kg})](4\ \text{J}\cdot\text{K}^{-1}\cdot\text{g}^{-1})} = 30\ \text{K}\cdot\text{day}^{-1}$$

The mass of water needed to maintain the body temperature is

$$m = \frac{1.0\times10^7\ \text{J}\cdot\text{day}^{-1}}{(44\ \text{kJ}\cdot\text{mol}^{-1})[(10^3\ \text{J})/(1\ \text{kJ})][(1\ \text{mol})/(18\ \text{g})]} = 4100\ \text{g}$$

(assuming all the heat is dissipated by evaporation).

4.19 Hot carbon reacts with steam to produce an equimolar mixture of CO(g) and H_2(g) known as water gas. What is the energy released as water gas is used as fuel?

$$CO(g) + \tfrac{1}{2}O_2(g) \rightarrow CO_2(g) \qquad \Delta_rH° = -282.984\ \text{kJ}$$
$$H_2(g) + \tfrac{1}{2}O_2(g) \rightarrow H_2O(g) \qquad \Delta_rH° = -241.818\ \text{kJ}$$

▮ For the equimolar mixture, the energy is $\Delta_rH° = -282.984\ \text{kJ} + (-241.818\ \text{kJ}) = -524.802\ \text{kJ}$.

4.20 The "water gas shift" reaction converts the CO(g) in water gas to CO_2(g) and H_2(g) by reacting additional steam with the water gas at high temperature. $CO(g) + H_2O(g) \rightarrow CO_2(g) + H_2(g)$ What is the heat value of this fuel, which contains only combustible H_2(g), compared to the original water gas fuel?

▮ The original fuel would generate 282.984 kJ of heat as the CO(g) burned. The new fuel will generate only 241.818 kJ of heat as the H_2(g) burns. Thus the heat value of the new fuel is less.

4.21 An equilibrium will be established between gaseous *cis*-1,2-dichloroethene and gaseous *trans*-1,2-dichloroethene in which 29% of the cis isomer is converted to the trans isomer. What is the energy involved in the reaction?

$$cis\text{-}CHCl{=}CHCl(g) \rightarrow trans\text{-}CHCl{=}CHCl(g) \qquad \Delta_r H° = 2.38 \text{ kJ}$$

▮ The $\Delta_r H°$ value assumes the reaction of 1 mol of the cis isomer. If only 29% of the isomer reacts, then the amount of energy is

$$\Delta_r H° = (0.29)(2.38 \text{ kJ}) = 0.69 \text{ kJ}$$

4.22 Ethanol can undergo decomposition to form two sets of products:

$$C_2H_5OH(g) \rightarrow \begin{cases} C_2H_4(g) + H_2O(g) & \Delta_r H° = 45.54 \text{ kJ} \\ CH_3CHO(g) + H_2(g) & \Delta_r H° = 68.91 \text{ kJ} \end{cases}$$

If the molar ratio of C_2H_4 to CH_3CHO is 8:1 in a set of product gases, determine the energy involved in the decomposition process.

▮ The 8:1 molar ratio implies that 8/9 of the C_2H_5OH will undergo the first reaction and 1/9 will undergo the second reaction. Thus, for 1 mol of C_2H_5OH reacting,

$$\Delta_r H° = \tfrac{8}{9}(45.54 \text{ kJ}) + \tfrac{1}{9}(68.91 \text{ kJ}) = 48.14 \text{ kJ}$$

4.23 Find $\Delta_r H° - \Delta_r U°$ for $4KClO_3(s) \rightarrow 3KClO_3(s) + KCl(s)$.

▮ The only major contributions to $\Delta_r H° - \Delta_r U°$ will be from gaseous reactants and products. Thus *(3.5)* gives

$$\Delta_r H° - \Delta_r U° = RT\,\Delta\nu_{i,g} = 0$$

4.24 Only at extremely high pressures does $\Delta_r H° - \Delta_r U°$ for condensed state reactions become significantly different than 0. Determine the pressure at which $\Delta_r H° - \Delta_r U°$ is equal to -1 kJ for the reaction

$$\text{C(graphite)} \rightarrow \text{C(diamond)}$$

For graphite, $\rho = 2.25 \text{ g} \cdot \text{cm}^{-3}$ and for diamond, $\rho = 3.51 \text{ g} \cdot \text{cm}^{-3}$.

▮ For this process, using *(3.2)* gives $\Delta_r H° - \Delta_r U° = P\,\Delta V$. Substituting the data and solving for P gives

$$-1000 \text{ J} = P(12.01 \text{ g})\left(\frac{1}{3.51 \text{ g} \cdot \text{cm}^{-3}} - \frac{1}{2.25 \text{ g} \cdot \text{cm}^{-3}}\right)\left(\frac{10^{-3} \text{ L}}{1 \text{ cm}^3}\right)\left(\frac{10^2 \text{ J}}{1 \text{ L} \cdot \text{bar}}\right)$$

$$P = 5200 \text{ bar}$$

4.25 Calculate $\Delta_r U°$ for

$$H_2(g) + \tfrac{1}{2}O_2(g) \rightarrow H_2O(l) \qquad \Delta_r H° = -285.830 \text{ kJ}$$

▮ For this reaction,

$$\Delta\nu_{i,g} = 0 - (1 \text{ mol} + \tfrac{1}{2} \text{ mol}) = -\tfrac{3}{2} \text{ mol}$$

Solving *(3.5)* for $\Delta_r U°$ and substituting the data gives

$$\Delta_r U° = -285.830 \text{ kJ} - (8.314 \times 10^{-3} \text{ J} \cdot \text{K}^{-1} \cdot \text{mol}^{-1})(298.15 \text{ K})(-\tfrac{3}{2} \text{ mol})$$
$$= -282.112 \text{ kJ}$$

4.26 Calculate $\Delta_r H°$ for

$$C_6H_5COOH(s) + \tfrac{15}{2}O_2(g) \rightarrow 7CO_2(g) + 3H_2O(l)$$

given $\Delta_r U° = -6316 \text{ cal} \cdot (\text{g benzoic acid})^{-1}$.

▮ For the thermochemical equation,

$$\Delta_r U° = (-6316 \text{ cal} \cdot \text{g}^{-1})\left(\frac{4.184 \text{ J}}{1 \text{ cal}}\right)\left(\frac{122.12 \text{ g}}{1 \text{ mol}}\right)\left(\frac{10^{-3} \text{ kJ}}{1 \text{ J}}\right) = -3227 \text{ kJ}$$

For this reaction

$$\Delta\nu_{i,g} = 7 \text{ mol} - \tfrac{15}{2} \text{ mol} = -\tfrac{1}{2} \text{ mol}$$

Solving *(3.5)* for $\Delta_r H°$ and substituting the data gives

$$\Delta_r H° = -3227 \text{ kJ} + (8.314 \times 10^{-3} \text{ J} \cdot \text{K}^{-1} \cdot \text{mol}^{-1})(298.15 \text{ K})(-\tfrac{1}{2} \text{ mol})$$
$$= -3228 \text{ kJ}$$

4.27 For the decomposition of $O_3(g)$ to form $O_2(g)$, calculate ΔH, ΔU, q, and w.

$$2O_3(g) \rightarrow 3O_2(g) \qquad \Delta_r H^\circ = -285.4\ \text{kJ}$$

▌ Because the reaction is carried out under constant-pressure conditions,

$$q = \Delta_r H^\circ = -285.4\ \text{kJ}$$

Substituting $\Delta \nu_{i,g} = +1\ \text{mol}$ into *(3.5)* gives

$$\Delta_r U^\circ = -285.4\ \text{kJ} - (8.314 \times 10^{-3}\ \text{J} \cdot \text{K}^{-1} \cdot \text{mol}^{-1})(298.15\ \text{K})(1\ \text{mol})$$
$$= -287.9\ \text{kJ}$$

Solving *(3.47)* for w and substituting the data gives

$$w = -287.9\ \text{kJ} - (-285.4\ \text{kJ}) = -2.5\ \text{kJ}$$

Note that w could also be calculated using

$$w = -P_{\text{ext}}\,\Delta V = -(1.00\ \text{bar})\,\Delta(nRT/P)$$
$$= -(1.00\ \text{bar})\frac{(8.314\ \text{J} \cdot \text{K}^{-1} \cdot \text{mol}^{-1})(298\ \text{K})\,\Delta n_g}{1.00\ \text{bar}}$$
$$= -(1.00\ \text{bar})\frac{(8.314\ \text{J} \cdot \text{K}^{-1} \cdot \text{mol}^{-1})(298\ \text{K})(3\ \text{mol} - 2\ \text{mol})}{1.00\ \text{bar}} = -2479\ \text{J}$$

4.28 Calculate $\Delta_r H^\circ$, $\Delta_r U^\circ$, q, and w for the reaction of 1.00 g of Zn placed in excess HCl.

$$\text{Zn(s)} + 2\text{HCl(aq)} \rightarrow \text{ZnCl}_2\text{(aq)} + \text{H}_2\text{(g)} \qquad \Delta_r H^\circ = -153.87\ \text{kJ}$$

▌ For each mole of Zn that reacts,

$$q = \Delta_r H^\circ = -153.87\ \text{kJ} \qquad \Delta \nu_{i,g} = +1\ \text{mol}$$
$$\Delta_r U^\circ = -153.87\ \text{kJ} - (8.314 \times 10^{-3}\ \text{kJ} \cdot \text{K}^{-1} \cdot \text{mol}^{-1})(298.15\ \text{K})(1\ \text{mol})$$
$$= -156.35\ \text{kJ}$$
$$w = -156.35\ \text{kJ} - (-153.87\ \text{kJ}) = -2.48\ \text{kJ}$$

Converting to the basis of 10.0 g of Zn,

$$q = \Delta_r H^\circ = \frac{-153.87\ \text{kJ}}{1\ \text{mol Zn}}(10.0\ \text{g Zn})\frac{1\ \text{mol Zn}}{65.39\ \text{g}} = -23.53\ \text{kJ}$$
$$\Delta_r U^\circ = \frac{-156.35\ \text{kJ}}{1\ \text{mol Zn}}(10.0\ \text{g Zn})\frac{1\ \text{mol Zn}}{65.39\ \text{g}} = -23.91\ \text{kJ}$$
$$w = \frac{-2.48\ \text{kJ}}{1\ \text{mol Zn}}(10.0\ \text{g Zn})\frac{1\ \text{mol Zn}}{65.39\ \text{g}} = -0.38\ \text{kJ}$$

4.29 What is $\Delta_r H^\circ_{373}$ for the reaction

$$\text{O}_3\text{(g)} + \text{OH(g)} \rightarrow \text{H(g)} + 2\text{O}_2\text{(g)} \qquad \Delta_r H^\circ_{298} = 36.3\ \text{kJ}$$

given $C_{P,\text{m}}/(\text{J} \cdot \text{K}^{-1} \cdot \text{mol}^{-1}) = 39.29$ for O_3, 29.355 for O_2, 20.784 for H, and 29.886 for OH?

▌ The special case of *(3.23)* for a chemical reaction is

$$\Delta_r H^\circ(T_2) = \Delta_r H^\circ(T_1) + \int_{T_1}^{T_2} \Delta_r C^\circ_P\, dT \qquad (4.1)$$

where

$$\Delta_r C^\circ_P = \sum_i \nu_i C^\circ_{P,\text{m},i} \qquad (4.2)$$

where ν_i, the stoichiometric coefficients in the chemical equation (ν_i is positive for products and negative for reactants), have the units of mol. For this reaction, using *(4.2)* gives

$$\Delta_r C^\circ_P = [(1\ \text{mol})(20.784\ \text{J} \cdot \text{K}^{-1} \cdot \text{mol}^{-1}) + (2\ \text{mol})(29.355\ \text{J} \cdot \text{K}^{-1} \cdot \text{mol}^{-1})$$
$$- [(1\ \text{mol})(39.20\ \text{J} \cdot \text{K}^{-1} \cdot \text{mol}^{-1}) + (1\ \text{mol})(29.886\ \text{J} \cdot \text{K}^{-1} \cdot \text{mol}^{-1})]$$
$$= 10.41\ \text{J} \cdot \text{K}^{-1}$$

Substituting the data into *(4.1)* gives

$$\Delta_r H^\circ_{373} = 36.3\ \text{kJ} + (10.41\ \text{J}\cdot\text{K}^{-1}\cdot\text{mol}^{-1})(373\ \text{K} - 298\ \text{K})\frac{10^{-3}\ \text{kJ}}{1\ \text{J}}$$
$$= 37.1\ \text{kJ}$$

4.30 Determine $\Delta_r H^\circ$ for the reaction

$$N_2(g) + 3H_2(g) \rightarrow 2NH_3(g) \qquad \Delta_r H^\circ = -92.22\ \text{kJ}$$

as a function of temperature. Prepare a plot of $\Delta_r H^\circ$ against T between 298 K and 1000 K. For these gases, $C_{P,m}/(\text{J}\cdot\text{K}^{-1}\cdot\text{mol}^{-1}) = 27.565 + 5.230 \times 10^{-3}[T/(\text{K})] - 0.04 \times 10^{-7}[T/(\text{K})]^2$ for N_2, $28.894 - 0.836 \times 10^{-3}[T/(\text{K})] + 20.17 \times 10^{-7}[T/(\text{K})]^2$ for H_2, and $25.895 + 32.999 \times 10^{-3}[T/(\text{K})] - 30.46 \times 10^{-7}[T/(\text{K})]^2$ for NH_3.

▌ Substituting the empirical heat capacity expressions given by *(3.19)* into *(4.1)* and integrating gives

$$\Delta_r H^\circ(T_2) = \Delta_r H^\circ(T_1) + (\Delta a)(T_2 - T_1) + \tfrac{1}{2}(\Delta b)(T_2^2 - T_1^2) + \tfrac{1}{3}(\Delta c)(T_2^3 - T_1^3)$$
$$+ \tfrac{1}{4}(\Delta d)(T_2^4 - T_1^4) + \cdots - (\Delta c')(T_2^{-1} - T_1^{-1}) \qquad (4.3a)$$

where Δa, Δb, etc. are defined by *(4.2)*. If J is defined as

$$J = \Delta_r H^\circ_{298} - (\Delta a)(298\ \text{K}) - \tfrac{1}{2}(\Delta b)(298\ \text{K})^2 - \tfrac{1}{3}(\Delta c)(298\ \text{K})^3 - \tfrac{1}{4}(\Delta d)(298\ \text{K})^4 - \cdots + (\Delta c')(298\ \text{K})^{-1} \qquad (4.4)$$

then

$$\Delta_r H^\circ_T = J + (\Delta a)T + \tfrac{1}{2}(\Delta b)T^2 + \tfrac{1}{3}(\Delta c)T^3 + \tfrac{1}{4}(\Delta d)T^4 + \cdots - (\Delta c')T^{-1} \qquad (4.3b)$$

For the reaction, using *(4.2)* gives

$$\Delta a = (2\ \text{mol})(25.895\ \text{J}\cdot\text{K}^{-1}\cdot\text{mol}^{-1}) - [(1\ \text{mol})(27.565\ \text{J}\cdot\text{K}^{-1}\cdot\text{mol}^{-1}) + (3\ \text{mol})(28.894\ \text{J}\cdot\text{K}^{-1}\cdot\text{mol}^{-1})]$$
$$= -62.457\ \text{J}\cdot\text{K}^{-1}$$
$$\Delta b/(10^{-3}) = (2\ \text{mol})(32.999\ \text{J}\cdot\text{K}^{-2}\cdot\text{mol}^{-1}) - [(1\ \text{mol})(5.230\ \text{J}\cdot\text{K}^{-2}\cdot\text{mol}^{-1}) + (3\ \text{mol})(-0.836\ \text{J}\cdot\text{K}^{-2}\cdot\text{mol}^{-1})]$$
$$= 63.276\ \text{J}\cdot\text{K}^{-2}$$
$$\Delta c/(10^{-7}) = (2\ \text{mol})(-30.46\ \text{J}\cdot\text{K}^{-3}\cdot\text{mol}^{-1}) - [(1\ \text{mol})(-0.04\ \text{J}\cdot\text{K}^{-3}\cdot\text{mol}^{-1}) + (3\ \text{mol})(20.17\ \text{J}\cdot\text{K}^{-3}\cdot\text{mol}^{-1})]$$
$$= -121.39\ \text{J}\cdot\text{K}^{-3}$$

and substituting the data into *(4.4)* gives

$$J = -92.22\ \text{kJ} - (-62.457\ \text{J}\cdot\text{K}^{-1})\frac{10^{-3}\ \text{kJ}}{1\ \text{J}}(298\ \text{K})$$
$$-\tfrac{1}{2}(63.276 \times 10^{-3}\ \text{J}\cdot\text{K}^{-2})\frac{10^{-3}\ \text{kJ}}{1\ \text{J}}(298\ \text{K})^2 - \tfrac{1}{3}(-121.39 \times 10^{-7}\ \text{J}\cdot\text{K}^{-3})\frac{10^{-3}\ \text{kJ}}{1\ \text{J}}(298\ \text{K})^3$$
$$= -76.30\ \text{kJ}$$

Thus at any temperature

$$\Delta_r H^\circ_T/(\text{kJ}) = -76.30 - (6.2457 \times 10^{-2})[T/(\text{K})] + (3.1638 \times 10^{-5})[T/(\text{K})]^2 - (1.2139 \times 10^{-8})[T/(\text{K})]^3$$

As a sample calculation, at $T = 500\ \text{K}$,

$$\Delta_r H^\circ_{500}/(\text{kJ}) = -76.30 - (6.2457 \times 10^{-2})(500) + (3.1638 \times 10^{-5})(500)^2 - (1.2139 \times 10^{-8})(500)^3$$
$$= -101.14\ \text{kJ}$$

This plot is shown in Fig. 4-3.

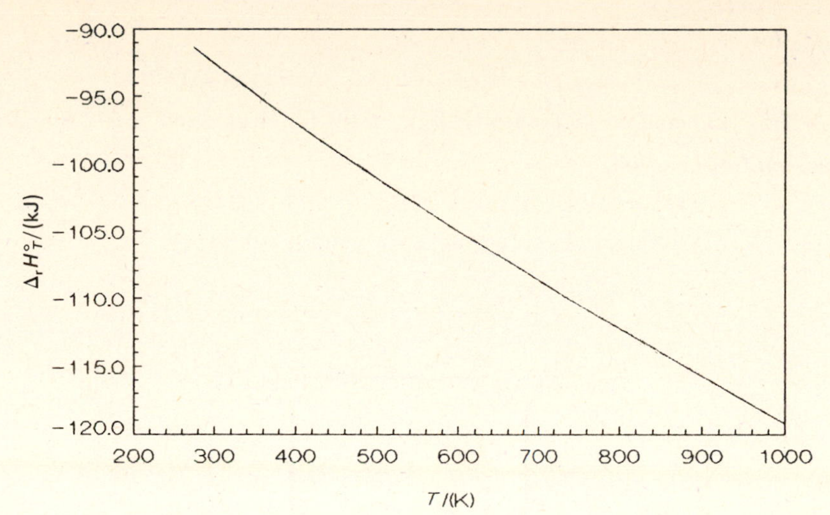

Fig. 4-3

4.31 Determine $\Delta_r H^\circ$ for the reaction

$$N_2(g) + 2O_2(g) \rightarrow N_2O_4(g) \qquad \Delta_r H^\circ = 9.079\ \text{kJ}$$

as a function of temperature. For these gases, $C_{P,m}/(\text{J}\cdot\text{K}^{-1}\cdot\text{mol}^{-1}) = 27.565 + 5.230\times 10^{-3}[T/(\text{K})] - 0.04\times 10^{-7}[T/(\text{K})]^2$ for N_2, $25.849 + 12.98\times 10^{-3}[T/(\text{K})] - 38.62\times 10^{-7}[T/(\text{K})]^2$ for O_2, and $37.49 + 0.156[T/(\text{K})] - 7.5\times 10^{-5}[T/(\text{K})]^2$ for N_2O_4. Find the temperature at which $\Delta_r H^\circ$ is a minimum.

▌ For the reaction, using *(4.2)* gives

$$\begin{aligned}\Delta a &= (1\ \text{mol})(37.49\ \text{J}\cdot\text{K}^{-1}\cdot\text{mol}^{-1}) - [(1\ \text{mol})(27.565\ \text{J}\cdot\text{K}^{-1}\cdot\text{mol}^{-1}) + (2\ \text{mol})(25.849\ \text{J}\cdot\text{K}^{-1}\cdot\text{mol}^{-1})]\\ &= -41.77\ \text{J}\cdot\text{K}^{-1}\end{aligned}$$

$$\begin{aligned}\Delta b &= (1\ \text{mol})(0.156\ \text{J}\cdot\text{K}^{-2}\cdot\text{mol}^{-1})\\ &\quad - [(1\ \text{mol})(5.230\times 10^{-3}\ \text{J}\cdot\text{K}^{-2}\cdot\text{mol}^{-1}) + (2\ \text{mol})(12.98\times 10^{-3}\ \text{J}\cdot\text{K}^{-2}\cdot\text{mol}^{-1})]\\ &= 0.125\ \text{J}\cdot\text{K}^{-2}\end{aligned}$$

$$\begin{aligned}\Delta c &= (1\ \text{mol})(-7.5\times 10^{-5}\ \text{J}\cdot\text{K}^{-3}\cdot\text{mol}^{-1})\\ &\quad - [(1\ \text{mol})(-0.04\times 10^{-7}\ \text{J}\cdot\text{K}^{-3}\cdot\text{mol}^{-1}) + (2\ \text{mol})(-38.62\times 10^{-7}\ \text{J}\cdot\text{K}^{-3}\cdot\text{mol}^{-1})]\\ &= -6.7\times 10^{-5}\ \text{J}\cdot\text{K}^{-3}\end{aligned}$$

and substituting the data into *(4.4)* gives

$$\begin{aligned}J &= 9.079\ \text{kJ} - (-41.77\ \text{J}\cdot\text{K}^{-1})\frac{10^{-3}\ \text{kJ}}{1\ \text{J}}(298\ \text{K})\\ &\quad - \tfrac{1}{2}(0.125\ \text{J}\cdot\text{K}^{-2})\frac{10^{-3}\ \text{kJ}}{1\ \text{J}}(298\ \text{K})^2\\ &\quad - \tfrac{1}{3}(-6.7\times 10^{-5}\ \text{J}\cdot\text{K}^{-3})\frac{10^{-3}\ \text{kJ}}{1\ \text{J}}(298\ \text{K})^3\\ &= 16.569\ \text{kJ}\end{aligned}$$

Thus at any temperature

$$\Delta_r H^\circ_T/(\text{kJ}) = 16.569 - (4.177\times 10^{-2})[T/(\text{K})] + (6.25\times 10^{-5})[T/(\text{K})]^2 - (2.2\times 10^{-8})[T/(\text{K})]^3$$

Differentiating the expression for $\Delta_r H^\circ_T$ with respect to T, setting the result equal to zero, and solving for T gives the minimum as

$$\frac{\partial \Delta_r H^\circ_T}{\partial T} = -4.177\times 10^{-2} + 2(6.25\times 10^{-5})[T/(\text{K})] - 3(2.2\times 10^{-8})[T/(\text{K})]^2 = 0$$

$$T = 433\ \text{K}$$

At this temperature,

$$\begin{aligned}\Delta_r H^\circ_{433}/(\text{kJ}) &= 16.569 - (4.177\times 10^{-2})(433) + (6.25\times 10^{-5})(433)^2 + (2.2\times 10^{-8})(433)^3\\ &= 8.415\end{aligned}$$

4.32 Calculate $\Delta_r H_0^\circ$ for

$$HCl(g) \rightarrow H(g) + Cl(g) \qquad \Delta_r H_{298}^\circ = 431.961\ \text{kJ}$$

given $(H_T^\circ - H_0^\circ)/(\text{kJ}\cdot\text{mol}^{-1}) = 8.644$ for HCl(g), 6.197 for H(g), and 6.272 for Cl(g) at 298 K.

▮ As can be seen from Fig. 4-4,

$$\begin{aligned}\Delta H_0^\circ &= (H_T^\circ - H_0^\circ)_{HCl} + \Delta_r H_{298}^\circ - (H_T^\circ - H_0^\circ)_H - (H_T^\circ - H_0^\circ)_{Cl} \\ &= 8.644\ \text{kJ} + 431.961\ \text{kJ} - 6.197\ \text{kJ} - 6.272\ \text{kJ} \\ &= 428.136\ \text{kJ}\end{aligned}$$

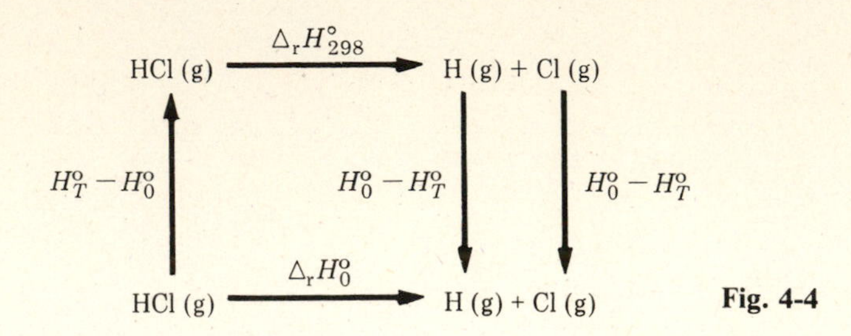

Fig. 4-4

4.33 For liquid water at 25 °C,

$$H_2(g) + \tfrac{1}{2}O_2(g) \rightarrow H_2O(l) \qquad \Delta_r H_{298}^\circ = -285.830\ \text{kJ}$$

Given $C_{P,m}/(\text{J}\cdot\text{K}^{-1}\cdot\text{mol}^{-1}) = 75.291$ for $H_2O(l)$ and 33.577 for $H_2O(g)$ and that the molar heat of vaporization of liquid water at 100 °C is $40.883\ \text{kJ}\cdot\text{mol}^{-1}$, find $\Delta_r H_{298}^\circ$ for $H_2(g) + \frac{1}{2}O_2(g) \rightarrow H_2O(g)$.

▮ Figure 4-5 shows

$$\begin{aligned}\Delta_r H_{298}^\circ &= \Delta H_1 + \Delta H_2 + \Delta H_3 + \Delta H_4 \\ &= -285.830\ \text{kJ} + (1\ \text{mol})(75.291\ \text{J}\cdot\text{K}^{-1}\cdot\text{mol}^{-1})(373\ \text{K} - 298\ \text{K})\frac{10^{-3}\ \text{kJ}}{1\ \text{J}} \\ &\quad + (1\ \text{mol})(40.883\ \text{kJ}\cdot\text{mol}^{-1}) \\ &\quad + (1\ \text{mol})(33.577\ \text{J}\cdot\text{K}^{-1}\cdot\text{mol}^{-1})(298\ \text{K} - 373\ \text{K})\frac{10^{-3}\ \text{kJ}}{1\ \text{J}} \\ &= -241.818\ \text{kJ}\end{aligned}$$

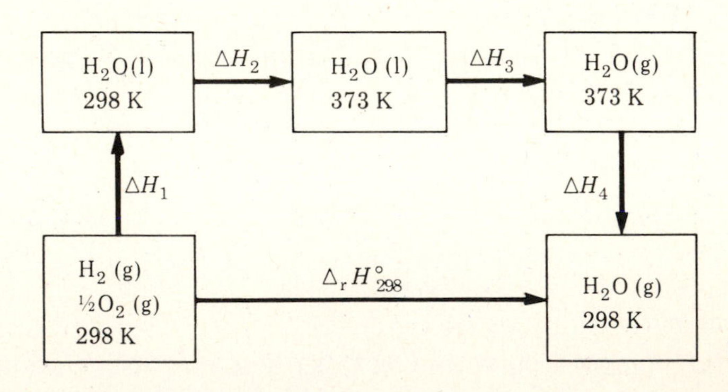

Fig. 4-5

4.34 Between 831 K and 1000 K, $\Delta_r H_T^\circ/(\text{kJ}) = -88.07 + 2.12 \times 10^{-5}[T/(\text{K})]^2$ for the reaction $Ag(s) + \frac{1}{2}I_2(g) \rightarrow AgI(l)$. Find $\Delta_r C_P^\circ$ for this reaction at 1000 K.

▮ Using *(3.21)* gives

$$\begin{aligned}\Delta_r C_P^\circ &= \left(\frac{\partial \Delta_r H_T^\circ}{\partial T}\right)_P = \left(\frac{\partial[-88.07 \times 10^3\ \text{J} + (0.0212\ \text{J}\cdot\text{K}^{-2})T^2]}{\partial T}\right)_P \\ &= (0.0424\ \text{J}\cdot\text{K}^{-2})T\end{aligned}$$

At 1000 K, $\Delta_r C_P^\circ = (0.0424\ \text{J}\cdot\text{K}^{-2})(1000\ \text{K}) = 42.4\ \text{J}\cdot\text{K}^{-1}$.

4.35 Exactly 25 mL of 0.500 M HCl and 25 mL of 0.500 M NaOH were mixed in a solution calorimeter, and a temperature increase of 3.015 K was observed. Determine the calorimeter constant given that the specific heat and the density of the resulting 0.250 M NaCl solution are $4.100\ \mathrm{J \cdot K^{-1} \cdot g^{-1}}$ and $1.0100\ \mathrm{g \cdot mL^{-1}}$, respectively. The thermochemical equation for the neutralization process may be written as

$$\mathrm{HCl\ (in\ 100\ H_2O) + NaOH\ (in\ 100\ H_2O) \rightarrow NaCl\ (in\ 201\ H_2O)} \qquad \Delta_r H° = -57.182\ \mathrm{kJ}$$

▮ Both reactants are being added in stoichiometric amounts. The amount of energy released by the reaction is

$$\Delta H = (25.00\ \mathrm{mL})\frac{10^{-3}\ \mathrm{L}}{1\ \mathrm{mL}}\left(\frac{0.500\ \mathrm{mol\ HCl}}{1\ \mathrm{L}}\right)\left(\frac{-57.182\ \mathrm{kJ}}{1\ \mathrm{mol\ HCl}}\right)\left(\frac{10^3\ \mathrm{J}}{1\ \mathrm{kJ}}\right) = -715\ \mathrm{J}$$

For the calorimeter and products of reaction, *(3.26)* can be written as

$$C_P(\text{calorimeter}) = \frac{715\ \mathrm{J}}{3.015\ \mathrm{K}} - (50.00\ \mathrm{mL})(1.0100\ \mathrm{g \cdot mL^{-1}})(4.100\ \mathrm{J \cdot K^{-1} \cdot g^{-1}})$$
$$= 30.1\ \mathrm{J \cdot K^{-1}}$$

4.36 The calorimeter described in Problem 4.35 was used to determine $\Delta_r H°$ for the reaction

$$\mathrm{CH_3COOH\ (in\ 100\ H_2O) + NaOH\ (in\ 100\ H_2O) \rightarrow NaCH_3COO\ (in\ 201\ H_2O)}$$

Exactly 25 mL of 0.500 M NaOH and 25.00 mL of 0.500 M CH_3COOH were mixed in the calorimeter, and a 2.934-K temperature increase was observed. Given $\rho = 1.0103\ \mathrm{g \cdot mL^{-1}}$ and $c_P = 4.113\ \mathrm{J \cdot K^{-1} \cdot g^{-1}}$ for the resulting $NaCH_3COO$ solution, determine $\Delta_r H°$.

▮ Both reactants are being added in stoichiometric amounts. The amount of reaction is

$$(25.00\ \mathrm{mL})\frac{10^{-3}\ \mathrm{L}}{1\ \mathrm{mL}}\left(\frac{0.500\ \mathrm{mol\ HCl}}{1\ \mathrm{L}}\right) = 1.25 \times 10^{-2}\ \mathrm{mol\ HCl}$$

The energy generated can be calculated from *(3.26)* as

$$\Delta H = [C_P(\text{calorimeter}) + mc_P]\,\Delta T$$
$$= [30.1\ \mathrm{J \cdot K^{-1}} + (50.00\ \mathrm{mL})(1.0103\ \mathrm{g \cdot mL^{-1}})(4.113\ \mathrm{J \cdot K^{-1} \cdot g^{-1}})](2.934\ \mathrm{K})$$
$$= 697.9\ \mathrm{J}$$

The heat of reaction is

$$\frac{(-697.9\ \mathrm{J})[(10^{-3}\ \mathrm{kJ})/(1\ \mathrm{J})]}{1.25 \times 10^{-2}\ \mathrm{mol\ HCl}} = -55.8\ \mathrm{kJ \cdot (mol\ HCl)^{-1}}$$

Thus $\Delta_r H° = -55.8\ \mathrm{kJ}$ for the above thermochemical equation.

4.37 An adiabatic bomb calorimeter is commonly used to measure heats of combustion of substances under constant-volume conditions. A 0.5421-g sample of benzoic acid was burned in a calorimeter with a resulting temperature increase of 2.680 K. What is the calorimeter constant?

$$\mathrm{C_6H_5COOH(s) + \tfrac{15}{2}O_2(g) \rightarrow 7CO_2(g) + 3H_2O(l)} \qquad \Delta_r U° = -3227\ \mathrm{kJ}$$

▮ The amount of heat released is

$$(0.5421\ \mathrm{g\ C_6H_5COOH})\left(\frac{1\ \mathrm{mol}}{122.12\ \mathrm{g \cdot mol^{-1}}}\right)\left(\frac{-3227\ \mathrm{kJ}}{1\ \mathrm{mol\ C_6H_5COOH}}\right) = -14.32\ \mathrm{kJ}$$

A bomb calorimeter is designed so that $C_P(\text{contents}) \ll C_P(\text{calorimeter})$. Thus *(3.26)* gives

$$C_P(\text{calorimeter}) = \frac{14.32\ \mathrm{kJ}}{2.680\ \mathrm{K}} = 5.343\ \mathrm{kJ \cdot K^{-1}}$$

4.38 The calorimeter described in Problem 4.37 was used to determine $\Delta_r U°$ for the combustion of naphthalene, $C_{10}H_8(s)$. The burning of a 0.6521-g sample of naphthalene caused a 1.173-K increase in temperature. What is $\Delta_r U°$ for $\mathrm{C_{10}H_8(s) + 12O_2(g) \rightarrow 10CO_2(g) + 4H_2O(l)}$?

▮ The amount of reaction is

$$\frac{0.6521\ \mathrm{g\ C_{10}H_8}}{128.19\ \mathrm{g \cdot mol^{-1}}} = 5.087 \times 10^{-3}\ \mathrm{mol\ C_{10}H_8}$$

The energy generated can be calculated from *(3.26)* as

$$\Delta H = (5.343\ \text{kJ}\cdot\text{K}^{-1})(1.173\ \text{K}) = 6.267\ \text{kJ}$$

The heat of reaction is

$$\frac{-6.267\ \text{kJ}}{5.087\times 10^{-3}\ \text{mol}\ C_{10}H_8} = -1232\ \text{kJ}\cdot(\text{mol}\ C_{10}H_8)^{-1}$$

Thus $\Delta_r U^\circ = -1232$ kJ for the above thermochemical equation.

4.39 As a 0.1000-mol sample of solid NH_4Cl was dissolved in 50.0 mol of water, the temperature of the solution decreased. A small electrical immersion heater restored the temperature of the system by passing 0.1246 A from a 15.03-V power supply for a period of 809 s. Determine $\Delta_r H^\circ$ for

$$NH_4Cl(s) \xrightarrow{500\ H_2O} NH_4Cl\ (\text{in } 500\ H_2O)$$

▮ The amount of energy absorbed by the endothermic reaction is

$$\Delta H = VIt = (15.03\ \text{V})(0.1246\ \text{A})(809\ \text{s})\left(\frac{1\ \text{kg}\cdot\text{m}^2\cdot\text{s}^{-3}\cdot\text{A}^{-1}}{1\ \text{V}}\right)\left(\frac{1\ \text{J}}{1\ \text{kg}\cdot\text{m}^2\cdot\text{s}^{-2}}\right)$$

$$= 1515\ \text{J}$$

Both reactants were added in stoichiometric amounts, so scaling the above answer by 10 gives $\Delta_r H^\circ = 15.15$ kJ.

4.2 CALCULATIONS INVOLVING THERMOCHEMICAL EQUATIONS

4.40 Use the following thermochemical equations

$$2ClO_2(g) + O_3(g) \rightarrow Cl_2O_7(g) \qquad \Delta_r H^\circ = -75.7\ \text{kJ}$$

$$O_3(g) \rightarrow O_2(g) + O(g) \qquad \Delta_r H^\circ = 106.7\ \text{kJ}$$

$$2ClO_3(g) + O(g) \rightarrow Cl_2O_7(g) \qquad \Delta_r H^\circ = -287\ \text{kJ}$$

$$O_2(g) \rightarrow 2O(g) \qquad \Delta_r H^\circ = 498.340\ \text{kJ}$$

to determine $\Delta_r H^\circ$ for $ClO_2(g) + O(g) \rightarrow ClO_3(g)$.

▮ Because $\Delta_r H^\circ$ is a state function, it can be calculated from a given sequence of thermochemical equations that can be combined algebraically to give the desired equation (*Hess's law*). In this case, multiplying both the first equation and the reverse of the third equation by 1/2 and adding gives

$$\tfrac{1}{2}[2ClO_2(g) + O_3(g) \rightarrow Cl_2O_7(g) \qquad \Delta_r H^\circ = -75.7\ \text{kJ}]$$
$$\tfrac{1}{2}[Cl_2O_7(g) \rightarrow 2ClO_3(g) + O(g) \qquad \Delta_r H^\circ = 287\ \text{kJ}]$$
$$\overline{ClO_2(g) + \tfrac{1}{2}O_3(g) \rightarrow ClO_3(g) + \tfrac{1}{2}O(g) \qquad \Delta_r H^\circ = 106\ \text{kJ}}$$

which has the "major" reactants and products—the ClO_2 and ClO_3—in place. To eliminate the O_3 from the reactants side, add one-half the reverse of the second equation:

$$ClO_2(g) + \tfrac{1}{2}O_3(g) \rightarrow ClO_3(g) + \tfrac{1}{2}O(g) \qquad \Delta_r H^\circ = 106\ \text{kJ}$$
$$\tfrac{1}{2}[O_2(g) + O(g) \rightarrow O_3(g) \qquad \Delta_r H^\circ = -106.7\ \text{kJ}]$$
$$\overline{ClO_2(g) + \tfrac{1}{2}O_2(g) \rightarrow ClO_3(g) \qquad \Delta_r H^\circ = 53\ \text{kJ}}$$

The desired equation can be obtained by adding one-half the reverse of the fourth equation:

$$ClO_2(g) + \tfrac{1}{2}O_2(g) \rightarrow ClO_3(g) \qquad \Delta_r H^\circ = 53\ \text{kJ}$$
$$\tfrac{1}{2}[2O(g) \rightarrow O_2(g) \qquad \Delta_r H^\circ = -498.340\ \text{kJ}]$$
$$\overline{ClO_2(g) + O(g) \rightarrow ClO_3(g) \qquad \Delta_r H^\circ = -196\ \text{kJ}}$$

If there were additional thermochemical equations given in the statement of the problem, they would not be used.

4.41 Use the following thermochemical equations:

$$\tfrac{1}{2}Cl_2(g) + O_2(g) \rightarrow ClO_2(g) \qquad \Delta_r H^\circ = 102.5\ \text{kJ}$$

$$\tfrac{1}{2}Cl_2(g) + \tfrac{3}{2}O_2(g) \rightarrow ClO_3(g) \qquad \Delta_r H^\circ = 155\ \text{kJ}$$

$$Cl_2(g) + \tfrac{7}{2}O_2(g) \rightarrow Cl_2O_7(g) \qquad \Delta_r H^\circ = 272.0\ \text{kJ}$$

to determine $\Delta_r H^\circ$ for $3ClO_3(g) \rightarrow Cl_2O_7(g) + ClO_2(g)$.

▮ Usually Hess's law problems are done in one step instead of a series of steps as illustrated in Problem 4.40. For the desired equation,

$$\begin{array}{lr} 3[ClO_3(g) \rightarrow \tfrac{1}{2}Cl_2(g) + \tfrac{3}{2}O_2(g) & \Delta_r H^\circ = -155\ \text{kJ}] \\ Cl_2(g) + \tfrac{7}{2}O_2(g) \rightarrow Cl_2O_7(g) & \Delta_r H^\circ = 272.0\ \text{kJ} \\ \tfrac{1}{2}Cl_2(g) + O_2(g) \rightarrow ClO_2(g) & \Delta_r H^\circ = 102.5\ \text{kJ} \\ \hline 3ClO_3(g) \rightarrow Cl_2O_7(g) + ClO_2(g) & \Delta_r H^\circ = -91\ \text{kJ} \end{array}$$

4.42 Use the following thermochemical equations:

$$\begin{array}{lr} Ag(s) \rightarrow Ag^+(aq) + e^- & \Delta_r H^\circ = 105.579\ \text{kJ} \\ \tfrac{1}{2}Cl_2(g) + e^- \rightarrow Cl^-(aq) & \Delta_r H^\circ = -167.159\ \text{kJ} \\ Br_2(l) + 2e^- \rightarrow 2Br^-(aq) & \Delta_r H^\circ = -243.10\ \text{kJ} \\ AgCl(s) \rightarrow Ag^+(aq) + Cl^-(aq) & \Delta_r H^\circ = 65.488\ \text{kJ} \\ AgBr(s) \rightarrow Ag^+(aq) + Br^-(aq) & \Delta_r H^\circ = 84.40\ \text{kJ} \end{array}$$

to determine which of the thermochemical equations:

$$Ag(s) + \tfrac{1}{2}Cl_2(g) \rightarrow AgCl(s)$$

$$Ag(s) + \tfrac{1}{2}Br_2(l) \rightarrow AgBr(s)$$

is more exothermic.

▮ Using Hess's law for each desired thermochemical equation gives

$$\begin{array}{lr} Ag(s) \rightarrow Ag^+(aq) + e^- & \Delta_r H^\circ = 105.579\ \text{kJ} \\ \tfrac{1}{2}Cl_2(g) + e^- \rightarrow Cl^-(aq) & \Delta_r H^\circ = -167.159\ \text{kJ} \\ Ag^+(aq) + Cl^-(aq) \rightarrow AgCl(s) & \Delta_r H^\circ = -65.488\ \text{kJ} \\ \hline Ag(s) + \tfrac{1}{2}Cl_2(g) \rightarrow AgCl(s) & \Delta_r H^\circ = -127.068\ \text{kJ} \end{array}$$

$$\begin{array}{lr} Ag(s) \rightarrow Ag^+(aq) + e^- & \Delta_r H^\circ = 105.579\ \text{kJ} \\ \tfrac{1}{2}[Br_2(l) + 2e^- \rightarrow 2Br^-(aq) & \Delta_r H^\circ = -243.10\ \text{kJ}] \\ Ag^+(aq) + Br^-(aq) \rightarrow AgBr(s) & \Delta_r H^\circ = -84.40\ \text{kJ} \\ \hline Ag(s) + \tfrac{1}{2}Br_2(l) \rightarrow AgBr(s) & \Delta_r H^\circ = -100.37\ \text{kJ} \end{array}$$

The silver–chlorine reaction is more exothermic.

4.43 Write the chemical equation describing the formation of $Na_2SO_4(s)$ at 25 °C. Repeat this question for 100 °C.

▮ The general format of a formation equation is to list the elements in their standard states on the reactants' side of the chemical equation and to list 1 mol of the substance in its standard state as the only product on the right-hand side of the equation. At 25 °C, Na is a solid metal, S exists in the rhombic crystalline form, and O is a diatomic gaseous molecule. Thus the chemical equation would be

$$2Na(s) + S(\text{rhombic}) + 2O_2(g) \rightarrow Na_2SO_4(s)$$

At 100 °C, the standard states of two of the reactants have changed: Na is a molten metal and S exists in the monoclinic crystalline form. Thus

$$2Na(l) + S(\text{monoclinic}) + 2O_2(g) \rightarrow Na_2SO_4(s)$$

4.44 At 25 °C, the enthalpy of formation ($\Delta_f H^\circ_{298}$) for any element in its standard state is zero. What is the value for $O_2(g)$ at 350 K, given $C_{P,m} = 29.6\ \text{J} \cdot \text{K}^{-1} \cdot \text{mol}^{-1}$?

▮ Even though the thermal enthalpy has increased by

$$H^\circ_{350} - H^\circ_{298} = (29.6\ \text{J} \cdot \text{K}^{-1} \cdot \text{mol}^{-1})(350\ \text{K} - 298\ \text{K}) = 1540\ \text{J}$$

the heat of formation is still defined as 0.

4.45 Determine $\Delta_f H^\circ_{298}$ for gaseous water. See Problem 4.33 for additional data.

▮ The thermochemical equation and the value of $\Delta_r H^\circ_{298}$ calculated in Problem 4.33 corresponds to $\Delta_f H^\circ_{298}$ for gaseous water.

4.46 Determine $\Delta_r H^\circ$ for $3ClO_3(g) \rightarrow Cl_2O_7(g) + ClO_2(g)$ using the heat of formation data given in Table 4-1.

Table 4-1

	$\Delta_f H^\circ_{298}/(kJ \cdot mol^{-1})$		$\Delta_f H^\circ_{298}/(kJ \cdot mol^{-1})$
$O(g)$	249.170	$S^{2+}(g)$	3 543.4
$O_3(g)$	142.7	$S^{3+}(g)$	6 908.2
$H(g)$	217.965	$S^{4+}(g)$	11 478.8
$H^+(g)$	1 536.202	$S^{5+}(g)$	18 497
$H^-(g)$	138.99	$S^{6+}(g)$	26 995
$OH^-(aq)$	−229.994	$S^{7+}(g)$	54 107
$H_2O(g)$	−241.818	$HS^-(aq)$	−17.6
$H_2O(l)$	−285.830	$H_2S(aq)$	−39.7
$HDO(l)$	−289.888	$NH_3(g)$	−46.11
$D_2O(l)$	−294.600	$C(g)$	716.682
$H_2O_2(g)$	−136.31	$CO_2(g)$	−393.509
$H_2O_2(l)$	−187.78	$CS_2(g)$	117.36
$F^-(s)$	−255.39	$Ag^+(aq)$	105.579
$Cl^-(g)$	−233.13	$AgBr_3^{2-}(aq)$	−284.5
$Cl^-(aq)$	−167.159	Co(β-solid)	0.46
$HCl(g)$	−92.307	*cis*-$[Co(NH_3)_4(NO_2)_2]I(s)$	−745.6
$HCl(aq)$	−167.159	*trans*-$[Co(NH_3)_4(NO_2)]_2I(s)$	−744.8
$ClO_2(g)$	102.5	$K^+(g)$	514.26
$ClO_3(g)$	155	$KOH(aq)$	−482.37
$Cl_2O_7(g)$	272.0	$KF(s)$	−567.27
$Br^-(g)$	−219.07	$KCl(s)$	−436.747
$Br^-(aq)$	−121.55	$KCl(aq)$	−419.53
$I^-(g)$	−197	$KBr(s)$	−393.793
$S(g)$	278.805	$KI(s)$	−327.900
$S^+(g)$	1 284.111		

▮ Enthalpy of formation data can be combined to calculate $\Delta_r H^\circ$ using a special case of Hess's law,

$$\Delta_r H^\circ_T = \sum_i \nu_i \, \Delta_f H^\circ_{T,i} \tag{4.5}$$

where the ν_i are the stoichiometric coefficients in the desired equation (ν_i is positive for products and negative for reactants) and have the units of mol. For the desired equation,

$$\begin{aligned}\Delta_r H^\circ &= [(1\text{ mol})\,\Delta_f H^\circ(Cl_2O_7) + (1\text{ mol})\,\Delta_f H^\circ(ClO_2)] - [(3\text{ mol})\,\Delta_f H^\circ(ClO_3)] \\ &= [(1\text{ mol})(272.0\text{ kJ}\cdot\text{mol}^{-1}) + (1\text{ mol})(102.5\text{ kJ}\cdot\text{mol}^{-1})] - [(3\text{ mol})(155\text{ kJ}\cdot\text{mol}^{-1})] \\ &= -91\text{ kJ}\end{aligned}$$

Note that this is the same as the answer determined in Problem 4.41 using the complete Hess law calculations.

4.47 The stable form of Co(s) at 25 °C is the α form (a hexagonal crystal). Determine ΔH° for the equation Co(α solid) → Co(β solid).

▮ Recalling that $\Delta_f H^\circ = 0$ for the stable form of an element, substituting the data from Table 4-1 into *(4.5)* gives

$$\begin{aligned}\Delta H^\circ &= [(1\text{ mol})\,\Delta_f H^\circ(\beta)] - [(1\text{ mol})\,\Delta_f H^\circ(\alpha)] \\ &= [(1\text{ mol})(0.46\text{ kJ}\cdot\text{mol}^{-1})] - [(1\text{ mol})(0)] \\ &= 0.46\text{ kJ}\end{aligned}$$

4.48 Use the data of Table 4-1 to determine the enthalpy of reaction for $H_2O(l) + D_2O(l) \rightarrow 2HDO(l)$, where D represents deuterium, ^2H.

▮ Using *(4.5)* gives

$$\Delta_r H° = [(2\text{ mol})\,\Delta_f H°(\text{HDO})] - [(1\text{ mol})\,\Delta_f H°(\text{H}_2\text{O}) + (1\text{ mol})\,\Delta_f H°(\text{D}_2\text{O})]$$
$$= [(2\text{ mol})(-289.888\text{ kJ}\cdot\text{mol}^{-1})] - [(1\text{ mol})(-285.830\text{ kJ}\cdot\text{mol}^{-1}) + (1\text{ mol})(-294.600\text{ kJ}\cdot\text{mol}^{-1})]$$
$$= 0.654\text{ kJ}$$

4.49 Determine $\Delta_f H°$ for O(g) given

$$\text{O}_3(\text{g}) \rightarrow \text{O}_2(\text{g}) + \text{O}(\text{g}) \qquad \Delta_r H° = 106.5\text{ kJ}$$

▮ Substituting the data given in Table 4-1 into *(4.5)* and solving for $\Delta_f H°(\text{O})$ gives

$$\Delta_r H° = [(1\text{ mol})\,\Delta_f H°(\text{O}_2) + (1\text{ mol})\,\Delta_f H°(\text{O})] - [(1\text{ mol})\,\Delta_f H°(\text{O}_3)]$$
$$106.5\text{ kJ} = [(1\text{ mol})(0) + (1\text{ mol})\,\Delta_f H°(\text{O})] - [(1\text{ mol})(142.7\text{ kJ}\cdot\text{mol}^{-1})]$$
$$\Delta_f H°(\text{O}) = 249.2\text{ kJ}\cdot\text{mol}^{-1}$$

4.50 Determine $\Delta_f H°$ for $\text{Cl}^-(\text{aq})$.

▮ For aqueous solutions of strong electrolytes (salts, strong acids, etc.) the symbols HCl(aq) and $\text{H}^+(\text{aq}) + \text{Cl}^-(\text{aq})$ mean the same thing. Thus, for the equation $\text{HCl(aq)} \rightarrow \text{H}^+(\text{aq}) + \text{Cl}^-(\text{aq})$, $\Delta_r H = 0$. Substituting $\Delta_f H° = 0$ for $\text{H}^+(\text{aq})$ into *(4.5)* gives

$$\Delta_r H° = [(1\text{ mol})\,\Delta_f H°(\text{H}^+) + (1\text{ mol})\,\Delta_f H°(\text{Cl}^-)] - [(1\text{ mol})\,\Delta_f H°(\text{HCl})]$$
$$0 = [(1\text{ mol})(0) + (1\text{ mol})\,\Delta_f H°(\text{Cl}^-)] - [(1\text{ mol})(-167.159\text{ kJ}\cdot\text{K}^{-1}\cdot\text{mol}^{-1})]$$

Solving gives $\Delta_f H°(\text{Cl}^-) = -167.159\text{ kJ}\cdot\text{mol}^{-1}$.

4.51 Determine $\Delta_f H°$ for $\text{K}^+(\text{aq})$.

▮ For the equation $\text{KCl(aq)} \rightarrow \text{K}^+(\text{aq}) + \text{Cl}^-(\text{aq})$, $\Delta_r H° = 0$. Substituting the data from Table 4-1 into *(4.5)* gives

$$0 = [(1\text{ mol})\,\Delta_f H°(\text{K}^+) + (1\text{ mol})\,\Delta_f H°(\text{Cl}^-)] - [(1\text{ mol})\,\Delta_f H°(\text{KCl})]$$
$$0 = [(1\text{ mol})\,\Delta_f H°(\text{K}^+) + (1\text{ mol})(-167.159\text{ kJ}\cdot\text{mol}^{-1})] - [(1\text{ mol})(-419.53\text{ kJ}\cdot\text{mol}^{-1})]$$

Solving gives $\Delta_f H°(\text{K}^+) = -252.37\text{ kJ}\cdot\text{mol}^{-1}$.

4.52 Calculate $\Delta_r H°$ for $\text{Cl}_2(\text{g}) + 2\text{Br}^-(\text{aq}) \rightarrow \text{Br}_2(\text{l}) + 2\text{Cl}^-(\text{aq})$.

▮ Substituting the data of Table 4-1 into *(4.5)* gives

$$\Delta_r H° = [(1\text{ mol})\,\Delta_f H°(\text{Br}_2) + (2\text{ mol})\,\Delta_f H°(\text{Cl}^-)] - [(1\text{ mol})\,\Delta_f H°(\text{Cl}_2) + (2\text{ mol})\,\Delta_f H°(\text{Br}^-)]$$
$$= [(1\text{ mol})(0) + (2\text{ mol})(-167.159\text{ kJ}\cdot\text{mol}^{-1})] - [(1\text{ mol})(0) + (2\text{ mol})(-121.55\text{ kJ}\cdot\text{mol}^{-1})]$$
$$= -91.22\text{ kJ}$$

4.53 Write the chemical equation describing the combustion of liquid methanol.

▮ The general format of a combustion equation is to list 1 mol of the substance in its standard state and sufficient $\text{O}_2(\text{g})$ to balance the equation as the only reactants and to list the combustion products in their standard states on the products' side of the equation. Thus the desired equation would be

$$\text{CH}_3\text{OH(l)} + \tfrac{3}{2}\text{O}_2(\text{g}) \rightarrow \text{CO}_2(\text{g}) + 2\text{H}_2\text{O(l)}$$

4.54 Interpret the thermochemical equation

$$\text{H}_2(\text{g}) + \tfrac{1}{2}\text{O}_2(\text{g}) \rightarrow \text{H}_2\text{O(l)} \qquad \Delta_r H° = -285.830\text{ kJ}$$

▮ This equation is in the correct format to describe the formation of liquid water, so $\Delta_f H° = -285.830\text{ kJ}\cdot\text{mol}^{-1}$ for $\text{H}_2\text{O(l)}$. This equation is also in the correct format to describe the combusion of hydrogen, so $\Delta_c H° = -285.830\text{ kJ mol}^{-1}$ for $\text{H}_2(\text{g})$.

4.55 Using the enthalpy of combustion data given for acetylene in Table 4-2, determine the enthalpy of formation of $\text{C}_2\text{H}_2(\text{g})$.

Table 4-2

	$\Delta_c H^\circ_{298}/(\text{kJ} \cdot \text{mol}^{-1})$
$H_2(g)$	−285.830
C(graphite)	−393.509
CO(g)	−282.984
$CH_4(g)$	−890.36
$C_2H_2(g)$	−1299.57
$C_2H_4(g)$	−1410.93
cis-CHCOOH=CHCOOH(s)(maleic acid)	−1355.16
trans-CHCOOH=CHCOOH(s)(fumaric acid)	−1334.65
B(s)	−636.39
$B_5H_9(l)$	−4510.84

▮ Using data from Table 4-2 and Hess's law gives

$$\begin{array}{ll} 2CO_2 + H_2O(l) \rightarrow \frac{5}{2}O_2(g) + C_2H_2(g) & \Delta_r H^\circ = 1299.57\ \text{kJ} \\ H_2(g) + \frac{1}{2}O_2(g) \rightarrow H_2O(l) & \Delta_r H^\circ = -285.830\ \text{kJ} \\ (2)[C(s) + O_2(g) \rightarrow CO_2(g) & \Delta_r H^\circ = -393.509\ \text{kJ}] \\ \hline 2C(s) + H_2(g) \rightarrow C_2H_2(g) & \Delta_f H^\circ = 226.72\ \text{kJ} \end{array}$$

4.56 Determine $\Delta_r H^\circ$ for $C_2H_2(g) + H_2(g) \rightarrow C_2H_4(g)$ using enthalpy of combustion data.

▮ Enthalpy of combustion data can be combined to calculate $\Delta_r H^\circ$ using a special case of Hess's law:

$$\Delta_r H^\circ_T = -\sum_i \nu_i\, \Delta_c H^\circ_{T,i} \tag{4.6}$$

where the ν_i have been defined in Problem 4.46. For the desired equation,

$$\begin{aligned} \Delta_r H^\circ &= -[(1\ \text{mol})\,\Delta_c H^\circ(C_2H_4)] + [(1\ \text{mol})\,\Delta_c H^\circ(C_2H_2) + (1\ \text{mol})\,\Delta_c H^\circ(H_2)] \\ &= -[(1\ \text{mol})(-1410.93\ \text{kJ} \cdot \text{mol}^{-1})] + [(1\ \text{mol})(-1299.5\ \text{kJ} \cdot \text{mol}^{-1}) + (1\ \text{mol})(-285.830\ \text{kJ} \cdot \text{mol}^{-1})] \\ &= -174.47\ \text{kJ} \end{aligned}$$

4.57 Using the data given in Table 4-2, determine $\Delta_f H^\circ$ for $B_5H_9(l)$.

▮ The formation equation is $5B(s) + \frac{9}{2}H_2(g) \rightarrow B_5H_9(l)$. Substituting the combustion enthalpy data into *(4.6)* gives

$$\begin{aligned} \Delta_f H^\circ &= -[(1\ \text{mol})\,\Delta_c H^\circ(B_5H_9)] + [(5\ \text{mol})\,\Delta_c H^\circ(B) + (\tfrac{9}{2}\ \text{mol})\,\Delta_c H^\circ(H_2)] \\ &= -[(1\ \text{mol})(-4510.84\ \text{kJ} \cdot \text{mol}^{-1})] + [(5\ \text{mol})(-636.39\ \text{kJ} \cdot \text{mol}^{-1}) + (\tfrac{9}{2}\ \text{mol})(-285.830\ \text{kJ} \cdot \text{mol}^{-1})] \\ &= 42.66\ \text{kJ} \end{aligned}$$

4.58 Although methane can be used as a fuel directly, it can be reacted with steam to generate a water-gas type of fuel:

$$CH_4(g) + H_2O(g) \rightarrow CO(g) + 3H_2(g)$$

Which generates more heat burning—pure CH_4 or the CO–H_2 mixture?

▮ The enthalpy of combustion of 1 mol of methane is −890.36 kJ. The enthalpy of combustion of the mixture will be

$$\begin{aligned} \Delta_c H^\circ &= (1\ \text{mol})\,\Delta_c H^\circ(CO) + (3\ \text{mol})\,\Delta_c H^\circ(H_2) \\ &= (1\ \text{mol})(-282.984\ \text{kJ} \cdot \text{mol}^{-1}) + (3\ \text{mol})(-285.830\ \text{kJ} \cdot \text{mol}^{-1}) \\ &= -1140.474\ \text{kJ} \end{aligned}$$

The mixture is the better fuel.

4.59 Determine the bond enthalpy for an oxygen-oxygen double bond. Use the data given in Table 4-1.

▮ The bond enthalpy (energy) is the energy needed to break 1 mol of that type of bond into atomic fragments with all species in the gaseous state. For the O=O bond, this corresponds to the thermochemical equation

$$O{=}O(g) \rightarrow 2O(g) \qquad \Delta_r H^\circ = \text{BE}$$

Using *(4.5)* and the data given in Table 4-1,

$$\begin{aligned} \text{BE} &= [(2\text{ mol})\,\Delta_f H^\circ(\text{O})] - [(1\text{ mol})\,\Delta_f H^\circ(\text{O}_2)] \\ &= [(2\text{ mol})(249.170\text{ kJ}\cdot\text{mol}^{-1})] - [(1\text{ mol})(0)] \\ &= 498.340\text{ kJ} \end{aligned}$$

This value is very similar to that given in Table 4-3 for O=O.

Table 4-3. BE/(kJ · mol^{-1})

	C	N	S	O	I	Br	Cl	F	H
H—	414	389	368	464	297	368	431	569	435
F—	490	280	343	213	280	285	255	159	
Cl—	326	201	272	205	209	218	243		
Br—	272	163	209	—	176	192			
I—	218	—	—	—	151				
O—	326	230	423	142					
O=	803[a]	590[b]	523	498					
O≡	1075	—	—						
S—	289	—	247						
S=	582	—							
N—	285	159							
N=	515	473							
N≡	858	946							
C—	331								
C=	590[c]								
C≡	812								

[a] 728 if —C(—)=O.
[b] 406 if $-NO_2$; 368 if $-NO_3$.
[c] 506 if alternating — and =.

4.60 Use the following thermochemical equations to determine the average bond enthalpy for a C—H bond.

$$CH_4(g) \rightarrow CH_3(g) + H(g) \qquad \Delta_r H^\circ = 438.47\text{ kJ}$$

$$CH_3(g) \rightarrow CH_2(g) + H(g) \qquad \Delta_r H^\circ = 462.65\text{ kJ}$$

$$CH_2(g) \rightarrow CH(g) + H(g) \qquad \Delta_r H^\circ = 423.40\text{ kJ}$$

$$CH(g) \rightarrow C(g) + H(g) \qquad \Delta_r H^\circ = 338.85\text{ kJ}$$

▮ The four equations can be added using Hess's law to give

$$CH_4(g) \rightarrow C(g) + 4H(g) \qquad \Delta_r H^\circ = 1663.37\text{ kJ}$$

The average bond enthalpy is

$$\overline{\text{BE}}(\text{C—H}) = \frac{1663.37\text{ kJ}}{4\text{ mol}} = 415.84\text{ kJ}\cdot\text{mol}^{-1}$$

Because the actual C—H bond energy depends on the chemical environment, values of $\overline{\text{BE}}_i$ are usually reported to only the units place and represent averages determined for several molecules.

4.61 Use the data given in Table 4-1 to calculate the average C=S bond enthalpy.

▮ For the equation $S{=}C{=}S(g) \rightarrow C(g) + 2S(g)$, the average bond enthalpy will be $\Delta_r H^\circ/2$. Substituting the data into *(4.5)* and solving gives

$$\Delta_r H° = [(1\text{ mol})\,\Delta_f H°(\text{C}) + (2\text{ mol})\,\Delta_f H°(\text{S})] - (1\text{ mol})\,\Delta_f H°(\text{CS}_2)$$
$$= [(1\text{ mol})(716.682\text{ kJ}\cdot\text{mol}^{-1}) + (2\text{ mol})(278.805\text{ kJ}\cdot\text{mol}^{-1})] - (1\text{ mol})(117.36\text{ kJ}\cdot\text{mol}^{-1})$$
$$= 1156.93\text{ kJ}$$

$$\text{BE(C=S)} = \frac{1156.93\text{ kJ}}{2\text{ mol}} = 578.47\text{ kJ}\cdot\text{mol}^{-1}$$

4.62 Use the data given in Table 4-1 to determine the average bond enthalpy in ozone, O_3. Briefly discuss the significance of the value.

▌ For the equation $O_3(g) \rightarrow 3O(g)$, the average bond enthalpy will be $\Delta_r H°/2$. Substituting the data into *(4.5)* and solving gives

$$\Delta_r H° = (3\text{ mol})\,\Delta_f H°(\text{O}) - (1\text{ mol})\,\Delta_f H°(\text{O}_3)$$
$$= (3\text{ mol})(249.170\text{ kJ}\cdot\text{mol}^{-1}) - (1\text{ mol})(142.7\text{ kJ}\cdot\text{mol}^{-1})$$
$$= 604.8\text{ kJ}$$

$$\overline{\text{BE}} = \frac{604.8\text{ kJ}}{2\text{ mol}} = 302.4\text{ kJ}\cdot\text{mol}^{-1}$$

This value differs significantly from 142 $\text{kJ}\cdot\text{mol}^{-1}$ for O—O and 498 $\text{kJ}\cdot\text{mol}^{-1}$ for O=O. The molecular structure of O_3 is considered to be an example of resonance (or delocalized π bonding) in which the bonding between oxygen atoms is described as an average of single and double bonding.

4.63 Use the following thermochemical equation to estimate $\overline{\text{BE}}$(C—O).

```
   H
   |
H—C—O—H(g) → C(g) + O(g) + 4H(g)      ΔrH° = 2038.37 kJ
   |
   H
```

▌ The value of $\Delta_r H°$ is simply the sum of the bond enthalpies. Using the values for $\overline{\text{BE}}$(C—H) and $\overline{\text{BE}}$(O—H) given in Table 4-3 gives

$$\Delta_r H° = (3\text{ mol})\,\overline{\text{BE}}\,(\text{C—H}) + (1\text{ mol})\,\overline{\text{BE}}\,(\text{C—O}) + (1\text{ mol})\,\overline{\text{BE}}\,(\text{O—H})$$
$$2038.37\text{ kJ} = (3\text{ mol})(414\text{ kJ}\cdot\text{mol}^{-1}) + (1\text{ mol})\,\overline{\text{BE}}\,(\text{C—O}) + (1\text{ mol})(464\text{ kJ}\cdot\text{mol}^{-1})$$
$$\overline{\text{BE}}\,(\text{C—O}) = 332\text{ kJ}\cdot\text{mol}^{-1}$$

This value of the average bond energy is in good agreement with that given in Table 4-3.

4.64 At absolute zero,

$$\text{C}\equiv\text{O(g)} \rightarrow \text{C(g)} + \text{O(g)} \qquad \text{BE} = 1071.79\text{ kJ}$$

If $\Delta_f H_0°/(\text{kJ}\cdot\text{mol}^{-1}) = -113.801$ for CO(g) and 246.785 for O(g), determine $\Delta_f H_0°$ for C(g).

▌ Substituting the data into *(4.5)* and solving for $\Delta_f H_0°(\text{C})$ gives

$$\text{BE} = [(1\text{ mol})\,\Delta_f H_0°(\text{C}) + (1\text{ mol})\,\Delta_f H_0°(\text{O})] - (1\text{ mol})\,\Delta_f H_0°(\text{CO})$$
$$1071.79\text{ kJ} = [(1\text{ mol})\,\Delta_f H_0°(\text{C}) + (1\text{ mol})(246.785\text{ kJ}\cdot\text{mol}^{-1})] - (1\text{ mol})(-113.801\text{ kJ}\cdot\text{mol}^{-1})$$
$$\Delta_f H_0° = 711.20\text{ kJ}\cdot\text{mol}^{-1}$$

4.65 Use the average bond enthalpy data given in Table 4-3 to estimate the enthalpy of hydrogenation of ethene.

▌ The chemical equation for this process is

```
   H  H                       H  H
   |  |                       |  |
H—C=C—H(g) + H—H(g) → H—C—C—H(g)
                              |  |
                              H  H
```

Bond enthalpy data can be combined to calculate $\Delta_r H°$ using a special case of Hess's law:

$$\Delta_r H_T° = -\sum_i^{\text{products}} n_i\,\overline{\text{BE}}_i + \sum_j^{\text{reactants}} n_j\,\overline{\text{BE}}_j \qquad (4.7)$$

where n_i and n_j represent the numbers of moles of bonds involved in the reaction. For the hydrogenation reaction, substituting the data from Table 4-3 into *(4.7)* gives

$$\begin{aligned}\Delta_r H^\circ &= -[(6\text{ mol})\,\overline{\text{BE}}\,(\text{C—H}) + (1\text{ mol})\,\overline{\text{BE}}\,(\text{C—C})]\\ &\quad + [(4\text{ mol})\,\overline{\text{BE}}\,(\text{C—H}) + (1\text{ mol})\,\overline{\text{BE}}\,(\text{C=C}) + (1\text{ mol})\,\overline{\text{BE}}\,(\text{H—H})]\\ &= -[(6\text{ mol})(414\text{ kJ}\cdot\text{mol}^{-1}) + (1\text{ mol})(331\text{ kJ}\cdot\text{mol}^{-1})]\\ &\quad + [(4\text{ mol})(414\text{ kJ}\cdot\text{mol}^{-1}) + (1\text{ mol})(590\text{ kJ}\cdot\text{mol}^{-1}) + (1\text{ mol})(435\text{ kJ}\cdot\text{mol}^{-1})]\\ &= -134\text{ kJ}\end{aligned}$$

The actual value is −136.94 kJ.

4.66 The enthalpy of combustion of formaldehyde to form gaseous water is $-527\text{ kJ}\cdot\text{mol}^{-1}$. Determine the C=O bond enthalpy in this molecule.

▮ For the equation

$$\text{H—}\overset{\overset{\displaystyle\text{O}}{\|}}{\text{C}}\text{—H(g)} + \text{O=O(g)} \rightarrow \text{O=C=O(g)} + \text{H—O—H(g)}$$

substituting the data from Table 4-3 into *(4.7)* gives

$$\begin{aligned}\Delta_r H^\circ &= -[(2\text{ mol})\,\overline{\text{BE}}\,(\text{C=O}) + (2\text{ mol})\,\overline{\text{BE}}\,(\text{O—H})]\\ &\quad + [(2\text{ mol})\,\overline{\text{BE}}\,(\text{C—H}) + (1\text{ mol})\,\overline{\text{BE}}\,(\text{C=O}) + (1\text{ mol})\,\overline{\text{BE}}\,(\text{O=O})]\\ -527\text{ kJ} &= -[(2\text{ mol})(803\text{ kJ}\cdot\text{mol}^{-1}) + (2\text{ mol})(464\text{ kJ}\cdot\text{mol}^{-1})]\\ &\quad + [(2\text{ mol})(414\text{ kJ}\cdot\text{mol}^{-1}) + (1\text{ mol})\text{ BE (C=O)} + (1\text{ mol})(498\text{ kJ}\cdot\text{mol}^{-1})]\\ \text{BE (C=O)} &= 681\text{ kJ}\cdot\text{mol}^{-1}\end{aligned}$$

4.67 Use average bond enthalpy data to calculate the value of $\Delta_f H^\circ$ for $H_2O_2(l)$. The enthalpy of vaporization of hydrogen peroxide is $51.47\text{ kJ}\cdot\text{mol}^{-1}$ at 25 °C. Compare the calculated value to the actual value listed in Table 4-1.

▮ The use of *(4.7)* is limited to gaseous reactants and products. For the equation

$$\text{H—H(g)} + \text{O=O(g)} \rightarrow \text{H—O—O—H(g)} \qquad \Delta_f H^\circ$$

substituting the data from Table 4-3 into *(4.7)* gives

$$\begin{aligned}\Delta_f H^\circ &= -[(2\text{ mol})\,\overline{\text{BE}}\,(\text{O—H}) + (1\text{ mol})\,\overline{\text{BE}}\,(\text{O—O})] + [(1\text{ mol})\,\overline{\text{BE}}\,(\text{H—H}) + (1\text{ mol})\,\overline{\text{BE}}\,(\text{O=O})]\\ &= -[(2\text{ mol})(464\text{ kJ}\cdot\text{mol}^{-1}) + (1\text{ mol})(142\text{ kJ}\cdot\text{mol}^{-1})]\\ &\quad + [(1\text{ mol})(435\text{ kJ}\cdot\text{mol}^{-1}) + (1\text{ mol})(498\text{ kJ}\cdot\text{mol}^{-1})]\\ &= -137\text{ kJ}\end{aligned}$$

Combining this value with the vaporization data using Hess's law gives

$$\begin{array}{rr} H_2(g) + O_2(g) \rightarrow H_2O_2(g) & \Delta_f H^\circ = -137\text{ kJ}\\ H_2O_2(g) \rightarrow H_2O_2(l) & \Delta_{vap} H^\circ = -51.47\text{ kJ}\\ \hline H_2(g) + O_2(g) \rightarrow H_2O_2(l) & \Delta_f H^\circ = -188\text{ kJ}\end{array}$$

These values agree exactly within the rules of significant figures.

4.68 Assume benzene to consist of six C—H bonds, three C=C bonds, and three C—C bonds. Use the data in Table 4-3 to calculate the enthalpy combustion for gaseous benzene. The actual value is $-3225\text{ kJ}\cdot\text{mol}^{-1}$. The discrepancy in the values is too large to be explained on the basis of average bond enthalpies. What is the cause of the discrepancy?

▮ For the equation

```
        H
        |
        C
      //  \
  H—C      C—H
    |      ||      (g) + 15/2 O=O(g) → 6O=C=O(g) + 3H—O—H(g)
  H—C      C—H
      \\  /
        C
        |
        H
```

substituting the data into *(4.7)* gives

$$\Delta H_c^\circ = -[(12\text{ mol})\,\overline{\text{BE}}\,(\text{C=O}) + (6\text{ mol})\,\overline{\text{BE}}\,(\text{O—H})] + [(\tfrac{15}{2}\text{ mol})\,\overline{\text{BE}}\,(\text{O=O}) + (6\text{ mol})\,\overline{\text{BE}}\,(\text{C—H})$$
$$+ (3\text{ mol})\,\overline{\text{BE}}\,(\text{C—C}) + (3\text{ mol})\,\overline{\text{BE}}\,(\text{C=C})]$$
$$= -[(12\text{ mol})(803\text{ kJ}\cdot\text{mol}^{-1}) + (6\text{ mol})(464\text{ kJ}\cdot\text{mol}^{-1})]$$
$$+ [(\tfrac{15}{2}\text{ mol})(498\text{ kJ}\cdot\text{mol}^{-1}) + (6\text{ mol})(414\text{ kJ}\cdot\text{mol}^{-1})$$
$$+ (3\text{ mol})(331\text{ kJ}\cdot\text{mol}^{-1}) + (3\text{ mol})(590\text{ kJ}\cdot\text{mol}^{-1})]$$
$$= -3438\text{ kJ}$$

The difference in these values, 213 kJ mol^{-1}, is known as the *resonance energy* and is the result of the delocalized π bonding in the benzene ring. This value is similar to $3(590\text{ kJ}\cdot\text{mol}^{-1} - 506\text{ kJ}\cdot\text{mol}^{-1}) = 252\text{ kJ}$, which can be calculated using the two entries for the C=O average bond enthalpy in Table 4-3. Comparison of hydrogenation reaction calculations (see Problem 4.65) gives the resonance energy a value of about 165 kJ · mol^{-1}.

4.69 Given

$$O_3(g) + OH(g) \rightarrow H(g) + 2O_2(g) \qquad \Delta_r U_0^\circ = 31.95\text{ kJ}$$

and $\bar{\nu}_i/(\text{cm}^{-1}) = 1110$, 705, and 1043 for O_3; 3735.21 for OH; and 1580.1932 for O_2, confirm the value of $\Delta_r H_{373}^\circ$ determined in Problem 4.29.

I For a chemical reaction,

$$\Delta_r H^\circ(\text{thermal}) = \sum_i \nu_i H^\circ(\text{thermal}) \tag{4.8a}$$

$$\Delta_r E^\circ(\text{thermal}) = \sum_i \nu_i E^\circ(\text{thermal}) \tag{4.8b}$$

where the ν_i have been defined in Problem 4.46. For O_3,

$$x_1 = 1.4388(1110)/373 = 4.243$$

Likewise, $x_2 = 2.72$ and $x_3 = 4.023$ for O_3, $x = 14.408$ for OH, and $x = 6.095$ for O_2. Using *(3.12)* and *(3.10)* for O_3 gives

$$H^\circ(\text{thermal, }O_3) = 4(8.314\text{ J}\cdot\text{K}^{-1}\cdot\text{mol}^{-1})(373\text{ K})$$
$$+ (8.314\text{ J}\cdot\text{K}^{-1}\cdot\text{mol}^{-1})(373\text{ K})\left(\frac{4.243}{e^{4.243}-1} + \frac{2.72}{e^{2.72}-1} + \frac{4.023}{e^{4.023}-1}\right)$$
$$= 13\,419\text{ J}\cdot\text{mol}^{-1}$$

Using *(3.12)* and *(3.8)* for OH and O_2 gives

$$H^\circ(\text{thermal, OH}) = \tfrac{7}{2}(8.314\text{ J}\cdot\text{K}^{-1}\cdot\text{mol}^{-1})(373\text{ K}) + \frac{8.314(373)(14.408)}{e^{14.408}-1}$$
$$= 10\,854\text{ J}\cdot\text{mol}^{-1}$$

$$H^\circ(\text{thermal, }O_2) = \tfrac{7}{2}(8.314)(373) + \frac{8.314(373)(6.095)}{e^{6.095}-1} = 10\,897\text{ J}\cdot\text{mol}^{-1}$$

Likewise, using *(3.12)* and *(3.7)* gives

$$H^\circ(\text{thermal, H}) = \tfrac{5}{2}(8.314\text{ J}\cdot\text{K}^{-1}\cdot\text{mol}^{-1})(373\text{ K}) = 7753\text{ J}\cdot\text{mol}^{-1}$$

Using *(4.8a)* gives

$$\Delta_r H^\circ(\text{thermal}) = [(1\text{ mol})H^\circ(\text{thermal, H}) + (2\text{ mol})H^\circ(\text{thermal, }O_2)]$$
$$- [(1\text{ mol})H^\circ(\text{thermal, }O_3) + (1\text{ mol})H^\circ(\text{thermal, OH})]$$
$$= [(1\text{ mol})(7.753\text{ kJ}\cdot\text{mol}^{-1}) + (2\text{ mol})(10.897\text{ kJ}\cdot\text{mol}^{-1})]$$
$$- [(1\text{ mol})(13.419\text{ kJ}\cdot\text{mol}^{-1}) + (1\text{ mol})(10.854\text{ kJ}\cdot\text{mol}^{-1})]$$
$$= 5.274\text{ kJ}$$

At any temperature,

$$\Delta_r H_T^\circ = \Delta_r U_0^\circ + \Delta_r H^\circ(\text{thermal}) \tag{4.9a}$$

$$\Delta_r U_T^\circ = \Delta_r U_0^\circ + \Delta_r E^\circ(\text{thermal}) \tag{4.9b}$$

Substituting into *(4.9a)* gives

$$\Delta_r H^\circ_{373} = 31.95\ \text{kJ} + 5.274\ \text{kJ} = 37.22\ \text{kJ}$$

This value agrees to within 0.3% with the value determined in Problem 4.29.

4.70 Confirm the value of $\Delta_f H^\circ_{500}$ for $NH_3(g)$ calculated in Problem 4.30 given $\bar{\nu}_i/(\text{cm}^{-1}) = 2358.583$ for N_2; 4403.566 for H_2; and 3506, 1022, 3577 (doubly degenerate), and 1691 (doubly degenerate) for NH_3.

▌ From Table 4-1, the thermochemical equation can be written

$$\tfrac{1}{2}N_2(g) + \tfrac{3}{2}H_2(g) \rightarrow NH_3(g) \qquad \Delta_f H^\circ_{298} = -46.11\ \text{kJ}$$

For N_2 at 298 K,

$$x = (1.4388)(2358.583)/298 = 11.38$$

Likewise, $x = 21.25$ for H_2 and $x_1 = 16.20$, $x_2 = 4.932$, $x_3 = 17.26$, and $x_4 = 8.160$. Using *(3.12)* and *(3.10)* for NH_3 at 298 K gives

$$H^\circ(\text{thermal}, NH_3) = 4(8.314\ \text{J}\cdot\text{K}^{-1}\cdot\text{mol}^{-1})(298\ \text{K}) + (8.314\ \text{J}\cdot\text{K}^{-1}\cdot\text{mol}^{-1})(298\ \text{K}) \times\left(\frac{16.20}{e^{16.20}-1} + \frac{4.932}{e^{4.932}-1} + \frac{2(17.26)}{e^{17.26}-1} + \frac{2(8.160)}{e^{8.160}-1}\right) = 10\,016\ \text{J}\cdot\text{mol}^{-1}$$

Using *(3.12)* and *(3.8)* for N_2 and H_2 at 298 K gives

$$H^\circ(\text{thermal}, N_2) = \tfrac{7}{2}(8.314\ \text{J}\cdot\text{K}^{-1}\cdot\text{mol}^{-1})(298\ \text{K}) + \frac{(8.314\ \text{J}\cdot\text{K}^{-1}\cdot\text{mol}^{-1})(298\ \text{K})(11.38)}{e^{11.38}-1} = 8677\ \text{J}\cdot\text{mol}^{-1}$$

$$H^\circ(\text{thermal}, H_2) = \tfrac{7}{2}(8.314)(298) + \frac{8.314(298)(21.25)}{e^{21.25}-1} = 8676\ \text{J}\cdot\text{mol}^{-1}$$

Substituting these results at 298 K into *(4.8a)* gives

$$\begin{aligned}\Delta_r H^\circ(\text{thermal}) &= (1\ \text{mol})H^\circ(\text{thermal}, NH_3) - [(\tfrac{1}{2}\ \text{mol})H^\circ(\text{thermal}, N_2) + (\tfrac{3}{2}\ \text{mol})H^\circ(\text{thermal}, H_2)] \\ &= (1\ \text{mol})(10.016\ \text{kJ}\cdot\text{mol}^{-1}) - [(\tfrac{1}{2}\ \text{mol})(8.677\ \text{kJ}\cdot\text{mol}^{-1}) + (\tfrac{3}{2}\ \text{mol})(8.676\ \text{kJ}\cdot\text{mol}^{-1})] \\ &= -7.337\ \text{kJ}\end{aligned}$$

The value of $\Delta_r U^\circ_0$ for the reaction can be determined using *(4.9a)*:

$$\Delta_r U^\circ_0 = -46.11\ \text{kJ} - (-7.337\ \text{kJ}) = -38.77\ \text{kJ}$$

Now, repeating the calculations for x_i for the substances at 500 K gives $x = 6.787$ for N_2; 12.67 for H_2; and $x_1 = 10.09$, $x_2 = 2.941$, $x_3 = 10.29$, and $x_4 = 4.866$ for NH_3. The respective thermal enthalpy values are $17.629\ \text{kJ}\cdot(\text{mol}\ NH_3)^{-1}$, $14.582\ \text{kJ}\cdot(\text{mol}\ N_2)^{-1}$, and $14.550\ \text{kJ}\cdot(\text{mol}\ H_2)^{-1}$. Using *(4.8a)* at 500 K gives

$$\begin{aligned}\Delta_r H^\circ(\text{thermal}) &= (1\ \text{mol})(17.629\ \text{kJ}\cdot\text{mol}^{-1}) - [(\tfrac{1}{2}\ \text{mol})(14.582\ \text{kJ}\cdot\text{mol}^{-1}) + (\tfrac{3}{2}\ \text{mol})(14.550\ \text{kJ}\cdot\text{mol}^{-1})] \\ &= -11.487\ \text{kJ}\end{aligned}$$

The value of $\Delta_f H^\circ_{500}$ will be given by *(4.9a)* as

$$\Delta_f H^\circ_{500} = -38.77\ \text{kJ} + (-11.487\ \text{kJ}) = -50.26\ \text{kJ}$$

The value calculated in Problem 4.30 is $(-101.14\ \text{kJ})/(2\ \text{mol}) = -50.57\ \text{kJ}\cdot\text{mol}$, which differs by 0.6%.

4.71 Prepare a plot of the bond enthalpy for O=O from 0 K to 2000 K given

$$O{=}O(g) \rightarrow 2O(g) \qquad \Delta_r H^\circ_0 = 493.570\ \text{kJ}$$

For $O_2(g)$, $\nu = 1580.1932\ \text{cm}^{-1}$, and for atomic oxygen, electronic energy data are $\varepsilon/(\text{cm}^{-1}) = 0\ (g_0 = 5)$, $158.265\ (g_1 = 3)$, $226.977\ (g_2 = 1)$, $15\,867.862\ (g_3 = 5)$, and $33\,792.583\ (g_4 = 1)$.

▌ As a sample calculation, consider $T = 298$ K. The thermal enthalpy for $O_2(g)$ will be given by *(3.12)* and *(3.8)* as

$$H^\circ(\text{thermal}, O_2) = \tfrac{7}{2}(8.314\ \text{J}\cdot\text{K}^{-1}\cdot\text{mol}^{-1})(298\ \text{K}) + \frac{(8.314\ \text{J}\cdot\text{K}^{-1}\cdot\text{mol}^{-1})(298\ \text{K})(7.626)}{e^{7.626}-1} = 8686\ \text{J}\cdot\text{mol}^{-1}$$

where the value of x was calculated by the usual method. For atomic oxygen, using *(3.12)* and *(3.7)* for the translational contribution and the expression from Table 3-2 for the electronic contribution gives

$$H°(\text{thermal, O}) = \tfrac{5}{2}(8.314)(298) + (8.314\ \text{J}\cdot\text{K}^{-1}\cdot\text{mol}^{-1})(298\ \text{K})$$
$$\times\left[\frac{5(0)\,e^{-0} + 3(0.7637)\,e^{-0.7637} + 1(1.0953)\,e^{-1.0953} + 5(76.574)\,e^{-76.574} + 1(163.075)\,e^{-163.075}}{5\,e^{-0} + 3\,e^{-0.7637} + 1\,e^{-1.0953} + 5\,e^{-76.574} + 1\,e^{-163.075}}\right]$$
$$= 6725\ \text{J}\cdot\text{mol}^{-1}$$

where the values of ε_j/kT were calculated using $1.4388\varepsilon_j/[T/(\text{K})]$. Substituting into *(4.8a)* gives

$$\Delta_r H°(\text{thermal}) = (2\ \text{mol})H°(\text{thermal, O}) - (1\ \text{mol})H°(\text{thermal, O}_2)$$
$$= (2\ \text{mol})(6.725\ \text{kJ}\cdot\text{mol}^{-1}) - (1\ \text{mol})(8.686\ \text{kJ}\cdot\text{mol}^{-1})$$
$$= 4.764\ \text{kJ}$$

and using *(4.9a)* gives

$$\text{BE(O=O)} = 493.570\ \text{kJ} + 4.764\ \text{kJ} = 498.334\ \text{kJ}$$

The plot is shown in Fig. 4-6.

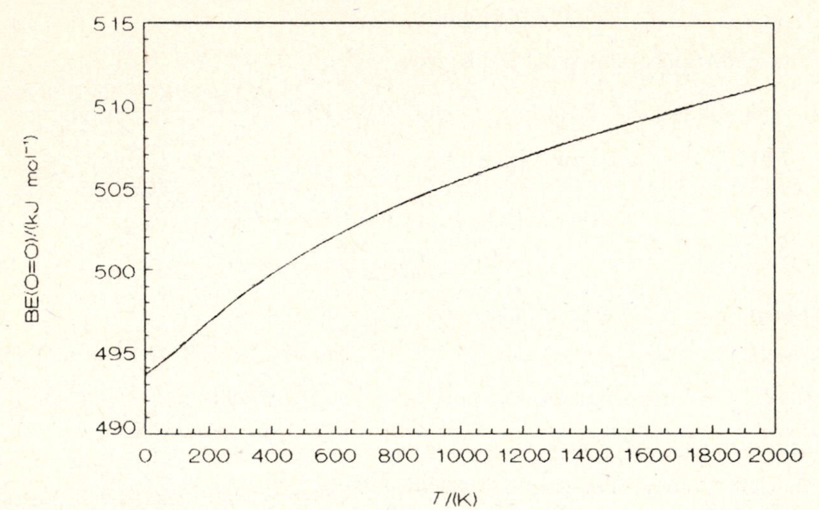

Fig. 4-6

4.3 APPLICATIONS TO SELECTED CHEMICAL REACTIONS

4.72 Using the enthalpy formation data given in Table 4-1 for H(g), H^+(g), and H^-(g), calculate the ionization energy and the electron affinity of H. Which of these is exothermic?

▮ The thermochemical equation for the ionization of a hydrogen atom is

$$\text{H(g)} \rightarrow \text{H}^+\text{(g)} + \text{e}^- \qquad \Delta_r H° = I$$

where I is the *ionization energy*. Using *(4.5)* gives

$$I = (1\ \text{mol})\,\Delta_f H°(\text{H}^+) - (1\ \text{mol})\,\Delta_f H°(\text{H})$$
$$= (1\ \text{mol})(1536.202\ \text{kJ}\cdot\text{mol}^{-1}) - (1\ \text{mol})(217.965\ \text{kJ}\cdot\text{mol}^{-1})$$
$$= 1318.237\ \text{kJ}$$

The thermochemical equation for the gain of an electron by a hydrogen atom is

$$\text{H(g)} + \text{e}^- \rightarrow \text{H}^-\text{(g)} \qquad \Delta_r H° = E_A(\text{H})$$

where E_A is the *electron affinity*. Using *(4.5)* gives

$$E_A(\text{H}) = (1\ \text{mol})\,\Delta_f H°(\text{H}^-) - (1\ \text{mol})\,\Delta_f H°(\text{H})$$
$$= (1\ \text{mol})(138.99\ \text{kJ}\cdot\text{mol}^{-1}) - (1\ \text{mol})(217.965\ \text{kJ}\cdot\text{mol}^{-1})$$
$$= -78.98\ \text{kJ}$$

The electron affinity is exothermic.

4.73 Use the data in Table 4-1 to calculate the first through seventh ionization energies for sulfur. Divide each value by the charge on the ion formed to eliminate the Coulombic effect, and interpret the results.

▮ As a sample calculation, consider the seventh ionization energy

$$S^{6+}(g) \rightarrow S^{7+}(g) + e^- \qquad \Delta_r H^\circ = I(S^{6+})$$

where *(4.5)* gives

$$\begin{aligned} I(S^{6+}) &= (1\ \text{mol})\ \Delta_f H^\circ(S^{7+}) - (1\ \text{mol})\ \Delta_f H^\circ(S^{6+}) \\ &= (1\ \text{mol})(54\,107\ \text{kJ}\cdot\text{mol}^{-1}) - (1\ \text{mol})(26\,995\ \text{kJ}\cdot\text{mol}^{-1}) \\ &= 27\,112\ \text{kJ} \end{aligned}$$

The respective values are $I(S) = 1005.306\ \text{kJ}\cdot\text{mol}^{-1}$, $I(S^+) = 2259.3\ \text{kJ}\cdot\text{mol}^{-1}$, $I(S^{2+}) = 3364.8\ \text{kJ}\cdot\text{mol}^{-1}$, $I(S^{3+}) = 4570.6\ \text{kJ}\cdot\text{mol}^{-1}$, $I(S^{4+}) = 7018\ \text{kJ}\cdot\text{mol}^{-1}$, and $I(S^{5+}) = 8498\ \text{kJ}\cdot\text{mol}^{-1}$. Dividing each value by the charge of the ion formed, for example,

$$\frac{I(S^{6+})}{Z} = \frac{27\,112\ \text{kJ}\cdot\text{mol}^{-1}}{7} = 3873\ \text{kJ}\cdot\text{mol}^{-1}$$

gives $I(S)/Z = 1005.306\ \text{kJ}\cdot\text{mol}^{-1}$, $I(S^+)/Z = 1129.7\ \text{kJ}\cdot\text{mol}^{-1}$, $I(S^{2+})/Z = 1121.6\ \text{kJ}\cdot\text{mol}^{-1}$, $I(S^{3+})/Z = 1142.65\ \text{kJ}\cdot\text{mol}^{-1}$, $I(S^{4+})/Z = 1404\ \text{kJ}\cdot\text{mol}^{-1}$, and $I(S^{5+})/Z = 1416\ \text{kJ}\cdot\text{mol}^{-1}$. The values of $I(S)/Z$, $I(S^+)/Z$, $I(S^{2+})/Z$, and $I(S^{3+})/Z$ are all similar because in each case a $3p$ electron is being ionized. Note that the value of $I(S^+)/Z$ is slightly higher because of the stability of the half-filled $3p$ subshell. The values of $I(S^{4+})/Z$ and $I(S^{5+})/Z$ are similar because a $3s$ electron is being ionized. Because the average position of the $3s$ electrons is closer to the nucleus than that of the $3p$ electrons, these ionization values are a little higher. The value of $I(S^{6+})/Z$ is considerably greater than the others because the electron being removed is a $2p$ electron.

4.74 At 25 °C, $I(\text{Na}) = 502.04\ \text{kJ}\cdot\text{mol}^{-1}$ and $E_A(\text{Cl}) = -354.81\ \text{kJ}\cdot\text{mol}^{-1}$. Determine $\Delta_r H^\circ$ for the equation

$$Na(g) + Cl(g) \rightarrow Na^+(g) + Cl^-(g)$$

Is this a favorable energy change?

▮ Writing the thermochemical equations for each of the processes and using Hess's law gives

$$\begin{array}{rl} Na(g) \rightarrow Na^+(g) + e^- & \Delta_r H^\circ = 502.04\ \text{kJ} \\ Cl(g) + e^- \rightarrow Cl^-(g) & \Delta_r H^\circ = -354.81\ \text{kJ} \\ \hline Na(g) + Cl(g) \rightarrow Na^+(g) + Cl^-(g) & \Delta_r H^\circ = 147.23\ \text{kJ} \end{array}$$

This is an unfavorable energy change.

4.75 Use the data of Table 4-1 to calculate the lattice energy for KCl(s). Is this an endothermic or exothermic process?

▮ The thermochemical equation for the *lattice energy of an ionic solid* is

$$M_aX_b(s) \rightarrow aM^{b+}(g) + bX^{a-}(g) \qquad \Delta_r H_T^\circ = \Delta_{\text{lat}} H_T^\circ$$

For KCl(s), the equation becomes $KCl(s) \rightarrow K^+ + Cl^-$. Substituting the data into *(4.5)* gives

$$\begin{aligned} \Delta_{\text{lat}} H^\circ &= [(1\ \text{mol})\ \Delta_f H^\circ(K^+) + (1\ \text{mol})\ \Delta_f H^\circ(Cl^-)] - (1\ \text{mol})\ \Delta_f H^\circ(KCl) \\ &= [(1\ \text{mol})(514.26\ \text{kJ}\cdot\text{mol}^{-1}) + (1\ \text{mol})(-233.13\ \text{kJ}\cdot\text{mol}^{-1})] - (1\ \text{mol})(-436.747\ \text{kJ}\cdot\text{mol}^{-1}) \\ &= 717.88\ \text{kJ} \end{aligned}$$

This is an endothermic process.

4.76 Calculate $\Delta_{\text{lat}} H_T^\circ$ for KF, KBr, and KI. Compare these values to that for KCl determined in Problem 4.75. What is the trend in the values as the interionic distance increases?

▮ As a sample calculation, consider KF. Using *(4.5)* and the data from Table 4-1 gives

$$\begin{aligned} \Delta_{\text{lat}} H^\circ &= [(1\ \text{mol})\ \Delta_f H^\circ(K^+) + (1\ \text{mol})\ \Delta_f H^\circ(F^-)] - (1\ \text{mol})\ \Delta_f H^\circ(KF) \\ &= [(1\ \text{mol})(514.26\ \text{kJ}\cdot\text{mol}^{-1}) + (1\ \text{mol})(-255.39\ \text{kJ}\cdot\text{mol}^{-1})] - (1\ \text{mol})(-567.27\ \text{kJ}\cdot\text{mol}^{-1}) \\ &= 826.4\ \text{kJ} \end{aligned}$$

Likewise, $\Delta_{\text{lat}} H^\circ/(\text{kJ}\cdot\text{mol}^{-1}) = 688.99$ for KBr and 645.16 for KI. All of these values are similar in magnitude to the value for KCl ($\Delta_{\text{lat}} H^\circ = 717.88\ \text{kJ}\cdot\text{mol}^{-1}$), which means that a considerable amount of energy is needed to destroy the respective crystal. The interionic distance increases as the anion increases in size and the lattice energy decreases.

4.77 The lattice energy of NaCl is 787.38 kJ · mol^{-1}. Using the results of Problem 4.74, calculate $\Delta_r H°$ for the equation $Na(g) + Cl(g) \rightarrow NaCl(s)$. Is this a favorable energy change?

▌ Writing the thermochemical equation for lattice energy and using Hess's law gives

$$\begin{array}{rl} Na(g) + Cl(g) \rightarrow Na^+(g) + Cl^-(g) & \Delta_r H° = 147.23 \text{ kJ} \\ Na^+(g) + Cl^-(g) \rightarrow NaCl(s) & \Delta_r H° = -787.38 \text{ kJ} \\ \hline Na(s) + Cl(g) \rightarrow NaCl(s) & \Delta_r H° = -640.15 \text{ kJ} \end{array}$$

This a favorable energy change.

4.78 Use the data of Table 4-1 to calculate $\Delta_r H°$ for

$$KOH(aq) + HCl(aq) \rightarrow KCl(aq) + H_2O(l)$$

▌ The *enthalpy of neutralization* can be found by substituting the data into *(4.5)*, giving

$$\begin{aligned} \Delta_{neut} H° &= [(1 \text{ mol})\, \Delta_f H°(KCl) + (1 \text{ mol})\, \Delta_f H°(H_2O)] \\ &\quad - [(1 \text{ mol})\, \Delta_f H°(KOH) - (1 \text{ mol})\, \Delta_f H°(HCl)] \\ &= [(1 \text{ mol})(-419.53 \text{ kJ} \cdot \text{mol}^{-1}) + (1 \text{ mol})(-285.830 \text{ kJ} \cdot \text{mol}^{-1})] \\ &\quad - [(1 \text{ mol})(-482.37 \text{ kJ} \cdot \text{mol}^{-1}) + (1 \text{ mol})(-167.159 \text{ kJ} \cdot \text{mol}^{-1})] \\ &= -55.83 \text{ kJ} \end{aligned}$$

4.79 Show that for strong acids and bases the enthalpy of neutralization can be determined by calculating $\Delta_r H°$ for $H^+(aq) + OH^-(aq) \rightarrow H_2O(l)$. Why does this net ionic equation give the same result as in Problem 4.78?

▌ Substituting the data from Table 4-1 into *(4.5)* gives

$$\begin{aligned} \Delta_{neut} H° &= (1 \text{ mol})\, \Delta_f H°(H_2O) - [(1 \text{ mol})\, \Delta_f H°(H^+) + (1 \text{ mol})\, \Delta_f H°(OH^-)] \\ &= (1 \text{ mol})(-285.830 \text{ kJ} \cdot \text{mol}^{-1}) + [(1 \text{ mol})(0) + (1 \text{ mol})(-229.994 \text{ kJ} \cdot \text{mol}^{-1})] \\ &= -55.836 \text{ kJ} \end{aligned}$$

This is the same numerical value as determined in Problem 4.78. For strong electrolytes, the symbolism $HCl(aq)$ and $H^+(aq) + Cl^-(aq)$ mean the same thing (see Problem 4.50). Thus, eliminating the spectator ions from the equation given in Problem 4.78 (and their value of $\Delta_r H°$) makes no difference in the calculation.

4.80 The enthalpy of neutralization for the first proton of $H_2S(aq)$ is -33.7 kJ mol^{-1}. Calculate the first acid ionization energy for $H_2S(aq)$.

▌ Writing the thermochemical equation for the neutralization reaction and combining it with the thermochemical equations from Problem 4.79 using Hess's law gives

$$\begin{array}{rl} H_2S(aq) + OH^-(aq) \rightarrow HS^-(aq) + H_2O(l) & \Delta_{neut} H° = -33.7 \text{ kJ} \\ H_2O(l) \rightarrow OH^-(aq) + H^+(aq) & \Delta_r H° = 55.836 \text{ kJ} \\ \hline H_2S(aq) \rightarrow HS^-(aq) + H^+(aq) & \Delta_r H° = 22.1 \text{ kJ} \end{array}$$

4.81 The enthalpy of neutralization of $HS^-(aq)$ is -5.1 kJ · mol^{-1}. Calculate the enthalpy of formation of $S^{2-}(aq)$.

▌ For the thermochemical equation

$$HS^-(aq) + OH^-(aq) \rightarrow S^{2-}(aq) + H_2O(l) \qquad \Delta_{neut} H° = -5.1 \text{ kJ}$$

substituting the data from Table 4-1 into *(4.5)* and solving gives

$$\begin{aligned} \Delta_{neut} H° &= [(1 \text{ mol})\, \Delta_f H°(S^{2-}) + (1 \text{ mol})\, \Delta_f H°(H_2O)] - [(1 \text{ mol})\, \Delta_f H°(HS^-) + (1 \text{ mol})\, \Delta_f H°(OH^-)] \\ -5.1 \text{ kJ} &= [(1 \text{ mol})\, \Delta_f H°(S^{2-}) + (1 \text{ mol})(-285.830 \text{ kJ} \cdot \text{mol}^{-1})] \\ &\quad - [(1 \text{ mol})(-17.6 \text{ kJ} \cdot \text{mol}^{-1}) + (1 \text{ mol})(-229.994 \text{ kJ} \cdot \text{mol}^{-1})] \\ \Delta_f H°(S^{2-}) &= 33.1 \text{ kJ} \cdot \text{mol}^{-1} \end{aligned}$$

4.82 Using the data given in Table 4-1, calculate the enthalpy of solution for hydrogen chloride.

▌ The *integral enthalpy of solution* for HCl will be $\Delta_r H°_T$ for

$$HCl(g) \xrightarrow{H_2O(l)} HCl(aq)$$

Substituting the data from Table 4-1 into *(4.5)* gives

$$\Delta_{soln}H° = (1\ \text{mol})\ \Delta_f H°(\text{aq}) - (1\ \text{mol})\ \Delta_f H°(\text{g})$$
$$= (1\ \text{mol})(-167.159\ \text{kJ}\cdot\text{mol}^{-1}) - (1\ \text{mol})(-92.307\ \text{kJ}\cdot\text{mol}^{-1})$$
$$= -74.852\ \text{kJ}$$

4.83 Use the following data to calculate $\Delta_{soln}H°$ for KNO_3 at various concentrations. Prepare a plot of the results against $n(H_2O)$.

$n(H_2O)/(\text{mol})$	$KNO_3(s)$	15	20	25	30	40	50
$\Delta_f H°/(\text{kJ}\cdot\text{mol}^{-1})$	−494.63	−467.265	−466.068	−465.206	−464.558	−463.633	−462.997

$n(H_2O)/(\text{mol})$	75	100	150	200	300	400	500
$\Delta_f H°/(\text{kJ}\cdot\text{mol}^{-1})$	−462.027	−461.487	−460.884	−460.558	−460.215	−460.043	−459.939

▌ The integral enthalpy of solution at a given concentration corresponds to the value of $\Delta_r H°$ for $KNO_3(s) + nH_2O \rightarrow KNO_3$ (in nH_2O). As a sample calculation, consider the data given for $n(H_2O) = 20$ mol.

$$\Delta_{soln}H° = (1\ \text{mol})\ \Delta_f H°(\text{aq}) - (1\ \text{mol})\ \Delta_f H°(\text{s})$$
$$= (1\ \text{mol})(-466.068\ \text{kJ}\cdot\text{mol}^{-1}) - (1\ \text{mol})(-494.63\ \text{kJ}\cdot\text{mol}^{-1})$$
$$= 28.56\ \text{kJ}$$

The plot is shown in Fig. 4-7.

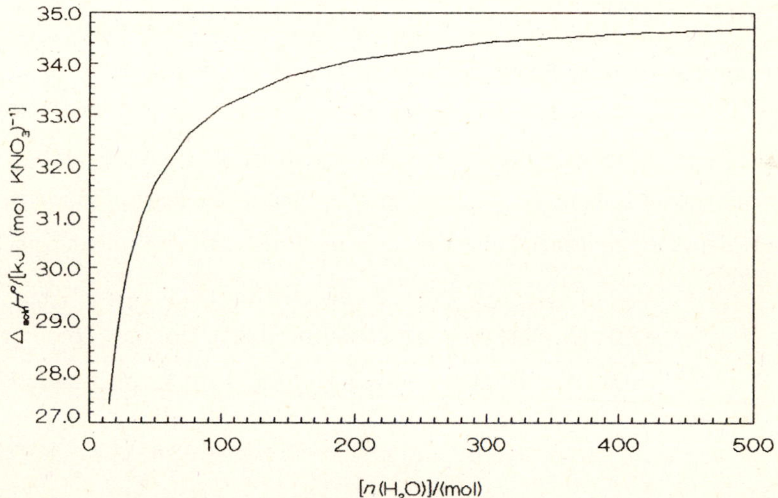

Fig. 4-7

4.84 Using the results of Problem 4.83, prepare a plot of the enthalpy of solution for KNO_3 against the molality of the solution. What is the major difference in the independent variable for the plots?

▌ As a sample calculation of the conversion of concentrations, consider the solution for which $n(H_2O) =$ 20 mol. Recalling that molality (m) is the amount of solute per kilogram of solvent,

$$m = \frac{1.000\ \text{mol KNO}_3}{(20\ \text{mol})\left(\dfrac{18.15\ \text{g H}_2\text{O}}{1\ \text{mol H}_2\text{O}}\right)\left(\dfrac{10^{-3}\ \text{kg}}{1\ \text{g}}\right)} = 2.775\ (\text{mol KNO}_3)\cdot\text{kg}^{-1}$$

The enthalpy of solution is

$$\Delta_{soln}H° = [28.56\ \text{kJ}\cdot(\text{mol KNO}_3)^{-1}][2.775\ (\text{mol KNO}_3)\cdot\text{kg}^{-1}] = 79.25\ \text{kJ}\cdot\text{kg}^{-1}$$

The plot is shown in Fig. 4-8. Both this and the plot in Fig. 4-7 are enthalpy of solution versus concentration. In

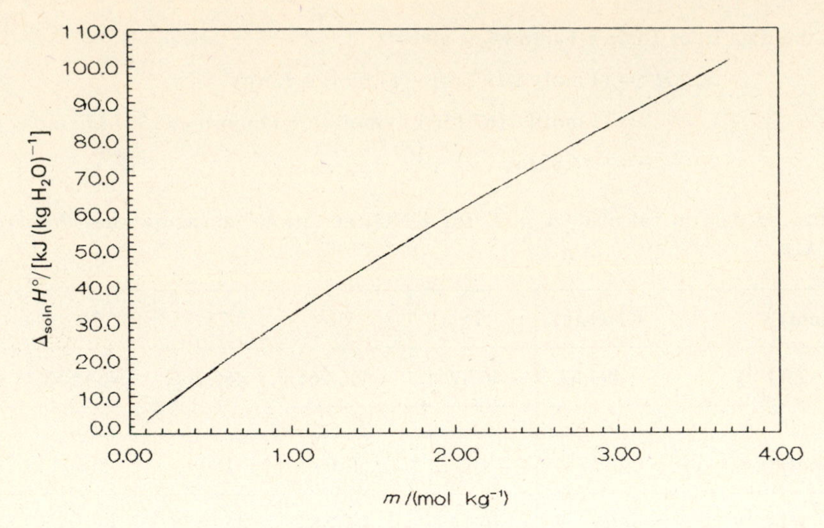

Fig. 4-8

Fig. 4-7 the amount of solute is held constant (1 mol) while the amount of solute changes, and in Fig. 4-8 the amount of solvent is held constant (1 kg or 55.51 mol) while the amount of solvent changes.

4.85 The enthalpy of solution of $Al(NO_3)_3 \cdot 9H_2O(s)$ is $2602\ kJ \cdot mol^{-1}$, and that of $Al(NO_3)_3 \cdot 6H_2O$ is $1695\ kJ \cdot mol^{-1}$ to form identical solutions. Calculate $\Delta_r H°$ for

$$Al(NO_3)_3 \cdot 6H_2O + 3H_2O(l) \rightarrow Al(NO_3)_3 \cdot 9H_2O(s)$$

▮ Writing the thermochemical equations for the dissolution processes and using Hess's law gives

$$\begin{array}{rlr} Al(NO_3)_3 \cdot 6H_2O(s) + (n+3)H_2O(l) & \rightarrow Al(NO_3)_3\ [\text{in } (n+9)H_2O] & \Delta_{soln}H° = 1695\ kJ \\ Al(NO_3)_3\ [\text{in } (n+9)H_2O] & \rightarrow Al(NO_3)_3 \cdot 9H_2O(s) + nH_2O & \Delta_r H_0° = -2602\ kJ \\ \hline Al(NO_3)_3 \cdot 6H_2O(s) + 3H_2O(s) & \rightarrow Al(NO_3)_3 \cdot 9H_2O(s) & \Delta_r H° = -907\ kJ \end{array}$$

4.86 Consider the dissolution process of methanol in water to consist of two steps:

$$CH_3OH(l) \rightarrow CH_3OH(g) \qquad \Delta_{vap}H° = 38.00\ kJ$$

$$CH_3OH(g) + nH_2O(l) \rightarrow CH_3OH\ (\text{in } nH_2O) \qquad \Delta_{solvation}H°$$

Given that the enthalpy of solution is $-7.27\ kJ \cdot mol^{-1}$, find the *enthalpy of solvation.*

▮ Adding the two given thermochemical equations using Hess's law gives the thermochemical equation for the dissolution process:

$$\begin{array}{rl} CH_3OH(l) \rightarrow CH_3OH(g) & \Delta_{vap}H° = 38.00\ kJ \\ CH_3OH(g) + nH_2O(l) \rightarrow CH_3OH\ (\text{in } nH_2O) & \Delta_{solvation}H° \\ \hline CH_3OH(l) + nH_2O(l) \rightarrow CH_3OH\ (\text{in } nH_2O) & \Delta_{soln}H° = -7.27\ kJ \end{array}$$

Thus,

$$\Delta_{solvation}H° = \Delta_{soln}H° - \Delta_{vap}H° = -7.27\ kJ - 38.00\ kJ = -45.27\ kJ$$

4.87 The dissolution of an ionic salt such as NaCl in water can be described in terms of the following thermochemical processes:

$$NaCl(s) \rightarrow Na^+(g) + Cl^-(g) \qquad \Delta_{lat}H° = 787.38\ kJ$$

$$Na^+(g) + Cl^-(g) + nH_2O(l) \rightarrow Na^+\ (\text{in } nH_2O) + Cl^-\ (\text{in } nH_2O) \qquad \sum \Delta_{solvation}H°(i)$$

The enthalpy of solvation of NaCl is $3.88\ kJ \cdot mol^{-1}$. Calculate the sum of the solvation enthalpies. A table of hydration enthalpies of ions lists $\Delta_{solvation}H°/(kJ \cdot mol^{-1}) = -406$ for Na^+ and -363 for Cl^-, with a typical error of $\pm 12\ kJ \cdot mol^{-1}$. Do the values from this table agree with your results?

▮ Using Hess's law on the above equations gives

$$\Delta_{soln}H° = \Delta_{lat}H° + \sum \Delta_{solvation}H°(i)$$

Substituting the values and solving for the sum of the hydration enthalpies gives

$$\sum \Delta_{solvation}H°(i) = 3.88\ kJ - 787.38\ kJ = -783.50\ kJ$$

The sum calculated from the individual values is

$$\sum \Delta_{solvation} H^\circ(i) = -406 \text{ kJ} + (-363 \text{ kJ}) = -769 \text{ kJ}$$

The two values agree within experimental error.

4.88 Use the data given in Problem 4.83 to calculate the *enthalpy of dilution* of an aqueous solution of KNO_3:

$$KNO_3 \text{ (in } 50H_2O) + 25H_2O(l) \rightarrow KNO_3 \text{ (in } 75H_2O)$$

Is this an endothermic or an exothermic process?

▌ Substituting the data into *(4.5)* gives

$$\begin{aligned}\Delta_{dil} H^\circ &= (1 \text{ mol})\, \Delta_f H^\circ(\text{diluted}) - (1 \text{ mol})\, \Delta_f H^\circ(\text{concentrated}) \\ &= (1 \text{ mol})(-462.027 \text{ kJ} \cdot \text{mol}^{-1}) - (1 \text{ mol})(-462.997 \text{ kJ} \cdot \text{mol}^{-1}) \\ &= 0.970 \text{ kJ}\end{aligned}$$

This is an endothermic process.

4.89 Determine $\Delta_r H^\circ$ for the complexation of Ag^+ by Br^-:

$$Ag^+(g) + 3Br^-(aq) \rightarrow AgBr_3^{2-}(aq)$$

▌ Substituting the data of Table 4-1 into *(4.5)* gives

$$\begin{aligned}\Delta_r H^\circ &= (1 \text{ mol})\, \Delta_f H^\circ(AgBr_3^{2-}) - [(1 \text{ mol})\, \Delta_f H^\circ(Ag^+) + (3 \text{ mol})\, \Delta_f H^\circ(Br^-)] \\ &= (1 \text{ mol})(-284.5 \text{ kJ} \cdot \text{mol}^{-1}) - [(1 \text{ mol})(105.579 \text{ kJ} \cdot \text{mol}^{-1}) + (3 \text{ mol})(-121.55 \text{ kJ} \cdot \text{mol}^{-1})] \\ &= -25.4 \text{ kJ}\end{aligned}$$

4.90 Calculate the enthalpy change for the cis-trans isomerization of $[Co(NH_3)_4(NO_2)_2]I$.

▌ Using the data given in Table 4-1, the enthalpy change for the *isomerization* process

$$cis\text{-}[Co(NH_3)_4(NO_2)_2]I(s) \rightarrow trans\text{-}[Co(NH_3)_4(NO_2)_2]I(s)$$

will be given by *(4.5)* as

$$\begin{aligned}\Delta_r H^\circ &= (1 \text{ mol})\, \Delta_f H^\circ(\text{trans}) - (1 \text{ mol})\, \Delta_f H^\circ(\text{cis}) \\ &= (1 \text{ mol})(-744.8 \text{ kJ} \cdot \text{mol}^{-1}) - (1 \text{ mol})(-745.6 \text{ kJ} \cdot \text{mol}^{-1}) \\ &= 0.8 \text{ kJ}\end{aligned}$$

4.91 Using the data given in Table 4-2, determine the enthalpy change for the cis-trans isomerism of CHCOOH=CHCOOH(s).

▌ For the reaction

```
COOH     COOH     H           COOH
    \   /           \        /
     C=C      →      C=C
    /   \           /   \
   H     H       COOH    H
      cis              trans
```

substituting the data into *(4.5)* gives

$$\begin{aligned}\Delta_r H^\circ &= -(1 \text{ mol})\, \Delta_c H^\circ(\text{trans}) + (1 \text{ mol})\, \Delta_c H^\circ(\text{cis}) \\ &= -(1 \text{ mol})(-1334.65 \text{ kJ} \cdot \text{mol}^{-1}) + (1 \text{ mol})(-1355.16 \text{ kJ} \cdot \text{mol}^{-1}) \\ &= -20.51 \text{ kJ}\end{aligned}$$

4.4 APPLICATIONS TO PHYSICAL CHANGES

4.92 Determine the heat of vaporization for H_2O_2 at 25 °C using the data of Table 4-1.

▌ For the vaporization process $H_2O_2(l) \rightarrow H_2O_2(g)$, using *(4.5)* gives

$$\begin{aligned}\Delta_{vap} H^\circ &= (1 \text{ mol})\, \Delta_f H^\circ(g) - (1 \text{ mol})\, \Delta_f H^\circ(l) \\ &= (1 \text{ mol})(-136.31 \text{ kJ} \cdot \text{mol}^{-1}) - (1 \text{ mol})(-187.78 \text{ kJ} \cdot \text{mol}^{-1}) \\ &= 51.47 \text{ kJ}\end{aligned}$$

The heat of vaporization is $51.47 \text{ kJ} \cdot \text{mol}^{-1}$ for H_2O_2 at 25 °C.

4.93 The heat of combustion of methane to produce gaseous water at 25 °C is $-802.34\ \text{kJ}\cdot\text{mol}^{-1}$. Determine the heat of vaporization of water at 25 °C.

▮ Using the data from Table 4-2 and Hess's law gives

$$\begin{array}{lr} CH_4(g) + 2O_2(g) \rightarrow CO_2(g) + 2H_2O(g) & \Delta_c H^\circ = -802\ 34\ \text{kJ} \\ CO_2(g) + 2H_2O(l) \rightarrow CH_4(g) + 2O_2(g) & \Delta_c H^\circ = 890.36\ \text{kJ} \\ \hline 2H_2O(l) \rightarrow 2H_2O(g) & \Delta_r H^\circ = 88.02\ \text{kJ} \end{array}$$

The value of $\Delta_{\text{vap}}H^\circ$ is $44.01\ \text{kJ}\cdot\text{mol}^{-1}$.

4.94 The solid-I form of BeF_2 melts at 825 K. Given $(H^\circ_T - H^\circ_{298})/(\text{kJ}\cdot\text{mol}^{-1}) = 35.530$ for the solid form and 40.287 for the liquid, determine the enthalpy of fusion, $\Delta_{\text{fus}}H^\circ_T$.

▮ The equation representing the fusion process is $BeF_2(\text{solid-I}) \rightarrow BeF_1(l)$. Substituting the data into *(3.15)* gives

$$\begin{aligned} \Delta_{\text{fus}}H^\circ &= (1\ \text{mol})(H^\circ_T - H^\circ_{298})_l - (1\ \text{mol})(H^\circ_T - H^\circ_{298})_s \\ &= (1\ \text{mol})(40.287\ \text{kJ}\cdot\text{mol}^{-1}) - (1\ \text{mol})(35.530\ \text{kJ}\cdot\text{mol}^{-1}) \\ &= 4.757\ \text{kJ} \end{aligned}$$

4.95 At 25 °C, $\Delta_{\text{vap}}H^\circ = 42.59\ \text{kJ}\cdot\text{mol}^{-1}$ for ethanol. Determine $\Delta_{\text{vap}}H^\circ_{273}$ given $C_{P,m}/(\text{J}\cdot\text{K}^{-1}\cdot\text{mol}^{-1}) = 111.46$ for the liquid and 65.44 for the gas.

▮ The vaporization process can be represented by the equation $CH_3CH_2OH(l) \rightarrow CH_3CH_2OH(g)$. For the process, *(3.12)* gives

$$\Delta_{\text{vap}}H^\circ_{T_2} = \Delta_{\text{vap}}H^\circ_{T_1} + \int_{T_1}^{T_2} \Delta_{\text{vap}}C^\circ_P\, dT \tag{4.10}$$

$$\begin{aligned} &= \Delta_{\text{vap}}H^\circ_{T_1} + n[C_{P,m}(g) - C_{P,m}(l)](T_2 - T_1) \\ &= 42.59\ \text{kJ} + (1\ \text{mol})(65.44\ \text{J}\cdot\text{K}^{-1}\cdot\text{mol}^{-1} - 111.46\ \text{J}\cdot\text{K}^{-1}\cdot\text{mol}^{-1})(273\ \text{K} - 298\ \text{K})[(10^{-3}\ \text{kJ})/(1\ \text{J})] \\ &= 43.74\ \text{kJ} \end{aligned}$$

4.96 The *Kirchhoff equation*, *(4.10)*, is valid for phase changes at constant pressure, such as vaporization and sublimation processes. However, condensed-phase changes are described by the *Planck equation*

$$\frac{d(\Delta_{\text{I}\rightarrow\text{II}}H^\circ)}{dT} = \Delta_{\text{I}\rightarrow\text{II}}C^\circ_P + \frac{\Delta_{\text{I}\rightarrow\text{II}}H^\circ}{T} - \Delta_{\text{I}\rightarrow\text{II}}H^\circ_T \frac{\partial \ln(\Delta_{\text{I}\rightarrow\text{II}}V)}{\partial T} \tag{4.11}$$

At 0 °C, $\Delta_{\text{fus}}H^\circ = 6.0095\ \text{kJ}\cdot\text{mol}^{-1}$, $\Delta_{\text{fus}}C_{P,m} = 37.28\ \text{J}\cdot\text{K}^{-1}\cdot\text{mol}^{-1}$, $V_m/(\text{cm}^3\cdot\text{mol}^{-1}) = 19.65$ for ice and 18.02 for liquid water, and $\alpha/(10^{-5}\ \text{K}^{-1}) = 11.0$ for ice and -6.805 for liquid water. Find $d(\Delta_{\text{fus}}H)/dT$ for water.

▮ Rearranging *(1.11)* for $\partial(\Delta_{\text{fus}}V_m)/\partial T$ and substituting the data gives

$$\begin{aligned} \frac{\partial(\Delta_{\text{fus}}V_m)}{\partial T} &= \alpha_l V_{m,l} - \alpha_s V_{m,s} \\ &= (-6.805\times10^{-5}\ \text{K}^{-1})(18.02\ \text{cm}^3\cdot\text{mol}^{-1}) - (11.0\times10^{-5}\ \text{K}^{-1})(19.65\ \text{cm}^3\cdot\text{mol}^{-1}) \\ &= -3.388\times10^{-3}\ \text{cm}^3\cdot\text{K}^{-1}\cdot\text{mol}^{-1} \end{aligned}$$

Recognizing that $\partial(\ln \Delta V)/\partial T = (1/\Delta V)(\partial \Delta V/\partial T)$, *(4.11)* gives

$$\begin{aligned} \frac{\partial(\Delta_{\text{fus}}H^\circ_m)}{\partial T} &= (37.28\ \text{J}\cdot\text{K}^{-1}\cdot\text{mol}^{-1}) + \frac{(6.0095\ \text{kJ}\cdot\text{mol}^{-1})[(10^3\ \text{J})/(1\ \text{kJ})]}{273.15\ \text{K}} \\ &\quad - \frac{(6.0095\ \text{kJ}\cdot\text{mol}^{-1})[(10^3\ \text{J})/(1\ \text{kJ})]}{-1.63\ \text{cm}^3\cdot\text{K}^{-1}\cdot\text{mol}^{-1}}(-3.388\times10^{-3}\ \text{cm}^3\cdot\text{K}^{-1}\cdot\text{mol}^{-1}) \\ &= 46.8\ \text{J}\cdot\text{K}^{-1}\cdot\text{mol}^{-1} \end{aligned}$$

4.97 Use the results of Problem 4.96 to determine $\Delta_f H^\circ$ for ice at 25 °C.

▌ Over this small temperature range, $\Delta_{fus}H^\circ$ at 25 °C will be given by

$$\Delta_{fus}H^\circ_{298} = \Delta_{fus}H^\circ_{273} + \int_{273\text{ K}}^{298\text{ K}} \frac{d(\Delta_{fus}H)}{dT}\,dT$$

$$= 6.0095\text{ kJ}\cdot\text{mol}^{-1} + (46.8\text{ J}\cdot\text{K}^{-1}\cdot\text{mol}^{-1})(298\text{ K} - 273\text{ K})[(10^{-3}\text{ kJ})/(1\text{ J})]$$

$$= 7.180\text{ kJ}\cdot\text{mol}^{-1}$$

Thus for the fusion process $H_2O(s) \rightarrow H_2O(l)$, using *(4.5)* and the data from Table 4-1 gives

$$\Delta_{fus}H^\circ = (1\text{ mol})\,\Delta_f H^\circ(l) - (1\text{ mol})\,\Delta_f H^\circ(s)$$

$$7.180\text{ kJ} = (1\text{ mol})(-285.830\text{ kJ}\cdot\text{mol}^{-1}) - (1\text{ mol})\,\Delta_f H^\circ(s)$$

$$\Delta_f H^\circ(s) = -293.010\text{ kJ}\cdot\text{mol}^{-1}$$

4.98 An approximate relation between the enthalpy of vaporization and the normal boiling point of a liquid (at 1 atm) is given by *Trouton's rule*,

$$\Delta_{vap}H^\circ_{T,m} \approx (88\text{ J}\cdot\text{K}^{-1}\cdot\text{mol}^{-1})\,T_{bp} \qquad (4.12)$$

provided the substance does not contain hydrogen bonding or form dimers. Use this relationship to predict the boiling point of C_2F_3Cl given $\Delta_{vap}H^\circ_{T,m} = 21.9\text{ kJ}\cdot\text{mol}^{-1}$.

▌ Solving *(4.12)* for T_{bp} and substituting the data gives

$$T_{bp} = \frac{(21.9\text{ kJ}\cdot\text{mol}^{-1})[(10^3\text{ J})/(1\text{ kJ})]}{88\text{ J}\cdot\text{K}^{-1}\cdot\text{mol}^{-1}} = 250\text{ K}$$

This value agrees very well with the experimental value of 245.3 K.

4.99 Calculate the value of the Trouton's rule constant for methanol given $\Delta_{vap}H^\circ_{T,m} = 35.3\text{ kJ}\cdot\text{mol}^{-1}$ at 337.9 K. Why is this value significantly different than that given in *(4.12)*?

▌ The constant will be given by

$$\text{Constant} = \frac{(35.3\text{ kJ}\cdot\text{mol}^{-1})[(10^3\text{ J})/(1\text{ kJ})]}{337.9\text{ K}} = 104\text{ J}\cdot\text{K}^{-1}\cdot\text{mol}^{-1}$$

Methanol has extensive hydrogen bonding in the liquid state, which makes the value of $\Delta_{vap}H_{T,m}$ greater for most liquids.

4.100 An approximate relation between the enthalpy of fusion and the melting point of a metallic element is given by

$$\Delta_{fus}H^\circ_T \approx (9.2\text{ J}\cdot\text{K}^{-1}\cdot\text{mol}^{-1})\,T_{mp} \qquad (4.13)$$

Determine the enthalpy of fusion for silver given that the melting point is 961 °C.

▌ Substituting the data into *(4.13)* gives

$$\Delta_{fus}H^\circ_T = (9.2\text{ J}\cdot\text{K}^{-1}\cdot\text{mol}^{-1})[(961 + 273)\text{ K}] = 11\,400\text{ J}\cdot\text{mol}^{-1}$$

This agrees very well with the experimental value of $11.3\text{ kJ}\cdot\text{mol}^{-1}$.

4.101 The heat capacity of solid NaCl from 500 K to 1074 K is given by

$$52.996\text{ J}\cdot\text{K}^{-1}\cdot\text{mol}^{-1} - (7.86\times10^{-3}\text{ J}\cdot\text{K}^{-2}\cdot\text{mol}^{-1})T + (1.97\times10^{-5}\text{ J}\cdot\text{K}^{-3}\cdot\text{mol}^{-1})T^2$$

and that of liquid NaCl from 1074 K to 1500 K is given by

$$125.637\text{ J}\cdot\text{K}^{-1}\cdot\text{mol}^{-1} - (8.187\times10^{-2}\text{ J}\cdot\text{K}^{-2}\cdot\text{mol}^{-1})T + (2.85\times10^{-5}\text{ J}\cdot\text{K}^{-3}\cdot\text{mol}^{-1})T^2$$

If $\Delta_{fus}H^\circ_{T,m} = 28.158\text{ kJ}\cdot\text{mol}^{-1}$ at 1074 K, determine $H^\circ_{1500} - H^\circ_{500}$ for NaCl.

▌ The desired quantity is simply the enthalpy change for heating NaCl from 500 K to 1500 K. This, in general, will be given by

$$\Delta H^\circ = \sum_i^{\text{phases}} \int C^\circ_{P,i}\,dT + \sum_j^{\text{transitions}} \Delta_j H^\circ_T \qquad (4.14)$$

In this case,

$$\Delta H^\circ_m = \int_{500\text{ K}}^{1074\text{ K}} C^\circ_{P,m}(\text{s})\,dT + \Delta_{\text{fus}}H^\circ_{T,m} + \int_{1074\text{ K}}^{1500\text{ K}} C^\circ_{P,m}(\text{l})\,dT$$

$$= (52.996\text{ J}\cdot\text{K}^{-1}\cdot\text{mol}^{-1})(1074\text{ K} - 500\text{ K})$$

$$-\tfrac{1}{2}(7.86\times10^{-3}\text{ J}\cdot\text{K}^{-2}\cdot\text{mol}^{-1})[(1074\text{ K})^2 - (500\text{ K})^2]$$

$$+\tfrac{1}{3}(1.97\times10^{-5}\text{ J}\cdot\text{K}^{-3}\cdot\text{mol}^{-1})[(1074\text{ K})^3 - (500\text{ K})^3]$$

$$+(28.158\text{ kJ}\cdot\text{mol}^{-1})\frac{10^3\text{ J}}{1\text{ kJ}} + (125.637\text{ J}\cdot\text{K}^{-1}\cdot\text{mol}^{-1})(1500\text{ K} - 1074\text{ K})$$

$$-\tfrac{1}{2}(8.187\times10^{-2}\text{ J}\cdot\text{K}^{-2}\cdot\text{mol}^{-1})[(1500\text{ K})^2 - (1074\text{ K})^2]$$

$$+\tfrac{1}{3}(2.85\times10^{-5}\text{ J}\cdot\text{K}^{-3}\cdot\text{mol}^{-1})[(1500\text{ K})^3 - (1074\text{ K})^3]$$

$$= 91\,270\text{ J}\cdot\text{mol}^{-1}$$

Thus $H^\circ_{1500} - H^\circ_{500} = 91.270\text{ kJ}\cdot\text{mol}^{-1}$.

4.102 Solid BeF_2 undergoes a solid-solid phase transformation at 500 K. To heat 1-mol sample of BeF_2 from 300 K to 800 K at constant pressure required 33.570 kJ. Calculate $\Delta_{\text{II}\to\text{I}}H^\circ_T$ given $C^\circ_{P,m} = 38.738\text{ J}\cdot\text{K}^{-1}\cdot\text{mol}^{-1} - (1.18\times10^{-3}\text{ J}\cdot\text{K}^{-2}\cdot\text{mol}^{-1})T + (1.516\times10^{-4}\text{ J}\cdot\text{K}^{-3}\cdot\text{mol}^{-1})T^2$ for solid-II and $47.364\text{ J}\cdot\text{K}^{-1}\cdot\text{mol}^{-1} + (3.347\times10^{-2}\text{ J}\cdot\text{K}^{-2}\cdot\text{mol}^{-1})T$ for solid-I.

▮ For this process, *(4.14)* gives

$$\Delta H^\circ_m = \int_{300\text{ K}}^{500\text{ K}} C^\circ_{P,m}(\text{solid-II})\,dT + \Delta_{\text{II}\to\text{I}}H^\circ_{T,m} + \int_{500\text{ K}}^{800\text{ K}} C^\circ_{P,m}(\text{solid-I})\,dT$$

$$(33.570\text{ kJ}\cdot\text{mol}^{-1})(1\text{ mol})\frac{10^3\text{ J}}{1\text{ kJ}} = (38.738\text{ J}\cdot\text{K}^{-1}\cdot\text{mol}^{-1})(500\text{ K} - 300\text{ K})$$

$$-\tfrac{1}{2}(1.18\times10^{-3}\text{ J}\cdot\text{K}^{-2}\cdot\text{mol}^{-1})[(500\text{ K})^2 - (300\text{ K})^2]$$

$$+\tfrac{1}{3}(1.516\times10^{-4}\text{ J}\cdot\text{K}^{-3}\cdot\text{mol}^{-1})[(500\text{ K})^3 - (300\text{ K})^3]$$

$$+\Delta_{\text{II}\to\text{I}}H^\circ_{T,m} + (47.364\text{ J}\cdot\text{K}^{-1}\cdot\text{mol}^{-1})(800\text{ K} - 500\text{ K})$$

$$+\tfrac{1}{2}(3.347\times10^{-2}\text{ J}\cdot\text{K}^{-2}\cdot\text{mol}^{-1})[(800\text{ K})^2 - (500\text{ K})^2]$$

$$\Delta_{\text{II}\to\text{I}}H^\circ_{T,m} = 229\text{ J}\cdot\text{mol}^{-1}$$

4.103 At the triple point of molecular oxygen, 54.351 K and 1.138 torr, the enthalpy of fusion is $444.8\text{ J}\cdot\text{mol}^{-1}$, the density of the solid is $42.46\text{ mol}\cdot\text{cm}^{-3}$, and the density of the liquid is $40.83\text{ mol}\cdot\text{cm}^{-3}$. Calculate dP/dT under these conditions.

▮ For a phase transformation, the *Clapeyron equation* gives

$$\frac{dP}{dT} = \frac{\Delta_{\text{I}\to\text{II}}H}{T\,\Delta_{\text{I}\to\text{II}}V} \qquad (4.15)$$

In this case, for 1 mol,

$$\Delta_{\text{fus}}V = V_l - V_s = (1\text{ mol})\left(\frac{1}{40.83\text{ mol}\cdot\text{cm}^{-3}} - \frac{1}{42.46\text{ mol}\cdot\text{cm}^{-3}}\right)\frac{10^{-3}\text{ L}}{1\text{ cm}^3}$$

$$= 9.40\times10^{-7}\text{ L}$$

Substituting into *(4.15)* gives

$$\frac{dP}{dT} = \frac{(444.8\text{ J})[(1\text{ L}\cdot\text{bar})/(10^2\text{ J})]}{(54.351\text{ K})(9.40\times10^{-7}\text{ L})} = 8.71\times10^4\text{ bar}\cdot\text{K}^{-1}$$

4.104 Calculate the freezing point of water at a depth of 10.0 cm below the surface of the water. The density of the water is $0.9998\text{ g}\cdot\text{cm}^{-3}$. See Problem 4.96 for additional data.

▮ For small temperature-pressure changes, *(4.15)* can be integrated to

$$\frac{\Delta P}{\Delta T} = \frac{\Delta_{\text{I}\to\text{II}}H}{T_{\text{I}\to\text{II}}\,\Delta_{\text{I}\to\text{II}}V} \qquad (4.16a)$$

where $T_{I \to II}$ is the temperature for the transition under the original pressure-temperature conditions. The change in pressure is given by *(1.4)* as

$$\Delta P = (0.9998\ \text{g} \cdot \text{cm}^{-3}) \frac{10^{-3}\ \text{kg}}{1\ \text{g}} \left(\frac{10^2\ \text{cm}}{1\ \text{m}}\right)^3 (9.81\ \text{m} \cdot \text{s}^{-2})(10.0\ \text{cm}) \frac{10^{-2}\ \text{m}}{1\ \text{cm}} \left(\frac{1\ \text{Pa}}{1\ \text{kg} \cdot \text{m}^{-1} \cdot \text{s}^{-2}}\right)\left(\frac{10^{-5}\ \text{bar}}{1\ \text{Pa}}\right)$$

$$= 9.81 \times 10^{-3}\ \text{bar}$$

and the volume change for 1 mol is given by

$$\Delta_{\text{fus}} V = (1\ \text{mol})(18.02\ \text{cm}^3 \cdot \text{mol}^{-1} - 19.65\ \text{cm}^3 \cdot \text{mol}^{-1}) \frac{10^{-3}\ \text{L}}{1\ \text{cm}^3}$$

$$= -1.63 \times 10^{-3}\ \text{L}$$

Solving *(4.16a)* for ΔT and substituting the data gives

$$\Delta T = \frac{(9.81 \times 10^{-3}\ \text{bar})(273\ \text{K})(-1.63 \times 10^{-3}\ \text{L})[(10^2\ \text{J})/(1\ \text{L} \cdot \text{bar})]}{(6.0095\ \text{kJ})[(10^3\ \text{J})/(1\ \text{kJ})]}$$

$$= -7.26 \times 10^{-5}\ \text{K}$$

4.105 The ice skate of a 210-lb hockey player is in contact with the ice over a 2.5 mm × 15 cm area. Determine the freezing point of the ice under the blade of the skate. See Problem 4.104 for additional data.

▌ For larger pressure-temperature changes, *(4.15)* can be integrated to

$$\Delta P = \frac{\Delta_{I \to II} H}{\Delta_{I \to II} V} \ln \frac{T_2}{T_1} \tag{4.16b}$$

The pressure under the skate blade is given by *(1.4)* as

$$P = \frac{(210\ \text{lb})[(0.454\ \text{kg})/(1\ \text{lb})](9.81\ \text{m} \cdot \text{s}^{-2})[(1\ \text{Pa})/(1\ \text{kg} \cdot \text{m}^{-1} \cdot \text{s}^{-2})][(10^{-5}\ \text{bar})/(1\ \text{Pa})]}{(2.5\ \text{mm})(15\ \text{cm})[(10^{-3}\ \text{m})/(1\ \text{mm})][(10^{-2}\ \text{m})/(1\ \text{cm})]}$$

$$= 25\ \text{bar}$$

Substituting the data for 1 mol into *(4.16b)* and solving for T_2 gives

$$(25 - 1)\ \text{bar} = \frac{(6.0095\ \text{kJ})[(10^3\ \text{J})/(1\ \text{kJ})]}{(-1.63 \times 10^{-3}\ \text{L})[(10^2\ \text{J})/(1\ \text{L} \cdot \text{bar})]} \ln \frac{T_2}{273.15\ \text{K}}$$

$$\ln \frac{T_2}{273.15\ \text{K}} = -6.5 \times 10^{-4}$$

$$T_2 = 272.97\ \text{K} = -0.18\ °\text{C}$$

4.106 A barometer on a space probe was designed to measure the atmospheric pressure on the surface of a planet by measuring the melting point of mercury. The normal melting point of mercury is −38.87 °C, and on the surface of the planet being investigated the melting point increased to −37.91 °C. Calculate the pressure of the atmosphere on this planet given that $\Delta_{\text{fus}} H_{\text{m}} = 2330\ \text{J} \cdot \text{mol}^{-1}$, $\rho = 14.193\ \text{g} \cdot \text{cm}^{-3}$ for the solid, and $\rho = 13.690\ \text{g} \cdot \text{cm}^{-3}$ for the liquid.

▌ The change of volume for 1 mol is

$$\Delta_{\text{fus}} V = (200.59\ \text{g})\left(\frac{1}{13.690\ \text{mol} \cdot \text{cm}^{-3}} - \frac{1}{14.193\ \text{mol} \cdot \text{cm}^{-3}}\right) \frac{10^{-3}\ \text{L}}{1\ \text{cm}^3}$$

$$= 5.19 \times 10^{-4}\ \text{L}$$

Substituting the data for 1 mol into *(4.16b)* gives

$$\Delta P = \frac{(2330\ \text{J})[(1\ \text{L} \cdot \text{bar})/(10^2\ \text{J})]}{5.19 \times 10^{-4}\ \text{L}} \ln \frac{235.24\ \text{K}}{234.28\ \text{K}} = 184\ \text{bar}$$

The atmospheric pressure on the surface of the planet is 184 bar.

4.107 Assume that graphite can be changed to diamond at 298 K by increasing the pressure to 1.5×10^4 bar. Given $\Delta H_{\text{m}} = 1895\ \text{J} \cdot \text{mol}^{-1}$ and $\Delta V_{\text{m}} = -1.92 \times 10^{-3}\ \text{L} \cdot \text{mol}^{-1}$ for this process, determine the temperature at which graphite can be changed to diamond at 1 bar.

▮ For this solid-solid phase transition, using *(4.16b)* for 1 mol gives

$$(1 - 1.5 \times 10^4)\ \text{bar} = \frac{(1895\ \text{J})[(1\ \text{L}\cdot\text{bar})/(10^2\ \text{J})]}{-1.92 \times 10^{-3}\ \text{L}} \ln \frac{T_2}{298\ \text{K}}$$

$$\ln \frac{T_2}{298\ \text{K}} = 1.5 \qquad T_2 = 1400\ \text{K}$$

4.108 At the normal boiling point of water, $\Delta_{\text{vap}}H^\circ_{\text{m}} = 40.656\ \text{kJ}\cdot\text{mol}^{-1}$. Determine an equation for dT/dP for water at the boiling point assuming that the vapor acts ideally.

▮ For the vaporization of 1 mol of liquid (or solid),

$$\Delta V_{\text{m}} = V_{\text{g,m}} - V_{\text{l,m}} \approx V_{\text{g,m}} = RT/P$$

which upon substituting into *(4.15)* gives the *Clausius-Clapeyron equation*

$$\frac{dP}{dT} = \frac{(\Delta_{\text{vap}}H^\circ \text{ or } \Delta_{\text{sub}}H^\circ)P}{RT^2} \tag{4.17}$$

Inverting *(4.17)* and substituting the data gives

$$\frac{dT}{dP} = \frac{(8.314\ \text{J}\cdot\text{K}^{-1}\cdot\text{mol}^{-1})(373\ \text{K})^2}{(40.656\ \text{kJ}\cdot\text{mol}^{-1})[(10^3\ \text{J})/(1\ \text{kJ})](1\ \text{bar})} = 28.5\ \text{K}\cdot\text{bar}^{-1}$$

4.109 What will be the boiling point of water at exactly 1 bar?

▮ For small pressure-temperature changes,

$$\frac{\Delta T}{\Delta P} \approx \frac{dT}{dP} = 28.5\ \text{K}\cdot\text{bar}^{-1}$$

(see Problem 4.108). Substituting the normal boiling point data (373.15 K, 1 atm) gives

$$T - 373.15\ \text{K} = (28.5\ \text{K}\cdot\text{bar}^{-1})\left[1.000\,00\ \text{bar} - (1.000\ \text{atm})\frac{1.013\,25\ \text{bar}}{1\ \text{atm}}\right]$$

$$T = 372.77\ \text{K} = 99.62\ ^\circ\text{C}$$

4.110 Using Trouton's rule, derive an expression for $\Delta T/\Delta P$ for the vaporization of typical liquids. What is the correction factor for a liquid that has a normal boiling point (1 atm) of 200.0 °C when expressing its "standard" (1 bar) boiling point?

▮ For small pressure-temperature changes, substituting *(4.12)* into *(4.17)* gives

$$\frac{\Delta T}{\Delta P} \approx \frac{dP}{dT} = \frac{RT^2}{(\Delta_{\text{vap}}H)P} = \frac{RT}{[(\Delta_{\text{vap}}H)/T]P} = \frac{RT}{(88\ \text{J}\cdot\text{K}^{-1}\cdot\text{mol}^{-1})P} \tag{4.18}$$

$$= \frac{(8.314\ \text{J}\cdot\text{K}^{-1}\cdot\text{mol}^{-1})T}{(88\ \text{J}\cdot\text{K}^{-1}\cdot\text{mol}^{-1})P} = 0.094\frac{T}{P}$$

Solving *(4.18)*, sometimes called *Craft's rule*, for ΔT and substituting gives

$$\Delta T = 0.094\frac{(573.2\ \text{K})(1.000\,00\ \text{bar} - 1.013\,25\ \text{bar})}{1.013\,25\ \text{bar}} = -0.70\ \text{K} = -0.70\ ^\circ\text{C}$$

4.111 The vapor pressure of solid acetylene as a function of temperature is given in a handbook as

$$\log[P/(\text{torr})] = \frac{-52.34B}{T/(\text{K})} + C$$

where $B = 21.914$ and $C = 8.933$. Determine the enthalpy of vaporization.

▮ The integral form of the *Clausius-Clapeyron equation*, *(4.17)*, can be written as

$$\ln P = \frac{-(\Delta_{\text{vap}}H \text{ or } \Delta_{\text{sub}}H)}{R}\left(\frac{1}{T}\right) + k \tag{4.19a}$$

Equating the $1/T$ terms, converting from natural to common logarithms, and substituting the data gives

$$-\Delta_{\text{sub}}H/2.303R = -52.23B$$

$$\Delta_{\text{sub}}H = 2.303(8.314\ \text{J}\cdot\text{K}^{-1}\cdot\text{mol}^{-1})(-52.23)(21.914)\frac{10^{-3}\ \text{kJ}}{1\ \text{J}}$$

$$= 21.92\ \text{kJ}\cdot\text{mol}^{-1}$$

4.112 Use the following temperature–vapor pressure data to determine $\Delta_{\text{sub}}H$, $\Delta_{\text{vap}}H$, and $\Delta_{\text{fus}}H$ for water at 0 °C:

Ice

T/(°C)	−2.0	−1.5	−1.0	−0.5	−0.0
P/(torr)	3.880	4.045	4.217	4.395	4.579

Water

T/(°C)	0.0	0.5	1.0	1.5	2.0
P/(torr)	4.579	4.750	4.926	5.107	5.294

▮ According to *(4.19a)*, a plot of ln P against T^{-1} will be linear and will have a slope equal to $-(\Delta_{\text{vap}}H$ or $\Delta_{\text{sub}}H)/R$. The slopes of the lines in Fig. 4-9 are −6137 K for ice and −5451 K for liquid water, giving

$$\Delta_{\text{vap}}H = -(8.314\ \text{J}\cdot\text{K}^{-1}\cdot\text{mol}^{-1})(-5451\ \text{K})\frac{10^{-3}\ \text{kJ}}{1\ \text{J}} = 45.32\ \text{kJ}\cdot\text{mol}^{-1}$$

$$\Delta_{\text{sub}}H = -8.314(-6137)(10^{-3}) = 51.03\ \text{kJ}\cdot\text{mol}^{-1}$$

The enthalpy change for fusion is

$$\Delta_{\text{fus}}H = \Delta_{\text{sub}}H - \Delta_{\text{vap}}H = 51.03\ \text{kJ}\cdot\text{mol}^{-1} - 45.32\ \text{kJ}\cdot\text{mol}^{-1} = 5.71\ \text{kJ}\cdot\text{mol}^{-1}$$

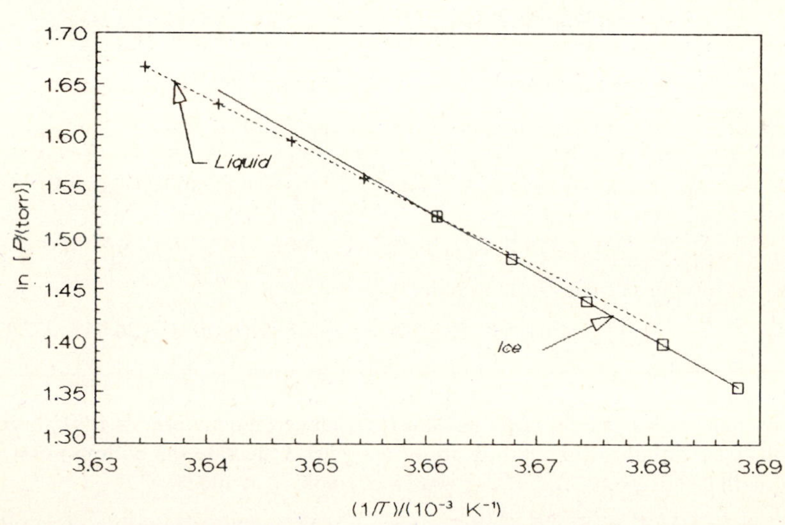

Fig. 4-9

4.113 At 100.0 °C the vapor pressure of liquid mercury is 0.2729 torr, and at 300.0 °C the value is 246.80 torr. Determine the enthalpy of vaporization of mercury.

▮ Assuming $\Delta_{\text{vap}}H$ to be constant over this temperature range, *(4.19a)* can be written as

$$\ln\frac{P_2}{P_1} = \frac{-\Delta_{\text{vap}}H}{R}\left(\frac{1}{T_2} - \frac{1}{T_1}\right) \qquad (4.19b)$$

Solving *(4.19b)* for $\Delta_{\text{vap}}H$ and substituting the data gives

$$\Delta_{\text{vap}}H = \frac{-(8.314\ \text{J}\cdot\text{K}^{-1}\cdot\text{mol}^{-1})[(10^{-3}\ \text{kJ})/(1\ \text{J})]\ln[(246.80\ \text{torr})/(0.2729\ \text{torr})]}{1/(573.2\ \text{K}) - 1/(373.2\ \text{K})}$$

$$= 60.54\ \text{kJ}\cdot\text{mol}^{-1}$$

4.114 Use the results of Problem 4.113 to predict the vapor pressure of Hg at 200.0 °C.

▮ Substituting either set of given pressure-temperature data into *(4.19b)* and solving for P_2 gives

$$\ln \frac{P_2}{246.80\ \text{torr}} = \frac{-(60.54\ \text{kJ}\cdot\text{mol}^{-1})[(10^3\ \text{J})/(1\ \text{kJ})]}{8.314\ \text{J}\cdot\text{K}^{-1}\cdot\text{mol}^{-1}}\left(\frac{1}{473.2\ \text{K}} - \frac{1}{573.2\ \text{K}}\right) = -2.684$$

$$P_2 = 16.8\ \text{torr}$$

This pressure agrees fairly well with the experimental value of 17.287 torr.

4.115 Use the results of Problem 4.113 to predict the normal boiling point of Hg.

▮ The normal boiling point of Hg will be the temperature at which the vapor pressure is exactly 760 torr. Substituting the data into *(4.19b)* and solving for T_2 gives

$$\ln \frac{760\ \text{torr}}{246.80\ \text{torr}} = \frac{-(60.54\ \text{kJ}\cdot\text{mol}^{-1})[(10^3\ \text{J})/(1\ \text{kJ})]}{8.314\ \text{J}\cdot\text{K}^{-1}\cdot\text{mol}^{-1}}\left(\frac{1}{T_2} - \frac{1}{573.2\ \text{K}}\right)$$

$$T_2 = 628.9\ \text{K} = 355.0\ ^\circ\text{C}$$

This temperature agrees fairly well with the experimental value of 356.57 °C.

4.116 Over large temperature ranges the *Kirchhoff equation* is used to describe the vapor pressure-temperature behavior of a substance.

$$\ln P = -\frac{\Delta_{\text{vap}}H \text{ or } \Delta_{\text{sub}}H}{R}\left(\frac{1}{T}\right) + \frac{\Delta_{\text{vap}}C_P \text{ or } \Delta_{\text{sub}}C_P}{R}(\ln T) + k \qquad (4.20)$$

Will a plot of $\ln P$ against T^{-1} show positive or negative deviation from the straight line predicted by *(4.19a)*?

▮ The difference between the values will be equal to the value of $[(\Delta_{\text{vap}}C_P \text{ or } \Delta_{\text{sub}}C_P)/R] \ln T$. Because the heat capacity of the liquid or solid is greater than that for the gas for most substances, $\Delta_{\text{vap}}C_P$ or $\Delta_{\text{sub}}C_P$ will be negative. Thus the plot of *(4.20)* will show negative deviations from the straight line plot of *(4.19a)*.

4.117 The vapor pressure of liquid ammonia from −78 °C to 132 °C is given by

$$\ln[P/(\text{atm})] = \frac{-3323}{T/(\text{K})} + (-1.860)\ln[T/(\text{K})] + 24.04$$

Determine $\Delta_{\text{vap}}H_{\text{m}}$ and $\Delta_{\text{vap}}C_{P,\text{m}}$. What is the vapor pressure at 25 °C?

▮ Comparison of the terms of the empirical equation and *(4.20)* gives

$$\Delta_{\text{vap}}H_{\text{m}} = -(8.314\ \text{J}\cdot\text{K}^{-1}\cdot\text{mol}^{-1})\frac{10^{-3}\ \text{kJ}}{1\ \text{J}}(-3323\ \text{K}) = 27.63\ \text{kJ}\cdot\text{mol}^{-1}$$

$$\Delta_{\text{vap}}C_{P,\text{m}} = (8.314\ \text{J}\cdot\text{K}^{-1}\cdot\text{mol}^{-1})(-1.860) = -15.46\ \text{J}\cdot\text{K}^{-1}\cdot\text{mol}^{-1}$$

At 25 °C, substituting into the empirical equation gives

$$\ln[P/(\text{atm})] = -3323/298 + (-1.860)(\ln 298) + 24.04 = 2.29$$

$$P = 9.87\ \text{atm}$$

4.118 Combine the barometric law, *(1.5)*, with the Clausius-Clapeyron equation, *(4.19b)*, to derive an equation relating the boiling point of a liquid to the altitude above sea level. Calculate the boiling point of water into air at 15 °C at an altitude of 10 000 ft given $\Delta_{\text{vap}}H_{\text{m}} = 40.656\ \text{kJ}\cdot\text{mol}^{-1}$ at 100.00 °C.

▮ Assigning $P_2 = P$, $T_2 = T$, $P_1 = P_0$, and $T_1 = T_0$ and substituting *(1.5)* into *(4.19b)* gives

$$\ln\frac{P}{P_0} = \ln e^{-gMh/RT_{\text{air}}} = \frac{-gMh}{RT_{\text{air}}} = -\frac{\Delta_{\text{vap}}H}{R}\left(\frac{1}{T} - \frac{1}{T_0}\right)$$

Solving for $1/T$ and substituting the data gives

$$\frac{1}{T} = \frac{1}{T_0} + \frac{gMh}{\Delta_{\text{vap}}H\,T_{\text{air}}}$$

$$= \frac{1}{373.15\ \text{K}} + \frac{(9.81\ \text{m}\cdot\text{s}^{-2})(29\times10^{-3}\ \text{kg}\cdot\text{mol}^{-1})(10\,000\ \text{ft})[(0.3048\ \text{m})/(1\ \text{ft})]}{(40.656\ \text{kJ}\cdot\text{mol}^{-1})[(10^3\ \text{J})/(1\ \text{kJ})](288\ \text{K})}$$

$$= 2.7539\times10^{-3}\ \text{K}^{-1}$$

$$T = 363.12\ \text{K} = 89.97\ ^\circ\text{C}$$

4.119 Prepare a graph of $\Delta_f H_T^\circ$ for N_2O_4 as a function of temperature from the following data (JANAF Thermochemical Tables):

$T/(\mathrm{K})$	$\dfrac{\Delta_f H_T^\circ(\mathrm{s})}{(\mathrm{kJ \cdot mol^{-1}})}$	$\dfrac{\Delta_f H_T^\circ(\mathrm{l})}{(\mathrm{kJ \cdot mol^{-1}})}$	$\dfrac{\Delta_f H_T^\circ(\mathrm{g})}{(\mathrm{kJ \cdot mol^{-1}})}$
0	−30.399		18.718
100	−35.874	−27.487	14.011
200	−36.947	−24.199	10.749
300	−34.982	−19.463	9.060
400	−29.882	−13.139	8.523
500	−22.746	−5.213	8.769

The melting point of N_2O_4 is 262 K. Determine $\Delta_{\mathrm{fus}}H_{262}^\circ$ and $\Delta_{\mathrm{vap}}H_{262}^\circ$. Also determine $\Delta_{\mathrm{vap}}H_{400}^\circ$.

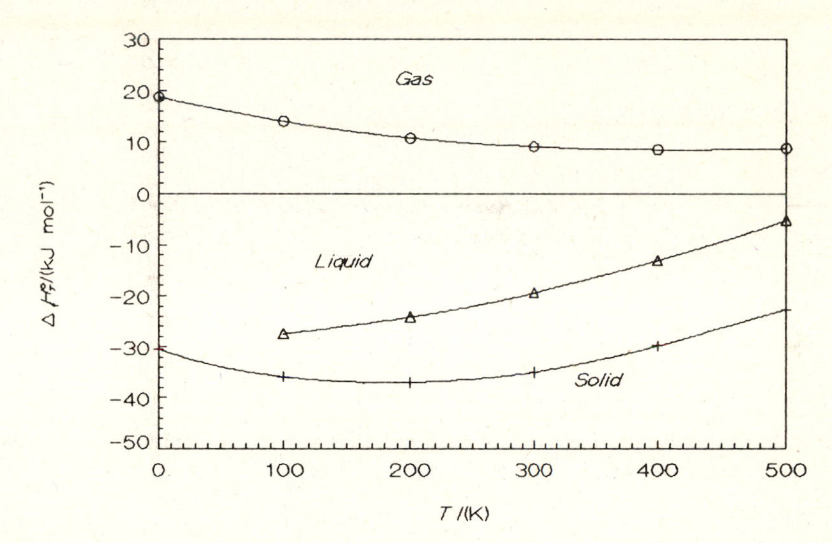

Fig. 4-10

▮ The data are plotted in Fig. 4-10. The vertical distance between the respective curves will give the values of the enthalpy changes for the phase transitions. From the graph,

$$\Delta_{\mathrm{fus}}H_{262}^\circ = -21.4\ \mathrm{kJ \cdot mol^{-1}} - (-36.1\ \mathrm{kJ \cdot mol^{-1}}) = 14.7\ \mathrm{kJ \cdot mol^{-1}}$$

$$\Delta_{\mathrm{vap}}H_{262}^\circ = 9.5\ \mathrm{kJ \cdot mol^{-1}} - (-21.4\ \mathrm{kJ \cdot mol^{-1}}) = 30.9\ \mathrm{kJ \cdot mol^{-1}}$$

$$\Delta_{\mathrm{vap}}H_{400}^\circ = 8.523\ \mathrm{kJ \cdot mol^{-1}} - (-13.139\ \mathrm{kJ \cdot mol^{-1}}) = 21.662\ \mathrm{kJ \cdot mol^{-1}}$$

CHAPTER 5
Entropy

5.1 THE SECOND LAW OF THERMODYNAMICS—HEAT ENGINES

5.1 A Carnot engine operates between T_h, 25 °C, and T_c, 0 °C, using 1.00 mol of an ideal monatomic gas. The initial pressure–volume conditions are 1.000 bar and 24.8 L. During the isothermal expansion step, the volume changes to 50.0 L. Calculate q, w, and ΔU for each step of the cycle and for the overall process.

▮ The first step of a Carnot cycle is a reversible isothermal expansion. Substituting the data into *(3.50)* gives

$$\Delta U_1 = 0 \qquad q_1 = -w_1$$

$$w_1 = -nRT_h \ln\frac{V_2}{V_1} = -(1.00\ \text{mol})(8.314\ \text{J}\cdot\text{K}^{-1}\cdot\text{mol}^{-1})(298\ \text{K})\ln\frac{50.0\ \text{L}}{24.8\ \text{L}}$$
$$= -1740\ \text{J}$$

The second step of a Carnot cycle is a reversible adiabatic expansion. Substituting $C_{V,m} = \frac{3}{2}R$ and the data into *(3.56b)* gives

$$V_3 = V_2\left(\frac{T_h}{T_c}\right)^{C_{V,m}/R} = (50.0\ \text{L})\left(\frac{298\ \text{K}}{273\ \text{K}}\right)^{3/2} = 57.0\ \text{L}$$

Using *(3.57)* gives

$$q_2 = 0$$

$$w_2 = \Delta U_2 = \int_{298\ \text{K}}^{273\ \text{K}} nC_{V,m}\,dT$$
$$= (1.00\ \text{mol})(\tfrac{3}{2})(8.314\ \text{J}\cdot\text{K}^{-1}\cdot\text{mol}^{-1})(273\ \text{K} - 298\ \text{K}) = -310\ \text{J}$$

The third step of a Carnot cycle is a reversible isothermal compression. The volume is given by

$$V_4 = V_3\frac{V_1}{V_2} = (57.0\ \text{L})\frac{24.8\ \text{L}}{50.0\ \text{L}} = 28.3\ \text{L}$$

Substituting the data into *(3.50)* gives

$$\Delta U_3 = 0 \qquad q_3 = -w_3$$

$$w_3 = -nRT_c \ln\frac{V_4}{V_3} = -(1.00\ \text{mol})(8.314\ \text{J}\cdot\text{K}^{-1}\cdot\text{mol}^{-1})(273\ \text{K})\ln\frac{28.3\ \text{L}}{57.0\ \text{L}}$$
$$= 1590\ \text{J}$$

The fourth step of a Carnot cycle is a reversible adiabatic compression back to the initial state of the system. Using *(3.57)* gives

$$q_4 = 0$$

$$w_4 = \Delta U_4 = \int_{273\ \text{K}}^{298\ \text{K}} nC_{V,m}\,dT$$
$$= (1.00\ \text{mol})(\tfrac{3}{2})(8.314\ \text{J}\cdot\text{K}^{-1}\cdot\text{mol}^{-1})(298\ \text{K} - 273\ \text{K}) = 310\ \text{J}$$

For the overall process,

$$q = q_1 + q_2 + q_3 + q_4 = 1740\ \text{J} + 0 + (-1590\ \text{J}) + 0 = 150\ \text{J}$$

$$w = w_1 + w_2 + w_3 + w_4 = -1740\ \text{J} + (-310) + 1590\ \text{J} + 310\ \text{J} = -150\ \text{J}$$

$$\Delta U = \Delta U_1 + \Delta U_2 + \Delta U_3 + \Delta U_4 = 0 + (-310\ \text{J}) + 0 + 310\ \text{J} = 0$$

5.2 Prepare a sketch showing the two heat reservoirs, the Carnot engine, and the "flow" of heat and work for the process described in Problem 5.1.

▮ The two heat reservoirs are represented by rectangles and the engine by a circle in Fig. 5-1. The heat flow from the reservoir at T_h to the engine, q_h, is equal to the value of q_1; the heat flow from the engine to the reservoir at T_c, q_c, is equal to q_3; and the work produced by the engine is equal to the value of w for the overall cycle.

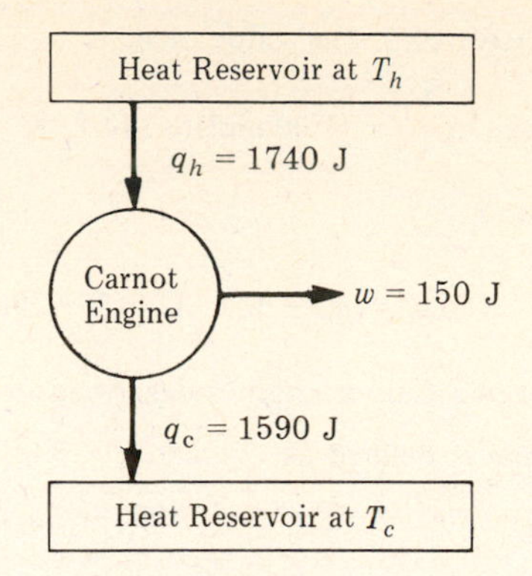

Fig. 5-1

5.3 Prepare a P-V plot for the Carnot cycle described in Problem 5.1. Determine the area enclosed by the figure. What does this area represent?

▮ The plot is shown in Fig. 5-2. The curves for the reversible isothermal steps (steps 1 and 3) were calculated using *(1.7)*, and the curves for the reversible adiabatic steps (steps 2 and 4) were calculated using *(3.56b)* and

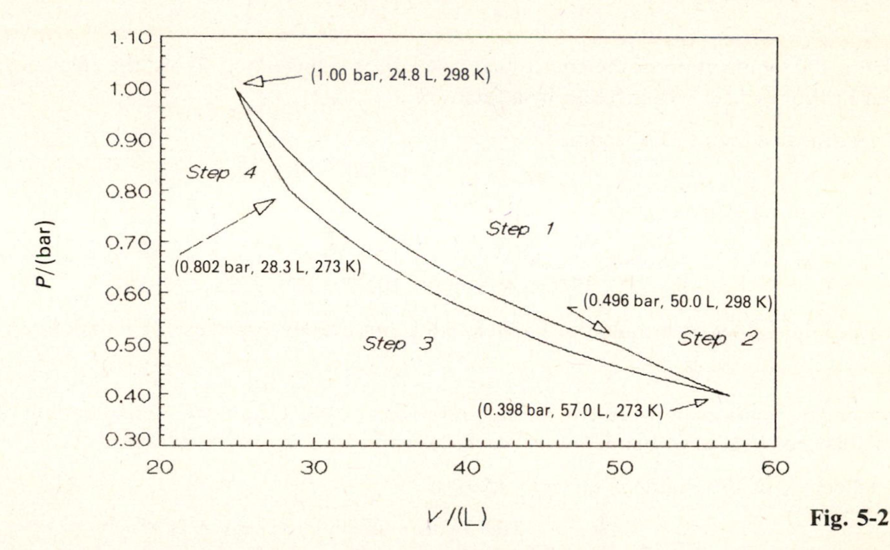

Fig. 5-2

(1.18) at various temperatures between 298 K and 273 K. By graphical integration, the area under the curves for steps 1 and 2 is 20.49 L · bar, and the area under the curves for steps 3 and 4 is 19.01 L · bar. The difference in these values (the area enclosed by the figure) is the work produced by the engine

$$w = -(20.49\ \text{L}\cdot\text{bar} - 19.01\ \text{L}\cdot\text{bar})\frac{10^2\ \text{J}}{1\ \text{L}\cdot\text{bar}} = -148\ \text{J}$$

5.4 Suppose the ideal monatomic gas in the Carnot engine described in Problem 5.1 was replaced by an ideal diatomic gas. What will be the work generated by the engine?

▮ The net work of the engine is given by

$$w = -nR(T_h - T_c)\ln\frac{V_2}{V_1} \tag{5.1}$$

This equation is not a function $C_{V,m}$, so the work generated will be the same for the replacement gas as for the original gas.

5.5 The Carnot engine described in Problem 5.1 was redesigned so that the gas expands isothermally to 50.0 L and adiabatically to −25 °C. By what factor has the work output from the engine increased? What are the values of q_h and q_c in Fig. 5-1 for this engine?

▮ Substituting the data into *(5.1)* gives

$$w = -(1.00\ \text{mol})(8.314\ \text{J}\cdot\text{K}^{-1}\cdot\text{mol}^{-1})(298\ \text{K} - 248\ \text{K})\ln\frac{50.0\ \text{L}}{24.8\ \text{L}} = -290\ \text{J}$$

The work has increased by a factor of 2. The value of q_h is

$$q_h = nRT_h \ln\frac{V_2}{V_1} = (1.00 \text{ mol})(8.314 \text{ J}\cdot\text{K}^{-1}\cdot\text{mol}^{-1})(298 \text{ K}) \ln\frac{50.0 \text{ L}}{24.8 \text{ L}}$$

$$= 1740 \text{ J}$$

and

$$q_c = q_h + w = 1740 \text{ J} - 290 \text{ J} = 1450 \text{ J}$$

5.6 What are the efficiencies of the two Carnot engines described in Problems 5.1 and 5.5?

▮ The *efficiency of a heat engine* is defined as

$$\varepsilon = 100(-w/q_h) \tag{5.2}$$

For the first engine,

$$\varepsilon = 100\frac{-(-150 \text{ J})}{1740 \text{ J}} = 8.6\%$$

and for the second engine,

$$\varepsilon = 100\frac{-(-290 \text{ J})}{1740 \text{ J}} = 16\%$$

5.7 Show that if the temperature of the cooler heat reservoir is changed to −25 °C, the efficiency of the Carnot engine described in Problem 5.1 will increase by a factor of 2.

▮ For a Carnot engine, *(5.2)* becomes

$$\varepsilon_{\text{Carnot}} = 100(T_h - T_c)/T_h \tag{5.3}$$

Taking a ratio of efficiencies gives

$$\frac{\varepsilon_{\text{Carnot}}(0\ ^\circ\text{C reservoir})}{\varepsilon_{\text{Carnot}}(-25\ ^\circ\text{C reservoir})} = \frac{100(298 \text{ K} - 273 \text{ K})/(298 \text{ K})}{100(298 \text{ K} - 248 \text{ K})/(298 \text{ K})} = 2$$

(The ratio of the values determined in Problem 5.6 is not exactly 2 because of rounding errors in the values of q and w.)

5.8 A Carnot engine operates between two thermal reservoirs at 100 °C and 0 °C. What amount of work can it produce for each 1000 J of heat absorbed from the warm reservoir?

▮ The efficiency of the engine is given by *(5.3)* as

$$\varepsilon_{\text{Carnot}} = 100(373 \text{ K} - 273 \text{ K})/(373 \text{ K}) = 26.8\%$$

Solving *(5.2)* for w and substituting the data gives

$$w = -(26.8)(1000 \text{ J})/100 = -268 \text{ J}$$

5.9 A Carnot engine is to operate with at least a 15% efficiency rating. The warm heat reservoir will be boiling water at 100 °C, and the cool heat reservoir will be cool stream water. What is the maximum temperature of the cooling water that will keep the efficiency rating above 15%?

▮ Substituting into *(5.3)* and solving for T_c gives

$$15 < 100(373 \text{ K} - T_c)/(373 \text{ K}) \qquad T_c > 317 \text{ K} = 44\ ^\circ\text{C}$$

5.10 A Carnot engine operates between two thermal reservoirs at 100 °C and 0 °C. During the operation of the engine, 25% of the work produced is dissipated by friction. How much heat must the warm reservoir supply in order for the engine to produce 1000 J of useful work?

▮ The total work that must be generated by the engine is

$$w_{\text{total}} = w_{\text{useful}} + w_{\text{friction}} = 1000 \text{ J} + 0.25w_{\text{total}}$$

$$w_{\text{total}} = 1330 \text{ J}$$

Equating the expressions for efficiency, *(5.2)* and *(5.3)*, and solving for q_h gives

$$\varepsilon = 100\frac{373 \text{ K} - 273 \text{ K}}{373 \text{ K}} = 100\frac{1330 \text{ J}}{q_h}$$

$$q_h = 4960 \text{ J}$$

5.11 Consider a hypothetical internal combustion engine that operates as efficiently as a Carnot engine between heat reservoirs at 2000 K and 1000 K. If $q_h = 1.5 \times 10^8$ J (the typical energy available from a gallon of gasoline), how far off the ground can the engine lift a 1500-kg object?

▮ Substituting the data into *(5.3)* gives

$$\varepsilon_{\text{Carnot}} = 100\frac{2000\text{ K} - 1000\text{ K}}{2000\text{ K}} = 50\%$$

Solving *(5.2)* for w and substituting the data gives

$$w = -(50)(1.5 \times 10^8\text{ J})/100 = -8 \times 10^7\text{ J}$$

The work that will be done on the object is given by mgl (see Table 3-5). Solving for l and substituting the data gives

$$l = \frac{w}{mg} = \frac{(8 \times 10^7\text{ J})[(1\text{ kg}\cdot\text{m}^2\cdot\text{s}^{-2})/(1\text{ J})]}{(1500\text{ kg})(9.81\text{ m}\cdot\text{s}^{-2})} = 5000\text{ m}$$

5.12 Assume that the human body is a heat engine. For a food intake equivalent to $1.0 \times 10^7\text{ J}\cdot\text{day}^{-1}$, a human can work for $6\text{ h}\cdot\text{day}^{-1}$ at a rate equivalent to 0.04 hp. What is the efficiency of this system?

▮ The total work produced in 1 day is

$$w = (1\text{ day})(6\text{ h}\cdot\text{day}^{-1})(0.04\text{ hp})\frac{746\text{ W}}{1\text{ hp}}\left(\frac{1\text{ J}\cdot\text{s}^{-1}}{1\text{ W}}\right)\left(\frac{3600\text{ s}}{1\text{ h}}\right) = -6 \times 10^5\text{ J}$$

Substituting into *(5.2)* gives

$$\varepsilon = 100\frac{-(-6 \times 10^5\text{ J})}{1.0 \times 10^7\text{ J}} = 6\%$$

5.13 Consider 1.00 mol of liquid water at 100 °C to be the warm heat reservoir for a Carnot engine. As heat is drawn from this reservoir by the engine, the temperature of the heat reservoir drops. How much work can be obtained from this reservoir as it cools to the temperature of the cool heat reservoir at 0 °C? Assume $C_{P,\text{m}} = 75\text{ J}\cdot\text{K}^{-1}\cdot\text{mol}^{-1}$ for water over this temperature range. What is the efficiency of this process?

▮ The efficiency of the system at any time during the process is given by *(5.3)* as

$$\varepsilon = 100(T - 273\text{ K})/T$$

For the transfer of a very small amount of heat from the water and the production of a very small amount of work by the engine, *(5.2)* can be written as

$$đw = \frac{\varepsilon}{100}\, đq_h$$

Substituting *(3.23)* and integrating gives

$$đw = \frac{100(T - 273\text{ K})/T}{100} nC_{P,\text{m}}\,dT = nC_{P,\text{m}}\left(1 - \frac{273\text{ K}}{T}\right)dT$$

$$\begin{aligned} w &= nC_{P,\text{m}}\int_{373\text{ K}}^{273\text{ K}}\left(1 - \frac{273\text{ K}}{T}\right)dT \\ &= (1.00\text{ mol})(75\text{ J}\cdot\text{K}^{-1}\cdot\text{mol}^{-1})\{(273\text{ K} - 373\text{ K}) - (273\text{ K})\ln[(273\text{ K})/(373\text{ K})]\} \\ &= -1100\text{ J} \end{aligned}$$

The efficiency of the process is given by *(5.2)* as

$$\varepsilon = 100\frac{-(-1100\text{ J})}{1.00(75\text{ J}\cdot\text{K}^{-1}\cdot\text{mol}^{-1})(100\text{ K})} = 15\%$$

(Compare this to the results of Problem 5.8, when $\varepsilon = 26.8\%$ for a similar engine operating between heat reservoirs at the same temperatures that do not change temperature during the process.)

5.14 What happens to the efficiency of a Carnot engine as T_h increases? Under what conditions will the efficiency of an engine be 100%?

▮ The high-temperature limit of *(5.3)* is

$$\lim_{T_h\to\infty}(\varepsilon_{\text{Carnot}}) = \lim_{T_h\to\infty}\left(100\frac{T_h - T_c}{T_h}\right) = 100\%$$

This result shows that by absorbing heat at extremely high temperatures, a Carnot engine will increase in efficiency. Likewise, the low-temperature limit of *(5.3)* is

$$\lim_{T_c \to 0} (\varepsilon_{\text{Carnot}}) = \lim_{T_c \to 0} \left(100 \frac{T_h - T_c}{T_h} \right) = 100\%$$

This result shows that by releasing heat at extremely low temperatures, the efficiency of an engine will increase. Although high efficiences can be reached, the *second law of thermodynamics* forbids the attainment of 0 K, so 100% efficiency cannot be achieved.

5.15 An ideal gas engine was constructed to run on the following cycle: **(*a*)** step 1, isobaric heating from P_1, V_1, T_1 to P_1, V_2, T_2; **(*b*)** step 2, isochoric cooling from P_1, V_2, T_2 to P_3, V_2, T_3; and **(*c*)** step 3, adiabatic compression from P_3, V_2, T_3 to P_1, V_1, T_1. Assume that 1.00 mol of an ideal monatomic gas starting at 298 K, 1.000 bar, and 24.8 L undergoes this cycle and that the engine is operating between heat reservoirs at 601 K and 186 K. Calculate q_h, w, and ε for the cycle. Compare ε to that for a Carnot engine operating between these same heat reservoirs.

▮ For the first steps, the volume can be determined using *(1.10)*, giving

$$V_2 = V_1 \frac{T_2}{T_1} = 24.8\ \text{L} \frac{601\ \text{K}}{298\ \text{K}} = 50.0\ \text{L}$$

Using *(3.54)* for this step gives

$$w_1 = -P_1(V_2 - V_1) = -(1.000\ \text{bar})(50.0\ \text{L} - 24.8\ \text{L}) \frac{10^2\ \text{J}}{1\ \text{L} \cdot \text{bar}} = -2520\ \text{J}$$

$$q_1 = nC_{V,m}(T_2 - T_1) + P_1(V_2 - V_1)$$

$$= (1.000\ \text{mol})(\tfrac{3}{2})(8.314\ \text{J} \cdot \text{K}^{-1} \cdot \text{mol}^{-1})(601\ \text{K} - 298\ \text{K}) + (1.000\ \text{bar})(50.0\ \text{L} - 24.8\ \text{L}) \frac{10^2\ \text{J}}{1\ \text{L} \cdot \text{bar}} = 6300\ \text{J}$$

For the second step, using *(3.53)* gives

$$w_2 = 0$$

$$q_2 = nC_{V,m}(T_3 - T_2) = (1.000\ \text{mol})(\tfrac{3}{2})(8.314\ \text{J} \cdot \text{K}^{-1} \cdot \text{mol}^{-1})(186\ \text{K} - 601\ \text{K})$$

$$= -5180\ \text{J}$$

For the third step, using *(3.57)* gives

$$q_3 = 0$$

$$w_3 = nC_{V,m}(T_1 - T_3)$$

$$= (1.000\ \text{mol})(\tfrac{3}{2})(8.314\ \text{J} \cdot \text{K}^{-1} \cdot \text{mol}^{-1})(298\ \text{K} - 186\ \text{K}) = 1400\ \text{J}$$

For the entire process,

$$q_h = q_1 = 6300\ \text{J}$$

$$w_{\text{total}} = w_1 + w_2 + w_3 = -2520\ \text{J} + 0 + 1400\ \text{J} = -1120\ \text{J}$$

Substituting these results into *(5.2)* gives

$$\varepsilon = 100 \frac{-(-1120\ \text{J})}{6300\ \text{J}} = 17.8\%$$

For a Carnot engine operating between these same heat reservoirs, *(5.3)* gives

$$\varepsilon_{\text{Carnot}} = 100 \frac{601\ \text{K} - 186\ \text{K}}{601\ \text{K}} = 69.1\%$$

These values of efficiency illustrate another statement of the *second law of thermodynamics* in that no engine can operate more efficiently than a Carnot engine.

5.16 Define the performance factor for the refrigerator shown in Fig. 5-3. Using the results of Problem 5.1, derive the equation for a Carnot refrigerator.

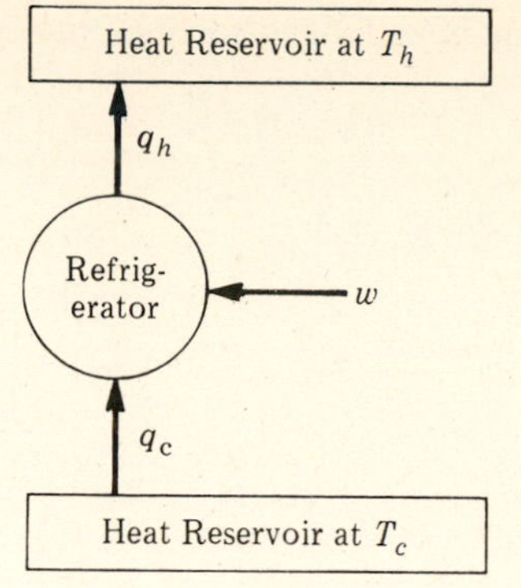

Fig. 5-3

▮ The *performance factor* is defined as

$$\varepsilon_{\text{ref}} = 100w/q_c \tag{5.4}$$

Substituting *(5.1)* and the expression derived for $q_c = q_3$ into *(5.4)* gives

$$\begin{aligned}\varepsilon_{\text{Carnot ref}} &= 100\frac{-nR(T_h - T_c)\ln(V_2/V_1)}{nRT_c\ln(V_4/V_1)} \\ &= 100\frac{-nR(T_h - T_c)\ln(V_2/V_1)}{-nRT_c\ln(V_2/V_1)} \\ &= 100(T_h - T_c)/T_c\end{aligned} \tag{5.5}$$

5.17 Consider an ideal refrigerator operating between 0 °C and 25 °C. The refrigerator is to produce 1.0 g of ice each second at 0 °C from water at 0 °C. How much work must be done? The molar heat of fusion of water is $6.0095\ \text{kJ}\cdot\text{mol}^{-1}$.

▮ The heat needed to be removed from the water each second is

$$q_c = (1\ \text{s})(1.0\ \text{g}\cdot\text{s}^{-1})\frac{1\ \text{mol}}{18\ \text{g}}(-6.0095\ \text{kJ}\cdot\text{mol}^{-1})\frac{10^3\ \text{J}}{1\ \text{kJ}} = 330\ \text{J}$$

Equating *(5.4)* and *(5.5)* and substituting the data gives

$$\varepsilon_{\text{ref}} = 100\frac{w}{330\ \text{J}} = 100\frac{298\ \text{K} - 273\ \text{K}}{273\ \text{K}}$$

$$w = 30\ \text{J}$$

This is equivalent to

$$P = (30\ \text{J}\cdot\text{s}^{-1})\frac{1\ \text{W}}{1\ \text{J}\cdot\text{s}^{-1}} = 30\ \text{W}$$

or 0.040 hp.

5.18 Suppose the water being introduced into the ice maker described in Problem 5.17 is originally at 25 °C and the refrigerator must cool the water as well as freeze it. How much total work must be done? The molar heat capacity of water is $75.3\ \text{J}\cdot\text{K}^{-1}\cdot\text{mol}^{-1}$.

▮ The performance factor of the system at any time during the cooling process is given by *(5.5)* as

$$\varepsilon_{\text{ref}} = 100(298\ \text{K} - T)/T$$

For the transfer of a very small amount of heat from the water and the intake of a very small amount of work by the refrigerator, *(5.4)* can be written as

$$đw = \frac{\varepsilon_{\text{ref}}}{100}\,đq_c$$

Substituting *(3.23)* and integrating gives for the cooling process for each second

$$đw = \frac{100(298\text{ K} - T)/T}{100}(-nC_{P,\text{m}}\,dT) = -nC_{P,\text{m}}\left(\frac{298\text{ K}}{T} - 1\right)dT$$

$$w = -nC_{P,\text{m}}\int_{298\text{ K}}^{273\text{ K}}\left(\frac{298\text{ K}}{T} - 1\right)dT$$

$$= -(1\text{ s})(1.0\text{ g}\cdot\text{s}^{-1})\frac{1\text{ mol}}{18\text{ g}}(75.3\text{ J}\cdot\text{K}^{-1}\cdot\text{mol}^{-1})\left[(298\text{ K})\ln\frac{273\text{ K}}{298\text{ K}} - (273\text{ K} - 298\text{ K})\right]$$

$$= 4.6\text{ J}$$

The total work required is

$$w_{\text{total}} = w_{\text{cooling}} + w_{\text{freezing}} = 4.6\text{ J} + 30\text{ J} = 35\text{ J}$$

The power requirement is

$$P = (35\text{ J}\cdot\text{s}^{-1})\frac{1\text{ W}}{1\text{ J}\cdot\text{s}^{-1}} = 35\text{ W} = 0.047\text{ hp}$$

5.19 Determine the performance factor of an ideal heat pump used for heating a building in the winter (outside temperature = 0 °C, inside temperature = 23 °C) and for cooling in the summer (outside temperature = 35 °C, inside temperature = 25 °C). Assume that the building gains or loses 1.0 kJ · s^{-1} for each kelvin difference between the inside and outside temperatures. Calculate the relative amounts of work needed to maintain the inside temperature during the winter and the summer.

I During the summer, the heat pump acts as a refrigerator (an air conditioner). The performance factor is given by *(5.5)* as

$$\varepsilon_{\text{ref}} = 100\frac{308\text{ K} - 298\text{ K}}{298\text{ K}} = 3.4\%$$

The heat pump must remove from the building each second

$$q_{\text{c}} = (1\text{ s})(1.0\text{ kJ}\cdot\text{s}^{-1}\cdot\text{K}^{-1})(10\text{ K}) = 10\text{ kJ}$$

Using *(5.4)* gives

$$w_{\text{summer}} = \frac{3.4}{100}(10\text{ kJ}) = 0.34\text{ kJ}$$

During the winter, the performance of the heat pump is given by *(5.5)* as

$$\varepsilon_{\text{ref}} = 100\frac{296\text{ K} - 273\text{ K}}{273\text{ K}} = 8.4\%$$

The heat pump must supply to the building each second

$$q_{\text{h}} = (1\text{ s})(1.0\text{ kJ}\cdot\text{s}^{-1}\cdot\text{K}^{-1})(23\text{ K}) = 23\text{ kJ}$$

For this process,

$$q_{\text{c}} = q_{\text{h}} - w = 23\text{ kJ} - w$$

Using *(5.4)* gives

$$8.4\% = 100w_{\text{winter}}/(23\text{ kJ} - w_{\text{winter}})$$

$$w_{\text{winter}} = 1.8\text{ kJ}$$

The ratio of the work needed during the seasons is

$$w_{\text{winter}}/w_{\text{summer}} = (1.8\text{ kJ})/(0.34\text{ kJ}) = 5.3$$

5.2 THE SECOND LAW OF THERMODYNAMICS—ENTROPY CALCULATIONS

5.20 Exactly 100 J of heat was transferred reversibly to a block of gold at 25.00 °C from a thermal reservoir at 25.01 °C, and then exactly 100 J of heat was absorbed reversibly from the block of gold by a thermal reservoir at 24.99 °C. What is ΔS(system) for this process?

▮ For each step the entropy change of the gold block for the *reversible isothermal transfer of heat* will be given by

$$\Delta S(\text{system}) = q_{\text{rev}}/T \tag{5.6}$$

Substituting the data into *(5.6)* gives

$$\Delta S(\text{system})_1 = \frac{100\ \text{J}}{298.15\ \text{K}} = 0.335\ 40\ \text{J}\cdot\text{K}^{-1}$$

$$\Delta S(\text{system})_2 = \frac{-100\ \text{J}}{298.15\ \text{K}} = -0.335\ 40\ \text{J}\cdot\text{K}^{-1}$$

For the process,

$$\Delta S(\text{system}) = \Delta S(\text{system})_1 + \Delta S(\text{system})_2$$
$$= 0.335\ 40\ \text{J}\cdot\text{K}^{-1} + (-0.335\ 40\ \text{J}\cdot\text{K}^{-1}) = 0$$

Entropy is a state function, because the state of the system is unchanged for the process, $\Delta S(\text{system}) = 0$.

5.21 What will be $\Delta S(\text{system})$ for steps 2 and 4 in the Carnot cycle described in Problem 5.1?

▮ These steps represent the reversible adiabatic expansion and compression steps for which $q_{\text{rev}} = 0$. Substituting into *(5.6)* gives a *reversible adiabatic process,*

$$\Delta S(\text{system}) = 0 \tag{5.7}$$

5.22 Determine $\Delta S^\circ_{\text{vap}}(\text{system})$ for iron at the boiling point of 3133 K given $\Delta_{\text{vap}}H^\circ_{\text{m}} = 349\ \text{kJ}\cdot\text{mol}^{-1}$.

▮ For a *reversible isothermal phase transition, (5.6)* gives

$$\Delta_{\text{I}\to\text{II}}S^\circ_{\text{m}}(\text{system}) = \frac{q_{\text{rev}}}{T} = \frac{\Delta_{\text{I}\to\text{II}}H^\circ}{T} \tag{5.8}$$

Substituting the data into *(5.8)* gives

$$\Delta_{\text{vap}}S^\circ_{\text{m}}(\text{system}) = \frac{(349\ \text{kJ}\cdot\text{mol}^{-1})[(10^3\ \text{J})/(1\ \text{kJ})]}{3133\ \text{K}} = 111\ \text{J}\cdot\text{K}^{-1}\cdot\text{mol}^{-1}$$

This is a relatively large entropy change because the system is going from a state of rather high order to one of rather high disorder.

5.23 Iron undergoes two solid-solid transitions as it is heated from 298 K to 1800 K. Given $\Delta H^\circ = 0.900\ \text{kJ}\cdot\text{mol}^{-1}$ at 1184 K for the alpha-gamma transition and $\Delta H^\circ = 0.837\ \text{kJ}\cdot\text{mol}^{-1}$ at 1665 K for the gamma-delta transition, calculate $\Delta S^\circ(\text{system})$ for each change. Comment on the changes in randomness.

▮ Substituting the data into *(5.8)* gives

$$\Delta_{\alpha\to\gamma}S^\circ_{\text{m}}(\text{system}) = \frac{(0.900\ \text{kJ}\cdot\text{mol}^{-1})[(10^3\ \text{J})/(1\ \text{kJ})]}{1184\ \text{K}} = 0.760\ \text{J}\cdot\text{K}^{-1}\cdot\text{mol}^{-1}$$

$$\Delta_{\gamma\to\delta}S^\circ_{\text{m}}(\text{system}) = \frac{(0.837\ \text{kJ}\cdot\text{mol}^{-1})[(10^3\ \text{J})/(1\ \text{kJ})]}{1665\ \text{K}} = 0.503\ \text{J}\cdot\text{K}^{-1}\cdot\text{mol}^{-1}$$

These relatively small values of $\Delta S(\text{system})$ indicate that a very small increase in randomness occurs during these condensed-phase changes.

5.24 Predict the enthalpy of fusion of iron at 1809 K and the enthalpy of vaporization at 3133 K.

▮ Both Trouton's rules, *(4.12)*, describing vaporization and *(4.13)* describing fusion, are based on *(5.8)* and can be restated as

$$\Delta_{\text{vap}}S^\circ_{\text{m}}(\text{system}) \approx 88\ \text{J}\cdot\text{K}^{-1}\cdot\text{mol}^{-1}$$

$$\Delta_{\text{fus}}S^\circ_{\text{m}}(\text{system}) \approx 9.2\ \text{J}\cdot\text{K}^{-1}\cdot\text{mol}^{-1}$$

Solving *(5.8)* for $\Delta_{\text{I}\to\text{II}}H^\circ$ and substituting the data gives

$$\Delta_{\text{vap}}H^\circ_{\text{m}} = (88\ \text{J}\cdot\text{K}^{-1}\cdot\text{mol}^{-1})(3133\ \text{K})\frac{10^{-3}\ \text{kJ}}{1\ \text{J}} = 280\ \text{kJ}\cdot\text{mol}^{-1}$$

$$\Delta_{\text{fus}}H^\circ_{\text{m}} = (9.2\ \text{J}\cdot\text{K}^{-1}\cdot\text{mol}^{-1})(1809\ \text{K})\frac{10^{-3}\ \text{kJ}}{1\ \text{J}} = 17\ \text{kJ}\cdot\text{mol}^{-1}$$

The predicted value of $\Delta_{\text{vap}}H^\circ_{\text{m}}$ is too low compared to the actual value of $349\ \text{kJ}\cdot\text{mol}^{-1}$ (see Problem 5.22), and the predicted value of $\Delta_{\text{fus}}H^\circ_{\text{m}}$ is too high compared to the actual value of $13.807\ \text{kJ}\cdot\text{mol}^{-1}$.

5.25 Calculate ΔS(system) for reversibly cooling an ideal monatomic gas from 298 K to 273 K under constant-volume conditions. Is this an increase or a decrease in the randomness of the system?

▌ For a system undergoing a *reversible expansion*, the entropy change will be given by

$$\Delta S(\text{system}) = \int_{T_1}^{T_2} \frac{C_V}{T}\, dT + \int_{V_1}^{V_2} \left(\frac{\partial P}{\partial T}\right)_V dV \qquad (5.9a)$$

$$\Delta S(\text{system}) = \int_{T_1}^{T_2} \frac{C_P}{T}\, dT - \int_{P_1}^{P_2} \left(\frac{\partial V}{\partial T}\right)_P dP \qquad (5.9b)$$

Under constant-volume conditions, $dV = 0$. Substituting $C_{V,\text{m}} = \frac{3}{2}R$ into *(5.9a)* gives

$$\Delta S_\text{m}(\text{system}) = \int_{298\text{ K}}^{273\text{ K}} \frac{\frac{3}{2}R}{T}\, dT + 0 = \tfrac{3}{2}R \ln \frac{273\text{ K}}{298\text{ K}}$$

$$= \tfrac{3}{2}(8.314\text{ J}\cdot\text{K}^{-1}\cdot\text{mol}^{-1}) \ln \frac{273\text{ K}}{298\text{ K}} = -1.09\text{ J}\cdot\text{K}^{-1}\cdot\text{mol}^{-1}$$

The negative entropy change means that the randomness of the system has decreased.

5.26 Repeat the calculations of Problem 5.25 for a reversible isobaric cooling of the gas. Compare the results.

▌ Under constant-pressure conditions, $dP = 0$. Substituting $C_{P,\text{m}} = \frac{5}{2}R$ into *(5.9b)* gives

$$\Delta S_\text{m}(\text{system}) = \int_{298\text{ K}}^{273\text{ K}} \frac{\frac{5}{2}R}{T}\, dT + 0 = \tfrac{5}{2}(8.314\text{ J}\cdot\text{K}^{-1}\cdot\text{mol}^{-1}) \ln \frac{273\text{ K}}{298\text{ K}}$$

$$= -1.82\text{ J}\cdot\text{K}^{-1}\cdot\text{mol}^{-1}$$

The entropy change for the isobaric process is larger because there has been a change in volume in this process.

5.27 Consider the reversible expansion of 2.000 mol of oxygen from 298 K and 24.8 L to 700 K and 38.8 L. Assuming ideal gas behavior, calculate ΔS(system) for this process. For $O_2(g)$, $C_{P,\text{m}} = 29.355\text{ J}\cdot\text{K}^{-1}\cdot\text{mol}^{-1}$.

▌ For the *reversible expansion of 1 mol of an ideal gas, (5.9)* becomes

$$\Delta S_\text{m}(\text{system}) = \int_{T_1}^{T_2} \frac{C_{V,\text{m}}}{T}\, dT + R \ln \frac{V_2}{V_1} \qquad (5.10a)$$

$$\Delta S_\text{m}(\text{system}) = \int_{T_1}^{T_2} \frac{C_{P,\text{m}}}{T}\, dT - R \ln \frac{P_2}{P_1} \qquad (5.10b)$$

Substituting the data into *(5.10a)* gives for 1 mol

$$\Delta S_\text{m}(\text{system}) = \int_{298\text{ K}}^{700\text{ K}} \frac{29.355\text{ J}\cdot\text{K}^{-1}\cdot\text{mol}^{-1} - R}{T}\, dT + R \ln \frac{38.8\text{ L}}{24.8\text{ L}}$$

$$= (29.355\text{ J}\cdot\text{K}^{-1}\cdot\text{mol}^{-1} - 8.314\text{ J}\cdot\text{K}^{-1}\cdot\text{mol}^{-1}) \ln \frac{700\text{ K}}{298\text{ K}} + (8.314\text{ J}\cdot\text{K}^{-1}\cdot\text{mol}^{-1}) \ln \frac{38.8\text{ L}}{24.8\text{ L}}$$

$$= 21.7\text{ J}\cdot\text{K}^{-1}\cdot\text{mol}^{-1}$$

For the 2-mol sample,

$$\Delta S(\text{system}) = (2.000\text{ mol})(21.7\text{ J}\cdot\text{K}^{-1}\cdot\text{mol}^{-1}) = 43.4\text{ J}\cdot\text{K}^{-1}$$

5.28 Repeat the calculations of Problem 5.27 using $C_{P,\text{m}}/(\text{J}\cdot\text{K}^{-1}\cdot\text{mol}^{-1}) = 25.849 + (12.98 \times 10^{-3})[T/(\text{K})] - (38.62 \times 10^{-7})[T/(\text{K})]^2$. Compare the results.

▮ Substituting the empirical heat capacity expression into *(5.10a)* gives

$$\Delta S_m(\text{system}) = \int_{298\text{ K}}^{700\text{ K}} \frac{\left[\begin{array}{c}(25.849\text{ J}\cdot\text{K}^{-1}\cdot\text{mol}^{-1} - 8.314\text{ J}\cdot\text{K}^{-1}\cdot\text{mol}^{-1})\\ +(12.98\times10^{-3}\text{ J}\cdot\text{K}^{-2}\cdot\text{mol}^{-1})T\\ -(38.62\times10^{-7}\text{ J}\cdot\text{K}^{-3}\cdot\text{mol}^{-1})T^2\end{array}\right]}{T}\,dT + R\ln\frac{38.8\text{ L}}{24.8\text{ L}}$$

$$\begin{aligned}&= (17.535\text{ J}\cdot\text{K}^{-1}\cdot\text{mol}^{-1})\ln[(700\text{ K})/(298\text{ K})]\\ &\quad + (12.98\times10^{-3}\text{ J}\cdot\text{K}^{-2}\cdot\text{mol}^{-1})(700\text{ K} - 298\text{ K})\\ &\quad - \tfrac{1}{2}(38.62\times10^{-7}\text{ J}\cdot\text{K}^{-3}\cdot\text{mol}^{-1})[(700\text{ K})^2 - (298\text{ K})^2]\\ &\quad + (8.314\text{ J}\cdot\text{K}^{-1}\cdot\text{mol}^{-1})\ln[(38.8\text{ L})/(24.8\text{ L})]\\ &= 23.1\text{ J}\cdot\text{K}^{-1}\cdot\text{mol}^{-1}\end{aligned}$$

For the 2-mol sample,

$$\Delta S(\text{system}) = (2.000\text{ mol})(23.1\text{ J}\cdot\text{K}^{-1}\cdot\text{mol}^{-1}) = 46.2\text{ J}\cdot\text{K}^{-1}$$

The simpler approach of Problem 5.27 introduces a −6% error.

5.29 What correction factor is needed to change entropy values listed in handbooks at 25 °C and 1 atm to values at 25 °C and 1 bar for ideal gases?

▮ For this isothermal change, $dT = 0$. Substituting into *(5.10b)* gives

$$\begin{aligned}\Delta S_m(\text{system}) &= 0 - (8.314\text{ J}\cdot\text{K}^{-1}\cdot\text{mol}^{-1})\ln\frac{1\text{ bar}}{(1\text{ atm})[(1.013\ 25\text{ bar})/(1\text{ atm})]}\\ &= 0.109\ 44\text{ J}\cdot\text{K}^{-1}\cdot\text{mol}^{-1}\end{aligned}$$

5.30 Consider a reversible isentropic expansion of 1.000 mol of an ideal gas from 25 °C to 75 °C. If the initial pressure was 1.000 bar, determine the final pressure.

▮ For an isentropic process, $\Delta S(\text{system}) = 0$. Substituting into *(5.10b)* and solving for P_2 gives for 1 mol

$$0 = \int_{298\text{ K}}^{348\text{ K}} \frac{\frac{5}{2}R}{T}\,dT - R\ln\frac{P_2}{1.000\text{ bar}}$$

$$\ln\frac{P_2}{1.000\text{ bar}} = \tfrac{5}{2}\ln\frac{348\text{ K}}{298\text{ K}} = 0.388$$

$$P_2 = 1.474\text{ bar}$$

5.31 Repeat the calculations of Problem 5.28 assuming that O_2 obeys the van der Waals equation of state. Compare the results.

▮ For the *reversible expansion of a van der Waals gas, (5.9a)* becomes

$$\Delta S(\text{system}) = \int_{T_1}^{T_2} \frac{C_V}{T}\,dT + nR\ln\frac{V_2 - nb}{V_1 - nb} \tag{5.11}$$

Substituting the results of Problem 5.28 and the data from Table 1-3 into *(5.11)* gives

$$\begin{aligned}\Delta S(\text{system}) &= (2.000\text{ mol})(19.4\text{ J}\cdot\text{K}^{-1}\cdot\text{mol}^{-1})\\ &\quad + (2.000\text{ mol})(8.314\text{ J}\cdot\text{K}^{-1}\cdot\text{mol}^{-1})\ln\frac{38.8\text{ L} - (2.000\text{ mol})(0.0319\text{ L}\cdot\text{mol}^{-1})}{24.8\text{ L} - (2.000\text{ mol})(0.0319\text{ L}\cdot\text{mol}^{-1})}\\ &= 46.3\text{ J}\cdot\text{K}^{-1}\end{aligned}$$

The results agree to within 0.2%.

5.32 Derive an equation for $(\partial S/\partial V)_T$ for a real gas obeying the Dieterici equation of state *(1.53)*. Use the relationship $(\partial S/\partial V)_T = (\partial P/\partial T)_V$.

Taking the derivative of *(1.53)* with respect to T gives

$$\left(\frac{\partial S}{\partial V}\right)_T = \left(\frac{\partial P}{\partial T}\right)_V = \left[\frac{\partial}{\partial T}\left(\frac{RT}{V_m - b}e^{-a/V_mRT}\right)\right]_V$$

$$= \frac{R}{V_m - b}e^{-a/V_mRT} + \frac{RT}{V_m - b}\left(\frac{a}{VRT^2}\right)e^{-a/V_mRT}$$

$$= \frac{RT}{V_m - b}e^{-a/V_mRT}\left(\frac{1}{T} + \frac{a}{V_mRT^2}\right) = P\frac{V_mRT + a}{V_mRT^2}$$

[The use of this result in *(5.9a)* would be very difficult.]

5.33 Calculate ΔS(system) for steps 1 and 3 in the Carnot cycle described in Problem 5.1. Combine these results with these of Problem 5.21 to prepare a plot of entropy against temperature for the process. Determine the area of the figure in the plot and identify its significance.

Step 1 is a reversible isothermal expansion for which *(5.10a)* gives

$$\Delta S(\text{system})_1 = (1.00\ \text{mol})(8.314\ \text{J}\cdot\text{K}^{-1}\cdot\text{mol}^{-1})\ln\frac{50.0\ \text{L}}{24.8\ \text{L}} = 5.83\ \text{J}\cdot\text{K}^{-1}$$

and step 3 is a reversible isothermal compression for which *(5.10a)* gives

$$\Delta S(\text{system})_3 = (1.00\ \text{mol})(8.314\ \text{J}\cdot\text{K}^{-1}\cdot\text{mol}^{-1})\ln\frac{28.3\ \text{L}}{57.0\ \text{L}} = -5.82\ \text{J}\cdot\text{K}^{-1}$$

Assuming that the system has an entropy of S at the beginning of the cycle, the coordinates of the first point are (298 K, S J · K^{-1}). As a result of the entropy change of step 1, the coordinates of the second point are (298 K, (S + 5.83) J · K^{-1}). Only the temperature changes during step 2, so the coordinates of the third point are (273 K, (S + 5.83) J · K^{-1}). During step 3, the entropy decreases by the same amount that it increased in step 1, so the coordinates of the fourth point are (273 K, S J · K^{-1}). Step 4 restores the system to the initial conditions. These points are plotted in Fig. 5-4. The area of the rectangle is

$$\text{Area} = (5.83\ \text{J}\cdot\text{K}^{-1})(298\ \text{K} - 273\ \text{K}) = 146\ \text{J}$$

which is equal to q for the overall process ($= -w$).

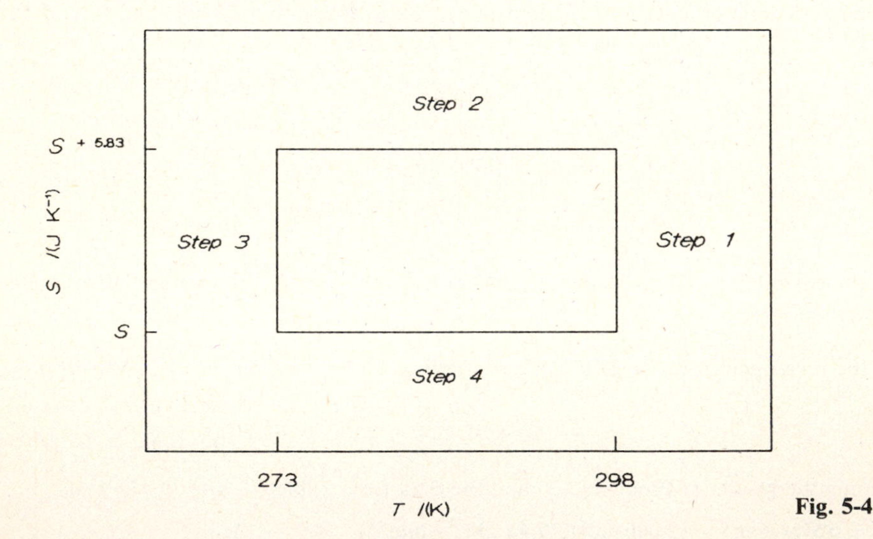

Fig. 5-4

5.34 A flask containing 1.00 mol of N_2 at 4.00 bar and 298 K was connected to a flask containing 1.00 mol of N_2 at 2.00 bar and 298 K. The gases were allowed to mix isothermally. What was the entropy change for the system?

The respective volumes of the flasks are given by *(1.18)* as

$$V_a = \frac{(1.00\ \text{mol})(0.083\,14\ \text{L}\cdot\text{bar}\cdot\text{K}^{-1}\cdot\text{mol}^{-1})(298\ \text{K})}{4.00\ \text{bar}} = 6.19\ \text{L}$$

$$V_b = \frac{1.00(0.083\,14)(298)}{2.00} = 12.39\ \text{L}$$

and the total volume of the system is

$$V = V_a + V_b = 6.19\ \text{L} + 12.39\ \text{L} = 18.58\ \text{L}$$

The final pressure of the gas after opening the stopcock is given by *(1.18)* as

$$P = \frac{(2.00\ \text{mol})(0.083\ 14\ \text{L}\cdot\text{bar}\cdot\text{K}^{-1}\cdot\text{mol}^{-1})(298\ \text{K})}{18.58\ \text{L}} = 2.67\ \text{bar}$$

The entropy change for each gas is calculated assuming the gas to expand reversibly from its original pressure to the final pressure. Using *(5.10b)* gives

$$\Delta S(\text{system})_a = -(1.00\ \text{mol})(8.314\ \text{J}\cdot\text{K}^{-1}\cdot\text{mol}^{-1})\ln\frac{2.67\ \text{bar}}{4.00\ \text{bar}} = 3.36\ \text{J}\cdot\text{K}^{-1}$$

$$\Delta S(\text{system})_b = -(1.00\ \text{mol})(8.314\ \text{J}\cdot\text{K}^{-1}\cdot\text{mol}^{-1})\ln\frac{2.67\ \text{bar}}{2.00\ \text{bar}} = -2.40\ \text{J}\cdot\text{K}^{-1}$$

The entropy change for the system is

$$\Delta S(\text{system}) = 3.36\ \text{J}\cdot\text{K}^{-1} + (-2.40\ \text{J}\cdot\text{K}^{-1}) = 0.96\ \text{J}\cdot\text{K}^{-1}$$

5.35 Consider the series of adiabats and isotherms shown in Fig. 5-5. Show that $\Delta S(\text{system})$ is a state function by calculating $\Delta S(\text{system})$ for the following paths: **(*a*)** *a-b-d-f*, **(*b*)** *a-c-d-f*, and **(*c*)** *a-c-e-f*.

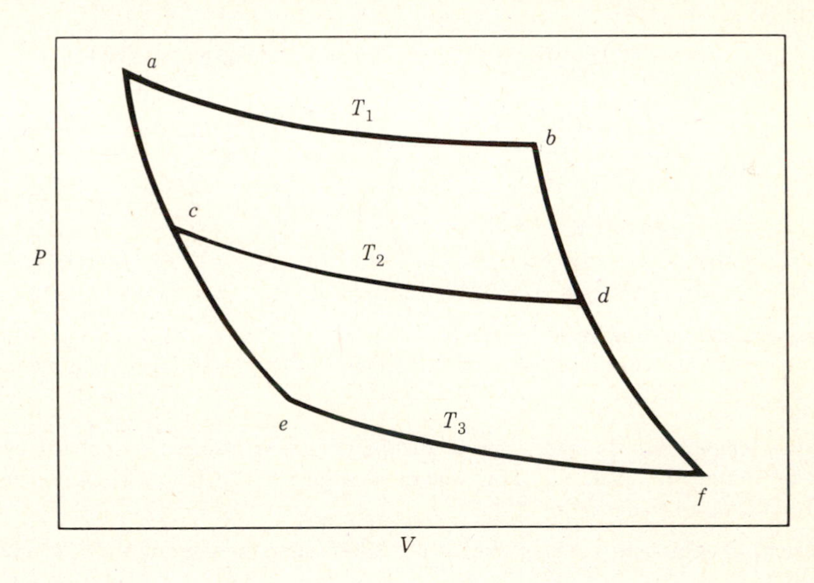

Fig. 5-5

▮ Along any of the adiabats, *(5.7)* gives $\Delta S(\text{system}) = 0$. For the isothermal paths, *(5.10a)* gives

$$\Delta S(\text{system})_{a\text{-}b} = R\ln\frac{V_b}{V_a} \qquad \Delta S(\text{system})_{c\text{-}d} = R\ln\frac{V_d}{V_c} \qquad \Delta S(\text{system})_{e\text{-}f} = R\ln\frac{V_f}{V_e}$$

The relationship between the volumes along the adiabats is given by *(3.56b)* as

$$V_c = V_a\left(\frac{T_1}{T_2}\right)^{C_V/R} \qquad V_d = V_b\left(\frac{T_1}{T_2}\right)^{C_V/R}$$

$$V_e = V_a\left(\frac{T_1}{T_3}\right)^{C_V/R} \qquad V_f = V_b\left(\frac{T_1}{T_3}\right)^{C_V/R}$$

From these relationships,

$$\frac{V_d}{V_c} = \frac{V_b}{V_a} \qquad \frac{V_f}{V_e} = \frac{V_b}{V_a}$$

which upon substituting into the expressions for $\Delta S(\text{system})$ gives

$$\Delta S(\text{system})_{a\text{-}b} = \Delta S(\text{system})_{c\text{-}d} = \Delta S(\text{system})_{e\text{-}f} = R\ln\frac{V_b}{V_a}$$

The value of $\Delta S(\text{system})$ for each path is

(a) $\Delta S(\text{system})_{a\text{-}b\text{-}d\text{-}f} = \Delta S(\text{system})_{a\text{-}b} + 0 + 0 = \Delta S(\text{system})_{a\text{-}b}$
(b) $\Delta S(\text{system})_{a\text{-}c\text{-}d\text{-}f} = 0 + \Delta S(\text{system})_{c\text{-}d} + 0 = \Delta S(\text{system})_{a\text{-}b}$
(c) $\Delta S(\text{system})_{a\text{-}c\text{-}e\text{-}f} = 0 + 0 + \Delta S(\text{system})_{e\text{-}f} = \Delta S(\text{system})_{a\text{-}b}$

All three values are equal—they are path-independent (a state function).

5.36 The pressure change for the sample of oxygen in Problem 5.27 was from 2.000 bar to 3.000 bar. Consider a 2-mol sample of NaCl undergoing the same pressure-temperature change as the oxygen. Calculate $\Delta S(\text{system})$ for this process given $C_{P,\text{m}}/(\text{J}\cdot\text{K}^{-1}\cdot\text{mol}^{-1}) = 45.94 + (16.32\times10^{-3})[T/(\text{K})]$, $V_\text{m} = 27.0\ \text{mL}\cdot\text{mol}^{-1}$, and $\alpha = 1.21\times10^{-4}\ \text{K}^{-1}$. Compare the results for the two substances.

▌ For the *reversible expansion of 1 mol of a condensed state, (5.9)* becomes

$$\Delta S_\text{m}(\text{system}) = \int_{T_1}^{T_2} \frac{C_{V,\text{m}}}{T}\,dT + \int_{V_1}^{V_2} \frac{\alpha}{\kappa}\,dV \tag{5.12a}$$

$$\Delta S_\text{m}(\text{system}) = \int_{T_1}^{T_2} \frac{C_{P,\text{m}}}{T}\,dT - \int_{P_1}^{P_2} \alpha V_\text{m}\,dP \tag{5.12b}$$

Substituting the data for 1 mol into *(5.12b)* and integrating gives

$$\begin{aligned}\Delta S_\text{m}(\text{system}) &= \int_{298\ \text{K}}^{700\ \text{K}} \frac{45.94\ \text{J}\cdot\text{K}^{-1}\cdot\text{mol}^{-1} + (16.32\times10^{-3}\ \text{J}\cdot\text{K}^{-2}\cdot\text{mol}^{-1})T}{T}\,dT \\ &\quad - (1.21\times10^{-4}\ \text{K}^{-1})(27.0\ \text{mL}\cdot\text{mol}^{-1})(3.000\ \text{bar} - 2.000\ \text{bar})\frac{10^{-3}\ \text{L}}{1\ \text{mL}}\left(\frac{10^2\ \text{J}}{1\ \text{L}\cdot\text{bar}}\right) \\ &= (45.94\ \text{J}\cdot\text{K}^{-1}\cdot\text{mol}^{-1})\ln\frac{700\ \text{K}}{298\ \text{K}} + (16.32\times10^{-3}\ \text{J}\cdot\text{K}^{-2}\cdot\text{mol}^{-1})(700\ \text{K} - 298\ \text{K}) \\ &\quad - 3.27\times10^{-4}\ \text{J}\cdot\text{K}^{-1}\cdot\text{mol}^{-1} \\ &= 45.8\ \text{J}\cdot\text{K}^{-1}\cdot\text{mol}^{-1}\end{aligned}$$

For the 2-mol sample,

$$\Delta S(\text{system}) = (2.000\ \text{mol})(45.8\ \text{J}\cdot\text{K}^{-1}\cdot\text{mol}^{-1}) = 91.6\ \text{J}\cdot\text{K}^{-1}$$

The entropy change for the solid is greater than that for the gas. For the gas, both the temperature and pressure-volume change contributions were important, but for the solid, only the temperature contribution was significant.

5.37 Calculate the temperature change when a 10.0-bar pressure change is applied reversibly and adiabatically to 1 mol of water at 25.00 °C. For water, $\alpha = 2.35\times10^{-4}\ \text{K}^{-1}$, $V_\text{m} = 18.0\ \text{mL}\cdot\text{mol}^{-1}$, and $C_{P,\text{m}} = 75.3\ \text{J}\cdot\text{K}^{-1}\cdot\text{mol}^{-1}$.

▌ For an isentropic process, $\Delta S(\text{system}) = 0$. Substituting the data into *(5.12b)* gives for 1 mol

$$0 = (75.3\ \text{J}\cdot\text{K}^{-1}\cdot\text{mol}^{-1})\ln\frac{T_2}{298.15\ \text{K}} - (2.35\times10^{-4}\ \text{K}^{-1})(18.0\ \text{mL}\cdot\text{mol}^{-1})\frac{10^{-3}\ \text{L}}{1\ \text{mL}}(10.0\ \text{bar})\frac{10^2\ \text{J}}{1\ \text{L}\cdot\text{bar}}$$

$$\ln\frac{T_2}{298.15\ \text{K}} = 5.62\times10^{-5} \qquad T_2 = 298.17\ \text{K}$$

The temperature change is 0.02 K.

5.38 Derive an equation for $(\partial T/\partial P)_S$ for a condensed state. Assuming $(\partial T/\partial P)_S \approx \Delta T/\Delta P$, repeat the calculations of Problem 5.37 for water.

▌ Setting the differential form of *(5.12b)* equal to zero for the isentropic process and rearranging gives for 1 mol

$$dS_\text{m} = (C_{P,\text{m}}/T)\,dT - \alpha V_\text{m}\,dP = 0$$

$$\left(\frac{\partial T}{dP}\right)_S = \frac{\alpha V_\text{m} T}{C_{P,\text{m}}}$$

Substituting the data gives

$$\begin{aligned}\Delta T &\approx \frac{(2.35\times10^{-4}\ \text{K}^{-1})(18.0\ \text{mL}\cdot\text{mol}^{-1})[(10^{-3}\ \text{L})/(1\ \text{mL})](298.15\ \text{K})(10.0\ \text{bar})}{(75.3\ \text{J}\cdot\text{K}^{-1}\cdot\text{mol}^{-1})[(1\ \text{L}\cdot\text{bar})/(10^2\ \text{J})]} \\ &= 1.67\times10^{-2}\ \text{K}\end{aligned}$$

5.39 Two 1-mol blocks of aluminum, one at 100 °C and the other at 0 °C, were brought into thermal contact. What was the overall entropy change of the aluminum blocks? For Al, $C_{P,m} = 24.35\ \mathrm{J \cdot K^{-1} \cdot mol^{-1}}$.

▮ The final temperature in this case will be $(T_{\mathrm{hot}} + T_{\mathrm{cold}})/2$. Assuming that each block can transfer heat reversibly, *(5.12b)* gives

$$\Delta S(\mathrm{system})_{\mathrm{hot}} = nC_{P,\mathrm{m}} \ln \frac{T_{\mathrm{final}}}{T_{\mathrm{hot}}}$$

$$\Delta S(\mathrm{system})_{\mathrm{cold}} = nC_{P,\mathrm{m}} \ln \frac{T_{\mathrm{final}}}{T_{\mathrm{cold}}}$$

Combining these results gives

$$\begin{aligned}\Delta S(\mathrm{system}) &= nC_{P,\mathrm{m}} \ln \frac{T_{\mathrm{final}}}{T_{\mathrm{hot}}} + nC_{P,\mathrm{m}} \ln \frac{T_{\mathrm{final}}}{T_{\mathrm{cold}}} = nC_{P,\mathrm{m}} \ln \frac{T_{\mathrm{final}}^2}{T_{\mathrm{hot}} T_{\mathrm{cold}}} \\ &= nC_{P,\mathrm{m}} \ln \frac{[(T_{\mathrm{hot}} + T_{\mathrm{cold}})/2]^2}{T_{\mathrm{hot}} T_{\mathrm{cold}}} = nC_{P,\mathrm{m}} \ln \frac{(T_{\mathrm{hot}} + T_{\mathrm{cold}})^2}{4 T_{\mathrm{hot}} T_{\mathrm{cold}}} \\ &= (1.00\ \mathrm{mol})(24.35\ \mathrm{J \cdot K^{-1} \cdot mol^{-1}}) \ln \frac{(373\ \mathrm{K} + 273\ \mathrm{K})^2}{4(373\ \mathrm{K})(273\ \mathrm{K})} \\ &= 0.59\ \mathrm{J \cdot K^{-1}}\end{aligned}$$

5.40 A 25.00-g sample of water at 25.0 °C was mixed reversibly and adiabatically with a 75.00-g sample of water at 75.0 °C. Calculate the entropy change for each sample of water and for the combined system. For $H_2O(l)$, $C_{P,m} = 75.3\ \mathrm{J \cdot K^{-1} \cdot mol^{-1}}$.

▮ The final temperature of the system can be determined using *(3.23)* and the law of conservation of energy:

$$(25.00\ \mathrm{g}) \frac{1\ \mathrm{mol}}{18.015\ \mathrm{g}} (75.3\ \mathrm{J \cdot K^{-1} \cdot mol^{-1}})(T - 298.2\ \mathrm{K}) + (75.00\ \mathrm{g}) \frac{1\ \mathrm{mol}}{18.015\ \mathrm{g}} (75.3\ \mathrm{J \cdot K^{-1} \cdot mol^{-1}})(T - 348.3\ \mathrm{K}) = 0$$

$$T = 335.7\ \mathrm{K}$$

Using *(5.12b)* for each water sample gives

$$\begin{aligned}\Delta S(\mathrm{system})_{\mathrm{hot}} &= (75.00\ \mathrm{g}) \frac{1\ \mathrm{mol}}{18.015\ \mathrm{g}} (75.3\ \mathrm{J \cdot K^{-1} \cdot mol^{-1}}) \ln \frac{335.7\ \mathrm{K}}{348.2\ \mathrm{K}} \\ &= -11.46\ \mathrm{J \cdot K^{-1}}\end{aligned}$$

$$\Delta S(\mathrm{system})_{\mathrm{cold}} = (25.00\ \mathrm{g}) \frac{1\ \mathrm{mol}}{18.015\ \mathrm{g}} (75.3) \ln \frac{335.7\ \mathrm{K}}{298.2\ \mathrm{K}} = 12.38\ \mathrm{J \cdot K^{-1}}$$

The entropy change for the combined system is

$$\begin{aligned}\Delta S(\mathrm{system}) &= \Delta S(\mathrm{system})_{\mathrm{hot}} + \Delta S(\mathrm{system})_{\mathrm{cold}} \\ &= -11.46\ \mathrm{J \cdot K^{-1}} + 12.38\ \mathrm{J \cdot K^{-1}} = 0.92\ \mathrm{J \cdot K^{-1}}\end{aligned}$$

5.41 Exactly 100 g of ice at 0.00 °C is mixed reversibly and adiabatically with exactly 100 g of liquid water at 100.00 °C. Determine the entropy change for each sample of water and for the combined system. For H_2O, $\Delta_{\mathrm{fus}}H = 6.0095\ \mathrm{kJ \cdot mol^{-1}}$ and $C_{P,m} = 75.3\ \mathrm{J \cdot K^{-1} \cdot mol^{-1}}$.

▮ The final temperature of the system can be determined using *(3.23)*, *(3.55)*, and the law of conservation of energy:

$$\begin{aligned}&(100.00\ \mathrm{g}) \frac{1\ \mathrm{mol}}{18.015\ \mathrm{g}} (6.0095\ \mathrm{kJ \cdot mol^{-1}}) \frac{10^3\ \mathrm{J}}{1\ \mathrm{kJ}} \\ &+ (100.00\ \mathrm{g}) \frac{1\ \mathrm{mol}}{18.015\ \mathrm{g}} (75.3\ \mathrm{J \cdot K^{-1} \cdot mol^{-1}})(T - 273.15\ \mathrm{K}) \\ &+ (100.00\ \mathrm{g}) \frac{1\ \mathrm{mol}}{18.015\ \mathrm{g}} (75.3\ \mathrm{J \cdot K^{-1} \cdot mol^{-1}})(T - 373.15\ \mathrm{K}) = 0\end{aligned}$$

$$T = 283.25\ \mathrm{K}$$

The entropy change for the cold water is given by *(5.8)* and *(5.12b)* as

$$\Delta S(\text{system})_{\text{cold}} = \frac{(100.0\ \text{g})[(1\ \text{mol})/(18.015\ \text{g})](6.0095\ \text{kJ}\cdot\text{mol}^{-1})[(10^3\ \text{J})/(1\ \text{kJ})]}{273.15\ \text{K}}$$

$$+ (100.0\ \text{g})\frac{1\ \text{mol}}{18.015\ \text{g}}(75.3\ \text{J}\cdot\text{K}^{-1}\cdot\text{mol}^{-1})\ln\frac{283.25\ \text{K}}{273.15\ \text{K}}$$

$$= 137.3\ \text{J}\cdot\text{K}^{-1}$$

and for the hot water by *(5.12b)* as

$$\Delta S(\text{system})_{\text{hot}} = (100.0\ \text{g})\frac{1\ \text{mol}}{18.015\ \text{g}}(75.3\ \text{J}\cdot\text{K}^{-1}\cdot\text{mol}^{-1})\ln\frac{283.25\ \text{K}}{273.15\ \text{K}}$$

$$= -115.2\ \text{J}\cdot\text{K}^{-1}$$

The entropy change for the combined system is

$$\Delta S(\text{system}) = \Delta S(\text{system})_{\text{hot}} + \Delta S(\text{system})_{\text{cold}}$$

$$= -115.2\ \text{J}\cdot\text{K}^{-1} + 137.3\ \text{J}\cdot\text{K}^{-1} = 22.1\ \text{J}\cdot\text{K}^{-1}$$

5.42 Determine $\Delta S(\text{system})$ for 100.0 g of supercooled liquid water at −10.0 °C freezing to form ice at −10.0 °C. For water, $\Delta_{\text{fus}}H_{273} = 6.0095\ \text{kJ}\cdot\text{K}^{-1}\cdot\text{mol}^{-1}$ and $C_{P,\text{m}}/(\text{J}\cdot\text{K}^{-1}\cdot\text{mol}^{-1}) = 75.3$ for liquid water and 38.0 for ice.

▮ The process must be carried out using a reversible path to calculate $\Delta S(\text{system})$. Because $\Delta S(\text{system})$ is a state function, the value determined for the reversible path will be equal to the value for the actual irreversible path. Consider the following three steps: (1) reversibly heating the liquid from −10.0 °C to 0.0 °C, (2) reversibly freezing the liquid at 0.0 °C, and (3) reversibly cooling the ice from 0.0 °C to −10.0 °C. Using *(5.8)* and *(5.12b)* gives

$$\Delta S(\text{system})_1 = (100.00\ \text{g})\frac{1\ \text{mol}}{18.015\ \text{g}}(75.3\ \text{J}\cdot\text{K}^{-1}\cdot\text{mol}^{-1})\ln\frac{273.2\ \text{K}}{263.2\ \text{K}}$$

$$= 15.59\ \text{J}\cdot\text{K}^{-1}$$

$$\Delta S(\text{system})_2 = \frac{(100.00\ \text{g})[(1\ \text{mol})/(18.015\ \text{g})](-6.0095\ \text{J}\cdot\text{K}^{-1}\cdot\text{mol}^{-1})[(10^3\ \text{J})/(1\ \text{kJ})]}{273.2\ \text{K}}$$

$$= -122.10\ \text{J}\cdot\text{K}^{-1}$$

$$\Delta S(\text{system})_3 = (100.00\ \text{g})\frac{1\ \text{mol}}{18.015\ \text{g}}(38.0\ \text{J}\cdot\text{K}^{-1}\cdot\text{mol}^{-1})\ln\frac{263.2\ \text{K}}{273.2\ \text{K}}$$

$$= -7.87\ \text{J}\cdot\text{K}^{-1}$$

For the entire process,

$$\Delta S(\text{system}) = \Delta S(\text{system})_1 + \Delta S(\text{system})_2 + \Delta S(\text{system})_3$$

$$= 15.59\ \text{J}\cdot\text{K}^{-1} + (-122.10\ \text{J}\cdot\text{K}^{-1}) + (-7.87\ \text{J}\cdot\text{K}^{-1})$$

$$= -114.38\ \text{J}\cdot\text{K}^{-1}$$

5.43 What is the entropy change for the system for producing a 1-mol sample of artificial air consisting of 79% N_2 and 21% O_2?

▮ The entropy change for *reversible isothermal mixing* is given by

$$\Delta_{\text{mix}}S(\text{system}) = -R\sum_i n_i \ln x_i \tag{5.13}$$

For the 1-mol sample, $n_i = x_i$. Substituting $x(N_2) = 0.79$ and $x(O_2) = 0.21$ into *(5.13)* gives

$$\Delta_{\text{mix}}S(\text{system}) = -(8.314\ \text{J}\cdot\text{K}^{-1}\cdot\text{mol}^{-1})[(0.79\ \text{mol})\ln 0.79 + (0.21\ \text{mol})\ln 0.21]$$

$$= 4.27\ \text{J}\cdot\text{K}^{-1}$$

5.44 Determine the composition of a binary mixture for which $\Delta_{\text{mix}}S(\text{system})$ is a maximum.

▮ For a binary mixture, the mole fractions will be x_i and $1 - x_i$. Differentiating *(5.13)* with respect to x_i, setting the result equal to zero, and solving for x_i gives

$$\frac{\partial[\Delta_{\text{mix}}S(\text{system})]}{\partial x_i} = \frac{\partial\{-R[x_i \ln x_i + (1 - x_i)\ln(1 - x_i)]\}}{\partial x_i}$$

$$= -R\left[\ln x_i + \frac{x_i}{x_i} - \ln(1 - x_i) - \frac{1 - x_i}{1 - x_i}\right] = 0$$

$$\ln x_i = \ln(1 - x_i) \qquad x_i = 0.5$$

Preparing an equimolar mixture will give the largest entropy change.

5.45 Hydrogen has two isotopes, 99.985% H and 0.015% D (where D represents deuterium, ^2H). Calculate the entropy change for preparing molecular hydrogen by mixing these isotopes.

▮ There are three possible combinations of isotopes: H_2, HD, and D_2. The mole fraction of each of these is

$$x(H_2) = (0.999\,85)(0.999\,85) = 0.999\,70$$

$$x(HD) = 2(0.999\,85)(0.000\,15) = 3.0 \times 10^{-4}$$

$$x(D_2) = (0.000\,15)(0.000\,15) = 2.3 \times 10^{-8}$$

Using *(5.13)* gives

$$\Delta_{\text{mix}}S(\text{system}) = -(8.314\ \text{J}\cdot\text{K}^{-1}\cdot\text{mol}^{-1})[0.999\,70 \ln 0.999\,70$$
$$+ (3.0 \times 10^{-4})\ln(3.0 \times 10^{-4}) + (2.3 \times 10^{-8})\ln(2.3 \times 10^{-8})]$$
$$= 0.0227\ \text{J}\cdot\text{K}^{-1}\cdot\text{mol}^{-1}$$

5.46 Repeat the calculations of Problem 5.34 assuming that the second flask contains Ar instead of N_2.

▮ In addition to the value of ΔS(system) calculated in Problem 5.34, there will be a contribution for mixing given by *(5.13)* as

$$\Delta_{\text{mix}}S(\text{system}) = -(8.314\ \text{J}\cdot\text{K}^{-1}\cdot\text{mol}^{-1})[(1.00\ \text{mol})\ln 0.500 + (1.00\ \text{mol})\ln 0.500]$$
$$= 11.53\ \text{J}\cdot\text{K}^{-1}$$

The total entropy change for the system is

$$\Delta S(\text{system}) = 0.96\ \text{J}\cdot\text{K}^{-1} + 11.53\ \text{J}\cdot\text{K}^{-1} = 12.49\ \text{J}\cdot\text{K}^{-1}$$

5.47 Calculate the entropy change for the thermal reservoirs in the surroundings and for the thermodynamic universe for the two processes described in Problem 5.20. Comment.

▮ The *entropy change for the surroundings* is given by

$$\Delta S(\text{surroundings}) = q(\text{surroundings})/T(\text{surroundings}) \tag{5.14}$$

and the *entropy change for the universe* as

$$\Delta S(\text{universe}) = \Delta S(\text{system}) + \Delta S(\text{surroundings}) \tag{5.15}$$

For the two processes,

$$\Delta S(\text{surroundings})_1 = \frac{-100\ \text{J}}{298.16\ \text{K}} = 0.335\,39\ \text{J}\cdot\text{K}^{-1}$$

$$\Delta S(\text{surroundings})_2 = \frac{100\ \text{J}}{298.14\ \text{K}} = 0.335\,41\ \text{J}\cdot\text{K}^{-1}$$

$$\Delta S(\text{universe})_1 = 0.335\,40\ \text{J}\cdot\text{K}^{-1} + (-0.335\,39\ \text{J}\cdot\text{K}^{-1}) = 1 \times 10^{-5}\ \text{J}\cdot\text{K}^{-1}$$

$$\Delta S(\text{universe})_2 = -0.335\,40\ \text{J}\cdot\text{K}^{-1} + 0.335\,41\ \text{J}\cdot\text{K}^{-1} = 1 \times 10^{-5}\ \text{J}\cdot\text{K}^{-1}$$

Note that ΔS(universe) is positive for these spontaneous processes. The very low values are the result of the processes being very close to reversible.

5.48 What will be ΔS(surroundings) and ΔS(universe) for each of the steps and for the entire process in the Carnot cycle described in Problem 5.1?

▮ For the first step,

$$q(\text{surroundings})_1 = -q(\text{system})_1 = -1740\ \text{J}$$

which upon substitution into *(5.14)* gives

$$\Delta S(\text{surroundings})_1 = \frac{-1740\ \text{J}}{298\ \text{K}} = -5.84\ \text{J}\cdot\text{K}^{-1}$$

Combining this with the result of Problem 5.33 gives

$$\Delta S(\text{universe})_1 = 5.83\ \text{J}\cdot\text{K}^{-1} + (-5.84\ \text{J}\cdot\text{K}^{-1}) = -0.01\ \text{J}\cdot\text{K}^{-1}$$

This small difference results from rounding errors. For *reversible processes,*

$$\Delta S(\text{universe}) = 0 \qquad (5.16a)$$

$$\Delta S(\text{system}) = -\Delta S(\text{surroundings}) \qquad (5.16b)$$

For the second step,

$$q(\text{surroundings})_2 = q(\text{system})_2 = 0$$

$$\Delta S(\text{system})_2 = \Delta S(\text{surroundings})_2 = \Delta S(\text{universe})_2 = 0$$

For the third step,

$$q(\text{surroundings})_3 = -q(\text{system})_3 = 1590\ \text{J}$$

$$\Delta S(\text{surroundings})_3 = (1590\ \text{J})/(273\ \text{K}) = 5.82\ \text{J}\cdot\text{K}^{-1}$$

$$\Delta S(\text{universe})_3 = -5.82\ \text{J}\cdot\text{K}^{-1} + 5.82\ \text{J}\cdot\text{K}^{-1} = 0$$

and for the fourth step,

$$q(\text{surroundings})_4 = q(\text{system})_4 = 0$$

$$\Delta S(\text{system})_4 = \Delta S(\text{surroundings})_4 = \Delta S(\text{universe})_4 = 0$$

For the overall process,

$$\Delta S(\text{surroundings}) = \Delta S(\text{surroundings})_1 + \Delta S(\text{surroundings})_2 + \Delta S(\text{surroundings})_3 + \Delta S(\text{surroundings})_4$$

$$= -5.84\ \text{J}\cdot\text{K}^{-1} + 0 + 5.82\ \text{J}\cdot\text{K}^{-1} + 0 = -0.02\ \text{J}\cdot\text{K}^{-1}$$

$$\Delta S(\text{universe}) = \Delta S(\text{universe})_1 + \Delta S(\text{universe})_2 + \Delta S(\text{universe})_3 + \Delta S(\text{universe})_4$$

$$= -0.01\ \text{J}\cdot\text{K}^{-1} + 0 + 0 + 0 = -0.01\ \text{J}\cdot\text{K}^{-1}$$

Because of rounding errors, these small values are all equal to zero. For a reversible cyclic process,

$$\Delta S(\text{system}) = \Delta S(\text{surroundings}) = \Delta S(\text{universe}) = 0$$

5.49 Calculate ΔS(surroundings) and ΔS(universe) for the process described in Problem 5.22 if it is done reversibly and if it is done by heating the iron in an oven at 3500 K.

▌ For the reversible process, *(5.16)* gives

$$\Delta S(\text{surroundings}) = -111\ \text{J}\cdot\text{K}^{-1}\cdot\text{mol}^{-1} \qquad \Delta S(\text{universe}) = 0$$

For the irreversible process,

$$q(\text{surroundings}) = (-349\ \text{kJ}\cdot\text{mol}^{-1})\frac{10^3\ \text{J}}{1\ \text{kJ}} = -349\ 000\ \text{J}$$

Using *(5.14)* and *(5.15)* gives

$$\Delta S(\text{surroundings}) = (-349\ 800\ \text{J})/(3500\ \text{K}) = -1.0\times 10^2\ \text{J}\cdot\text{K}^{-1}\cdot\text{mol}^{-1}$$

$$\Delta S(\text{universe}) = 111\ \text{J}\cdot\text{K}^{-1}\cdot\text{mol}^{-1} + (-1.0\times 10^2\ \text{J}\cdot\text{K}^{-1}\cdot\text{mol}^{-1}) = 11\ \text{J}\cdot\text{K}^{-1}\cdot\text{mol}^{-1}$$

5.50 Calculate ΔS(surroundings) and ΔS(universe) for the process described in Problem 5.25 if it is done reversibly and if it is done by placing the gas in an ice-water bath at 273 K.

▌ For the reversible process, *(5.16)* gives

$$\Delta S(\text{surroundings}) = 1.09\ \text{J}\cdot\text{K}^{-1}\cdot\text{mol}^{-1} \qquad \Delta S(\text{universe}) = 0$$

For the irreversible process,

$$q(\text{surroundings}) = -\int_{298}^{273} \tfrac{3}{2}R\,dT = -\tfrac{3}{2}(8.314\ \text{J}\cdot\text{K}^{-1}\cdot\text{mol}^{-1})(273\ \text{K} - 298\ \text{K})$$

$$= 312\ \text{J}\cdot\text{mol}^{-1}$$

Using *(5.14)* and *(5.15)* gives

$$\Delta S(\text{surroundings}) = (312\ \text{J}\cdot\text{mol}^{-1})/(273\ \text{K}) = 1.14\ \text{J}\cdot\text{K}^{-1}\cdot\text{mol}^{-1}$$

$$\Delta S(\text{universe}) = -1.09\ \text{J}\cdot\text{K}^{-1}\cdot\text{mol}^{-1} + 1.14\ \text{J}\cdot\text{K}^{-1}\cdot\text{mol}^{-1} = 0.05\ \text{J}\cdot\text{K}^{-1}\cdot\text{mol}^{-1}$$

5.51 Calculate ΔS(surroundings) and ΔS(universe) for the process described in Problem 5.34.

▌ In this case, q(surroundings) = 0, which gives ΔS(surroundings) = 0 and ΔS(universe) = $0.96\ \text{J}\cdot\text{K}^{-1}\cdot\text{mol}^{-1}$.

5.52 Calculate ΔS(surroundings) and ΔS(universe) for the process described in Problem 5.40 assuming that both samples transfer heat to a heat reservoir at 335.7 K.

▌ For the two processes,

$$q(\text{surroundings})_{\text{cold}} = -(25.00\ \text{g})\frac{1\ \text{mol}}{18.015\ \text{g}}(75.3\ \text{J}\cdot\text{K}^{-1}\cdot\text{mol}^{-1})(335.7\ \text{K} - 298.2\ \text{K})$$
$$= -3919\ \text{J}$$

$$q(\text{surroundings})_{\text{hot}} = -(75.00\ \text{g})\frac{1\ \text{mol}}{18.015\ \text{g}}(75.3\ \text{J}\cdot\text{K}^{-1}\cdot\text{mol}^{-1})(335.7\ \text{K} - 348.2\ \text{K})$$
$$= 3919\ \text{J}$$

Using *(5.14)* and substituting the results for ΔS(system) from Problem 5.40 into *(5.15)* gives

$$\Delta S(\text{surroundings})_{\text{cold}} = \frac{-3919\ \text{J}}{335.7\ \text{K}} = -11.67\ \text{J}\cdot\text{K}^{-1}\cdot\text{mol}^{-1}$$

$$\Delta S(\text{universe})_{\text{cold}} = 12.38\ \text{J}\cdot\text{K}^{-1} + (-11.67\ \text{J}\cdot\text{K}^{-1}) = 0.71\ \text{J}\cdot\text{K}^{-1}$$

$$\Delta S(\text{surroundings})_{\text{hot}} = \frac{3919\ \text{J}}{335.7\ \text{K}} = 11.67\ \text{J}\cdot\text{K}^{-1}\cdot\text{mol}^{-1}$$

$$\Delta S(\text{universe})_{\text{hot}} = -11.46\ \text{J}\cdot\text{K}^{-1} + 11.67\ \text{J}\cdot\text{K}^{-1} = 0.21\ \text{J}\cdot\text{K}^{-1}$$

For the entire process,

$$\Delta S(\text{surroundings}) = \Delta S(\text{surroundings})_{\text{cold}} + \Delta S(\text{surroundings})_{\text{hot}}$$
$$= -11.67\ \text{J}\cdot\text{K}^{-1} + 11.67\ \text{J}\cdot\text{K}^{-1} = 0$$

$$\Delta S(\text{universe}) = \Delta S(\text{universe})_{\text{cold}} + \Delta S(\text{universe})_{\text{hot}}$$
$$= 0.71\ \text{J}\cdot\text{K}^{-1} + 0.21\ \text{J}\cdot\text{K}^{-1} = 0.92\ \text{J}\cdot\text{K}^{-1}$$

5.53 Calculate ΔS(surroundings) and ΔS(universe) for the process described in Problem 5.42 assuming the sample to be in contact with a heat reservoir at −25 °C.

▌ For the three steps,

$$q(\text{surroundings}) = -(100.0\ \text{g})\frac{1\ \text{mol}}{18.015\ \text{g}}\left[(75.3\ \text{J}\cdot\text{K}^{-1}\cdot\text{mol}^{-1})(273.2\ \text{K} - 263.2\ \text{K})\right.$$
$$+ (-6.0095\ \text{kJ}\cdot\text{mol}^{-1})\frac{10^3\ \text{J}}{1\ \text{kJ}}$$
$$\left. + (38.0\ \text{J}\cdot\text{K}^{-1}\cdot\text{mol}^{-1})(263.2\ \text{K} - 273.2\ \text{K})\right]$$
$$= 31\,300\ \text{J}$$

Using *(5.14)* and substituting the results of Problem 5.42 into *(5.15)* gives

$$\Delta S(\text{surroundings}) = (31\,300\ \text{J})/(284\ \text{K}) = 126\ \text{J}\cdot\text{K}^{-1}$$

$$\Delta S(\text{universe}) = -114.38\ \text{J}\cdot\text{K}^{-1}\cdot\text{mol}^{-1} + 126\ \text{J}\cdot\text{K}^{-1} = 12\ \text{J}\cdot\text{K}^{-1}$$

5.54 Calculate ΔS(surroundings) and ΔS(universe) for the process described in Problem 5.43.

▌ Because the process is isothermal, q(surroundings) = 0, giving

$$\Delta S(\text{surroundings}) = 0 \quad \text{and} \quad \Delta S(\text{system}) = \Delta S(\text{universe}) = 4.27\ \text{J}\cdot\text{K}^{-1}$$

5.3 THE THIRD LAW OF THERMODYNAMICS

5.55 What is the value of $S^\circ_{0,\mathrm{m}}$ for CO and CO_2?

▮ At absolute zero, the *third law of thermodynamics* gives the absolute entropy of a pure substance in a perfect crystal as

$$S^\circ_0 = nR \ln \Omega \qquad (5.17)$$

where Ω is the number of unique orientations of the molecules, etc., in the crystal. For CO the molecules can be oriented as either C≡O or O≡C, giving $\Omega(\mathrm{CO}) = 2$; and for CO_2 the molecule can be oriented only in one way (O=C=O), giving $\Omega(\mathrm{CO_2}) = 1$. Substituting these values into *(5.17)* gives

$$S^\circ_0(\mathrm{CO})_\mathrm{m} = (8.314\ \mathrm{J\cdot K^{-1}\cdot mol^{-1}}) \ln 2 = 5.763\ \mathrm{J\cdot K^{-1}\cdot mol^{-1}}$$

$$S^\circ_0(\mathrm{CO_2})_\mathrm{m} = 8.314 \ln 1 = 0$$

5.56 At 298 K, $S^\circ_{298,\mathrm{m}} = 223.066\ \mathrm{J\cdot K^{-1}\cdot mol^{-1}}$ for $Cl_2(g)$. Determine $S^\circ_{1500,\mathrm{m}}$ given

$$C^\circ_{P,\mathrm{m}}/(\mathrm{J\cdot K^{-1}\cdot mol^{-1}}) = 31.284 + (10.144\times10^{-3})[T/(\mathrm{K})] - (40.38\times10^{-7})[T/(\mathrm{K})]^2$$

▮ Under these conditions Cl_2 will act as an ideal gas. At 1500 K, $S^\circ_{1500,\mathrm{m}} = S^\circ_{298,\mathrm{m}} + \Delta S^\circ_\mathrm{m}(\text{system})$, where $\Delta S^\circ(\text{system})$ is given by *(5.10b)* for heating the gas from 298 K to 1500 K. Substituting the data gives

$$S^\circ_{1500,\mathrm{m}} = 223.066\ \mathrm{J\cdot K^{-1}\cdot mol^{-1}} + \int_{298\ \mathrm{K}}^{1500\ \mathrm{K}} \frac{\left[\begin{array}{r}31.284\ \mathrm{J\cdot K^{-1}\cdot mol^{-1}} + (10.144\times10^{-3}\ \mathrm{J\cdot K^{-2}\cdot mol^{-1}})T\\ - (40.38\times10^{-7}\ \mathrm{J\cdot K^{-3}\cdot mol^{-1}})T^2\end{array}\right]}{T}\, dT$$

$$= 223.066\ \mathrm{J\cdot K^{-1}\cdot mol^{-1}} + (31.284\ \mathrm{J\cdot K^{-1}\cdot mol^{-1}}) \ln\frac{1500\ \mathrm{K}}{298\ \mathrm{K}}$$

$$+ (10.144\times10^{-3}\ \mathrm{J\cdot K^{-2}\cdot mol^{-1}})(1500\ \mathrm{K} - 298\ \mathrm{K})$$

$$- \tfrac{1}{2}(40.38\times10^{-7}\ \mathrm{J\cdot K^{-3}\cdot mol^{-1}})[(1500\ \mathrm{K})^2 - (298\ \mathrm{K})^2]$$

$$= 281.437\ \mathrm{J\cdot K^{-1}\cdot mol^{-1}}$$

5.57 Determine $S_{1500,\mathrm{m}}$ for $Cl_2(g)$ at 0.500 bar. See Problem 5.56 for additional data.

▮ The entropy will increase because of the lower pressure. Including the $-R\ln(P_2/P_1)$ term from *(5.10b)* in the results of Problem 5.56 gives

$$S_{1500,\mathrm{m}} = S^\circ_{298,\mathrm{m}} + \int_{298\ \mathrm{K}}^{1500\ \mathrm{K}} \frac{C_{P,\mathrm{m}}}{T}\, dT - R\ln\frac{P_2}{P_1}$$

$$= 281.437\ \mathrm{J\cdot K^{-1}\cdot mol^{-1}} - (8.314\ \mathrm{J\cdot K^{-1}\cdot mol^{-1}}) \ln\frac{0.500\ \mathrm{bar}}{1.000\ \mathrm{bar}}$$

$$= 287.200\ \mathrm{J\cdot K^{-1}\cdot mol^{-1}}$$

5.58 For K_2SO_4(α-solid), $S^\circ_{298,\mathrm{m}} = 175.544\ \mathrm{J\cdot K^{-1}\cdot mol^{-1}}$. Determine $S^\circ_{1000,\mathrm{m}}$ for the β-solid given the following data: $C^\circ_{P,\mathrm{m}}/(\mathrm{J\cdot K^{-1}\cdot mol^{-1}}) = 89.07 + 0.148\,68\ [T/(\mathrm{K})] - 1.314\times10^{-5}[T/(\mathrm{K})]^2$ for the α-solid and $115.21 + 7.953\times10^{-2}[T/(\mathrm{K})]$ for the β-solid, and $\Delta_{\alpha\to\beta}H^\circ_{857} = 8.452\ \mathrm{kJ\cdot mol^{-1}}$.

▮ Adding the two heating contributions and the phase-change contribution to $S^\circ_{298,\mathrm{m}}$ gives

$$S^\circ_{1000,\mathrm{m}} = 175.544\ \mathrm{J\cdot K^{-1}\cdot mol^{-1}} + \int_{298\ \mathrm{K}}^{857\ \mathrm{K}} \frac{\left[\begin{array}{r}89.07\ \mathrm{J\cdot K^{-1}\cdot mol^{-1}} + (0.148\,68\ \mathrm{J\cdot K^{-2}\cdot mol^{-1}})T\\ - (1.314\times10^{-5}\ \mathrm{J\cdot K^{-3}\cdot mol^{-1}})T^2\end{array}\right]}{T}\, dT$$

$$+ \frac{(8.452\ \mathrm{J\cdot K^{-1}\cdot mol^{-1}})[(10^3\ \mathrm{J})/(1\ \mathrm{kJ})]}{857\ \mathrm{K}}$$

$$+ \int_{857\ \mathrm{K}}^{1000\ \mathrm{K}} \frac{115.21\ \mathrm{J\cdot K^{-1}\cdot mol^{-1}} + (7.953\times10^{-2}\ \mathrm{J\cdot K^{-2}\cdot mol^{-1}})T}{T}\, dT$$

$$= 175.544\ \mathrm{J\cdot K^{-1}\cdot mol^{-1}} + (89.07\ \mathrm{J\cdot K^{-1}\cdot mol^{-1}}) \ln[(857\ \mathrm{K})/(298\ \mathrm{K})]$$

$$+ (0.148\,68\ \mathrm{J\cdot K^{-2}\cdot mol^{-1}})(857\ \mathrm{K} - 298\ \mathrm{K}) - \tfrac{1}{2}(1.314\times10^{-5}\ \mathrm{J\cdot K^{-3}\cdot mol^{-1}})[(857\ \mathrm{K})^2 - (298\ \mathrm{K})^2]$$

$$+ 9.86\ \mathrm{J\cdot K^{-1}\cdot mol^{-1}} + (115.21\ \mathrm{J\cdot K^{-1}\cdot mol^{-1}}) \ln[(1000\ \mathrm{K})/(857\ \mathrm{K})]$$

$$+ (7.953\times10^{-2}\ \mathrm{J\cdot K^{-2}\cdot mol^{-1}})(1000\ \mathrm{K} - 857\ \mathrm{K})$$

$$= 387.49\ \mathrm{J\cdot K^{-1}\cdot mol^{-1}}$$

5.59 Prepare a plot of $S^\circ_{T,\mathrm{m}}$ for graphite from 298 K to 2000 K given the following data: $S^\circ_{298,\mathrm{m}} = 5.740\ \mathrm{J \cdot K^{-1} \cdot mol^{-1}}$ and

$$C^\circ_{P,\mathrm{m}}/(\mathrm{J \cdot K^{-1} \cdot mol^{-1}}) = 16.71 + (4.77 \times 10^{-3})[T/(\mathrm{K})] - (8.54 \times 10^{5})[T/(\mathrm{K})]^{-2}$$

▮ At any temperature, the molar entropy will be given by

$$S^\circ_{T,\mathrm{m}} = S^\circ_{298,\mathrm{m}} + \int_{298\ \mathrm{K}}^{T} \frac{\begin{bmatrix} 16.71\ \mathrm{J \cdot K^{-1} \cdot mol^{-1}} + (4.77 \times 10^{-3}\ \mathrm{J \cdot K^{-2} \cdot mol^{-1}})T \\ - (8.54 \times 10^{-5}\ \mathrm{J \cdot K \cdot mol^{-1}})T^{-2} \end{bmatrix}}{T}\, dT$$

As an example, consider $T = 500$ K.

$$\begin{aligned} S^\circ_{500,\mathrm{m}} &= 5.740\ \mathrm{J \cdot K^{-1} \cdot mol^{-1}} + (16.71\ \mathrm{J \cdot K^{-1} \cdot mol^{-1}}) \ln [(500\ \mathrm{K})/(298\ \mathrm{K})] \\ &\quad + (4.77 \times 10^{-3}\ \mathrm{J \cdot K^{-2} \cdot mol^{-2}})(500\ \mathrm{K} - 298\ \mathrm{K}) \\ &\quad + \tfrac{1}{2}(8.54 \times 10^{5}\ \mathrm{J \cdot K \cdot mol^{-1}})[(500\ \mathrm{K})^{-2} - (298\ \mathrm{K})^{-2}] \\ &= 12.247\ \mathrm{J \cdot K^{-1} \cdot mol^{-1}} \end{aligned}$$

The plot is shown in Fig. 5-6.

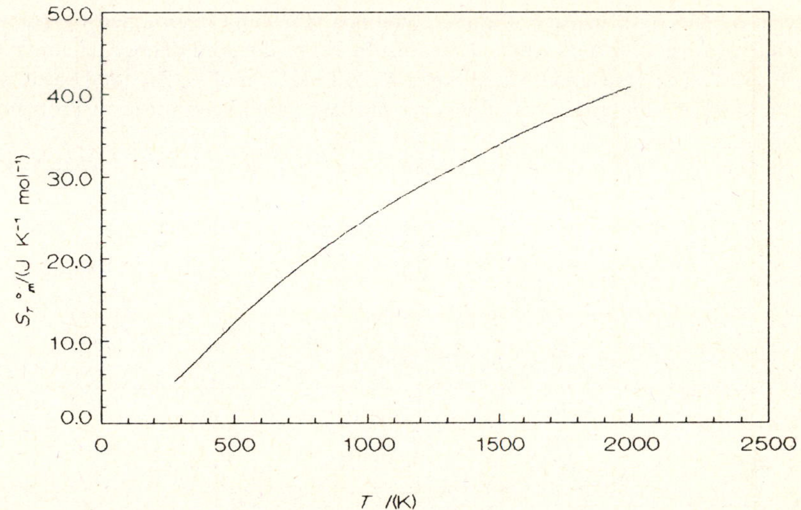

Fig. 5-6

5.60 Given $S^\circ_{298,\mathrm{m}} = 152.231\ \mathrm{J \cdot K^{-1} \cdot mol^{-1}}$ for $Br_2(\mathrm{l})$, find $S^\circ_{298,\mathrm{m}}$ for $Br_2(\mathrm{g})$. Assume $C^\circ_{P,\mathrm{m}}/(\mathrm{J \cdot K^{-1} \cdot mol^{-1}}) = 75.488$ for the liquid and 36.190 for the gas. The boiling point of bromine at 1 bar is 332.503 K, and $\Delta_{\mathrm{vap}} H^\circ_{\mathrm{m}} = 29.563\ \mathrm{kJ \cdot mol^{-1}}$.

▮ Using a three-step process in which the liquid is heated to the boiling point, the liquid is vaporized, and the gas is cooled to 298 K gives

$$\begin{aligned} S^\circ_{298,\mathrm{m}}(\mathrm{g}) &= S^\circ_{298,\mathrm{m}}(\mathrm{l}) + \Delta S(\mathrm{system})_1 + \Delta S(\mathrm{system})_2 + \Delta S(\mathrm{system})_3 \\ &= 152.231\ \mathrm{J \cdot K^{-1} \cdot mol^{-1}} + (75.488\ \mathrm{J \cdot K^{-1} \cdot mol^{-1}}) \ln \frac{332.503\ \mathrm{K}}{298.15\ \mathrm{K}} \\ &\quad + \frac{(29.563\ \mathrm{kJ \cdot mol^{-1}})[(10^3\ \mathrm{J})/(1\ \mathrm{kJ})]}{332.503\ \mathrm{K}} + (36.190\ \mathrm{J \cdot K^{-1} \cdot mol^{-1}}) \ln \frac{298.15\ \mathrm{K}}{332.503\ \mathrm{K}} \\ &= 245.428\ \mathrm{J \cdot K^{-1} \cdot mol^{-1}} \end{aligned}$$

5.61 At 10.00 K, $C^\circ_{P,\mathrm{m}} = 0.431\ \mathrm{J \cdot K^{-1} \cdot mol^{-1}}$ for gold. Calculate $S^\circ_{10,\mathrm{m}}$.

▮ At such low temperatures, using *(3.29)* and *(3.40)* for metals gives $C^\circ_{P,\mathrm{m}} = C^\circ_{V,\mathrm{m}} = \alpha T^3$. Substituting into *(5.12b)* gives

$$S^\circ_{T,\mathrm{low}} = \Delta S = \int_0^{T_{\mathrm{low}}} \frac{\alpha T^3}{T}\, dT = \frac{\alpha}{3}(T_{\mathrm{low}} - 0)^3 = \frac{C^\circ_{P,\mathrm{m}}}{3} \tag{5.18}$$

Thus,

$$S^\circ_{10,\mathrm{m}} = (0.431\ \mathrm{J \cdot K^{-1} \cdot mol^{-1}})/3 = 0.144\ \mathrm{J \cdot K^{-1} \cdot mol^{-1}}$$

5.62 Determine $S^\circ_{298,m}$ for gold given the following heat capacity data:

$T/(\mathrm{K})$	10	15	20	25	30	35	40	45	50	60
$C_{P,m}/(\mathrm{J\cdot K^{-1}\cdot mol^{-1}})$	0.431	1.474	3.187	5.245	7.375	9.395	11.22	12.86	14.29	16.59

$T/(\mathrm{K})$	70	80	90	100	125	150	175	200	250	298.15
$C_{P,m}/(\mathrm{J\cdot K^{-1}\cdot mol^{-1}})$	18.31	19.63	20.64	21.44	22.78	23.59	24.08	24.41	24.97	25.42

▌ The absolute entropy of a substance at any temperature can be calculated using

$$S^\circ_{T,m} = S^\circ_{0,m} + \sum_i^{\text{phases}} \int \frac{C^\circ_{P,m}(i)}{T}\,dT + \sum_i^{\text{transitions}} \frac{\Delta_j H^\circ_m}{T_j} \tag{5.19a}$$

$$S^\circ_{T,m} = S^\circ_{0,m} + \sum_i^{\text{phases}} \int C^\circ_{P,m}(i)\,d(\ln T) + \sum_j^{\text{transitions}} \frac{\Delta_j H^\circ_m}{T_j} \tag{5.19b}$$

The second term on the right in each of these equations is usually determined by graphical interpretation of a plot of $C^\circ_{P,m}(i)/T$ against T or a plot $C^\circ_{P,m}(i)$ against $\ln T$. For the gold data given above, a plot of $C^\circ_{P,m}T$ against T is given in Fig. 5-7. The area under the curve is $47.26\ \mathrm{J\cdot K^{-1}\cdot mol^{-1}}$. The total value of the integral in *(5.19a)* from 0 K to 298 K will be the area under the curve plus the value determined in Problem 5.61.

$$\int_0^{298} \frac{C^\circ_{P,m}(i)}{T}\,dT = 0.144\ \mathrm{J\cdot K^{-1}\cdot mol^{-1}} + 47.26\ \mathrm{J\cdot K^{-1}\cdot mol^{-1}} = 47.40\ \mathrm{J\cdot K^{-1}\cdot mol^{-1}}$$

For Au, $S^\circ_0 = 0$, and because there are no phase changes over this temperature interval, *(5.19a)* gives

$$S^\circ_{298,m} = 0 + 47.40\ \mathrm{J\cdot K^{-1}\cdot mol^{-1}} + 0 = 47.40\ \mathrm{J\cdot K^{-1}\cdot mol^{-1}}$$

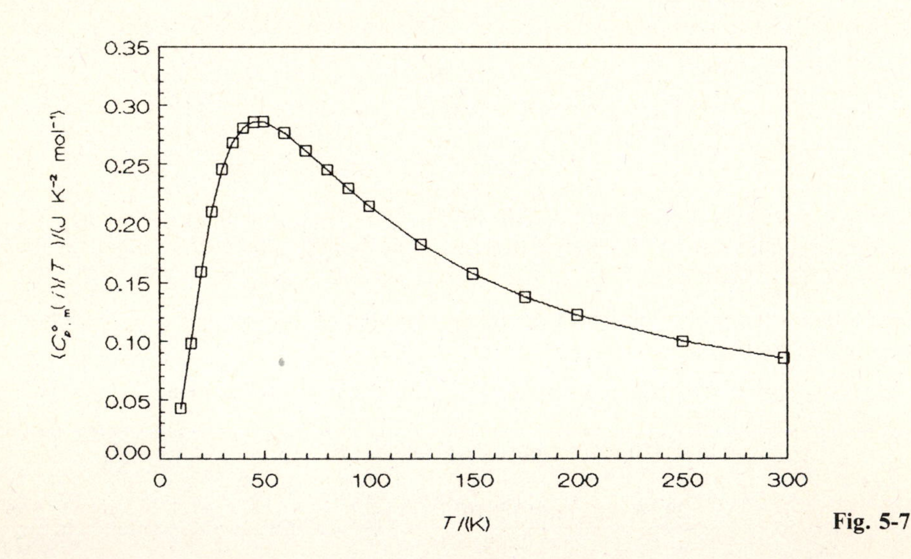

Fig. 5-7

5.63 Suppose the value of $S^\circ_{10,m}$ determined in Problem 5.61 was 25% in error. What would be the resulting error in $S^\circ_{298,m}$?

▌ A 25% error in the value of Problem 5.61 is $(0.144\ \mathrm{J\cdot K^{-1}\cdot mol^{-1}})(0.25) = 0.036\ \mathrm{J\cdot K^{-1}\cdot mol^{-1}}$, which corresponds to an error of

$$\frac{0.036\ \mathrm{J\cdot K^{-1}\cdot mol^{-1}}}{47.40\ \mathrm{J\cdot K^{-1}\cdot mol^{-1}}} \times 100 = 0.076\%$$

in the value of $S^\circ_{298,m}$ determined in Problem 5.62.

5.64 Determine $S^\circ_{1000,m}$ for NaOH given the following heat capacity data:

Solid-I

$T/(K)$	100	200	300	400	500	572
$C_{P,m}/(J\cdot K^{-1}\cdot mol^{-1})$	27.744	49.580	59.664	64.936	75.157	85.887

$$\Delta_{I\to II}H^\circ_{572} = 7.196\ kJ\cdot mol^{-1}$$

$$C^\circ_{P,m} = 86.023\ J\cdot K^{-1}\cdot mol^{-1} \qquad \text{for } 572\ K \le T \le 596\ K$$

$$\Delta_{fus}H^\circ_{596} = 6.611\ kJ\cdot mol^{-1}$$

Liquid

$T/(K)$	596	600	700	800	900	1000
$C_{P,m}/(J\cdot K^{-1}\cdot mol^{-1})$	86.111	86.065	85.479	84.893	84.303	83.722

$$S^\circ_{100,m} = 15.513\ J\cdot K^{-1}\cdot mol^{-1}$$

■ A plot of $C_{P,m}(i)$ against $\ln T$ for the data is shown in Fig. 5-8. The area under the curve is 140.87 $J\cdot K^{-1}\cdot mol^{-1}$. Using *(5.19b)* gives

$$S^\circ_{1000,m} = 15.513\ J\cdot K^{-1}\cdot mol^{-1} + 140.87\ J\cdot K^{-1}\cdot mol^{-1} + \frac{(7.196\ kJ\cdot mol^{-1})[(10^3\ J)/(1\ kJ)]}{572\ K} + \frac{(6.611\ kJ\cdot mol^{-1})[(10^3\ J)/(1\ kJ)]}{596\ K}$$

$$= 180.06\ J\cdot K^{-1}\cdot mol^{-1}$$

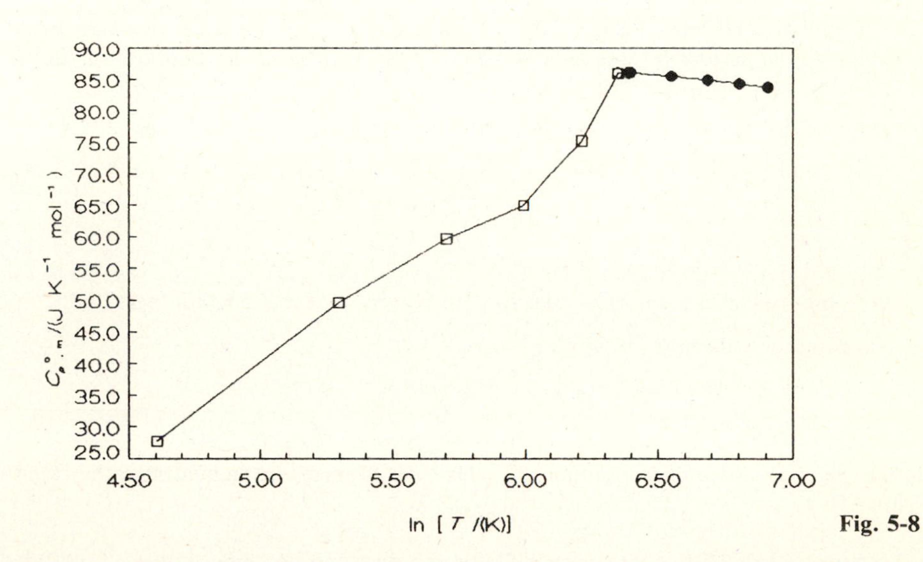

Fig. 5-8

5.4 ΔS FOR CHEMICAL REACTIONS

5.65 Should $\Delta_r S^\circ$ be large and positive, large and negative, or small for the following chemical reaction?

$$NaHCO_3(s) + HCl(aq) \rightarrow NaCl(aq) + H_2O(l) + CO_2(g)$$

Confirm your prediction by calculating $\Delta_r S^\circ$ for the equation using the data given in Table 5-1.

■ The molar absolute entropy of gases is considerably larger than that of the condensed phases. Because there is the net increase of 1 mol of gas in the equation, the value of $\Delta_r S^\circ$ would be predicted to be large and positive. For a chemical equation,

$$\Delta_r S^\circ_T = \sum_i \nu_i S^\circ_{T,i} \qquad (5.20)$$

where ν_i has been defined in Problem 4.46. Substituting the data from Table 5-1 into *(5.20)* gives

$$\begin{aligned}\Delta_r S^\circ &= [(1\text{ mol})S^\circ(\text{NaCl}) + (1\text{ mol})S^\circ(\text{H}_2\text{O}) + (1\text{ mol})S^\circ(\text{CO}_2)] - [(1\text{ mol})S^\circ(\text{NaHCO}_3) + (1\text{ mol})S^\circ(\text{HCl})] \\ &= [(1\text{ mol})(115.5\text{ J}\cdot\text{K}^{-1}\cdot\text{mol}^{-1}) + (1\text{ mol})(69.91\text{ J}\cdot\text{K}^{-1}\cdot\text{mol}^{-1}) + (1\text{ mol})(213.74\text{ J}\cdot\text{K}^{-1}\cdot\text{mol}^{-1})] \\ &\quad - [(1\text{ mol})(101.7\text{ J}\cdot\text{K}^{-1}\cdot\text{mol}^{-1}) + (1\text{ mol})(56.5\text{ J}\cdot\text{K}^{-1}\cdot\text{mol}^{-1})] \\ &= 240.95\text{ J}\cdot\text{K}^{-1}\end{aligned}$$

Table 5-1

	S°_{298} $(\text{J}\cdot\text{K}^{-1}\cdot\text{mol}^{-1})$		S°_{298} $(\text{J}\cdot\text{K}^{-1}\cdot\text{mol}^{-1})$
$O_2(g)$	205.138	C(graphite)	5.740
$O_3(g)$	238.93	$CO_2(g)$	213.74
H(g)	114.713	$CH_4(g)$	186.264
$H^+(aq)$	0	Hg(l)	76.02
$H_2(g)$	130.684	$Hg^{2+}(aq)$	−32.2
OH(g)	183.745	$Hg_2^{2+}(aq)$	84.5
$H_2O(l)$	69.91	Fe(s)	27.28
HCl(aq)	56.5	$Fe_2O_3(s)$	87.40
$N_2(g)$	191.61	NaCl(aq)	115.5
$NH_3(g)$	192.45	$NaHCO_3(s)$	101.7
$N_2O_4(g)$	304.29		

5.66 Repeat Problem 5.65 for the formation of 1 mol of $Fe_2O_3(s)$.

▌ The chemical equation $2\text{Fe(s)} + \frac{3}{2}\text{O}_2\text{(g)} \rightarrow \text{Fe}_2\text{O}_3\text{(s)}$ shows that a net decrease of 1.5 mol of gas occurs in the reaction, so the predicted value of $\Delta_f S^\circ$ would be large and negative. Substituting the data from Table 5-1 into *(5.20)* gives for the entropy of formation

$$\begin{aligned}\Delta_f S^\circ &= (1\text{ mol})S^\circ(\text{Fe}_2\text{O}_3) - [(2\text{ mol})S^\circ(\text{Fe}) + (\tfrac{3}{2}\text{ mol})S^\circ(\text{O}_2)] \\ &= (1\text{ mol})(87.40\text{ J}\cdot\text{K}^{-1}\cdot\text{mol}^{-1}) - [(2\text{ mol})(27.28\text{ J}\cdot\text{K}^{-1}\cdot\text{mol}^{-1}) + (\tfrac{3}{2}\text{ mol})(205.138\text{ J}\cdot\text{K}^{-1}\cdot\text{mol}^{-1})] \\ &= -274.87\text{ J}\cdot\text{K}^{-1}\end{aligned}$$

5.67 Repeat the calculations of Problem 5.66 at 1000 K given $S^\circ_{1000}/(\text{J}\cdot\text{K}^{-1}\cdot\text{mol}^{-1}) = 66.672$ for Fe(s), 252.882 for $Fe_2O_3(s)$, and 243.578 for $O_2(g)$. Has $\Delta_f S^\circ$ increased or decreased with increased temperature?

▌ Substituting the data into *(5.20)* gives

$$\begin{aligned}\Delta_f S^\circ_{1000} &= (1\text{ mol})(252.882\text{ J}\cdot\text{K}^{-1}\cdot\text{mol}^{-1}) - [(2\text{ mol})(66.672\text{ J}\cdot\text{K}^{-1}\cdot\text{mol}^{-1}) + (\tfrac{3}{2}\text{ mol})(243.578\text{ J}\cdot\text{K}^{-1}\cdot\text{mol}^{-1})] \\ &= -245.829\text{ J}\cdot\text{K}^{-1}\end{aligned}$$

Although the individual molar absolute entropies have all increased significantly, the value of $\Delta_f S^\circ$ has decreased by about 11%.

5.68 Use the data in Table 5-1 to calculate $\Delta_r S^\circ$ for the disproportionation of $Hg_2^{2+}(aq)$. Why is the value of $\Delta_r S^\circ$ negative even though $\Delta\nu_i > 0$?

▌ For the equation $\text{Hg}_2^{2+}\text{(aq)} \rightarrow \text{Hg(l)} + \text{Hg}^{2+}\text{(aq)}$, using *(5.20)* gives

$$\begin{aligned}\Delta_r S^\circ &= [(1\text{ mol})S^\circ(\text{Hg}) + (1\text{ mol})S^\circ(\text{Hg}^{2+})] - (1\text{ mol})S^\circ(\text{Hg}_2^{2+}) \\ &= [(1\text{ mol})(76.02\text{ J}\cdot\text{K}^{-1}\cdot\text{mol}^{-1}) + (1\text{ mol})(-32.2\text{ J}\cdot\text{K}^{-1}\cdot\text{mol}^{-1})] + (1\text{ mol})(84.5\text{ J}\cdot\text{K}^{-1}\cdot\text{mol}^{-1}) \\ &= -40.7\text{ J}\cdot\text{K}^{-1}\end{aligned}$$

The aqueous Hg_2^{2+} cation contains a considerable amount of disorder compared to the monatomic aqueous Hg^{2+} cation and the liquid Hg. Thus the chemical reaction produces a decrease in entropy.

5.69 Calculate $\Delta_r S^\circ$ for $\text{NO}_2\text{(g)} + \text{NO(g)} \rightarrow \text{N}_2\text{O}_3\text{(g)}$ given $\Delta_f S^\circ/(\text{J}\cdot\text{K}^{-1}\cdot\text{mol}^{-1}) = -60.88$ for $NO_2(g)$, 12.39 for NO(g), and −187.04 for $N_2O_3(g)$.

▮ Molar values of absolute entropies of formation can be combined to calculate $\Delta_r S^\circ$ using a special case of Hess's law:

$$\Delta_r S_T^\circ = \sum_i \nu_i \Delta_f S_{T,i}^\circ \tag{5.21}$$

where ν_i has been defined in Problem 4.46. Substituting the data gives

$$\begin{aligned}\Delta_r S^\circ &= (1\ \text{mol})\,\Delta_f S^\circ(N_2O_3) - [(1\ \text{mol})\,\Delta_f S^\circ(NO_2) + (1\ \text{mol})\,\Delta_f S^\circ(NO)]\\ &= (1\ \text{mol})(-187.04\ \text{J}\cdot\text{K}^{-1}\cdot\text{mol}^{-1}) - [(1\ \text{mol})(-60.88\ \text{J}\cdot\text{K}^{-1}\cdot\text{mol}^{-1}) + (1\ \text{mol})(12.39\ \text{J}\cdot\text{K}^{-1}\cdot\text{mol}^{-1})]\\ &= -138.55\ \text{J}\cdot\text{K}^{-1}\end{aligned}$$

5.70 The saturated Weston cell, which can be represented by the chemical reaction

$$\text{Cd(amalgam)} + Hg_2SO_4(s) + \tfrac{8}{3}H_2O(l) \rightarrow CdSO_4\cdot\tfrac{8}{3}H_2O(s) + 2Hg(l)$$

is used to produce reference potentials. The potential of the cell is given by

$$E^\circ/(\text{V}) = 1.018\,410 - (4.93\times10^{-5}\ \text{K}^{-1})X - (8.0\times10^{-7}\ \text{K}^{-2})X^2 + (1\times10^{-8}\ \text{K}^{-3})X^3$$

where $X = T - 298\ \text{K}$ over the temperature range of 278 K to 323 K. Calculate $\Delta_r S_{300}^\circ$ for this reaction.

▮ The relation between $\Delta_r S$ and E is given by

$$\Delta_r S = nF\frac{dE}{dT} \tag{5.22}$$

where $n = 2\ \text{mol}$ for this equation. Applying the chain rule of derivatives, *(5.22)* gives

$$\begin{aligned}\Delta_r S^\circ &= nF\frac{dE^\circ}{dX}\frac{dX}{dT}\\ &= nF[(-4.93\times10^{-5}\ \text{V}\cdot\text{K}^{-1}) - 2(8.0\times10^{-7}\ \text{V}\cdot\text{K}^{-2})X + 3(1\times10^{-8}\ \text{V}\cdot\text{K}^{-3})X^2]\frac{d(T-298\ \text{K})}{dT}\\ &= (2\ \text{mol})(96\,484.56\ \text{C}\cdot\text{mol}^{-1})[(-4.93\times10^{-5}\ \text{V}\cdot\text{K}^{-1}) - 2(8.0\times10^{-7}\ \text{V}\cdot\text{K}^{-2})(300\ \text{K} - 298\ \text{K})\\ &\qquad +3(1\times10^{-8}\ \text{V}\cdot\text{K}^{-3})(300\ \text{K} - 298\ \text{K})^2](1)\frac{1\ \text{A}\cdot\text{s}}{1\ \text{C}}\left(\frac{1\ \text{J}}{1\ \text{V}\cdot\text{A}\cdot\text{s}}\right)\\ &= -10.11\ \text{J}\cdot\text{K}^{-1}\end{aligned}$$

5.71 Calculate ΔS(system), ΔS(surroundings), and ΔS(universe) for the chemical equation

$$\text{C(graphite)} + 2H_2(g) \rightarrow CH_4(g) \qquad \Delta_f H_{298}^\circ = -74.81\ \text{kJ}$$

▮ The value of ΔS(system) will be equal to $\Delta_f S^\circ$. Using the data in Table 5-1 and *(5.20)* gives

$$\begin{aligned}\Delta S(\text{system}) &= \Delta_f S^\circ = (1\ \text{mol})S^\circ(CH_4) - [(1\ \text{mol})S^\circ(C) + (2\ \text{mol})S^\circ(H_2)]\\ &= (1\ \text{mol})(186.264\ \text{J}\cdot\text{K}^{-1}\cdot\text{mol}^{-1})\\ &\quad - [(1\ \text{mol})(5.740\ \text{J}\cdot\text{K}^{-1}\cdot\text{mol}^{-1}) + (2\ \text{mol})(130.684\ \text{J}\cdot\text{K}^{-1}\cdot\text{mol}^{-1})]\\ &= -80.844\ \text{J}\cdot\text{K}^{-1}\end{aligned}$$

For the chemical reaction, $q(\text{surroundings}) = -\Delta_f H_{298}^\circ$. Substituting into *(5.14)* gives

$$\Delta S(\text{surroundings}) = \frac{-(-74.81\ \text{kJ})[(10^3\ \text{J})/(1\ \text{kJ})]}{298.15\ \text{K}} = 250.9\ \text{J}\cdot\text{K}^{-1}$$

Substituting both of these values into *(5.15)* gives

$$\Delta S(\text{universe}) = -80.444\ \text{J}\cdot\text{K}^{-1} + 250.9\ \text{J}\cdot\text{K}^{-1} = 170.1\ \text{J}\cdot\text{K}^{-1}$$

5.72 What is $\Delta_r S_{373}^\circ$ for the reaction $O_3(g) + OH(g) \rightarrow H(g) + 2O_2(g)$? See Problem 4.29 for additional data.

▮ Substituting the data from Table 5-1 into *(5.20)* gives

$$\begin{aligned}\Delta_r S_{298}^\circ &= [(1\ \text{mol})S^\circ(H) + (2\ \text{mol})S^\circ(O_2)] - [(1\ \text{mol})S^\circ(O_3) + (1\ \text{mol})S^\circ(OH)]\\ &= [(1\ \text{mol})(114.713\ \text{J}\cdot\text{K}^{-1}\cdot\text{mol}^{-1}) + (2\ \text{mol})(205.138\ \text{J}\cdot\text{K}^{-1}\cdot\text{mol}^{-1})]\\ &\quad - [(1\ \text{mol})(238.93\ \text{J}\cdot\text{K}^{-1}\cdot\text{mol}^{-1}) + (1\ \text{mol})(183.745\ \text{J}\cdot\text{K}^{-1}\cdot\text{mol}^{-1})]\\ &= 102.31\ \text{J}\cdot\text{K}^{-1}\end{aligned}$$

The temperature dependence of $\Delta_r S_T^\circ$ is given by

$$\Delta_r S^\circ(T_2) = \Delta_r S^\circ(T_1) + \int_{T_1}^{T_2} \frac{\Delta_r C_P^\circ}{T}\,dT \qquad (5.23)$$

For this reaction, $\Delta_r C_P^\circ = 10.41\ \mathrm{J\cdot K^{-1}}$ (see Problem 4.29). Substituting into *(5.23)* and integrating gives

$$\Delta_r S_{373}^\circ = 102.31\ \mathrm{J\cdot K^{-1}} + (10.41\ \mathrm{J\cdot K^{-1}})\ln\frac{373\ \mathrm{K}}{298\ \mathrm{K}} = 104.65\ \mathrm{J\cdot K^{-1}}$$

5.73 Determine $\Delta_r S^\circ$ for the reaction $N_2(g) + 3H_2(g) \rightarrow 2NH_3(g)$ as a function of temperature. Prepare a plot of $\Delta_r S^\circ$ against T between 298 K and 1000 K. See Problem 4.30 for additional data.

▮ Substituting the data from Table 5-1 into *(5.20)* gives

$$\begin{aligned}\Delta_r S_{298}^\circ &= (2\ \mathrm{mol})S^\circ(NH_3) - [(1\ \mathrm{mol})S^\circ(N_2) + (3\ \mathrm{mol})S^\circ(H_2)]\\ &= (2\ \mathrm{mol})(192.45\ \mathrm{J\cdot K^{-1}\cdot mol^{-1}}) - [(1\ \mathrm{mol})(191.61\ \mathrm{J\cdot K^{-1}\cdot mol^{-1}}) + (3\ \mathrm{mol})(130.684\ \mathrm{J\cdot K^{-1}\cdot mol^{-1}})]\\ &= -198.76\ \mathrm{J\cdot K^{-1}}\end{aligned}$$

Substituting the empirical heat capacity expressions given by *(3.19)* and *(5.23)* and integrating gives

$$\begin{aligned}\Delta_r S^\circ(T_2) &= \Delta_r S^\circ(T_1) + (\Delta a)\ln(T_2/T_1) + (\Delta b)(T_2 - T_1) + \tfrac{1}{2}(\Delta c)(T_2^2 - T_1^2)\\ &\quad + \tfrac{1}{3}(\Delta d)(T_2^3 - T_1^3) + \cdots - \tfrac{1}{2}(\Delta c')(T_2^{-2} - T_1^{-2})\end{aligned} \qquad (5.24a)$$

If L is defined as

$$\begin{aligned}L &= \Delta_r S_{298}^\circ - (\Delta a)\ln(298\ \mathrm{K}) - (\Delta b)(298\ \mathrm{K}) - \tfrac{1}{2}(\Delta c)(298\ \mathrm{K})^2\\ &\quad - \tfrac{1}{3}(\Delta d)(298\ \mathrm{K})^3 - \cdots + \tfrac{1}{2}(\Delta c')(298\ \mathrm{K})^{-2}\end{aligned} \qquad (5.25)$$

then

$$\Delta_r S_T^\circ = L + (\Delta a)\ln T + (\Delta b)T + \tfrac{1}{2}(\Delta c)T^2 + \tfrac{1}{3}(\Delta d)T^3 + \cdots - \tfrac{1}{2}(\Delta c')T^{-2} \qquad (5.24b)$$

For the reaction, $\Delta a = -62.457\ \mathrm{J\cdot K^{-1}}$, $\Delta b = 63.276 \times 10^{-3}\ \mathrm{J\cdot K^{-2}}$, and $\Delta c = -121.39 \times 10^{-7}\ \mathrm{J\cdot K^{-3}}$ (see Problem 4.30). Substituting into *(5.25)* gives

$$\begin{aligned}L &= -198.76\ \mathrm{J\cdot K^{-1}} - (-62.457\ \mathrm{J\cdot K^{-1}})\ln(298\ \mathrm{K}) - (63.276 \times 10^{-3}\ \mathrm{J\cdot K^{-2}})(298\ \mathrm{K})\\ &\quad - \tfrac{1}{2}(-121.39 \times 10^{-7}\ \mathrm{J\cdot K^{-3}})(298\ \mathrm{K})^2\\ &= 138.77\ \mathrm{J\cdot K^{-1}}\end{aligned}$$

Thus at any temperature *(5.24b)* gives

$$\Delta_r S_T^\circ/(\mathrm{J\cdot K^{-1}}) = 138.77 + (-62.457)\ln[T/(\mathrm{K})] + (63.276 \times 10^{-3})[T/(\mathrm{K})] + \tfrac{1}{2}(-121.39 \times 10^{-7})[T/(\mathrm{K})]^2$$

As a sample calculation, at $T = 500\ \mathrm{K}$,

$$\begin{aligned}\Delta_r S_{500}^\circ/(\mathrm{J\cdot K^{-1}}) &= 138.77 + (-62.457)\ln(500) + (63.276 \times 10^{-3})(500) + \tfrac{1}{2}(-121.39 \times 10^{-7})(500)^2\\ &= -219.26\ \mathrm{J\cdot K^{-1}}\end{aligned}$$

The graph is shown in Fig. 5-9.

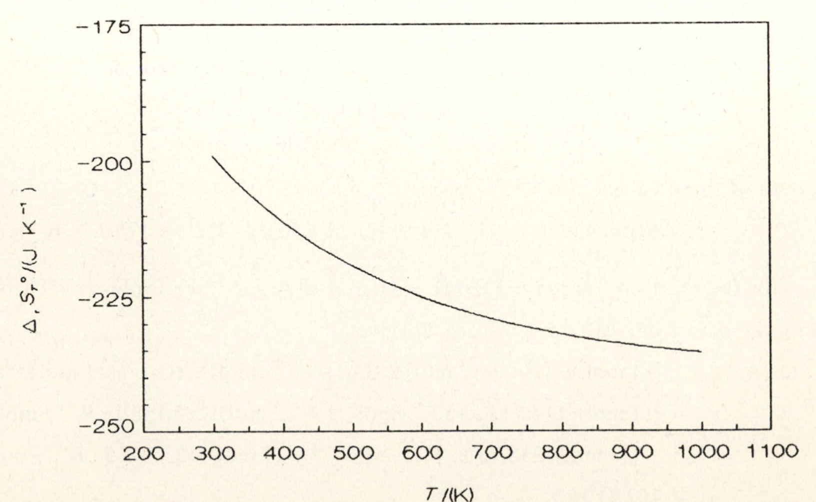

Fig. 5-9

5.74 Determine $\Delta_r S°$ for the reaction $N_2(g) + 2O_2(g) \rightarrow N_2O_4(g)$ ($\Delta_r S° = -297.60\ \text{J}\cdot\text{K}^{-1}$) as a function of temperature. See Problem 4.31 for additional data. Find the temperature at which $\Delta_r S°_T$ is a minimum.

▮ For this reaction, $\Delta a = -41.77\ \text{J}\cdot\text{K}^{-1}$, $\Delta b = 0.125\ \text{J}\cdot\text{K}^{-2}$, and $\Delta c = -6.7 \times 10^{-5}\ \text{J}\cdot\text{K}^{-3}$ (see Problem 4.31). Substituting into *(5.25)* gives

$$L = -297.60\ \text{J}\cdot\text{K}^{-1} - (-41.77\ \text{J}\cdot\text{K}^{-1})\ln(298\ \text{K}) - (0.125\ \text{J}\cdot\text{K}^{-2})(298\ \text{K}) - \tfrac{1}{2}(-6.7\times10^{-5}\ \text{J}\cdot\text{K}^{-3})(298\ \text{K})^2$$
$$= -93.91\ \text{J}\cdot\text{K}^{-1}$$

Thus at any temperature *(5.24b)* gives

$$\Delta_r S°_T/(\text{J}\cdot\text{K}^{-1}) = -93.91 + (-41.77)\ln[T/(\text{K})] + (0.125)[T/(\text{K})] + \tfrac{1}{2}(-6.7\times10^{-5})[T/(\text{K})]^2$$

Differentiating the expression for $\Delta_r S°_T$ with respect to T, setting the result equal to zero, and solving for T gives the minimum as

$$\frac{\partial[\Delta_r S°_T]}{\partial T} = \frac{-41.77}{T/(\text{K})} + 0.125 + (-6.7\times10^{-5})[T/(\text{K})] = 0$$

$$T = 436\ \text{K}$$

At this temperature,

$$\Delta_r S°_{433}/(\text{J}\cdot\text{K}^{-1}) = -93.91 + (-41.77)\ln 436 + (0.125)(436) + \tfrac{1}{2}(-6.7\times10^{-5})(436)^2$$
$$= -299.64$$

Note that the minimum of $\Delta_r S°_T$ occurs at very nearly the same temperature as the minimum of $\Delta_r H°_T$.

5.75 Calculate $\Delta_r S$ for the reaction $N_2(g) + 2O_2(g) \rightarrow N_2O_4(g)$, assuming it to be performed at 0.500 bar.

▮ Under standard conditions, substituting the data from Table 5-1 in *(5.20)* gives

$$\Delta_r S° = (1\ \text{mol})S°(N_2O_4) - [(1\ \text{mol})S°(N_2) + (2\ \text{mol})S°(O_2)]$$
$$= (1\ \text{mol})(304.29\ \text{J}\cdot\text{K}^{-1}\cdot\text{mol}^{-1}) - [(1\ \text{mol})(191.61\ \text{J}\cdot\text{K}^{-1}\cdot\text{mol}^{-1}) + (2\ \text{mol})(205.138\ \text{J}\cdot\text{K}^{-1}\cdot\text{mol}^{-1})]$$
$$= -297.60\ \text{J}\cdot\text{K}^{-1}$$

Neglecting the pressure contributions from *(5.12b)*, *(5.10b)* gives

$$\Delta_r S_T = \Delta_r S°_T - R\,\Delta n_g \ln[P/(\text{bar})] \qquad (5.26)$$

where Δn_g has been defined in Problem 3.8. In this case,

$$\Delta n_g = 1\ \text{mol} - (1\ \text{mol} + 2\ \text{mol}) = -2\ \text{mol}$$

Substituting into *(5.26)* gives

$$\Delta_r S_T = -297.60\ \text{J}\cdot\text{K}^{-1} - (8.314\ \text{J}\cdot\text{K}^{-1}\cdot\text{mol}^{-1})(-2\ \text{mol})\ln 0.500$$
$$= -309.13\ \text{J}\cdot\text{K}^{-1}$$

CHAPTER 6
Free Energy

6.1 FREE ENERGY

6.1 Determine ΔG° and ΔA° for the chemical equation $C(s) + CO_2(g) \rightarrow 2CO(g)$, given $\Delta_r H^\circ = 172.459\ \text{kJ}$, $\Delta_r U^\circ = 169.981\ \text{kJ}$, and $\Delta_r S^\circ = 175.87\ \text{J}\cdot\text{K}^{-1}$. Is this reaction spontaneous?

▮ The definitions of the two free energy functions are

$$G = H - TS \tag{6.1a}$$

$$A = U - TS \tag{6.1b}$$

Under constant-temperature conditions (like a chemical reaction),

$$\Delta G^\circ = \Delta H^\circ - T\,\Delta S^\circ \tag{6.2a}$$

$$= 172.459\ \text{kJ} - (298.15\ \text{K})(175.87\ \text{J}\cdot\text{K}^{-1})\frac{10^{-3}\ \text{kJ}}{1\ \text{J}} = 120.023\ \text{kJ}$$

$$\Delta A^\circ = \Delta U^\circ - T\,\Delta S^\circ \tag{6.2b}$$

$$= 169.981 - (298.15)(175.87)(10^{-3}) = 117.545\ \text{kJ}$$

The values of the *Gibbs free energy change* (or Gibbs energy) is positive, implying that the process is not spontaneous under constant-pressure conditions (at 1 bar). Likewise, the *Helmholtz free energy change* is positive, implying that the process is not spontaneous under constant-volume conditions.

6.2 Calculate $\Delta G - \Delta A$ for heating 1.00 mol of Zn(s) from 25 °C to 98 °C at 1.00 bar. The Zn undergoes a volume change from $9.16\ \text{mL}\cdot\text{mol}^{-1}$ to $9.22\ \text{mL}\cdot\text{mol}^{-1}$ during this temperature increase.

▮ Subtracting *(6.1b)* from *(6.1a)* and substituting *(3.1a)* gives

$$G - A = (H - TS) - (U - TS) = H - U = PV \tag{6.3a}$$

For the process carried out under isobaric conditions,

$$\Delta G - \Delta A = \Delta(PV) = P\,\Delta V \tag{6.3b}$$

$$= \left[(1.00\ \text{bar})\frac{10^5\ \text{Pa}}{1\ \text{bar}}\right]\left[(1.00\ \text{mol})(9.22\ \text{mL}\cdot\text{mol}^{-1} - 9.16\ \text{mL}\cdot\text{mol}^{-1})\frac{1\ \text{cm}^3}{1\ \text{mL}}\left(\frac{10^{-6}\ \text{m}^3}{1\ \text{cm}^3}\right)\left(\frac{1\ \text{J}}{1\ \text{m}^3\cdot\text{Pa}}\right)\right]$$

$$= 6\times 10^{-3}\ \text{J}$$

6.3 Repeat the calculations of Problem 6.2 for 1.00 mol of an ideal gas undergoing the same temperature increase.

▮ For an ideal gas, substituting *(1.18)* into *(6.3a)* gives

$$\Delta G - \Delta A = \Delta(PV) = \Delta(nRT) \tag{6.4}$$

$$= (1.00\ \text{mol})(8.314\ \text{J}\cdot\text{K}^{-1}\cdot\text{mol}^{-1})(371\ \text{K} - 298\ \text{K}) = 607\ \text{J}$$

6.4 Calculate $\Delta G^\circ - \Delta A^\circ$ for the chemical equation $C(s) + CO_2(g) \rightarrow 2CO(g)$. Confirm your results by using the values determined in Problem 6.1.

▮ If the contributions to the $\Delta(PV)$ terms in *(6.3)* are neglected for all substances except gases,

$$\Delta G - \Delta A = RT\,\Delta n_g = RT\,\Delta\nu_{i,g} \tag{6.5}$$

where Δn_g and $\Delta\nu_{i,g}$ were defined in Problem 3.8. For the chemical reaction being considered,

$$\Delta n_g = \nu(CO) + \nu(CO_2) = 2\ \text{mol} + (-1\ \text{mol}) = 1\ \text{mol}$$

Substituting into *(6.5)* gives

$$\Delta G^\circ - \Delta A^\circ = (8.314\ \text{J}\cdot\text{K}^{-1}\cdot\text{mol}^{-1})(298.15\ \text{K})(1\ \text{mol}) = 2479\ \text{J}$$

From Problem 6.1,

$$\Delta G^\circ - \Delta A^\circ = 120.023\ \text{kJ} - 117.545\ \text{kJ} = 2.478\ \text{kJ}$$

These values agree within the rules of significant figures.

6.5 The thermochemical equation $2Cu^+(aq) \rightarrow Cu(s) + Cu^{2+}(aq)$ $(\Delta H^\circ = -78.57\ kJ)$ was studied electrochemically. Determine ΔS° for the reaction given $E^\circ = 0.368\ V$. What is the maximum useful work that can be obtained from this reaction?

▮ The maximum useful work is equal to ΔG°. Values of ΔG° can be calculated from experimental electrochemical measurements using

$$\Delta G = -nFE \qquad (6.6)$$

Substituting into *(6.6)* gives

$$\Delta G^\circ = -(1\ mol)(9.648 \times 10^4\ J \cdot mol^{-1} \cdot V^{-1})(0.368\ V)\frac{10^{-3}\ kJ}{1\ J} = -35.5\ kJ$$

Rearranging *(6.2a)* and substituting the data gives

$$\Delta S^\circ = \frac{[-78.57\ kJ - (-35.5\ kJ)][(10^3\ kJ)/(1\ kJ)]}{298.15\ K} = -144\ J \cdot K^{-1}$$

6.6 At the melting point of KF, $S^\circ_m/(J \cdot K^{-1} \cdot mol^{-1}) = 163.582$ for the liquid and 139.536 for the solid and $\Delta_{fus}H^\circ_{1131,m} = 27.196\ kJ \cdot mol^{-1}$. Determine $\Delta_{fus}G^\circ_{1131}$.

▮ Using *(5.20)* for the reaction $KF(s) \rightarrow KF(l)$ gives

$$\Delta_{fus}S^\circ_{1131} = (1\ mol)(163.582\ J \cdot K^{-1} \cdot mol^{-1}) - (1\ mol)(139.536\ J \cdot K^{-1} \cdot mol^{-1})$$
$$= 24.046\ J \cdot K^{-1}$$

Substituting into *(6.2a)* gives

$$\Delta_{fus}G^\circ_{1131} = (27.196\ kJ)\frac{10^3\ J}{1\ kJ} - (1131\ K)(24.046\ J \cdot K^{-1}) = 0$$

For a phase change performed under standard-state conditions, $\Delta G^\circ = 0$.

6.7 At 25 °C, $\Delta_f H^\circ_m/(kJ \cdot mol^{-1}) = 117.36$ for $CS_2(g)$ and 89.70 for $CS_2(l)$ and $S^\circ_m/(J \cdot K^{-1} \cdot mol^{-1}) = 237.84$ for $CS_2(g)$ and 151.34 for $CS_2(l)$. Calculate $\Delta_{vap}G^\circ$. Which form of CS_2 is more stable at 25 °C?

▮ For the equation $CS_2(l) \rightarrow CS_2(g)$, using *(4.5)* and *(5.20)* gives

$$\Delta_{vap}H^\circ = (1\ mol)\ \Delta_f H^\circ(g) - (1\ mol)\ \Delta_f H^\circ(l)$$
$$= (1\ mol)(117.36\ kJ \cdot mol^{-1}) - (1\ mol)(89.70\ kJ \cdot mol^{-1}) = 27.66\ kJ$$
$$\Delta_{vap}S^\circ = (1\ mol)S^\circ(g) - (1\ mol)S^\circ(l)$$
$$= (1\ mol)(237.84\ kJ \cdot mol^{-1}) - (1\ mol)(151.34\ kJ \cdot mol^{-1}) = 86.50\ J \cdot K^{-1}$$

Substituting these results into *(6.2a)* gives

$$\Delta_{vap}G^\circ = 27.66\ kJ - (298.15\ K)(86.50\ J \cdot K^{-1})\frac{10^{-3}\ kJ}{1\ J} = 1.87\ kJ$$

Based on the positive value of $\Delta_{vap}G^\circ$, vaporization is not spontaneous under standard-state conditions, and so the liquid state is more stable at 25 °C.

6.8 At 1000 K, $\Delta_r G^\circ = 38.634\ kJ$ and $\Delta_r S^\circ = 126.172\ J \cdot K^{-1}$ for the equation $F_2(g) \rightarrow 2F(g)$. Calculate $\Delta_r H^\circ_{1000}$. Is this reaction spontaneous at 1000 K under standard-state conditions?

At 4000 K, $\Delta_r G^\circ = -350.006\ kJ$ and $\Delta_r S^\circ = 131.885\ J \cdot K^{-1}$. Calculate $\Delta_r H^\circ_{4000}$. What has happened to the spontaneity of the reaction? What is the controlling driving force for the reaction in each case?

▮ At 1000 K, the reaction is not spontaneous, as indicated by the positive value of the $\Delta_r G^\circ$. Solving *(6.2a)* for $\Delta_r H^\circ$ and substituting the data gives

$$\Delta_r H^\circ_{1000} = 38.634\ kJ + (1000\ K)(126.172\ J \cdot K^{-1})\frac{10^{-3}\ kJ}{1\ J} = 164.806\ kJ$$

The unfavorable endothermic heat of reaction is the controlling factor at 1000 K. At 4000 K, the reaction is spontaneous, as indicated by the negative value of $\Delta_r G^\circ$. Repeating the calculations gives

$$\Delta_r H^\circ_{4000} = -350.006\ kJ + (4000\ K)(131.885\ J \cdot K^{-1})\frac{10^{-3}\ kJ}{1\ J} = 177.534\ kJ$$

The favorable increase in entropy is the controlling factor at 4000 K.

6.9 The maximum free energy change for an ideal ternary mixture occurs for $x_i = 0.333$. Determine $\Delta_{mix}G°$ at 25 °C for preparing 1 mol of solution.

▮ For an ideal solution, $\Delta_{mix}H° = 0$. Substituting *(5.13)* into *(6.2a)* gives for *isothermal mixing*

$$\Delta_{mix}G° = 0 - T(-R\sum_i n_i \ln x_i) = RT\sum_i n_i \ln x_i \tag{6.7}$$

Recognizing that for 1 mol of solution, $n_i = x_i$, using *(6.7)* gives

$$\Delta_{mix}G° = (8.314\ \text{J}\cdot\text{K}^{-1}\cdot\text{mol}^{-1})(298.15\ \text{K})[0.333\ln 0.333 + 0.333\ln 0.333 + 0.333\ln 0.333]$$
$$= -2723\ \text{J}$$

6.10 Determine ΔG for compressing a 1.00-mol sample of O_2 at 0 °C from 1.00 bar to 10.00 bar using the following P-V data:

P/(bar)	1.000	1.250	1.500	1.750	2.000	2.500	3.000	3.500
V_m/(L · mol^{-1})	22.711	18.164	15.133	12.968	11.344	9.071	7.555	6.473

P/(bar)	4.000	5.000	6.000	7.000	8.000	9.000	10.000
V_m/(L · mol^{-1})	5.661	4.524	3.767	3.225	2.820	2.504	2.251

▮ The pressure dependence of G is given by

$$\left(\frac{\partial G}{\partial P}\right)_T = V \tag{6.8}$$

Integrating *(6.8)* implies that ΔG will be given by the area under the curve of a plot of V_m against P. Plotting the data (see Fig. 6-1) gives

$$\Delta G_m = \int_{P_1}^{P_2} V_m\,dP = (52.2\ \text{L}\cdot\text{bar}\cdot\text{mol}^{-1})\frac{10^2\ \text{J}}{1\ \text{L}\cdot\text{bar}} = 5220\ \text{J}$$

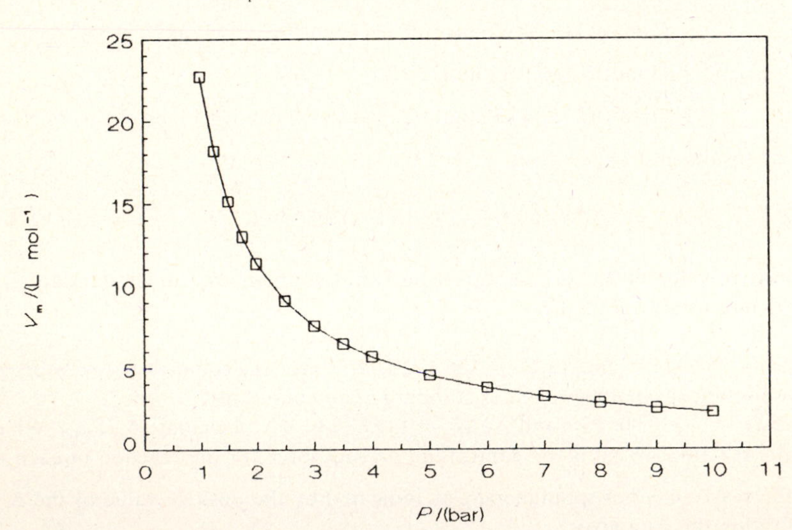

Fig. 6-1

6.11 A 1-mol sample of O_2 at 1.00 bar at 0 °C was compressed from 22.71 L to 2.25 L under isothermal conditions. Calculate ΔG for this process assuming ideal gas behavior.

▮ The *pressure-volume dependence of* ΔG under isothermal conditions for an ideal gas is given by

$$\Delta G = -nRT\ln(V_2/V_1) = nRT\ln(P_2/P_1) \tag{6.9}$$

Substituting the data gives

$$\Delta G = -(1.00\ \text{mol})(8.314\ \text{J}\cdot\text{K}^{-1}\cdot\text{mol}^{-1})(273.15\ \text{K})\ln\frac{2.25\ \text{L}}{22.71\ \text{L}} = 5250\ \text{J}$$

6.12 Repeat the calculations of Problem 6.11 assuming that O_2 obeys the van der Waals equation of state. Compare the results of the two calculations.

▌ For a van der Waals gas, *(6.8)* becomes

$$\Delta G = -nRT \ln \frac{V_2 - nb}{V_1 - nb} + n^2bRT \ln \left(\frac{1}{V_2 - nb} - \frac{1}{V_1 - nb} \right) - 2an^2 \left(\frac{1}{V_2} - \frac{1}{V_1} \right) \tag{6.10}$$

Substituting the data from Table 1-3 gives

$$\Delta G = -(1.00\ \text{mol})(8.314\ \text{J}\cdot\text{K}^{-1}\cdot\text{mol}^{-1})(273.15\ \text{K}) \ln \frac{2.25\ \text{L} - (1.00\ \text{mol})(0.0319\ \text{L}\cdot\text{mol}^{-1})}{22.71\ \text{L} - (1.00\ \text{mol})(0.0319\ \text{L}\cdot\text{mol}^{-1})}$$

$$+ (1.00\ \text{mol})^2(0.0319\ \text{L}\cdot\text{mol}^{-1})(8.314\ \text{J}\cdot\text{K}^{-1}\cdot\text{mol}^{-1})(273.15\ \text{K})$$

$$\times \left(\frac{1}{2.25\ \text{L} - (1.00\ \text{mol})(0.031\,83\ \text{L}\cdot\text{mol}^{-1})} - \frac{1}{22.71\ \text{L} - (1.00\ \text{mol})(0.031\,83\ \text{L}\cdot\text{mol}^{-1})} \right)$$

$$- 2(1.380\ \text{L}^2\cdot\text{bar}\cdot\text{mol}^{-2})(1.00\ \text{mol})^2 \left(\frac{1}{2.25} - \frac{1}{22.71} \right) \frac{10^2\ \text{J}}{1\ \text{L}\cdot\text{bar}}$$

$$= 5.20 \times 10^3\ \text{J}$$

The two values differ by 1%.

6.13 For a real gas that can be described by the pressure virial equation of state, *(1.32)*,

$$G(\text{ideal}) - G(\text{real}) = -B_pP - \tfrac{1}{2}C_pP^2 \tag{6.11}$$

Calculate $G(\text{ideal}) - G(\text{real})$ for 1.00 mol of O_2 at 1.00 bar and 0 °C and at 10.00 bar and 0 °C. Confirm the calculations of Problem 6.12.

▌ At 1.00 bar, using the virial coefficients given in Table 1-3,

$$G(\text{ideal}) - G(\text{real}) = (1.00\ \text{mol})[-(-2.777 \times 10^{-2}\ \text{L}\cdot\text{mol}^{-1})(1.00\ \text{bar})$$

$$- \tfrac{1}{2}(1.1073 \times 10^{-4}\ \text{L}\cdot\text{bar}^{-1}\cdot\text{mol}^{-1})(1.00\ \text{bar})^2] \frac{10^2\ \text{J}}{1\ \text{L}\cdot\text{bar}}$$

$$= 2.771\ \text{J}$$

and at 10.00 bar,

$$G(\text{ideal}) - G(\text{real}) = (1.00)[-(-2.777 \times 10^{-2})(10.00) - \tfrac{1}{2}(1.1073 \times 10^{-4})(10.00)^2](10^2)$$

$$= 27.22\ \text{J}$$

For the compression described in Problem 6.12, subtracting the above results gives

$$\Delta[G(\text{ideal}) - G(\text{real})] = 27.22\ \text{J} - 2.771\ \text{J} = 24.45\ \text{J}$$

From the results of Problems 6.11 and 6.12,

$$\Delta[G(\text{ideal}) - G(\text{real})] = 5250\ \text{J} - 5200\ \text{J} = 50\ \text{J}$$

These results are very similar.

6.14 The pressure on a 1.00-mol sample of NaCl was changed from 1 bar to 1000 bar. What is the ΔG for this process? For NaCl at 25 °C, $V_m = 27.0 \times 10^{-6}\ \text{m}^3\cdot\text{mol}^{-1}$.

▌ For condensed states, ΔG for an isothermal expansion is given by

$$\Delta G_m = V_m\,\Delta P \tag{6.12a}$$

Substituting the data gives

$$\Delta G_m = (27.0 \times 10^{-6}\ \text{m}^3\cdot\text{mol}^{-1})(1000\ \text{bar} - 1\ \text{bar}) \frac{10^5\ \text{Pa}}{1\ \text{bar}} \left(\frac{1\ \text{J}}{1\ \text{Pa}\cdot\text{m}^3} \right) \left(\frac{10^{-3}\ \text{kJ}}{1\ \text{J}} \right)$$

$$= 2.70\ \text{kJ}\cdot\text{mol}^{-1}$$

6.15 A more complete equation relating ΔG to pressure changes for condensed states is

$$\Delta G_m = V_m\,\Delta P(1 - \tfrac{1}{2}\kappa\,\Delta P) \tag{6.12b}$$

if $\kappa\,\Delta P \ll 1$. Given $\kappa = 4.2 \times 10^{-11}\ \text{Pa}^{-1}$ for NaCl, calculate the size of the error made in Problem 6.14 by neglecting the second term in *(6.12b)*.

▮ The error will be represented by the deviation from unity by the $1 - \frac{1}{2}\kappa\,\Delta P$ term. Substituting the data gives

$$\tfrac{1}{2}(4.2 \times 10^{-11}\ \mathrm{Pa}^{-1})(1000\ \mathrm{bar} - 1\ \mathrm{bar})[(10^5\ \mathrm{Pa})/(1\ \mathrm{bar})] = 2.1 \times 10^{-3}$$

The error in the answer of Problem 6.14 is 0.21%.

6.16 An ideal monatomic gas expands from 1.000 bar, 24.8 L, and 298 K into an evacuated container. The final volume is 49.6 L. Calculate q, w, ΔU, ΔH, ΔS, ΔG, and ΔA for the process.

▮ Because this is an ideal gas, $\Delta T = 0$. Thus *(3.48)* and *(3.49)* give $\Delta U = \Delta H = 0$. Because this a free expansion, $P_{ext} = 0$. Thus,

$$w = \int_{V_1}^{V_2} P_{ext}\, dV = 0$$

Using *(3.47)* gives $q = \Delta U - w = 0$. Assuming that the expansion can be done reversibly, *(5.10a)* gives

$$\Delta S_m = (8.314\ \mathrm{J\cdot K^{-1}\cdot mol^{-1}})\ln\frac{49.6\ \mathrm{L}}{24.8\ \mathrm{L}} = 5.763\ \mathrm{J\cdot K^{-1}\cdot mol^{-1}}$$

Substituting the above results into *(6.2)* gives

$$\Delta G_m = 0 - (298\ \mathrm{K})(5.763\ \mathrm{J\cdot K^{-1}\cdot mol^{-1}}) = -1718\ \mathrm{J\cdot mol^{-1}}$$

$$\Delta A_m = 0 - (298\ \mathrm{K})(5.763\ \mathrm{J\cdot K^{-1}\cdot mol^{-1}}) = -1718\ \mathrm{J\cdot mol^{-1}}$$

6.17 The system in Problem 6.16 is restored to the original conditions by an isothermal compression using a constant external pressure of 1.001 bar. Repeat the calculations for the restoring process and for the overall cycle process.

▮ For the restoring process (an isothermal isobaric process), *(3.52)* gives

$$\Delta U = \Delta H = 0$$

$$w = -q = -(1.001\ \mathrm{bar})(24.8\ \mathrm{L} - 49.6\ \mathrm{L})\frac{10^2\ \mathrm{J}}{1\ \mathrm{L\cdot bar}} = 2480\ \mathrm{J}$$

Assuming that the compression can be done reversibly, *(5.10a)* gives

$$\Delta S_m = (8.314\ \mathrm{J\cdot K^{-1}\cdot mol^{-1}})\ln\frac{24.8\ \mathrm{L}}{49.6\ \mathrm{L}} = -5.763\ \mathrm{J\cdot K^{-1}\cdot mol^{-1}}$$

Substituting the above results into *(6.2)* gives

$$\Delta G_m = \Delta A_m = 0 - (298\ \mathrm{K})(-5.763\ \mathrm{J\cdot K^{-1}\cdot mol^{-1}}) = 1718\ \mathrm{J\cdot mol^{-1}}$$

For the overall process, adding the above results to those from Problem 6.16 gives

$$\Delta U = \Delta H = \Delta S = \Delta G = \Delta A = 0$$

which is expected for state functions in a cyclic process and

$$w = 0 + 2480\ \mathrm{J} = 2480\ \mathrm{J}$$

$$q = 0 + (-2480)\ \mathrm{J} = -2480\ \mathrm{J}$$

6.18 At 25 °C, $\Delta_f G^\circ_m = -33.56\ \mathrm{kJ\cdot mol^{-1}}$ for $H_2S(g)$. Calculate $\Delta_f G_m$ at 0.500 bar and 10.00 bar. At what pressure will $\Delta_f G_m$ become equal to 0?

▮ Substituting $P_1 = 1\ \mathrm{bar}$ into *(6.9)* gives

$$\Delta_f G_m = \Delta_f G^\circ_m + RT\ln[P/(\mathrm{bar})] \qquad (6.13)$$

$$\Delta_f G_m = -33.56\ \mathrm{kJ\cdot mol^{-1}} + (8.314\ \mathrm{J\cdot K^{-1}\cdot mol^{-1}})\frac{10^{-3}\ \mathrm{kJ}}{1\ \mathrm{J}}(298.15\ \mathrm{K})\ln[P/(\mathrm{bar})]$$

The respective values of $\Delta_f G_m$ are

$$\Delta_f G_m(P = 0.500\ \mathrm{bar}) = -33.56\ \mathrm{kJ\cdot mol^{-1}} + (8.314\ \mathrm{J\cdot K^{-1}\cdot mol^{-1}})\frac{10^{-3}\ \mathrm{kJ}}{1\ \mathrm{J}}(298.15\ \mathrm{K})\ln 0.500$$

$$= -35.28\ \mathrm{kJ\cdot mol^{-1}}$$

$$\Delta_f G_m(P = 10.00\ \mathrm{bar}) = -33.56 + (8.314)(10^{-3})(298.15)\ln 10.00$$

$$= -27.85\ \mathrm{kJ\cdot mol^{-1}}$$

Setting *(6.13)* equal to zero, substituting the data, and solving for P gives

$$0 = -33.56\ \text{kJ}\cdot\text{mol}^{-1} + (8.314\ \text{J}\cdot\text{K}^{-1}\cdot\text{mol}^{-1})\frac{10^{-3}\ \text{kJ}}{1\ \text{J}}(298.15\ \text{K})\ln[P/(\text{bar})]$$

$$\ln[P/(\text{bar})] = 13.54 \qquad P = 7.6\times 10^5\ \text{bar}$$

6.19 The density of graphite is $2.25\ \text{g}\cdot\text{cm}^{-3}$, and that of diamond is $3.51\ \text{g}\cdot\text{cm}^{-3}$. If $\Delta G^\circ_{298} = 2.900\ \text{kJ}$ for C(graphite) → C(diamond), determine the pressure at which graphite and diamond are in equilibrium at 25 °C.

▌ Integrating *(6.8)* gives

$$\Delta G = \Delta G^\circ + \int_{P_1}^{P_2} \Delta V\, dP = \Delta G^\circ + \Delta V\,\Delta P \qquad (6.14)$$

Setting *(6.14)* equal to 0 for the equilibrium conditions and substituting the data gives

$$0 = (2.900\ \text{kJ}\cdot\text{mol}^{-1})\frac{10^3\ \text{J}}{1\ \text{kJ}} + (12.01\ \text{g}\cdot\text{mol}^{-1})\left(\frac{1}{3.51\ \text{g}\cdot\text{cm}^{-3}} - \frac{1}{2.25\ \text{g}\cdot\text{cm}^{-3}}\right)\frac{10^{-3}\ \text{L}}{1\ \text{cm}^3}(P - 1\ \text{bar})\frac{10^2\ \text{J}}{1\ \text{L}\cdot\text{bar}}$$

$$P = 15\,100\ \text{bar}$$

6.20 The vapor pressure of liquid water at −15 °C is 1.436 torr, and that of ice at −15 °C is 1.241 torr. Determine $\Delta_{\text{fus}}G_{\text{m}}$ for water at −15 °C.

▌ Consider the following three-step process to describe the melting process:

$$H_2O(\text{s}, 1.241\ \text{torr}) \rightarrow H_2O(\text{g}, 1.241\ \text{torr}) \rightarrow H_2O(\text{g}, 1.436\ \text{torr}) \rightarrow H_2O(\text{l}, 1.436\ \text{torr})$$

Because the first and third steps can be done reversibly under equilibrium conditions, $\Delta G_{\text{m},1} = \Delta G_{\text{m},3} = 0$. Using *(6.9)* for the second step gives

$$\Delta G_{\text{m},2} = (8.314\ \text{J}\cdot\text{K}^{-1}\cdot\text{mol}^{-1})(258\ \text{K})\ln\frac{1.436\ \text{torr}}{1.241\ \text{torr}} = 313\ \text{J}\cdot\text{mol}^{-1}$$

Adding the three contributions gives

$$\Delta_{\text{fus}}G_{273,\text{m}} = 0 + 313\ \text{J}\cdot\text{mol}^{-1} + 0 = 313\ \text{J}\cdot\text{mol}^{-1}$$

6.21 For the vaporization of hydrogen peroxide at 25 °C, $\Delta_{\text{vap}}G^\circ = 14.78\ \text{kJ}\cdot\text{mol}^{-1}$. Determine the vapor pressure of H_2O_2 at 25 °C.

▌ Consider the following two-step process to describe the vaporization process:

$$H_2O_2(\text{l}, P) \rightarrow H_2O_2(\text{g}, P) \rightarrow H_2O_2(\text{g}, 1\ \text{bar})$$

Because the first step can be done under equilibrium conditions, $\Delta G_{\text{m},1} = 0$. Using *(6.9)* for the second step gives

$$\Delta G_{\text{m},2} = (8.314\ \text{J}\cdot\text{K}^{-1}\cdot\text{mol}^{-1})(258.15\ \text{K})\ln\frac{1\ \text{bar}}{P}$$

Adding the contributions and solving for P gives

$$(14.78\ \text{kJ}\cdot\text{mol}^{-1})\frac{10^3\ \text{J}}{1\ \text{kJ}} = 0 + (8.314\ \text{J}\cdot\text{K}^{-1}\cdot\text{mol}^{-1})(298.15\ \text{K})\ln\frac{1\ \text{bar}}{P}$$

$$\ln[P/(\text{bar})] = -5.962 \qquad P = 2.57\times 10^{-3}\ \text{bar}$$

6.22 The vapor pressure of water at 25 °C is 23.756 torr. What is the difference in the values of $\Delta_f G^\circ_{\text{m}}$ for liquid and gaseous water at this temperature? Calculate $\Delta_f G^\circ_{\text{m}}$ for H_2O(g) given $\Delta_f G^\circ_{\text{m}} = -237.129\ \text{kJ}\cdot\text{mol}^{-1}$ for H_2O(l).

▌ Consider the following two-step process to describe the vaporization process:

$$H_2O(\text{l}, P) \rightarrow H_2O(\text{g}, P) \rightarrow H_2O(\text{g}, 1\ \text{bar})$$

Because the first step can be done under reversible equilibrium conditions, $\Delta G_{\text{m},1} = 0$. Using *(6.9)* for the second step gives

$$\Delta G_{\text{m},2} = (8.314\ \text{J}\cdot\text{K}^{-1}\cdot\text{mol}^{-1})(258.15\ \text{K})\frac{10^{-3}\ \text{kJ}}{1\ \text{J}}\ln\left[\frac{1\ \text{bar}}{(23.756\ \text{torr})[(1\ \text{atm})/(760\ \text{torr})][(1.013\,25\ \text{bar})/(1\ \text{atm})]}\right]$$

$$= 8.558\ \text{kJ}\cdot\text{mol}^{-1}$$

The difference between the values of $\Delta_f G^\circ_m$ is

$$\Delta_f G^\circ_m(g) - \Delta_f G^\circ_m(l) = 0 + 8.558\ \text{kJ} \cdot \text{mol}^{-1} = 8.558\ \text{kJ} \cdot \text{mol}^{-1}$$

giving

$$\Delta_f G^\circ_m(g) = -237.129\ \text{kJ} \cdot \text{mol}^{-1} + 8.558\ \text{kJ} \cdot \text{mol}^{-1} = -228.571\ \text{kJ} \cdot \text{mol}^{-1}$$

6.23 At $-1.00\ ^\circ\text{C}$, $\Delta G^\circ = -21.9\ \text{J}$, and at $1.00\ ^\circ\text{C}$, $\Delta G^\circ = 22.1\ \text{J}$ for the freezing of water. Calculate $\Delta_{\text{fus}} S^\circ_m$ at $0.00\ ^\circ\text{C}$.

▮ The temperature dependence of the free energy in terms of entropy is given by one of the forms of the *Gibbs–Helmholtz equation,*

$$\left(\frac{\partial G}{\partial T}\right)_P = -S \qquad (6.15a)$$

Assuming that $\Delta(\Delta G^\circ)/\Delta T \approx \partial(\Delta G)/\partial T$, *(6.15a)* gives for the freezing process

$$\Delta_{\text{cryst}} S^\circ_m = -\frac{22.1\ \text{J} \cdot \text{mol}^{-1} - (-21.9\ \text{J} \cdot \text{mol}^{-1})}{274.15\ \text{K} - 272.15\ \text{K}} = -22\ \text{J} \cdot \text{K}^{-1} \cdot \text{mol}^{-1}$$

Thus $\Delta_{\text{fus}} S^\circ_m = -\Delta_{\text{cryst}} S^\circ_m = 22\ \text{J} \cdot \text{K}^{-1} \cdot \text{mol}^{-1}$.

6.24 Use the data given in Problem 6.23 to determine $\Delta_{\text{fus}} H^\circ_m$.

▮ The temperature dependence of the free energy in terms of enthalpy is given by

$$\left(\frac{\partial(G/T)}{\partial T}\right)_P = -\frac{H}{T^2} \qquad (6.15b)$$

$$\left(\frac{\partial(G/T)}{\partial(1/T)}\right)_P = H \qquad (6.15c)$$

Assuming that $[\partial(G/T)/\partial(1/T)]_P \approx \Delta(\Delta G^\circ/T)/\Delta(1/T)$, *(6.15c)* gives for the freezing process

$$\Delta_{\text{cryst}} H^\circ_m = -\frac{(22.1\ \text{J} \cdot \text{mol}^{-1})/(274.15\ \text{K}) - (-21.9\ \text{J} \cdot \text{mol}^{-1})/(272.15\ \text{K})}{1/(274.15\ \text{K}) - 1/(272.15\ \text{K})}$$

$$= -6010\ \text{J} \cdot \text{mol}^{-1}$$

6.25 Repeat the calculations of Problem 6.19 at $100\ ^\circ\text{C}$ given $C_{P,m}/(\text{J} \cdot \text{K}^{-1} \cdot \text{mol}^{-1}) = 6.113$ for diamond and 8.527 for graphite, and $\Delta_r H^\circ_{298} = 1895\ \text{J}$. Assume that the densities and heat capacities do not change over this temperature range.

▮ For the reaction given in Problem 6.19,

$$\Delta_r C_P = (1\ \text{mol})(6.113\ \text{J} \cdot \text{K}^{-1} \cdot \text{mol}^{-1}) - (1\ \text{mol})(8.527\ \text{J} \cdot \text{K}^{-1} \cdot \text{mol}^{-1})$$

$$= -2.414\ \text{J} \cdot \text{K}^{-1}$$

At any temperature *(4.1)* gives

$$\Delta_r H^\circ_T = \Delta_r H^\circ_{298} + \Delta_r C_P(T - 298\ \text{K})$$

$$= 1895\ \text{J} + (-2.414\ \text{J} \cdot \text{K}^{-1})(T - 298\ \text{K})$$

$$= 2614\ \text{J} - (2.141\ \text{J} \cdot \text{K}^{-1})T$$

Substituting into *(6.15b)* and integrating gives

$$\frac{\Delta_r G^\circ_{373}}{373\ \text{K}} = \frac{\Delta_r G^\circ_{298}}{298\ \text{K}} - \int_{298\ \text{K}}^{373\ \text{K}} \frac{2614\ \text{J} - (2.414\ \text{J} \cdot \text{K}^{-1})T}{T^2}\, dT$$

$$= \frac{(2.900\ \text{kJ})[(10^3\ \text{J})/(1\ \text{kJ})]}{298\ \text{K}} + (2614\ \text{J})\left(\frac{1}{373\ \text{K}} - \frac{1}{298\ \text{K}}\right) + (2.414\ \text{J} \cdot \text{K}^{-1}) \ln[(373\ \text{K})/(298\ \text{K})]$$

$$= 8.510\ \text{J} \cdot \text{K}^{-1}$$

$$\Delta_r G^\circ_{373} = (373\ \text{K})(8.510\ \text{J} \cdot \text{K}^{-1}) = 3170\ \text{J}$$

Repeating the calculations for P in Problem 6.19 gives

$$0 = (3170\ \text{J} \cdot \text{mol}^{-1}) + (12.01\ \text{g} \cdot \text{mol}^{-1})\left(\frac{1}{3.15\ \text{g} \cdot \text{cm}^{-3}} - \frac{1}{2.25\ \text{g} \cdot \text{cm}^{-3}}\right)\left(\frac{10^{-3}\ \text{L}}{1\ \text{cm}^3}\right)(P - 1\ \text{bar})\left(\frac{10^2\ \text{J}}{1\ \text{L} \cdot \text{bar}}\right)$$

$$P = 16\,500\ \text{bar}$$

6.26 At 25.00 °C, $\Delta_f H_m^\circ = -92.307\ \text{kJ}\cdot\text{mol}^{-1}$ and $\Delta_f G_m^\circ = -95.299\ \text{kJ}\cdot\text{mol}^{-1}$. Assuming that $\Delta_f H_m^\circ$ is temperature-independent between 298.15 K and 300.00 K, determine $\Delta_f G_{300,m}^\circ$ for HCl.

▌ Integrating *(6.15c)* and substituting the data gives

$$\frac{\Delta_f G_{300,m}^\circ}{300.00\ \text{K}} = \frac{\Delta_f G_{298,m}^\circ}{298.15\ \text{K}} + \int_{298.15\ \text{K}}^{300.00\ \text{K}} \Delta_f H_m^\circ\, d\left(\frac{1}{T}\right)$$

$$= \frac{-95.299\ \text{kJ}\cdot\text{mol}^{-1}}{298.15\ \text{K}} + (-92.307\ \text{kJ}\cdot\text{mol}^{-1})\left(\frac{1}{300.00\ \text{K}} - \frac{1}{298.15\ \text{K}}\right)$$

$$= -0.317\,73\ \text{kJ}\cdot\text{K}^{-1}\cdot\text{mol}^{-1}$$

$$\Delta_f G_{300,m}^\circ = (300.00\ \text{K})(-0.317\,73\ \text{kJ}\cdot\text{K}^{-1}\cdot\text{mol}^{-1}) = -95.319\ \text{kJ}\cdot\text{mol}^{-1}$$

6.27 For the equation

$$4KClO_3(s) \rightarrow 3KClO_4(s) + KCl(s)$$

$\Delta_r H^\circ = -144.08\ \text{kJ}$ and $\Delta_r S^\circ = -36.8\ \text{J}\cdot\text{K}^{-1}$. Calculate $\Delta_r G^\circ$. Using the data given in Table 6-1, confirm the value of $\Delta_r G^\circ$. Is this reaction spontaneous?

Table 6-1

	$\Delta_f G_{298}^\circ$ (kJ · mol⁻¹)		$\Delta_f G_{298}^\circ$ (kJ · mol⁻¹)
$O_3(g)$	163.2	HDO(l)	−241.857
H(g)	203.247	$D_2O(l)$	−243.439
OH(g)	34.23	$Cl^-(aq)$	−131.228
$OH^-(g)$	−138.698	$Br^-(aq)$	−103.96
$OH^-(aq)$	−157.244	PbO(s, red)	−188.93
$OH^+(g)$	1306.437	NaCl(s)	−384.138
$HO_2(g)$	22.6	NaCl(aq)	−393.133
$H_2O(g)$	−228.572	KCl(s)	−409.14
$H_2O(l)$	−237.129	$KClO_3(s)$	−296.25
$H_2O_2(g)$	−105.57	$KClO_4(s)$	−303.09

▌ Substituting the data into *(6.2a)* gives

$$\Delta_r G^\circ = -144.08\ \text{kJ} - (298.15\ \text{K})(-36.8\ \text{J}\cdot\text{K}^{-1})\frac{10^{-3}\ \text{kJ}}{1\ \text{J}} = -133.11\ \text{kJ}$$

Data on free energy of formation can be combined to calculate $\Delta_r G^\circ$ using a special case of Hess's law,

$$\Delta_r G_T^\circ = \sum_i \nu_i\, \Delta_f G_{T,i}^\circ \qquad (6.16)$$

where ν_i was defined in Problem 4.46. Substituting the data from Table 6-1 into *(6.16)* gives

$$\Delta_r G^\circ = [(3\ \text{mol})\,\Delta_f G^\circ(KClO_4) + (1\ \text{mol})\,\Delta_f G^\circ(KCl)] - (4\ \text{mol})\,\Delta_f G^\circ(KClO_3)$$

$$= [(3\ \text{mol})(-303.09\ \text{kJ}\cdot\text{mol}^{-1}) + (1\ \text{mol})(-409.14\ \text{kJ}\cdot\text{mol}^{-1})] - (4\ \text{mol})(-296.25\ \text{kJ}\cdot\text{mol}^{-1})$$

$$= -133.41\ \text{kJ}$$

Both results indicate that the reaction is thermodynamically spontaneous at 25 °C under standard-state conditions.

6.28 As sodium chloride dissolves in water, the resulting solution is slightly cool to the touch as indicated by the thermochemical equation

$$NaCl(s) \xrightarrow{H_2O} NaCl(aq) \qquad \Delta_{soln} H^\circ = 3.88\ \text{kJ}$$

What is the sign of $\Delta_{soln} S^\circ$? Calculate $\Delta_{soln} S^\circ$ using the data in Table 6-1.

▌ The dissolution process is spontaneous. If $\Delta_{soln} H^\circ$ is unfavorable (endothermic), then $\Delta_{soln} S^\circ$ must be favorable or positive. Substituting the data from Table 6-1 into *(6.16)* gives

$$\Delta_{soln} G^\circ = (1\ \text{mol})\,\Delta_f G^\circ(aq) - (1\ \text{mol})\,\Delta_f G_T^\circ(s)$$

$$= (1\ \text{mol})(-393.133\ \text{kJ}\cdot\text{mol}^{-1}) - (1\ \text{mol})(-384.138\ \text{kJ}\cdot\text{mol}^{-1})$$

$$= -8.995\ \text{kJ}$$

Solving *(6.2a)* for ΔS and substituting the data gives

$$\Delta_{\text{soln}}S^\circ = \frac{[3.88\ \text{kJ} - (-8.995\ \text{kJ})][(10^3\ \text{J})/(1\ \text{kJ})]}{298.15\ \text{K}} = 43.18\ \text{J}\cdot\text{K}^{-1}$$

6.29 What is the standard-state free energy change for $H_2O(l) + D_2O(l) \rightarrow 2HDO(l)$? Use the data from Table 6-1. Is this reaction spontaneous?

▌ Using *(6.16)* gives

$$\begin{aligned}\Delta_r G^\circ &= (2\ \text{mol})\ \Delta_f G^\circ(\text{HDO}) - [(1\ \text{mol})\ \Delta_f G^\circ(\text{H}_2\text{O}) + (1\ \text{mol})\ \Delta_f G^\circ(\text{D}_2\text{O})]\\ &= (2\ \text{mol})(-241.857\ \text{kJ}\cdot\text{mol}^{-1}) - [(1\ \text{mol})(-237.129\ \text{kJ}\cdot\text{mol}^{-1}) + (1\ \text{mol})(-243.439\ \text{kJ}\cdot\text{mol}^{-1})]\\ &= -3.146\ \text{kJ}\end{aligned}$$

This reaction is spontaneous under standard-state conditions.

6.30 Using the data given in Table 6-1, calculate $\Delta_r G^\circ$ for $Cl_2(g) + 2Br^-(aq) \rightarrow Br_2(l) + 2Cl^-$. Will chlorine oxidize bromide ion in aqueous solution?

▌ The free energy of any element in its standard state is zero. Substituting the data for the ions into *(6.16)* gives

$$\begin{aligned}\Delta_r G^\circ &= [(1\ \text{mol})\ \Delta_f G^\circ(\text{Br}_2) + (2\ \text{mol})\ \Delta_f G^\circ(\text{Cl}^-)] - [(1\ \text{mol})\ \Delta_f G^\circ(\text{Cl}_2) + (2\ \text{mol})\ \Delta_f G^\circ(\text{Br}^-)]\\ &= [(1\ \text{mol})(0) + (2\ \text{mol})(-131.228\ \text{kJ}\cdot\text{mol}^{-1})] - [(1\ \text{mol})(0) + (2\ \text{mol})(-103.96\ \text{kJ}\cdot\text{mol}^{-1})]\\ &= -54.54\ \text{kJ}\end{aligned}$$

6.31 Use the thermochemical equation for the phase transition

$$\text{PbO(s, yellow)} \rightarrow \text{PbO(s, red)} \qquad \Delta_r G^\circ = -1.04\ \text{kJ}$$

and the data from Table 6-1 to determine $\Delta_f G_m^\circ$ for PbO(s, yellow).

▌ Using *(6.16)* for the above thermochemical equation gives

$$\begin{aligned}\Delta_r G^\circ &= (1\ \text{mol})\ \Delta_f G^\circ(\text{red}) - (1\ \text{mol})\ \Delta_f G^\circ(\text{yellow})\\ -1.04\ \text{kJ} &= (1\ \text{mol})(-188.93\ \text{kJ}\cdot\text{mol}^{-1}) - (1\ \text{mol})\ \Delta_f G^\circ(\text{yellow})\\ \Delta_f G_m^\circ(\text{yellow}) &= -187.89\ \text{kJ}\cdot\text{mol}^{-1}\end{aligned}$$

6.32 Based on thermodynamic considerations, which set of products will be most favored?

$$2\text{OH(g)} \rightarrow \begin{cases} \text{H}_2\text{O}_2\text{(g)} & (a)\\ \text{H}_2\text{O(g)} + \tfrac{1}{2}\text{O}_2\text{(g)} & (b)\\ \text{OH}^-\text{(g)} + \text{OH}^+\text{(g)} & (c)\\ \text{H}_2\text{(g)} + \text{O}_2\text{(g)} & (d)\\ \text{HO}_2\text{(g)} + \tfrac{1}{2}\text{H}_2\text{(g)} & (e) \end{cases}$$

▌ The equation with the largest negative value of $\Delta_r G^\circ$ will be preferred. Substituting the data from Table 6-1 into *(6.16)* for each reaction gives

$$\begin{aligned}\Delta_r G_a^\circ &= (1\ \text{mol})\ \Delta_f G^\circ(\text{H}_2\text{O}_2) - (2\ \text{mol})\ \Delta_f G^\circ(\text{OH})\\ &= (1\ \text{mol})(-105.57\ \text{kJ}\cdot\text{mol}^{-1}) - (2\ \text{mol})(34.23\ \text{kJ}\cdot\text{mol}^{-1}) = -174.03\ \text{kJ}\\ \Delta_r G_b^\circ &= [(1\ \text{mol})\ \Delta_f G^\circ(\text{H}_2\text{O}) + (\tfrac{1}{2}\ \text{mol})\ \Delta_f G^\circ(\text{O}_2)] - (2\ \text{mol})\ \Delta_f G^\circ(\text{OH})\\ &= [(1\ \text{mol})(-228.572\ \text{kJ}\cdot\text{mol}^{-1}) + (\tfrac{1}{2}\ \text{mol})(0)] - (2\ \text{mol})(34.23\ \text{kJ}\cdot\text{mol}^{-1})\\ &= -297.03\ \text{kJ}\\ \Delta_r G_c^\circ &= [(1\ \text{mol})\ \Delta_f G^\circ(\text{OH}^-) + (1\ \text{mol})\ \Delta_f G^\circ(\text{OH}^+)] - (2\ \text{mol})\ \Delta_f G^\circ(\text{OH})\\ &= [(1\ \text{mol})(-138.698\ \text{kJ}\cdot\text{mol}^{-1}) + (1\ \text{mol})(1306.437\ \text{kJ}\cdot\text{mol}^{-1})] - (2\ \text{mol})(34.23\ \text{kJ}\cdot\text{mol}^{-1})\\ &= 1099.28\ \text{kJ}\\ \Delta_r G_d^\circ &= [(1\ \text{mol})\ \Delta_f G^\circ(\text{H}_2) + (1\ \text{mol})\ \Delta_f G^\circ(\text{O}_2)] - (2\ \text{mol})\ \Delta_f G^\circ(\text{OH})\\ &= [(1\ \text{mol})(0) + (1\ \text{mol})(0)] - (2\ \text{mol})(34.23\ \text{kJ}\cdot\text{mol}^{-1}) = -68.46\ \text{kJ}\end{aligned}$$

$$\Delta_r G_e^\circ = [(1\ \text{mol})\,\Delta_f G^\circ(HO_2) + (\tfrac{1}{2}\ \text{mol})\,\Delta_f G^\circ(H_2)] - (2\ \text{mol})\,\Delta_f G^\circ(OH)$$

$$= [(1\ \text{mol})(22.6\ \text{kJ}\cdot\text{mol}^{-1}) + (\tfrac{1}{2}\ \text{mol})(0)] - (2\ \text{mol})(34.23\ \text{kJ}\cdot\text{mol}^{-1})$$

$$= -45.9\ \text{kJ}$$

Reaction **(b)** has the largest negative value of $\Delta_r G^\circ$. Thus the major products should be $H_2O(g) + \frac{1}{2}O_2(g)$.

6.33 A proposed mechanism for the reaction $I^-(aq) + ClO^-(aq) \rightarrow IO^-(aq) + Cl^-(aq)$ consists of the following three steps:

$$ClO^-(aq) + H_2O(l) \rightarrow HClO(aq) + OH^-(aq) \qquad \Delta_r G^\circ = 36.8\ \text{kJ}$$

$$I^-(aq) + HClO(aq) \rightarrow HIO(aq) + Cl^-(aq) \qquad \Delta_r G^\circ = -98.9\ \text{kJ}$$

$$OH^-(aq) + HIO(aq) \rightarrow H_2O(l) + IO^-(aq) \qquad \Delta_r G^\circ = -19.3\ \text{kJ}$$

Why does the first reaction occur even though it is thermodynamically nonspontaneous?

▌ The reaction is part of an overall process that is spontaneous. Using Hess's law gives

$$\begin{array}{ll} ClO^-(aq) + H_2O(l) \rightarrow HClO(aq) + OH^-(aq) & \Delta_r G^\circ = 36.8\ \text{kJ} \\ I^-(aq) + HClO(aq) \rightarrow HIO(aq) + Cl^-(aq) & \Delta_r G^\circ = -98.9\ \text{kJ} \\ OH^-(aq) + HIO(aq) \rightarrow H_2O(l) + IO^-(aq) & \Delta_r G^\circ = -19.3\ \text{kJ} \\ \hline I^-(aq) + ClO^-(aq) \rightarrow IO^-(aq) + Cl^-(aq) & \Delta_r G^\circ = -81.4\ \text{kJ} \end{array}$$

6.34 From 500 K to 1000 K, $\Delta_f H_m^\circ = -80.5\ \text{kJ}\cdot\text{mol}^{-1}$ and is relatively temperature-independent. Given $\Delta_f G_{500,m}^\circ = -19.788\ \text{kJ}\cdot\text{mol}^{-1}$, find $\Delta_f G_{1000,m}^\circ$.

▌ Assuming a constant value of $\Delta_f H^\circ$ and substituting the data into the integral form of *(6.15b)* gives

$$\frac{\Delta_r G_{1000,m}^\circ}{1000\ \text{K}} = \frac{-19.788\ \text{kJ}\cdot\text{mol}^{-1}}{500\ \text{K}} + (-80.5\ \text{kJ}\cdot\text{mol}^{-1})\left[\frac{1}{1000\ \text{K}} - \frac{1}{500\ \text{K}}\right]$$

$$= 4.09 \times 10^{-2}\ \text{kJ}\cdot\text{K}^{-1}\cdot\text{mol}^{-1}$$

$$\Delta_r G_{1000,m}^\circ = (1000\ \text{K})(4.09 \times 10^{-2}\ \text{kJ}\cdot\text{K}^{-1}\cdot\text{mol}^{-1}) = 40.9\ \text{kJ}\cdot\text{mol}^{-1}$$

6.35 For "quick and dirty" calculations of $\Delta_r G_T^\circ$ over small temperature changes, *(6.2a)* is often used assuming $\Delta_r H_T^\circ$ and $\Delta_r S_T^\circ$ are constant. Given $\Delta_f H_{298,m}^\circ = -928.8\ \text{kJ}\cdot\text{mol}^{-1}$ and $\Delta_r G_{298,m}^\circ = -824.9\ \text{kJ}\cdot\text{mol}^{-1}$, calculate the approximate value of $\Delta_f G_{200,m}^\circ$.

▌ The entropy change of the reaction is found using *(6.2a)* at 298 K as

$$\Delta_f S_m^\circ = \frac{-928.8\ \text{kJ}\cdot\text{mol}^{-1} - (-824.9\ \text{kJ}\cdot\text{mol}^{-1})}{298.15\ \text{K}} = -0.3485\ \text{kJ}\cdot\text{K}^{-1}\cdot\text{mol}^{-1}$$

Substituting the data into *(6.2a)* at 200 K gives

$$\Delta_f G_{200,m}^\circ = -928.8\ \text{kJ}\cdot\text{mol}^{-1} - (200\ \text{K})(-0.3485\ \text{kJ}\cdot\text{K}^{-1}\cdot\text{mol}^{-1}) = -859.1\ \text{kJ}\cdot\text{mol}^{-1}$$

6.36 What is $\Delta_r G_{373}^\circ$ for the reaction $O_3(g) + OH(g) \rightarrow H(g) + 2O_2(g)$? See Problems 4.29 and 5.72 for additional data.

▌ Substituting the data from Table 6-1 into *(6.16)* gives

$$\Delta_r G_{298}^\circ = [(1\ \text{mol})\,\Delta_f G^\circ(H) + (2\ \text{mol})\,\Delta_f G^\circ(O_2)] - [(1\ \text{mol})\,\Delta_f G^\circ(O_3) + (1\ \text{mol})\,\Delta_f G^\circ(OH)]$$

$$= [(1\ \text{mol})(203.247\ \text{kJ}\cdot\text{mol}^{-1}) + (2\ \text{mol})(0)] - [(1\ \text{mol})(163.2\ \text{kJ}\cdot\text{mol}^{-1}) + (1\ \text{mol})(34.23\ \text{kJ}\cdot\text{mol}^{-1})]$$

$$= 5.8\ \text{kJ}$$

The temperature dependence of $\Delta_r G_T^\circ$ is given by a special case of *(6.15)* as

$$\Delta_r G_T^\circ = J + KT - (\Delta a)T\ln T - \tfrac{1}{2}(\Delta b)T^2 - \tfrac{1}{6}(\Delta c)T^3 - \tfrac{1}{12}(\Delta d)T^4 - \cdots - \tfrac{1}{2}(\Delta c')T^{-1} \tag{6.17a}$$

where J is defined by *(4.4)* and

$$K = \frac{\Delta_r G_{298}^\circ - J}{298\ \text{K}} + (\Delta a)\ln(298\ \text{K}) + \tfrac{1}{2}(\Delta b)(298\ \text{K}) + \tfrac{1}{6}(\Delta c)(298\ \text{K})^2$$

$$+ \tfrac{1}{12}(\Delta d)(298\ \text{K})^3 + \cdots + \tfrac{1}{2}(\Delta c')(298\ \text{K})^{-2} \tag{6.17b}$$

In this case, $\Delta_r C_P^\circ = \Delta a = 10.41\ \mathrm{J \cdot K^{-1}}$ (see Problem 4.29). Substituting into *(4.4)* gives

$$J = \Delta_r H_{298}^\circ - (\Delta a)(298\ \mathrm{K}) = (36.3\ \mathrm{kJ})\frac{10^3\ \mathrm{J}}{1\ \mathrm{kJ}} - (10.41\ \mathrm{J \cdot K^{-1}})(298\ \mathrm{K}) = 33\,200\ \mathrm{J}$$

Using *(6.17b)* gives

$$K = \frac{(5.8\ \mathrm{kJ})[(10^3\ \mathrm{J})/(1\ \mathrm{kJ})] - 33\,200\ \mathrm{J}}{298\ \mathrm{K}} + (10.41\ \mathrm{J \cdot K^{-1}}) \ln (298\ \mathrm{K}) = -32.6\ \mathrm{J \cdot K^{-1}}$$

Substituting into *(6.17a)* gives

$$\Delta_r G_{373}^\circ = 33\,200\ \mathrm{J} + (-32.6\ \mathrm{J \cdot K^{-1}})(373\ \mathrm{K}) - (10.41\ \mathrm{J \cdot K^{-1}})(373\ \mathrm{K}) \ln (373\ \mathrm{K})$$
$$= -2000\ \mathrm{J} = -2.0\ \mathrm{kJ}$$

6.37 In Problem 6.36, the value of $\Delta_r G_T^\circ$ changed from positive at 298 K to negative at 373 K. Above what temperature will the reaction be thermodynamically spontaneous?

▌ Setting the expression for $\Delta_r G_T^\circ$ equal to zero and solving for T gives

$$\Delta_r G_T^\circ = 0 = 33\,200\ \mathrm{J} + (-32.6\ \mathrm{J \cdot K^{-1}})T - (10.41\ \mathrm{J \cdot K^{-1}})T \ln T$$
$$T = 354\ \mathrm{K}$$

6.38 Determine $\Delta_r G^\circ$ for the reaction

$$N_2(g) + 3H_2(g) \rightarrow 2NH_3(g) \qquad \Delta_r G_{298}^\circ = -32.90\ \mathrm{kJ}$$

as a function of temperature. Prepare a plot of $\Delta_r G^\circ$ against T between 298 K and 1000 K. See Problems 4.30 and 5.73 for additional data. At what temperature does the formation of ammonia under standard-state conditions become nonspontaneous?

▌ For the reaction, $\Delta a = -62.457\ \mathrm{J \cdot K^{-1}}$, $\Delta b = 63.276 \times 10^{-3}\ \mathrm{J \cdot K^{-2}}$, $\Delta c = -121.39 \times 10^{-7}\ \mathrm{J \cdot K^{-3}}$, and $J = -76.30\ \mathrm{kJ}$ (see Problem 4.30). Substituting into *(6.17b)* gives

$$K = \frac{[-32.90\ \mathrm{kJ} - (-76.30\ \mathrm{kJ})][(10^3\ \mathrm{J})/(1\ \mathrm{kJ})]}{298\ \mathrm{K}} + (-62.457\ \mathrm{J \cdot K^{-1}}) \ln (298\ \mathrm{K})$$
$$+ \tfrac{1}{2}(63.276 \times 10^{-3}\ \mathrm{J \cdot K^{-2}})(298\ \mathrm{K}) + \tfrac{1}{6}(-121.39 \times 10^{-7}\ \mathrm{J \cdot K^{-3}})(298\ \mathrm{K})^2$$
$$= -201.0\ \mathrm{J \cdot K^{-1}}$$

Thus at any temperature, *(6.17a)* gives

$$\Delta_r G_T^\circ/(\mathrm{kJ}) = -76.30 - (0.2010)[T/(\mathrm{K})] + (6.2457 \times 10^{-2})[T/(\mathrm{K})] \ln [T/(\mathrm{K})]$$
$$- (3.1638 \times 10^{-5})[T/(\mathrm{K})]^2 + (2.0232 \times 10^{-9})[T/(\mathrm{K})]^3$$

As a sample calculation, at $T = 500$ K,

$$\Delta_r G_{500}^\circ/(\mathrm{kJ}) = -76.30 - (0.2010)(500) + (6.2457 \times 10^{-2})(500) \ln (500)$$
$$- (3.1638 \times 10^{-5})(500)^2 + (2.0232 \times 10^{-9})(500)^3$$
$$= 9.62\ \mathrm{kJ}$$

The graph is shown in Fig. 6-2. From the graph, at $T > 456$ K, the formation of NH_3 is no longer spontaneous.

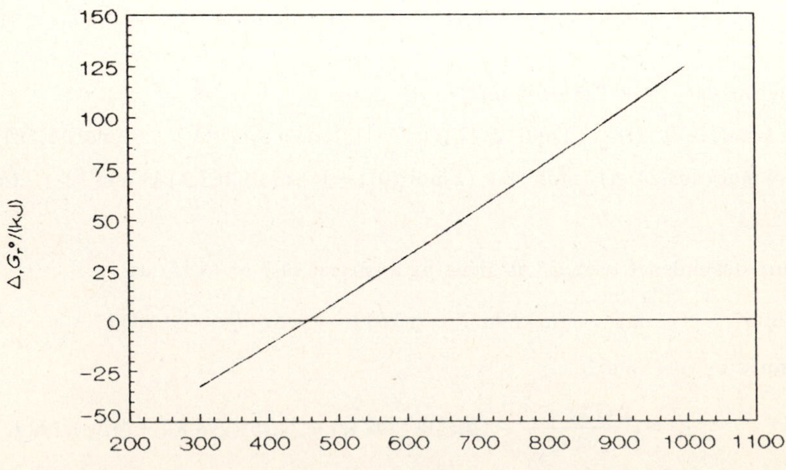

Fig. 6-2

6.39 Determine $\Delta_r G°$ for the reaction

$$N_2(g) + 2O_2(g) \rightarrow N_2O_4(g) \qquad \Delta_r G° = 97.787 \text{ kJ}$$

as a function of temperature. See Problems 4.31 and 5.74 for additional data. Prepare a plot of $\Delta_f H_T^\circ$, $-T\,\Delta_f S_T^\circ$, and $\Delta_f G_T^\circ$ against T for $100 < T/(\text{K}) \leq 1000$. Which factor—the enthalpy of formation or the entropy of formation—is more critical to the value of the free energy of formation?

▮ For the reaction, $\Delta a = -41.77\ \text{J}\cdot\text{K}^{-1}$, $\Delta b = 0.125\ \text{J}\cdot\text{K}^{-2}$, $\Delta c = -6.7\times10^{-5}\ \text{J}\cdot\text{K}^{-3}$, and $J =$ 16.569 kJ (see Problem 4.31). Substituting into *(6.17b)* gives

$$K = \frac{(97.787\text{ kJ} - 16.569\text{ kJ})[(10^3\text{ J})/(1\text{ kJ})]}{298\text{ K}} + (-41.77\ \text{J}\cdot\text{K}^{-1})\ln(298\text{ K})$$

$$+\tfrac{1}{2}(0.125\ \text{J}\cdot\text{K}^{-2})(298\text{ K}) + \tfrac{1}{6}(-6.7\times10^{-5}\ \text{J}\cdot\text{K}^{-3})(298\text{ K})^2$$

$$= 52.060\ \text{J}\cdot\text{K}^{-1}$$

Thus at any temperature, *(6.17a)* gives

$$\Delta_r G_T^\circ/(\text{kJ}) = 16.569 + (0.052\,060)[T/(\text{K})] + (4.177\times10^{-2})[T/(\text{K})]\ln[T/(\text{K})]$$

$$-(6.25\times10^{-5})[T/(\text{K})]^2 + (1.1\times10^{-8})[T/(\text{K})]^3$$

Plots of $\Delta_f H_T^\circ$ using the equation determined in Problem 4.31, of $-T\,\Delta_f S_T^\circ$ using the equation determined in Problem 5.74, and of $\Delta_f G_T^\circ$ using the above equation appear in Fig. 6-3. Note that although $\Delta_f H_T^\circ$ and $\Delta_f S_T^\circ$ both have a minimum near 435 K (see Problems 4.31 and 5.74), there is no minimum in either the $-T\,\Delta_f S_T^\circ$ curve or the $\Delta_f G_T^\circ$ curve. The unfavorable entropy change is the major contribution to the nonspontaneous value $\Delta_f G_T^\circ$.

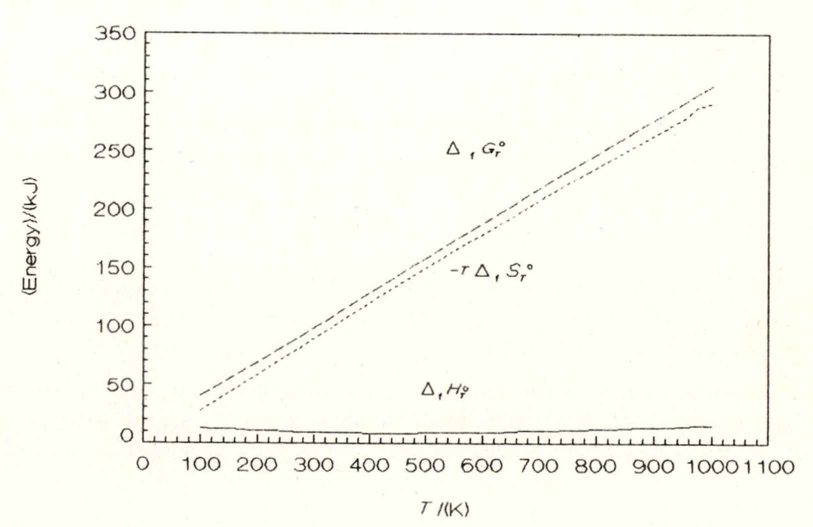

Fig. 6-3

6.40 For the formation of $SF_2(g)$,

$$\Delta_f G_{T,m}^\circ/(\text{kJ}\cdot\text{mol}^{-1}) = -335.8 + (0.9863)[T/(\text{K})] - (0.1637)[T/(\text{K})]\ln[T/(\text{K})] + (1.780\times10^{-4})[T/(\text{K})]^2$$

over the temperature range of 300 K to 1000 K. Determine the temperature at which the formation of SF_2 is most spontaneous.

▮ Differentiating the expression for $\Delta_f G_{T,m}^\circ$ with respect to T, setting the result equal to zero, and solving for T gives

$$\frac{\partial(\Delta_f G_{T,m}^\circ)}{\partial T} = \frac{\partial\{-335.8 + (0.9863)[T/(\text{K})] - (0.1637)[T/(\text{K})]\ln[T/(\text{K})] + (1.780\times10^{-4})[T/(\text{K})]^2\}}{\partial T}$$

$$= 0.9863 - (0.1637)\{\ln[T/(\text{K})] + 1\} + (2)(1.780\times10^{-4})[T/(\text{K})] = 0$$

$$T = 705\text{ K}$$

6.41 Prepare a plot of $\Delta_f G_{T,m}^\circ$ against T for each of the following gaseous oxides of sulfur.

$T/(\mathrm{K})$	0	500	1000	1500	2000
$\Delta_f G^\circ_{T,m}(SO)/(kJ \cdot mol^{-1})$	5.028	−37.391	−64.382	−66.854	−69.294
$\Delta_f G^\circ_{T,m}(S_2O)/(kJ \cdot mol^{-1})$	−58.859	−122.290	−179.507	−217.275	−258.325
$\Delta_f G^\circ_{T,m}(SO_2)/(kJ \cdot mol^{-1})$	−294.299	−300.871	−288.725	−252.239	−215.929
$\Delta_f G^\circ_{T,m}(SO_3)/(kJ \cdot mol^{-1})$	−390.025	−352.668	−293.639	−211.247	−129.768

$T/(\mathrm{K})$	2500	3000	3500	4000	4500
$\Delta_f G^\circ_{T,m}(SO)/(kJ \cdot mol^{-1})$	−71.708	−74.111	−76.512	−78.908	−81.291
$\Delta_f G^\circ_{T,m}(S_2O)/(kJ \cdot mol^{-1})$	−301.726	−346.922	−393.557	−441.382	−490.215
$\Delta_f G^\circ_{T,m}(SO_2)/(kJ \cdot mol^{-1})$	−179.675	−143.383	−106.996	−70.484	−33.829
$\Delta_f G^\circ_{T,m}(SO_3)/(kJ \cdot mol^{-1})$	−48.855	31.748	112.210	192.643	273.120

$T/(\mathrm{K})$	5000	5500	6000
$\Delta_f G^\circ_{T,m}(SO)/(kJ \cdot mol^{-1})$	−83.648	−85.965	−88.235
$\Delta_f G^\circ_{T,m}(S_2O)/(kJ \cdot mol^{-1})$	−539.913	−590.360	−641.455
$\Delta_f G^\circ_{T,m}(SO_2)/(kJ \cdot mol^{-1})$	2.981	39.960	77.128
$\Delta_f G^\circ_{T,m}(SO_3)/(kJ \cdot mol^{-1})$	353.700	434.427	515.359

At what temperature is the formation of $SO_2(g)$ and of $SO_3(g)$ equally thermodynamically favorable? Above what temperature is $S_2O(g)$ the preferred product of the reaction between S and O_2? Over what temperature range is $SO_2(g)$ the preferred product? At high temperatures, will $SO_2(g)$ be stable?

▮ The plot is shown in Fig. 6-4. The curves for $SO_2(g)$ and $SO_3(g)$ cross near $T = 1000$ K. At $T >$ 1700 K, $S_2O(g)$ is the preferred product. Between 1000 K and 1700 K, $SO_2(g)$ is the preferred product. At high temperatures the values of $\Delta_f G^\circ_{T,m}$ for $SO_2(g)$ becomes positive, and so $SO_2(g)$ is expected to decompose.

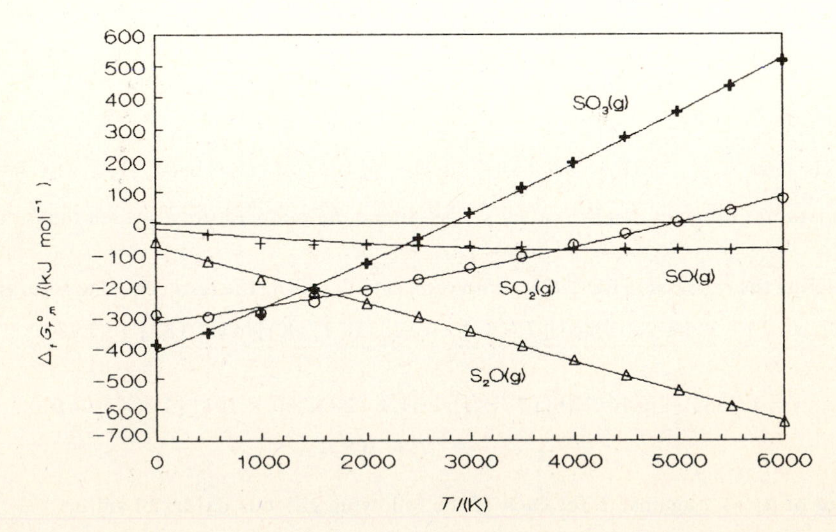

Fig. 6-4

6.42 At 500 K, $[(G_T^\circ - H_{298}^\circ)/T] = -243.688\ \mathrm{J\cdot K^{-1}\cdot mol^{-1}}$ for $O_3(g)$. Calculate the value of $(G_T^\circ - H_0^\circ)/T$ at 500 K. For $O_3(g)$, $H_T - H_{298} = -10.351\ \mathrm{kJ\cdot mol^{-1}}$ at 0 K.

▮ The two different *free energy functions* (or *phi, Planck,* or *Giauque functions*) are related by

$$\frac{G_T^\circ - H_0^\circ}{T} = \frac{G_T^\circ - H_0^\circ + H_{298}^\circ - H_{298}^\circ}{T} = \frac{G_T^\circ - H_{298}^\circ}{T} + \frac{H_{298}^\circ - H_0^\circ}{T} \tag{6.18}$$

$$= -243.688\ \mathrm{J\cdot K^{-1}\cdot mol^{-1}} + \frac{-(-10.351\ \mathrm{kJ\cdot mol^{-1}})[(10^3\ \mathrm{J})/(1\ \mathrm{kJ})]}{500\ \mathrm{K}}$$

$$= -222.986\ \mathrm{J\cdot K^{-1}\cdot mol^{-1}}$$

6.43 Use the following data to predict $(G_T^\circ - H_{298}^\circ)/T$ for $CCl_4(g)$ at 1425 K.

$T/(\mathrm{K})$	1000	1200	1400	1600	1800	2000
$[(G_T^\circ - H_{298}^\circ)/T]/(\mathrm{J\cdot K^{-1}\cdot mol^{-1}})$	−357.595	−370.798	−382.757	−393.632	−403.577	−412.725

▮ As can be seen from Fig. 6-5, a plot of $(G_T^\circ - H_{298}^\circ)/T$ against T is nearly linear. Usually table values are linearly interpolated, giving

$$\left(\frac{G_T^\circ - H_{298}^\circ}{T}\right)_{1425\ \mathrm{K}} = \left(\frac{G_T^\circ - H_{298}^\circ}{T}\right)_{1400\ \mathrm{K}} + \frac{1425\ \mathrm{K} - 1400\ \mathrm{K}}{1600\ \mathrm{K} - 1400\ \mathrm{K}}\left[\left(\frac{G_T^\circ - H_{298}^\circ}{T}\right)_{1600\ \mathrm{K}} - \left(\frac{G_T^\circ - H_{298}^\circ}{T}\right)_{1400\ \mathrm{K}}\right]$$

$$= -382.257\ \mathrm{J\cdot K^{-1}\cdot mol^{-1}} + \frac{1425\ \mathrm{K} - 1400\ \mathrm{K}}{1600\ \mathrm{K} - 1400\ \mathrm{K}}[-393.632\ \mathrm{J\cdot K^{-1}\cdot mol^{-1}} - (-382.757\ \mathrm{J\cdot K^{-1}\cdot mol^{-1}})]$$

$$= -384.116\ \mathrm{J\cdot K^{-1}\cdot mol^{-1}}$$

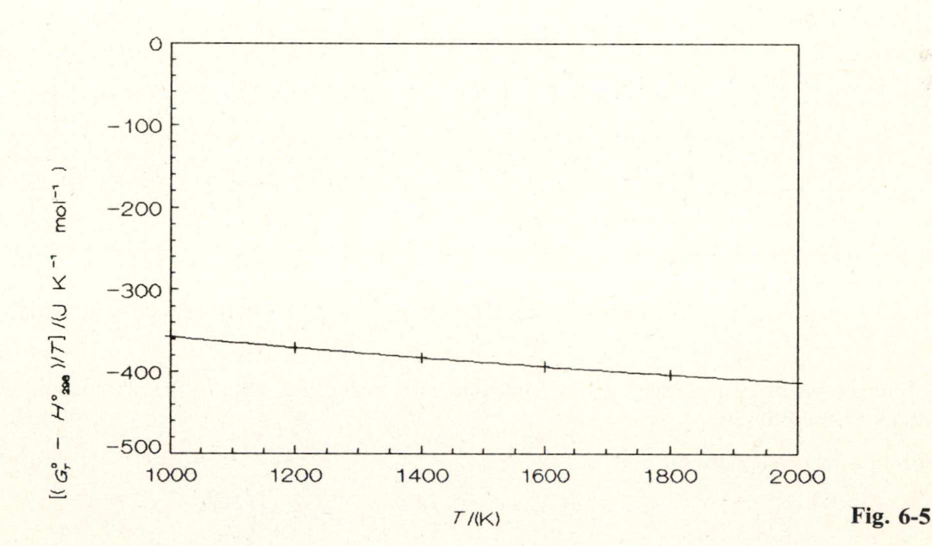

Fig. 6-5

6.44 One of the reactions important in ozone depletion is

$$Cl(g) + O_3(g) \rightarrow ClO(g) + O_2(g) \qquad \Delta H_0^\circ = -163.668\ \mathrm{kJ}$$

Given $[(G_T^\circ - H_0^\circ)/T]/(\mathrm{J\cdot K^{-1}\cdot mol^{-1}}) = -155.164$ for Cl(g), −222.986 for $O_3(g)$, −212.638 for ClO(g), and −191.158 for $O_2(g)$ at 500 K, determine $\Delta_r G_{500}^\circ$.

▮ For the reaction

$$\frac{\Delta(G_T^\circ - H_0^\circ)}{T} = \sum_i \nu_i \left(\frac{G_T^\circ - H_0^\circ}{T}\right)_i \tag{6.19}$$

$$= \left[(1\ \text{mol})\left(\frac{G_T^\circ - H_0^\circ}{T}\right)_{\text{Cl}} + (1\ \text{mol})\left(\frac{G_T^\circ - H_0^\circ}{T}\right)_{\text{O}_2}\right]$$

$$- \left[(1\ \text{mol})\left(\frac{G_T^\circ - H_0^\circ}{T}\right)_{\text{Cl}} + (1\ \text{mol})\left(\frac{G_T^\circ - H_0^\circ}{T}\right)_{\text{O}_3}\right]$$

$$= [(1\ \text{mol})(-212.638\ \text{J}\cdot\text{K}^{-1}\cdot\text{mol}^{-1}) + (1\ \text{mol})(-191.158\ \text{J}\cdot\text{K}^{-1}\cdot\text{mol}^{-1})]$$

$$- [(1\ \text{mol})(-155.164\ \text{J}\cdot\text{K}^{-1}\cdot\text{mol}^{-1}) + (1\ \text{mol})(-222.986\ \text{J}\cdot\text{K}^{-1}\cdot\text{mol}^{-1})]$$

$$= -25.646\ \text{J}\cdot\text{K}^{-1}$$

and

$$\Delta_r G_T^\circ = T[\Delta(G_T^\circ - H_0^\circ)/T] + \Delta_r H_0^\circ \tag{6.20}$$

$$= (500\ \text{K})(-25.646\ \text{J}\cdot\text{K}^{-1})\frac{10^{-3}\ \text{kJ}}{1\ \text{J}} + (-163.668\ \text{kJ}) = -176.491\ \text{kJ}$$

6.45 Elemental silicon can be prepared by

$$\text{SiO}_2(\text{l}) + 2\text{C(s)} \xrightarrow{3300\ \text{K}} \text{Si(l)} + 2\text{CO(g)} \qquad \Delta_r H_{298}^\circ = 730.077\ \text{kJ}$$

Given $[(G_T^\circ - H_{298}^\circ)/T]/(\text{J}\cdot\text{K}^{-1}\cdot\text{mol}^{-1}) = -145.191$ for $SiO_2(l)$, -32.756 for C(s), -85.103 for Si(l), and -245.435 for CO(g), calculate $\Delta_r G_{3300}^\circ$.

▮ For the reaction

$$\frac{\Delta(G_T^\circ - H_{298}^\circ)}{T} = \sum_i \nu_i \left(\frac{G_T^\circ - H_{298}^\circ}{T}\right)_i \tag{6.21}$$

$$= \left[(1\ \text{mol})\left(\frac{G_T^\circ - H_{298}^\circ}{T}\right)_{\text{Si}} + (2\ \text{mol})\left(\frac{G_T^\circ - H_{298}^\circ}{T}\right)_{\text{CO}}\right]$$

$$- \left[(1\ \text{mol})\left(\frac{G_T^\circ - H_{298}^\circ}{T}\right)_{\text{SiO}_2} + (2\ \text{mol})\left(\frac{G_T^\circ - H_{298}^\circ}{T}\right)_{\text{C}}\right]$$

$$= [(1\ \text{mol})(-85.103\ \text{J}\cdot\text{K}^{-1}\cdot\text{mol}^{-1}) + (2\ \text{mol})(-245.435\ \text{J}\cdot\text{K}^{-1}\cdot\text{mol}^{-1})]$$

$$- [(1\ \text{mol})(-145.191\ \text{J}\cdot\text{K}^{-1}\cdot\text{mol}^{-1}) + (2\ \text{mol})(-32.756\ \text{J}\cdot\text{K}^{-1}\cdot\text{mol}^{-1})]$$

$$= -365.270\ \text{J}\cdot\text{K}^{-1}$$

and

$$\Delta_r G_T^\circ = T\frac{\Delta(G_T^\circ - H_{298}^\circ)}{T} + \Delta_r H_{298}^\circ \tag{6.22}$$

$$= (3300\ \text{K})(-365.270\ \text{J}\cdot\text{K}^{-1})\frac{10^{-3}\ \text{kJ}}{1\ \text{J}} + 730.077\ \text{kJ} = -475.314\ \text{kJ}$$

6.46 Two solid phases of a substance are in equilibrium with each other. What is the relationship between the chemical potentials of the phases?

▮ For an open system where $G = G(P, T, n_i)$,

$$dG = \left(\frac{\partial G}{\partial T}\right)_{P,n_i} dT + \left(\frac{\partial G}{\partial P}\right)_{T,n_i} dP + \sum_i \left(\frac{\partial G}{\partial n_i}\right)_{P,T,n_{j\neq i}} dn_i$$

$$= \left(\frac{\partial G}{\partial T}\right)_{P,n_i} dT + \left(\frac{\partial G}{\partial P}\right)_{T,n_i} dP + \sum_i \mu_i\, dn_i \tag{6.23}$$

where the symbol $n_{j\neq i}$ means that the amounts of all substances other than substance i are held constant and the *chemical potential* is defined as

$$\mu_i = \left(\frac{\partial G}{\partial n_i}\right)_{P,T,n_{j\neq i}} \tag{6.24}$$

For a system in which an equilibrium has been established between two phases, $dT = dP = 0$ and $dn_i = -dn_j$. Substituting into *(6.23)* gives

$$0 = 0 + 0 + \mu_i\,dn_i + \mu_j\,dn_j = \mu_i\,dn_i + \mu_j(-dn_i)$$

$$\mu_i = \mu_j$$

6.47 Prepare a plot of $(\mu - \mu°)/RT$ against P for a pure ideal gas. Determine $\mu - \mu°$ at 5.5 bar and 273 K.

▌ For a pure gas, *(6.24)* gives

$$\mu = G/n = G_m \tag{6.25}$$

Using *(6.25)* and *(6.9)* gives

$$\frac{\mu - \mu°}{RT} = \frac{G_m - G_m^\circ}{RT} = \frac{RT\ln(P/P°)}{RT} = \ln[P/(\text{bar})]$$

A plot of $\ln[P/(\text{bar})]$ against P is given in Fig. 6-6. From the plot, $(\mu - \mu°)/RT = 1.70$ at 5.5 bar, giving

$$\mu - \mu° = (8.314\ \text{J}\cdot\text{K}^{-1}\cdot\text{mol}^{-1})(273\ \text{K})(1.70) = 3870\ \text{J}\cdot\text{mol}^{-1}$$

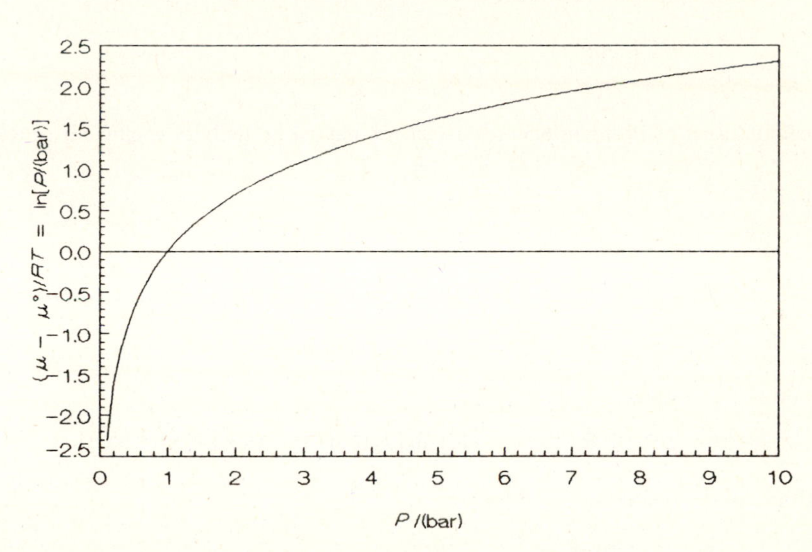

Fig. 6-6

6.2 ACTIVITIES

6.48 What is the free energy change as the activity of 1.00 mol of a substance changes from 5.25 to 1.25 at 25 °C?

▌ The free energy is related to the *activity* (a) by

$$\Delta G = G_2 - G_1 = nRT\ln(a_2/a_1) \tag{6.26}$$

Substituting the data gives

$$\Delta G = (1.00\ \text{mol})(8.314\ \text{J}\cdot\text{K}^{-1}\cdot\text{mol}^{-1})(298\ \text{K})\ln\frac{1.25}{5.25} = -3560\ \text{J}$$

6.49 What is the activity of a substance at 25 °C such that $G_m - G_m^\circ = 0.100\ \text{kJ}\cdot\text{mol}^{-1}$?

▌ Substituting $G_1 = G°$ and $a_1 = 1$ into *(6.26)* gives

$$G - G° = nRT\ln a \tag{6.27}$$

Substituting the data and solving for a gives

$$(0.100\ \text{kJ})\frac{10^3\ \text{J}}{1\ \text{kJ}} = (1\ \text{mol})(8.314\ \text{J}\cdot\text{K}^{-1}\cdot\text{mol}^{-1})(298\ \text{K})\ln a$$

$$\ln a = 0.0403 \qquad a = 1.041$$

6.50 Find the pressure of an ideal gas at which $G_m - G_m^\circ = 2.500\ \text{kJ}\cdot\text{mol}^{-1}$ at 25 °C.

▌ For an ideal gas,

$$a = P/(\text{bar}) \tag{6.28}$$

Substituting into *(6.27)* gives

$$G - G^\circ = nRT \ln [P/(\text{bar})] \tag{6.29}$$

Substituting the data and solving for P gives

$$(2.500\ \text{kJ})\frac{10^3\ \text{J}}{1\ \text{kJ}} = (1.00\ \text{mol})(8.314\ \text{J}\cdot\text{K}^{-1}\cdot\text{mol}^{-1})(298.15\ \text{K}) \ln [P/(\text{bar})]$$

$$\ln [P/(\text{bar})] = 1.008 \qquad P = 2.74\ \text{bar}$$

6.51 What is $G_m - G_m^\circ$ for an ideal gas at a pressure of 1.0 torr and 875 °C?

▮ Substituting into *(6.29)* gives

$$G_m - G_m^\circ = (8.314\ \text{J}\cdot\text{K}^{-1}\cdot\text{mol}^{-1})(1148\ \text{K}) \ln \left[\frac{(1.0\ \text{torr})[(1\ \text{atm})/(760\ \text{torr})][(1.013\ 25\ \text{bar})/(1\ \text{atm})]}{\text{bar}}\right]$$

$$= -63\ 000\ \text{J}\cdot\text{mol}^{-1}$$

6.52 Repeat the calculations of Problem 6.50 for a real gas having an activity coefficient of 0.95.

▮ For a real gas,

$$a = f/(\text{bar}) \tag{6.30}$$

where the *fugacity* (f) is defined in terms of the *activity coefficient* (or *fugacity coefficient*) as

$$f = \gamma P \tag{6.31}$$

Substituting into *(6.27)* gives

$$G - G^\circ = nRT \ln [f/(\text{bar})] = nRT \ln [\gamma P/(\text{bar})] \tag{6.32}$$

Substituting the data and solving for P gives

$$(2.500\ \text{kJ})\frac{10^3\ \text{J}}{1\ \text{kJ}} = (1.00\ \text{mol})(8.314\ \text{J}\cdot\text{K}^{-1}\cdot\text{mol}^{-1})(298.15\ \text{K}) \ln [\gamma P/(\text{bar})]$$

$$\ln [\gamma P/(\text{bar})] = 1.008 \qquad P = \frac{2.74\ \text{bar}}{\gamma} = \frac{2.74\ \text{bar}}{0.95} = 2.88\ \text{bar}$$

6.53 A 1.00-mol sample of $CCl_2F_2(g)$ at 425 K expanded from 75.2 bar to 12.5 bar. What was the free energy change? Assume real gas behavior and that γ is given by Fig. 6-7. For CCl_2F_2, $T_c = 111.5\ °C$ and $P_c = 40.1$ bar.

▮ The reduced temperature and pressures are given by *(1.30)* and *(1.31)* as

$$T_r = (425\ \text{K})/(384.7\ \text{K}) = 1.10$$

$$P_{r,1} = \frac{75.2\ \text{bar}}{40.1\ \text{bar}} = 1.88 \qquad P_{r,2} = \frac{12.5\ \text{bar}}{40.1\ \text{bar}} = 0.312$$

From the plot of γ against P_r shown in Fig. 6-7, $\gamma_1 = 0.57$ and $\gamma_2 = 0.94$. Substituting into *(6.31)* gives

$$f_2 = (0.94)(12.5\ \text{bar}) = 11.8\ \text{bar}$$

$$f_1 = (0.57)(75.2\ \text{bar}) = 43\ \text{bar}$$

The free energy change will be given by

$$\Delta G = G_2 - G_1 = nRT \ln (f_2/f_1) \tag{6.33}$$

$$= (1.00)(8.314\ \text{J}\cdot\text{K}^{-1}\cdot\text{mol}^{-1})(425\ \text{K}) \ln \frac{11.8\ \text{bar}}{43\ \text{bar}} = -4570\ \text{J}$$

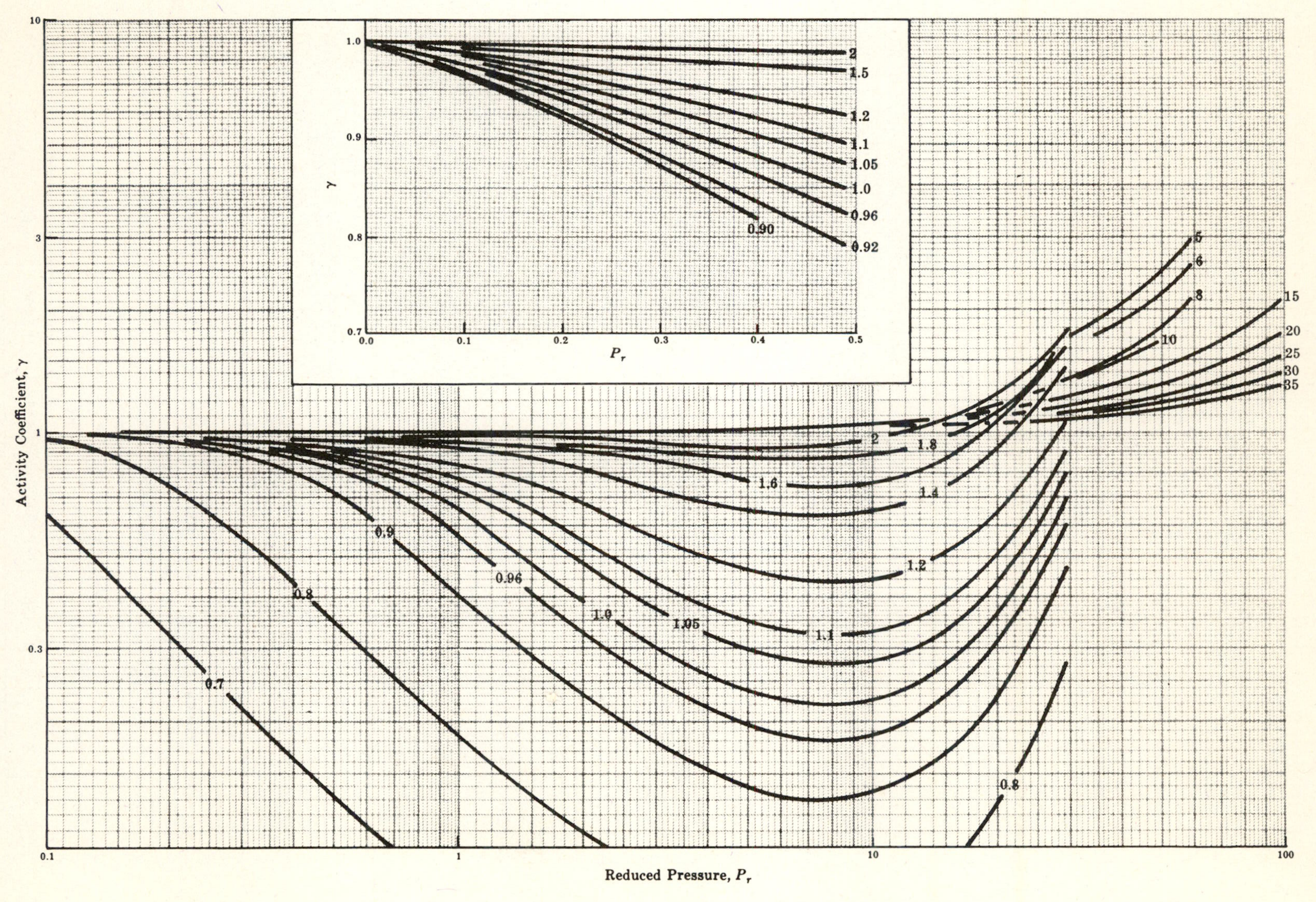
Activity Coefficient, γ
Reduced Pressure, P_r
10
3
1
0.3
0.1
1
10
100
0.7
0.8
0.9
0.96
1.0
1.05
1.1
1.2
1.4
1.6
1.8
2
5
6
8
10
15
20
25
30
35
γ
P_r
1.0
0.9
0.8
0.7
0.0
0.1
0.2
0.3
0.4
0.5
2
1.5
1.2
1.1
1.05
1.0
0.96
0.92
0.90

Fig. 6-7

6.54 Determine γ and f for $O_2(g)$ at 1000 bar and 0 °C from the following data:

P/(bar)	1	100	200	300	400	500	600	700	800	900	1000
V_m/(L · mol^{-1})	22.710 79	0.209 82	0.104 47	0.071 50	0.060 06	0.052 68	0.047 07	0.044 27	0.043 02	0.040 35	0.038 43

▌ For a real gas,

$$RT \ln \gamma = \int_0^P \left(V_m - \frac{RT}{P} \right) dP \tag{6.34}$$

The value of the integral is evaluated by determining the area under the curve of a plot of $V_m - RT/P$ against P. As a sample calculation, for $P = 600$ bar,

$$V_m - \frac{RT}{P} = 0.047\,07 \text{ L} \cdot \text{mol}^{-1} - \frac{(0.083\,14 \text{ L} \cdot \text{bar} \cdot \text{K}^{-1} \cdot \text{mol}^{-1})(273.15 \text{ K})}{600 \text{ bar}}$$

$$= 9.22 \times 10^{-3} \text{ L} \cdot \text{mol}^{-1}$$

The plot is shown in Fig. 6-8. The area under the curve is 3.43 L · bar · mol^{-1}, giving

$$\ln \gamma = \frac{3.43 \text{ L} \cdot \text{bar} \cdot \text{mol}^{-1}}{(0.083\,14 \text{ L} \cdot \text{bar} \cdot \text{K}^{-1} \cdot \text{mol}^{-1})(273.15 \text{ K})} = 0.151$$

$$\gamma = 1.16$$

Substituting the results into *(6.31)* gives

$$f = (1.16)(1000 \text{ bar}) = 1160 \text{ bar}$$

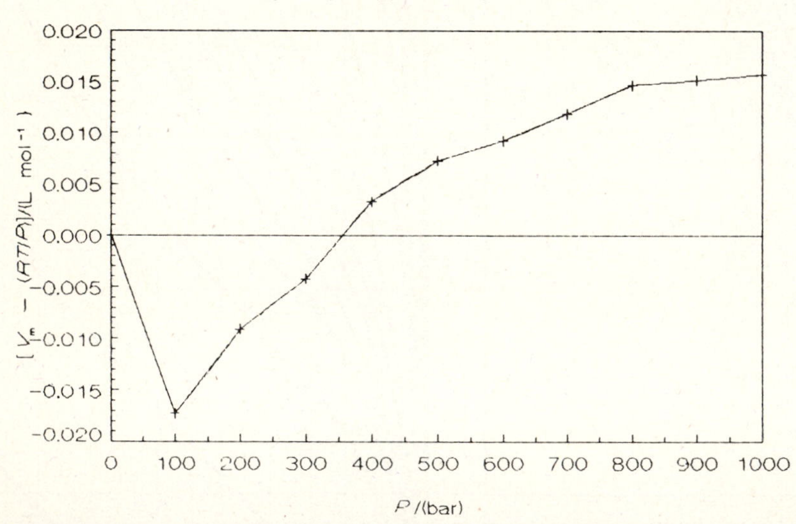

Fig. 6-8

6.55 Determine γ and f for $O_2(g)$ at 1000 bar and 0 °C for the following data:

P/(bar)	1	100	200	300	400	500	600	700	800	900	1000
Z	1	0.9239	0.9200	0.9445	1.0579	1.1599	1.2436	1.3647	1.5156	1.5990	1.6922

For $O_2(g)$, $P_c = 50.8$ bar.

▮ For a real gas,

$$\ln \gamma = \int_{P_1}^{P_2} (Z - 1)\, d \ln P_r \tag{6.35}$$

The value of the integral is evaluated by determining the area under the curve of a plot of $Z - 1$ against $\ln P_r$. This plot appears in Fig. 6-9. The area under the curve is 0.033, giving $\gamma = 1.03$. Substituting the results into *(6.31)* gives

$$f = (1.03)(1000 \text{ bar}) = 1030 \text{ bar}$$

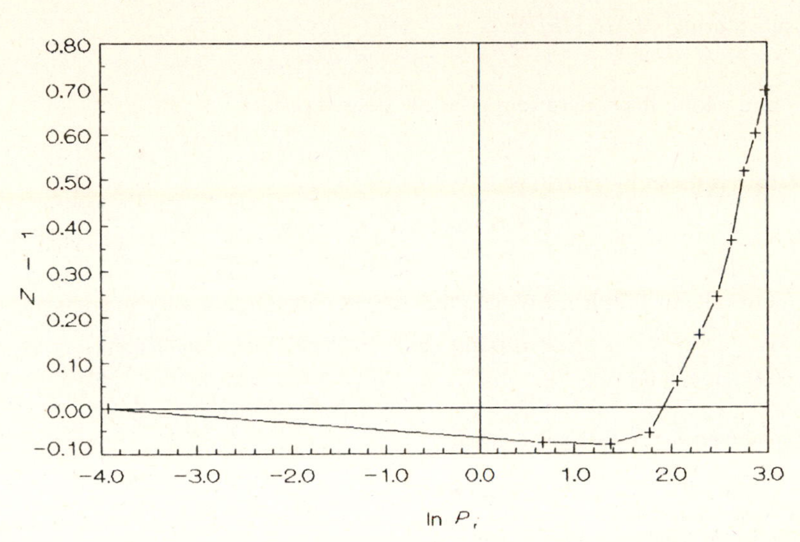

Fig. 6-9

6.56 Derive an equation for determining γ for a real gas for which *(1.32)* is valid. Given $B_p = -2.228 \times 10^{-2}$ L and $C_p = 4.84 \times 10^{-5}$ L·bar^{-1} for $O_2(g)$ at 0 °C for $0 \le P \le 100$ bar, determine $G_{m,\text{real}} - G_{m,\text{ideal}}$ at $P = 10.000$ bar for $O_2(g)$.

▮ Substituting *(1.32)* into *(6.34)* gives

$$RT \ln \gamma = \int_0^P \left(\frac{RT + B_pP + C_pP^2 + D_pP^3 + \cdots}{P} - \frac{RT}{P} \right) dP$$

$$= B_pP + \tfrac{1}{2}C_pP^2 + \tfrac{1}{3}D_pP^3 + \cdots \tag{6.36}$$

For an ideal gas, *(6.31)* gives $f_{\text{ideal}} = \gamma P = P$, which upon substituting into *(6.33)* gives

$$G_{m,\text{real}} - G_{m,\text{ideal}} = RT \ln \frac{f}{P} = RT \ln \gamma = B_pP + \tfrac{1}{2}C_pP^2 + \tfrac{1}{3}D_pP^3 + \cdots$$

$$= [(-2.228 \times 10^{-2} \text{ L})(10.000 \text{ bar}) + \tfrac{1}{2}(4.84 \times 10^{-5} \text{ L}\cdot\text{bar}^{-1})(10.000 \text{ bar})^2][(10^2 \text{ J})/(1 \text{ L}\cdot\text{bar})]$$

$$= -22.04 \text{ J}$$

6.57 At low pressure and high temperature, the real gas behavior of many substances can be described by $Z = 1 + b$. Derive an equation for determining f. Given $b = 4.00 \times 10^{-4}$ bar^{-1} at 200 °C for H_2, calculate f at $P = 20.00$ bar.

▮ Solving *(1.28)* for PV_m and substituting the expression for Z gives

$$PV_m = ZRT = RT(1 + bP) = RT + B_pP$$

where $B_p = bRT$. Recognizing that this is a special case of *(1.32)*, using *(6.36)* gives

$$RT \ln \gamma = B_pP = bRTP \qquad \ln \gamma = bP \qquad \gamma = e^{bP}$$

Substituting into *(6.31)* gives

$$f = Pe^{bP} \tag{6.37a}$$

$$= (20.00 \text{ bar})\, e^{(4.00 \times 10^{-4} \text{ bar}^{-1})(20.00 \text{ bar})} = 20.16 \text{ bar}$$

6.58 Show that the geometric mean of the ideal pressure and the fugacity is approximately equal to the real pressure of a gas that can be described by the equation of state given in Problem 6.57. At $P = 20.00\ \text{bar}$ and 473 K, $V_m = 1.981\ \text{L}\cdot\text{mol}^{-1}$. Show that the result is reasonably valid.

▌ Using a series expansion for the exponential in *(6.37a)* gives

$$f = Pe^{bP} + P(1 + bP + \cdots) = P[1 + (Z - 1) + \cdots] = PZ \qquad (6.37b)$$

From *(1.28)*

$$Z = \frac{PV_m}{RT} = \frac{P}{RT/V_m} = \frac{P}{P_{\text{ideal}}}$$

which upon substituting into *(6.37b)* gives

$$f = P(P/P_{\text{ideal}})$$

Solving for P and taking the square root gives the desired result:

$$P = (fP_{\text{ideal}})^{1/2}$$

The ideal pressure is given by *(1.18)* as

$$P_{\text{ideal}} = \frac{(0.083\,14\ \text{L}\cdot\text{bar}\cdot\text{K}^{-1}\cdot\text{mol}^{-1})(473\ \text{K})}{1.981\ \text{L}\cdot\text{mol}^{-1}} = 19.85\ \text{bar}$$

Substituting the results of Problem 6.57 into the derived equation gives

$$P = [(20.16\ \text{bar})(19.85\ \text{bar})]^{1/2} = 20.00\ \text{bar}$$

which is in good agreement.

6.59 The fugacity for a van der Waals gas is given by

$$\ln[f/(\text{bar})] = \ln\frac{nRT}{V - nb} + \frac{nb}{V - nb} - \frac{2an}{RTV} \qquad (6.38a)$$

At what pressure is the standard state of oxygen at 0 °C?

▌ The standard state of a real gas is defined as $f = 1\ \text{bar}$. Substituting the data from Table 1-3 into *(6.38a)* for exactly 1 mol gives

$$\ln 1 = \ln\left[\frac{(1\ \text{mol})(0.083\,14\ \text{L}\cdot\text{bar}\cdot\text{K}^{-1}\cdot\text{mol}^{-1})(273.15\ \text{K})}{V - (1\ \text{mol})(0.0319\ \text{L}\cdot\text{mol}^{-1})}\right]$$

$$+ \frac{(1\ \text{mol})(0.0319\ \text{L}\cdot\text{mol}^{-1})}{V - (1\ \text{mol})(0.0319\ \text{L}\cdot\text{mol}^{-1})} - \frac{2(1.138\ \text{L}^2\cdot\text{bar}\cdot\text{mol}^{-1})(1\ \text{mol})}{(0.083\,14\ \text{L}\cdot\text{bar}\cdot\text{K}^{-1}\cdot\text{mol}^{-1})(273.15\ \text{K})V}$$

$$0 = \ln\frac{22.71}{[V/(\text{L})] - 0.0319} + \frac{0.0319}{V/(\text{L}) - 0.0319} - \frac{0.1215}{V/(\text{L})}$$

Solving the equation gives $V = 22.65\ \text{L}$. Substituting into *(1.44c)* gives

$$P = \frac{(1\ \text{mol})(0.083\,14\ \text{L}\cdot\text{bar}\cdot\text{K}^{-1}\cdot\text{mol}^{-1})(273.15\ \text{K})}{22.65\ \text{L} - (1\ \text{mol})(0.0319\ \text{L}\cdot\text{mol}^{-1})} - \frac{(1.380\ \text{L}^2\cdot\text{bar}\cdot\text{mol}^{-2})(1\ \text{mol})^2}{(22.65\ \text{L})^2}$$

$$= 1.0014\ \text{bar}$$

6.60 To a good approximation, *(1.44a)* can be written as

$$\frac{PV_m}{RT} = 1 + \left(b - \frac{a}{RT}\right)\frac{P}{RT}$$

Derive an expression for f using the above approximation. Given $a = 0.2476\ \text{L}^2\cdot\text{bar}\cdot\text{mol}^{-2}$ and $b = 0.026\,61\ \text{L}\cdot\text{mol}^{-1}$, determine the pressure corresponding to the standard state of H_2 at 0 °C. Why is this value less than unity?

▌ Solving the approximation for V_m, substituting into *(6.34)*, and integrating gives

$$RT\ln\gamma = \int_0^P\left[\frac{RT}{P} + \left(b - \frac{a}{RT}\right) - \frac{RT}{P}\right]dP = \left(b - \frac{a}{RT}\right)P$$

Substituting *(6.31)* and rearranging gives

$$\ln[f/(\text{bar})] = \ln[P/(\text{bar})] + \left(b - \frac{a}{RT}\right)\frac{P}{RT} \qquad (6.38b)$$

Letting $f = 1$ bar and substituting the data gives

$$\ln 1 = \ln[P/(\text{bar})] + \left(0.026\,61\ \text{L}\cdot\text{mol}^{-1} - \frac{0.2476\ \text{L}^2\cdot\text{bar}\cdot\text{mol}^{-2}}{(0.083\,14\ \text{L}\cdot\text{bar}\cdot\text{K}^{-1}\cdot\text{mol}^{-1})(273.15\ \text{K})}\right) \times \left[\frac{P}{(0.083\,14\ \text{L}\cdot\text{bar}\cdot\text{K}^{-1}\cdot\text{mol}^{-1})(273.15\ \text{K})}\right]$$

Solving gives $P = 0.9993$ bar. Hydrogen shows positive deviation from ideality, and so the standard fugacity will be at a pressure less than 1 bar.

6.61 Prepare a plot of fugacity against pressure for $O_2(g)$, $H_2(g)$, and an ideal gas. Assume that *(6.38b)* is valid for the real gases.

▮ For the ideal gas, $f = P$. The values of a and b for O_2 can be found in Table 1-3, and those for H_2 are given in Problem 6.60. As a sample calculation, consider $P = 10.00$ bar:

$$f(\text{ideal}) = 10.00\ \text{bar}$$

$$\ln[f(H_2)] = \ln 10.00 + \left(0.026\,61\ \text{L}\cdot\text{mol}^{-1} - \frac{0.2476\ \text{L}^2\cdot\text{bar}\cdot\text{mol}^{-2}}{(0.083\,14\ \text{L}\cdot\text{bar}\cdot\text{K}^{-1}\cdot\text{mol}^{-1})(273.15\ \text{K})}\right) \times \left(\frac{10.00\ \text{bar}}{(0.083\,14\ \text{L}\cdot\text{bar}\cdot\text{K}^{-1}\cdot\text{mol}^{-1})(273.15\ \text{K})}\right) = 2.3095$$

$$f(H_2) = 10.07\ \text{bar}$$

$$\ln[f(O_2)] = \ln 10.00 + \left(0.0319 - \frac{1.380}{(0.083\,14)(273.15)}\right)\left(\frac{10.00}{(0.083\,14)(273.15)}\right) = 2.2899$$

$$f(O_2) = 9.87\ \text{bar}$$

The plot is shown in Fig. 6-10.

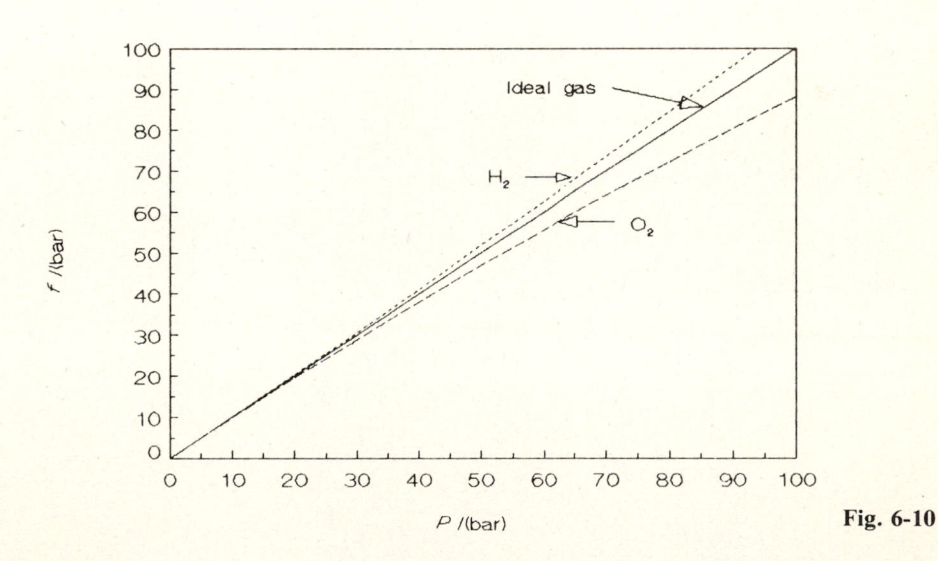

Fig. 6-10

6.62 A real gas can be described by the equation $P(V_m - b) = RT$. Derive an equation for the fugacity of this gas.

▮ Solving the expression for V_m and substituting into *(6.34)* gives

$$RT \ln \gamma = \int_0^P \left(\frac{RT}{P} + b - \frac{RT}{P}\right) dP = \int_0^P b\, dP = bP$$

$$\gamma = e^{bP/RT}$$

Substituting this result into *(6.31)* and solving for f gives $f = Pe^{bP/RT}$.

6.63 For a gas that obeys the Berthelot equation of state, *(1.56)*, the fugacity can be determined from the critical constants using

$$\ln \gamma = \frac{9T_c}{128P_cT}\left(1 - \frac{6T_c^2}{T^2}\right)P \tag{6.39}$$

Determine $\mu - \mu^\circ$ for a 1.00-mol sample of O_2 at 10.0 bar and 0 °C. For O_2, $T_c = -118.4\ °C$ and $P_c =$ 50.8 bar.

▌ From *(6.8)*,

$$\mu - \mu^\circ = \int_{P^\circ}^{P} V_m\, dP = RT \ln \frac{P}{P^\circ} + \frac{9RT_c}{128P_c}\left(1 - \frac{6T_c^2}{T^2}\right)(P - P^\circ)$$

$$= (8.314\ \mathrm{J \cdot K^{-1} \cdot mol^{-1}})(273\ \mathrm{K}) \ln \frac{10.0\ \mathrm{bar}}{1.00\ \mathrm{bar}}$$

$$+ \frac{9(8.314\ \mathrm{J \cdot K^{-1} \cdot mol^{-1}})(154.8\ \mathrm{K})}{128(50.8\ \mathrm{bar})}\left(1 - \frac{6(154.8\ \mathrm{K})^2}{(273\ \mathrm{K})^2}\right)(10.0\ \mathrm{bar} - 1.00\ \mathrm{bar})$$

$$= 5210\ \mathrm{J \cdot mol^{-1}}$$

6.64 Use *(6.38b)* to derive equations for $\mu - \mu^\circ$, $S - S^\circ$, and $H - H^\circ$ for a van der Waals gas.

▌ From *(6.33)* and *(6.38b)*,

$$G_m - G_m^\circ = \mu - \mu^\circ = RT \ln f = RT \ln [P/(\mathrm{bar})] + (b - a/RT)P$$

Using *(6.15a)* gives

$$S_m - S_m^\circ = -\left[\frac{\partial(G_m - G_m^\circ)}{\partial T}\right]_P = -R \ln [P/(\mathrm{bar})] - \frac{a}{RT^2}P$$

Using *(6.2a)* gives

$$H_m - H_m^\circ = RT \ln [P/(\mathrm{bar})] + \left(b - \frac{a}{RT}\right)P + T\left(-R \ln [P/(\mathrm{bar})] - \frac{a}{RT^2}P\right)$$

$$= (b - 2a/RT)P$$

6.65 Calculate the activity of ice at 0 °C and 1.0 torr. For ice, $V_m^\circ = 19.6 \times 10^{-6}\ \mathrm{m^3 \cdot mol^{-1}}$ and $\kappa = 12 \times 10^{-11}\ \mathrm{Pa^{-1}}$.

▌ For the condensed states of matter, *(6.8)* and *(6.27)* give

$$\ln a = \frac{1}{RT}\int_{1\ \mathrm{bar}}^{P} V_m\, dP \tag{6.40}$$

Substituting $V_m = V_m^\circ(1 - \kappa P)$ and integrating gives

$$\ln a = \frac{V_m^\circ}{RT}\left\{(P - 1\ \mathrm{bar}) - \frac{\kappa}{2}[P^2 - (1\ \mathrm{bar})^2]\right\} \tag{6.41}$$

For $1.0\ \mathrm{torr} = 133\ \mathrm{Pa}$,

$$\ln a = \frac{19.6 \times 10^{-6}\ \mathrm{m^3 \cdot mol^{-1}}}{(8.314\ \mathrm{Pa \cdot m^3 \cdot K^{-1} \cdot mol^{-1}})(273\ \mathrm{K})}$$

$$\times \left\{(133\ \mathrm{Pa} - 1 \times 10^5\ \mathrm{Pa}) - \frac{12 \times 10^{-11}\ \mathrm{Pa^{-1}}}{2}[(133\ \mathrm{Pa})^2 - (1 \times 10^5\ \mathrm{Pa})^2]\right\}$$

$$= -8.62 \times 10^{-4}$$

$$a = 0.999\ 138$$

6.66 A student omitted the κP term in the derivation of *(6.41)*. How much of an error was introduced in the value of a for ice at 0 °C and 1.0 torr?

▌ The abbreviated form of *(6.41)* is

$$\ln a = \frac{V_m^\circ}{RT}(P - 1\ \mathrm{bar}) \frac{19.6 \times 10^{-6}\ \mathrm{m^3 \cdot mol^{-1}}}{(8.314\ \mathrm{Pa \cdot m^3 \cdot K^{-1} \cdot mol^{-1}})(273\ \mathrm{K})}(133\ \mathrm{Pa} - 1 \times 10^5\ \mathrm{Pa})$$

$$= -8.62 \times 10^{-4}$$

$$a = 0.999\ 138$$

There was no significant error introduced.

6.67 For an ideal gas, $\Delta G_{m,425} = -4570\ J$ for an expansion from 75.2 bar to 12.5 bar. What is the corresponding $\Delta G_{m,273}$ for ice? See Problem 6.65 for additional data.

▮ Substituting the data into *(6.41)* gives

$$\ln a = \frac{(19.6 \times 10^{-6}\ m^3 \cdot mol^{-1})[(10^5\ Pa)/(1\ bar)]}{(8.314\ Pa \cdot m^3 \cdot K^{-1} \cdot mol^{-1})(273\ K)}$$
$$\times \left\{(75.2\ bar - 1\ bar) - \frac{(12 \times 10^{-11}\ Pa^{-1})[(10^5\ Pa)/(1\ bar)]}{2}[(75.2\ bar)^2 - (1\ bar)^2]\right\}$$
$$= 0.0640$$

$$a = 1.066$$

$$\ln a = \frac{(19.6 \times 10^{-6})(10^5)}{(8.314)(273)}\left\{(12.5 - 1) - \frac{(12 \times 10^{-11})(10^5)}{2}[(12.5)^2 - (1)^2]\right\}$$
$$= 0.009\,93$$

$$a = 1.010$$

Substituting into *(6.26)* gives

$$\Delta G_{m,273} = (8.314\ J \cdot K^{-1} \cdot mol^{-1})(273\ K) \ln \frac{1.010}{1.066} = -122\ J$$

6.68 Calculate the ionic strength of a 0.020 M solution of $CaCl_2$.

▮ The *ionic strength* (I) of a solution of an electrolyte is given by

$$I = \tfrac{1}{2} \sum_i^{\text{ions}} C_i z_i^2 \tag{6.42}$$

where z_i is the charge on the ion i. The *formal* (or *analytical*) *concentration* of the solution (C) is related to the ionic concentrations by

$$C_i = \nu_i C \tag{6.43a}$$

where ν_i is the number of moles of ions produced by 1 mol of the electrolyte. For $CaCl_2$,

$$C(Ca^{2+}) = (1)(0.020\ M) = 0.020\ M \qquad C(Cl^-) = (2)(0.020\ M) = 0.040\ M$$

which upon substituting into *(6.42)* gives

$$I = \tfrac{1}{2}[C(Ca^{2+})z(Ca^{2+})^2 + C(NO_3^-)z(NO_3^-)^2]$$
$$= \tfrac{1}{2}[(0.020\ M)(+2)^2 + (0.040\ M)(-1)^2]$$
$$= 0.060\ M$$

This result agrees with that shown in Tables 6-2 for a 2:1 electrolyte:

$$I = 3C = 3(0.020\ M) = 0.060\ M$$

Table 6-2

strong electrolyte type	example	ionic strength I	mean ionic concentration $C_\pm$	solute activity a
1:1	NaCl	C	C	$y_\pm^2 C^2$
2:1	$CaCl_2$	$3C$	$4^{1/3}C$	$4y_\pm^3 C^3$
1:2	Na_2SO_4	$3C$	$4^{1/3}C$	$4y_\pm^3 C^3$
2:2	$CuSO_4$	$4C$	C	$y_\pm^2 C^2$
3:1	$FeCl_3$	$6C$	$27^{1/4}C$	$27y_\pm^4 C^4$
1:3	Na_3PO_4	$6C$	$27^{1/4}C$	$27y_\pm^4 C^4$
3:2	$Al_2(SO_4)_3$	$15C$	$108^{1/5}C$	$108y_\pm^5 C^5$
2:3	$Ca_3(PO_4)_2$	$15C$	$108^{1/5}C$	$108y_\pm^5 C^5$
3:3	$AlPO_4$	$9C$	C	$y_\pm^2 C^2$

Note: Molalities (m) and $\gamma_\pm$ values may be substituted for molarities (C) and $y_\pm$ values.

6.69 Calculate the ionic strength of a solution containing 0.0010 M Na_2SO_4 and 0.0030 M $NaNO_3$ at 25 °C.

▌ The respective ionic concentrations are given by *(6.43a)* as

$$C(Na^+) = 2(0.0010\text{ M}) + (0.0030\text{ M}) = 0.0050\text{ M}$$
$$C(SO_4^{2-}) = 1(0.0010\text{ M}) = 0.0010\text{ M}$$
$$C(NO_3^-) = 1(0.0030\text{ M}) = 0.0030\text{ M}$$

Substituting into *(6.42)* gives

$$I = \tfrac{1}{2}[C(Na^+)z(Na^+)^2 + C(SO_4^{2-})z(SO_4^{2-})^2 + C(NO_3^-)z(NO_3^-)^2]$$
$$= \tfrac{1}{2}[(0.0050\text{ M})(+1)^2 + (0.0010\text{ M})(-2)^2 + (0.0030\text{ M})(-1)^2]$$
$$= 0.0060\text{ M}$$

This is the same as the result determined by adding the contributions from Table 6-2.

$$I = I(Na_2SO_4) + I(NaNO_3) = 3C(Na_2SO_4) + C(NaNO_3)$$
$$= 3(0.0010\text{ M}) + 0.0030\text{ M} = 0.0060\text{ M}$$

6.70 Determine the ionic strength for a weak electrolyte such as 0.025 M CH_3COOH, which is only 4.1% ionized.

▌ For a weak electrolyte *(6.43)* becomes

$$C_i = \nu_i \alpha C \tag{6.43b}$$

where α is the degree of ionization for the weak electrolyte. Substituting the data into *(6.43b)* gives

$$C(H^+) = C(CH_3COO^-) = 1(0.041)(0.025\text{ M}) = 0.0010\text{ M}$$

Substituting into *(6.42)* gives

$$I = \tfrac{1}{2}[C(H^+)z(H^+)^2 + C(CH_3OO^-)z(CH_3COO^-)^2]$$
$$= \tfrac{1}{2}[(0.0010\text{ M})(+1)^2 + (0.0010\text{ M})(-1)^2] = 0.0010\text{ M}$$

6.71 Determine the mean ionic concentration of the solutions given in Problems 6.68 and 6.70.

▌ The *mean ionic concentration* ($C_\pm$) is defined as

$$C_\pm = (C_+^{\nu_+} C_-^{\nu_-})^{1/\nu} \tag{6.44}$$

where $\nu = \nu_+ + \nu_-$. Substituting the results of Problems 6.68 and 6.70 into *(6.44)* gives

$$C_\pm(CaCl_2) = [C(Ca^{2+})C(Cl^-)^2]^{1/3} = [(0.0020\text{ M})(0.0040\text{ M})^2]^{1/3} = 0.032\text{ M}$$
$$C_\pm(CH_3COOH) = [C(H^+)C(CH_3COO^-)]^{1/2} = [(0.0010\text{ M})(0.0010\text{ M})]^{1/2} = 0.0010\text{ M}$$

6.72 What concentration of KNO_3 is required to produce a solution having the same ionic strength as 0.010 M $CuSO_4$? What concentration of KNO_3 is required to produce a solution having the same mean ionic concentration as 0.010 M $CuSO_4$?

▌ The two electrolytes are classified as 1:1 and 2:2, respectively. From Table 6-2, $I(KNO_3) = C(KNO_3)$ and $I(CuSO_4) = 4C(CuSO_4)$. Thus,

$$C(KNO_3) = 4C(CuSO_4) = 4(0.010\text{ M}) = 0.040\text{ M}$$

in order for both solutions to have the same ionic strength. Likewise from Table 6-2, $C_\pm(KNO_3) = C(KNO_3)$ and $C_\pm(CuSO_4) = C(CuSO_4)$. Thus,

$$C(KNO_3) = C(CuSO_4) = 0.010\text{ M}$$

in order for both solutions to have the same mean ionic concentration.

6.73 The mean ionic activity coefficient of a 0.494 M aqueous solution of $CaCl_2$ is 0.516. Determine the mean ionic activity.

▌ The *mean ionic activity* ($a_\pm$) is given by

$$a_\pm = y_\pm[C_\pm/(\text{M})] \tag{6.45a}$$

where $y_\pm$ is the *mean ionic activity coefficient* given by

$$y_\pm = (y_+^{\nu_+} y_-^{\nu_-})^{1/\nu} \tag{6.46}$$

For this solution of a 2:1 electrolyte, Table 6-2 gives

$$C_\pm = 4^{1/3}C = 4^{1/3}(0.494\ \text{M}) = 0.784\ \text{M}$$

Substituting into *(6.45)* gives $a_\pm = (0.516)(0.784) = 0.405$.

6.74 The mean ionic activity coefficient considered in Problem 6.73 was based on the molarity concentration scale. Determine the mean ionic activity coefficient and the mean ionic activity on the molality concentration scale for the 0.494 M aqueous solution of $CaCl_2$. For this solution, $\rho = 1.042 \times 10^3\ \text{kg}\cdot\text{m}^3$ and $m = 0.500\ m$. Compare the results of $a_\pm$.

▌ The relation between the mean activity coefficients on the molality scale ($\gamma_\pm$) and the molarity scale is given by

$$\gamma_\pm = y_\pm \frac{C}{\rho_{\text{solvent}} m(10^{-3})} \tag{6.47}$$

where the density is in units of $\text{kg}\cdot\text{m}^{-3}$. Substituting the data gives

$$\gamma_\pm = 0.516 \frac{0.494}{(1.00 \times 10^3)(0.500)(10^{-3})} = 0.510$$

On the molality basis,

$$a_\pm = \gamma_\pm[m_\pm/(\text{m})] \tag{6.45b}$$

In this case $m_\pm = 4^{1/3}m$, which upon substituting into *(6.45b)* gives

$$a_\pm = (0.510)(4^{1/3})(0.500) = 0.405$$

The values of $a_\pm$ determined above and in Problem 6.73 are the same.

6.75 Calculate the activity of $CaCl_2$ in the solution described in Problem 6.73.

▌ The activity of a strong electrolyte $A_{\nu_+}B_{\nu_-}$ in solution is given by

$$a = a_+^{\nu_+}a_-^{\nu_-} \tag{6.48}$$

where

$$a_\pm = (a_+^{\nu_+}a_-^{\nu_-})^{1/\nu} \tag{6.49}$$

For the $CaCl_2$ solution, $a(CaCl_2) = a_+^{\nu} = (0.405)^3 = 0.0664$.

6.76 Calculate $\Delta\mu$ for diluting a solution of HCl at 25 °C from 0.010 m to 0.0010 m assuming $\gamma_\pm = 0.964$ at 0.0010 m and 0.887 at 0.010 m.

▌ The activity of the HCl in each solution is given by Table 6-2 as

$$a(0.010\ \text{m}) = (0.887)^2(0.010)^2 = 7.9 \times 10^{-5}$$

$$a(0.0010\ \text{m}) = (0.964)^2(0.0010)^2 = 9.3 \times 10^{-7}$$

Substituting into *(6.26)* gives

$$\Delta G = \Delta\mu = (8.314\ \text{J}\cdot\text{K}^{-1}\cdot\text{mol}^{-1})(298\ \text{K}) \ln\frac{9.3 \times 10^{-7}}{7.9 \times 10^{-5}} = -11\,000\ \text{J}\cdot\text{mol}^{-1}$$

6.77 Confirm the values of $\gamma_\pm$ given for HCl in Problem 6.76 assuming that the Debye–Hückel limiting law is valid.

▌ The mean activity coefficient can be calculated from the *Debye–Hückel theory* using

$$\log y_\pm = -\frac{z_+z_-AI^{1/2}}{1 + aBI^{1/2}} + bI \tag{6.50}$$

where A and B are parameters dependent on the solvent, a and b are parameters dependent on the solute, and z_i is the absolute value of the ionic valence. For aqueous solutions at 25 °C with concentrations of $I \le 10^{-2}$ M, *(6.50)* reduces to the *Debye–Hückel limiting law* given by

$$\log y_\pm = -z_+z_-(0.5116)I^{1/2} \tag{6.51}$$

where I can be expressed in terms of either molarity or molality. For the HCl solutions, Table 6-2 gives $I = m$. Thus

$$\log[\gamma_\pm(0.010\text{ m})] = -(1)(1)(0.5116)(0.10)^{1/2} = -5.2 \times 10^{-2}$$

$$\gamma_\pm(0.010\text{ m}) = 0.887$$

$$\log[\gamma_\pm(0.0010\text{ m})] = -(1)(1)(0.5116)(0.0010)^{1/2} = -1.6 \times 10^{-2}$$

$$\gamma_\pm(0.0010\text{ m}) = 0.964$$

6.78 Calculate $y(Na^+)$ in the solution described in Problem 6.69.

▌ The ionic strength of the solution is 0.0060 M (see Problem 6.69). Because this value is less than 10^{-2} M, the activity coefficient of the ion can be calculated using a special case of *(6.51)*:

$$\log y_i = -z_i^2(0.5116)I^{1/2} = -(1)^2(0.5116)(0.0060)^{1/2} = -0.040 \qquad (6.52)$$

$$y(Na^+) = 0.912$$

6.79 Calculate γ_+, γ_-, and $\gamma_\pm$ for a 0.0025 m solution of $CaCl_2$ at 25 °C.

▌ From Table 6-2, $I = 3m = 3(0.0025\text{ m}) = 0.0075$ m. Substituting into *(6.51)* and *(6.52)* and taking antilogarithms gives

$$\log \gamma_\pm = -(2)(1)(0.5116)(0.0075)^{1/2} = -0.089$$

$$\gamma_\pm = 0.815$$

$$\log \gamma_+ = -(2)^2(0.5116)(0.0075)^{1/2} = -0.177$$

$$\gamma_+ = 0.665$$

$$\log \gamma_- = -(1)^2(0.5116)(0.0075)^{1/2} = -0.044$$

$$\gamma_- = 0.904$$

As a check, substituting the values of γ_+ and γ_- into *(6.46)* gives

$$\gamma_\pm = [(0.605)(0.904)^2]^{1/3} = 0.816$$

6.80 Show

$$\gamma(ZnSO_4)^2 = \frac{\gamma(ZnCl_2)^3\gamma(Na_2SO_4)^3}{\gamma(NaCl)^4}$$

provided that all mean activity coefficients are determined at the same ionic strength. Calculate $\gamma_\pm(ZnSO_4)$ given $\gamma_\pm = 0.515$ for 0.1 m $ZnCl_2$, 0.445 for 0.1 m $NaSO_4$, and 0.710 for 0.3 m NaCl.

▌ Using *(6.46)* for each of the electrolytes gives

$$\gamma(ZnSO_4)^2 = \frac{\gamma(Zn^{2+})\gamma(Cl^-)^2\gamma(Na^+)^2\gamma(SO_4^{2-})}{[\gamma(Na^+)\gamma(Cl^-)]^2} = \gamma(Na^+)^2\gamma(SO_4^{2-})$$

Substituting the data gives

$$\gamma(ZnSO_4)^2 = \frac{(0.515)^3(0.445)^3}{(0.710)^4} = 0.0474$$

Taking the square root gives $\gamma(ZnSO_4) = 0.218$. The observed value is 0.177.

6.81 Derive a general equation for the radius of the ionic atmosphere as a function of ionic strength. Evaluate this radius for 0.010 m Na_2SO_4. For water at 25 °C, $\rho_{solv} = 997\text{ kg}\cdot\text{m}^{-3}$ and the dielectric constant is 78.54.

▌ The desired radius is given by

$$r = 1/\kappa \qquad (6.53a)$$

where

$$\kappa^2 = \frac{8\pi e^2 L^2 \rho_{solv}}{(4\pi\varepsilon_0)(\varepsilon/\varepsilon_0)RT} I \qquad (6.53b)$$

where e is the charge on an electron ($1.602\,189\,2 \times 10^{-19}$ C), $4\pi\varepsilon_0$ is the permittivity constant ($1.112\,650\,056 \times$

$10^{-10}\ C^2\cdot N^{-1}\cdot m^{-2}$), and $\varepsilon/\varepsilon_0$ is the dielectric constant. Substituting the constants into *(6.53b)* gives

$$\kappa^2 = \frac{8\pi(1.602\times10^{-19}\ C)^2(6.022\times10^{23}\ mol^{-1})^2(997\ kg\cdot m^{-3})}{(1.113\times10^{-10}\ C^2\cdot N^{-1}\cdot m^{-2})(78.54)(8.314\ J\cdot K^{-1}\cdot mol^{-1})(298\ K)[(1\ N\cdot m)/(1\ J)]}I$$

$$= (1.076\times10^{19}\ mol^{-1}\cdot kg\cdot m^{-2})I$$

Using *(6.53a)* gives

$$r = \left(\frac{1}{(1.076\times10^{19}\ mol^{-1}\cdot kg\cdot m^{-2})I}\right)^{1/2} = \frac{3.049\times10^{-10}\ m}{[I/(mol\cdot kg^{-1})]^{1/2}}$$

For a 0.010-m solution of Na_2SO_4, Table 6-2 gives

$$I = 3m = 3(0.010\ m) = 0.030\ mol\cdot kg^{-1}$$

Substituting gives

$$r = \frac{3.049\times10^{-10}\ m}{(0.030)^{1/2}} = 1.76\times10^{-9}\ m = 1.76\ nm$$

6.82 The Debye–Hückel constant A in *(6.50)* is given by

$$A = \frac{e^3L^2}{2.303(4\pi\varepsilon_0)(\varepsilon/\varepsilon_0)RT}\left(\frac{2\pi\rho_{solv}}{4\pi\varepsilon_0(\varepsilon/\varepsilon_0)RT}\right)^{1/2} \qquad (6.54)$$

Determine A for solutions of strong electrolytes in ethanol at 25 °C. For CH_3CH_2OH, $\rho = 785.1\ kg\cdot m^{-3}$ and $\varepsilon/\varepsilon_0 = 24.25$.

▌ Substituting the data into *(6.54)* gives

$$A = \frac{(1.602\times10^{-19}\ C)^3(6.022\times10^{23}\ mol^{-1})^2}{(2.303)(1.113\times10^{-10}\ C^2\cdot N^{-1}\cdot m^{-2})(24.25)(8.314\ J\cdot K^{-1}\cdot mol^{-1})(298\ K)[(1\ N\cdot m)/(1\ J)]}$$
$$\times\left(\frac{2\pi(785.1\ kg\cdot m^{-3})}{(1.113\times10^{-10}\ C^2\cdot N^{-1}\cdot m^{-2})(24.25)(8.314\ J\cdot K^{-1}\cdot mol^{-1})(298\ K)[(1\ N\cdot m)/(1\ J)]}\right)^{1/2}$$
$$= 2.627\ kg^{1/2}\cdot mol^{1/2}$$

6.83 Prepare a plot of ln $\gamma_\pm$ against $m^{1/2}$ for $CaCl_2$ at 25 °C.

$m/(mol\cdot kg^{-1})$	0.000	0.005	0.010	0.020	0.050	0.100	0.200	0.500	1.000	2.000
$\gamma_\pm$	1.000	0.789	0.732	0.669	0.584	0.524	0.491	0.510	0.725	1.554

Also plot the curves predicted by *(6.50)* and *(6.51)* given $A = 0.5116$, $B = 0.329\times10^{10}\ m^{-1}$, and $a = 5.24\times10^{-10}\ m$.

▌ For $CaCl_2$, $I = 3m$ (see Table 6-2). Substituting into *(6.50)* and *(6.51)* gives

$$\log\gamma_\pm = -\frac{(2)(1)(0.5116)(3m)^{1/2}}{1+(5.24\times10^{-10}\ m)(0.329\times10^{10}\ m^{-1})(3m)^{1/2}} = -\frac{1.772m^{1/2}}{1+2.99m^{1/2}}$$

$$\log\gamma_\pm = -(2)(1)(0.5116)(3m)^{1/2} = -1.772m^{1/2}$$

As a sample calculation, for $m = 0.050$ m,

$$\log\gamma_\pm = -\frac{1.772(0.050)^{1/2}}{1+2.99(0.050)^{1/2}} = -0.237$$

$$\log\gamma_\pm = -1.772(0.050)^{1/2} = -0.396$$

The graph is shown in Fig. 6-11. Note that the values calculated using *(6.51)* agree with the experimental values only at very low concentrations and that the values calculated using *(6.50)* agree with the experimental values over a larger concentration range.

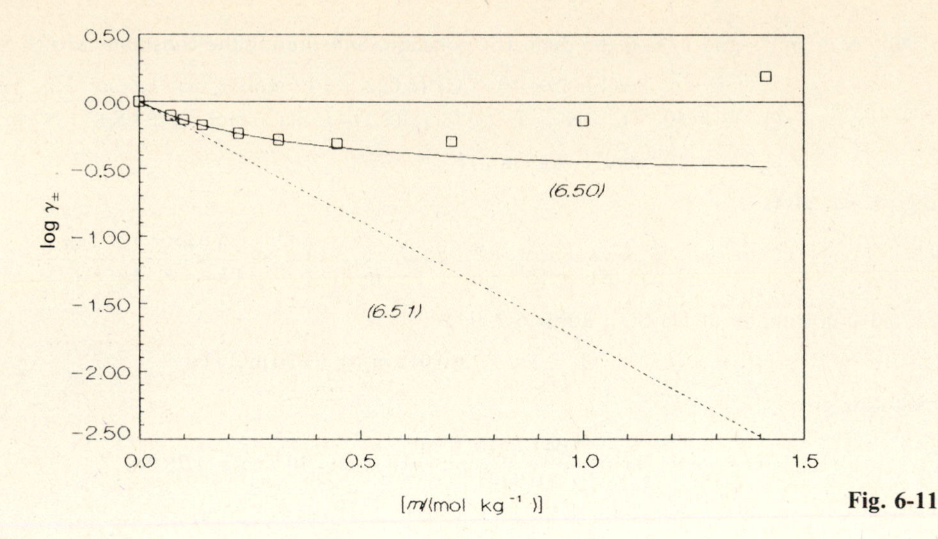

Fig. 6-11

6.84 What is ΔG for the reaction

$$N_2(g) + 3H_2(g) \rightarrow 2NH_3(g) \qquad \Delta_r G^\circ_{298} = -32.90\ \text{kJ}$$

if it is carried out under pressure conditions such that $a(NH_3) = 0.2$, $a(N_2) = a(H_2) = 10.0$? Is this reaction more or less favorable under these conditions?

▮ The relation between $\Delta_r G^\circ$ and $\Delta_r G$ is given by

$$\Delta_r G_T = \Delta_r G^\circ_T + RT \ln Q \tag{6.55}$$

where Q is the *reaction quotient* defined as

$$Q = \prod_i a_{I_i}^{\nu_i} \tag{6.56}$$

where ν_i has been defined in Problem 4.46 and I_i is the chemical formula for the reactant or product. For this reaction, *(6.56)* gives

$$Q = \frac{[a(NH_3)]^2}{a(N_2)[a(H_2)]^3}$$

which upon substituting into *(6.55)* gives

$$\Delta_r G_T = -32.90\ \text{kJ} + (8.314\ \text{J}\cdot\text{K}^{-1}\cdot\text{mol}^{-1})\frac{10^{-3}\ \text{kJ}}{1\ \text{J}}(298.15\ \text{K})(\text{mol}) \ln \frac{(0.2)^2}{(10.0)(10.0)^3}$$

$$= -63.71\ \text{kJ}$$

Under these nonstandard state conditions, the reaction has become more favorable. Note: Although the activities used in *(6.55)* and *(6.56)* are dimensionless, each stoichiometric coefficient (ν_i) has the unit of mole, which will generate the unit of mol shown in front of the logarithm term above.

6.85 Calculate $\Delta_r G$ at 25 °C for $H_2O(l) \rightarrow H^+(aq) + OH^-(aq)$, given $C(H^+) = C(OH^-) = 1 \times 10^{-7}$ M.

▮ Substituting the data from Table 6-1 into *(6.16)* gives

$$\Delta_r G^\circ = [(1\ \text{mol})\ \Delta_f G^\circ(H^+) + (1\ \text{mol})\ \Delta_f G^\circ(OH^-)] - [(1\ \text{mol})\ \Delta_f G^\circ(H_2O)]$$

$$= [(1\ \text{mol})(0) + (1\ \text{mol})(-157.244\ \text{kJ}\cdot\text{mol}^{-1})] - [(1\ \text{mol})(-237.129\ \text{kJ}\cdot\text{mol}^{-1})]$$

$$= 79.885\ \text{kJ}$$

Assuming $a(H_2O) = 1$, *(6.56)* gives

$$Q = \frac{a(H^+)a(OH^-)}{a(H_2O)} = \frac{\gamma(H^+)C(H^+)\gamma(OH^-)C(OH^-)}{1} = \gamma_\pm^2 C(H^+)^2$$

For this dilute solution, $\gamma_\pm = 1$. Substituting into *(6.55)* gives

$$\Delta_r G = 79.885\ \text{kJ} + (8.314\ \text{J}\cdot\text{K}^{-1}\cdot\text{mol}^{-1})\frac{10^{-3}\ \text{kJ}}{1\ \text{J}}(298.15\ \text{K})(\text{mol}) \ln [(1)^2(1 \times 10^{-7})^2]$$

$$= 0\ \text{kJ}$$

6.86 For the reaction $D_2O(l) \rightarrow D_2O(g)$ $(\Delta_r G^\circ_{298} = 8.904\ \text{kJ})$, what is the vapor pressure of D_2O at 25 °C?

▮ For an equilibrium process, $\Delta G = 0$. Assuming ideal behavior, $a(g) = [P/(\text{bar})]$ and $a(l) = 1$. Substituting into *(6.56)* gives

$$Q = a(g)/a(l) = [P/(\text{bar})]$$

which upon substituting into *(6.55)* and solving for P gives

$$0 = 8.904\ \text{kJ} + (8.314\ \text{J}\cdot\text{K}^{-1}\cdot\text{mol}^{-1})\frac{10^{-3}\ \text{kJ}}{1\ \text{J}}(298.15\ \text{K})(\text{mol})\ln[P/(\text{bar})]$$

$$\ln P = -3.592 \qquad P = 2.75 \times 10^{-2}\ \text{bar}$$

6.87 At 25 °C, the pressure of water vapor over an equilibrium mixture of $CuSO_4{\cdot}5H_2O(s)$ and $CuSO_4{\cdot}3H_2O(s)$ is 7.8 torr. Calculate $\Delta_r G^\circ$ for

$$CuSO_4{\cdot}5H_2O(s) \rightarrow CuSO_4{\cdot}3H_2O(s) + 2H_2O(g)$$

▮ For this reaction, $a(CuSO_4{\cdot}5H_2O) = a(CuSO_4{\cdot}3H_2O) = 1$ and *(6.28)* gives

$$a(H_2) = (7.8\ \text{torr})\frac{1\ \text{atm}}{760\ \text{torr}}\left(\frac{1.013\,25\ \text{bar}}{1\ \text{atm}}\right) = 0.0104\ \text{bar}$$

Substituting $\Delta G = 0$ and the data into *(6.55)* gives

$$0 = \Delta_r G^\circ + (8.314\ \text{J}\cdot\text{K}^{-1}\cdot\text{mol}^{-1})\frac{10^{-3}\ \text{kJ}}{1\ \text{J}}(298.15\ \text{K})(\text{mol})\ln(0.0104)^2$$

$$\Delta_r G^\circ = 22.64\ \text{kJ}$$

6.88 Prepare a plot of $\Delta_r G_{500}$ against the *progress* (or *advancement*) *of reaction* ξ for

$$Br_2(g) + Cl_2(g) \rightarrow 2BrCl(g) \qquad \Delta_r G^\circ_{500} = -7.388\ \text{kJ}$$

Assume that the original reaction mixture contains 1 mol each of Br_2 and Cl_2. Discuss the plot.

▮ Letting ξ represent the amount of Br_2 (and Cl_2) that reacts, the amounts of substances present at any time will be $(1 - \xi)$ mol for both Br_2 and Cl_2 and 2ξ mol for BrCl. The total number of moles will be $(1 - \xi)$ mol + $(1 - \xi)$ mol + 2ξ mol = 2 mol, giving the mole fractions as

$$x(Br_2) = x(Cl_2) = (1 - \xi)/2 \qquad x(BrCl) = 2\xi/2 = \xi$$

and partial pressures as

$$P(Br_2) = P(Cl_2) = \frac{1 - \xi}{2}(1\ \text{bar}) \qquad P(BrCl) = \xi(1\ \text{bar})$$

The reaction quotient for this reaction is given by *(6.56)* as

$$Q = \frac{a(BrCl)^2}{a(Br_2)a(Cl_2)} = \frac{P(BrCl)^2}{P(Br_2)P(Cl_2)}$$

$$= \frac{[\xi(1\ \text{bar})]^2}{[\frac{1}{2}(1 - \xi)(1\ \text{bar})][\frac{1}{2}(1 - \xi)(1\ \text{bar})]} = \frac{4\xi^2}{(1 - \xi)^2}$$

where the activities of the gases are assumed to be equal to the partial pressures. Substituting into *(6.55)* gives

$$\Delta_r G_{500} = -7.388\ \text{kJ} + (8.314\ \text{J}\cdot\text{K}^{-1}\cdot\text{mol}^{-1})\frac{10^{-3}\ \text{kJ}}{1\ \text{J}}(500\ \text{K})(\text{mol})\ln\frac{4\xi^2}{(1 - \xi)^2}$$

As a sample calculation, consider $\xi = 0.50$:

$$\Delta_r G_{500} = -7.388\ \text{kJ} + (8.314\ \text{J}\cdot\text{K}^{-1}\cdot\text{mol}^{-1})\frac{10^{-3}\ \text{kJ}}{1\ \text{J}}(500\ \text{K})(\text{mol})\ln\frac{4(0.050)^2}{(1 - 0.50)^2}$$

$$= -1.625\ \text{kJ}$$

The plot is shown in Fig. 6-12. At the beginning of the reaction $\xi = 0$ and the value of $\Delta_r G$ is always negative, implying that the reaction will occur. As ξ reaches the equilibrium value, in this case $\xi = 0.55$, $\Delta_r G$ approaches zero and the net reaction stops (dynamic equilibrium is established). For values of ξ greater than the equilibrium value, $\Delta_r G$ will be positive and further reaction will not occur. (For reactions with larger values of $\Delta_r G$, plotting $\Delta_r G$ against $\ln \xi$ will generate the same type of curve.)

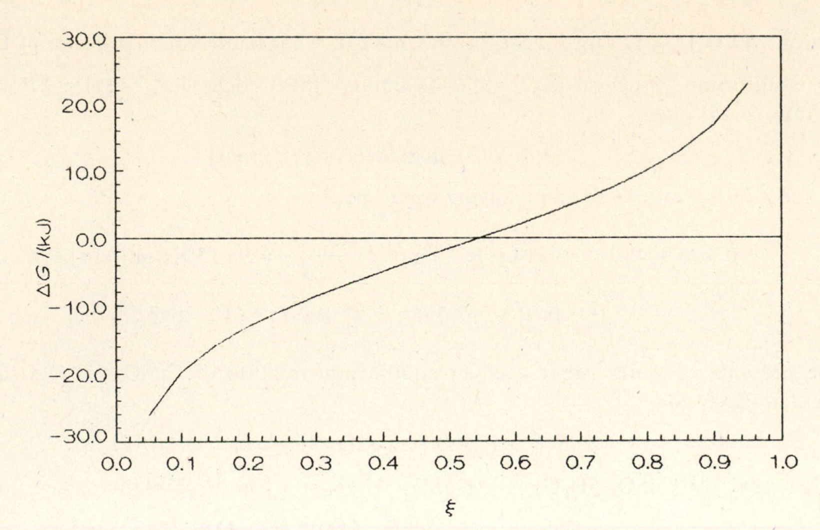

Fig. 6-12

6.89 Prepare a plot of $G - \mu°(\mathrm{Br_2}) - \mu°(\mathrm{Cl_2})$ against ξ for the reaction described in Problem 6.88. Discuss the plot.

▮ For each substance, *(6.27)* gives

$$\mu_i = \mu_i^\circ + RT \ln a_i = \mu_i^\circ + RT \ln P_i$$

assuming the activities of the gases to be equal to the partial pressures. Using *(6.23)* and the results of Problem 6.88 gives

$$\begin{aligned} G = \sum n_i\mu_i &= (1-\xi)\mu(\mathrm{Br_2}) + (1-\xi)\mu(\mathrm{Cl_2}) + 2\xi\mu(\mathrm{BrCl}) \\ &= (1-\xi)\left(\mu°(\mathrm{Br_2}) + RT\ln\frac{(1-\xi)(1\text{ bar})}{2}\right) \\ &\quad + (1-\xi)\left(\mu°(\mathrm{Cl_2}) + RT\ln\frac{(1-\xi)(1\text{ bar})}{2}\right) \\ &\quad + 2\xi\{\mu°(\mathrm{BrCl}) + RT\ln[\xi(1\text{ bar})]\} \\ &= \mu°(\mathrm{Br_2}) + \mu°(\mathrm{Cl_2}) + \xi[2\mu°(\mathrm{BrCl}) - \mu°(\mathrm{Br_2}) - \mu°(\mathrm{Cl_2})] \\ &\quad + 2RT\left[(1-\xi)\ln\frac{(1-\xi)(1\text{ bar})}{2} + \xi\ln[\xi(1\text{ bar})]\right] \\ &= \mu°(\mathrm{Br_2}) + \mu°(\mathrm{Cl_2}) + \xi\,\Delta_r G° + 2RT\left[(1-\xi)\ln\frac{1-\xi}{2} + \xi\ln\xi\right] \end{aligned}$$

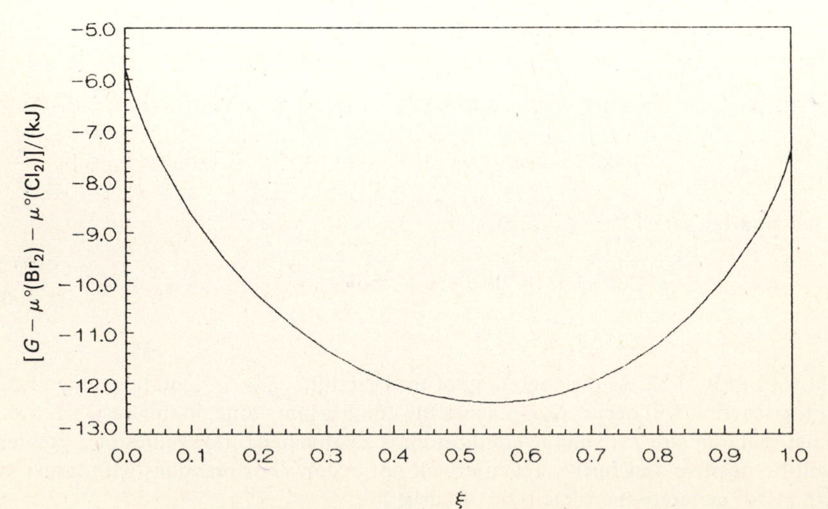

Fig. 6-13

As a sample calculation, consider $\xi = 0.50$:

$$G - \mu°(\text{Br}_2) - \mu°(\text{Cl}_2) = (0.50)(-7.388\ \text{kJ}) + 2(8.314\ \text{J}\cdot\text{K}^{-1}\cdot\text{mol}^{-1})\frac{10^{-3}\ \text{kJ}}{1\ \text{J}}(500\ \text{K})(\text{mol})$$

$$\times\left[(1-0.50)\ln\frac{1-0.50}{2} + (0.50)\ln 0.50\right]$$

$$= -12.339\ \text{kJ}$$

The plot is shown in Fig. 6-13. The value at $\xi = 0$ is equal to $2RT\ln(1/2)$, and the value at $\xi = 1$ is equal to $\Delta_r G°_{500}$. The curve reaches a minimum at $\xi = 0.55$ where equilibrium is established (see Problem 6.88). (For reactions with larger values of $\Delta_r G°$ plotting $G - \sum \mu_i°$ against $\ln \xi$ will generate the same type of curve.)

6.3 THERMODYNAMIC RELATIONS

6.90 Derive the Maxwell relation for $(\partial T/\partial V)_S$.

Table 6-3

$$dU = T\,dS - P\,dV \qquad dG = -S\,dT + V\,dP$$

$$dH = T\,dS + V\,dP \qquad dA = -S\,dT - P\,dV$$

$$\left(\frac{\partial x}{\partial y}\right)_w = \left(\frac{\partial x}{\partial z}\right)_w\left(\frac{\partial z}{\partial y}\right)_w \qquad \left(\frac{\partial x}{\partial y}\right)_z\left(\frac{\partial y}{\partial z}\right)_x\left(\frac{\partial z}{\partial x}\right)_y = 1$$

$$\left(\frac{\partial z}{\partial x}\right)_y = \frac{1}{(\partial x/\partial z)_y}$$

$$\left(\frac{\partial H}{\partial T}\right)_P = C_P \qquad \left(\frac{\partial E}{\partial T}\right)_V = C_V \qquad \left(\frac{\partial T}{\partial P}\right)_H = \mu$$

$$\alpha = \frac{1}{V}\left(\frac{\partial V}{\partial T}\right)_P \qquad \kappa = -\frac{1}{V}\left(\frac{\partial V}{\partial P}\right)_T \qquad \gamma = C_P/C_V$$

▮ From Table 6-3, $dU = T\,dS - P\,dV$. At constant V,

$$\left(\frac{\partial U}{\partial S}\right)_V = T$$

and at constant S,

$$\left(\frac{\partial U}{\partial V}\right)_S = -P$$

Differentiating the first equation with respect to V at constant S, and the second equation with respect to S at constant V, gives

$$\left[\frac{\partial}{\partial V}\left(\frac{\partial U}{\partial S}\right)_V\right]_S = \left(\frac{\partial T}{\partial V}\right)_S \qquad \left[\frac{\partial}{\partial S}\left(\frac{\partial U}{\partial V}\right)_S\right]_V = \left(\frac{\partial(-P)}{\partial S}\right)_V$$

But U is a state function, and one of the properties of such a function is that the order of differentiation is not important. Hence, the above second derivatives are equal, which gives

$$\left(\frac{\partial T}{\partial V}\right)_S = -\left(\frac{\partial P}{\partial S}\right)_U$$

6.91 A general format for the *Maxwell relations* is given by

$$\left(\frac{\partial W}{\partial X}\right)_Z = \pm\left(\frac{\partial Y}{\partial Z}\right)_X \qquad (6.57)$$

where the cross products XY and WZ must be work terms, i.e., PV and ST, and the negative sign is chosen if W and X or Y and Z are T and V. Use this equation to determine the Maxwell relation for $(\partial T/\partial V)_S$.

▌ Letting $W = T$ and $X = V$ in *(6.57)*, then $Y = P$ and $Z = S$. Choosing the negative sign because W and X are T and V gives

$$\left(\frac{\partial T}{\partial V}\right)_S = -\left(\frac{\partial P}{\partial S}\right)_V$$

Note that this is the same result as in Problem 6.90.

6.92 Show that $C_P = T(\partial S/\partial T)_P$ and $C_V = T(\partial S/\partial T)_V$.

▌ Under constant-pressure conditions, $đq = dH = C_P\,dT$. Substituting into *(5.6)* gives

$$dS = \frac{đq}{T} = \frac{C_P\,dT}{T} \qquad \left(\frac{\partial S}{\partial T}\right)_P = \frac{C_P}{T}$$

$$C_P = T\left(\frac{\partial S}{\partial T}\right)_P \tag{6.58a}$$

Likewise, under constant-volume conditions, $đq = dU = C_V\,dT$, giving

$$C_V = T\left(\frac{\partial S}{\partial T}\right)_V \tag{6.58b}$$

6.93 Show that $(\partial U/\partial V)_S = (\partial A/\partial V)_T$.

▌ From the equation for dU and dA in Table 6-3,

$$\left(\frac{\partial U}{\partial V}\right)_S = P \qquad \left(\frac{\partial A}{\partial V}\right)_T = P$$

Equating these results gives the desired equation.

6.94 Show that $C_P = -T(\partial^2 G/\partial T^2)_P$.

▌ From the equation for dG in Table 6-3,

$$\left(\frac{\partial G}{\partial T}\right)_P = -S$$

Taking a second derivative with respect to T and substituting *(6.58a)* gives

$$\left(\frac{\partial^2 G}{\partial T^2}\right)_P = -\left(\frac{\partial S}{\partial T}\right)_P = -\frac{C_P}{T}$$

Rearranging gives the desired result.

6.95 Show that $(\partial C_P/\partial P)_T = -T(\partial^2 V/\partial T^2)_P$.

▌ Using the definition for C_P, the equation for dH given in Table 6-3, and *(6.57)* gives

$$\left(\frac{\partial C_P}{\partial P}\right)_T = \left[\frac{\partial}{\partial P}\left(\frac{\partial H}{\partial T}\right)_P\right]_T = \left[\frac{\partial}{\partial T}\left(\frac{\partial H}{\partial P}\right)_T\right]_P = \left(\frac{\partial}{\partial T}\left[T\left(\frac{\partial S}{\partial P}\right)_T + V\right]\right)_P = \left(\frac{\partial}{\partial T}\left[T\left(-\frac{\partial V}{\partial T}\right)_P + V\right]\right)_P$$

$$= -T\left(\frac{\partial^2 V}{\partial T^2}\right)_P + \left(-\frac{\partial V}{\partial T}\right)_P + \left(\frac{\partial V}{\partial T}\right)_P = -T\left(\frac{\partial^2 V}{\partial T^2}\right)_P$$

6.96 Show that $(\partial\alpha/\partial P)_T = -(\partial\kappa/\partial T)_P$.

▌ Using the definition of α and κ given in Table 6-3 gives

$$\left(\frac{\partial\alpha}{\partial P}\right)_T = \left[\frac{\partial}{\partial P}\left(\frac{1}{V}\right)\left(\frac{\partial V}{\partial T}\right)_P\right]_T = \frac{1}{V}\left[\frac{\partial}{\partial P}\left(\frac{\partial V}{\partial T}\right)_P\right]_T = \frac{1}{V}\left[\frac{\partial}{\partial T}\left(\frac{\partial V}{\partial P}\right)_T\right]_P = \left[\frac{\partial}{\partial T}\left(\frac{1}{V}\right)\left(\frac{\partial V}{\partial P}\right)_T\right]_P$$

$$= \left[\frac{\partial(-\kappa)}{\partial T}\right]_P = -\left(\frac{\partial\kappa}{\partial T}\right)_P$$

6.97 Show that $(\partial S/\partial P)_V = \kappa C_V/\alpha T$.

▌ Using the properties of derivatives given in Table 6-3, *(6.58b)*, and the definitions of α and κ gives

$$\left(\frac{\partial S}{\partial P}\right)_V = \left(\frac{\partial S}{\partial T}\right)_V\left(\frac{\partial T}{\partial P}\right)_V = \frac{C_V}{T}\left(\frac{\partial T}{\partial P}\right)_V = \frac{C_V}{T}\left(\frac{-(\partial V/\partial P)_T}{(\partial V/\partial T)_P}\right) = \frac{C_V}{T}\left(\frac{-(1/V)(\partial V/\partial P)_T}{(1/V)(\partial V/\partial T)_P}\right) = C_V\kappa/T\alpha$$

6.98 Show that $(\partial S/\partial V)_P = C_P/\alpha TV$.

Using the properties of derivatives given in Table 6-3, *(6.58a)*, and the definition of α gives

$$\left(\frac{\partial S}{\partial P}\right)_V = \left(\frac{\partial S}{\partial T}\right)_P\left(\frac{\partial T}{\partial V}\right)_P = \frac{C_P}{T}\left(\frac{\partial T}{\partial V}\right)_P = \frac{C_P}{T}\frac{1}{(\partial V/\partial T)_P} = \frac{C_P}{T}\frac{1}{V(1/V)(\partial V/\partial T)_P} = C_P/TV\alpha$$

6.99 The isentropic expansion is defined as

$$\alpha_S = -\frac{1}{V}\left(\frac{\partial V}{\partial T}\right)_S$$

Show that $\alpha_S\alpha = -C_V\kappa/TV$.

Substituting the definitions, using the properties of derivatives given in Table 6-3, using *(6.58b)*, and substituting the definition of the compressibility gives

$$\alpha_S\alpha = -\frac{1}{V}\left(\frac{\partial V}{\partial T}\right)_S\left(\frac{1}{V}\right)\left(\frac{\partial V}{\partial T}\right)_P = \frac{1}{V^2}\left(\frac{\partial V}{\partial T}\right)_P\frac{-(\partial S/\partial T)_V}{(\partial S/\partial T)_T} = -\frac{1}{V^2}\left(\frac{\partial V}{\partial T}\right)_P\left(\frac{\partial V}{\partial S}\right)_T\frac{C_V}{T}$$

$$= -\frac{1}{V^2}\left(-\frac{\partial S}{\partial P}\right)_T\left(\frac{\partial V}{\partial S}\right)_T\frac{C_V}{T} = \frac{1}{V^2}\left(\frac{\partial V}{\partial P}\right)_T\frac{C_V}{T} = \frac{-\kappa C_V}{VT}$$

6.100 The isentropic compressibility is defined as

$$\kappa_S = -\frac{1}{V}\left(\frac{\partial V}{\partial P}\right)_S$$

Show that $\kappa_S/\kappa = 1/\gamma$.

Substituting the definitions, using the properties of derivatives given in Table 6-3, and substituting *(6.58)* gives

$$\frac{\kappa_S}{\kappa} = \frac{-(1/V)(\partial V/\partial P)_S}{-(1/V)(\partial V/\partial P)_T} = \frac{(\partial V/\partial P)_S}{(\partial V/\partial P)_T} = \frac{-(\partial S/\partial P)_V/(\partial S/\partial V)_P}{-(\partial T/\partial P)_V/(\partial T/\partial V)_P}$$

$$= \frac{(\partial S/\partial P)_V(\partial P/\partial T)_V}{(\partial S/\partial V)_P(\partial V/\partial T)_P} = \frac{(\partial S/\partial T)_V}{(\partial S/\partial T)_P} = \frac{C_V/T}{C_P/T} = \frac{C_V}{C_P} = \frac{1}{\gamma}$$

6.101 Show that $\kappa - \kappa_S = \alpha^2VT/C_P$.

Substituting *(3.29)* into the result of Problem 6.100 and rearranging gives

$$\kappa_S = \kappa\frac{C_V}{C_P} = \frac{\kappa}{C_P}\left(C_P - \frac{\alpha^2VT}{\kappa}\right) = \kappa - \frac{\alpha^2VT}{C_P}$$

$$\kappa - \kappa_S = \alpha^2VT/C_P$$

6.102 Show that $\mu = (\partial H/\partial n_i)_{S,P,n_{i\neq j}} = (\partial A/\partial n_i)_{T,V,n_{i\neq j}} = (\partial U/\partial n_i)_{S,V,n_{i\neq j}}$.

For open systems, the expressions for dU, dH, and dA given in Table 6-3 become

$$dU = T\,dS - P\,dV + \sum_i\left(\frac{\partial U}{\partial n}\right)_{S,V,n_{i\neq j}}dn_i$$

$$dH = T\,dS + V\,dP + \sum_i\left(\frac{\partial H}{\partial n}\right)_{S,P,n_{i\neq j}}dn_i$$

$$dA = -S\,dT - P\,dV + \sum_i\left(\frac{\partial A}{\partial n}\right)_{T,V,n_{i\neq j}}dn_i$$

Upon comparison to *(6.23)* and *(6.24)*, μ is defined by the derivatives following the summation signs.

CHAPTER 7
Chemical Equilibrium

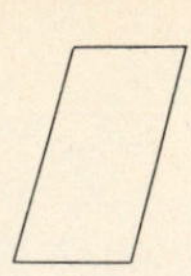

7.1 EQUILIBRIUM CONSTANTS

7.1 For the chemical equation $O_3(g) + OH(g) \rightarrow H(g) + 2O_2(g)$, $\Delta_r G° = 5.8$ kJ at 298 K, 0 J at 354 K, and −2.0 kJ at 373 K. Calculate the value of the equilibrium constant for this reaction at each of these temperatures.

▌ Under equilibrium conditions, $\Delta G = 0$ and $Q = K$. Thus *(6.55)* becomes

$$\ln K = \frac{-\Delta_r G°}{RT} \tag{7.1a}$$

or

$$K = e^{-\Delta_r G°/RT} \tag{7.1b}$$

Substituting the respective values of $\Delta_r G°$ and temperatures gives

$$K_{298} = \exp\left(-\frac{(5.8\ \text{kJ})[(10^3\ \text{J})/(1\ \text{kJ})]}{(8.314\ \text{J}\cdot\text{K}^{-1}\cdot\text{mol}^{-1})(298\ \text{K})(\text{mol})}\right) = 0.096$$

$$K_{354} = e^{-(0)(10^3)/(8.314)(354)} = 1.0$$

$$K_{373} = e^{-(-2.0)(10^3)/(8.314)(373)} = 1.9$$

7.2 Compare the relative partial pressures of the gases under the three different equilibrium systems described in Problem 7.1.

▌ The equilibrium constant expression for this reaction is given by *(6.56)* as

$$K = \prod_i a_i^{\nu_i} \tag{7.2}$$

$$K = \frac{a(H)a(O_2)^2}{a(O_3)a(OH)}$$

Assuming ideal gas behavior, *(6.28)* gives $a_i = P_i/(\text{bar})$. Thus the equilibrium constant expression becomes

$$K_P = \frac{[P(H)/(\text{bar})][P(O_2)/(\text{bar})]^2}{[P(O_3)/(\text{bar})][P(OH)/(\text{bar})]}$$

where the subscript notation on K refers to pressure. At 298 K, where the reaction is not spontaneous,

$$K = \frac{[P(H)/(\text{bar})][P(O_2)/(\text{bar})]^2}{[P(O_3)/(\text{bar})][P(OH)/(\text{bar})]} = 0.096 < 1$$

$$[P(H)/(\text{bar})][P(O_2)/(\text{bar})]^2 < [P(O_3)/(\text{bar})][P(OH)/(\text{bar})]$$

Likewise at 354 K,

$$[P(H)/(\text{bar})][P(O_2)/(\text{bar})]^2 = [P(O_3)/(\text{bar})][P(OH)/(\text{bar})]$$

and at 373 K,

$$[P(H)/(\text{bar})][P(O_2)/(\text{bar})]^2 > [P(O_3)/(\text{bar})][P(OH)/(\text{bar})]$$

7.3 Calculate the value of K for the equation

$$CO(g) + \tfrac{1}{2}O_2(g) \rightarrow CO_2(g) \qquad \Delta_r G°_{298} = -257.191\ \text{kJ}$$

What is the relation of the value of K for the equation

$$2CO(g) + O_2(g) \rightarrow 2CO_2(g)$$

to that for the first reaction?

▌ For the first reaction, the equilibrium constant K_1 is given by *(7.1b)* as

$$K_1 = \exp\left(-\frac{(257.191\ \text{kJ})[(10^3\ \text{J})/(1\ \text{kJ})]}{(8.314\ \text{J}\cdot\text{K}^{-1}\cdot\text{mol}^{-1})(298.15\ \text{K})(\text{mol})}\right) = 1.14 \times 10^{45}$$

The second equation is simply double the first. Thus $\Delta_r G^\circ_{298} = -514.382$ kJ. Substituting into *(7.1b)* gives

$$K_2 = e^{-(-514.382)(10^3)/(8.314)(298.15)} = 1.31 \times 10^{90}$$

The general rule is that if two equations describing the same reaction are related by a multiplication factor of k, then the equilibrium constants are related by the power of k. In this case,

$$K_2 = 1.31 \times 10^{90} = (1.14 \times 10^{45})^2 = K_1^2$$

7.4 Calculate the value of K at 298 K for the equation

$$CO_2(g) \rightarrow CO(g) + \tfrac{1}{2}O_2(g) \qquad \Delta_r G^\circ_{298} = 257.191 \text{ kJ}$$

What is the relation of the value of K for this equation to that for K_1 for the first equation in Problem 7.3?

▮ Using *(7.1b)* gives

$$K = \exp\left(-\frac{(257.191 \text{ kJ})[(10^3 \text{ J})/(1 \text{ kJ})]}{(8.314 \text{ J} \cdot \text{K}^{-1} \cdot \text{mol}^{-1})(298.15 \text{ K})(\text{mol})}\right) = 8.75 \times 10^{-46}$$

This chemical equation is simply the reverse of the first equation in Problem 7.3. The general rule is that if two equations describing the same reaction are the reverse of each other, then the equilibrium constants are inverses of each other (a special case of the general rule given in Problem 7.4 where $k = -1$). In this case,

$$K = 8.75 \times 10^{-46} = \frac{1}{1.14 \times 10^{45}} = \frac{1}{K_1}$$

7.5 Calculate the value of K at 298 K for each of the following equations:

$$CO(g) \rightarrow \tfrac{1}{2}O_2(g) + C(\text{graphite}) \qquad \Delta_r G^\circ_{298} = 137.168 \text{ kJ}$$

$$C(\text{graphite}) + O_2(g) \rightarrow CO_2(g) \qquad \Delta_r G^\circ_{298} = -394.359 \text{ kJ}$$

What is the relation of these equilibrium constants and K_1 for the first equation in Problem 7.3?

▮ Using *(7.1b)* for each equation gives

$$K' = \exp\left(-\frac{(137.168 \text{ kJ})[(10^3 \text{ J})/(1 \text{ kJ})]}{(8.314 \text{ J} \cdot \text{K}^{-1} \cdot \text{mol}^{-1})(298.15 \text{ K})(\text{mol})}\right) = 9.31 \times 10^{-25}$$

$$K'' = \exp[-(-394.359)(10^3)/(8.314)(298.15)] = 1.23 \times 10^{69}$$

The chemical equation in Problem 7.3 is the sum of these two equations. The general rule is that if two equations can be added using Hess's law to give another equation, the equilibrium constant of the latter will be given by the product of the equilibrium constants of the two equations. In this case,

$$K'K'' = (9.31 \times 10^{-25})(1.23 \times 10^{69}) = 1.14 \times 10^{45} = K_1$$

7.6 Use the following sets of equilibrium partial pressure data at 731 K to calculate K for the equation $H_2(g) + I_2(g) \rightarrow 2HI(g)$.

	$P(H_2)/(\text{bar})$	$P(I_2)/(\text{bar})$	$P(HI)/(\text{bar})$
Trial 1	0.276 18	0.064 38	0.9387
Trial 2	0.100 27	0.103 06	0.7176

▮ For this equation, *(7.2)* gives

$$K = \frac{a(HI)^2}{a(H_2)a(I_2)} = \frac{[P(HI)/(\text{bar})]^2}{[P(H_2)/(\text{bar})][P(I_2)/(\text{bar})]}$$

assuming ideal gas behavior. Substituting the two sets of data gives

$$K = \frac{(0.9387)^2}{(0.276\,18)(0.064\,38)} = 49.56 \qquad K = \frac{(0.7176)^2}{(0.100\,27)(0.103\,06)} = 49.83$$

The average value is 49.70.

7.7 Use the results of Problem 7.6 to calculate $\Delta_f G^\circ_{m,731}$ for HI.

▮ Rearranging *(7.1a)* and substituting the data gives

$$\Delta_r G^\circ_{731} = -(8.314\ \text{J}\cdot\text{K}^{-1}\cdot\text{mol}^{-1})\frac{10^{-3}\ \text{kJ}}{1\ \text{J}}(731\ \text{K})(\text{mol})\ \ln 49.70 = -23.74\ \text{kJ}$$

Dividing this result by 2 mol gives $\Delta_f G^\circ_{m,731} = -11.37\ \text{kJ}\cdot\text{mol}^{-1}$.

7.8 At 400 °C and 350 bar, a 1:3 mixture of nitrogen and hydrogen reacts to form an equilibrium mixture containing $x(NH_3) = 0.50$. Assuming ideal gas behavior, calculate K for $N_2(g) + 3H_2(g) \rightarrow 2NH_3(g)$.

▮ At equilibrium,

$$x(NH_3) = 0.50 \qquad x(NH_3) + x(N_2) + x(H_2) = 1.00 \qquad x(H_2) = 3x(N_2)$$

Solving simultaneously gives $x(NH_3) = 0.50$, $x(N_2) = 0.13$, and $x(H_2) = 0.38$. The partial pressures are given by *(1.23)* as

$$P(NH_3) = (0.50)(350\ \text{bar}) = 175\ \text{bar}$$
$$P(N_2) = (0.13)(350) = 46\ \text{bar}$$
$$P(H_2) = (0.38)(350) = 137\ \text{bar}$$

Using *(7.2)* and substituting the data gives

$$K = \frac{a(NH_3)^2}{a(N_2)a(H_2)^3} = \frac{[P(NH_3)/(\text{bar})]^2}{[P(N_2)/(\text{bar})][P(H_2)/(\text{bar})]^3} = \frac{(175)^2}{46(137)^3} = 2.6 \times 10^{-4}$$

7.9 What is the value of K for the equation $Cl_2(g) \rightarrow 2Cl(g)$ when the system contains equal numbers of Cl atoms and Cl_2 molecules? For Cl_2, this equilibrium condition is reached at 1990 K. Calculate $\Delta_r G^\circ_{1990}$.

▮ Equal numbers of atoms and molecules means equal numbers of moles of atoms and molecules, and hence $x(Cl_2) = x(Cl) = 0.50$. Using *(7.2)* and *(1.23)* gives

$$K = \frac{a(Cl)^2}{a(Cl_2)} = \frac{[P(Cl)/(\text{bar})]^2}{[P(Cl_2)/(\text{bar})]} = \frac{\{x(Cl)[P_t/(\text{bar})]\}^2}{x(Cl_2)[P_t/(\text{bar})]} = \frac{[(0.50)(1.00)]^2}{0.50(1.00)} = 0.50$$

Rearranging *(7.1a)* and substituting the data gives

$$\Delta_r G^\circ_{1990} = -(8.314\ \text{J}\cdot\text{K}^{-1}\cdot\text{mol}^{-1})\frac{10^{-3}\ \text{kJ}}{1\ \text{J}}(1990\ \text{K})(\text{mol})\ \ln 0.50 = 11.5\ \text{kJ}$$

7.10 Calculate K for the equation

```
H         Cl    H          H
 \       /       \        /
  C=C        →     C=C
 /       \       /        \
Cl         H    Cl          Cl
```

given $\Delta_f G^\circ/(\text{kJ}\cdot\text{mol}^{-1}) = 27.34$ for *trans*-1,2-dichloroethene and 22.11 for *cis*-1,2-dichloroethene. Assuming the activities of the isomers to be given by the respective concentrations, what is the ratio of the concentrations of the cis isomer to the trans isomer at equilibrium?

▮ For the equation, *(6.16)* gives

$$\Delta_r G^\circ = (1\ \text{mol})\ \Delta_f G^\circ(\text{cis}) - (1\ \text{mol})\ \Delta_f G^\circ(\text{trans})$$
$$= (1\ \text{mol})(22.11\ \text{kJ}\cdot\text{mol}^{-1}) - (1\ \text{mol})(27.34\ \text{kJ}\cdot\text{mol}^{-1}) = -5.23\ \text{kJ}$$

Substituting into *(7.1b)* gives

$$K = \exp\left(-\frac{(-5.23\ \text{kJ})[(10^3\ \text{J})/(1\ \text{kJ})]}{(8.314\ \text{J}\cdot\text{K}^{-1}\cdot\text{mol}^{-1})(298\ \text{K})(\text{mol})}\right) = 8.2$$

The equilibrium constant for this isomerism reaction is given by *(7.2)* as

$$K = \frac{a(\text{cis})}{a(\text{trans})} = \frac{[\text{cis}]}{[\text{trans}]} = 8.2$$

The square bracket notation refers to the equilibrium concentration expressed in molarity, and the symbol for the equilibrium constant is often written as K_C.

7.11 The disproportionation of $Cu^+(aq)$ was studied electrochemically:

$$2Cu^+(aq) \rightarrow Cu(s) + Cu^{2+}(aq) \qquad E° = 0.368\ V$$

Write the equilibrium constant expression for this equation and calculate K.

▮ Applying *(7.2)* to this equation gives

$$K = a(Cu)a(Cu^{2+})/a(Cu^+)^2$$

Assuming the activity of the solid Cu to be unity and the activities of the ions to be given by *(6.45)*, the expression for K becomes

$$K = \frac{(1)y(Cu^{2+})[Cu^{2+}]}{\{y(Cu^+)[Cu^+]\}^2} = \frac{y(Cu^{2+})}{y(Cu^+)^2}\frac{[Cu^{2+}]}{[Cu^+]^2}$$

Substituting *(6.6)* into *(7.1)* gives

$$K = e^{nFE°/RT}$$

$$= \exp\left(\frac{(1\ mol)(9.648 \times 10^4\ J \cdot mol^{-1} \cdot V^{-1})(0.368\ V)}{(8.314\ J \cdot K^{-1} \cdot mol^{-1})(298\ K)(mol)}\right)$$

$$= 1.7 \times 10^6$$

7.12 The pressure dependence of the equilibrium conditions for the equation

$$\tfrac{1}{2}N_2(g) + \tfrac{3}{2}H_2(g) \rightarrow NH_3(g)$$

was studied by measuring the mole fraction of NH_3 produced at various pressures for 1:3 mixtures of N_2 and H_2 at 500 °C. Use the following data to prepare a plot of K_P against P. Why is K_P pressure-dependent?

P/(bar)	100	200	300	400	500	600	700	800	900	1000
x(NH₃)	0.10	0.18	0.25	0.32	0.37	0.42	0.46	0.50	0.53	0.56

▮ Many equilibrium problems are solved most easily by setting up a small table that includes the chemical equation, initial reaction conditions, stoichiometric changes as a result of reaction, and final equilibrium conditions. Letting n be the amount of NH_3 formed by the reaction, the table in this case becomes

equation	$\tfrac{1}{2}N_2(g)$ +	$\tfrac{3}{2}H_2(g)$ →	$NH_3(g)$
$n_{initial}$/(mol)	1	3	
n_{change}/(mol)	$-\tfrac{1}{2}n$	$-\tfrac{3}{2}n$	$+n$
$n_{equilibrium}$/(mol)	$1 - \tfrac{1}{2}n$	$3 - \tfrac{3}{2}n$	n

The total number of moles present at equilibrium is

$$n_t = n(H_2) + n(N_2) + n(NH_3) = (3 - \tfrac{3}{2}n) + (1 - \tfrac{1}{2}n) + n = 4 - n$$

In each case the mole fraction of NH_3 is given as data. The relationship between $x(NH_3)$ and n is

$$x(NH_3) = n/(4 - n)$$

which upon solving for n gives

$$n = \frac{4x(NH_3)}{1 + x(NH_3)}$$

The mole fractions of H_2 and N_2 in terms of $x(NH_3)$ are

$$x(N_2) = \frac{1 - \tfrac{1}{2}n}{4 - n} = \frac{1 - \tfrac{1}{2}[4x(NH_3)]/[1 + x(NH_3)]}{4 - [4x(NH_3)]/[1 + x(NH_3)]} = \frac{1 - x(NH_3)}{4}$$

$$x(H_2) = \frac{3 - \tfrac{3}{2}n}{4 - n} = \frac{3 - \tfrac{3}{2}[4x(NH_3)]/[1 + x(NH_3)]}{4 - [4x(NH_3)]/[1 + x(NH_3)]} = \frac{3 - 3x(NH_3)}{4}$$

Using *(1.23)* gives the partial pressures as

$$P(\mathrm{NH_3}) = x(\mathrm{NH_3})P_t \qquad P(\mathrm{N_2}) = \frac{1 - x(\mathrm{NH_3})}{4}P_t \qquad P(\mathrm{H_2}) = \frac{3 - 3x(\mathrm{NH_3})}{4}P_t$$

Substituting these into *(7.2)* gives

$$K_P = \frac{P(\mathrm{NH_3})/(\mathrm{bar})}{[P(\mathrm{N_2})/(\mathrm{bar})]^{1/2}[P(\mathrm{H_2})/(\mathrm{bar})]^{3/2}}$$

$$= \frac{x(\mathrm{NH_3})[P_t/(\mathrm{bar})]}{\{\frac{1}{4}[1 - x(\mathrm{NH_3})][P_t/(\mathrm{bar})]\}^{1/2}\{\frac{1}{4}[3 - 3x(\mathrm{NH_3})][P_t/(\mathrm{bar})]\}^{3/2}}$$

$$= \frac{16x(\mathrm{NH_3})}{[(1 - x(\mathrm{NH_3})]^{1/2}[3 - 3x(\mathrm{NH_3})]^{3/2}P_t}$$

As a sample calculation, consider $P = 100$ bar:

$$K_P = \frac{16(0.10)}{(1 - 0.10)^{1/2}[3 - 3(0.10)]^{3/2}(100)} = 3.8 \times 10^{-3}$$

The plot is shown in Fig. 7-1. As the pressure increases, the gases begin to deviate from ideality and the assumption that $a_i = P_i$ is no longer valid.

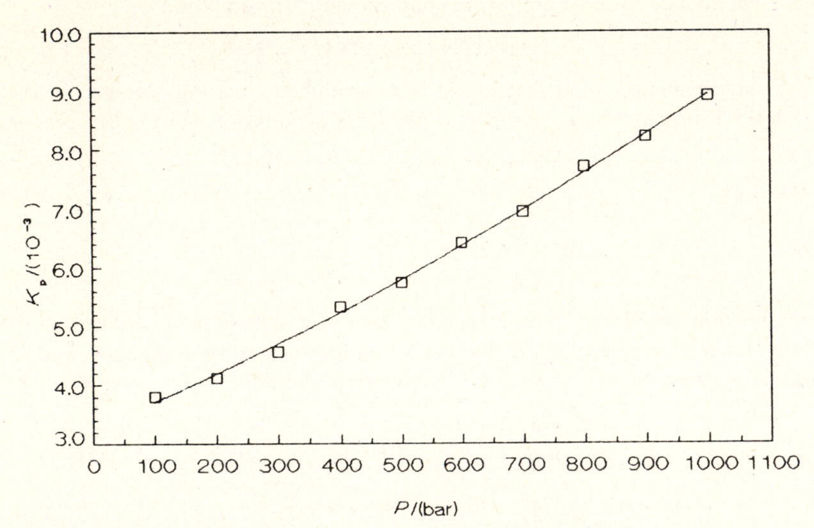

Fig. 7-1

7.13 A sample of 1,3-butadiene was allowed to come into equilibrium with a mixture of 1,2-butadiene, 1-butyne, and 2-butyne at 25 °C. Given $\Delta_f G°/(\mathrm{kJ \cdot mol^{-1}}) = 152.4$, 201.7, 203.0, and 187.1, respectively, calculate the ratio of the partial pressures. Assume ideal gas behavior.

▮ For the equation

```
  H  H  H  H            H     H  H
  |  |  |  |            |     |  |
H—C=C—C=C—H  →  H—C=C=C—C—H
                              |
                              H
```

(6.16) gives

$$\Delta_r G° = (1\ \mathrm{mol})\,\Delta_f G°(\text{1,2-butadiene}) - (1\ \mathrm{mol})\,\Delta_f G°(\text{1,3-butadiene})$$

$$= (1\ \mathrm{mol})(201.7\ \mathrm{kJ \cdot mol^{-1}}) - (1\ \mathrm{mol})(152.4\ \mathrm{kJ \cdot mol^{-1}}) = 49.3\ \mathrm{kJ}$$

Using *(7.1b)* gives

$$K = \exp\left(-\frac{(49.3\ \mathrm{kJ})[(10^3\ \mathrm{J})/(1\ \mathrm{kJ})]}{(8.314\ \mathrm{J \cdot K^{-1} \cdot mol^{-1}})(289\ \mathrm{K})(\mathrm{mol})}\right) = 2 \times 10^{-9}$$

Substituting into *(7.2)* gives

$$K = \frac{a(\text{1,2-butadiene})}{a(\text{1,3-butadiene})} = \frac{P(\text{1,2-butadiene})/(\mathrm{bar})}{P(\text{1,3-butadiene})/(\mathrm{bar})} = 2 \times 10^{-9}$$

Likewise for the equation

```
  H  H  H  H                   H  H
  |  |  |  |                   |  |
H—C=C—C=C—H → H - C≡C—C—C—H
                               |  |
                               H  H
```

$\Delta_r G^\circ = 50.6\ \text{kJ}$, $K = 1 \times 10^{-9}$, and

$$\frac{P(\text{1-butyne})/(\text{bar})}{P(\text{1,3-butadiene})/(\text{bar})} = 1 \times 10^{-9}$$

and for the equation

```
  H  H  H  H          H        H
  |  |  |  |          |        |
H—C=C—C=C—H → H—C—C≡C—C—H
                      |        |
                      H        H
```

$\Delta_r G^\circ = 34.7\ \text{kJ}$, $K = 8 \times 10^{-7}$, and

$$\frac{P(\text{2-butyne})/(\text{bar})}{P(\text{1,3-butadiene})/(\text{bar})} = 8 \times 10^{-7}$$

7.14 In Problems 6.88 and 6.89, the reaction between $Br_2(g)$ and $Cl_2(g)$ continued until $\xi = 0.55$. Show that this value of ξ represents the equilibrium conditions of the reaction.

▌ The value of Q at $\xi = 0.55$ is

$$Q = 4\xi^2/(1 - \xi)^2 = 4(0.55)^2/(1 - \xi)^2 = 6.0$$

Substituting $\Delta_r G^\circ_{500} = -7.388\ \text{kJ}$ into *(7.1b)* gives

$$K = \exp\left(-\frac{(-7.388\ \text{kJ})[(10^3\ \text{J})/(1\ \text{kJ})]}{(8.314\ \text{J} \cdot \text{K}^{-1} \cdot \text{mol}^{-1})(500\ \text{K})(\text{mol})}\right) = 5.91$$

These values of Q and K are essentially the same, indicating equilibrium.

7.15 Often the criterion "$Q > K$ for chemical reaction means a nonspontaneous reaction and $Q < K$ means a spontaneous reaction" is used to predict the spontaneity of a reaction instead of calculating ΔG using *(6.55)*. Show that the statement is true for the reaction considered in Problems 6.88 and 6.89.

▌ For values of $\xi < 0.55$, the forward reaction will be spontaneous as can be seen from Figs. 6-12 and 6-13. The value of Q for $\xi < 0.55$ will be

$$Q = 4(<0.55)^2/[1 - (<0.55)]^2 < 5.91$$

which agrees with the statement. For values of $\xi > 0.55$, the forward reaction will not be spontaneous (see Figs. 6-12 and 6-13). The value of Q for $\xi > 0.55$ will be

$$Q = 4(>0.55)^2/[1 - (>0.55)]^2 < 5.91$$

which also agrees with the statement.

7.16 Show that the maximum yield of product at equilibrium occurs when stoichiometric amounts of reactants are used. Use the reaction considered in Problems 6.88 and 6.89 and assume ideal gas behavior.

▌ Let r be the ratio of $P(Cl_2)$ to $P(Br_2)$. Thus, $P(Cl_2) = rP(Br_2)$. Using *(1.22)* gives

$$P_t = P(Cl_2) + P(Br_2) + P(BrCl) = rP(Br_2) + P(Br_2) + P(BrCl)$$
$$= (r + 1)P(Br_2) + P(BrCl)$$

Solving for $P(Br_2)$ gives

$$P(Br_2) = \frac{P_t - P(BrCl)}{r + 1} \qquad P(Cl_2) = r\left(\frac{P_t - P(BrCl)}{r + 1}\right)$$

Writing the equilibrium constant expression using *(7.3)* gives

$$K_P = \frac{P(BrCl)^2}{P(Cl_2)P(Br_2)} = \frac{P(BrCl)^2}{\left[r\left(\frac{P_t - P(BrCl)}{r + 1}\right)\right]\left(\frac{P_t - P(BrCl)}{r + 1}\right)}$$
$$= \frac{P(BrCl)^2}{r\{[P_t - P(BrCl)]/(r + 1)\}^2}$$

Taking the logarithm of each side of the equation and differentiating with respect to r gives

$$\ln K_P = 2\ln[P(\mathrm{BrCl})] - \ln r - 2\ln[P_t - P(\mathrm{BrCl})] + 2\ln(r+1)$$

$$0 = 2\frac{1}{P(\mathrm{BrCl})}\left(\frac{\partial P(\mathrm{BrCl})}{\partial r}\right)_{P,T} - \frac{1}{r} - 2\left(\frac{1}{P_t - P(\mathrm{BrCl})}\right)\left(\frac{-\partial P(\mathrm{BrCl})}{\partial r}\right)_{P,T} + 2\frac{1}{r+1}$$

Assigning $[\partial P(\mathrm{BrCl})/\partial r]_{P,T} = 0$ and solving for r gives

$$0 = 0 - \frac{1}{r} - 0 + 2\frac{1}{r+1} \qquad r = 1$$

Stoichiometric amounts of Cl_2 and Br_2 generate the maximum yield of BrCl.

7.17 A student made a 25% random error in the evaluation of an equilibrium constant. What is the random error in the value of $\Delta_r G^\circ_{298}$ that the student reported?

▮ If $y = y(x_1, x_2, \ldots)$, then the random error in y will be given by

$$\rho(y) = \left[\sum_i \left(\frac{\partial y}{\partial x_i}\right)^2 \rho(x_i)^2\right]^{1/2} \tag{7.3}$$

where $\rho(x_i)$ is the random error in the variable x_i. Solving *(7.1a)* for $\Delta_r G^\circ$ and using *(7.3)* gives

$$\Delta_r G^\circ = -RT\ln K$$

$$\rho(\Delta_r G^\circ) = \left[\left(\frac{\partial(\Delta_r G^\circ)}{\partial K}\right)^2 \rho(K)^2\right]^{1/2} = \left[\left(-RT\frac{1}{K}\right)^2 \rho(K)^2\right]^{1/2} = \frac{RT}{K}\rho(K)$$

Substituting $\rho(K) = \pm 0.25K$ gives

$$\rho(\Delta_r G^\circ) = \frac{(8.314\ \mathrm{J\cdot K^{-1}\cdot mol^{-1}})(298\ \mathrm{K})(\mathrm{mol})}{K}(\pm 0.25)K = \pm 618\ \mathrm{J}$$

For most reactions where $\Delta_r G^\circ$ is on the order of tens or hundreds of kilojoules, this is an insignificant error.

7.18 What is the resulting error in K if a 25% random error is made in the measurement of $\Delta_r G^\circ_{298}$?

▮ Rearranging the equation derived in Problem 7.17 and substituting $\rho(\Delta_r G^\circ) \pm (0.25)\Delta_r G^\circ$ gives the random error as

$$\rho(K) = \frac{K}{RT}\rho(\Delta_r G^\circ) = \frac{K}{RT}(0.25)\,\Delta_r G^\circ = \pm(0.25)K\ln K$$

The value of $\rho(K)$ is a function of K; for example, for $K = 1\times 10^{-10}$, $\rho(K) = \pm 6\times 10^{-10}$; for $K = 1\times 10^{-2}$, $\rho(K) = \pm 1\times 10^{-2}$; for $K = 1$, $\rho(K) = 0$; for $K = 1\times 10^{2}$, $\rho(K) = \pm 1\times 10^{2}$; and for $K = 1\times 10^{10}$, $\rho(K) = 6\times 10^{10}$. The respective relative errors $[\rho(K)/K]$ are ±600%, ±100%, 0, ±100%, and ±600%. Note that errors made in measuring $\Delta_r G^\circ$ values introduce large random errors in the values of K.

7.19 The equilibrium constant (K_w) for the equation $H_2O(l) \rightarrow H^+(aq) + OH^-(aq)$ was measured at several temperatures. Determine $\Delta_r H^\circ_{298}$ and $\Delta_r S^\circ_{298}$ for the reaction using the following data:

$T/(^\circ\mathrm{C})$	0	25	40	60
K_w	1.15×10^{-15}	1.00×10^{-14}	2.95×10^{-14}	9.55×10^{-14}

▮ The temperature dependence of the equilibrium constant is given by the *van't Hoff equation*

$$\ln\frac{K_{T_2}}{K_{T_1}} = \frac{1}{R}\int_{T_1}^{T_2}\frac{\Delta_r H^\circ}{T^2}\,dT \tag{7.4}$$

A special form of *(7.4)* that is valid for small temperature ranges when $\Delta_r H^\circ$ and $\Delta_r S^\circ$ are constant is

$$\ln K = -\Delta_r H^\circ/RT + \Delta_r S^\circ/R \tag{7.5a}$$

which implies that a plot of $\ln K$ against T^{-1} will be linear with a slope of $-\Delta_r H^\circ/R$ and an intercept of $\Delta_r S^\circ/R$

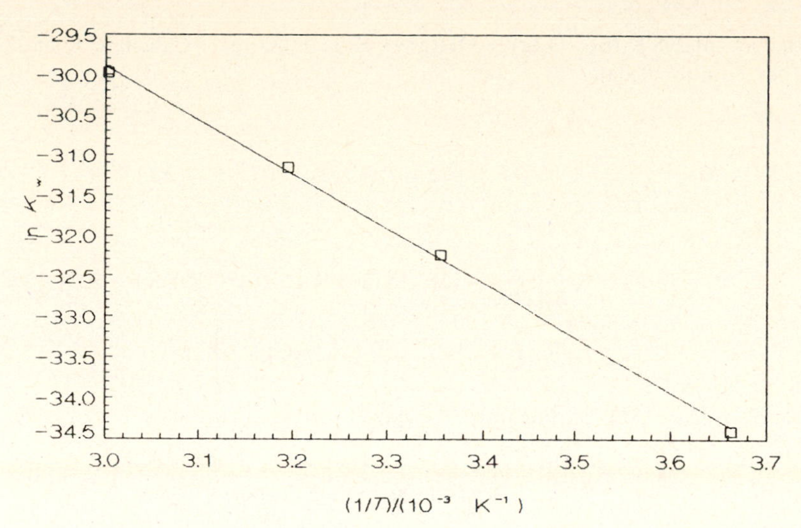

Fig. 7-2

(see Fig. 7-2). For this plot, the slope is −6720 K and the intercept is −9.72, giving

$$\Delta_r H^\circ = -(8.314\ \mathrm{J \cdot K^{-1} \cdot mol^{-1}})\frac{10^{-3}\ \mathrm{kJ}}{1\ \mathrm{J}}(-6720\ \mathrm{K})(\mathrm{mol}) = 55.9\ \mathrm{kJ}$$

$$\Delta_r S^\circ = (8.314\ \mathrm{J \cdot K^{-1} \cdot mol^{-1}})(-9.72)(\mathrm{mol}) = -80.8\ \mathrm{J \cdot K^{-1}}$$

7.20 At 25 °C, $K = 1.35 \times 10^{28}$ for

$$\alpha,\beta\text{-D-glucose(aq)} \rightarrow 2\ \text{lactic acid(aq)} \qquad \Delta_r H^\circ_{298} = -108.66\ \mathrm{kJ}$$

Calculate K at the physiological temperature of 37.0 °C assuming $\Delta_r H^\circ$ to be constant over this temperature range.

▌ Values of K at two different temperatures are related to $\Delta_r H^\circ$ by a special form of *(7.5a)*:

$$\ln \frac{K_{T_2}}{K_{T_1}} = \frac{-\Delta_r H^\circ}{R}\left(\frac{1}{T_2} - \frac{1}{T_1}\right) \tag{7.5b}$$

Substituting the data into *(7.5b)* gives

$$\ln \frac{K_{T_2}}{1.35 \times 10^{28}} = \frac{-(-108.66\ \mathrm{kJ})[(10^3\ \mathrm{J})/(1\ \mathrm{kJ})]}{(8.314\ \mathrm{J \cdot K^{-1} \cdot mol^{-1}})(\mathrm{mol})}\left(\frac{1}{310.2\ \mathrm{K}} - \frac{1}{298.2\ \mathrm{K}}\right) = -1.70$$

$$\frac{K_{T_2}}{1.35 \times 10^{28}} = 0.18 \qquad K_{T_2} = 2.4 \times 10^{27}$$

7.21 For the equation $Br_2(g) \rightarrow 2Br(g)$, $K = 38.4$ at 2400 K and $K = 84.7$ at 2600 K. Determine $\Delta_r G^\circ$, $\Delta_r H^\circ$, $\Delta_r S^\circ$, and K for this reaction at 2500 K.

▌ Solving *(7.5b)* for $\Delta_r H^\circ$ and substituting the data gives

$$\Delta_r H^\circ_{2500} = \frac{-(8.314\ \mathrm{J \cdot K^{-1} \cdot mol^{-1}})[(10^{-3}\ \mathrm{kJ})/(1\ \mathrm{J})](\mathrm{mol})\ln(84.7/28.4)}{(2600\ \mathrm{K})^{-1} - (2400\ \mathrm{K})^{-1}} = 205\ \mathrm{kJ}$$

Substituting either set of data and the value of $\Delta_r H^\circ$ into *(7.5b)* gives

$$\ln \frac{K_{T_2}}{38.4} = \frac{-(205\ \mathrm{kJ})[(10^3\ \mathrm{J})/(1\ \mathrm{kJ})]}{(8.314\ \mathrm{J \cdot K^{-1} \cdot mol^{-1}})(\mathrm{mol})}\left(\frac{1}{2500\ \mathrm{K}} - \frac{1}{2400\ \mathrm{K}}\right) = 0.411$$

$$K_{2500} = 57.9$$

Substituting into *(7.1a)* gives

$$\Delta_r G^\circ_{2500} = -(8.314\ \mathrm{J \cdot K^{-1} \cdot mol^{-1}})\frac{10^{-3}\ \mathrm{kJ}}{1\ \mathrm{J}}(2500\ \mathrm{K})(\mathrm{mol})\ln 57.9$$

$$= -84.4\ \mathrm{kJ}$$

Rearranging *(6.2a)* and substituting the data gives

$$\Delta_r S^\circ_{2500} = \frac{[205\ \mathrm{kJ} - (-84.4\ \mathrm{kJ})][(10^3\ \mathrm{J})/(1\ \mathrm{kJ})]}{2500\ \mathrm{K}} = 116\ \mathrm{J \cdot K^{-1}}$$

7.22 Given $K = 0.096$ at 298 K for $O_3(g) + OH(g) \rightarrow H(g) + 2O_2(g)$. Calculate K at 373 K. See Problems 4.29, 5.72, and 6.36 for additional data.

▮ A special case of *(7.4)* is

$$\ln K = M + \frac{1}{R}\left[\frac{-J}{T} + (\Delta a)T + \tfrac{1}{2}(\Delta b)T + \tfrac{1}{6}(\Delta c)\ln T^2 + \tfrac{1}{12}(\Delta d)T^3 + \cdots + \tfrac{1}{2}(\Delta c')T^{-2}\right] \tag{7.6}$$

where

$$M = \ln K_{298} + \frac{1}{R}\left[\frac{J}{298} - (\Delta a)\ln(298\ \text{K}) - \tfrac{1}{2}(\Delta b)(298\ \text{K}) - \tfrac{1}{6}(\Delta c)(298\ \text{K})^2 - \tfrac{1}{12}(\Delta d)(298\ \text{K})^3 - \cdots - \tfrac{1}{2}(\Delta c')(298\ \text{K})^{-2}\right] \tag{7.7}$$

In this case $\Delta_r G_P^\circ = \Delta a$. Substituting into *(7.7)* gives

$$M = \ln 0.096 + \frac{1}{(8.314\ \text{J}\cdot\text{K}^{-1}\cdot\text{mol}^{-1})(\text{mol})}\left(\frac{33\,200\ \text{J}}{298\ \text{K}}\right) = 11.05$$

Using *(7.6)* gives

$$\ln K = 11.05 + \frac{1}{(8.314\ \text{J}\cdot\text{K}^{-1}\cdot\text{mol}^{-1})(\text{mol})}\left(\frac{-33\,200\ \text{J}}{373\ \text{K}}\right) = 0.34$$

$$K = 1.4$$

7.23 In Problem 7.22, the value of K changed from 0.096 at 298 K to 1.4 at 373 K. Above what temperature will the reaction become thermodynamically spontaneous?

▮ The reaction will become spontaneous at temperatures above which $K > 1$ or $\ln K > 0$. Substituting into the expression for $\ln K$ determined in Problem 7.22 and solving for T gives

$$0 < 11.05 + \frac{1}{(8.314\ \text{J}\cdot\text{K}^{-1}\cdot\text{mol}^{-1})(\text{mol})}\left(\frac{-33\,200\ \text{J}}{T}\right)$$

$$T > 361\ \text{K}$$

This is in good agreement with the results of Problem 6.37.

7.24 Determine $\ln K$ for the equation $N_2(g) + 3H_2(g) \rightarrow 2NH_3(g)$ as a function of temperature. Prepare a plot of $\ln K$ against T^{-1} between 298 K and 1000 K. See Problems 4.30, 5.73, and 6.38 for additional data. Also, show a plot based on *(7.5)*. At what temperature does the formation of ammonia become spontaneous under standard-state conditions?

▮ Using $\Delta_r G_{298}^\circ = -32.90$ kJ, *(7.1b)* gives

$$K_{298} = \exp\left(-\frac{(-32.90\ \text{kJ})[(10^3\ \text{J})/(1\ \text{kJ})]}{(8.314\ \text{J}\cdot\text{K}^{-1}\cdot\text{mol}^{-1})(298.15\ \text{K})(\text{mol})}\right) = 5.8 \times 10^5$$

For the reaction, $\Delta a = -62.457\ \text{J}\cdot\text{K}^{-1}$, $\Delta b = 63.276 \times 10^{-3}\ \text{J}\cdot\text{K}^{-2}$, $\Delta c = -121.39 \times 10^{-7}\ \text{J}\cdot\text{K}^{-3}$, and $J = -76.30$ kJ. Substituting into *(7.7)* gives

$$M = \ln(5.8 \times 10^5) + \frac{1}{(8.314\ \text{J}\cdot\text{K}^{-1}\cdot\text{mol}^{-1})(\text{mol})} \times \left[\frac{(-76.30\ \text{kJ})[(10^3\ \text{J})/(1\ \text{kJ})]}{298\ \text{K}} - (-62.457\ \text{J}\cdot\text{K}^{-1})\ln(298\ \text{K}) - \tfrac{1}{2}(63.276 \times 10^{-3}\ \text{J}\cdot\text{K}^{-2})(298\ \text{K}) - \tfrac{1}{6}(-121.39 \times 10^{-7}\ \text{J}\cdot\text{K}^{-3})(298\ \text{K})^2\right]$$

$$= 24.18$$

Thus at any temperature, *(7.6)* gives

$$\ln K = 24.18 + \frac{1}{8.314}\left(\frac{76.30 \times 10^3}{T/(\text{K})} - 62.457\ln[T/(\text{K})] + \tfrac{1}{2}(63.276 \times 10^{-3})[T/(\text{K})] - \tfrac{1}{6}(121.39 \times 10^{-7})[T/(K)]^2\right)$$

As a sample calculation, at $T = 500$ K,

$$\ln K = 24.18 + \frac{1}{8.314}\left[\frac{76.30 \times 10^3}{500} - (62.457)\ln(500) + \tfrac{1}{2}(63.276 \times 10^{-3})(500) - \tfrac{1}{6}(121.39 \times 10^{-7})(500)^2\right]$$
$$= -2.31$$

The plot is shown in Fig. 7-3. The plot based on *(7.5)* is a straight line that begins at $\ln(5.8 \times 10^5) = 13.27$ and has a slope given by

$$\text{Slope} = \frac{-\Delta_r H^\circ_{298}}{R} = \frac{-(92.22\ \text{kJ})[(10^3\ \text{J})/(1\ \text{kJ})]}{(8.314\ \text{J} \cdot \text{K}^{-1} \cdot \text{mol}^{-1})(\text{mol})} = 11\ 090\ \text{K}$$

From the graph, at $T^{-1} < 2.20 \times 10^{-3}\ \text{K}^{-1}$ or $T > 455$ K, the formation of NH_3 is no longer spontaneous.

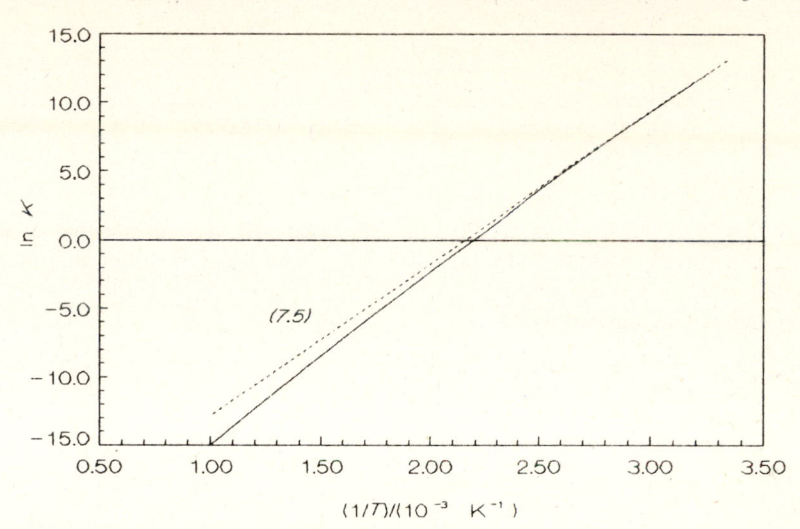

Fig. 7-3

7.25 A plot of $\Delta_f G^\circ_T$ against T for BO_2(g) appears in Fig. 7-4. Explain the two discontinuities. Predict the sign of each of the following thermodynamic properties at the temperatures indicated.

$T/(\text{K})$	$\Delta_r G^\circ$	$[\partial(\Delta_r G^\circ)/\partial T]_P$	$\Delta_r S^\circ$	$\Delta_r H^\circ$	$[\partial(\Delta_r S^\circ)/\partial T]_P$
1000					
3000					
5000					

$T/(\text{K})$	$[\partial(\Delta_r H^\circ)/\partial T]_P$	$\Delta_r C^\circ_P$	$\ln K$	$[\partial(\ln K)/\partial T]_P$
1000				
3000				
5000				

▌ The discontinuities occur at the temperatures at which boron melts (2350 K) and boils (4140 K). For $\Delta_r G^\circ$, the sign corresponds to the value of $\Delta_r G^\circ$ on the plot. In all three cases, it is negative. For $[\partial(\Delta_r G^\circ)/\partial T]_P$, the sign corresponds to the slope of the curve at that temperature. The respective signs are −, +, and +. For $\Delta_r S^\circ$, *(6.15a)* gives $\Delta_r S^\circ = -[\partial(\Delta_r G^\circ)/\partial T]_P$, and the sign will be opposite that of the slope: +, −, and −. For $\Delta_r H^\circ$, *(6.2a)* gives $\Delta_r H^\circ = \Delta_r G^\circ + T\,\Delta_r S^\circ$, and the sign will be that of the numerically larger quantity, $\Delta_r G^\circ$ or $T\,\Delta_r S^\circ$. In this case, $T\,\Delta_r S^\circ$ is not very large until very high temperatures are reached. Thus $\Delta_r H^\circ$ will be − in all three cases.

For $[\partial(\Delta_r S^\circ)/\partial T]_P = -[\partial^2(\Delta_r G^\circ)/\partial T^2]_P$, the sign will be opposite that of the second derivative of $\Delta_r G^\circ$. It is very difficult to tell from the plot what the signs of this parameter are. For $[\partial(\Delta_r H^\circ)/\partial T]_P = T[\partial(\Delta_r S^\circ)/\partial T]_P$, the sign will be that of $[\partial(\Delta_r S^\circ)/\partial T]_P$. Again, these signs cannot be determined from the plot. For $\Delta_r C^\circ_P = T[\partial(\Delta_r S^\circ)/\partial T]_P$, the sign will that of $[\partial(\Delta_r S^\circ)/\partial T]_P$, which cannot be determined from the plot.

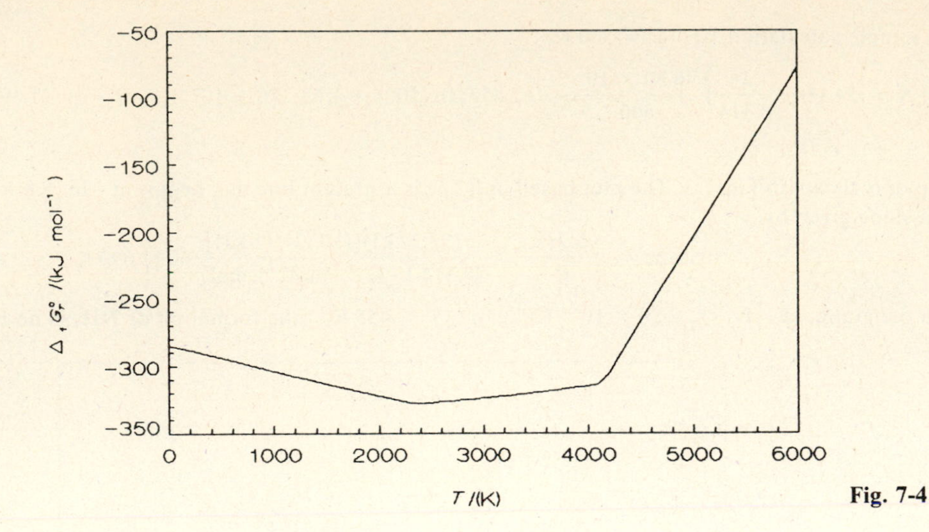

Fig. 7-4

For ln K, *(7.1a)* gives $\ln K = -\Delta_r G°/RT$, and the sign will be opposite that of $\Delta_r G°$: +, +, and +. For $[\partial(\ln K)/\partial T]_P$, *(7.4)* gives $[\partial(\ln K)/\partial T]_P = \Delta_r H°/RT^2$, and the sign will be that of $\Delta_r H°$. In this case: −, −, and −.

The answers can be summarized as

$T/(\mathrm{K})$	$\Delta_r G°$	$[\partial(\Delta_r G°)/\partial T]_P$	$\Delta_r S°$	$\Delta_r H°$	$[\partial(\Delta_r S°)/\partial T]_P$
1000	−	−	+	−	?
3000	−	+	−	−	?
5000	−	+	−	−	?

$T/(\mathrm{K})$	$[\partial(\Delta_r H°)/\partial T]_P$	$\Delta_r C_P°$	ln K	$[\partial(\ln K)/\partial T]_P$
1000	?	?	+	−
3000	?	?	+	−
5000	?	?	+	−

7.2 EQUILIBRIUM AND GASES

7.26 A flask containing $F_2(g)$ at 2.00 bar and 1000 K is allowed to equilibrate with F(g):

$$F_2(g) \rightarrow 2F(g) \qquad K = 9.59 \times 10^{-3}$$

What are the partial pressures of the gases once equilibrium has been reached? Assume ideal gas behavior.

▮ Because the partial pressures of the gases are directly proportional to the respective amounts of substances ($P_i = n_i RT/V$), the equilibrium table presented in Problem 7.12 can be written in terms of P_i. Letting x represent the amount of F_2 that dissociates gives

equation	$F_2(g)$ →	$2F(g)$
$P_{i,\mathrm{initial}}/(\mathrm{bar})$	2.00	
$P_{i,\mathrm{change}}/(\mathrm{bar})$	$-x$	$+2x$
$P_{i,\mathrm{equilibrium}}/(\mathrm{bar})$	$2.00 - x$	$2x$

Ideal gas behavior implies that $K = K_P$. For this reaction *(7.2)* gives

$$K_P = \frac{[P(\mathrm{F})/(\mathrm{bar})]^2}{[P(\mathrm{F_2})/(\mathrm{bar})]} = \frac{(2x)^2}{2.00 - x} = 9.59 \times 10^{-3}$$

Solving gives $x = 6.81 \times 10^{-2}$. Thus,

$$P(\mathrm{F_2}) = 2.00\ \mathrm{bar} - 0.0681\ \mathrm{bar} = 1.93\ \mathrm{bar}$$

$$P(\mathrm{F}) = 2(0.0681\ \mathrm{bar}) = 0.136\ \mathrm{bar}$$

7.27 A flask containing $F_2(g)$ at a concentration of $0.0240\ \mathrm{mol \cdot L^{-1}}$ is allowed to equilibrate with F(g):

$$\mathrm{F_2(g) \rightarrow 2F(g)} \qquad K_{1000} = 9.59 \times 10^{-3}$$

What are the equilibrium concentrations of the gases once equilibrium has been reached? Assume ideal gas behavior.

▌ Because the concentrations of the gases are directly proportional to the respective amounts of substances ($C_i = n_i/V$), the equilibrium table can be written in terms of C_i. Letting x represent the amount of F_2 that dissociates gives

equation	$F_2(g)$ →	2F(g)
$C_{i,\mathrm{initial}}/(\mathrm{mol \cdot L^{-1}})$	0.0240	
$C_{i,\mathrm{change}}/(\mathrm{mol \cdot L^{-1}})$	$-x$	$+2x$
$C_{i,\mathrm{equilibrium}}/(\mathrm{mol \cdot L^{-1}})$	$0.0240 - x$	$2x$

Ideal gas behavior implies that $K = K_P$. The relation between K_P and K_C, the concentration equilibrium constant, is

$$K_P = (RT)^{\Delta n_g} K_C \tag{7.8}$$

where Δn_g was defined in Problem 3.8. Substituting the data into *(7.8)* and solving for K_C gives

$$K_C = \frac{9.59 \times 10^{-3}}{[(0.083\,14\ \mathrm{L \cdot bar \cdot K^{-1} \cdot mol^{-1}})(1000\ \mathrm{K})]^1} = 1.153 \times 10^{-4}$$

For this reaction *(7.2)* gives

$$K_C = \frac{[\mathrm{F}]^2}{[\mathrm{F_2}]} = \frac{(2x)^2}{0.0240 - x} = 1.153 \times 10^{-4}$$

where $[i]$ represents the equilibrium concentration expressed in molarity. Solving gives $x = 8.17 \times 10^{-4}\ \mathrm{mol \cdot L^{-1}}$. Thus

$$[\mathrm{F_2}] = 0.0240\ \mathrm{mol \cdot L^{-1}} - 8.17 \times 10^{-4}\ \mathrm{mol \cdot L^{-1}} = 0.0232\ \mathrm{mol \cdot L^{-1}}$$

$$[\mathrm{F}] = 2(8.17 \times 10^{-4}\ \mathrm{mol \cdot L^{-1}}) = 1.63 \times 10^{-3}\ \mathrm{mol \cdot L^{-1}}$$

7.28 A flask containing 0.0600 mol of $F_2(g)$ is allowed to equilibrate with F(g):

$$\mathrm{F_2(g) \rightarrow 2F(g)} \qquad K_{1000} = 9.59 \times 10^{-3}$$

What are the mole fractions of the gases once equilibrium has been reached? The total pressure of the gases at equilibrium is 2.07 bar. Assume ideal gas behavior.

▌ Letting x respresent the amount of F_2 that dissociates gives

equation	$F_2(g)$ →	2F(g)
$n_{i,\mathrm{initial}}/(\mathrm{mol})$	0.0600	
$n_{i,\mathrm{change}}/(\mathrm{mol})$	$-x$	$+2x$
$n_{i,\mathrm{equilibrium}}/(\mathrm{mol})$	$0.0600 - x$	$2x$

The total number of moles and the respective mole fractions are

$$n_t = (0.0600 - x) + 2x = 0.0600 + x$$

$$x(F_2) = \frac{0.0600 - x}{0.0600 + x} \qquad x(F) = \frac{2x}{0.0600 + x}$$

Ideal gas behavior implies that $K = K_P$. The relation between K_P and K_x, the mole fraction equilibrium constant, is

$$K_P = P^{\Delta n_g} K_x \tag{7.9}$$

where P is the total pressure of the gases (including any inert gases that may be present) and Δn_g was defined in Problem 3.8. Substituting the data into *(7.9)* and solving for K_x gives

$$K_x = (9.59 \times 10^{-3})/(2.07)^1 = 4.63 \times 10^{-3}$$

For this reaction *(7.2)* gives

$$K_x = \frac{x(F)^2}{x(F_2)} = \frac{[2x/(0.0600 + x)]^2}{(0.0600 - x)/(0.0600 + x)} = \frac{4x^2}{(0.0600 - x)(0.0600 + x)}$$

$$= 4x^2/(3.60 \times 10^{-3} - x^2) = 4.63 \times 10^{-3}$$

Solving gives $x = 2.04 \times 10^{-3}$ mol. Thus,

$$x(F_2) = \frac{0.0600 \text{ mol} - 0.002\,04 \text{ mol}}{0.0600 \text{ mol} + 0.002\,04 \text{ mol}} = 0.934$$

$$x(F) = \frac{2(0.002\,04 \text{ mol})}{0.0600 \text{ mol} + 0.002\,04 \text{ mol}} = 0.0658$$

7.29 What is the degree of dissociation of $F_2(g)$ into F(g) at 1000 K? See Problems 7.26–7.28 for further data.

▮ The degree of dissociation (α) is given by the amount of F_2 that dissociates divided by the amount of F_2 that could dissociate. From Problems 7.26–7.28,

$$\alpha = \frac{0.0681 \text{ bar}}{2.00 \text{ bar}} = \frac{8.17 \times 10^{-4} \text{ mol} \cdot \text{L}^{-1}}{0.0240 \text{ mol} \cdot \text{L}^{-1}} = \frac{2.04 \times 10^{-3} \text{ mol}}{0.0600 \text{ mol}} = 0.0340$$

7.30 Derive the relationship between the equilibrium values of ξ and K_P for the dissociation of $F_2(g)$:

$$F_2(g) \rightarrow 2F(g) \qquad K_{298} = 1.49 \times 10^{-22}$$

Assume ideal gas behavior.

▮ The value of ξ_e represents the amount of F_2 that dissociates in order to establish equilibrium. The equilibrium table becomes

equation	$F_2(g)$ →	2F(g)
$n_{i,\text{initial}}$/(mol)	1	
$n_{i,\text{change}}$/(mol)	$-\xi_e$	$+2\xi_e$
$n_{i,\text{equilibrium}}$/(mol)	$1 - \xi_e$	$2\xi_e$

The total number of moles and the respective mole fractions are

$$n_t = (1 - \xi_e) + 2\xi_e = 1 + \xi_e$$

$$x(F_2) = \frac{1 - \xi_e}{1 + \xi_e} \qquad x(F) = \frac{2\xi_e}{1 + \xi_e}$$

The partial pressures are

$$P(F_2) = \frac{1 - \xi_e}{1 + \xi_e} P \qquad \text{and} \qquad P(F) = \frac{2\xi_e}{1 + \xi_e} P$$

which upon substitution into *(7.2)* give

$$K_P = \frac{[P(\mathrm{F})/(\mathrm{bar})]^2}{[P(\mathrm{F_2})/(\mathrm{bar})]}$$

$$= \frac{\{[2\xi_e/(1+\xi_e)]P\}^2}{[(1-\xi_e)/(1+\xi_e)]P} = \frac{4\xi_e P}{(1+\xi_e)(1-\xi_e)} = \frac{4\xi_e P}{1-\xi_e^2}$$

Because the value of K is very small at 298 K, $1 - \xi_e^2 \approx 1$ and $\xi_e = K_P/4P$.

7.31 At what pressure will $\alpha = 0.25$ for the dissociation of $F_2(g)$ at 1000 K? See Problem 7.30 for further data. Assume ideal gas behavior.

▮ Substituting $\alpha = \xi_e$ into the results of Problem 7.30 gives

$$K_P = \frac{4\alpha P}{1-\alpha^2} = 9.59 \times 10^{-3}$$

Solving for P and substituting $\alpha = 0.25$ gives

$$P = \frac{(9.59 \times 10^{-3})[1-(0.25)^2]}{4(0.25)} = 0.0090\ \mathrm{bar}$$

(Note that the approximate solution $K_P = 4\xi P$ is not valid in this case because K_P is much larger at 1000 K than it is at 298 K.)

7.32 The average molar mass of an equilibrium mixture of $F_2(g)$ and $F(g)$ at 1000 K was $36.739\ \mathrm{g \cdot mol^{-1}}$. Calculate K_x for the equation $F_2(g) \rightarrow 2F(g)$.

▮ For this binary mixture, *(1.26)* gives

$$x(\mathrm{F})(18.9984\ \mathrm{g \cdot mol^{-1}}) + x(\mathrm{F_2})(37.9968\ \mathrm{g \cdot mol^{-1}}) = 36.739\ \mathrm{g \cdot mol^{-1}}$$

$$x(\mathrm{F}) + x(\mathrm{F_2}) = 1$$

Solving the simultaneous equations gives $x(\mathrm{F}) = 0.0662$ and $x(\mathrm{F_2}) = 0.9338$. Substituting into *(7.2)* gives

$$K_x = \frac{x(\mathrm{F})^2}{x(\mathrm{F_2})} = \frac{(0.0662)^2}{0.9338} = 4.69 \times 10^{-3}$$

7.33 A mixture of air at 1.00 bar and 2000 K was passed through an electric arc to produce nitric oxide.

$$N_2(g) + O_2(g) \rightarrow 2NO(g) \qquad K_{2000} = 4.00 \times 10^{-4}$$

What are the partial pressures of the gases once equilibrium has been established? What fraction of the O_2 is converted? Assume ideal gas behavior.

▮ Assuming the initial pressures of N_2 and O_2 to be 0.80 bar and 0.20 bar, respectively, the equilibrium table from this reaction is

equation	$N_2(g)$ +	$O_2(g)$ →	$2NO(g)$
$P_{i,\mathrm{initial}}/(\mathrm{bar})$	0.80	0.20	
$P_{i,\mathrm{change}}/(\mathrm{bar})$	$-x$	$-x$	$2x$
$P_{i,\mathrm{equilibrium}}/(\mathrm{bar})$	$0.80 - x$	$0.20 - x$	$2x$

where x represents the amounts of N_2 and of O_2 that combine. Ideal gas behavior implies $K = K_P$; thus *(7.2)* gives

$$K_P = \frac{[P(\mathrm{NO})/(\mathrm{bar})]^2}{[P(\mathrm{N_2})/(\mathrm{bar})]/[P(\mathrm{O_2})/(\mathrm{bar})]} = \frac{(2x)^2}{(0.80-x)(0.20-x)} = 4.00 \times 10^{-4}$$

Solving gives $x = 4.0 \times 10^{-3}$ bar. The partial pressures are

$$P(\mathrm{N_2}) = 0.80\ \mathrm{bar} - 0.0040\ \mathrm{bar} = 0.80\ \mathrm{bar}$$

$$P(\mathrm{O_2}) = 0.20\ \mathrm{bar} - 0.0040\ \mathrm{bar} = 0.20\ \mathrm{bar}$$

$$P(\mathrm{NO}) = 2(0.0040\ \mathrm{bar}) = 0.0080\ \mathrm{bar}$$

The fraction of O_2 converted is

$$\alpha(O_2) = \frac{0.0040\ \text{bar}}{0.20\ \text{bar}} = 0.020$$

7.34 The process described in Problem 7.33 was carried out at an initial pressure of 2.0 kbar. What will be the partial pressures of the gases once equilibrium calculations have been established? The critical temperatures and pressures for the gases are −93 °C and 65 bar for NO, −147 °C and 33.9 bar for N_2, and −118.4 °C and 50.8 bar for O_2. Has the fraction of O_2 converted changed under these higher-pressure conditions?

▮ Using *(1.30)* and *(1.31)* for each gas gives

$$P_r(\text{NO}) = \frac{(2.0\ \text{kbar})[(10^3\ \text{bar})/(1\ \text{kbar})]}{65\ \text{bar}} = 31$$

$$T_r(\text{NO}) = \frac{2000\ \text{K}}{(-93 + 273)\text{K}} = 11.1$$

$P_r(N_2) = 59$, $T_r(N_2) = 15.9$, $P_r(O_2) = 39$, and $T_r(O_2) = 12.9$. From Fig. 6-7, $\gamma(\text{NO}) = 1.21$, $\gamma(N_2) = 1.50$, and $\gamma(O_2) = 1.39$. The activity coefficient equilibrium constant is given by *(7.2)* for this equation as

$$K_\gamma = \frac{\gamma(\text{NO})^2}{\gamma(N_2)\gamma(O_2)} = \frac{(1.21)^2}{(1.50)(1.39)} = 0.702$$

The relation among K, K_P, and K_γ is

$$K = K_P K_\gamma \qquad (7.10)$$

Solving *(7.10)* for K_P and substituting the data gives

$$K_P = (4.00 \times 10^{-4})/0.702 = 5.70 \times 10^{-4}$$

The initial partial pressures of $P(N_2)$ and $P(O_2)$ are 1.6 kbar and 0.40 kbar. From Problem 7.33,

$$K_P = \frac{(2x)^2}{(1.6 - x)(0.40 - x)} = 5.70 \times 10^{-4}$$

Solving gives $x = 9.4$ bar. The partial pressures are

$$P(N_2) = 1.6\ \text{kbar} - 9.4\ \text{bar} = 1.6\ \text{kbar}$$

$$P(O_2) = 0.40\ \text{kbar} - 9.4\ \text{bar} = 0.40\ \text{kbar}$$

$$P(\text{NO}) = 2(9.4\ \text{bar}) = 18.8\ \text{bar}$$

Under these conditions,

$$\alpha(O_2) = \frac{9.4\ \text{bar}}{(0.40\ \text{kbar})[(10^3\ \text{bar})/(1\ \text{kbar})]} = 0.024$$

The fraction of O_2 that has reacted has increased slightly under the higher-pressure conditions.

7.35 A 0.100-mol sample of NO_2 was placed in a 10.0-L container and heated to 750 K. The total pressure of the equilibrium mixture as a result of the decomposition $2NO_2(g) \rightarrow 2NO(g) + O_2(g)$ was 0.827 bar. What is the value of K_P at this temperature?

▮ The initial pressure of NO_2 is given by *(1.18)* as

$$P(NO_2)_{\text{initial}} = \frac{(0.100\ \text{mol})(0.083\,14\ \text{L}\cdot\text{bar}\cdot\text{K}^{-1}\cdot\text{mol}^{-1})(750\ \text{K})}{10.0\ \text{L}} = 0.624\ \text{bar}$$

Letting x be the amount of O_2 formed, the equilibrium table is

equation	$2NO_2(g)$ +	$2NO(g)$ →	$O_2(g)$
$P_{i,\text{initial}}$/(bar)	0.624		
$P_{i,\text{change}}$/(bar)	$-2x$	$+2x$	$+x$
$P_{i,\text{equilibrium}}$/(bar)	$0.624 - 2x$	$2x$	x

The total pressure of the gases is given by *(1.22)* as

$$P_t = P(NO_2) + P(NO) + P(O_2) = (0.624 - 2x)\text{ bar} + 2x\text{ bar} + x\text{ bar}$$
$$= (0.624 + x)\text{ bar} = 0.827\text{ bar}$$

Solving gives $x = 0.203$ bar. The partial pressures are

$$P(NO_2) = 0.624\text{ bar} - 2(0.203\text{ bar}) = 0.218\text{ bar}$$
$$P(N_2) = 2(0.203\text{ bar}) = 0.406\text{ bar}$$
$$P(O_2) = 0.203\text{ bar}$$

which upon substitution into *(7.2)* gives

$$K_P = \frac{[P(NO)/(\text{bar})]^2[P(O_2)/(\text{bar})]}{[P(NO_2)/(\text{bar})]^2} = \frac{(0.406)^2(0.203)}{(0.218)^2} = 0.704$$

7.36 A 0.0100-mol sample of NO_2 was placed in the container described in Problem 7.35, and equilibrium was established at 750 K. What is the degree of dissociation? Compare the value to that for the 0.100-mol NO_2 sample in Problem 7.35.

▮ The initial concentration of NO_2 was

$$C(NO_2)_{\text{initial}} = \frac{0.0100\text{ mol}}{10.0\text{ L}} = 0.001\,00\text{ mol}\cdot\text{L}^{-1}$$

Letting x represent the amount of O_2 formed, the equilibrium table is

equation	$2NO_2(g)$ +	$2NO(g)$ →	$O_2(g)$
$C_{i,\text{initial}}/(\text{mol}\cdot\text{L}^{-1})$	0.001 00		
$C_{i,\text{change}}/(\text{mol}\cdot\text{L}^{-1})$	$-2x$	$+2x$	$+x$
$C_{i,\text{equilibrium}}/(\text{mol}\cdot\text{L}^{-1})$	$0.001\,00 - 2x$	$2x$	x

For this equation, $\Delta n_g = +1$ mol. Solving *(7.8)* for K_C and substituting the data gives

$$K_C = \frac{0.704}{[(0.083\,14\text{ L}\cdot\text{bar}\cdot\text{K}^{-1}\cdot\text{mol}^{-1})(750\text{ K})]^1} = 0.0113$$

Substituting into *(7.2)* and solving for x gives

$$K_C = \frac{[NO]^2[O_2]}{[NO_2]^2} = \frac{(2x)^2(x)}{(0.001\,00 - 2x)^2} = 0.0013$$
$$x = 4.19 \times 10^{-4}\text{ mol}\cdot\text{L}^{-1}$$

The fraction of NO_2 that undergoes dissociation is

$$\alpha = \frac{2(4.19 \times 10^{-4}\text{ mol}\cdot\text{L}^{-1})}{0.001\,00\text{ mol}\cdot\text{L}^{-1}} = 0.838$$

This fraction is greater than that at the higher pressure in Problem 7.35, where

$$\alpha = 2(0.203\text{ bar})/0.624\text{ bar} = 0.651$$

7.37 Repeat the calculation of Problem 7.36 for a 0.0100-mol sample of NO_2 mixed with 0.100 mol of an inert gas. The total pressure of the equilibrium mixture is 0.712 bar. Assume ideal gas behavior. What is the effect of the inert gas?

▮ The initial pressure of NO_2 and the partial pressure of the N_2 at equilibrium are given by *(1.18)* as

$$P(NO_2)_{\text{initial}} = \frac{(0.0100\text{ mol})(0.083\,14\text{ L}\cdot\text{bar}\cdot\text{K}^{-1}\cdot\text{mol}^{-1})(750\text{ K})}{10.0\text{ L}} = 0.0624\text{ bar}$$

$$P(N_2) = (0.100)(0.083\,14)(750)/10.0 = 0.624\text{ bar}$$

Letting x be the amount of O_2 formed, the equilibrium table is

equation	$2NO_2(g)$ +	$2NO(g)$ →	$O_2(g)$
$P_{i,\text{initial}}/(\text{bar})$	0.0624		
$P_{i,\text{change}}/(\text{bar})$	$-2x$	$+2x$	x
$P_{i,\text{equilibrium}}/(\text{bar})$	$0.0624 - 2x$	$2x$	x

The sum of the partial pressures of the gases involved in the reaction is given by *(1.22)* as

$$P_t = P(NO_2) + P(NO) + P(O_2) = (0.0624 - 2x)\ \text{bar} + 2x\ \text{bar} + x\ \text{bar}$$

$$= (0.0624 + x)\ \text{bar} = 0.712\ \text{bar} - 0.624\ \text{bar} = 0.088\ \text{bar}$$

Solving gives $x = 0.206$ bar. The fraction of dissociation is

$$\alpha = 2(0.026\ \text{bar})/(0.0624\ \text{bar}) = 0.83$$

As long as the gases act ideally, the pressure of the inert gas has no effect on α.

7.38 What amount of NO_2 must be placed in the container described in Problem 7.35 to obtain an equilibrium concentration of O_2 of $0.100\ \text{mol} \cdot \text{L}^{-1}$?

▮ Letting x represent the original amount of NO_2 and y the amount of O_2 formed, the equilibrium table is

equation	$2NO_2(g)$ +	$2NO(g)$ →	$O_2(g)$
$C_{i,\text{initial}}/(\text{mol} \cdot \text{L}^{-1})$	x		
$C_{i,\text{change}}/(\text{mol} \cdot \text{L}^{-1})$	$-2y$	$+2y$	$+y$
$C_{i,\text{equilibrium}}/(\text{mol} \cdot \text{L}^{-1})$	$x - 2y$	$2y$	y

Substituting the above results with $y = 0.100\ \text{mol} \cdot \text{L}^{-1}$ into the expression for K_C determined in Problem 7.36 gives

$$K_C = \frac{(2y)^2 y}{(x - 2y)^2} = \frac{[2(0.100)]^2(0.100)}{[x - 2(0.100)]^2} = 0.0113$$

Solving gives $x = 0.795\ \text{mol} \cdot \text{L}^{-1}$. This corresponds to

$$n(NO_2)_{\text{initial}} = (10.0\ \text{L})(0.785\ \text{mol} \cdot \text{L}^{-1}) = 7.95\ \text{mol}$$

7.39 What amount of NO_2 must be placed in the container described in Problem 7.35 to obtain an equilibrium concentration of NO_2 of $0.100\ \text{mol} \cdot \text{L}^{-1}$?

▮ The equilibrium table for this problem will be the same as that given in Problem 7.38, where $x - 2y = 0.100\ \text{mol} \cdot \text{L}^{-1}$. Substituting into the expression for K_C gives

$$K_C = \frac{(2y)^2 y}{(x - 2y)^2} = \frac{(2y)^2 y}{(0.100)^2} = 0.0113$$

Solving gives $y = 0.0305\ \text{mol} \cdot \text{L}^{-1}$. Thus,

$$x = 0.100\ \text{mol} \cdot \text{L}^{-1} + 2(0.0305\ \text{mol} \cdot \text{L}^{-1}) = 0.161\ \text{mol} \cdot \text{L}^{-1}$$

$$n(NO_2)_{\text{initial}} = (10.0\ \text{L})(0.161\ \text{mol} \cdot \text{L}^{-1}) = 1.61\ \text{mol}$$

7.40 A mixture containing 0.100 bar NO_2, 0.100 bar O_2, and 0.100 bar NO was allowed to equilibrate at 750 K. What are the equilibrium partial pressures? See Problem 7.35 for additional data. Assume ideal gas behavior.

▮ The value of the reaction coordinate is given by *(6.56)* as

$$Q_P = (0.100)^2(0.100)/(0.100)^2 = 0.100$$

Because $Q_P < K_P$, additional NO_2 will decompose to form NO and O_2. Letting x represent the amount of additional O_2 formed, the equilibrium table is

equation	$2NO_2(g)$ +	$2NO(g)$ →	$O_2(g)$
$P_{i,\text{initial}}/(\text{bar})$	0.100	0.100	0.100
$P_{i,\text{change}}/(\text{bar})$	$-2x$	$+2x$	$+x$
$P_{i,\text{equilibrium}}/(\text{bar})$	$0.100 - 2x$	$0.100 + 2x$	$0.100 + x$

Substituting into the expression for K_P and solving for x gives

$$K_P = \frac{(0.100 + 2x)^2(0.100 + x)}{(0.100 - 2x)^2} = 0.704$$

$$x = 0.0207 \text{ bar}$$

The equilibrium partial pressures are

$$P(NO_2) = 0.100 \text{ bar} - 2(0.0207 \text{ bar}) = 0.059 \text{ bar}$$

$$P(NO) = 0.100 \text{ bar} + 2(0.0207 \text{ bar}) = 0.141 \text{ bar}$$

$$P(O_2) = 0.100 \text{ bar} + 0.0207 \text{ bar} = 0.121 \text{ bar}$$

7.41 A mixture containing 1.25 bar of hydrogen and 0.75 bar of acetylene reacted to form ethene, and equilibrium was established at 25 °C. What are the partial pressures of the gases at equilibrium?

$$C_2H_2(g) + H_2(g) \rightarrow C_2H_4(g) \qquad K_P = 5.14 \times 10^{24}$$

▌ For a reaction that has a large value of K, usually the equilibrium table is modified to include a step in which the reaction is assumed to go to completion before the changes resulting from equilibrium are considered. Letting x represent the amount of acetylene present at equilibrium, the equilibrium table is

equation	$C_2H_2(g)$ +	$H_2(g)$ →	$C_2H_4(g)$
$P_{i,\text{initial}}/(\text{bar})$	0.75	1.25	
$P_{i,\text{reaction}}/(\text{bar})$	-0.75	-0.75	$+0.75$
$P_{i,\text{change}}/(\text{bar})$	$+x$	$+x$	$-x$
$P_{i,\text{equilibrium}}/(\text{bar})$	x	$0.50 + x$	$0.75 - x$

Substituting these results into *(7.2)* gives

$$K_P = \frac{0.75 - x}{x(0.50 + x)} = 5.14 \times 10^{24}$$

Solving gives $x = 2.9 \times 10^{-25}$ bar. The partial pressures are

$$P(C_2H_2) = 2.9 \times 10^{-25} \text{ bar}$$

$$P(H_2) = 0.50 \text{ bar} + 2.9 \times 10^{-25} \text{ bar} = 0.50 \text{ bar}$$

$$P(C_2H_4) = 0.75 \text{ bar} - 2.9 \times 10^{-25} \text{ bar} = 0.75 \text{ bar}$$

7.42 At a pressure of 1.00 bar, an equilibrium exists at 2000 K between 0.250 mol of $Br_2(g)$, 0.750 mol of $F_2(g)$, and 0.497 mol of $BrF_3(g)$. What will be the amounts of each gas after the pressure on the system has been increased to 2.00 bar and equilibrium at 2000 K reestablished?

$$Br_2(g) + 3F_2(g) \rightarrow 2BrF_3(g)$$

▮ Under the initial equilibrium conditions, the total amount of substance and the mole fraction were

$$n_t = 0.250\text{ mol} + 0.750\text{ mol} + 0.497\text{ mol} = 1.497\text{ mol}$$

$$x(Br_2) = (0.250\text{ mol})/(1.497\text{ mol}) = 0.167$$

$$x(F_2) = (0.750\text{ mol})/(1.497\text{ mol}) = 0.501$$

$$x(BrF_3) = (0.497\text{ mol})/(1.497\text{ mol}) = 0.332$$

The value of K_P as expressed using *(7.2)* is

$$K_P = \frac{[P(BrF_3)/(\text{bar})]^2}{[P(Br_2)/(\text{bar})][P(F_2)/(\text{bar})]^3} = \frac{[(0.332)(1.00)]^2}{[(0.167)(1.00)][(0.501)(1.00)]^3}$$
$$= 5.25$$

After the pressure has increased to 2.00 bar, the reaction quotient given by *(6.56)* is

$$Q_P = \frac{[(0.332)(2.00)]^2}{[(0.167)(2.00)][(0.501)(2.00)]^3} = 1.31$$

Because $Q_P < K_P$, additional Br_2 will react with F_2 to form more BrF_3 until equilibrium conditions are again established. Letting x represent the amount of Br_2 that reacts, the equilibrium table is

equation	$Br_2(g)$ +	$3F_2(g)$ →	$2BrF_3(g)$
$n_{i,\text{initial}}/(\text{mol})$	0.250	0.750	0.497
$n_{i,\text{change}}/(\text{mol})$	$-x$	$-3x$	$+2x$
$n_{i,\text{equilibrium}}/(\text{mol})$	$0.250 - x$	$0.750 - 3x$	$0.497 + 2x$

The total amount of substance and mole fractions are

$$n_t = (0.250 - x)\text{ mol} + (0.750\text{ mol} - 3x)\text{ mol} + (0.497 + 2x)\text{ mol} = (1.497 - 2x)\text{ mol}$$

$$x(Br_2) = \frac{(0.250 - x)\text{ mol}}{(1.497 - 2x)\text{ mol}}$$

$$x(F_2) = \frac{(0.750 - 3x)\text{ mol}}{(1.497 - 2x)\text{ mol}}$$

$$x(BrF_3) = \frac{(0.497 + 2x)\text{ mol}}{(1.497 - 2x)\text{ mol}}$$

Substituting into *(7.2)* and solving for x gives

$$K_P = \frac{\{[(0.497 + 2x)/(1.497 - 2x)]2.00\}^2}{\{[(0.250 - x)/(1.497 - 2x)]2.00\}\{[(0.750 - 3x)/(1.497 - 2x)]2.00\}^3} = 5.25$$

$$x = 0.0609\text{ mol}$$

The amounts of substances under the new equilibrium conditions will be

$$n(Br_2) = 0.250\text{ mol} - 0.0609\text{ mol} = 0.189\text{ mol}$$

$$n(F_2) = 0.750\text{ mol} - 3(0.0609\text{ mol}) = 0.567\text{ mol}$$

$$n(BrF_3) = 0.497\text{ mol} + 2(0.0609\text{ mol}) = 0.619\text{ mol}$$

7.43 Derive the relation between $\ln K_C$ and $\Delta_r A°$ for a chemical reaction involving ideal gases.

▮ Taking logarithms of both sides of *(7.8)* and substituting *(7.1a)* and *(6.5)* gives

$$\ln K_P = \Delta n_g \ln RT + \ln K_C$$

$$\frac{-\Delta_r G°}{RT} = \Delta n_g \ln RT + \ln K_C$$

$$-\Delta_r G° = RT\,\Delta n_g \ln RT + RT \ln K_C$$

$$-\Delta_r A^\circ - RT\,\Delta n_g = RT\,\Delta n_g \ln RT + RT \ln K_C$$

$$\ln K_C = \frac{-\Delta_r A^\circ}{RT} - \Delta n_g(1 + \ln RT)$$

7.44 Derive the relation for the temperature dependence of $\ln K_C$ for a reaction involving ideal gases.

❚ Substituting *(7.8)* into *(7.4)*, differentiating with respect to T, substituting *(3.5)*, and solving for $\partial(\ln K_C)/\partial T$ gives

$$\frac{\partial(\ln K_P)}{\partial T} = \frac{\partial\{\ln[(RT)^{\Delta n_g} K_C]\}}{\partial T} = \frac{\partial(\Delta n_g \ln R + \Delta n_g \ln T + \ln K_C)}{\partial T} = \frac{\Delta n_g}{T} + \frac{\partial(\ln K_C)}{\partial T} = \frac{\Delta_r H^\circ}{RT^2}$$

$$\frac{\partial(\ln K_C)}{\partial T} = \frac{\Delta_r H^\circ}{RT^2} - \frac{\Delta n_g}{T} = \frac{\Delta_r H^\circ}{RT^2} - \frac{RT\,\Delta n_g}{RT^2} = \frac{\Delta_r H^\circ - RT\,\Delta n_g}{RT^2} = \frac{\Delta_r U^\circ}{RT^2}$$

7.45 Derive the relation for the pressure dependence of $\ln K_\gamma$ in terms of $\ln K_P$.

❚ Taking logarithms of both sides of *(7.10)*, differentiating with respect to P, and rearranging gives the desired relation:

$$\ln K = \ln K_\gamma + \ln K_P$$

$$\left(\frac{\partial(\ln K)}{\partial P}\right)_T = 0 = \left(\frac{\partial(\ln K_\gamma)}{\partial P}\right)_T + \left(\frac{\partial(\ln K_P)}{\partial P}\right)_T$$

$$\left(\frac{\partial(\ln K_\gamma)}{\partial P}\right)_T = -\left(\frac{\partial(\ln K_P)}{\partial P}\right)_T$$

7.46 Find $[\partial(\ln K_x)/\partial T]_P$ and $[\partial(\ln K_x)/\partial P]_T$ for a reaction involving ideal gases.

❚ Taking logarithms of both sides of *(7.9)*, solving for $\ln K_x$, differentiating with respect to T, and substituting into *(7.4)* gives

$$\ln K_P = \Delta n_g \ln P + \ln K_x$$

$$\ln K_x = \ln K_P - \Delta n_g \ln P$$

$$\left(\frac{\partial(\ln K_x)}{\partial T}\right)_P = -\left(\frac{\partial(\ln K_P)}{\partial T}\right)_P - 0 = \frac{\Delta_r H^\circ}{RT^2}$$

Differentiating the above expression for $\ln K_x$ with respect to P gives

$$\left(\frac{\partial(\ln K_x)}{\partial P}\right)_T = 0 - \frac{\Delta n_g}{P} = \frac{-\Delta n_g}{P}$$

7.47 For the equation $CaCO_3(s) \rightarrow CaO(s) + CO_2(g)$, $K_{1000} = 0.059$. Exactly 10 g of $CaCO_3$ is placed in a 10.0-L container at 1000 K. After equilibrium is reached, what mass of $CaCO_3$ remains?

❚ Assuming the activities of the pure solids to be unity and assuming ideal gas behavior for CO_2, *(7.2)* gives

$$K = a(\mathrm{CaO})a(\mathrm{CO_2})/a(\mathrm{CaCO_3}) = [P(\mathrm{CO_2})/(\mathrm{bar})] = 0.059$$

Substituting into *(1.18)* gives

$$n = \frac{(0.059\ \mathrm{bar})(10.0\ \mathrm{L})}{(0.083\,14\ \mathrm{L\cdot bar\cdot K^{-1}\cdot mol^{-1}})(1000\ \mathrm{K})} = 0.0071\ \mathrm{mol\ CO_2}$$

The mass of $CaCO_3$ remaining is

$$m = 10.00\ \mathrm{g} - (0.071\ \mathrm{mol\ CO_2})\frac{1\ \mathrm{mol\ CaCO_3}}{1\ \mathrm{mol\ CO_2}}(100.09\ \mathrm{g\cdot mol^{-1}}) = 9.29\ \mathrm{g\ CaCO_3}$$

7.48 The vapor pressure of water over an equilibrium mixture of $Na_2SO_4\cdot 10H_2O(s)$ and $Na_2SO_4(s)$ is 19.1 torr at 25 °C. Determine $\Delta_r G^\circ$ for

$$Na_2SO_4\cdot 10H_2O(s) \rightarrow Na_2SO_4(s) + 10H_2O(g)$$

❚ Assuming the activities of the pure solids to be unity and assuming ideal gas behavior for H_2O, *(7.2)* gives

$$K = a(\mathrm{Na_2SO_4})a(\mathrm{H_2O})^{10}/a(\mathrm{Na_2SO_4\cdot 10H_2O}) = [P(\mathrm{H_2O})/(\mathrm{bar})]^{10}$$

Substituting the data gives

$$K = \left[(19.1\ \text{torr})\frac{1\ \text{atm}}{760\ \text{torr}}\left(\frac{1.013\,25\ \text{bar}}{1\ \text{atm}}\right)\right]^{10} = 1.15 \times 10^{-16}$$

Rearranging *(7.1a)* and substituting the data gives

$$\Delta_r G^\circ = -(8.314\ \text{J} \cdot \text{K}^{-1} \cdot \text{mol}^{-1})\frac{10^{-3}\ \text{kJ}}{1\ \text{J}}(298.15\ \text{K})(\text{mol})\ln(1.15 \times 10^{-16})$$

$$= 90.98\ \text{kJ}$$

7.49 The vapor pressure of H_2 over LiH(s) is 0.278 mbar at 800 K and 6.19 mbar at 900 K. Determine $\Delta_r H^\circ_{850}$ for $LiH(s) \rightarrow Li(l) + \frac{1}{2}H_2(g)$.

▮ Assuming the activities of the pure solid and pure liquid to be unity and assuming ideal gas behavior for H_2, *(7.2)* gives

$$K = a(\text{Li})a(\text{H}_2)^{1/2}/a(\text{LiH}) = [P(\text{H}_2)/(\text{bar})]^{1/2}$$

Substituting the data gives

$$K_{800} = (2.78 \times 10^{-4})^{1/2} = 1.67 \times 10^{-2}$$

$$K_{900} = (6.19 \times 10^{-3})^{1/2} = 7.87 \times 10^{-2}$$

Rearranging *(7.5b)* and substituting the data gives

$$\Delta_r H^\circ_{850} = \frac{-(8.314\ \text{J} \cdot \text{K}^{-1} \cdot \text{mol}^{-1})[(10^{-3}\ \text{kJ})/(1\ \text{J})](\text{mol})\ln[(7.87 \times 10^{-2})/(1.67 \times 10^{-2})]}{(900\ \text{K})^{-1} - (800\ \text{K})^{-1}}$$

$$= 92.8\ \text{kJ}$$

7.50 Hot copper turnings can be used as an "oxygen getter" for inert gas supplies by slowly passing the gas over the turnings at 600 K:

$$2\text{Cu(s)} + \tfrac{1}{2}\text{O}_2\text{(g)} \rightarrow \text{Cu}_2\text{O(s)} \qquad \Delta_f G^\circ = -124.944\ \text{kJ}$$

How many molecules of O_2 are left in 1 L of a gas supply after equilibrium has been reached?

▮ Using *(7.1a)* gives

$$\ln K = \frac{-(-124.944\ \text{kJ})[(10^3\ \text{J})/(1\ \text{kJ})]}{(8.314\ \text{J} \cdot \text{K}^{-1} \cdot \text{mol}^{-1})(600\ \text{K})(\text{mol})} = 25.05$$

$$K = 7.5 \times 10^{10}$$

Assuming the activities of the pure solids to be unity and assuming ideal gas behavior for O_2, *(7.2)* gives

$$K = \frac{a(\text{Cu}_2\text{O})}{a(\text{Cu})^2 a(\text{O}_2)^{1/2}} = \frac{1}{[P(\text{O}_2)/(\text{bar})]^{1/2}} = 7.5 \times 10^{10}$$

$$P(\text{O}_2) = 1.8 \times 10^{-22}\ \text{bar}$$

Substituting into *(1.21)* gives

$$N^* = \frac{(1.8 \times 10^{-22}\ \text{bar})(6.022 \times 10^{23}\ \text{mol}^{-1})}{(0.08314\ \text{L} \cdot \text{bar} \cdot \text{K}^{-1} \cdot \text{mol}^{-1})(600\ \text{K})} = 2.2\ \text{L}^{-1}$$

7.51 Determine the effect on the equilibrium for the equation

$$\text{Br}_2\text{(g)} + 3\text{F}_2\text{(g)} \rightarrow 2\text{BrF}_3\text{(g)}$$

caused by **(*a*)** increasing $x(F_2)$, **(*b*)** decreasing $x(Br_2)$, **(*c*)** increasing $x(BrF_3)$, **(*d*)** increasing the total pressure a small amount, **(*e*)** increasing the temperature given that $\Delta_r H^\circ$ is negative, and **(*f*)** adding a small amount of an inert gas.

▮ For a gaseous reaction, using *(6.2a)*, *(7.1b)*, *(7.9)*, and *(7.10)* gives the mole fraction of one of the products (x_R^r) as

$$x_R^r = \frac{x_L^l x_M^m \cdots}{x_S^s \cdots} K_\gamma^{-1} P^{-\Delta n_g} e^{-\Delta_r H^\circ/RT} e^{\Delta_r S^\circ/R} \qquad (7.11)$$

where L, M, ... are reactants and S, ... are products. For the formation of BrF_3,

$$x(\text{BrF}_3)^2 = x(\text{Br}_2)x(\text{F}_2)^3 K_\gamma^{-1} P^2 e^{-\Delta_r H^\circ/RT} e^{\Delta_r S^\circ/R}$$

According to *(7.11)*, **(a)** increasing $x(F_2)$ will cause $x(BrF_3)$ to increase [and $x(Br_2)$ to decrease]; **(b)** decreasing $x(Br_2)$ will cause $x(BrF_3)$ to decrease [and $x(F_2)$ to increase]; and **(c)** increasing $x(BrF_3)$ will cause both $x(Br_2)$ and $x(F_2)$ to increase. A small increase in total pressure will not result in a significant change in K_y, so **(d)** increasing the pressure will cause $x(BrF_3)$ to increase. The value of $e^{-\Delta_r H°/RT}$ is greater than unity since $\Delta_r H°$ is negative, so **(e)** increasing the temperature will decrease the value of the exponential, which will cause $x(BrF_3)$ to decrease. As long as the gases are ideal, **(f)** adding a small amount of an inert gas will not change $x(BrF_3)$.

7.3 EQUILIBRIUM IN AQUEOUS SOLUTIONS

7.52 What is the pH of a 0.010 M solution of hydroiodic acid?

▌ According to the Brønsted-Lowry theory, an equilibrium is established between an acid (A) and its conjugate base (B) represented by either

$$A(aq) + H_2O(l) \rightleftarrows H_3O^+(aq) + B(aq) \quad \text{or} \quad A(aq) \rightleftarrows H^+(aq) + B(aq) \tag{7.12}$$

where the equilibrium concentrations of the acid and its conjugate base are

$$[A] = C_A - ([H^+] - [OH]) \tag{7.13a}$$

$$[B] = C_B + ([H^+] - [OH]) \tag{7.13b}$$

In aqueous solution, HI is a "strong" acid that is essentially 100% ionized. Thus $[A] \approx 0$ and *(7.12a)* gives

$$[H^+] = C_A + [OH^-] = C_A + K_w/[H^+] \tag{7.14}$$

where the *water ionization equilibrium constant* expression is given by

$$K_w = [H^+][OH^-] = 1.00 \times 10^{-14} \tag{7.15}$$

at 25 °C. In this case $C_A \gg [OH^-]$. Using *(7.14)* gives $[H^+] = 0.010$ M. The pH is

$$pH = -\log[H^+] \tag{7.16}$$

$$pH = -\log 0.010 = 2.00$$

7.53 What is the pH of a 1.00×10^{-8} M solution of NaOH?

▌ For a strong base, $[B] = 0$. Thus *(7.13b)* becomes

$$0 = C_B + [H^+] - [OH^-] = C_B + \frac{K_w}{[OH^-]} - [OH^-] \tag{7.17}$$

In this case, C_B is not considerably greater than $[H^+]$, and so the quadratic equation must be solved. Substituting the data gives

$$0 = 1.00 \times 10^{-8} + \frac{1.00 \times 10^{-14}}{[OH^-]} - [OH^-]$$

$$[OH^-] = 1.05 \times 10^{-7}\ M$$

The pOH of the solution is

$$pOH = -\log[OH^-] \tag{7.18}$$

$$pOH = -\log(1.05 \times 10^{-7}) = 6.979$$

and the pH is

$$pH = pK_w - pOH \tag{7.19}$$

$$pH = 14.000 - 6.979 = 7.021$$

7.54 Find the pH of a 0.015 M solution of cyanic acid given $K_a = 3.3 \times 10^{-4}$ for HNCO.

▌ The thermodynamic equilibrium constant for *(7.12)* is

$$K_a = \frac{a(H^+)a(B)}{a(A)} = \frac{y(H^+)y(B)}{y(A)}\left(\frac{[H^+][B]}{[A]}\right) \tag{7.20}$$

when K_a is the *acid ionization* (or *acid dissociation*) *constant.* In most calculations, the activity coefficients are ignored, and *(7.20)* becomes

$$K_a = [H^+][B]/[A] \tag{7.21}$$

Substitution of *(7.13)* into *(7.21)* gives

$$K_a = \frac{[H^+](C_B + [H^+] - [OH^-])}{C_A - [H^+] + [OH^-]} \tag{7.22}$$

For a solution of a weak acid, $C_B = 0$. Thus *(7.22)* becomes

$$K_a = \frac{[H^+]([H^+] - [OH^-])}{C_A - [H^+] + [OH^-]} \tag{7.23a}$$

This is a cubic equation in $[H^+]$. However, if $[H^+] \gg [OH^-]$ (which is usually true of an acid unless $K_a \ll 10^{-15}$), then

$$K_a = \frac{[H^+]^2}{C_A - [H^+]} \tag{7.23b}$$

which is only a quadratic equation in $[H^+]$, and if $C_A > 10^4 K_a$, then

$$K_a = [H^+]^2/C_A \tag{7.23c}$$

For the cyanic acid solution of interest, *(7.23b)* can be used, giving

$$3.3 \times 10^{-4} = \frac{[H^+]^2}{0.015 - [H^+]}$$

$$[H^+] = 2.1 \times 10^{-3}\ \text{M}$$

The pH is given by *(7.16)* as $\text{pH} = -\log(2.1 \times 10^{-3}) = 2.68$.

7.55 Find the pH of a 1.0×10^{-3} M aqueous solution of phenol given $K_a = 1.28 \times 10^{-10}$.

▌ The low value of K_a indicates that probably *(7.23a)* should be used to solve for $[H^+]$. Substituting *(7.15)* for $[OH^-]$ and arranging gives

$$[H^+]^3 + K_a[H^+]^2 - (K_aC_A + K_w)[H^+] - K_wK_a = 0$$

Substituting the data and solving for $[H^+]$ gives

$$[H^+]^3 + (1.28 \times 10^{-10})[H^+]^2 - \{(1.28 \times 10^{-10})(1.0 \times 10^{-3}) + (1.00 \times 10^{-14})\}[H^+]$$
$$- (1.00 \times 10^{-14})(1.28 \times 10^{-10}) = 0$$

$$[H^+] = 3.7 \times 10^{-7}\ \text{M}$$

The pH is given by *(7.16)* as $\text{pH} = -\log(3.7 \times 10^{-7}) = 6.43$.

7.56 What is the pH of a 0.010 M solution of ammonia given $K_b = 1.6 \times 10^{-5}$ for NH_3?

▌ For the equilibrium involving a weak base, using *(7.22)* with $C_A = 0$ gives

$$K_a = \frac{[H^+](C_B + [H^+] - [OH^-])}{-[H^+] + [OH^-]} \tag{7.24a}$$

Usually *(7.24a)* is written using the *base ionization* (or *base dissociation*) *constant* given by

$$K_w = K_aK_b \tag{7.25}$$

Solving *(7.24a)* requires the solution of a cubic equation in $[OH^-]$. However, if $[OH^-] \gg [H^+]$, then *(7.24a)* can be simplified to

$$K_a = [H^+](C_B - [OH^-])/[OH^-] \tag{7.24b}$$

Finally, if $C_B > 10^4 K_a$, *(7.24b)* can be simplified to

$$K_a = [H^+]C_B/[OH^-] \tag{7.24c}$$

For the ammonia solution being considered, *(7.24b)* can be used, giving

$$K_a = \frac{[H^+](C_B - [OH^-])}{[OH^-]} = \frac{K_w}{K_b} = \frac{(K_w/[OH^-])(C_B - [OH^-])}{[OH^-]}$$

$$K_b = \frac{[OH^-]^2}{C_B - [OH^-]} = \frac{[OH^-]^2}{0.010 - [OH^-]} = 1.6 \times 10^{-5}$$

Solving gives $[OH^-] = 3.9 \times 10^{-4}$ M. Using *(7.18)* and *(7.19)* gives

$$pOH = -\log(3.9 \times 10^{-4}) = 3.41$$

$$pH = 14.00 - 3.41 = 10.59$$

7.57 What is the pH of a buffer solution prepared using 0.010 M formic acid and 0.010 M Na(HCOO)? For HCOOH, $K_a = 1.772 \times 10^{-4}$.

▌ This buffer solution contains a weak acid and its conjugate base. For an acidic solution where $[H^+] \gg [OH^-]$, *(7.22)* becomes

$$K_a = \frac{[H^+](C_B + [H^+])}{C_A - [H^+]} \tag{7.26a}$$

If C_B and $C_A \gg [H^+]$, then

$$K_a = [H^+]C_B/C_A \tag{7.26b}$$

which in the logarithm form is known as the *Henderson-Hasselbalch equation*

$$pH = pK_a + \log(C_B/C_A) \tag{7.26c}$$

In this case C_B and C_A will not be considerably greater than $[H^+]$, so using *(7.26a)* gives

$$1.772 \times 10^{-4} = \frac{[H^+](0.010 + [H^+])}{0.010 - [H^+]}$$

The solution is $[H^+] = 1.7 \times 10^{-4}$, which corresponds to $pH = -\log(1.7 \times 10^{-4}) = 3.77$.

7.58 Determine the pH of the buffer solution described in Problem 7.57 after the addition of 0.001 mol of NaOH to 1.00 L of the buffer. Compare this change in pH to that for addition of 0.001 mol of NaOH to 1.00 L of pure water.

▌ The 0.001 mol of NaOH reacts with the un-ionized HCOOH to produce additional Na(HCOO). Substituting

$$C_A = \frac{[(0.010\ \text{mol})/(1\ \text{L})](1.00\ \text{L}) - (0.001\ \text{mol NaOH})[(1\ \text{mol HCOOH})/(1\ \text{mol NaOH})]}{1.00\ \text{L}}$$

$$= 0.009\ \text{mol} \cdot \text{L}^{-1}$$

$$C_B = \frac{[(0.010\ \text{mol})/(1\ \text{L})](1.00\ \text{L}) + (0.001\ \text{mol NaOH})[(1\ \text{mol HCOO}^-)/(1\ \text{mol NaOH})]}{1.00\ \text{L}}$$

$$= 0.011\ \text{mol} \cdot \text{L}^{-1}$$

into *(7.26a)* gives

$$1.772 \times 10^{-4} = \frac{[H^+](0.011 + [H^+])}{0.009 - [H^+]}$$

and solving gives $[H^+] = 1.4 \times 10^{-4}$ M, which corresponds to

$$pH = -\log(1.4 \times 10^{-4}) = 3.85$$

This is a change of only 0.08 pH unit compared to that of pure water, which would change from 7.0 to 11.0 for the same process.

7.59 The *buffer capacity* (β) for a weak acid-conjugate base buffer is defined as the number of moles of strong acid or base needed to change the pH of 1 L of solution by 1 pH unit, where

$$\beta = \frac{d[A]}{d\text{pH}} = \frac{2.303(C_A + C_B)K_a[H^+]}{([H^+] + K_a)^2} \tag{7.27}$$

Under what conditions will a buffer best resist a change in pH?

▌ From *(7.27)* it can be seen that β is directly proportional to the concentrations of the weak acid and the conjugate base. Differentiating *(7.27)* with respect to $[H^+]$, setting the result equal to zero, and solving for $[H^+]$ gives

$$\frac{\partial \beta}{\partial [H^+]} = 2.303(C_A + C_B)K_a\left[\frac{1}{([H^+] + K_a)^2} + \frac{-2[H^+]}{([H^+] + K_a)^3}\right] = 0$$

$$1 - \frac{2[H^+]}{[H^+] + K_a} = 0 \qquad [H^+] = K_a \qquad pH = pK_a$$

The best buffering action will occur at $pH = pK_a$.

7.60 Prepare a plot of the buffer capacity against pH for the buffer described in Problem 7.57. If the effective buffer range is considered to include values of β greater than 0.25 the maximum value of β, find the effective buffer range. What does this result imply about C_A/C_B?

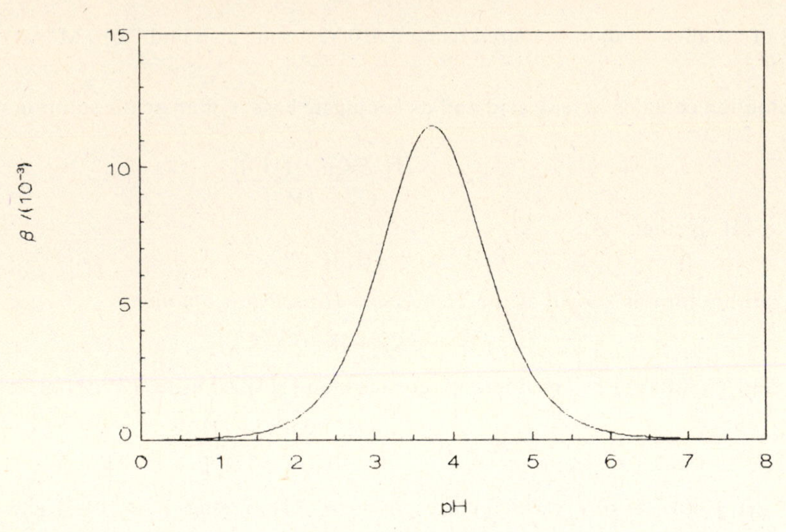

Fig. 7-5

A plot of β against pH is shown in Fig. 7-5. As a sample calculation, consider pH = 4.00 or $[H^+] = 1.0 \times 10^{-4}\ \text{mol}\cdot\text{L}^{-1}$:

$$\beta = \frac{2.303(0.010 + 0.010)(1.772 \times 10^{-4})(1.0 \times 10^{-4})}{(1.0 \times 10^{-4} + 1.772 \times 10^{-4})^2} = 0.011$$

Substituting $[H^+] = K_a$ into *(7.27)* gives the maximum value of β as

$$\beta_{\text{max}} = \frac{2.303(C_A + C_B)K_aK_a}{(K_a + K_a)^2} = \frac{2.303(C_A + C_B)}{4}$$

Setting *(7.27)* equal to $0.25\beta_{\text{max}}$ and solving for $[H^+]$ gives

$$\frac{2.303(C_A + C_B)K_a[H^+]}{([H^+] + K_a)^2} = 0.25\frac{2.303(C_A + C_B)}{4}$$

$$[H^+]^2 - 14K_a[H^+] + K_a^2 = 0$$

$$[H^+] = 0.07K_a \qquad \text{pH} = \text{p}K_a + 1.2$$

This value, ± 1 pH unit, is similar to that observed in Fig. 7-5. Using *(7.26c)* indicates that the ratio of C_A/C_B will vary from 0.1 to 10 over this pH range.

7.61 What is the pH of a buffer solution consisting of 0.10 M NH_3 and 0.15 M NH_4NO_3? For NH_3, $K_b = 1.6 \times 10^{-5}$ at 25 °C.

This buffer solution contains a weak base and its conjugate acid. For a basic solution where $[OH^-] \gg [H^+]$, *(7.22)* becomes

$$K_a = \frac{[H^+](C_B - [OH^-])}{C_A + [OH^-]} = \frac{K_w}{K_b} = \frac{(K_w/[OH^-])(C_B - [OH^-])}{C_A + [OH^-]} \qquad (7.28)$$

Substituting the data into *(7.28)* and solving for $[OH^-]$ gives

$$K_b = \frac{[OH^-](C_A + [OH^-])}{(C_B - [OH^-])} = 1.6 \times 10^{-5} = \frac{[OH^-](0.15 + [OH^-])}{0.10 - [OH^-]}$$

$$[OH^-] = 1.1 \times 10^{-5}$$

Substituting into *(7.18)* and *(7.19)* gives

$$\text{pOH} = -\log(1.1 \times 10^{-5}) = 4.96$$

$$\text{pH} = 14.00 - 4.96 = 9.04$$

7.62 What is the pH of a 0.25 M solution of $NaHCO_3$? For H_2CO_3 at 25 °C, $K_{a_1} = 4.5 \times 10^{-7}$ and $K_{a_2} = 4.8 \times 10^{-11}$.

For a solution of an *ampholyte* (or *amphoteric ion*) with the formula HA^-,

$$[H^+] = \left(\frac{K_{a_1}(K_{a_2}[HA^-] + K_w)}{K_{a_1} + [HA^-]}\right)^{1/2} \tag{7.29a}$$

where $[HA^-]$ is usually assumed to be equal to the formal concentration of the salt. In this case, $K_{a_2}[HCO_3^-] > K_w$ and $[HCO_3^-] > K_{a_1}$, and *(7.29a)* simplifies to

$$[H^+] = (K_{a_1}K_{a_2})^{1/2} \tag{7.29b}$$

$$[H^+] = [(4.5 \times 10^{-7})(4.8 \times 10^{-11})]^{1/2} = 4.6 \times 10^{-9}\ M$$

Using *(7.16)* gives $pH = -\log(4.6 \times 10^{-9}) = 8.34$.

7.63 What is the pH of a 0.010 M glycine solution? For glycine, $K_{a_1} = 4.47 \times 10^{-3}$ and $K_{a_2} = 1.67 \times 10^{-10}$ at 25 °C.

The equilibrium of the zwitterion can be written as

$$\underset{HGly^+}{NH_3^+-CH_2-COOH} \overset{K_{a_1}}{\rightleftharpoons} \underset{Gly}{NH_3^+-CH_2-COO^-} \overset{K_{a_2}}{\rightleftharpoons} \underset{Gly^-}{NH_2-CH_2-COO^-}$$

This is just a special case of the equilibrium theory developed for ampholytes where $[H_2A] = [HGly^+]$, $[HA] = [Gly]$, and $[A^-] = [Gly^-]$. Substituting the data into *(7.29a)* gives

$$[H^+] = \left(\frac{(4.47 \times 10^{-3})[(1.67 \times 10^{-10})(0.010) + 1.00 \times 10^{-14}]}{4.47 \times 10^{-3} + 0.010}\right)^{1/2}$$

$$= 7.2 \times 10^{-7}\ M$$

and using *(7.16)* gives $pH = -\log(7.2 \times 10^{-7}) = 6.14$.

7.64 What is the pH of a 0.100 M NH_4Cl solution given $K_b = 1.6 \times 10^{-5}$ for NH_3?

An acidic solution will be formed as the NH_4^+ undergoes hydrolysis. Substituting *(7.25)* into *(7.23c)* gives

$$[H^+] \approx [(K_w/K_b)C_A]^{1/2} \tag{7.23d}$$

Substituting the data into *(7.23d)* and using *(7.16)* gives

$$[H^+] = \left(\frac{1.00 \times 10^{-14}}{1.6 \times 10^{-5}}(0.100)\right)^{1/2} = 7.9 \times 10^{-6}\ M$$

$$pH = -\log(7.9 \times 10^{-6}) = 5.10$$

7.65 What is the pH of a 0.100 M solution of Na_2SO_4? For H_2SO_4, $K_{a_2} = 1.0 \times 10^{-2}$.

A basic solution will be found as the anion undergoes hydrolysis. Substituting *(7.25)* into *(7.24c)* gives

$$[OH^-] \approx [(K_w/K_a)C_B]^{1/2} \tag{7.24d}$$

Substituting the data into *(7.24d)* and using *(7.18)* and *(7.19)* gives

$$[OH^-] = \left(\frac{1.00 \times 10^{-14}}{1.0 \times 10^{-2}}(0.100)\right)^{1/2} = 3.2 \times 10^{-7}\ M$$

$$pOH = -\log(3.2 \times 10^{-7}) = 6.49$$

$$pH = 14.00 - 6.49 = 7.51$$

7.66 What is the pH of a 0.010 M solution of $NH_4(HCOO)$ given $K_a = 1.772 \times 10^{-4}$ for HCOOH and $K_b = 1.6 \times 10^{-5}$ for NH_3?

Because both the anion and cation hydrolyze, *(7.29a)* becomes

$$[H^+] = \left(\frac{K_wK_a(C + K_b)}{K_b(C + K_a)}\right)^{1/2} \tag{7.29c}$$

provided $K_aK_b \gg K_w$. In this case, $C \gg K_a$ and K_b, so

$$[H^+] = (K_w K_a / K_b)^{1/2} \tag{7.29d}$$

$$[H^+] = \left[\frac{(1.00 \times 10^{-14})(1.772 \times 10^{-4})}{1.6 \times 10^{-5}}\right]^{1/2} = 3.3 \times 10^{-7}\ M$$

Using *(7.16)* gives $pH = -\log(3.3 \times 10^{-7}) = 6.48$.

7.67 A 50.0-mL aliquot of 0.010 M solution of HCOOH was titrated with 0.100 M NaOH. Predict the pH of the solution **(a)** at the beginning of the reaction, **(b)** at the half-equivalence point, **(c)** at the equivalence point, and **(d)** after 10.0 mL of base has been added. For HCOOH, $K_a = 1.772 \times 10^{-4}$.

(a) At the beginning of the reaction, *(7.23c)* gives

$$pH = \frac{pK_a - \log C_A}{2} = \frac{-\log(1.772 \times 10^{-4}) - \log 0.010}{2} = 2.88$$

(b) At the half-equivalence point, $C_A \approx C_B$ and *(7.26b)* gives

$$pH = pK_a = -\log(1.772 \times 10^{-4}) = 3.7515$$

(c) At the equivalence point, hydrolysis of the anion will occur. The value of C_B is

$$C_B = C_{A,0}\frac{V_A}{V_{soln}} = (0.010\ M)\frac{50.0\ mL}{55.0\ mL} = 0.0091\ M$$

Using *(7.24d)* gives

$$pH = \frac{pK_w + pK_a + \log C_B}{2} = \frac{14.00 - \log(1.772 \times 10^{-4}) + \log(0.0091)}{2} = 7.86$$

(d) In excess base, the concentration of NaOH is

$$C_B = \frac{C_{B,0}V_B - C_{A,0}V_A}{V_{soln}} = \frac{(0.100\ M)(10.0\ mL) - (0.010\ M)(50.0\ mL)}{60.0\ mL} = 0.0083\ M$$

Substituting into *(7.17)* gives

$$pH = pK_w + \log C_B = 14.00 + \log 0.0083 = 11.92$$

7.68 Consider the titration of a 0.010 M solution of phthalic acid with 0.010 M NaOH. Predict the pH **(a)** at the beginning of the reaction, **(b)** at the first half-equivalence point, **(c)** at the first equivalence point, **(d)** at the second half-equivalence point, and **(e)** at the second equivalence point. For $C_6H_6(COOH)_2$, $K_{a_1} = 1.3 \times 10^{-3}$ and $K_{a_2} = 3.9 \times 10^{-6}$.

For each of the points *(7.29a)* gives

(a) $$pH = \frac{pK_{a_1} - \log C_A}{2} = \frac{-\log(1.3 \times 10^{-3}) - \log(0.010)}{2} = 2.44$$

(b) $$pH = pK_{a_1} = -\log(1.3 \times 10^{-3}) = 2.89$$

(c) $$pH = \frac{pK_{a_1} + pK_{a_2}}{2} = \frac{-\log(1.3 \times 10^{-3}) - \log(3.9 \times 10^{-6})}{2} = 4.15$$

(d) $$pH = pK_{a_2} = -\log(3.9 \times 10^{-6}) = 5.41$$

(e) $$pH = \frac{pK_w + pK_{a_2} + \log C_B}{2} = \frac{14.00 - \log(3.9 \times 10^{-6}) + \log(0.010/3)}{2}$$

$$= 8.47$$

7.69 The pH of the half-equivalence point of the titration of valine, $(CH_3)_2CHCHNH_2COOH$, is 2.286 with HCl and 9.719 with NaOH. Calculate the *isoelectric point* of valine—the pH at which the dipolar ion does not migrate in an electrical field.

The equilibrium of the zwitterion can be written as

$$\begin{array}{c}NH_3^+\\|\\HCCH(CH_3)_2\\|\\COOH\end{array} \xrightleftharpoons{pH'=2.286} \begin{array}{c}NH_3^+\\|\\HCCH(CH_3)_2\\|\\COO^-\end{array} \xrightleftharpoons{pH''=9.719} \begin{array}{c}NH_2\\|\\HCCH(CH_3)_2\\|\\COO^-\end{array}$$

where

$$pH = \frac{pH' + pH''}{2} = \frac{2.286 + 9.179}{2} = 6.003$$

7.70 What is the degree of ionization in a 0.100 M aqueous solution of lactic acid, $CH_3CHOHCOOH$, given $K_a = 1.374 \times 10^{-4}$? Do not assume ideal behavior.

▌ If the solution were acting ideally, *(7.23b)* would give the value of $[H^+]$ as

$$1.374 \times 10^{-4} = [H^+]^2/(0.100 - [H^+])$$

$$[H^+] = 3.64 \times 10^{-3}\ M$$

The ionic strength of the solution of the 1:1 electrolyte (see Table 6-2) is

$$I = C = 3.64 \times 10^{-3}\ M$$

which upon substitution into *(6.51)* gives

$$\log y_\pm = -(1)(1)(0.5116)(3.64 \times 10^{-3})^{1/2} = -0.0309$$

$$y_\pm = 0.931$$

Assuming $y(A) = 1$, using *(7.20)* gives

$$1.374 \times 10^{-4} = \frac{(0.931)^2}{1}\left(\frac{[H^+]}{0.100 - [H^+]}\right)$$

Solving gives $[H^+] = 3.90 \times 10^{-3}$ M. Repeating the process of calculating I, $y_\pm$, and $[H^+]$ gives a final answer of $[H^+] = 3.91 \times 10^{-3}$ M. The fraction of lactic acid molecules undergoing ionization is

$$\alpha = \frac{3.91 \times 10^{-3}\ M}{0.100\ M} = 0.0391$$

7.71 The ionization constant for acetic acid has been measured at 25 °C as a function of ionic strength. Use the following data to determine K_a.

$I/(M)$	0.009 51	0.024 03	0.041 75	0.097 81	0.160 95	0.179 94
$K'/(10^{-5})$	1.752	1.747	1.743	1.734	1.724	1.726

For each concentration, calculate $y_\pm$ and prepare a plot of $\log y_\pm$ against $I^{1/2}$.

▌ Assuming $y(A) = 1$, using *(7.20)* gives

$$K_a = \frac{y(H^+)y(CH_3COO^-)}{y(CH_3COOH)} K' = y_\pm^2 K'$$

Taking logarithms of both sides, rearranging, and substituting *(6.51)* gives

$$\log K' = \log K_a - 2 \log y_\pm = \log K_a + 2(1)(1)(0.5116)I^{1/2}$$

Thus a plot of $\log K'$ against $I^{1/2}$ should be linear and have an intercept equal to $\log K_a$. From Fig. 7-6, the

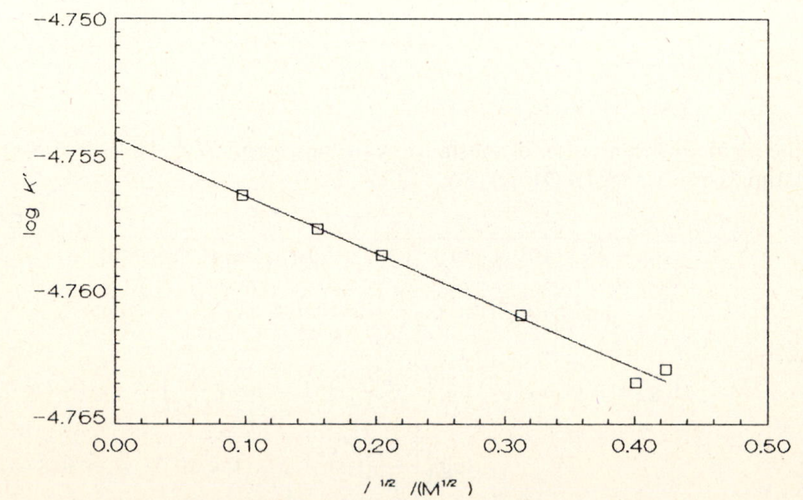

Fig. 7-6

intercept is $\log K_a = -4.7544$, giving $K_a = 1.760 \times 10^{-5}$. From the above relationship,

$$y_{\pm} = (K_a/K')^{1/2}$$

As a sample calculation, for $I = 0.179\,94$ M,

$$y_{\pm} = \left(\frac{1.760 \times 10^{-5}}{1.726 \times 10^{-5}}\right)^{1/2} = 1.010$$

The plot of $\log y_{\pm}$ is shown in Fig. 7-7.

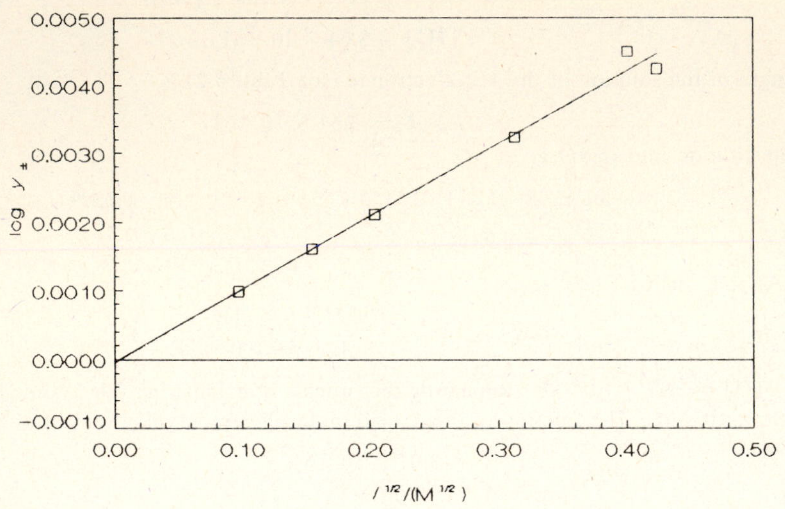

Fig. 7-7

7.72 For the stepwise formation of the thiosulfate complexes of Ag^+, $K_{f1} = 8 \times 10^8$ and $K_{f2} = 4 \times 10^4$. Prepare a plot of the fraction of each ion (α), given by $[i]/C(Ag^+)$, against $\log [S_2O_3^{2-}]$. What is the concentration of each of the ions at $[S_2O_3^{2-}] = 1 \times 10^{-8}$ M if $C(Ag^+) = 0.100$ M?

I For the following stepwise formation reactions and equilibrium constant expressions,

$$\begin{aligned} A + B &\rightleftarrows AB & K_{f1} &= [AB]/[A][B] \\ AB + B &\rightleftarrows AB_2 & K_{f2} &= [AB_2]/[AB][B] \\ &\vdots \\ AB_{n-1} + B &\rightleftarrows AB_n & K_{fn} &= [AB_n]/[AB_{n-1}][B] \end{aligned} \tag{7.30}$$

the equilibrium concentrations of the various species at a given value of [B] are

$$\begin{aligned} [A] &= \frac{C_A}{1 + K_{f1}[B] + K_{f1}K_{f2}[B]^2 + \cdots + K_{f1}K_{f2}\cdots K_{fn}[B]^n} \\ [AB] &= K_{f1}[B][A] \\ [AB_2] &= K_{f1}K_{f2}[B]^2[A] \\ &\vdots \\ [AB_n] &= K_{f1}K_{f2}\cdots K_{fn}[B]^n[A] \end{aligned} \tag{7.31}$$

where C_A is the formal concentration of substance A. In this case, $A = Ag^+$, $B = S_2O_3^{2-}$, and $n = 2$. As a sample calculation, consider $\log [S_2O_3^{2-}] = -6$:

$$[Ag^+] = \frac{C(Ag^+)}{1 + (8 \times 10^8)(1 \times 10^{-6}) + (8 \times 10^8)(4 \times 10^4)(1 \times 10^{-6})^2} = \frac{C(Ag^+)}{833}$$

$$\log [\alpha(Ag^+)] = \log \left(\frac{[Ag^+]}{C(Ag^+)}\right) = \log \frac{1}{833} = -2.921$$

$$\log [\alpha(AgS_2O_3^-)] = \log \left(\frac{[AgS_2O_3^-]}{C(Ag^+)}\right) = \log \left(\frac{K_{f1}}{833}\right) + \log [S_2O_3^{2-}]$$

$$= \log \left(\frac{8 \times 10^8}{833}\right) + \log (1 \times 10^{-6}) = -0.018$$

$$\log [\alpha(Ag(S_2O_3)_2^{3-})] = \log\left(\frac{[Ag(S_2O_3)_2^{3-}]}{C(Ag^+)}\right) = \log\left(\frac{K_{f1}K_{f2}}{833}\right) + 2\log[S_2O_3^{2-}]$$

$$= \log\left(\frac{(8\times 10^8)(4\times 10^4)}{833}\right) + 2\log(1\times 10^{-6}) = -1.415$$

The plot is shown in Fig. 7-8. From the graph, $\alpha(Ag^+) = -0.95$, $\alpha(AgS_2O_3^-) = -0.05$, and $\alpha(Ag(S_2O_3)_2^{3-}) = -3.45$, giving

$$[Ag^+] = C(Ag^+)10^{\log[\alpha(Ag^+)]} = (0.100\ M)10^{-0.95} = 0.011\ M$$

$$[AgS_2O_3^-] = C(Ag^+)10^{\log[\alpha(AgS_2O_3^-)]} = (0.100\ M)10^{-0.05} = 0.089\ M$$

$$[Ag(S_2O_3)_2^{3-}] = C(Ag^+)10^{\log[\alpha(Ag(S_2O_3)_2^{3-}]} = (0.100\ M)10^{-3.45} = 3\times 10^{-5}\ M$$

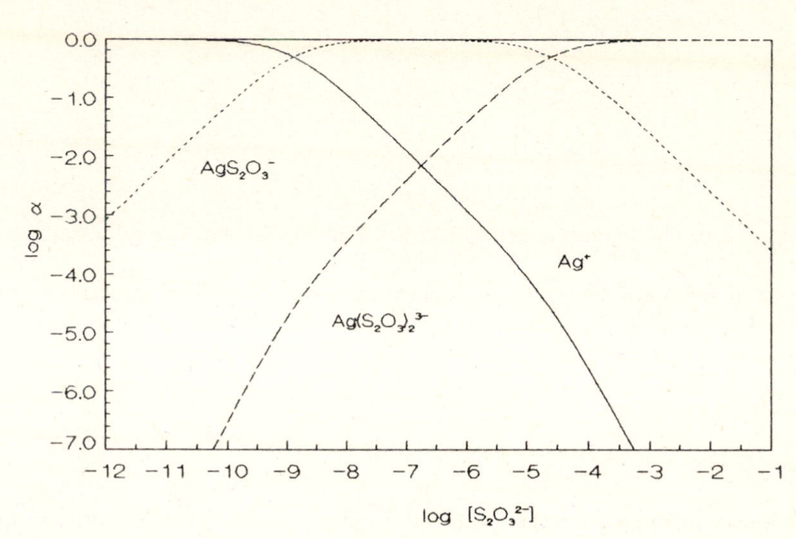

Fig. 7-8

7.73 Given $K_{a_1} = 1\times 10^{-2}$, $K_{a_2} = 2.1\times 10^{-3}$, $K_{a_3} = 6.9\times 10^{-7}$, and $K_{a_4} = 5.5\times 10^{-11}$ for ethylenediaminetetraacetic acid [EDTA or $(HOOCCH_2)_2NCH_2CH_2N(CH_2COOH)_2$], find $[H_4A]$, $[H_3A^-]$, $[H_2A^{2-}]$, $[HA^{3-}]$, and $[A^{4-}]$ at pH = 9.00 in a 0.100 M EDTA solution.

▮ Equations *(7.31)* are applicable to the stepwise ionization of a polyprotic acid where B is H^+, A is the anion of the acid, and

$$K_{f1} = \frac{1}{K_{an}} \qquad K_{f2} = \frac{1}{K_{a(n-1)}} \qquad \cdots \qquad K_{fn} = \frac{1}{K_{a_1}} \tag{7.32}$$

The concentration $[H^+]$ in the solution is found by using the antilogarithm form of *(7.16)*:

$$[H^+] = 10^{-pH} = 10^{-9.00} = 1.0\times 10^{-9}\ M$$

and $C_A = 0.100$ M. Using *(7.32)* with $K_{f1} = 1.8\times 10^{10}$, $K_{2f} = 1.4\times 10^6$, $K_{f3} = 480$, and $K_{f4} = 100$ gives

$$[A^{4-}] = \frac{0.100}{\left[\begin{array}{l} 1 + (1.8\times 10^{10})(1.0\times 10^{-9}) + (1.8\times 10^{10})(1.4\times 10^6)(1.0\times 10^{-9})^2 \\ \quad + (1.8\times 10^{10})(1.4\times 10^6)(480)(1.0\times 10^{-9})^3 \\ \quad + (1.8\times 10^{10})(1.4\times 10^6)(480)(100)(1.0\times 10^{-9})^4 \end{array}\right]}$$

$$= 0.0053\ M$$

$$[HA^{3-}] = (1.8\times 10^{10})(1.0\times 10^{-9})(0.0053) = 0.095\ M$$

$$[H_2A^{2-}] = (1.8\times 10^{10})(1.4\times 10^6)(1.0\times 10^{-9})^2(0.0053) = 1.3\times 10^{-4}\ M$$

$$[H_3A^-] = (1.8\times 10^{10})(1.4\times 10^6)(480)(1.0\times 10^{-9})^3(0.0053) = 6.4\times 10^{-11}\ M$$

$$[H_4A] = (1.8\times 10^{10})(1.4\times 10^6)(480)(100)(1.0\times 10^{-9})^4(0.0053) = 6.4\times 10^{-18}\ M$$

This type of calculation can be repeated for various values of pH, and a plot of the fraction of concentration (α) as a function of pH can be made (see Fig. 7-9).

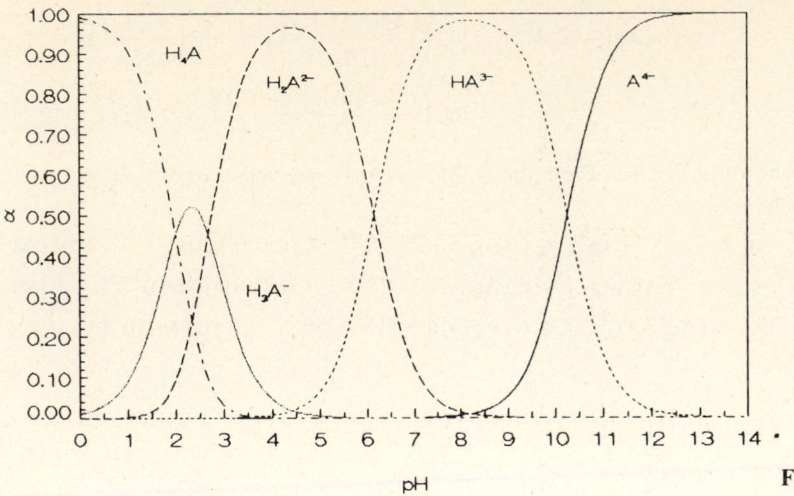

Fig. 7-9

7.74 What is the pH of a 0.100 M solution of EDTA? See Problem 7.73 for the values of K_a.

▮ The only two major contributions to the concentration of H^+ are from the first two ionization steps. Letting x represent the concentration of H_4A that ionizes and y the concentration of H_2A^{2-} formed, the equilibrium table approach described in Problem 7.12 gives $[H_4A] = (0.100 - x)$ M, $[H_3A^-] = (x - y)$ M, $[H_2A^{2-}] = y$ M and $[H^+] = (x + y)$ M. Substituting into *(7.23b)* gives

$$\frac{(x + y)(x - y)}{0.100 - x} = 1 \times 10^{-2} \qquad \frac{(x + y)(y)}{x - y} = 2.1 \times 10^{-3}$$

These simultaneous equations can be solved by an iterative process: (1) the approximation $x + y \approx x - y \approx x$ is made in the second equation, giving $y = 2.1 \times 10^{-3}$ M; (2) substitution of $y = 2.1 \times 10^{-3}$ M into the first gives $x = 2.7 \times 10^{-2}$ M; (3) substitution of this value of x into the second equation gives $y = 1.8 \times 10^{-3}$ M; (4) substitution of this value of y into the first equation gives $x = 2.7 \times 10^{-2}$ M. No further iterations are needed. The pH of the solution is given by *(7.16)* as

$$[H^+] = 2.7 \times 10^{-2}\ \text{M} + 1.8 \times 10^{-3}\ \text{M} = 2.9 \times 10^{-2}\ \text{M}$$

$$\text{pH} = -\log(2.9 \times 10^{-2}) = 1.54$$

7.75 Prepare a plot of $\bar{\nu}$, the average number of protons per molecule, against pH for EDTA. See Problem 7.73 for further data.

▮ The *average number of ligands* bound to the A atom in *(7.30)* is

$$\bar{\nu} = \frac{K_{f1}[B] + 2K_{f1}K_{f2}[B]^2 + \cdots + nK_{f1}K_{f2}\cdots K_{fn}[B]^n}{1 + K_{f1}[B] + K_{f1}K_{f2}[B]^2 + \cdots + K_{f1}K_{f2}\cdots K_{fn}[B]^n} \qquad (7.33)$$

In this case, $[B] = [H^+] = 10^{-\text{pH}}$. As a sample calculation, consider $\text{pH} = 9.00$:

$$\bar{\nu} = \frac{\left[\begin{aligned}&(1.8 \times 10^{10})(1.0 \times 10^{-9}) + 2(1.8 \times 10^{10})(1.4 \times 10^{6})(1.0 \times 10^{-9})^2 \\ &\quad + 3(1.8 \times 10^{10})(1.4 \times 10^{6})(480)(1.0 \times 10^{-9})^3 \\ &\quad\quad + 4(1.8 \times 10^{10})(1.4 \times 10^{6})(480)(100)(1.0 \times 10^{-9})^4\end{aligned}\right]}{\left[\begin{aligned}&1 + (1.8 \times 10^{10})(1.0 \times 10^{-9}) + (1.8 \times 10^{10})(1.4 \times 10^{6})(1.0 \times 10^{-9})^2 \\ &\quad + (1.8 \times 10^{10})(1.4 \times 10^{6})(480)(1.0 \times 10^{-9})^3 \\ &\quad\quad + (1.8 \times 10^{10})(1.4 \times 10^{6})(480)(100)(1.0 \times 10^{-9})^4\end{aligned}\right]}$$

$$= 0.95$$

The graph is shown in Fig. 7-10.

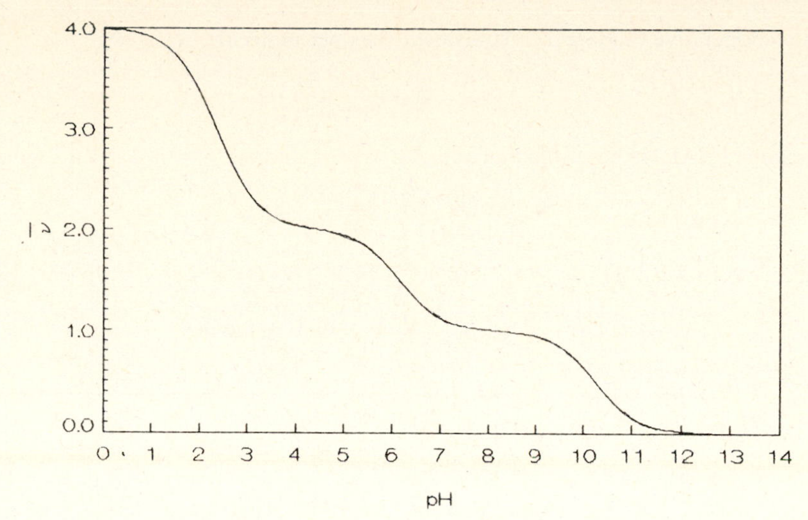

Fig. 7-10

7.76 What is the average number of O_2 molecules bound to myoglobin at a partial pressure of 0.20 bar? The pressure of O_2 required to bind 50% of myoglobin molecules is 0.037 bar.

▌ Each myoglobin will bind a maximum of one O_2 molecule. Substituting $[\mathrm{Mb}] = [\mathrm{A}]$, $[\mathrm{MbO_2}] = [\mathrm{AB}]$, and $P(\mathrm{O_2}) = [\mathrm{A}]$ into *(7.30)* gives

$$K_{f1} = \frac{[\mathrm{MbO_2}]}{[\mathrm{Mb}][P(\mathrm{O_2})/(\mathrm{bar})]}$$

If 50% of the molecules are bound, then $[\mathrm{MbO_2}] = [\mathrm{Mb}]$ and

$$K_{f1} = \frac{[\mathrm{MbO_2}]}{[\mathrm{MbO_2}](0.037)} = 27$$

The average number of molecules will be given by *(7.33)* as

$$\bar{\nu} = \frac{(27)(0.20)}{1 + (27)(0.20)} = 0.84$$

7.77 Each myoglobin molecule will bind a maximum of either one O_2 or one CO molecule. Given $K_{f1} = 1.1 \times 10^4\ \mathrm{bar}^{-1}$ for CO in the absence of O_2 and $27\ \mathrm{bar}^{-1}$ for O_2, find the partial pressure of CO in air necessary to bind with 25% of the myoglobin molecules.

▌ The *fractional saturation* (θ) is related to the average number of ligands attached to A by

$$\theta = \bar{\nu}/n \qquad (7.34)$$

where n is the number of available sites per molecule. In this case, $n = 1$, and so *(7.34)* and *(7.33)* give in the presence of air

$$\theta = \frac{[P(\mathrm{CO})/(\mathrm{bar})]K_{f1}(\mathrm{CO})}{1 + [P(\mathrm{CO})/(\mathrm{bar})]K_{f1}(\mathrm{CO}) + [P(\mathrm{O_2})/(\mathrm{bar})]K_{f1}(\mathrm{O_2})}$$

Letting $P(\mathrm{O_2}) = 0.20\ \mathrm{bar}$ for air, substituting the data, and solving for $P(\mathrm{CO})$ gives

$$0.25 = \frac{P(\mathrm{CO})(1.1 \times 10^4)}{1 + P(\mathrm{CO})(1.1 \times 10^4) + (0.20\ \mathrm{bar})(27)}$$

$$P(\mathrm{CO}) = 1.9 \times 10^{-4}\ \mathrm{bar}$$

7.78 Use the following data for the binding of NADH to lactate dehydrogenase to determine the number of binding sites per molecule (n) and the intrinsic formation constant (K_f):

[NADH]/(10^{-7} M)	0.49	0.71	0.99	2.4	3.6	4.8	7.8	18	56
$\bar{\nu}$	0.47	0.58	0.71	1.52	1.98	2.21	2.73	3.50	3.92

I For the noncooperative binding at n equivalent sites per molecule,

$$\bar{\nu} = \frac{n[\mathrm{B}]}{[\mathrm{B}] + 1/K_\mathrm{f}} \tag{7.35}$$

A plot of $\bar{\nu}$ against [NADH] has a limiting value of $\bar{\nu}$ of n, and at $\nu = n/2$ *(7.35)* gives

$$\frac{n}{2} = \frac{n[\mathrm{NADH}]}{[\mathrm{NADH}] + 1/K_\mathrm{f}} \qquad K_\mathrm{f} = \frac{1}{[\mathrm{NADH}]}$$

From the plot given in Fig. 7-11, $n = 4$, and at $n = 2$, $[\mathrm{NADH}] = 3.7 \times 10^{-7}$ M, giving

$$K_\mathrm{f} = \frac{1}{3.7 \times 10^{-7}\ \mathrm{M}} = 2.7 \times 10^6$$

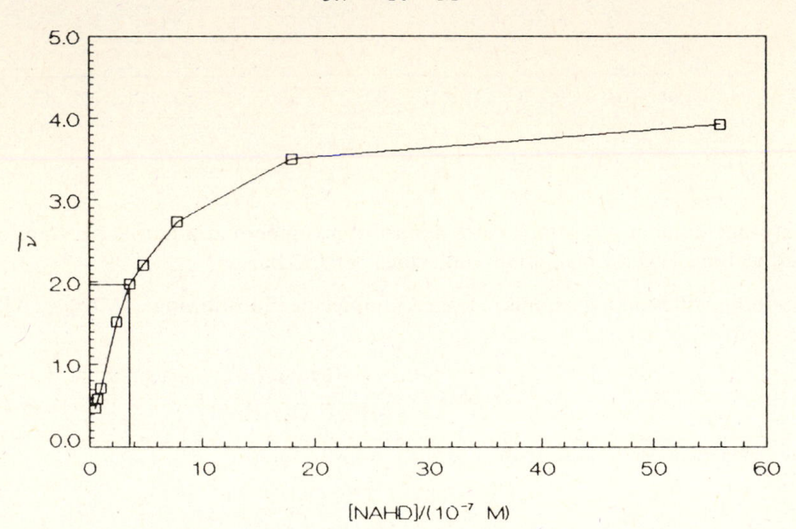

Fig. 7-11

7.79 Prepare a Hughes-Klotz plot and a Scatchard plot of the data given in Problem 7.78, and determine n and K_f.

I One of the linear forms of *(7.35)* is

$$\frac{1}{\bar{\nu}} = \frac{1}{nK_\mathrm{f}}\frac{1}{[\mathrm{B}]} + \frac{1}{n}$$

Thus a plot of $1/\bar{\nu}$ against $1/[\mathrm{B}]$, a *Hughes-Klotz plot*, will be linear and have an intercept equal to $1/n$ and a slope equal to $1/nK_\mathrm{f}$. From Fig. 7-12,

$$n = \frac{1}{\text{intercept}} = \frac{1}{0.26} = 4$$

$$K_\mathrm{f} = \frac{1}{n(\text{slope})} = \frac{1}{(4)(0.98 \times 10^{-7})} = 2.6 \times 10^6$$

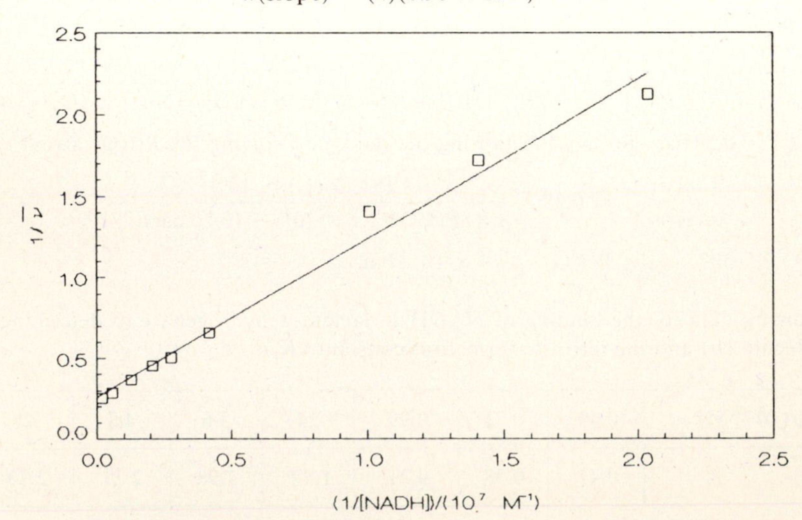

Fig. 7-12

Another of the linear forms of *(7.35)* is

$$\bar{\nu}/[\mathrm{B}] = -K_f\bar{\nu} + nK_f$$

Thus a plot of $\bar{\nu}/[\mathrm{B}]$ against $\bar{\nu}$, a *Scatchard plot*, will be linear and have a slope of $-K_f$ and an intercept of nK_f. The x intercept will be n. From Fig. 7-13,

$$K_f = -\text{slope} = -(-0.23 \times 10^7) = 2.3 \times 10^6$$

$$n = \frac{\text{intercept}}{K_f} = \frac{0.97 \times 10^7}{2.3 \times 10^6} = 4$$

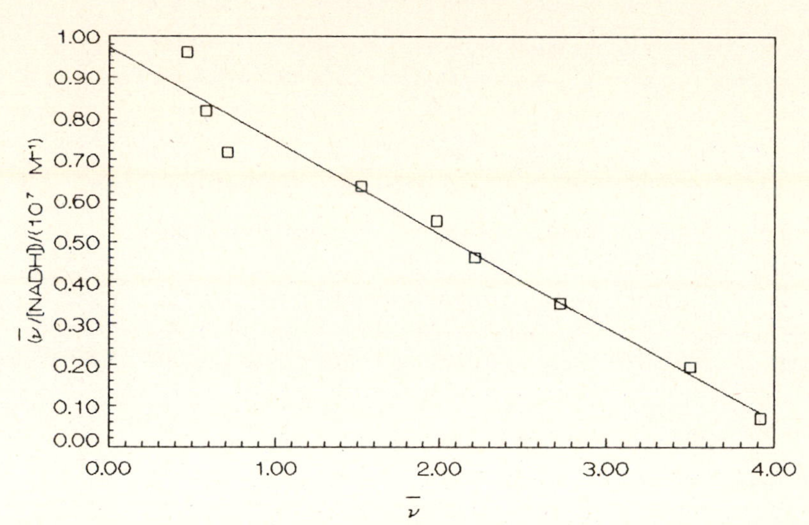

Fig. 7-13

7.80 Prepare a Hill plot for the data given in Problem 7.78 to show that the binding is noncooperative, and determine K_f.

▮ A *Hill plot* is prepared by plotting $\log[\theta/(1-\theta)] = \log[\bar{\nu}/(n-\bar{\nu})]$ against $\log[\mathrm{B}]$. From *(7.35)*,

$$\log\frac{\bar{\nu}}{n-\bar{\nu}} = \log\frac{\theta}{1-\theta} = \log K_f + \log[\mathrm{B}]$$

Thus the plot should be linear with a slope of unity and an intercept of $\log K_f$. The slope of the plot shown in Fig. 7-14 is 1.22, and the intercept is 7.9, giving

$$\log K_f = \text{intercept} = 7.9 \qquad \text{and} \qquad K_f = 8 \times 10^7$$

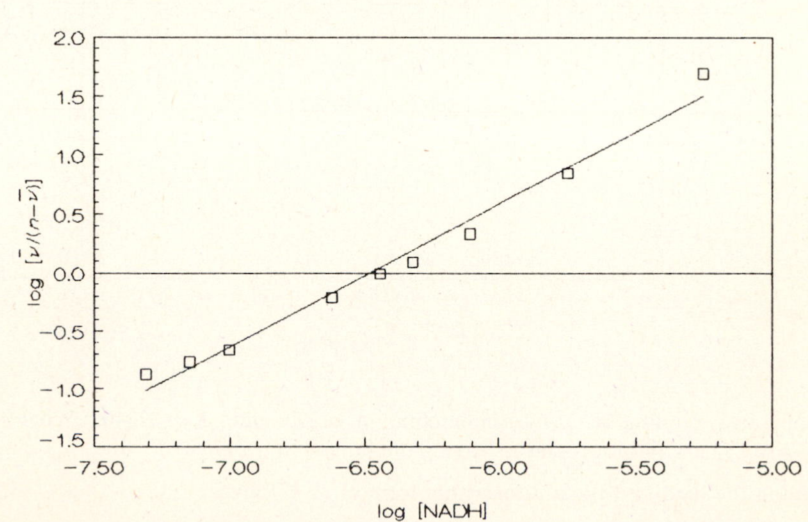

Fig. 7-14

7.81 Using the value of $K_f = 2.5 \times 10^6$ for the noncooperative binding of NADH to lactate dehydrogenase, determine the value of the formation constants for each step.

▮ The intrinsic formation constant is related to each of the formation constants for each step by

$$K_{f,i} = \frac{n - i + 1}{i} K_f \tag{7.36}$$

Substituting values of i gives

$$K_{f1} = \frac{4 - 1 + 1}{1}(2.5 \times 10^6) = 1.0 \times 10^7$$

$$K_{f2} = \frac{4 - 2 + 1}{2}(2.5 \times 10^6) = 3.8 \times 10^6$$

$$K_{f3} = \frac{4 - 3 + 1}{3}(2.5 \times 10^6) = 1.7 \times 10^6$$

$$K_{f4} = \frac{4 - 4 + 1}{4}(2.5 \times 10^6) = 6.3 \times 10^5$$

7.82 Prepare a Hill plot for the cooperative binding of O_2 by squid hemocyanin:

θ	0.0030	0.0192	0.0839	0.329	0.517	0.734	0.834	0.875	0.894	0.913
$P(O_2)/(\text{bar})$	0.0015	0.0103	0.0423	0.1340	0.1823	0.2709	0.4360	0.6037	0.7558	0.9822

What is the value of the Hill coefficient?

▮ For cooperative binding,

$$\log \frac{\bar{\nu}}{n - \bar{\nu}} = \log \frac{\theta}{1 - \theta} = \log K_f + n \log [\text{B}] \tag{7.37}$$

Thus a plot of $\log[\theta/(1 - \theta)]$ against $\log[\text{B}]$ will be linear with a slope equal to n, the Hill coefficient. From the plot shown in Fig. 7-15, $n = 2.45$ in the center of the curve. At high and low values of $\log[P(O_2)]$, the slope is nearly unity because the binding is noncooperative under these conditions.

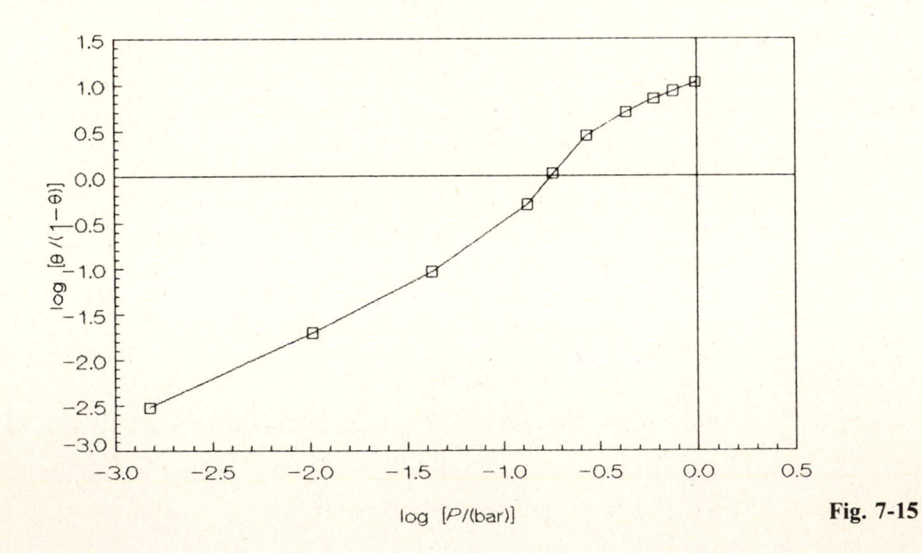

Fig. 7-15

7.83 For the cooperative binding of O_2 by hemoglobin, $n = 2.7$ and $K_f = 7340$. What is the fractional saturation for $P(O_2) = 0.15$ bar in blood that is leaving the lungs?

▮ Substituting the data into the antilogarithm form of *(7.37)* gives

$$\frac{\theta}{1 - \theta} = K_f[P(O_2)/(\text{bar})]^n = (7340)(0.15)^{2.7} = 44$$

Solving gives $\theta = 0.98$.

7.84 The solubility of NaCl in water at 25 °C is 5.33 M. Calculate the mean ionic activity coefficient for NaCl in a saturated solution given

$$NaCl(s) \rightarrow Na^+(aq) + Cl^-(aq) \qquad \Delta_r G° = -8.995 \text{ kJ}$$

▮ For the equilibrium established by a solid salt $A_{\nu_+}B_{\nu_-}$ and its ions in solution,

$$K_{sp} = (y_\pm)^\nu [A^{(Z_+)+}]^{\nu_+}[B^{(Z_-)-}]^{\nu_-} \tag{7.38}$$

where K_{sp} is called the *solubility product constant.* Using *(7.1b)* gives

$$K_{sp} = \exp\left(-\frac{(-8.995\text{ kJ})[(10^3\text{ J})/(1\text{ kJ})]}{(8.314\text{ J}\cdot\text{K}^{-1}\cdot\text{mol}^{-1})(298\text{ K})(\text{mol})}\right) = 37.7$$

For a 1 : 1 electrolyte, Table 6-2 gives $K_{sp} = y_\pm^2 C^2$. Substituting the data and solving for $y_\pm$ gives

$$37.7 = y_\pm^3 (5.33)^2 \qquad y_\pm = 1.15$$

7.85 Calculate the concentration of Hg^{2+} in **(*a*)** pure water, **(*b*)** 0.0010 M KI, and **(*c*)** 0.0010 M $NaNO_3$. $K_{sp} = 3.00 \times 10^{-26}$ for HgI_2.

▮ For the 2 : 1 electrolyte, Table 6-2 gives $K_{sp} = 4y_\pm^3 C^3$.

(*a*) In pure water, assume that $y_\pm = 1$. Thus,

$$C = \left(\frac{3.00 \times 10^{-26}}{4(1)^3}\right)^{1/3} = 1.96 \times 10^{-9}\text{ M}$$

From Table 6-2, $I = 3C = 3(1.96 \times 10^{-9}) = 5.88 \times 10^{-9}$ M. Substituting into *(6.51)* gives

$$\log y_\pm = -(1)(1)(0.5116)(5.88 \times 10^{-9})^{1/2} = -3.92 \times 10^{-5}$$

$$y_\pm = 0.9999$$

which confirms the assumption. Thus $C = 1.96 \times 10^{-9}$ M.

(*b*) In 0.0010 M KI, Table 6-2 gives $I = C(\text{KI}) + 3C(HgI_2) \approx C(\text{KI}) = 0.0010$ M.

$$\log y_\pm = -(1)(1)(0.5116)(0.0010)^{1/2} = -0.016$$

$$y_\pm = 0.963$$

In this case, the added iodide ion reduces the solubility (the *common ion effect*) where $C_+ = C$ and $C_- = 2C + 0.0010$ M, giving

$$3.00 \times 10^{-26} = (0.963)^3(C)(2C + 0.0010)^2$$

$$C = 3.4 \times 10^{-20}\text{ M}$$

(*c*) In 0.0010 M $NaNO_3$, Table 6-2 gives $I = C(NaNO_3) + 3C(HgI_2) \approx C(\text{KI}) = 0.0010$ M.

$$\log y_\pm = -(1)(1)(0.5116)(0.0010)^{1/2} = -0.016$$

$$y_\pm = 0.963$$

Thus the solubility is increased slightly (the *diverse ion effect*) to

$$C = \left(\frac{3 \times 10^{-26}}{4(0.963)^3}\right)^{1/3} = 2.03 \times 10^{-9}\text{ M}$$

7.86 Calculate the concentration of Pb^{2+} in **(*a*)** pure water (neglecting hydrolysis of the anion) and **(*b*)** in an acidic solution having pH = 4.0. For $PbSO_4$, $K_{sp} = 2.20 \times 10^{-8}$, and for H_2SO_4, $K_{a_2} = 1.0 \times 10^{-2}$. Assume $y_\pm = 1$.

▮ **(*a*)** For the 2 : 2 electrolyte, Table 6-2 gives $K_{sp} = y_\pm^2 C^2$. Substituting the data and solving for C gives

$$C = \left(\frac{2.2 \times 10^{-8}}{1^2}\right)^{1/3} = 1.48 \times 10^{-4}\text{ M}$$

(*b*) For the solution in which hydrolysis of the anion occurs,

$$C_+ = \left[\frac{K_{sp}}{y_\pm^2}\left(1 + \frac{[H^+]}{K_{a_2}}\right)\right]^{1/2} = \left[\frac{2.20 \times 10^{-8}}{1^2}\left(1 + \frac{1.0 \times 10^{-4}}{1.0 \times 10^{-2}}\right)\right]^{1/2}$$

$$= 1.49 \times 10^{-4}\text{ M}$$

7.87 Given $K_{sp} = 1.8 \times 10^{-10}$, $K_{f1} = 1.1 \times 10^3$, $K_{f2} = 1.0 \times 10^2$, $K_{f3} = 1$, and $K_{f4} = 1.8$ for AgCl, find the equilibrium concentrations of Ag^+, AgCl(aq), $AgCl_2^-$, $AgCl_3^{2-}$, and $AgCl_4^{3-}$ at $[Cl^-] = 0.100$ M.

▮ The expressions for the various equilibrium concentrations given by *(7.31)* must be modified slightly because of the additional solubility product equilibrium. For a 1:1 electrolyte, *(7.38)* gives

$$[Ag^+] = \frac{K_{sp}}{[Cl^-]} = \frac{1.8 \times 10^{-10}}{0.100} = 1.8 \times 10^{-9}\ \text{M}$$

The remainder of the expressions in *(7.31)* become

$$[AgCl] = K_{f1}[Cl^-][Ag^+] = K_{f1}K_{sp} = (1.1 \times 10^3)(1.8 \times 10^{-10}) = 2.0 \times 10^{-7}\ \text{M}$$

$$[AgCl_2^-] = K_{f1}K_{f2}[Cl^-]^2[Ag^+] = K_{f1}K_{f2}K_{sp}[Cl^-]$$

$$= (1.1 \times 10^3)(1.0 \times 10^2)(1.8 \times 10^{-10})(0.100) = 2.0 \times 10^{-6}\ \text{M}$$

$$[AgCl_3^{2-}] = K_{f1}K_{f2}K_{f3}[Cl^-]^3[Ag^+] = K_{f1}K_{f2}K_{f3}K_{sp}[Cl^-]^2$$

$$= (1.1 \times 10^3)(1.0 \times 10^2)(1)(1.8 \times 10^{-10})(0.100)^2 = 2.0 \times 10^{-7}\ \text{M}$$

$$[AgCl_4^{3-}] = K_{f1}K_{f2}K_{f3}K_{f4}[Cl^-]^4[Ag^+] = K_{f1}K_{f2}K_{f3}K_{f4}K_{sp}[Cl^-]^3$$

$$= (1.1 \times 10^3)(1.0 \times 10^2)(1)(1.8)(1.8 \times 10^{-10})(0.100)^3 = 3.6 \times 10^{-8}\ \text{M}$$

These calculations are easily repeated for any value of $[Cl^-]$ and can be plotted in the form of $[i]$ or $\log [i]$ against $\log [Cl^-]$ (see Fig. 7-16).

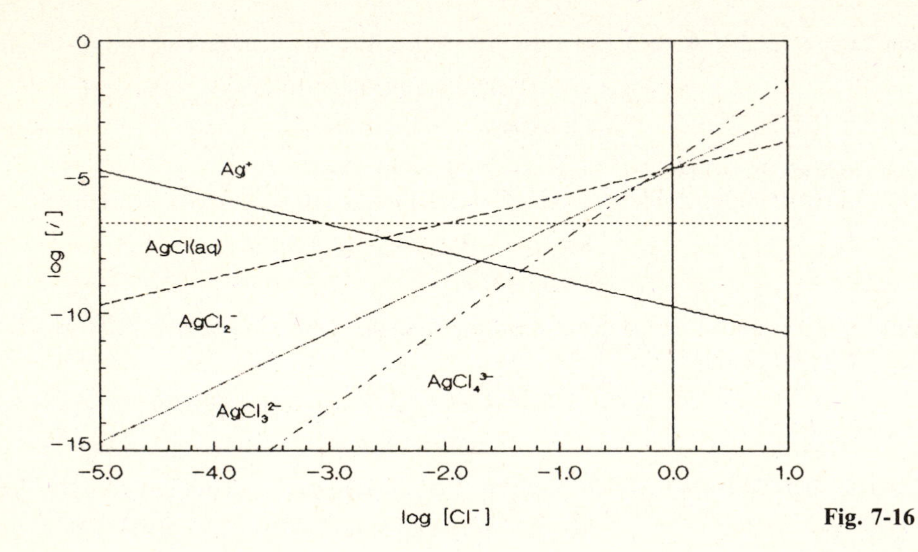

Fig. 7-16

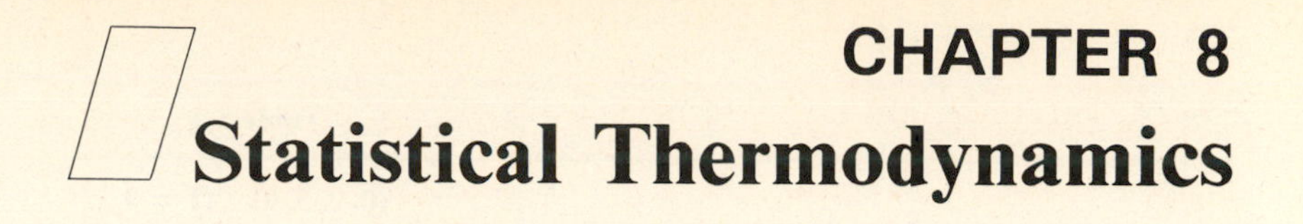

CHAPTER 8
Statistical Thermodynamics

8.1 ENSEMBLES

8.1 Determine the possible distributions among the energy states in an ensemble of systems consisting of five particles such that $E_t/kT = 20$. Assume that the energy of each particle is restricted to $E_i/kT = 0, 2, 4,$ or 6.

The distribution $(0, 0, 5, 0)$ means that there are no particles in the first state contributing $E_1/kT = 0$ to the total energy, none in the second state contributing $E_2/kT = 2$, five particles in the third state contributing $E_3/kT = 4$, and none in the fourth state contributing $E_4/kT = 6$. The total energy in this distribution is

$$E_t/kT = (0)(0) + (0)(2) + (5)(4) + (0)(6) = 20$$

The other distributions are $(0, 1, 3, 1)$, $(0, 2, 1, 2)$, $(1, 0, 2, 2)$, and $(1, 1, 0, 3)$.

8.2 Find the number of ensemble states corresponding to each of the acceptable distributions in Problem 8.1.

▌ The number of ensemble states (Ω) corresponding to a given distribution of N_s systems is given by

$$\Omega = \frac{N_s}{\prod_i N_i!} = \frac{N_s![\prod_j g_j^{N_j}]}{\prod_j N_j!} \tag{8.1}$$

where i refers to energy states and j refers to energy levels of degeneracy g_j. For each of the distributions,

$$\Omega(0, 0, 5, 0) = 5!/0!0!5!0! = 1$$

$$\Omega(0, 1, 3, 1) = 5!/0!1!3!1! = 20$$

$$\Omega(0, 2, 1, 2) = 5!/0!2!1!2! = 30$$

$$\Omega(1, 0, 2, 2) = 5!/1!0!2!2! = 30$$

$$\Omega(1, 1, 0, 3) = 5!/1!1!0!3! = 20$$

8.3 Prepare a table showing how the distributions determined in Problem 8.1 are realized for the five distinguishable particles A, B, C, D, and E. Compare these results to the values of Ω calculated in Problem 8.2.

▌ See Table 8-1. For each of the distributions the number of arrangements of the particles agrees with the respective value of Ω.

8.4 What is the probability of finding system C in state 1 for the distribution $(1, 0, 2, 2)$? Compare the numerical answer to the actual count of the number of times C appears in the state 1 column of Table 8-1.

▌ The probability of finding a chosen system in the ith state in a given distribution is given by

$$P_i(N_1, N_2, \ldots) = N_i/N_s \tag{8.2}$$

Substituting $N_i = 1$ and $N_s = 5$ for the distribution gives $P_1(1, 0, 2, 2) = \frac{1}{5}$. System C appears in the state 1 column 6 times out of the total of 30 entries, giving 6/30 or 1/5.

8.5 What is the overall probability of finding system C in state 1 for the distributions listed in Problem 8.1? Compare the numerical answer to the actual count of the number of times system C appears in the state 1 column of Table 8-1.

▌ The overall probability of finding a chosen system in the ith state in a given distribution is given by

$$P_i = \frac{\sum N_i \Omega}{N_s \sum \Omega} \tag{8.3}$$

Substituting into *(8.3)* gives

$$P_1 = \frac{(0)(1) + (0)(20) + (0)(30) + (1)(30) + (1)(20)}{5(1 + 20 + 30 + 30 + 20)} = 0.099$$

System C appears in the state 1 column 10 times out of the total of 101 entries, giving 10/101 or 0.099.

Table 8-1

(0, 0, 5, 0) $\Omega = 1$

State 1	State 2	State 3	State 4
		A B C D E	

(0, 1, 3, 1) $\Omega = 20$

State 1	State 2	State 3	State 4
	A	B C D	E
	A	B C E	D
	A	B D E	C
	A	C D E	B
	B	A C D	E
	B	A C E	D
	B	A D E	C
	B	C D E	A
	C	A B D	E
	C	A B E	D
	C	A D E	B
	C	B D E	A
	D	A B C	E
	D	A B E	C
	D	A C E	B
	D	B C E	A
	E	A B C	D
	E	A B D	C
	E	A C D	B
	E	B C D	A

(1, 1, 0, 3) $\Omega = 20$

State 1	State 2	State 3	State 4
A	B		C D E
A	C		B D E
A	D		B C E
A	E		B C D
B	A		C D E
B	C		A D E
B	D		A C E
B	E		A C D
C	A		B D E
C	B		A D E
C	D		A B E
C	E		A B D
D	A		B C E
D	B		A C E
D	C		A B E
D	E		A B C
E	A		B C D
E	B		A C D
E	C		A B D
E	D		A B C

(0, 2, 1, 2) $\Omega = 30$

State 1	State 2	State 3	State 4
	A B	C	D E
	A B	D	C E
	A B	E	C D
	A C	B	D E
	A C	D	B E
	A C	E	B D
	A D	B	C E
	A D	C	B E
	A D	E	B C
	A E	B	C D
	A E	C	B D
	A E	D	B C
	B C	A	D E
	B C	D	A E
	B C	E	A D
	B D	A	C E
	B D	C	A E
	B D	E	A C
	B E	A	C D
	B E	C	A D
	B E	D	A C
	C D	A	B E
	C D	B	A E
	C D	E	A B
	C E	A	B D
	C E	B	A D
	C E	D	A B
	D E	A	B C
	D E	B	A C
	D E	C	A B

(1, 0, 2, 2) $\Omega = 30$

State 1	State 2	State 3	State 4
A		B C	D E
A		B D	C E
A		B E	C D
A		C D	B E
A		C E	B D
A		D E	B C
B		A C	D E
B		A D	C E
B		A E	C D
B		C D	A E
B		C E	A D
B		D E	A C
C		A B	D E
C		A D	B E
C		A E	B D
C		B D	A E
C		B E	A D
C		D E	A B
D		A B	C E
D		A C	B E
D		A E	B C
D		B C	A E
D		B E	A C
D		C E	A B
E		A B	C D
E		A C	B D
E		A D	B C
E		B C	A D
E		B D	A C
E		C D	A B

8.6 Consider the placement of three indistinguishable molecules into two distinguishable energy levels with no restrictions on their placement. Assuming $g_1 = 1$ and $g_2 = 1$, determine the probability of finding each distribution.

▮ The number of ways that N indistinguishable objects can be placed into n distinguishable locations is given by

$$t = (N + n - 1)!/N!(n - 1)! \qquad (8.4)$$

Substituting $N = 3$ and $n = 2$ into *(8.4)* gives the number of distributions as

$$t = (3 + 2 - 1)!/3!(2 - 1)! = 4$$

The four distributions can be represented by

	Level 1	Level 2	
(3, 0)	XXX		$\Omega(3,0) = \dfrac{3!(1^3 1^0)}{3!0!} = 1$
(2, 1)	XX	X	$\Omega(2,1) = \dfrac{3!(1^2 1^1)}{2!1!} = 3$
(1, 2)	X	XX	$\Omega(1,2) = \dfrac{3!(1^1 1^2)}{1!2!} = 3$
(0, 3)		XXX	$\Omega(0,3) = \dfrac{3!(1^0 1^3)}{0!3!} = 1$

where the values of Ω were calculated using *(8.1)* and represent the number of times out of the total that the particular distribution will occur. (These values of Ω can be confirmed by determining the number of possible arrangements of three distinguishable particles A, B, and C.) The total number of occurrences is $1 + 3 + 3 + 1 = 8$, the same result as determined using the equation to determine the number of ways that N distinguishable objects can be placed into n distinguishable locations,

$$t = n^N \qquad (8.5)$$

which in this case is $t = 2^3 = 8$. The probability of each distribution is

$$P(3,0) = \tfrac{1}{8} = 0.125 \qquad P(2,1) = \tfrac{3}{8} = 0.375$$

$$P(1,2) = \tfrac{3}{8} = 0.375 \qquad P(0,3) = \tfrac{1}{8} = 0.125$$

8.7 Repeat the calculations of Problem 8.6 assuming $g_1 = 2$ and $g_2 = 1$. Compare the results.

▮ As in Problem 8.6, the number of distributions is 4 and may be represented by

	Level 1	Level 2	
(3, 0)	XXX		$\Omega = \dfrac{3!(2^3 1^0)}{3!0!} = 8$
(2, 1)	XX	X	$\Omega = \dfrac{3!(2^2 1^1)}{2!1!} = 12$
(1, 2)	X	XX	$\Omega = \dfrac{3!(2^1 1^2)}{1!2!} = 6$
(0, 3)		XXX	$\Omega = \dfrac{3!(2^0 1^3)}{0!3!} = 1$

The total number of occurrences is $8 + 12 + 6 + 1 = 27$, giving

$$P(3,0) = \tfrac{8}{27} = 0.296 \qquad P(2,1) = \tfrac{12}{27} = 0.444$$

$$P(1,2) = \tfrac{6}{27} = 0.222 \qquad P(3,0) = \tfrac{1}{27} = 0.037$$

The symmetric probabilities found in Problem 8.6 are no longer valid—the probability of the distributions containing more molecules in the level with the higher degeneracy is greater than that of the levels with the smaller degeneracy.

8.8 Consider 20 molecules divided equally between four energy states. What is the number of ensemble states for this distribution? What is the value of Ω if one molecule is removed from one state and added to another?

▮ Using *(8.1)* for each distribution gives

$$\Omega(5,5,5,5) = \frac{20!}{5!\,5!\,5!\,5!} = 1.17 \times 10^{10}$$

$$\Omega(5,4,6,5) = \frac{20!}{5!\,4!\,6!\,5!} = 9.78 \times 10^{9}$$

8.9 Consider two flasks of equal volume connected by a stopcock. Suppose one molecule is in the system. What is the probability of the molecule being in a given flask? What is the probability of all the molecules being in a given flask for a system containing two molecules and for a system containing 100 molecules?

▮ The individual probability of a given molecule being a given flask is 1/2. Thus for the one-molecule system, $P = 1/2$. The combined or overall probability of finding N molecules in the given flask is given by

$$P = (1/2)^{N}$$

which gives

$$P = (1/2)^2 = 0.25 \qquad P = (1/2)^{100} = 7.89 \times 10^{-31}$$

for the large systems.

8.10 A system consists of three energy states—a ground state ($E_0 = 0$), a first excited state ($E_1 = 2kT$), and a second excited state ($E_2 = 6kT$). Determine the partition function. Repeat the calculations if the system consists of three energy levels—a ground level ($E_0 = 0$, $g_0 = 1$), a first excited level ($E_1 = 2kT$, $g_1 = 3$), and a second excited level ($E_2 = 6kT$, $g_2 = 2$).

▮ The *partition function* (Q) is defined as

$$Q = \sum_i e^{-E_i/kT} = \sum_j g_j\, e^{-E_j/kT} \qquad (8.6)$$

For each system, *(8.6)* gives

$$Q = e^{-0/kT} + e^{-2kT/kT} + e^{-6kT/kT} = 1.138$$

$$Q = (1)\, e^{-0/kT} + (3)\, e^{-2kT/kT} + (2)\, e^{-6kT/kT} = 1.411$$

8.11 What is the overall probability for the second state in the first system described in Problem 8.10? What is the probability for the second level in the second system described in Problem 8.10?

▮ In terms of the partition function, *(8.3)* becomes

$$P_i = e^{-E_i/kT}/Q = g_i\, e^{-E_i/kT}/Q \qquad (8.7)$$

For each system *(8.7)* gives

$$P_2 = e^{-2kT/kT}/1.138 = 0.119$$

$$P_2 = 3\, e^{-2kT/kT}/1.411 = 0.288$$

8.12 What is the ratio of the number of particles in each of the states in the first system and levels in the second system described in Problem 8.10?

▮ The ratio of the numbers of particles in each state or level is given by the Boltzmann distribution law, *(2.22)*,

$$\frac{N_i}{N_j} = e^{-(E_i-E_j)/kT} = \frac{g_i}{g_j}\, e^{-(E_i-E_j)/kT} \qquad (2.22)$$

For the first system,

$$\frac{N_2}{N_0} = e^{-(6kT-0)/kT} = 2.48 \times 10^{-3}$$

$$\frac{N_1}{N_0} = e^{-(2kT-0)/kT} = 0.135$$

and for the second system,

$$\frac{N_2}{N_0} = \frac{2}{1} e^{-(6kT-0)/kT} = 4.96 \times 10^{-3}$$

$$\frac{N_1}{N_0} = \frac{3}{1} e^{-(2kT-0)/kT} = 0.406$$

8.13 Assume that each of the systems described in Problem 8.10 contains 1 mol of particles. How many particles are in each state of the first system and in each level of the second system?

▮ For the first system, the total number of particles will be

$$6.022 \times 10^{23} = N_0 + N_1 + N_2 = N_0 + 0.135N_0 + 2.48 \times 10^{-3} N_0 = 1.137N_0$$

where the expressions for N_1 and N_2 were determined in Problem 8.12. Solving gives $N_0 = 5.296 \times 10^{23}$ particles. Thus,

$$N_1 = (0.135)(5.296 \times 10^{23}) = 7.15 \times 10^{22} \text{ particles}$$

$$N_2 = (2.48 \times 10^{-3})(5.296 \times 10^{23}) = 1.31 \times 10^{21} \text{ particles}$$

Likewise for the second system,

$$6.022 \times 10^{23} = N_0 + 0.406N_0 + 4.96 \times 10^{-3}N_0 = 1.411N_0$$

$$N_0 = 4.268 \times 10^{23} \text{ particles}$$

$$N_1 = (0.406)(4.268 \times 10^{23}) = 1.73 \times 10^{23} \text{ particles}$$

$$N_2 = (4.96 \times 10^{-3})(4.268 \times 10^{23}) = 2.12 \times 10^{21} \text{ particles}$$

8.14 What is the high-temperature limit of *(2.22)*?

▮ In terms of states,

$$\lim_{T\to\infty} \frac{N_i}{N_j} = \lim_{T\to\infty} e^{(-E_i-E_j)/kT} = 1$$

Thus all the energy states will be equally populated. In terms of levels,

$$\lim_{T\to\infty} \frac{N_i}{N_j} = \lim_{T\to\infty} \frac{g_i}{g_j} e^{-(E_i-E_j)/kT} = \frac{g_i}{g_j}(1) = \frac{g_i}{g_j}$$

Thus all the energy levels will be occupied in ratios of the degeneracies.

8.15 A system consisting of two energy states separated by 2×10^{-23} J has a ratio of particles in each state of 51/49. What is the temperature of the system?

▮ Substituting into *(2.22)* and solving for T gives

$$\tfrac{51}{49} = e^{-(2\times10^{-23}\,\text{J})/(1.381\times10^{-23}\,\text{J}\cdot\text{K}^{-1})T} \qquad T = -36\text{ K}$$

8.16 Typical values of energy differences are 10^{-42} J for translational states, 10^{-23} J for rotational states, and 10^{-20} J for vibrational states. Calculate N_2/N_1 for each type of energy at 100 K, 298 K, and 1000 K. Discuss the results.

▮ For the translational states, *(2.22)* gives

$$\left(\frac{N_2}{N_1}\right)_{100} = \exp\left(-\frac{10^{-42}\text{ J}}{(1.381 \times 10^{-23}\text{ J}\cdot\text{K}^{-1})(100\text{ K})}\right) = 1$$

$$\left(\frac{N_2}{N_1}\right)_{298} = \exp\left(-\frac{10^{-23}}{(1.381 \times 10^{-23})(298)}\right) = 1$$

$$\left(\frac{N_2}{N_1}\right)_{1000} = \exp\left(-\frac{10^{-20}}{(1.381 \times 10^{-23})(1000)}\right) = 1$$

For translational motion, the molecules in a system are evenly distributed over all the states and can be treated classically. For the rotational states,

$$\left(\frac{N_2}{N_1}\right)_{100} = \exp\left(-\frac{10^{-23}}{(1.381 \times 10^{-23})(100)}\right) = 0.993$$

$$\left(\frac{N_2}{N_1}\right)_{298} = \exp\left(-\frac{10^{-23}}{(1.381 \times 10^{-23})(298)}\right) = 0.998$$

$$\left(\frac{N_2}{N_1}\right)_{1000} = \exp\left(-\frac{10^{-23}}{(1.381 \times 10^{-23})(1000)}\right) = 0.999$$

For rotational motion, the molecules in a system are evenly distributed over all the states and can be treated classically (unless at very low temperatures). For the vibrational states,

$$\left(\frac{N_2}{N_1}\right)_{100} = \exp\left(-\frac{10^{-20}}{(1.381 \times 10^{-23})(100)}\right) = 0.0007$$

$$\left(\frac{N_2}{N_1}\right)_{298} = \exp\left(-\frac{10^{-20}}{(1.381 \times 10^{-23})(298)}\right) = 0.088$$

$$\left(\frac{N_2}{N_1}\right)_{1000} = \exp\left(-\frac{10^{-20}}{(1.381 \times 10^{-23})(1000)}\right) = 0.485$$

For vibrational motion, most of the molecules in a system are in the lowest state, even at high temperatures, and cannot be treated classically.

8.2 IDEAL GAS PARTITION FUNCTIONS

8.17 Translational motion of an ideal gas can be described in terms of a particle in a three-dimensional cubic box where the energy of a state is given by

$$\varepsilon_{\text{trans}} = (n_x^2 + n_y^2 + n_z^2)\frac{h^2}{8ma^2} \tag{8.8}$$

where h is Planck's constant, m is the molecular mass, a is the length of the cube, and n_i is a quantum number. Assuming $n_x = n_y = n_z = n$, find n for a nitrogen molecule at 298 K in a cube having $a = 0.2916$ m.

▮ Setting the average translational kinetic energy given by *(2.19a)* equal to $\varepsilon_{\text{trans}}$ given by *(8.8)* and solving for n gives

$$\bar{\varepsilon}_{\text{trans}} = \tfrac{3}{2}kT = (n^2 + n^2 + n^2)\frac{h^2}{8ma^2} = (3n^2)\frac{h^2}{8ma^2}$$

$$n = \left(\frac{4(1.381 \times 10^{-23}\ \text{J}\cdot\text{K}^{-1})(298\ \text{K})\dfrac{28.01 \times 10^{-3}\ \text{kg}\cdot\text{mol}^{-1}}{6.022 \times 10^{23}\ \text{mol}^{-1}}(0.2916\ \text{m})^2}{(6.626 \times 10^{-34}\ \text{J}\cdot\text{s})^2[(1\ \text{kg}\cdot\text{m}^2\cdot\text{s}^{-2})/(1\ \text{J})]}\right)^{1/2}$$

$$= 1.218 \times 10^{10}$$

8.18 Calculate $\varepsilon_{\text{trans}}(1, 1, 1)$ and $\varepsilon_{\text{trans}}(1, 1, 2)$ for a nitrogen molecule in a cube having $a = 0.2196$ m. Compare $\Delta\varepsilon_{\text{trans}}$ to $\bar{\varepsilon}_{\text{trans}}$ at 298 K.

▮ Substituting into *(8.8)* gives

$$\varepsilon_{\text{trans}}(1, 1, 1) = (1^2 + 1^2 + 1^2) \times \frac{(6.626 \times 10^{-34}\ \text{J}\cdot\text{s})^2[(1\ \text{kg}\cdot\text{m}^2\cdot\text{s}^{-2})/(1\ \text{J})]}{8[(28.01 \times 10^{-3}\ \text{kg}\cdot\text{mol}^{-1})/(6.022 \times 10^{23}\ \text{mol}^{-1})](0.2916\ \text{m})^2}$$

$$= 4.163 \times 10^{-41}\ \text{J}$$

$$\varepsilon_{\text{trans}}(1, 1, 2) = (1^2 + 1^2 + 2^2)\frac{(6.626 \times 10^{-34})^2}{8[(28.01 \times 10^{-3})/(6.022 \times 10^{23})](0.2916)^2}$$

$$= 8.326 \times 10^{-41}\ \text{J}$$

The average translational energy is given by *(2.19a)* as

$$\bar{\varepsilon}_{\text{trans}} = \tfrac{3}{2}(1.381 \times 10^{-23}\ \text{J}\cdot\text{K}^{-1})(298\ \text{K}) = 6.17 \times 10^{-21}\ \text{J}$$

giving

$$\frac{\Delta\varepsilon_{\text{trans}}}{\bar{\varepsilon}_{\text{trans}}} = \frac{8.326 \times 10^{-41}\ \text{J} - 4.163 \times 10^{-41}\ \text{J}}{6.17 \times 10^{-21}\ \text{J}} = 6.74 \times 10^{-21}$$

The translational energy states are very closely spaced compared to the value of the average translational energy.

8.19 Calculate the value of the molecular translational partition function for nitrogen at 298 K in a 24.79-L container (the standard molar volume of an ideal gas under these conditions).

∎ For an ideal gas,

$$q_{\text{trans}} = (2\pi mkT)^{3/2}\frac{V}{h^3} = (2\pi mkT)^{3/2}\frac{nRT}{h^3P} \tag{8.9}$$

Substituting into *(8.9)* gives

$$q_{\text{trans}} = \{2\pi[(28.01 \times 10^{-3}\ \text{kg}\cdot\text{mol}^{-1})/(6.022 \times 10^{23}\ \text{mol}^{-1})]$$
$$\times(1.381 \times 10^{-23}\ \text{J}\cdot\text{K}^{-1})(298\ \text{K})\}^{3/2} \times \frac{(24.79\ \text{L})[(10^{-3}\ \text{m}^3)/(1\ \text{L})]}{[(1\ \text{kg}\cdot\text{m}^2\cdot\text{s}^{-2})/(1\ \text{J})]^{3/2}(6.626 \times 10^{-34}\ \text{J}\cdot\text{s})^3}$$
$$= 3.555 \times 10^{30}$$

8.20 How will the value of q_{trans} change with an increase in T, a decrease of P, or an increase in V? What is the value of q_{trans} for N(g) and for N_3(g) under the same conditions as N_2 given in Problem 8.19?

∎ From *(8.9)*, q_{trans} is directly proportional to $T^{5/2}$, inversely proportional to P, and directly proportional to V. Thus any of the three given changes will cause an increase in q_{trans}. From *(8.9)*, q_{trans} is directly proportional to $m^{3/2}$, so $q_{\text{trans}}(\text{N}) = 1.258 \times 10^{30}$ and $q_{\text{trans}}(\text{N}_3) = 6.533 \times 10^{30}$.

8.21 At temperatures below which the value of q_{trans} is less than 10, the translational energy should no longer be treated classically. What is this temperature for the system described in Problem 8.19?

∎ Substituting $q_{\text{trans}} = 10$ into *(8.9)* and solving for T gives

$$10 = \{2\pi[(28.01 \times 10^{-3}\ \text{kg}\cdot\text{mol}^{-1})/(6.022 \times 10^{23}\ \text{mol}^{-1})](1.381 \times 10^{-23}\ \text{J}\cdot\text{K}^{-1})T\}^{3/2}$$
$$\times\frac{(24.79\ \text{L})[(10^{-3}\ \text{m}^3)/(1\ \text{L})]}{[(1\ \text{kg}\cdot\text{m}^2\cdot\text{s}^{-2})/(1\ \text{J})]^{3/2}(6.626 \times 10^{-34}\ \text{J}\cdot\text{s})^3}$$
$$T = 5.938 \times 10^{-18}\ \text{K}$$

8.22 Calculate ε_{rot} for nitrogen at 298 K for $J = 0$ and $J = 2$. Compare $\Delta\varepsilon_{\text{rot}}$ to $\bar{\varepsilon}_{\text{rot}}$ at 298 K. The moment of inertia is $1.407 \times 10^{-46}\ \text{kg}\cdot\text{m}^2$ for N_2.

∎ For a rigid rotator,

$$\varepsilon_{\text{rot}} = J(J+1)(h^2/8\pi^2 I) \tag{8.10}$$

and I is the moment of inertia and J is the rotational quantum number. Substituting into *(8.10)* gives

$$\varepsilon_{\text{rot}}(J=0) = (0)(0+1)\frac{(6.626 \times 10^{-34}\ \text{J}\cdot\text{s})^2[(1\ \text{kg}\cdot\text{m}^2\cdot\text{s}^{-2})/(1\ \text{J})]}{8\pi^2(1.407 \times 10^{-46}\ \text{kg}\cdot\text{m}^2)} = 0\ \text{J}$$

$$\varepsilon_{\text{rot}}(J=2) = (2)(2+1)\frac{(6.626 \times 10^{-34})^2}{8\pi^2(1.407 \times 10^{-46})} = 2.371 \times 10^{-22}\ \text{J}$$

The average rotational energy is given by $\frac{1}{2}kT$ for each degree of freedom, giving

$$\bar{\varepsilon}_{\text{rot}} = \tfrac{1}{2}(1.381 \times 10^{-23}\ \text{J}\cdot\text{K}^{-1})(298\ \text{K}) = 2.06 \times 10^{-21}\ \text{J}$$

The ratio of $\Delta\varepsilon_{\text{rot}}$ to $\bar{\varepsilon}_{\text{rot}}$ is

$$\frac{\Delta\varepsilon_{\text{rot}}}{\bar{\varepsilon}_{\text{rot}}} = \frac{2.371 \times 10^{-22}\ \text{J} - 0\ \text{J}}{2.06 \times 10^{-21}\ \text{J}} = 0.115$$

The rotational energy levels are relatively close compared to the values of the average rotational energy (unless at low temperatures).

8.23 Calculate the value of the molecular rotational partition function for N_2(g) at 298 K. The moment of inertia is $1.407 \times 10^{-46}\ \text{kg}\cdot\text{m}^2$, and the symmetry number is 2 for N_2.

∎ For a diatomic or linear polyatomic molecule,

$$q_{\text{rot}} = 8\pi^2 kT/h^2\sigma \tag{8.11}$$

where σ is the *symmetry number* for the molecule. Substituting into *(8.11)* gives

$$q_{\text{rot}} = \frac{8\pi^2(1.407 \times 10^{-46}\ \text{kg}\cdot\text{m}^2)(1.381 \times 10^{-23}\ \text{J}\cdot\text{K}^{-1})(298\ \text{K})}{(6.626 \times 10^{-34}\ \text{J}\cdot\text{s})^2(2)[(1\ \text{kg}\cdot\text{m}^2\cdot\text{s}^{-2})/(1\ \text{J})]} = 52.08$$

8.24 How will the value of q_{rot} change with an increase of T, a decrease in P, or an increase in V? What is the value of q_{rot} for N(g) and for N_3(g) under the same conditions as N_2 given in Problem 8.23? For N_3, $I = 6.4936 \times 10^{-46}\ \text{kg} \cdot \text{m}^2$ and $\sigma = 2$.

▮ From *(8.11)*, q_{rot} is directly proportional to T but does not depend on P or V. Thus an increase in T will cause an increase in q_{rot}. From *(8.11)*, q_{rot} is directly proportional to I, so $q_{rot}(N_3) = 240.3$. There are no rotational degrees of freedom for an atom.

8.25 Calculate the value of the molecular rotational partition function for O_3(g) at 298 K. The symmetry number is 2, and the rotational constants are 3.553 81 cm^{-1}, 0.445 30 cm^{-1}, and 0.394 77 cm^{-1} for O_3.

▮ For a nonlinear polyatomic molecule,

$$q_{rot} = \frac{8\pi^2(8\pi^3 I_x I_y I_z)^{1/2}(kT)^{3/2}}{h^3\sigma} \tag{8.12}$$

where I_x, I_y, and I_z are the three principal moments of inertia for the molecule. The spectroscopic data given in cm^{-1} can be converted to moments of inertia given by kg · m^2 by dividing the spectroscopic value into 2.7993×10^{-46}, giving

$$I = \frac{2.7993 \times 10^{-46}}{3.553\,81} = 7.8769 \times 10^{-47}\ \text{kg} \cdot \text{m}^2$$

$$I = \frac{2.7993 \times 10^{-46}}{0.445\,30} = 6.2852 \times 10^{-46}\ \text{kg} \cdot \text{m}^2$$

$$I = \frac{2.7993 \times 10^{-46}}{0.394\,77} = 7.0909 \times 10^{-46}\ \text{kg} \cdot \text{m}^2$$

Substituting into *(8.12)* gives

$$q_{rot} = \frac{\left[\begin{array}{l} 8\pi^2[8\pi^3(7.8769 \times 10^{-47}\ \text{kg} \cdot \text{m}^2)(6.2852 \times 10^{-46}\ \text{kg} \cdot \text{m}^2) \\ \times (7.0909 \times 10^{-46}\ \text{kg} \cdot \text{m}^2)]^{1/2}[(1.381 \times 10^{-23}\ \text{J} \cdot \text{K}^{-1})(298\ \text{K})]^{3/2} \end{array}\right]}{(6.626 \times 10^{-34}\ \text{J} \cdot \text{s})^3(2)[(1\ \text{kg} \cdot \text{m}^2 \cdot \text{s}^{-2})/(1\ \text{J})]^{3/2}}$$

$$= 3344$$

8.26 The symmetry number in *(8.11)* represents the number of indistinguishable positions into which the molecule can be placed by simple rigid rotations. Determine σ for asymmetric linear molecules (e.g., HCl), symmetric linear molecules (e.g., N_2), N_3, O_3, H_2O, N_2O, BrF_5, CCl_3F, and CCl_4.

▮ There is only one position for a heteronuclear diatomic molecule like HCl and for an asymmetric linear molecule like NNO, giving $\sigma = 1$. There are two positions related by rotation for a homogeneous diatomic molecule like NN and for a symmetric linear molecule like NNN, giving $\sigma = 2$. For symmetric nonlinear molecules like OOO and HOH, there are two positions related by rotation, giving $\sigma = 2$. The BrF_5 molecule is in the shape of a square pyramid, giving $\sigma = 4$, and the CCl_3F molecule is in the shape of a trigonal pyramid, giving $\sigma = 3$. The CCl_4 molecule is a regular tetrahedron, for which $\sigma = 12$.

8.27 For the system described in Problem 8.23, at what temperature will $q_{rot} = 10$ as calculated using *(8.11)*? What is the actual value of q_{rot} using *(8.6)*?

▮ Solving *(8.11)* for T and substituting the data gives

$$T = \frac{(10)(6.626 \times 10^{-34}\ \text{J} \cdot \text{s})^2(2)[(1\ \text{kg} \cdot \text{m}^2 \cdot \text{s}^{-2})/(1\ \text{J})]}{8\pi^2(1.407 \times 10^{-46}\ \text{kg} \cdot \text{m}^2)(1.381 \times 10^{-23}\ \text{J} \cdot \text{K}^{-1})} = 57.25\ \text{K}$$

The degeneracy of each rotational energy level is $2J + 1$. Substituting *(8.10)* into *(8.6)* gives

$$q_{rot} = \sum (2J + 1)\, e^{-J(J+1)h^2/8\pi^2 IkT}$$

$$= [2(0) + 1]\, e^{-0(0+1)(0.05000)} + [2(1) + 1]\, e^{-1(1+1)(0.05000)} + \cdots$$

$$= 20.34$$

where roughly 20 terms are needed before q_{rot} changes insignificantly. Recalling that $\sigma = 2$ for N_2, $q_{rot} = 10.17$.

8.28 Calculate ε_{vib} for nitrogen at 298 K for $v = 0$ and $v = 1$. Compare $\Delta\varepsilon_{vib}$ to ε_{vib} at 298 K. For N_2, $\bar{\nu} = 2354.999$ cm^{-1}.

▮ For a simple harmonic oscillator,

$$\varepsilon_{\text{vib}} = (v + \tfrac{1}{2})h\nu \qquad (8.13)$$

where ν is the frequency of the vibrational motion. The value of ν is

$$\nu = c\bar{\nu} = (2.9979 \times 10^{10}\ \text{cm} \cdot \text{s}^{-1})(2354.999\ \text{cm}^{-1}) = 7.0601 \times 10^{13}\ \text{s}^{-1}$$

Substituting into *(8.13)* gives

$$\varepsilon_{\text{vib}}(v = 0) = (0 + \tfrac{1}{2})(6.626 \times 10^{-34}\ \text{J} \cdot \text{s})(7.0601 \times 10^{13}\ \text{s}^{-1}) = 2.339 \times 10^{-20}\ \text{J}$$

$$\varepsilon_{\text{vib}}(v = 1) = (1 + \tfrac{1}{2})(6.626 \times 10^{-34})(7.0601 \times 10^{13}) = 7.017 \times 10^{-20}\ \text{J}$$

Dividing $\Delta\varepsilon_{\text{vib}}$ by $\varepsilon_{\text{vib}}(v = 0)$ gives

$$\frac{\Delta\varepsilon_{\text{vib}}}{\varepsilon_{\text{vib}}(v = 0)} = \frac{7.017 \times 10^{-20}\ \text{J} - 2.339 \times 10^{-20}\ \text{J}}{2.339 \times 10^{-20}\ \text{J}} = 2.000$$

The vibrational energy states are relatively far apart compared to the values of the vibrational energy (unless at very high temperatures).

8.29 Calculate the value of the molecular vibrational partition function for $N_2(g)$ at 298 K. For N_2, $\bar{\nu} = 2754.999\ \text{cm}^{-1}$.

▮ For each degree of vibrational freedom,

$$q_{\text{vib}} = \frac{1}{1 - e^{-x}} \qquad (8.14a)$$

where x was defined in Table 3-2. For N_2,

$$x = (1.4388)(2354.999)/298 = 11.365$$

which upon substituting into *(8.14a)* gives

$$q_{\text{vib}} = 1/(1 - e^{-11.365}) = 1.000\,012$$

8.30 How will the value of q_{vib} change with an increase in T? At what temperature will $q_{\text{vib}} = 10$ for N_2? For N_2, $\bar{\nu} = 2354.999\ \text{cm}^{-1}$.

▮ Setting $q_{\text{vib}} = 10$ in *(8.14a)* and solving for x gives

$$10 = 1/(1 - e^{-x}) \qquad x = 0.1054$$

which corresponds to

$$T = \frac{(1.4388)(2354.999)}{0.1054} = 32\,150\ \text{K}$$

8.31 Which molecule will have the greater value of q_{vib} at a given temperature—one with a high value of $\bar{\nu}$ or one with a low value of $\bar{\nu}$? Confirm your prediction by calculating q_{vib} for $I_2(g)$ at 298 K, and compare the answer to that given for N_2 in Problem 8.29. For I_2, $\bar{\nu} = 214.5481\ \text{cm}^{-1}$.

▮ For a lower value of $\bar{\nu}$, the corresponding value of x will be lower. The value of e^{-x} will be larger and $1 - e^{-x}$ will be smaller. Thus q_{vib} is predicted to be greater for a smaller value of $\bar{\nu}$. For I_2,

$$x = (1.4388)(214.5481)/298 = 1.035$$

$$q_{\text{vib}} = 1/(1 - e^{-1.035}) = 1.551$$

As $\bar{\nu}$ decreased from $2354.999\ \text{cm}^{-1}$ to $214.5481\ \text{cm}^{-1}$, q_{vib} increased from 1.000 012 to 1.551.

8.32 Calculate the value of the molecular vibrational partition function for $N_3(g)$ at 298 K. For N_3, $\bar{\nu}_1 = 1800\ \text{cm}^{-1}$; $\bar{\nu}_2 = 500\ \text{cm}^{-1}$, $g_2 = 2$; and $\bar{\nu}_3 = 1200\ \text{cm}^{-1}$.

▮ For a polyatomic molecule,

$$q_{\text{vib}} = \prod_{i=1}^{3\Lambda-5 \text{ or } 3\Lambda-6} \frac{1}{1 - e^{-x_i}} \qquad (8.14b)$$

The values of x_i are

$$x_1 = (1.4288)(1800)/298 = 8.7$$

$x_2 = x_3 = 2.4$, and $x_4 = 5.8$. Substituting into *(8.14b)* gives

$$q_{\text{vib}} = \frac{1}{1 - e^{-8.7}}\left(\frac{1}{1 - e^{-2.4}}\right)^2\left(\frac{1}{1 - e^{-5.8}}\right) = 1.213$$

8.33 Which molecule, O_3 or N_3, will have a larger product of $q_{rot}q_{vib}$ at the same temperature? Confirm your answer by using the results of Problems 8.24, 8.25, and 8.32. For $O_3(g)$, $\bar{\nu}/(\text{cm}^{-1}) = 1110$, 705, and 1043.

▮ The values of q_{rot} are considerably greater than those of q_{vib}. Thus, O_3 with three rotational degrees of freedom and three vibrational degrees of freedom will have a greater value of $q_{rot}q_{vib}$ than N_3, which has only two rotational degrees of freedom and four vibrational degrees of freedom. For O_3, the values of x are 5.36, 3.40, and 5.03, respectively, at 298 K. Using *(8.14b)* gives

$$q_{vib} = \frac{1}{1 - e^{-5.36}}\left(\frac{1}{1 - e^{-3.40}}\right)^2\left(\frac{1}{1 - e^{-5.03}}\right) = 1.046$$

The respective products of partition functions are

$$q_{rot}q_{vib}(N_3) = (240.3)(1.213) = 291.5$$

$$q_{rot}q_{vib}(O_3) = (3344)(1.046) = 3498$$

which confirms the prediction.

8.34 Calculate the number of molecules in a 1-mol sample of N_2 at 298 K having $v = 1$ and $v = 2$. Repeat the calculations at 1000 K, and compare the results.

▮ At 298 K, $x = 11.365$ (see Problem 8.29). The number of molecules in the ground state is given by

$$N_0 = N(1 - e^{-x}) \qquad (8.15a)$$

where N is the total number of molecules present. Substituting into *(8.15a)* gives

$$N_0 = (6.022\,045 \times 10^{23})(1 - e^{-11.365}) = 6.021\,975 \times 10^{23}$$

The number of molecules in a given excited state is given by

$$N_v = N_0\, e^{-vx} \qquad (8.15b)$$

Substituting gives

$$N_1 = (6.021\,975 \times 10^{23})\, e^{-1(11.365)} = 6.982 \times 10^{18}$$

$$N_2 = (6.021\,975 \times 10^{23})\, e^{-2(11.365)} = 8.095 \times 10^{13}$$

At 1000 K, $x = 3.388$, giving $N_0 = 5.819 \times 10^{23}$, $N_1 = 1.965 \times 10^{22}$, and $N_2 = 6.638 \times 10^{20}$, a significant increase in the population of the excited states.

8.35 What is the ratio of the number of molecules with $v = 1$ and $J = 2$ to those with $v = 2$ and $J = 6$ for N_2 at 1000 K? For N_2, $\nu = 7.06 \times 10^{13}\ \text{s}^{-1}$ and $I = 1.407 \times 10^{-46}\ \text{kg} \cdot \text{m}^2$.

▮ Using *(8.10)* and *(8.13)* gives

$$\varepsilon(v = 1, J = 2) = (1 + \tfrac{1}{2})(6.626 \times 10^{-34}\ \text{J} \cdot \text{s})(7.06 \times 10^{13}\ \text{s}^{-1})$$

$$+(2)(2 + 1)\frac{(6.626 \times 10^{-34}\ \text{J} \cdot \text{s})^2[(1\ \text{kg} \cdot \text{m}^2 \cdot \text{s}^{-2})/(1\ \text{J})]}{8\pi^2(1.407 \times 10^{-46}\ \text{kg} \cdot \text{m}^2)}$$

$$= 7.04 \times 10^{-20}\ \text{J}$$

$$\varepsilon(v = 2, J = 6) = (2 + \tfrac{1}{2})(6.626 \times 10^{-34})(7.06 \times 10^{13}) + (6)(6 + 1)\frac{(6.626 \times 10^{-34})^2}{8\pi^2(1.407 \times 10^{-46})}$$

$$= 1.186 \times 10^{-19}\ \text{J}$$

Substituting into *(2.22)* gives

$$\frac{N(v = 1, J = 2)}{N(v = 2, J = 6)} = \frac{[2(2) + 1]\exp[-(7.04 \times 10^{-20}\ \text{J})/(1.381 \times 10^{-23}\ \text{J} \cdot \text{K}^{-1})(1000\ \text{K})]}{[2(6) + 1]\exp[-(1.186 \times 10^{-19}\ \text{J})/(1.381 \times 10^{-23}\ \text{J} \cdot \text{K}^{-1})(1000\ \text{K})]}$$

$$= 12.6$$

8.36 Calculate the molecular electronic partition function for atomic nitrogen at 298 K given the following electronic data:

level	0	1	2	3	4
$\varepsilon_j/(\text{cm}^{-1})$	0	19 224.464	19 233.177	28 838.920	28 839.306
g_j	4	6	4	2	4

▮ The energies of each level are calculated using

$$\varepsilon = h\nu = hc\bar{\nu} = (6.626 \times 10^{-34}\ \text{J}\cdot\text{s})(2.9979 \times 10^{10}\ \text{cm}\cdot\text{s}^{-1})(0\ \text{cm}^{-1}) = 0\ \text{J}$$

Likewise, $\varepsilon_1 = 3.819 \times 10^{-19}\ \text{J}$, $\varepsilon_2 = 3.821 \times 10^{-19}\ \text{J}$, $\varepsilon_3 = 5.729 \times 10^{-19}\ \text{J}$, and $\varepsilon_4 = 5.729 \times 10^{-19}\ \text{J}$. Substituting into *(8.6)* gives

$$\begin{aligned} q_{\text{elec}} &= 4\exp\left(-\frac{0}{(1.381 \times 10^{-23}\ \text{J}\cdot\text{K}^{-1})(298\ \text{K})}\right) + 6\exp\left(-\frac{3.819 \times 10^{-19}}{(1.381 \times 10^{-23})(298)}\right) \\ &\quad + 4\exp\left(-\frac{3.821 \times 10^{-19}}{(1.381 \times 10^{-23}\ \text{J}\cdot\text{K}^{-1})(298)}\right) + 2\exp\left(-\frac{5.729 \times 10^{-19}}{(1.381 \times 10^{-23})(298)}\right) \\ &\quad + 4\exp\left(-\frac{5.729 \times 10^{-19}}{(1.381 \times 10^{-23})(298)}\right) \\ &= 4 \end{aligned}$$

None of the excited states makes a significant contribution to q_{elec} at 298 K.

8.37 What electronic energy spacing will contribute 1×10^{-6} to q_{elec} at 1000 K? Express the answer in units of J, $\text{J}\cdot\text{mol}^{-1}$, and cm^{-1}.

▮ Choosing a generous value of $g = 10$, substituting into *(8.6)*, and solving for ε gives

$$1 \times 10^{-6} = 10\exp[-\varepsilon/(1.381 \times 10^{-23}\ \text{J}\cdot\text{K}^{-1})(1000\ \text{K})]$$

$$\varepsilon = 2 \times 10^{-19}\ \text{J}$$

Expressing this value in the other units gives

$$\nu = \frac{\varepsilon}{hc} = \frac{2 \times 10^{-19}\ \text{J}}{(6.626 \times 10^{-34}\ \text{J}\cdot\text{s})(2.9979 \times 10^{10}\ \text{cm}\cdot\text{s}^{-1})} = 1 \times 10^{4}\ \text{cm}^{-1}$$

$$E = L\varepsilon = (6.022 \times 10^{23}\ \text{mol}^{-1})(2 \times 10^{-19}\ \text{J}) = 1 \times 10^{5}\ \text{J}\cdot\text{mol}^{-1}$$

8.38 Calculate q_{elec} at 298 K for N_2 given $\bar{\nu}_j/(\text{cm}^{-1}) = 49\,754.78\ (g_0 = 3)$, $59\,306.81\ (g_1 = 6)$, and $59\,380\ (g_2 = 6)$.

▮ Because $\bar{\nu}_j > 1 \times 10^{4}\ \text{cm}^{-1}$, the contribution of the excited states will be insignificant at 298 K (see Problem 8.37). Thus *(8.6)* gives

$$q_{\text{elec}} = 3\,e^{-0/kT} = 3$$

8.39 Prepare a plot of q_{elec} for N_3 between 0 and 1000 K given $\bar{\nu}_j/(\text{cm}^{-1}) = 0\ (g_0 = 2)$, $71.3\ (g_1 = 2)$, and $36\,811\ (g_2 = 2)$.

▮ The respective energies are

$$\varepsilon_0 = (6.626 \times 10^{-34}\ \text{J}\cdot\text{s})(2.9979 \times 10^{10}\ \text{cm}\cdot\text{s}^{-1})(0\ \text{cm}^{-1}) = 0\ \text{J}$$

$$\varepsilon_1 = (6.626 \times 10^{-34})(2.9979 \times 10^{10})(71.3) = 1.42 \times 10^{-21}\ \text{J}$$

$$\varepsilon_2 = (6.626 \times 10^{-34})(2.9979 \times 10^{10})(36\,811) = 7.31 \times 10^{-19}\ \text{J}$$

At any given temperature, *(8.6)* gives

$$\begin{aligned} q_{\text{elec}} &= 2\exp[-0/(1.381 \times 10^{-23}\ \text{J}\cdot\text{K}^{-1})T] + 2\exp[-(1.42 \times 10^{-21}\ \text{J})/(1.381 \times 10^{-23}\ \text{J}\cdot\text{K}^{-1})T] \\ &\quad +2\exp[-(7.31 \times 10^{-19}\ \text{J})/(1.381 \times 10^{-23}\ \text{J}\cdot\text{K}^{-1})T] \end{aligned}$$

As a sample calculation, consider $T = 298\ \text{K}$:

$$q_{\text{elec}} = 2 + 2\exp\left(-\frac{1.42 \times 10^{-21}}{(1.381 \times 10^{-23})(298)}\right) + 2\exp\left(-\frac{7.31 \times 10^{-19}}{(1.381 \times 10^{-23})(298)}\right) = 3.417$$

The plot is shown in Fig. 8-1.

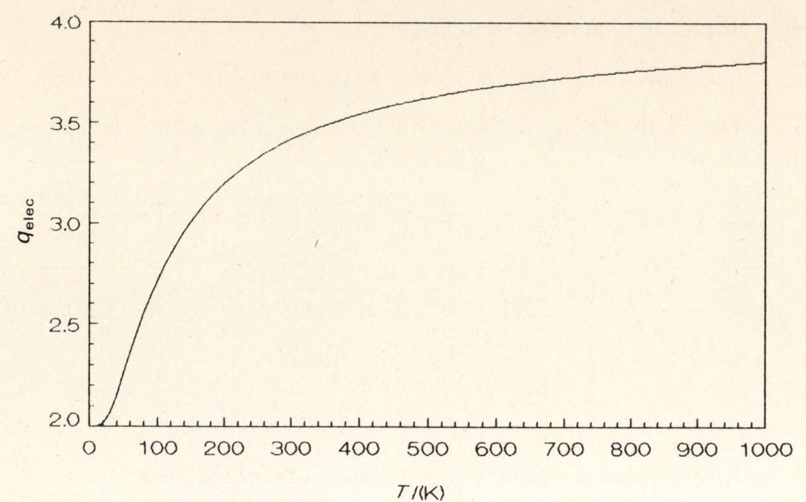

Fig. 8-1

8.40 Calculate the fraction of N_2 molecules at 298 K that are in the first excited level. See Problem 8.36 for additional data.

▌ Substituting $\varepsilon_0 = 0$ J, $g_0 = 4$, $\varepsilon_1 = 3.819 \times 10^{-19}$ J, and $g_1 = 6$ into *(2.22)* gives

$$\frac{N_1}{N_0} = \tfrac{6}{4}\exp\left(-\frac{3.819 \times 10^{-19}\ \text{J}}{(1.381 \times 10^{-23}\ \text{J}\cdot\text{K}^{-1})(298\ \text{K})}\right) = 7.67 \times 10^{-41}$$

Assuming $N_0 \approx N$, then $N_1/N = 7.67 \times 10^{-41}$.

8.41 Nuclear energy levels are separated by energies on the order of 1 MeV. What is the nuclear partition function?

▌ Converting the energy to J (see Table 3-1),

$$\varepsilon = (1\ \text{MeV})\frac{10^6\ \text{eV}}{1\ \text{MeV}}\left(\frac{1.602 \times 10^{-19}\ \text{J}}{1\ \text{eV}}\right) = 2 \times 10^{-13}\ \text{J}$$

The exponential in *(8.6)* at 298 K is

$$\exp\left(-\frac{2 \times 10^{-13}\ \text{J}}{(1.381 \times 10^{-23}\ \text{J}\cdot\text{K}^{-1})(298\ \text{K})}\right) = e^{-5\times10^7} = 10^{-2\times10^7} \approx 0$$

Thus *(8.6)* gives $q_{\text{nuc}} = g_0$.

8.42 Calculate $\ln Q$, the logarithm of the canonical ensemble partition function, for N(g), N_2(g), and N_3(g) at 298 K. See Problems 8.19, 8.20, 8.24, 8.29, 8.32, 8.36, 8.38, and 8.39 for additional data.

▌ The ensemble partition function is given in terms of the molecular contributions by

$$Q = (q_{\text{nuc}}q_{\text{elec}}q_{\text{vib}}q_{\text{rot}})^N(q_{\text{trans}}^N/N!) \tag{8.16}$$

For ordinary chemical processes, it is customary to omit q_{nuc} in *(8.16)*. From Problems 8.20 and 8.36, *(8.16)* gives, for 1 mol of N(g),

$$\begin{aligned}\ln[Q(\text{N})] &= L(\ln q_{\text{elec}} + \ln q_{\text{trans}}) - \ln(L!)\\ &= (6.022 \times 10^{23})[\ln 4 + \ln(1.258 \times 10^{30})] - [(6.022 \times 10^{23})\ln(6.022 \times 10^{23}) - 6.022 \times 10^{23}]\\ &= 1.020 \times 10^{25}\end{aligned}$$

where *Stirling's approximation*

$$\ln(N!) = N\ln N - N \tag{8.17}$$

was used to evaluate $\ln(L!)$. From Problems 8.19, 8.29, and 8.38, *(8.16)* gives, for 1 mol of N_2(g),

$$\begin{aligned}\ln[Q(\text{N}_2)] &= L(\ln q_{\text{elec}} + \ln q_{\text{vib}} + \ln q_{\text{rot}} + \ln q_{\text{trans}}) - \ln(L!)\\ &= (6.022 \times 10^{23})[\ln 3 + \ln 1.000\,012 + \ln 52.08 + \ln(3.555 \times 10^{30})] - 3.2377 \times 10^{25}\\ &= 1.303 \times 10^{25}\end{aligned}$$

From Problems 8.20, 8.24, 8.32, and 8.39, *(8.16)* gives for 1 mol of $N_3(g)$

$$\ln[Q(N_3)] = L(\ln q_{elec} + \ln q_{vib} + \ln q_{rot} + \ln q_{trans}) - \ln(L!)$$
$$= (6.022 \times 10^{23})[\ln 3.417 + \ln 1.213 + \ln 240.3 + \ln(6.533 \times 10^{30})] - 3.2377 \times 10^{25}$$
$$= 1.452 \times 10^{25}$$

8.3 APPLICATION TO THERMODYNAMICS INVOLVING IDEAL GASES

8.43 Derive an expression for the pressure of an ideal gas from *(8.16)*.

▌ The relationship between pressure and Q is

$$P = kT(\partial \ln Q/\partial V)_{T,N} \tag{8.18}$$

Substituting *(8.16)* gives

$$P = kT\left(\frac{\partial}{\partial V}[N(\ln q_{nuc} + \ln q_{elec} + \ln q_{vib} + \ln q_{trans}) - \ln N!]\right)_{T,N}$$
$$= kTN\left[0 + 0 + 0 + \left(\frac{\partial}{\partial V}\ln\left[(2\pi mkT)^{3/2}\frac{V}{h^3}\right]\right)_{T,N} - 0\right]$$
$$= kTN(1/V) \tag{8.19}$$

This result is the same as *(1.18)*, the ideal gas law.

8.44 Calculate the Helmholtz free energy for $N(g)$, $N_2(g)$, and $N_3(g)$ at 298 K. See Problem 8.42 for additional data.

▌ The relationship between A and Q is

$$A = -kT \ln Q \tag{8.20}$$

Substituting the results of Problem 8.42 into *(8.20)* gives

$$A(N) = -(1.381 \times 10^{-23}\ \text{J}\cdot\text{K}^{-1})(298\ \text{K})(1.020 \times 10^{25}) = -41\,990\ \text{J}$$
$$A(N_2) = -(1.381 \times 10^{-23})(298)(1.303 \times 10^{25}) = -53\,640\ \text{J}$$
$$A(N_3) = -(1.381 \times 10^{-23})(298)(1.452 \times 10^{25}) = -59\,770\ \text{J}$$

8.45 Given $N_3(g) \rightarrow N_2(g) + N(g)$ ($\Delta E° = 53.851$ kJ), use the results of Problem 8.44 to calculate $\Delta A°_{298}$ and $\Delta G°_{298}$ for this reaction.

▌ For the chemical equation,

$$\Delta A° = [(1\ \text{mol})A°(N_2) + (1\ \text{mol})A°(N)] - (1\ \text{mol})A°(N_3)$$
$$= [(1\ \text{mol})(-53.64\ \text{kJ}\cdot\text{mol}^{-1}) + (1\ \text{mol})(-41.99\ \text{kJ}\cdot\text{mol}^{-1})] - (1\ \text{mol})(-59.77\ \text{kJ}\cdot\text{mol}^{-1})$$
$$= -35.86\ \text{kJ}$$

giving

$$\Delta A°_{298} = \Delta E°_0 + \Delta A° = 53.851\ \text{kJ} - 35.86\ \text{kJ} = 18.00\ \text{kJ}$$

Using $\Delta n_g = +1$, *(6.5)* gives

$$\Delta G°_{298} = 18.00\ \text{kJ} + (8.314 \times 10^{-3}\ \text{kJ}\cdot\text{K}^{-1}\cdot\text{mol}^{-1})(298\ \text{K})(1\ \text{mol}) = 20.48\ \text{kJ}$$

8.46 Prepare a plot of the translational contribution to the standard-state entropy of an ideal gas as a function of molar mass. Prepare isotherms at 100 K, 200 K, 500 K, 750 K, and 1000 K.

▌ The translational contribution to the entropy is given by the *Sakur-Tetrode equation*,

$$S°(\text{trans}) = R\left[\tfrac{5}{2} + \ln\left(\frac{(2\pi mkT)^{3/2}V°}{h^3 L}\right)\right] = R\left[\tfrac{5}{2} + \ln\left(\frac{(2\pi mkT)^{3/2}RT}{h^3 LP}\right)\right] \tag{8.21a}$$

Upon substitution of the constants, *(8.21a)* can be written as

$$S°(\text{trans})/(\text{J}\cdot\text{K}^{-1}\cdot\text{mol}^{-1}) = 12.471\,62 \ln[M/(\text{g}\cdot\text{mol}^{-1})] + 20.7860 \ln[T/(\text{K})] - 9.685\,21 \tag{8.21b}$$

As a sample calculation, consider the results for $N_3(g)$, $M = 42.0201\ \text{g}\cdot\text{mol}^{-1}$ at 298 K:

$$S°(\text{trans})/(\text{J}\cdot\text{K}^{-1}\cdot\text{mol}^{-1}) = 12.471\,62 \ln 42.0201 + 20.7860 \ln 298.15 - 9.685\,21$$
$$= 155.37$$

The plot is shown in Fig. 8-2.

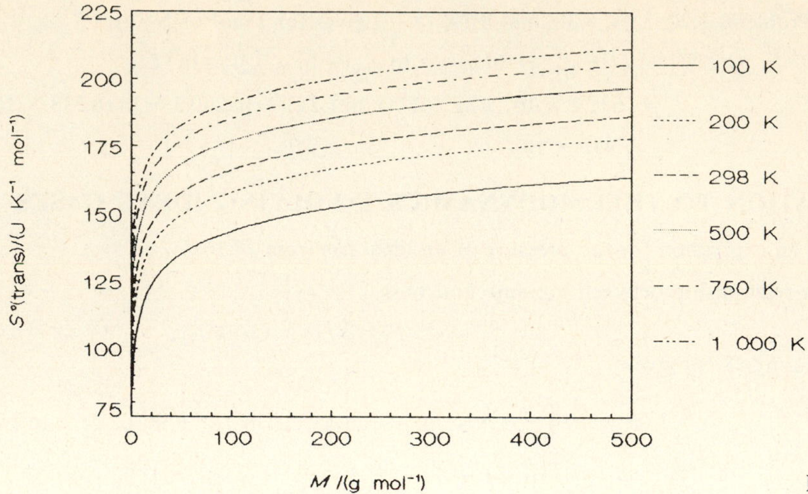

Fig. 8-2

8.47 Prepare a plot of the rotational contribution to the standard-state entropy of a linear polyatomic or diatomic ideal gas as a function of the moment of inertia. Prepare isotherms at the same temperatures listed in Problem 8.46.

I The rotational contribution to the entropy is given by

$$S°(\text{rot}) = R[1 + \ln(8\pi^2 IkT/h^2\sigma)] \qquad (8.22a)$$

$$S°(\text{rot}) = R\left(\tfrac{3}{2} + \ln\frac{8\pi^2(8\pi^3 I_x I_y I_z)^{1/2}(kT)^{3/2}}{h^2\sigma}\right) \qquad (8.23a)$$

for a linear polyatomic or diatomic gas and for a nonlinear polyatomic gas, respectively. Upon substitution of the constants, *(8.22a)* and *(8.23a)* can be written as

$$S°(\text{rot})/(\text{J}\cdot\text{K}^{-1}\cdot\text{mol}^{-1}) = 8.314\,41 \ln\frac{[I/(10^{-47}\,\text{kg}\cdot\text{m}^2)][T/(\text{K})]}{\sigma} - 22.377\,86 \qquad (8.22b)$$

$$S°(\text{rot})/(\text{J}\cdot\text{K}^{-1}\cdot\text{mol}^{-1}) = 229.5910 + 12.471\,62 \ln[T/(\text{K})] - 8.314\,41 \ln\sigma + 4.157\,21 \ln[I_x I_y I_z/(10^{-114}\,\text{kg}^3\cdot\text{m}^6)] \qquad (8.23b)$$

As a sample calculation, consider the results for a heteronuclear linear molecule with $I = 6.4963 \times 10^{-46}\,\text{kg}\cdot\text{m}^2$ at 298 K:

$$S°(\text{rot})/(\text{J}\cdot\text{K}^{-1}\cdot\text{mol}^{-1}) = 8.314\,41 \ln\frac{(64.936)(298.15)}{1} - 22.377\,86 = 59.694$$

[The value of $S°(\text{rot})$ for a symmetric linear molecule is given by the above value less $R \ln 2$.] The plot is shown in Fig. 8-3.

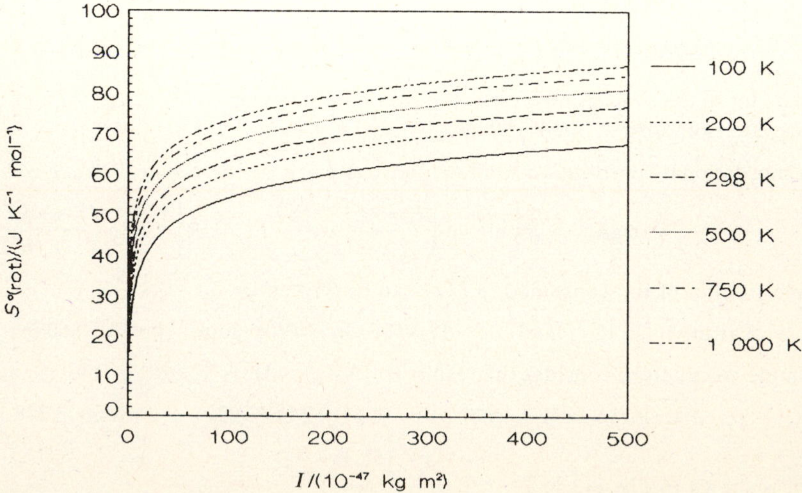

Fig. 8-3

8.48 Prepare a plot of the vibrational contribution to the standard-state entropy of an ideal gas for each degree of freedom as a function of the vibrational wavenumber, $\bar{\nu}$. Prepare isotherms at the same temperatures listed in Problem 8.46.

▌ The vibrational contribution to the entropy is given by

$$S°(\text{vib}) = R \sum_{i=1}^{3A-5 \text{ or } 3A-6} \left(\frac{x_i}{e^{x_i} - 1} - \ln(1 - e^{-x_i}) \right) \tag{8.24}$$

As a sample calculation, consider $\bar{\nu} = 500 \text{ cm}^{-1}$ at 298 K:

$$x = (1.4388)(500)/298 = 2.41$$

$$S°(\text{vib})/(\text{J} \cdot \text{K}^{-1} \cdot \text{mol}^{-1}) = (8.314)\left(\frac{2.41}{e^{2.41} - 1} - \ln(1 - e^{-2.41}) \right) = 2.76$$

The graph is shown in Fig. 8-4.

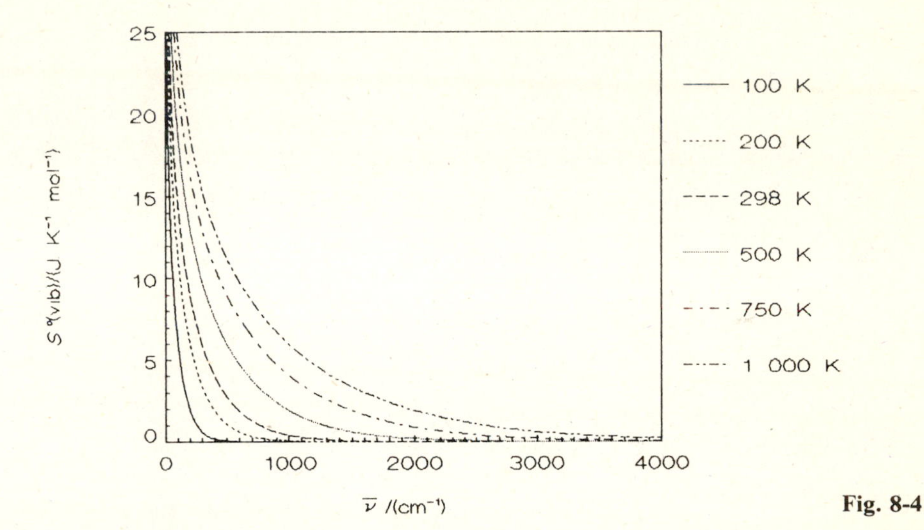

Fig. 8-4

8.49 Calculate the electronic contribution to the standard-state entropy of $N_3(g)$ at 298 K. See Problem 8.39 for additional data.

▌ The electronic contribution to the entropy is given by

$$S°(\text{elec}) = R\left(\ln q_{\text{elec}} + \frac{\sum g_j(\varepsilon_j/kT)\, e^{-\varepsilon_j/kT}}{q_{\text{elec}}} \right) \tag{8.25}$$

Substituting the values from Problem 8.39 gives

$$S°(\text{elec}) = (8.314 \text{ J} \cdot \text{K}^{-1} \cdot \text{mol}^{-1})$$

$$\times \left(\ln 3.417 + \frac{\begin{bmatrix} 2\left(\frac{0}{(1.381 \times 10^{-23} \text{ J} \cdot \text{K}^{-1})(298 \text{ K})}\right) \exp\left(-\frac{0}{(1.381 \times 10^{-23} \text{ J} \cdot \text{K}^{-1})(298 \text{ K})}\right) \\ + 2\left(\frac{1.42 \times 10^{-21} \text{ J}}{(1.381 \times 10^{-23} \text{ J} \cdot \text{K}^{-1})(298 \text{ K})}\right) \exp\left(-\frac{1.42 \times 10^{-21} \text{ J} \cdot \text{K}^{-1}}{(1.381 \times 10^{-23} \text{ J} \cdot \text{K}^{-1})(298 \text{ K})}\right) \\ + 2\left(\frac{7.31 \times 10^{-19} \text{ J}}{(1.381 \times 10^{-23} \text{ J} \cdot \text{K}^{-1})(298 \text{ K})}\right) \exp\left(-\frac{7.31 \times 10^{-19} \text{ J} \cdot \text{K}^{-1}}{(1.381 \times 10^{-23} \text{ J} \cdot \text{K}^{-1})(298 \text{ K})}\right) \end{bmatrix}}{3.417} \right)$$

$$= 14.28 \text{ J} \cdot \text{K}^{-1} \cdot \text{mol}^{-1}$$

8.50 Calculate $\Delta_r S°_{298}$ for the equation $N_3(g) \rightarrow N_2(g) + N(g)$. See Problems 8.23, 8.24, 8.28, 8.32, 8.36, 8.38, and 8.39 for additional data.

▮ From Figs. 8-2, 8-3, and 8-4 and Problem 8.49,

$$S°(N_3) = S°(\text{trans}) + S°(\text{rot}) + S°(\text{vib}) + S°(\text{elec})$$
$$= 155\ \text{J}\cdot\text{K}^{-1}\cdot\text{mol}^{-1} + [58\ \text{J}\cdot\text{K}^{-1}\cdot\text{mol}^{-1} - (8.314\ \text{J}\cdot\text{K}^{-1}\cdot\text{mol}^{-1})\ln 2]$$
$$+[2(3\ \text{J}\cdot\text{K}^{-1}\cdot\text{mol}^{-1}) + 0 + 0] + 14\ \text{J}\cdot\text{K}^{-1}\cdot\text{mol}^{-1}$$
$$= 227\ \text{J}\cdot\text{K}^{-1}\cdot\text{mol}^{-1}$$

$$S°(N_2) = 150\ \text{J}\cdot\text{K}^{-1}\cdot\text{mol}^{-1} + [46\ \text{J}\cdot\text{K}^{-1}\cdot\text{mol}^{-1} - (8.314\ \text{J}\cdot\text{K}^{-1}\cdot\text{mol}^{-1})\ln 2] + 0 + (8.314\ \text{J}\cdot\text{K}^{-1}\cdot\text{mol}^{-1})\ln 1$$
$$= 190\ \text{J}\cdot\text{K}^{-1}\cdot\text{mol}^{-1}$$

$$S°(N) = 140\ \text{J}\cdot\text{K}^{-1}\cdot\text{mol}^{-1} + (8.314\ \text{J}\cdot\text{K}^{-1}\cdot\text{mol}^{-1})\ln 4$$
$$= 152\ \text{J}\cdot\text{K}^{-1}\cdot\text{mol}^{-1}$$

For N_2 and N, the electronic contribution was calculated using *(8.25)*, neglecting the higher energy levels. Substituting into *(5.20)* gives

$$\Delta_r S° = [(1\ \text{mol})S°(N_2) + (1\ \text{mol})S°(N)] - (1\ \text{mol})S°(N_3)$$
$$= [(1\ \text{mol})(190\ \text{J}\cdot\text{K}^{-1}\cdot\text{mol}^{-1}) + (1\ \text{mol})(152\ \text{J}\cdot\text{K}^{-1}\cdot\text{mol}^{-1})] - (1\ \text{mol})(227\ \text{J}\cdot\text{K}^{-1}\cdot\text{mol}^{-1})$$
$$= 115\ \text{J}\cdot\text{K}^{-1}$$

8.51 Calculate $\Delta S°$ for cooling 1.00 mol of Ne from 298 K to 273 K under constant-volume conditions.

▮ The only contributions to the molar entropy of Ne is translational motion. Using *(8.21a)* gives

$$\Delta S° = \Delta S°(\text{trans}) = R\left(\tfrac{5}{2} + \ln\frac{(2\pi mkT_2)^{3/2}V}{h^3L}\right) - R\left(\tfrac{5}{2} + \ln\frac{(2\pi mkT_1)^{3/2}V}{h^3L}\right)$$
$$= R\ln\left(\frac{T_2}{T_1}\right)^{3/2} = \tfrac{3}{2}R\ln\frac{T_2}{T_1} = \tfrac{3}{2}(8.314\ \text{J}\cdot\text{K}^{-1}\cdot\text{mol}^{-1})\ln\frac{273\ \text{K}}{298\ \text{K}}$$
$$= -1.09\ \text{J}\cdot\text{K}^{-1}\cdot\text{mol}^{-1}$$

This is the same answer as determined in Problem 5.25.

8.52 Calculate $\Delta S°$ for cooling 1.00 mol of Ne from 298 K to 273 K under constant-pressure conditions.

▮ Repeating the calculations of Problem 8.51 gives

$$\Delta S° = \Delta S°(\text{trans}) = R\left(\tfrac{5}{2} + \ln\frac{(2\pi mkT_2)^{3/2}RT_2}{h^3LP}\right) - R\left(\tfrac{5}{2} + \ln\frac{(2\pi mkT_1)^{3/2}RT_1}{h^3LP}\right)$$
$$= R\ln\left(\frac{T_2}{T_1}\right)^{3/2} = \tfrac{5}{2}R\ln\frac{T_2}{T_1} = \tfrac{5}{2}(8.314\ \text{J}\cdot\text{K}^{-1}\cdot\text{mol}^{-1})\ln\frac{273\ \text{K}}{298\ \text{K}}$$
$$= -1.82\ \text{J}\cdot\text{K}^{-1}\cdot\text{mol}^{-1}$$

This is the same answer as determined in Problem 5.26.

8.53 Calculate ΔS for the expansion of 1.00 mol of $O_2(g)$ from 298 K and 24.8 L to 700 K and 38.8 L. For O_2, $\bar{\nu} = 1580.1932\ \text{cm}^{-1}$, $I = 1.947 \times 10^{-46}\ \text{kg}\cdot\text{m}^2$, and $g_0 = 3$.

▮ Under the initial conditions, using *(8.21a)*, *(8.22b)*, *(8.24)*, and *(8.25)* gives

$S(\text{trans})/(\text{J}\cdot\text{K}^{-1}\cdot\text{mol}^{-1})$

$$= (8.314)\left[\tfrac{5}{2} + \ln\left(\frac{\left[2\pi\dfrac{32.00\times10^{-3}\ \text{kg}\cdot\text{mol}^{-1}}{6.022\times10^{23}\ \text{mol}^{-1}}(1.381\times10^{-23}\ \text{J}\cdot\text{K}^{-1})(298\ \text{K})\right]^{3/2}(24.8\times10^{-3}\ \text{m}^3)}{(6.626\times10^{-34}\ \text{J}\cdot\text{s})^3(6.022\times10^{23}\ \text{mol}^{-1})[(1\ \text{kg}\cdot\text{m}^2\cdot\text{s}^{-2})/(1\ \text{J})]^{3/2}}\right)\right]$$
$$= 152.1$$

$$S(\text{rot})/(\text{J}\cdot\text{K}^{-1}\cdot\text{mol}^{-1}) = 8.314\ln\frac{(19.47)(298)}{2} - 22.378 = 43.92$$

$$x = \frac{(1.4388)(1580.1932)}{298} = 7.63$$

$$S(\text{vib})/(\text{J}\cdot\text{K}^{-1}\cdot\text{mol}^{-1}) = 8.314\left[\frac{7.63}{e^{7.63}-1} - \ln(1-e^{-7.63})\right] = 0.03$$

$$S(\text{elec})/(\text{J}\cdot\text{K}^{-1}\cdot\text{mol}^{-1}) = 8.314\ln 3 = 9.13$$

giving

$$S_1 = 152.1\ \text{J}\cdot\text{K}^{-1}\cdot\text{mol}^{-1} + 43.92\ \text{J}\cdot\text{K}^{-1}\cdot\text{mol}^{-1} + 0.03\ \text{J}\cdot\text{K}^{-1}\cdot\text{mol}^{-1} + 9.13\ \text{J}\cdot\text{K}^{-1}\cdot\text{mol}^{-1}$$
$$= 205.2\ \text{J}\cdot\text{K}^{-1}\cdot\text{mol}^{-1}$$

Likewise under the final conditions, $S(\text{trans}) = 166.4\ \text{J}\cdot\text{K}^{-1}\cdot\text{mol}^{-1}$, $S(\text{rot}) = 51.01\ \text{J}\cdot\text{K}^{-1}\cdot\text{mol}^{-1}$, $S(\text{vib}) = 1.42\ \text{J}\cdot\text{K}^{-1}\cdot\text{mol}^{-1}$, and $S(\text{elec}) = 9.13\ \text{J}\cdot\text{K}^{-1}\cdot\text{mol}^{-1}$, giving

$$S_2 = 166.4\ \text{J}\cdot\text{K}^{-1}\cdot\text{mol}^{-1} + 51.01\ \text{J}\cdot\text{K}^{-1}\cdot\text{mol}^{-1} + 1.42\ \text{J}\cdot\text{K}^{-1}\cdot\text{mol}^{-1} + 9.13\ \text{J}\cdot\text{K}^{-1}\cdot\text{mol}^{-1}$$
$$= 228.0\ \text{J}\cdot\text{K}^{-1}\cdot\text{mol}^{-1}$$

The entropy change for the expansion is

$$\Delta S = S_2 - S_1 = 228.0\ \text{J}\cdot\text{K}^{-1}\cdot\text{mol}^{-1} - 205.2\ \text{J}\cdot\text{K}^{-1}\cdot\text{mol}^{-1} = 22.8\ \text{J}\cdot\text{K}^{-1}\cdot\text{mol}^{-1}$$

which agrees favorably with the value of $21.7\ \text{J}\cdot\text{K}^{-1}\cdot\text{mol}^{-1}$ determined in Problem 5.27.

8.54 Calculate $\Delta_r S°$ for $\text{Na(g)} \rightarrow \text{Na}^+\text{(g)} + e^-\text{(g)}$, given $M/(\text{g}\cdot\text{mol}^{-1}) = 22.989\,77$, $22.989\,22$, and 5.4858×10^{-4} and $g_0 = 2$, 0, and 2, respectively.

▌ For Na, using *(8.21b)* and *(8.25)*,

$$S°(\text{Na})/(\text{J}\cdot\text{K}^{-1}\cdot\text{mol}^{-1}) = 12.471\,62\ln 22.989\,77 + 20.7860\ln 298.15 - 9.685\,21 + 8.314\,41\ln 2$$
$$= 153.607$$

Likewise, $S°(\text{Na}^+) = 147.844\ \text{J}\cdot\text{K}^{-1}\cdot\text{mol}^{-1}$ and $S°(e^-) = 20.869\ \text{J}\cdot\text{K}^{-1}\cdot\text{mol}^{-1}$. Substituting into *(5.20)* gives

$$\Delta_r S° = [(1\ \text{mol})S°(\text{Na}^+) + (1\ \text{mol})S°(e^-)] - (1\ \text{mol})S°(\text{Na})$$
$$= [(1\ \text{mol})(147.844\ \text{J}\cdot\text{K}^{-1}\cdot\text{mol}^{-1}) + (1\ \text{mol})(20.869\ \text{J}\cdot\text{K}^{-1}\cdot\text{mol}^{-1})] - (1\ \text{mol})(153.607\ \text{J}\cdot\text{K}^{-1}\cdot\text{mol}^{-1})$$
$$= 15.106\ \text{J}\cdot\text{K}^{-1}\cdot\text{mol}^{-1}$$

8.55 What is the approximate value of $\Delta_r S°$ for

$$^{35}\text{Cl}^{35}\text{Cl(g)} + {}^{37}\text{Cl}^{37}\text{Cl(g)} \rightarrow 2\,{}^{35}\text{Cl}^{37}\text{Cl(g)}$$

assuming that any differences in the molar masses, moments of inertia, and vibrational energy levels are negligible for the isotopes?

▌ The major contribution to $\Delta_r S°$ will be from the σ term in *(8.22)*. Using *(5.20)* gives

$$\Delta_r S° = (2\ \text{mol})S°(^{35}\text{Cl}^{37}\text{Cl}) - [(1\ \text{mol})S°(^{35}\text{Cl}^{35}\text{Cl}) + (1\ \text{mol})S°(^{37}\text{Cl}^{37}\text{Cl})]$$
$$= (2\ \text{mol})[-R\ln\sigma(^{35}\text{Cl}^{37}\text{Cl})] - \{(1\ \text{mol})[-R\ln\sigma(^{35}\text{Cl}^{35}\text{Cl})] + (1\ \text{mol})[-R\ln\sigma(^{37}\text{Cl}^{37}\text{Cl})]\}$$
$$= -(8.314\ \text{J}\cdot\text{K}^{-1}\cdot\text{mol}^{-1})[2\ln 1 - 1\ln 2 - 1\ln 2]$$
$$= 11.53\ \text{J}\cdot\text{K}^{-1}\cdot\text{mol}^{-1}$$

8.56 The entropy of CO(g) was determined to be $193.3\ \text{J}\cdot\text{K}^{-1}\cdot\text{mol}^{-1}$ by adding the following contributions: Debye extrapolation from 0 to 11.70 K, graphical integration of C_P–T data from 11.70 K to 61.55 K, solid-solid phase transition at 61.55 K, graphical integration of C_P–T data from 61.55 K to 68.09 K, fusion at 68.09 K, graphical integration of C_P–T data from 68.09 K to 81.61 K, vaporization at 81.61 K, real gas behavior correction, and graphical integration of C_P–T data from 81.61 K to 298.15 K. The value calculated from statistical mechanics is $197.674\ \text{J}\cdot\text{K}^{-1}\cdot\text{mol}^{-1}$. Explain the difference between these values.

▌ The difference corresponds to the omission of the $S_0°$ contribution in the thermodynamic calculations given by *(5.17)* as

$$S_0° = (8.314\ \text{J}\cdot\text{K}^{-1}\cdot\text{mol}^{-1})\ln 2 = 5.763\ \text{J}\cdot\text{K}^{-1}\cdot\text{mol}^{-1}$$

This term is slightly larger than the actual difference of $4.4\ \text{J}\cdot\text{K}^{-1}\cdot\text{mol}^{-1}$ because the orientation of the CO molecules in the crystal is not perfectly random, resulting from a small dipole moment.

8.57 Calculate the free energy function for $N_3(g)$ at 298 K. See Problems 8.24, 8.32, and 8.39 for additional data.

▮ The translational contribution to $(G_T^\circ - E_0^\circ)/T = (G_T^\circ - H_0^\circ)/T$ is given by

$$[(G_T^\circ - E_0^\circ)/T]_{\text{trans}} = -R\ln[(2\pi mkT)^{3/2}RT/h^3L]$$
$$= \{30.473 - 12.471\,62\ln[M/(\text{g}\cdot\text{mol}^{-1})] - 20.786\,03\ln[T/(\text{K})]\}\ \text{J}\cdot\text{K}^{-1}\cdot\text{mol}^{-1} \tag{8.26}$$

Substituting the data into *(8.26)* gives

$$[(G_T^\circ - E_0^\circ)/T]_{\text{trans}} = (30.473 - 12.471\,62\ln 42.0201 - 20.786\,03\ln 298.15)\ \text{J}\cdot\text{K}^{-1}\cdot\text{mol}^{-1}$$
$$= -134.58\ \text{J}\cdot\text{K}^{-1}\cdot\text{mol}^{-1}$$

The rotational contribution is given by

$$[(G_T^\circ - E_0^\circ)/T]_{\text{rot}} = -R\ln(8\pi^2 IkT/h^2\sigma)$$
$$= \{30.728 + 8.314\,41\ln\{\sigma/[T/(\text{K})]\} - 8.314\,41\ln[I/(10^{-47}\ \text{kg}\cdot\text{m}^2)]\}\ \text{J}\cdot\text{K}^{-1}\cdot\text{mol}^{-1} \tag{8.27a}$$

$$[(G_T^\circ - E_0^\circ)/T]_{\text{rot}} = -R\ln[8\pi^2(8\pi^3 I_xI_yI_z)^{1/2}(kT)^{3/2}/h^3\sigma]$$
$$= \{-217.119 - 4.157\,21\ln[I_xI_yI_z/(10^{-114}\ \text{kg}^3\cdot\text{m}^6)]$$
$$- 12.471\,62\ln[T/(\text{K})] + 8.314\,41\ln\sigma\}\ \text{J}\cdot\text{K}^{-1}\cdot\text{mol}^{-1} \tag{8.27b}$$

where *(8.27a)* is used for a linear polyatomic or diatomic molecule and *(8.27b)* is used for a nonlinear polyatomic molecule. Substituting the data of Problem 8.24 into *(8.27a)* gives

$$\left(\frac{G_T^\circ - E_0^\circ}{T}\right)_{\text{rot}} = \left(30.728 + 8.314\,41\ln\frac{2}{298.15} - 8.314\,41\ln 64.936\right)\ \text{J}\cdot\text{K}^{-1}\cdot\text{mol}^{-1}$$
$$= -45.580\ \text{J}\cdot\text{K}^{-1}\cdot\text{mol}^{-1}$$

The vibrational contribution is given by

$$[(G_T^\circ - E_0^\circ)/T]_{\text{vib}} = -R\ln(1 - e^{-x}) \tag{8.28}$$

for each degree of vibrational freedom. Substituting the results of Problem 8.32 into *(8.28)* gives

$$[(G_T^\circ - E_0^\circ)/T]_{\text{vib}} = (8.314\ \text{J}\cdot\text{K}^{-1}\cdot\text{mol}^{-1})[\ln(1 - e^{-8.7}) + 2\ln(1 - e^{-2.4}) + \ln(1 - e^{-5.8})]$$
$$= -1.61\ \text{J}\cdot\text{K}^{-1}\cdot\text{mol}^{-1}$$

The electronic contribution is given by

$$[(G_T^\circ - E_0^\circ)/T]_{\text{elec}} = -R\ln q_{\text{elec}} \tag{8.29}$$

Substituting the results of Problem 8.39 into *(8.29)* gives

$$[(G_T^\circ - E_0^\circ)/T]_{\text{elec}} = -(8.314\ \text{J}\cdot\text{K}^{-1}\cdot\text{mol}^{-1})\ln 3.417 = -10.22\ \text{J}\cdot\text{K}^{-1}\cdot\text{mol}^{-1}$$

Adding the contributions gives

$$[(G_T^\circ - E_0^\circ)/T] = -134.58\ \text{J}\cdot\text{K}^{-1}\cdot\text{mol}^{-1} + (-45.580\ \text{J}\cdot\text{K}^{-1}\cdot\text{mol}^{-1})$$
$$+(-1.61\ \text{J}\cdot\text{K}^{-1}\cdot\text{mol}^{-1}) + (-10.22\ \text{J}\cdot\text{K}^{-1}\cdot\text{mol}^{-1})$$
$$= -191.99\ \text{J}\cdot\text{K}^{-1}\cdot\text{mol}^{-1}$$

8.58 Consider a hypothetical chemical reaction A → B, where A consists of the two energy levels and B consists of the one energy level shown in Fig. 8-5. Prepare a plot of K as a function of temperature.

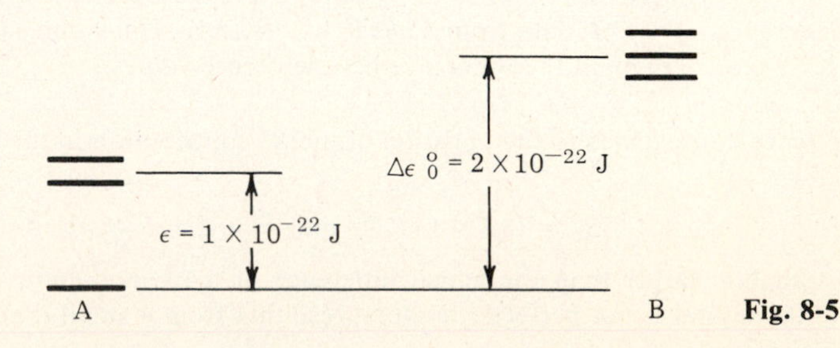

Fig. 8-5

▮ Combining *(8.26)*-*(8.29)* gives

$$[(G_T^\circ - E_0^\circ)/T] = -R \ln\left(\frac{q_{\text{trans}}}{N} q_{\text{rot}} q_{\text{vib}} q_{\text{elec}}\right) \qquad (8.30)$$

The system shown in Fig. 8-5 is similar to that for electronic calculations. Using *(8.6)* gives

$$q(\text{A}) = 1\, e^{-0/kT} + 2\, e^{-(1\times 10^{-22}\,\text{J})/kT} = 1 + 2\, e^{-(1\times 10^{-22}\,\text{J})/kT}$$

$$q(\text{B}) = 3\, e^{-0/kT} = 3$$

Substituting into *(8.30)* gives

$$[(G_T^\circ - E_0^\circ)/T]_\text{A} = -R \ln (1 + 2\, e^{-(1\times 10^{-22}\,\text{J})/kT})$$

$$[(G_T^\circ - E_0^\circ)/T]_\text{B} = -R \ln 3$$

Using *(6.19)* gives

$$\Delta(G_T^\circ - E_0^\circ)/T = \{(1\text{ mol})[(G_T^\circ - E_0^\circ)/T]_\text{B}\} - \{(1\text{ mol})[(G_T^\circ - E_0^\circ)/T]_\text{A}\}$$
$$= (1\text{ mol})(-R \ln 3) - (1\text{ mol})[-R \ln (1 + 2\, e^{-(1\times 10^{-22}\,\text{J})/kT})]$$

Substituting into *(6.20)* gives

$$\Delta_r G_T^\circ = -RT(1\text{ mol})[\ln 3 - \ln (1 + 2\, e^{-(1\times 10^{-23}\,J/kT)})] + (6.022 \times 10^{23}\text{ mol}^{-1})(2 \times 10^{-22}\text{ J})$$

and using *(7.1b)* gives

$$K = \exp\left(\frac{\ln 3 - \ln (1 + 2\ e^{-(1\times 10^{-22}\,\text{J})/kT}) - (6.022 \times 10^{23}\text{ mol}^{-1})(2 \times 10^{-22}\text{ J})}{(8.314\text{ J}\cdot\text{K}^{-1}\cdot\text{mol}^{-1})T}\right)$$

As a sample calculation, consider $T = 298$ K:

$$K = \exp\left(\frac{\ln 3 - \ln \{1 + 2 \exp [-(1 \times 10^{-22}\text{ J})/(1.381 \times 10^{-23}\text{ J}\cdot\text{K}^{-1})(298\text{ K})]\} - (6.022 \times 10^{23}\text{ mol}^{-1})(2 \times 10^{-22}\text{ J})}{(8.314\text{ J}\cdot\text{K}^{-1}\cdot\text{mol}^{-1})(298\text{ K})}\right)$$
$$= 0.968$$

The graph is shown in Fig. 8-6.

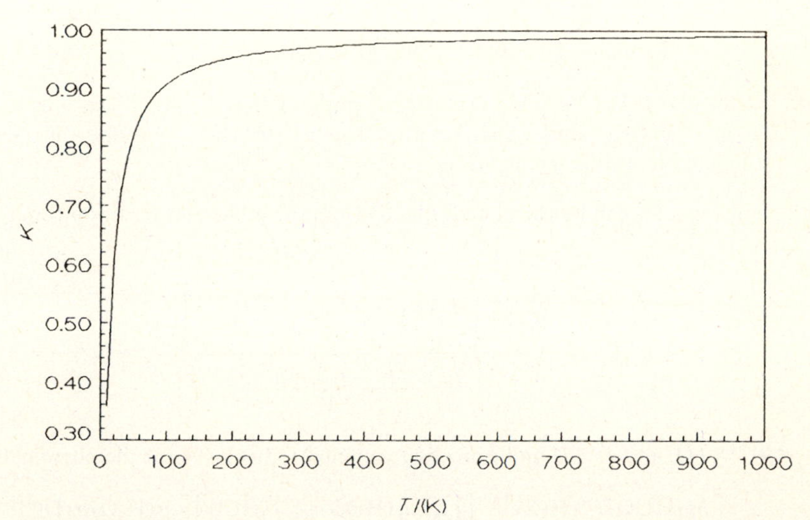

Fig. 8-6

8.59 Calculate K_{1000} for $2\text{Na(g)} \rightleftarrows \text{Na}_2\text{(g)}$ ($\Delta_r E_0^\circ = -70.538$ kJ), given $g_0 = 2$ for Na(g) and $g_0 = 2$, $\bar{\nu} = 159.11\text{ cm}^{-1}$, and $I = 1.809 \times 10^{-45}\text{ kg}\cdot\text{m}^2$ for $\text{Na}_2\text{(g)}$.

▮ Using *(8.26)*, *(8.27a)*, *(8.28)*, and *(8.29)* gives

$$[(G_T^\circ - E_0^\circ)/T]_{\text{trans,Na}} = [30.473 - 12.471\,62 \ln 22.989\,77 - 20.786\,03 \ln (1000)]\text{ J}\cdot\text{K}^{-1}\cdot\text{mol}^{-1}$$
$$= -152.211\text{ J}\cdot\text{K}^{-1}\cdot\text{mol}^{-1}$$

$$[(G_T^\circ - E_0^\circ)/T]_{\text{trans,Na}_2} = [30.473 - 12.471\,62 \ln 45.795 - 20.786\,03 \ln (1000)]\text{ J}\cdot\text{K}^{-1}\cdot\text{mol}^{-1}$$
$$= -160.805\text{ J}\cdot\text{K}^{-1}\cdot\text{mol}^{-1}$$

$$[(G_T^\circ - E_0^\circ)/T]_{\text{rot,Na}_2} = [30.728 - 8.314\,41 \ln(2/1000) - 8.314\,41 \ln 180.9]\ \text{J}\cdot\text{K}^{-1}\cdot\text{mol}^{-1}$$
$$= -64.161\ \text{J}\cdot\text{K}^{-1}\cdot\text{mol}^{-1}$$
$$x = \frac{(1.4388)(159.11)}{1000} = 0.228\,93$$
$$[(G_T^\circ - E_0^\circ)/T]_{\text{vib,Na}_2} = (8.314\ \text{J}\cdot\text{K}^{-1}\cdot\text{mol}^{-1}) \ln(1 - e^{-0.22893}) = -13.192\ \text{J}\cdot\text{K}^{-1}\cdot\text{mol}^{-1}$$
$$[(G_T^\circ - E_0^\circ)/T]_{\text{elec,Na}} = -(8.314\ \text{J}\cdot\text{K}^{-1}\cdot\text{mol}^{-1}) \ln 2 = -5.763\ \text{J}\cdot\text{K}^{-1}\cdot\text{mol}^{-1}$$
$$[(G_T^\circ - E_0^\circ)/T]_{\text{elec,Na}_2} = -(8.314\ \text{J}\cdot\text{K}^{-1}\cdot\text{mol}^{-1}) \ln 1 = 0$$

Summarizing the contributions gives

$$[(G_T^\circ - E_0^\circ)/T]_{\text{Na}} = -152.211\ \text{J}\cdot\text{K}^{-1}\cdot\text{mol}^{-1} + (-5.763\ \text{J}\cdot\text{K}^{-1}\cdot\text{mol}^{-1})$$
$$= -157.974\ \text{J}\cdot\text{K}^{-1}\cdot\text{mol}^{-1}$$
$$[(G_T^\circ - E_0^\circ)/T]_{\text{Na}_2} = -160.805\ \text{J}\cdot\text{K}^{-1}\cdot\text{mol}^{-1} + (-64.161\ \text{J}\cdot\text{K}^{-1}\cdot\text{mol}^{-1}) + (-13.192\ \text{J}\cdot\text{K}^{-1}\cdot\text{mol}^{-1}) + 0$$
$$= -238.158\ \text{J}\cdot\text{K}^{-1}\cdot\text{mol}^{-1}$$

For the chemical reaction, using *(6.19)* and *(6.20)* gives

$$\frac{\Delta(G_T^\circ - E_0^\circ)}{T} = (1\ \text{mol})\left(\frac{G_T^\circ - E_0^\circ}{T}\right)_{\text{Na}_2} - (2\ \text{mol})\left(\frac{G_T^\circ - E_0^\circ}{T}\right)_{\text{Na}}$$
$$= (1\ \text{mol})(-238.158\ \text{J}\cdot\text{K}^{-1}\cdot\text{mol}^{-1}) - (2\ \text{mol})(-157.974\ \text{J}\cdot\text{K}^{-1}\cdot\text{mol}^{-1})$$
$$= 77.790\ \text{J}\cdot\text{K}^{-1}$$
$$\Delta_r G_{1000}^\circ = (1000\ \text{K})(77.790\ \text{J}\cdot\text{K}^{-1})\frac{10^{-3}\ \text{kJ}}{1\ \text{J}} + (-70.538\ \text{kJ}) = 7.252\ \text{kJ}$$

Substituting into *(7.1b)* gives

$$K = \exp\left(-\frac{(7.252\ \text{kJ})[(10^3\ \text{J})/(1\ \text{kJ})]}{(8.314\ \text{J}\cdot\text{K}^{-1}\cdot\text{mol}^{-1})(1000\ \text{K})(\text{mol})}\right) = 0.418$$

8.60 Calculate K_C at 298 K for the equation

$$H_2O(g) + DCl(g) \rightleftarrows HDO(g) + HCl(g) \qquad \Delta_r E_0^\circ = -2.510\ \text{kJ}$$

given $I_xI_yI_z = 5.8410 \times 10^{-141}\ \text{kg}^3\cdot\text{m}^6$ for H_2O and $1.595\,84 \times 10^{-140}\ \text{kg}^3\cdot\text{m}^6$ for HDO and $I = 5.1416 \times 10^{-47}\ \text{kg}\cdot\text{m}^2$ for DCl and $2.6437 \times 10^{-47}\ \text{kg}\cdot\text{m}^2$ for HCl. Assume the electronic and vibrational contributions are negligible at this temperature.

▌ The equilibrium constant can be calculated directly from the molecular partition functions of the reactants and products by

$$K_C = \frac{\prod_i^{\text{products}} \left[\dfrac{q_{\text{trans},i}q_{\text{rot},i}q_{\text{vib},i}q_{\text{elec},i}q_{\text{nuc},i}}{V}\right]^{\nu_i}}{\prod_j^{\text{reactants}} \left[\dfrac{q_{\text{trans},j}q_{\text{rot},j}q_{\text{vib},j}q_{\text{elec},j}q_{\text{nuc},j}}{V}\right]^{\nu_j}} e^{-\Delta E_0^\circ/RT} \tag{8.31}$$

Substituting *(8.9)*, *(8.11)*, and *(8.12)* and canceling common terms gives for the chemical equation

$$K_C = \left[\frac{M(\text{HDO})M(\text{HCl})}{M(\text{H}_2\text{O})M(\text{DCl})}\right]^{3/2}\left[\frac{[I_xI_yI_z(\text{HDO})]^{1/2}I(\text{HCl})}{[I_xI_yI_z(\text{H}_2\text{O})]^{1/2}I(\text{DCl})}\right]\left[\frac{\sigma(\text{H}_2\text{O})\sigma(\text{DCl})}{\sigma(\text{HDO})\sigma(\text{HCl})}\right] e^{-\Delta_r E_0^\circ/RT}$$

Substituting the data gives

$$K_C = \left[\frac{(19.021\ \text{g}\cdot\text{mol}^{-1})(36.461\ \text{g}\cdot\text{mol}^{-1})}{(18.015\ \text{g}\cdot\text{mol}^{-1})(37.467\ \text{g}\cdot\text{mol}^{-1})}\right]^{3/2}\left[\frac{(1.595\,84 \times 10^{-140}\ \text{kg}^3\cdot\text{m}^6)^{1/2}(2.6437 \times 10^{-47}\ \text{kg}\cdot\text{m}^2)}{(5.8410 \times 10^{-141}\ \text{kg}^3\cdot\text{m}^6)^{1/2}(5.1416 \times 10^{-47}\ \text{kg}\cdot\text{m}^2)}\right]\left[\frac{(2)(1)}{(1)(1)}\right]$$
$$\times \exp\left(-\frac{(-2.510\ \text{kJ})[(10^3\ \text{J})/(1\ \text{kJ})]}{(8.314\ \text{J}\cdot\text{K}^{-1}\cdot\text{mol}^{-1})(298.15\ \text{K})(\text{mol})}\right)$$
$$= 4.873$$

8.4 MONATOMIC CRYSTALS

8.61 Calculate $E_T^\circ - E_0^\circ$ for silver at 298 K using both the Einstein characteristic temperature ($\Theta_E = 161$ K) and the Debye characteristic temperature ($\Theta_D = 208$ K).

▌ The thermal energy for a monatomic crystal is given by the Einstein theory as

$$E_T^\circ - E_0^\circ = \frac{3R\Theta_E}{e^{\Theta_E/T} - 1} \tag{8.32}$$

Substituting the value of Θ_E into *(8.32)* gives

$$E_{298}^\circ - E_0^\circ = \frac{3(8.314\ \mathrm{J\cdot K^{-1}\cdot mol^{-1}})(161\ \mathrm{K})}{e^{161\ \mathrm{K}/298\ \mathrm{K}} - 1} = 5610\ \mathrm{J\cdot mol^{-1}}$$

The thermal energy according to the Debye theory is

$$E_T^\circ - E_0^\circ = 3RT\left[3\left(\frac{T}{\Theta_D}\right)^3 \int_0^{\Theta_D/T} \frac{x^3\,dx}{e^x - 1}\right] = 3RT\vartheta(\Theta_D)_E \tag{8.33}$$

where $\vartheta(\Theta_D)_E$ represents the Debye internal energy function (see Fig. 8-7). For Ag,

$$\Theta_D/T = (208\ \mathrm{K})/(298\ \mathrm{K}) = 0.70$$

and $\vartheta(\Theta_D)_E = 0.77$ from Fig. 8-7. Thus *(8.33)* gives

$$E_{298}^\circ - E_0^\circ = 3(8.314\ \mathrm{J\cdot K^{-1}\cdot mol^{-1}})(298\ \mathrm{K})(0.77) = 5700\ \mathrm{J\cdot mol^{-1}}.$$

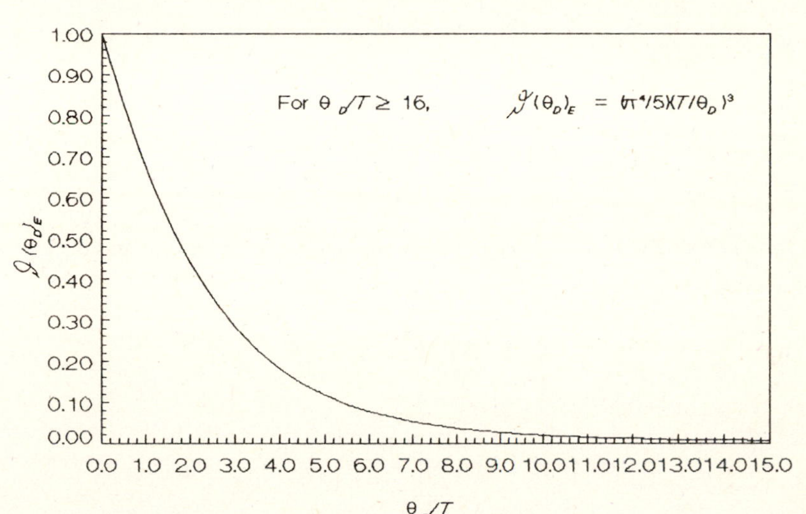

Fig. 8-7

8.62 What are the high- and low-temperature limits of *(8.32)*?

▌ At very high temperatures, Θ_E/T will approach 0 and $e^{\Theta_E/T}$ will approach 1. Using a series expansion for $e^{\Theta_E/T}$ under these conditions gives

$$\lim_{T\to\infty}(E_T^\circ - E_0^\circ) = \lim_{T\to\infty}\left(\frac{3R\Theta_E}{e^{\Theta_E/T} - 1}\right) = \frac{3R\Theta_E}{(1 + \Theta_E/T + \cdots) - 1} = 3RT$$

At very low temperatures, Θ_E/T will become very large and $e^{\Theta_E/T}$ will become very large:

$$\lim_{T\to 0}(E_T^\circ - E_0^\circ) = \lim_{T\to 0}\left[\frac{3R\Theta_E}{e^{\Theta_E/T} - 1}\right] = \frac{3R\Theta_E}{e^{\Theta_E/T}} = 0$$

8.63 Calculate $A_T^\circ - E_0^\circ$ for silver at 298 K using both the Einstein theory ($\Theta_E = 161$ K) and the Debye theory ($\Theta_D = 208$ K).

▌ The thermal Helmholtz free energy is given by the Einstein theory as

$$A_T^\circ - E_0^\circ = 3RT\ln(1 - e^{-\Theta_E/T}) \tag{8.34}$$

Substituting the value of Θ_E into *(8.34)* gives

$$A_{298}^\circ - E_0^\circ = 3(8.314\ \mathrm{J\cdot K^{-1}\cdot mol^{-1}})(298\ \mathrm{K})\ln(1 - e^{-(161\ \mathrm{K})/(298\ \mathrm{K})}) = -6500\ \mathrm{J\cdot mol^{-1}}$$

The thermal Helmholtz free energy according to the Debye theory is

$$A_T^\circ - E_0^\circ = -3RT\left[3\left(\frac{T}{\Theta_D}\right)^3 \int_0^{\Theta_D/T} x^2 \ln(1 - e^{-x})\,dx\right] = -3RT\vartheta(\Theta_D)_A \tag{8.35}$$

where $\vartheta(\Theta_D)_A$ represents the Debye Helmholtz energy function (see Fig. 8-8). For Ag,

$$\Theta_D/T = (208\text{ K})/(298\text{ K}) = 0.70$$

and $\vartheta(\Theta_D)_A = 0.90$ from Fig. 8-8. Thus *(8.35)* gives

$$A_{298}^\circ - E_0^\circ = -3(8.314\text{ J}\cdot\text{K}^{-1}\cdot\text{mol}^{-1})(298\text{ K})(0.90) = -6700\text{ J}\cdot\text{mol}^{-1}$$

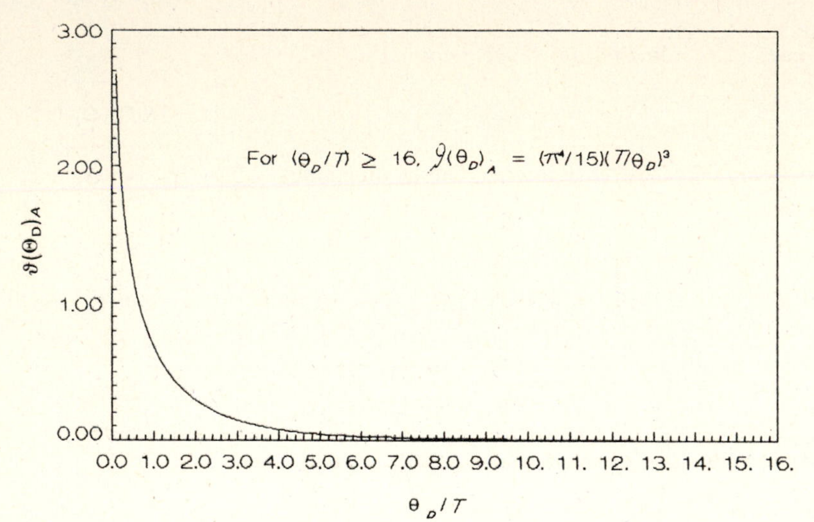

Fig. 8-8

8.64 Calculate S_T° for silver at 298 K using both the Einstein theory ($\Theta_E = 161$ K) and the Debye theory ($\Theta_D = 208$ K).

▮ The entropy is given by the Einstein theory as

$$S_T^\circ = 3R\left(\frac{\Theta_E/T}{e^{\Theta_E/T} - 1} - \ln(1 - e^{-\Theta_E/T})\right) \tag{8.36}$$

Substituting the value of Θ_E into *(8.36)* gives

$$S_{298}^\circ = 3(8.314\text{ J}\cdot\text{K}^{-1}\cdot\text{mol}^{-1})\left[\frac{(161\text{ K})/(298\text{ K})}{e^{(161\text{ K})/(298\text{ K})} - 1} - \ln(1 - e^{-(161\text{ K})/(298\text{ K})})\right]$$

$$= 40.6\text{ J}\cdot\text{K}^{-1}\cdot\text{mol}^{-1}$$

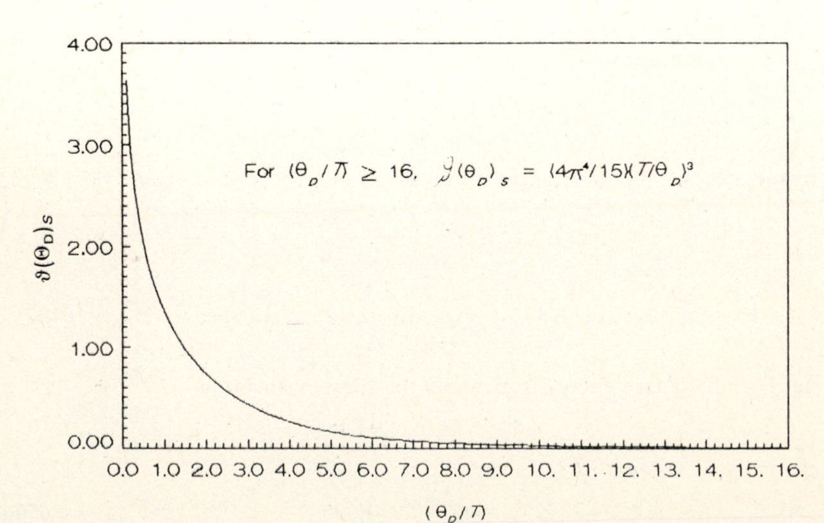

Fig. 8-9

The entropy according to the Debye theory is

$$S_T^\circ = 3R\left[3\left(\frac{T}{\Theta_D}\right)^3 \int_0^{\Theta_D/T}\left(\frac{x^3}{e^x - 1} - x^2 \ln(1 - e^{-x})\right) dx\right] = 3RT\vartheta(\Theta_D)_S \tag{8.37}$$

where $\vartheta(\Theta_D)_S$ represents the Debye entropy function (see Fig. 8-9). For Ag,

$$\Theta_D/T = (208\ \text{K})/(298\ \text{K}) = 0.70$$

and $\vartheta(\Theta_D)_S = 1.70$ from Fig. 8-9. Thus *(8.37)* gives

$$S_{298}^\circ = 3(8.314\ \text{J}\cdot\text{K}^{-1}\cdot\text{mol}^{-1})(1.70) = 42.4\ \text{J}\cdot\text{K}^{-1}\cdot\text{mol}^{-1}$$

(The tabulated entropy of Ag at 298 K is $42.55\ \text{J}\cdot\text{K}^{-1}\cdot\text{mol}^{-1}$.)

CHAPTER 9
Electrochemistry

9.1 OXIDATION–REDUCTION

9.1 Complete and balance the following oxidation-reduction equation:

$$H_2S(aq) + NO_3^-(aq) \rightarrow S(s) + NO(g)$$

How many moles of electrons are involved in the reaction?

▮ The reaction is occurring in an acidic solution as indicated by the presence of H_2S. (If the solution were alkaline, the reactant would be written in the form of HS^- or S^{2-} depending on the pH of the solution.) In an acidic aqueous solution, $H_2O(l)$ and $H^+(aq)$ may be added as needed to balance the equation. The individual half-reactions are

$$H_2S(aq) \rightarrow S(s) \qquad \text{and} \qquad NO_3^-(aq) \rightarrow NO(g)$$

The first half-reaction can be balanced "by inspection" by adding two $H^+(aq)$ to the right-hand side, giving

$$H_2S(aq) \rightarrow S(s) + 2H^+(aq)$$

To balance the second half-reaction, $xH^+(aq)$ and $yH_2O(l)$ are added to the equation:

$$NO_3^-(aq) + xH^+(aq) + yH_2O(l) \rightarrow NO(g)$$

and counts of hydrogen atoms and oxygen items give

$$3 + y = 1$$

$$x + 2y = 0$$

Solving these equations gives $x = 4$ and $y = -2$. The half-reaction becomes

$$NO_3^-(aq) + 4H^+(aq) \rightarrow NO(g) + 2H_2O(l)$$

Balancing each half-reaction electrically by adding electrons gives

$$H_2S(aq) \rightarrow S(s) + 2H^+(aq) + 2e^-$$

$$NO_3^-(aq) + 4H^+(aq) + 3e^- \rightarrow NO(g) + 2H_2O(l)$$

Because the number of electrons used in the reduction process must be equal to the number generated by the oxidation process, the first half-reaction is multiplied by 3 and the second half-reaction is multiplied by 2 before they are added to give the final balanced equation:

$$(3)[H_2S(aq) \rightarrow S(s) + 2H^+(aq) + 2e^-]$$

$$(2)[NO_3^-(aq) + 4H^+(aq) + 3e^- \rightarrow NO(g) + 2H_2O(l)]$$

$$3H_2S(aq) + 2NO_3^-(aq) + 2H^+(aq) \rightarrow 3S(s) + 2NO(g) + 4H_2O(l)$$

The number of electrons common to both sides of the above equation is 6, so $n = 6$ mol.

9.2 Complete and balance the following oxidation-reduction equation:

$$Mn(OH)_2(s) + H_2O_2(aq) \rightarrow MnO_2(s)$$

How many moles of electrons are involved in the reaction?

▮ The reaction is occurring in an alkaline solution as indicated by the formulas of the manganese compounds. In a basic aqueous solution, $H_2O(l)$ and $OH^-(aq)$ may be added as needed to balance the equation. The individual half-reactions are

$$Mn(OH)_2(s) \rightarrow MnO_2(s) \qquad \text{and} \qquad H_2O_2(aq) \rightarrow OH^-(aq)$$

The first half-reaction can be balanced by adding $xOH^-(aq)$ and $yH_2O(l)$ to the equation:

$$Mn(OH)_2(s) + xOH^-(aq) + yH_2O(l) \rightarrow MnO_2(s)$$

and counting the numbers of hydrogen atoms and oxygen atoms to give

$$2 + x + y = 2$$

$$2 + x + 2y = 0$$

Solving these equations gives $x = 2$ and $y = -2$. The half-reaction becomes

$$Mn(OH)_2(s) + 2OH^-(aq) \rightarrow MnO_2(s) + 2H_2O(l)$$

The second half-reaction can be balanced "by inspection":

$$H_2O_2(aq) \rightarrow 2OH^-(aq)$$

[If both $OH^-(aq)$ and $H_2O(l)$ were added and the simultaneous equation approach used, the result would be to add $2OH^-(aq)$ to the right-hand side of the equation and to include no $H_2O(l)$.] Balancing each half-reaction electrically by adding electrons gives

$$Mn(OH)_2(s) + 2OH^-(aq) \rightarrow MnO_2(s) + 2H_2O(l) + 2e^-$$

$$H_2O_2(aq) + 2e^- \rightarrow 2OH^-(aq)$$

Because the number of electrons generated by the first half-reaction is equal to the number required by the second half-reaction, the half-reactions are added to give

$$Mn(OH)_2(s) + H_2O_2(aq) \rightarrow MnO_2(s) + 2H_2O(l)$$

The number of electrons common to both sides of the above equation is 2, so $n = 2$ mol.

9.3 For the reaction described in Problem 9.2, $\Delta_r G^\circ_{298} = -190.4$ kJ. What is the cell potential that this reaction could produce if carried out electrochemically?

▌ Solving *(6.6)* for E and substituting the data gives

$$E = \frac{-\Delta_r G^\circ}{nF} = \frac{-(-190.4\text{ kJ})[(10^3\text{ J})/(1\text{ kJ})]}{(2\text{ mol e}^-)[96\,484.56\text{ J}\cdot\text{V}^{-1}\cdot(\text{mol e}^-)^{-1}]} = 0.987\text{ V}$$

9.4 Determine the standard reduction potential for $K^+(aq)$ given $\Delta_f G^\circ_{298} = -283.27\text{ kJ}\cdot\text{mol}^{-1}$ for $K^+(aq)$.

▌ For the reduction of $K^+(aq)$, $K^+(aq) + e^- \rightarrow K(s)$, using *(6.16)* gives

$$\begin{aligned}\Delta_r G^\circ &= [(1\text{ mol})\,\Delta_f G^\circ(K)] - [(1\text{ mol})\,\Delta_f G^\circ(K^+) + (1\text{ mol})\,\Delta_f G^\circ(e^-)]\\ &= [(1\text{ mol})(0)] - [(1\text{ mol})(-283.27\text{ kJ}\cdot\text{mol}^{-1}) + (1\text{ mol})(0)]\\ &= 283.27\text{ kJ}\end{aligned}$$

Substituting into *(6.6)* gives

$$E = \frac{-(283.27\text{ kJ})[(10^3\text{ J})/(1\text{ kJ})]}{(2\text{ mol e}^-)[96\,484.56\text{ J}\cdot\text{V}^{-1}\cdot(\text{mol e}^-)^{-1}]} = -2.9359\text{ V}$$

9.5 A pair of inert electrodes were placed in an aqueous solution containing $AgNO_3$ and $Cu(NO_3)_2$. The potential applied to the cell was increased from 0 V until a chemical reaction was observed to occur. What was this chemical reaction? Assume reversible conditions.

▌ The possible oxidation reactions might include the following:

$$2H_2O(l) \rightarrow O_2(g) + 4H^+(aq) + 4e^- \qquad E^\circ = -1.229\text{ V}$$

$$2H_2O(l) \rightarrow H_2O_2(aq) + 2H^+(aq) + 2e^- \qquad E^\circ = -1.776\text{ V}$$

Recalling that a negative value of $\Delta_r G^\circ$ represents a spontaneous reaction, the reaction with the more positive value of E° will be the preferred reaction:

$$2H_2O(l) \rightarrow O_2(g) + 4H^+(aq) + 4e^-$$

Likewise, the possible reduction reactions might include the following:

$$Ag^+(aq) + e^- \rightarrow Ag(s) \qquad E^\circ = 0.7991\text{ V}$$

$$Cu^{2+}(aq) + 2e^- \rightarrow Cu(s) \qquad E^\circ = 0.337\text{ V}$$

$$2H^+(aq) + 2e^- \rightarrow H_2(g) \qquad E^\circ = 0.0000\text{ V}$$

Using the same reasoning as above, the preferred reaction will be

$$Ag^+(aq) + e^- \rightarrow Ag(s)$$

Combining the two half-reactions gives

$$4Ag^+(aq) + 2H_2O(l) \rightarrow 4Ag(s) + O_2(g) + 4H^+(aq)$$

9.6 During an electrochemical experiment, 0.2773 g of Ag was transferred from one electrode to the other electrode in a coulometer. What electric charge passed through the circuit?

▮ The amount of silver transferred was

$$n = \frac{m}{M} = \frac{0.2773\ \text{g}}{107.868\ \text{g} \cdot \text{mol}^{-1}} = 2.571 \times 10^{-3}\ \text{mol Ag}$$

From the half-reaction $Ag(s) \rightarrow Ag^+(aq) + e^-$, using simple stoichiometry gives

$$(2.571 \times 10^{-3}\ \text{mol Ag})\frac{1\ \text{mol e}^-}{1\ \text{mol Ag}} = 2.571 \times 10^{-3}\ \text{mol e}^-$$

Multiplying by Faraday's constant gives

$$Q = (2.571 \times 10^{-3}\ \text{mol e}^-)[96\,484.56\ \text{C} \cdot (\text{mol e}^-)^{-1}] = 248.1\ \text{C}$$

where Q is the quantity of electricity.

9.7 An electrochemical cell was based on the oxidation-reduction reaction described in Problem 9.2. During the operation of this cell for 45 s, 0.136 g of MnO_2 was produced. What is the average electric current produced by the cell?

▮ The amount of MnO_2 produced was

$$n = \frac{m}{M} = \frac{0.136\ \text{g}}{86.94\ \text{g} \cdot \text{mol}^{-1}} = 1.56 \times 10^{-3}\ \text{mol MnO}_2$$

This corresponds to

$$(1.56 \times 10^{-3}\ \text{mol MnO}_2)\frac{2\ \text{mol e}^-}{1\ \text{mol MnO}_2} = 3.12 \times 10^{-3}\ \text{mol e}^-$$

$$Q = (3.12 \times 10^{-3}\ \text{mol e}^-)[96\,484.56\ \text{C} \cdot (\text{mol e}^-)^{-1}] = 301\ \text{C}$$

The electric current is

$$I = Q/t \tag{9.1}$$

$$I = \frac{(301\ \text{C})[(1\ \text{A} \cdot \text{s})/(1\ \text{C})]}{45\ \text{s}} = 6.7\ \text{A}$$

9.8 A student designed an experiment to determine the electronic charge, e. A solution of $I^-(aq)$ was electrolyzed for a period of 6.244 h using a current of 0.042 92 A to produce 1.269 g of I_2. What is the experimental value of e?

▮ The number of electrons is

$$N = \left[\frac{1.269\ \text{g I}_2}{253.810\ \text{g} \cdot \text{mol}^{-1}}\right]\left[\frac{2\ \text{mol e}^-}{1\ \text{mol I}_2}\right](6.022\,045 \times 10^{23}\ \text{mol}^{-1}) = 6.022 \times 10^{21}$$

The charge passed through the circuit is given by *(9.1)* as

$$Q = (0.042\,92\ \text{A})(6.244\ \text{h})\frac{3600\ \text{s}}{1\ \text{h}}\left(\frac{1\ \text{C}}{1\ \text{A} \cdot \text{s}}\right) = 964.8\ \text{C}$$

The electronic charge is

$$e = \frac{Q}{N} = \frac{964.8\ \text{C}}{6.022 \times 10^{21}} = 1.602 \times 10^{-19}\ \text{C}$$

9.9 What amount of work is needed to move an electron against a potential difference of 1.000 V? What amount of work is needed for 1.00 mol of electrons?

▮ The electrical work equivalent to 1 eV is given by Table 3-5 as

$$w = \Delta V Q = (1.000\ \text{V})(1.602 \times 10^{-19}\ \text{C})\frac{1\ \text{J}}{1\ \text{C}\cdot\text{V}} = 1.602 \times 10^{-19}\ \text{J}$$

$$w = (6.022 \times 10^{23}\ \text{mol}^{-1})(1.602 \times 10^{-19}\ \text{J}) = 96\,470\ \text{J}\cdot\text{mol}^{-1}$$

The last value is equal to the Faraday constant, F.

9.10 What is the force between two electrons in a vacuum separated by 0.100 nm?

▮ The force is given by *Coulomb's law* as

$$F = \frac{-Q_1 Q_2}{4\pi\varepsilon_0(\varepsilon/\varepsilon_0)r^2} \tag{9.2}$$

where Q_i represents the charge on each species, $4\pi\varepsilon_0$ is the permittivity constant ($1.112\,650\,056 \times 10^{-10}\ \text{C}^2\cdot\text{N}^{-1}\cdot\text{m}^{-2}$), $\varepsilon/\varepsilon_0$ is the dielectric constant of the substance in which the charges are present, and r is the distance between the charges. For a vacuum, $\varepsilon/\varepsilon_0 = 1$. Substituting into *(9.2)* gives

$$F = \frac{-(-1.602 \times 10^{-19}\ \text{C})(-1.602 \times 10^{-19}\ \text{C})}{(1.113 \times 10^{-10}\ \text{C}^2\cdot\text{N}^{-1}\cdot\text{m}^{-2})(1)\{(0.100\ \text{nm})[(10^{-9}\ \text{m})/(1\ \text{nm})]\}^2}$$

$$= -2.31 \times 10^{-8}\ \text{N}$$

The negative sign on the value indicates that this is a repulsive force.

9.11 What is the electrical work needed to bring the two electrons described in Problem 9.10 from infinity to the separation of 0.100 nm?

▮ The integrated form of *(9.2)* is

$$w = \frac{Q_1 Q_2}{4\pi\varepsilon_0(\varepsilon/\varepsilon_0)r} \tag{9.3}$$

Substituting the data gives

$$w = \frac{(-1.602 \times 10^{-19}\ \text{C})(-1.602 \times 10^{-19}\ \text{C})[(1\ \text{J})/(1\ \text{N}\cdot\text{m})]}{(1.113 \times 10^{-10}\ \text{C}^2\cdot\text{N}^{-1}\cdot\text{m}^{-2})(1)(0.100\ \text{nm})[(10^{-9}\ \text{m})/(1\ \text{nm})]}$$

$$= 2.31 \times 10^{-18}\ \text{J}$$

The positive sign on the value indicates that work is required to bring the electrons together.

9.12 At what distance will the electric field strength be equal to $5.8 \times 10^9\ \text{V m}^{-1}$ around a K^+ ion in a vacuum?

▮ The electric field strength is given by

$$E = \frac{Q}{4\pi\varepsilon_0(\varepsilon/\varepsilon_0)r^2} \tag{9.4}$$

Solving *(9.4)* for r and substituting the data gives

$$r = \left(\frac{(1.602 \times 10^{-19}\ \text{C})[(1\ \text{C}\cdot\text{V})/(1\ \text{J})][(1\ \text{J})/(1\ \text{N}\cdot\text{m})]}{(1.113 \times 10^{-10}\ \text{C}^2\cdot\text{N}^{-1}\cdot\text{m}^{-2})(1)(5.8 \times 10^9\ \text{V}\cdot\text{m}^{-1})}\right)^{1/2}$$

$$= 5.0 \times 10^{-10}\ \text{m} = 0.50\ \text{nm}$$

9.13 What is the potential difference between two points in a vacuum separated by a distance of 0.30 nm and 0.25 nm from a K^+ ion? Repeat the calculation for two points in water assuming the charge to be shielded. At 25 °C, $\varepsilon/\varepsilon_0 = 78.85$ for water and the Debye–Hückel theory gives the shielding radius as 5 nm for dilute aqueous solutions.

▮ The Coulombic potential is given by

$$\phi = \frac{Q}{4\pi\varepsilon_0(\varepsilon/\varepsilon_0)r} \tag{9.5}$$

Substituting into *(9.5)* gives the potential difference as

$$\Delta\phi = \frac{Q}{4\pi\varepsilon_0(\varepsilon/\varepsilon_0)}\left(\frac{1}{r_2} - \frac{1}{r_1}\right)$$

$$= \frac{(1.602 \times 10^{-19}\ \mathrm{C})[(1\ \mathrm{C \cdot V})/(1\ \mathrm{J})][(1\ \mathrm{J})/(1\ \mathrm{N \cdot m})]}{(1.113 \times 10^{-10}\ \mathrm{C^2 \cdot N^{-1} \cdot m^{-2}})(1)}\left(\frac{1}{0.25\ \mathrm{nm}} - \frac{1}{0.30\ \mathrm{nm}}\right)\left(\frac{10^9\ \mathrm{nm}}{1\ \mathrm{m}}\right)$$

$$= 0.96\ \mathrm{V}$$

In aqueous solution the right-hand side of *(9.5)* is multiplied by an additional e^{-r/r_D} term to correct for the shielding of the solvent. Thus for the aqueous solution,

$$\Delta\phi = \frac{1.602 \times 10^{-19}}{(1.113 \times 10^{-10})(78.85)}\left[\frac{e^{-0.25/5}}{0.25} - \frac{e^{-0.30/5}}{0.30}\right](10^9) = 0.012\ \mathrm{V}$$

9.2 CONDUCTIVITY

9.14 The resistivity of aluminum is $2.824 \times 10^{-8}\ \Omega \cdot \mathrm{m}$. Calculate the potential drop across a piece of aluminum wire that is 2.0 mm in diameter and 1.00 m long. Assume the current is 1.25 A.

▌ The resistance (R) is related to the resistivity (ρ), the length (l), and the area (A) by

$$R = \rho(l/A) \tag{9.6}$$

Substituting the data gives

$$R = (2.824 \times 10^{-8}\ \Omega \cdot \mathrm{m})\frac{1.00\ \mathrm{m}}{\pi(1.0 \times 10^{-3}\ \mathrm{m})^2} = 9.0 \times 10^{-3}\ \Omega$$

The relationship between the electric potential difference (ΔV), the electric current (I), and the resistance is given by *Ohm's law* as

$$\Delta V = IR \tag{9.7}$$

Substituting the data into *(9.7)* gives

$$\Delta V = (1.25\ \mathrm{A})(9.0 \times 10^{-3}\ \Omega)\frac{1\ \mathrm{V}}{1\ \mathrm{A \cdot \Omega}} = 1.13 \times 10^{-2}\ \mathrm{V}$$

9.15 What is the conductance and conductivity of the aluminum wire described in Problem 9.14?

▌ The *conductance* (L) is defined as

$$L = 1/R \tag{9.8}$$

and the *conductivity*, or *specific conductance* (κ), is defined as

$$\kappa = 1/\rho \tag{9.9}$$

Substituting the results of Problem 9.14 into *(9.8)* and *(9.9)* gives

$$L = \frac{1}{9.0 \times 10^{-3}\ \Omega} = 111\ \Omega^{-1} = 111\ \mathrm{S}$$

$$\kappa = \frac{1}{2.824 \times 10^{-8}\ \Omega \cdot \mathrm{m}} = 3.541 \times 10^7\ \Omega^{-1} \cdot \mathrm{m}^{-1} = 3.541 \times 10^7\ \mathrm{S \cdot m^{-1}}$$

where the unit of *siemens* ($1\ \mathrm{S} = 1\ \Omega^{-1} = 1\ \mathrm{A \cdot V^{-1}} = 1\ \mathrm{m^{-2} \cdot kg^{-1} \cdot s^3 \cdot A^2}$) is commonly used for conductivity measurements.

9.16 A conductance cell containing fresh laboratory triply distilled water had a resistance of 426 kΩ, and when it was filled with a calibration solution of KCl it had a resistance of 2171 Ω. Determine the cell constant (A/l) given that $\kappa = 0.140\,88\ \mathrm{S \cdot m^{-1}}$ for the calibration solution (roughly 0.01 m).

▌ Using *(9.8)* for the water and for the calibration solution gives

$$L(\mathrm{H_2O}) = \frac{1}{(426\ \mathrm{k\Omega})[(10^3\ \Omega)/(1\ \mathrm{k\Omega})]} = 2.35 \times 10^{-6}\ \mathrm{S}$$

$$L(\mathrm{soln}) = \frac{1}{2171\ \Omega} = 4.606 \times 10^{-4}\ \mathrm{S}$$

The conductance of the electrolyte is

$$L(\mathrm{KCl}) = L(\mathrm{soln}) - L(\mathrm{H_2O}) = 4.606 \times 10^{-4}\,\mathrm{S} - 2.35 \times 10^{-6}\,\mathrm{S}$$
$$= 4.583 \times 10^{-4}\,\mathrm{S}$$

The equation equivalent to *(9.6)* for conductance is

$$L = K[A/l] \tag{9.10}$$

Solving for A/l and substituting the data gives

$$\frac{A}{l} = \frac{4.583 \times 10^{-4}\,\mathrm{S}}{0.140\,88\,\mathrm{S \cdot m^{-1}}} = 3.253 \times 10^{-3}\,\mathrm{m}$$

9.17 The conductance cell described in Problem 9.16 was used to determine the conductivity and molar conductance of a 8.42×10^{-4} M KCl solution. Given $L(\mathrm{KCl}) = 4.034 \times 10^{-5}\,\mathrm{S}$ after making the correction for the conductance of the water, calculate κ and Λ for this solution.

▮ Rearranging *(9.10)* and substituting the data gives

$$\kappa = L\frac{l}{A} = (4.034 \times 10^{-5}\,\mathrm{S})\frac{1}{3.253 \times 10^{-3}\,\mathrm{m}} = 1.240 \times 10^{-2}\,\mathrm{S \cdot m^{-1}}$$

The *molar conductance* can be calculated using

$$\Lambda/(\mathrm{S \cdot m^2 \cdot mol^{-1}}) = \frac{10^{-3}[\kappa/(\mathrm{S \cdot m^{-1}})]}{C/(\mathrm{mol \cdot L^{-1}})} \tag{9.11}$$

Substituting the data gives

$$\Lambda = \frac{10^{-3}(1.240 \times 10^{-2})}{8.42 \times 10^{-4}}\,\mathrm{S \cdot m^2 \cdot mol^{-1}} = 0.014\,73\,\mathrm{S \cdot m^2 \cdot mol^{-1}}$$

9.18 Using $\Lambda = 0.0147\,\mathrm{S \cdot m^2 \cdot mol^{-1}}$, determine the concentration of a KCl solution that has $L(\mathrm{soln}) = 10L(\mathrm{H_2O})$, where $\kappa = 5.7 \times 10^{-6}\,\mathrm{S \cdot m^{-1}}$ for pure water.

▮ The conductance of the pure water is given by *(9.10)* as $L(\mathrm{H_2O}) = \kappa(\mathrm{H_2O})(A/l)$, giving

$$L(\mathrm{KCl}) = L(\mathrm{soln}) - L(\mathrm{H_2O}) = 10L(\mathrm{H_2O}) - L(\mathrm{H_2O}) = 9L(\mathrm{H_2}) = 9\kappa(\mathrm{H_2O})(A/l)$$

Solving *(9.10)* for $\kappa(\mathrm{KCl})$ and substituting the data gives

$$\kappa(\mathrm{KCl}) = 9\kappa(\mathrm{H_2O})(A/l)(l/A) = 9\kappa(\mathrm{H_2O})$$

Solving *(9.11)* for C and substituting the data gives

$$C = \frac{9(5.7 \times 10^{-6})(10^{-3})}{0.0147}\,\mathrm{mol \cdot L^{-1}} = 3.5 \times 10^{-6}\,\mathrm{M}$$

9.19 An old handbook lists the equivalent conductance of a 0.01 N solution of $CaCl_2$ as 120.36 mho · cm². Convert this value to the molar conductance in units of $\mathrm{S \cdot m^2 \cdot mol^{-1}}$.

▮ The relationship between equivalents and moles [and between normality (N) and molarity] for electrolytes is determined by the electrical charge (z):

$$z = \nu_+ z_+ = \nu_- z_- \tag{9.12}$$

where z_+ and z_- are the absolute values of the valences on the ions formed. For $CaCl_2$,

$$z = (1)(2) = (2)(1) = 2\,\mathrm{equiv \cdot mol^{-1}}$$

The equivalent and molar conductances are related by

$$\Lambda = z\Lambda_e \tag{9.13}$$

$$\Lambda = (2\,\mathrm{equiv \cdot mol^{-1}})(120.36\,\mathrm{mho \cdot cm^2 \cdot equiv^{-1}})\frac{1\,\Omega^{-1}}{1\,\mathrm{mho}}\left(\frac{1\,\mathrm{S}}{1\,\Omega^{-1}}\right)\left(\frac{1\,\mathrm{m}}{10^2\,\mathrm{cm}}\right)^2$$
$$= 2.4072 \times 10^{-2}\,\mathrm{S \cdot m^2 \cdot mol^{-1}}$$

9.20 Use the following concentration-molar conductance data for KCl at 25 °C to determine Λ_0, the molar conductance at infinite dilution.

$C/(10^{-4}\ \text{M})$	0.326	1.045	2.657	3.522	6.090
$\Lambda/(10^{-2}\ \text{S}\cdot\text{m}^2\cdot\text{mol}^{-1})$	1.4937	1.4895	1.4842	1.4816	1.4756

$C/(10^{-4}\ \text{M})$	9.286	14.080	20.291	27.848	32.827
$\Lambda/(10^{-2}\ \text{S}\cdot\text{m}^2\cdot\text{mol}^{-1})$	1.4711	1.4650	1.4576	1.4504	1.4468

Empirically, a plot of Λ against $C^{1/2}$ (a *Kohlrausch plot*) for dilute solutions of a strong electrolyte will be linear with an intercept equal to Λ_0. From Fig. 9-1, the intercept gives $\Lambda_0 = 0.014\,988\ \text{S}\cdot\text{m}^2\cdot\text{mol}^{-1}$, and the slope of the straight line is $-0.009\,11\ \text{S}\cdot\text{m}^2\cdot\text{mol}^{-1}\cdot\text{M}^{-1/2}$.

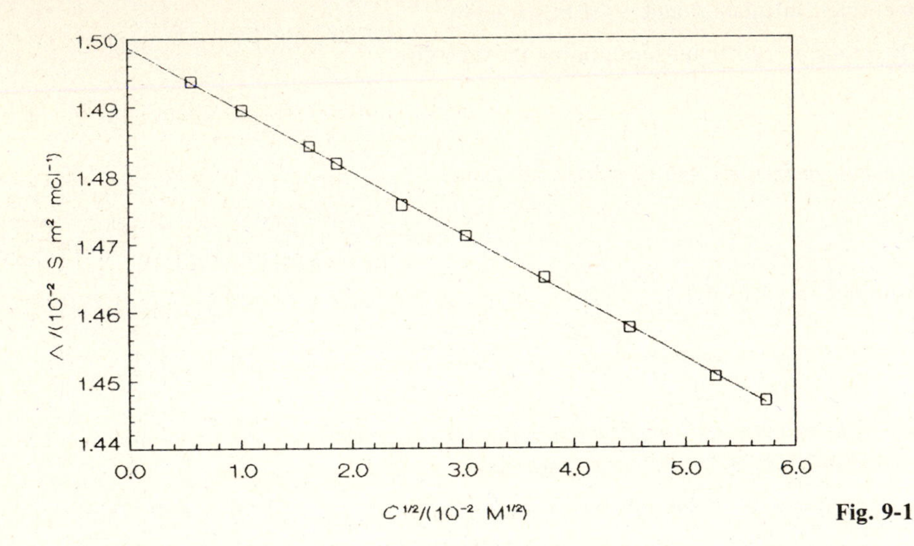

Fig. 9-1

9.21 Compare the slope of the line determined in Problem 9.20 to that predicted by the Debye-Hückel-Onsager equation.

For dilute solutions of strong electrolytes,

$$\Lambda = \Lambda_0 - (A + B\Lambda_0)C^{1/2} \tag{9.14}$$

where for a 1 : 1 electrolyte in H_2O at 25 °C,

$$A = \frac{ZeF}{3\pi\eta}\left(\frac{2z^2eL}{\varepsilon kT}\right)^{1/2} = 6.020 \times 10^{-3}\ \text{S}\cdot\text{m}^2\cdot\text{mol}^{-1}\cdot\text{M}^{-1/2}$$

$$B = \frac{e^2z^2(0.59)}{24\pi\varepsilon kT}\left(\frac{2z^2e^2L}{\pi\varepsilon kT}\right)^{1/2} = 0.229\ \text{M}^{-1/2}$$

Thus *(9.14)* predicts the slope of a plot of Λ against $C^{1/2}$ to be

$$\begin{aligned}\text{Slope} &= -(A + B\Lambda_0)\\ &= -[6.020 \times 10^{-3}\ \text{S}\cdot\text{m}^2\cdot\text{mol}^{-1}\cdot\text{M}^{-1/2} + (0.229\ \text{M}^{-1/2})(0.014\,988\ \text{S}\cdot\text{m}^2\cdot\text{mol}^{-1})]\\ &= -0.009\,45\ \text{S}\cdot\text{m}^2\cdot\text{mol}^{-1}\cdot\text{M}^{-1/2}\end{aligned}$$

which agrees favorably with the value of $-0.009\,11\ \text{S}\cdot\text{m}^2\cdot\text{mol}^{-1}\cdot\text{M}^{-1/2}$ determined in Problem 9.20.

9.22 At higher concentrations the molar conductivity of KCl(aq) is given by

$$\frac{\Lambda + AC^{1/2}}{1 - BC^{1/2}} = \Lambda_0 + (1.419 \times 10^{-2}\ \text{S}\cdot\text{m}^2\cdot\text{mol}^{-1}\cdot\text{M}^{-1})C + (2.924 \times 10^{-3}\ \text{S}\cdot\text{m}^2\cdot\text{mol}^{-1}\cdot\text{M}^{-1})C\log[C/(\text{M})]$$

$$- (1.806 \times 10^{-2}\ \text{S}\cdot\text{m}^2\cdot\text{mol}^{-1}\cdot\text{M}^{-2})C^2$$

Calculate Λ at $C = 0.100\ \text{M}$ and compare the value to the experimental value of $1.2896 \times 10^{-2}\ \text{S}\cdot\text{m}^2\cdot\text{mol}^{-1}$. See Problem 9.21 for additional data.

▮ Substituting the data gives

$$\frac{\Lambda + (6.020 \times 10^{-3}\ \mathrm{S \cdot m^2 \cdot mol^{-1} \cdot M^{-1/2}})(0.100\ \mathrm{M})^{1/2}}{1 - (0.229\ \mathrm{M^{-1/2}})(0.100\ \mathrm{M})^{1/2}}$$

$$= 1.4988 \times 10^{-2}\ \mathrm{S \cdot m^2 \cdot mol^{-1}} + (1.419 \times 10^{-2}\ \mathrm{S \cdot m^2 \cdot mol^{-1} \cdot M^{-1}})(0.100\ \mathrm{M})$$

$$+ (2.924 \times 10^{-3}\ \mathrm{S \cdot m^2 \cdot mol^{-1} \cdot M^{-1}})(0.100\ \mathrm{M}) \log 0.100 - (1.806 \times 10^{-2}\ \mathrm{S \cdot m^2 \cdot mol^{-1} \cdot M^{-2}})(0.100\ \mathrm{M})^2$$

Solving gives $\Lambda = 1.2876 \times 10^{-2}\ \mathrm{S \cdot m^2 \cdot mol^{-1}}$, a difference of -0.16%.

9.23 The ionic molar conductances at infinite dilution are $7.352 \times 10^{-3}\ \mathrm{S \cdot m^2 \cdot mol^{-1}}$ for K^+ and $7.634 \times 10^{-3}\ \mathrm{S \cdot m^2 \cdot mol^{-1}}$ for Cl^-. Calculate Λ_0 for KCl from these data.

▮ The molar conductance is related to the ionic molar conductance (λ_i) by

$$\Lambda_0 = \nu_+ \lambda_{0,+} + \nu_- \lambda_{0,-} \tag{9.15}$$

Substituting the data gives

$$\Lambda_0(\mathrm{KCl}) = (1)(7.352 \times 10^{-3}\ \mathrm{S \cdot m^2 \cdot mol^{-1}}) + (1)(7.634 \times 10^{-3}\ \mathrm{S \cdot m^2 \cdot mol^{-1}})$$

$$= 1.4986 \times 10^{-2}\ \mathrm{S \cdot m^2 \cdot mol^{-1}}$$

9.24 Using $\Lambda_0/(10^{-2}\ \mathrm{S \cdot m^2 \cdot mol^{-1}}) = 4.261$ for HCl, 1.4986 for KCl, and 1.132 for $K(CH_2ClCOO)$, calculate the value of the molar conductance at infinite dilution for monochloroacetic acid.

▮ At infinite dilution,

$$\Lambda_0(\mathrm{WZ}) = \Lambda_0(\mathrm{WX}) + \Lambda_0(\mathrm{YZ}) - \Lambda_0(\mathrm{YX}) \tag{9.16}$$

Substituting into *(9.16)* gives

$$\Lambda_0(\mathrm{CH_2ClCOOH}) = \Lambda_0(\mathrm{HCl}) + \Lambda_0(\mathrm{K(CH_2ClCOO)}) - \Lambda_0(\mathrm{KCl})$$

$$= 4.2616 \times 10^{-2}\ \mathrm{S \cdot m^2 \cdot mol^{-1}} + 1.132 \times 10^{-2}\ \mathrm{S \cdot m^2 \cdot mol^{-1}} - 1.4986 \times 10^{-2}\ \mathrm{S \cdot m^2 \cdot mol^{-1}}$$

$$= 3.895 \times 10^{-2}\ \mathrm{S \cdot m^2 \cdot mol^{-1}}$$

9.25 Prepare a plot of Λ against $C^{1/2}$ for $CH_2ClCOOH(aq)$.

$C/(10^{-3}\ \mathrm{M})$	0.110	0.303	0.590	1.323	2.821	3.812	7.462	14.043	20.179
$\Lambda/(10^{-2}\ \mathrm{S \cdot m^2 \cdot mol^{-1}})$	3.6210	3.2892	2.9558	2.4615	1.9714	1.7798	1.3985	1.0900	0.9383

Using the results of Problems 9.21 and 9.24, show that *(9.14)* is valid at infinite dilution.

▮ The plot is shown in Fig. 9-2. The straight line in the figure is a plot of *(9.14)*, where

$$\Lambda/(\mathrm{S \cdot m^2 \cdot mol^{-1}}) = 3.895 \times 10^{-2} - [6.020 \times 10^{-3}\ \mathrm{M^{-1/2}} + (0.229)(3.895 \times 10^{-2})\ \mathrm{M^{-1/2}}]C^{1/2}$$

$$= 3.895 \times 10^{-2} - 1.494 \times 10^{-2}[C/(\mathrm{M})]^{1/2}$$

Note that extrapolation to $C = 0$ is very difficult to do for a weak electrolyte.

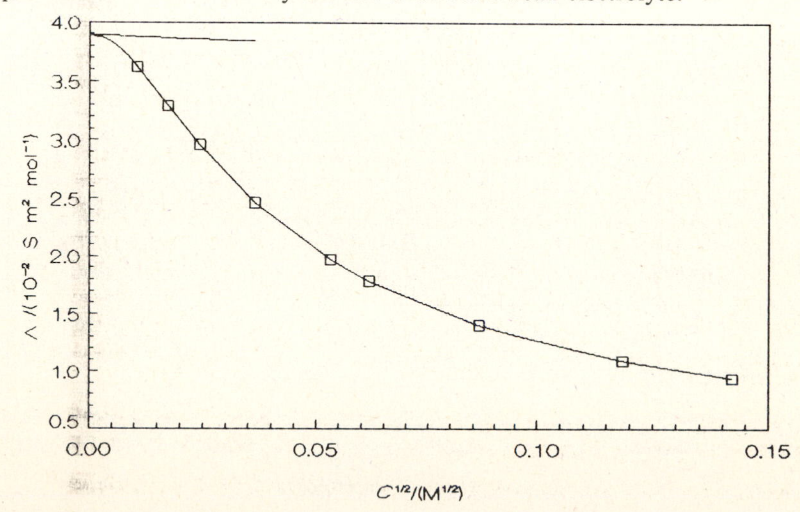

Fig. 9-2

9.26 Determine the apparent value of K_a for $CH_2ClCOOH(aq)$ at $C = 1.323 \times 10^{-3}$ M. See Problem 9.25 for additional data.

▮ The fraction of $CH_2ClCOOH$ that is ionized is

$$\alpha = \Lambda/\Lambda_0 \qquad (9.17)$$

where the usual symbol α is used instead of ξ. Substituting the data gives

$$\alpha = \frac{2.4615 \times 10^{-2}\ \mathrm{S \cdot m^2 \cdot mol^{-1}}}{3.895 \times 10^{-2}\ \mathrm{S \cdot m^2 \cdot mol^{-1}}} = 0.6320$$

Setting up the usual equilibrium table, see Problem 7.12,

equation	$CH_2ClCOOH(aq)$ ⇄	$CH_2ClCOO^-(aq)$ +	$H^+(aq)$
$C_{initial}/(M)$	C		
$C_{change}/(M)$	$-\alpha C$	$+\alpha C$	$+\alpha C$
$C_{equilibrium}/(M)$	$(1-\alpha)C$	αC	αC

Using *(7.20)* gives

$$K_a = \frac{y_\pm^2}{y(CH_2ClCOOH)}\left(\frac{[H^+][CH_2ClCOO^-]}{[CH_2ClCOOH]}\right) = \frac{y_\pm^2}{y(CH_2ClCOOH)}K'_a$$

where K'_a is the apparent value of K_a. Substituting the data gives

$$K'_a = \frac{(\alpha C)(\alpha C)}{(1-\alpha)C} = \frac{\alpha^2 C}{1-\alpha} = \frac{(0.6320)^2(1.323 \times 10^{-3})}{1-0.6320} = 1.436 \times 10^{-3}$$

9.27 Determine $y_\pm$ for $CH_2ClCOOH(aq)$ at $C = 0.020\,179$ M. See Problems 9.25 and 9.26 for additional data.

▮ Assuming $y(CH_2ClCOOH) = 1$, from Problem 9.26, $K_a = y_\pm^2 K'_a$. Taking logarithms of both sides gives

$$\log K_a = 2 \log y_\pm + \log K'_a$$

As the solutions become more dilute, the $\log y_\pm$ term can be replaced by *(6.51)*, giving

$$\log K_a = 2[-(1)(1)(0.5116)I^{1/2}] + \log K'_a$$

Substituting $I = \alpha C$ (see Table 6-2) and rearranging gives

$$\log K'_a = \log K_a + 1.0232(\alpha C)^{1/2}$$

Thus a plot of $\log K'_a$ against $(\alpha C)^{1/2}$ will be linear with a theoretical slope of 1.0232 and an intercept equal to $\log K_a$. Following the procedure in Problem 9.26, values of K'_a and αC were determined and are shown in Fig. 9-3.

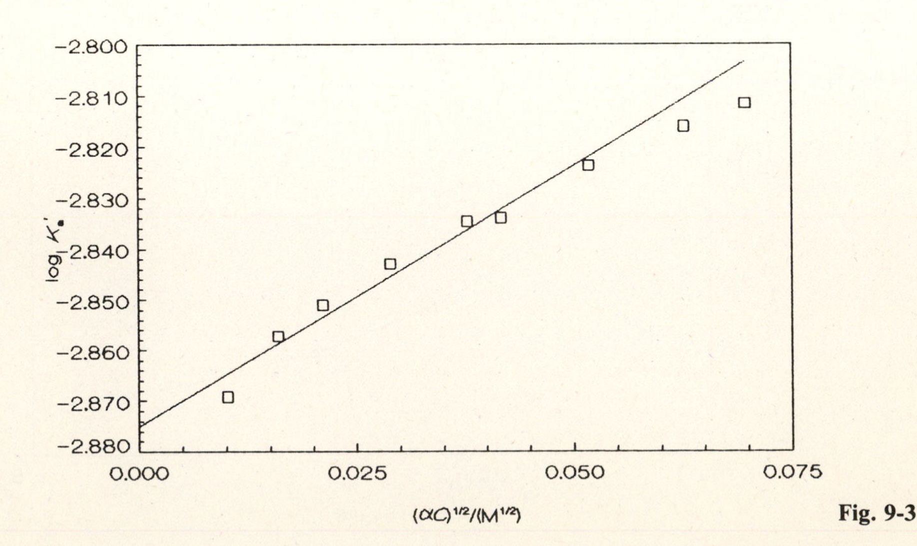

Fig. 9-3

The slope is 1.028, which agrees well with the theoretical value, and the intercept is −2.875, giving

$$\log K_a = -2.875 \qquad K_a = 1.33 \times 10^{-3}$$

From the above, for $C = 0.020\,179$ M,

$$\log y_{\pm} = \frac{\log K_a - \log K'_a}{2} = \frac{-2.875 - (-2.812)}{2} = -0.032$$

$$y_{\pm} = 0.929$$

9.28 What will be the value of Λ for a 0.0010 M aqueous solution of ammonia? For NH_3, $K_b = 1.6 \times 10^{-5}$ and $\Lambda_0 = 2.38 \times 10^{-2}\ S \cdot m^2 \cdot mol^{-1}$. Assume $y_{\pm} = y(NH_3) = 1$.

▌ Following the procedure in Problem 7.56 to determine $[OH^-]$ gives

$$K_b = \frac{[OH^-]^2}{0.0010 - [OH^-]} = 1.6 \times 10^{-5}$$

Solving gives $[OH^-] = 1.2 \times 10^{-4}$ M. The value of α is

$$\alpha = (1.2 \times 10^{-4}\ M)/(0.0010\ M) = 0.12$$

Rearranging *(9.17)* and substituting the data gives

$$\Lambda = (0.12)(2.38 \times 10^{-2}\ S \cdot m^2 \cdot mol^{-1}) = 2.9 \times 10^{-3}\ S \cdot m^2 \cdot mol^{-1}$$

9.29 At 25 °C, $\lambda_0(H^+) = 3.4982 \times 10^{-2}\ S \cdot m^2 \cdot mol^{-1}$ and $\lambda_0(OH^-) = 1.98 \times 10^{-2}\ S \cdot m^2 \cdot mol^{-1}$. Determine $\Lambda_0(H_2O)$. Given $\kappa = 5.7 \times 10^{-6}\ S \cdot m^{-1}$ for H_2O, determine K_w.

▌ Substituting into *(9.15)* gives

$$\Lambda_0 = (1)(3.4982 \times 10^{-2}\ S \cdot m^2 \cdot mol^{-1}) + (1)(1.98 \times 10^{-2}\ S \cdot m^2 \cdot mol^{-1})$$
$$= 5.48 \times 10^{-2}\ S \cdot m^2 \cdot mol^{-1}$$

The "concentration" of water is

$$C = \frac{(0.999\,87\ g \cdot mL^{-1})[(10^3\ mL)/(1\ L)]}{18.015\ g \cdot mol^{-1}} = 55.50\ mol \cdot L^{-1}$$

Substituting into *(9.11)* gives

$$\Lambda/(S \cdot m^{-2} \cdot mol^{-1}) = 10^{-3}(5.7 \times 10^{-6})/55.50 = 1.0 \times 10^{-10}$$

Substituting into *(9.17)* gives

$$\alpha = \frac{1.0 \times 10^{-10}\ S \cdot m^2 \cdot mol^{-1}}{5.48 \times 10^{-2}\ S \cdot m^2 \cdot mol^{-1}} = 1.8 \times 10^{-9}$$

For the ionization of water,

$$[H^+] = [OH^-] = \alpha C = (1.8 \times 10^{-9})(55.50\ mol \cdot L^{-1}) = 1.0 \times 10^{-7}\ M$$

which upon substituting into *(7.15)* gives

$$K_w = [H^+][OH^-] = (1.0 \times 10^{-7})^2 = 1.0 \times 10^{-14}$$

9.30 Determine K_{sp} for AgCl given $\kappa = 1.802 \times 10^{-4}\ S \cdot m^{-1}$. For the ions, $\lambda_{0,i}/(10^{-2}\ S \cdot m^2 \cdot mol^{-1}) = 0.6192$ for Ag^+ and 0.7634 for Cl^-.

▌ For AgCl, *(9.15)* gives

$$\Lambda_0 = (1)(0.6192 \times 10^{-2}\ S \cdot m^2 \cdot mol^{-1}) + (1)(0.7634 \times 10^{-2}\ S \cdot m^2 \cdot mol^{-1})$$
$$= 1.3826 \times 10^{-2}\ S \cdot m^2 \cdot mol^{-1}$$

Substituting into *(9.14)* gives

$$\Lambda/(S \cdot m^2 \cdot mol^{-1}) = 1.3826 \times 10^{-2} - [6.020 \times 10^{-3} + (0.229)(1.3826 \times 10^{-2})][C/(M)]^{1/2}$$
$$= 1.3826 \times 10^{-2} - (9.186 \times 10^{-3})[C/(M)]^{1/2}$$

Because C is quite small for a slightly soluble salt such as AgCl, as a first approximation assume $\Lambda = \Lambda_0$. Rearranging *(9.11)* and substituting gives

$$C = \frac{10^{-3}(1.802 \times 10^{-4})}{1.3826 \times 10^{-2}} = 1.303 \times 10^{-5}\ M$$

Substituting this value into the expression for Λ gives

$$\Lambda = [1.3826 \times 10^{-2} - (9.186 \times 10^{-3})(1.303 \times 10^{-5})^{1/2}]\ \mathrm{S \cdot m^2 \cdot mol^{-1}}$$
$$= 1.3793 \times 10^{-2}\ \mathrm{S \cdot m^2 \cdot mol^{-1}}$$

Because the value of Λ has changed slightly, the iterative process is repeated until C does not change:

$$C = \frac{10^{-3}(1.802 \times 10^{-4})}{1.3793 \times 10^{-2}} = 1.306 \times 10^{-5}\ \mathrm{M}$$

$$\Lambda = [1.3826 \times 10^{-2} - (9.186 \times 10^{-3})(1.306 \times 10^{-5})^{1/2}]$$
$$= 1.3793 \times 10^{-2}\ \mathrm{S \cdot m^2 \cdot mol^{-1}}$$

Recognizing that $[Ag^+] = [Cl^-] = C$ and that at these concentrations $y_\pm = 1$, *(7.38)* gives

$$K_{sp} = (1)^2(1.306 \times 10^{-5})(1.306 \times 10^{-5}) = 1.706 \times 10^{-10}$$

9.31 The salinity of a brine was measured using the conductivity cell described in Problem 9.16. The resistance of the brine was 1235 Ω. Assuming the conduction as the result of NaCl ($\Lambda_0 = 1.2645 \times 10^{-2}\ \mathrm{S \cdot m^2 \cdot mol^{-1}}$), find the concentration of NaCl. At higher concentrations the molar conductivity of NaCl(aq) is given by

$$\frac{\Lambda + AC^{1/2}}{1 - BC^{1/2}} = \Lambda_0 + (9.579 \times 10^{-3}\ \mathrm{S \cdot m^2 \cdot mol^{-1} \cdot M^{-1}})C - (6.529 \times 10^{-3}\ \mathrm{S \cdot m^2 \cdot mol^{-1} \cdot M^{-2}})C^2$$

▮ Substituting into *(9.8)* and *(9.10)* gives

$$L = 1/(1235\ \Omega) = 8.10 \times 10^{-4}\ \mathrm{S}$$

$$\kappa = \frac{8.10 \times 10^{-4}\ \mathrm{S}}{3.253 \times 10^{-3}\ \mathrm{m}} = 0.249\ \mathrm{S \cdot m^{-1}}$$

The same iterative process described in Problem 9.30 using the above equation and *(9.11)* is used to find C:

$$C = \frac{10^{-3}(0.249)}{1.2645 \times 10^{-2}} = 1.97 \times 10^{-2}\ \mathrm{M}$$

$$\Lambda = [1.2645 \times 10^{-2}\ \mathrm{S \cdot m^2 \cdot mol^{-1}} + (9.579 \times 10^{-3}\ \mathrm{S \cdot m^2 \cdot mol^{-1} \cdot M^{-1}})(1.97 \times 10^{-2}\ \mathrm{M})$$
$$- (6.529 \times 10^{-3}\ \mathrm{S \cdot m^2 \cdot mol^{-1} \cdot M^{-2}})(1.97 \times 10^{-2}\ \mathrm{M})^2][1 - (0.229\ \mathrm{M^{-1/2}})(1.97 \times 10^{-2}\ \mathrm{M})^{1/2}]$$
$$- (6.020 \times 10^{-3}\ \mathrm{S \cdot m^2 \cdot mol^{-1} \cdot M^{-1/2}})(1.97 \times 10^{-2}\ \mathrm{M})^{1/2}$$
$$= 1.1569 \times 10^{-2}\ \mathrm{S \cdot m^2 \cdot mol^{-1}}$$

Repeating the calculations gives $C = 2.16 \times 10^{-2}$ M.

9.32 Consider the hypothetical Hittorf cell having inert electrodes shown in Fig. 9-4a. Each of the compartments contains 5 mol of electrolyte as represented by + and − signs. Construct a diagram showing the arrangement of the ions after passing 3 mol of electrons, assuming negligible migration. Construct a diagram showing the arrangement of the ions after ionic migration with the transport number of the anion equal to twice the transport number of the anion (i.e., $t_- = 2t_+$).

▮ The discharge of 3 mol e⁻ at the electrode requires three of the − signs in the left side of the cell to be removed at the electrode, leaving two in that compartment, and three of the + signs in the right side of the cell to be removed at the electrode, leaving two in that compartment (see Fig. 9-4b). For the passing of 3 mol e⁻ through the solution, one + sign moves to the right for every two − signs moving to the left across each boundary, because $t_- = 2t_+$. This leaves 4 mol in the left side, 7 mol in the middle, and 3 mol in the right side (see Fig. 9-4c).

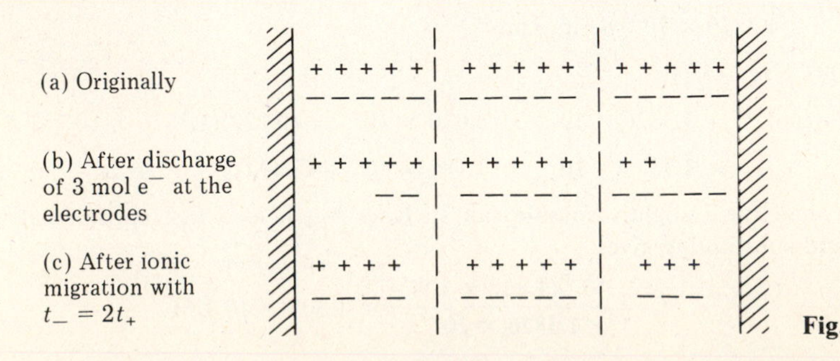

Fig. 9-4

9.33 Using the data shown in Figs. 9-4a and 9-4c, determine t_+ and t_-.

▮ The transport number is given by

$$t_i = \frac{|z_i(n_{i,0} - n_{i,f} \pm n_{i,e})|}{n_e} \tag{9.18}$$

where in the left portion of the cell, the original number of moles of electrolyte present ($n_{i,0}$) was 5 mol, the final number present ($n_{i,f}$) is 4 mol, the number of moles of + involved in the electrode reaction ($n_{+,e}$) was 0 mol, the number of moles of − involved in the electrode reaction ($n_{-,e}$) was −3 mol (where the minus sign is used to represent removal), and the number of moles of electrons passed through the cell (n_e) was 3 mol. Substituting the data into *(9.18)* for each ion gives

$$t_+ = \frac{|(1)(5\text{ mol} - 4\text{ mol} \pm 0\text{ mol})|}{3\text{ mol}} = 0.333$$

$$t_- = \frac{|(1)(5\text{ mol} - 4\text{ mol} - 3\text{ mol})|}{3\text{ mol}} = 0.667$$

Analysis of the right portion of the cell using $n_{i,0} = 5$ mol, $n_{i,f} = 3$ mol, $n_{+,e} = -3$ mol, $n_{-,e} = 0$ mol, and $n_e = 3$ mol also gives $t_+ = 0.333$ and $t_- = 0.667$.

9.34 A Hittorf cell was used to determine the transport number of K^+ ion in KCl(aq). The middle compartment contained a 7.1474 mass % solution of KCl. After the passing of 2.3024×10^{-2} mol e^- through the cell, the anode solution weighing 121.41 g contained a 6.5099 mass % solution of KCl. Calculate $t(K^+)$.

▮ The masses of KCl and H_2O in the anode solution after the experiment were

$$m_f(\text{KCl}) = (121.41\text{ g})(0.065\,099) = 7.904\text{ g}$$

$$m(\text{H}_2\text{O}) = (121.41\text{ g})(1 - 0.065\,099) = 113.51\text{ g}$$

Assuming that the mass of the water in the compartment remained constant, the mass of KCl originally in the compartment was

$$m_0(\text{KCl}) = 0.071\,474 \frac{113.51\text{ g}}{1 - 0.071\,474} = 8.738\text{ g}$$

The various values of n for *(9.18)* are

$$n_e = 2.3034 \times 10^{-2}\text{ mol}$$

$$n_0(\text{K}^+) = \frac{8.738\text{ g}}{74.551\text{ g}\cdot\text{mol}^{-1}} = 0.117\,21\text{ mol}$$

$$n_f(\text{K}^+) = \frac{7.904\text{ g}}{74.551\text{ g}\cdot\text{mol}^{-1}} = 0.106\,02\text{ mol}$$

$$n_e(\text{K}^+) = 0\text{ mol}$$

Substituting into *(9.18)* gives

$$t(\text{K}^+) = \frac{|(1)(0.117\,21\text{ mol} - 0.106\,02\text{ mol} \pm 0)|}{2.3024 \times 10^{-2}\text{ mol}} = 0.4860$$

9.35 As an independent check on the value of $t(K^+)$ determined in Problem 9.34, the cathode solution was analyzed to determine $t(Cl^-)$. If the mass of the solution was 125.66 g and contained 7.7668% mass KCl, calculate $t(Cl^-)$. How does this value confirm the value of $t(K^+)$?

▮ As in Problem 9.34,

$$m_f(\text{KCl}) = (125.66\text{ g})(0.077\,668) = 9.760\text{ g}$$

$$m(\text{H}_2\text{O}) = (125.66\text{ g})(1 - 0.077\,668) = 115.90\text{ g}$$

$$m_0(\text{KCl}) = (0.071\,474)\frac{115.90\text{ g}}{1 - 0.071\,474} = 8.921\text{ g}$$

giving

$$n_0(\text{Cl}^-) = \frac{8.921\text{ g}}{74.551\text{ g}\cdot\text{mol}^{-1}} = 0.119\,66\text{ mol}$$

$$n_f(\text{Cl}^-) = \frac{9.760\text{ g}}{74.551\text{ g}\cdot\text{mol}^{-1}} = 0.130\,92\text{ mol}$$

$$n_e(\text{Cl}^-) = +2.3024 \times 10^{-2}\text{ mol}$$

Substituting into *(9.18)* gives

$$r(Cl^-) = \frac{|(1)(0.119\,66\text{ mol} - 0.130\,92\text{ mol} + 0.023\,024\text{ mol})|}{2.3024 \times 10^{-2}\text{ mol}} = 0.5109$$

Because the transport number of an ion represents the fraction of current carried by that ion,

$$t_+ + t_- = 1 \tag{9.19}$$

Adding the results of Problem 9.34 to the value of $t(Cl^-)$ determined above gives

$$0.4860 + 0.5109 = 0.9969 \approx 1$$

9.36 In a moving boundary experiment to determine $t(Na^+)$, the boundary between a 0.020 00 M solution of NaCl and $CdCl_2$(aq) moved a distance equivalent to 0.8910 cm³ during 2757 s as a current of 1.6001 mA was passed through the cell. Calculate $t(Na^+)$.

▮ The apparent transport number is given by

$$t_i = \frac{z_i F(10^3\text{ mol})[C_i/(\text{M})]}{I}\frac{dV}{dt} \tag{9.20}$$

Assuming $dV/dt = V/t$, substituting the data into *(9.20)* gives

$$t(Na^+) = \frac{(1)(9.648\,456 \times 10^4\text{ C}\cdot\text{mol}^{-1})(10^3\text{ mol})(0.020\,00)}{(1.6001 \times 10^{-3}\text{ A})[(1\text{ C})/(1\text{ A}\cdot\text{s})]}\frac{(0.8910\text{ cm}^3)[(10^{-6}\text{ m}^3)/(1\text{ cm}^3)]}{2757\text{ s}}$$

$$= 0.3897$$

9.37 Using the data given in Problem 9.36, determine the ionic mobility of Na^+.

▮ The *ionic mobility* (u_i) is given by

$$u_i = \frac{1}{t(dE/dl)} = \frac{V\kappa}{It} \tag{9.21}$$

where dE/dl is the electric field strength and l is the length. For the NaCl solution, Λ is given by the relationship given in Problem 9.31 as

$$\begin{aligned}\Lambda &= [1.2645 \times 10^{-2}\text{ S}\cdot\text{m}^2\cdot\text{mol}^{-1} + (9.579 \times 10^{-3}\text{ S}\cdot\text{m}^2\cdot\text{mol}^{-1}\cdot\text{M}^{-1})(0.020\,00\text{ M})\\ &\quad - (6.529 \times 10^{-3}\text{ S}\cdot\text{m}^2\cdot\text{mol}^{-1}\cdot\text{M}^{-2})(0.020\,00\text{ M})^2][1 - (0.229\text{ M}^{-1/2})\\ &\quad \times (0.020\,00\text{ M})^{1/2}] - (6.020 \times 10^{-3}\text{ S}\cdot\text{m}^2\cdot\text{mol}^{-1}\cdot\text{M}^{-1/2})(0.020\,00\text{ M})^{1/2}\\ &= 1.1567 \times 10^{-2}\text{ S}\cdot\text{m}^2\cdot\text{mol}^{-1}\end{aligned}$$

Rearranging *(9.11)* and substituting gives

$$\kappa = (1.1567 \times 10^{-2})(0.020\,00)(10^3)\text{ S}\cdot\text{m}^{-1} = 0.2313\text{ S}\cdot\text{m}^{-1}$$

Substituting into *(9.21)* gives

$$u(K^+) = \frac{(0.8910\text{ cm}^3)[(10^{-6}\text{ m}^3)/(1\text{ cm}^3)](0.2313\text{ S}\cdot\text{m}^{-1})[(1\,\Omega^{-1})/(1\text{ S})]}{(1.6001 \times 10^{-3}\text{ A})(2757\text{ s})[(1\text{ V})/(1\text{ A}\cdot\Omega)]}$$

$$= 4.666 \times 10^{-8}\text{ m}^2\cdot\text{V}^{-1}\cdot\text{s}^{-1}$$

9.38 Using the data of Problems 9.36 and 9.37, calculate $u(Cl^-)$.

▮ The relation between the ionic mobilities and the transport number is

$$t_i = \frac{z_i\nu_i u_i}{z_+\nu_+u_+ + z_-\nu_-u_-} \tag{9.22}$$

Solving *(9.22)* for u_- and substituting the data gives

$$\begin{aligned}u(Cl^-) &= \frac{z(Na^+)\nu(Na^+)u(Na^+)}{z(Cl^-)\nu(Cl^-)}\left(\frac{1}{t(Na^+)} - 1\right)\\ &= \frac{(1)(1)(4.666 \times 10^{-8}\text{ m}^2\cdot\text{V}^{-1}\cdot\text{s}^{-1})}{(1)(1)}\left(\frac{1}{0.3897} - 1\right)\\ &= 7.307 \times 10^{-8}\text{ m}^2\cdot\text{V}^{-1}\cdot\text{s}^{-1}\end{aligned}$$

9.39 Using the results of Problems 9.37 and 9.38, calculate λ_i for the ions.

▮ The relation between the ionic molar conductance and ionic mobility is

$$\lambda_i = z_i F u_i \tag{9.23}$$

Substituting into *(9.23)* gives

$$\lambda(\mathrm{Na^+}) = (1)(9.648\,456 \times 10^4\ \mathrm{C \cdot mol^{-1}})(4.666 \times 10^{-8}\ \mathrm{m^2 \cdot V^{-1} \cdot s^{-1}})\frac{1\ \mathrm{A \cdot s}}{1\ \mathrm{C}}\left(\frac{1\ \mathrm{V}}{1\ \mathrm{A \cdot \Omega}}\right)\left(\frac{1\ \mathrm{S}}{1\ \Omega^{-1}}\right)$$

$$= 4.502 \times 10^{-3}\ \mathrm{S \cdot m^2 \cdot mol^{-1}}$$

$$\lambda(\mathrm{Cl^-}) = (1)(9.648\,456 \times 10^4)(7.307 \times 10^{-8}) = 7.050 \times 10^{-3}\ \mathrm{S \cdot m^2 \cdot mol^{-1}}$$

9.40 At infinite dilution, the ionic molar conductance of $K^+(aq)$ is $7.352 \times 10^{-3}\ \mathrm{S \cdot m^2 \cdot mol^{-1}}$. Determine the drift speed of $K^+(aq)$ under an electric field strength of $1.52\ \mathrm{V \cdot cm^{-1}}$.

▮ The relation between the drift speed and ionic mobility is

$$v_i = u_i \frac{dE}{dl} \tag{9.24}$$

Rearranging *(9.23)* and substituting the data gives

$$u(\mathrm{K^+}) = \frac{7.352 \times 10^{-3}\ \mathrm{S \cdot m^2 \cdot mol^{-1}}}{(1)(9.648\,456 \times 10^4\ \mathrm{C \cdot mol^{-1}})[(1\ \mathrm{A \cdot s})/(1\ \mathrm{C})][(1\ \mathrm{V})/(1\ \mathrm{A \cdot \Omega})][(1\ \mathrm{s})/(1\ \Omega^{-1})]}$$

$$= 7.620 \times 10^{-8}\ \mathrm{m^2 \cdot V^{-1} \cdot s^{-1}}$$

Substituting into *(9.24)* gives

$$v_i = (7.620 \times 10^{-8}\ \mathrm{m^2 \cdot V^{-1} \cdot s^{-1}})(1.52\ \mathrm{V \cdot cm^{-1}})\frac{10^2\ \mathrm{cm}}{1\ \mathrm{m}} = 1.16 \times 10^{-5}\ \mathrm{m}$$

9.41 What is the radius of the K^+ ion in the solution described in Problem 9.40? The viscosity of water is $8.904 \times 10^{-4}\ \mathrm{Pa \cdot s}$.

▮ The relation between ionic mobility, viscosity, and radius is given by

$$r = z_i e / 6\pi\eta u_i \tag{9.25}$$

Substituting the data into *(9.25)* gives

$$r = \frac{(1)(1.602 \times 10^{-19}\ \mathrm{C})[(1\ \mathrm{Pa})/(1\ \mathrm{kg \cdot m^{-1} \cdot s^{-2}})][(1\ \mathrm{A \cdot s})/(1\ \mathrm{C})][(1\ \mathrm{kg \cdot m^2 \cdot s^{-3} \cdot A^{-1}})/(1\ \mathrm{V})]}{6\pi(8.904 \times 10^{-4}\ \mathrm{Pa \cdot s})(7.620 \times 10^{-8}\ \mathrm{m^2 \cdot V^{-1} \cdot s^{-1}})}$$

$$= 1.253 \times 10^{-10}\ \mathrm{m}$$

9.42 Using $\lambda_0(\mathrm{Ca^{2+}}) = 1.1900 \times 10^{-2}\ \mathrm{S \cdot m^2 \cdot mol^{-1}}$ and $\lambda_0(\mathrm{Cl^-}) = 7.634 \times 10^{-3}\ \mathrm{S \cdot m^2 \cdot mol^{-1}}$, calculate the diffusion coefficient of each ion at 25 °C.

▮ The relation between the diffusion coefficient and the ionic mobility is given by the *Einstein relation*

$$D_i^\infty = \frac{u_{0,i} RT}{z_i F} = \frac{RT\lambda_{0,i}}{F^2 z_i^2} \tag{9.26}$$

Substituting into *(9.26)* for each ion gives

$$D^\infty(\mathrm{Ca^{2+}}) = \frac{(8.314\ \mathrm{Pa \cdot m^2 \cdot K^{-1} \cdot mol^{-1}})(298\ \mathrm{K})(1.1900 \times 10^{-2}\ \mathrm{S \cdot m^2 \cdot mol^{-1}})[(1\ \mathrm{kg \cdot m^{-1} \cdot s^{-2}})/(1\ \mathrm{Pa})]}{(9.648\,456 \times 10^4\ \mathrm{C \cdot mol^{-1}})^2(2)^2[(1\ \mathrm{A \cdot s})/(1\ \mathrm{C})]^2[(1\ \mathrm{s})/(1\ \mathrm{A^2 \cdot s^3 \cdot kg^{-1} \cdot m^{-2}})]}$$

$$= 7.922 \times 10^{-10}\ \mathrm{m^2 \cdot s^{-1}}$$

$$D^\infty(\mathrm{Cl^-}) = \frac{(8.314)(298)(7.634 \times 10^{-3})}{(9.648\,456 \times 10^4)^2(1)^2} = 2.033 \times 10^{-9}\ \mathrm{m^2 \cdot s^{-1}}$$

9.43 Using the results of Problem 9.42, calculate $D^\infty(\mathrm{CaCl_2})$ at 25 °C.

▮ The ionic contribution to D^∞ can be made using

$$\frac{\nu}{D^\infty} = \frac{\nu_+}{D_+^\infty} + \frac{\nu_-}{D_-^\infty} = \frac{F^2}{RT}\left(\frac{\nu_+ z_+^2}{\lambda_{0,+}} + \frac{\nu_- z_-^2}{\lambda_{0,-}}\right) \tag{9.27}$$

Substituting the values of D_i^∞ determined in Problem 9.42 into *(9.27)* gives

$$\frac{3}{D^\infty(\mathrm{CaCl_2})} = \frac{1}{7.922 \times 10^{-10}\ \mathrm{m^2 \cdot s^{-1}}} + \frac{2}{2.033 \times 10^{-9}\ \mathrm{m^2 \cdot s^{-1}}}$$

Solving gives $D^\infty(\mathrm{CaCl_2}) = 1.336 \times 10^{-9}\ \mathrm{m^2 \cdot s^{-1}}$.

9.44 Using $D^\infty(\mathrm{H^+}) = 9.32 \times 10^{-9}\ \mathrm{m^2 \cdot s^{-1}}$ and $D^\infty(\mathrm{OH^-}) = 5.28 \times 10^{-9}\ \mathrm{m^2 \cdot s^{-1}}$ at 25 °C, calculate the value of Λ_0 for water.

▮ The molar conductivity is related to the diffusion coefficients by the *Einstein–Nernst equation*,

$$\Lambda_0 = \frac{F^2}{RT}(\nu_+ z_+^2 D_+^\infty + \nu_- z_-^2 D_-^\infty) \tag{9.28}$$

Substituting the data gives

$$\Lambda_0 = \frac{(9.648 \times 10^4\ \mathrm{C \cdot mol^{-1}})^2[(1\ \mathrm{A \cdot s})/(1\ \mathrm{C})]^2[(1\ \mathrm{s})/(1\ \mathrm{A^2 \cdot s^{-3} \cdot kg^{-1} \cdot m^{-2}})]}{(8.314\ \mathrm{Pa \cdot m^3 \cdot K^{-1} \cdot mol^{-1}})(298\ \mathrm{K})[(1\ \mathrm{kg \cdot m^{-1} \cdot s^{-2}})/(1\ \mathrm{Pa})]}$$
$$\times [(1)(1)^2(9.32 \times 10^{-9}\ \mathrm{m^2 \cdot s^{-1}}) + (1)(1)^2(5.28 \times 10^{-9})]$$
$$= 5.48 \times 10^{-2}\ \mathrm{S \cdot m^2 \cdot mol^{-1}}$$

9.45 Using the data given in Problem 9.42, calculate $t(\mathrm{Ca^{2+}})$.

▮ The relationship between the ionic molar conductance and transport number is

$$t_i = \frac{\lambda_i/z_i}{\lambda_+/z_+ + \lambda_-/z_-} \tag{9.29}$$

Substituting the data gives

$$t(\mathrm{Ca^{2+}}) = \frac{(1.1900 \times 10^{-2}\ \mathrm{S \cdot m^2 \cdot mol^{-1}})/2}{(1.1900 \times 10^{-2}\ \mathrm{S \cdot m^2 \cdot mol^{-1}})/2 + (7.634 \times 10^{-3}\ \mathrm{S \cdot m^2 \cdot mol^{-1}})/1} = 0.4380$$

9.46 Calculate the value of the ionic molar conductance for $\mathrm{Ca^{2+}}$ in a 0.10 M $\mathrm{CaCl_2}$ solution given that $\Lambda = 2.0492 \times 10^{-2}\ \mathrm{S \cdot m^2 \cdot mol^{-1}}$ and $t(\mathrm{Ca^{2+}}) = 0.4060$.

▮ The relationship among the transport number, the ionic molar conductance, and the molar conductance is

$$\nu_i \lambda_i = t_i \Lambda \tag{9.30}$$

Rearranging for λ_i and substituting the data gives

$$\lambda(\mathrm{Ca^{2+}}) = \frac{(0.4060)(2.0492 \times 10^{-2}\ \mathrm{S \cdot m^2 \cdot mol^{-1}})}{2} = 4.160 \times 10^{-3}\ \mathrm{S \cdot m^2 \cdot mol^{-1}}$$

9.3 ELECTROCHEMICAL CELLS

9.47 Prepare a sketch and write the half-reaction and overall cell reaction for the electrochemical cell designated by

$$\mathrm{Pt|Ag(s)|AgCl(s)|Cl^-}\ (a = 1)|\mathrm{Cl_2}\ (a = 1)|\mathrm{C(graphite)|Pt}$$

▮ The information concerning the anode (the site of oxidation) is written at the left of the cell designation and corresponds to the half-reaction

$$\mathrm{Ag(s) + Cl^-(aq) \rightarrow AgCl(s) + e^-}$$

and the information at the right of the cell designation concerns the cathode (the site of reduction) and corresponds to the half-reaction

$$\mathrm{Cl_2(g) + 2e^- \rightarrow 2Cl^-(aq)}$$

Multiplying the oxidation half-reaction by 2 and adding the half-reactions gives the overall cell reaction as $\mathrm{2Ag(s) + Cl_2(g) \rightarrow 2AgCl(s)}$ with $n = 2$ mol.

The sketches shown in Fig. 9-5 can be used to draw a sketch of the cell. In this case, the anode compartment is a metal precipitate half-cell (see Fig. 9-5e), and the cathode compartment is a gas half-cell (see Fig. 9-5a). The vertical lines in the cell designation refer to the simple phase boundaries between the substances, and so combining the two half sketches gives the cell shown in Fig. 9-6.

9.48 Write the overall chemical equation and the cell notation for the electrochemical cell shown in Fig. 9-7.

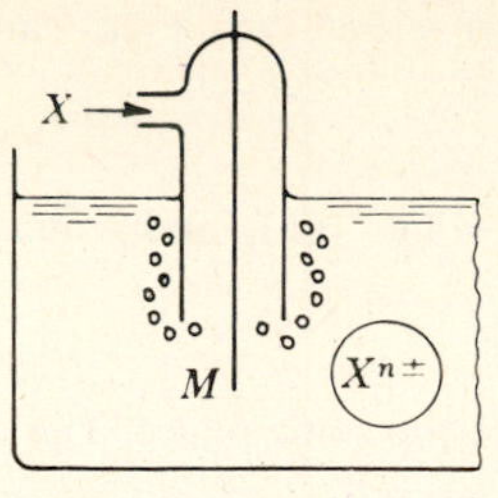

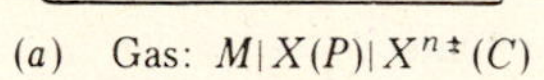
(a) Gas: $M|X(P)|X^{n\pm}(C)$

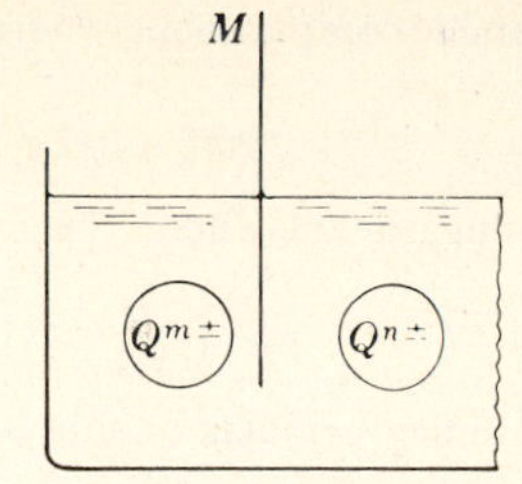

(b) Redox: $M\,|\,Q^{m\pm}\,(C), Q^{n\pm}\,(C)$

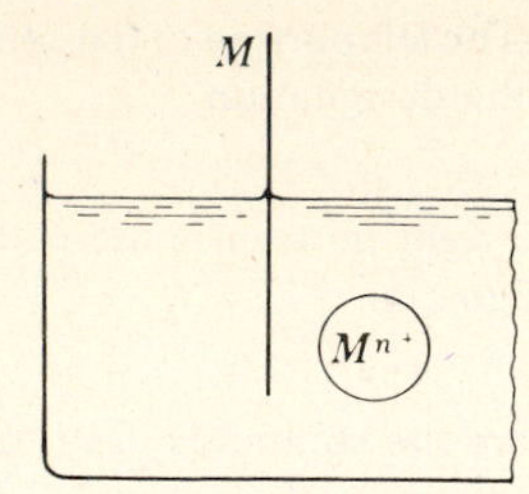

(c) Metal: $M|M^{n+}(C)$

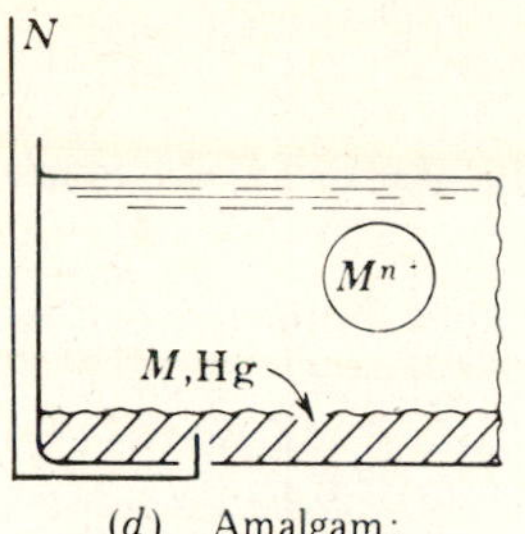

(d) Amalgam: $N|M(\mathrm{Hg}, C)|M^{n+}(C)$

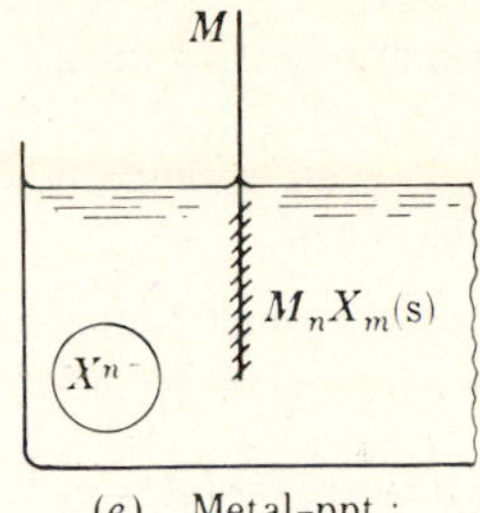

(e) Metal-ppt.: $M|M_nX_m(\mathrm{s})|X^{n-}(C)$

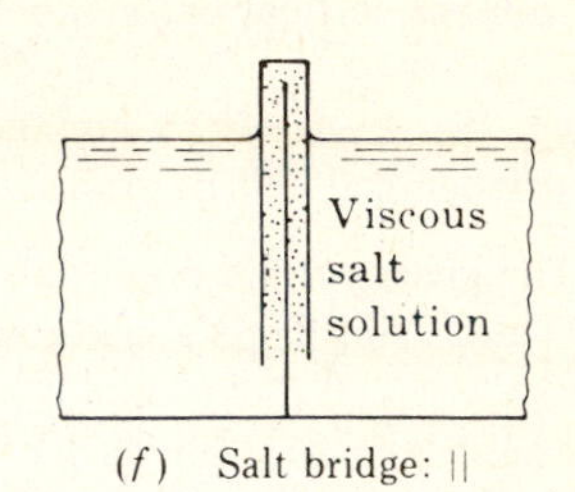

(f) Salt bridge: ||

Fig. 9-5

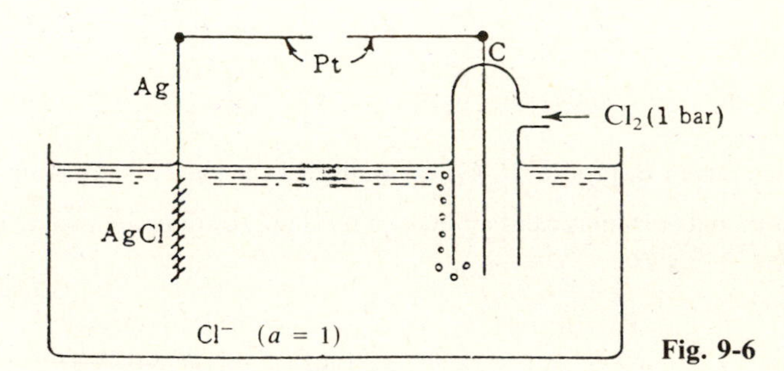

Fig. 9-6

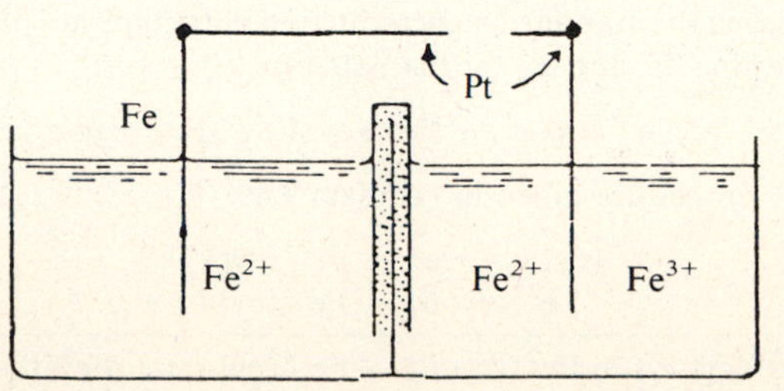

Fig. 9-7

▮ The left portion of the cell is the anode compartment, where $Fe(s) \rightarrow Fe^{2+}(aq) + 2e^-$ and would be represented by the designation

$$Pt|Fe(s)|Fe^{2+}(aq)$$

The right portion is the cathode compartment where $Fe^{3+} + e^- \rightarrow Fe^{2+}(aq)$ and would be represented by the designation

$$Fe^{2+}(aq),\ Fe^{3+}(aq)|Pt$$

where the comma is used to separate the formulas of substances in the same phase. The overall equation is

$$Fe(s) + 2Fe^{3+}(aq) \rightarrow 3Fe^{2+}(aq)$$

which can be represented by

$$Pt|Fe(s)|Fe^{2+}(aq)||Fe^{2+}(aq),\ Fe^{3+}(aq)|Pt$$

The double vertical bar symbol is used to represent the salt bridge.

9.49 Using $E^0 = -0.4402$ V for the Fe^{2+}/Fe couple and 0.771 V for the Fe^{3+}/Fe^{2+} couple, determine $E°$ for the electrochemical cell described in Problem 9.48.

▮ The given couple potentials are standard-state reduction potentials for the two half-reactions. Substituting into *(6.6)* and using Hess's law gives

$$\begin{array}{rl} & \Delta_r G° \\ Fe(s) \rightarrow 2Fe^{2+}(aq) + 2e^- & -(2)F(0.4402\text{ V}) \\ 2[Fe^{3+}(aq) + e^- \rightarrow Fe^{2+}(aq)] & (2)[-(1)F(0.771\text{ V})] \\ \hline Fe(s) + 2Fe^{3+}(aq) \rightarrow 3Fe^{2+}(aq) & -2F(0.4402\text{ V} + 0.771\text{ V}) \end{array}$$

where the sign on the potential for the Fe^{2+}/Fe couple was changed to represent the potential for the oxidation half-reaction. Solving *(6.6)* for $E°$ and substituting the above result gives

$$E° = \frac{-2F(0.4402\text{ V} + 0.771\text{ V})}{-2F} = 1.211\text{ V}$$

Note that this process is always equivalent to the simple addition of a potential for the reduction half-reaction and a potential for the oxidation half-reaction

$$\begin{array}{rl} Fe(s) \rightarrow 2Fe^{2+}(aq) + 2e^- & E° = 0.4402\text{ V} \\ 2[Fe^{3+}(aq) + e^- \rightarrow Fe^{2+}(aq)] & E° = 0.771\text{ V} \\ \hline Fe(s) + 2Fe^{3+}(aq) \rightarrow 3Fe^{2+}(aq) & E° = 1.211\text{ V} \end{array}$$

9.50 Using the data given in Problem 9.49, determine $E°$ for the Fe^{3+}/Fe couple.

▮ The desired half-reaction can be obtained by simply adding the two given reduction half-reactions. Proceeding as in Problem 9.49 gives

$$\begin{array}{rl} & \Delta_r G° \\ Fe^{2+}(aq) + 2e^- \rightarrow Fe(s) & -(2)F(-0.4402\text{ V}) \\ Fe^{3+}(aq) + e^- \rightarrow Fe^{2+}(aq) & -(1)F(0.771\text{ V}) \\ \hline Fe^{3+}(aq) + 3e^- \rightarrow Fe(s) & -F[(2)(-0.4402\text{ V}) + (1)(0.771\text{ V})] \end{array}$$

Substituting this result into *(6.6)* gives

$$E° = \frac{-F[(2)(-0.4402\text{ V}) + (1)(0.771\text{ V})]}{-3F} = -0.036\text{ V}$$

Note that potentials for half-reactions cannot simply be added to give the potential for another half-reaction.

9.51 Given $E' = 0.08$ V for the $Fe^{3+}(\text{cyt } b)/Fe^{2+}(\text{cyt } b)$ couple and $E' = 0.22$ V for the $Fe^{3+}(\text{cyt } c_1)/Fe^{2+}(\text{cyt } c_1)$ couple, where E' represents the standard-state reduction potentials at pH = 7.0 at 25 °C and cyt is an abbreviation for cytochromes. Determine E' and K for the equation

$$Fe^{3+}(\text{cyt } c_1) + Fe^{2+}(\text{cyt } b) \rightarrow Fe^{2+}(\text{cyt } c_1) + Fe^{3+}(\text{cyt } b)$$

▮ Following the short procedure given in Problem 9.49,

$$\begin{array}{rl} Fe^{3+}(\text{cyt } c_1) + e^- \rightarrow Fe^{2+}(\text{cyt } c_1) & E' = 0.22\text{ V} \\ Fe^{2+}(\text{cyt } b) \rightarrow Fe^{3+}(\text{cyt } b) + e^- & E' = -0.08\text{ V} \\ \hline Fe^{3+}(\text{cyt } c_1) + Fe^{2+}(\text{cyt } b) \rightarrow Fe^{2+}(\text{cyt } c_1) + Fe^{3+}(\text{cyt } b) & E' = 0.14\text{ V} \end{array}$$

The relation between the standard state potential and the equilibrium constant can be found by combining *(7.1b)* and *(6.6)* to give

$$K = e^{nFE°/RT} \tag{9.31}$$

Substituting into *(9.31)* gives

$$K = \exp\left(\frac{(1\ \text{mol})(9.648 \times 10^4\ \text{J}\cdot\text{V}^{-1}\cdot\text{mol}^{-1})(0.14\ \text{V})}{(8.314\ \text{J}\cdot\text{K}^{-1}\cdot\text{mol}^{-1})(298\ \text{K})(\text{mol})}\right) = 230$$

9.52 Given $E° = -0.268$ V for the $PbCl_2/Pb$ couple and -0.126 V for the Pb^{2+}/Pb couple, determine K_{sp} for $PbCl_2$ at 25 °C.

▮ The desired chemical equation for the solubility of a slightly soluble salt can be obtained by adding the reduction half-reaction involving the $PbCl_2/Pb$ couple to the oxidation half-reaction involving the Pb^{2+}/Pb couple. Proceeding as in Problem 9.49 gives

	$\Delta_r G°$
$PbCl(s) + 2e^- \rightarrow Pb(s) + 2Cl^-(aq)$	$-(2)F(-0.268\ V)$
$Pb(s) \rightarrow Pb^{2+}(aq) + 2e^-$	$-(2)F(0.126\ V)$
$PbCl_2(s) \rightarrow Pb^{2+}(aq) + 2Cl^-(aq)$	$-(2)F(-0.268\ V + 0.126\ V)$

Substituting into *(9.31)* gives

$$K = \exp\left[\frac{(2\ \text{mol})(9.648 \times 10^4\ \text{J}\cdot\text{V}^{-1}\cdot\text{mol}^{-1})(-0.268\ \text{V} + 0.126\ \text{V})}{(8.314\ \text{J}\cdot\text{K}^{-1}\cdot\text{mol}^{-1})(298\ \text{K})(\text{mol})}\right] = 1.6 \times 10^{-5}$$

9.53 Given $E° = 0.0000$ V for the H^+/H_2 couple and -0.8281 V for the $H_2O/H_2, OH^-$ couple. Determine K_w at 25 °C.

▮ Proceeding as in Problem 9.52 gives

	$\Delta_r G°$
$H_2(g) \rightarrow 2H^+(aq) + 2e^-$	$-(2)F(0.0000\ V)$
$2H_2O(l) + 2e^- \rightarrow H_2(g) + 2OH^-(aq)$	$-(2)F(-0.8281\ V)$
$2H_2O(l) \rightarrow 2H^+(aq) + 2OH^-(aq)$	$-(2)F(0.8281\ V)$

The desired chemical reaction and corresponding $\Delta_r G°$ is

$$H_2O(l) \rightarrow H^+ + OH^-(aq) \qquad \Delta_r G° = -\frac{(2)F(-0.8281\ \text{V})}{2}$$

Substituting into *(9.31)* gives

$$K = \exp\left(\frac{(1\ \text{mol})(9.648 \times 10^4\ \text{J}\cdot\text{V}^{-1}\cdot\text{mol}^{-1})(-0.8281\ \text{V})}{(8.314\ \text{J}\cdot\text{K}^{-1}\cdot\text{mol}^{-1})(298.15\ \text{K})(\text{mol})}\right) = 1.01 \times 10^{-14}$$

9.54 The temperature dependence of the standard state cell potential for the lead storage cell is

$$E°/(\text{V}) = 2.1191 + 1.62 \times 10^{-4}[T/(°\text{C})] + 8.5 \times 10^{-7}[T/(°\text{C})]^2$$

Calculate $\Delta_r G°$, $\Delta_r S°$, $\Delta_r H°$, and $\Delta_r C_P°$ at 25 °C for the reaction

$$Pb(s) + PbO_2(s) + 2H_2SO_4(aq) \rightarrow 2PbSO_4(s) + 2H_2O(l)$$

▮ Using *(6.6)* gives

$$\begin{aligned}\Delta_r G° &= -(2\ \text{mol})(9.648 \times 10^4\ \text{J}\cdot\text{V}^{-1}\cdot\text{mol}^{-1}) \\ &\quad \times [2.1191 + (1.62 \times 10^{-4})(25) + (8.5 \times 10^{-7})(25)^2]\ \text{V}\,[(10^{-3}\ \text{kJ})/(1\ \text{J})] \\ &= -409.8\ \text{kJ}\end{aligned}$$

Combining *(6.6)* and *(6.15a)* gives

$$\Delta_r S° = nF\left(\frac{\partial E°}{\partial T}\right)_P \tag{9.32}$$

Using the chain rule of differentiation gives

$$\begin{aligned}\Delta_r S° &= nF\left(\frac{\partial E°}{\partial[T/(°\text{C})]}\right)\left(\frac{\partial[T/(°\text{C})]}{\partial T}\right) \\ &= nF\{1.62 \times 10^{-4} + (8.5 \times 10^{-7})(2)[T/(°\text{C})]\}\ \text{V}\cdot\text{K}^{-1}(1) \\ &= (2\ \text{mol})(9.648 \times 10^4\ \text{J}\cdot\text{V}^{-1}\cdot\text{mol}^{-1})\{1.62 \times 10^{-4} + (8.5 \times 10^{-7})(2)(25)\}\ \text{V}\cdot\text{K}^{-1} \\ &= 39.5\ \text{J}\cdot\text{K}^{-1}\end{aligned}$$

Combining *(6.6)* and *(6.2a)* gives

$$\Delta_r H^\circ = -nF\left[E^\circ - T\left(\frac{\partial E^\circ}{\partial T}\right)_P\right] \tag{9.33}$$

Substituting into *(9.33)* gives

$$\begin{aligned}\Delta_r H^\circ &= -(2\text{ mol})(9.648\,456 \times 10^4\text{ J}\cdot\text{V}^{-1}\cdot\text{mol}^{-1})\{[2.1191 + (1.62 \times 10^{-4})(25) + (8.5 \times 10^{-7})(25^2)]\text{ V}\\ &\quad -(298\text{ K})[1.62 \times 10^{-4} + (8.5 \times 10^{-7})(2)(25)]\text{ V}\cdot\text{K}^{-1}\}[(10^{-3}\text{ kJ})/(1\text{ J})]\\ &= -398.0\text{ kJ}\end{aligned}$$

Combining *(3.21)* and *(9.33)* gives

$$\Delta_r C_P^\circ = nFT\frac{\partial^2 E^\circ}{\partial T^2} \tag{9.34}$$

Using the chain rule of differentiation gives

$$\begin{aligned}\Delta_r C_P^\circ &= nFT\frac{\partial^2 E^\circ}{\partial[T/(^\circ\text{C})]^2}\left(\frac{\partial[T/(^\circ\text{C})]}{\partial T}\right)^2\\ &= (2\text{ mol})(9.648 \times 10^4\text{ J}\cdot\text{V}^{-1}\cdot\text{mol}^{-1})[(8.5 \times 10^{-7})(2)\text{ V}\cdot\text{K}^{-1}](1)\\ &= 33\text{ J}\cdot\text{K}^{-1}\end{aligned}$$

9.55 The standard reduction potential for $Ni^{2+}(aq)$ is -0.250 V. What is the half-cell potential if $a(Ni^{2+}) = 0.015$?

▌ Combining *(6.6)* and *(6.55)* gives the *Nernst equation*

$$E = E^\circ - \frac{RT}{nF}\ln Q \tag{9.35}$$

At 25 °C, *(9.35)* becomes

$$E = E^\circ - \frac{0.059\,157\text{ V}\cdot\text{mol}}{n}\log Q \tag{9.36}$$

For the reduction half-reaction, $Ni^{2+}(aq) + 2e^- \rightarrow Ni(s)$, using *(6.56)* gives

$$Q = \frac{a(\text{Ni})}{a(\text{Ni}^{2+})} = \frac{1}{0.015} = 67$$

Substituting into *(9.36)* gives

$$E = -0.250\text{ V} - \frac{0.059\,157\text{ V}\cdot\text{mol}}{2\text{ mol}}\log 67 = -0.304\text{ V}$$

9.56 Calculate the potential of the following cell:

$$\text{Cu}|\text{Mn(s)}|\text{MnCl}_2(0.0010\text{ M}),\ \text{HCl}(0.010\text{ M})|\text{O}_2(0.25\text{ bar})|\text{Pt}|\text{Cu}$$

given $E^\circ = -1.185$ V for the Mn^{2+}/Mn couple and 1.229 V for the $O_2/H_2O, H^+$ couple.

▌ The chemical equation and standard-state potential are

$$\begin{array}{rl} 2[\text{Mn(s)} \rightarrow \text{Mn}^{2+}(\text{aq}) + 2e^-] & E^\circ = 1.185\text{ V}\\ \text{O}_2(\text{g}) + 4\text{H}^+(\text{aq}) + 4e^- \rightarrow 2\text{H}_2\text{O(l)} & E^\circ = 1.229\text{ V}\\ \hline 2\text{Mn(s)} + \text{O}_2(\text{g}) + 4\text{H}^+(\text{aq}) \rightarrow 2\text{Mn}^{2+}(\text{aq}) + 2\text{H}_2\text{O(l)} & E^\circ = 2.414\text{ V}\end{array}$$

For the chemical reaction, using *(6.56)* gives

$$Q = \frac{a(\text{Mn}^{2+})^2 a(\text{H}_2\text{O})^2}{a(\text{Mn})^2 a(\text{O}_2) a(\text{H}^+)^4} = \frac{y(\text{Mn}^{2+})^2}{y(\text{H}^+)^4}\,\frac{[C(\text{Mn}^{2+})/(\text{M})]^2}{[P(\text{O}_2)/(\text{bar})][C(\text{H}^+)/(\text{M})]^4}$$

where the oxygen was assumed to be acting ideally. At these dilute concentrations, y_i can be given by *(6.52)*. From Table 6-2, $I(\text{MnCl}_2) = 3C(\text{MnCl}_2)$ and $I(\text{HCl}) = C(\text{HCl})$. Substituting into *(6.52)* gives

$$\begin{aligned}\log[y(\text{Mn}^{2+})] &= -(2)^2(0.5116)[(3)(0.0010)]^{1/2} = -0.11\\ y(\text{Mn}^{2+}) &= 0.77\\ \log[y(\text{H}^+)] &= -(1)^2(0.5116)(0.010)^{1/2} = -0.051\\ y(\text{H}^+) &= 0.89\end{aligned}$$

The value of the reaction quotient is

$$Q = \frac{(0.77)^2}{(0.89)^4}\frac{(0.0010)^2}{(0.25)(0.010)^4} = 380$$

Substituting into *(9.36)* gives

$$E = 2.414\ \text{V} - \frac{0.059\,157\ \text{V}\cdot\text{mol}}{4\ \text{mol}}\log 380 = 2.452\ \text{V}$$

9.57 Use the following data for the cell

$$\text{Pt}|\text{H}_2(a = 1)|\text{H}^+(C)|\text{AgCl(s)}|\text{Ag(s)}|\text{Pt}$$

to determine the standard reduction potential for the AgCl/Ag couple:

$m/(\text{mol}\cdot\text{kg}^{-1})$	0.003 215	0.004 488	0.005 619	0.007 311	0.009 138	0.011 195	0.013 407
$E/(\text{V})$	0.520 70	0.504 01	0.492 74	0.479 65	0.468 77	0.458 78	0.449 91

▮ The chemical reaction corresponding to the cell rotation is

$$\tfrac{1}{2}[\text{H}_2(\text{g}) \rightarrow 2\text{H}^+(\text{aq}) + 2\text{e}^-]$$
$$\text{AgCl(s)} + \text{e}^- \rightarrow \text{Ag(s)} + \text{Cl}^-(\text{aq})$$
$$\overline{\text{AgCl(s)} + \tfrac{1}{2}\text{H}_2(\text{g}) \rightarrow \text{Ag(s)} + \text{H}^+(\text{aq}) + \text{Cl}^-(\text{aq})}$$

Using *(6.56)* and *(9.36)* gives

$$Q = \frac{a(\text{Ag})a(\text{H}^+)a(\text{Cl}^-)}{a(\text{AgCl})a(\text{H}_2)^{1/2}} = \gamma_\pm^2[m(\text{HCl})/(\text{m})]^2$$

$$E = E^\circ - \frac{0.059\,157\ \text{V}\cdot\text{mol}^{-1}}{1\ \text{mol}}\log\{\gamma_\pm^2[m(\text{HCl})/(\text{m})]^2\}$$

Rearranging the equation for E so that all the measured quantities are on the left side gives

$$E + 2(0.059\,157\ \text{V})\log[m(\text{HCl})/(\text{m})] = E^\circ - 2(0.059\,157\ \text{V})\log\gamma_\pm$$

As the concentration approaches zero, *(6.51)* can be used to describe the log $\gamma_\pm$ term giving

$$E + (0.118\,314\ \text{V})\log[m(\text{HCl})/(\text{m})] = E^\circ - (0.118\,314\ \text{V})\{-(1)(1)(0.5116)[m/(\text{m})]^{1/2}\}$$
$$= E^\circ + (0.060\,53)[m/(\text{m})]^{1/2}\ \text{V}$$

This result indicates that a plot of $E + (0.118\,314\ \text{V})\log[m(\text{HCl})/(\text{m})]$ against $m^{1/2}$ will be linear as the concentration approaches zero and will have an intercept equal to E° and a theoretical slope of 0.060 53. The plot appears in Fig. 9-8. The intercept is 0.2235 V, and the limiting slope is 0.040. Because $E^\circ = 0$ V (by definition) for the H^+/H_2 couple, $E^\circ(\text{AgCl/Ag}) = E^\circ = 0.2235$ V.

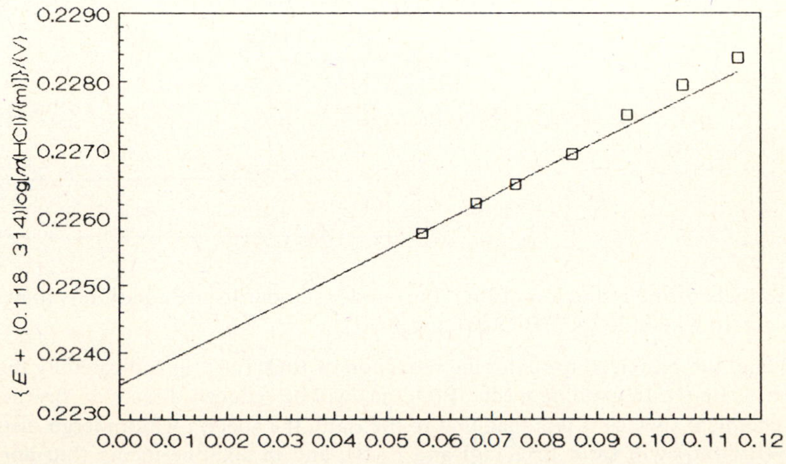

Fig. 9-8

9.58 Using the results of Problem 9.57, calculate $\gamma_{\pm}(\mathrm{HCl})$ at 0.013 407 m.

▮ Rearranging the equation in Problem 9.57 relating E, $E°$, $\gamma_{\pm}^2$, and $m(\mathrm{HCl})$ and substituting the data gives

$$\log \gamma_{\pm} = \frac{E° - E}{0.118\,314\ \mathrm{V}} - \log[m/(\mathrm{m})]$$

$$= \frac{(0.2235\ \mathrm{V}) - (0.449\,91\ \mathrm{V})}{0.118\,314\ \mathrm{V}} - \log 0.013\,407 = -0.041$$

Taking antilogarithms gives $\gamma_{\pm} = 0.910$.

9.59 The cell

$$\mathrm{Ag(s)|AgCl(s)|HCl(0.100\ M)|glass|H^+}(C)\mathrm{|KCl(sat)|Hg_2Cl_2(s)|Hg(l)}$$

is used to measure pH. Given $E = 0.2873$ V for AgCl/Ag, $E = 0.2415$ V for $\mathrm{Hg_2Cl_2/Hg}$, and $\gamma_{\pm} = 0.796$ for 0.1 M HCl, determine the theoretical value for the cell potential as a function of pH.

▮ The $\mathrm{H^+(aq)}$ ions pass through the membrane, and the above cell is equivalent to the following two cells connected in series:

$$\mathrm{Ag(s)|AgCl(s)|HCl(0.1\ M)|H_2}(a = 1)\mathrm{|Pt\cdots Pt|H_2}(a = 1)\mathrm{|H^+}(C)\mathrm{|KCl(sat)|Hg_2Cl_2(s)|Hg(l)]}$$

The corresponding chemical equations are

$$\mathrm{2Ag(s) + 2Cl^-(0.1\ M) + 2H^+(0.1\ M) \rightarrow 2AgCl(s) + H_2(g)}$$

$$E = E° - \frac{0.059\,157\ \mathrm{V \cdot mol}}{2\ \mathrm{mol}} \log \frac{1}{a(\mathrm{Cl^-})^2 a(\mathrm{H^+})^2}$$

$$= -0.2873\ \mathrm{V} + (0.059\,157\ \mathrm{V}) \log\{a[\mathrm{H^+(0.1\ M)}]\}$$

and

$$\mathrm{Hg_2Cl_2(s) + H_2(g) \rightarrow 2Hg(l) + 2Cl^-(sat\ KCl) + 2H^+}(C)$$

$$E = E° - \frac{0.059\,157\ \mathrm{V \cdot mol}}{2\ \mathrm{mol}} \log[a(\mathrm{Cl^-})^2 a(\mathrm{H^+})^2]$$

$$= 0.2415\ \mathrm{V} - (0.059\,157\ \mathrm{V}) \log\{a[\mathrm{H^+}(C)]\}$$

The total cell potential is

$$E = -0.2873\ \mathrm{V} + (0.059\,157\ \mathrm{V}) \log[(0.100)(0.796)] + 0.2415\ \mathrm{V} - (0.059\,157\ \mathrm{V}) \log\{a[\mathrm{H^+}(C)]\}$$

$$= -0.0458\ \mathrm{V} - 0.0650\ \mathrm{V} + (0.059\,157\ \mathrm{V})\ \mathrm{pH}$$

$$= -0.1108\ \mathrm{V} + (0.059\,157\ \mathrm{V})\ \mathrm{pH}$$

9.60 Inorganic chemists often use Latimer diagrams to discuss the reactivity, stability, and possible products of a reaction. Using the following *Latimer diagrams* for bromine

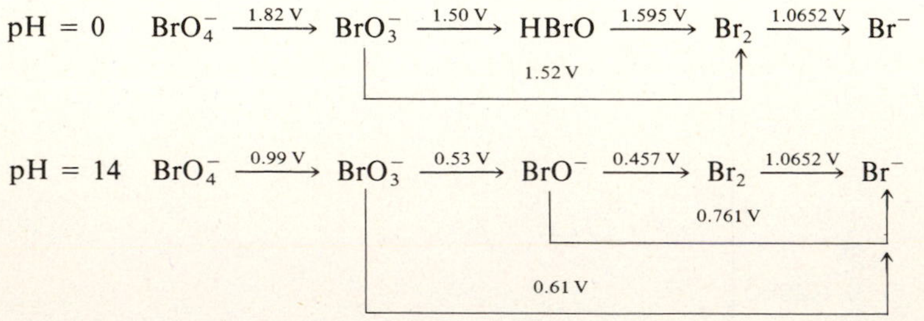

$$\mathrm{pH} = 0 \quad \mathrm{BrO_4^-} \xrightarrow{1.82\ \mathrm{V}} \mathrm{BrO_3^-} \xrightarrow{1.50\ \mathrm{V}} \mathrm{HBrO} \xrightarrow{1.595\ \mathrm{V}} \mathrm{Br_2} \xrightarrow{1.0652\ \mathrm{V}} \mathrm{Br^-}$$

($\mathrm{BrO_3^-} \rightarrow \mathrm{Br_2}$: 1.52 V)

$$\mathrm{pH} = 14 \quad \mathrm{BrO_4^-} \xrightarrow{0.99\ \mathrm{V}} \mathrm{BrO_3^-} \xrightarrow{0.53\ \mathrm{V}} \mathrm{BrO^-} \xrightarrow{0.457\ \mathrm{V}} \mathrm{Br_2} \xrightarrow{1.0652\ \mathrm{V}} \mathrm{Br^-}$$

($\mathrm{BrO^-} \rightarrow \mathrm{Br^-}$: 0.761 V; $\mathrm{BrO_3^-} \rightarrow \mathrm{Br^-}$: 0.61 V)

compare the products of the reduction of $\mathrm{BrO_3^-(aq)}$ under standard-state conditions in acidic and alkaline media. Also briefly discuss the stability of HBrO(aq) and $\mathrm{Br_2(l)}$.

▮ In acidic media, the preferred route for the reduction of $\mathrm{BrO_3^-(aq)}$ (as indicated by the higher potential) is the direct formation of $\mathrm{Br_2(l)}$. In alkaline media, $\mathrm{BrO_3^-(aq)}$ will be reduced directly to $\mathrm{Br^-(aq)}$. If the potential to the left of a given chemical species is less than that to the right, the species will undergo disproportionation. Thus in acidic media, HBrO(aq) will form $\mathrm{BrO_3^-(aq)}$ and $\mathrm{Br_2(l)}$, and in alkaline media (but not in acidic media) $\mathrm{Br_2(l)}$ will form $\mathrm{BrO^-(aq)}$ and $\mathrm{Br^-(aq)}$.

9.61 Using the data given in Problem 9.60, prepare a Frost diagram for bromine. Use this diagram to predict the stability of HBrO(aq).

▮ A *Frost diagram* is a plot of $\Delta G°/F$ for a species relative to the free element against the oxidation state of that species. (The ordinate is sometimes called the "volt equivalent" because it is determined by multiplying a potential by the number of moles of e^- involved.) As sample calculations, consider the data given for $Br_2(l)$, HBrO(aq), and $BrO_3^-(aq)$:

$$BrO_3^- \xrightarrow{1.50\text{ V}} HBrO \xrightarrow{1.595\text{ V}} Br_2$$

The point for $Br_2(l)$ will be at oxidation state = 0 and $\Delta_r G°/F = 0$. The point for HBrO(aq) will be at oxidation state = +1 and

$$\Delta G°/F = 0 + (1\text{ mol e}^-)(1.595\text{ V}) = 1.595\text{ V}\cdot(\text{mol e}^-)$$

The point for BrO_3^- will be at oxidation state = +5 and

$$\Delta G°/F = 1.595\text{ V}\cdot(\text{mol e}^-) + (4\text{ mol e}^-)(1.50\text{ V}) = 7.60\text{ V}\cdot(\text{mol e}^-)$$

The diagram is shown in Fig. 9-9. A disproportionation reaction will occur whenever the value of $\Delta G°/F$ of a species is above a straight line joining the adjacent higher and lower oxidation states. A straight line between $Br_2(l)$ and $BrO_3^-(aq)$ (see Fig. 9-9) shows that the point for HBrO(aq) lies above it, so HBrO(aq) would be expected to form $Br_2(l)$ and $BrO_3^-(aq)$.

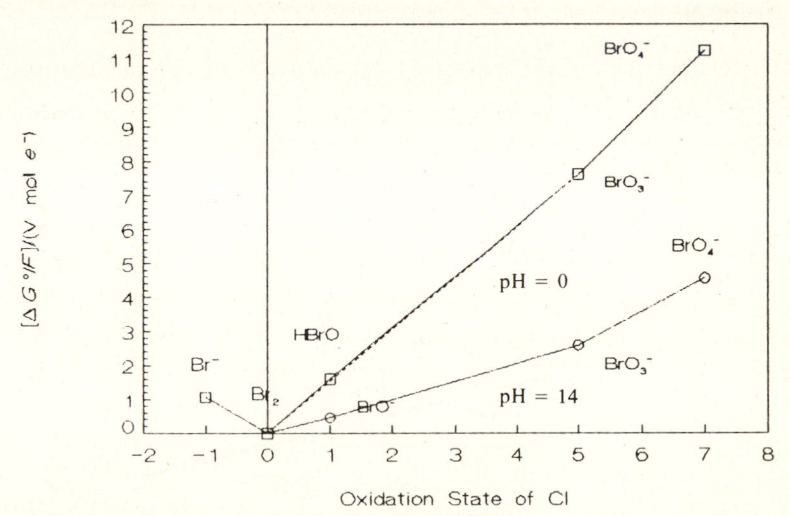

Fig. 9-9

9.62 Latimer and Frost diagrams are useful only for considering chemical reactions under conditions of pH = 0 and pH = 14. Using the data below, construct the Pourbaix potential-pH diagram for fluorine.

$$F_2(g) + 2e^- \rightarrow 2F^-(aq) \qquad E° = 2.87\text{ V}$$

$$F_2(g) + 2H^+(aq) + 2e^- \rightarrow 2HF(aq) \qquad E° = 3.06\text{ V}$$

$$HF(aq) \rightleftarrows H^+(aq) + F^-(aq) \qquad K_a = 6.5 \times 10^{-4}$$

▮ A *Pourbaix diagram* consists of a plot of E' against pH, where E' represents the standard-state potential for all substances at unit activity except $H^+(aq)$. A vertical line representing the ionization of HF(aq) will appear at

$$K_a = \frac{[H^+][F^-]}{[HF]} = \frac{[H^+](1)}{1} = 6.5 \times 10^{-4}$$

$$[H^+] = 6.5 \times 10^{-4} \qquad pH = -\log(6.5 \times 10^{-4}) = 3.19$$

For pH ≥ 3.19, a horizontal line will appear at $E' = 2.87$ V for the reduction of $F_2(g)$ to $F^-(aq)$. For pH ≤ 3.19, the reduction of $F_2(g)$ to HF(aq) will be described by

$$E' = E° - \frac{0.059\,157\text{ V}\cdot\text{mol}}{2\text{ mol}}\log\frac{[HF]^2}{P(F_2)[H^+]^2}$$

$$= E° - \frac{0.059\,157\text{ V}\cdot\text{mol}}{2\text{ mol}}\log\frac{(1)^2}{(1)[H^+]^2}$$

$$= 3.06\text{ V} - \frac{0.059\,157\text{ V}\cdot\text{mol}}{2\text{ mol}}2(-\log[H^+]) = 3.06\text{ V} - (0.59\,157\text{ V})(\text{pH})$$

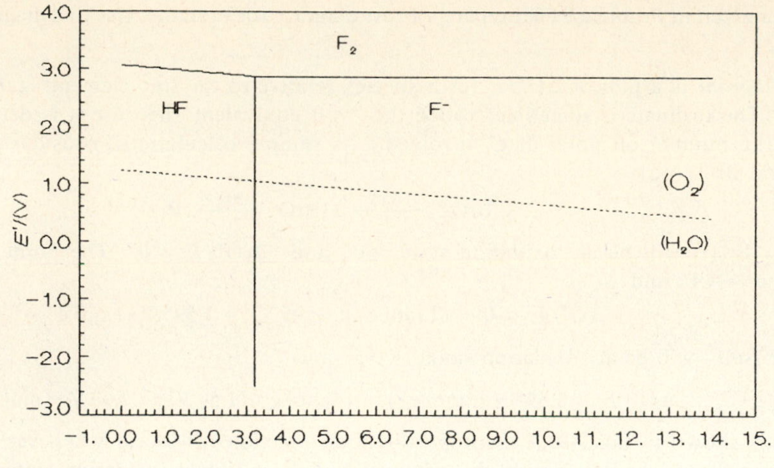

Fig. 9-10

The Pourbaix plot for fluorine appears in Fig. 9-10. [Also shown in the figure is a plot for the $O_2(g) + 4H^+(aq) + 4e^- \rightarrow 2H_2O(l)$ half-reaction, where $E° = 1.229\ V - (0.059\ 157\ V)pH$.]

9.63 Using Fig. 9-10, discuss the stability of a system of $F_2(g)$ and $H_2O(l)$ under standard-state conditions.

▌ If two Pourbaix diagrams are superimposed, information concerning the reactivity of the systems under standard-state conditions can be obtained. Consider the hypothetical superimposed diagrams for substances A and B shown in Fig. 9-11. If a mixture of the oxidized form of A is mixed with the reduced form of B at a specified pH, A will undergo reduction and B will be oxidized. From Fig. 9-10, such a situation exists with $F_2(g)$ and $H_2O(l)$, so $F_2(g)$ will react with $H_2O(l)$ to produce $O_2(g)$ and $F^-(aq)$ at $pH \geq 3.19$ or $O_2(g)$ and $HF(aq)$ at $pH \leq 3.19$. The figure shows that $F^-(aq)$ will not undergo any oxidation-reduction reactions with $O_2(g)$ or with $H_2O(l)$.

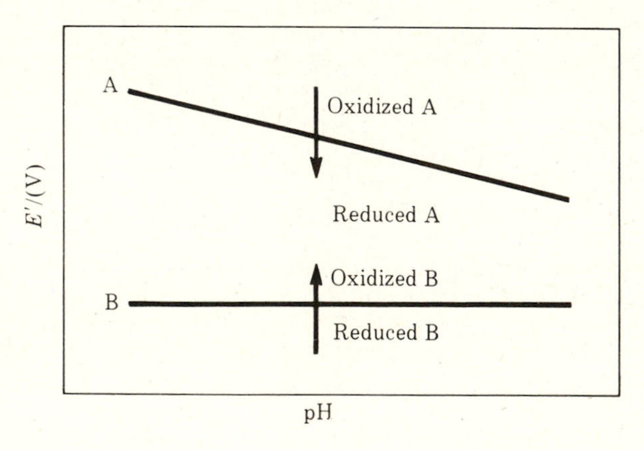

Fig. 9-11

9.64 Determine the exchange current density and the transfer coefficient for an iron electrode from the following data for the overpotential of a H^+/H_2 cathode at 25 °C:

$j/(A \cdot m^{-2})$	10	100	1000	10 000
$\eta/(V)$	0.40	0.53	0.64	0.77

▌ The relation between the current density (j) and the overpotential (η) for a one-electron exchange process is given by the *Butler–Volmer equation,*

$$j = j_e[e^{-\alpha\eta F/RT} - e^{(1-\alpha)\eta F/RT}] \tag{9.37}$$

where j_e is the exchange current density and α is the transfer coefficient (or symmetry factor). When the overpotential is large ($\eta > 75$ mV), *(9.37)* becomes

$$j = \begin{cases} j_e\, e^{(1-\alpha)\eta F/RT} & \text{(cathode)} \quad (9.38a) \\ j_e\, e^{-\alpha\eta F/RT} & \text{(anode)} \quad (9.38b) \end{cases}$$

where *(9.38a)* describes a cathode and *(9.38b)* describes an anode. Taking logarithms of both sides of *(9.38a)* gives

$$\ln j = \ln j_e + \frac{(1-\alpha)F}{RT}\eta$$

A plot of $\ln j$ against η (a *Tafel plot*) will be linear with an intercept of $\ln j_e$ and an intercept of $(1-\alpha)F/RT$. From Fig. 9-12 the slope is 18.9 V^{-1} and the intercept is −5.27, giving $j_e = 5.1 \times 10^{-3}\ A\cdot m^{-2}$ and

$$\alpha = 1 - \frac{(\text{slope})RT}{F} = 1 - \frac{(18.9\ V^{-1})(8.314\ J\cdot K^{-1}\cdot mol^{-1})(298\ K)}{9.648\times 10^4\ J\cdot V^{-1}\cdot mol^{-1}} = 0.514$$

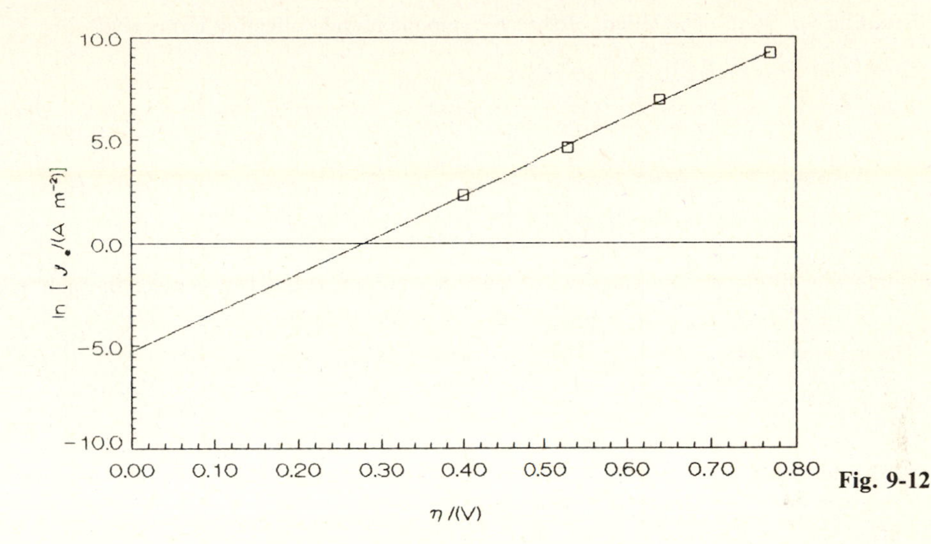

Fig. 9-12

9.65 Using the results of Problem 9.64, determine the current flow through an iron electrode of total area of 5 cm^2 given $\eta = 15$ mV.

▮ When the overpotential is small ($\eta < 75$ mV), *(9.37)* becomes

$$j = \frac{-j_e\eta F}{RT} \tag{9.39}$$

Substituting the data gives

$$j = \frac{-(5.1\times 10^{-3}\ A\cdot m^{-2})(15\ mV)[(10^{-3}\ V)/(1\ mV)](9.648\times 10^4\ J\cdot V^{-1}\cdot mol^{-1})}{(8.314\ J\cdot K^{-1}\cdot mol^{-1})(298\ K)}$$

$$= -3.0\times 10^{-3}\ A\cdot m^{-2}$$

The current will be

$$I = jA = (-3.0\times 10^{-3}\ A\cdot m^{-2})(5\ cm^2)[(10^{-4}\ m^2)/(1\ cm^2)]$$

$$= -6.0\times 10^{-7}\ A = -0.60\ \mu A$$

9.66 The work function for Pt is 5.65 eV, and the work function for Cu is 4.65 eV. What happens as a piece of Cu is placed in contact with a piece of Pt?

▮ Because the work function of Cu is less than that of Pt, a net flow of electrons will occur from the Cu to the Pt until a junction potential ($\Delta\phi$) equal to

$$\Delta\phi = 5.65\ V - 4.65\ V = 1.00\ V$$

is produced.

9.67 Write the expression for the interface potential for the $Cr_2O_7^{2-}/Cr^{3+}$ couple. Combine the expression with $2F\,\Delta\phi(Zn/Zn^{2+}) = \mu(Zn^{2+}) - \mu(Zn) + 2\mu(e^-)$ for the oxidation of Zn to obtain the expression for the electrochemical cell

$$Pt|Zn(s)|Zn^{2+}(aq)||Cr^{3+}(aq), Cr_2O_7^{2-}(aq), H^+(aq)|Pt$$

▮ Writing *(6.6)* in terms of chemical potentials for the half-reaction

$$Cr_2O_7^{2-}(aq) + 14H^+(aq) + 6e^- \rightarrow 2Cr^{3+}(aq) + 7H_2O(l)$$

gives

$$6F\,\Delta\phi(Cr_2O_7^{2-}, Cr^{3+}) = 2\mu(Cr^{3+}) + 7\mu(H_2O) - \mu(Cr_2O_7^{2-}) - 14\mu(H^+) - 6\mu(e^-)$$

Multiplying the expression for the oxidation of Zn by 3 and adding gives

$$6F[\Delta\phi(Zn/Zn^{2+}) + \Delta\phi(Cr_2O_7^{2-}, Cr^{3+})] = 3\mu(Zn^{2+}) - 3\mu(Cr^{3+}) + 6\mu(e^-) + 2\mu(Cr^{3+}) + 7\mu(H_2O) - \mu(Cr_2O_7^{2-}) - 14\mu(H^+) - 6\mu(e^-)$$

Substituting $\mu_i = \mu_i^\circ + RT\ln a_i$ for each component and collecting terms gives

$$E = \Delta\phi(Zn/Zn^{2+}) + \Delta\phi(Cr_2O_7^{2-}, Cr^{3+})$$

$$= \frac{3\mu(Zn^{2+}) - 3\mu(Zn) + 2\mu^\circ(Cr^{3+}) + 7\mu(H_2O) - \mu(Cr_2O_7^{2-}) - 14\mu(H^+)}{6F} + \frac{RT}{6F}\ln\frac{a(Zn^{2+})^3 a(Cr^{3+})^2 a(H_2O)^7}{a(Zn)^3 a(Cr_2O_7^{2-}) a(H^+)^{14}}$$

$$= E^\circ + \frac{RT}{6F}\ln\frac{a(Zn^{2+})^3 a(Cr^{3+})^2}{a(Zn)^3 a(Cr_2O_7^{2-}) a(H^+)^{14}}$$

9.68 Calculate the diffusion potential for a 0.0100 M solution of NaCl in contact through a liquid junction with a 0.0050 M solution of NaCl at 25 °C. The average transport number for Na^+ is 0.39.

▮ The general equation for a liquid junction potential (E_J) is

$$dE_J = \frac{-RT}{F}\sum\frac{t_i}{z_i}\,d\ln a_i \qquad (9.40)$$

where the summation is performed over both ions. For the same electrolyte on both sides of the liquid junction, *(9.40)* becomes

$$E_J = \frac{-RT}{F}\left[\frac{t_+}{z_+} + \frac{t_-}{z_-}\right]\ln\frac{a_{\pm,2}}{a_{\pm,1}} \qquad (9.41)$$

Assuming *(6.51)* to be valid

$$\log y_\pm = -(1)(1)(0.5116)(0.0050)^{1/2} = -0.036$$

$$y_\pm = 0.920$$

for the 0.0050 M solution and $\log y_\pm = -0.051$ and $y_\pm = 0.889$ for the 0.0100 M solution. Substituting *(9.19)* and the data into *(9.41)* gives

$$E_J = \frac{-(8.314\ \mathrm{J \cdot K^{-1} \cdot mol^{-1}})(298\ \mathrm{K})}{9.648 \times 10^4\ \mathrm{J \cdot V^{-1} \cdot mol^{-1}}}\left(\frac{0.39}{1} + \frac{1 - 0.39}{-1}\right)\ln\frac{(0.920)(0.0050)}{(0.889)(0.0100)}$$

$$= -0.0037\ \mathrm{V}$$

9.69 Calculate the liquid junction potential for a 0.010 M solution of HCl in contact with a 0.010 M solution of NaCl at 25 °C. The molar conductances are $1.1851 \times 10^{-2}\ \mathrm{S \cdot m^2 \cdot mol^{-1}}$ for NaCl and $4.1200 \times 10^{-2}\ \mathrm{S \cdot m^2 \cdot mol^{-1}}$ for HCl.

▮ For 1:1 electrolytes at the same concentration having an ion in common, *(9.40)* becomes

$$E_J = \frac{RT}{F}\ln\frac{\Lambda_2}{\Lambda_1} \qquad (9.42)$$

Substituting the data into *(9.42)* gives

$$E_J = \frac{(8.314\ \mathrm{J \cdot K^{-1} \cdot mol^{-1}})(298\ \mathrm{K})}{9.648 \times 10^4\ \mathrm{J \cdot V^{-1} \cdot mol^{-1}}}\ln\frac{4.1200 \times 10^{-2}\ \mathrm{S \cdot m^2 \cdot mol^{-1}}}{1.1851 \times 10^{-2}\ \mathrm{S \cdot m^2 \cdot mol^{-1}}} = 0.032\,01\ \mathrm{V}$$

9.70 Determine the cell potential for the following concentration cell at 25 °C:

$$Pt|H_2(1.00\ \mathrm{bar})|HCl(aq)|H_2(0.50\ \mathrm{bar})|Pt$$

▮ The overall chemical equation and standard state potential is

$$H_2(1.00\ \mathrm{bar}) \rightarrow H_2(0.50\ \mathrm{bar}) \qquad (E^\circ = 0)$$

with $n = 2 \text{ mol e}^-$. This cell has no liquid junction, so *(9.36)* is used, giving

$$E = 0 - \frac{0.059\ 157\ \text{V}\cdot\text{mol}}{2\ \text{mol}} \log\frac{0.50\ \text{bar}}{1.00\ \text{bar}} = 8.9 \times 10^{-3}\ \text{V}$$

9.71 Compare the cell potential at 25 °C for the cells

$$\text{Ag(s)}|\text{AgCl(s)}|\text{HCl(0.1000 M)}||\text{HCl(0.0100 M)}|\text{AgCl(s)}|\text{Ag(s)}$$

and

$$\text{Ag(s)}|\text{AgCl(s)}|\text{HCl(0.1000 M)}|\text{HCl(0.0100 M)}|\text{AgCl(s)}|\text{Ag(s)}$$

given $y_\pm = 0.798$ and $t(\text{H}^+) = 0.8314$ for 0.1000 M HCl and $y_\pm = 0.906$ and $t(\text{H}^+) = 0.8251$ for 0.0100 M HCl.

▮ The presence of the salt bridge in the first cell is to eliminate the liquid junction potential. Using *(9.36)* gives

$$E = -\frac{0.059\ 157\ \text{V}\cdot\text{mol}}{1\ \text{mol}} \log\frac{(0.906)^2(0.0100)^2}{(0.798)^2(0.1000)^2} = 0.112\ \text{V}$$

The second cell has a liquid junction present. For this cell, *(9.36)* becomes

$$E = -\frac{0.059\ 157\ \text{V}\cdot\text{mol}}{n} t_i \log Q \qquad (9.43)$$

where t_i is the average transport number of the spectator ion—in this case, $\text{H}^+\text{(aq)}$. Substituting into *(9.43)* gives

$$E = -\frac{0.059\ 157\ \text{V}\cdot\text{mol}}{1\ \text{mol}}\left(\frac{0.8314 + 0.8251}{2}\right) \log\frac{(0.906)^2(0.0100)^2}{(0.798)^2(0.1000)^2} = 0.093\ \text{V}$$

The difference between these values is the liquid junction potential

$$E_J = 0.112\ \text{V} - 0.093\ \text{V} = 0.019\ \text{V}$$

which can also be calculated using *(9.41)*.

9.72 For the thermocell

$$\text{Cl}_2\text{(g)}\ \underset{T_1}{|}\ \text{AgCl(l)}\ \underset{T_2}{|}\ \text{Cl}_2\text{(g)}$$

operating between 800 K and 1200 K, $dE/dT = -7.0 \times 10^{-4}\ \text{V}\cdot\text{K}^{-1}$. Find E for this cell operating and $T_1 = 850\ \text{K}$ and $T_2 = 1050\ \text{K}$.

▮ The potential of a thermocell is given by

$$E = -\int_{T_1}^{T_2} \left(\frac{dE}{dT}\right) dT \qquad (9.44)$$

Substituting the data and integrating gives

$$E = -(-7.0 \times 10^{-4}\ \text{V}\cdot\text{K}^{-1})(1050\ \text{K} - 850\ \text{K}) = 0.14\ \text{V}$$

CHAPTER 10
Heterogeneous Equilibria

10.1 PHASE RULE

10.1 Consider a gaseous mixture of $^{35}Cl_2O$ and $^{37}Cl_2O$. Determine the number of **(a)** phases, **(b)** components, and **(c)** degrees of freedom in the system.

I **(a)** A *phase* is the portion of the system that is submacroscopically homogeneous and is separated from other such portions by definite physical boundaries. In this case, the number of phases (p) is 1, a gas phase.

(b) The *number of components* is the minimum number of independently variable chemical species necessary to describe the composition of each phase. In this case, the number of components (c) is two—$^{35}Cl_2O$ and $^{37}Cl_2O$. Even though a third substance, $^{35}Cl^{37}ClO$, is present, it is not considered a component because its presence and amount would be easily calculated using standard equilbrium techniques from the equation

$$^{35}Cl_2O(g) + {}^{37}Cl_2O(g) \rightleftharpoons 2\,{}^{35}Cl^{37}ClO(g)$$

(c) The *number of degrees of freedom* is the minimum number of intensive variables needed to fix the values of all the remaining intensive variables. It is given by the *Gibbs phase rule* as

$$f = c - p + 2 \tag{10.1}$$

where f is the number of degrees of freedom. In this case, $f = 2 - 1 + 2 = 3$. The variables chosen might be T, P, and $x(^{35}Cl_2O)$.

10.2 Consider a system consisting of solid, liquid, and gaseous acetone at equilibrium. Determine p, c, and f for this system.

I The system consists of three phases (pure solid acetone, pure liquid acetone, and pure gaseous acetone) and one component (acetone). Substituting into *(10.1)* gives $f = 1 - 3 + 2 = 0$. A single component at its triple point is commonly called an *invariant system*.

10.3 Determine p, c, and f for a system consisting of an aqueous solution of acetone in equilibrium with crushed ice.

I The number of phases in the system is two (solid and aqueous solution), and the number of components is two (water and acetone). Substituting into *(10.1)* gives

$$f = 2 - 2 + 2 = 2$$

The variables chosen to describe the system might be T and P.

10.4 Determine p, c, and f for an aqueous solution of HCl in equilibrium with its vapor. Repeat the problem for an aqueous solution of HCl and HI.

I In both cases, $p = 2$ (vapor and aqueous solution). For the HCl solution, $c = 2$ (HCl and H_2O), and for the mixture of acids, $c = 3$ (HCl, HI, and H_2O). Substituting into *(10.1)* gives

$$f(\text{HCl}) = 2 - 2 + 2 = 2 \qquad f(\text{HCl, HI}) = 3 - 2 + 2 = 3$$

Any two variables such as T and P, or T and $x(H_2O)$ could be chosen to describe the HCl solution, and any three variables such as T, P, and $x(\text{HCl})$ could be chosen to describe the HCl-HI mixture.

10.5 What are the chemical species present in a dilute aqueous solution of H_2SO_4? Determine p, c, and f for this system.

I The important chemical species are $H^+(aq)$, $HSO_4^-(aq)$, $SO_4^{2-}(aq)$, and $OH^-(aq)$. These species are related by several equilibrium expressions, and the number of components is two (H_2O and H_2SO_4). The number of phases is one. Substituting into *(10.1)* gives

$$f = 2 - 1 + 2 = 3$$

The variables T, P, and $x(H_2SO_4)$ can be used to describe the system.

10.6 An equimolar mixture of nicotine and water will separate into two layers, each consisting of a saturated solution. Determine p, c, and f for such a system. Repeat the problem for such a mixture in equilibrium with a vapor phase.

▮ For both systems, $c = 2$ (nicotine and H_2O). For the first system, $p = 2$ (H_2O saturated with nicotine and nicotine saturated with H_2O), and for the second system, $p = 3$ (H_2O saturated with nicotine, nicotine saturated with H_2O, and a gaseous mixture of nicotine and H_2O). Substituting into *(10.1)* gives

$$f = 2 - 2 + 2 = 2$$

for the two-phase system and

$$f = 2 - 3 + 1 = 1$$

for the three-phase system. For the *divariant system* ($f = 2$), the chosen variables could be T and P, and for the *univariant system* ($f = 1$), the chosen variable could be T.

10.7 Consider a magic catalyst that allows solid C to react with $O_2(g)$ to form gaseous CO, CO_2, C_3O_2, C_5O_2, and $C_{12}O_9$. Determine c, p, and f for this system once equilibrium has been reached.

▮ The system consists of two phases (gaseous mixture and solid C) and of two components (C and O_2). All of the other species present are related to these through various equilibrium expressions. Substituting into *(10.1)* gives

$$f = 2 - 2 + 2 = 2$$

The chosen variables could be T and P.

10.8 Consider the chemical equation

$$CuSO_4{\cdot}5H_2O(s) \rightarrow CuSO_4(s) + 5H_2O(g)$$

Assuming that all of the hydrated salt decomposes, find c, p, and f for the system. Repeat this problem for an equilibrium system containing all three chemical species.

▮ In both cases, $c = 2$. In the first case, the choice of components might be $CuSO_4$ and H_2O, and in the second case, the choice might be $CuSO_4{\cdot}5H_2O$ and $CuSO_4$ (H_2O is related by the value of K for the equation). In the first case, $p = 2$ ($CuSO_4$ and H_2O), and in the second case, $p = 3$ ($CuSO_4{\cdot}5H_2O$, $CuSO_4$, and H_2O). Substituting into *(10.1)* gives

$$f(\text{decomposition}) = 2 - 2 + 2 = 2 \qquad f(\text{equilibrium}) = 2 - 3 + 2 = 1$$

The choices of variables might be T and P for the first system and T for the second system.

10.9 Consider an aqueous solution of $AlCl_3$. Determine c, p, and f for this system. Repeat the problem for a concentrated solution in which $Al(OH)_3(s)$ has precipitated from the solution.

▮ For both systems, $c = 2$ (H_2O and $AlCl_3$). For the simple solution, $p = 1$ (aqueous $AlCl_3$), and for the heterogeneous system, $p = 2$ [$Al(OH)_3(s)$ and aqueous $AlCl_3$ with excess $H^+(aq)$ and $Cl^-(aq)$ from the hydrolysis]. Substituting into *(10.1)* gives

$$f(\text{homogeneous system}) = 2 - 1 + 2 = 3$$

$$f(\text{heterogeneous system}) = 2 - 2 + 2 = 2$$

Possible choices of variables would be T, P, and $x(AlCl_3)$ for the homogeneous system and T and P for the heterogeneous system.

10.10 Determine p, c, and f for the equilibrium system $FeO(s) + CO(g) \rightleftharpoons CO_2(g) + Fe(s)$.

▮ Only three components are needed to describe the system (FeO, Fe, and either CO or CO_2). The number of phases is three (solid FeO, solid Fe, and the gaseous mixture of CO and CO_2). Substituting into *(10.1)* gives

$$f = 3 - 3 + 2 = 2$$

The variables T and P could be chosen to describe the system.

10.2 PHASE DIAGRAMS FOR ONE-COMPONENT SYSTEMS

10.11 What is the maximum number of phases that can be in equilibrium in a one-component system?

▮ Substituting $f = 0$ and $c = 1$ into *(10.1)* and solving for p gives

$$0 = 1 - p + 2 \qquad p = 3$$

No more than three phases can be in equilibrium.

10.12 A 0.25-g sample of CCl_4 was introduced into a 0.25-L closed system at 23.0 °C. If the vapor pressure of CCl_4 is 100.0 torr at this temperature, describe the system once equilibrium has been reached.

▮ Assuming ideal gas behavior, *(1.18)* gives

$$n = \frac{(100.0\ \text{torr})[(1\ \text{atm})/(760\ \text{torr})](0.25\ \text{L})}{(0.0821\ \text{L}\cdot\text{atm}\cdot\text{K}^{-1}\cdot\text{mol}^{-1})(298\ \text{K})} = 1.4\times10^{-3}\ \text{mol}$$

or $m = 0.22$ g. If liquid CCl_4 was introduced, 0.22 g would vaporize and establish a liquid–vapor equilibrium with 0.03 g of liquid remaining, or if gaseous CCl_4 was introduced, 0.03 g would condense and establish a liquid–vapor equilibrium with 0.22 g of vapor.

10.13 Use the following data to prepare a phase diagram for O_2:

	triple point	**boiling point**	**melting point**	**critical point**	**vapor pressure of solid**
***T*/(K)**	54.4	90.2	54.8	154.6	54.1
***P*/(bar)**	0.0015	1.0133	1.0133	50.43	0.0013

	vapor pressure of liquid									
***T*/(K)**	60	70	80	90	100	110	120	130	140	150
***P*/(bar)**	0.0071	0.062	0.301	0.994	2.542	5.434	10.216	17.478	27.865	42.190

▮ The phase diagram for a one-component system is a plot of P against T. The line separating the solid–vapor areas not visible in the figure is generated by plotting the vapor pressure data of the solid from the lowest temperature to the triple point. The line separating the liquid–vapor areas is generated by plotting the vapor pressure data of the liquid from the triple point to the critical point. The nearly vertical line separating the liquid–solid areas is generated by plotting the melting point data to the triple point, and the vertical line representing the critical isotherm runs from the critical point to higher pressures. The term *vapor* is often used to describe the gaseous form of a substance at temperatures below the critical temperature. The phase diagram is shown in Fig. 10-1.

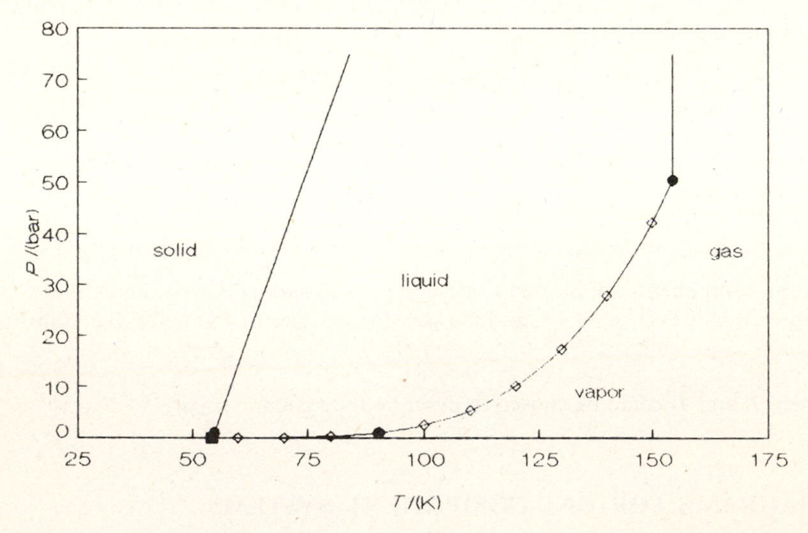

Fig. 10-1

10.14 Describe the changes in the carbon system shown in Fig. 10-2 for **(*a*)** isobarically heating from point *a* and **(*b*)** isothermally compressing from point *a*.

▮ **(*a*)** The graphite will simply increase in temperature until it reaches 3900 K, at which point it sublimes. The resulting carbon vapor will increase in temperature.

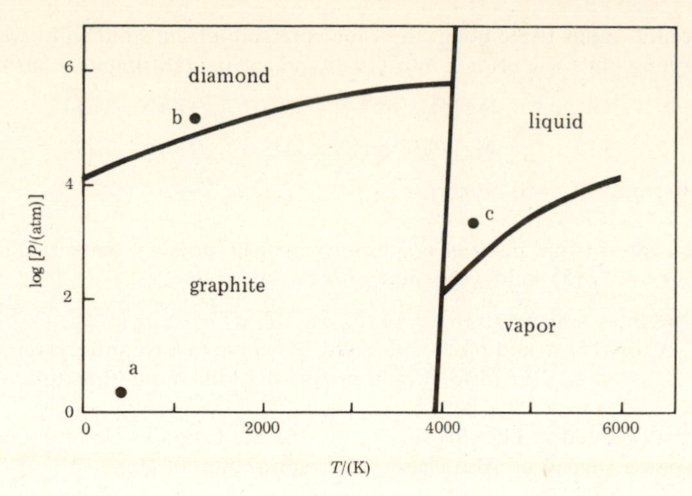

Fig. 10-2

(*b*) The graphite will decrease in volume slightly as the pressure is increased until it reaches 2×10^4 atm. At this pressure the crystalline graphite will undergo a solid–solid phase change to diamond (provided sufficient time is allowed). As the pressure is increased further, the diamond will be compressed.

10.15 Describe the changes in the carbon system shown in Fig. 10-2 for (*a*) isobarically heating from point *b* and (*b*) isothermally decreasing the pressure from point *c*.

▮ (*a*) The diamond will increase in temperature until it reaches 2000 K, at which point it will undergo a phase change to graphite (if sufficient time is allowed). The graphite will increase in temperature until it reaches 4000 K, at which point it will melt. Once the sample has melted, the liquid formed will increase in temperature.

(*b*) The liquid carbon will expand slightly as the pressure is reduced. Once the pressure is 300 atm, the liquid will vaporize. As the pressure is further reduced, the volume of the gas increases.

10.16 The triple point of CO_2 is 5 bar and −57 °C, the critical point is 75 bar and 31 °C, and the sublimation temperature at 1 bar is −78 °C. What will be the physical state of CO_2 at (*a*) 0.5 bar and − 70 °C, (*b*) 25 bar and 25 °C, and (*c*) 70 bar and 25 °C?

▮ Although a very rough P–T phase diagram could be prepared from the given data, it would not be accurate enough to answer question (*c*). An alternative phase diagram consisting of a plot of ln P against $1/T$ can easily be prepared using straight lines based on *(4.19a)*; see Fig. 10-3. The three data points are shown on the phase diagram, giving (*a*) vapor, (*b*) vapor, and (*c*) liquid.

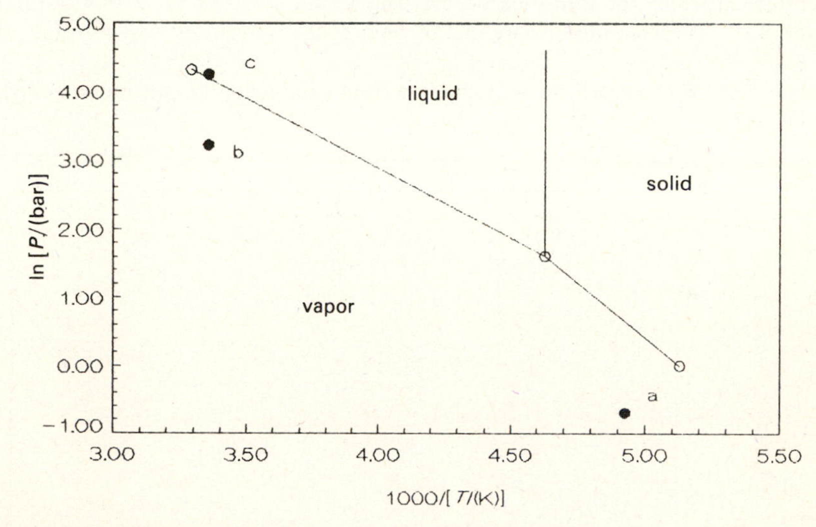

Fig. 10-3

10.17 The vapor pressure of Cd is given by

$$\log [P/(\text{torr})] = a/[T/(\text{K})] - b \tag{10.2}$$

where $a = 5693$ and $b = 8.564$ for solid Cd from 150 °C to 321 °C and $a = 5218$ and $b = 7.897$ for liquid Cd from 500 °C to 840 °C. Calculate the triple point for Cd from these data.

▮ At the solid–liquid–vapor triple point, the vapor pressure of the solid will be equal to the vapor pressure of the liquid. Substituting both sets of data into *(10.2)* and solving the simultaneous equations

$$\log [P_{tp}/(\text{torr})] = 5693/[T_{tp}/(\text{K})] - 8.564$$

$$\log [P_{tp}/(\text{torr})] = 5218/[T_{tp}/(\text{K})] - 7.897$$

gives $T_{tp} = 712\ \text{K}$ and $P_{tp} = 0.270\ \text{torr}$.

10.18 At the solid–liquid–vapor triple point of a substance, which curve will have the greater slope: **(*a*)** solid–vapor line or liquid–vapor line? **(*b*)** solid–vapor line or solid–liquid line?

▮ The slope of the lines will be given by *(4.15)* as $dP/dT = \Delta_{I \to II}H/\Delta_{I \to II}V$. **(*a*)** Because $\Delta_{sub}H > \Delta_{vap}H$ and $\Delta_{sub}V \approx \Delta_{vap}V$, *(4.15)* would predict the solid–gas curve to have the greater slope. **(*b*)** Although $\Delta_{fus}H < \Delta_{sub}H$, because $\Delta_{fus}V \ll \Delta_{sub}V$, *(4.15)* would predict the solid–liquid line to have the greater slope.

10.19 Using the C_P°–T data plotted in Fig. 10-4 for Hg, classify the fusion of Hg as either a first-order, a second-order, or a higher-order phase transition. Also classify the vaporization of Hg.

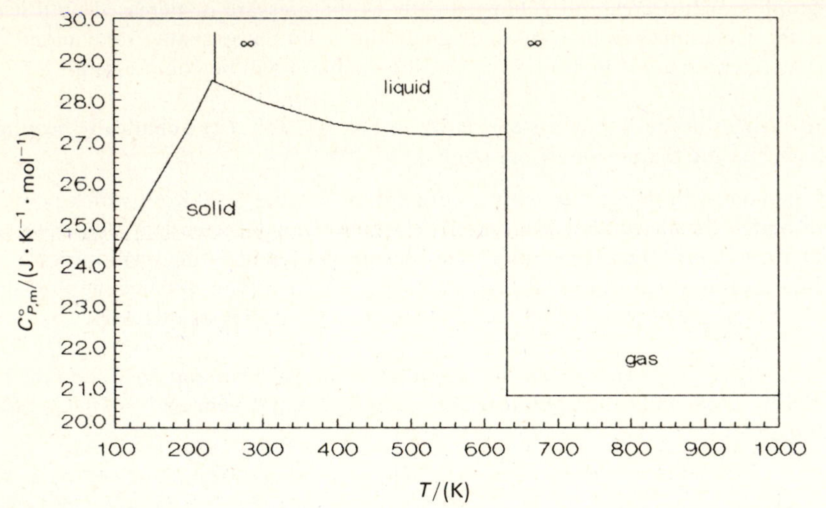

Fig. 10-4

▮ A *first-order phase transition* is characterized by $\Delta_{I \to II}H^\circ \neq 0$, $\Delta_{I \to II}S^\circ \neq 0$, $\Delta_{I \to II}V \neq 0$, and $C_{P,m}^\circ = \infty$ at the temperature at which the transition occurs. The reason $C_{P,m}^\circ = \infty$ is that during the transition $dT = 0$ but $đq = dH \neq 0$. These conditions are met by both of the transitions shown in Fig. 10-4.

10.20 Repeat Problem 10.19 for "white" Sn as it changes from conducting to superconducting at 3.72 K. See Fig. 10-5 for C_P°–T data.

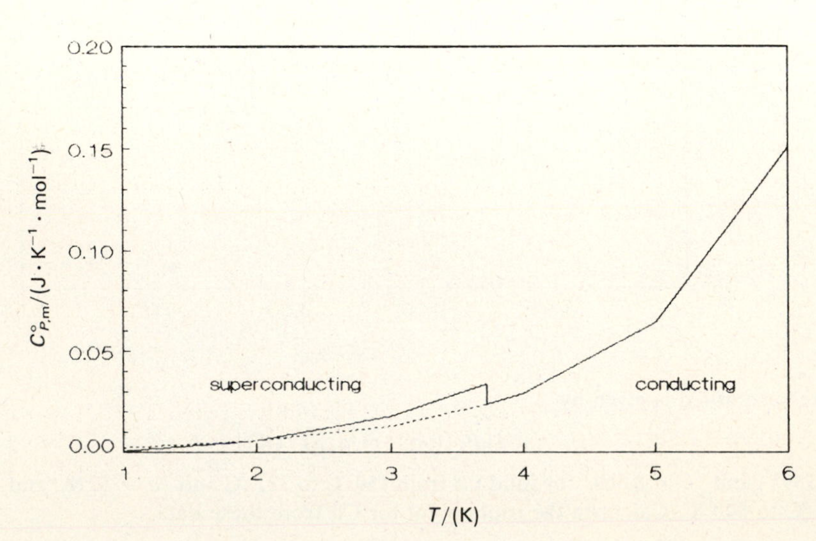

Fig. 10-5

▮ A *second-order phase transition* is characterized by $\Delta_{I\to II}H^\circ = \Delta_{I\to II}S^\circ = \Delta_{I\to II}V = 0$ and $C^\circ_{P,m} \neq \infty$. Although the $C^\circ_{P,m}$–T plot for Sn is discontinuous, all the values remain finite, and this change is an example of a second-order transition.

10.21 Derive an expression for dP/dT for a first-order phase transition.

▮ For the equilibrium between two phases undergoing a first-order transition $\mu_I = \mu_{II}$, giving $d\mu_I = d\mu_{II}$. Substituting the total differentials for $d\mu_i$ gives

$$\left(\frac{\partial \mu_I}{\partial P}\right)_T dP + \left(\frac{\partial \mu_I}{\partial T}\right)_P dT = \left(\frac{\partial \mu_{II}}{\partial P}\right)_T dP + \left(\frac{\partial \mu_{II}}{\partial T}\right)_P dT$$

Solving for dP/dT gives

$$\frac{dP}{dT} = \frac{-[(\partial \mu_{II}/\partial T)_P - (\partial \mu_I/\partial T)_P]}{(\partial \mu_{II}/\partial P)_T - (\partial \mu_I/\partial P)_T}$$

Substituting *(6.8)*, *(6.15a)*, and *(5.8)* gives

$$\frac{dP}{dT} = \frac{S_{II} - S_I}{V_{II} - V_I} = \frac{\Delta_{I\to II}S}{\Delta_{I\to II}V} = \frac{\Delta_{I\to II}H}{T(\Delta_{I\to II}V)}$$

10.22 Derive an expression for dP/dT for a second-order phase transition.

▮ As described in Problem 10.20, for the equilibrium between two phases undergoing a second-order transition, $V_I = V_{II}$, giving $dV_I = dV_{II}$. Substituting the total differentials for dV_i gives

$$\left(\frac{\partial V_I}{\partial P}\right)_T dP + \left(\frac{\partial V_I}{\partial T}\right)_P dT = \left(\frac{\partial V_{II}}{\partial P}\right)_T dP + \left(\frac{\partial V_{II}}{\partial T}\right)_P dT$$

Solving for dP/dT gives

$$\frac{dP}{dT} = \frac{-[(\partial V_{II}/\partial T)_P - (\partial V_I/\partial T)_P]}{(\partial V_{II}/\partial P)_T - (\partial V_I/\partial P)_T}$$

Substituting *(1.8)* and *(1.11)* gives the *Ehrenfest equation*

$$\frac{dP}{dT} = \frac{\alpha_{II} - \alpha_I}{\kappa_{II} - \kappa_I} = \frac{\Delta_{I\to II}\alpha}{\Delta_{I\to II}\kappa} \qquad (10.3)$$

10.3 PHASE DIAGRAMS FOR TWO-COMPONENT SYSTEMS

10.23 Use the following data to prepare a liquid–liquid phase diagram for the formic acid–benzene system.

$T/(^\circ C)$	25	30	40	50	60	70	73.2
$w(C_6H_6)$ **in formic acid-rich phase**	0.141	0.151	0.178	0.210	0.257	0.351	0.518
$w(C_6H_6)$ **in benzene-rich phase**	0.904	0.896	0.870	0.834	0.779	0.686	0.518

where w_i is the mass fraction of component *i*.

▮ The data indicate that below 73.2 °C the liquids are only partially miscible and form two layers once equilibrium is reached—a saturated solution of C_6H_6 in HCOOH (the formic acid-rich phase) and a saturated solution of HCOOH in C_6H_6 (the benzene-rich phase). The plot of T as a function of $w(C_6H_6)$ is shown in Fig. 10-6. The respective data points are connected in the figure by horizontal *tie lines*.

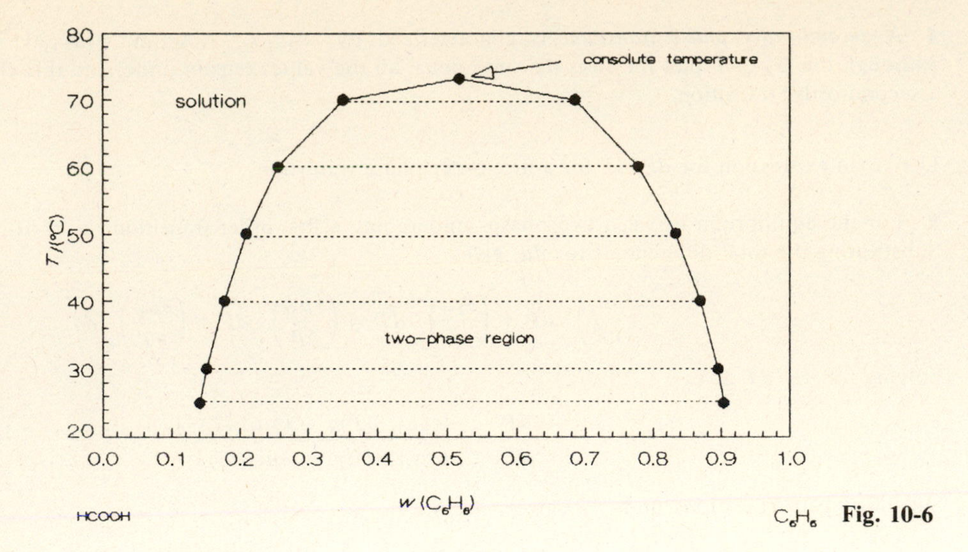

Fig. 10-6

10.24 Describe the equilibrium conditions at 50 °C for a mixture containing 50.0 g of C_6H_6 and 50.0 g of HCOOH. See Problem 10.23 for additional data.

▌ The system will consist of two phases. The composition of the formic acid-rich phase will be $w_1(C_6H_6) = 0.210$, and the composition of the benzene-rich phase will be $w_2(C_6H_6) = 0.834$. The respective mass of each phase is related to the total mass (m_0) by

$$m_1 + m_2 = m_0 \tag{10.4}$$

and the ratio of the masses of each phase is given by

$$\frac{m_1}{m_2} = \frac{w_{i,2} - w_{i,0}}{w_{i,0} - w_{i,1}} \tag{10.5}$$

Substituting the data into *(10.4)* and *(10.5)* gives the two simultaneous equations

$$m_1 + m_2 = 100.0 \text{ g}$$

$$\frac{m_1}{m_2} = \frac{0.834 - 0.500}{0.500 - 0.210}$$

which upon solving give $m_1 = 53.5$ g and $m_2 = 46.5$ g.

10.25 Using Fig. 10-7, describe what will happen as 1000 g of 2,6-dimethylpyridine (C_7H_9N) is added in 1-g aliquots to a 100-g sample of water at 25 °C. What will be observed for the addition process if it is carried out at 125 °C?

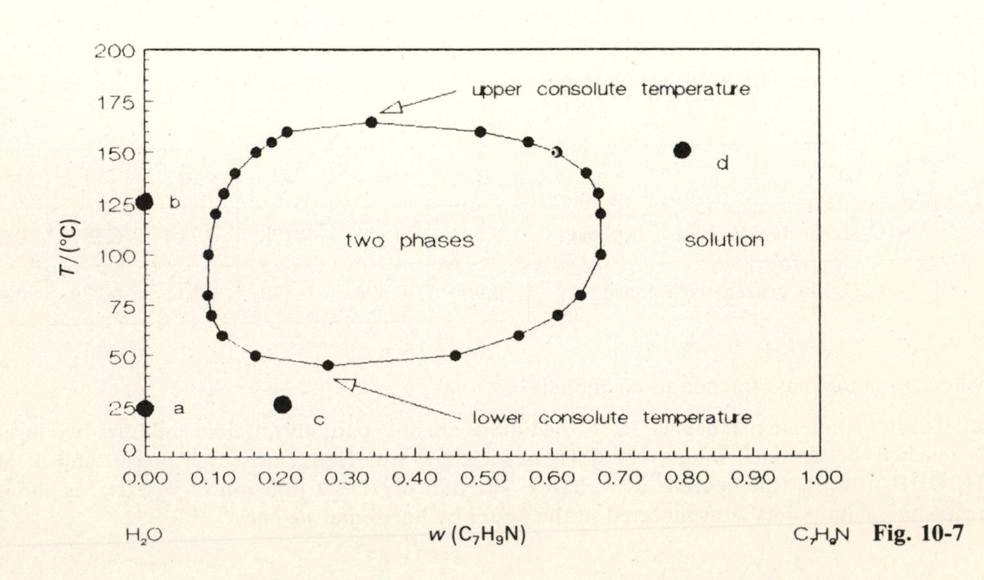

Fig. 10-7

▮ Moving horizontally from point *a* shows that the added C_7H_9N will simply dissolve in the H_2O, forming a solution. Likewise, moving horizontally from point *b* shows that the initial aliquots of C_7H_9N will dissolve in the water to form a solution, but once the concentration of the solution reaches $w(C_7H_9N) = 0.12$, a second phase will appear. The concentration of this second phase is $w(C_7H_9N) = 0.67$. As more C_7H_9N is added, these two phases remain in equilibrium with no changes in composition but in mass proportions given by *(10.5)*. After the addition of nearly 200 g of C_7H_9N, the system becomes a solution once more, with the C_7H_9N essentially acting as the solvent and the H_2O acting as the solute. Further addition of C_7H_9N simply results in a solution in which $w(C_7H_9N)$ is increasing.

10.26 Describe what will happen to a system of H_2O-C_7H_9N at point *c* in Fig. 10-7 as the system is heated.

▮ Initially the temperature of the solution will increase. Once the system reaches 50 °C, the system changes into a two-phase system—a saturated solution of H_2O in C_7H_9N and a saturated solution of C_7H_9N in H_2O. As the temperature increases, the composition and the relative amounts of the saturated solutions change. Once the system reaches a temperature of 160 °C, the two components are completely miscible and a one-phase system [$w(C_7H_9N) = 0.20$] results. The temperature of the solution will increase until liquid–vapor equilibrium becomes important.

10.27 Point *c* in Fig. 10-7 is located between the upper and lower consolute temperatures. Does this mean that a mixture of 80 g of C_7H_9N and 20 g of H_2O at 150 °C will form a two-phase system?

▮ No, point *d* is located in the one-phase region of the diagram.

10.28 A sample of N_2O_4 is decomposed to form N_2 and O_2, and the resulting gaseous mixture is cooled to 70 K. As the temperature of the liquid mixture is increased to the boiling point, describe the liquid–vapor equilibrium using Fig. 10-8.

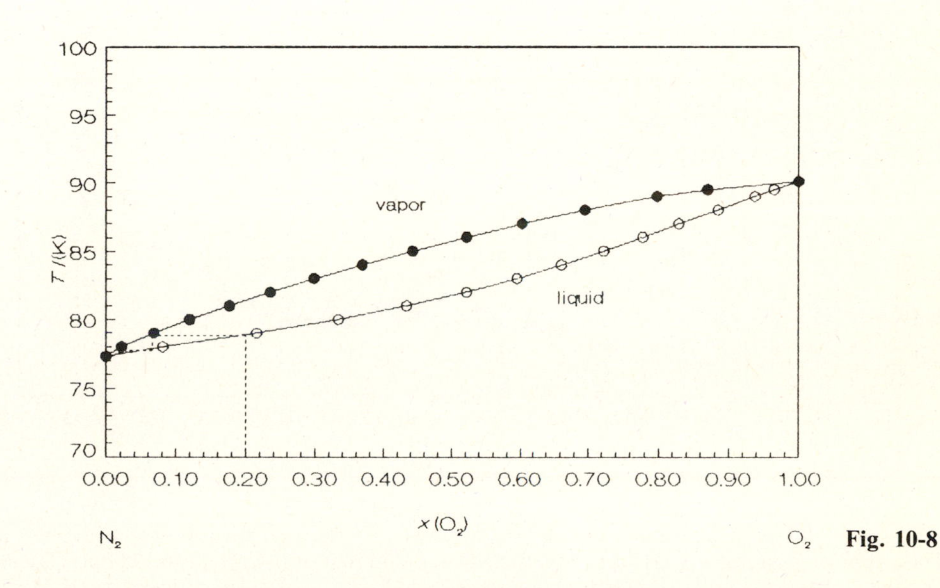

Fig. 10-8

▮ The decomposition mixture will produce a mixture of gases in the ratio of 2 mol of O_2 to 1 mol of N_2, or $x(O_2) = 0.67$. A horizontal tie line would indicate that a liquid mixture with composition of $x(O_2) = 0.67$ will establish an equilibrium with a vapor of composition $x(O_2)_{vap} = 0.37$ at 84 K. The vapor is richer in the more volatile component—N_2.

10.29 How many theoretical equivalent plates are needed to distill N_2 from air?

▮ Assuming $x(O_2) = 0.20$ for air, Fig. 10-8 shows that liquid air will establish a liquid–vapor equilibrium with a gas of composition $x(O_2)_{vap} = 0.07$ at 79 K as seen by the horizontal dashed tie line. If this vapor is cooled at 78 K, a new liquid–vapor equilibrium is established with a gas of composition $x(O_2)_{vap} = 0.01$. Cooling this vapor and allowing one more liquid–vapor equilibrium to be established will generate gaseous N_2 as the product. Each of those liquid–vapor equilibrium steps is known as a *theoretical equivalent plate* (TEP), and three TEPs are needed for this process.

10.30 Use the following temperature–composition data

$T/(°C)$	78.3	76.6	75.5	73.9	72.8	72.1	71.8
$x(CH_3COOCH_2CH_3)_{liq}$	0.000	0.050	0.100	0.200	0.300	0.400	0.500
$x(CH_3COOCH_2CH_3)_{vap}$	0.000	0.102	0.187	0.305	0.389	0.457	0.516

$T/(°C)$	71.8	71.9	72.2	73.0	74.7	76.0	77.1
$x(CH_3COOCH_2CH_3)_{liq}$	0.540	0.600	0.700	0.800	0.900	0.950	1.000
$x(CH_3COOCH_2CH_3)_{vap}$	0.540	0.576	0.644	0.726	0.837	0.914	1.000

to prepare a liquid-vapor phase diagram for the ethanol-ethyl acetate system at 1 atm. Is it possible to distill pure ethyl acetate from a mixture containing $x(CH_3COOCH_2CH_3) = 0.25$?

▌ The phase diagram is shown in Fig. 10-9. The data points are related by horizontal tie lines (not shown). The distillation process would increase the concentration of ethyl acetate until the *azeotrope* at $x(CH_3COOCH_2CH_3) = 0.540$ is reached. At this point, the compositions of the liquid and vapor are identical, so pure ethyl acetate cannot be produced by this distillation process.

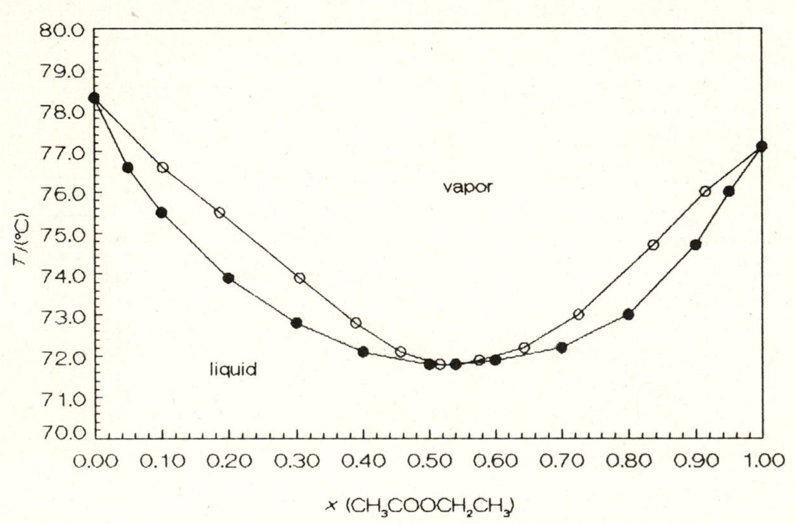

Fig. 10-9

10.31 The vapor pressure of n-C_5H_{12} at 25 °C is 0.674 bar, and that of n-C_6H_{14} is 0.198 bar. Calculate the partial pressure of each substance and the total pressure over a solution containing $x(n\text{-}C_5H_{12}) = 0.25$. Assume that these substances form an ideal solution.

▌ The partial pressure of each substance over an ideal solution is given by *Raoult's law* as

$$P_i = P_i^\circ x_i \tag{10.6}$$

where P_i° is the vapor pressure of the pure substance. Substituting the data into *(10.6)* gives

$$P(n\text{-}C_5H_{12}) = (0.674 \text{ bar})(0.25) = 0.17 \text{ bar}$$

$$P(n\text{-}C_6H_{14}) = (0.198)(0.75) = 0.15 \text{ bar}$$

Substituting into *(1.22)* gives

$$P_t = 0.17 \text{ bar} + 0.15 \text{ bar} = 0.32 \text{ bar}$$

10.32 What is the composition of the vapor phase in equilibrium with the solution described in Problem 10.31?

▮ Substituting the results of Problem 10.31 into *(1.23)* gives

$$x(n\text{-}C_5H_{12})_{vap} = \frac{0.17\text{ bar}}{0.32\text{ bar}} = 0.53$$

for the vapor.

10.33 Prepare a pressure-composition diagram for the n-C_5H_{12}–n-C_6H_{14} system described in Problem 10.31.

▮ The partial pressure of each component will be represented by a straight line given by *(10.6)* as

$$P(n\text{-}C_5H_{12}) = x(n\text{-}C_5H_{12})P°(n\text{-}C_5H_{12}) = [1 - x(n\text{-}C_6H_{14})]P°(n\text{-}C_5H_{12})$$

$$P(n\text{-}C_6H_{14}) = x(n\text{-}C_6H_{14})P°(n\text{-}C_6H_{14})$$

and the total pressure above the solution is represented by a straight line given by *(1.22)* as

$$P_t = [1 - x(n\text{-}C_6H_{14})]P°(n\text{-}C_5H_{12}) + x(n\text{-}C_6H_{14})P°(n\text{-}C_6H_{14})$$

$$= P°(n\text{-}C_5H_{12}) + [P°(n\text{-}C_6H_{14}) - P°(n\text{-}C_5H_{12})]x(n\text{-}C_6H_{14})$$

As a sample calculation, consider $x(n\text{-}C_6H_{14}) = 0.750$.

$$P(n\text{-}C_5H_{12}) = (1 - 0.750)(0.674\text{ bar}) = 0.169\text{ bar}$$

$$P(n\text{-}C_6H_{14}) = (0.750)(0.198) = 0.149\text{ bar}$$

$$P_t = 0.674\text{ bar} + (0.198\text{ bar} - 0.674\text{ bar})(0.750) = 0.317\text{ bar}$$

The plots are shown in Fig. 10-10.

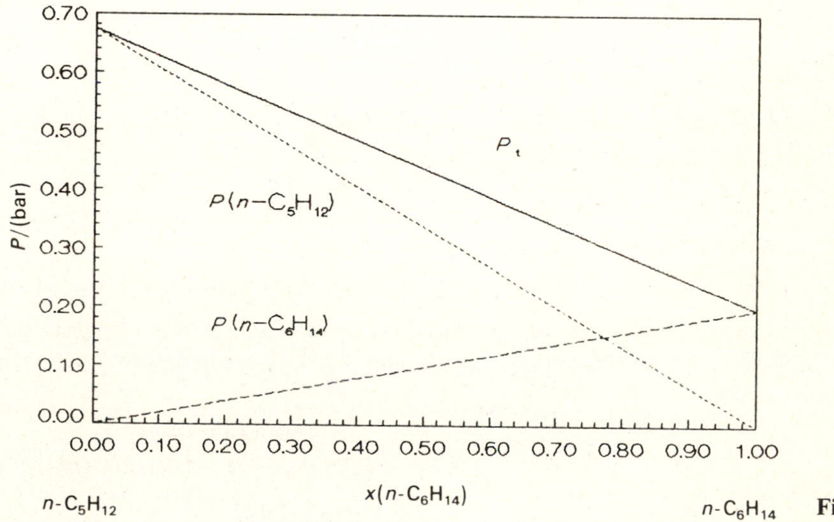

Fig. 10-10

10.34 Prepare a plot of the bubble point line and the dew point line for the n-C_5H_{12}–n-C_6H_{14} system described in Problem 10.31.

▮ The *bubble point line* or liquid line is the expression for P_t determined in Problem 10.33:

$$P_{liq} = P°(n\text{-}C_5H_{12}) + [P°(n\text{-}C_6H_{14}) - P°(n\text{-}C_5H_{12})]x(n\text{-}C_6H_{14})$$

The composition of the vapor is given by *(1.23)* as

$$x(n\text{-}C_6H_{14})_{vap} = \frac{P(n\text{-}C_6H_{14})}{P_t}$$

$$= \frac{x(n\text{-}C_6H_{14})P°(n\text{-}C_6H_{14})}{P°(n\text{-}C_5H_{12}) + [P°(n\text{-}C_6H_{14}) - P°(n\text{-}C_5H_{12})]x(n\text{-}C_6H_{14})}$$

Solving for $x(n\text{-}C_6H_{14})$ and substituting into the expression for P_t gives

$$P_{vap} = \frac{P°(n\text{-}C_5H_{12})P°(n\text{-}C_6H_{14})}{P°(n\text{-}C_6H_{14}) + [P°(n\text{-}C_5H_{12}) - P°(n\text{-}C_6H_{14})]x(n\text{-}C_6H_{14})_{vap}}$$

which is the equation for the *dew point line* or vapor line. As a sample calculation, consider $x(n\text{-}C_6H_{14}) = 0.750$:

$$P_{\text{liq}} = (0.674) + (0.198 \text{ bar} - 0.674 \text{ bar})(0.750) = 0.317 \text{ bar}$$

$$P_{\text{vap}} = \frac{(0.674 \text{ bar})(0.198 \text{ bar})}{0.198 \text{ bar} + (0.674 \text{ bar} - 0.198 \text{ bar})(0.750)} = 0.240 \text{ bar}$$

The plots are shown in Fig. 10-11.

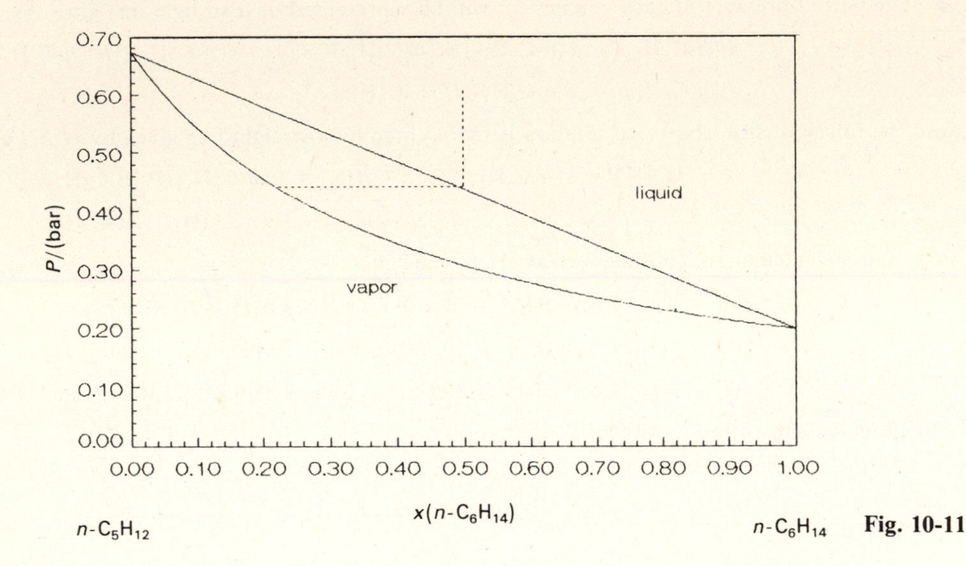

Fig. 10-11

10.35 The pressure over an equimolar mixture of $n\text{-}C_5H_{12}$ and $n\text{-}C_6H_{14}$ at 25 °C is reduced until the solution begins to boil. At what pressure will the solution begin to boil? What is the composition of the first bubble of vapor? See Problem 10.31 for further data.

▌ Substituting $x(n\text{-}C_6H_{14})$ into the expression for the bubble point line determined in Problem 10.35 gives the pressure at which boiling will begin as

$$P_{\text{liq}} = 0.674 \text{ bar} + (0.198 \text{ bar} - 0.674 \text{ bar})(0.500) = 0.436 \text{ bar}$$

This result is shown graphically in Fig. 10-11, where the vertical line intersects the liquid line. Substituting $P_{\text{vap}} = 0.436$ bar into the expression for the dew point line determined in Problem 10.35 and solving for $x(n\text{-}C_6H_{14})_{\text{vap}}$ gives

$$0.436 \text{ bar} = \frac{(0.674 \text{ bar})(0.198 \text{ bar})}{0.198 \text{ bar} + (0.674 \text{ bar} - 0.198 \text{ bar})x(n\text{-}C_6H_{14})}$$

$$x(n\text{-}C_6H_{14}) = 0.227$$

This result is shown graphically in Fig. 10-11, where the horizontal tie line intersects the vapor line.

10.36 The liquid–vapor phase diagram for two partially miscible liquids is shown in Fig. 10-12. Determine the number of phases, and describe the system in each of the numbered areas in the phase diagram.

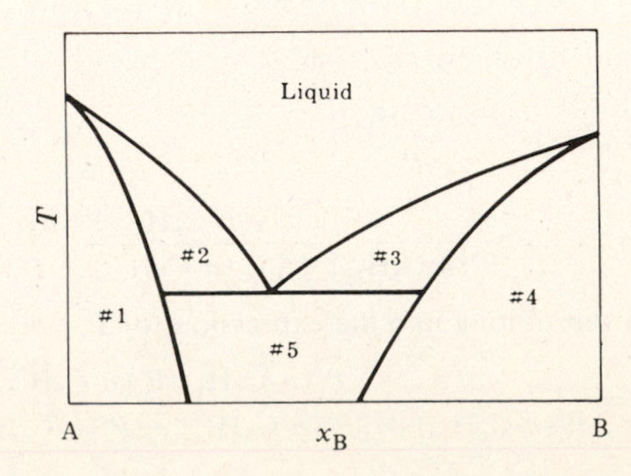

Fig. 10-12

▮ Areas 1, 3, and 5 are one-phase regions consisting of a liquid solution of B in A, a liquid solution of A in B, and a gaseous solution, respectively. Areas 2 and 4 are two-phase regions consisting of vapor in equilibrium with a liquid solution of A saturated with B and of vapor in equilibrium with a liquid solution of B saturated with A, respectively. Area 6 is also a two-phase region consisting of two liquid solutions—A saturated with B and B saturated with A.

10.37 The vapor pressure of toluene, C_7H_8, is 40.0 torr at 31.8 °C, and that of water at the same temperature is 35.3 torr. Assuming that these liquids are completely immiscible, determine the vapor pressure over a mixture of these two liquids at 31.8 °C.

▮ The contribution to the total vapor pressure of the mixture by each liquid is equal to the vapor pressure of the pure liquid. Substituting the data into *(1.22)* gives

$$P_t = P(C_7H_8) + P(H_2O) = 40.0 \text{ torr} + 35.3 \text{ torr} = 75.3 \text{ torr}$$

10.38 What is the composition of the vapor over the C_7H_8–H_2O mixture described in Problem 10.37 given that $x(C_7H_8) = 0.25$? What is the composition given that $x(C_7H_8) = 0.75$?

▮ The partial pressures and the total vapor pressure over the mixture are independent of the amounts of liquids. Substituting the data of Problem 10.37 into *(1.23)* gives for both solutions

$$x(C_7H_8) = \frac{40.0 \text{ torr}}{75.3 \text{ torr}} = 0.531 \qquad x(H_2O) = 1 - 0.531 = 0.469$$

10.39 Figure 10-13 is a Clausius–Clapeyron plot, see *(4.19a)*, of the vapor pressure of toluene, the vapor pressure of water, and the total vapor pressure above a mixture of the immiscible liquids. At what temperature would it be possible to purify toluene by steam distillation at 760 torr?

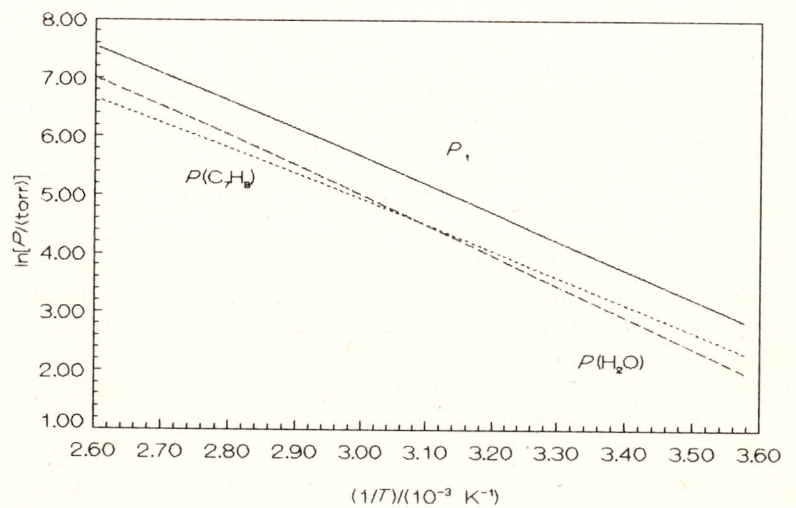

Fig. 10-13

▮ The mixture will boil at the temperature at which the total vapor pressure is equal to 760 torr. From the graph, $1/T = 2.785 \times 10^{-3}\ \text{K}^{-1}$ for $\ln 760 = 6.633$, and the individual pressures are $\ln[P(C_7H_8)] = 5.82$ and $\ln[P(H_2O)] = 6.05$, giving

$$T = \frac{1}{2.785 \times 10^{-3}\ \text{K}^{-1}} = 359.1 \text{ K} = 85.9 \text{ °C}$$

$$P(C_7H_8) = e^{5.82} \text{ torr} = 337 \text{ torr}$$

$$P(H_2O) = e^{6.05} \text{ torr} = 424 \text{ torr}$$

10.40 What mass of toluene will be distilled for each 100.0 g of water that vaporizes in the system described in Problem 10.39?

▮ The relation between vapor pressures, mole fractions, masses, and molar masses of liquids for steam distillation is

$$\frac{P_1^\circ}{P_2^\circ} = \frac{x_1}{x_2} = \frac{m_1 M_2}{m_2 M_1} \qquad (10.7)$$

Solving *(10.7)* for $m(C_7H_8)$ and substituting the data gives

$$m(C_7H_8) = (100.0\ g)\frac{(92.2\ g \cdot mol^{-1})(337\ torr)}{(18.0\ g \cdot mol^{-1})(424\ torr)} = 407\ g$$

10.41 The solid-liquid phase diagram for the W–Mo system is shown in Fig. 10-14. Describe the physical state of a mixture of W and Mo at 3000 °C given $x(Mo) = 0.20$. What is the physical state for $x(Mo) = 0.65$? What is the state for $x(Mo) = 0.90$?

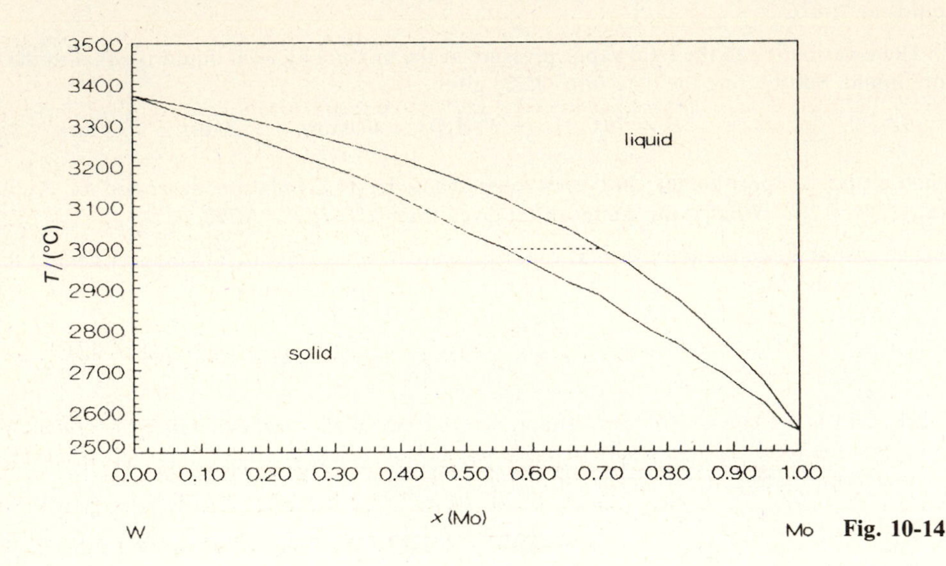

Fig. 10-14

▮ The phase diagram shows two one-phase regions and one two-phase region. Mixtures of W and Mo are completely soluble in each other in both the solid and liquid phases. At 3000 °C and $x(Mo) = 0.20$, the equilibrium mixture will consist of a solid solution of W and Mo. At $x(Mo) = 0.90$, the equilibrium mixture will consist of a liquid solution of W and Mo. At $x(Mo) = 0.65$, the system will consist of two phases—a liquid solution with $x(Mo) = 0.70$ and a solid solution with $x(Mo) = 0.55$—in equilibrium. This is illustrated by the horizontal tie line shown in the figure.

10.42 A series of Ni–Mn mixtures were prepared and allowed to reach equilibrium at various temperatures. Use the following data to prepare a phase diagram for the Ni–Mn system:

Mn-Rich Mixtures

$T/(°C)$	1260	1200	1150	1100	1050	1000
$w(Ni)_{sol}$	0.00	0.04	0.08	0.13	0.22	0.45
$w(Ni)_{liq}$	0.00	0.07	0.12	0.18	0.29	0.45

Ni-Rich Mixtures

$T/(°C)$	1050	1100	1150	1200	1250	1300	1350	1400	1450
$w(Ni)_{sol}$	0.58	0.64	0.70	0.75	0.80	0.85	0.90	0.96	1.00
$w(Ni)_{liq}$	0.54	0.62	0.68	0.73	0.78	0.83	0.88	0.94	1.00

where w_i is the mass fraction. In each case the liquid phase was a solution of Ni and Mn and the solid phase was a solution of Ni and Mn.

▮ The data are plotted in Fig. 10-15. Each set of compositions is connected by a horizontal tie line.

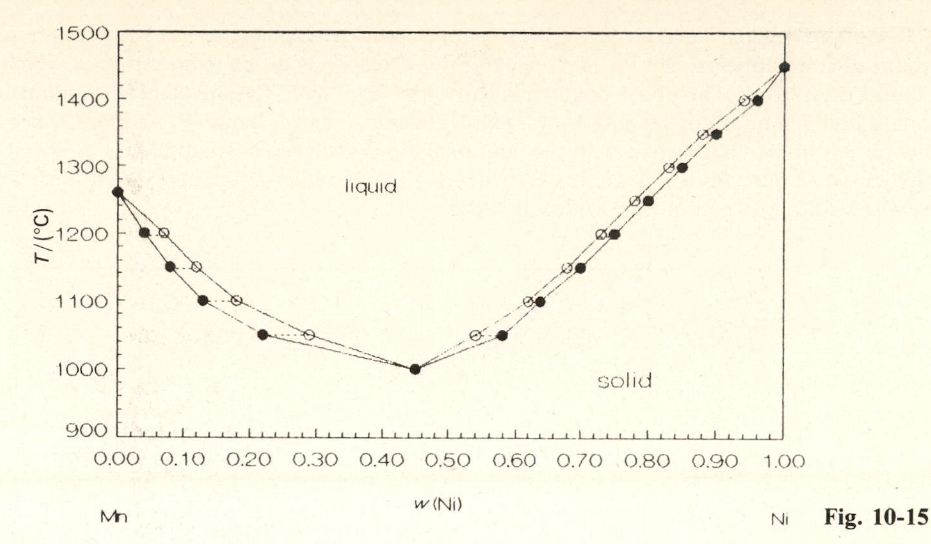

Fig. 10-15

10.43 What is the composition of the solid solution in equilibrium with a liquid mixture of Ni and Mn at $w(\text{Ni}) = 0.45$ and $1000\,^{\circ}\text{C}$?

▌ The phase diagram shown in Fig. 10-15 shows that the composition of the solid solution will be the same as the composition of the liquid solution.

10.44 The solid–liquid phase diagram for the Pb–Bi system is shown in Fig. 10-16. Describe the physical state of the system in the various numbered areas.

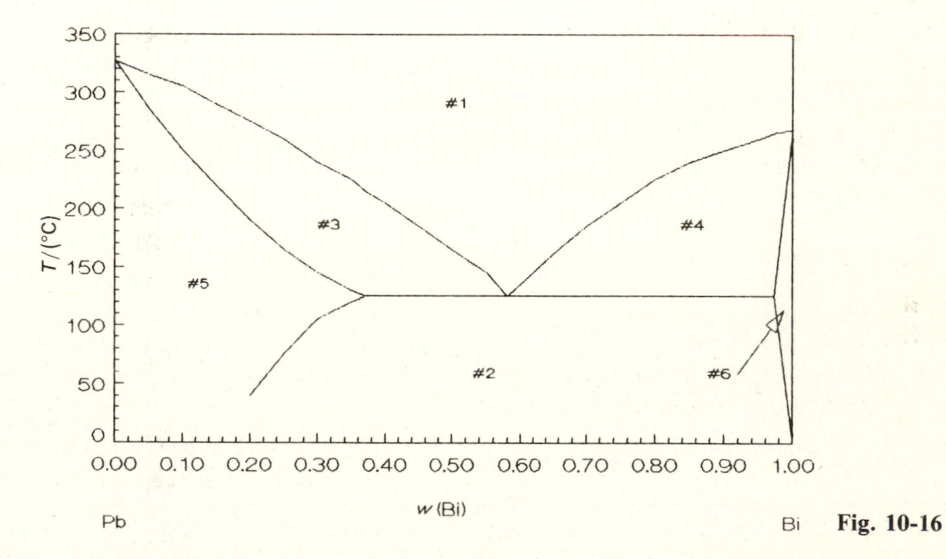

Fig. 10-16

▌ There are three one-phase regions and three two-phase regions shown in the diagram. Area 1 is the single-phase liquid solution of Pb and Bi. Area 5 is a one-phase solid solution of Bi in Pb, and area 6 is a one-phase solid solution of Pb in Bi. The diagram indicates that these metals are not completely miscible. Area 2 is a two-phase region in which a saturated solid solution of Bi in Pb is in equilibrium with a saturated solid solution of Pb in Bi. Area 3 consists of a liquid solution in equilibrium with a solid saturated solution of Bi in Pb. Area 4 consists of a liquid solution in equilibrium with a saturated solution of Pb in Bi.

10.45 Use the following cooling curve data to prepare a liquid–solid phase diagram for the Pb–As system:

w (Pb)	pure As	0.10	0.20	0.30	0.40	0.50	0.60	0.70	0.80	0.90	pure Pb
T/(°C) liquidus curve	—	830	790	750	700	650	600	530	455	360	—
T/(°C) solidus curve	852	285?	290	288	287	289	290	286	288	289?	327

▮ The higher temperature given for each composition corresponds to the temperature at which the first small amount of solid appears. The second temperature corresponds to the temperature at which the last small amount of liquid disappears. The phase diagram is shown in Fig. 10-17. The area labeled "liquid" is a one-phase area in which a liquid solution of Pb and As will exist. The area labeled "liquid + As(s)" is a two-phase region in which solid pure As is in equilibrium with the liquid solution. The area labeled "liquid + Pb(s)" is a two-phase region in which solid pure Pb is in equilibrium with the liquid solution. The remaining area is a two-phase region in which solid Pb is in equilibrium with solid As.

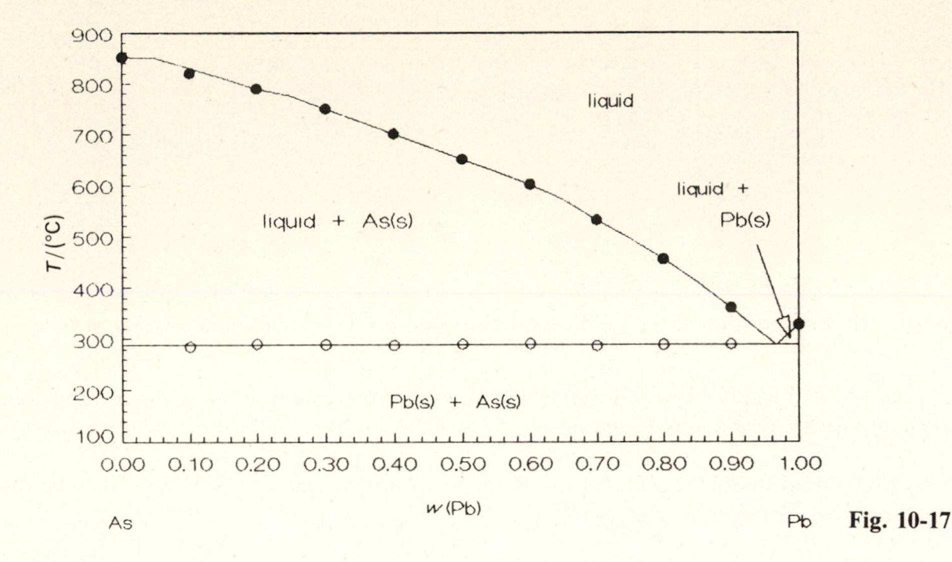

Fig. 10-17

10.46 What is the maximum mass of pure As that can be obtained by the crystallization of 100.0 g of a Pb-As mixture with $w(\text{Pb}) = 0.500$? See Problem 10.45 for further data.

▮ The *eutectic point* for the Pb-As system is 288 °C, and $w(\text{Pb}) = 0.967$ (see Fig. 10-17). If the original liquid mixture is cooled to a temperature just slightly above the eutectic temperature, the maximum amount of solid pure As will be obtained. Substituting into *(10.4)* and *(10.5)* gives

$$m(\text{As}) + m(\text{eutectic}) = 100.0\ \text{g} \qquad \frac{m(\text{As})}{m(\text{eutectic})} = \frac{0.967 - 0.500}{0.500 - 0.000}$$

Solving these simultaneous equations gives $m(\text{As}) = 48.3$ g.

10.47 Pure silver melts at 962 °C and has an enthalpy of fusion equal to $11.30\ \text{kJ}\cdot\text{mol}^{-1}$. What is the freezing point of a Pb-Ag solution of concentration $w(\text{Ag}) = 0.90$?

▮ The relation between the concentration and melting point of a solution is given by

$$\ln x_i = \frac{\Delta_{\text{fus}} H_i}{R}\left(\frac{1}{T_i^\circ} - \frac{1}{T_i}\right) \tag{10.8}$$

where T_i° is the melting point of the pure substance and T_i is the temperature at which the substance begins to freeze from the solution. The masses of metals present in 100.0 g of mixture are $m(\text{Ag}) = 90.0$ g and $m(\text{Pb}) = 10.0$ g, giving

$$x(\text{Ag}) = \frac{(90.0\ \text{g})/(107.9\ \text{g}\cdot\text{mol}^{-1})}{(90.0\ \text{g})/(107.9\ \text{g}\cdot\text{mol}^{-1}) + (10.0\ \text{g})/(207.2\ \text{g}\cdot\text{mol}^{-1})} = 0.945$$

Solving *(10.8)* for T_i and substituting the data gives

$$\frac{1}{T} = \frac{1}{1235\ \text{K}} - \frac{(8.314\ \text{J}\cdot\text{K}^{-1}\cdot\text{mol}^{-1})\ln 0.945}{(11.30\ \text{kJ}\cdot\text{mol}^{-1})[(10^3\ \text{J})/(1\ \text{kJ})]} = 8.51\times 10^{-4}$$

$$T = 1175\ \text{K} = 902\ ^\circ\text{C}$$

The observed value is 930 °C.

10.48 Silver and lead form a simple eutectic system. Given that $\Delta_{\text{fus}} H_{601} = 5.10\ \text{kJ}\cdot\text{mol}^{-1}$ for Pb, predict the eutectic temperature and composition. See Problem 10.47 for additional data.

▮ Substituting the data for Ag and Pb into *(10.8)* gives

$$\ln x(\mathrm{Ag}) = \frac{(11.30\ \mathrm{kJ \cdot mol^{-1}})[(10^3\ \mathrm{J})/(1\ \mathrm{kJ})]}{8.314\ \mathrm{J \cdot K^{-1} \cdot mol^{-1}}}\left(\frac{1}{1235\ \mathrm{K}} - \frac{1}{T_i}\right)$$

$$\ln [1 - x(\mathrm{Ag})] = \frac{(5.10\ \mathrm{kJ \cdot mol^{-1}})[(10^3\ \mathrm{J})/(1\ \mathrm{kJ})]}{8.314\ \mathrm{J \cdot K^{-1} \cdot mol^{-1}}}\left(\frac{1}{601\ \mathrm{K}} - \frac{1}{T_i}\right)$$

Solving the simultaneous equations gives $T_i = 496\ \mathrm{K} = 223\ ^\circ\mathrm{C}$ and $x(\mathrm{Ag}) = 0.19$. The observed values are $T_i = 305\ ^\circ\mathrm{C}$ and $x(\mathrm{Ag}) = 0.04$.

10.49 The solid-liquid phase diagram for the Mg-Zn system is shown in Fig. 10-18. Describe the physical state of the system in the various numbered areas.

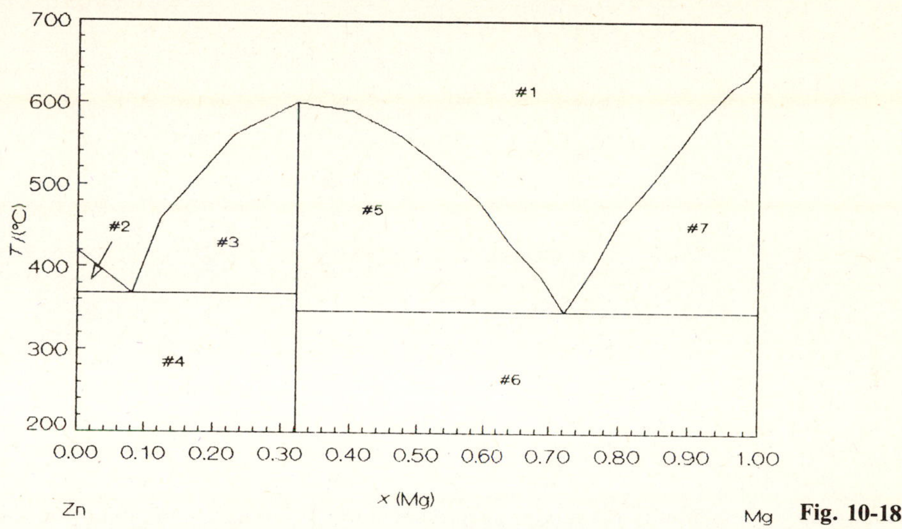

Fig. 10-18

▮ The metals form a single-phase system of a solution of Mg and Zn in area 1. Area 2 represents a two-phase region between the liquid and solid Zn, and area 7 represents a two-phase region between the liquid and solid Mg. The vertical line at $x(\mathrm{Mg}) \approx 0.33$ represents the formation of a *congruent melting compound.* The empirical formula of this compound is $Mg_{0.33}Zn_{0.67}$ or $MgZn_2$. Both areas 3 and 5 represent two-phase regions between the liquid and solid $MgZn_2$. Area 4 is a two-phase region between solid Zn and solid $MgZn_2$, and area 6 is a two-phase region between solid Mg and solid $MgZn_2$.

10.50 The solid-liquid phase diagram for the Mg-Ni system is shown in Fig. 10-19. Describe what will happen as a sample of Mg_2Ni is heated from 400 °C.

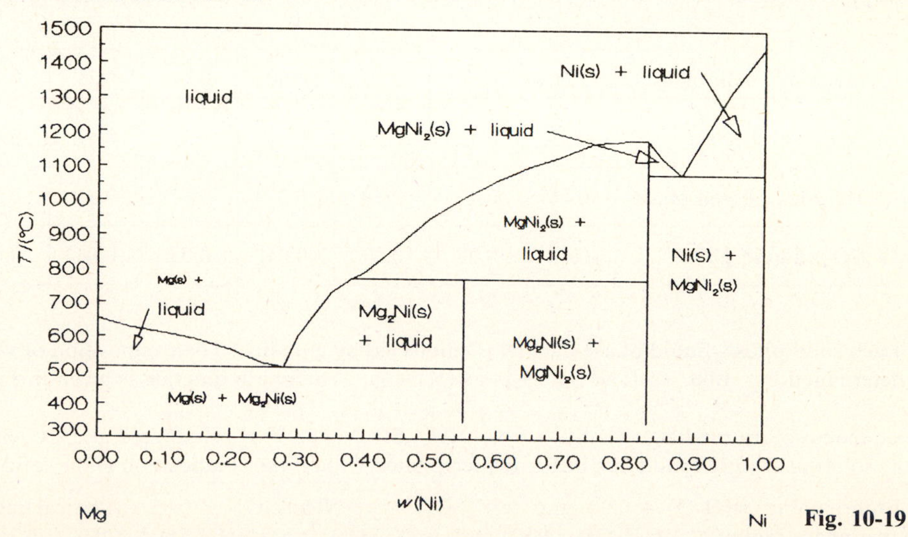

Fig. 10-19

▮ The phase diagram shows that Mg_2Ni is an *incongruent melting compound* because at 770 °C it decomposes to form solid $MgNi_2$ and a liquid phase of composition $w(\mathrm{Ni}) = 0.38$.

10.4 PHASE DIAGRAMS FOR THREE-COMPONENT SYSTEMS

10.51 What is the maximum number of degrees of freedom allowed for a three-component system? How are phase diagrams for these systems prepared?

▌ For a three-component system, *(10.1)* gives $f = 3 - p + 2 = 5 - p$. The minimum number of phases is one, so the maximum value of f is 4. Usually the phase diagram is prepared by plotting the composition on triangular graph paper (two independent degrees of freedom) for isothermal and isobaric conditions (see Fig. 10-20).

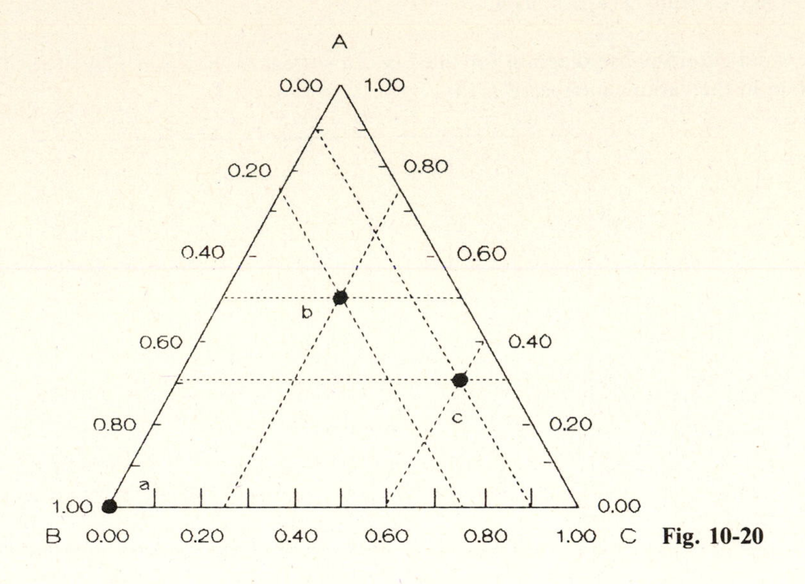

Fig. 10-20

10.52 What is the composition of the mixtures represented by **(*a*)** point *a*, **(*b*)** point *b*, and **(*c*)** point *c* in Fig. 10-20?

▌ The composition (mass fraction or mole fraction) for each component is read on the scale that increases from the opposite side of the triangle (=0.00) to the respective vertex (=1.00). For component A the scale on the right side of the triangle is used, for component B the scale on the left side, and for component C the scale along the bottom.

(*a*) Point *a* is pure B $[x(\mathrm{A}) = 0.00,\ x(\mathrm{B}) = 1.00,\ x(\mathrm{C}) = 0.00]$.

(*b*) Point *b* is $x(\mathrm{A}) = 0.50,\ x(\mathrm{B}) = 0.25,\ x(\mathrm{C}) = 0.25$.

(*c*) Point *c* is $x(\mathrm{A}) = 0.30,\ x(\mathrm{B}) = 0.10,\ x(\mathrm{C}) = 0.60$.

Note that $\sum x_i = 1.00$ in each case.

10.53 Using the following data for the K_2SO_4-$(NH_4)_2SO_4$-H_2O system at 25 °C, prepare the phase diagram for this system:

$w[(NH_4)_2SO_4]$, solid phase	1.00	0.88	0.72	0.60	0.47	0.31	0.16	0.06	0.00
$w(K_2SO_4)$, solid phase	0.00	0.12	0.28	0.40	0.53	0.69	0.84	0.94	1.00
$w[(NH_4)_2SO_4]$, liquid phase	0.56	0.41	0.39	0.37	0.35	0.31	0.22	0.11	0.00
$w(K_2SO_4)$, liquid phase	0.00	0.02	0.03	0.04	0.05	0.06	0.07	0.10	0.11

▌ Each solid phase–liquid phase data set is connected by a tie line. The composition of water present can always be determined by $1.00 - w[(NH_4)_2SO_4] - w(K_2SO_4)$. The phase diagram is given in Fig. 10-21.

10.54 An aqueous solution with $w(K_2SO_4) = w[(NH_4)_2SO_4] = 0.05$ is evaporated at 25 °C. What is the composition of the solution at which solid first begins to precipitate? What is the composition of the solid formed? See Fig. 10-21.

▌ From a point $w(H_2O) = 0.90$ and $w(K_2SO_4) = w[(NH_4)_2SO_4] = 0.05$, a vertical line is drawn and intersects the two-phase region at $w(H_2O) = 0.82$ and $w(K_2SO_4) = w[(NH_4)_2SO_4] = 0.09$. A tie line is drawn through the two-phase region to give the composition of the solid phase as $w(H_2O) = 0.00$, $w[(NH_4)_2SO_4] = 0.04$, and $w(K_2SO_4) = 0.96$. See Fig. 10-21.

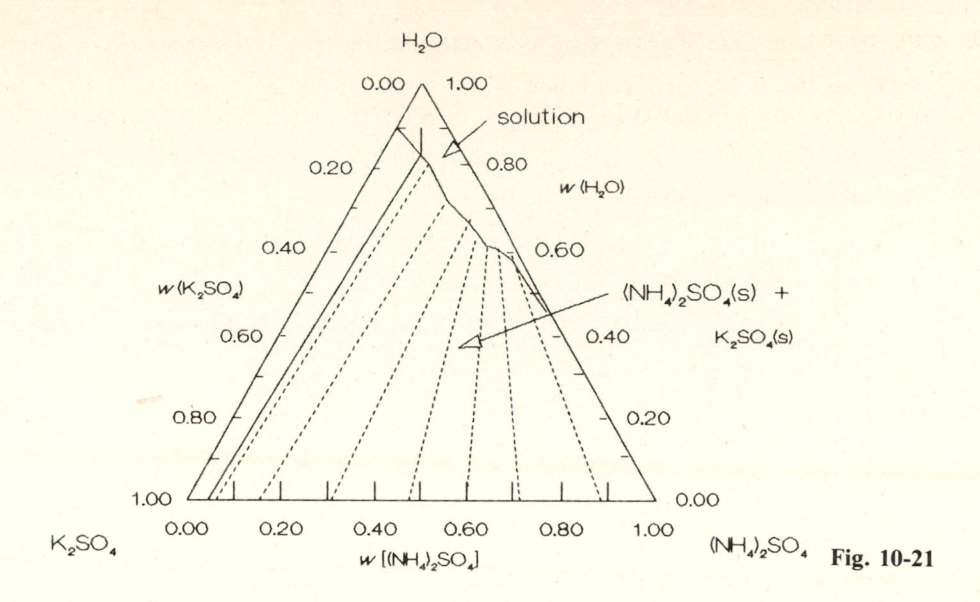

Fig. 10-21

10.55 Figure 10-22 shows the phase diagram for the $Pb(NO_3)_2$–$NaNO_3$–H_2O system at 25 °C. Label the areas.

▌ There are two two-phase areas, one one-phase area, and one three-phase area on the diagram. Area 1 is a liquid aqueous solution of $Pb(NO_3)_2$ and $NaNO_3$. Area 2 represents an equilibrium between solid $Pb(NO_3)_2$ and an aqueous solution saturated with $Pb(NO_3)_2$ and $NaNO_3$. Area 3 represents an equilibrium between solid $NaNO_3$ and an aqueous solution saturated with $Pb(NO_3)_2$ and $NaNO_3$. Area 4 represents an equilibrium between solid $NaNO_3$, solid $Pb(NO_3)_2$, and saturated aqueous solution with $w(NaNO_3) = 0.43$, $w[Pb(NO_3)_2] = 0.17$, and $w(H_2O) = 0.40$.

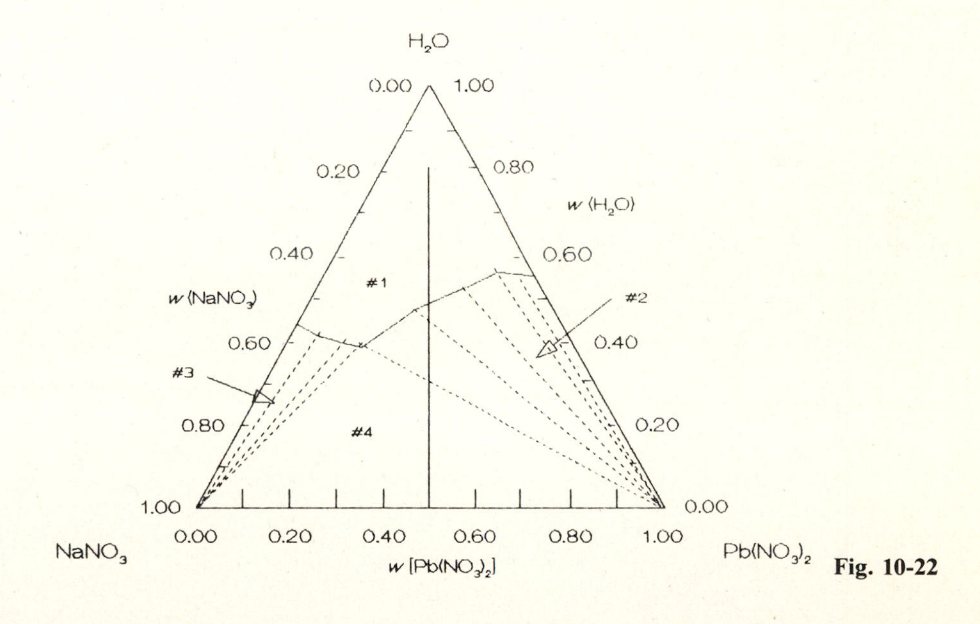

Fig. 10-22

10.56 Exactly 50 g each of $Pb(NO_3)_2$ and $NaNO_3$ are mixed, and 1-g aliquots of water are added. Describe the system as the water is added. See Fig. 10-22.

▌ The physical state of the system will be described by the vertical line shown in Fig. 10-22. As long as the amount of water added keeps the composition of the system in area 4, the system will consist of solid $NaNO_3$, solid $Pb(NO_3)_2$, and saturated aqueous solution with $w(NaNO_3) = 0.43$, $w[Pb(NO_3)_2] = 0.17$, and $w(H_2O) = 0.40$. Once sufficient water has been added to bring the system into area 2, the $NaNO_3$ will be completely dissolved, leaving solid $Pb(NO_3)_2$ in equilibrium with an aqueous solution saturated with $NaNO_3$ and $Pb(NO_3)_2$. As more water is added, once the system reaches area 1, the remaining $Pb(NO_3)_2$ dissolves and an aqueous solution of $Pb(NO_3)_2$ and $NaNO_3$ is formed. As more water is added, the solution becomes more and more dilute.

10.57 For the system defined in Problem 10.56, determine the mass of water needed to dissolve all of the salts.

▮ From Fig. 10-22, the composition of the system at the point where the system becomes a single phase is $w(H_2O) = 0.50$, $w(NaNO_3) = 0.25$, and $w[Pb(NH_3)_2] = 0.25$. Recognizing that

$$w(NaNO_3)m(soln) = m(NaNO_3)$$

and substituting the data gives

$$0.25m(soln) = 50.0 \text{ g} \qquad m(soln) = 200 \text{ g}$$

The mass of water is

$$m(H_2O) = 200 \text{ g soln} - 50.0 \text{ g } NaNO_3 - 50.0 \text{ g } Pb(NO_3)_2 = 100 \text{ g } H_2O$$

CHAPTER 11
Solutions

11.1 CONCENTRATIONS

11.1 Identify the solvent and solute in each of the following hydrochloric acid solutions: **(*a*)** $m(\text{HCl}) = 3.0\text{ g}$, $m(\text{H}_2\text{O}) = 1.00\text{ kg}$; **(*b*)** $V[\text{HCl(g)}] = 2.0\text{ L}$ under room conditions, $V(\text{H}_2\text{O}) = 1.0\text{ L}$; and **(*c*)** $x(\text{HCl}) = 1.5 \times 10^{-3}$.

The *solvent* is often considered to be the component of the solution present in the largest mass, volume, or amount of substance, and the *solutes* are the other components present. In solution **(*a*)** the solvent is easily identified as H_2O and the solute is HCl on the basis of mass. In solution **(*c*)** the solvent is chosen as H_2O and the solute is HCl on the basis of the relative amounts of substance. In solution **(*b*)**, water is still considered to be the solvent even through the volume of the gaseous HCl is twice as large as that of H_2O. The reason is that the H_2O determines the final physical state of the solution—a liquid.

11.2 Show that the three solutions described in Problem 11.1 are really the same solution.

Assuming that a sample of the solution contains 3.0 g of HCl and 1.00 kg of H_2O, corresponding to solution **(*a*)**, the amounts of each substance are

$$n(\text{HCl}) = \frac{3.0\text{ g HCl}}{36.46\text{ kg}\cdot\text{mol}^{-1}} = 0.082\text{ mol HCl}$$

$$n(\text{H}_2\text{O}) = \frac{(1.00\text{ kg H}_2\text{O})[(10^3\text{ g})/(1\text{ kg})]}{18.0152\text{ g}\cdot\text{mol}^{-1}} = 55.5\text{ mol H}_2\text{O}$$

giving

$$x(\text{HCl}) = \frac{0.082\text{ mol}}{0.082\text{ mol} + 55.5\text{ mol}} = 1.5 \times 10^{-3}$$

which corresponds to solution **(*c*)**. Assuming ideal gas behavior for HCl, *(1.18)* gives

$$V(\text{HCl}) = \frac{(0.082\text{ mol})(0.083\ 14\text{ L}\cdot\text{bar}\cdot\text{K}^{-1}\cdot\text{mol}^{-1})(298\text{ K})}{1.00\text{ bar}} = 2.0\text{ L}$$

and assuming the density of water to be $1.00\text{ g}\cdot\text{mL}^{-1}$,

$$V(\text{H}_2\text{O}) = \frac{(1.00\text{ kg})[(10^3\text{ g})/(1\text{ kg})][(1\text{ L})/(10^3\text{ mL})]}{1.00\text{ g}\cdot\text{mL}^{-1}} = 1.00\text{ L}$$

which corresponds to solution **(*b*)**.

11.3 Express the concentration of a 5.00 mass % solution of $Na_2S_2O_3$ in terms of **(*a*)** molarity, **(*b*)** molality, and **(*c*)** mole fraction of solute. The density of the solution at 25 °C is $1.040\text{ g}\cdot\text{mL}^{-1}$.

Assume that a sample of the solution contains 5.00 g of $Na_2S_2O_3$ and 95.00 g of H_2O. The amounts of substance are

$$n(\text{Na}_2\text{S}_2\text{O}_3) = \frac{5.00\text{ g Na}_2\text{S}_2\text{O}_3}{158.13\text{ g}\cdot\text{mol}^{-1}} = 0.0316\text{ mol Na}_2\text{S}_2\text{O}_3$$

$$n(\text{H}_2\text{O}) = \frac{95.00\text{ g H}_2\text{O}}{18.015\text{ g}\cdot\text{mol}^{-1}} = 5.273\text{ mol H}_2\text{O}$$

and the volume of the solution is

$$V = \frac{100.00\text{ g}}{(1.040\text{ g}\cdot\text{mL}^{-1})[(10^3\text{ mL})/(1\text{ L})]} = 0.096\ 15\text{ L}$$

(*a*) The *molarity* (C) is

$$C/(\text{mol}\cdot\text{L}^{-1}) = \frac{[n(\text{solute})/(\text{mol})]}{[V(\text{solution})/(\text{L})]} = \frac{0.0136}{0.096\ 15} = 0.329 \qquad (11.1)$$

or $C = 0.329\text{ mol/L}$ or 0.329 M.

(b) The *molality* (m) is

$$m/(\text{mol}\cdot\text{kg}^{-1}) = \frac{[n(\text{solute})/(\text{mol})]}{[m(\text{solvent})/(\text{kg})]} = \frac{0.0316}{0.095\,00} = 0.333 \qquad (11.2)$$

or $m = 0.333\ \text{mol}\cdot\text{kg}^{-1}$ or 0.333 m.

(c) The mole fraction of solute is

$$x(\text{Na}_2\text{S}_2\text{O}_3) = \frac{0.0316\ \text{mol}}{0.0316\ \text{mol} + 5.273\ \text{mol}} = 5.96 \times 10^{-3}$$

11.4 The only source of $Na_2S_2O_3$ for preparing the solution described in Problem 11.3 was $Na_2S_2O_3 \cdot 5H_2O$. How could 1.000 kg of the solution be prepared?

▮ The mass of the $Na_2S_2O_3$ needed is

$$m(\text{Na}_2\text{S}_2\text{O}_3) = \frac{5.00\ \text{g Na}_2\text{S}_2\text{O}_3}{100.00\ \text{g solution}}(1.000\ \text{kg})\left(\frac{10^3\ \text{g}}{1\ \text{kg}}\right) = 50.0\ \text{g Na}_2\text{S}_2\text{O}_3$$

which corresponds to

$$m(\text{Na}_2\text{S}_2\text{O}_3\cdot 5\text{H}_2\text{O}) = (50.0\ \text{g Na}_2\text{S}_2\text{O}_3)\frac{248.21\ \text{g Na}_2\text{S}_2\text{O}_3\cdot 5\text{H}_2\text{O mol}^{-1}}{158.13\ \text{g Na}_2\text{S}_2\text{O}_3\ \text{mol}^{-1}}$$

$$= 78.5\ \text{g Na}_2\text{S}_2\text{O}_3\cdot 5\text{H}_2\text{O}$$

The solution can be prepared by adding 921.5 g of H_2O to 78.5 g of $Na_2S_2O_3 \cdot 5H_2O$.

11.5 A handbook gives the concentration of water in a 2.772 M aqueous solution of NaOH as $c(H_2O) = 998.0\ g \cdot L^{-1}$. What are the molality and mole fraction of NaOH in this solution?

▮ The relation between m and C is

$$m/(\text{mol}\cdot\text{kg}^{-1}) = \frac{1000[C/(\text{mol}\cdot\text{L}^{-1})]}{c(\text{H}_2\text{O})/(\text{g}\cdot\text{L}^{-1})} \qquad (11.3)$$

Substituting the data into *(11.3)* gives

$$m/(\text{mol}\cdot\text{kg}^{-1}) = \frac{(1000)(2.772)}{998.0} = 2.778$$

The relation between m and x(solute) is

$$x(\text{solute}) = \frac{[m/(\text{mol}\cdot\text{kg}^{-1})][M(\text{solvent})/(\text{g}\cdot\text{mol}^{-1})]}{1000 + [m/(\text{mol}\cdot\text{kg}^{-1})][M(\text{solvent})/(\text{g}\cdot\text{mol}^{-1})]} \qquad (11.4)$$

Substituting the data into *(11.4)* gives

$$x(\text{NaOH}) = \frac{(2.778)(18.015)}{1000 + (2.778)(18.015)} = 4.766 \times 10^{-2}$$

11.6 Aqueous solutions of ethanol are often described in terms of "proof"—double the alcoholic percentage by volume at 60 °F. For example, a 50.00 proof solution contains 25.00 vol % ethanol. Given that the specific gravity of such a solution is 0.970 97, determine the composition of ethanol in terms of mass percent. The specific gravity of ethanol at 60 °F is 0.7939.

▮ Assuming that a 100.00-mL sample of solution gives $V(C_2H_5OH) = 25.00\ mL$, the mass of alcohol is

$$m(\text{C}_2\text{H}_5\text{OH}) = (25.00\ \text{mL})(0.7939\ \text{g}\cdot\text{mL}^{-1}) = 19.85\ \text{g}$$

The mass of the solution sample is

$$m(\text{solution}) = (100.00\ \text{mL})(0.970\,97\ \text{g}\cdot\text{mL}^{-1}) = 97.097\ \text{g}$$

giving the mass fraction as

$$w(\text{C}_2\text{H}_5\text{OH}) = \frac{19.85\ \text{g}}{97.097\ \text{g}} = 0.2044$$

or 20.44 mass %.

11.7 A 25.00-mL aliquot of 1.00 M NaOH and a 10.00-mL aliquot of 0.50 M NaCl were mixed. What are the concentrations of the compounds in the resulting solution? Assume that the volume of the resulting solution is 35.00 mL.

▮ The concentration–volume relationship for dilution is

$$C_1V_1 = C_2V_2 \tag{11.5}$$

where the subscript 1 refers to the initial conditions and the subscript 2 refers to the final conditions. Substituting the data for each compound into *(11.5)* gives

$$C(\text{NaOH}) = (1.00\ \text{mol}\cdot\text{L}^{-1})[(25.00\ \text{mL})/(35.00\ \text{mL})] = 0.71\ \text{mol}\cdot\text{L}^{-1}$$

$$C(\text{NaCl}) = (0.50)(10.00/35.00) = 0.14\ \text{mol}\cdot\text{L}^{-1}$$

11.8 What are the concentrations of the ions in the resulting solution described in Problem 11.7?

▮ The concentrations of each of the ions will be equal to the formal concentrations because both compounds are strong electrolytes. Thus

$$C(\text{OH}^-) = 0.71\ \text{mol}\cdot\text{L}^{-1}$$

$$C(\text{Cl}^-) = 0.14\ \text{mol}\cdot\text{L}^{-1}$$

$$C(\text{Na}^+) = 0.71\ \text{mol}\cdot\text{L}^{-1} + 0.14\ \text{mol}\cdot\text{L}^{-1} = 0.85\ \text{mol}\cdot\text{L}^{-1}$$

11.9 Commercial "concentrated" nitric acid is 15.6 M. How would a stockroom attendant prepare 10.0 L of laboratory "dilute" nitric acid that is 6.0 M?

▮ The volume of concentrated acid needed is given by *(11.5)* as

$$V(\text{conc}) = (10.0\ \text{L})\frac{6.0\ \text{mol}\cdot\text{L}^{-1}}{15.6\ \text{mol}\cdot\text{L}^{-1}} = 3.8\ \text{L}$$

The attendant will dilute a 3.8-L sample of the concentrated acid to 10.0 L by adding water.

11.10 Determine the concentration of CO(g) in water at 25 °C and 0.010 atm given that the Henry's law constant for this system is 5.80×10^4 atm.

▮ The mole fraction of gas in solution is related to the partial pressure of the gas above the solution by *Henry's law*,

$$P_2 = x_2K_2 \tag{11.6}$$

where K_2 is the Henry's law constant and the subscript 2 refers to the solute. Solving *(11.6)* for $x(\text{CO})$ and substituting the data gives

$$x(\text{CO}) = \frac{0.010\ \text{atm}}{5.80 \times 10^4\ \text{atm}} = 1.7 \times 10^{-7}$$

11.11 What is the ratio of the solubilities of N_2 and O_2 in H_2O at 100 °C given $K_2/(\text{atm}) = 12.6 \times 10^4$ for N_2 and 7.01×10^4 for O_2? Assume that $P(N_2) = 0.80$ atm and $P(O_2) = 0.20$ atm.

▮ The ratio of the mole fractions is given by *(11.6)* as

$$\frac{x(N_2)}{x(O_2)} = \frac{P(N_2)/K(N_2)}{P(O_2)/K(O_2)} = \frac{(0.80\ \text{atm})/(12.6 \times 10^4\ \text{atm})}{(0.20\ \text{atm})/(7.01 \times 10^4\ \text{atm})} = 2.2$$

11.12 Use the following partial pressure-concentration data at 23 °C for solutions of H_2(g) in water to determine $K(H_2)$.

$P(H_2)/(\text{atm})$	1.447	2.632	3.947	5.263	6.579	7.895	9.211	10.789
$x(H_2)/(10^{-5})$	1.87	3.39	5.08	6.74	8.34	9.87	11.29	12.83

▮ A plot of $P(H_2)$ against $x(H_2)$ is predicted to be linear by *(11.6)* and to have a slope equal to $K(H_2)$. The plot is shown in Fig. 11-1. The slope of the straight line at lower pressures gives $K(H_2) = 7.76 \times 10^4$ atm.

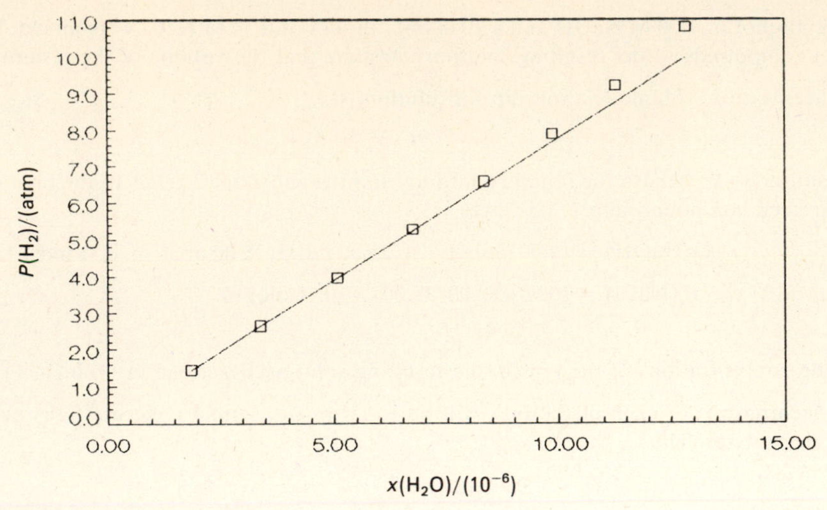

Fig. 11-1

11.13 At 0 °C, $K(CO_2) = 728$ atm, and at 60 °C, $K(CO_2) = 3410$ atm. What is the concentration of CO_2 in water at 0 °C given $P(CO_2) = 0.10$ atm? If the solution is heated to 60 °C, what will be the pressure needed to keep the CO_2 in solution?

▮ Solving *(11.6)* for $x(CO_2)$ and substituting the data gives

$$x(CO_2) = \frac{0.10\text{ atm}}{728\text{ atm}} = 1.4 \times 10^{-4}$$

At 60 °C,

$$P(CO_2) = (1.4 \times 10^{-4})(3410\text{ atm}) = 0.48\text{ atm}$$

11.14 Use the following data to determine the value of $\Delta_{soln}H°$ for $O_2(g)$ in water.

$T/(°C)$	0	10	20	30
$K(O_2)/(10^4\text{ bar})$	2.55	3.27	4.01	4.75

▮ The dissolution process can be considered to be an equilibrium established between the gas and the dissolved solute:

$$O_2(g) \rightleftharpoons O_2(aq)$$

Substituting *(6.28)* and *(11.6)* into *(7.2)* gives

$$K = \frac{a[O_2(aq)]}{a[O_2(g)]} = \frac{x[O_2(aq)]}{P/(\text{bar})} = \frac{1}{K(O_2)}$$

where it was assumed that $a \approx x$ for a solute (see Sec. 11.2). From *(7.5a)*, a plot of $\ln[1/K(O_2)]$ against T^{-1} will be linear and will have a slope equal to $-\Delta_{soln}H°/R$. The data are plotted in Fig. 11-2. The slope is 1.72×10^3 K,

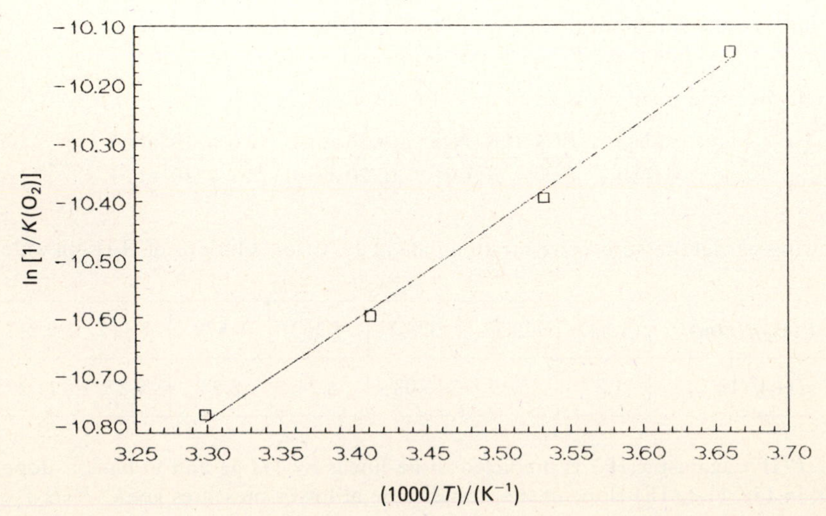

Fig. 11-2

giving the enthalpy of solution as

$$\Delta_{\text{soln}}H^\circ = -(8.314\ \text{J}\cdot\text{K}^{-1}\cdot\text{mol}^{-1})(1.72\times10^3\ \text{K})\frac{10^{-3}\ \text{kJ}}{1\ \text{J}} = 14.3\ \text{kJ}\cdot\text{mol}^{-1}$$

11.15 What is the relation between the volume of gas that dissolves in a fixed volume of solution and the partial pressure of the gas?

▮ The partial pressure of the gas is given by *(11.6)* as

$$P_2 = x_2K_2 = (n_2/n_t)K_2$$

Substituting *(1.18)* for P_2 gives

$$\frac{n_2RT}{V} = \frac{n_2}{n_t}K_2$$

and solving for V gives

$$V = n_tRT/K_2 \qquad (11.7)$$

The volume is independent of the partial pressure, provided the volume is measured at that pressure. (This is an alternative statement of Henry's law.)

11.16 The *Bunsen absorption coefficient* (α) is the volume of gas measured at STP that dissolves in a unit volume of solvent.

$$\alpha = V(\text{gas})/V(\text{solvent}) \qquad (11.8)$$

Calculate the value of α for N_2 in water at 25 °C given $K(N_2) = 8.65\times10^4$ atm.

▮ For dilute solutions, the value of n_t in *(11.7)* will be essentially equal to $n(H_2O)$. Letting V in *(11.7)* represent the volume of gas that dissolves in 1 L (=1000 g) of solvent, the absorption coefficient is

$$\alpha = \frac{[(1000\ \text{g})/(18.02\ \text{g}\cdot\text{mol}^{-1})](0.0821\ \text{L}\cdot\text{atm}\cdot\text{K}^{-1}\cdot\text{mol}^{-1})(273\ \text{K})/(8.65\times10^4\ \text{atm})}{1\ \text{L}}$$

$$= 1.44\times10^{-2}$$

11.17 The Bunsen absorption coefficient at 25 °C for methane is 3.01×10^{-2}, and for ethane it is 4.12×10^{-2}. What volume of each gas, measured at STP, will dissolve in 1.00 L of H_2O?

▮ Solving *(11.8)* for V(gas) and substituting the data gives

$$V(CH_4) = (3.01\times10^{-2})(1.00\ \text{L})[(10^3\ \text{mL})/(1\ \text{L})] = 30.1\ \text{mL}$$

$$V(C_2H_6) = (4.12\times10^{-2})(1.00)(10^3) = 41.2\ \text{mL}$$

11.18 What is the mole ratio of C_2H_6 to CH_4 in an aqueous solution at 25 °C if the gas above the solution consists of an equimolar mixture of these gases at 1 atm?

▮ For partial pressures other than 1 atm, *(11.8)* becomes $V(\text{gas}) = \alpha V(\text{solvent})P$. Substituting the data gives

$$V(CH_4) = (3.01\times10^{-2})(1.00\ \text{L})[(10^3\ \text{mL})/(1\ \text{L})](0.500) = 15.1\ \text{mL}$$

$$V(C_2H_6) = (4.12\times10^{-2})(1.00)(10^3)(0.500) = 20.6\ \text{mL}$$

Taking a ratio of these volumes gives

$$\frac{n(C_2H_6)}{n(CH_4)} = \frac{V(C_2H_6)}{V(CH_4)} = \frac{20.6\ \text{mL}}{15.1\ \text{mL}} = 1.36$$

11.19 An aqueous solution of $FeCl_3$ in 5 M HCl is shaken with twice its volume of isopropyl alcohol (also containing 5 M HCl), and 88% of the iron ion is extracted. What is the distribution coefficient for the compound?

▮ For simple extraction, the distribution coefficient (k_d) is given by

$$k_d = c_2'/c_2 \qquad (11.9)$$

where c_2 and c_2' are the concentrations of the solute in the layers. Substituting the data into *(11.9)* gives

$$k_d = \frac{0.88/2V}{0.12/V} = 3.7$$

11.20 How many extraction steps would be needed to reduce the concentration of $FeCl_3$ to 0.1% of its original concentration in the system described in Problem 11.19?

▮ The fraction of unextracted solute remaining after n extractions is given by

$$f_n = \left(1 + k_d \frac{V'}{V}\right)^{-n} \tag{11.10}$$

Solving *(11.10)* for n and substituting the data gives

$$n = \frac{-\ln 0.001}{\ln[1 + 3.7(2V/V)]} = 3.2$$

Using four extractions will reduce the concentration to less than 0.1% of its original value.

11.2 THERMODYNAMIC PROPERTIES OF SOLUTIONS

11.21 Compare the strengths of the solute-solute, solvent-solvent, and solute-solvent intermolecular forces for solutions of CH_3OH in H_2O and of CH_3OH in C_6H_6 given $\Delta_f H°/(\text{kJ}\cdot\text{mol}^{-1}) = -238.66$ for $CH_3OH(l)$, -242.781 for CH_3OH (in $4\,H_2O$), and -235.174 for CH_3OH (in $4\,C_6H_6$). Are either of these solutions ideal?

▮ Both dissolution processes can be represented by the equation

$$CH_3OH(l) + 4(\text{solvent}) \rightarrow CH_3OH(\text{in 4 solvent})$$

Substituting the data into *(4.5)* gives

$$\Delta_{soln}H°(H_2O) = -242.781\ \text{kJ}\cdot\text{mol}^{-1} - (-238.66\ \text{kJ}\cdot\text{mol}^{-1}) = -4.12\ \text{kJ}\cdot\text{mol}^{-1}$$

$$\Delta_{soln}H°(C_6H_6) = -235.174 - (-238.66) = 3.49\ \text{kJ}\cdot\text{mol}^{-1}$$

The exothermic process means that the solvent-solute intermolecular forces are stronger than the sum of the intermolecular forces in the pure components, and the endothermic process means that the sum of the intermolecular forces in the pure components is stronger than the solvent-solute forces in the solution. Neither of these solutions is ideal because $\Delta_{soln}H° \neq 0$.

11.22 A "100 proof" solution of ethanol in water consists of 50.00 mL of $C_2H_5OH(l)$ and 50.00 mL of $H_2O(l)$ mixed at 15.56 °C. The density of the solution is $0.9344\ \text{g}\cdot\text{mL}^{-1}$, that of pure H_2O is $1.0000\ \text{g}\cdot\text{mL}^{-1}$, and that of pure C_2H_5OH is $0.7939\ \text{g}\cdot\text{mL}^{-1}$. Is this solution ideal?

▮ The masses of the pure components are

$$m(C_2H_5OH) = (50.00\ \text{mL})(0.7939\ \text{g}\cdot\text{mL}^{-1}) = 39.70\ \text{g}$$

$$m(H_2O) = (50.00)(1.0000) = 50.00\ \text{g}$$

giving $m(\text{soln}) = 89.70\ \text{g}$. The volume of the solution is

$$V(\text{soln}) = \frac{89.70}{0.9344\ \text{g}\cdot\text{mL}^{-1}} = 96.00\ \text{mL}$$

This solution is not ideal because $V(\text{soln}) \neq V(\text{solvent}) + V(\text{solute})$.

11.23 The vapor pressure of water over an aqueous solution of NH_3 at 70 °C is 15.5 torr. Given that the concentration of the solution is $x(NH_3) = 0.150$, determine whether or not the solution is ideal. The vapor pressure of pure $H_2O(l)$ at 70 °C is 233.7 torr.

▮ For an ideal solution, the vapor pressure of the components will be given by *(10.6)*. For H_2O, the predicted vapor pressure would be

$$P(H_2O) = (1 - 0.150)(233.7\ \text{torr}) = 198.7\ \text{torr}$$

Because the actual vapor pressure of H_2O is considerably less than that predicted by *(10.6)*, the solution is not ideal.

11.24 At 39.9 °C, $P°/(\text{torr}) = 54.7$ for $H_2O(l)$ and 260.7 for $CH_3OH(l)$. Assuming ideal behavior, prepare a plot of the partial pressures and the total vapor pressure of the solution as a function of concentration.

▮ For an ideal solution, the partial pressure for each component will be given by *(10.6)*, and the total vapor pressure will be given by *(1.22)*. As a sample calculation, consider $x(CH_3OH) = 0.470$.

$$P(CH_3OH) = (0.470)(260.7 \text{ torr}) = 122.5 \text{ torr}$$

$$P(H_2O) = (1 - 0.470)(54.7) = 29.0 \text{ torr}$$

$$P_t = 122.5 \text{ torr} + 29.0 \text{ torr} = 151.5 \text{ torr}$$

The plots are shown by the dashed straight lines in Fig. 11-3.

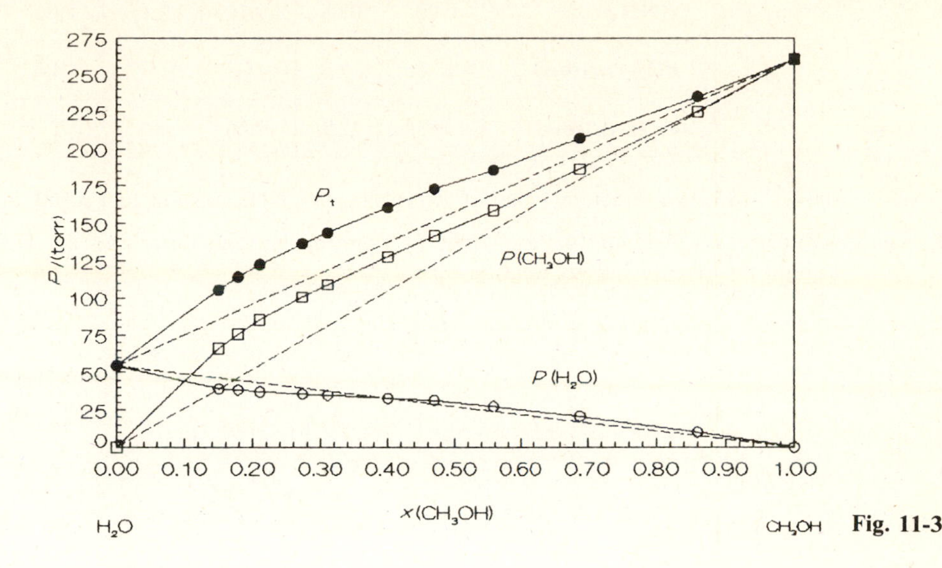

Fig. 11-3

11.25 Prepare a plot of the partial pressures and the total vapor pressure of the solution as a function of concentration for the CH_3OH–H_2O system at 39.9 °C. Use the following table.

$x(CH_3OH)$	0	0.1499	0.1785	0.2107	0.2731	0.3106
$P(H_2O)$/(torr)	54.7	39.2	38.5	37.2	35.8	34.9
$P(CH_3OH)$/(torr)	0.0	66.1	75.5	85.2	100.6	108.8

$x(CH_3OH)$	0.401	0.470	0.558	0.689	0.860	1.000
$P(H_2O)$/(torr)	32.8	31.5	27.3	20.7	10.1	0.0
$P(CH_3OH)$/(torr)	127.7	141.6	158.4	186.6	225.2	260.7

Does this system show positive or negative deviations from ideality?

▌ The data are plotted as the solid lines in Fig. 11-3. The total vapor pressure of the solution was calculated using *(1.22)*. Comparison of the actual vapor pressure curves to those calculated using *(10.6)* in the figure shows that CH_3OH shows positive deviations from ideality over the entire range of composition, H_2O shows slight negative deviations in the water-rich solutions and slight positive deviations in the methanol-rich solutions, and the solution shows positive deviations over the entire range of composition.

11.26 Using Fig. 11-3, determine the values of the Henry's law constants for both components.

▌ As $x_i \rightarrow 0$, the general shape of the partial pressure curve for a solute is linear with a slope equal to K_2 as predicted by *(11.6)*, rather than having a slope equal to P_0° as predicted by *(10.6)*. From the figure,

$$K(H_2O) = \frac{10.1 \text{ torr} - 0.0 \text{ torr}}{1.000 - 0.860} = 72.1 \text{ torr}$$

$$K(CH_3OH) = \frac{66.1 - 0.0}{0.1499 - 0} = 441 \text{ torr}$$

11.27 Using Fig. 11-3, describe the general shape of the partial pressure curves as $x_i \to 1$.

▮ In both cases, the actual curve for the solute asymptotically approaches the curve predicted by *(10.6)*.

11.28 Prepare a partial pressure–composition curve for the NH_3–H_2O system at 70 °F using the following data:

$x(NH_3)$	0.00	0.05	0.10	0.15	0.20
$P(H_2O)/(\text{torr})$	18.6	17.6	16.5	15.5	14.5
$P(NH_3)/(\text{torr})$	0.0	42.9	78.6	134.5	221.3

Given $K(NH_3) = 858$ torr, include a plot of Henry's law for NH_3. Also include a plot of Raoult's law for H_2O.

▮ The plots are shown in Fig. 11-4. Also shown is a plot of the total vapor pressure of the solution. Note that H_2O shows slight positive deviations from ideality.

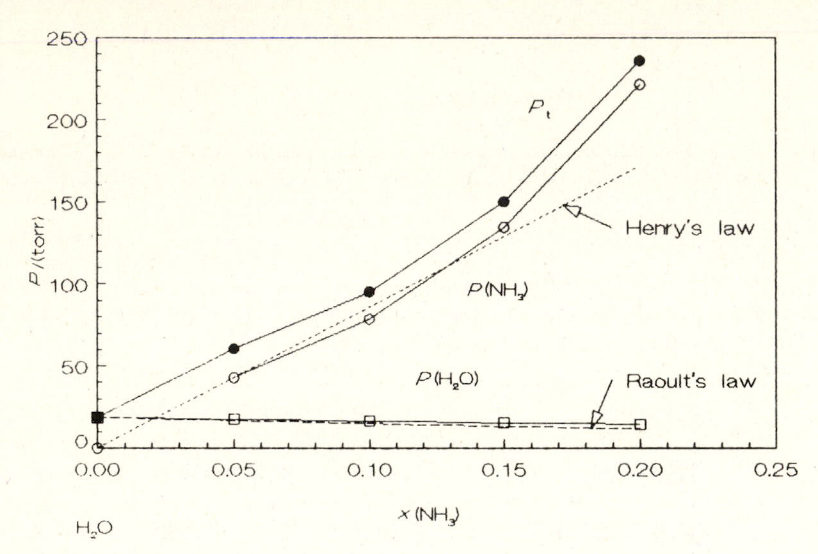

Fig. 11-4

11.29 Assuming that C_6H_6 and $C_6H_5CH_3$ form ideal solutions, calculate ΔG, ΔH, and ΔS at 25 °C for the addition of 1.00 mol of C_6H_6 to an infinitely large sample of solution with $x(C_6H_6) = 0.35$. The vapor pressure of C_6H_6 at 25 °C is 0.153 bar.

▮ For an ideal solution, $\Delta_{\text{soln}}H = 0$. The value of $\Delta_{\text{soln}}G$ for one component is given by *(6.27)* as

$$\Delta_{\text{soln}}G_i = nRT \ln \{x_i[P_i^\circ/(\text{bar})]\} \tag{11.11}$$

Substituting the data gives

$$\Delta_{\text{soln}}G = (1.00\ \text{mol})(8.314\ \text{J}\cdot\text{K}^{-1}\cdot\text{mol}^{-1})(298\ \text{K}) \ln [(0.35)(0.153)] = -7300\ \text{J}$$

Solving *(6.2a)* for $\Delta_{\text{soln}}S$ gives

$$\Delta_{\text{soln}}S = \frac{0 - (-7300\ \text{J})}{298\ \text{K}} = 24\ \text{J}\cdot\text{K}^{-1}$$

11.30 At 298 K, $\Delta_f G^\circ/(\text{kJ}\cdot\text{mol}^{-1}) = -394.359$ for $CO_2(g)$ and -385.98 for $CO_2(aq)$. Calculate the solubility of CO_2 in water at a gaseous pressure of 1.00 bar.

▮ If the standard state for the solution is chosen as a 1 m solution, *(6.27)* gives

$$\Delta_{\text{soln}}G_{\text{m}}^\circ = -RT \ln [m/(\text{m})] \tag{11.12}$$

where *(6.16)* gives

$$\Delta_{\text{soln}}G_{\text{m}}^\circ = [(1\ \text{mol})(-385.98\ \text{kJ}\cdot\text{mol}^{-1})] - [(1\ \text{mol})(-394.359\ \text{kJ}\cdot\text{mol}^{-1})] = 8.38\ \text{kJ}$$

Substituting into *(11.12)* and solving for the molality gives

$$(8.38\ \text{kJ})[(10^3\ \text{J})/(1\ \text{kJ})] = -(8.314\ \text{J}\cdot\text{K}^{-1}\cdot\text{mol}^{-1})(298\ \text{K}) \ln [m/(\text{m})]$$

$$m = 0.0340\ \text{mol}\cdot\text{kg}^{-1}$$

11.31 The solubility of N_2 in H_2O at 25 °C is 6.4×10^{-4} mol · kg^{-1} at a gaseous pressure of 1.00 bar. Calculate $\Delta_{soln}G°$ and $\Delta_f G°$ for N_2(aq).

▮ Substituting the data into *(11.12)* gives

$$\Delta_{soln}G^\circ_m = -(8.314\ \mathrm{J \cdot K^{-1} \cdot mol^{-1}})[(1\ \mathrm{kJ})/(10^3\ \mathrm{J})](298\ \mathrm{K}) \ln (6.4 \times 10^{-4}) = 18.2\ \mathrm{kJ \cdot mol^{-1}}$$

For the dissolution process, *(6.16)* gives

$$18.2\ \mathrm{kJ} = [(1\ \mathrm{mol})\Delta_f G^\circ(\mathrm{aq\ N_2})] - [(1\ \mathrm{mol})(0)]$$

Solving gives $\Delta_f G° = 18.2\ \mathrm{kJ \cdot mol^{-1}}$ for N_2(aq).

11.32 At 25 °C, $S°/(\mathrm{J \cdot K^{-1} \cdot mol^{-1}}) = 269.2$ for C_6H_6(l) and 319.7 for $C_6H_5CH_3$. What is the value of $S°$ for a solution prepared by mixing 0.50 mol and C_6H_6 with 1.50 mol $C_6H_5CH_3$?

▮ For the mixture,

$$S^\circ = n(C_6H_6)S^\circ(C_6H_6) + n(C_6H_5CH_3)S^\circ(C_6H_5CH_3) + \Delta_{soln}S$$

where $\Delta_{soln}S$ is given by *(5.13)*. Substituting the data gives

$$\begin{aligned} S^\circ &= (0.50\ \mathrm{mol})(269.2\ \mathrm{J \cdot K^{-1} \cdot mol^{-1}}) + (1.50\ \mathrm{mol})(319.7\ \mathrm{J \cdot K^{-1} \cdot mol^{-1}}) \\ &\quad - (8.314\ \mathrm{J \cdot K^{-1} \cdot mol^{-1}})[(0.50\ \mathrm{mol})(\ln 0.25) + (1.50\ \mathrm{mol})(\ln 0.75)] \\ &= 624\ \mathrm{J \cdot K^{-1}} \end{aligned}$$

11.33 Prepare a plot of $\Delta_{soln}H$, $\Delta_{soln}G$, and $T\,\Delta_{soln}S$ against mole fraction for preparing 1.00 mol of an ideal solution at 25 °C.

▮ For an ideal solution, $\Delta_{soln}H_m = 0$. As a sample calculation, consider $n_1/(\mathrm{mol}) = x_1 = 0.25$ and $n_2/(\mathrm{mol}) = x_2 = 0.75$. Substituting into *(5.13)* and *(6.7)* gives

$$T\,\Delta_{soln}S_m = -(8.314\ \mathrm{J \cdot K^{-1} \cdot mol^{-1}})(298\ \mathrm{K})[(0.25\ \mathrm{mol})(\ln 0.25) + (0.75\ \mathrm{mol})(\ln 0.75)] = 1394\ \mathrm{J}$$

$$\Delta_{soln}G^\circ_m = (8.314)(298)[(0.25)(\ln 0.25) + (0.75)(\ln 0.75)] = -1394\ \mathrm{J}$$

The plots are shown in Fig. 11-5.

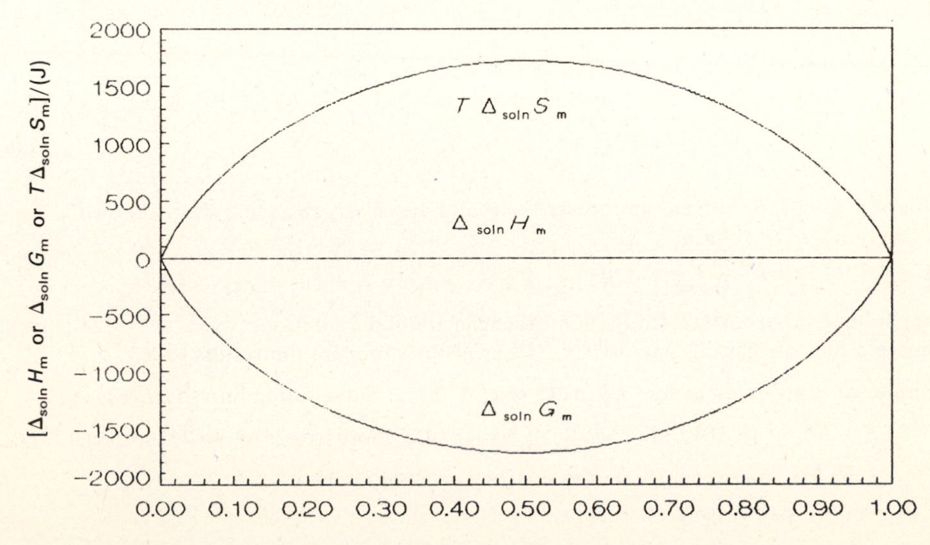

Fig. 11-5

11.34 Use the following data to determine $\Delta_{soln}H°$ for preparing 1.00 mol of solution of CCl_4(l) in C_6H_6(l) at 25 °C:

$n(C_6H_6)$/(mol)	CCl_4(l)	0.1	0.25	0.50	1	1.5	2
$\Delta_f H°_m$/(kJ · mol^{-1})	−135.44	−135.394	−135.344	−135.281	−135.206	−135.156	−135.122

$n(C_6H_6)$/(mol)	3	4	5	6	8	10
$\Delta_f H°_m$/(kJ · mol^{-1})	−135.080	−135.051	−135.030	−135.018	−134.997	−134.980

Prepare a plot of $\Delta_{soln}H°$ against $x(C_6H_6)$ and compare it to Fig. 11-5.

▌ As a sample calculation, consider the solution for which $n(C_6H_6) = 0.25$ mol. For the dissolution process *(4.5)* gives

$$\Delta_{soln}H° = [(1\ \text{mol})(-135.344\ \text{kJ}\cdot\text{mol}^{-1}) - (1\ \text{mol})(-135.44\ \text{kJ}\cdot\text{mol}^{-1})][(10^3\ \text{J})/(1\ \text{kJ})] = 96\ \text{J}$$

For 1.00 mol of solution,

$$x(C_6H_6) = \frac{0.25\ \text{mol}}{1.00\ \text{mol} + 0.25\ \text{mol}} = 0.200$$

$$\Delta_{soln}H°_m = \frac{96\ \text{J}}{1.00\ \text{mol} + 0.25\ \text{mol}} = 77\ \text{J}\cdot\text{mol}^{-1}$$

The plot is shown in Fig. 11-6. Note that $\Delta_{soln}H°_m$ for this solution is not zero as it is for an ideal solution.

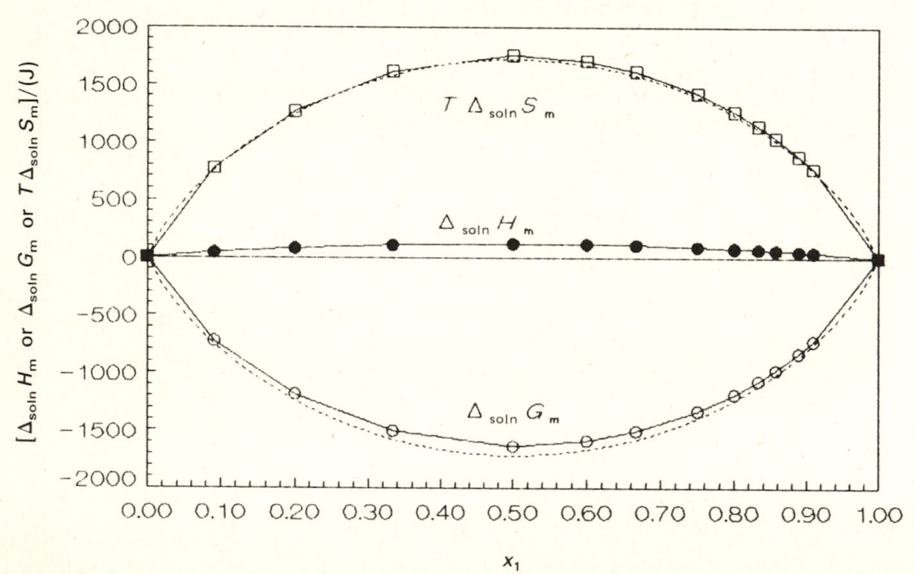

Fig. 11-6

11.35 A real solution in which the solvent and solute molecules are nearly the same size is known as a *regular solution*. For such a solution *(6.7)* becomes

$$\Delta_{soln}G° = RT[n_1(\ln x_1) + n_2(\ln x_2)] + (n_1 + n_2)x_1x_2w \qquad (11.13)$$

where w is a parameter that corrects for the nonideal behavior of the solution. Given $w = 324$ J for the CCl_4–C_6H_6 system, prepare a plot of $\Delta_{soln}G°_m$ against $x(C_6H_6)$ at 25 °C. Compare the results to Fig. 11-5.

▌ As a sample calculation, consider $n_1/(\text{mol}) = x_1 = 0.25$. Substituting into *(11.13)* gives

$$\begin{aligned}\Delta_{soln}G°_m &= (8.314\ \text{J}\cdot\text{K}^{-1}\cdot\text{mol}^{-1})(298\ \text{K})[(0.25\ \text{mol})(\ln 0.25) + (0.75\ \text{mol})(\ln 0.75)] \\ &\quad + (0.25\ \text{mol} + 0.75\ \text{mol})(0.25)(0.75)(324\ \text{J}) \\ &= -1333\ \text{J}\end{aligned}$$

The plot is shown in Fig. 11-6 along with the plot of *(6.7)*. The curve for the real solution shows that the dissolution process is not quite as favorable as that for the ideal solution. The difference between these plots, resulting from the $(n_1 + n_2)x_1x_2w$ term in *(11.13)*, is known as the *excess free energy of mixing.*

11.36 Prepare a plot of $T\,\Delta_{soln}S^\circ_m$ for the CCl_4-C_6H_6 system using the results of Problems 11.34 and 11.35. Compare the results to Fig. 11-5.

▌ As a sample calculation for $n(C_6H_6) = 1.00$ mol, $\Delta_{soln}H^\circ_m = 106$ J and $\Delta_{soln}G^\circ_m = -1637$ J. Substituting into *(6.2a)* gives

$$T\,\Delta_{soln}S^\circ_m = 117\text{ J} - (-1637\text{ J}) = 1754\text{ J}$$

The plot is shown in Fig. 11-6 along with the plot of *(5.13)*. The curve for the real solution shows that the entropy change for the dissolution process is a little more favorable than that for the ideal solution.

11.37 Using the data given in Problem 11.25, calculate the activity coefficient for both components at $x(CH_3OH) = 0.2107$. Prepare a plot of the activity coefficients as a function of concentration. Does this system show positive or negative deviations from ideality?

▌ For a solution composed of two volatile components that are completely miscible, the activity coefficient of each component is given by

$$f_i = P_i/x_iP^\circ_i \tag{11.14}$$

Substituting the data given in Problem 11.25 into *(11.14)* gives

$$f(H_2O) = \frac{37.2\text{ torr}}{(1 - 0.2107)(54.7\text{ torr})} = 0.862$$

$$f(CH_3OH) = \frac{85.2}{(0.2107)(260.7)} = 1.551$$

The plot is shown in Fig. 11-7. At all concentrations, $f(CH_3OH) > 1$, so CH_3OH shows positive deviations from ideality. In the water-rich solutions, $f(H_2O) < 1$, so H_2O shows negative deviations; and in the methanol-rich solutions, $f(H_2O) > 1$, so H_2O shows positive deviations in the solutions.

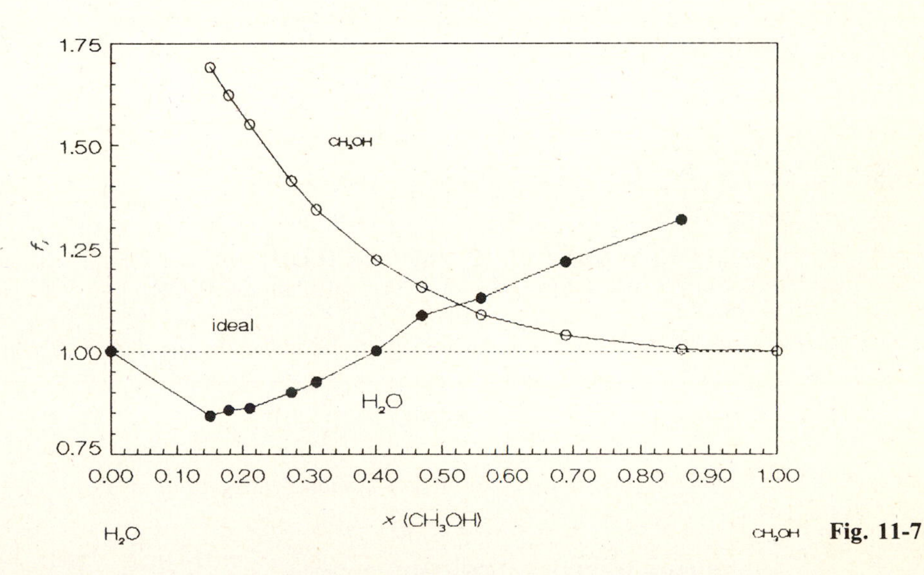

Fig. 11-7

11.38 Using the results of Problem 11.37, prepare a plot of the activities of the components as a function of concentration. Does this system show positive or negative deviations from ideality?

▌ The activity of each component is given by

$$a_i = f_ix_i \tag{11.15}$$

As a sample calculation, consider $x(CH_3OH) = 0.2107$. Substituting into *(11.15)* gives

$$a(H_2O) = (0.862)(1 - 0.2107) = 0.6804$$

$$a(CH_3OH) = (1.551)(0.2107) = 0.3268$$

The graph is shown in Fig. 11-8. For an ideal solution, *(11.15)* gives $a_i = x_i$. These ideal lines are shown in

Fig. 11-8 by dashed lines. The figure shows that for all concentrations $a(CH_3OH)$ is greater than $x(CH_3OH)$, so CH_3OH shows positive deviations from ideality. In the water-rich solution, $a(H_2O)$ is less than $x(H_2O)$, so H_2O shows negative deviations in these solutions, and in the methanol-rich solutions, $a(H_2O)$ is greater than $x(H_2O)$, so H_2O shows positive deviations in these solutions.

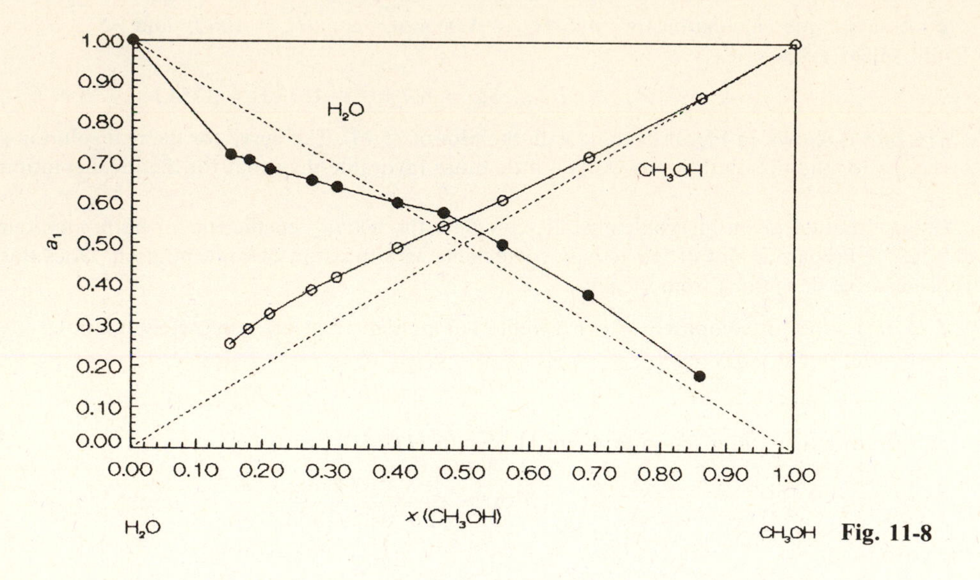

Fig. 11-8

11.39 Using the data given in Problem 11.28, determine the activity coefficient and the activity of NH_3 at $x(NH_3) = 0.15$.

For a gaseous solute, the activity coefficient is given by

$$f_2 = P_2/K_2x_2 \tag{11.16}$$

Substituting the data into *(11.16)* gives

$$f(NH_3) = \frac{134.5 \text{ torr}}{(858 \text{ torr})(0.15)} = 1.045$$

Substituting into *(11.15)* gives $a(NH_3) = (1.045)(0.15) = 0.1568$.

11.40 At $x(CH_3OH) = 0.1499$, $a(H_2O) = 0.717$, and $a(CH_3OH) = 0.254$ for the system described in Problem 11.25. If $a(H_2O) = 0.704$ at $x(CH_3OH) = 0.1785$, calculate $a(CH_3OH)$.

The activity of the solute is related to the activity of the solvent by the *Gibbs-Duhem equation.*

$$d(\ln a_2) = -\frac{x_1}{x_2}\, d(\ln a_1) \tag{11.17}$$

where *(11.17)* is usually evaluated by performing a graphical integration of a plot of x_1/x_2 against $\ln a_1$. The values to plot are

$$x(H_2O) = 0.8501 \qquad x(CH_3OH) = 0.1499 \qquad \frac{x(H_2O)}{x(CH_3OH)} = \frac{0.8501}{0.1499} = 5.671$$

$$\ln[a(H_2O)] = \ln 0.717 = 0.1499$$

$$\ln[a(CH_3OH)] = \ln 0.254 = -1.370$$

for the first set of data and $x(H_2O) = 0.8215$, $x(CH_3OH) = 0.1785$, $x(H_2O)/x(CH_3OH) = 4.602$, and $\ln[a(H_2O)] = -0.351$ for the second set of data. The graph appears in Fig. 11-9, and the area of the enclosed figure is -0.093, giving

$$\ln[a(CH_3OH)] = -1.370 - (-0.093) = -1.277$$

or $a(CH_3OH) = 0.279$. (The value calculated in Problem 11.38 is 0.290.)

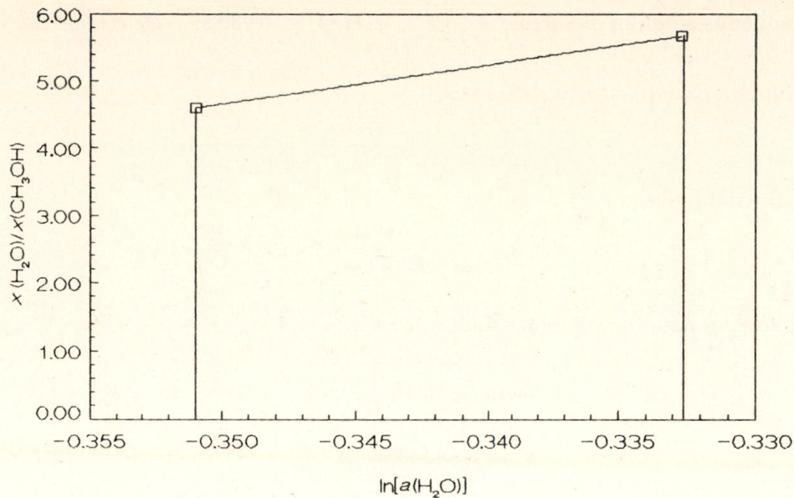

Fig. 11-9

11.41 The change in the chemical potential for a component in a regular solution ($\Delta\mu_i$) can be determined from *(11.13)* as

$$\Delta\mu_i = RT \ln x_i + (1 - x_i)^2 w \qquad (11.18)$$

Calculate the activity coefficients for the CCl_4-C_6H_6 system described in Problem 11.35, and prepare a plot of γ_i against $x(C_6H_6)$.

▮ Upon a term-by-term comparison of *(11.18)* with *(6.27)*,

$$\Delta\mu_i = RT \ln a_i = RT \ln f_i x_i = RT \ln x_i + RT \ln f_i$$

the equation for the activity coefficient is

$$RT \ln f_i = (1 - x_i)^2 w \qquad (11.19)$$

As a sample calculation, consider $x(C_6H_6) = 0.25$.

$$\ln [f(C_6H_6)] = \frac{(1 - 0.25)^2(324 \text{ J} \cdot \text{mol}^{-1})}{(8.314 \text{ J} \cdot \text{K}^{-1} \cdot \text{mol}^{-1})(298 \text{ K})} = 0.0735$$

$$\ln [f(CCl_4)] = \frac{(0.25)^2(324)}{(8.314)(298)} = 0.008\,17$$

giving $f(C_6H_6) = 1.076$ and $f(CCl_4) = 1.0082$. The results are plotted in Fig. 11-10.

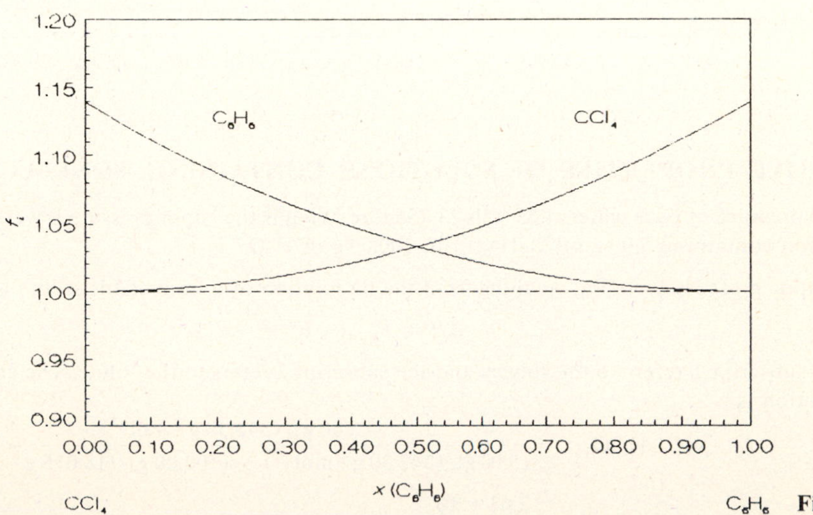

Fig. 11-10

11.42 In a 1.00 m aqueous solution of sucrose at 25 °C, $a(H_2O) = 0.980\ 59$. Determine the value of the molal osmotic coefficient.

▮ The *molal osmotic coefficient* is defined as

$$\phi = \frac{-1000}{[M(H_2O)/(g \cdot mol^{-1})][m/(m)]} \ln [a(H_2O)] \tag{11.20}$$

Substituting the data gives

$$\phi = \frac{-1000}{(18.015)(1.00)} \ln 0.980\ 59 = 1.088$$

11.43 Using the following data for aqueous solutions of LiCl at 25 °C,

$m/(m)$	0.00	0.01	0.05	0.10
ϕ	1.000	0.950	0.948	0.939

determine $\gamma_\pm$ for LiCl at 0.10 m.

▮ The relation between ϕ and the mean activity coefficient is

$$\ln \gamma_\pm = (\phi - 1) + \int_0^m \frac{\phi - 1}{m}\, dm \tag{11.21}$$

The integral is evaluated graphically (see Fig. 11-11) as -0.187, giving

$$\ln \gamma_\pm = (0.939 - 1) + (-0.187) = -0.248$$

Taking antilogarithms gives $\gamma_\pm = 0.780$. Experimentally, the value is 0.790.

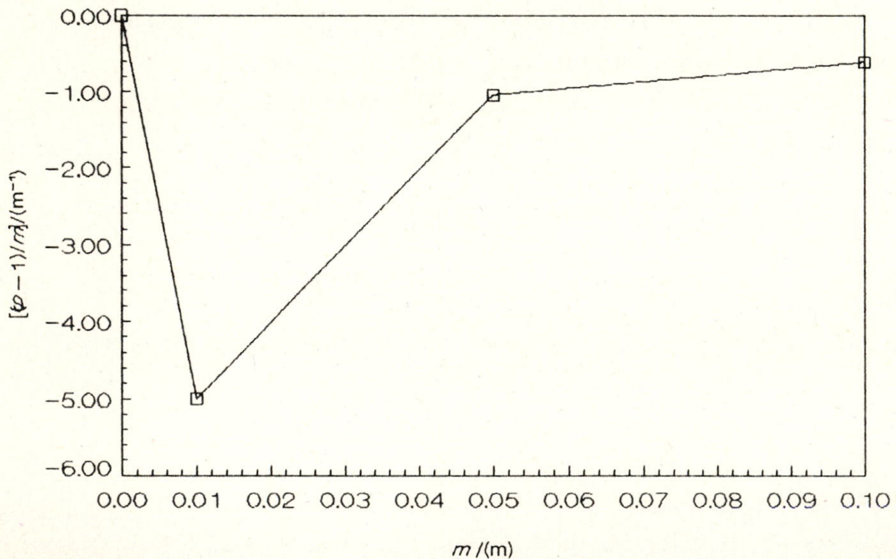

Fig. 11-11

11.3 COLLIGATIVE PROPERTIES OF SOLUTIONS CONTAINING NONELECTROLYTE SOLUTES

11.44 The vapor pressure of pure water at 25 °C is 23.756 torr. What is the vapor pressure lowering and the vapor pressure of a solution containing 5.00 g of $C_{12}H_{22}O_{11}$ in 100.00 g of H_2O?

▮ The vapor pressure lowering resulting from the dissolution of a nonvolatile solute is given by

$$\Delta P = P_{soln} - P_1^\circ = -P_1^\circ x_2 \tag{11.22}$$

where the subscript 1 refers to the solvent and the subscript 2 refers to the solute. The concentration of $C_{12}H_{22}O_{11}$ in the solution is

$$x(C_{12}H_{22}O_{11}) = \frac{(5.00\ g)/(342.30\ g \cdot mol^{-1})}{(5.00\ g)/(342.30\ g \cdot mol^{-1}) + (100.00\ g)/(18.015\ g \cdot mol^{-1})}$$

$$= 2.62 \times 10^{-3}$$

Substituting into *(11.22)* gives

$$\Delta P = -(23.756 \text{ torr})(2.62 \times 10^{-3}) = -0.0622 \text{ torr}$$

$$P_{\text{soln}} = P_1^\circ + \Delta P = 23.756 \text{ torr} + (-0.0622 \text{ torr}) = 23.694 \text{ torr}$$

11.45 A beaker containing 0.010 mol of $C_{12}H_{22}O_{11}$ in 100.00 g of H_2O and a beaker containing 0.020 mol of $C_{12}H_{22}O_{11}$ in 100.00 g of H_2O are placed in a chamber and allowed to equilibrate. What is the concentration of $C_{12}H_{22}O_{11}$ in the resulting solutions?

▮ Water vapor will be transferred from the more dilute solution to the more concentrated solution until both solutions reach the same concentration. Letting $n(H_2O)$ represent the amount of water that is transferred,

$$\frac{0.010 \text{ mol}}{0.010 \text{ mol} + (100.00 \text{ g})/(18.015 \text{ g}\cdot\text{mol}^{-1}) - n(H_2O)} = \frac{0.020 \text{ mol}}{0.020 \text{ mol} + (100.00 \text{ g})/(18.015 \text{ g}\cdot\text{mol}^{-1}) + n(H_2O)}$$

Solving gives $n(H_2O) = 1.850$ mol. The final concentration of the solutions is

$$x(C_{12}H_{22}O_{11}) = \frac{0.010 \text{ mol}}{0.010 \text{ mol} + (100.00 \text{ mol})/(18.015 \text{ g}\cdot\text{mol}^{-1}) - 1.850 \text{ mol}} = 0.002\,69$$

11.46 Repeat Problem 11.45, assuming that the second beaker contains 0.020 mol of urea.

▮ None of the calculations involved in Problem 11.45 depend on the identity of the solute and therefore will not change for this system. Thus $x(C_{12}H_{22}O_{11}) = 0.002\,69$ in the first beaker and $x(H_2NCONH_2) = 0.002\,69$ in the second beaker.

11.47 What will be the vapor pressure lowering at 25 °C for a solution containing 0.010 mol of $C_{12}H_{22}O_{11}$ and 0.015 mol of H_2NCONH_2 dissolved in 100.00 g of H_2O? At 25 °C, $P°(H_2O) = 23.756$ torr.

▮ The total mole fraction of solute is

$$x_2 = \frac{0.010 \text{ mol} + 0.015 \text{ mol}}{0.010 \text{ mol} + 0.015 \text{ mol} + (100.00 \text{ g})/(18.015 \text{ g}\cdot\text{mol}^{-1})} = 4.5 \times 10^{-3}$$

Substituting into *(11.22)* gives

$$\Delta P = -(23.756 \text{ torr})(4.5 \times 10^{-3}) = 0.11 \text{ torr}$$

11.48 What mass of a solute having a molar mass of 345 g · mol^{-1} is needed to decrease the vapor pressure of 100.0 g of H_2O at 25 °C by 1.00 torr? At 25 °C, $P°(H_2O) = 23.756$ torr.

▮ Solving *(11.22)* for x_2 and substituting the data gives

$$x_2 = -(-1.00 \text{ torr})/(23.756 \text{ torr}) = 0.0421$$

Solving

$$0.0421 = \frac{m_2/(345 \text{ g}\cdot\text{mol}^{-1})}{m_2/(345 \text{ g}\cdot\text{mol}^{-1}) + (100.0 \text{ g})/(18.015 \text{ g}\cdot\text{mol}^{-1})}$$

for m_2 gives $m_2 = 84.2$ g.

11.49 What is the boiling point of a solution of 1.00 g of naphthalene dissolved in 100.0 g of toluene? The normal boiling point of toluene is 110.7 °C, and $K_{x,\text{bp}} = 36.1$ K for toluene.

▮ The mole fraction of solute is

$$x(C_{10}H_8) = \frac{(1.00 \text{ g})/(128.18 \text{ g}\cdot\text{mol}^{-1})}{(1.00 \text{ g})/(128.18 \text{ g}\cdot\text{mol}^{-1}) + (100.0 \text{ g})/(92.14 \text{ g}\cdot\text{mol}^{-1})} = 7.14 \times 10^{-3}$$

The boiling point of the solution of a nonvolatile solute is given by

$$T_{\text{bp,soln}} = T_{\text{bp,1}} + K_{x,\text{bp}}x_2 \qquad (11.23)$$

where $K_{x,\text{bp}}$ is the mole fraction boiling point elevation constant. Substituting the data into *(11.23)* gives

$$T_{\text{bp,soln}} = 383.9 \text{ K} + (36.1 \text{ K})(7.14 \times 10^{-3}) = 384.2 \text{ K} = 111.0 \text{ °C}$$

11.50 Using $K_{x,\text{bp}} = 36.1$ K for toluene, calculate the enthalpy of vaporization of toluene. The normal boiling point of toluene is 110.7 °C.

▮ The relation between the boiling point constant and the enthalpy of vaporization is given by

$$K_{x,\mathrm{bp}} = RT_{\mathrm{bp},1}^2/\Delta_{\mathrm{vap}}H_1 \tag{11.24}$$

Solving *(11.24)* for $\Delta_{\mathrm{vap}}H_1$ and substituting the data gives

$$\Delta_{\mathrm{vap}}H(\text{toluene}) = \frac{(8.314\ \mathrm{J\cdot K^{-1}\cdot mol^{-1}})(383.9\ \mathrm{K})^2[(10^{-3}\ \mathrm{kJ})/(1\ \mathrm{J})]}{36.1\ \mathrm{K}} = 33.9\ \mathrm{kJ\cdot mol^{-1}}$$

11.51 What must be the molality of an aqueous solution that has a boiling point elevation of 1.00 K? For H_2O, $K_{\mathrm{bp}} = 0.512\ \mathrm{K\cdot kg\cdot mol^{-1}}$.

▮ In terms of molality, *(11.23)* becomes

$$T_{\mathrm{bp,soln}} = T_{\mathrm{bp},1} + K_{\mathrm{bp}}m \tag{11.25}$$

Solving for m and substituting the data gives

$$m = \frac{T_{\mathrm{bp,soln}} - T_{\mathrm{bp},1}}{K_{\mathrm{bp}}} = \frac{\Delta T_{\mathrm{bp}}}{K_{\mathrm{bp}}} = \frac{1.00\ \mathrm{K}}{0.512\ \mathrm{K\cdot kg\cdot mol^{-1}}} = 1.95\ \mathrm{mol\cdot kg^{-1}}$$

11.52 Calculate the value of K_{bp} for benzene given that $\Delta_{\mathrm{vap}}H = 94.3\ \mathrm{cal\cdot g^{-1}}$ and that the normal boiling point is 80.15 °C.

▮ In terms of molality, *(11.24)* becomes

$$K_{\mathrm{bp}} = \frac{RT_{\mathrm{bp},1}^2 M_1}{[(10^3\ \mathrm{g})/(1\ \mathrm{kg})]\,\Delta_{\mathrm{vap}}H_1} \tag{11.26}$$

Substituting the data into *(11.26)* gives

$$K_{\mathrm{bp}} = \frac{(8.314\ \mathrm{J\cdot K^{-1}\cdot mol^{-1}})(353.30\ \mathrm{K})^2(78.11\ \mathrm{g\cdot mol^{-1}})}{[(10^3\ \mathrm{g})/(1\ \mathrm{kg})](94.3\ \mathrm{cal\cdot g^{-1}})[(4.184\ \mathrm{J})/(1\ \mathrm{cal})](78.11\ \mathrm{g\cdot mol^{-1}})} = 2.63\ \mathrm{K\cdot kg\cdot mol^{-1}}$$

11.53 For water at 1.00 atm, $K_{\mathrm{bp}} = 0.512\ \mathrm{K\cdot kg\cdot mol^{-1}}$. Determine K_{bp} at 1.00 bar. The normal boiling point of water is 99.975 °C.

▮ The pressure dependence of K_{bp} is given by

$$K_{\mathrm{bp}} = \frac{K_{\mathrm{bp}}(1\ \mathrm{atm})}{1 - [(10^3\ \mathrm{g})/(1\ \mathrm{kg})][K_{\mathrm{bp}}(1\ \mathrm{atm})/M_1T]\ln[P/(1\ \mathrm{atm})]} \tag{11.27}$$

Substituting the data gives

$$K_{\mathrm{bp}} = \frac{0.512\ \mathrm{K\cdot kg\cdot mol^{-1}}}{1 - \dfrac{10^3\ \mathrm{g}}{1\ \mathrm{kg}}\left(\dfrac{0.512\ \mathrm{K\cdot kg\cdot mol^{-1}}}{(18.105\ \mathrm{g\cdot mol^{-1}})(373.125\ \mathrm{K})}\right)\ln\left(\dfrac{(1.00\ \mathrm{bar})[(1\ \mathrm{atm})/(1.013\,25\ \mathrm{bar})]}{1\ \mathrm{atm}}\right)}$$
$$= 0.512\ \mathrm{K\cdot kg\cdot mol^{-1}}$$

There is an insignificant change in the value of K_{bp} for this change in standard states.

11.54 A handbook lists the freezing point constant for benzene as $5.12\ \mathrm{K\cdot kg\cdot mol^{-1}}$. What is the value of $K_{x,\mathrm{fp}}$?

▮ Equations similar to *(11.24)* and *(11.26)* relate $K_{x,\mathrm{fp}}$ and K_{fp} through the enthalpy of fusion. Taking a ratio gives

$$\frac{K_{x,\mathrm{fp}}}{K_{\mathrm{fp}}} = \frac{RT_{\mathrm{fp},1}^2/\Delta_{\mathrm{fus}}H_1}{RT_{\mathrm{fp},1}^2 M_1/\{[(10^3\ \mathrm{g})/(1\ \mathrm{kg})]\,\Delta_{\mathrm{fus}}H_1\}} = \frac{(10^3\ \mathrm{g})/(1\ \mathrm{kg})}{M_1} \tag{11.28}$$

for dilute solutions. Solving for $K_{x,\mathrm{fp}}$ and substituting the data gives

$$K_{x,\mathrm{fp}} = \left[\frac{(10^3\ \mathrm{g})/(1\ \mathrm{kg})}{78.11\ \mathrm{g\cdot mol^{-1}}}\right](5.12\ \mathrm{K\cdot kg\cdot mol^{-1}}) = 65.5\ \mathrm{K}$$

11.55 A 1.00-g sample of acetic acid was dissolved in 100.0 g of C_6H_6, and a freezing point depression of 0.45 K was observed. What is the apparent molar mass of CH_3COOH in this solution? For C_6H_6, $K_{\mathrm{fp}} = 5.12\ \mathrm{K\cdot kg\cdot mol^{-1}}$.

▮ The freezing point of a solution is given by

$$T_{\mathrm{fp,soln}} = T_{\mathrm{fp},1} - K_{\mathrm{fp}}m \tag{11.29}$$

Solving for the molality and substituting the data gives

$$m = \frac{-(T_{fp,soln} - T_{fp,1})}{K_{fp}} = \frac{-(-0.45\ K)}{5.12\ K \cdot kg \cdot mol^{-1}} = 0.088\ mol \cdot kg^{-1}$$

Solving *(11.2)* for n(solute) gives

$$n(\text{solute})/(\text{mol}) = (0.088)(0.1000) = 8.8 \times 10^{-3}$$

which corresponds to a molar mass of solute of

$$M(\text{solute}) = \frac{1.00\ g}{8.8 \times 10^{-3}\ mol} = 114\ g \cdot mol^{-1}$$

This value of molar mass is nearly twice the value determined from the formula. This implies that nearly all of the acetic acid molecules are dimerized in the nonpolar solvent as a result of hydrogen bonding taking place at the —COOH function group.

11.56 At 0 °C, $K_2/(\text{bar}) = 2.58 \times 10^4$ for O_2 and 5.36×10^4 for N_2. What is the freezing point of water that is saturated with air at a pressure of 1.013 bar? For H_2O, $K_{fp} = 1.86\ K \cdot kg \cdot mol^{-1}$.

▮ Solving *(11.28)* for $K_{x,fp}$ and substituting the data gives

$$K_{x,fp} = \frac{[(10^3\ g)/(1\ kg)](1.86\ K \cdot kg \cdot mol^{-1})}{18.015\ g \cdot mol^{-1}} = 103\ K$$

In terms of mole fraction, *(11.29)* becomes

$$T_{fp,soln} = T_{fp,1} - K_{x,fp}x_2 \qquad (11.30)$$

Solving *(11.6)* for x_2 and substituting the data gives for each gas

$$x(N_2) = \frac{(1.013\ bar)(0.80)}{5.36 \times 10^4\ bar} = 1.5 \times 10^{-5}$$

$$x(O_2) = \frac{(1.013)(0.20)}{2.58 \times 10^4} = 7.9 \times 10^{-6}$$

For this extremely dilute solution, the contribution to the freezing point lowering by each gas can be calculated using *(11.30)*, and the overall effect will be the sum of the contributions. Solving *(11.30)* for ΔT_{fp} and substituting the data gives

$$\Delta T_{fp}(N_2) = T_{fp,soln} - T_{fp,1} = -K_{x,fp}x(N_2) = -(103\ K)(1.5 \times 10^{-5}) = -1.5 \times 10^{-3}\ K$$

$$\Delta T_{fp}(O_2) = (103\ K)(7.9 \times 10^{-6}) = -8.1 \times 10^{-4}\ K$$

The overall change in the freezing point is -2.3×10^{-3} K, giving

$$T_{fp,soln} = 273.1500\ K - 2.3 \times 10^{-3}\ K = 273.1477\ K$$

11.57 A handbook lists the freezing point depression of a 0.436 m aqueous solution of sucrose as −0.850 °C. Calculate the osmotic coefficient given $K_{fp} = 1.86\ K \cdot kg \cdot mol^{-1}$.

▮ The relation between the osmotic coefficient of a nonelectrolyte and the freezing point depression is

$$\phi = -\Delta T_{fp}/mK_{fp} \qquad (11.31)$$

Substituting the data gives

$$\phi = \frac{-(-0.850\ K)}{(0.436\ mol \cdot kg^{-1})(1.86\ K \cdot kg \cdot mol^{-1})} = 1.05$$

11.58 Use the following data to determine $\gamma(C_{12}H_{22}O_{11})$ in a 1.00 m aqueous solution.

$m/(mol \cdot kg^{-1})$	0.0151	0.0598	0.1055	0.1541	0.2201	0.2711	0.3243	0.4365	0.5564
$[-\Delta T_{fp}/K_{fp}]/(mol \cdot kg^{-1})$	0.0150	0.0600	0.107	0.156	0.225	0.279	0.336	0.457	0.590

$m/(mol \cdot kg^{-1})$	0.6858	0.8246	1.1360	1.2524	1.3753	1.5054	1.7909	2.1154
$[-\Delta T_{fp}/K_{fp}]/(mol \cdot kg^{-1})$	0.736	0.897	1.275	1.421	1.582	1.757	2.160	2.652

▮ The values of $\ln[\gamma(C_{12}H_{22}O_{11})]$ can be calculated by the graphical integration of a plot of $(\phi - 1)/m$ against m, see *(11.21)*. As can be seen from *(11.31)*, the respective values of ϕ can be determined by dividing the value of $(-\Delta T_{fp}/K_{fp})$ by the respective value of m. The plot of $(\phi - 1)/m$ is shown in Fig. 11-12. The area under the curve between $m = 0$ and $m = 1.00\ \text{m}$ is 0.091 69, which upon substitution into *(11.21)* gives

$$\ln[\gamma(C_{12}H_{22}O_{11})] = (1.097 - 1) + 0.091\ 69 = 0.189$$

Taking antilogarithms gives $\gamma(C_{12}H_{22}O_{11}) = 1.208$.

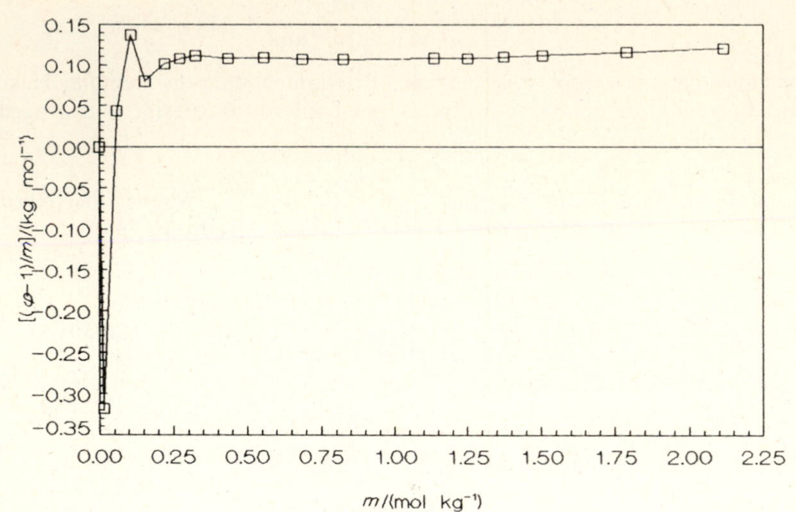

Fig. 11-12

11.59 In the sucrose-water system described in Problem 11.58, the value of ϕ at $m = 1.0000\ \text{mol}\cdot\text{kg}^{-1}$ is 1.097. Determine the activity of water in this solution.

▮ The activity of the solvent is given by

$$\ln a_1 = \frac{-M_1\,\Delta T_{fp}[(10^{-3}\ \text{kg})/(1\ \text{g})]}{K_{fp}} = -M_1 m\phi\frac{10^{-3}\ \text{kg}}{1\ \text{g}} \tag{11.32}$$

Substituting the data and taking antilogarithms gives

$$\ln[a(H_2O)] = -(18.015\ \text{g}\cdot\text{mol}^{-1})(1.000\ \text{mol}\cdot\text{kg}^{-1})(1.097)[(10^{-3}\ \text{kg})/(1\ \text{g})] = -0.019\ 76$$

$$a(H_2O) = 0.980\ 43$$

11.60 Use the following freezing point data for aqueous solutions to determine the molar mass of maltose:

w(maltose)	0.0100	0.0200	0.0300	0.0400	0.0500
$-\Delta T_{fp}/(\text{K})$	0.055	0.112	0.169	0.229	0.290

For H_2O, $K_{fp} = 1.86\ \text{K}\cdot\text{kg}\cdot\text{mol}^{-1}$.

▮ Combining *(11.2)* and *(11.29)* gives

$$M_2/(\text{g}\cdot\text{mol}^{-1}) = \frac{[m_2/(\text{g})][K_{fp}/(\text{K}\cdot\text{kg}\cdot\text{mol}^{-1})]}{[-\Delta T_{fp}/(\text{K})][m_1/(\text{kg})]} \tag{11.33}$$

As a sample calculation, consider $w(\text{maltose}) = 0.0500$. Assuming 100.0 g of solution, $m(\text{maltose}) = 5.00\ \text{g}$ and $m(H_2O) = 0.0950\ \text{kg}$. Substituting into *(11.33)* gives

$$M(\text{maltose})/(\text{g}\cdot\text{mol}^{-1}) = \frac{(5.00)(1.86)}{(0.290)(0.0950)} = 338$$

A plot of $M(\text{maltose})$ against $w(\text{maltose})$ is shown in Fig. 11-13. The extrapolated value is $342\ \text{g}\cdot\text{mol}^{-1}$.

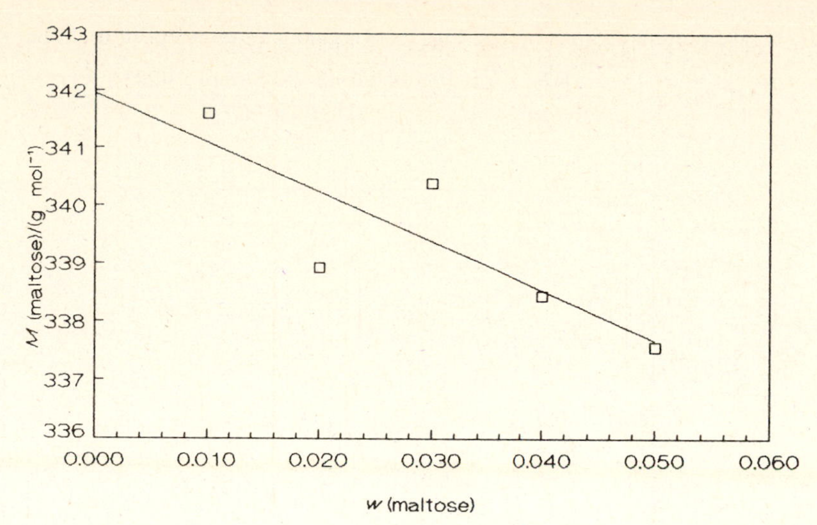

Fig. 11-13

11.61 The osmotic pressure (Π) of a solution of a nonvolatile nonelectrolyte is given by

$$\Pi = -\frac{RT}{V_{m,1}} \ln \frac{P_{soln}}{P_1^\circ} \tag{11.34}$$

Assuming ideal behavior, predict Π for an aqueous glucose solution at 10.2 °C that contains 5.00×10^{-4} mol of glucose per gram of solution. The density of water is $0.9997\ g \cdot cm^{-3}$.

▌ Assuming ideal behavior, *(11.34)* can be written as

$$\Pi = -(RT/V_{m,1}) \ln x_1 \tag{11.35}$$

Each gram of solution contains

$$m(C_6H_{12}O_6) = (5.00 \times 10^{-4}\ mol)(180.16\ g \cdot mol^{-1}) = 0.0901\ g$$

$$m(H_2O) = 1.0000\ g - 0.0901\ g = 0.9099\ g$$

The mole fraction of water in the solution is

$$x(H_2O) = \frac{(0.9099\ g)/(18.015\ g \cdot mol^{-1})}{5.00 \times 10^{-4}\ mol + (0.9099\ g)/(18.015\ g \cdot mol^{-1})} = 0.9902$$

which upon substituting into *(11.35)* gives

$$\Pi = -\frac{(0.083\,14\ bar \cdot L \cdot K^{-1} \cdot mol^{-1})(283.4\ K)}{[(18.015\ g \cdot mol^{-1})/(0.9997\ g \cdot cm^{-3})][(10^{-3}\ L)/(1\ cm^3)]} \ln 0.9902 = 12.88\ bar$$

This prediction differs by 10.1% from the observed value of 11.70 bar.

11.62 What is the osmotic pressure of a 0.109 M aqueous solution of glycerol at 25 °C? Assume that the solution is ideal.

▌ For a dilute, ideal solution,

$$\Pi = CRT \tag{11.36}$$

where C is the concentration expressed in either molarity or molality. Substituting the data gives

$$\Pi = (0.109\ mol \cdot L^{-1})(0.083\,14\ bar \cdot L \cdot K^{-1} \cdot mol^{-1})(298\ K) = 2.70\ bar$$

11.63 Use the following data to determine the molar mass of glucose at 10.2 °C.

$C'/[(g\ C_6H_{12}O_6) \cdot L^{-1}]$	18.1	93.1	193.0
Π/(bar)	2.42	11.70	24.12

▌ For real solutions, *(11.36)* can be written as

$$\Pi/C' = RT/M_2 + bC' \tag{11.37}$$

Values of Π/C' are plotted against C' (see Fig. 11-14), giving an intercept of $0.133\ \text{bar}\cdot\text{L}\cdot\text{g}^{-1}$. From *(11.37)*,

$$M(\text{glucose}) = \frac{RT}{\text{intercept}} = \frac{(0.083\,14\ \text{bar}\cdot\text{L}\cdot\text{K}^{-1}\cdot\text{mol}^{-1})(283.4\ \text{K})}{0.133\ \text{bar}\cdot\text{L}\cdot\text{g}^{-1}} = 177\ \text{g}\cdot\text{mol}^{-1}$$

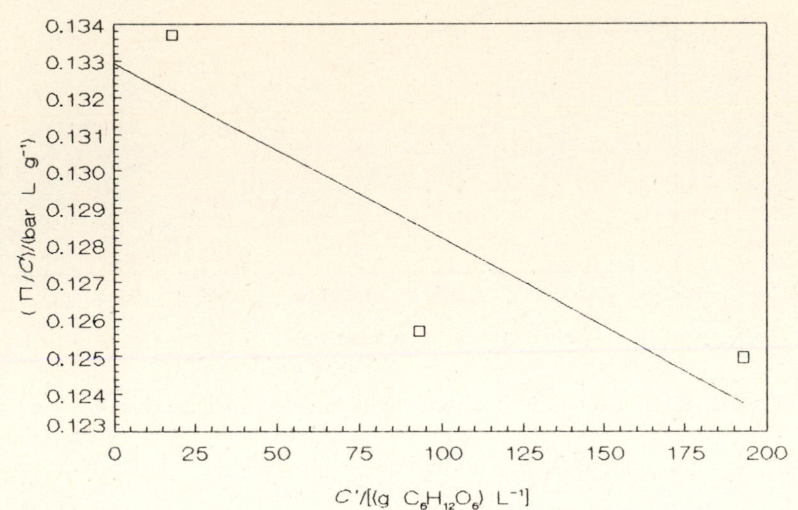

Fig. 11-14

11.64 Determine the osmotic coefficient for the aqueous solution described in Problem 11.61.

▮ The osmotic coefficient is related to the osmotic pressure by

$$\phi = n_1 V_{m,1}\Pi/n_2RT$$

Substituting the data gives

$$\phi = \frac{[(0.9099\ \text{g})/(18.015\ \text{g}\cdot\text{mol}^{-1})][(18.015\ \text{g}\cdot\text{mol}^{-1})/(0.9997\ \text{g}\cdot\text{cm}^{-3})](11.70\ \text{bar})}{(5.00\times10^{-4}\ \text{mol})(0.083\,14\ \text{bar}\cdot\text{L}\cdot\text{K}^{-1}\cdot\text{mol}^{-1})(283.4\ \text{K})[(10^3\ \text{cm}^3)/(1\ \text{L})]} = 0.904$$

11.4 SOLUTIONS OF ELECTROLYTES

11.65 Determine the van't Hoff factor for a 0.475 m aqueous solution of $CaCl_2$ given $\Delta T_{fp} = -2.345$ K. For water, $K_{fp} = 1.86\ \text{kg}\cdot\text{mol}^{-1}$.

▮ For a solution of an electrolyte, *(11.29)* becomes

$$T_{fp,soln} = T_{fp,1} - iK_{fp}m \tag{11.38}$$

where i is the *van't Hoff factor*. Solving *(11.38)* for i and substituting the data gives

$$i = \frac{-(-2.345\ \text{K})}{(1.86\ \text{K}\cdot\text{kg}\cdot\text{mol}^{-1})(0.475\ \text{mol}\cdot\text{kg}^{-1})} = 2.65$$

11.66 Use the following data for aqueous $BaCl_2$ solutions to find the relation between i and ν, the total number of ions produced by a strong electrolyte in solution.

$m/(\text{mol}\cdot\text{kg}^{-1})$	0.001 00	0.005 00	0.010 00	0.020 00	0.050 00	0.100 00
$-\Delta T_{fp}/(\text{K})$	0.005 30	0.025 60	0.050 34	0.098 76	0.239 8	0.469 8

For water, $K_{fp} = 1.86\ \text{K}\cdot\text{kg}\cdot\text{mol}^{-1}$.

▮ Values of i calculated using *(11.38)* are plotted against m in Fig. 11-15. The extrapolated intercept is $i \to \nu = 3$, the total number of cations and anions produced by the electrolyte.

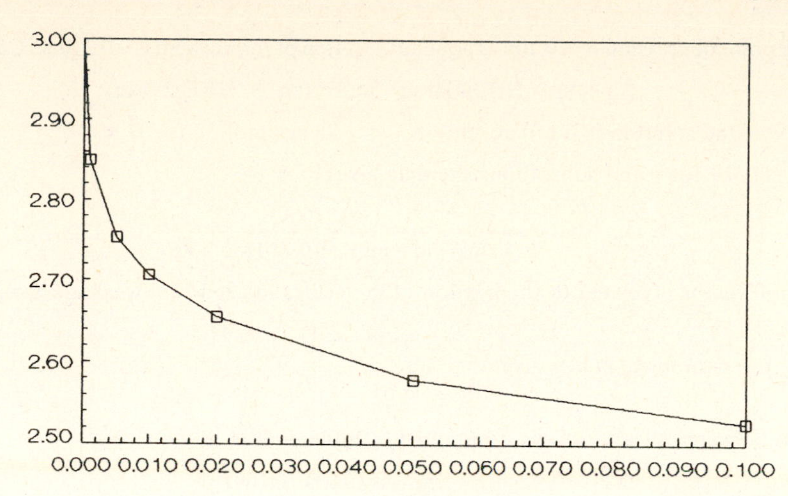

Fig. 11-15

11.67 The freezing point depression of a 1.00×10^{-3} m solution of $K_xFe(CN)_6$ is -7.10×10^{-3} K. Determine x given $K_{fp} = 1.86\ K \cdot kg \cdot mol^{-1}$ for H_2O.

▮ Solving *(11.38)* for i and substituting the data gives

$$i = \frac{-(-7.10 \times 10^{-3}\ K)}{(1.00 \times 10^{-3}\ mol \cdot kg^{-1})(1.86\ K \cdot kg \cdot mol^{-1})} = 3.82$$

Assuming that this value of i implies $\nu = 4$, the formula of the coordination compound is $K_3[Fe(CN)_6]$, or $x = 3$.

11.68 What is the value of the osmotic coefficient for the solution described in Problem 11.67?

▮ The relation between i and ϕ is

$$i = \phi\nu \qquad (11.39)$$

Solving *(11.39)* for ϕ and substituting the data gives $\phi = 3.82/4 = 0.955$.

11.69 The average osmotic pressure of human blood is 7.8 bar at 37 °C. What is the concentration of an aqueous NaCl solution that could be used in the bloodstream? Assume ideal solution behavior.

▮ For a strong electrolyte, *(11.36)* becomes

$$\Pi = iCRT \qquad (11.40)$$

Assuming $i = \nu = 2$, the concentration is given by *(11.40)* as

$$C = \frac{7.8\ bar}{(2)(0.083\ 14\ L \cdot bar \cdot K^{-1} \cdot mol^{-1})(310\ K)} = 0.15\ mol \cdot L^{-1}$$

11.70 Calculate the osmotic pressure for a 0.010 m aqueous solution of NaBr at 25 °C assuming **(*a*)** and ideal solution and **(*b*)** $\gamma_\pm = 0.903$.

▮ Assume that 0.010 m = 0.010 M for this dilute solution. **(*a*)** Using *(11.40)* for the ideal solution gives

$$\Pi = (2)(0.010\ mol \cdot L^{-1})(0.083\ 14\ L \cdot bar \cdot K^{-1} \cdot mol^{-1})(298\ K) = 0.50\ bar$$

(*b*) For a real solution of a 1:1 electrolyte, *(11.40)* can be written as

$$\Pi = RT[\gamma_\pm^2 m^2]^{1/2}\nu \qquad (11.41)$$

Substituting the data into *(11.41)* gives

$$\Pi = (0.083\ 14)(298)[(0.0903)^2(0.010)^2]^{1/2}(2) = 0.45\ bar$$

11.71 The freezing point depression of a 0.001 00 M aqueous solution of HCl is -3.690×10^{-3} K. What is the osmotic pressure of the solution at 25 °C? For H_2O, $K_{fp} = 1.86\ K \cdot kg \cdot mol^{-1}$.

▮ For this dilute solution, assume 0.001 00 M = 0.001 00 m. Solving *(11.38)* for i and substituting the data gives

$$i = \frac{-(-3.601 \times 10^{-3}\ K)}{(1.86\ K \cdot kg \cdot mol^{-1})(0.001\ 00\ mol \cdot kg^{-1})} = 1.94$$

The osmotic pressure is given by *(11.40)* as

$$\Pi = (1.94)(0.001\ 00\ mol \cdot L^{-1})(0.083\ 14\ L \cdot bar \cdot K^{-1} \cdot mol^{-1})(298\ K) = 0.0481\ bar$$

11.72 The freezing point depression of a 0.109 M aqueous solution of formic acid is −0.210 K. Calculate K'_a for the equation

$$HCOOH(aq) \rightleftharpoons H^+(aq) + HCOO^-(aq)$$

The molality of the solution is 0.110 m, and $K_{fp} = 1.86\ \text{kg} \cdot \text{mol}^{-1}$ for H_2O.

▌ Solving *(11.38)* for i and substituting the data gives

$$i = \frac{-(-0.210\ \text{K})}{(1.86\ \text{K} \cdot \text{kg} \cdot \text{mol}^{-1})(0.110\ \text{mol} \cdot \text{kg}^{-1})} = 1.03$$

The van't Hoff factor is related to the fraction of molecules ionized in a weak electrolyte by

$$\alpha = (i - 1)/(\nu - 1) \tag{11.42}$$

Substituting the data into *(11.42)* gives

$$\alpha = (1.03 - 1)/(2 - 1) = 0.03$$

The relation between K'_a and α was determined in Problem 9.26 as

$$K'_a = \frac{\alpha^2 C}{1 - \alpha} = \frac{(0.03)^2(0.109)}{1 - 0.03} = 1 \times 10^{-4}$$

The accepted value is 1.772×10^{-4}.

11.5 PARTIAL MOLAR QUANTITIES

11.73 The volume of aqueous NaOH solutions containing 1000.00 g of H_2O for $0 \le m/(\text{mol} \cdot \text{kg}^{-1}) \le 2$ is given by

$$V/(\text{cm}^3) = 1001.56 - 4.35[m/(\text{mol} \cdot \text{kg}^{-1})] + 1.74[m/(\text{mol} \cdot \text{kg}^{-1})]^2$$

Determine the partial molar volume of NaOH in a 1.000 m solution.

▌ If X represents an extensive property of a system, the *partial molar quantity* is defined as

$$\bar{X}_i = \left(\frac{\partial X}{\partial n_i}\right)_{T,P,n_{j \neq i}} \tag{11.43}$$

Because the concentration is given as molality,

$$n(H_2O) = \frac{1000.00\ \text{g}}{18.015\ \text{g} \cdot \text{mol}^{-1}} = 55.509\ \text{mol}$$

and is constant—as required by *(11.43)*—and $n(\text{NaOH}) = m$. Thus,

$$V(\text{NaOH})/(\text{cm}^3 \cdot \text{mol}^{-1}) = \left(\frac{\partial[V/(\text{cm}^3)]}{\partial n(\text{NaOH})}\right)_{T,P,n(H_2O)} = -4.35 + 3.48\,[m/(\text{mol} \cdot \text{kg}^{-1})]$$

At 1.00 m,

$$\bar{V}(\text{NaOH})/(\text{cm}^3 \cdot \text{mol}^{-1}) = -4.35 + (3.48)(1.000) = -0.87$$

11.74 Using the data given in Problem 11.73, determine the partial molar volume of H_2O for $m = 1.000\ \text{mol} \cdot \text{kg}^{-1}$.

▌ The relation between the extensive property of the system and the partial molar quantities is

$$X = \sum_i^{\text{components}} \bar{X}_i n_i \tag{11.44}$$

At $m = 1.000$ m,

$$V/(\text{cm}^3) = 1001.56 - (4.35)(1.000) + (1.74)(1.000)^2 = 998.95$$

Substituting into *(11.44)* and solving for $\bar{V}(H_2O)$ gives

$$998.95\ \text{cm}^3 = (-0.87\ \text{cm}^3 \cdot \text{mol}^{-1})(1.000\ \text{mol}) + \bar{V}(H_2O)(55.509\ \text{mol})$$

$$\bar{V}(H_2O) = 18.012\ \text{cm}^3$$

11.75 Use the following volume-concentration data to determine $\bar{V}(\text{NaOH})$ at $m = 1.000\ \text{mol} \cdot \text{kg}^{-1}$.

$m/(\text{mol} \cdot \text{kg}^{-1})$	0.773	0.907	1.042	1.178	1.316
$V/(\text{cm}^3)$	999.15	999.01	998.91	998.87	998.89

where V is the volume of solution containing 1000.00 g of H_2O.

▌ A plot of V against m is shown in Fig. 11-16. The value of $\bar{V}(\text{NaOH})$ is predicted by *(11.43)* to be equal to the slope of the tangent to the curve at $m = 1.000\ \text{mol}\cdot\text{kg}^{-1}$. For the graph, $\bar{V}(\text{NaOH}) = -0.62\ \text{cm}^3\cdot\text{mol}^{-1}$.

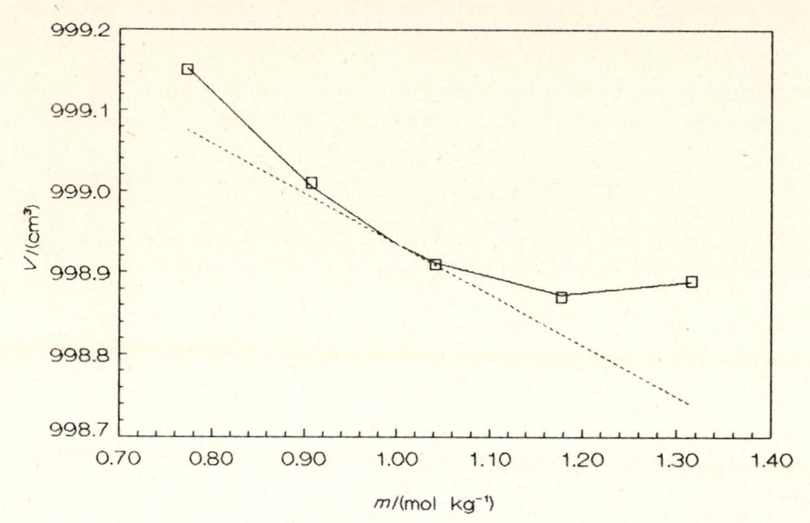

Fig. 11-16

11.76 Using the data given in Problem 11.75, calculate the apparent molar volume for NaOH for each solution and use these results to determine $\bar{V}(\text{NaOH})$ at $m = 1.000\ \text{mol}\cdot\text{kg}^{-1}$. The density of water is $0.9982\ \text{g}\cdot\text{cm}^{-3}$, and $V = 998.94\ \text{cm}^3$ at $m = 1.000\ \text{mol}\cdot\text{kg}^{-1}$.

▌ The *apparent molar quantity* is defined as

$$\phi_i = (X - n_1 X^{\circ}_{m,1})/n_i \tag{11.45}$$

where $X^{\circ}_{m,1}$ is the molar value for the pure solvent. As a sample calculation, consider $m = 1.000\ \text{mol}\cdot\text{kg}^{-1}$.

$$\phi(\text{NaOH}) = \frac{998.94\ \text{cm}^3 - (55.509\ \text{mol})[(18.015\ \text{g}\cdot\text{mol}^{-1})/(0.9982\ \text{g}\cdot\text{cm}^3)]}{1.000\ \text{mol}} = -2.86\ \text{cm}^3\cdot\text{mol}^{-1}$$

The relation between the partial molar quantity and ϕ_i is

$$\bar{x}_i = \phi_i + n_i\frac{\partial\phi_i}{\partial n_i} = \phi_i + \frac{\partial\phi_i}{\partial\{\ln[n_i/(\text{mol})]\}} \tag{11.46}$$

The $\partial\phi_i/\partial(\ln n_i)$ term can be evaluated by determining the slope of a plot of ϕ_i against $\ln n_i$ at the appropriate point. From the plot of $\phi(\text{NaOH})$ against $\ln[n(\text{NaOH})]$ shown in Fig. 11-17, the slope is $2.27\ \text{cm}^3\cdot\text{mol}^{-1}$, giving

$$V(\text{NaOH}) = -2.86\ \text{cm}^3\cdot\text{mol}^{-1} + 2.27\ \text{cm}^3\cdot\text{mol}^{-1} = -0.59\ \text{cm}^3\cdot\text{mol}^{-1}$$

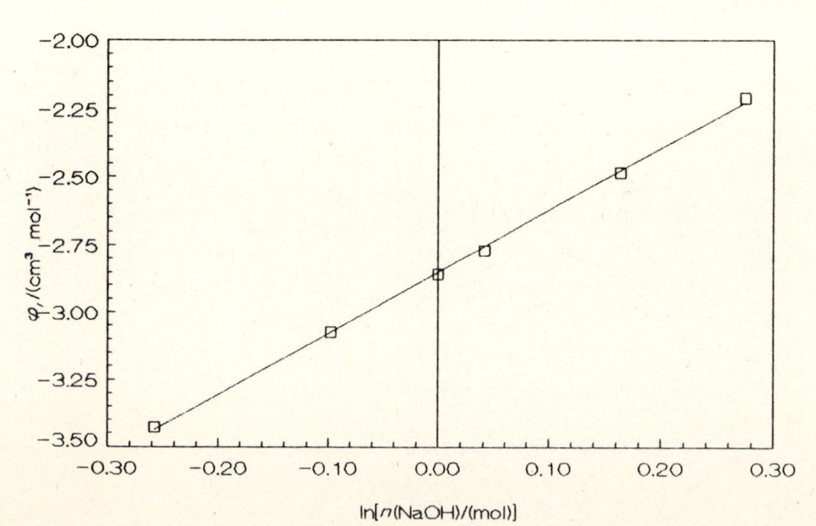

Fig. 11-17

11.77 Use the data given in Problem 11.75 to determine $\bar{V}(\text{NaOH})$ and $\bar{V}(\text{H}_2\text{O})$ at $m = 1.000\ \text{mol}\cdot\text{kg}^{-1}$, using the method of intercepts.

▮ The *method of intercepts* is based on preparing a plot of X', where

$$X' = X/(n_1 + n_2) \tag{11.47}$$

against x_2 and drawing a tangent to the curve at the desired concentration. The intercepts of the tangent on the axes gives the respective values of $\bar{X}_i$. As a sample calculation, for $m = 1.000\ \text{mol}\cdot\text{kg}^{-1}$,

$$V' = \frac{998.94\ \text{cm}^3}{55.509\ \text{mol} + 1.000\ \text{mol}} = 17.678\ \text{cm}^3\cdot\text{mol}^{-1}$$

The graph is shown in Fig. 11-18. The intercept on the $x(\text{NaOH}) = 0$ axis gives $\bar{V}(\text{H}_2\text{O}) = 18.005\ \text{cm}^3\cdot\text{mol}^{-1}$, and an extrapolation to the $x(\text{NaOH}) = 1$ axis gives $-0.48\ \text{cm}^3\ \text{mol}^{-1}$.

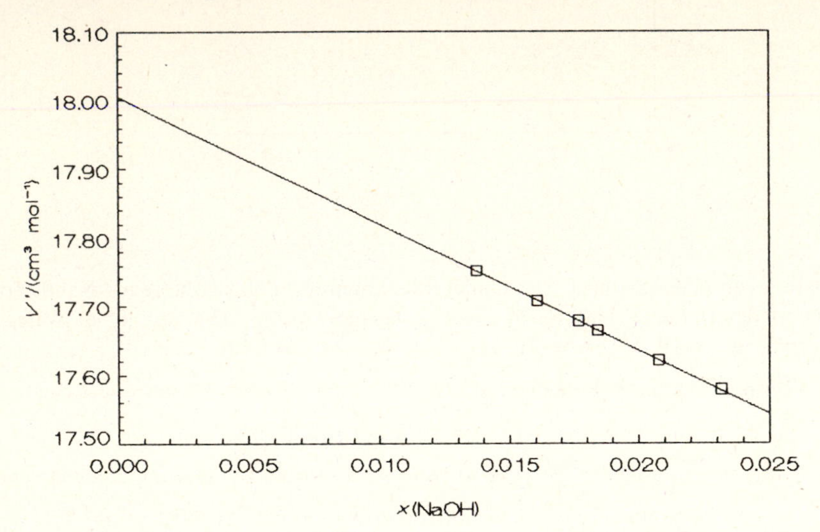

Fig. 11-18

11.78 A 1.00-mol sample of $\text{CH}_3\text{CH}_2\text{OH}(\text{l})$ was added to an infinitely large sample of a $\text{CH}_3\text{CH}_2\text{OH}$–$\text{H}_2\text{O}$ solution with $x(\text{CH}_3\text{CH}_2\text{OH}) = 0.090$. What was the change in volume for the process? For this solution, $\bar{V}(\text{H}_2\text{O}) = 18.1\ \text{cm}^3\cdot\text{mol}^{-1}$ and $\bar{V}(\text{CH}_3\text{CH}_2\text{OH}) = 52.2\ \text{cm}^3\cdot\text{mol}^{-1}$.

▮ For this process, $\Delta V = \bar{V}(\text{CH}_3\text{CH}_2\text{OH}) = 52.2\ \text{cm}^3$.

11.79 Using the data given in Problem 11.78, determine the volumes of the pure liquids needed to prepare $100.0\ \text{cm}^3$ of a solution with $x(\text{CH}_3\text{CH}_2\text{OH}) = 0.090$. The densities of the pure liquids are $0.7893\ \text{g}\cdot\text{cm}^{-3}$ for $\text{CH}_3\text{CH}_2\text{OH}$ and $0.9970\ \text{g}\cdot\text{cm}^{-3}$ for H_2O.

▮ For this solution, *(11.44)* gives

$$100.0\ \text{cm}^3 = n(\text{H}_2\text{O})(18.1\ \text{cm}^3\cdot\text{mol}^{-1}) + n(\text{CH}_3\text{CH}_2\text{OH})(52.2\ \text{cm}^3\cdot\text{mol}^{-1})$$

$$0.090 = n(\text{CH}_3\text{CH}_2\text{OH})/[n(\text{CH}_3\text{CH}_2\text{OH}) + n(\text{H}_2\text{O})]$$

Solving these equations gives $n(\text{CH}_3\text{CH}_2\text{OH}) = 0.425\ \text{mol}$ and $n(\text{H}_2\text{O}) = 4.297\ \text{mol}$. The volumes of the liquids needed are

$$V(\text{CH}_3\text{CH}_2\text{OH}) = \frac{(0.425)(46.07\ \text{g}\cdot\text{mol}^{-1})}{0.7893\ \text{g}\cdot\text{cm}^3} = 24.8\ \text{cm}^3$$

$$V(\text{H}_2\text{O}) = \frac{(4.297)(18.015)}{0.9970} = 77.6\ \text{cm}^3$$

CHAPTER 12
Rates of Chemical Reactions

12.1 RATE EQUATIONS FOR SIMPLE REACTIONS

12.1 For the general chemical equation

$$\sum_i \nu_i I_i \tag{12.1}$$

where ν_i is the stoichiometric coefficient and I is the chemical formula of substance i, write the expression for the extent of reaction, the rate of reaction, the rate of change of i, and the rate of change of the concentration of i.

▮ The *extent of reaction* is given by

$$d\xi = \frac{1}{\nu_i}\, dn_i \tag{12.2}$$

and the *rate of reaction* is given by

$$\frac{d\xi}{dt} = \frac{1}{\nu_i}\frac{dn_i}{dt} \tag{12.3}$$

The last term in *(12.3)*, dn_i/dt, is the *rate of change of i*, and the *rate of change of the concentration of i* is dC_i/dt.

12.2 Express the rate of reaction for

$$H_3AsO_4(aq) + 2H^+(aq) + 3I^-(aq) \rightarrow H_3AsO_3(aq) + I_3^-(aq) + H_2O(l)$$

▮ Substituting into *(12.3)* gives

$$\frac{d\xi}{dt} = \frac{dn(H_3AsO_3)}{dt} = \frac{dn(I_3^-)}{dt} = \frac{dn(H_2O)}{dt} = -\frac{dn(H_3AsO_4)}{dt} = -\frac{1}{2}\frac{dn(H^+)}{dt} = -\frac{1}{3}\frac{dn(I^-)}{dt}$$

12.3 What is the relation among the various rates of concentration for the following equation?

$$2N_2O_5(g) \rightarrow 2N_2O_4(g) + O_2(g)$$

▮ For this reaction,

$$\frac{1}{2}\frac{dC(N_2O_4)}{dt} = -\frac{1}{2}\frac{dC(N_2O_5)}{dt} = \frac{dC(O_2)}{dt}$$

12.4 For gaseous reactions, the rate is often described in terms of dP_i/dt instead of dC_i/dt or dn_i/dt. What is the relation among these three expressions?

▮ Substituting *(1.18)* into dP_i/dt gives

$$\frac{dP_i}{dt} = \frac{d(n_iRT/V)}{dt} = \frac{1}{VRT}\frac{dn_i}{dt}$$

Recognizing that $C_i = n_i/V$ gives

$$\frac{dC_i}{dt} = \frac{d(n_i/V)}{dt} = \frac{1}{V}\frac{dn_i}{dt}$$

Comparing both results gives

$$\frac{dP_i}{dt} = \frac{1}{VRT}\frac{dn_i}{dt} = \frac{1}{RT}\frac{dC_i}{dt}$$

12.5 During the initial stages of the reaction $H_2(g) + Br_2(g) \rightarrow 2HBr(g)$ the rate law is

$$\frac{dC(HBr)}{dt} = kC(H_2)C(Br_2)^{1/2}$$

where k is the rate constant. What is the order of the reaction with respect to each component and the overall order?

▮ The *order of reaction* is equal to the exponent on the concentration term. Thus the reaction is first-order with respect to H_2 and one-half-order with respect to Br_2. The overall order is $1 + \frac{1}{2} = \frac{3}{2}$.

12.6 Is the reaction described in Problem 12.5 an elementary reaction?

▮ If the reaction were elementary, the reaction rate would involve one molecule of H_2 reacting with one molecule of Br_2 to give the bimolecular rate law

$$\frac{dC(\text{HBr})}{dt} = kC(\text{H}_2)C(\text{Br}_2)$$

Because this is not the observed rate law, the reaction is not elementary.

12.7 The rate law for the equation $CH_3NC(g) \rightarrow CH_3CN(g)$ is first-order in CH_3NC at high pressures and is second-order in CH_3NC at low pressures. Write the rate law for each case.

▮ At high pressures the rate law is

$$-\frac{dC(\text{CH}_3\text{NC})}{dt} = k_1C(\text{CH}_3\text{NC})$$

and at low pressures the rate law is

$$-\frac{dC(\text{CH}_3\text{NC})}{dt} = k_2C(\text{CH}_3\text{NC})^2$$

12.8 The reaction $2H(g) + Ar(g) \rightarrow H_2(g) + Ar(g)$ is elementary with $k = 2.2 \times 10^8\ \text{mol}^{-2} \cdot \text{L}^2 \cdot \text{s}^{-1}$ at 3000 K. Write the rate law for this process.

▮ The order of reaction for each reactant will be equal to the stoichiometric coefficient for an elementary reaction. Thus

$$\frac{dC(\text{H}_2)}{dt} = (2.2 \times 10^8\ \text{mol}^{-2} \cdot \text{L}^2 \cdot \text{s}^{-1})C(\text{H})^2C(\text{Ar})$$

12.9 Confirm the units on the rate constant given in Problem 12.8.

▮ In general, the units for k are $C^{1-(\text{overall order})}t^{-1}$. In this case the units are

$$(\text{mol} \cdot \text{L}^{-1})^{1-3}(\text{s})^{-1} = \text{mol}^{-2} \cdot \text{L}^2 \cdot \text{s}^{-1}$$

12.10 At 3000 K, $-dC(\text{H}_2)/dt = kC(\text{H}_2)C(\text{Ar})$, where $k = 2.2 \times 10^4\ \text{mol}^{-1} \cdot \text{L} \cdot \text{s}^{-1}$ for the equation

$$\text{H}_2(g) + \text{Ar}(g) \rightarrow 2\text{H}(g) + \text{Ar}(g)$$

Calculate $-dC(\text{H}_2)/dt$ given $C(\text{H}_2) = 4.1 \times 10^{-3}\ \text{mol} \cdot \text{L}^{-1}$ and $C(\text{Ar}) = 4.1 \times 10^{-4}\ \text{mol} \cdot \text{L}^{-1}$.

▮ Substituting the concentrations into the rate law gives

$$-\frac{dC(\text{H}_2)}{dt} = (2.2 \times 10^4\ \text{mol}^{-1} \cdot \text{L} \cdot \text{s}^{-1})(4.1 \times 10^{-3}\ \text{mol} \cdot \text{L}^{-1})(4.1 \times 10^{-4}\ \text{mol} \cdot \text{L}^{-1})$$

$$= 0.037\ \text{mol} \cdot \text{L}^{-1} \cdot \text{s}^{-1}$$

12.11 What concentration of Ar is needed to double the reaction rate for the system described in Problem 12.10?

▮ Solving the rate law given in Problem 12.10 for $C(\text{Ar})$ and substituting the data gives

$$C(\text{Ar}) = \frac{(2)(0.037\ \text{mol} \cdot \text{L}^{-1} \cdot \text{s}^{-1})}{(2.2 \times 10^4\ \text{mol}^{-1} \cdot \text{L} \cdot \text{s}^{-1})(4.1 \times 10^{-3}\ \text{mol} \cdot \text{L}^{-1})} = 8.2 \times 10^{-4}\ \text{mol} \cdot \text{L}^{-1}$$

12.12 Derive the integrated rate law for an nth-order reaction for the chemical equation $\text{A} \rightarrow \text{products}$, where $n \neq 1$. Repeat the derivation for $n = 1$.

▮ Integrating the rate law

$$-\frac{dC(\text{A})}{dt} = kC(\text{A})^n$$

between $C(\mathrm{A}) = C(\mathrm{A})_0$ at $t = 0$ and $C(\mathrm{A})$ at t gives

$$[C(\mathrm{A})]^{1-n} - [C(\mathrm{A})_0]^{1-n} = -(1-n)kt \tag{12.4}$$

The integration of $-dC(\mathrm{A})/dt = kC(\mathrm{A})^1$ gives

$$\ln[C(\mathrm{A})] - \ln[C(\mathrm{A})_0] = -kt \tag{12.5}$$

12.13 Derive equations for t_α, the time required for $\xi = 1 - \alpha$, for $n \neq 1$ and for $n = 1$.

▌ Multiplying *(12.4)* by $(1-n)[C(\mathrm{A})_0]^{1-n}$ gives

$$\left(\frac{C(\mathrm{A})}{C(\mathrm{A})_0}\right)^{1-n} = 1 + [C(\mathrm{A})_0]^{n-1}(n-1)kt$$

Substituting $\alpha = C(\mathrm{A})/C(\mathrm{A})_0$ and solving for t_α gives

$$t_\alpha = \frac{\alpha^{1-n} - 1}{[C(\mathrm{A})_0]^{n-1}(n-1)k} \tag{12.6}$$

Likewise, for *(12.5)*, substituting $\alpha = C(\mathrm{A})/C(\mathrm{A})_0$ and solving for t_α gives

$$t_\alpha = (-\ln \alpha)/k \tag{12.7}$$

12.14 A chemical reaction is known to be zeroth-order with $k = 5 \times 10^{-8}\ \mathrm{mol \cdot L^{-1} \cdot s^{-1}}$. **(a)** How long does it take for $C(\mathrm{A})$ to decrease from $4 \times 10^{-4}\ \mathrm{mol \cdot L^{-1}}$ to $2 \times 10^{-2}\ \mathrm{mol \cdot L^{-1}}$? **(b)** How long does it take for $C(\mathrm{A})$ to decrease from $2 \times 10^{-2}\ \mathrm{mol \cdot L^{-1}}$ to $1 \times 10^{-2}\ \mathrm{mol \cdot L^{-1}}$?

▌ Both of these changes are half-life periods. Substituting $n = 0$ and $\alpha = 0.5$ into *(12.6)* gives

$$t_{1/2} = 0.5C(\mathrm{A})_0/k$$

Substituting the data for both changes gives

(a)
$$t_{1/2} = \frac{(0.5)(4 \times 10^{-4}\ \mathrm{mol \cdot L^{-1}})}{5 \times 10^{-8}\ \mathrm{mol \cdot L^{-1} \cdot s^{-1}}} = 4 \times 10^3\ \mathrm{s}$$

(b)
$$t_{1/2} = \frac{(0.5)(2 \times 10^{-4})}{5 \times 10^{-8}} = 2 \times 10^3\ \mathrm{s}$$

12.15 Show that any property that is directly proportional to concentration can be used to confirm that a reaction is first-order and gives the correct value of k.

▌ Substituting $x = bC(\mathrm{A})$, where x is the property being measured and b is the proportionality constant,

$$\ln \frac{x/b}{x_0/b} = \ln \frac{x}{x_0} = -kt$$

12.16 Use the following time-concentration data for the equation α-mannose(aq) → β-mannose(aq) to confirm that the reaction is first-order.

t/(s)	0	900	2700	4500	6300	8100	10 500	12 900	15 600	18 000	∞
S	3.55	3.90	4.85	5.70	6.25	6.95	7.80	8.65	9.50	10.15	19.3

where S is proportional to the concentration of β-mannose as measured in a saccharimeter. Determine the value of k.

▌ Substituting $C(\alpha\text{-mannose})_0 = S_\infty - S_0$ and $C(\alpha\text{-mannose}) = S_\infty - S$ into *(12.5)* gives

$$\ln(S_\infty - S) = \ln(S_\infty - S_0) - kt$$

Thus a plot of $\ln(S_\infty - S)$ against t will be linear for this first-order reaction and will have a slope equal to $-k$. The plot is shown in Fig. 12-1, giving

$$k = -\mathrm{slope} = -(-3.02 \times 10^{-5}\ \mathrm{s^{-1}}) = 3.02 \times 10^{-5}\ \mathrm{s^{-1}}$$

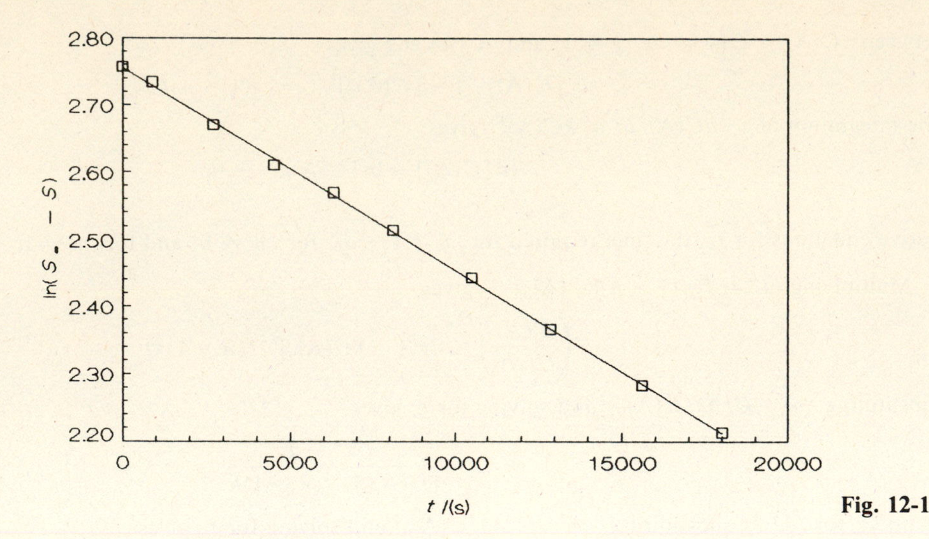

Fig. 12-1

12.17 The inversion of sucrose was studied at 40 °C for a solution containing 17.1 mass % sucrose and 0.02 M HCl. Confirm that the reaction is first-order, and determine k using the following data.

t/(s)	0	1188	2383	3561	3781	4650	6085	10 897	13 578
fraction of sucrose remaining	1.000	0.975	0.950	0.924	0.920	0.900	0.870	0.780	0.734

t/(s)	17 941	22 281	29 412	36 522	48 698	66 036	69 150	89 184	161 784
fraction of sucrose remaining	0.660	0.613	0.507	0.430	0.324	0.218	0.200	0.126	0.020

▮ The data represent the values of $C(\text{sucrose})/C(\text{sucrose})_0$ for various times. For a first-order reaction, *(12.5)* indicates that a plot of $\ln[C(\text{sucrose})/C(\text{sucrose})_0]$ against t will be linear and have a slope equal to $-k$. From Fig. 12-2,

$$k = -\text{slope} = -(-2.39 \times 10^{-5}\ \text{s}^{-1}) = 2.39 \times 10^{-5}\ \text{s}^{-1}$$

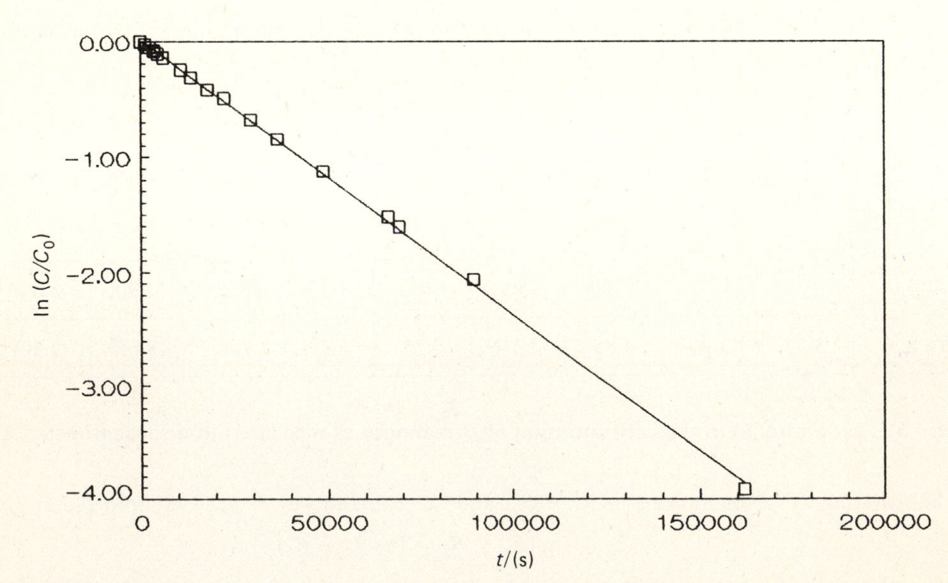

Fig. 12-2

12.18 Use the data given in Problem 12.17 to graphically determine $t_{1/2}$ for the reaction. Compare the result to that predicted by *(12.7)*.

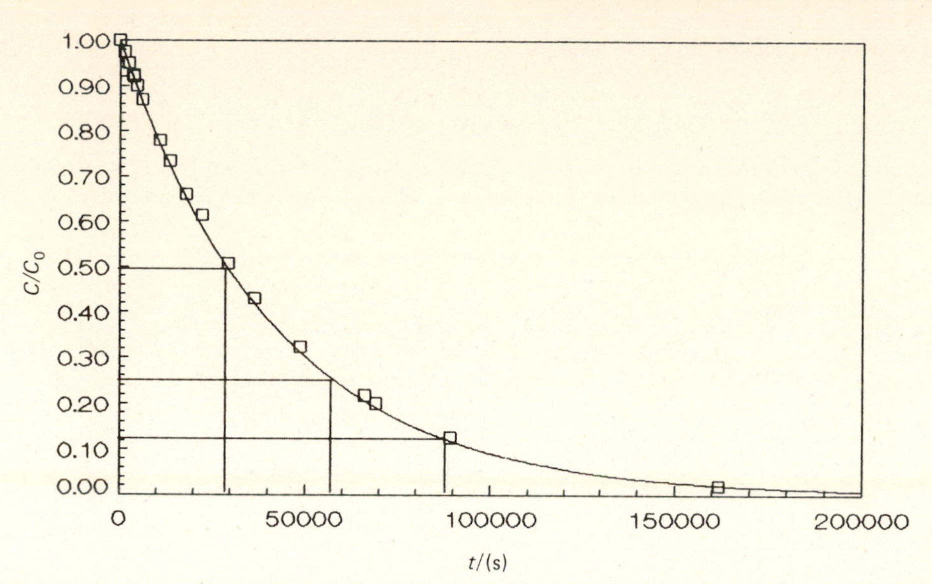

Fig. 12-3

▌ The $C(\text{sucrose})/C(\text{sucrose})_0$ data are plotted against t in Fig. 12-3. The time required for the fraction to decrease from 1.000 to 0.500 is 29 000 s, the time from 0.500 to 0.250 is 28 000 s, and the time from 0.250 to 0.125 is 31 000 s. The average half-life is 29 000 s. Substituting $k = 2.3 \times 10^{-5}\ \text{s}^{-1}$ into *(12.7)* gives

$$t_{1/2} = \frac{-\ln 0.5}{2.39 \times 10^{-5}\ \text{s}^{-1}} = 2.90 \times 10^4\ \text{s}$$

which is in good agreement with the average graphical value.

12.19 What is the ratio of $t_{1/2}$ to $t_{1/3}$ for a first-order reaction?

▌ Substituting into *(12.7)* gives

$$\frac{t_{1/2}}{t_{1/3}} = \frac{(-\ln 0.500)/k}{(-\ln 0.333)/k} = 0.630$$

12.20 Calculate the average life for a reactant undergoing a first-order reaction.

▌ The *average lifetime* for reactant A is given by

$$\bar{t} = \frac{\int_0^\infty C(\text{A})\,dt}{C(\text{A})_0}$$

Substituting the exponential form of *(12.5)* and integrating gives

$$\bar{t} = \frac{\int_0^\infty C(\text{A})_0\,e^{-kt}\,dt}{C(\text{A})_0} = \int_0^\infty e^{-kt}\,dt = \frac{1}{k}$$

12.21 Consider the chemical equation $\text{A(g)} \rightarrow n\text{B(g)}$, which is first-order. Derive an equation for the total pressure of the system as a function of time.

▌ At any time, *(12.5)* gives $P(\text{A}) = P(\text{A})_0\,e^{-kt}$. The partial pressure of B is given by

$$P(\text{B}) = n[P(\text{A})_0 - P(\text{A})] = n[P(\text{A})_0 - P(\text{A})_0\,e^{-kt}] = nP(\text{A})_0(1 - e^{-kt})$$

where $P(\text{A})_0 - P(\text{A})$ represents the amount of A that has reacted. Substituting into *(1.22)* gives

$$P_t = P(\text{A})_0\,e^{-kt} + nP(\text{A})_0(1 - e^{-kt}) = P(\text{A})_0[n + (1 - n)\,e^{-kt}]$$

Note that if $n = 1$, the total pressure remains constant.

12.22 On the basis of the chemical equation $2\text{N}_2\text{O}_5\text{(g)} \rightarrow 2\text{N}_2\text{O}_4\text{(g)} + \text{O}_2\text{(g)}$ a student assumed that the reaction would be second-order. Using the following data, determine whether or not the student's assumption was correct.

t/(s)	0	1200	2400	3600	4800	6000	7200	8400	9600	10 800	12 000	13 200
$P(\text{N}_2\text{O}_5)$/(torr)	268.7	247.2	236.2	227.1	217.8	209.5	201.8	193.2	185.8	178.1	164.9	152.4

▮ For a second-order reaction, *(12.4)* gives

$$\frac{1}{C(\text{A})} = \frac{1}{C(\text{A})_0} + kt$$

This equation implies that a plot of $[C(\text{A})]^{-1}$ against t will be linear. From Fig. 12-4, a plot of $1/P$ against t (see Problem 12.4), it can be seen that this is not the case. The reaction is not second-order.

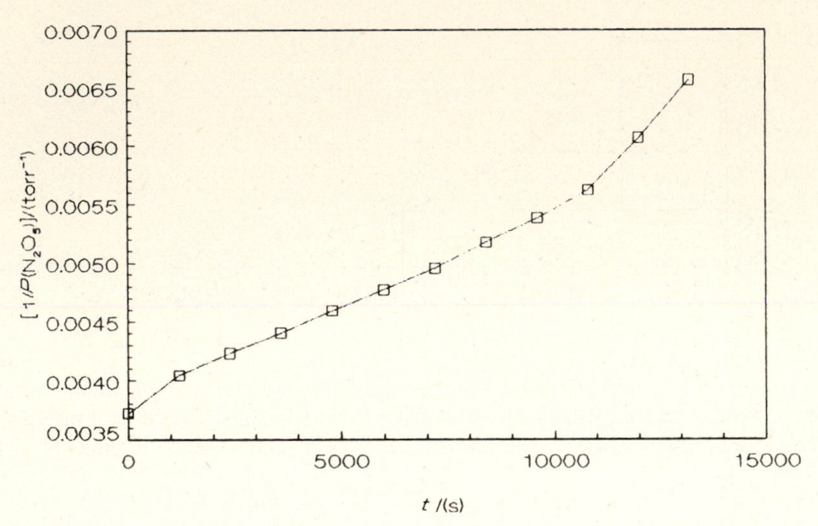

Fig. 12-4

12.23 For the bimolecular equation

$$O(g) + OH(g) \rightarrow O_2(g) + H(g)$$

the rate constant is $k = 1.3 \times 10^{10}\ \text{L} \cdot \text{mol}^{-1} \cdot \text{s}^{-1}$. Compare the initial value of $dC(\text{O})/dt$ for $C(\text{O})_0 = C(\text{OH})_0 = 1.0 \times 10^{-4}\ \text{mol} \cdot \text{L}^{-1}$ to that for $C(\text{O})_0 = 1.0 \times 10^{-4}\ \text{mol} \cdot \text{L}^{-1}$ and $C(\text{OH})_0 = 2.0 \times 10^{-4}\ \text{mol} \cdot \text{L}^{-1}$.

▮ The rate of change of the concentration of O(g) is

$$\frac{dC(\text{O})}{dt} = -kC(\text{O})C(\text{OH})$$

For the reaction in which both initial concentrations are the same,

$$\frac{dC(\text{O})}{dt} = -(1.3 \times 10^{10}\ \text{L} \cdot \text{mol}^{-1} \cdot \text{s}^{-1})(1.0 \times 10^{-4}\ \text{mol} \cdot \text{L}^{-1})(1.0 \times 10^{-4}\ \text{mol} \cdot \text{L}^{-1})$$

$$= -130\ \text{mol} \cdot \text{L}^{-1} \cdot \text{s}^{-1}$$

and for the reaction in which $C(\text{OH})_0$ is $2C(\text{O})_0$,

$$\frac{dC(\text{O})}{dt} = -(1.3 \times 10^{10})(1.0 \times 10^{-4})(2 \times 10^{-4}) = -260\ \text{mol} \cdot \text{L}^{-1} \cdot \text{s}^{-1}$$

Doubling the initial concentration of one of the components increases the initial reaction rate by a factor of 2.

12.24 What is the concentration of O(g) at $t = 1.0 \times 10^{-6}\ \text{s}$ for the system described in Problem 12.23?

▮ The integrated rate law for the reaction in which the initial concentrations were identical is given by *(12.4)* as

$$\frac{1}{C(\text{O})} = \frac{1}{C(\text{O})_0} + kt = \frac{1}{1.0 \times 10^{-4}\ \text{mol} \cdot \text{L}^{-1}} + (1.3 \times 10^{10}\ \text{L} \cdot \text{mol}^{-1} \cdot \text{s}^{-1})(1.0 \times 10^{-6}\ \text{s})$$

$$= 2.3 \times 10^{4}\ \text{L} \cdot \text{mol}^{-1}$$

giving $C(\text{O}) = 4.3 \times 10^{-5}\ \text{mol} \cdot \text{L}^{-1}$. The integrated rate law for $C(\text{O})_0 \neq C(\text{OH})$ is

$$\frac{1}{C(\text{O})_0 - C(\text{OH})_0} \ln \frac{C(\text{OH})_0 C(\text{O})}{C(\text{O})_0 C(\text{OH})} = kt$$

At any time during the reaction, $C(\text{O}) = 1.0 \times 10^{-4}\ \text{mol} \cdot \text{L}^{-1} - x$ and $C(\text{OH}) = 2.0 \times 10^{-4}\ \text{mol} \cdot \text{L}^{-1} - x$, where x represents the concentration of each that has reacted. Substituting into the integrated rate law and solving gives

$$\frac{1}{1.0 \times 10^{-4}\,\text{mol}\cdot\text{L}^{-1} - 2.0 \times 10^{-4}\,\text{mol}\cdot\text{L}^{-1}} \ln \frac{(2.0 \times 10^{-4}\,\text{mol}\cdot\text{L}^{-1})(1.0 \times 10^{-4}\,\text{mol}\cdot\text{L}^{-1} - x)}{(1.0 \times 10^{-4}\,\text{mol}\cdot\text{L}^{-1})(2.0 \times 10^{-4}\,\text{mol}\cdot\text{L}^{-1} - x)}$$

$$= (1.3 \times 10^{10}\,\text{L}\cdot\text{mol}^{-1}\cdot\text{s}^{-1})(1.0 \times 10^{-6}\,\text{s})$$

$$\ln \frac{2(1.0 \times 10^{-4} - x)}{2.0 \times 10^{-4} - x} = -1.3 \qquad \frac{2(1.0 \times 10^{-4} - x)}{2.0 \times 10^{-4} - x} = 0.27$$

$$x = 8.4 \times 10^{-5}\,\text{mol}\cdot\text{L}^{-1}$$

Thus $C(\text{O}) = 1.0 \times 10^{-4}\,\text{mol}\cdot\text{L}^{-1} - 8.4 \times 10^{-5}\,\text{mol}\cdot\text{L}^{-1} = 1.6 \times 10^{-5}\,\text{mol}\cdot\text{L}^{-1}$.

12.25 How does the rate of formation of $H_2(g)$ change as the pressure of Ar(g) is doubled in the system described in Problem 12.8? Suppose $P(\text{H})_0 \ll P(\text{Ar})_0$, so that $P(\text{Ar}) \approx P(\text{Ar})_0$ throughout the reaction. What is the corresponding rate equation?

▮ Because the reaction is first-order in the concentration of Ar, doubling the amount of Ar present will double the reaction rate. If $C(\text{Ar})$ is essentially constant, the pseudo-order rate equation is

$$\frac{dC(\text{H}_2)}{dt} = kC(\text{H})^2$$

where $k = (2.2 \times 10^8\,\text{mol}^2\cdot\text{L}^{-2}\cdot\text{s}^{-1})C(\text{Ar})$.

12.26 Use the following data:

$t/(\text{s})$	0	300	600	900	1500	2100	2700	3300	3900	4500	∞
$V(\text{O}_2)/(\text{cm}^3)$	0.00	7.50	14.00	19.65	28.80	35.80	41.20	45.20	48.30	50.60	57.90

to show that the equation

$$2\text{H}_2\text{O}_2(\text{aq}) \xrightarrow{\text{I}^-\,(0.0200\,\text{M})} 2\text{H}_2\text{O}(\text{l}) + \text{O}_2(\text{g})$$

is pseudo-first-order in the concentration of H_2O_2. The actual rate law is

$$\frac{dC(\text{H}_2\text{O}_2)}{dt} = -kC(\text{I}^-)C(\text{H}_2\text{O}_2)$$

Determine k.

▮ Using the results of Problem 12.15, the volume of O_2 produced is related to the concentration of unreacted H_2O_2 by

$$C(\text{H}_2\text{O}_2) = b[V(\text{O}_2)_\infty - V(\text{O}_2)]$$

Thus a plot of $\ln[V(\text{O}_2)_\infty - V(\text{O}_2)]$ against t will be linear if the reaction is pseudo-first-order, and the slope will be equal to $-kC(\text{I}^-)$. From the plot shown in Fig. 12-5,

$$k = \frac{-\text{slope}}{C(\text{I}^-)} = \frac{-(-4.60 \times 10^{-4}\,\text{s}^{-1})}{0.0200\,\text{mol}\cdot\text{L}^{-1}} = 0.0230\,\text{L}\cdot\text{mol}^{-1}\cdot\text{s}^{-1}$$

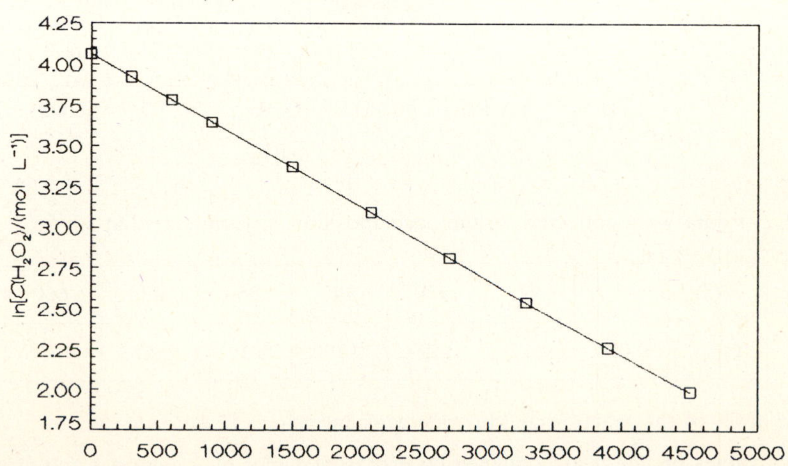

Fig. 12-5

12.2 DETERMINATION OF REACTION ORDER AND RATE CONSTANTS

12.27 Use the following rates of decomposition of acetaldehyde

$[dP(CH_3CHO)/dt]/(\text{torr} \cdot s^{-1})$	0.1422	0.1248	0.1123	0.0983	0.0857	0.0782
ξ	0.000	0.050	0.100	0.150	0.200	0.250

$[dP(CH_3CHO)/dt]/(\text{torr} \cdot s^{-1})$	0.0718	0.0625	0.0518	0.0445	0.0382
ξ	0.300	0.350	0.400	0.450	0.500

to determine the order of reaction for

$$CH_3\overset{\overset{O}{\|}}{C}H(g) \rightarrow \text{products}$$

▌ For the general rate law

$$\pm\frac{dC_i}{dt} = kC_i^n$$

taking logarithms gives

$$\log\left[\pm\frac{dC_i}{dt}\right] = \log k + n \log C_i \qquad (12.8)$$

A plot of $\log[dP(CH_3CHO)/dt]$ against $\log(1 - \xi)$ appears in Fig. 12-6. The slope gives $n = 1.86$.

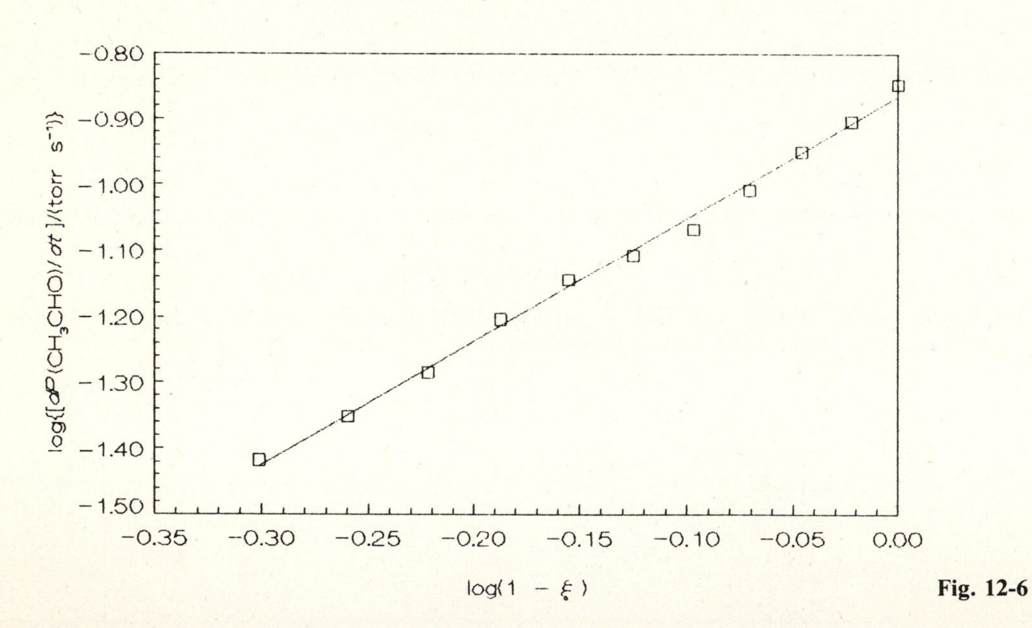

Fig. 12-6

12.28 The following ξ–t data were collected for the decomposition of formic acid at 50 °C:

$t/(s)$	0	60	120	240	360	480	660	840
ξ	0.000	0.120	0.223	0.390	0.521	0.623	0.735	0.810

Determine the order of the reaction and the rate constant using *(12.8)*.

▌ The various values of $d\xi/dt$ are obtained by preparing a plot of ξ against t and graphically determining the tangent to the curve at the various values of t (see Fig. 12-7). (For clarity, only two tangents are shown in the

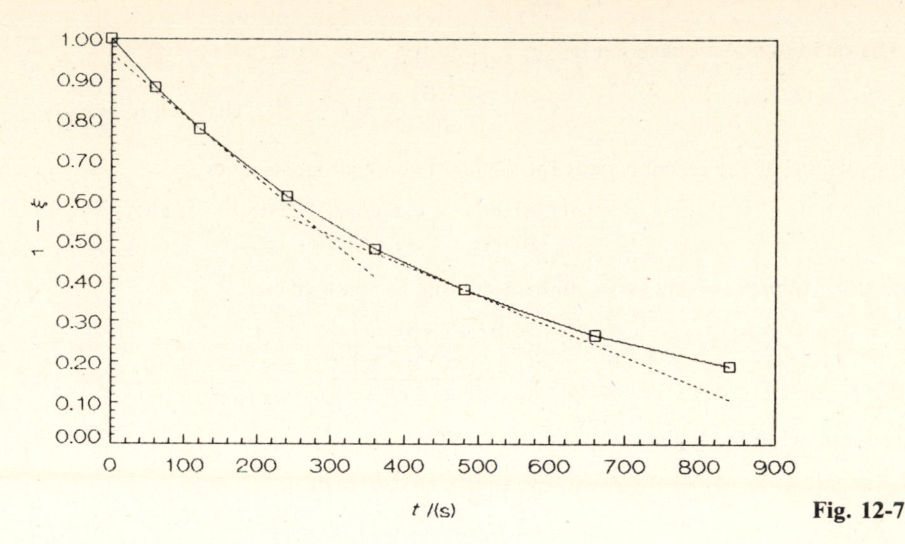

Fig. 12-7

figure.) The respective values of the rates are

ξ	1.120	0.223	0.390	0.521	0.623	0.735
$(d\xi/dt)/(10^{-3}\ s^{-1})$	1.742	1.538	1.208	0.948	0.746	0.525

The plot of $\log(d\xi/dt)$ against $\log(1-\xi)$ appears in Fig. 12-8, giving $n = 1.000$ from the slope and $\log k = -2.703$ or $k = 1.98 \times 10^{-3}\ s^{-1}$ from the intercept.

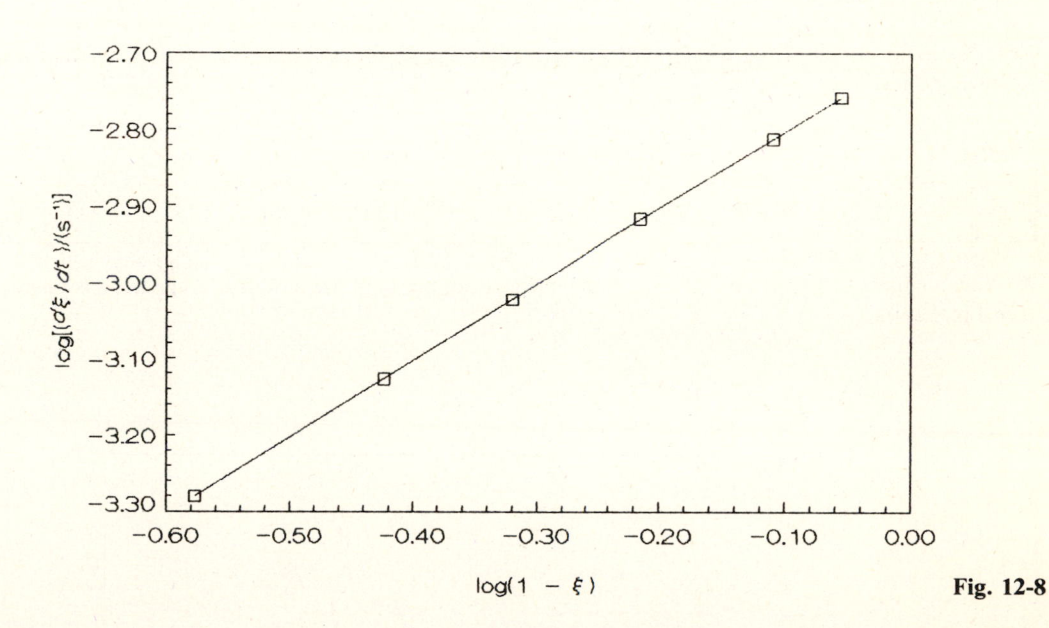

Fig. 12-8

12.29 The initial rates of reaction for the equation $2A + B \rightarrow$ products were determined under various initial concentrations of reactants. Use the following data to determine the order of reaction for each component.

experiment	$C(A)_0/(M)$	$C(B)_0/(M)$	$-[dC(B)/dt]/(mol \cdot L^{-1} \cdot s^{-1})$
1	0.10	0.10	0.25
2	0.20	0.10	0.50
3	0.10	0.20	0.25

▮ The rate law for the reaction is

$$-\frac{dC(\mathrm{B})}{dt} = k[C(\mathrm{A})]^a[C(\mathrm{B})]^b$$

Taking a ratio of the reaction rates for the first two experiments gives

$$\frac{-[dC(\mathrm{B})/dt]_1}{-[dC(\mathrm{B})/dt]_2} = \frac{k[C(\mathrm{A})_1]^a[C(\mathrm{B})_1]^b}{k[C(\mathrm{A})_2]^a[C(\mathrm{B})_2]^b} = \left(\frac{C(\mathrm{A})_1}{C(\mathrm{A})_2}\right)^a$$

Taking logarithms, solving for a, and substituting the data gives

$$a = \frac{\log\left[\dfrac{-[dC(\mathrm{B})/dt]_1}{-[dC(\mathrm{B})/dt]_2}\right]}{\log[C(\mathrm{A})_1/C(\mathrm{A})_2]} = \frac{\log(0.25/0.50)}{\log(0.10/0.20)} = 1$$

Likewise, solving for b gives

$$b = \frac{\log\left[\dfrac{-[dC(\mathrm{B})/dt)]_1}{-[dC(\mathrm{B})/dt)]_3}\right]}{\log[C(\mathrm{B})_1/C(\mathrm{B})_3]} = \frac{\log(0.25/0.25)}{\log(0.10/0.20)} = 0$$

The reaction is first-order with respect to A and zeroth-order with respect to B.

12.30 For the reaction $A + 2B \rightarrow$ products, the reaction rate was halved as the concentration of A was doubled. What is the order of reaction with respect to A?

▮ As in Problem 12.29,

$$a = \frac{\log(\mathrm{rate}_2/\mathrm{rate}_1)}{\log[C(\mathrm{A})_2/C(\mathrm{A})_1]} = \frac{\log[0.5\,\mathrm{rate}_1/\mathrm{rate}_1]}{\log[2C(\mathrm{A})_1/C(\mathrm{A})_1]} = -1$$

The reaction is negative first-order with respect to A.

12.31 Determine the order of reaction for the dimerization equation $2CH_3OC_6H_4CNO(CCl_4) \rightarrow$ products using the following data.

t/(s)	0	3600	7200	12 900	19 500	33 900	56 520	64 800	72 720	81 480	91 080
ξ	1.000	0.909	0.833	0.735	0.673	0.527	0.391	0.353	0.334	0.315	0.297

▮ Various plots of *(12.4)* and *(12.5)* could be prepared, but only the plot of $1/\xi$ against t is linear, giving $n = 2$ (see Fig. 12-9).

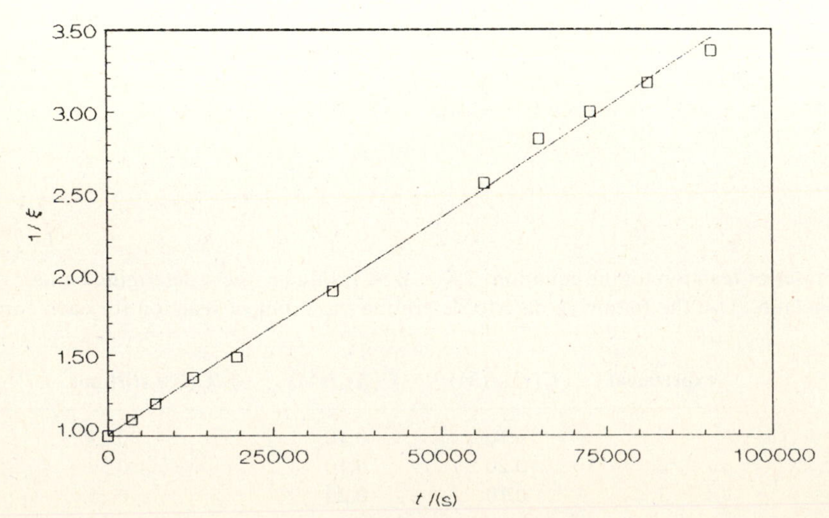

Fig. 12-9

12.32 The total pressure of the system at 279.2 °C for the equation $SO_2Cl_2(g) \rightarrow SO_2(g) + Cl_2(g)$ was observed as a function of time. Determine the reaction order and the rate constant from the following data.

t/(s)	204	942	1686	2466	3270	4098	4944	5778	6642	7500	8400	9354	10 308	∞
P_t/(torr)	325	335	345	355	365	375	385	395	405	415	425	435	445	594.2

▮ At any time during the reaction, $P_t = P(SO_2Cl_2) + P(SO_2) + P(Cl_2)$. Letting x represent the amount of SO_2Cl_2 that reacts,

$$P(SO_2) = P(Cl_2) = x \qquad \text{and} \qquad P(SO_2Cl_2) = P(SO_2Cl_2)_0 - x.$$

Thus

$$P_t = P(SO_2Cl_2)_0 - x + x + x = P(SO_2Cl_2)_0 + x$$

Solving for x gives $x = P_t - P(SO_2Cl_2)_0$ and

$$P(SO_2Cl_2) = P(SO_2Cl_2)_0 - [P_t - P(SO_2Cl_2)_0] = 2P(SO_2Cl_2)_0 - P_t$$

The value of $2P(SO_2Cl_2)_0$ is equal to the data entry for $t = \infty$. Thus,

$$P(SO_2Cl_2) = 594.2\text{ torr} - P_t$$

Various plots of *(12.4)* and *(12.5)* could be prepared, but only the plot of ln $[P(SO_2Cl_2)]$ against t is linear, giving $n = 1$ (see Fig. 12-10). From the slope,

$$k = -\text{slope} = -(-5.81 \times 10^{-5}\text{ s}^{-1}) = 5.81 \times 10^{-5}\text{ s}^{-1}$$

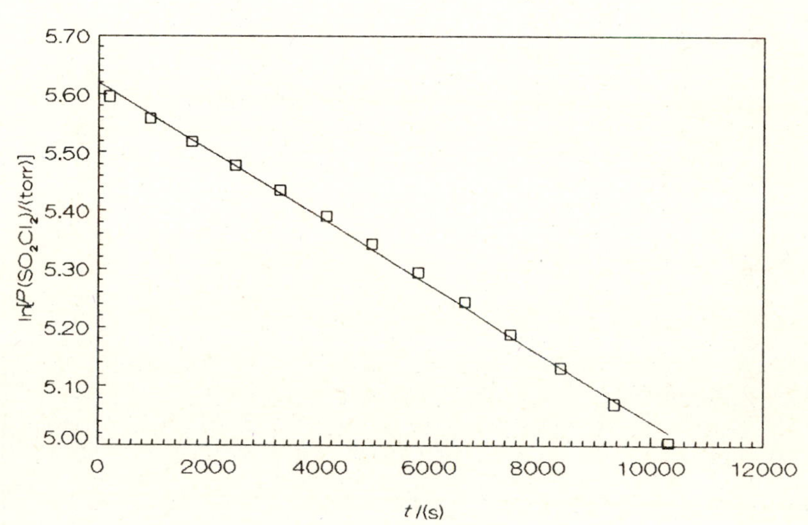

Fig. 12-10

12.33 For the data given in Problem 12.32, use *(12.5)* to calculate values of k for consecutive data to confirm the reaction order and the value of k determined in Problem 12.32.

▮ Substituting $P_{t,\infty} - P_i$ at t_i into *(12.5)* and solving for k gives

$$k = \frac{1}{t_2 - t_1} \ln \frac{P_{t,\infty} - P_1}{P_{t,\infty} - P_2}$$

As a sample calculation, consider the first two data.

$$k = \frac{1}{942\text{ s} - 204\text{ s}} \ln \frac{594.2\text{ torr} - 325\text{ torr}}{594.2\text{ torr} - 335\text{ torr}} = 5.13 \times 10^{-5}\text{ s}^{-1}$$

The range of values of k is $5.13 \times 10^{-5}\text{ s}^{-1}$ to 7.64×10^{-5}, and the average value is $6.11 \times 10^{-5}\text{ s}^{-1}$.

12.34 Use the technique introduced in Problem 12.33 to confirm the results of Problem 12.28 for the decomposition of formic acid.

I The expression for k is

$$k = \frac{1}{t_2 - t_1} \ln \frac{(1-\xi)_1}{(1-\xi)_2}$$

As a sample calculation, consider the first two data.

$$k = \frac{1}{60\ \text{s} - 0\ \text{s}} \ln \frac{1 - 0.000}{1 - 0.120} = 2.13 \times 10^{-3}\ \text{s}^{-1}$$

The range of values of k is $2.13 \times 10^{-3}\ \text{s}^{-1}$ to $1.86 \times 10^{-3}\ \text{s}^{-1}$, and the average value is $2.01 \times 10^{-3}\ \text{s}^{-1}$.

12.35 Use the following half-life data for the decomposition of $N_2O(g)$ at 1030 K to determine the order of reaction and the rate constant.

$t_{1/2}/(\text{s})$	212	255	470	860
$P(N_2O)_0/(\text{torr})$	360	290	139	52.5

I The data show that there is an inverse relation between $t_{1/2}$ and $P(N_2O)_0$; thus the reaction order is greater than first-order. Substituting $\alpha = 0.5$ into *(12.6)* and taking logarithms gives

$$\log t_{1/2} = \log \frac{2^{n-1} - 1}{(n-1)k} - (n-1) \log C_{i,0} \qquad (12.9)$$

A plot of $\log t_{1/2}$ against $\log P(N_2O)_0$ will be linear and the slope will be equal to $-(n-1)$ and the intercept equal to $\log [(2^{n-1} - 1)/(n-1)k]$. From Fig. 12-11,

$$n = 1 - (\text{slope}) = 1 - (-0.73) = 1.73$$

$$k = \frac{2^{n-1} - 1}{(n-1) \log^{-1} (\text{intercept})} = \frac{2^{1.73-1} - 1}{(1.73 - 1) \log^{-1} 4.203} = 5.65 \times 10^{-5}$$

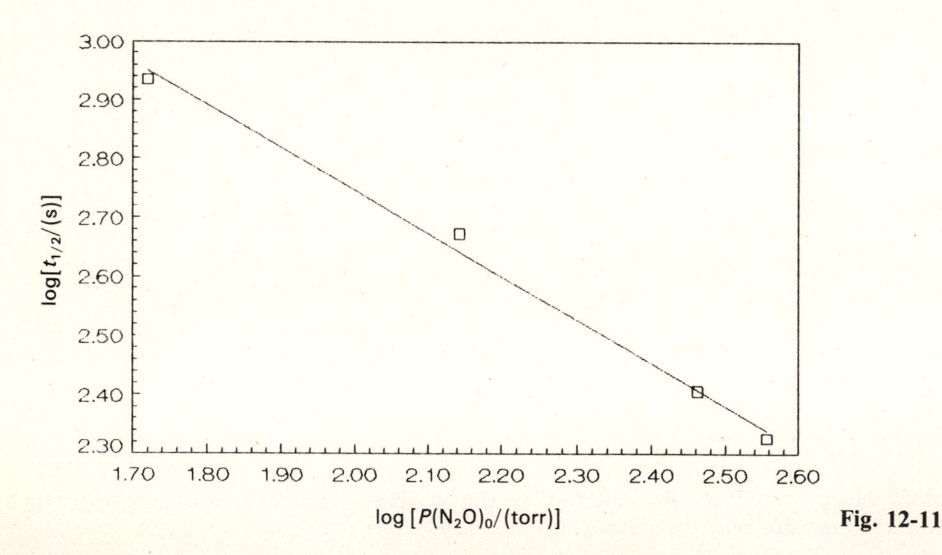

Fig. 12-11

12.36 Use the half-life data given in Problem 12.18 to confirm that the reaction is first-order.

I A constant value of $t_{1/2}$ occurs only for first-order reactions, see *(12.7)*.

12.37 For the reaction between $NO(g)$ and $H_2(g)$ at 1100 K, the following half-life data were obtained for $P(NO)_0 = P(H_2)_0$.

$t_{1/2}/(\text{s})$	81	102	140	224
$P_0/(\text{torr})$	354	341	288	202

Determine the overall order of the reaction.

▮ Because the initial partial pressures are identical,

$$-\frac{d(\text{NO})}{dt} = kP(\text{NO})^a P(\text{H}_2)^b = kP(\text{NO})^n$$

where $n = a + b$. Preparing the plot corresponding to *(12.9)* (see Fig. 12-12) gives

$$n = 1 - \text{slope} = 1 - (-1.7) = 2.7$$

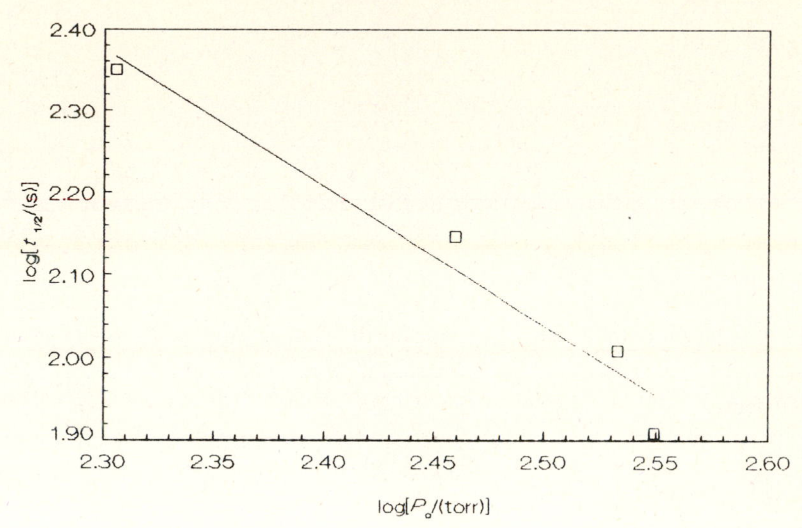

Fig. 12-12

12.38 The half-life for a given reaction was doubled as the initial concentration of a reactant was doubled. What is the order of reaction for this component?

▮ Substituting the data into *(12.9)* gives

$$\log t_{1/2} = \log \frac{2^{n-1} - 1}{(n - 1)k} - (n - 1) \log C_{i,0}$$

$$\log 2t_{1/2} = \log \frac{2^{n-1} - 1}{(n - 1)k} - (n - 1) \log 2C_{i,0}$$

and solving by subtraction gives $n = 0$.

12.39 The time required for ξ to reach 0.90 for the reaction discussed in Problem 12.16 is 76 200 s. Use Fig. 12-13 to determine the order of reaction.

▮ Values of $\xi = 1 - (S_\infty - S)/(S_\infty - S_0)$ are plotted against $t/t_{0.90}$ as solid circles in Fig. 12-13. These points lie along the $n = 1$ curve.

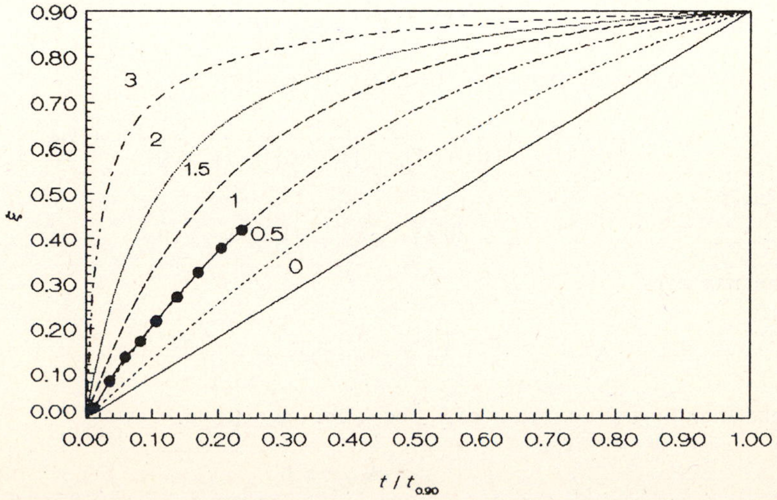

Fig. 12-13

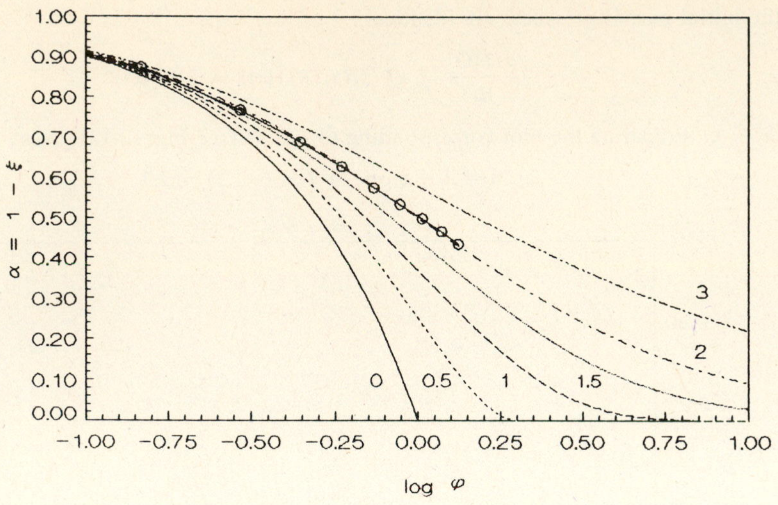

Fig. 12-14

12.40 Use Fig. 12-14 to determine the order of reaction for the dimerization of butadiene at 600 K.

P(butadiene)/(torr)	632	552	485	435	396	362	336	314	294	274
t/(s)	0	600	1200	1800	2400	3000	3600	4200	4800	5400

▮ Figure 12-14 is known as a *Powell plot* where $\alpha = 1 - \xi$ and $\phi = k[C(\text{A})_0]^{n-1}t$. The reaction order can be determined by plotting experimental values of α and $\log t$ on the same scale as the master plot and moving the plot of experimental values horizontally (keeping the $\log t$ and $\log \phi$ axes together) until a match is obtained. As a sample calculation, consider $t = 5400$ s.

$$\alpha = \frac{P(\text{butadiene})}{P(\text{butadiene})_0} = \frac{274\ \text{torr}}{632\ \text{torr}} = 0.434$$

$$\log t = \log 5400 = 3.732$$

The scale chosen for the $\log t$ values is 2.000 to 4.000 so that it matches with the $\log \phi$ scale. The experimental plot is represented by the open circles in Fig. 12-14 and matches the curve for $n = 2$.

12.41 Confirm the value of $\alpha = 0.622$ at $\phi = 0.794$ for $n = 3$ plotted in Fig. 12-14.

▮ For a third-order reaction, *(12.4)* gives

$$1/[C(\text{A})]^2 = 1/[C(\text{A})_0]^2 + 2kt$$

Rearranging the definition of ϕ given in Problem 12.40 gives, for a third-order reaction,

$$kt = \frac{1}{\phi[C(\text{A})_0]^2}$$

which upon substitution into the expression for $1/[C(\text{A})]^2$ gives

$$\frac{1}{[C(\text{A})]^2} = \frac{1}{[C(\text{A})_0]^2} + 2\frac{\phi}{[C(\text{A})_0]^2} = \frac{1}{[C(\text{A})_0]^2}(1 + 2\phi)$$

Solving for α gives

$$\alpha = C(\text{A})/C(\text{A})_0 = [1/(1 + 2\phi)]^{1/2}$$

Substituting the data gives

$$\alpha = \left(\frac{1}{1 + (2)(0.794)}\right)^{1/2} = 0.622$$

12.42 Express the relaxation time as a function of k_1 and k_2 for the reaction

$$\text{A} \underset{k_2}{\overset{k_1}{\rightleftharpoons}} \text{B}$$

The restoration process for a reaction returning to equilibrium after a disturbance is always first-order for a small displacement from equilibrium. Thus *(12.5)* gives

$$\Delta C_i = \Delta C_{i,0}\, e^{-1/\tau} \qquad (12.10)$$

where τ is the relaxation time. Upon differentiation,

$$\frac{d(\Delta C_i)}{dt} = -\Delta C_i \frac{1}{\tau} \qquad (12.11)$$

For the chemical reaction, the rate expression is

$$\frac{dC(\mathrm{B})}{dt} = k_1 C(\mathrm{A}) - k_2 C(\mathrm{B})$$

and at equilibrium $k_1 C(\mathrm{A}) - k_2 C(\mathrm{B}) = 0$. For a small increase in $C(\mathrm{B})$, $\Delta C(\mathrm{B})$,

$$C(\mathrm{A}) = [\mathrm{A}] - \Delta C(\mathrm{B}) \qquad C(\mathrm{B}) = [\mathrm{B}] + \Delta C(\mathrm{B})$$

Substituting into the rate equation gives

$$\begin{aligned} \frac{dC(\mathrm{B})}{dt} &= k_1([\mathrm{A}] - \Delta C[\mathrm{B}]) - k_2([\mathrm{B}] + \Delta C[\mathrm{B}]) \\ &= (k_1[\mathrm{A}] - k_2[\mathrm{B}]) - (k_1 + k_2)\,\Delta C[\mathrm{B}] \\ &= 0 - (k_1 - k_2)\,\Delta C[\mathrm{B}] \end{aligned}$$

Upon comparison with *(12.11)*, $1/\tau = k_1 + k_2$, or

$$\tau = (k_1 + k_2)^{-1} \qquad (12.12)$$

12.43 Given $\tau = 23\ \mu\mathrm{s}$ and $K = 1.0 \times 10^3$, determine k_1 and k_2 for the reaction described in Problem 12.42.

At equilibrium, $k_1[\mathrm{A}] - k_2[\mathrm{B}] = 0$, giving

$$K = [\mathrm{B}]/[\mathrm{A}] = k_1/k_2$$

Substituting the data into the above expression for K and into *(12.12)* gives

$$1.0 \times 10^3 = \frac{k_1}{k_2} \qquad \frac{1}{23 \times 10^{-6}\ \mathrm{s}} = k_1 + k_2$$

Solving these simultaneous equations gives $k_1 = 4.3 \times 10^4\ \mathrm{s}^{-1}$ and $k_2 = 43\ \mathrm{s}^{-1}$.

12.44 For the study of the reaction

$$\mathrm{A + B} \underset{k_2}{\overset{k_1}{\rightleftharpoons}} \mathrm{C}$$

the relaxation time is

$$\tau = \{k_2 + k_1([\mathrm{A}] + [\mathrm{B}])\}^{-1} \qquad (12.13)$$

Determine k_1 and k_2 given $t = 7.6\ \mu\mathrm{s}$ in water at $\mathrm{pH} = 6.00$.

Under equilibrium conditions,

$$K = \frac{[\mathrm{H_2O}]}{[\mathrm{H^+}][\mathrm{OH^-}]} = \frac{55.6}{(1.0 \times 10^{-6})(1.0 \times 10^{-8})} = 5.6 \times 10^{15}$$

where $K = k_1/k_2$. Substituting the data into the expression for K and for τ gives

$$7.6 \times 10^{-6}\ \mathrm{s} = \{k_2 + k_1[(1.0 \times 10^{-6}) + (1.0 \times 10^{-8})]\}^{-1} \qquad 5.6 \times 10^{15} = k_1/k_2$$

Solving these simultaneous equations gives $k_1 = 1.3 \times 10^{11}\ \mathrm{L \cdot mol^{-1} \cdot s^{-1}}$ and $k_2 = 2.3 \times 10^{-5}\ \mathrm{s}^{-1}$.

12.45 For the reaction

$$\mathrm{H^+(aq) + C_6H_5COO^-(aq)} \underset{k_2}{\overset{k_1}{\rightleftharpoons}} \mathrm{C_6H_5COOH(aq)}$$

$k_1 = 3.5 \times 10^{10}\ \mathrm{L \cdot mol^{-1} \cdot s^{-1}}$, $k_2 = 2.2 \times 10^6\ \mathrm{s}^{-1}$, and $K = 6.6 \times 10^{-5}$. Calculate the relaxation time for a 0.010 M solution of benzoic acid.

Using the procedure outlined in Problem 7.54, $C(H^+) = C(C_6H_5COO^-) = 7.8 \times 10^{-4}\ mol \cdot L^{-1}$. Substituting into *(12.13)* gives

$$\tau = [2.2 \times 10^6\ s^{-1} + (3.5 \times 10^{10}\ L \cdot mol^{-1} \cdot s^{-1})(7.4 \times 10^{-4}\ mol \cdot L^{-1} + 7.4 \times 10^{-4}\ mol \cdot L^{-1})]^{-1}$$

$$= 1.9 \times 10^{-8}\ s = 19\ ns$$

12.46 For the reaction

$$HF(aq) \underset{k_2}{\overset{k_1}{\rightleftharpoons}} H^+(aq) + F^-(aq)$$

$\tau = 0.63$ ns for a 0.100 M solution of HF ($[H^+] = [F^-] = 7.7 \times 10^{-3}\ mol \cdot L^{-1}$) and 2.04 ns for a 0.010 M solution of HF ($[H^+] = [F^-] = 2.2 \times 10^{-3}\ mol \cdot L^{-1}$). Determine k_1, k_2, and K for this reaction.

For the ionization reaction, *(12.13)* gives $\tau = \{k_1 + k_2([H^+] + [F^-])\}^{-1}$. This equation implies that a plot of $1/\tau$ against $([H^+] + [F^-])$ would be linear with an intercept of k_1 and a slope of k_2. However, for only two sets of data, it is simpler to solve the simultaneous equations

$$0.63 \times 10^{-9}\ s = [k_1 + k_2(7.7 \times 10^{-3}\ mol \cdot L^{-1} + 7.7 \times 10^{-3}\ mol \cdot L^{-1})]^{-1}$$

$$2.04 \times 10^{-9}\ s = [k_1 + k_2(2.2 \times 10^{-3} + 2.2 \times 10^{-3})]^{-1}$$

giving $k_1 = 5 \times 10^7\ s^{-1}$ and $k_2 = 1.00 \times 10^{11}\ L \cdot mol^{-1} \cdot s^{-1}$. For this process,

$$K = \frac{k_1}{k_2} = \frac{5 \times 10^7}{1.00 \times 10^{11}} = 5 \times 10^{-4}$$

which agrees favorably with the thermodynamic value of 6.5×10^{-4}.

12.3 RATE EQUATIONS FOR COMPLEX REACTIONS

12.47 Consider the following proposed mechanism

$$A \underset{k_2}{\overset{k_1}{\rightleftharpoons}} B \qquad B + A \xrightarrow{k_3} C$$

for the overall chemical equation $2A \rightarrow C$. Write the rate expressions for each species. Assuming that $C(B)$ is essentially constant, rewrite the rate expressions for $C(A)$ and $C(C)$ so they do not contain the $C(B)$ term. Under what conditions will the reaction be pseudo-first-order to pseudo-second-order?

The rate expressions are

$$\frac{-dC(A)}{dt} = k_1C(A) - k_2C(B) + k_3C(B)C(A)$$

$$\frac{dC(B)}{dt} = k_1C(A) - k_2C(B) - k_2C(A)C(B)$$

$$\frac{dC(C)}{dt} = k_3C(A)C(B)$$

Using the *steady-state approximation* for $C(B)$ $[dC(B)/dt = 0]$ and solving for $C(B)$ gives

$$\frac{dC(B)}{dt} = k_1C(A) - k_2C(B) - k_3C(A)C(B) = 0$$

$$C(B) = k_1C(A)/[k_2 + k_3C(A)]$$

Substituting this expression for $C(B)$ into the expressions for $-dC(A)/dt$ and $dC(C)/dt$ gives

$$\frac{-dC(A)}{dt} = k_1C(A) - k_2\frac{k_1C(A)}{k_2 + k_3C(A)} + k_3\frac{k_1C(A)}{k_2 + k_3C(A)}C(A) = \frac{2k_1k_3[C(A)]^2}{k_2 + k_3C(A)}$$

$$\frac{dC(C)}{dt} = k_3C(A)\frac{k_1C(A)}{k_2 + k_3C(A)} = \frac{k_1k_3[C(A)]^2}{k_2 + k_3C(A)}$$

Note that $-dC(A)/dt = 2[dC(C)/dt]$. If $k_3C(A) \gg k_2$, then

$$\frac{dC(C)}{dt} \approx \frac{k_1k_3[C(A)]^2}{k_3[C(A)]^2} = k_1C(A)$$

and if $k_3C(\mathrm{A}) \ll k_2$, then

$$\frac{dC(\mathrm{C})}{dt} \approx \frac{k_1k_3[C(\mathrm{A})]^2}{k_2}$$

12.48 Consider the following proposed mechanism

$$\mathrm{A} \underset{k_2}{\overset{k_1}{\rightleftharpoons}} \mathrm{B} \qquad \mathrm{B} + \mathrm{C} \xrightarrow{k_3} \mathrm{D}$$

for the overall chemical equation $\mathrm{A} + \mathrm{C} \rightarrow \mathrm{D}$. Assuming B to be an intermediate described by the steady-state approximation, write the rate expression for $C(\mathrm{A})$.

▮ Applying the steady-state approximation to the rate expression for $C(\mathrm{B})$ gives

$$\frac{dC(\mathrm{B})}{dt} = k_1C(\mathrm{A}) - k_2C(\mathrm{B}) - k_3C(\mathrm{B})C(\mathrm{C}) = 0$$

$$C(\mathrm{B}) = k_1C(\mathrm{A})/[k_2 + k_3C(\mathrm{C})]$$

Writing the rate expression for $C(\mathrm{A})$ and substituting the expression for $C(\mathrm{B})$ gives

$$\frac{-dC(\mathrm{A})}{dt} = k_1C(\mathrm{A}) - k_2C(\mathrm{B}) = k_1C(\mathrm{A}) - k_2\frac{k_1C(\mathrm{A})}{k_2 + k_3C(\mathrm{C})} = \frac{k_1k_3C(\mathrm{A})C(\mathrm{C})}{k_2 + k_3C(\mathrm{C})}$$

12.49 Consider the following proposed mechanism:

$$\mathrm{A_2} \rightleftarrows 2\mathrm{A} \qquad (K_1)$$

$$\mathrm{A} + \mathrm{B} \rightleftarrows \mathrm{C} \qquad (K_2)$$

$$\mathrm{A_2} + \mathrm{C} \xrightarrow{k} \mathrm{D} + \mathrm{A}$$

for the overall chemical equation $\mathrm{A_2} + \mathrm{B} \rightarrow \mathrm{D}$. Assuming that the equilibria are rapidly established in the first two steps, write the rate expression for $C(\mathrm{D})$.

▮ From the equilibrium steps

$$C(\mathrm{A}) = [K_1C(\mathrm{A_2})]^{1/2}$$

$$C(\mathrm{C}) = K_2C(\mathrm{A})C(\mathrm{B}) = K_2[K_1C(\mathrm{A_2})]^{1/2}C(\mathrm{B}) = K_1^{1/2}K_2[C(\mathrm{A_2})]^{1/2}C(\mathrm{B})$$

Substituting these expressions for $C(\mathrm{A})$ and $C(\mathrm{C})$ into the rate expression for $C(\mathrm{D})$ gives

$$\frac{dC(\mathrm{D})}{dt} = kC(\mathrm{A_2})C(\mathrm{C}) = kC(\mathrm{A_2})K_1^{1/2}K_2[C(\mathrm{A})]^{1/2}C(\mathrm{B}) = kK_1^{1/2}K_2[C(\mathrm{A_2})]^{3/2}C(\mathrm{B})$$

12.50 Consider the following proposed mechanism:

$$\mathrm{A_2} \rightleftarrows 2\mathrm{A} \qquad (K)$$

$$\mathrm{A} + \mathrm{B_2} \xrightarrow{k_1} \mathrm{C} + \mathrm{B} \qquad \mathrm{B} + \mathrm{A_2} \xrightarrow{k_2} \mathrm{C} + \mathrm{A}$$

for the overall chemical equation $\mathrm{A_2} + \mathrm{B_2} \rightarrow 2\mathrm{C}$. Assuming that the equilibrium is rapidly established and the steady-state approximation is valid for $C(\mathrm{B})$, write the rate expression for $C(\mathrm{C})$.

▮ From the equilibrium step $C(\mathrm{A}) = K^{1/2}[C(\mathrm{A_2})]^{1/2}$. Applying the steady-state approximation for $C(\mathrm{B})$ gives

$$\frac{dC(\mathrm{B})}{dt} = k_1C(\mathrm{A})C(\mathrm{B_2}) - k_2C(\mathrm{B})C(\mathrm{A_2}) = 0$$

$$C(\mathrm{B}) = \frac{k_1C(\mathrm{A})C(\mathrm{B_2})}{k_2C(\mathrm{A_2})} = \frac{k_1K^{1/2}[C(\mathrm{A_2})]^{1/2}C(\mathrm{B})}{k_2C(\mathrm{A_2})}$$

The rate expression for $C(\mathrm{C})$ is

$$\frac{dC(\mathrm{C})}{dt} = k_1C(\mathrm{A})C(\mathrm{B_2}) + k_2C(\mathrm{B})C(\mathrm{A_2})$$

$$= k_1K^{1/2}[C(\mathrm{A_2})]^{1/2}C(\mathrm{B_2}) + k_2\frac{k_1K^{1/2}[C(\mathrm{A_2})]^{1/2}C(\mathrm{B_2})}{k_2C(\mathrm{A_2})C(\mathrm{A_2})}$$

$$= 2k_1K^{1/2}[C(\mathrm{A_2})]^{1/2}C(\mathrm{B_2})$$

12.51 Consider the following proposed mechanism:

$$A \xrightarrow{k_1} B + C \qquad A + C \xrightarrow{k_2} B + D$$

for the overall chemical equation $A \rightarrow B + \frac{1}{2}D$. Assuming that the steady-state approximation is valid for $C(C)$, write the rate expression for $C(D)$.

▮ Applying the steady-state approximation for $C(C)$ gives

$$\frac{dC(C)}{dt} = k_1C(A) - k_2C(A)C(C) = 0 \qquad C(C) = k_1/k_2$$

The rate expression for $C(D)$ is

$$\frac{dC(D)}{dt} = k_2C(A)C(C) = k_2C(A)\frac{k_1}{k_2} = k_1C(A)$$

12.52 Consider a second proposed mechanism:

$$A \rightleftarrows B + C \qquad (K)$$

$$A + C \xrightarrow{k} B + D$$

for the overall chemical equation described in Problem 12.51. Assuming that the equilibrium is rapidly established, write the rate expression for $C(D)$. According to this mechanism, how does increasing $C(B)$ affect the reaction rate?

▮ From the equilibrium step, $C(C) = KC(A)/C(B)$. The rate expression for $C(D)$ is

$$\frac{dC(D)}{dt} = kC(A)C(C) = kC(A)\frac{KC(A)}{C(B)} = kK[C(A)]^2[C(B)]^{-1}$$

Increasing $C(B)$ will decrease the reaction rate.

12.53 Consider the four following proposed mechanisms for the overall chemical equation $A_2 + B_2 \rightarrow 2AB$.

mechanism 1	mechanism 2	mechanism 3	mechanism 4
$A_2 \xrightarrow{k_1} 2A$	$A_2 + B_2 \xrightarrow{k_1} A_2B_2$	$A_2 \rightleftarrows 2A \quad (K_1)$	$A_2 \rightleftarrows 2A \quad (K)$
$A + B_2 \xrightarrow{k_2} AB + B$	$A_2B_2 \xrightarrow{k_2} 2AB$	$B_2 \rightleftarrows 2B \quad (K_2)$	$B_2 \xrightarrow{k_1} 2B$
$A + B \xrightarrow{k_3} AB$	$k_1 \ll k_2$	$A + B \xrightarrow{k} AB$	$A + B \xrightarrow{k_2} AB$
$k_1 \ll k_2 \approx k_3$			$k_1 \ll k_2$

Write the rate expression for $C(AB)$ for each mechanism.

▮ For the first mechanism, the relative values of the rate constants indicate that the first step is the *rate-determining step*. Thus

$$\frac{dC(AB)}{dt} = -2\frac{dC(A_2)}{dt} = -2[-k_1C(A_2)] = 2k_1C(A_2)$$

Likewise for the second mechanism,

$$\frac{dC(B)}{dt} = -2\frac{dC(A_2)}{dt} = -2[-k_1C(A_2)C(B_2)] = 2k_1C(A_2)C(B_2)$$

The concentrations of the intermediates in the third mechanism are $C(A) = K_1^{1/2}[C(A_2)]^{1/2}$ and $C(B) = K_2^{1/2}[C(B_2)]^{1/2}$. The rate expression for $C(AB)$ is

$$\frac{dC(AB)}{dt} = kC(A)C(B) = kK_1^{1/2}[C(A_2)]^{1/2}K_2^{1/2}[C(B_2)]^{1/2} = kK_1^{1/2}K_2^{1/2}[C(A_2)]^{1/2}[C(B_2)]^{1/2}$$

For the fourth mechanism, the rate determining step is the second step, giving

$$\frac{dC(AB)}{dt} = -2\frac{dC(B)}{dt} = -2[-k_1C(B_2)] = 2k_1C(B_2)$$

12.54 Consider the two following proposed mechanisms for the decomposition of ozone, $2O_3(g) \rightarrow 3O_2(g)$.

mechanism 1	mechanism 2
$O_3 \xrightarrow{k_1} O_2 + O\cdot$	$O_3 \xrightarrow{k_1} O_2 + O\cdot$
$O\cdot + O_3 \xrightarrow{k_2} 2O_2$	$O\cdot + O_2 \xrightarrow{k_2} O_3$
	$O\cdot + O_3 \xrightarrow{k_3} 2O_2$

Which mechanism predicts a first-order reaction during the early stages of the decomposition?

▮ Applying the steady-state approximation to $C(O\cdot)$ in the first mechanism gives

$$\frac{dC(O\cdot)}{dt} = k_1 C(O_3) - k_2 C(O\cdot)C(O_3) = 0 \qquad C(O\cdot) = k_1/k_2$$

The rates of decomposition of O_3 and of formation of O_2 are

$$\frac{-dC(O_3)}{dt} = k_1 C(O_3) + k_2 C(O\cdot)C(O_3) = k_2 C(O_3) + k_2 \frac{k_1}{k_2} C(O_3) = 2k_1 C(O_3)$$

$$\frac{dC(O_2)}{dt} = 2k_2 C(O\cdot)C(O_3) + k_1 C(O_3) = 3k_1 C(O_3)$$

This mechanism correctly predicts the first-order nature of the reaction. Applying the steady-state approximation to $C(O\cdot)$ in the second mechanism gives

$$\frac{dC(O\cdot)}{dt} = k_1 C(O_3) - k_2 C(O\cdot)C(O_2) - k_3 C(O\cdot)C(O_3) = 0$$

$$C(O\cdot) = \frac{k_1 C(O_3)}{k_2 C(O_2) + k_3 C(O_3)}$$

The rates of decomposition of O_3 and of formation of O_2 are

$$\frac{-dC(O_3)}{dt} = k_1 C(O_3) - k_2 C(O\cdot)C(O_2) + k_3 C(O\cdot)C(O_3)$$

$$= k_1 C(O_3) - k_2 \frac{k_1 C(O_3)}{k_2 C(O_2) + k_3 C(O_3)} C(O_2) + k_3 \frac{k_1 C(O_3)}{k_2 C(O_2) + k_2 C(O_3)} C(O_3)$$

$$= \frac{2k_1 k_3 [C(O_3)]^2}{k_2 C(O_2) + k_3 C(O_3)}$$

$$\frac{dC(O_2)}{dt} = k_1 C(O_3) - k_2 C(O\cdot)C(O_2) + 2k_3 C(O\cdot)C(O_3) = \frac{3k_1 k_3 [C(O_3)]^2}{k_2 C(O_2) + k_3 C(O_3)}$$

Under experimental conditions such that $k_2 C(O_2) \ll k_3 C(O_3)$, this mechanism also predicts the first-order nature of the reaction

$$\frac{dC(O_2)}{dt} \approx \frac{3k_1 k_3 [C(O_3)]^2}{k_3 C(O_3)} = 3k_1 C(O_3)$$

12.55 Consider the three following proposed mechanisms for the overall equation $2A + B \rightarrow 2C$.

mechanism 1	mechanism 2	mechanism 3
$2A + B \xrightarrow{k} 2C$	$2A \rightleftarrows A_2 \quad (K)$	$2A \underset{k_2}{\overset{k_1}{\rightleftharpoons}} A_2$
	$A_2 + B \xrightarrow{k} 2C$	$A_2 + B \xrightarrow{k_3} 2C$

Which mechanism predicts a third-order reaction?

▮ The first mechanism is a simple termolecular reaction, giving

$$\frac{dC(C)}{dt} = 2k[C(A)]^2 C(B)$$

The equilibrium step in the second mechanism gives $C(A_2) = K[C(A)]^2$, which upon substitution into the expression for $dC(C)/dt$ gives

$$\frac{dC(C)}{dt} = 2kC(A_2)C(B) = 2kK[C(A)]^2C(B)$$

Applying the steady-state approximation to $C(A_2)$ in the third mechanism gives

$$\frac{dC(A_2)}{dt} = k_1[C(A)]^2 - k_2C(A_2) - k_3C(A_2)C(B) = 0$$

$$C(A_2) = k_1[C(A)]^2/[k_2 + k_3C(B)]$$

The rate expression of $C(C)$ is

$$\frac{dC(C)}{dt} = 2k_3C(A_2)C(B) = \frac{2k_1k_3[C(A)]^2C(B)}{k_2 + k_3C(B)}$$

Both the first and second proposed mechanisms predict a third-order reaction. Likewise, if $k_2 \gg k_3C(B)$, the third mechanism also predicts a third-order reaction.

12.56 Consider the following proposed mechanism

$$2A \rightleftarrows A_2 \qquad (K_1)$$

$$A + B \rightleftarrows C \qquad (K_2)$$

$$A_2 + C \xrightarrow{k} D + 2A$$

to describe the overall equation $A + C \rightarrow D$. What is the reaction order predicted by this proposed mechanism?

▌ The two equilibrium steps give

$$C(A_2) = K_1[C(A)]^2 \qquad C(C) = K_2C(A)C(B)$$

which upon substitution into the rate expression for $C(D)$ gives

$$\frac{dC(D)}{dt} = kC(A_2)C(C) = kK_1[C(A)]^2K_2C(A)C(B) = kK_1K_2[C(A)]^3C(B)$$

This mechanism predicts a fourth-order reaction.

12.57 The *Lindemann mechanism* used to describe the overall chemical equation $A \rightarrow$ products consists of

$$2A \underset{k_{-2}}{\overset{k_2}{\rightleftharpoons}} A^* + A \qquad A^* \xrightarrow{k_1} \text{products}$$

where A^* represents an excited molecule. Assuming the steady-state approximation for $C(A^*)$, derive the rate expression for C(products). Under what conditions is this a pseudo-first-order or a pseudo-second-order reaction?

▌ Applying the steady-state approximation to $C(A^*)$ gives

$$\frac{dC(A^*)}{dt} = k_2[C(A)]^2 - k_{-2}C(A^*)C(A) - k_1C(A^*) = 0$$

$$C(A^*) = \frac{k_2[C(A)]^2}{k_1 + k_{-2}C(A)}$$

which upon substitution into the rate expression for C(products) gives

$$\frac{dC(\text{products})}{dt} = k_1C(A^*) = \frac{k_1k_2[C(A)]^2}{k_1 + k_{-2}C(A)} \qquad (12.14)$$

Under conditions such that $k_1 \gg k_{-2}C(A)$ the reaction is pseudo-second-order with a rate constant of k_2, and under conditions such that $k_1 \ll k_{-2}C(A)$ the reaction is pseudo-first-order with a rate constant of k_1k_2/k_{-2}.

12.58 Unless the pressure of reactant A becomes too small, the accuracy of *(12.14)* can be tested by writing

$$\frac{dC(\text{products})}{dt} = \frac{k_1k_2C(A)}{k_1 + k_{-2}C(A)}C(A) = k_{\text{uni}}C(A)$$

where k_{uni} is the pseudo-first-order rate constant. Using the following data for the decomposition of azomethane at 330 °C, determine k_2 and k_1k_2/k_{-2}.

$P(CH_3NNCH_3)/(\text{torr})$	393	235	144	56.5	33.3	14.4	7.5
$k_{uni}/(10^{-3}\ s^{-1})$	2.82	2.82	2.65	2.13	1.76	1.31	1.10

▮ The definition of k_{uni} can be put into linear form by inverting:

$$\frac{1}{k_{uni}} = \frac{k_1 + k_{-2}C(A)}{k_1 k_2 C(A)} = \frac{1}{k_2 C(A)} + \frac{k_{-2}}{k_1 k_2}$$

This result implies that a plot of $1/k_{uni}$ against $1/C(A)$ will be linear with a slope equal to $1/k_2$ and an intercept equal to k_{-2}/k_1k_2. From the plot shown in Fig. 12-15 using pressures (see Problem 12.4),

$$k_2 = (\text{slope})^{-1} = (8060\ \text{torr} \cdot \text{s})^{-1} = 1.24 \times 10^{-4}\ \text{torr}^{-1} \cdot \text{s}^{-1}$$

$$k_1 k_2 / k_{-2} = (\text{intercept})^{-1} = (326\ \text{s})^{-1} = 3.07 \times 10^{-3}\ \text{s}^{-1}$$

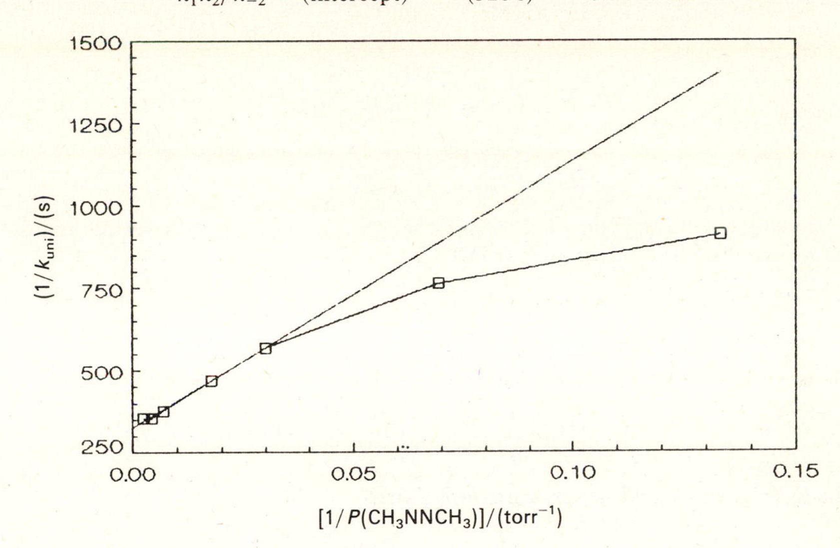

Fig. 12-15

12.59 Consider the following mechanism used to describe the overall chemical equation A → products, where M is an inert molecule:

$$A + M \underset{k_{-2}}{\overset{k_2}{\rightleftharpoons}} A^* + M \qquad A^* \xrightarrow{k_1} \text{products}$$

Assuming the steady-state approximation for $C(A^*)$, derive the rate expression for $C(\text{products})$. Compare the results to those in Problem 12.57.

▮ Applying the steady-state approximation to $C(A^*)$ gives

$$\frac{dC(A^*)}{dt} = k_2 C(A)C(M) - k_{-2}C(A^*)C(M) - k_1 C(A^*) = 0$$

$$C(A^*) = k_2 C(A)C(M)/[k_1 + k_{-2}C(M)]$$

which upon substitution into the rate expression for $C(\text{products})$ gives

$$\frac{dC(\text{products})}{dt} = k_1 C(A^*) = \frac{k_1 k_2 C(M)}{k_1 + k_{-2}C(M)} C(A)$$

This result is similar to *(12.14)* if the terms containing $C(M)$ are replaced by $C(A)$. This mechanism predicts a first-order reaction.

12.60 Consider the following proposed mechanism for the decomposition of ozone (see Problem 12.54).

mechanism 3

$$O_3 + M \underset{k_{-2}}{\overset{k_2}{\rightleftharpoons}} O_2 + O\cdot + M$$

$$O\cdot + O_3 \xrightarrow{k_2'} 2O_2$$

Does this mechanism predict a first-order reaction during the early stages of the decomposition? What is the pseudo-order during the latter stages of the reaction?

▮ Applying the steady-state approximation for $C(\text{O}\cdot)$ gives

$$\frac{dC(\text{O}\cdot)}{dt} = k_2C(\text{O}_3)C(\text{M}) - k_{-2}C(\text{O}_2)C(\text{O}\cdot)C(\text{M}) - k_2'C(\text{O}\cdot)C(\text{O}_3) = 0$$

$$C(\text{O}\cdot) = \frac{k_2C(\text{O}_3)C(\text{M})}{k_2'C(\text{O}_3) + k_{-2}C(\text{O}_2)C(\text{M})} = \frac{k_2C(\text{O}_3)}{k_2'C(\text{O}_3)/C(\text{M}) + k_2C(\text{O}_2)}$$

which upon substitution into the rate expression for $C(\text{O}_3)$ gives

$$\frac{-dC(\text{O}_3)}{dt} = k_2C(\text{O}_3)C(\text{M}) - k_{-2}C(\text{O}_2)C(\text{O}\cdot)C(\text{M}) + k_2'C(\text{O}\cdot)C(\text{O}_3)$$

$$= k_2C(\text{O}_3)C(\text{M}) - k_{-2}C(\text{O}_2)\frac{k_2C(\text{O}_3)}{k_2'C(\text{O}_3)/C(\text{M}) + k_{-2}C(\text{O}_2)}C(\text{M})$$

$$+ k_2'\frac{k_2C(\text{O}_3)}{k_2'C(\text{O}_3)/C(\text{M}) + k_{-2}C(\text{O}_2)}C(\text{O}_3)$$

$$= \frac{2k_2k_2'[C(\text{O}_3)]^2}{k_2'C(\text{O}_3)/C(\text{M}) + k_{-2}C(\text{O}_2)}$$

During the early stages of the reaction, $k_2'C(\text{O}_3)/C(\text{M}) \gg k_{-2}C(\text{O}_2)$, and this result predicts the pseudo-first-order reaction with respect to the concentration of O_3.

$$\frac{-dC(\text{O}_3)}{dt} = 2k_2C(\text{M})C(\text{O}_3)$$

During the latter stages of the reaction, $k_2'C(\text{O}_3)/C(\text{M}) \ll k_{-2}C(\text{O}_2)$, and the rate expression becomes

$$\frac{-dC(\text{O}_3)}{dt} = \frac{2k_2k_2'}{k_{-2}}[C(\text{O}_3)]^2[C(\text{O}_2)]^{-1}$$

which is second-order in O_3 and negative first-order in O_2.

12.61 Consider the following two proposed mechanisms for the overall equation $\text{A} + \text{B}_2 \rightarrow \text{C}$.

mechanism 1

$$\text{B}_2 + \text{M} \rightleftharpoons 2\text{B} + \text{M} \qquad (K_1)$$

$$\text{B} + \text{A} + \text{M} \rightleftharpoons \text{AB} + \text{M} \qquad (K_2)$$

$$\text{AB} + \text{B}_2 \xrightarrow{k} \text{C} + \text{B}$$

mechanism 2

$$\text{B}_2 + \text{M} \xrightarrow{k_1} 2\text{B} + \text{M}$$

$$\text{B} + \text{A} \xrightarrow{k_2} \text{AB}$$

$$\text{AB} + \text{B}_2 \xrightarrow{k_3} \text{C} + \text{B}$$

$$2\text{B} \xrightarrow{k_4} \text{B}_2$$

Identify the chain initiation, propagation, and termination steps in each mechanism. Do both mechanisms give the same expression for $dC(\text{C})/dt$?

▮ In the first mechanism, the forward reaction of the first step is the initiation step, the reverse reaction of the first step is the termination step, and the second and third steps are the propagation steps. In the second mechanism, the first step is the initiation step, the fourth step is the termination step, and the second and third steps are the propagation steps.

For the first mechanism, the equilibrium reactions give

$$C(\text{B}) = K_1^{1/2}[C(\text{B}_2)]^{1/2}$$

$$C(\text{AB}) = K_2C(\text{B})C(\text{A}) = K_1^{1/2}K_2C(\text{A})[C(\text{B}_2)]^{1/2}$$

which upon substitution into the rate expression for $C(\text{C})$ gives

$$\frac{dC(\text{C})}{dt} = kC(\text{AB})C(\text{B}_2) = kK_1^{1/2}K_2C(\text{A})[C(\text{B}_2)]^{3/2}$$

For the second mechanism, applying the steady-state approximation for $C(\mathrm{B})$ and $C(\mathrm{AB})$ gives

$$\frac{dC(\mathrm{B})}{dt} = 2k_1C(\mathrm{B_2})C(\mathrm{M}) - k_2C(\mathrm{B})C(\mathrm{A}) + k_3C(\mathrm{AB})C(\mathrm{B_2}) - 2k_4[C(\mathrm{B})]^2 = 0$$

$$\frac{dC(\mathrm{AB})}{dt} = k_2C(\mathrm{B})C(\mathrm{A}) - k_3C(\mathrm{AB})C(\mathrm{B_2}) = 0$$

Solving these equations simultaneously give

$$C(\mathrm{B}) = \frac{k_1^{1/2}}{k_4^{1/2}}[C(\mathrm{M})]^{1/2}[C(\mathrm{B_2})]^{1/2} \qquad C(\mathrm{AB}) = \frac{k_1^{1/2}k_2}{k_3k_4^{1/2}}\frac{C(\mathrm{A})[C(\mathrm{M})]^{1/2}}{[C(\mathrm{B_2})]^{1/2}}$$

which upon substitution into the rate expression for $C(\mathrm{C})$ gives

$$\frac{dC(\mathrm{C})}{dt} = k_3C(\mathrm{AB})C(\mathrm{B_2}) = \frac{k_1^{1/2}k_2}{k_4^{1/2}}C(\mathrm{A})[C(\mathrm{B_2})]^{1/2}[C(\mathrm{M})]^{1/2}$$

The different termination steps give different rate equations.

12.62 Consider the following three proposed mechanisms for the overall equation $2\mathrm{A} \rightarrow \mathrm{A_2}$.

mechanism 1	**mechanism 2**	**mechanism 3**
$2\mathrm{A} \underset{k_{-1}}{\overset{k_2}{\rightleftharpoons}} \mathrm{A_2^*}$	$\mathrm{A + M \rightleftarrows AM} \quad (K)$	$2\mathrm{A} \xrightarrow{k} \mathrm{A_2}$
$\mathrm{A_2^*} + \mathrm{M} \xrightarrow{k_2'} \mathrm{A_2 + M}$	$\mathrm{AM + A} \xrightarrow{k} \mathrm{A_2 + M}$	

Which of these mechanisms predicts a second-order reaction in $C(\mathrm{A})$? How could one or more of these mechanisms be eliminated?

▮ For the first mechanism, applying the steady-state approximation for $C(\mathrm{A_2^*})$ gives

$$\frac{dC(\mathrm{A_2^*})}{dt} = k_2[C(\mathrm{A})]^2 - k_{-1}C(\mathrm{A_2^*}) - k_2'C(\mathrm{A_2^*})C(\mathrm{M}) = 0$$

$$C(\mathrm{A_2^*}) = \frac{k_2[C(\mathrm{A})]^2}{k_{-1} + k_2'C(\mathrm{M})}$$

which upon substitution into the expression for $dC(\mathrm{A_2})/dt$ gives

$$\frac{dC(\mathrm{A_2})}{dt} = k_2'C(\mathrm{A_2^*})C(\mathrm{M}) = \frac{k_2k_2'C(\mathrm{M})}{k_{-1} + k_2'C(\mathrm{M})}[C(\mathrm{A})]^2 = k_{\mathrm{obs}}[C(\mathrm{A})]^2$$

which is second-order in $C(\mathrm{A})$. This mechanism would be acceptable if k_{obs} would show a complicated dependence on $C(\mathrm{M})$.

For the second mechanism, the equilibrium reaction gives $C(\mathrm{AM}) = KC(\mathrm{A})C(\mathrm{M})$, which upon substitution into the rate expression for $C(\mathrm{A_2})$ gives

$$\frac{dC(\mathrm{A_2})}{dt} = kC(\mathrm{AM})C(\mathrm{A}) = kKC(\mathrm{M})[C(\mathrm{A})]^2 = k_{\mathrm{obs}}[C(\mathrm{A})]^2$$

which is second-order in $C(\mathrm{A})$. This mechanism would be acceptable if k_{obs} would show a linear dependence on $C(\mathrm{M})$.

For the third mechanism, $dC(\mathrm{A_2})/dt = k[C(\mathrm{A})]^2$, which is second-order in $C(\mathrm{A})$. This mechanism would be acceptable if k were independent of $C(\mathrm{M})$.

12.63 The rate equation for the overall equation

$$\mathrm{H_3AsO_4(aq) + 2H^+(aq) + 3I^-(aq)} \underset{k_r}{\overset{k_f}{\rightleftharpoons}} \mathrm{H_3AsO_3(aq) + I_3^-(aq) + H_2O(l)}$$

is

$$\frac{dC(\mathrm{I_3^-})}{dt} = k_fC(\mathrm{H_3AsO_4})C(\mathrm{I^-})C(\mathrm{H^+}) - k_r\frac{C(\mathrm{H_3AsO_3})C(\mathrm{I_3^-})}{[C(\mathrm{I^-})]^2C(\mathrm{H^+})}$$

where $k_f = 4.7 \times 10^{-4}\ \mathrm{L^2 \cdot mol^{-2} \cdot min^{-1}}$ and $k_r = 3 \times 10^{-3}\ \mathrm{mol^2 \cdot L^{-2} \cdot min^{-1}}$. Determine the equilibrium constant for this reaction from the rate constants.

▌ Once equilibrium has been reached,

$$\frac{dC(I_3^-)}{dt} = k_f C(H_3AsO_4)C(I^-)C(H^+) - k_r \frac{C(H_3AsO_3)C(I_3^-)}{[C(I^-)]^2 C(H^+)} = 0$$

giving

$$K = \frac{[H_3AsO_3][I_3^-]}{[H_3AsO_4][I^-]^3[H^+]^2} = \frac{k_f}{k_r} = \frac{4.7 \times 10^{-4}}{3 \times 10^{-3}} = 0.2$$

12.64 For the reaction

$$H_2(g) + Ar(g) \underset{k_r}{\overset{k_f}{\rightleftharpoons}} 2H(g) + Ar(g)$$

$k_f = 2.2 \times 10^4\ L \cdot mol^{-1} \cdot s^{-1}$ and $K_C = 1.02 \times 10^{-4}$ at 3000 K. What is the value of k_r?

▌ Recognizing that at equilibrium (see Problem 12.63)

$$K_C = k_f / k_r \qquad (12.15)$$

Solving *(12.15)* for k_r and substituting the data gives

$$k_r = \frac{2.2 \times 10^4\ L \cdot mol^{-1} \cdot s^{-1}}{1.02 \times 10^{-4}\ mol \cdot L^{-1}} = 2.2 \times 10^8\ L^2 \cdot mol^{-2} \cdot s^{-1}$$

12.65 For the reaction

$$A \underset{k_{-1}}{\overset{k_1}{\rightleftharpoons}} B$$

the integrated rate law is

$$\ln \frac{C(A) - [A]}{C(A)_0 - [A]} = -(k_1 + k_{-1})t \qquad (12.16)$$

Describe the limiting case where $k_1 \gg k_{-1}$.

▌ In this limiting case, the reaction will essentially be irreversible and $[A] = 0$. Thus

$$\ln \frac{C(A) - 0}{C(A)_0 - 0} = -(k_1 + 0)t$$

which is identical to *(12.5)*.

12.66 An alternative form of *(12.16)* is

$$\frac{C(A)}{C(A)_0} = \frac{1}{1 + K} + \frac{K}{1 + K} e^{-kt}$$

where $k = k_1 + k_{-1}$. Prepare a plot of $C(A)/C(A)_0$ against kt for values of $K = 0.10, 1.00, 10.00$, and ∞. Briefly compare the curves.

▌ The plot is shown in Fig. 12-16. All the curves exponentially approach the value of $(1 + K)^{-1}$. The curve for $K = \infty$ is the usual exponential decay for the irreversible case.

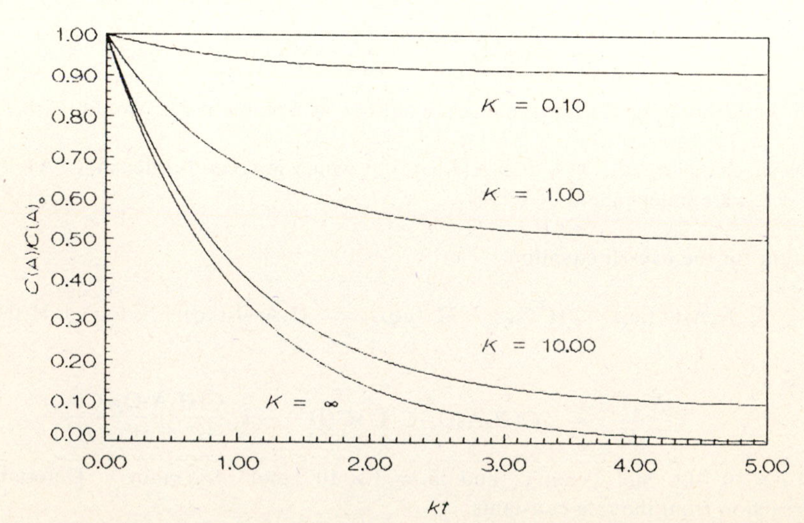

Fig. 12-16

12.67 For the reaction

$$A \underset{k_{-2}}{\overset{k_1}{\rightleftharpoons}} B + C$$

the integrated rate law is

$$\ln \frac{[C(A)_0]^2 - [A]C(A)}{C(A)_0\{C(A) - [A]\}} = k_1 \frac{C(A)_0 + [A]}{C(A)_0 - [A]} t$$

where $C(B)_0 = C(C)_0 = 0$. Given $k_1 = 5 \times 10^5\ s^{-1}$ for the ionization of NH_3, determine the time required for $C(OH^-)$ to reach $[OH^-]/2$ in a 0.0100 M solution of NH_3.

▌ In this solution, $[OH^-] = 3.9 \times 10^{-4}\ mol \cdot L^{-1}$ (see Problem 7.56). Under these conditions,

$$[NH_3] = 0.0100\ mol \cdot L^{-1} - 0.000\,39\ mol \cdot L^{-1} = 0.0096\ mol \cdot L^{-1}$$

$$C(NH_3) = 0.0100 - (0.000\,39/2) = 0.0098\ mol \cdot L^{-1}$$

Solving the above equation for t and substituting the data gives

$$t = \frac{\ln \dfrac{(0.0100\ mol \cdot L^{-1})^2 - (0.0096\ mol \cdot L^{-1})(0.0098\ mol \cdot L^{-1})}{(0.0100\ mol \cdot L^{-1})(0.0098\ mol \cdot L^{-1} - 0.0096\ mol \cdot L^{-1})}}{(5 \times 10^5\ s^{-1}) \dfrac{0.0100\ mol \cdot L^{-1} + 0.0096\ mol \cdot L^{-1}}{0.0100\ mol \cdot L^{-1} - 0.0096\ mol \cdot L^{-1}}} = 4 \times 10^{-8}\ s$$

12.68 For the reaction

$$[Cr(H_2O)_4Cl_2]^+(aq) \xrightarrow{k_1} [Cr(H_2O)_5Cl]^{2+}(aq) \xrightarrow{k_2} [Cr(H_2O)_6]^{3+}(aq)$$

$k_1 = 1.78 \times 10^{-3}\ s^{-1}$ and $k_2 = 5.8 \times 10^{-5}\ s^{-1}$ for $C([Cr(H_2O)_4Cl_2]^+)_0 = 0.0174\ mol \cdot L^{-1}$ at 0 °C. Prepare a plot of the concentration of the three complexes as a function of time.

▌ For two consecutive reactions the respective concentrations at time t are given by

$$C(A) = C(A)_0\, e^{-k_1 t} \qquad (12.17)$$

$$C(B) = C(A)_0 \frac{k_1}{k_2 - k_1}(e^{-k_1 t} - e^{-k_2 t}) \qquad (12.18)$$

$$C(C) = C(A)_0\left(1 - \frac{k_2\, e^{-k_2 t} - k_1\, e^{-k_2 t}}{k_2 - k_1}\right) \qquad (12.19)$$

As a sample calculation, consider $t = 500$ s.

$$C([Cr(H_2O)_4Cl_2]^+) = (0.0174\ mol \cdot L^{-1})\, e^{-(1.78\times10^{-3}\ s^{-1})(500\ s)} = 0.0071\ mol \cdot L^{-1}$$

$$C([Cr(H_2O)_5Cl]^{2+}) = (0.0174) \frac{1.78 \times 10^{-3}}{5.8 \times 10^{-5} - 1.78 \times 10^{-3}}(e^{-(1.78\times10^{-3})(500)} - e^{-(5.8\times10^{-5})(500)})$$
$$= 0.0101\ mol \cdot L^{-1}$$

$$C([Cr(H_2O)_6]^{3+}) = (0.0174)\left[1 - \frac{(5.8 \times 10^{-5})\, e^{-(1.78\times10^{-3})(500)} - (1.78 \times 10^{-3})\, e^{-(5.8\times10^{-5})(500)}}{(5.8 \times 10^{-5}) - (1.78 \times 10^{-3})}\right]$$
$$= 0.0002\ mol \cdot L^{-1}$$

The plots are shown in Fig. 12-17.

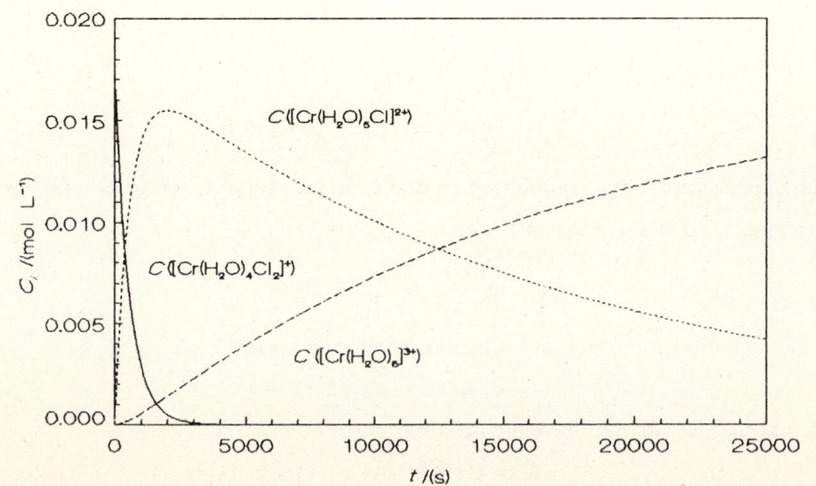

Fig. 12-17

12.69 Calculate the value of t at which the value of $C([Cr(H_2O)_5Cl]^{2+})$ is a maximum. See Problem 12.68 for additional data. What is the maximum value of $C([Cr(H_2O)_5Cl]^{2+})$? Compare the results to those shown in Fig. 12-17.

▮ Setting the derivative of *(12.18)* with respect to t equal to zero gives

$$\frac{d[C(B)/C(A)_0]}{dt} = \frac{k_1}{k_2 - k_1}(-k_1 e^{-k_1 t} + k_2 e^{-k_2 t}) = 0$$

$$t = \frac{\ln(k_2/k_1)}{k_2 - k_1} = \frac{\ln[(5.8 \times 10^{-5}\ \text{s}^{-1})/(1.78 \times 10^{-3}\ \text{s}^{-1})]}{5.8 \times 10^{-5}\ \text{s}^{-1} - 1.78 \times 10^{-3}\ \text{s}^{-1}} = 1990\ \text{s}$$

Substituting into *(12.18)* gives

$$C([Cr(H_2O)_5Cl]^{2+}) = \frac{(1.78 \times 10^{-3}\ \text{s}^{-1})(0.0174\ \text{mol}\cdot\text{L}^{-1})}{5.8 \times 10^{-5}\ \text{s}^{-1} - 1.78 \times 10^{-3}\ \text{s}^{-1}}[e^{-(1.78\times10^{-3}\ \text{s}^{-1})(1990\ \text{s})} - e^{-(5.8\times10^{-5}\ \text{s}^{-1})(1990\ \text{s})}]$$

$$= 0.0155\ \text{mol}\cdot\text{L}^{-1}$$

These values are in good agreement with the curve shown for $C([Cr(H_2O)_5Cl]^{2+})$ in Fig. 12-17.

12.70 Prepare plots of *(12.17)–(12.19)* against $k_1 t$, where $k_2/k_1 = 0.10$, 1.00, and 10.00. Briefly discuss the curves.

▮ For the case where $k_2 = k_1 = k$, *(12.18)* and *(12.19)* are undefined. For this special case,

$$C(B) = C(A)_0 kt\, e^{-kt} \qquad C(C) = C(A)_0[1 - (1 + kt)\, e^{-kt}]$$

The plots are shown in Fig. 12-18. The curve for $C(A)$ is the same for all three cases. For the $k_2/k_1 = 0.10$ set of curves, $C(B)$ increases as $C(A)$ decreases and reaches a rather high concentration before a large amount of C is produced. For the $k_2/k_1 = 1.00$ set of curves, $C(B)$ increases as $C(A)$ decreases, but B begins to react before $C(B)$ reaches a high value. For the $k_2/k_1 = 10.00$ set of curves, $C(B)$ is quite low and is nearly insignificant in the system.

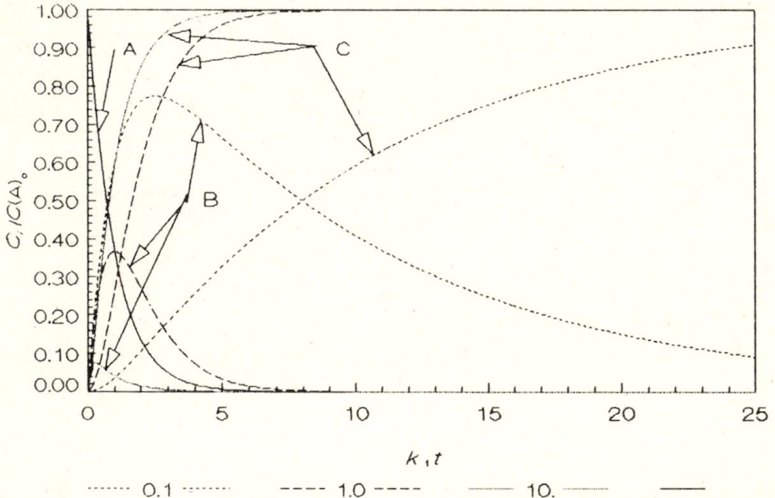

Fig. 12-18

12.71 A proposed mechanism for the reaction A → B is

$$A \underset{k_{-1}}{\overset{k_1}{\rightleftharpoons}} C \qquad C \underset{k'_{-1}}{\overset{k'_1}{\rightleftharpoons}} B$$

Determine an expression for the equilibrium constant for the overall reaction in terms of the rate constants.

▮ At equilibrium, *(12.15)* gives for each step

$$K = \frac{C(C)}{C(A)} = \frac{k_1}{k_{-1}} \qquad K' = \frac{C(B)}{C(C)} = \frac{k'_1}{k'_{-1}}$$

Solving the first equation for $C(C)$ and substituting into the second equation gives

$$C(B)/C(A)(k_1/k_{-1}) = k'_1/k'_{-1}$$

Recognizing that the overall equilibrium constant is given by $C(B)/C(A)$,

$$K = C(B)/C(A) = (k'_1/k'_{-1})(k_1/k_{-1})$$

12.72 Consider a mechanism in which a single reactant produces several products by the following parallel first-order reactions:

$$A \xrightarrow{k_1} B \qquad A \xrightarrow{k_1'} C \qquad A \xrightarrow{k_1''} D$$

Write the rate expression for $C(A)$, and determine the relative concentrations of the products at time t given $C(B)_0 = C(C)_0 = C(D)_0 = 0$.

▮ The rate of disappearance of A is given by

$$\frac{-dC(A)}{dt} = k_1C(A) + k_1'C(A) + k_1''C(A) = (k_1 + k_1' + k_1'')C(A) = kC(A) \tag{12.20}$$

where $k = k_1 + k_1' + k_1''$. At time t, *(12.17)* gives $C(A) = C(A)_0 e^{-kt}$.
For product B, $dC(B)/dt = k_1C(A) = k_1C(A)_0 e^{-kt}$, which integrates to

$$C(B) = C(B)_0 + \frac{k_1C(A)_0}{k}(1 - e^{-kt}) \tag{12.21}$$

Similar expressions can be written for $C(C)$ and $C(D)$. Taking ratios of these expressions gives

$$C(B):C(C):C(D) = k_1:k_1':k_1''$$

12.73 If $k_1 = 100.0\ \mathrm{s^{-1}}$ and $k_1' = 1.00\ \mathrm{s^{-1}}$, what is the half-life of A in Problem 12.72?

▮ From Problem 12.72, $k = k_1 + k_1' = 100.0\ \mathrm{s^{-1}} + 10.0\ \mathrm{s^{-1}} = 110.0\ \mathrm{s^{-1}}$. Substituting into *(12.7)* gives

$$t_{1/2} = \frac{-\ln 0.5}{110.0\ \mathrm{s^{-1}}} = 6.30 \times 10^{-3}\ \mathrm{s}$$

12.74 What are the concentrations of B and C at $t = 1.0$ ms in the system described in Problem 12.73? Assume $C(B)_0 = C(C)_0 = 0$ and $C(A)_0 = 0.100\ \mathrm{mol \cdot L^{-1}}$.

▮ Using *(12.21)* for each product gives

$$C(B) = 0 + \frac{(100.0\ \mathrm{s^{-1}})(0.100\ \mathrm{mol \cdot L^{-1}})}{110.0\ \mathrm{s^{-1}}}[1 - e^{-(110.0\ \mathrm{s^{-1}})(0.0010\ \mathrm{s})}] = 9.5 \times 10^{-3}\ \mathrm{mol \cdot L^{-1}}$$

$$C(C) = 0 + \frac{(10.0)(0.100)}{110.0}[1 - e^{-(110.0)(0.0010)}] = 9.5 \times 10^{-4}\ \mathrm{mol \cdot L^{-1}}$$

Note that $C(B)/C(C) = 10/1$ as predicted by the results of Problem 12.72.

12.75 Consider a mechanism in which a single reactant produces several products by the following parallel reactions:

$$A \xrightarrow{k_1} B \qquad 2A \xrightarrow{k_2} C + D$$

Write the rate expression for $C(A)$. Derive the integrated rate law for this mechanism assuming $C(B)_0 = C(A)_0 = C(D)_0 = 0$.

▮ The rate of disappearance of A is given by $-dC(A)/dt = k_1C(A) + 2k_2[C(A)]^2$. Rearranging this expression gives

$$\frac{dC(A)}{k_1C(A) + 2k_2[C(A)]^2} = -dt$$

The integral on the left side of the equation is given by

$$\int \frac{dx}{x(a + bx)} = -\frac{1}{a}\ln\frac{a + bx}{x}$$

Evaluating between the limits of $C(A) = C(A)_0$ at $t = 0$ and $C(A) = C(A)$ at $t = t$ gives

$$-\frac{1}{k_1}\ln\frac{k_1 + 2k_2C(A)}{C(A)} + \frac{1}{k_1}\ln\frac{k_1 + 2k_2C(A)_0}{C(A)_0} = -(t - 0)$$

$$\ln\frac{C(A)_0[k_1 + 2k_2C(A)]}{C(A)[k_1 + 2k_2C(A)_0]} = k_1t$$

12.76 Consider the two following proposed parallel mechanisms for the overall equation $A + B \rightarrow C + D$.

mechanism 1	mechanism 2
$A \xrightarrow{k_1} D + E$	$A + B \xrightarrow{k_2} C + D$
$B + E \xrightarrow{k_1'} C$	

where $k_1' \gg k_1$. Write the rate expression for $-dC_A/dt$.

▮ The rate expression is

$$\frac{-dC(A)}{dt} = k_1C(A) + k_2C(A)C(B) = [k_1 + k_2C(B)]C(A)$$

12.77 A mixture of two different substances, A and B, undergoes parallel first-order reactions to form a common product, C.

$$A \xrightarrow{k_1} C \qquad B \xrightarrow{k_1'} C$$

Write an expression for the concentration of C as a function of t.

▮ At any time t, $C(C) = [C(A)_0 - C(A)] + [C(B)_0 - C(B)]$. Substituting *(12.17)* gives

$$C(C) = [C(A)_0 - C(A)_0\,e^{-k_1t}] + [C(B)_0 - C(B)_0\,e^{-k_1't}] \tag{12.22}$$

12.78 Describe the behavior of *(12.22)* if $k_1' = k_1$ and if $k_1' \neq k_1$.

▮ The logarithm form of *(12.22)* is

$$\ln[C(A)_0 + C(B)_0 - C(C)] = \ln[C(A)_0\,e^{-k_1t} + C(B)_0\,e^{-k_1't}] \tag{12.23}$$

If $k_1' = k_1 = k$, then *(12.23)* gives

$$\ln[C(A)_0 + C(B)_0 - C(C)] = \ln\{[C(A)_0 + C(B)_0]\,e^{-kt}\} = \ln[C(A)_0 + C(B)_0] - kt$$

A plot of $\ln[C(A)_0 + C(B)_0 - C(C)]$ against t will be linear with a slope equal to $-k$. If $k_1' \neq k_1$, then the reactant involved in the reaction with the greater k reacts more rapidly, and the contribution from this reaction decreases as t increases. Thus, *(12.23)* predicts a nonlinear relationship between $\ln[C(A)_0 + C(B)_0 - C(C)]$ and t until only the slower reaction remains significant, at which point the plot becomes linear. From the linear portion of the curve, the value of the smaller k can be determined, and, in turn, from the initial portion of the curve, the value of the greater k can be determined.

12.79 Consider the following data for the hydrolysis of diethyl-t-butylcarbinyl chloride in which the mechanism given in Problem 12.77 is thought to be applicable.

$C(C)/(\text{mol}\cdot\text{L}^{-1})$	0	0.0106	0.0179	0.0283	0.0350	0.0397	0.0431
$t/(\text{s})$	0	900	1800	3600	5400	7200	9000

$C(C)/(\text{mol}\cdot\text{L}^{-1})$	0.0460	0.0482	0.0498	0.0525	0.0542	0.0580
$t/(\text{s})$	10 800	12 600	14 400	18 000	21 600	∞

Determine the values of k_1 and k_1'.

▮ The value of $C(A)_0 + C(B)_0 - C(C)$ in *(12.23)* can be determined from the data by setting it equal to $C(C)_\infty - C(C)$. A plot of $\ln[C(C)_\infty - C(C)]$ against t is shown in Fig. 12-19. For $t > 11\,000$ s, the plot is linear with a slope of $-1.08 \times 10^{-4}\,\text{s}^{-1}$ and an intercept of -3.259 as indicated by the solid line. Letting k_1' represent the slower reaction, then

$$k' = -\text{slope} = -(-1.08 \times 10^{-4}\,\text{s}^{-1}) = 1.08 \times 10^{-4}\,\text{s}^{-1}$$

$$C(B)_0 = e^{-3.259} = 0.0384\ \text{mol}\cdot\text{L}^{-1}$$

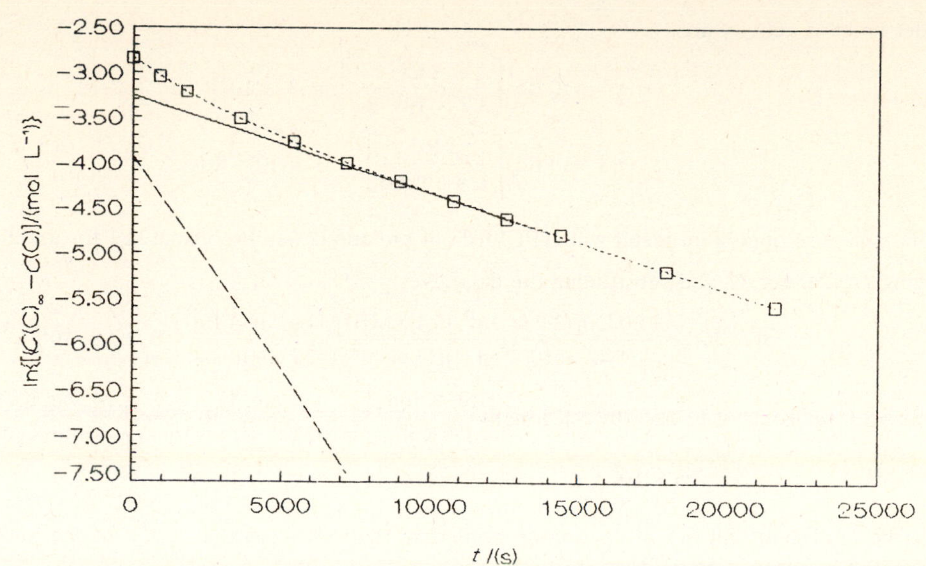

Fig. 12-19

Rearranging *(12.22)* and taking logarithms gives for the initial portion of the curve

$$\ln[C(\mathrm{A})_0 + C(\mathrm{B})_0 - C(\mathrm{C}) - C(\mathrm{B})_0 e^{-k_1' t}] = \ln[C(\mathrm{C})_\infty - C(\mathrm{C}) - C(\mathrm{B})_0 e^{-k_1' t}]$$
$$= \ln[C(\mathrm{A})_0] - k_1 t$$

The dashed line shown in Fig. 12-19 is a plot of this function. The plot is linear with a slope of $-4.78 \times 10^{-4}\ \mathrm{s^{-1}}$ and an intercept of -3.927, giving

$$k_1 = -\text{slope} = -(-4.78 \times 10^{-4}\ \mathrm{s^{-1}}) = 4.78 \times 10^{-4}\ \mathrm{s^{-1}}$$
$$C(\mathrm{A})_0 = e^{-3.927} = 0.0197\ \mathrm{mol \cdot L^{-1}}$$

Using the results for the initial concentration and rate constants, the dotted line shown in Fig. 12-19 was determined using *(12.23)* and is in good agreement with the experimental data.

12.4 RADIOACTIVE DECAY

12.80 What is the decay constant for the β^- emission of $^3\mathrm{H}$ given $t_{1/2} = 12.26$ yr?

▌ The decay constant (λ) is the first-order rate constant for radioactive decay processes. Solving *(12.27)* for λ and substituting the data gives

$$\lambda = \frac{-\ln 0.5}{12.26\ \mathrm{yr}} = 0.056\,54\ \mathrm{yr^{-1}}$$

12.81 What is the activity of a 1.00-g sample of KCl? Naturally occurring potassium consists of 0.01% $^{40}\mathrm{K}$, which has a half-life of 1.28×10^9 yr.

▌ The decay constant is

$$\lambda = \frac{-\ln 0.5}{1.28 \times 10^9\ \mathrm{yr}} = 5.42 \times 10^{-10}\ \mathrm{yr^{-1}}$$

In a 1.00-g sample of KCl,

$$N(^{40}\mathrm{K}) = (1.00\ \mathrm{g\ KCl})\left(\frac{1\ \mathrm{mol\ KCl}}{74.56\ \mathrm{g\ KCl}}\right)\left(\frac{6.022 \times 10^{23}\ \mathrm{K\ ions}}{1\ \mathrm{mol\ KCl}}\right)\left(\frac{1\ ^{40}\mathrm{K\ ion}}{10^4\ \mathrm{K\ ions}}\right) = 8 \times 10^{17}$$

The activity (A) is given by

$$A = \lambda N = (5.42 \times 10^{-10}\ \mathrm{yr^{-1}})(8 \times 10^{17}) = 4 \times 10^8\ \mathrm{yr^{-1}} \qquad (12.24)$$

12.82 Express the activity of the KCl sample described in Problem 12.81 in units of becquerels, curies, and rutherfords.

▌ The SI unit becquerel is defined as $1\ \mathrm{Bq} = 1\ \mathrm{s^{-1}}$. Thus

$$A = (4 \times 10^8\ \mathrm{yr^{-1}})\left[\frac{1\ \mathrm{yr}}{3.1536 \times 10^7\ \mathrm{s}}\right] = 10\ \mathrm{s^{-1}} = 10\ \mathrm{Bq}$$

The older units of activity are

$$A = (10\ \text{Bq})\left[\frac{1\ \text{Ci}}{3.7 \times 10^{10}\ \text{Bq}}\right] = 3 \times 10^{-10}\ \text{Ci}$$

$$A = (10\ \text{Bq})\left[\frac{1\ \text{Rd}}{1 \times 10^{6}\ \text{Bq}}\right] = 1 \times 10^{-5}\ \text{Rd}$$

12.83 What mass of pure potassium metal will emit 1.0 Ci of radiation? See Problem 12.81 for additional data.

▌ Solving *(12.24)* for N and substituting the data gives

$$N(^{40}\text{K}) = \frac{(1.0\ \text{Ci})[(3.7 \times 10^{10}\ \text{Bq})/(1\ \text{Ci})][(1\ \text{s}^{-1})/(1\ \text{Bq})]}{(5.42 \times 10^{-10}\ \text{yr}^{-1})[(1\ \text{yr})/(3.1536 \times 10^{7}\ \text{s})]} = 2.2 \times 10^{27}$$

$$m(\text{KCl}) = (2.2 \times 10^{27}\ {}^{40}\text{K ions})\left(\frac{10^4\ \text{K ions}}{1\ {}^{40}\text{K ion}}\right)\left(\frac{1\ \text{mol KCl}}{6.022 \times 10^{23}\ \text{K ions}}\right)\left(\frac{74.56\ \text{g KCl}}{1\ \text{mol KCl}}\right)$$

$$= 2.7 \times 10^{9}\ \text{g}$$

12.84 What is the minimum half-life of an isotope needed so that not more than 0.1% of the nuclei undergo decay during a 3.0-h laboratory period?

▌ Solving *(12.17)* for λ and substituting the data gives

$$\lambda = \frac{-\ln 0.999}{3.0\ \text{h}} = 3.3 \times 10^{-4}\ \text{h}^{-1}$$

which upon substitution into *(12.7)* gives

$$t_{1/2} = \frac{-\ln 0.5}{3.3 \times 10^{-4}\ \text{h}^{-1}} = 2100\ \text{h} = 88\ \text{days}$$

12.85 What fraction of tritium atoms in a sample remains after 100.0 yr? If the original number of tritium atoms was 1.50×10^{18}, what is the number remaining? See Problem 12.80 for additional data.

▌ Substituting the data into *(12.17)* gives

$$\frac{N}{N_0} = \exp[-(0.056\,54\ \text{yr}^{-1})(100.0\ \text{yr})] = 3.50 \times 10^{-3}$$

which is equivalent to

$$N = (1.50 \times 10^{18})(3.50 \times 10^{-3}) = 5.26 \times 10^{15}$$

12.86 The nuclide ^{8}Be undergoes α emission with $t_{1/2} = 2 \times 10^{-16}$ s. What volume of He will be generated at 25 °C and 1.0 bar from 1.0 pg of ^{8}Be undergoing decay for 1.0 as? (1 attosecond $= 10^{-18}$ s.)

▌ The nuclear equation for the decay is ${}^8_4\text{Be} \rightarrow 2\,{}^4_2\text{He}$. Solving *(12.7)* for λ and substituting the data gives

$$\lambda = \frac{-\ln 0.5}{2 \times 10^{-16}\ \text{s}} = 3 \times 10^{15}\ \text{s}^{-1}$$

The number of ^{8}Be nuclei that have undergone decay is $N_0 - N$. Substituting *(12.17)* and the data gives

$$N_0 - N = N_0(1 - e^{-\lambda t})$$

$$= \left[(1.0 \times 10^{-12}\ \text{g})\left(\frac{1\ \text{mol}}{8.00\ \text{g}}\right)\left(\frac{6.022 \times 10^{23}}{1\ \text{mol}}\right)\right]\{1 - \exp[-(3 \times 10^{15}\ \text{s}^{-1})(1.0 \times 10^{-18}\ \text{s})]\}$$

$$= 2.3 \times 10^{8}$$

The number of He atoms formed is $2(2.3 \times 10^8) = 4.6 \times 10^8$ or 7.6×10^{-16} mol. Solving *(1.18)* for V and substituting the data gives

$$V = \frac{(7.6 \times 10^{-16}\ \text{mol})(8.314\ \text{Pa} \cdot \text{m}^3 \cdot \text{K}^{-1} \cdot \text{mol}^{-1})(298\ \text{K})}{(1.0\ \text{bar})[(1 \times 10^{5}\ \text{Pa})/(1\ \text{bar})]} = 1.9 \times 10^{-17}\ \text{m}^3$$

12.87 Isotopes of oxygen with mass number less than 16 undergo β^+ emission. Assuming an equimolar mixture of ^{14}O and ^{15}O, find the ratio of the nuclides at the end of 1.00 h given $t_{1/2}/(\text{s}) = 71.0$ for ^{14}O and 124 for ^{15}O.

▮ For each nuclide *(12.7)* gives

$$\lambda(^{14}O) = \frac{-\ln 0.5}{71.0\ s} = 9.76 \times 10^{-3}\ s^{-1}$$

$$\lambda(^{15}O) = \frac{-\ln 0.5}{124} = 5.59 \times 10^{-3}\ s^{-1}$$

For each isotope, N will be given by *(12.17)*. Taking a ratio and substituting the data gives

$$\frac{N(^{14}O)}{N(^{15}O)} = \frac{\exp[-(9.76 \times 10^{-3}\ s^{-1})(3600\ s)]}{\exp[-(5.59 \times 10^{-3}\ s^{-1})(3600\ s)]} = 3.02 \times 10^{-7}$$

12.88 At what time will the ratio of ^{14}O nuclei to ^{15}O nuclei be equal to 0.250 in the system described in Problem 12.87?

▮ Substituting the data into the expression for $N(^{14}O)/N(^{15}O)$ and solving for t gives

$$\frac{N(^{14}O)}{N(^{15}O)} = \frac{e^{-(9.76\times10^{-3}\ s^{-1})t}}{e^{-(5.59\times10^{-3}\ s^{-1})t}} = 0.250 \qquad t = 332\ s$$

12.89 The nuclide ^{227}Ac undergoes β^- emission (98.6%) or α emission (1.4%) with a half-life of 21.6 g. Determine $\lambda(\alpha)$ and $\lambda(\beta^-)$.

▮ The overall decay constant is given by *(12.7)* as

$$\lambda = \frac{-\ln 0.5}{21.6\ yr} = 0.0321\ yr^{-1}$$

For the parallel modes of decay, *(12.20)* gives

$$\lambda = \lambda(\alpha) + \lambda(\beta^-) = 0.0321\ yr^{-1}$$

and from the results of Problem 12.72

$$\lambda(\alpha)/\lambda(\beta^-) = 0.014/0.986$$

Solving these simultaneous equations gives $\lambda(\alpha) = 4.5 \times 10^{-4}\ yr^{-1}$ and $\lambda(\beta^-) = 0.0317\ yr^{-1}$.

12.90 A portion of the "thorium" decay series consists of the following decay processes:

$$^{228}Ac \xrightarrow{-\beta^-} {}^{228}Th \xrightarrow{-\alpha} {}^{224}Ra \xrightarrow{-\alpha} {}^{220}Rn \xrightarrow{-\alpha} {}^{216}Po$$

where $t_{1/2} = 6.13$ h, 1.913 yr, 3.64 d, and 55 s, respectively. What is the $N(^{228}Ac)/N(^{228}Th)$ ratio at $t = 2.0 \times 10^5$ s?

▮ Substituting the data into *(12.7)* gives

$$\lambda(^{228}Ac) = \frac{-\ln 0.5}{(6.13\ h)[(3600\ s)/(1\ h)]} = 3.14 \times 10^{-5}\ s^{-1}$$

$$\lambda(^{228}Th) = 1.148 \times 10^{-8}\ s^{-1}$$

$$\lambda(^{224}Ra) = 2.20 \times 10^{-6}\ s^{-1}$$

$$\lambda(^{220}Rn) = 1.26 \times 10^{-2}\ s^{-1}$$

This decay is an example of *disequilibrium* in which $t_{1/2}(\text{parent}) < t_{1/2}(\text{daughter})$ or $\lambda(\text{parent}) > \lambda(\text{daughter})$. For this case of consecutive first-order reactions, using *(12.17)* and *(12.18)* gives

$$N(^{228}Th) = \frac{(3.14 \times 10^{-5}\ s^{-1})N(^{228}Ac)_0}{1.148 \times 10^{-8}\ s^{-1} - 3.14 \times 10^{-5}\ s^{-1}}$$
$$\times \{\exp[-(3.14 \times 10^{-5}\ s^{-1})(2.0 \times 10^5\ s)] - \exp[-(1.148 \times 10^{-8}\ s^{-1})(2.0 \times 10^5\ s)]\}$$
$$= (0.996)N(^{228}Ac)_0$$

$$N(^{228}Ac) = N(^{228}Ac)_0 \exp[-(3.14 \times 10^{-5}\ s^{-1})(2.0 \times 10^5\ s)] = (1.87 \times 10^{-3})N(^{228}Ac)_0$$

The desired ratio is

$$\frac{N(^{228}Ac)}{N(^{228}Th)} = \frac{(1.87 \times 10^{-3})N(^{228}Ac)_0}{0.996N(^{228}Ac)_0} = 1.88 \times 10^{-3}$$

12.91 For the decay of ^{228}Ac to ^{228}Th described in Problem 12.90, determine the time for the radioactive daughter to reach its maximum activity.

▮ From Problem 12.69,

$$t = \frac{\ln[\lambda(^{228}\text{Th})/\lambda(^{228}\text{Ac})]}{\lambda(^{228}\text{Th}) - \lambda(^{228}\text{Ac})} = \frac{\ln[(1.148 \times 10^{-8}\ \text{s}^{-1})/(3.14 \times 10^{-5}\ \text{s}^{-1})]}{1.148 \times 10^{-8}\ \text{s}^{-1} - 3.14 \times 10^{-5}\ \text{s}^{-1}} = 2.52 \times 10^5\ \text{s}$$

12.92 Consider the ^{228}Th-^{224}Ra decay step described in Problem 12.90. Determine the $N(^{228}\text{Th})/N(^{224}\text{Ra})$ ratio once *transient equilibrium* in which $t_{1/2}(\text{parent}) > t_{1/2}(\text{daughter})$ or $\lambda(\text{parent}) < \lambda(\text{daughter})$ has been established.

▮ For this special case, *(12.18)* becomes

$$\frac{N(\text{parent})}{N(\text{daughter})} \approx \frac{\lambda(\text{daughter}) - \lambda(\text{parent})}{\lambda(\text{parent})}$$

which upon substituting the data gives

$$\frac{N(^{228}\text{Th})}{N(^{224}\text{Ra})} = \frac{2.20 \times 10^{-6}\ \text{s}^{-1} - 1.148 \times 10^{-8}\ \text{s}^{-1}}{1.148 \times 10^{-8}\ \text{s}^{-1}} = 191$$

12.93 Consider the ^{224}Ra-^{220}Rn decay step described in Problem 12.90. Determine the $N(^{224}\text{Th})/N(^{220}\text{Ra})$ ratio once *secular equilibrium* in which $t_{1/2}(\text{parent}) \gg t_{1/2}(\text{daughter})$ or $\lambda(\text{parent}) \ll \lambda(\text{daughter})$ has been established.

▮ For this special case *(12.18)* becomes

$$\frac{N(\text{parent})}{N(\text{daughter})} \approx \frac{\lambda(\text{daughter})}{\lambda(\text{parent})}$$

Substituting the data gives

$$\frac{N(^{224}\text{Ra})}{N(^{220}\text{Rn})} = \frac{1.26 \times 10^{-2}\ \text{s}^{-1}}{2.20 \times 10^{-6}\ \text{s}^{-1}} = 5730$$

12.94 What is the limiting value of the time required for the radioactive daughter to reach its maximum activity as the value of $t_{1/2}(\text{parent})/t_{1/2}(\text{daughter})$ increases?

▮ The limiting case occurs when the value of $\lambda(\text{daughter})/\lambda(\text{parent})$ approaches infinity. In this special case of secular equilibrium

$$\lim_{t_{1/2}(\text{parent}) \gg t_{1/2}(\text{daughter})} t = \lim_{t_{1/2}(\text{parent}) \gg t_{1/2}(\text{daughter})} \left[\frac{\log[\lambda(\text{daughter})/\lambda(\text{parent})]}{\lambda(\text{daughter}) - \lambda(\text{parent})}\right] \to \infty$$

12.95 One of the products from the thermal neutron fission of ^{235}U is ^{144}Xe, which is the parent of the following decay series:

$$^{144}\text{Xe} \xrightarrow{-\beta^-} {}^{144}\text{Cs} \xrightarrow{-\beta^-} {}^{144}\text{Ba} \xrightarrow{-\beta^-} {}^{144}\text{La} \xrightarrow{-\beta^-} {}^{144}\text{Ce} \to \cdots$$

where $t_{1/2}/(\text{s}) = 1.2\ \text{s}, 1.0\ \text{s}, 11.9\ \text{s}$, and $40\ \text{s}$, respectively. Prepare a plot of $N_i/N(^{144}\text{Xe})_0$ for the various nuclides for $0 \le t/(\text{s}) \le 100$.

▮ Using *(12.7)* for the nuclides gives

$$\lambda(^{144}\text{Xe}) = \frac{-\ln 0.5}{1.2\ \text{s}} = 0.58\ \text{s}^{-1} \qquad \lambda(^{144}\text{Cs}) = 0.69\ \text{s}^{-1}$$

$$\lambda(^{144}\text{Ba}) = 0.0582\ \text{s}^{-1} \qquad \lambda(^{144}\text{La}) = 0.017\ \text{s}^{-1}$$

The number of nuclei present at time t for each nuclide is given by

$$N(1) = N(1)_0 e^{-\lambda_1 t} \tag{12.25}$$

$$N(2) = N(1)_0 \lambda_1 \left[\frac{e^{-\lambda_1 t}}{\lambda_2 - \lambda_1} + \frac{e^{-\lambda_2 t}}{\lambda_1 - \lambda_2}\right] \tag{12.26}$$

$$N(3) = N(1)_0 \lambda_1 \lambda_2 \left[\frac{e^{-\lambda_1 t}}{(\lambda_2 - \lambda_1)(\lambda_3 - \lambda_1)} + \frac{e^{-\lambda_2 t}}{(\lambda_1 - \lambda_2)(\lambda_3 - \lambda_2)} + \frac{e^{-\lambda_3 t}}{(\lambda_1 - \lambda_3)(\lambda_2 - \lambda_3)}\right] \tag{12.27}$$

$$N(4) = N(1)_0 \lambda_1 \lambda_2 \lambda_3 \left[\frac{e^{-\lambda_1 t}}{(\lambda_2 - \lambda_1)(\lambda_3 - \lambda_1)(\lambda_4 - \lambda_1)} + \frac{e^{-\lambda_2 t}}{(\lambda_1 - \lambda_2)(\lambda_3 - \lambda_2)(\lambda_4 - \lambda_2)} + \frac{e^{-\lambda_3 t}}{(\lambda_1 - \lambda_3)(\lambda_2 - \lambda_3)(\lambda_4 - \lambda_3)} + \frac{e^{-\lambda_4 t}}{(\lambda_1 - \lambda_4)(\lambda_2 - \lambda_4)(\lambda_3 - \lambda_4)}\right] \tag{12.28}$$

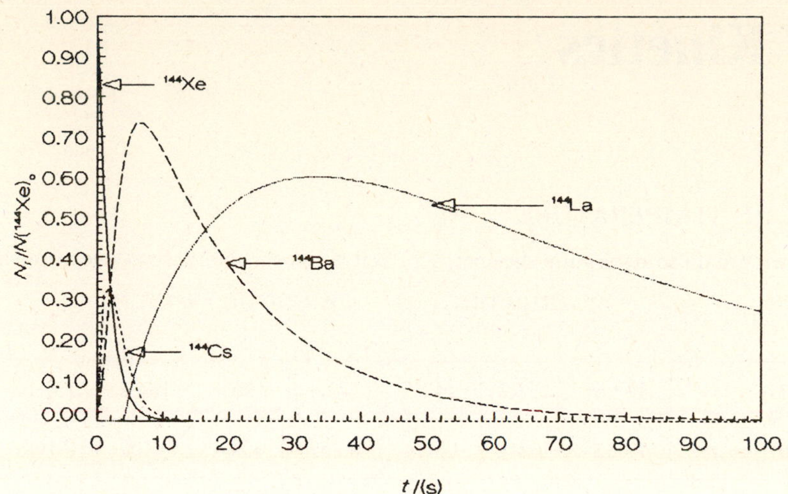

Fig. 12-20

The plot is shown in Fig. 12-20. Note that because all species are radioactive, the values of N_i all decrease as t increases—unlike component C in *(12.19)*, which increases as t increases.

12.96 Assuming $A_0 = 12.6\ \text{min}^{-1}\cdot\text{g}^{-1}$ and $t_{1/2} = 5730\ \text{yr}$ for ^{14}C, predict the ^{14}C activity for a piece of cloth manufactured 1990 years ago.

I For ^{14}C, $\lambda = 1.21 \times 10^{-4}\ \text{yr}^{-1}$. Solving *(12.17)* for A and substituting the data gives

$$A = (12.6\ \text{min}^{-1}\cdot\text{g}^{-1})\exp[-(1.21\times10^{-4}\ \text{yr}^{-1})(1990\ \text{yr})] = 9.9\ \text{min}^{-1}\cdot\text{g}^{-1}$$

12.97 If the Shroud of Turin has a ^{14}C activity of $11.5\ \text{min}^{-1}\cdot\text{g}^{-1}$, estimate the age of this piece of cloth. See Problem 12.96 for additional data.

I Solving *(12.17)* for t and substituting the data gives

$$t = \frac{-\ln[(11.5\ \text{min}^{-1}\cdot\text{g}^{-1})/(12.6\ \text{min}^{-1}\cdot\text{g}^{-1})]}{1.21\times10^{-4}\ \text{yr}^{-1}} = 755\ \text{yr}$$

12.98 Material from a stony meteorite had an $N(^{207}Pb)/N(^{206}Pb)$ ratio equal to 0.589. Using

$$\frac{N(^{207}\text{Pb})}{N(^{206}\text{Pb})} = (7.25\times10^{-3})\frac{e^{\lambda_{235}t}-1}{e^{\lambda_{238}t}-1}$$

where $\lambda_{235} = 9.8\times10^{-10}\ \text{yr}^{-1}$ and $\lambda_{238} = 1.54\times10^{-10}\ \text{yr}^{-1}$, determine the age of the meteorite.

I Substituting the data and solving for t gives

$$0.589 = 7.25\times10^{-3}\frac{e^{(9.8\times10^{-10}\ \text{yr}^{-1})t}-1}{e^{(1.54\times10^{-10}\ \text{yr}^{-1})t}-1}$$

$$t = 4.37\times10^{9}\ \text{yr}$$

12.99 A granite pebble from Southern Zimbabwe had an $N(^{40}Ar)/N(^{40}K)$ ratio of 0.524. Determine the age of the rock.

I The age of a system determined by the potassium-argon method can be calculated using

$$t = \frac{1}{\lambda_{EC}+\lambda_{\beta^-}}\ln\left[1+\frac{(\lambda_{EC}+\lambda_{\beta^-})}{\lambda_{EC}}\left(\frac{N(^{40}\text{Ar})}{N(^{40}\text{K})}\right)\right]$$

where $\lambda_{EC} = 5.85\times10^{-11}\ \text{yr}^{-1}$ and $\lambda_{\beta^-} = 4.72\times10^{-10}\ \text{yr}^{-1}$. Substituting the data gives

$$t = \frac{1}{5.85\times10^{-11}\ \text{yr}^{-1}+4.72\times10^{-10}\ \text{yr}^{-1}}\ln\left[1+\frac{5.85\times10^{-11}\ \text{yr}^{-1}+4.72\times10^{-10}\ \text{yr}^{-1}}{5.85\times10^{-11}\ \text{yr}^{-1}}(0.524)\right]$$

$$= 3.30\times10^{9}\ \text{yr}$$

CHAPTER 13
Reaction Kinetics

13.1 INFLUENCE OF TEMPERATURE

13.1 Use the following data to determine the energy of activation for the thermal decomposition of malonic acid,

$$CH_2(COOH)_2(g) \rightarrow CH_3COOH(g) + CO_2(g)$$

$T/(°C)$	153.6	153.2	143.5	142.3	136.4	134.2	133.6	129.4	125.9
$k/(10^{-3}\ s^{-1})$	1.083	1.045	0.410	0.367	0.208	0.169	0.160	0.107	0.0763

▌ According to the *Arrhenius theory*, the rate constant is related to the activation energy (E_a) by

$$k = A\,e^{-E_a/RT} \tag{13.1a}$$

or

$$\ln k = \ln A - E_a/RT \tag{13.1b}$$

where A is the *pre-exponential factor*. Thus a plot of $\ln k$ against T^{-1} will be linear with a slope equal to $-(E_a/R)$. From Fig. 13-1,

$$E_a = -R(\text{slope}) = -(8.314\ \text{J}\cdot\text{K}^{-1}\cdot\text{mol}^{-1})(-16\,400\ \text{K})\left(\frac{10^{-3}\ \text{kJ}}{1\ \text{J}}\right) = 136\ \text{kJ}\cdot\text{mol}^{-1}$$

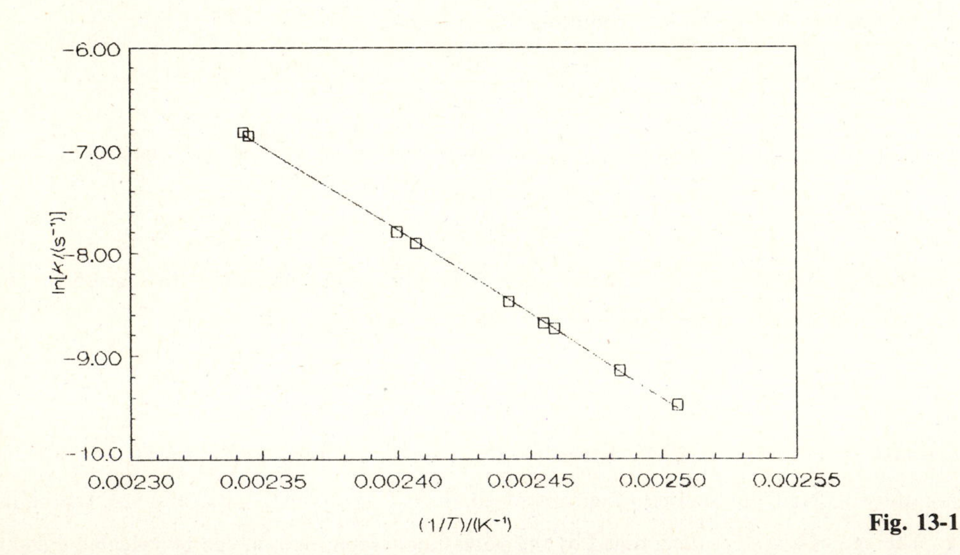

Fig. 13-1

13.2 The rate equation for the chemical reaction $CCl_3COOH \rightarrow CHCl_3 + CO_2$ is first-order with $k = 6.7 \times 10^{-7}\ s^{-1}$ at 25 °C and $1.33 \times 10^{-5}\ s^{-1}$ at 45 °C. Determine the energy of activation for this reaction.

▌ For only two sets of rate constant–temperature data, *(13.1b)* can be written as

$$\ln\frac{k_2}{k_1} = \frac{-E_a}{R}\left(\frac{1}{T_2} - \frac{1}{T_1}\right) \tag{13.2}$$

Solving *(13.2)* for E_a and substituting the data gives

$$E_a = \frac{-(8.314\ \text{J}\cdot\text{K}^{-1}\cdot\text{mol}^{-1})[(10^{-3}\ \text{kJ})/(1\ \text{J})]\ln[(1.33\times10^{-5}\ \text{s}^{-1})/(6.7\times10^{-7}\ \text{s}^{-1})]}{1/(318\ \text{K}) - 1/(298\ \text{K})} = 118\ \text{kJ}\cdot\text{mol}^{-1}$$

13.3 For the displacement reaction,

$$[Co(NH_3)_5Cl]^{2+}(aq) + H_2O(l) \rightarrow [Co(NH_3)_5(H_2O)]^{3+}(aq) + Cl^-(aq)$$

the rate constant is given by

$$\ln [k/(\text{min}^{-1})] = \frac{-11\,067\ \text{K}}{T} + 31.330$$

Evaluate k at 25.00 °C.

▌ Substituting the data and taking logarithms gives

$$\ln [k/(\text{min}^{-1})] = \frac{-11\,067\ \text{K}}{298.15\ \text{K}} + 31.330 = -5.789$$

$$k = 3.06 \times 10^{-3}\ \text{min}^{-1} = 5.10 \times 10^{5}\ \text{s}^{-1}$$

13.4 Using the data given in Problem 13.3, determine E_a and A for the chemical reaction.

▌ Upon comparison of the terms of *(13.1b)* with those given in the empirical equation for ln k,

$$\ln A = 31.330 \qquad A = 4.04 \times 10^{13}\ \text{min}^{-1} = 6.73 \times 10^{11}\ \text{s}^{-1}$$

and

$$-E_a/RT = (-11\,067\ \text{K})/T \qquad E_a = (8.314\ \text{J}\cdot\text{K}^{-1}\cdot\text{mol}^{-1})[(10^{-3}\ \text{kJ})/(1\ \text{J})](11\,067) = 92.02\ \text{kJ}\cdot\text{mol}^{-1}$$

13.5 Compare the values of $t_{1/2}$ at 25.00 °C and 30.00 °C for the reaction described in Problem 13.3.

▌ At 30.00 °C,

$$\ln [k/(\text{min}^{-1})] = (-11\,067\ \text{K})/(303.15\ \text{K}) + 31.330 = -5.177$$

$$k = 5.65 \times 10^{-3}\ \text{min}^{-1} = 9.41 \times 10^{-5}\ \text{s}^{-1}$$

Substituting the values of k at 25.00 °C (see Problem 13.3) and 30.00 °C into *(12.7)* gives

$$t_{1/2}(25.00\ °\text{C}) = \frac{-\ln 0.5}{5.10 \times 10^{-5}\ \text{s}^{-1}} = 1.36 \times 10^4\ \text{s}$$

$$t_{1/2}(30.00\ °\text{C}) = \frac{-\ln 0.5}{9.41 \times 10^{-5}\ \text{s}^{-1}} = 7.37 \times 10^3\ \text{s}$$

Taking a ratio gives

$$\frac{t_{1/2}(25.00\ °\text{C})}{t_{1/2}(30.00\ °\text{C})} = \frac{1.36 \times 10^4\ \text{s}}{7.37 \times 10^3\ \text{s}} = 1.85$$

13.6 Compare the respective times required for 95% conversion of $[Co(NH_3)_5Cl]^{2+}(aq)$ to $[Co(NH_3)_5(H_2O)]^{3+}(aq)$ at 25.00 °C and 30.00 °C for the reaction described in Problem 13.3.

▌ From Problems 13.3 and 13.5, $k(25.00\ °\text{C}) = 5.10 \times 10^{-5}\ \text{s}^{-1}$ and $k(30.00\ °\text{C}) = 9.41 \times 10^{-5}\ \text{s}^{-1}$. Substituting $C(\text{A})/C(\text{A})_0 = 0.05$ into *(12.17)* and solving for t gives

$$t(25.00\ °\text{C}) = (-\ln 0.05)/(5.10 \times 10^{-5}\ \text{s}^{-1}) = 5.87 \times 10^4\ \text{s}$$

$$t(30.00\ °\text{C}) = (-\ln 0.05)/(9.41 \times 10^{-5}\ \text{s}^{-1}) = 3.18 \times 10^4\ \text{s}$$

Taking a ratio gives

$$\frac{t(25.00\ °\text{C})}{t(30.00\ °\text{C})} = \frac{5.87 \times 10^4\ \text{s}}{3.18 \times 10^4\ \text{s}} = 1.85$$

Note that this is the same ratio as for the half-lives (see Problem 13.5).

13.7 At what temperature will the reaction rate be double that at 25.00 °C for the reaction described in Problem 13.3?

▌ At 25.00 °C, $k = 3.06 \times 10^{-3}\ \text{min}^{-1}$ (see Problem 13.3). Rearranging the expression for the rate constant and substituting the data gives

$$T = \frac{-11\,067\ \text{K}}{\ln [(2)(3.06 \times 10^{-3})] - 31.330} = 303.82\ \text{K} = 30.67\ °\text{C}$$

13.8 Consider the following data for the decomposition of azomethane,

$$H_3CNNCH_3(g) \rightarrow N_2(g) + C_2H_6(g)$$

T = 298.4 °C

t/(s)	0	600	1200	1980	2760	3900
$P(H_3CNNCH_3)$/(torr)	430.8	371.8	313.6	251.9	205.2	155.1

T = 320.4 °C

t/(s)	0	180	360	540	720	1020	1500
$P(H_3CNNCH_3)$/(torr)	212.3	161.7	130.3	102.0	80.6	55.5	31.5

Determine the Arrhenius constants for this reaction.

▌ A plot of ln P against t is linear for each set of data (see Fig. 13-2). The reaction is first-order at both temperatures. From *(12.5)*,

$$k(298.4\ ^\circ\text{C}) = -\text{slope} = -(2.65 \times 10^{-4}\ \text{s}^{-1}) = 2.65 \times 10^{-4}\ \text{s}^{-1}$$

and $k(320.4\ ^\circ\text{C}) = 1.27 \times 10^{-3}\ \text{s}^{-1}$. Solving *(13.2)* for E_a and substituting the data gives

$$E_a = \frac{-(8.314\ \text{J} \cdot \text{K}^{-1} \cdot \text{mol}^{-1}) \ln[(1.27 \times 10^{-3}\ \text{s}^{-1})/(2.65 \times 10^{-4}\ \text{s}^{-1})]}{[1/(593.6\ \text{K}) - 1/(571.6\ \text{K})]} = 2.01 \times 10^5\ \text{J} \cdot \text{mol}^{-1}$$

Solving *(13.1a)* for A and substituting the data gives

$$A = \frac{2.65 \times 10^{-4}\ \text{s}^{-1}}{\exp[-(2.01 \times 10^5\ \text{J} \cdot \text{mol}^{-1})/(8.314\ \text{J} \cdot \text{K}^{-1} \cdot \text{mol}^{-1})(571.6\ \text{K})]} = 6.18 \times 10^{14}\ \text{s}^{-1}$$

For this reaction,

$$k = (6.18 \times 10^{14}\ \text{s}^{-1}) \exp[-(2.01 \times 10^5\ \text{J} \cdot \text{mol}^{-1})/RT]$$

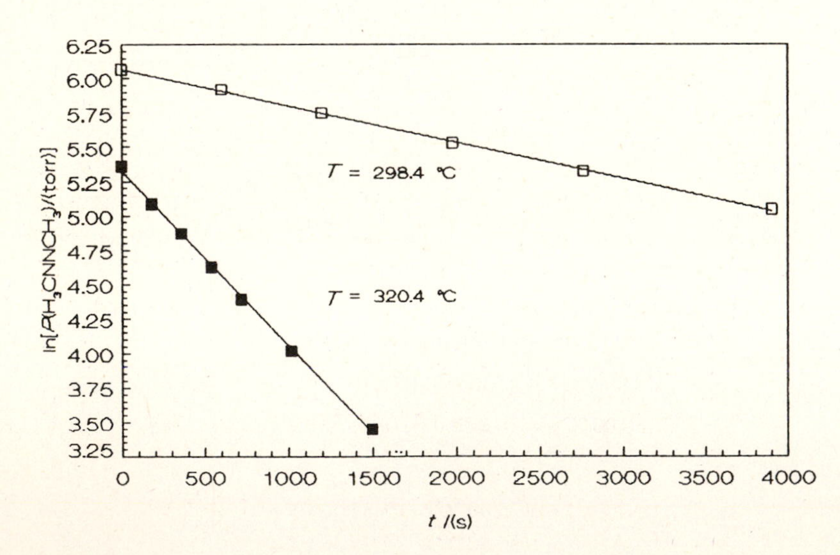

Fig. 13-2

13.9 What is the limiting value of k as predicted by *(13.1a)* as T increases?

▌ Taking the limit of *(13.1a)* as $T \rightarrow \infty$ gives

$$\lim_{T \rightarrow \infty} (k) = \lim_{T \rightarrow \infty} (A\, e^{-E_a/RT}) = A$$

13.10 Which reaction will have the greater temperature dependence for the rate constant—one with a small value of E_a or one with a large value of E_a?

▮ The temperature dependence of k can be found by taking the derivative of *(13.1a)* with respect to T, giving

$$\frac{dk}{dT} = A\,e^{-E_a/RT}\,\frac{E_a}{RT^2} = k\,\frac{E_a}{RT^2} \qquad (13.3)$$

For positive values of E_a, the temperature dependence will be greater for reactions with large values of E_a.

13.11 The rate constant for the reaction

$$COCl_2(g) \rightarrow CO(g) + Cl_2(g) \qquad \Delta_r U_{700} = 103.1\ \text{kJ}$$

is given by

$$\log\,[k/(\text{min}^{-1})] = \frac{-11\,420\ \text{K}}{T} + 15.154$$

Prepare a reaction coordinate diagram describing this reaction.

▮ A *reaction coordinate diagram* is a plot of E against the progress of the reaction similar to Fig. 4-1 except that the energy levels are connected by an energy hump represented by E_a that must be surmounted by the reactants before the products are formed. The value of E_a for this reaction is

$$E_a = (\ln 10)(8.314\ \text{J}\cdot\text{K}^{-1}\cdot\text{mol}^{-1})(11\,420\ \text{K})\,\frac{10^{-3}\ \text{kJ}}{1\ \text{J}} = 218.6\ \text{kJ}\cdot\text{mol}^{-1}$$

The diagram is shown in Fig. 13-3.

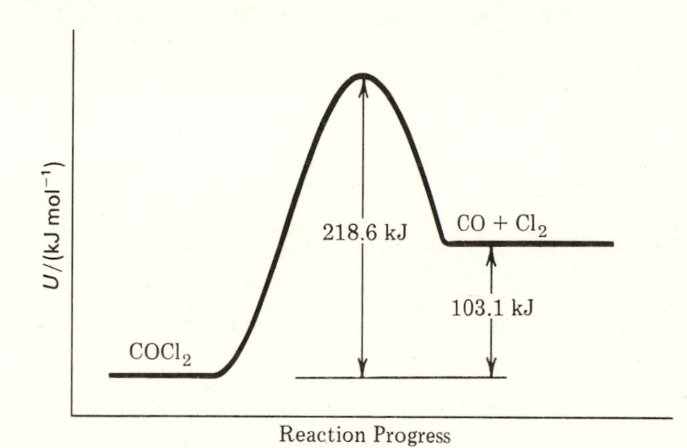

Fig. 13-3

13.12 What is the activation energy for the reverse reaction for the system described in Problem 13.11?

▮ The activation energies for the forward and reverse reactions are related to $\Delta_r U$ by

$$\Delta_r U = E_a - E_{a,r} \qquad (13.4)$$

Solving *(13.4)* for $E_{a,r}$ and substituting the data from Problem 13.11 gives

$$E_{a,r} = 218.6\ \text{kJ}\cdot\text{mol}^{-1} - 103.1\ \text{kJ}\cdot\text{mol}^{-1} = 115.5\ \text{kJ}\cdot\text{mol}^{-1}$$

13.13 A chain mechanism for the thermal decomposition of organic compounds is

$$A_1 \xrightarrow{k_1} R_1 + A_2 \qquad E_a = 320\ \text{kJ}\cdot\text{mol}^{-1}$$

$$R_1 + A_1 \xrightarrow{k_2} R_1H + R_2 \qquad E_a = 40\ \text{kJ}\cdot\text{mol}^{-1}$$

$$R_2 \xrightarrow{k_3} R_1 + A_3 \qquad E_a = 140\ \text{kJ}\cdot\text{mol}^{-1}$$

$$R_1 + R_2 \xrightarrow{k_4} A_4 \qquad E_a = 0\ \text{kJ}\cdot\text{mol}^{-1}$$

where R represents a highly reactive radical and A represents a stable molecule. An analysis of the mechanism gives the rate equation for a long-chain organic compound as

$$\frac{-dC(A_1)}{dt} = \left(\frac{k_1 k_2 k_3}{2k_4}\right)^{1/2} C(A_1) = k_{obs}C(A_1)$$

where k_{obs} is the observed rate constant. What is the value of the activation energy for the overall reaction?

▮ Substituting *(13.1a)* for each rate constant gives

$$A_{obs}\, e^{-E_{a,obs}/RT} = \left(\frac{A_1\, e^{-E_{a,1}/RT} A_2\, e^{-E_{a,2}/RT} A_3\, e^{-E_{a,3}/RT}}{2A_4\, e^{-E_{a,4}/RT}}\right)^{1/2}$$

Taking logarithms and collecting terms gives

$$RT \ln A_{obs} - E_{a,obs} = \frac{RT}{2} \ln\left(\frac{A_1 A_2 A_3}{2A_4}\right) - \tfrac{1}{2}(E_{a,1} + E_{a,2} + E_{a,3} - E_{a,4})$$

Upon comparison of terms,

$$E_{a,obs} = \tfrac{1}{2}(E_{a,1} + E_{a,2} + E_{a,3} - E_{a,4})$$
$$= \tfrac{1}{2}(320\ \text{kJ}\cdot\text{mol}^{-1} + 40\ \text{kJ}\cdot\text{mol}^{-1} + 140\ \text{kJ}\cdot\text{mol}^{-1} - 0) = 250\ \text{kJ}\cdot\text{mol}^{-1}$$

13.14 A chain mechanism for the reaction between $H_2(g)$ and $Br_2(g)$ is

$$Br_2 + M \xrightarrow{k_1} 2Br + M \qquad E_{a,1} = 192.9\ \text{kJ}\cdot\text{mol}^{-1}$$

$$2Br + M \xrightarrow{k_2} Br_2 + M \qquad E_{a,2} = 0\ \text{kJ}\cdot\text{mol}^{-1}$$

$$Br + H_2 \xrightarrow{k_3} HBr + H \qquad E_{a,3} = 73.6\ \text{kJ}\cdot\text{mol}^{-1}$$

$$HBr + H \xrightarrow{k_4} Br + H_2 \qquad E_{a,4} = 3.8\ \text{kJ}\cdot\text{mol}^{-1}$$

$$H + Br_2 \xrightarrow{k_5} HBr + Br \qquad E_{a,5} = 3.8\ \text{kJ}\cdot\text{mol}^{-1}$$

where M refers to any gaseous species. An analysis of the mechanism gives the rate equation as

$$\frac{dC(\text{HBr})}{dt} = \frac{2k_3(k_1/k_2)^{1/2}C(\text{H}_2)[C(\text{Br}_2)]^{1/2}}{1 + k_4 C(\text{HBr})/k_5 C(\text{Br}_2)} = \frac{2kC(\text{H}_2)[C(\text{Br}_2)]^{1/2}}{1 + k'C(\text{HBr})/C(\text{Br}_2)}$$

where k and k' are the observed rate constants. What is the value of the activation energy for the overall reaction $H_2(g) + Br_2(g) \rightarrow 2HBr(g)$? What is the temperature dependence of k'?

▮ Substituting *(13.1a)* for the rate constants k', k_4, and k_5 and taking logarithms gives

$$k' = k_4/k_5$$
$$A'\, e^{-E_a'/RT} = A_4\, e^{-E_{a,4}/RT}/A_5\, e^{-E_{a,5}/RT}$$
$$RT \ln A' - E_a' = RT \ln(A_4/A_5) - E_{a,4} + E_{a,5}$$
$$E_a' = E_{a,4} - E_{a,5} = 3.8\ \text{kJ}\cdot\text{mol}^{-1} - 3.8\ \text{kJ}\cdot\text{mol}^{-1} = 0.0\ \text{kJ}\cdot\text{mol}^{-1}$$

Because E_a' is zero, k' should be relatively independent of T. (The value of k' increases from 0.116 at 25 °C to 0.122 at 300 °C.)

With the significant temperature dependence appearing only in the numerator,

$$k = k_3(k_1/k_2)^{1/2}$$
$$A\, e^{-E_a/RT} = A_3\, e^{-E_{a,3}/RT}\left[\frac{A_1\, e^{-E_{a,1}/RT}}{A_2\, e^{-E_{a,2}/RT}}\right]^{1/2}$$
$$RT \ln A - E_a = RT \ln[A_3(A_1/A_2)^{1/2}] - E_{a,3} + \tfrac{1}{2}(-E_{a,1} + E_{a,2})$$
$$E_a = E_{a,3} + \tfrac{1}{2}(E_{a,1} - E_{a,2}) = 73.6\ \text{kJ}\cdot\text{mol}^{-1} + \tfrac{1}{2}(192.9\ \text{kJ}\cdot\text{mol}^{-1} - 0\ \text{kJ}\cdot\text{mol}^{-1})$$
$$= 170.1\ \text{kJ}\cdot\text{mol}^{-1}$$

(This value agrees well with the observed value of $E_a = 166\ \text{kJ}\cdot\text{mol}^{-1}$.)

13.15 Consider the following parallel reactions involving A:

$$A \xrightarrow{k_1} B \qquad A \xrightarrow{k_2} C$$

where

$$\frac{-dC(\text{A})}{dt} = (k_1 + k_2)C(\text{A}) = k_{obs}C(\text{A})$$

and k_{obs} is the observed rate constant. What is the value of the activation energy for this system?

▮ Solving *(13.3)* for E_a gives

$$E_a = RT^2 \frac{1}{k}\frac{dk}{dT} = RT^2 \frac{d \ln k}{dT}$$

Substituting,

$$E_{a,obs} = RT^2 \frac{d \ln (k_1 + k_2)}{dT} = RT^2 \frac{1}{k_1 + k_2}\frac{d(k_1 + k_2)}{dT}$$

$$= RT^2 \frac{1}{k_1 + k_2}\left(\frac{dk_1}{dT} + \frac{dk_2}{dT}\right) = RT^2 \frac{1}{k_1 + k_2}\left(k_1 \frac{E_{a,1}}{RT^2} + k_2 \frac{E_{a,2}}{RT^2}\right)$$

$$= \frac{k_1 E_{a,1} + k_2 E_{a,2}}{k_1 + k_2}$$

13.2 REACTION RATE THEORY

13.16 What fraction of bimolecular collisions involve molecules with energies greater than 100 kJ · mol^{-1} (a typical value of E_a for chemical reactions) at 298 K? Repeat the calculations for 1000 K.

▮ As long as $E_a \gg RT$, *(2.22)* can be used to calculate the fraction of effective collisions. Substituting the data into *(2.22)* gives

$$\frac{N(\text{effective})}{N(\text{total})} = \exp\left[-\frac{(100\ \text{kJ}\cdot\text{mol}^{-1})[(10^3\ \text{J})/(1\ \text{kJ})]}{(8.314\ \text{J}\cdot\text{K}^{-1}\cdot\text{mol}^{-1})(298\ \text{K})}\right] = 3 \times 10^{-18}$$

at 298 K and $N(\text{effective})/N(\text{total}) = 6 \times 10^{-6}$ at 1000 K.

13.17 Consider the bimolecular collision

$$NO(g) + Cl_2(g) \rightarrow NOCl(g) + Cl(g)$$

with $\sigma_{12} = 3.5$ Å. Determine the value of the pre-exponential factor for this reaction as a function of T given that the value of the steric factor is 0.014.

▮ For bimolecular collisions between unlike molecules, the expression for the rate law is

$$\frac{-dC_i}{dt} = p\left(\frac{8\pi RT(M_1 + M_2)}{M_1 M_2}\right)^{1/2} \sigma_{12}^2 (10^3\ \text{L})\, e^{-E_a/RT} C_1 C_2$$

$$= (2.753 \times 10^{29}) p\sigma_{12}^2 \left(\frac{T(M_1 + M_2)}{M_1 M_2}\right)^{1/2} e^{-E_a/RT} C_1 C_2 \qquad (13.5)$$

where p is the steric factor and C_i is in units of molarity, M_i is in units of grams per mole, and σ is in units of meters. Comparison of *(13.5)* with *(13.1a)* gives

$$A = (2.753 \times 10^{29}) p\sigma_{12}^2 \left(\frac{T(M_1 + M_2)}{M_1 M_2}\right)^{1/2} \qquad (13.6)$$

Substituting the data into *(13.6)* gives

$$A = (2.753 \times 10^{29})(0.014)(3.5 \times 10^{-10})^2 \left(\frac{T(30.0 + 70.9)}{(30.0 + 70.9)}\right)^{1/2}$$

$$= (1.03 \times 10^8) T^{1/2}\ \text{L}\cdot\text{mol}^{-1}\cdot\text{s}^{-1}$$

The observed value is $(1.0 \times 10^8) T^{1/2}$ L · mol^{-1} · s^{-1}.

13.18 Consider the bimolecular collision $2NOCl(g) \rightarrow 2NO(g) + Cl_2(g)$ with $\sigma = 3.5$ Å. Predict the value of the pre-exponential factor for this reaction as a function of T. If the observed value is given by $\log (A/T^{1/2}) = 9.51$, calculate the value of the steric factor.

▮ For bimolecular collisions between like molecules, the expression for the rate law is

$$\frac{-dC_i}{dt} = 4p\left(\frac{\pi RT}{M}\right)^{1/2} \sigma^2 (10^3\ L)\, e^{-E_a/RT} C_i^2$$

$$= (3.893 \times 10^{29}) p\sigma^2 (T/M)^{1/2} e^{-E_a/RT} C_i^2 \qquad (13.7)$$

where the terms and units were discussed in Problem 13.17. Comparison of *(13.7)* with *(13.1a)* gives

$$A = (3.893 \times 10^{29}) p\sigma^2 (T/M)^{1/2} \qquad (13.8)$$

Substituting the data into *(13.8)* gives

$$A = (3.893 \times 10^{29})p(3.5 \times 10^{-10})^2\left(\frac{T}{65.5}\right)^{1/2} = (5.89 \times 10^9)pT^{1/2}\ \mathrm{L \cdot mol^{-1} \cdot s^{-1}}$$

The observed value of A is

$$A = 10^{9.51}T^{1/2} = (3.2 \times 10^9)T^{1/2}\ \mathrm{L \cdot mol^{-1} \cdot s^{-1}}$$

which gives

$$p = \frac{(3.2 \times 10^9)T^{1/2}\ \mathrm{L \cdot mol^{-1} \cdot s^{-1}}}{(5.89 \times 10^9)T^{1/2}\ \mathrm{L \cdot mol^{-1} \cdot s^{-1}}} = 0.54$$

13.19 For the bimolecular collision

$$2HI(g) \rightarrow H_2(g) + I_2(g)$$

$E_a = 183\ \mathrm{kJ \cdot mol^{-1}}$, $\sigma = 3.5$ Å, and $p = 0.44$. Predict the value of the rate constant as a function of T.

▌ Comparison of *(13.7)* and *(13.1)* gives

$$k = (3.893 \times 10^{29})p\sigma^2(T/M)^{1/2}\, e^{-E_a/RT}$$

Substituting the data gives

$$k = (3.893 \times 10^{29})(0.44)(3.5 \times 10^{-10})^2(T/127.9)^{1/2} \exp\left[-\left(\frac{(183\ \mathrm{kJ \cdot mol^{-1}})[(10^3\ \mathrm{J})/(1\ \mathrm{kJ})]}{(8.314\ \mathrm{J \cdot K^{-1} \cdot mol^{-1}})T}\right)\right]$$

$$= (1.86 \times 10^9)T^{1/2}\, e^{-(22000/T)}$$

13.20 Continue the analysis of Problem 13.9 in terms of *(13.5)*.

▌ The limit of k as T increases was shown to be equal to A in Problem 13.9. The Arrhenius theory assumes that A is a constant and so predicts that k approaches a finite value as T increases. However, *(13.5)* showed that A is a function of $T^{1/2}$, so as T increases so does A. Thus collision theory predicts that k will not approach a constant value but rather will continue to increase as T increases.

13.21 Using the results of Problem 13.19, prepare an Arrhenius plot for $300 \le T/(\mathrm{K}) \le 1500$. For comparison, prepare a plot of *(13.1b)* with the value of A determined at 900 K. Compare the two plots.

▌ From the results of Problem 13.19, the temperature dependence of ln k is given by

$$\ln k = \ln(1.86 \times 10^9) + \frac{\ln T}{2} - \frac{22\,000}{T} = 21.344 + \frac{\ln T}{2} - \frac{22\,000}{T}$$

A plot of this equation is shown in Fig. 13-4 as a solid line. A plot of

$$\ln k = \ln(1.86 \times 10^9) + \frac{\ln 900}{2} - \frac{22\,000}{T} = 24.745 - \frac{22\,000}{T}$$

is also shown in Fig. 13-4 as a dashed line. As can be seen from the plots, the actual Arrhenius plot has a slightly larger slope that the plot of *(13.1b)*, which would give a slightly greater value for E_a.

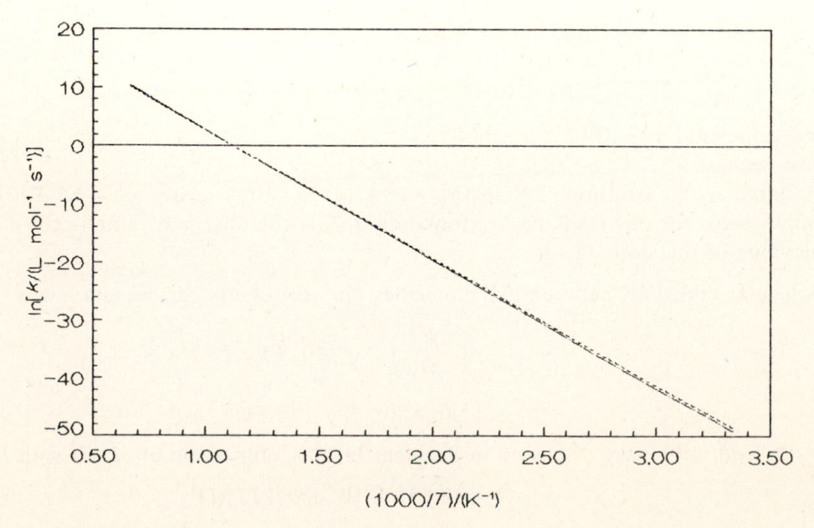

Fig. 13-4

13.22 For the bimolecular collision

$$Br(g) + CH_4(g) \rightarrow HBr(g) + CH_3(g)$$

$A = (1.48 \times 10^9)T^{1/2}\ L\cdot mol^{-1}\cdot s^{-1}$ and $p = 0.21$. Calculate the collision cross section for the reaction.

▮ The *collision cross section* is defined as the product $\pi\sigma_{12}^2$. Solving *(13.6)* for $\pi\sigma_{12}^2$ and substituting the data gives

$$\sigma_{12}^2 = \frac{(1.48 \times 10^9)T^{1/2}}{(2.753 \times 10^{29})(0.21)}\left(\frac{(79.9)(16.0)}{T^{1/2}(79.9 + 16.0)}\right)^{1/2} = 9.3 \times 10^{-20}\ m^2$$

The collision cross section is $2.94 \times 10^{-19}\ m^2$.

13.23 The rate constant for a nonionic reaction that occurs by a bimolecular collision in dilute solution is given by

$$k_D = aRT/3\eta \qquad (13.9)$$

where $a = 4$ for collisions between like molecules, $a = 8$ for collisions between unlike molecules, and η is the coefficient of viscosity of the solvent. Calculate k_D for a reaction between unlike molecules in water at 25 °C. For water, $\eta = 0.900 \times 10^{-3}\ Pa\cdot s$ at 25 °C.

▮ Substituting the data into *(13.9)* gives

$$k_D = \frac{8(8.314\ J\cdot K^{-1}\cdot mol^{-1})[(1\ Pa\cdot m^3)/(1\ J)][(10^3\ L)/(1\ m^3)](298\ K)}{3(0.900 \times 10^{-3}\ Pa\cdot s)} = 7.34 \times 10^9\ L\cdot mol^{-1}\cdot s^{-1}$$

13.24 How much faster or slower will the reaction described in Problem 13.23 be if the solvent was changed to ethanol? For CH_3CH_2OH, $\eta = 1.200 \times 10^{-3}\ Pa\cdot s$ at 25 °C.

▮ Taking a ratio of *(13.9)* for the two rate constants gives

$$\frac{k_D(CH_3CH_2OH)}{k_D(H_2O)} = \frac{\eta(H_2O)}{\eta(CH_3CH_2OH)} = \frac{0.900 \times 10^{-3}\ Pa\cdot s}{1.200 \times 10^{-3}\ Pa\cdot s} = 0.750$$

The rate reaction in ethanol will be 75.0% of that in water.

13.25 What is the range of values of k_D for ionic reactions involving bimolecular collisions between unlike ions in water at 25 °C? For H_2O, $\eta = 0.900 \times 10^{-3}\ Pa\cdot s$ at 25 °C.

Table 13-1

ionic charges	+2 and +2 or −2 and −2	+2 and +1 or −2 and −1	+1 and +1 or −1 and −1	none
b	0.019	0.17	0.45	1

ionic charges	+1 and −1	+2 and −1 or −2 and +1	+3 and −1 or −3 and +1	+2 and −2
b	1.9	3.0	4.3	5.7

▮ For ionic reactions, *(13.9)* is multiplied by the constant b found in Table 13-1. From Problem 13.23, the minimum value of k_D will be for like-charged divalent ions, where

$$k_D(\text{min}) = (0.019)(7.34 \times 10^9\ L\cdot mol^{-1}\cdot s^{-1}) = 1.4 \times 10^8\ L\cdot mol\cdot s^{-1}$$

and the maximum value of k_D will be for oppositely charged divalent ions, where

$$k_D(\text{max}) = (5.7)(7.34 \times 10^9) = 4.2 \times 10^{10}\ L\cdot mol^{-1}\cdot s^{-1}$$

13.26 The rate constant for the bimolecular collision reaction

$$H_3O^+(aq) + OH^-(aq) \rightarrow 2H_2O(l)$$

is $1.3 \times 10^{11}\ L\cdot mol^{-1}\cdot s^{-1}$. Determine $r(H_3O^+) + r(OH^-)$ for this diffusion-controlled reaction given $D(H_3O^+) = 9.31 \times 10^{-9}\ m^2\cdot s^{-1}$ and $D(OH^-) = 5.30 \times 10^{-9}\ m^2\cdot s^{-1}$ at 25 °C. Interpret the results.

■ In terms of diffusion coefficients, *(13.9)* for unlike molecules becomes

$$k_D = 4\pi L(D_1 + D_2)(r_1 + r_2)b$$

where values of b can be obtained from Table 13-1. Solving for $(r_1 + r_2)$ and substituting the data gives

$$r(H_3O^+) + r(OH^-) = \frac{(1.3 \times 10^{11}\ L \cdot mol^{-1} \cdot s^{-1})[(1\ m^3)/(10^3\ L)]}{4\pi(6.022 \times 10^{23}\ mol^{-1})(9.31 \times 10^{-9}\ m^2 \cdot s^{-1} + 5.30 \times 10^{-9}\ m^2 \cdot s^{-1})(1.9)}$$

$$= 6.2 \times 10^{-10}\ m$$

This sum of radii is too large for simple H_3O^+ ions interacting with OH^- ions. A plausible explanation is that the actual species in solution are $[H(H_2O)_4]^+$ and $[OH(H_2O)_3]^-$.

13.27 The fundamental reactions involved in the *transition state theory* are

$$A + B \underset{}{\overset{K_C}{\rightleftharpoons}} (A - B)^\ddagger \xrightarrow{k_1} \text{products} \qquad (13.10)$$

where the symbol ‡ refers to the *transition state* (or *activated complex*). Briefly discuss the number of degrees of freedom for each of the species.

■ If either of the reactants, A or B, is an atomic species, then each will have $3(1) = 3$ translational degrees of freedom (see Problems 3.9-3.11). If either of the reactants is a linear molecule, the reactant molecule will have 3 translational degrees, 2 rotational degrees, and $3\Lambda - 5$ internal (usually vibrational) degrees of freedom, where Λ is the number of atoms in the molecule. If either of the reactants is a nonlinear molecule, the reactant molecule will have 3 translational degrees, 3 rotational degrees, and $3\Lambda - 6$ internal degrees of freedom. If the activated complex is linear, it will have 3 translational, 2 rotational, and $3\Lambda(A) + 3\Lambda(B) - 6$ internal degrees of freedom, where the extra internal degree of freedom describes the vibration of the bond holding the activated complex together. If the activated complex is nonlinear, it will have 3 translational, 3 rotational, and $3\Lambda(A) + 3\Lambda(B) - 7$ internal degrees of freedom.

13.28 The transition state theory predicts the rate law as

$$\frac{-dC_i}{dt} = w(10^3\ L)\left(\frac{kT}{h}\right)\left(\frac{Q_\ddagger}{Q(A)Q(B)}\right) e^{-\Delta U_0^\circ/RT} C_A C_B \qquad (13.11)$$

where w is the transmission coefficient (usually set equal to unity) and Q_i is the molecular partition function (with $V = 1\ m^3$). Write out the contribution to Q_i for the formation of a linear activated complex from an atomic reactant and a diatomic molecule.

■ For the atomic reactant, the three translational degrees of freedom give

$$Q(A) = q_{trans}$$

and for the diatomic reactant, the three translational, two rotational, and one vibrational degrees of freedom give

$$Q(B) = q_{trans} q_{rot} q_{vib}$$

For a linear activated complex consisting of three atoms, the three translational, two rotational, and three vibrational degrees of freedom give

$$Q_\ddagger = q_{trans} q_{rot} q_{vib} q'_{vib} q''_{vib}$$

13.29 Determine the order of magnitude of the molecular contributions to the partition function for a typical molecule involved in a chemical reaction at 298 K having $M \approx 1\ g \cdot mol^{-1}$, $I \approx 10^{-46}\ kg \cdot m^2$, and $\bar{\nu} \approx 10^3\ cm^{-1}$.

■ The molecular mass for the molecule is

$$m = \frac{(1\ g \cdot mol^{-1})[(10^{-3}\ kg)/(1\ g)]}{6.022 \times 10^{23}\ mol^{-1}} = 2 \times 10^{-27}\ kg$$

Substituting into *(8.9)* gives

$$q_{trans} \approx \frac{\{2\pi(2 \times 10^{-27}\ kg)(1.381 \times 10^{-23}\ J \cdot K^{-1})(298\ K)[(1\ kg \cdot m^2 \cdot s^{-2})/(1\ J)]\}^{3/2}(1\ m^3)}{\{(6.626 \times 10^{-34}\ J \cdot s)[(1\ kg \cdot m^2 \cdot s^{-2})/(1\ J)]\}^3} = 10^{30}$$

for three degrees of translational motion. For each degree of freedom, $q_{trans} = 1 \times 10^{10}$. Substituting into *(8.11)* gives

$$q_{rot} \approx \frac{8\pi^2(10^{-46}\ kg \cdot m^2)(1.381 \times 10^{-23}\ J \cdot K^{-1})(298\ K)}{(6.626 \times 10^{-34}\ J \cdot s)^2[(1\ kg \cdot m^2 \cdot s^{-2})/(1\ J)](1)} = 10^2$$

for each degree of rotational motion. Using the definition of x given in Table 3-2,

$$x = \frac{(1.4388)(10^3\ \mathrm{cm^{-1}})}{298\ \mathrm{K}} = 5$$

which, upon substitution into *(8.14a)*, gives

$$q_{vib} = 1/(1 - e^{-5}) = 1$$

for each degree of vibrational freedom.

13.30 Using the results of Problems 13.28 and 13.29, estimate the order of magnitude of the $Q_\ddagger/Q(\mathrm{A})Q(\mathrm{B})$ term in *(13.11)* for the formation of a linear activated complex from an atomic reactant and a diatomic molecule.

▌ Substituting $q_{trans} = 10^{30}$, $q_{rot} = 10^2$, and $q_{vib} = 1$ into the expression for Q_i determined in Problem 13.28 gives

$$\frac{Q_\ddagger}{Q(\mathrm{A})Q(\mathrm{B})} = \frac{(10^{30})(10^2)(1)^3}{(10^{30})[(10^{30})(10^2)(1)]} = 10^{-30}$$

13.31 If the temperature exponent is 1/2 for each translational and each rotational degree of freedom and $\theta = 0$ to 1 for each vibrational degree of freedom, predict the temperature exponent in A for the formation of a linear activated complex from an atomic reactant and a diatomic molecule.

▌ From Problem 13.28,

$$Q(\mathrm{A}) \propto (T^{1/2})^3 = T^{3/2}$$

$$Q(\mathrm{B}) \propto (T^{1/2})^3(T^{1/2})^2(T^\theta) = T^{5/2+\theta}$$

$$Q_\ddagger \propto (T^{1/2})^3(T^{1/2})^2(T^\theta)^3 = T^{5/2+3\theta}$$

Comparison of *(13.11)* with *(13.1a)* gives

$$A/(\mathrm{L\cdot mol^{-1}\cdot s^{-1}}) = w(10^3\ L)\left(\frac{kT}{h}\right)\left(\frac{Q_\ddagger}{Q(\mathrm{A})Q(\mathrm{B})}\right) \qquad (13.12)$$

$$A \propto T\frac{T^{5/2+3\theta}}{(T^{3/2})(T^{5/2+\theta})} = T^{-1/2+2\theta}$$

If $\theta = 0$, then the minimum exponent is $-1/2$, and if $\theta = 1$, the predicted maximum exponent is 3/2. However, the maximum exponent is 1/2 (see Table 13-2), so the range of exponents for the reaction is $-1/2$ to 1/2.

Table 13-2

reactant A	+ reactant B	= activated complex $(AB)^\ddagger$	exponent of T in A	$A/(\mathrm{L\cdot mol^{-1}\cdot s^{-1}})$ at 25 °C	approximate p
atom	atom	linear	$\frac{1}{2}$	10^{12}	1
atom	linear molecule	linear	$-\frac{1}{2}$ to $\frac{1}{2}$	10^{10}	10^{-2}
atom	linear molecule	nonlinear	0 to $\frac{1}{2}$	10^{11}	10^{-1}
atom	nonlinear molecule	nonlinear	$-\frac{1}{2}$ to $\frac{1}{2}$	10^{10}	10^{-2}
linear molecule	linear molecule	linear	$-\frac{3}{2}$ to $\frac{1}{2}$	10^{7}	10^{-4}
linear molecule	linear molecule	nonlinear	-1 to $\frac{1}{2}$	10^{8}	10^{-3}
linear molecule	nonlinear molecule	nonlinear	$-\frac{3}{2}$ to $\frac{1}{2}$	10^{7}	10^{-4}
nonlinear molecule	nonlinear molecule	nonlinear	-2 to $\frac{1}{2}$	10^{7}	10^{-5}

13.32 Using the results of Problems 13.30 and 13.31, estimate the order of magnitude for A for the formation of a linear activated complex from an atomic reactant and a diatomic molecule.

▌ Substituting the results of Problem 13.30 into *(13.12)* gives

$$A/(\mathrm{L\cdot mol^{-1}\cdot s^{-1}}) = w(10^3)(6\times10^{23})\left[\frac{(1.381\times10^{-23}\ \mathrm{J\cdot K^{-1}})(298\ \mathrm{K})}{6.626\times10^{-34}\ \mathrm{J\cdot s}}\right](10^{-30}) = 10^9 w$$

This value is in good agreement with the entry of 10^{10} given in Table 13-2 for this reaction.

13.33 Estimate the order of magnitude for the value of the steric factor for the formation of a linear activated complex from an atomic reactant and a diatomic molecule.

▮ The value of p is formed by comparing the estimated value of $A = 10^9 w$ (see Problem 13.32) to that for the reaction of rigid spheres interacting to form a diatomic activated complex. For the latter reaction,

$$Q(\mathrm{A}) = Q(\mathrm{B}) = 10^{30} \qquad Q_{\ddagger} = (10^{30})(10^2) = 10^{32}$$

giving

$$A/(\mathrm{L\cdot mol^{-1}\cdot s^{-1}}) = pw(10^3)(6\times 10^{23})\frac{(1.381\times 10^{-23}\ \mathrm{J\cdot K^{-1}})(298\ \mathrm{K})}{6.626\times 10^{-34}\ \mathrm{J\cdot s}}\left(\frac{10^{32}}{(10^{30})(10^{30})}\right) = 10^{11}pw$$

Comparing the values of A gives

$$p = \frac{10^9 w\ \mathrm{L\cdot mol^{-1}\cdot s^{-1}}}{10^{11} w\ \mathrm{L\cdot mol^{-1}\cdot s^{-1}}} = 10^{-2}$$

which is in good agreement with the entry in Table 13-2.

13.34 For the equation

$$\mathrm{H(g) + H_2(g)} \overset{K_C}{\rightleftharpoons} (\mathrm{H-H-H})^{\ddagger} \xrightarrow{k_1} \mathrm{H_2(g) + H(g)}$$

the pre-exponential factor is given by

$$A = 10^{8.94}T^{1/2}\ \mathrm{L\cdot mol^{-1}\cdot s^{-1}}$$

Determine the shape of the activated complex using Table 13-2.

▮ At 298 K,

$$A = (10^{8.94})(298)^{1/2} = 1.5\times 10^{10}\ \mathrm{L\cdot mol^{-1}\cdot s^{-1}}$$

The value of $A = 10^{10}\ \mathrm{L\cdot mol^{-1}\cdot s^{-1}}$ and the value of 1/2 for the exponent of T in A best agree with the entries in Table 13-2 for the formation of a linear activated complex.

13.35 Predict the pre-exponential factor at 298 K for the reaction

$$\mathrm{Br(g) + Cl_2(g)} \overset{K_C}{\rightleftharpoons} (\mathrm{Br-Cl-Cl})^{\ddagger}(\mathrm{g}) \xrightarrow{k_1} \mathrm{BrCl(g) + Cl(g)}$$

using the transition state theory. Assume that the activated complex is linear with

$$I = 1.14\times 10^{-44}\ \mathrm{kg\cdot m^2} \qquad \bar\nu_1 = 190\ \mathrm{cm^{-1}} \qquad \bar\nu_2 = 110\ \mathrm{cm^{-1}}\ (g_2 = 2)$$

and $g_0 = 4$ for the ground electronic level. For $\mathrm{Cl_2(g)}$, $I = 1.15\times 10^{-45}\ \mathrm{kg\cdot m^2}$ and $\bar\nu = 550\ \mathrm{cm^{-1}}$, and for Br(g), $g_0 = 4$ for the ground electronic level.

▮ Using *(8.9)* gives

$$q_{\mathrm{trans}}(\mathrm{Br}) = \frac{\left[2\pi\dfrac{79.90\times 10^{-3}\ \mathrm{kg\cdot mol^{-1}}}{6.022\times 10^{23}\ \mathrm{mol^{-1}}}(1.381\times 10^{-23}\ \mathrm{J\cdot K^{-1}})(298\ \mathrm{K})\dfrac{1\ \mathrm{kg\cdot m^2\cdot s^{-2}}}{1\ \mathrm{J}}\right]^{3/2}(1\ \mathrm{m^3})}{\{(6.626\times 10^{-34}\ \mathrm{J\cdot s})[(1\ \mathrm{kg\cdot m^2\cdot s^{-2}})/(1\ \mathrm{J})]\}^3}$$

$$= 6.91\times 10^{32}$$

$$q_{\mathrm{trans}}(\mathrm{Cl_2}) = \frac{\{2\pi[(70.91\times 10^{-3})/(6.022\times 10^{23})](1.381\times 10^{-23})(298)\}^{3/2}(1)}{(6.626\times 10^{-34})^3} = 5.78\times 10^{32}$$

$$q_{\mathrm{trans}}(\mathrm{BrClCl}) = \frac{\{2\pi[(150.81\times 10^{-3})/(6.022\times 10^{23})](1.381\times 10^{-23})(298)\}^{3/2}(1)}{(6.626\times 10^{-34})^3} = 1.79\times 10^{33}$$

Using *(8.11)*,

$$q_{\mathrm{rot}}(\mathrm{Cl_2}) = \frac{8\pi^2(1.15\times 10^{-45}\ \mathrm{kg\cdot m^2})(1.381\times 10^{-23}\ \mathrm{J\cdot K^{-1}})(298\ \mathrm{K})}{(6.626\times 10^{-34}\ \mathrm{J\cdot s})^2(2)[(1\ \mathrm{kg\cdot m^2\cdot s^{-2}})/(1\ \mathrm{J})]} = 426$$

$$q_{\mathrm{rot}}(\mathrm{BrClCl}) = \frac{8\pi^2(1.14\times 10^{-44})(1.381\times 10^{-23})(298)}{(6.626\times 10^{-34})^2(1)} = 8440$$

Using *(8.14a)* gives

$$x(\mathrm{Cl_2}) = \frac{(1.4388)(550)}{298} = 2.65$$

$$q_{\mathrm{vib}}(\mathrm{Cl_2}) = \frac{1}{1 - e^{-2.65}} = 1.08$$

$$x(\mathrm{BrClCl})_1 = \frac{(1.4388)(190)}{298} = 0.92 \qquad x(\mathrm{BrClCl})_2 = \frac{(1.4388)(110)}{298} = 0.53$$

$$q_{\mathrm{vib}}(\mathrm{BrClCl}) = \left(\frac{1}{1 - e^{-0.92}}\right)\left(\frac{1}{1 - e^{-0.53}}\right)^2 = 9.82$$

The respective molecular partition functions are

$$Q(\mathrm{Br}) = q_{\mathrm{trans}}q_{\mathrm{elec}} = (6.91 \times 10^{32})(4) = 2.76 \times 10^{33}$$

$$Q(\mathrm{Cl_2}) = q_{\mathrm{trans}}q_{\mathrm{rot}}q_{\mathrm{vib}} = (5.78 \times 10^{32})(426)(1.08) = 2.66 \times 10^{35}$$

$$Q(\mathrm{BrClCl}) = q_{\mathrm{trans}}q_{\mathrm{rot}}q_{\mathrm{vib}}q_{\mathrm{elec}} = (1.79 \times 10^{33})(8440)(9.82)(4) = 5.93 \times 10^{38}$$

Substituting into *(13.12)* gives

$$A = w(10^3)(6.022 \times 10^{23})\left(\frac{(1.381 \times 10^{-23}\ \mathrm{J \cdot K^{-1}})(298\ \mathrm{K})}{6.626 \times 10^{-34}\ \mathrm{J \cdot s}}\right)\left(\frac{5.93 \times 10^{38}}{(2.76 \times 10^{33})(2.66 \times 10^{35})}\right)$$

$$= (3.02 \times 10^9)w\ \mathrm{L \cdot mol^{-1} \cdot s^{-1}}$$

which agrees well with the experimental value of $(4.5 \pm 2.0) \times 10^9\ \mathrm{L \cdot mol^{-1} \cdot s^{-1}}$.

13.36 For the reaction

$$\mathrm{H(g) + H_2(g)} \underset{}{\overset{K_C}{\rightleftharpoons}} (\mathrm{H - H - H})^\ddagger \xrightarrow{k_1} \mathrm{H_2(g) + H}$$

$E_a = 23\ \mathrm{kJ \cdot mol^{-1}}$ and $A = 1.5 \times 10^{10}\ \mathrm{L \cdot mol^{-1} \cdot s^{-1}}$ at 298 K. Determine $\Delta H^\circ_\ddagger$, the enthalpy change for the formation of the activated complex from the reactants.

▌ The relation between $\Delta H^\circ_\ddagger$ and E_a is

$$\Delta H^\circ_\ddagger = E_a - nRT \tag{13.13}$$

where $n = 1$ for condensed-phase reactions and n is the molecularity for gas-phase reactions. Substituting the data into *(13.13)* gives

$$\Delta H^\circ_\ddagger = 23\ \mathrm{kJ \cdot mol^{-1}} - (2)(8.314 \times 10^{-3}\ \mathrm{kJ \cdot mol^{-1} \cdot K^{-1}})(298\ \mathrm{K}) = 18\ \mathrm{kJ \cdot mol^{-1}}$$

13.37 Determine $\Delta S^\circ_\ddagger$ for the reaction described in Problem 13.36.

▌ The entropy change for the formation of an activated complex is given by

$$\Delta S^\circ_\ddagger = R[\ln(Ah/kT) - n] \tag{13.14}$$

Substituting the data given in Problem 13.36 gives

$$\Delta S^\circ_\ddagger = (8.314\ \mathrm{J \cdot K^{-1} \cdot mol^{-1}})\left[\ln\left(\frac{(1.5 \times 10^{10})(6.626 \times 10^{-34})}{(1.381 \times 10^{-23})(298)}\right) - 2\right] = -67\ \mathrm{J \cdot K^{-1} \cdot mol^{-1}}$$

13.38 Estimate $\Delta S^\circ_\ddagger$ for the reaction described in Problem 13.36 using typical values of $S^\circ(\mathrm{trans}) = 150\ \mathrm{J \cdot K^{-1} \cdot mol^{-1}}$, $S^\circ(\mathrm{rot}) = 30\ \mathrm{J \cdot K^{-1} \cdot mol^{-1}}$ for each degree of rotational freedom and $S^\circ(\mathrm{vib}) = 1\ \mathrm{J \cdot K^{-1} \cdot mol^{-1}}$ for each degree of vibrational freedom. Compare this estimate to the value of $\Delta S^\circ_\ddagger$ obtained in Problem 13.37.

▌ For each substance,

$$S^\circ(\mathrm{H}) = 150\ \mathrm{J \cdot K^{-1} \cdot mol^{-1}}$$

$$S^\circ(\mathrm{H_2}) = 150\ \mathrm{J \cdot K^{-1} \cdot mol^{-1}} + 2(30\ \mathrm{J \cdot K^{-1} \cdot mol^{-1}}) + 1(1\ \mathrm{J \cdot K^{-1} \cdot mol^{-1}}) = 211\ \mathrm{J \cdot K^{-1} \cdot mol^{-1}}$$

$$S^\circ(\mathrm{H_3}) = 150\ \mathrm{J \cdot K^{-1} \cdot mol^{-1}} + 2(30\ \mathrm{J \cdot K^{-1} \cdot mol^{-1}}) + 3(1\ \mathrm{J \cdot K^{-1} \cdot mol^{-1}}) = 213\ \mathrm{J \cdot K^{-1} \cdot mol^{-1}}$$

Substituting into *(5.20)* gives

$$\Delta S^\circ_\ddagger = 213\ \mathrm{J \cdot K^{-1} \cdot mol^{-1}} - (150\ \mathrm{J \cdot K^{-1} \cdot mol^{-1}} + 211\ \mathrm{J \cdot K^{-1} \cdot mol^{-1}}) = -148\ \mathrm{J \cdot K^{-1} \cdot mol^{-1}}$$

This value is roughly twice that of the experimental value.

13.39 Determine $\Delta G_{\ddagger}^{\circ}$ and $K_{\ddagger}$ for the reaction described in Problem 13.36.

▮ Substituting the results of Problem 13.36 and 13.37 into *(6.2a)* gives

$$\Delta G_{\ddagger}^{\circ} = 18\ \text{kJ}\cdot\text{mol}^{-1} - (298\ \text{K})(-67\ \text{J}\cdot\text{K}^{-1}\cdot\text{mol}^{-1})[(10^{-3}\ \text{kJ})/(1\ \text{J})] = 38\ \text{kJ}\cdot\text{mol}^{-1}$$

Substituting this value of $\Delta G_{\ddagger}^{\circ}$ into *(7.1b)* gives

$$K_{\ddagger} = \exp\left(-\frac{(38\ \text{kJ}\cdot\text{mol}^{-1})[(10^{3}\ \text{J})/(1\ \text{kJ})]}{(8.314\ \text{J}\cdot\text{K}^{-1}\cdot\text{mol}^{-1})(298\ \text{K})}\right) = 2.2\times10^{-7}$$

13.40 Using the results of Problem 13.39, determine the value of K_C for the reaction described in Problem 13.36. Assume a typical value for ν of $1\times10^{13}\ \text{s}^{-1}$.

▮ The relation between K_C and $K_{\ddagger}$ is given by

$$K_C = \frac{kT}{h\nu}K_{\ddagger} \tag{13.15}$$

where ν is the intramolecular vibrational frequency associated with the decomposition of the complex. Substituting the data into *(13.15)* gives

$$K_C = \frac{(1.381\times10^{-23}\ \text{J}\cdot\text{K}^{-1})(298\ \text{K})}{(6.626\times10^{-34}\ \text{J}\cdot\text{s})(1\times10^{13}\ \text{s}^{-1})}(2.2\times10^{-7}) = 1\times10^{-7}$$

13.41 At 30 °C, $\Delta V_{\ddagger} = -22.3\ \text{mL}\cdot\text{mol}^{-1}$ for the dimerization of cyclopentadiene. What is the ratio of the rate constant for this reaction carried out at 100 bar compared to the rate constant at 1 bar?

▮ The pressure dependence of the rate constant is given by

$$\frac{\partial \ln k_1}{\partial P} = \frac{-\Delta V_{\ddagger}}{RT} \tag{13.16}$$

Substituting the data into the integrated form of *(13.16)* gives

$$\ln\frac{k_1(P_2)}{k_1(P_1)} = \frac{-\Delta V_{\ddagger}}{RT}(P_2 - P_1) = \frac{-(-22.3\ \text{mL}\cdot\text{mol}^{-1})(100\ \text{bar} - 1\ \text{bar})[(10^{-3}\ \text{L})/(1\ \text{mL})]}{(0.083\,14\ \text{L}\cdot\text{bar}\cdot\text{K}^{-1}\cdot\text{mol}^{-1})(305\ \text{K})} = 8.7\times10^{-2}$$

Taking antilogarithms gives $k_1(P_2)/k_1(P_1) = 1.09$.

13.42 For a reaction involving ionic reactants described by the mechanism given by *(13.10)*,

$$\log k = \log k_1 K_C + 1.0232\, Z_A Z_B I^{1/2} \tag{13.17}$$

where I is the ionic strength of the solution. Given $\log k_1 K_C = -2.861$ for the reaction

$$S_2O_8^{2-}(aq) + 2I^-(aq) \rightarrow 2SO_4^{2-}(aq) + I_2(aq)$$

determine $k/(\text{L}\cdot\text{mol}^{-1}\cdot\text{s}^{-1})$ for a reaction mixture containing $C(\text{KI}) = 0.010\,000\ \text{M}$ and $C(Na_2S_2O_8) = 1.5\times10^{-4}\ \text{M}$.

▮ From Table 6-2, $I = 3(1.5\times10^{-4}\ \text{M}) + 1.000\times10^{-2}\ \text{M} = 1.045\times10^{-2}\ \text{M}$. Substituting the data into *(13.17)* gives

$$\log k = -2.861 + (1.0232)(-2)(-1)(1.045\times10^{-2})^{1/2} = -2.652$$

Taking antilogarithms gives $k = 2.23\times10^{-3}\ \text{L}\cdot\text{mol}^{-1}\cdot\text{s}^{-1}$.

13.43 How does the value of the rate constant in *(13.17)* depend on the signs of Z_A and Z_B?

▮ If the signs of Z_A and Z_B are both positive or both negative (as in Problem 13.42), the value of the rate constant increases as the ionic strength increases. If the signs of Z_A and Z_B are different, the value of the rate constant decreases as the ionic strength increases. This effect is known as the *primary salt effect*

13.3 HOMOGENEOUS CATALYSIS

13.44 A proposed mechanism for the catalyzed decomposition of aqueous hydrogen peroxide is

$$H_2O_2(aq) + I^-(aq) \xrightarrow{k_1} H_2O(l) + IO^-(aq)$$

$$H_2O_2(aq) + IO^-(aq) \xrightarrow{k_1'} H_2O(l) + O_2(g) + I^-(aq)$$

where $k_1' \gg k_1$. Derive the rate law for this reaction.

▮ The first reaction is the rate-determining step giving

$$\frac{-dC(H_2O_2)}{dt} = k_1 C(H_2O_2)C(I^-)$$

This rate law is considered in Problem 12.26.

13.45 For the equation

$$2H_2O_2(aq) \rightarrow 2H_2O(l) + O_2(g) \qquad \Delta_r U° = -191.78 \text{ kJ}$$

$E_a = 75.3 \text{ kJ} \cdot \text{mol}^{-1}$ for the uncatalyzed reaction and $E_a = 56.5 \text{ kJ} \cdot \text{mol}^{-1}$ for the I^--catalyzed reaction (see Problem 13.44). Prepare a reaction coordinate diagram describing this reaction.

▮ The diagram is shown in Fig. 13-5.

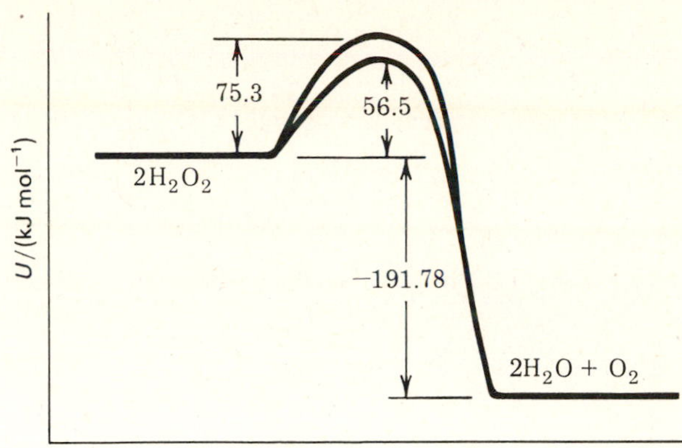

Fig. 13-5

13.46 What is the ratio of the rate constant for the I^--catalyzed decomposition of $H_2O_2(aq)$ to that for the uncatalyzed reaction at 298 K? See Problem 13.45 for additional data.

▮ Using *(13.1a)* gives

$$\frac{k(I^-)}{k(\text{uncat})} = \frac{A \exp[-(56.5 \times 10^3 \text{ J} \cdot \text{mol}^{-1})/(8.314 \text{ J} \cdot \text{K}^{-1} \cdot \text{mol}^{-1})(298 \text{ K})]}{A \exp[-(75.3 \times 10^3 \text{ J} \cdot \text{mol}^{-1})/(8.314 \text{ J} \cdot \text{K}^{-1} \cdot \text{mol}^{-1})(298 \text{ K})]} = 1970$$

13.47 By what factor will the rate constant for the catalyzed reverse reaction described in Problem 13.46 increase compared to that for the uncatalyzed reverse reaction?

▮ The activation energies for the reverse reaction are given by *(13.4)* as

$$E_{a,r}(I^-) = 56.5 \text{ kJ} \cdot \text{mol}^{-1} - (-191.78 \text{ kJ} \cdot \text{mol}^{-1}) = 248.3 \text{ kJ} \cdot \text{mol}^{-1}$$

$$E_{a,r}(\text{uncat}) = 75.3 \text{ kJ} \cdot \text{mol}^{-1} - (-191.78 \text{ kJ} \cdot \text{mol}^{-1}) = 267.1 \text{ kJ} \cdot \text{mol}^{-1}$$

Using *(13.1a)* gives

$$\frac{k_r(I^-)}{k_r(\text{uncat})} = \frac{A \exp[-(248.3 \times 10^3 \text{ J} \cdot \text{mol}^{-1})/(8.314 \text{ J} \cdot \text{K}^{-1} \cdot \text{mol}^{-1})(298 \text{ K})]}{A \exp[-(267.1 \times 10^3 \text{ J} \cdot \text{mol}^{-1})/(8.314 \text{ J} \cdot \text{K}^{-1} \cdot \text{mol}^{-1})(298 \text{ K})]} = 1970$$

The catalyst increases both the forward and reverse reactions by the same factor.

13.48 The chemical reaction

$$CH_3COOCH_3(aq) + H_2O(l) \rightarrow CH_3COOH(aq) + CH_3OH(aq)$$

is believed to be catalyzed by $H^+(aq)$ where the observed rate constant (k_{obs}) is related to the rate constant for the uncatalyzed reaction (k_{uncat}) by

$$k_{obs} = k_{uncat} C(\text{catalyst})^n \qquad (13.18)$$

Use the following data to determine k_{uncat} and n.

$k_{obs}/(10^{-4} \text{ s}^{-1})$	0.108	0.585	1.000	2.682	3.469
$C(HCl)/(\text{mol} \cdot \text{L}^{-1})$	0.1005	0.5024	0.8275	1.800	2.429

▌ Taking logarithms of both sides of *(13.18)* gives

$$\log k_{\text{obs}} = \log k_{\text{uncat}} + n \log [C(\text{H}^+)]$$

Thus a plot of log k_{obs} against log $[C(\text{HCl})]$ will be linear, with an intercept equal to log k_{uncat} and a slope equal to n. The data are plotted in Fig. 13-6, giving $n = 1.10 \approx 1$ and $k_{\text{uncat}} = 1.32 \times 10^{-4}\ \text{L} \cdot \text{mol}^{-1} \cdot \text{s}^{-1}$.

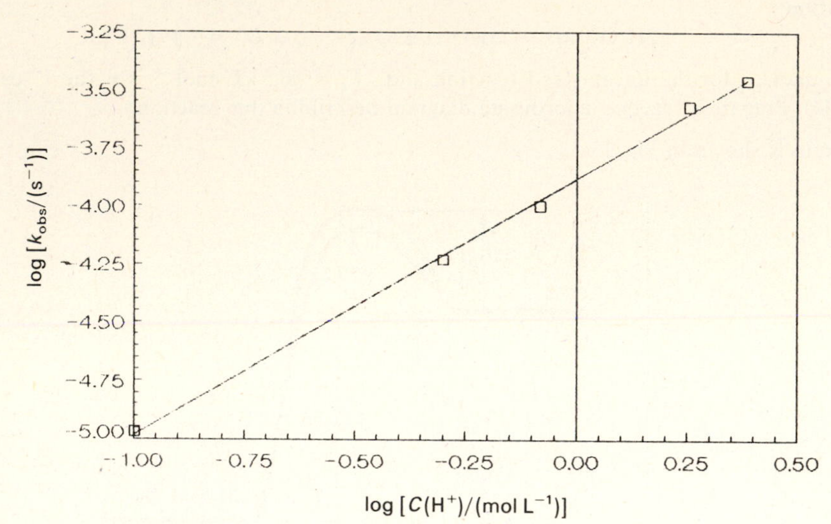

Fig. 13-6

13.49 For the acid-catalyzed reaction

```
  NH2                                      NH—C=O
  |                                        |    |
HN=C           O      (aq) → HN=C          |       (aq) + H2O(l)
  |            ||                |          |
H3C—N—CH2—C—OH          H3C—N——CH2
     creatine                       creatinine
```

the observed rate constant at 78 °C is given by *(13.18)*, where $k_{\text{uncat}} = 2.39 \times 10^{-4}\ \text{L} \cdot \text{mol}^{-1} \cdot \text{s}^{-1}$ and $n = 1$ for $C(\text{H}^+)$. Calculate the observed rate constants at 0.380 M HCl and at 0.760 M HCl and compare them.

▌ Substituting the data into *(13.18)* gives

$$k_{\text{obs}} = (2.39 \times 10^{-4}\ \text{L} \cdot \text{mol}^{-1} \cdot \text{s}^{-1})(0.380\ \text{mol} \cdot \text{L}^{-1}) = 9.08 \times 10^{5}\ \text{s}^{-1}$$

for the 0.380 M HCl solution and $k_{\text{obs}} = 1.82 \times 10^{-4}\ \text{s}^{-1}$ for the 0.760 M HCl. The ratio of the rate constants is 2:1, as expected for doubling the concentration of a reactant in a first-order reaction.

13.50 The chemical reaction

$$[\text{CrCl}_2(\text{H}_2\text{O})_4]^+(\text{aq}) + \text{H}_2\text{O}(\text{l}) \rightarrow [\text{CrCl}(\text{H}_2\text{O})_5]^{2+}(\text{aq}) + \text{Cl}^-(\text{aq})$$

is believed to be inhibited by H^+(aq), where the observed rate constant (k_{obs}) is given by

$$k_{\text{obs}} = k_{\text{uncat}} + k(\text{H}^+)/C(\text{H}^+)$$

Use the following data to determine k_{uncat} and $k(\text{H}^+)$.

C(HCl)/(10^{-3} mol · L^{-1})	0.200	0.861	1.005	4.196	8.000	9.953
k_{obs}/(10^{-3} s^{-1})	1.10	0.341	0.307	0.170	0.078	0.070

▌ A plot of k_{obs} against $[C(\text{H}^+)]^{-1}$ is shown in Fig. 13-7. The intercept gives $k_{\text{uncat}} = 6.9 \times 10^{-5}\ \text{s}^{-1}$, and the slope gives $k(\text{H}^+) = 2.09 \times 10^{-7}\ \text{mol} \cdot \text{L}^{-1} \cdot \text{s}^{-1}$.

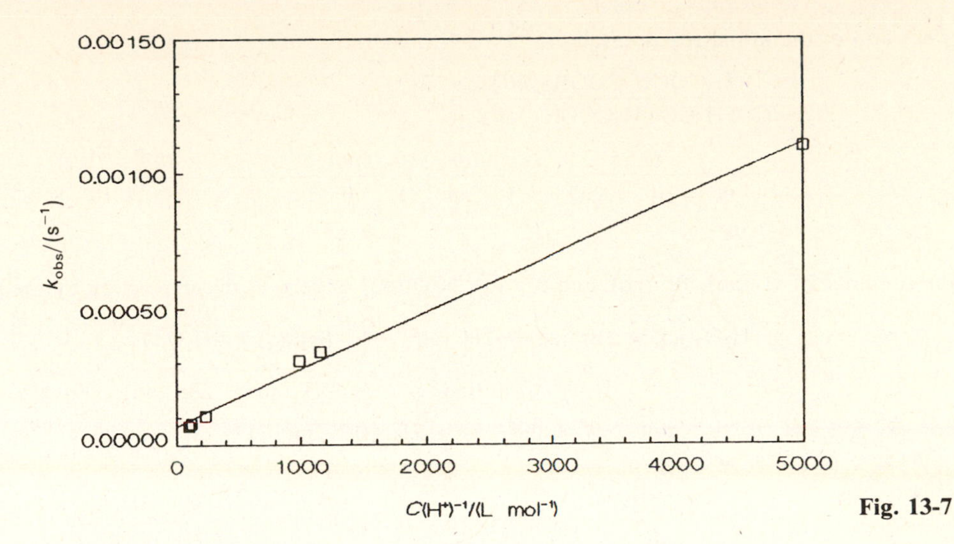

Fig. 13-7

13.51 The rate at which phenolphthalein changes color in 0.31 M KOH was studied as a function of the concentration of NaCl:

$C(\mathrm{NaCl})/(\mathrm{mol \cdot L^{-1}})$	0.000	0.116	0.194
$k'/(10^{-3}\ \mathrm{s^{-1}})$	1.746	2.006	2.176

The observed rate constant (k') for low concentrations of NaCl is given by $k' = k_{\mathrm{uncat}} + kC(\mathrm{NaCl})$. Use the above data to determine k_{uncat} and k.

▮ A plot of k' against $C(\mathrm{NaCl})$ is shown in Fig. 13-8. The intercept gives $k_{\mathrm{uncat}} = 1.747 \times 10^{-3}\ \mathrm{s^{-1}}$, and the slope gives $k = 2.22 \times 10^{-3}\ \mathrm{L \cdot mol^{-1} \cdot s^{-1}}$.

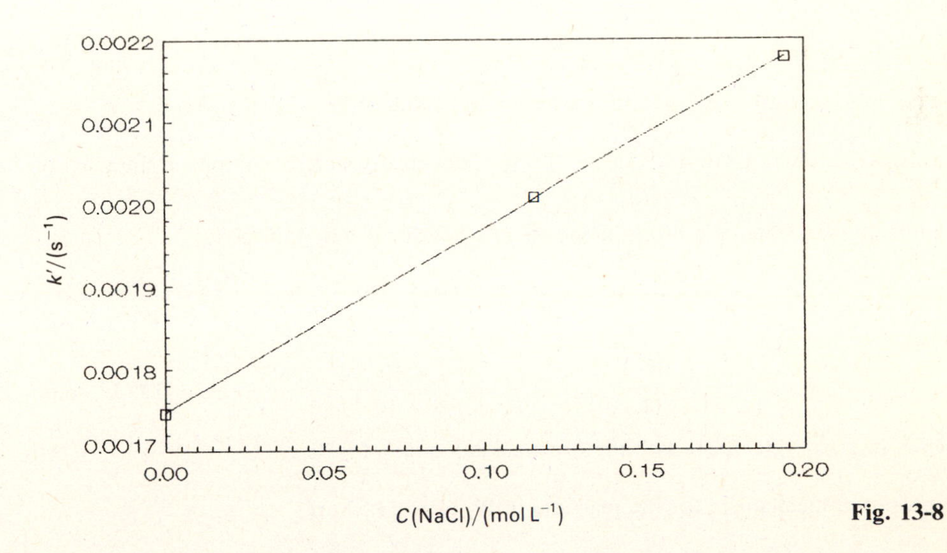

Fig. 13-8

13.52 The rate equation for the equation $\mathrm{CH_3COCH_2COOH(aq)} \rightarrow \mathrm{CH_3COCH_3(aq)} + \mathrm{CO_2(g)}$ is given by

$$\frac{-dC(\mathrm{CH_3COCH_2COOH})}{dt} = k_1 C(\mathrm{CH_3COCH_2COOH}) + k_2 C(\mathrm{CH_3COCH_2COO^-})$$

where $k_1 = 1.65 \times 10^{-5}\ \mathrm{s^{-1}}$ and $k_2 = 3 \times 10^{-7}\ \mathrm{s^{-1}}$ at 25 °C. Compare the initial reaction rates at $C(\mathrm{CH_3COCH_2COOH})_0 = 1.0 \times 10^{-3}$ M for an uncatalyzed reaction and at $C(\mathrm{CH_3COCH_2COO^-})_0 = 1.0 \times 10^{-3}$ M for a catalyzed reaction.

▌ The ratio of the initial reaction rates is

$$\frac{[-dC(CH_3COCH_2COOH)_0/dt]_{uncat}}{[-dC(CH_3COCH_2COOH)_0/dt]_{cat}}$$

$$= \frac{(1.65 \times 10^{-5}\ s^{-1})(1.0 \times 10^{-3}\ mol \cdot L^{-1}) + (3 \times 10^{-7}\ s^{-1})(0)}{(1.65 \times 10^{-5}\ s^{-1})(1.0 \times 10^{-3}\ mol \cdot L^{-1}) + (3 \times 10^{-7}\ s^{-1})(1.0 \times 10^{-3}\ mol \cdot L^{-1})}$$

$$= 0.98$$

13.53 In the presence of $H^+(aq)$, $Br^-(aq)$, and $Br_2(aq)$, $H_2O_2(aq)$ undergoes decomposition by the parallel reactions

$$H_2O_2(aq) + 2Br^-(aq) + 2H^+(aq) \xrightarrow{k_1} Br_2(aq) + 2H_2O(l)$$

$$H_2O_2(aq) + Br_2(aq) \xrightarrow{k_2} 2H^+(aq) + 2Br^-(aq) + O_2(g)$$

where $k_1 = 1.2 \times 10^{-4}\ L^2 \cdot mol^{-2} \cdot s^{-1}$ and $k_2 = 5.8 \times 10^{-4}\ mol \cdot L^{-1} \cdot s^{-1}$ at 25 °C. The rate laws for the simultaneous reactions are

$$\frac{-dC(H_2O_2)}{dt} = k_1 C(H_2O_2)C(H^+)C(Br^-) + k_2 \frac{C(H_2O_2)C(Br_2)}{C(H^+)C(Br^-)}$$

$$\frac{dC(Br_2)}{dt} = k_1 C(H_2O_2)C(H^+)C(Br^-) - k_2 \frac{C(H_2O_2)C(Br_2)}{C(H^+)C(Br^-)}$$

Assuming that the reaction rates are equal, determine the value of the observed rate constant (k_{obs}), where

$$\frac{-dC(H_2O_2)}{dt} = k_{obs} C(H_2O_2)C(H^+)C(Br^-)$$

▌ Addition of the rate laws for $-dC(H_2O_2)/dt$ and $dC(Br_2)/dt$ gives

$$\frac{-dC(H_2O_2)}{dt} + \frac{dC(Br_2)}{dt} = 2k_1 C(H_2O_2)C(H^+)C(Br^-)$$

Under steady-state conditions where $C(Br^-)$, $C(Br_2)$, and $C(H^+)$ are constant, $dC(Br_2)/dt = 0$, giving

$$\frac{-dC(H_2O_2)}{dt} = 2k_1 C(H_2O_2)C(H^+)C(Br^-)$$

Thus,

$$k_{obs} = 2k_1 = 2(1.2 \times 10^{-4}\ L^2 \cdot mol^{-2} \cdot s^{-1}) = 2.4 \times 10^{-4}\ L^2 \cdot mol^{-2} \cdot s^{-1}$$

which is in good agreement with the experimental value of $2.3 \times 10^{-4}\ L^2 \cdot mol^{-2} \cdot s^{-1}$.

13.54 What is the value of $C(Br_2)/C(H^+)^2C(Br^-)^2$ under steady-state conditions in the reaction described in Problem 13.53?

▌ Under steady-state conditions, using the rate law for $dC(Br_2)/dt$ gives

$$\frac{dC(Br_2)}{dt} = k_1 C(H_2O_2)C(H^+)C(Br^-) - k_2 \frac{C(H_2O_2)C(Br_2)}{C(H^+)C(Br^-)} = 0$$

giving

$$\frac{C(Br_2)}{[C(H^+)]^2[C(Br^-)]^2} = \frac{k_1}{k_2} = \frac{1.2 \times 10^{-4}\ L^2 \cdot mol^{-2} \cdot s^{-1}}{5.8 \times 10^{-4}\ mol \cdot L^{-1} \cdot s^{-1}} = 0.21\ L^3 \cdot mol^{-3}$$

which is in good agreement with the experimental value of 0.20.

13.55 The proposed mechanism for the reactions in Problem 13.53 are

$$H_2O_2(aq) + H^+(aq) + Br^-(aq) \xrightarrow{k'} HBrO(aq) + H_2O(l)$$

$$HBrO(aq) + H^+(aq) + Br^-(aq) \underset{(rapid)}{\rightleftharpoons} H_2O(l) + Br_2(aq)$$

and

$$H_2O_2(aq) + HBrO(aq) \xrightarrow{k''} H_2O(l) + H^+(aq) + Br^-(aq) + O_2(g)$$

$$Br_2(aq) + H_2O(l) \xrightarrow[(rapid)]{} HBrO(aq) + H^+(aq) + Br^-(aq)$$

Derive the expression for $-dC(H_2O_2)/dt$ given in Problem 13.53.

▌ From the two rate-determining steps,

$$\frac{-dC(H_2O_2)}{dt} = k'C(H_2O_2)C(H^+)C(Br^-) + k''C(H_2O_2)C(HBrO)$$

From the HBrO-Br^--Br_2 equilibrium, $K = C(Br_2)/C(HBrO)C(H^+)C(Br^-)$, so

$$\frac{-dC(H_2O_2)}{dt} = k'C(H_2O_2)C(H^+)C(Br^-) + \frac{k''C(H_2O_2)C(Br_2)}{KC(H^+)C(Br^-)}$$

which is the desired expression if $k' = k_1$ and $k''/K = k_2$.

13.56 Derive an equation for $C(B)$ in the autocatalytic reaction

$$A \xrightarrow{B} B + \cdots$$

where $-dC(A)/dt = kC(A)C(B)$. Assume that $C(B)_0 \neq 0$.

▌ Letting x represent the decrease in concentration of each reactant in time t gives

$$C(A) = C(A)_0 - x \qquad C(B) = C(B)_0 + x$$

$$\frac{-dC(A)}{dt} = \frac{-d[C(A)_0 - x]}{dt} = \frac{dx}{dt}$$

Using this notation, the rate equation becomes

$$\frac{dx}{dt} = k[C(A)_0 - x][C(B)_0 + x]$$

which upon integration gives

$$\ln\frac{C(A)_0[C(B)_0 + x]}{C(B)_0[C(A)_0 - x]} = \ln\frac{C(A)_0C(B)}{C(B)_0C(A)} = [C(A)_0 + C(B)_0]kt \tag{13.19}$$

Taking the antilogarithm of each side gives

$$\frac{C(B)}{C(A)} = \frac{C(B)_0}{C(A)_0}e^{[C(A)_0+C(B)_0]kt}$$

At any given time, $C(A) + C(B) = C(A)_0 + C(B)_0$. Solving this equation for $C(A)$ and substituting gives

$$\frac{C(B)}{C(A)_0 + C(B)_0 - C(B)} = \frac{C(B)_0}{C(A)_0}e^{[C(A)_0+C(B)_0]kt}$$

Solving for $C(B)$ gives the desired expression as

$$C(B) = \frac{C(A)_0 + C(B)_0}{[C(A)_0/C(B)_0]\,e^{-[C(A)_0+C(B)_0]kt} + 1} \tag{13.20}$$

13.57 For the chemical equation

$$CH_3COCH_3(aq) + I_2(aq) \xrightarrow{H^+} CH_3COCH_2I(aq) + H^+(aq) + I^-(aq)$$

the following data were collected for $C(CH_3COCH_3)_0 = 0.683$ M at 20 °C.

t/(h)	24	46	65
$x/(\text{mol}\cdot L^{-1})$	0.000 196	0.000 602	0.001 492

where x is the concentration of H^+(aq) formed. Show that the reaction is autocatalytic with the rate law

$$\frac{-dC(CH_3COCH_3)}{dt} = kC(CH_3COCH_3)C(H^+)$$

▌ Solving *(13.19)* for $\ln[C(B)]$ gives

$$\ln[C(B)] = \ln\frac{C(B)_0C(A)}{C(A)_0} + [C(A)_0 + C(B)_0]kt$$

Under the reaction conditions, $C(CH_3COCH_3) \gg x$, giving $C(CH_3COCH_3) \approx C(CH_3COCH_3)_0$, and $x \gg C(H^+)_0$, giving $C(H^+) \approx x$. Thus,

$$\ln x = \ln [C(H^+)_0] + [C(CH_3COCH_3)_0 + C(H^+)_0]kt$$

The plot of $\ln x$ against t shown in Fig. 13-9 is linear, indicating that the assumed rate law is valid. The intercept of the plot is −9.71, giving

$$C(H^+)_0 = e^{-9.71} = 6.1 \times 10^{-5}\ \text{mol} \cdot \text{L}^{-1}$$

and the slope is $0.050\ \text{h}^{-1}$, giving

$$k = \frac{\text{slope}}{C(CH_3COCH_3)_0 + C(H^+)_0} = \frac{0.050\ \text{h}^{-1}}{0.683\ \text{mol} \cdot \text{L}^{-1} + 6.1 \times 10^{-5}\ \text{mol} \cdot \text{L}^{-1}} = 7.3 \times 10^{-2}\ \text{L} \cdot \text{mol}^{-1} \cdot \text{h}^{-1}$$

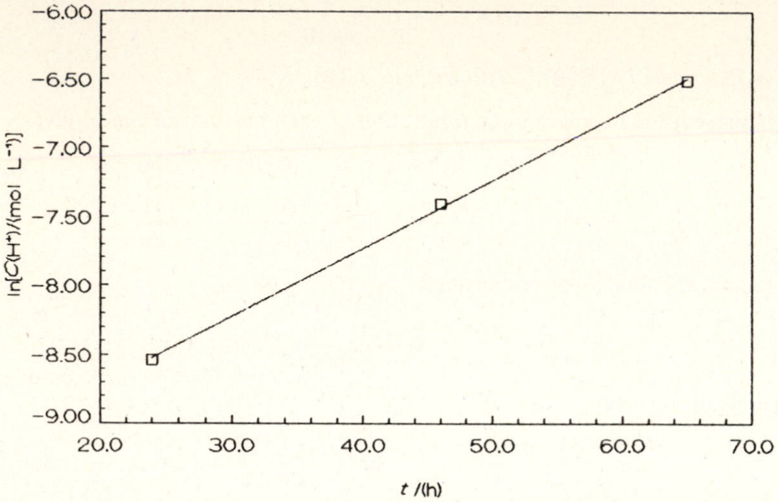

Fig. 13-9

13.58 Using the results of Problem 13.57, prepare a plot of $C(H^+)$ as a function of time from $0 \le t/(\text{h}) \le 400$.

▌ Substituting the results of Problem 13.57 into *(13.20)* gives

$$C(H^+) = \frac{0.683\ \text{mol} \cdot \text{L}^{-1} + 6.1 \times 10^{-5}\ \text{mol} \cdot \text{L}^{-1}}{\dfrac{0.683\ \text{mol} \cdot \text{L}^{-1}}{6.1 \times 10^{-5}\ \text{mol} \cdot \text{L}^{-1}} \exp\left\{-\left[\begin{array}{l}(0.683\ \text{mol} \cdot \text{L}^{-1} + 6.1 \times 10^{-5}\ \text{mol} \cdot \text{L}^{-1}) \\ \times (7.3 \times 10^{-2}\ \text{L} \cdot \text{mol}^{-1} \cdot \text{h}^{-1})(t/\text{h})\end{array}\right]\right\} + 1}$$

$$= \frac{0.683\ \text{mol} \cdot \text{L}^{-1}}{11\,200\, e^{-(0.050)(t/\text{h})} + 1}$$

As a sample calculation, consider $t = 200$ h.

$$C(H^+) = \frac{0.683\ \text{mol} \cdot \text{L}^{-1}}{11\,200\, e^{-(0.050)(200)} + 1} = 0.453\ \text{mol} \cdot \text{L}^{-1}$$

The plot is shown in Fig. 13-10.

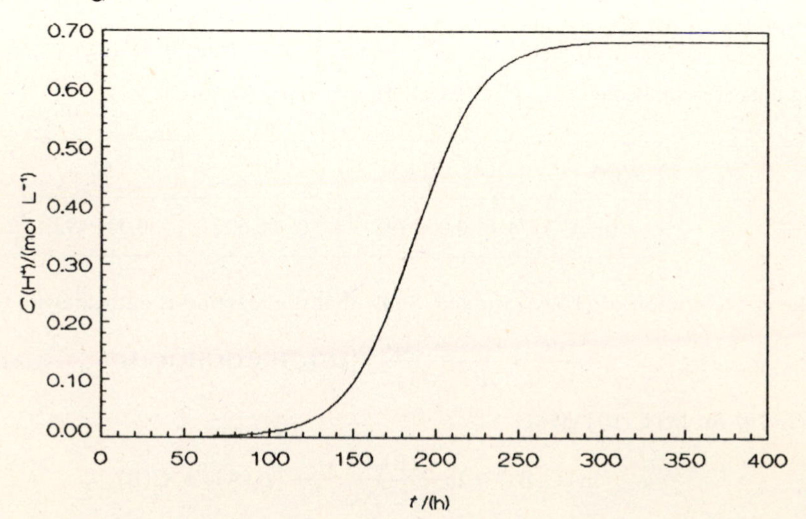

Fig. 13-10

13.59 The hydrolysis of fumarate ion to L-malate ion,

$$^{-}OOC{-}CH{=}CH{-}COO^{-}(aq) + H_2O(l) \rightarrow {}^{-}OOC{-}CHOH{-}CH_2{-}COO^{-}(aq)$$

is catalyzed by the enzyme fumarase. The proposed mechanism is

$$E + S \underset{k_{-1}}{\overset{k_1}{\rightleftharpoons}} X \underset{k_{-2}}{\overset{k_2}{\rightleftharpoons}} E + P \tag{13.21}$$

where E is the enzyme, S is the substrate, X is the enzyme-substrate complex, and P is the product. The rate expression is

$$\frac{dC(P)}{dt} = \frac{(V_S/K_S)C(S) - (V_P/K_P)C(P)}{1 + [C(S)/K_S] + [C(P)/K_P]} \tag{13.22}$$

where

$$V_S = k_2 C(E)_0 \tag{13.23}$$

$$V_P = k_{-1} C(E)_0 \tag{13.24}$$

$$K_S = (k_{-1} + k_2)/k_1 \tag{13.25}$$

$$K_P = (k_{-1} + k_2)/k_{-2} \tag{13.26}$$

and k_2 is known as the *turnover constant.* Describe the kinetics of the initial rate of this reaction for $C(S) \ll K_S$ and $C(S) \gg K_S$.

▌ The initial rate of the reaction is often represented by the system v where

$$v = \left(\frac{dC(P)}{dt}\right)_0$$

At the beginning of the reaction $C(P) = 0$, and *(13.22)* becomes the *Michaelis–Menten equation*

$$v = \frac{(V_S/K_S)C(S)}{1 + [C(S)/K_S]} = \frac{V_S/C(S)}{K_S + C(S)} \tag{13.27}$$

If $C(S) \ll K_S$, *(13.27)* becomes

$$v = V_S C(S)/K_S$$

a pseudo-first-order reaction in substrate, and if $C(S) \gg K_S$, *(13.26)* becomes

$$v = V_S C(S)/C(S) = V_S$$

a pseudo-zero-order reaction in substrate.

13.60 The hydrolysis of N-glutaryl-L-phenylalanine-p-nitroanilide (GPNA) by α-chymotrypsin (CT) to form p-nitroaniline and N-glutaryl-L-phenylalamine is described by *(13.21)*. Use the following data to determine V_S, K_S, and k_2 using a Lineweaver–Burke plot:

$$C(\text{CT}) = 4.0 \times 10^{-6}\ \text{mol} \cdot \text{L}^{-1}$$

$C(\text{GPNA})/(10^{-4}\ \text{mol} \cdot \text{L}^{-1})$	2.5	5.0	10.0	15.0
$v/(10^{-8}\ \text{mol} \cdot \text{L}^{-1} \cdot \text{s}^{-1})$	3.7	6.3	9.8	11.8

▌ The *Lineweaver–Burke equation* is a linear transformation of *(13.27)* where

$$\frac{1}{v} = \frac{K_S}{V_S}\left(\frac{1}{C(S)}\right) + \frac{1}{V_S} \tag{13.28a}$$

A plot of $1/v$ against $1/C(\text{GPNA})$ is shown in Fig. 13-11. From the intercept,

$$v_S = \frac{1}{\text{intercept}} = \frac{1}{0.469 \times 10^7\ \text{L} \cdot \text{s} \cdot \text{mol}^{-1}} = 2.13 \times 10^{-7}\ \text{mol} \cdot \text{L}^{-1} \cdot \text{s}^{-1}$$

and from the slope,

$$K_S = V_S(\text{slope}) = (2.13 \times 10^{-7}\ \text{mol} \cdot \text{L}^{-1} \cdot \text{s}^{-1})(5.58 \times 10^3\ \text{s}) = 1.19 \times 10^{-3}\ \text{mol} \cdot \text{L}^{-1}$$

Solving *(13.23)* for k_2 and substituting the data gives

$$k_2 = \frac{2.13 \times 10^{-7}\ \text{mol} \cdot \text{L}^{-1} \cdot \text{s}^{-1}}{4.0 \times 10^{-6}\ \text{mol} \cdot \text{L}^{-1}} = 0.053\ \text{s}^{-1}$$

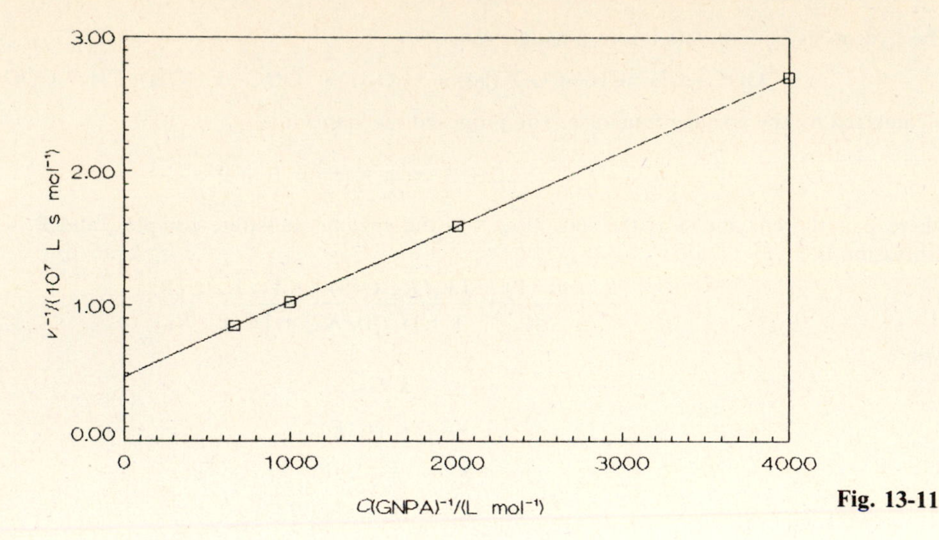

Fig. 13-11

13.61 Using the data given in Problem 13.60, determine V_S, K_S, and k_2 using a Hanes-Woolf plot.

▌ The *Hanes-Woolf equation* is a linear transformation of *(13.27)* where

$$\frac{C(\mathrm{S})}{v} = \frac{1}{V_S}C(\mathrm{S}) + \frac{K_S}{V_S} \tag{13.28b}$$

A plot of $C(\mathrm{GPNA})/v$ against $C(\mathrm{S})$ is shown in Fig. 13-12. From the slope,

$$V_S = \frac{1}{\text{slope}} = \frac{1}{4.74 \times 10^6\ \mathrm{L \cdot s \cdot mol^{-1}}} = 2.11 \times 10^{-7}\ \mathrm{mol \cdot L^{-1} \cdot s^{-1}}$$

and from the intercept,

$$K_S = V_S(\text{intercept}) = (2.11 \times 10^{-7}\ \mathrm{mol \cdot L^{-1} \cdot s^{-1}})(5.55 \times 10^3\ \mathrm{s}) = 1.17 \times 10^{-3}\ \mathrm{mol \cdot L^{-1}}$$

Solving *(13.23)* for k_2 and substituting the data gives

$$k_2 = \frac{2.11 \times 10^{-7}\ \mathrm{mol \cdot L^{-1} \cdot s^{-1}}}{4.0 \times 10^{-6}\ \mathrm{mol \cdot L^{-1}}} = 0.053\ \mathrm{s^{-1}}$$

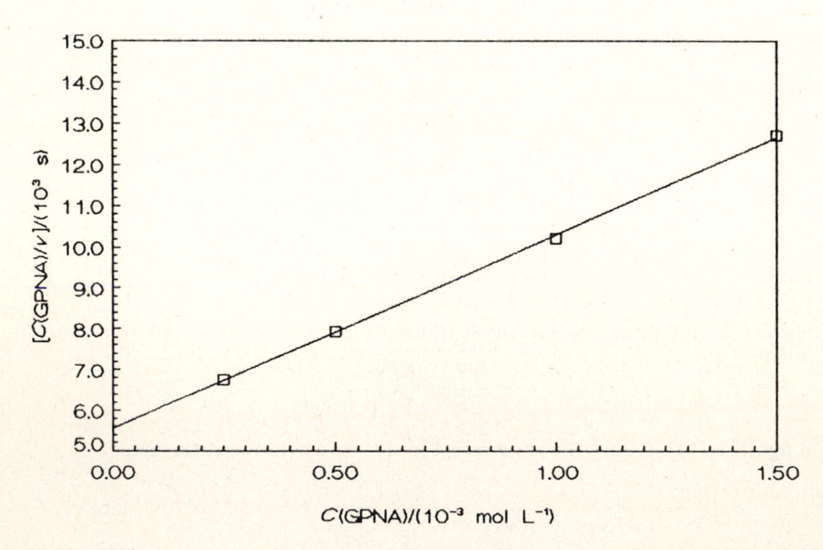

Fig. 13-12

13.62 Using the data given in Problem 13.60, determine V_S, K_S, and k_2 using an Eadie-Hofster plot.

▌ The *Eadie-Hofster equation* is a linear transformation of *(13.27)* where

$$v = -K_S\frac{v}{C(\mathrm{S})} + V_S \tag{13.28c}$$

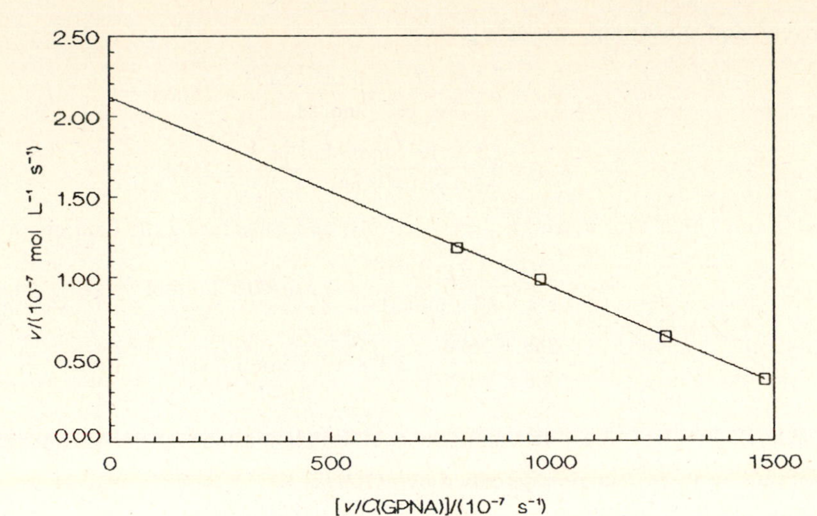

Fig. 13-13

A plot of v against $v/C(\text{GPNA})$ is shown in Fig. 13-13. From the intercept,

$$V_S = \text{intercept} = 2.12 \times 10^{-7}\ \text{mol} \cdot \text{L}^{-1} \cdot \text{s}^{-1}$$

and from the slope,

$$K_S = -(\text{slope}) = -(-1.18 \times 10^{-3}\ \text{mol} \cdot \text{L}^{-1}) = 1.18 \times 10^{-3}\ \text{mol} \cdot \text{L}^{-1}$$

Solving *(13.23)* for k_2 and substituting the data gives

$$k_2 = \frac{2.12 \times 10^{-7}\ \text{mol} \cdot \text{L}^{-1} \cdot \text{s}^{-1}}{4.0 \times 10^{-6}\ \text{mol} \cdot \text{L}^{-1}} = 0.053\ \text{s}^{-1}$$

13.63 For the reaction described in Problem 13.59, the following data were obtained for $C(\text{fumarase})_0 = 5.0 \times 10^{-10}\ \text{mol} \cdot \text{L}^{-1}$ at pH = 6, 25 °C, and $\mu = 0.01\ \text{mol} \cdot \text{L}^{-1}$.

$v/(\text{mol} \cdot \text{L}^{-1} \cdot \text{s}^{-1})$	1.09×10^{-7}	4.65×10^{-7}
$C(\text{fumarate ion})/(\text{mol} \cdot \text{L}^{-1})$	1.00×10^{-6}	1.00×10^{-5}

$v/(\text{mol} \cdot \text{L}^{-1} \cdot \text{s}^{-1})$	2.32×10^{-8}	8.90×10^{-8}
$C(\text{malate ion})/(\text{mol} \cdot \text{L}^{-1})$	1.00×10^{-6}	1.00×10^{-5}

Determine the values of k_1, k_{-1}, k_2, and k_{-2} for this reaction.

▌ Substituting the fumarate data into *(13.27)* gives

$$1.09 \times 10^{-7}\ \text{mol} \cdot \text{L}^{-1} \cdot \text{s}^{-1} = \frac{V_S(1.00 \times 10^{-6}\ \text{mol} \cdot \text{L}^{-1})}{K_S + 1.00 \times 10^{-6}\ \text{mol} \cdot \text{L}^{-1}}$$

$$4.65 \times 10^{-7} = \frac{V_S(1.00 \times 10^{-5})}{K_S + 1.00 \times 10^{-5}}$$

Solving these simultaneous equations gives $K_S = 5.7 \times 10^{-6}\ \text{mol} \cdot \text{L}^{-1}$ and $V_S = 7.3 \times 10^{-7}\ \text{mol} \cdot \text{L}^{-1} \cdot \text{s}^{-1}$. Likewise, for the malate data,

$$2.32 \times 10^{-8} = \frac{V_P(1.00 \times 10^{-6})}{K_P + 1.00 \times 10^{-6}}$$

$$8.90 \times 10^{-8} = \frac{V_P(1.00 \times 10^{-5})}{K_P + 1.00 \times 10^{-5}}$$

giving $K_P = 4.6 \times 10^{-6}\ \text{mol} \cdot \text{L}^{-1}$ and $V_P = 1.3 \times 10^{-7}\ \text{mol} \cdot \text{L}^{-1} \cdot \text{s}^{-1}$. Solving *(13.23)* and *(13.24)* for k_2 and

k_{-1}, respectively, and substituting the data gives

$$k_2 = \frac{7.3 \times 10^{-7}\ \text{mol}\cdot\text{L}^{-1}\cdot\text{s}^{-1}}{5.0 \times 10^{-10}\ \text{mol}\cdot\text{L}^{-1}} = 1500\ \text{s}^{-1}$$

$$k_{-1} = \frac{1.3 \times 10^{-7}\ \text{mol}\cdot\text{L}^{-1}\cdot\text{s}^{-1}}{5.0 \times 10^{-10}\ \text{mol}\cdot\text{L}^{-1}} = 260\ \text{s}^{-1}$$

Solving *(13.25)* and *(13.26)* for k_1 and k_{-2}, respectively, and substituting the data gives

$$k_1 = \frac{260\ \text{s}^{-1} + 1500\ \text{s}^{-1}}{5.7 \times 10^{-6}\ \text{mol}\cdot\text{L}^{-1}} = 2.6 \times 10^8\ \text{L}\cdot\text{mol}^{-1}\cdot\text{s}^{-1}$$

$$k_{-2} = \frac{260 + 1500}{4.6 \times 10^{-6}} = 3.3 \times 10^8\ \text{L}\cdot\text{mol}^{-1}\cdot\text{s}^{-1}$$

13.64 Using the results of Problem 13.63, determine the equilibrium constant for the overall reaction.

▌ For an enzyme-catalyzed reaction described by *(13.21)*,

$$K = \frac{V_S/K_S}{V_P/K_P} \tag{13.29}$$

Substituting the results of Problem 13.63 into *(13.29)* gives

$$K = \frac{(7.3 \times 10^{-7}\ \text{mol}\cdot\text{L}^{-1}\cdot\text{s}^{-1})/(5.7 \times 10^{-6}\ \text{mol}\cdot\text{L}^{-1})}{(1.3 \times 10^{-7}\ \text{mol}\cdot\text{L}^{-1}\cdot\text{s}^{-1})/(4.6 \times 10^{-6}\ \text{mol}\cdot\text{L}^{-1})} = 4.5$$

13.65 The catalysis of the hydrolysis of amino acids by various metallocarboxypeptidases can be inhibited by various substances. Consider the following data for the hydrolysis of benzoylglycylglycyl-L-phenylalanine (Bz-Gly-Gly-Phe) by 1.0×10^{-5} M zinc carboxypeptidase:

$C(\text{S})/(10^{-4}\ \text{mol}\cdot\text{L}^{-1})$	2.0	5.0	10.0
$v(\text{uninhibited})/(10^{-5}\ \text{mol}\cdot\text{L}^{-1}\cdot\text{s}^{-1})$	4.0	7.7	11.1
$v(\text{inhibited})/(10^{-5}\ \text{mol}\cdot\text{L}^{-1}\cdot\text{s}^{-1})$	1.4	2.7	3.9

where the inhibitor is 2.0×10^{-4} M β-phenylpropionic acid. Show that this is an example of noncompetitive inhibition.

▌ In *noncompetitive inhibition* the inhibitor does not attack the active site of the enzyme:

$$\begin{array}{ccccc} \text{E} + \text{S} & \rightleftharpoons & \text{X} & \rightleftharpoons & \text{E} + \text{P} \\ + & & + & & \\ \text{I} & & \text{I} & & \\ \upharpoonleft\downharpoonright & & \upharpoonleft\downharpoonright & & \\ \text{EI} + \text{S} & \rightleftharpoons & \text{XI} & & \end{array} \tag{13.30a}$$

For this mechanism, the Lineweaver–Burke equation is

$$\frac{1}{v} = \left[\frac{K_S}{V_S}\left(\frac{1}{C(\text{S})}\right) + \frac{1}{V_S}\right]\left(1 + \frac{C(\text{I})}{K_I}\right) \tag{13.31a}$$

where $C(\text{I})$ is the concentration of the inhibitor and K_I is the equilibrium constant for the inhibition process. The plots of $1/v$ against $1/C(\text{S})$ for both the uninhibited and inhibited processes are shown in Fig. 13-14. Comparison of *(13.31a)* with *(13.28a)* indicates that both the slope and the intercept for the uninhibited reaction will be increased by a factor of $[1 + C(\text{I})/K_I]$. The respective intercepts are $5.00 \times 10^3\ \text{L}\cdot\text{s}\cdot\text{mol}^{-1}$ and $1.427 \times 10^4\ \text{L}\cdot\text{s}\cdot\text{mol}^{-1}$, and the respective slopes are 4.00 s and 11.5 s, giving

$$1 + \frac{C(\text{I})}{K_I} = \frac{1.427 \times 10^4\ \text{L}\cdot\text{s}\cdot\text{mol}^{-1}}{5.00 \times 10^3\ \text{L}\cdot\text{s}\cdot\text{mol}^{-1}} = \frac{11.5\ \text{s}}{4.00\ \text{s}} = 2.86$$

Solving for K_I and substituting in the concentration of the inhibitor gives

$$K_I = \frac{2.0 \times 10^{-4}\ \text{mol}\cdot\text{L}^{-1}}{2.86 - 1} = 1.1 \times 10^{-4}\ \text{mol}\cdot\text{L}^{-1}$$

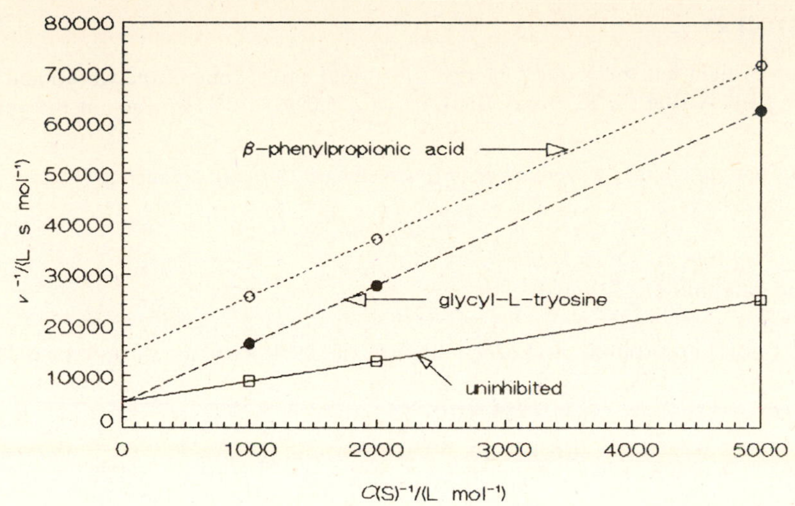

Fig. 13-14

13.66 Additional studies on the system described in Problem 13.65 with 2.0×10^{-4} M glycyl-L-tryosine as an inhibitor gave the following data:

$C(S)/(10^{-4}\ \text{mol} \cdot \text{L}^{-1})$	2.0	5.0	10.0
$v(\text{inhibited})/(10^{-5}\ \text{mol} \cdot \text{L}^{-1} \cdot \text{s}^{-1})$	1.6	3.6	6.1

Is this an example of competitive or uncompetitive inhibition?

▮ In *competitive inhibition* the inhibitor and the substrate compete for the same active site on the enzyme:

$$\begin{array}{l} E + S \rightleftharpoons X \rightleftharpoons E + P \\ + \\ I \\ \upharpoonleft\downharpoonright \\ EI \end{array} \tag{13.30b}$$

and in *uncompetitive inhibition* the inhibitor simply binds to the complex

$$\begin{array}{l} E + S \rightleftharpoons X \rightleftharpoons E + P \\ \qquad\quad + \\ \qquad\quad I \\ \qquad\quad \upharpoonleft\downharpoonright \\ \qquad\quad XI \end{array} \tag{13.30c}$$

The respective Lineweaver-Burke equations are

$$\frac{1}{v} = \frac{K_S}{V_S}\left(1 + \frac{C(I)}{K_I}\right)\left(\frac{1}{C(S)}\right) + \frac{1}{V_S} \tag{13.31b}$$

$$\frac{1}{v} = \frac{K_S}{V_S}\left(\frac{1}{C(S)}\right) + \frac{1}{V_S}\left(1 + \frac{C(I)}{K_I}\right) \tag{13.31c}$$

Comparison of *(13.31b)* with *(13.28a)* indicates that the slope of the uninhibited reaction will be increased by the factor of $[1 + C(I)/K_I]$ and that both reactions will have the same intercept. Comparison of *(13.31c)* with *(13.28a)* indicates that both reactions will have the same slope and that the intercept of the uninhibited reaction will be increased by a factor of $[1 + C(I)/K_I]$. The plot of $1/v$ against $1/C(S)$ is shown in Fig. 13-14. The plot for the inhibited reaction has the same intercept and a greater slope than the plot for the uninhibited reaction, indicating a competitive inhibition. The respective slopes of the plots are 4.00 s and 11.5 s, giving

$$K_I = \frac{2.0 \times 10^{-4}\ \text{mol} \cdot \text{L}^{-1}}{(11.5\ \text{s})/(4.00\ \text{s}) - 1} = 1.1 \times 10^{-4}\ \text{mol} \cdot \text{L}^{-1}$$

13.4 PHOTOCHEMISTRY

13.67 Which quantum of light has the greatest energy: $\bar{\nu} = 3651\ \text{cm}^{-1}$ (one of the vibrational absorption frequencies in H_2O), $\lambda = 1.544$ Å (the Cu K_α X-ray line), or $\nu = 5.090 \times 10^{14}$ Hz (one of the yellow lines in the visible Na spectrum)?

▮ The relation among energy, wavenumber ($\bar{\nu}$), wavelength (λ), and frequency (ν) is

$$E = h\nu = \frac{hc}{\lambda} = hc\bar{\nu} \qquad (13.32)$$

Substituting the data into *(13.22)* gives

$$E(\text{H}_2\text{O absorption}) = (6.626 \times 10^{-34}\ \text{J}\cdot\text{s})(2.9979 \times 10^{8}\ \text{m}\cdot\text{s}^{-1})(3651\ \text{cm}^{-1})\frac{100\ \text{cm}}{1\ \text{m}}$$

$$= 7.253 \times 10^{-20}\ \text{J}$$

$$E(\text{Cu K}_\alpha) = \frac{(6.626 \times 10^{-34}\ \text{J}\cdot\text{s})(2.9979 \times 10^{8}\ \text{m}\cdot\text{s}^{-1})}{(1.544\ \text{Å})[(10^{-10}\ \text{m})/(1\ \text{Å})]} = 1.287 \times 10^{-15}\ \text{J}$$

$$E(\text{Na}) = (6.626 \times 10^{-34}\ \text{J}\cdot\text{s})(5.090 \times 10^{14}\ \text{Hz})[(1\ \text{s}^{-1})/(1\ \text{Hz})] = 3.373 \times 10^{-19}\ \text{J}$$

The Cu K_α radiation has the greatest energy.

13.68 What is the wavelength of light absorbed by the C=O bond in acetone? Will a mercury light source operating at 254 nm be effective in breaking this bond?

▮ The $-\overset{|}{\text{C}}=\text{O}$ bond energy is 728 $\text{kJ}\cdot\text{mol}^{-1}$ (see Table 4-3). Solving *(13.32)* for λ and substituting the data gives

$$\lambda = \frac{(6.626 \times 10^{-34}\ \text{J}\cdot\text{s})(2.9979 \times 10^{8}\ \text{m}\cdot\text{s}^{-1})}{(728\ \text{kJ}\cdot\text{mol}^{-1})[(10^{3}\ \text{J})/(1\ \text{kJ})][(1\ \text{mol})/(6.022 \times 10^{23})]} = 1.64 \times 10^{-7}\ \text{m} = 164\ \text{nm}$$

Because λ and E are inversely related, the maximum wavelength needed is 164 nm. Thus the radiation of the mercury source will not be effective.

13.69 Assume that 1×10^{-17} J of light energy is needed by the interior of the human eye to "see" an object. How many photons of yellow light ($\lambda = 590$ nm) are needed to generate this minimum amount of energy?

▮ The energy of each photon is given by *(13.32)* as

$$E = \frac{(6.626 \times 10^{-34}\ \text{J}\cdot\text{s})(2.9979 \times 10^{8}\ \text{m}\cdot\text{s}^{-1})}{(590\ \text{nm})[(10^{-9}\ \text{m})/(1\ \text{nm})]} = 3.4 \times 10^{-19}\ \text{J}$$

The number of photons needed is

$$\frac{1 \times 10^{-17}\ \text{J}}{3.4 \times 10^{-19}\ \text{J/photon}} = 30\ \text{photons}$$

13.70 Consider the following mechanism for a photochemical reaction:

$$\text{A} + h\nu \xrightarrow{k_1} \text{A}^* \qquad \text{A}^* + \text{M} \xrightarrow{k_2} \text{A} + \text{M} \qquad \text{A}^* \xrightarrow{k_3} \text{B} + \text{C}$$

Derive an expression for the quantum yield (Φ).

▮ The rate of formation of A* in the first step is pseudo-zero-order in $C(\text{A})$ and directly proportional to the intensity of the light absorbed (I_a). Under steady-state conditions,

$$\frac{dC(\text{A}^*)}{dt} = k_1 I_a - k_2 C(\text{A}^*)C(\text{M}) - k_3 C(\text{A}^*) = 0$$

giving

$$C(\text{A}^*) = k_1 I_a/[k_3 + k_2 C(\text{M})]$$

The rate of formation of products is

$$\frac{dC(\text{B})}{dt} = k_3 C(\text{A}^*) = \frac{k_1 k_3 I_a}{k_3 + k_2 C(\text{M})}$$

The *quantum yield* is the number of molecules of reactant consumed or product generated per quantum absorbed.

Thus,

$$\Phi = \frac{dC(\mathrm{B})/dt}{I_a} = \frac{k_1 k_3 I_a/[k_3 + k_2 C(\mathrm{M})]}{I_a} = \frac{k_1 k_3}{k_3 + k_2 C(\mathrm{M})} \tag{13.33}$$

13.71 A popular actinometer for solution chemistry uses a solution of $K_3[Fe(C_2O_4)_3]$ in which the Fe^{3+} is reduced and the oxalate ion is oxidized photochemically. Assuming $\Phi = 1.24$ at 313 nm, calculate I_a needed to produce 1.3×10^{-5} mol of Fe^{2+} in a period of 36.5 min.

▌ The rate of formation of Fe^{2+} is

$$\frac{dn(\mathrm{Fe}^{2+})}{dt} = \frac{1.3 \times 10^{-5}\ \mathrm{mol}}{(36.5\ \mathrm{min})[(60\ \mathrm{s})/(1\ \mathrm{min})]} = 5.9 \times 10^{-9}\ \mathrm{mol \cdot s^{-1}}$$

Solving *(13.33)* for I_a and substituting the data gives

$$I_a = \frac{5.9 \times 10^{-9}\ \mathrm{mol \cdot s^{-1}}}{1.24} = 4.8 \times 10^{-9}\ \mathrm{mol \cdot s^{-1}}$$

13.72 A sample of CH_2CO was irradiated with the light source described in Problem 13.71 for a period of 15.2 min. If the quantum yield of C_2H_4 is 1.0 and that of CO is 2.0, determine the amount of each gas produced by the photochemical reaction.

▌ Solving *(13.33)* for the amount of product and substituting the data gives

$$n(\mathrm{CO}) = (2.0)(15.2\ \mathrm{min})[(60\ \mathrm{s})/(1\ \mathrm{min})](4.8 \times 10^{-9}\ \mathrm{mol \cdot s^{-1}}) = 8.8 \times 10^{-6}\ \mathrm{mol}$$

$$n(\mathrm{C_2H_4}) = (1.0)(15.2)(60)(4.8 \times 10^{-9}) = 4.4 \times 10^{-6}\ \mathrm{mol}$$

13.73 Aqueous solutions of $KMnO_4$ absorb light strongly at 522 nm. If $I/I_0 = 0.14$ for a 6.33×10^{-5} M solution in a 2.00-cm cell, determine the absorption coefficient of $KMnO_4$.

▌ The relation among the transmittance (I/I_0), the absorption coefficient (α), the concentration, and the cell path length (l) is given by the *Beer-Lambert-Bouguer law*

$$\ln (I_0/I) = (1000)[\alpha/(\mathrm{m^2 \cdot mol^{-1}})][C/(\mathrm{mol \cdot L^{-1}})][l/(\mathrm{m})] \tag{13.34}$$

Solving *(13.34)* for α and substituting the data gives

$$\alpha/(\mathrm{m^2 \cdot mol^{-1}}) = \frac{\ln (1/0.14)}{(1000)(6.33 \times 10^{-5})(2.00 \times 10^{-2})} = 1550$$

13.74 Using the results of Problem 13.73, predict the transmittance for a 1.58×10^{-5} M solution of $KMnO_4$ in the same cell.

▌ Substituting the data into *(13.34)* and taking antilogarithms gives

$$\ln (I_0/I) = (1000)(1550)(1.58 \times 10^{-5})(2.00 \times 10^{-2}) = 0.490$$

$$I_0/I = 1.63$$

Inverting to find the transmittance gives $I/I_0 = 1/1.63 = 0.61$.

13.75 The width of the absorption curve at half-height is approximately 100 nm for the $KMnO_4$ solution described in Problem 13.73. Estimate the value of the integrated absorption coefficient.

▌ The integrated absorption coefficient can be estimated by multiplying the absorption coefficient by the width of the band at half-height. The frequency range of the width is

$$\nu = \frac{c}{\lambda} = \frac{2.9979 \times 10^{8}\ \mathrm{m \cdot s^{-1}}}{(522 + 50\ \mathrm{nm})[(10^{-9}\ \mathrm{m})/(1\ \mathrm{nm})]} = 5.24 \times 10^{-14}\ \mathrm{s^{-1}}$$

$$\nu = \frac{2.9979 \times 10^{8}}{(522 - 50)(10^{-9})} = 6.35 \times 10^{14}\ \mathrm{s^{-1}}$$

$$\Delta\nu = 6.35 \times 10^{14}\ \mathrm{s^{-1}} - 5.24 \times 10^{14}\ \mathrm{s^{-1}} = 1.11 \times 10^{14}\ \mathrm{s^{-1}}$$

The estimated absorption coefficient is

$$\int \alpha\, d\nu = (1550\ \mathrm{m^2\ mol^{-1}})(1.11 \times 10^{14}\ \mathrm{s^{-1}}) = 1.72 \times 10^{17}\ \mathrm{m^2 \cdot mol^{-1} \cdot s^{-1}}$$

13.76 Determine the oscillator strength for the absorption described in Problem 13.73. Is this electronic transition "allowed" according to the oscillating electron model?

▮ The *oscillator strength* is the ratio of the experimental integrated absorption coefficient to the value of $\int \alpha\, d\nu = 1.598 \times 10^{18}\ \text{m}^2 \cdot \text{mol}^{-1} \cdot \text{s}^{-1}$. Using the results of Problem 13.75 gives

$$\text{Oscillator strength} = \frac{1.72 \times 10^{17}\ \text{m}^2 \cdot \text{mol}^{-1} \cdot \text{s}^{-1}}{1.598 \times 10^{18}\ \text{m}^2 \cdot \text{mol}^{-1} \cdot \text{s}^{-1}} = 0.108$$

Electronic transitions that have oscillator strengths near unity are like those allowed for the oscillating electron model, and those that have much smaller oscillator strengths (like the one for this system) suggest that this absorption is not completely allowed.

13.77 Assume that the emission of light from the absorption process described in Problem 13.73 is an example of spontaneous emission. Calculate the half-life for this emission.

▮ The rate equation describing spontaneous emission is

$$\left(\frac{dN_m}{dt}\right)_{\text{spont}} = A_{ml} N_m \tag{13.35}$$

where l and m describe the quantum states involved in the emission process and A_{ml} is the *Einstein coefficient* for spontaneous emission given by

$$A_{ml} = \frac{8\pi\nu^2}{c^2 L} \int \alpha\, d\nu \tag{13.36}$$

Recognizing that A_{ml} in *(13.35)* is similar to a first-order rate constant, *(12.7)* gives

$$t_{1/2} = \frac{-\ln 0.5}{A_{ml}} = \frac{1.49 \times 10^{39}\ \text{m}^2 \cdot \text{mol}^{-1} \cdot \text{s}^{-2}}{\nu^2 \int \alpha\, d\nu}$$

Substituting $\nu = 5.74 \times 10^{14}\ \text{s}^{-1}$ and the results of Problem 13.75 into the expression for $t_{1/2}$ gives

$$t_{1/2} = \frac{1.49 \times 10^{39}\ \text{m}^2 \cdot \text{mol}^{-1} \cdot \text{s}^{-2}}{(5.74 \times 10^{14}\ \text{s}^{-1})^2 (1.72 \times 10^{17}\ \text{m}^2 \cdot \text{mol}^{-1} \cdot \text{s}^{-1})} = 2.63 \times 10^{-8}\ \text{s}$$

13.78 Is the emission process described in Problem 13.77 an example of phosphorescence or of fluorescence?

▮ *Phosphorescence* corresponds to a transition between electronic states of different multiplicities (e.g., triplet excited state to a singlet ground state) and has typical lifetimes greater than 10^{-3} s. *Fluorescence* corresponds to a transition between electronic states of the same multiplicity and has typical lifetimes on the order of 10^{-8} s. Based on the results of Problem 13.77, this emission process is an example of fluorescence.

13.79 Consider the following mechanism describing the absorption and emission of light by Hg atoms in the presence of H_2:

Excitation: $\quad Hg + h\nu \longrightarrow Hg^*$

Fluorescence: $\quad Hg^* \xrightarrow{k_e} Hg + h\nu$

Quenching: $\quad Hg^* \xrightarrow{k_q} Hg + \text{thermal energy}$

$$Hg^* + H_2 \xrightarrow{k_2} Hg + 2H + \text{thermal energy}$$

Use the following lifetime data to determine k_e, k_q, and k_2:

$\tau/(10^{-7}\ \text{s})$	1.10	0.82	0.69	0.41	0.25
$C(H_2)/(10^{-5}\ \text{mol} \cdot \text{L}^{-1})$	0	1.0	2.0	5.0	10.0

▮ The net decay of Hg^* is

$$\frac{-dC(Hg^*)}{dt} = k_e C(Hg^*) + k_q C(Hg^*) + k_2 C(Hg^*) C(H_2) = [k_e + k_q + k_2 C(H_2)] C(Hg^*)$$

which is first-order in $C(Hg^*)$. The time required for the concentration to fall to $1/e$ of its initial value is represented by the symbol τ and is given by *(12.7)* as

$$\tau = \frac{-\ln(1/e)}{k_e + k_q + k_2C(H_2)} = \frac{1}{k_e + k_q + k_2C(H_2)} \tag{13.37}$$

This equation can be transformed into

$$1/\tau = 1/\tau_0 + k_2C(H_2)$$

where

$$\tau_0 = 1/(k_e + k_q)$$

Plotting values of $1/\tau$ against $C(H_2)$ gives the straight line shown in Fig. 13-15. From the plot,

$$k_2 = \text{slope} = 3.11 \times 10^{11}\ \text{L} \cdot \text{mol}^{-1} \cdot \text{s}^{-1}$$

$$k_e + k_q = \text{intercept} = 0.89 \times 10^7\ \text{s}^{-1}$$

(The individual values of k_e and k_q cannot be determined from the data given.)

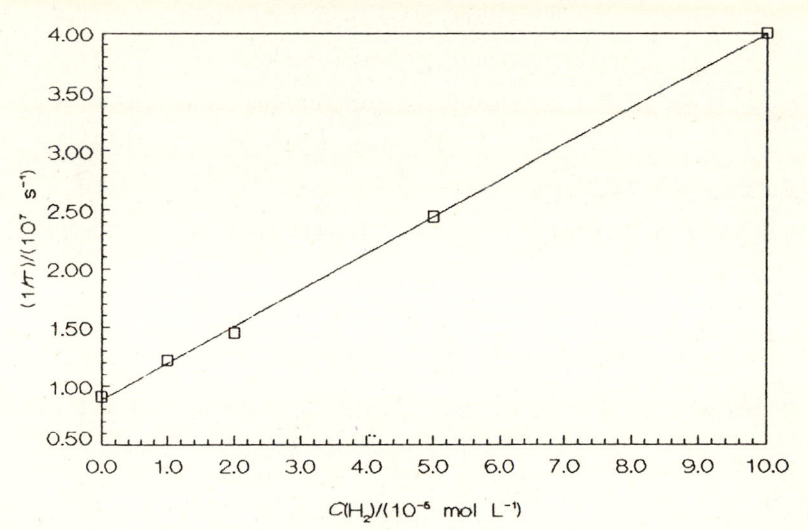

Fig. 13-15

13.80 Determine the value of the Stern–Volmer constant for the process described in Problem 13.79.

▌ The fluorescence signal (ϕ) at a concentration of quencher $C(Q)$ is related to the fluorescence signal in the absence of quencher (ϕ_0) and the rate constants given in Problem 13.79 by

$$\frac{\phi_0}{\phi} - 1 = \frac{k_2}{k_e + k_q} C(Q) = K_{SV}C(Q) \tag{13.38}$$

where K_{SV} is the *Stern–Volmer constant.* Substituting the results of Problem 13.79 into *(13.38)* and solving for K_{SV} gives

$$K_{SV} = \frac{k_2}{k_e + k_q} = \frac{3.11 \times 10^{11}\ \text{L} \cdot \text{mol}^{-1} \cdot \text{s}^{-1}}{0.89 \times 10^{-7}\ \text{s}^{-1}} = 3.49 \times 10^{18}\ \text{L} \cdot \text{mol}^{-1}$$

13.81 Determine the quenching cross section for the process described in Problem 13.79.

▌ The *quenching cross section* is given by

$$\sigma_q = \frac{k_2}{10^3 L}\left(\frac{\pi\mu}{8kT}\right)^{1/2} \tag{13.39}$$

where μ is given by *(2.23)*. For the Hg–H_2 system,

$$\mu = \frac{(2.02 \times 10^{-3}\ \text{kg} \cdot \text{mol}^{-1})(200.59 \times 10^{-3}\ \text{kg} \cdot \text{mol}^{-1})}{(2.02 \times 10^{-3}\ \text{kg} \cdot \text{mol}^{-1} + 200.59 \times 10^{-3}\ \text{kg} \cdot \text{mol}^{-1})(6.022 \times 10^{23}\ \text{mol}^{-1})} = 3.32 \times 10^{-27}\ \text{kg}$$

Substituting into *(13.39)* gives

$$\sigma_q = \frac{3.11 \times 10^{11}\ \text{L} \cdot \text{mol}^{-1} \cdot \text{s}^{-1}}{[(10^3\ \text{L})/(1\ \text{m}^3)](6.022 \times 10^{23}\ \text{mol}^{-1})}\left(\frac{\pi(3.32 \times 10^{-27}\ \text{kg})[(1\ \text{J})/(1\ \text{kg} \cdot \text{m}^2 \cdot \text{s}^{-2})]}{8(1.381 \times 10^{-23}\ \text{J} \cdot \text{K}^{-1})(298\ \text{K})}\right)^{1/2} = 2.91 \times 10^{-19}\ \text{m}^2$$

CHAPTER 14

Introduction to Quantum Mechanics

14.1 PRELIMINARIES

14.1 Determine the kinetic energy of a photoemitted electron from the surface of Cs if the wavelength of the incident light is 525 nm. The work function for Cs is 2.14 eV. What is the velocity of the electron?

▌ The relation between frequency (ν) and wavelength (λ) of light is given by

$$\nu\lambda = c \tag{14.1}$$

where c is the velocity of light. Solving *(14.1)* for ν and substituting the data gives

$$\nu = \frac{2.9979 \times 10^8 \text{ m} \cdot \text{s}^{-1}}{(525 \text{ nm})[(10^{-9} \text{ m})/(1 \text{ nm})]} = 5.71 \times 10^{14} \text{ s}^{-1}$$

The kinetic energy of the electron is related to the work function (Φ) and ν by

$$E = \tfrac{1}{2}mv^2 = h\nu - \Phi \tag{14.2}$$

Substituting the data into *(14.2)* gives

$$E = (6.626 \times 10^{-34} \text{ J} \cdot \text{s})(5.71 \times 10^{14} \text{ s}^{-1}) - (2.14 \text{ eV})[(1.602 \times 10^{-19} \text{ J})/(1 \text{ eV})] = 3.5 \times 10^{-20} \text{ J}$$

Solving *(14.2)* for v and substituting the data gives

$$v = \left(\frac{2(3.5 \times 10^{-20} \text{ J})(1 \text{ kg} \cdot \text{m}^2 \cdot \text{s}^{-2})/(1 \text{ J})}{9.11 \times 10^{-31} \text{ kg}}\right)^{1/2} = 2.8 \times 10^5 \text{ m} \cdot \text{s}^{-1}$$

14.2 The following data were collected for the photoelectric emission of an electron from Ca:

$\lambda/(\text{nm})$	253.6	313.2	365.0	404.7
$E/(\text{eV})$	1.95	0.98	0.50	0.14

Determine the work function and the value of Planck's constant.

▌ According to *(14.2)*, the intercept of a plot of E against ν will be equal to $-\Phi$, and the slope will be equal to h. From Fig. 14-1,

$$\Phi = -\text{intercept} = -(-4.59 \times 10^{-19} \text{ J}) = 4.59 \times 10^{-19} \text{ J} = 2.86 \text{ eV}$$

$$k = \text{slope} = 6.5 \times 10^{-34} \text{ J} \cdot \text{s}$$

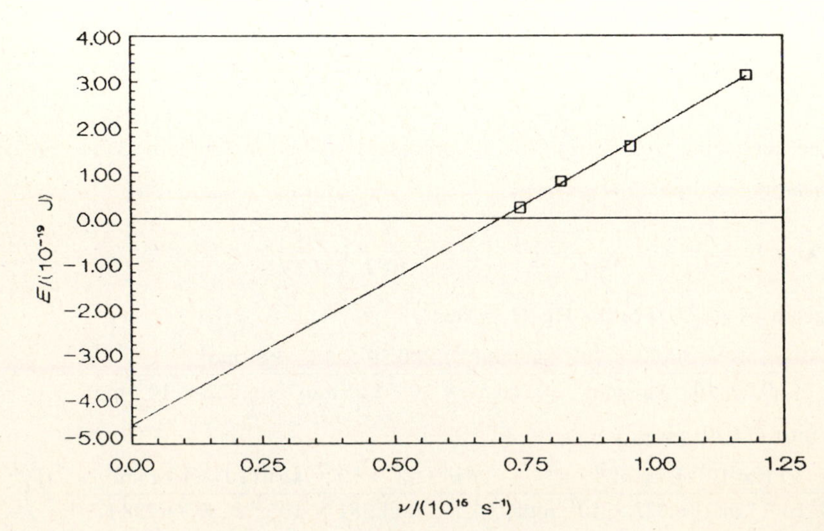

Fig. 14-1

14.3 What are the minimum frequency and maximum wavelength of light that will dislodge an electron from Ca? See Problem 14.2 for additional data.

▌ Setting $E \geq 0$ and solving *(14.2)* for ν gives

$$\nu \geq \frac{4.59 \times 10^{-19}\ \text{J}}{6.626 \times 10^{-34}\ \text{J} \cdot \text{s}} = 6.93 \times 10^{14}\ \text{s}^{-1}$$

which upon substitution into *(14.1)* gives

$$\nu \leq \frac{2.9979 \times 10^{8}\ \text{m} \cdot \text{s}^{-1}}{6.93 \times 10^{14}\ \text{s}^{-1}} = 4.33 \times 10^{-7} = 433\ \text{nm}$$

14.4 The *radiant energy density* (ρ_ν) for blackbody radiation is given by

$$\rho_\nu = \frac{8\pi h\nu^3}{c^3(e^{h\nu/kT} - 1)} \tag{14.3}$$

where $\rho_\nu\,d\nu$ is the energy per unit volume in the frequency range between ν and $\nu + d\nu$. Prepare a plot of the spectral concentration of the radiant exitance as a function of frequency for a blackbody at $T = 5776$ K (the temperature of the sun).

▌ The *spectral concentration of the radiant exitance* (M_ν) is the rate of emission of energy per unit area per unit time in the frequency range between ν and $\nu + d\nu$ and is given by

$$M_\nu = \frac{c}{4}\rho_\nu = \frac{2\pi h\nu^3}{c^2(e^{h\nu/kT} - 1)} \tag{14.4}$$

As a sample calculation, substituting $\nu = 3.50 \times 10^{14}\ \text{s}^{-1}$ into *(14.4)* gives

$$M_\nu = \frac{2\pi(6.626 \times 10^{-34}\ \text{J} \cdot \text{s})(3.50 \times 10^{14}\ \text{s}^{-1})^3}{(2.9979 \times 10^{8}\ \text{m} \cdot \text{s}^{-1})^2\left(\exp\left[\dfrac{(6.626 \times 10^{-34}\ \text{J} \cdot \text{s})(3.50 \times 10^{14}\ \text{s}^{-1})}{(1.381 \times 10^{-23}\ \text{J} \cdot \text{K}^{-1})(5776\ \text{K})}\right] - 1\right)} = 1.15 \times 10^{-7}\ \text{J} \cdot \text{m}^{-2}$$

The plot is shown in Fig. 14-2.

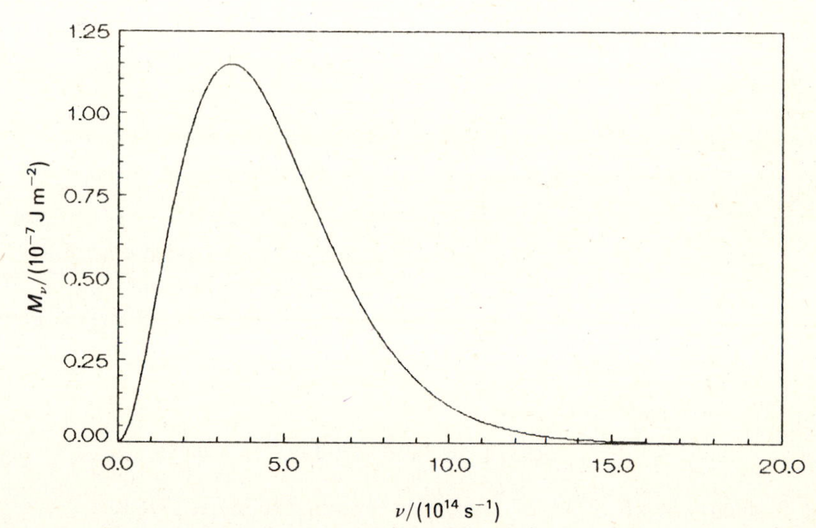

Fig. 14-2

14.5 Using *(14.3)* and *(14.4)*, derive expressions for ρ_λ and M_λ.

▌ The radiant energy density in terms of wavelength is represented by ρ_λ and is related to ρ_ν by

$$\rho_\lambda\,d\lambda = -\rho_\nu\,d\nu \tag{14.5}$$

where the negative sign accounts for the inverse relation between λ and ν. Solving *(14.5)* for ρ_λ and substituting *(14.1)* and *(14.3)* gives

$$\rho_\lambda = -\rho_\nu\frac{d\nu}{d\lambda} = \frac{-8\pi h\nu^3}{c^3(e^{h\nu/kT} - 1)}\,\frac{d(c/\lambda)}{d\lambda} = \frac{-8\pi h(c/\lambda)^3}{c^3(e^{h(c/\lambda)/kT} - 1)}\left(\frac{-c}{\lambda^2}\right) = \frac{8\pi hc}{\lambda^5(e^{hc/\lambda kT} - 1)} \tag{14.6}$$

As in Problem 14.4,

$$M_\lambda = \frac{c}{4}\rho_\lambda = \frac{2\pi hc^2}{\lambda^5(e^{hc/\lambda kT} - 1)} \tag{14.7}$$

14.6 Calculate the radiant exitance for a blackbody at $T = 5776$ K.

▌ The radiant exitance (M) is the total energy emitted per unit time per unit area and is given by

$$M = \int_0^\infty M_\lambda\, d\lambda = \int_0^\infty M_\nu\, d\nu \tag{14.8}$$

Substituting *(14.4)* into *(14.8)* gives

$$M = \frac{2\pi h}{c^2}\int_0^\infty \frac{\nu^3}{e^{h\nu/kT} - 1}\, d\nu = \frac{2\pi h}{c^2}\left(\frac{kT}{h}\right)^4 \int_0^\infty \frac{x^3}{e^x - 1}\, dx$$

where $x = h\nu/kT$. The integral is equal to $\pi^4/15$, giving

$$M = \frac{2\pi h}{c^2}\left(\frac{kT}{h}\right)^4\left(\frac{\pi^4}{15}\right) = \frac{2\pi^5 k^4}{15c^2h^3}T^4 = \sigma T^4 \tag{14.9}$$

where $\sigma = 5.670 \times 10^{-8}\ \mathrm{W \cdot m^{-2} \cdot K^{-4}}$ is the *Stefan–Boltzmann constant.* Substituting the data into *(14.9)* gives

$$M = (5.670 \times 10^{-8}\ \mathrm{W \cdot m^{-2} \cdot K^{-4}})(5776\ \mathrm{K})^4 = 6.311 \times 10^7\ \mathrm{W \cdot m^{-2}}$$

14.7 What is the radiant power of the sun given $r = 6.960 \times 10^8$ m?

▌ The radiant power (Φ) can be determined by

$$\Phi = MA \tag{14.10}$$

Substituting the results of Problem 14.6 into *(14.10)* gives

$$\Phi = (6.311 \times 10^7\ \mathrm{W \cdot m^{-2}})[4\pi(6.960 \times 10^8\ \mathrm{m})^2] = 3.842 \times 10^{26}\ \mathrm{W}$$

14.8 Determine the fraction of the radiant energy density emitted by the sun in the visible spectral range (400–700 nm).

▌ The total radiant energy density is given by

$$\rho_\lambda = \frac{4}{c}M = \frac{4(6.311 \times 10^{-7}\ \mathrm{W \cdot m^{-2}})[(1\ \mathrm{J})/(1\ \mathrm{W \cdot s})]}{(2.9979 \times 10^8\ \mathrm{m \cdot s^{-1}})} = 0.8420\ \mathrm{J \cdot m^{-3}}$$

The radiant energy density in the desired wavelength range can be approximated by

$$\begin{aligned}\rho_\lambda\, d\lambda \approx \rho_\lambda\, \Delta\lambda &= \frac{8\pi hc\, \Delta\lambda}{\bar{\lambda}^5(e^{hc/\bar{\lambda}kT} - 1)} \\ &= \frac{8\pi(6.626 \times 10^{-34}\ \mathrm{J \cdot s})(2.9979 \times 10^8\ \mathrm{m \cdot s^{-1}})(300 \times 10^{-9}\ \mathrm{m})}{(550 \times 10^{-9}\ \mathrm{m})^5\left(\exp\left[\dfrac{(6.626 \times 10^{-34}\ \mathrm{J \cdot s})(2.9979 \times 10^8\ \mathrm{m \cdot s^{-1}})}{(550 \times 10^{-9}\ \mathrm{m})(1.381 \times 10^{-23}\ \mathrm{J \cdot K^{-1}})(5776\ \mathrm{K})}\right] - 1\right)} \\ &= 0.325\ \mathrm{J \cdot m^{-3}}\end{aligned}$$

The fraction in the visible spectral range is

$$(0.325\ \mathrm{J \cdot m^{-3}})/(0.8420\ \mathrm{J \cdot m^{-3}}) = 0.386$$

14.9 Determine the frequency at which M_ν is a maximum for a blackbody at 5776 K.

▌ Differentiating *(14.4)* with respect to ν gives

$$\frac{dM_\nu}{d\nu} = \frac{2\pi h}{c^2}\left(\frac{3\nu^2}{(e^{h\nu/kT} - 1)} + \frac{-\nu^3 e^{h\nu/kT}(h/kT)}{(e^{h\nu/kT} - 1)^2}\right)$$

Setting the derivative equal to zero at $\nu = \nu_{\nu,\mathrm{max}}$ gives

$$\frac{2\pi h\nu_{\nu,\mathrm{max}}^2}{c^2(e^{h\nu_{\nu,\mathrm{max}}/kT} - 1)}\left(3 - \frac{x e^x}{e^x - 1}\right) = 0$$

where $x = h\nu_{\nu,\mathrm{max}}/kT$. Rearranging and solving the expression for x gives

$$e^x(x - 3) + 3 = 0 \qquad x = 2.821\,439$$

The expression for the frequency at which M_ν is a maximum is

$$\nu_{\nu,\max} = \frac{2.821\,439\,kT}{h} = (5.878\,89 \times 10^{10}\,\mathrm{s^{-1} \cdot K^{-1}})T \tag{14.11}$$

Substituting the data into *(14.11)* gives

$$\nu_{\nu,\max} = (5.879 \times 10^{10}\,\mathrm{s^{-1} \cdot K^{-1}})(5776\,\mathrm{K}) = 3.396 \times 10^{14}\,\mathrm{s^{-1}}$$

which agrees with the maximum of the curve shown in Fig. 14-2.

14.10 Determine the wavelength at which M_λ is a maximum for a blackbody at 5776 K. Compare this wavelength to the frequency determined in Problem 14.9 for the maximum of M_ν.

▌ Proceeding as in Problem 14.9 gives

$$\frac{dM_\lambda}{d\lambda} = 2\pi hc^2\left(\frac{-5}{\lambda^6(e^{hc/\lambda kT}-1)} + \frac{-e^{hc/\lambda kT}(-hc/\lambda^2 kT)}{\lambda^5(e^{hc/\lambda kT}-1)^2}\right)$$

$$\frac{2\pi hc^2}{\lambda^6_{\lambda,\max}(e^{hc/\lambda kT}-1)}\left(-5 + \frac{x\,e^x}{e^x - 1}\right) = 0$$

where $x = hC/\lambda_{\lambda,\max}kT$. Rearranging and solving the expression for x gives

$$e^x(x-5) + 5 = 0 \qquad x = 4.965\,114$$

The expression for the wavelength at which M_λ is a maximum is

$$\lambda_{\lambda,\max} = \frac{hc}{4.965\,114kT} = \frac{2.897\,79 \times 10^{-3}\,\mathrm{m \cdot K}}{T} \tag{14.12}$$

and is known as *Wein's displacement law*. Substituting the data into *(14.12)* gives

$$\lambda_{\lambda,\max} = \frac{2.897\,79 \times 10^{-3}\,\mathrm{m \cdot K}}{5776\,\mathrm{K}} = 5.017 \times 10^{-7}\,\mathrm{m} = 501.7\,\mathrm{nm}$$

Substituting into *(14.1)* gives

$$\nu_{\lambda,\max} = \frac{2.9979 \times 10^8\,\mathrm{m \cdot s^{-1}}}{(501.7\,\mathrm{nm})[(10^{-9}\,\mathrm{m})/(1\,\mathrm{nm})]} = 5.976 \times 10^{14}\,\mathrm{s^{-1}}$$

Note that $\nu_{\lambda,\max} \neq \nu_{\nu,\max}$ because of the way in which M_ν and M_λ are defined.

14.11 Compare the value of M_λ for a blackbody at 2000 K to that for a blackbody at 1500 K at $\lambda = 1.00\,\mu\mathrm{m}$.

▌ Substituting into *(14.7)* gives

$$\frac{M_\lambda(2000\,\mathrm{K})}{M_\lambda(1500\,\mathrm{K})} = \frac{2\pi hc^2/\lambda^5(e^{hc/\lambda k(2000\,\mathrm{K})}-1)}{2\pi hc^2/\lambda^5(e^{hc/\lambda k(1500\,\mathrm{K})}-1)} = \frac{e^{hc/\lambda k(1500\,\mathrm{K})}-1}{e^{hc/\lambda k(2000\,\mathrm{K})}-1}$$

$$= \frac{\exp\left(\dfrac{(6.626 \times 10^{-34}\,\mathrm{J \cdot s})(2.9979 \times 10^8\,\mathrm{m \cdot s^{-1}})}{(1.00 \times 10^{-6}\,\mathrm{m})(1.381 \times 10^{-23}\,\mathrm{J \cdot K^{-1}})(1500\,\mathrm{K})}\right) - 1}{\exp\left(\dfrac{(6.626 \times 10^{-34}\,\mathrm{J \cdot s})(2.9979 \times 10^8\,\mathrm{m \cdot s^{-1}})}{(1.00 \times 10^{-6}\,\mathrm{m})(1.381 \times 10^{-23}\,\mathrm{J \cdot K^{-1}})(2000\,\mathrm{K})}\right) - 1} = 11.0$$

14.12 Compare the maximum value of M_λ for a backbody at 2000 K to that for a blackbody at 1500 K.

▌ Substituting *(14.12)* into *(14.7)* gives

$$M_{\lambda,\max} = \frac{2\pi hc^2}{[hc/(4.965\,114)kT]^5(e^{4.965114}-1)} = \frac{2\pi(4.965\,114)(kT)^5}{h^4c^3(e^{4.965114}-1)}$$

Taking a ratio gives

$$\frac{M_{\lambda,\max}(2000\,\mathrm{K})}{M_{\lambda,\max}(1500\,\mathrm{K})} = \frac{2\pi(4.965\,114)k^5(2000\,\mathrm{K})^5/h^4c^3(e^{4.965114}-1)}{2\pi(4.965\,114)k^5(1500\,\mathrm{K})^5/h^4c^3(e^{4.965114}-1)} = \frac{(2000\,\mathrm{K})^5}{(1500\,\mathrm{K})^5} = 4.21$$

14.13 What is the de Broglie wavelength associated with a He atom at room temperature?

▌ Substituting into *(1.62)* gives

$$v_{\mathrm{rms}} = \left(\frac{3(8.314\,\mathrm{J \cdot K^{-1} \cdot mol^{-1}})[(1\,\mathrm{kg \cdot m^2 \cdot s^{-2}})/(1\,\mathrm{J})](298\,\mathrm{K})}{4.00 \times 10^{-3}\,\mathrm{kg}}\right)^{1/2} = 1360\,\mathrm{m \cdot s^{-1}}$$

The *de Broglie wavelength* is given by

$$\lambda = h/p = h/mv \tag{14.13}$$

where p is the linear momentum. Substituting into *(14.13)* gives

$$\lambda = \frac{(6.626 \times 10^{-34}\ \mathrm{J \cdot s})[(1\ \mathrm{kg \cdot m^2 \cdot s^{-2}})/(1\ \mathrm{J})]}{[(4.00 \times 10^{-3}\ \mathrm{kg \cdot mol^{-1}})/(6.022 \times 10^{23}\ \mathrm{mol^{-1}})](1360\ \mathrm{m \cdot s^{-1}})} = 7.34 \times 10^{-11}\ \mathrm{m}$$

This wavelength is slightly smaller than the atomic diameter of a He atom (0.05 nm).

14.14 What is the de Broglie wavelength associated with an electron accelerated by a voltage of 100 keV?

▮ The energy of the electron is

$$E = (100\ \mathrm{keV})\left(\frac{10^3\ \mathrm{eV}}{1\ \mathrm{keV}}\right)\left(\frac{1.602 \times 10^{-19}\ \mathrm{J}}{1\ \mathrm{eV}}\right) = 1.6 \times 10^{-14}\ \mathrm{J}$$

which upon substitution into *(15.9)* gives

$$v_{\mathrm{rms}} = \left(\frac{2(1.6 \times 10^{-14}\ \mathrm{J})[(1\ \mathrm{kg \cdot m^2 \cdot s^{-2}})/(1\ \mathrm{J})]}{9.11 \times 10^{-31}\ \mathrm{kg}}\right)^{1/2} = 1.9 \times 10^8\ \mathrm{m \cdot s^{-1}}$$

Because the calculated value of v_{rms} is $0.63c$, relativistic theory should be used, where

$$\begin{aligned} v_{\mathrm{rms}} &= c\left[1 - \left(\frac{mc^2}{E + mc^2}\right)^2\right]^{1/2} \\ &= (2.9979 \times 10^8\ \mathrm{m \cdot s^{-1}}) \\ &\quad \times \left[1 - \left(\frac{(9.11 \times 10^{-31}\ \mathrm{kg})(2.9979 \times 10^8\ \mathrm{m \cdot s^{-1}})^2}{(1.6 \times 10^{-14}\ \mathrm{J})\left[\dfrac{1\ \mathrm{kg \cdot m^2 \cdot s^{-2}}}{1\ \mathrm{J}}\right] + (9.11 \times 10^{-31}\ \mathrm{kg})(2.9979 \times 10^8\ \mathrm{m \cdot s^{-1}})^2}\right)^2\right]^{1/2} \\ &= 1.6 \times 10^8\ \mathrm{m \cdot s^{-1}} \end{aligned}$$

Substituting into *(14.13)* gives

$$\lambda = \frac{(6.626 \times 10^{-34}\ \mathrm{J \cdot s})[(1\ \mathrm{kg \cdot m^2 \cdot s^{-2}})/(1\ \mathrm{J})]}{(9.11 \times 10^{-31}\ \mathrm{kg})(1.6 \times 10^8\ \mathrm{m \cdot s^{-1}})} = 4.5 \times 10^{-12}\ \mathrm{m} = 4.5\ \mathrm{pm}$$

14.15 Yellow light with $\lambda = 590\ \mathrm{nm}$ passes through two slits, and the interference pattern is observed on a screen placed 1.00 m from the slits. The repeat distance of the pattern is 0.594 mm. Determine the distance between the slits.

▮ The repeat distance of the pattern (D) is related to the slit-screen distance (l) and the angle between the pattern from the axis of the system (θ) by

$$D = l \tan \theta \tag{14.14}$$

and the wavelength is related to the slit width (d) by

$$\lambda = d \sin \theta \tag{14.15}$$

Substituting the data into *(14.14)* and solving for θ gives

$$\theta = \tan^{-1}\left(\frac{(0.594\ \mathrm{mm})[(10^{-3}\ \mathrm{m})/(1\ \mathrm{mm})]}{1.00\ \mathrm{m}}\right) = 3.40 \times 10^{-2}$$

which upon substitution into *(14.15)* gives

$$d = \frac{(590\ \mathrm{nm})[(10^{-9}\ \mathrm{m})/(1\ \mathrm{nm})]}{\sin(3.40 \times 10^{-2})} = 9.9 \times 10^{-4}\ \mathrm{m} = 0.99\ \mathrm{mm}$$

14.16 A beam of electrons with a velocity of $1.00 \times 10^6\ \mathrm{m \cdot s^{-1}}$ passes through a slit of width 300 nm and forms a diffraction pattern on a screen placed 1.00 m from the slit. Determine the width of the primary diffraction pattern.

▮ The wavelength of the electrons is given by *(14.13)* as

$$\lambda = \frac{(6.626 \times 10^{-34}\ \mathrm{J \cdot s})[(1\ \mathrm{kg \cdot m^2 \cdot s^{-2}})/(1\ \mathrm{J})]}{(9.11 \times 10^{-31}\ \mathrm{kg})(1.00 \times 10^6\ \mathrm{m \cdot s^{-1}})} = 7.27 \times 10^{-10}\ \mathrm{m}$$

The equations describing diffraction through a slit are similar to *(14.14)* and *(14.15)* except that D is replaced by

$W/2$, where W is the width of the pattern. Solving *(14.15)* for θ and substituting the data gives

$$\theta = \tan^{-1}\left[\frac{7.27 \times 10^{-10}\,\text{m}}{(300\,\text{nm})[(10^{-9}\,\text{m})/(1\,\text{nm})]}\right] = 0.139$$

which upon substitution into *(14.14)* gives

$$W = 2(1.00\,\text{m})\tan 0.139 = 4.85 \times 10^{-3}\,\text{m} = 4.85\,\text{mm}$$

14.17 Suppose the uncertainty in the position of the electron described in Problem 14.14 was ±0.01 nm. What is the uncertainty in the momentum?

▌ For simultaneous measurements of position and momentum, the respective uncertainties are related by the *Heisenberg uncertainty principle*, where

$$\Delta p_x\, \Delta x \geq \hbar/2 \tag{14.16}$$

where $\hbar = h/2\pi$. Solving *(14.16)* for Δp_x and substituting the data gives

$$\Delta p_x \geq \frac{(1.0546 \times 10^{-34}\,\text{J}\cdot\text{s})[(1\,\text{kg}\cdot\text{m}^2\cdot\text{s}^{-2})/(1\,\text{J})]}{2(\pm 0.01\,\text{nm})[(10^{-9}\,\text{m})/(1\,\text{nm})]} = \pm 5 \times 10^{-24}\,\text{kg}\cdot\text{m}\cdot\text{s}^{-1}$$

Note that the uncertainty is ~3% of the momentum.

$$p_x = (9.11 \times 10^{-31}\,\text{kg})(1.6 \times 10^{8}\,\text{m}\cdot\text{s}^{-1}) = 1.5 \times 10^{-22}\,\text{kg}\cdot\text{m}\cdot\text{s}^{-1}$$

14.2 THE BOHR ATOM

14.18 Many lines of the Balmer series for atomic hydrogen appear in the visible range of the spectrum. What is the limiting wavelength for this series?

▌ The various series of lines in the spectra of atomic hydrogen are described by the *Rydberg equation*

$$\bar{\nu} = Z^2\mathcal{R}\left(\frac{1}{n_2^2} - \frac{1}{n_1^2}\right) \tag{14.17}$$

where Z is the atomic number, $n_1 > n_2$, and

$$\mathcal{R} = (109\,737.3177\,\text{cm}^{-1})\frac{m(\text{nucleus})}{m(\text{nucleus}) + m(\text{electron})} \tag{14.18}$$

Substituting data for H into *(14.18)* gives

$$\mathcal{R} = (109\,737.3177\,\text{cm}^{-1})\frac{1.672\,65 \times 10^{-27}\,\text{kg}}{1.672\,65 \times 10^{-27}\,\text{kg} + 9.109\,53 \times 10^{-31}\,\text{kg}} = 109\,678\,\text{cm}^{-1}$$

For the series limit of the Balmer series, $n_2 = 2$ and $n_1 = \infty$. Substituting into *(14.17)* gives

$$\bar{\nu} = (1)^2(109\,678\,\text{cm}^{-1})\left(\frac{1}{2^2} - \frac{1}{\infty^2}\right) = 27\,419.5\,\text{cm}^{-1}$$

Solving *(13.32)* for λ gives

$$\lambda = \frac{1}{\bar{\nu}} = \frac{(10^{-2}\,\text{m})/(1\,\text{cm})}{27\,419.5\,\text{cm}^{-1}} = 3.647\,04 \times 10^{-7}\,\text{m} = 364.704\,\text{nm}$$

14.19 In what part of the spectrum will the Lyman series ($n_2 = 1$) and the Paschen series ($n_2 = 3$) for atomic H appear?

▌ As in Problem 14.18, the series limits will be

$$\bar{\nu}(\text{Lyman}) = (1)^2(109\,678\,\text{cm}^{-1})\left(\frac{1}{1^2} - \frac{1}{\infty^2}\right) = 109\,678\,\text{cm}^{-1}$$

$$\bar{\nu}(\text{Paschen}) = (1)^2(109\,678)\left(\frac{1}{3^2} - \frac{1}{\infty^2}\right) = 12\,186.4\,\text{cm}^{-1}$$

$$\lambda(\text{Lyman}) = \frac{(10^{-2}\,\text{m})/(1\,\text{cm})}{109\,678\,\text{cm}^{-1}} = 9.1176 \times 10^{-8}\,\text{m} = 91.176\,\text{nm}$$

$$\lambda(\text{Paschen}) = \frac{10^{-2}}{12\,186.4} = 8.2059 \times 10^{-7}\,\text{m} = 820.59\,\text{nm}$$

which fall into the ultraviolet and near-infrared portions, respectively, of the spectrum.

14.20 Determine the ionization energy of atomic H.

▮ The ionization process $H(g) \rightarrow H^+(g) + e^-$ correspond to the $n_2 = 1$ to $n_1 = \infty$ transition in *(14.17)*. Substituting $\mathscr{R} = 109\,678\ \text{cm}^{-1}$ (see Problem 14.18) into *(14.17)* gives

$$\tilde{\nu} = (1)^2(109\,678\ \text{cm}^{-1})\left(\frac{1}{1^2} - \frac{1}{\infty^2}\right) = 109\,678\ \text{cm}^{-1}$$

Substituting into *(13.32)* gives

$$E = (6.626\,18 \times 10^{-34}\ \text{J}\cdot\text{s})(2.997\,925 \times 10^8\ \text{m}\cdot\text{s}^{-1})(109\,678\ \text{cm}^{-1})\frac{10^2\ \text{cm}}{1\ \text{m}} = 2.178\,73 \times 10^{-18}\ \text{J}$$

14.21 At what temperature will the translational kinetic energy of atomic H equal that for the $n_2 = 1$ to $n_1 = 2$ transition?

▮ Substituting $\mathscr{R} = 109\,678\ \text{cm}^{-1}$ (see Problem 14.18) into *(14.17)* gives

$$\tilde{\nu} = (1)^2(109\,678\ \text{cm}^{-1})\left(\frac{1}{1^2} - \frac{1}{2^2}\right) = 82\,259\ \text{cm}^{-1}$$

Substituting into *(13.32)* gives

$$E = (6.626 \times 10^{-34}\ \text{J}\cdot\text{s})(2.997\,925 \times 10^8\ \text{m}\cdot\text{s}^{-1})(82\,259\ \text{cm}^{-1})\frac{10^2\ \text{cm}}{1\ \text{m}} = 1.634\,05 \times 10^{-18}\ \text{J}$$

Solving *(1.59)* for T and substituting the data gives

$$T = \frac{2(1.634\,05 \times 10^{-18}\ \text{J})}{3(1.380\,66 \times 10^{-23}\ \text{J}\cdot\text{K}^{-1})} = 7.8902 \times 10^4\ \text{K}$$

14.22 What is the ratio of the number of atoms with $n_1 = 2$ to that with $n_2 = 1$ at 298.15 K?

▮ Substituting $E = 1.634\,05 \times 10^{-18}\ \text{J}$ (see Problem 14.21) into *(2.22)* gives

$$\frac{N(n_1 = 2)}{N(n_2 = 1)} = \frac{1}{1}\exp\left(-\frac{1.634\,05 \times 10^{-18}\ \text{J}}{(1.380\,66 \times 10^{-23}\ \text{J}\cdot\text{K}^{-1})(298.15\ \text{K})}\right) = e^{-396.96} = 4.2 \times 10^{-173}$$

14.23 What will be the difference in the wavelength of the $n_2 = 1$ to $n_1 = 2$ transition between H and deuterium? For D, $m(\text{nucleus}) = 3.343\,398 \times 10^{-27}$ kg.

▮ Substituting the results of Problem 14.21 into *(13.32)* gives

$$\lambda(\text{H}) = \frac{(10^{-2}\ \text{m})/(1\ \text{cm})}{82\,259\ \text{cm}^{-1}} = 1.215\,67 \times 10^{-7}\ \text{m} = 121.567\ \text{nm}$$

Substituting the data into *(14.18)* gives

$$\mathscr{R}(\text{D}) = (109\,737.3177\ \text{cm}^{-1})\frac{3.343\,398 \times 10^{-27}\ \text{kg}}{3.343\,398 \times 10^{-27}\ \text{kg} + 9.109\,53 \times 10^{-31}\ \text{kg}} = 109\,707\ \text{cm}^{-1}$$

Substituting into *(14.17)* gives

$$\tilde{\nu}(\text{D}) = (1)^2(109\,707\ \text{cm}^{-1})\left(\frac{1}{1^2} - \frac{1}{2^2}\right) = 82\,280\ \text{cm}^{-1}$$

which corresponds to

$$\lambda(\text{D}) = \frac{10^{-2}}{82\,280} = 1.215\,36 \times 10^{-7}\ \text{m} = 121.536\ \text{nm}$$

The line for the D transition will lie 0.031 nm below that for the H transition.

14.24 Compare the ionization energies of H and He^+.

▮ Assuming $\mathcal{R}(H) \approx \mathcal{R}(He^+)$, *(14.17)* gives

$$\frac{\bar{\nu}(He^+)}{\bar{\nu}(H)} = \frac{2^2\mathcal{R}(He^+)(1/1^2 - 1/\infty^2)}{1^2\mathcal{R}(H)(1/1^2 - 1/\infty^2)} = 4$$

which upon substituting into *(13.32)* gives

$$\frac{E(He^+)}{E(H)} = \frac{hc\nu(He^+)}{hc\nu(H)} = 4$$

The ionization energy of He^+ is nearly four times as great as that of H.

14.25 Calculate the first four Bohr radii of an atom of H.

▮ The *Bohr radii* are given by

$$r_n = n^2\frac{4\pi\varepsilon_0\hbar^2}{\mu e^2 Z} = \frac{(5.294\,653 \times 10^{-11}\text{ m})n^2}{Z} \tag{14.19}$$

where μ is given by *(2.23)*. Substituting into *(14.19)* gives

$$r_1 = \frac{(5.295 \times 10^{-11}\text{ m})(1)^2}{1} = 5.295 \times 10^{-11}\text{ m} = 0.052\,95\text{ nm}$$

Likewise, $r_2 = 0.211\,80$ nm, $r_3 = 0.476\,55$ nm, and $r_4 = 0.847\,20$ nm.

14.26 Find the relation between the Bohr radii for H and He^+. Are there any combinations of values for n for which the radii are equal?

▮ Assuming that μ for the electron is the same for both systems, taking a ratio of *(14.19)* gives

$$\frac{r_n(H)}{r_n(He^+)} = \frac{[n(H)]^2(4\pi\varepsilon_0)\hbar^2/\mu e^2 Z(H)}{[n(He^+)]^2(4\pi\varepsilon_0)\hbar^2/\mu e^2 Z(He^+)} = \frac{[n(H)]^2/1}{[n(He^+)]^2/2} = \frac{2[n(H)]^2}{[n(He^+)]^2}$$

If the radii are equal,

$$[n(He^+)]^2 = 2[n(H)]^2 \qquad [n(He^+)]^2 = 2^{1/2}n(H)$$

This result shows that integral values for $n(H)$ will give nonintegral values for $n(He^+)$. Because this cannot be true, there are no common radii in the systems.

14.27 The atomic radius of Li is 0.152 nm. Calculate the effective nuclear charge on electrons in the second Bohr orbit ($n = 2$).

▮ Solving *(14.19)* for effective nuclear charge (Z_{eff}) and substituting the data gives

$$Z_{eff} = \frac{(5.295 \times 10^{-11}\text{ m})(2)^2}{(0.152\text{ nm})[(10^{-9}\text{ m})/(1\text{ nm})]} = 1.39$$

14.28 Calculate the velocity of an electron in the ground state of an atom of H. What fraction of the velocity of light is this value? How long does it take for the electron to complete one revolution around the nucleus? How many times a second does the electron travel around the nucleus?

▮ According to the Bohr theory, the angular momentum is an integral multiple of $\hbar$:

$$mvr_n = n\hbar \tag{14.20}$$

Solving *(14.20)* for v and substituting $r_n = 5.295 \times 10^{-4}$ m (see Problem 14.25) gives

$$v = \frac{(1)(1.0546 \times 10^{-34}\text{ J}\cdot\text{s})[(1\text{ kg}\cdot\text{m}^2\cdot\text{s}^{-2})/(1\text{ J})]}{(9.11 \times 10^{-31}\text{ kg})(5.295 \times 10^{-11}\text{ m})} = 2.186 \times 10^6\text{ m}\cdot\text{s}^{-1}$$

This velocity, relative to c, is

$$\frac{v}{c} = \frac{2.186 \times 10^6\text{ m}\cdot\text{s}^{-1}}{2.9979 \times 10^8\text{ m}\cdot\text{s}^{-1}} = 7.29 \times 10^{-3}$$

The time required to travel the distance $2\pi r_n$ is

$$t = \frac{2\pi r_n}{v} = \frac{2\pi(5.295 \times 10^{-11}\text{ m})}{2.286 \times 10^6\text{ m}\cdot\text{s}^{-1}} = 1.522 \times 10^{-16}\text{ s}$$

The number of revolutions is

$$\nu = \frac{1}{t} = \frac{1}{1.522 \times 10^{-16}\ \text{s}} = 6.57 \times 10^{15}\ \text{s}^{-1}$$

14.29 Calculate the de Broglie wavelength of an electron in the ground state of an atom of H. Show that this wavelength is equal to the circumference of the orbit.

▮ Substituting $v = 2.186 \times 10^6\ \text{m} \cdot \text{s}^{-1}$ (see Problem 14.28) into *(14.13)* gives

$$\lambda = \frac{(6.626 \times 10^{-34}\ \text{J} \cdot \text{s})[(1\ \text{kg} \cdot \text{m}^2 \cdot \text{s}^{-2})/(1\ \text{J})]}{(9.11 \times 10^{-31}\ \text{kg})(2.186 \times 10^6\ \text{m} \cdot \text{s}^{-1})} = 3.33 \times 10^{-10}\ \text{m}$$

Taking a ratio of the circumference where $r_n = 5.295 \times 10^{-11}\ \text{m}$ (see Problem 14.25) to λ gives

$$\frac{2\pi(5.295 \times 10^{-11}\ \text{m})}{3.33 \times 10^{-10}\ \text{m}} = 0.999$$

(The difference from unity results in the round-off errors of the various calculations.)

14.30 The hypothetical element "positronium" consists of an electron moving in space around a nucleus consisting of a positron (a subatomic particle similar to the electron except possessing a positive charge). Using the Bohr theory, calculate the radius of the first order of the electron.

▮ Substituting $m(\text{nucleus}) = m(\text{electron})$ into *(2.23)* gives

$$\mu = \frac{m(\text{electron})m(\text{electron})}{m(\text{electron}) + m(\text{electron})} = \frac{m(\text{electron})}{2} = \frac{\mu(\text{H})}{2}$$

Substituting into *(14.19)* gives

$$r_n = n^2 \frac{4\pi\varepsilon_0\hbar^2}{(\mu/2)e^2 Z} = (1.0589 \times 10^{-10}\ \text{m})n^2$$

The radius of the first orbit is

$$r_1 = (1.0589 \times 10^{-10}\ \text{m})(1)^2 = 1.0589 \times 10^{-10}\ \text{m} = 0.105\,89\ \text{nm}$$

14.31 Calculate the energies of the first four Bohr orbits of an atom of H.

▮ The energies are given by

$$E_n = \frac{-\mu Z^2 e^4}{2\hbar^2(4\pi\varepsilon_0)^2}\left(\frac{1}{n^2}\right) = \frac{-(2.178\,720 \times 10^{-18}\ \text{J})Z^2}{n^2} \tag{14.21}$$

where the separated atom ($n = \infty$) is defined as the zero point of energy. Substituting into *(14.21)* gives

$$E_1 = \frac{-(2.179 \times 10^{-18}\ \text{J})(1)^2}{(1)^2} = -2.179 \times 10^{-18}\ \text{J}$$

Likewise, $E_2 = -5.448 \times 10^{-19}\ \text{J}$, $E_3 = -2.421 \times 10^{-19}\ \text{J}$, and $E_4 = -1.362 \times 10^{-19}\ \text{J}$.

14.32 Find the relation between the energies of the Bohr orbits for H and He^+. Is there any combination of values for n for which the energies are equal?

▮ Assuming that μ for the electron is the same for both systems, taking a ratio of *(14.21)* gives

$$\frac{E_n(\text{H})}{E_n(\text{He}^+)} = \frac{-\mu[Z(\text{H})]^2 e^4/2\hbar^2(4\pi\varepsilon_0)^2[n(\text{H})]^2}{-\mu[Z(\text{He}^+)]^2 e^4/2\hbar^2(4\pi\varepsilon_0)^2[n(\text{He}^+)]^2} = \frac{[n(\text{He}^+)]^2}{4[n(\text{H})]^2}$$

If the energies are equal,

$$[n(\text{He}^+)]^2 = 4[n(\text{H})]^2 \qquad [n(\text{He}^+)]^2 = 2n(\text{H})$$

This result shows that there are many common energies in the systems.

14.33 The ionization energy of Li is $520.2\ \text{kJ} \cdot \text{mol}^{-1}$. Calculate the effective nuclear charge on an electron in the second Bohr orbit.

▮ The ionization energy is equal to the negative of E_n. Solving *(14.21)* for the effective nuclear charge and substituting the data gives

$$Z_{\text{eff}} = \left[\frac{(-520.2\ \text{kJ} \cdot \text{mol}^{-1})(2)^2[(10^3\ \text{J})/(1\ \text{kJ})]}{-(2.179 \times 10^{-18}\ \text{J})(6.022 \times 10^{23}\ \text{mol}^{-1})}\right]^{1/2} = 1.26$$

which agrees fairly well with the results of Problem 14.27.

14.34 Using the Bohr theory, predict the electronic spectrum of positronium. See Problem 14.30 for additional data.

▌ Substituting $\mu = \mu(\mathrm{H})/2$ into *(14.21)* gives

$$E_n = \frac{-(\mu/2)Z^2e^4}{2\hbar^2(4\pi\varepsilon_0)^2}\left(\frac{1}{n^2}\right) = \frac{-1.089\,360 \times 10^{-18}\ \mathrm{J}}{n^2}$$

The energies of the first four orbits are $-1.089\,360 \times 10^{-18}$ J, $-0.272\,340 \times 10^{-18}$ J, $-0.121\,040 \times 10^{-18}$ J, and $-0.068\,085 \times 10^{-18}$ J. The energy changes corresponding to electronic transitions are

$$E_{1\to2} = E_2 - E_1 = (-0.272\,340 \times 10^{-18}\ \mathrm{J}) - (-1.089\,360 \times 10^{-18}\ \mathrm{J}) = 0.817\,020 \times 10^{-18}\ \mathrm{J}$$

Likewise, $E_{1\to3} = 0.968\,320 \times 10^{-18}$ J, $E_{1\to4} = 1.021\,275 \times 10^{-18}$ J, $E_{2\to3} = 0.151\,300 \times 10^{-18}$ J, $E_{2\to4} = 0.204\,255 \times 10^{-18}$ J, and $E_{3\to4} = 0.052\,955 \times 10^{-18}$ J. Solving *(13.32)* for $\bar{\nu}$ and substituting the values of E gives

$$\nu(1 \to 2) = \frac{(0.817\,020 \times 10^{-18}\ \mathrm{J})(10^{-2}\ \mathrm{m})/(1\ \mathrm{cm})}{(6.626\,18 \times 10^{-34}\ \mathrm{J\cdot s})(2.997\,925 \times 10^{8}\ \mathrm{m\cdot s^{-1}})} = 41\,129\ \mathrm{cm^{-1}}$$

Likewise, $\nu(1 \to 3) = 48\,746\ \mathrm{cm^{-1}}$, $\nu(1 \to 4) = 51\,411\ \mathrm{cm^{-1}}$, $\nu(2 \to 3) = 7616\ \mathrm{cm^{-1}}$, $\nu(2 \to 4) = 10\,282\ \mathrm{cm^{-1}}$, and $\nu(3 \to 4) = 2666\ \mathrm{cm^{-1}}$.

14.3 POSTULATES OF QUANTUM MECHANICS

14.35 Describe the following wave functions as symmetric (even), antisymmetric (odd), or neither (unsymmetric or asymmetric): ***(a)*** $\psi(\theta) = \cos\theta$; ***(b)*** $\psi(\theta) = \sin\theta\cos\theta$; ***(c)*** $\psi(x) = A\,e^{-x}$, where A is a constant; ***(d)*** $\psi(x) = x^n$, where n is odd; and ***(e)*** $\psi(x) = x + x^2$.

▌ A *symmetric wave function* is defined as $\psi(x) = \psi(-x)$, and an *antisymmetric wave function* is defined as $\psi(x) = -\psi(-x)$. Applying these definitions to the various wave functions gives

(a) $\psi(-\theta) = \cos(-\theta) = \cos\theta = \psi(\theta)$

(b) $\psi(-\theta) = \sin(-\theta)\cos(-\theta) = -\sin\theta\cos\theta = -\psi(\theta)$

(c) $\psi(-x) = A\,e^{-(-x)} = A\,e^{x} \neq \pm\psi(x)$

(d) $\psi(-x) = (-x)^n = -x^n = -\psi(x)$

(e) $\psi(-x) = (-x) + (-x)^2 = -x + x^2 \neq \pm\psi(x)$

The symmetric wave function is ***(a)***, the antisymmetric wave functions are ***(b)*** and ***(d)***, and the wave functions ***(c)*** and ***(e)*** are unsymmetric.

14.36 Identify which of the following wave functions are "acceptable" wave functions: ***(a)*** $\psi(x) = \pm x^2$; ***(b)*** $\psi(x) = Ax^2$, where A is a constant; ***(c)*** $\psi(\theta) = \cos\theta$; and ***(d)*** $\psi(x) = e^{-ax}$, where a is a constant.

▌ An *acceptable wave function* is one that is finite, continuous, single-valued, and square-integrable and that has continuous first and second derivatives. Applying this definition to the various wave functions gives

(a) Not acceptable, because $\pm x^2$ is not single-valued and approaches infinity as $x \to \infty$.
(b) Not acceptable, because Ax^2 approaches infinity as $x \to \infty$.
(c) Acceptable.
(d) Not acceptable, because e^{-ax} approaches infinity as $x \to -\infty$. If $a \geq 0$, then e^{-ax} is acceptable.

14.37 Determine $\psi^*\psi$ for the following wave functions: ***(a)*** $\psi(\theta) = \sin\theta + i\cos\theta$; ***(b)*** $\psi(x) = e^{iax}$; and ***(c)*** $\psi(x) = e^{-x^2}$, where $i = (-1)^{1/2}$.

▌ The symbol ψ^* represents the *complex conjugate* of ψ, which is formed by changing the sign of any term in ψ containing i. Applying this definition to the wave functions gives

(a) $$\psi^*(\theta)\psi(\theta) = (\sin\theta + i\cos\theta)^*(\sin\theta + i\cos\theta)$$
$$= (\sin\theta - i\cos\theta)(\sin\theta + i\cos\theta)$$
$$= \sin^2\theta - i^2\cos^2\theta = \sin^2\theta + \cos^2\theta = 1$$

(b) $\psi^*(x)\psi(x) = (e^{iax})^*(e^{iax}) = e^{-iax}\,e^{iax} = 1$

(c) $\psi^*(x)\psi(x) = (e^{-x^2})^*(e^{-x^2}) = e^{-x^2}\,e^{-x^2} = e^{-2x^2}$

14.38 For the wave function $\psi(\theta) = A\,e^{im\phi}$, where m is an integer, evaluate A so that the wave function is normalized.

▌ For a *normalized wave function*,

$$\int_{\text{all space}} \psi^*\psi\, dV = 1 \tag{14.22}$$

where dV is the volume element, which in three-dimensional space is given by

$$dV = dx\,dy\,dz = r^2 \sin\theta\, dr\, d\theta\, d\phi \tag{14.23}$$

where $0 \le r \le \infty$, $0 \le \theta \le \pi$, and $0 \le \phi \le 2\pi$. Substituting the given wave function into *(14.22)* and solving for A gives

$$\int_0^{2\pi} (A^*\,e^{-im\phi})(A\,e^{im\phi})\, d\theta = A^*A \int_0^{2\pi} d\phi = A^*A(2\pi) = 1$$

$$A^*A = 1/2\pi$$

Usually this is written as $A = 1/(2\pi)^{1/2}$.

14.39 Show that the wave functions $\psi_1(x) = \sin(n\pi x/a)$ and $\psi_2(x) = \cos(n\pi x/a)$, where n and a are constants, are orthogonal. The permitted values of x are $0 \le x \le a$.

▌ For *orthogonal wave functions*, ψ_i and ψ_j,

$$\int_{\text{all space}} \psi_i^*\psi_j\, dV = 0 \tag{14.24}$$

Substituting the given wave functions into *(14.24)* gives

$$\int_0^a \left(\sin\frac{n\pi x}{a}\right)^* \left(\cos\frac{n\pi x}{a}\right) dx = \int_0^a \sin\frac{n\pi x}{a}\cos\frac{n\pi x}{a}\, dx = \frac{a}{n\pi}\int_0^{n\pi} \sin u \cos u\, du = \frac{a}{n\pi}\frac{\sin^2 u}{2}\Bigg|_0^{n\pi}$$

$$= \frac{a}{2n\pi}(\sin^2 n\pi - \sin^2 0) = 0$$

where the following substitution of variables was made.

$$u = \frac{n\pi}{a}x \qquad dx = \frac{a}{n\pi}du$$

Note that the limits of integration were changed by the change in variables. Because the integral is equal to zero, the wave functions are orthogonal.

14.40 Show that the wave functions $\psi_1(\phi) = A\,e^{im\phi}$ and $\psi_2(\phi) = B\,e^{in\phi}$, where n and m are integers, are orthonormal.

▌ If ψ_i and ψ_j are normalized wave functions that obey *(14.24)*, they are known as *orthonormal wave functions*. From Problem 14.38, $A = B = 1/(2\pi)^{1/2}$. Substituting the given wave functions into *(14.24)* gives

$$\int_0^{2\pi} \left(\frac{1}{(2\pi)^{1/2}}e^{im\phi}\right)^* \left(\frac{1}{(2\pi)^{1/2}}e^{in\phi}\right) d\phi = \frac{1}{2\pi}\int_0^{2\pi} e^{i(n-m)\phi}\, d\phi$$

$$= \frac{1}{2\pi}\int_0^{2\pi} \cos[(n-m)\phi] + i\sin[(n-m)\phi]\, d\phi$$

$$= \frac{1}{2\pi}\left(\frac{\sin[(n-m)\phi]}{n-m} + \frac{-i\cos[(n-m)\phi]}{n-m}\right)\Bigg|_0^{2\pi}$$

$$= \frac{1}{2\pi(n-m)}[0 - 0 - i(1-1)] = 0$$

where *Euler's formula*

$$e^{\pm i\phi} = \cos\phi \pm i\sin\phi \tag{14.25}$$

was used to evaluate the integral. Because each wave function is normalized and because the above integral is equal to zero, the wave functions are orthonormal.

14.41 The operators for position and linear momentum are given by

$$\hat{x} = x \tag{14.26}$$

and

$$\hat{p}_x = \frac{\hbar}{i}\frac{\partial}{\partial x} \tag{14.27}$$

respectively. Determine the result of operating on the function $\psi(x) = A\sin(n\pi x/a)$, where A, n, and a are constants, with each operator.

▌ The results of performing the respective operations ($\hat{o}$) on the functions are

$$\hat{o}\psi(x) = \hat{x}\psi(x) = xA\sin(n\pi x/a)$$

$$\hat{o}\psi(x) = \hat{p}_x\psi(x) = \frac{\hbar}{i}\frac{\partial}{\partial x}A\sin\frac{n\pi x}{a} = \frac{\hbar}{i}A\frac{n\pi}{a}\cos\frac{n\pi x}{a}$$

14.42 Are the operators given by *(14.26)* and *(14.27)* linear or nonlinear?

▌ A *linear operator* has the following properties:

$$\hat{o}[a\psi_i(x) + b\psi_j(x)] = a\hat{o}\psi_i(x) + b\hat{o}\psi_j(x) \tag{14.28a}$$

$$\hat{o}[c\psi_i(x)] = c\hat{o}\psi_i(x) \tag{14.28b}$$

where a, b, and c are constants. Both operators are linear because

$$\hat{o}[a\psi_i(x) + b\psi_j(x)] = x[a\psi_i(x) + b\psi_j(x)] = xa\psi_i(x) + xb\psi_j(x) = ax\psi_i(x) + bx\psi_j(x) = a\hat{o}\psi_i(x) + b\hat{o}\psi_j(x)$$

$$\hat{o}[a\psi_i(x) + b\psi_j(x)] = \frac{\hbar}{i}\frac{\partial}{\partial x}[a\psi_i(x) + b\psi_j(x)] = \frac{\hbar}{i}\frac{\partial}{\partial x}a\psi_i(x) + \frac{\hbar}{i}\frac{\partial}{\partial x}b\psi_j(x)$$

$$= a\frac{\hbar}{i}\frac{\partial}{\partial x}a\psi_i(x) + b\frac{\hbar}{i}\frac{\partial}{\partial x}b\psi_j(x) = a\hat{o}\psi_i(x) + b\hat{o}\psi_j(x)$$

14.43 Are the operators given by *(14.26)* and *(14.27)* Hermitian?

▌ A *Hermitian operator* has the following property:

$$\int_{\text{all space}} \psi_i^*\hat{o}\psi_j\,dV = \int_{\text{all space}} \psi_j(\hat{o}\psi_i)^*\,dV \tag{14.29}$$

The position operator is a Hermitian operator because

$$\int_{\text{all space}} \psi_i^*\hat{o}\psi_j\,dV = \int_{\text{all space}} \psi_i^*(x)x\psi_j(x)\,dx = \int_{\text{all space}} \psi_j(x)x\psi_i^*(x)\,dx$$

$$= \int_{\text{all space}} \psi_j(x)[x\psi_i(x)]^*\,dx = \int_{\text{all space}} \psi_j(\hat{o}\psi_i)^*\,dV$$

where $x = x^*$ since x is real. The linear momentum operator is a Hermitian operator because

$$\int_{\text{all space}} \psi_i^*\hat{o}\psi_j\,dV = \int_{\text{all space}} \psi_i^*(x)\frac{\hbar}{i}\frac{\partial}{\partial x}\psi_j(x)\,dx = \frac{\hbar}{i}\int_{\text{all space}} \psi_i^*(x)\psi_j(x)\,dx - \int_{\text{all space}} \psi_j(x)\frac{\hbar}{i}\frac{\partial}{\partial x}\psi_i^*(x)\,dx$$

$$= 0 - \int_{\text{all space}} \psi_j(x)\left[\frac{\hbar}{-i}\frac{\partial}{\partial x}\psi_i(x)\right]^*\,dx$$

$$= \int_{\text{all space}} \psi_j(x)\left(\frac{\hbar}{i}\frac{\partial}{\partial x}\psi_i(x)\right)^*\,dx = \int_{\text{all space}} \psi_j(\hat{o}\psi_i)^*\,dV$$

where the integral was integrated by parts and $\psi_i(x)$ and $\psi_j(x)$ must both be zero at $x = \pm\infty$.

14.44 Using *(14.27)*, determine the operator $\hat{p}_x^2$. Does this operator follow the general rule that if $\hat{o}$ is Hermitian, then $\hat{o}^2$ will also be Hermitian?

▌ The notation $\hat{o}^n$ means to perform the operation $\hat{o}$ n successive times. In this case,

$$p_x^2\psi(x) = \left(\frac{\hbar}{i}\frac{\partial}{\partial x}\right)\left(\frac{\hbar}{i}\frac{\partial}{\partial x}\right)\psi(x) = -\hbar^2\frac{\partial^2}{\partial x^2}\psi(x)$$

giving

$$\hat{p}_x^2 = -\hbar^2\frac{\partial^2}{\partial x^2} \tag{14.30}$$

Since the operator $\hat{p}_x^2$ is real, the criterion of *(14.29)* is satisfied and the operator is Hermitian.

14.45 Do the operators given by *(14.26)* and *(14.27)* commute? Do these operators follow the general rule that operators having a nonzero commutator correspond to physical observables that cannot be determined simultaneously with no restriction in this precision?

▮ The *commutator* of two operators $\hat{o}_i$ and $\hat{o}_j$ is written as $[\hat{o}_i, \hat{o}_j]$, where

$$[\hat{o}_i, \hat{o}_j] = \hat{o}_i\hat{o}_j - \hat{o}_j\hat{o}_i \tag{14.31}$$

If $[\hat{o}_i, \hat{o}_j] = 0$ when operating on an arbitrary function, then the operators are said to *commute*. In this case the operators do not commute, because

$$[\hat{o}_i, \hat{o}_j]\psi = [\hat{x}, \hat{p}_x]\psi(x) = \left[x, \frac{\hbar}{i}\frac{\partial}{\partial x}\right]\psi(x) = x\frac{\hbar}{i}\frac{\partial}{\partial x}\psi(x) - \frac{\hbar}{i}\frac{\partial}{\partial x}[x\psi(x)]$$

$$= x\frac{\hbar}{i}\frac{\partial}{\partial x}\psi(x) - x\frac{\hbar}{i}\frac{\partial}{\partial x}\psi(x) - \psi(x)\frac{\hbar}{i}(1) = \frac{\hbar}{i}\psi(x) \neq 0$$

giving $[\hat{x}, \hat{p}_x] = \hbar/i$. The general rule is followed by these operators; see *(14.16)*.

14.46 Is $\psi(x) = A\sin(n\pi x/a)$, where A, n, and a are constants, an eigenfunction of either *(14.26)* or *(14.27)*? If so, what is the eigenvalue?

▮ If $\hat{o}$ and ψ are such that

$$\hat{o}\psi = o\psi \tag{14.32}$$

where o is a physically observable quantity, then o is known as an *eigenvalue* and ψ is an *eigenfunction*. For the position operator,

$$\hat{o}\psi = \hat{x}\psi(x) = x\psi(x) = o\psi$$

where ψ is an eigenfunction with an eigenvalue $o = x$. For the linear momentum operator,

$$\hat{o}\psi = \frac{\hbar}{i}\frac{\partial}{\partial x}\psi(x) = \frac{\hbar}{i}\frac{\partial}{\partial x}A\sin\frac{n\pi x}{a} = \frac{\hbar}{i}A\frac{n\pi}{a}\cos\frac{n\pi x}{a} \neq o\psi(x)$$

14.47 Show that a linear combination of two linear eigenfunctions having the same eigenvalue is an eigenfunction with the same eigenvalue.

▮ A *linear combination* is given by

$$\psi = a_1\psi_1 + a_2\psi_2 \tag{14.33}$$

where a_1 and a_2 are constants. Substituting *(14.33)* into *(14.32)* gives

$$\hat{o}\psi = \hat{o}(a_1\psi_1 + a_2\psi_2) = \hat{o}a_1\psi_1 + \hat{o}a_2\psi_2 = a_1\hat{o}\psi_1 + a_2\hat{o}\psi_2 = a_1o\psi_1 + a_2o\psi_2 = o[a_1\psi_1 + a_2\psi_2] = o\psi$$

where o is the eigenvalue common to the two linear eigenfunctions.

14.48 Determine the average position ($\langle x\rangle$) for the wave function $\psi(x) = A\sin(n\pi x/a)$, where A, n, and a are constants and $0 \le x \le a$.

▮ The *quantum-mechanical average* of a series of measurements (the *expectation value*) will be given by

$$\langle o\rangle = \frac{\int_{\text{all space}} \psi^*\hat{o}\psi\,dV}{\int_{\text{all space}} \psi^*\psi\,dV} \tag{14.34}$$

[The denominator in *(14.34)* does not appear if normalized wave functions are used.] Substituting into *(14.34)* gives

$$\langle x\rangle = \frac{\int_0^a \psi^*(x)\hat{x}\psi(x)\,dx}{\int_0^a \psi^*(x)\psi(x)\,dx} = \frac{\int_0^a \sin(n\pi x/a)x\sin(n\pi x/a)\,dx}{\int_0^a \sin(n\pi x/a)\sin(n\pi x/a)\,dx} = \frac{\int_0^a x\sin^2(n\pi x/a)\,dx}{\int_0^a \sin^2(n\pi x/a)\,dx}$$

$$= \frac{\left[\dfrac{x^2}{2} - \dfrac{x\sin(2n\pi x/a)}{4(n\pi x/a)} - \dfrac{\cos(2n\pi x/a)}{8(n\pi x/a)^2}\right]_0^a}{\left[\dfrac{x}{2} - \dfrac{\sin(2n\pi x/a)}{4(n\pi x/a)}\right]_0^a} = \frac{[a^2/4 - 0] - (0-0) - (1-1)}{[a/2 - 0] - (0-0)} = \frac{a}{2}$$

14.49 Show that if ψ is a solution to the time-independent Schrödinger wave equation, then $c\psi$ is also a solution where c is a constant.

■ The time-independent *Schrödinger wave equation* is given by

$$\mathscr{H}\psi = E\psi \tag{14.35}$$

where $\mathscr{H}$ is the *Hamiltonian operator* (sometimes represented by $\hat{H}$) and E is the total energy of the system. The Hamiltonian operator is the sum of the kinetic and potential energies of the system:

$$\mathscr{H} = -\frac{\hbar^2}{2}\sum_{i=1}^{N}\frac{1}{m_i}\nabla_i^2 + U \tag{14.36}$$

where the summation is performed for the number of particles in the system (N) and

$$\nabla^2 = \frac{\partial^2}{\partial x^2} + \frac{\partial^2}{\partial y^2} + \frac{\partial^2}{\partial z^2} \tag{14.37a}$$

$$= \frac{1}{r^2}\frac{\partial}{\partial r}\left(r^2\frac{\partial}{\partial r}\right) + \frac{1}{r^2 \sin\theta}\frac{\partial}{\partial\theta}\left(\sin\theta\frac{\partial}{\partial\theta}\right) + \frac{1}{r^2\sin^2\theta}\frac{\partial^2}{\partial\phi^2} \tag{14.37b}$$

Because $\mathscr{H}$ is linear, *(14.35)* gives $\mathscr{H}(c\psi) = c\mathscr{H}\psi = cE$, proving that $c\psi$ is also a solution to *(14.35)*.

14.4 APPLICATIONS TO TRANSLATIONAL, ROTATIONAL, AND VIBRATIONAL MOTION

14.50 Consider a *freely moving particle* ($U = 0$) in one dimension with mass m. Show that a wave function describing this system is

$$\psi(x) = A\,e^{\pm i(2mE)^{1/2}x/\hbar} \tag{14.38}$$

where $E \geq 0$ and A is a constant.

■ For this system, *(14.36)* becomes

$$\mathscr{H} = -\frac{\hbar^2}{2m}\frac{\partial^2}{\partial x^2}$$

Substituting into *(14.35)* gives

$$\mathscr{H}\psi(x) = -\frac{\hbar^2}{2m}\frac{\partial^2}{\partial x^2}A\,e^{\pm i(2mE)^{1/2}x/\hbar} = \mp A\frac{\hbar^2}{2m}\left(\frac{i(2mE)^{1/2}}{\hbar}\right)e^{\pm i(2mE)^{1/2}x/\hbar} = -A\frac{\hbar^2}{2m}\left(\frac{i^2(2mE)}{\hbar^2}\right)e^{\pm i(2mE)^{1/2}x/\hbar}$$

$$= EA\,e^{\pm i(2mE)^{1/2}x/\hbar} = E\psi(x)$$

14.51 For the system described in Problem 14.50, determine $\langle x\rangle$ and $\langle p_x\rangle$. Do these results agree with *(14.16)*?

■ Substituting $\hat{x} = x$ and *(14.38)* into *(14.34)* gives

$$\langle x\rangle = \frac{\int_{-\infty}^{\infty}(A\,e^{\pm i(2mE)^{1/2}x/\hbar})^* x(A\,e^{\pm i(2mE)^{1/2}x/\hbar})\,dx}{\int_{-\infty}^{\infty}(A\,e^{\pm i(2mE)^{1/2}x/\hbar})^*(A\,e^{\pm i(2mE)^{1/2}x/\hbar})\,dx} = \frac{\int_{-\infty}^{\infty}x\,dx}{\int_{-\infty}^{\infty}dx} = \frac{x^2/2|_{-\infty}^{\infty}}{x|_{-\infty}^{\infty}} = 0$$

where the rule of l'Hôpital was used to evaluate the limiting value of the fraction. This result implies that the particle will have an equal probability of being anywhere between $x = -\infty$ and $x = \infty$. Substituting *(14.27)* and *(14.38)* into *(14.34)* gives

$$\langle p_x\rangle = \frac{\int_{-\infty}^{\infty}(A\,e^{\pm i(2mE)^{1/2}x/\hbar})^*(\hbar/i)(\partial/\partial x)(A\,e^{\pm i(2mE)^{1/2}x/\hbar})\,dx}{\int_{-\infty}^{\infty}(A\,e^{\pm i(2mE)^{1/2}x/\hbar})^*(A\,e^{\pm i(2mE)^{1/2}x/\hbar})\,dx} = \frac{\pm(2mE)^{1/2}\int_{-\infty}^{\infty}dx}{\int_{-\infty}^{\infty}dx} = \pm(2mE)^{1/2}$$

This result implies that the momentum will be fixed for a given value of E. According to *(14.16)*, if p_x is fixed, then the uncertainty in x must be very large, as shown above.

14.52 Consider a particle with mass m moving freely in one dimension in which there is a potential energy well with a finite height V_0 and a width $0 \leq x \leq a$. Is there a greater probability of finding the particle over the potential energy well or outside the barriers of the well?

■ A sketch of the system is shown in Fig. 14-3. Using *(14.38)* for each of the three regions gives

$$\psi(x) = \begin{cases} A\,e^{\pm i[2m(E-V_0)]^{1/2}x/\hbar} & -\infty \leq x < 0 \\ B\,e^{\pm i(2mE)^{1/2}x/\hbar} & 0 \leq x \leq a \\ A\,e^{\pm i[2m(E-V_0)]^{1/2}x/\hbar} & a < x \leq \infty \end{cases}$$

In general, the exponential term in the wave function over the potential energy well will be greater than the exponential term outside the well, and because the probability of finding the particle is directly proportional to $\psi^*(x)\psi(x)$ there will be a greater probability of finding the particle over the well. As $E \to V_0$, this effect becomes significant because $\psi(x) \to 0$ outside the well, and as $E \to \infty$, this effect becomes insignificant.

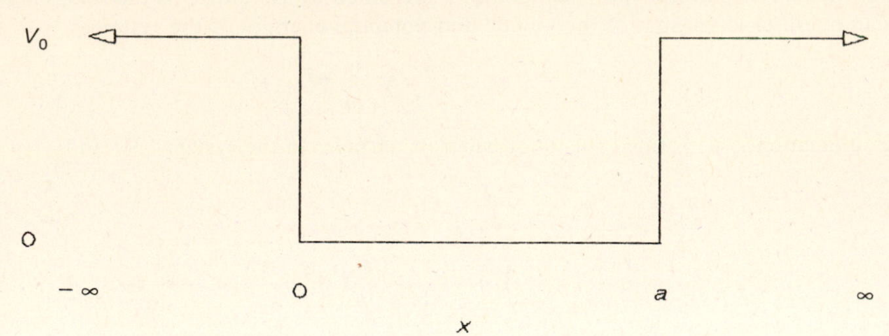

Fig. 14-3

14.53 Consider a system similar to that shown in Fig. 14-3 with

$$\psi(x) = \begin{cases} A\, e^{\pm i[2m(E-V_0)]^{1/2}x/\hbar} & -\infty \le x < 0 \\ B\, e^{\pm i(2mE)^{1/2}x/\hbar} & 0 \le x \le \infty \end{cases}$$

Calculate the fraction of the wave reflected at $x = 0$.

▮ In order for $\psi(x)$ and $\partial\psi(x)/\partial x$ to be continuous, $A \neq B$ and a reflected wave given by

$$\psi(x) = C\, e^{\pm i[2m(E-V_0)]^{1/2}x/\hbar}$$

will be present. The *reflection coefficient* (r) is

$$r = \frac{\psi^*(\text{reflected})\psi(\text{reflected})}{\psi^*(\text{incident})\psi(\text{incident})} = \frac{(C\, e^{\pm i[2m(E-V_0)]^{1/2}x/\hbar})^*(C\, e^{\pm i[2m(E-V_0)]^{1/2}x/\hbar})}{(A\, e^{\pm i[2m(E-V_0)]^{1/2}x/\hbar})^*(A\, e^{\pm i[2m(E-V_0)]^{1/2}x/\hbar})} = C^*C/A^*A$$

14.54 Consider a system similar to that shown in Fig. 14-3 with $E < V_0$. Write the wave function for this system and determine V_0 such that $E = 9.50$ eV for the lowest energy electron in a potential energy well with $a = 5.0$ nm.

▮ The wave function for a *particle in a potential energy well with finite walls* is

$$\psi(x) = \begin{cases} A\, e^{[2m(V_0-E)]^{1/2}x/\hbar} & -\infty \le x < 0 \\ B \sin[(2mE)^{1/2}x/\hbar] + C \cos[(2mE)^{1/2}x/\hbar] & 0 \le x \le a \\ D\, e^{-[2m(V_0-E)]^{1/2}x/\hbar} & a < x \le \infty \end{cases}$$

where the trigonometric form is commonly used for $\psi(x)$ within the potential energy well. These terms in $\psi(x)$ indicate that there is a small probability of finding the particle outside the well ("in the walls"). The wave function for the lowest energy state is symmetric and is given by

$$\left(\frac{E}{V_0 - E}\right)^{1/2} = \cot\frac{(2mE)^{1/2}a}{2\hbar}$$

Solving for V_0 and substituting the data gives

$$V_0 = (9.50\text{ eV})[(1.602\times10^{-19}\text{ J})/(1\text{ eV})]$$
$$\times\left(1 + \tan^2\frac{\left[2(9.11\times10^{-31}\text{ kg})(9.50\text{ eV})\left(\frac{1.602\times10^{-19}\text{ J}}{1\text{ eV}}\right)\right]^{1/2}(5.0\times10^{-9}\text{ m})}{2(1.0546\times10^{-34}\text{ J}\cdot\text{s})}\right)$$
$$= 2.555\times10^{-18}\text{ J} = 15.95\text{ eV}$$

14.55 An electron with $E = 9.50$ eV encounters a barrier 15.95 eV high and 5.0 nm wide. What is the probability that the electron will tunnel through the barrier?

▮ The probability that the electron passes through the barrier (D) is

$$D = \frac{1}{1 + (\frac{1}{4})(k_2/k_1 + k_1/k_2)^2 \sinh^2 k_2 a} \qquad (14.39a)$$

which becomes

$$D = \left(\frac{4}{k_2/k_1 + k_1/k_2}\right)^2 e^{-2k_2a} \tag{14.39b}$$

if $k_2a \gg 1$ and

$$k_1 = (2mE)^{1/2}/\hbar \tag{14.40}$$

$$k_2 = [2m(V_0 - E)]^{1/2}/\hbar \tag{14.41}$$

Substituting the data into *(14.40)* and *(14.41)* gives

$$k_1 = \frac{2(9.11 \times 10^{-31}\text{ kg})(9.50\text{ eV})[(1.602 \times 10^{-19}\text{ J})/(1\text{ eV})][(1\text{ kg}\cdot\text{m}^2\cdot\text{s}^{-2})/(1\text{ J})]}{(1.0546 \times 10^{-34}\text{ J}\cdot\text{s})[(1\text{ kg}\cdot\text{m}^2\cdot\text{s}^{-2})/(1\text{ J})]} = 1.58 \times 10^{10}\text{ m}^{-1}$$

$$k_2 = \frac{2(9.11 \times 10^{-31})(15.95 - 9.50)(1.602 \times 10^{-19})^{1/2}}{1.0546 \times 10^{-34}} = 1.30 \times 10^{10}\text{ m}^{-1}$$

Substituting into *(14.39b)* gives

$$D = \left(\frac{4}{\dfrac{1.30 \times 10^{10}\text{ m}^{-1}}{1.58 \times 10^{10}\text{ m}^{-1}} + \dfrac{1.58 \times 10^{10}\text{ m}^{-1}}{1.30 \times 10^{10}\text{ m}^{-1}}}\right)^2 \exp\left[-2(1.30 \times 10^{10}\text{ m}^{-1})(5.0 \times 10^{9}\text{ m})\right] = 1.34 \times 10^{-56}$$

14.56 What is the probability of an electron tunneling through a barrier as **(a)** $a \to \infty$ and **(b)** $(V_0 - E) \to 0$?

▌ **(a)** As the thickness of the barrier increases, the exponential term in *(14.39b)* decreases, giving $D \to 0$ as $a \to \infty$.

(b) As the energy of the electron approaches that of the height of the barrier, the value of k_2 as determined by *(14.41)* decreases and the exponential term in *(14.39b)* increases, giving $D \to 1$ as $(V_0 - E) \to 0$.

14.57 Consider a system similar to that shown in Fig. 14-3 with $V_0 = \infty$ and $E < V_0$. Write the wave function for this system, and determine the normalization constant. Show that $\psi_1(x)$ and $\psi_2(x)$ are orthogonal.

▌ The wave function for a *particle in a potential energy well with infinite walls* is

$$\psi_n(x) = \begin{cases} A \sin(n\pi x/a) & 0 \le x \le a \\ 0 & \text{otherwise} \end{cases} \tag{14.42}$$

where $n = 1, 2, \ldots$. Substituting *(14.42)* into *(14.22)* and solving for A gives

$$\int_0^a \left(A \sin\frac{n\pi x}{a}\right)^* \left(A \sin\frac{n\pi x}{a}\right) dx = A^*A \int_0^a \sin^2\frac{n\pi x}{a}\, dx$$

$$= A^*A\left(\frac{x}{2} - \frac{\sin(2n\pi x/a)}{4(n\pi/a)}\right)\Bigg|_0^a = A^*A\left(\frac{a}{2} - 0 - (0 - 0)\right) = A^*A(a/2) = 1$$

$$A^*A = 2/a \tag{14.43}$$

which is often written $A = (2/a)^{1/2}$. Substituting into *(14.24)* gives

$$\int_0^a \left(A \sin\frac{2\pi x}{a}\right)^* \left(A \sin\frac{\pi x}{a}\right) dx = A^*A \int_0^a \sin^2\frac{2\pi x}{a} \sin\frac{\pi x}{a}\, dx$$

$$= \frac{2}{a}\left(\frac{\sin[(2-1)\pi x/a]}{2(2-1)} - \frac{\sin[(2+1)\pi x/a]}{2(2+1)}\right)\Bigg|_0^a = 0$$

14.58 Using the results of Problem 14.57, derive $\langle E_n \rangle$.

▌ Substituting $\mathcal{H} = -(\hbar^2/2m)(\partial^2/\partial x^2)$ and *(14.42)* into *(14.34)* with the denominator of *(14.34)* set equal to unity gives

$$\langle E_n \rangle = \int_0^a \left(A \sin\frac{n\pi x}{a}\right)^* \left(-\frac{\hbar^2}{2m}\right)\frac{\partial^2}{\partial x^2}\left(A \sin\frac{n\pi x}{a}\right) dx = -\frac{\hbar^2}{2m} A^*A \int_0^a \sin\frac{n\pi x}{a} \frac{\partial}{\partial x}\left(\frac{n\pi}{a}\cos\frac{n\pi x}{a}\right) dx$$

$$= -\frac{\hbar^2}{2m}\left(\frac{n\pi}{a}\right) A^*A \int_0^a \left(\sin\frac{n\pi x}{a}\right)\left(\frac{n\pi}{a}\right)\left(-\sin\frac{n\pi x}{a}\right) dx = \frac{\hbar^2}{2m}\left(\frac{n\pi}{a}\right)^2 A^*A \int_0^a \sin^2\frac{n\pi x}{a}\, dx$$

$$= \frac{\hbar^2}{2m}\left(\frac{n\pi}{a}\right)^2 \left(\frac{2}{a}\right)\left(\frac{a}{2}\right) = n^2\hbar^2\pi^2/2ma^2 = n^2h^2/8ma^2 \tag{14.44}$$

where the integral was evaluated in Problem 14.57.

14.59 Consider an N_2 molecule confined to a 1.00-m potential energy well. What is the lowest energy state for this system? What is the quantum number of an "average" N_2 molecule in which the translational kinetic energy is $kT/2$ at $T = 298$ K?

I Substituting $n = 1$ into *(14.44)* gives

$$\langle E_1 \rangle = \frac{(1)^2(6.626 \times 10^{-34}\ \text{J}\cdot\text{s})^2[(1\ \text{kg}\cdot\text{m}^2\cdot\text{s}^{-2})/(1\ \text{J})]}{8[(28.0 \times 10^{-3}\ \text{kg}\cdot\text{mol}^{-1})/(6.022 \times 10^{23}\ \text{mol}^{-1})](1.00\ \text{m})^2} = 1.18 \times 10^{-42}\ \text{J}$$

Solving *(14.44)* for n and substituting the data gives

$$n = \left[\frac{8\dfrac{28.0 \times 10^{-3}\ \text{kg}\cdot\text{mol}^{-1}}{6.022 \times 10^{23}\ \text{mol}^{-1}}(1.00\ \text{m})^2(1.381 \times 10^{-23}\ \text{J}\cdot\text{K}^{-1})(298\ \text{K})/2}{(6.626 \times 10^{-34}\ \text{J}\cdot\text{s})^2[(1\ \text{kg}\cdot\text{m}^2\cdot\text{s}^{-2})/(1\ \text{J})]}\right]^{1/2} = 4.18 \times 10^{10}$$

14.60 Calculate the energy needed to promote an "average" N_2 molecule to the next quantum state in the system described in Problem 14.59.

I The energy difference between successive energy states is found using *(14.44)* to be

$$\Delta E_n = \frac{(n+1)^2h^2}{8ma^2} - \frac{n^2h^2}{8ma^2} = \frac{(2n+1)h^2}{8ma^2} \tag{14.45}$$

Substituting the results of Problem 14.59 into *(14.45)* gives

$$\Delta E_n = \frac{[2(4.18 \times 10^{10}) + 1](6.626 \times 10^{-34}\ \text{J}\cdot\text{s})^2[(1\ \text{kg}\cdot\text{m}^2\cdot\text{s}^{-2})/(1\ \text{J})]}{8[(28.0 \times 10^{-3}\ \text{kg}\cdot\text{mol}^{-1})/(6.022 \times 10^{23}\ \text{mol}^{-1})](1.00\ \text{m})^2} = 9.87 \times 10^{-32}\ \text{J}$$

14.61 For the system described in Problem 14.59, determine the number of energy states that are between $kT/2$ and $1.01kT/2$ at 298 K.

I The quantum number corresponding to $kT/2$ was determined to be $n = 4.18 \times 10^{10}$. Substituting $1.01kT/2$ into *(14.44)* gives

$$n = \left[\frac{8\dfrac{28.0 \times 10^{-3}\ \text{kg}\cdot\text{mol}^{-1}}{6.022 \times 10^{23}\ \text{mol}^{-1}}(1.00\ \text{m})^2(1.01)\dfrac{(1.381 \times 10^{-23}\ \text{J}\cdot\text{K}^{-1})(298\ \text{K})}{2}}{(6.626 \times 10^{-34}\ \text{J}\cdot\text{s})^2[(1\ \text{kg}\cdot\text{m}^2\cdot\text{s}^{-2})/(1\ \text{J})]}\right]^{1/2} = 4.20 \times 10^{10}$$

There are approximately 2×10^8 energy states in this small energy range.

14.62 Determine the lowest energy of an electron placed in a one-dimensional potential energy well with $a = 1 \times 10^{-15}$ m (a typical nuclear diameter).

I Substituting the data into *(14.44)* gives

$$\langle E_1 \rangle = \frac{(1)^2(6.626 \times 10^{-34}\ \text{J}\cdot\text{s})^2[(1\ \text{kg}\cdot\text{m}^2\cdot\text{s}^{-2})/(1\ \text{J})]}{8(9.11 \times 10^{-31}\ \text{kg})(1 \times 10^{-15}\ \text{m})^2} = 6 \times 10^{-8}\ \text{J}$$

This extremely high value, nearly 4×10^{11} eV, would imply that free electrons would not exist in the nucleus.

14.63 Determine $\langle x \rangle$ and $\langle x^2 \rangle$ for the system described in Problem 14.57.

I Substituting $\hat{x} = x$, $\hat{x}^2 = x^2$, and *(14.42)* into *(14.34)* with the denominator of *(14.34)* set equal to unity gives

$$\langle x \rangle = \int_0^a \left(A \sin\frac{n\pi x}{a}\right)^* x \left(A \sin\frac{n\pi x}{a}\right) dx = A^*A \int_0^a x \sin^2\frac{n\pi x}{a}\, dx$$

$$= \frac{2}{a}\left(\frac{x^2}{4} - \frac{x \sin(2n\pi x/a)}{4(n\pi/a)} - \frac{\cos(2n\pi x/a)}{8(n\pi/a)^2}\right)\Bigg|_0^a = \frac{2}{a}\left[\frac{a^2}{4} - 0 - (0 - 0) - \left(\frac{1}{8} - \frac{1}{8}\right)\right] = \frac{a}{2}$$

$$\langle x^2 \rangle = \int_0^a \left(A \sin\frac{n\pi x}{a}\right)^* x^2 \left(A \sin\frac{n\pi x}{a}\right) dx = A^*A \int_0^a x^2 \sin^2\frac{n\pi x}{a}\, dx$$

$$= \frac{2}{a}\left[\frac{x^3}{6} - \left(\frac{x^2}{4(n\pi/a)} - \frac{1}{8(n\pi/a)^3}\right)\sin\frac{2n\pi x}{a} - \frac{x\cos(2n\pi x/a)}{4(n\pi/a)^2}\right]\Bigg|_0^a$$

$$= \frac{2}{a}\left\{\frac{a^3}{6} - 0 - \left[\left(\frac{a^3}{4n\pi} - \frac{a^3}{8n^3\pi^3}\right)(0) - \left(0 - \frac{a^3}{8n^3\pi^3}\right)(0)\right] - \left(\frac{a^3(1)}{4n^2\pi^2} - 0\right)\right\}$$

$$= \frac{2}{a}\left(\frac{a^3}{6} - \frac{a^3}{4n^2\pi^2}\right) = a^2\left(\frac{1}{3} - \frac{1}{2n^2\pi^2}\right)$$

14.64 Determine $\langle p_x \rangle$ and $\langle p_x^2 \rangle$ for the system described in Problem 14.57.

▌ Substituting *(14.27)*, *(14.30)*, and *(14.42)* into *(14.34)* with the denominator of *(14.34)* set equal to unity gives

$$\langle p_x \rangle = \int_0^a \left(A \sin \frac{n\pi x}{a}\right)^* \frac{\hbar}{i} \frac{\partial}{\partial x} \left(A \sin \frac{n\pi x}{a}\right) dx = A^*A \frac{\hbar}{i}\left(\frac{n\pi}{a}\right) \int_0^a \sin \frac{n\pi x}{a} \cos \frac{n\pi x}{a}\, dx$$

$$= \frac{2}{a}\left(\frac{\hbar}{i}\right)\left(\frac{n\pi}{a}\right)\left(\frac{1}{2(n\pi x/a)}\right) \sin^2 \frac{n\pi x}{a}\bigg|_0^a = 0$$

$$\langle p_x^2 \rangle = \int_0^a \left(A \sin \frac{n\pi x}{a}\right)^* (-\hbar^2) \frac{\partial^2}{\partial x^2} \left(A \sin \frac{n\pi x}{a}\right) dx = A^*A(-\hbar^2) \int_0^a \sin \frac{n\pi x}{a} \frac{\partial}{\partial x} \frac{n\pi}{a} \cos \frac{n\pi x}{a}\, dx$$

$$= A^*A(-\hbar^2)\frac{n\pi}{a} \int_0^a \left(\sin \frac{n\pi x}{a}\right)\left(\frac{n\pi}{a}\right)\left(-\sin \frac{n\pi x}{a}\right) dx$$

$$= \frac{2}{a}\hbar^2\left(\frac{n\pi}{a}\right)^2 \int_0^a \sin^2 \frac{n\pi x}{a}\, dx = \frac{2}{a}\hbar^2\left(\frac{n\pi}{a}\right)^2\left(\frac{a}{2}\right) = \frac{h^2n^2}{4a^2}$$

14.65 Show that the particle in a one-dimensional potential energy well obeys the uncertainty principle for $n = 1$.

▌ The deviation of a physical quantity is given by

$$\Delta z = [\langle z^2 \rangle - \langle z \rangle^2]^{1/2} \tag{14.46}$$

Substituting the results of Problems 14.63 and 14.64 into *(14.46)* gives

$$\Delta p_x = \left(\frac{h^2(1)^2}{4a^2} - 0\right)^{1/2} = \frac{h}{2a}$$

$$\Delta x = \left[\frac{a^2}{3} - \frac{a^2}{2(1)^2\pi^2} - \left(\frac{a}{2}\right)^2\right]^{1/2} = 0.1808a$$

Multiplying Δp_x by Δx gives

$$\Delta p_x\, \Delta x = \frac{h}{2a}(0.1808a) = 0.180\pi\hbar = 0.5679\hbar$$

Note that the conditions of *(14.16)* are satisfied.

14.66 Determine the probability of finding the particle in a one-dimensional potential energy well in the middle half of the well $(a/4 \le x \le 3a/4)$. What is the classical limit of this probability?

▌ Substituting *(14.42)* into *(14.22)* and integrating over the limits on x gives the probability (P) as

$$P = \int_{a/4}^{3a/4} \left(A \sin \frac{n\pi x}{a}\right)^* \left(A \sin \frac{n\pi x}{a}\right) dx = A^*A \int_{a/4}^{3a/4} \sin^2 \frac{n\pi x}{a}\, dx = \frac{2}{a}\left(\frac{x}{2} - \frac{\sin(2n\pi x/a)}{4(n\pi/a)}\right)\bigg|_{a/4}^{3a/4}$$

$$= \frac{2}{a}\left(\frac{3a/4}{2} - \frac{a/4}{2} - \frac{\sin[2n\pi(3a/4)/a] - \sin[2n\pi(a/4)/a]}{4(n\pi/a)}\right) = \frac{1}{2} + \frac{1}{2n\pi}\left(\sin\frac{n\pi}{2} - \sin\frac{3n\pi}{2}\right)$$

$$= \frac{1}{2} + \frac{\sin(n\pi/2) - [-\sin(n\pi/2)]}{2n\pi} = \frac{1}{2} + \frac{\sin(n\pi/2)}{n\pi}$$

Substituting $n = 1, 2,$ and 10 gives $P = 0.5087, 0.5087,$ and 0.5003, respectively. As $n \to \infty$, P approaches the classical limit of 0.5.

14.67 Assume that the energies of the π electrons in a conjugated carbon bond system $CH_2{=}(CH{-}CH{=})_kCH_2$ can be described by *(14.44)* where

$$a = 2(k + 1)(0.141\ \text{nm})$$

The four π electrons in butadiene are placed in the $n = 1$ and $n = 2$ levels. Determine the wavelength of light needed to promote the highest-energy electron to the $n = 3$ level.

▌ The length of the potential energy well is

$$a = 2(1 + 1)(0.141\ \text{nm}) = 0.564\ \text{nm}$$

Substituting into *(14.45)* gives

$$\Delta E_n = \frac{[2(2) + 1](6.626 \times 10^{-34}\ \text{J}\cdot\text{s})^2[(1\ \text{kg}\cdot\text{m}^2\cdot\text{s}^{-2})/(1\ \text{J})]}{8(9.11 \times 10^{-31}\ \text{kg})(0.564 \times 10^{-9}\ \text{m})^2} = 9.47 \times 10^{-19}\ \text{J}$$

Solving *(13.32)* for λ and substituting gives

$$\lambda = \frac{(6.626 \times 10^{-34}\ \mathrm{J \cdot s})(2.9979 \times 10^{8}\ \mathrm{m \cdot s^{-1}})}{9.47 \times 10^{-19}\ \mathrm{J}} = 2.10 \times 10^{-7}\ \mathrm{m} = 210\ \mathrm{nm}$$

This value agrees well with the experimental value of 217 nm.

14.68 The wave function for a *particle in a three-dimensional potential energy well* is

$$\psi(x, y, z) = A \sin(n_x \pi x/a) \sin(n_y \pi y/b) \sin(n_z \pi z/c) \qquad (14.47)$$

where n_x, n_y, and n_z are quantum numbers and a, b, c are the dimensions of the well. Evaluate the normalization constant (A).

▮ Substituting *(14.47)* into *(14.22)* and solving for A gives

$$\int_0^a \int_0^b \int_0^c \left(A \sin\frac{n_x \pi x}{a} \sin\frac{n_y \pi y}{b} \sin\frac{n_z \pi z}{c}\right)^* \left(A \sin\frac{n_x \pi x}{a} \sin\frac{n_y \pi y}{b} \sin\frac{n_z \pi z}{c}\right) dx\, dy\, dz$$

$$= A^*A \int_0^a \int_0^b \sin^2\frac{n_x \pi x}{a} \sin^2\frac{n_y \pi y}{b}\, dx\, dy \left(\int_0^c \sin^2\frac{n_z \pi z}{c}\, dz\right) = A^*A \int_0^a \int_0^b \sin^2\frac{n_x \pi x}{a} \sin^2\frac{n_y \pi y}{b}\, dx\, dy \frac{c}{2}$$

$$= A^*A\frac{c}{2} \int_0^a \sin^2\frac{n_x \pi x}{a}\, dx \left[\int_0^b \sin^2\frac{n_y \pi y}{b}\, dy\right] = A^*A\frac{c}{2} \int_0^a \sin^2\frac{n_x \pi x}{a}\, dx \frac{b}{2}$$

$$= A^*A\left(\frac{b}{2}\right)\left(\frac{c}{2}\right) \int_0^a \sin^2\frac{n_x \pi x}{a}\, dx = A^*A\left(\frac{a}{2}\right)\left(\frac{b}{2}\right)\left(\frac{c}{2}\right)$$

$$A^*A = 8/abc \qquad (14.48)$$

which is often written $A = (8/abc)^{1/2}$.

14.69 Determine the eigenvalues of *(14.47)* for energy.

▮ Substituting *(14.36)*, using *(14.37a)* and *(14.47)*, into *(14.35)* gives

$$\mathcal{H}\psi(x, y, z) = \frac{-\hbar^2}{2m}\left(\frac{\partial^2}{\partial x^2} + \frac{\partial^2}{\partial y^2} + \frac{\partial^2}{\partial z^2}\right) A \sin\frac{n_x \pi x}{a} \sin\frac{n_y \pi y}{b} \sin\frac{n_z \pi z}{c}$$

$$= A\frac{-\hbar^2}{2m}\left(\sin\frac{n_y \pi y}{b} \sin\frac{n_z \pi z}{c} \frac{\partial^2}{\partial x^2} \sin\frac{n_x \pi x}{a}\right.$$

$$\left. + \sin\frac{n_x \pi x}{a} \sin\frac{n_z \pi z}{c} \frac{\partial^2}{\partial y^2} \sin\frac{n_y \pi y}{b} + \sin\frac{n_x \pi x}{a} \sin\frac{n_y \pi y}{b} \frac{\partial^2}{\partial z^2} \sin\frac{n_z \pi z}{c}\right)$$

$$= A\frac{-\hbar^2}{2m}\left[\sin\frac{n_y \pi y}{b}\left(\sin\frac{n_z \pi z}{c}\right)\left(\frac{-n_x^2 \pi^2}{a^2}\right) \sin\frac{n_x \pi x}{a}\right.$$

$$\left. + \sin\frac{n_x \pi x}{a}\left(\sin\frac{n_z \pi z}{c}\right)\left(\frac{-n_y^2 \pi^2}{b^2}\right) \sin\frac{n_y \pi y}{b} + \sin\frac{n_x \pi x}{a}\left(\sin\frac{n_y \pi y}{b}\right)\left(\frac{-n_z^2 \pi^2}{c^2}\right) \sin\frac{n_z \pi z}{c}\right]$$

$$= A\frac{\hbar^2 \pi^2}{2m} \sin\frac{n_x \pi x}{a} \sin\frac{n_y \pi y}{b} \sin\frac{n_z \pi z}{c}\left(\frac{n_x^2}{a^2} + \frac{n_y^2}{b^2} + \frac{n_z^2}{c^2}\right)$$

$$= \frac{\hbar^2 \pi^2}{2m}\left(\frac{n_x^2}{a^2} + \frac{n_y^2}{b^2} + \frac{n_z^2}{c^2}\right)\psi(x, y, z) = \frac{h^2}{8m}\left(\frac{n_x^2}{a^2} + \frac{n_y^2}{b^2} + \frac{n_z^2}{c^2}\right)\psi(x, y, z)$$

Thus,

$$E = \frac{h^2}{8m}\left(\frac{n_x^2}{a^2} + \frac{n_y^2}{b^2} + \frac{n_z^2}{c^2}\right) \qquad (14.49)$$

14.70 Assume that the system described in Problem 14.68 is cubic and that $E = 101h^2/8ma^2$. What is the degeneracy of this level?

▮ For a cube, *(14.49)* becomes

$$E = \frac{h^2}{8ma^2}(n_x^2 + n_y^2 + n_z^2) = \frac{h^2}{8mV^{2/3}}(n_x^2 + n_y^2 + n_z^2) \qquad (14.50)$$

where V is the volume of the cube. The solution to this problem is to find the sets of quantum numbers in *(14.50)* that will generate 101. By trial and error these are 10, 1, 0; 10, 0, 1; 1, 10, 0; 1, 0, 10; 0, 1, 10; 0, 10, 1; 9, 4, 2; 9, 2, 4;

2, 9, 4; 2, 4, 9; 4, 2, 9; 4, 9, 2; 8, 6, 1; 8, 1, 6; 1, 8, 6; 1, 6, 8; 6, 8, 1; 6, 1, 8; 7, 6, 4; 7, 4, 6; 6, 4, 7; 6, 7, 4; 4, 7, 6; and 4, 6, 7. There are 24 states in this energy level.

14.71 Determine the number of energy states less than $E = \frac{3}{2}kT$ at $T = 298$ K for an N_2 molecule in a container with $V = 24.8$ L.

▮ The number of states is given by

$$N = \frac{\pi V}{6h^3}(8mE)^{3/2} \tag{14.51}$$

Substituting the data into *(14.51)* gives

$$N = \frac{\pi(24.8\text{ L})[(10^{-3}\text{ m}^3)/(1\text{ L})]}{6(6.626\times 10^{-34}\text{ J}\cdot\text{s})^3}\left[8\,\frac{28.0\times 10^{-3}\text{ kg}\cdot\text{mol}^{-1}}{6.022\times 10^{23}\text{ mol}^{-1}}\left(\frac{3}{2}\right)(1.381\times 10^{-23}\text{ J}\cdot\text{K}^{-1})\left(\frac{1\text{ kg}\cdot\text{m}^2\cdot\text{s}^{-2}}{1\text{ J}}\right)(298\text{ K})\right]^{3/2}$$

$$= 4.91\times 10^{30}$$

14.72 Consider a *particle confined to a circle* of constant radius. Write the Hamiltonian operator and determine the wave function for this system.

▮ For this two-dimensional system, $\theta = \pi/2$, giving $\sin\theta = 1$ in *(14.37b)*. Using *(14.36)* with *(14.37b)* gives the Hamiltonian operator for this system as

$$\mathcal{H} = -\frac{\hbar^2}{2m}\left(\frac{1}{r^2(1)^2}\right)\frac{\partial^2}{\partial\phi^2} = -\frac{\hbar^2}{2mr^2}\frac{\partial^2}{\partial\phi^2}$$

Substituting into *(14.35)* gives

$$\mathcal{H}\psi(\phi) = -\frac{\hbar^2}{2mr^2}\frac{\partial^2}{\partial\phi^2}\psi(\phi) = E\psi(\phi)$$

The solution of this differential equation is

$$\psi(\phi) = A\,e^{\pm in\phi} \tag{14.52}$$

where $n = 0, 1, 2, \ldots$ and

$$E = \frac{n^2h^2}{8\pi^2mr^2} \tag{14.53}$$

[Alternative solutions are $\psi(\phi) = A\sin n\phi$ or $A\cos n\phi$.] Normalization of *(14.52)* gives

$$\int_0^{2\pi}(A\,e^{\pm in\phi})^*(A\,e^{\pm in\phi})\,d\phi = A^*A\int_0^{2\pi}d\phi = A^*A\phi\Big|_0^{2\pi} = A^*A(2\pi)$$

$$A^*A = 1/2\pi$$

which is commonly written as $A = 1/(2\pi)^{1/2}$.

14.73 Assume that the six π electrons in a benzene molecule can be treated as in Problem 14.72 with two electrons in the $n = 0$ level, four electrons in the $n = 2$ level, and $r = 0.130$ nm. What is the difference in energy between the third and fourth energy levels? What is the corresponding wavelength of light needed to promote an electron from $n = 1$ to $n = 2$?

▮ The difference in energies between two levels will be given by *(14.53)* as

$$\Delta E = \frac{(n+1)^2h^2}{8\pi^2mr^2} - \frac{n^2h^2}{8\pi^2mr^2} = \frac{(2n+1)h^2}{8\pi^2mr^2}$$

$$= \frac{[2(1)+1](6.626\times 10^{-34}\text{ J}\cdot\text{s})^2[(1\text{ kg}\cdot\text{m}^2\cdot\text{s}^{-2})/(1\text{ J})]}{8\pi^2(9.11\times 10^{-31}\text{ kg})(0.130\times 10^{-9}\text{ m})^2} = 1.08\times 10^{-18}\text{ J}$$

Solving *(13.32)* for λ and substituting gives

$$\lambda = \frac{(6.626\times 10^{-34}\text{ J}\cdot\text{s})(2.9979\times 10^8\text{ m}\cdot\text{s}^{-1})}{1.08\times 10^{-18}\text{ J}} = 1.83\times 10^{-7}\text{ m} = 183\text{ nm}$$

14.74 Write the Schrödinger wave equation for a *rigid-rotator* system having a reduced mass μ and moment of inertia I.

▮ The Hamiltonian operator is given by *(14.36)* using *(14.37b)* to describe the angular motion of the system. Substituting into *(14.35)* gives

$$-\frac{\hbar^2}{2\mu}\left[\frac{1}{r^2\sin\theta}\frac{\partial}{\partial\theta}\left[\sin\theta\frac{\partial}{\partial\theta}\right] + \frac{1}{r^2\sin^2\theta}\frac{\partial^2}{\partial\phi^2}\right]\psi(\theta,\phi) = E\psi(\theta,\phi)$$

14.75 The solution to the equation described in Problem 14.74 is

$$\psi_{J,m}(\theta, \phi) = \frac{1}{(2\pi)^{1/2}} \Theta_{J,m}(\theta)\, e^{\pm im\phi} \tag{14.54}$$

where

$$\Theta_{J,m}(\theta) = \left[\frac{(2J+1)(J-|m|)!}{2(J+|m|!)}\right]^{1/2} P_J^{|m|}(\cos\theta) \tag{14.55}$$

$$E_J = J(J+1)\hbar^2/2I \tag{14.56}$$

the *rotational quantum number* is given by $J = 0, 1, 2, \ldots,$ and the *associated Legendre functions* are given by

$$P_l^{|m|}(x) = \frac{1}{2^l l!}(1-x^2)^{|m|/2} \frac{d^{l+|m|}}{dx^{l+|m|}}(x^2-1)^l \tag{14.57}$$

What restrictions does *(14.55)* place on the values of m? What is the degeneracy of each rotational level?

▮ In order for the normalization constant in *(14.55)* to be defined, the maximum value of $|m|$ is J such that the $J - |m|$ term is greater than or equal to zero. Thus $m = 0, \pm 1, \pm 2, \ldots, \pm J$. For a given value of J, there are $2J+1$ permitted values of m, giving the degeneracy of each level as $g_J = 2J + 1$.

14.76 Derive the associated Legendre function $P_2^{|2|}(\cos\theta)$ listed in Table 14-1 using *(14.57)*.

Table 14-1. $P_l^{|m|}(\cos\theta)$

l	$m = 0$	$m = \pm 1$	$m = \pm 2$	$m = \pm 3$
0	1			
1	$\cos\theta$	$\sin\theta$		
2	$\frac{1}{2}[3\cos^2\theta - 1]$	$3\cos\theta\sin\theta$	$3\sin^2\theta$	
3	$\frac{1}{2}(5\cos^3\theta - 3\cos\theta)$	$\frac{3}{2}(5\cos^2\theta - 1)\sin\theta$	$15\cos\theta\sin^2\theta$	$15\sin^3\theta$

▮ Using *(14.57)* gives

$$P_2^{|2|}(x) = \frac{1}{2^2 2!}(1-x^2)^{2/2}\frac{d^{2+2}}{dx^{2+2}}(x^2-1)^2 = \tfrac{1}{8}(1-x^2)\frac{d^4}{dx^4}(x^4 - 2x^2 + 1) = \tfrac{1}{8}(1-x^2)\frac{d^3}{dx^3}(4x^3 - 4x)$$

$$= \tfrac{1}{2}(1-x^2)\frac{d^2}{dx^2}(3x^2 - 1) = \tfrac{1}{2}(1-x^2)\frac{d}{dx}(6x) = 3(1-x^2)$$

Substituting $x = \cos\theta$ gives

$$P_2^{|2|}(\cos\theta) = 3(1 - \cos^2\theta) = 3\sin^2\theta$$

14.77 The moment of inertia for N_2O is $6.649 \times 10^{-46}\ \mathrm{kg\cdot m^2}$. Determine the value of the rotational quantum number for a molecule with a rotational energy equal to $kT/2$ at $T = 298$ K.

▮ Solving *(14.56)* for J and substituting the data gives

$$J(J+1) = \frac{2(6.649\times10^{-46}\ \mathrm{kg\cdot m^2})(\frac{1}{2})(1.381\times10^{-23}\ \mathrm{J\cdot K^{-1}})(298\ \mathrm{K})[(1\ \mathrm{J})/(1\ \mathrm{kg\cdot m^2\cdot s^{-2}})]}{(1.0546\times10^{-34}\ \mathrm{J\cdot s})^2} = 246$$

$$J \approx 15$$

14.78 What is the general equation for the rotational energy spacing? Calculate ΔE for the $J = 15$ to $J = 16$ transition for the system described in Problem 14.77.

▮ Using *(14.56)* gives

$$\Delta E = \frac{(J+1)(J+2)\hbar^2}{2I} - \frac{J(J+1)\hbar^2}{2I} = \frac{(J+1)\hbar^2}{I} \tag{14.58}$$

Substituting the data into *(14.58)* gives

$$\Delta E = \frac{(15+1)(1.0546\times10^{-34}\ \mathrm{J\cdot s})^2}{(6.649\times10^{-46}\ \mathrm{kg\cdot m^2})[(1\ \mathrm{J})/(1\ \mathrm{kg\cdot m^2\cdot s^{-2}})]} = 2.676\times10^{-22}\ \mathrm{J}$$

14.79 Prepare a vector diagram to illustrate the relation between the total and component angular momenta for $l = 4$. What is the degeneracy of this rotational level?

▌ The vector length of the total angular momentum $(I\omega)$ is

$$I\omega = \hbar[l(l+1)]^{1/2} = \hbar[4(5)] = (20)^{1/2}\hbar$$

and the angular momentum component is given by $m\hbar$, where $m = 0, \pm 1, \pm 2, \ldots, \pm l$. In this case, the components will be $0, \pm\hbar, \pm 2\hbar, \pm 3\hbar$, and $\pm 4\hbar$. The vector diagram is shown in Fig. 14-4. The degeneracy is

$$g = 2l + 1 = 2(4) + 1 = 9$$

corresponding to the nine vectors shown. (The uncertainty principle does not permit exact knowledge of the positions of the vectors, and so the diagram is often represented by a series of concentric cones.)

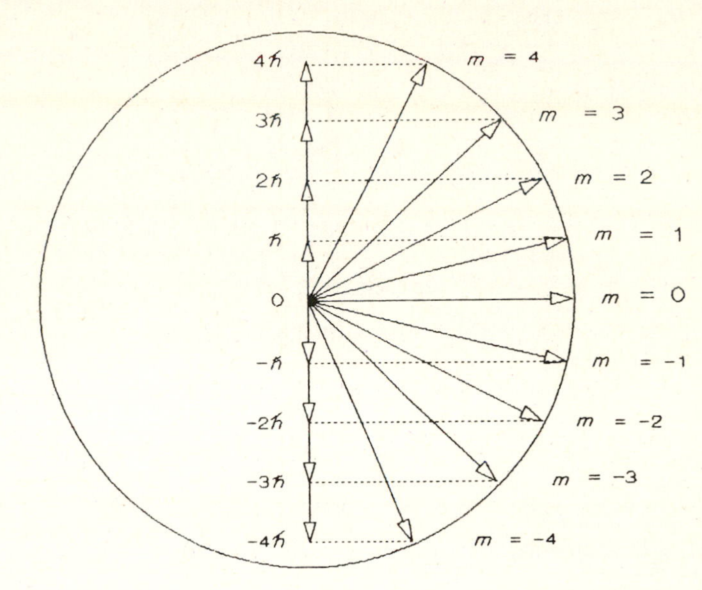

Fig. 14-4

14.80 Write the Schrödinger wave equation for a *simple harmonic oscillator* system having a reduced mass μ. The solution to the equation is

$$\psi_v = \frac{1}{(2^v v!)^{1/2}}\left(\frac{a}{\pi}\right)^{1/4} e^{-ax^2} H_v(a^{1/2}x) \tag{14.59}$$

where $a = 2\pi\nu_0\mu/\hbar$, the *vibrational quantum number* is given by $v = 0, 1, 2, \ldots$, ν_0 is the classical vibrational frequency, and the *Hermite polynomials* are given by

$$H_n(z) = (-1)^n e^{z^2} \frac{d^n}{dz^n} e^{-z^2} \tag{14.60}$$

Write the wave function for $v = 0$.

▌ The Hamiltonian operator is given by *(14.36)* using *(14.37a)*, giving

$$\left(-\frac{\hbar^2}{2\mu}\frac{d^2}{dx^2} + \frac{hx^2}{2}\right)\psi_v = E\psi_v$$

where $U(x) = kx^2/2$ (see Problem 3.75). Substituting $n = v = 0$ into *(14.60)* gives

$$H_0(a^{1/2}x) = H_0(z) = (-1)^0 e^{z^2} \frac{d^0}{dz^0} e^{-z^2} = 1$$

and *(14.59)* gives

$$\psi_0 = \frac{1}{[(2^0)(0!)]^{1/2}}\left(\frac{a}{\pi}\right)^{1/4} e^{-ax^2/2}(1) = \left(\frac{a}{\pi}\right)^{1/4} e^{-ax^2/2}$$

14.81 The Hermite polynomials in *(14.60)* can be determined using Table 14-2. Confirm the entry for $H_4(z)$ given in the table.

Table 14-2

	z^0	z^1	z^2	z^3	z^4	z^5	z^6	z^7	z^8
H_0	1								
H_1		2							
H_2	−2		4						
H_3		−12		8					
H_4	12		−48		16				
H_5		120		−160		32			
H_6	−120		720		−480		64		
H_7		−1680		3360		−1344		128	
H_8	1680		−13 440		13 440		−3580		256

▮ Using *(14.60)* gives

$$H_4(z) = (-1)^4 e^{z^2}\frac{d^4}{dz^4}e^{-z^2} = e^{z^2}\frac{d^3}{dz^3}[e^{-z^2}(-2z)] = e^{z^2}\frac{d^2}{dz^2}[-2\,e^{-z^2} - 2z\,e^{-z^2}(-2z)] = e^{z^2}\frac{d^2}{dz^2}[2(2z^2-1)\,e^{-z^2}]$$

$$= 2\,e^{z^2}\frac{d}{dz}[(4z)\,e^{-z^2} + (2z^2-1)(-2z)\,e^{-z^2}] = 4\,e^{z^2}\frac{d}{dz}[(3-2z^2)z\,e^{-z^2}]$$

$$= 4\,e^{z^2}[(-4z)z\,e^{-z^2} + (3-2z^2)\,e^{-z^2} + (3-2z^2)z(-2z)\,e^{-z^2}]$$

$$= 4[-4z + 3 - 2z^2 - 6z^2 + 4z^4] = 16z^4 - 48z^2 + 12$$

14.82 Using Table 14-2, show that the following recursion formulas are valid for $H_5(z)$:

$$\frac{dH_n(z)}{dz} = 2nH_{n-1}(z) \qquad (14.61)$$

$$H_n(z) - 2zH_{n-1}(z) + 2(n-1)H_{n-2}(z) = 0 \qquad (14.62)$$

▮ Substituting the expression for $H_5(z)$ from Table 14-2 into *(14.61)* gives

$$\frac{dH_5(z)}{dz} = \frac{d}{dz}(120z - 160z^3 + 32z^5) = 120 - 480z^2 + 160z^4 = 2(5)(12 - 48z^2 + 16z^4) = 2(5)H_4(z)$$

Substituting the expressions for $H_5(z)$, $H_4(z)$, and $H_3(z)$ into *(14.62)* gives

$$(120z - 160z^3 + 32z^5) - 2z(12 - 48z^2 + 16z^4) + 2(5-1)(-12z + 8z^3) = 0$$

14.83 A plot of $\psi_v^*\psi_v$ for $v = 8$ for a simple harmonic oscillator is shown in Fig. 14-5. How does the plot illustrate the correspondence principle?

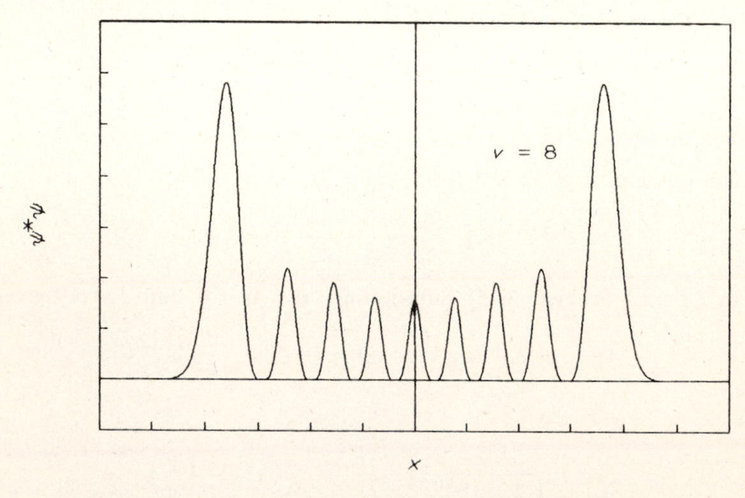

Fig. 14-5

▮ The value of $\psi^*\psi$ is proportional to the probability of locating the system at that value of x. The plot shows that the highest probability (even for this rather low value of v) is near the ends of the vibration as predicted by the classical theory.

14.84 The wave function for $v = 1$ for a simple harmonic oscillator is

$$\psi_1 = (2)^{1/2}\left(\frac{a^3}{\pi}\right)^{1/4} x\, e^{-ax^2/2}$$

Find the values of x such that $\psi^*\psi$ is a maximum.

▮ Differentiating $\psi^*\psi$ with respect to x and setting the result equal to zero gives

$$\frac{d(\psi^*\psi)}{dx} = 2\left(\frac{a^3}{\pi}\right)^{1/2}\frac{d}{dx}(x^2 e^{-ax^2}) = 2(a^3/\pi)^{1/2}[2x\, e^{-ax^2} + x^2 e^{-ax^2}(-2ax)] = 0$$

$$x = \pm(1/a)^{1/2} = \pm[\hbar/2\pi n \nu_0 \mu]^{1/2}$$

14.85 Determine $\langle x\rangle$ and $\langle x^2\rangle$ for $v = 0$ for a simple harmonic oscillator.

▮ Substituting the wave function found in Problem 14.80 and *(14.26)* into *(14.34)* with the denominator set to unity gives

$$\langle x\rangle = \int_{-\infty}^{\infty}\left[\left(\frac{a}{\pi}\right)^{1/4} e^{-ax^2/2}\right]^* x\left[\left(\frac{a}{\pi}\right)^{1/4} e^{-ax^2/2}\right] dx = \left(\frac{a}{\pi}\right)^{1/2}\int_{-\infty}^{\infty} x\, e^{-ax^2}\, dx = 0$$

Likewise, for $\hat{x}^2 = x^2$,

$$\langle x^2\rangle = \left(\frac{a}{\pi}\right)^{1/2}\int_{-\infty}^{\infty} x^2 e^{-ax^2}\, dx = \left(\frac{a}{\pi}\right)^{1/2} 2\int_{0}^{\infty} x^2 e^{-ax^2}\, dx = 2\left(\frac{a}{\pi}\right)^{1/2}\left(\frac{1}{4a}\right)\left(\frac{\pi}{a}\right)^{1/2} = \frac{1}{2a} = \frac{\hbar}{4\pi\nu_0\mu}$$

14.86 The vibrational force constant for the HI molecule is 314.14 $N \cdot m^{-1}$. Calculate the zero-point energy.

▮ The vibrational energy for a simple harmonic oscillator is given by

$$E = (v + \tfrac{1}{2})h\nu_0 \qquad (14.63)$$

where

$$\nu_0 = \frac{(k/\mu)^{1/2}}{2\pi} \qquad (14.64)$$

Using *(2.23)* gives

$$\mu = \frac{(1.0079\times10^{-3}\ \mathrm{kg\cdot mol^{-1}})(126.905\times10^{-3}\ \mathrm{kg\cdot mol^{-1}})}{(1.0079\times10^{-3}\ \mathrm{kg\cdot mol^{-1}} + 126.905\times10^{-3}\ \mathrm{kg\cdot mol^{-1}})(6.022\,045\times10^{23}\ \mathrm{mol^{-1}})} = 1.6605\times10^{-27}\ \mathrm{kg}$$

Substituting into *(14.64)* gives

$$\nu_0 = \left(\frac{(314.14\ \mathrm{N\cdot m^{-1}})[(1\ \mathrm{kg\cdot m\cdot s^{-2}})/(1\ \mathrm{N})]}{1.6605\times10^{-27}\ \mathrm{kg}}\right)^{1/2}\frac{1}{2\pi} = 6.9225\times10^{13}\ \mathrm{s^{-1}}$$

The zero-point energy corresponds to $v = 0$. Substituting into *(14.63)* gives

$$E = (0 + \tfrac{1}{2})(6.6262\times10^{-34}\ \mathrm{J\cdot s})(6.9225\times10^{13}\ \mathrm{s^{-1}}) = 2.2935\times10^{-20}\ \mathrm{J}$$

14.87 Calculate the probability that a simple harmonic oscillator in the $v = 0$ state will be outside its classical limits.

▮ At the classical limit of x, the energy given by *(14.63)* will equal the classical potential energy, giving

$$E = (0 + \tfrac{1}{2})h\nu_0 = kx_0/2$$

Solving for x_0 and substituting *(14.64)* gives

$$x_0 = \left(\frac{h\nu_0}{k}\right)^{1/2} = \left[\frac{h}{k}\left(\frac{(k/\mu)^{1/2}}{2\pi}\right)\right]^{1/2} = \left(\frac{\hbar}{(k\mu)^{1/2}}\right)^{1/2}$$

Substituting *(14.64)* into the definition of a gives

$$a = \frac{2\pi\nu_0\mu}{\hbar} = \frac{2\pi\mu}{\hbar}\left(\frac{(k/\mu)^{1/2}}{2\pi}\right) = \frac{(k\mu)^{1/2}}{\hbar} \qquad (14.65)$$

which upon substitution into the expression for x_0 gives

$$x_0 = (1/a)^{1/2} = a^{-1/2}$$

The probability of the system being within the classical limits is

$$P = \int_{-a^{-1/2}}^{a^{-1/2}} \psi^*\psi\,dx = 2\int_0^{a^{-1/2}} \psi^*\psi\,dx = 2\int_0^{a^{-1/2}} \left[\left(\frac{a}{\pi}\right)^{1/4} e^{-ax^2/2}\right]^* \left[\left(\frac{a}{\pi}\right)^{1/4} e^{-ax^2/2}\right] dx$$

$$= 2\left(\frac{a}{\pi}\right)^{1/2} \int_0^{a^{-1/2}} e^{-ax^2}\,dx = \frac{2}{\pi^{1/2}} \int_0^1 e^{-z^2}\,dz = \operatorname{erf}(1)$$

where $z^2 = ax^2$. From Table 2-1 and Fig. 2-1, $P = 0.8427$. The probability of the oscillator being outside the classical limits is

$$P = 1 - 0.8427 = 0.1573$$

14.5 APPROXIMATION METHODS

14.88 Use the trial wave function $\phi = \cos ax$, where a is a constant and $-\pi/2a \le x \le \pi/2a$, to calculate the energy of the first state of a simple harmonic oscillator. Does the result illustrate the variation theorem?

▌ Substituting $\mathcal{H} = -(\hbar^2/2\mu)(d^2/dx^2) + kx^2/2$ and the given wave function into *(14.34)* gives

$$E_\phi = \frac{\int_{-\pi/2a}^{\pi/2a} (\cos ax)^*[(-\hbar^2/2\mu)(d^2/dx^2) + kx^2/2](\cos ax)\,dx}{\int_{-\pi/2a}^{\pi/2a} (\cos ax)^*(\cos ax)\,dx}$$

$$= \frac{\dfrac{-\hbar^2}{2\mu}\displaystyle\int_{-\pi/2a}^{\pi/2a} \cos ax \frac{d^2}{dx^2}\cos ax\,dx + \frac{k}{2}\int_{-\pi/2a}^{\pi/2a} x^2\cos^2 ax\,dx}{\int_{-\pi/2a}^{\pi/2a} \cos^2 ax\,dx}$$

$$= \frac{\dfrac{-\hbar^2}{2\mu}a^2\displaystyle\int_{-\pi/2a}^{\pi/2a} \cos^2 ax\,dx + \frac{k}{2}\int_{-\pi/2a}^{\pi/2a} x^2\cos^2 ax\,dx}{\int_{-\pi/2a}^{\pi/2a} \cos^2 ax\,dx}$$

$$= \frac{\dfrac{\hbar^2}{2\mu}a^2\left(\dfrac{x}{2} + \dfrac{\sin 2ax}{4a}\right)\Bigg|_{-\pi/2a}^{\pi/2a} + \dfrac{k}{2}\left[\dfrac{x^3}{6} + \left(\dfrac{x^2}{4a} - \dfrac{1}{8a^3}\right)\sin 2ax + \dfrac{x\cos 2ax}{4a^2}\right]\Bigg|_{-\pi/2a}^{\pi/2a}}{[x/2 + (\sin 2ax)/4a]|_{-\pi/2a}^{\pi/2a}}$$

$$= \frac{\dfrac{\hbar^2}{2\mu}a^2\left(\dfrac{\pi/2a}{2} - \dfrac{-\pi/2a}{2} + 0 - 0\right) + \dfrac{k}{2}\left(\dfrac{(\pi/2a)^3}{6} - \dfrac{(-\pi/2a)^3}{6} + 0 - 0 + \dfrac{(\pi/2a)(-1)}{4a^2} - \dfrac{(-\pi/2a)(-1)}{4a^2}\right)}{(\pi/2a)/2 - (-\pi/2a)/2 + 0 - 0}$$

$$= \frac{\hbar^2 a^2}{2\mu} + \frac{k}{4a^2}\left(\frac{\pi^2}{6} - 1\right)$$

The expression for E_ϕ is minimized with respect to a by

$$\frac{\partial E_\phi}{\partial a} = \frac{2a\hbar^2}{2\mu} - \frac{2k}{4a^3}\left(\frac{\pi^2}{6} - 1\right) = 0$$

giving

$$a^2 = \left[\frac{k\mu}{2\hbar^2}\left(\frac{\pi^2}{6} - 1\right)\right]^{1/2}$$

The minimum energy is

$$E_{\min,\phi} = \frac{\hbar^2}{2\mu}\left[\frac{k\mu}{2\hbar^2}\left(\frac{\pi^2}{6} - 1\right)\right]^{1/2} + \frac{k(\pi^2/6 - 1)}{4[(k\mu/2\hbar^2)(\pi^2/6 - 1)]^{1/2}} = \hbar(k/\mu)^{1/2}(1/2^{1/2})(\pi^2/6 - 1)^{1/2} = 0.568h\nu_0$$

The *variation theorem* states that the energy calculated using the true wave function is related to the energy calculated using a trial wave function by

$$E \le E_{\min,\phi} \qquad (14.66)$$

In this case *(14.63)* gives the correct energy as $0.5h\nu_0$, so *(14.66)* is valid.

14.89 Use the trial wave function

$$\phi = c_1 x(a - x) + c_2 x^2(a - x)^2$$

where c_1 and c_2 are constants to calculate the energy of the first state of a one-dimensional potential energy well with infinite sides and width a. Does the result illustrate the variation theorem?

I If the trial wave function is constructed by taking a linear combination of m "basis functions" (μ_i)

$$\phi = \sum_{i=1}^{m} c_i \mu_i \tag{14.67}$$

the values of c_i that give the minimum energy satisfy the following set of equations:

$$\begin{aligned} c_1(H_{11} - ES_{11}) + c_2(H_{12} - ES_{12}) + \cdots + c_m(H_{1m} - ES_{1m}) &= 0 \\ c_1(H_{21} - ES_{21}) + c_2(H_{22} - ES_{22}) + \cdots + c_m(H_{2m} - ES_{2m}) &= 0 \\ \vdots \\ c_1(H_{m1} - ES_{m1}) + c_2(H_{m2} - ES_{m2}) + \cdots + c_m(H_{mm} - ES_{mm}) &= 0 \end{aligned} \tag{14.68}$$

where

$$H_{ij} = \int_{\text{all space}} \mu_i^* \mathcal{H} \mu_j \, dV \tag{14.69}$$

$$S_{ij} = \int_{\text{all space}} \mu_i^* \mu_j \, dV \tag{14.70}$$

The system *(14.68)* will have a nonzero solution only for those values of E that make the *secular determinant* of the coefficients vanish, that is,

$$\begin{vmatrix} H_{11} - ES_{11} & H_{12} - ES_{12} & \cdots & H_{1m} - ES_{1m} \\ H_{21} - ES_{21} & H_{22} - ES_{22} & \cdots & H_{2m} - ES_{2m} \\ \vdots & \vdots & & \vdots \\ H_{m1} - ES_{m1} & H_{m2} - ES_{m2} & \cdots & H_{mm} - ES_{mm} \end{vmatrix} = 0 \tag{14.71}$$

For the given trial wave function, $\mu_1 = x(a - x)$ and $\mu_2 = x^2(a - x)^2$. Substituting $\mathcal{H} = (-\hbar^2/2m)(d^2/dx^2)$ into *(14.69)* and *(14.70)* gives

$$H_{11} = \int_0^a \mu_1^* \mathcal{H} \mu_1 \, dx = -\frac{\hbar^2}{2m} \int_0^a \left[x(a - x)\right]^* \frac{d^2}{dx^2}[x(a - x)] \, dx$$

$$= -\frac{\hbar^2}{2m} \int_0^a x(a - x)(-2) \, dx = \frac{\hbar^2}{m} \left(\frac{a}{2}x^2 + \frac{1}{3}x^3 \right) \Bigg|_0^a = \frac{\hbar^2 a^3}{6m}$$

$$H_{12} = \int_0^a \mu_1^* \mathcal{H} \mu_2 \, dx = -\frac{\hbar^2}{2m} \int_0^a [x(a - x)]^* \frac{d^2}{dx^2}[x^2(a - x)^2] \, dx$$

$$= -\frac{\hbar^2}{2m} \int_0^a x(a - x)(-2a^2 - 12ax + 12x^2) \, dx = -\frac{\hbar^2}{2m} \int_0^a (a^3x - 7a^2x^2 + 12ax^3 - 6x^4) \, dx$$

$$= \frac{\hbar^2}{m} \left[\frac{a^3}{2}x^2 - \frac{7a^2}{3}x^3 + \frac{12a}{4}x^4 - \frac{6}{5}x^5 \right] \Bigg|_0^a = \frac{\hbar^2 a^5}{30m}$$

$$H_{21} = \cdots = H_{12}$$

$$H_{22} = \int_0^a \mu_2^* \mathcal{H} \mu_2 \, dx = -\frac{\hbar^2}{2m} \int_0^a [x^2(a - x)^2]^* \frac{d^2}{dx^2}[x^2(a - x)^2] \, dx$$

$$= -\frac{\hbar^2}{2m} \int_0^a x^2(a - x)^2(2a^2 - 12ax + 12ax^2) \, dx = -\frac{\hbar^2}{2m} \int_0^a (a^4x^2 - 8a^3x^3 + 19a^2x^4 - 18ax^5 + 6x^6) \, dx$$

$$= -\frac{\hbar^2}{2m} \left(\frac{a^4}{3}x^3 - \frac{8a^3}{4}x^4 - \frac{19a^2}{5}x^5 + \frac{18a}{6}x^6 - \frac{6}{7}x^7 \right) \Bigg|_0^a = \frac{\hbar^2 a^7}{105m}$$

$$S_{11} = \int_0^a \mu_1^* \mu_1 \, dx = \int_0^a [x(a - x)]^*[x(a - x)] \, dx = \int_0^a (a^2x^2 - 2ax^3 + x^4) \, dx$$

$$= \left(\frac{a^2}{3}x^3 - \frac{2a}{4}x^4 + \frac{1}{5}x^5 \right) \Bigg|_0^a = \frac{a^5}{30}$$

$$S_{12} = \int_0^a \mu_1^* \mu_2 \, dx = \int_0^a [x(a - x)]^*[x^2(a - x)^2] \, dx$$

$$= \int_0^a (a^3x^3 - 3a^2x^4 + 3ax^5 - x^6) \, dx = \left(\frac{a^3}{4}x^4 - \frac{3a^2}{5}x^5 + \frac{3a}{6}x^6 - \frac{1}{7}x^7 \right) \Bigg|_0^a = a^7/141$$

$$S_{21} = \cdots = S_{12}$$

$$S_{22} = \int_0^a \mu_2^* \mu_2 \, dx = \int_0^a [x^2(a-x)^2]^*[x^2(a-x)^2] \, dx = \int_0^a (a^4x^4 - 4a^3x^5 + 6a^2x^6 - 4ax^7 + x^8) \, dx$$

$$= \left(\frac{a^5}{5}x^5 - \frac{4a^3}{6}x^6 - \frac{6a^2}{7}x^7 - \frac{4a}{8}x^8 + \frac{1}{9}x^9 \right) \Bigg|_0^a = \frac{a^9}{630}$$

Substituting these results into *(14.71)* gives

$$\begin{vmatrix} \frac{\hbar^2 a^3}{6m} - E\frac{a^5}{30} & \frac{\hbar^2 a^5}{30m} - E\frac{a^7}{140} \\ \frac{\hbar^2 a^5}{30m} - E\frac{a^7}{140} & \frac{\hbar^2 a^7}{105m} - E\frac{a^9}{630} \end{vmatrix} = 0$$

Multiplying the elements of the determinant by 8820 and dividing by a^3 simplifies the determinant to

$$\begin{vmatrix} 1470\frac{\hbar^2}{m} - E(294a^2) & 294a^2\frac{\hbar^2}{m} - E(63a^4) \\ 294a^2\frac{\hbar^2}{m} - E(63a^4) & 84a^4\frac{\hbar^2}{m} - E(14a^6) \end{vmatrix} = 0$$

Expanding the determinant and dividing by 49 gives the quadratic equation

$$(3a^4)E^2 - (168a^2\hbar^2/m)E + 756\hbar^4/m^2 = 0$$

Solving for E gives

$$E = \frac{-(-168a^2\hbar^2/m) \pm [(-168a^2\hbar^2/m)^2 - 4(3a^4)(756\hbar^4/m^2)]^{1/2}}{2(3a^4)}$$

$$= \frac{\hbar^2}{ma^2}\left[\frac{168 \pm (19\,152)^{1/2}}{6}\right] = \frac{h^2}{(2\pi)^2 ma^2}(51.065 \text{ or } 4.934\,875) = 0.125\,001\,9\frac{h^2}{8ma^2} = 1.000\,015\frac{h^2}{8ma^2}$$

Comparison of this result with *(14.44)* shows that *(14.66)* is valid.

14.90 Calculate the first-order ground-state energy of an anharmonic oscillator having a potential energy given by

$$U(x) = \frac{kx^4}{2} + \frac{\alpha x^3}{6} + \frac{\beta x^4}{24}$$

▮ The Hamiltonian operator for this system is given by *(14.36)* and *(14.37a)* as

$$\mathcal{H} = -\frac{\hbar^2}{2\mu}\frac{d^2}{dx^2} + \frac{kx^2}{2} + \frac{\alpha x^3}{6} + \frac{\beta x^4}{24} = \mathcal{H}^{(0)} + \mathcal{H}^{(1)}$$

where the unperturbed Hamiltonian operator for the harmonic oscillator is

$$\mathcal{H}^{(0)} = -\frac{\hbar^2}{2\mu}\frac{d^2}{dx^2} + \frac{kx^2}{2}$$

(see Problem 14.80) and the perturbation is

$$\mathcal{H}^{(1)} = \frac{\alpha x^3}{6} + \frac{\beta x^4}{24}$$

The ground-state wave function (see Problem 14.80) is

$$\psi_0^{(0)} = \left(\frac{a}{\pi}\right)^{1/4} e^{-ax^2/2}$$

According to the *perturbation theory*, the first-order estimate for the energy is given by

$$E_n = E_n^{(0)} + H_{nn}^{(1)} \tag{14.72}$$

where $E_n^{(0)}$ is the energy for the unperturbed problem and $H_{nn}^{(1)}$ is given by *(14.69)* using $\mathcal{H}^{(1)}$ and $\psi_n^{(0)}$. Substituting into *(14.69)* gives

$$H_{00}^{(1)} = \int_{\text{all space}} \psi_0^{(0)*} \mathscr{H}^{(1)} \psi_0^{(0)}\, dV = \int_{-\infty}^{\infty} \left[\left(\frac{a}{\pi}\right)^{1/4} e^{-ax^2/2}\right]^* \left(\frac{\alpha x^3}{6} + \frac{\beta x^4}{24}\right) \left[\left(\frac{a}{\pi}\right)^{1/4} e^{-ax^2/2}\right] dx$$

$$= \left(\frac{a}{\pi}\right)^{1/2} \int_{-\infty}^{\infty} \left(\frac{\alpha x^3}{6} + \frac{\beta x^2}{24}\right) e^{-ax^2}\, dx = \left(\frac{a}{\pi}\right)^{1/2} \left[\frac{\alpha}{6} \int_{-\infty}^{\infty} x^3\, e^{-ax^2}\, dx - \frac{\beta}{24} \int_{-\infty}^{\infty} x^4\, e^{-ax^2}\, dx\right]$$

$$= \left(\frac{a}{\pi}\right)^{1/2} \left(0 + \frac{\beta}{24} 2 \int_{-\infty}^{\infty} x^4\, e^{-ax^2}\, dx\right) = \left(\frac{a}{\pi}\right)^{1/2} \left(\frac{\beta}{12}\right) \left(\frac{3\pi^{1/2}}{8a^{5/2}}\right) = \frac{\beta}{32a^2} = \frac{\hbar^2 \beta}{32k\mu}$$

where *(14.65)* was substituted for a. Substituting *(14.63)* and the result for $H_{00}^{(1)}$ into *(14.72)* gives

$$E_0 = \tfrac{1}{2} h\nu_0 + \frac{\hbar^2 \beta}{32k\mu}$$

CHAPTER 15
Atomic Structure and Spectroscopy

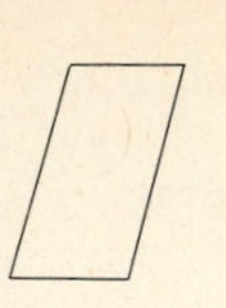

15.1 HYDROGENLIKE ATOMS

15.1 Write the Hamiltonian operator describing the motion of the electron in a hydrogenlike atom.

▌ A *hydrogenlike atom* consists of one electron moving around a nucleus of charge $+Ze$. Using *(14.36)* and *(14.37b)* gives

$$\mathscr{H} = -\frac{\hbar^2}{2\mu}\left[\frac{1}{r^2}\frac{\partial}{\partial r}\left(r^2\frac{\partial}{\partial r}\right) + \frac{1}{r^2 \sin\theta}\frac{\partial}{\partial\theta}\left(\sin\theta\frac{\partial}{\partial\theta}\right) + \frac{1}{r^2\sin^2\theta}\frac{\partial^2}{\partial\phi^2}\right] + \frac{-Ze^2}{4\pi\varepsilon_0 r} \tag{15.1}$$

where μ is the reduced mass ($\approx m_e$) and the $-Ze^2/4\pi\varepsilon_0 r$ term represents the Coulombic attraction between the nucleus and the electron.

15.2 Using Tables 15-1 and 15-2, write the complete wave function for an electron with the following quantum numbers: $n = 2,\ l = 1,\ m = +1$.

Table 15-1

n	l	$R_{n,l}(r)$ where $\rho = 2Zr/na_0$
1	0	$2(Z/a_0)^{3/2}\, e^{-\rho/2}$
2	0	$[1/2(2)^{1/2}](Z/a_0)^{3/2}(2-\rho)\, e^{-\rho/2}$
2	1	$[1/2(6)^{1/2}](Z/a_0)^{3/2}\rho\, e^{-\rho/2}$
3	0	$[2/81(3)^{1/2}](Z/a_0)^{3/2}(27 - 18\rho + 2\rho^2)\, e^{-\rho/2}$
3	1	$[4/81(6)^{1/2}](Z/a_0)^{3/2}(6-\rho)\rho\, e^{-\rho/2}$
3	2	$[4/81(30)^{1/2}](Z/a_0)^{3/2}\rho^2\, e^{-\rho/2}$

Table 15-2

l	m	$\Theta_{l,m}(\theta)$	$\Phi_m(\phi)$
0	0	$(2)^{1/2}/2$	$(1/2\pi)^{1/2}$
1	0	$[(6)^{1/2}/2]\cos\theta$	$(1/2\pi)^{1/2}$
1	+1	$[(3)^{1/2}/2]\sin\theta$	$(1/2\pi)^{1/2}\, e^{i\phi}$
1	−1	$[(3)^{1/2}/2]\sin\theta$	$(1/2\pi)^{1/2}\, e^{-i\phi}$
2	0	$[(15/2)^{1/2}/2][3\cos^2\theta - 1]$	$(1/2\pi)^{1/2}$
2	+1	$[(15)^{1/2}/2]\sin\theta\cos\theta$	$(1/2\pi)^{1/2}\, e^{i\phi}$
2	−1	$[(15)^{1/2}/2]\sin\theta\cos\theta$	$(1/2\pi)^{1/2}\, e^{-i\phi}$
2	+2	$[(15)^{1/2}/4]\sin^2\theta$	$(1/2\pi)^{1/2}\, e^{2i\phi}$
2	−2	$[(15)^{1/2}/4]\sin^2\theta$	$(1/2\pi)^{1/2}\, e^{-2i\phi}$

▌ The complete wave function for a hydrogenlike atom is given by

$$\psi(r,\theta,\phi) = R(r)\Theta(\theta)\Phi(\phi) \tag{15.2}$$

Substituting the radial wave function from Table 15-1 for $n = 2$ and $l = 1$ and the angular wave functions from Table 15-2 for $l = 1$ and $m = +1$ into *(15.2)* gives

$$\psi(r,\theta,\phi) = \frac{1}{2(6)^{1/2}}\left(\frac{Z}{a_0}\right)^{3/2}\rho\, e^{-\rho/2}\,\frac{3^{1/2}}{2}(\sin\theta)\,\frac{1}{(2\pi)^{1/2}}\,e^{i\phi} = \frac{1}{8(\pi)^{1/2}}\left(\frac{Z}{a_0}\right)^{3/2}\rho\, e^{-\rho/2}(\sin\theta)\, e^{i\phi}$$

15.3 Show that the wave function determined in Problem 15.2 is normalized.

▌ Substituting the wave function into *(14.22)*, using the spherical volume element given by *(14.23)*, and integrating gives

$$\int_{\text{all space}} \psi^* \psi \, dV = \int_0^\infty \int_0^\pi \int_0^{2\pi} \left[\frac{1}{8(\pi)^{1/2}} \left(\frac{Z}{a_0} \right)^{3/2} \rho \, e^{-\rho/2} (\sin\theta) \, e^{i\phi} \right]^*$$

$$\times \left[\frac{1}{8(\pi)^{1/2}} \left(\frac{Z}{a_0} \right)^{3/2} \rho \, e^{-\rho/2} (\sin\theta) \, e^{i\phi} \right] r^2 \sin\theta \, dr \, d\theta \, d\phi$$

$$= \frac{1}{64\pi} \left(\frac{Z}{a_0} \right)^3 \int_0^\infty \int_0^\pi \int_0^{2\pi} \rho^2 \, e^{-\rho} (\sin^3\theta) r^2 \, dr \, d\theta \, d\phi$$

$$= \frac{1}{64\pi} \left(\frac{Z}{a_0} \right)^5 \int_0^\infty \int_0^\pi r^4 \, e^{-Zr/a_0} \sin^3\theta \, dr \, d\theta \int_0^{2\pi} d\phi$$

$$= \frac{1}{32} \left(\frac{Z}{a_0} \right)^5 \int_0^\infty r^4 \, e^{-Zr/a_0} \, dr \int_0^\pi \sin^3\theta \, d\theta$$

$$= \frac{1}{32} \left(\frac{Z}{a_0} \right)^5 \int_0^\infty r^4 \, e^{-Zr/a_0} \, dr \left[-\tfrac{1}{3} (\cos\theta)(\sin^2\theta + 2) \right] \Big|_0^\pi$$

$$= -\frac{1}{96} \left(\frac{Z}{a_0} \right)^5 \int_0^\infty r^4 \, e^{-Zr/a_0} \, dr \left[(-1)(0+2) - (1)(0+2) \right]$$

$$= \frac{1}{24} \left(\frac{Z}{a_0} \right)^5 \int_0^\infty r^4 \, e^{-Zr/a_0} \, dr = \frac{1}{24} \left(\frac{Z}{a_0} \right)^5 \frac{4!}{(Z/a_0)^5} = 1$$

15.4 Show that the wave functions describing a 1s electron and a 2s electron are orthogonal.

$$\psi_{1s} = \left(\frac{Z}{a_0} \right)^{3/2} \frac{1}{\pi^{1/2}} e^{-\rho/2} = \left(\frac{Z}{a_0} \right)^{3/2} \frac{1}{\pi^{1/2}} e^{-Zr/a_0}$$

$$\psi_{2s} = \left(\frac{Z}{a_0} \right)^{3/2} \frac{1}{4(2\pi)^{1/2}} (2 - \rho) \, e^{-\rho/2} = \left(\frac{Z}{a_0} \right)^{3/2} \frac{1}{4(\pi)^{1/2}} \left(2 - \frac{Zr}{a_0} \right) e^{-Zr/2a_0}$$

▌ Substituting the wave functions into *(14.24)*, using the spherical volume element given by *(14.23)*, and integrating gives

$$\int_{\text{all space}} \psi_i^* \psi_j \, dV = \int_0^\infty \int_0^\pi \int_0^{2\pi} \left[\left(\frac{Z}{a_0} \right)^{3/2} \frac{1}{\pi^{1/2}} e^{-Zr/a_0} \right]^* \left[\left(\frac{Z}{a_0} \right)^{3/2} \frac{1}{4(2\pi)^{1/2}} \left(2 - \frac{Zr}{a_0} e^{-Zr/2a_0} \right) \right] r^2 \sin\theta \, dr \, d\theta \, d\phi$$

$$= \left(\frac{Z}{a_0} \right)^3 \frac{1}{4\pi(2)^{1/2}} \int_0^\infty \int_0^\pi \int_0^{2\pi} \left(2 - \frac{Zr}{a_0} \right) e^{-3Zr/2a_0} r^2 \sin\theta \, dr \, d\theta \, d\phi$$

$$= \left(\frac{Z}{a_0} \right)^3 \frac{1}{4\pi(2)^{1/2}} \int_0^\infty \int_0^\pi \left(2 - \frac{Zr}{a_0} \right) r^2 \, e^{-3Zr/2a_0} \sin\theta \, dr \, d\theta \int_0^{2\pi} d\phi$$

$$= \left(\frac{Z}{a_0} \right)^3 \frac{1}{2(2)^{1/2}} \int_0^\infty \left(2 - \frac{Zr}{a_0} \right) r^2 \, e^{-3Zr/2a_0} \, dr \int_0^\pi \sin\theta \, d\theta$$

$$= \left(\frac{Z}{a_0} \right)^3 \frac{1}{2(2)^{1/2}} \int_0^\infty \left(2 - \frac{Zr}{a_0} \right) r^2 \, e^{-3Zr/2a_0} \, dr \, (-\cos\theta) \Big|_0^\pi$$

$$= \left(\frac{Z}{a_0} \right)^3 \frac{1}{2(2)^{1/2}} \int_0^\infty \left(2 - \frac{Zr}{a_0} \right) r^2 \, e^{-3Zr/2a_0} \, dr \, (-1 - 1)$$

$$= \left(\frac{Z}{a_0} \right)^3 \frac{1}{2^{1/2}} \left[2 \int_0^\infty r^2 \, e^{-3Zr/2a_0} \, dr - \frac{Z}{a_0} \int_0^\infty r^3 \, e^{-3Zr/2a_0} \, dr \right]$$

$$= \left(\frac{Z}{a_0} \right)^3 \frac{1}{2^{1/2}} \left[2 \frac{2!}{(3Z/2a_0)^3} - \frac{Z}{a_0} \left(\frac{3!}{(3Z/2a_0)^4} \right) \right]$$

$$= \left(\frac{Z}{a_0} \right)^3 \frac{1}{2^{1/2}} \left[\frac{32}{27} \left(\frac{a_0}{Z} \right)^3 - \frac{Z}{a_0} \left(\frac{a_0}{Z} \right)^4 \left(\frac{96}{81} \right) \right] = 0$$

15.5 The angular wave function for a hydrogenlike atom is given by $Y(\theta, \phi) = \Theta(\theta)\Phi(\phi)$, where $\Theta(\theta)$ is given by *(14.55)* with the *azimuthal quantum number* (l) replacing J, $\Phi(\phi)$ is given by

$$\Phi(\phi) = \frac{1}{(2\pi)^{1/2}} e^{im\phi} \tag{15.3}$$

and m is the *magnetic quantum number.* Note that $m = 0, \pm 1, \pm 2, \ldots, \pm l$ (see Problem 14.75). Confirm the entry for $l = 1$ and $m = +1$ given in Table 15-2.

▮ From *(15.3)*,

$$\Phi(\phi) = \frac{1}{(2\pi)^{1/2}} e^{i(+1)\phi} = \frac{1}{(2\pi)^{1/2}} e^{i\phi}$$

which agrees with the entry for the $\Phi(\phi)$ wave function. From Table 14-1, $P_1^{|1|}(\cos\theta) = \sin\theta$. Substituting into *(14.55)* gives

$$\Theta(\theta)_{1,1} = \left[\frac{[2(1) + 1](1 - |1|)!}{2(1 + |1|!)}\right]^{1/2} \sin\theta = \left(\frac{3}{4}\right)^{1/2} \sin\theta = \frac{3^{1/2}}{2} \sin\theta$$

which agrees with the entry for the $\Theta(\theta)$ wave function.

15.6 Determine the magnitude of the orbital angular momentum and the component of angular momentum vector in the z direction for the angular wave function determined in Problem 15.5.

▮ The magnitude of the orbital angular momentum (L) is given by

$$L = [l(l + 1)]^{1/2}\hbar \tag{15.4}$$

Substituting $l = 1$ into *(15.4)* gives

$$L = [(1)(1 + 1)]^{1/2}\hbar = 2^{1/2}\hbar = 2^{1/2}(1.0546 \times 10^{-34}\ \text{J}\cdot\text{s}) = 1.4914 \times 10^{-34}\ \text{J}\cdot\text{s}$$

The possible values of the component of angular momentum vector in the z direction (L_z) are given by

$$L_z = m\hbar \tag{15.5}$$

Substituting $m = 0, \pm 1$ into *(15.5)* gives

$$L_z = \begin{cases} (+1)(1.0546 \times 10^{-34}\ \text{J}\cdot\text{s}) = 1.0546 \times 10^{-34}\ \text{J}\cdot\text{s} \\ (0)(1.0546 \times 10^{-34}) = 0 \\ (-1)(1.0546 \times 10^{-34}) = -1.0546 \times 10^{-34}\ \text{J}\cdot\text{s} \end{cases}$$

15.7 The angular wave functions for $l = 2$ and $m = \pm 1$ are

$$Y(\theta, \phi) = \frac{1}{2}\left(\frac{3}{2\pi}\right)^{1/2} (\sin\theta)\, e^{\pm i\phi}$$

Construct two real wave functions, $Y'(\theta, \phi)$ and $Y''(\theta, \phi)$, from these.

▮ The linear combinations

$$Y'(\theta, \phi) = 2^{1/2}\left[\frac{Y_+(\theta, \phi) + Y_-(\theta, \phi)}{2}\right] \tag{15.6a}$$

$$Y''(\theta, \phi) = 2^{1/2}\left[\frac{Y_+(\theta, \phi) - Y_-(\theta, \phi)}{2i}\right] \tag{15.6b}$$

will be real, where the subscripts on $Y(\theta, \phi)$ refer to the sign being used for the exponential term and $2^{1/2}$ is merely an additional normalization factor. Substituting into *(15.6)* and using *(14.25)* gives

$$Y'(\theta, \phi) = 2^{1/2}\frac{\frac{1}{2}(3/2\pi)^{1/2}(\sin\theta)\, e^{i\phi} + \frac{1}{2}(3/2\pi)^{1/2}(\sin\theta)\, e^{-i\phi}}{2} = (3/\pi)^{1/2}(\tfrac{1}{4}\sin\theta)[e^{i\phi} + e^{-i\phi}]$$

$$= (3/\pi)^{1/2}(\tfrac{1}{4}\sin\theta)[\cos\phi + i\sin\phi + \cos\phi - i\sin\phi] = \tfrac{1}{2}(3/\pi)^{1/2}\sin\theta\cos\phi \tag{15.7a}$$

$$Y''(\theta, \phi) = 2^{1/2}\frac{\frac{1}{2}(3/2\pi)^{1/2}(\sin\theta)\, e^{i\phi} - \frac{1}{2}(3/2\pi)^{1/2}(\sin\theta)\, e^{-i\phi}}{2i} = (3/\pi)^{1/2}(1/4i)(\sin\theta)(e^{i\phi} - e^{-i\phi})$$

$$= (3/\pi)^{1/2}(1/4i)(\sin\theta)(\cos\phi + i\sin\phi - \cos\phi + i\sin\phi) = \tfrac{1}{2}(3/\pi)^{1/2}\sin\theta\sin\phi \tag{15.7b}$$

15.8 Show that the wave functions given by *(15.7a)* and *(15.7b)* are orthonormal.

▌ Substituting *(15.7a)* and *(15.7b)* into *(14.22)* gives

$$\int_{\text{all space}} \psi^* \psi \, dV = \int_{\text{all space}} [Y'(\theta, \phi)]^*[Y''(\theta, \phi)] \, dV = \int_0^{2\pi} \int_0^{\pi} \left[\frac{1}{2}\left(\frac{3}{\pi}\right)^{1/2} \sin\theta \cos\phi\right]^*$$

$$\times \left[\frac{1}{2}\left(\frac{3}{\pi}\right)^{1/2} \sin\theta\cos\phi\right] \sin\theta \, d\theta \, d\phi = \frac{3}{4\pi}\int_0^{2\pi}\int_0^{\pi} \sin^3\theta \cos^2\phi \, d\theta \, d\phi$$

$$= \frac{3}{4\pi}\int_0^{\pi} \sin^3\theta \, d\theta \left(\int_0^{2\pi} \cos^2\phi \, d\phi\right) = \frac{3}{4\pi}\int_0^{\pi} \sin^3\theta \, d\theta \left(\frac{\phi}{2} + \frac{\sin 2\phi}{4}\right)\Bigg|_0^{2\pi}$$

$$= \frac{3}{4\pi}\int_0^{\pi} \sin^3\theta \, d\theta \left[\frac{2\pi}{2} - 0 + (0 - 0)\right] = \frac{3}{4}\int_0^{\pi} \sin^3\theta \, d\theta$$

$$= \frac{3}{4}\left(\frac{-1}{3}(\cos\phi)(\sin^2\phi + 2)\right)\Bigg|_0^{\pi} = \frac{-1}{4}[(-1)(0+2) - (1)(0+2)] = 1$$

$$\int_{\text{all space}} \psi^* \psi \, dV = \int_{\text{all space}} [Y''(\theta, \phi)]^* Y''(\theta, \phi) \, dV = \int_0^{2\pi} \int_0^{\pi} \left[\frac{1}{2}\left(\frac{3}{\pi}\right)^{1/2} \sin\theta \sin\phi\right]^*$$

$$\times \left[\frac{1}{2}\left(\frac{3}{\pi}\right)^{1/2} \sin\theta\sin\phi\right] \sin\theta \, d\theta \, d\phi = \frac{3}{4\pi}\int_0^{2\pi}\int_0^{\pi} \sin^3\theta \sin^2\phi \, d\theta \, d\phi$$

$$= \frac{3}{4\pi}\int_0^{\pi} \sin^3\theta \, d\theta \left[\int_0^{2\pi} \sin^2\phi \, d\phi\right] = \frac{3}{4\pi}\int_0^{\pi} \sin^3\theta \, d\theta \left(\frac{\phi}{2} + \frac{\sin 2\phi}{4}\right)\Bigg|_0^{2\pi}$$

$$= \frac{3}{4\pi}\int_0^{\pi} \sin^3\theta \, d\theta \left(\frac{2\pi}{2} - 0 + (0 - 0)\right) = \frac{3}{4}\int_0^{\pi} \sin^3\theta \, d\theta = 1$$

Both wave functions are normalized. Substituting into *(14.24)* gives

$$\int_{\text{all space}} \psi^* \psi \, dV = \int_{\text{all space}} [\psi''(\theta, \phi)]^*[\psi'(\theta, \phi)] \, dV = \int_0^{2\pi} \int_0^{\pi} \left[\frac{1}{2}\left(\frac{3}{\pi}\right)^{1/2} \sin\theta \sin\phi\right]^*$$

$$\times \left[\frac{1}{2}\left(\frac{3}{\pi}\right)^{1/2} \sin\theta\cos\phi\right] \sin\theta \, d\theta \, d\phi = \frac{3}{4\pi}\int_0^{2\pi}\int_0^{\pi} \sin^3\theta \sin\phi \cos\phi \, d\theta \, d\phi$$

$$= \frac{3}{4\pi}\int_0^{\pi} \sin^3\theta \, d\theta \left[\int_0^{2\pi} \sin\phi\cos\phi \, d\phi\right] = \frac{3}{4\pi}\int_0^{\pi} \sin^3\theta \, d\theta \left(\frac{1}{2}\sin^2\phi\right)\Bigg|_0^{2\pi}$$

$$= \frac{3}{8\pi}\int_0^{\pi} \sin^3\theta \, d\theta \, (0 - 0) = 0$$

The wave functions are orthogonal.

15.9 Prepare a plot of the wave function given by *(15.7b)* in the *yz* plane. On the same set of axes, prepare a plot of $Y''(\theta, \phi)^* Y''(\theta, \phi)$.

▌ In the *yz* plane, $\phi = 90°$ and $\phi = 270°$. As a sample calculation, consider $\phi = 90°$ and $\theta = 45°$.

$$Y''(\theta, \phi) = \frac{1}{2}\left(\frac{3}{\pi}\right)^{1/2} \sin 45° \sin 90° = 0.345$$

$$Y''(\theta, \phi)^* Y''(\theta, \phi) = \frac{3}{4\pi} \sin^2 45° \sin^2 90° = 0.119$$

The plots are shown in Fig. 15-1. For clarity, absolute values of $Y''(\theta, \phi)$ are shown. [Note that either polar graph paper, plotting values of θ and Y where $Y = |Y''(\theta, \phi)|$ or $Y''(\theta, \phi)^* Y''(\theta, \phi)$, or Cartesian graph paper, plotting values of $x = Y \sin\theta$ and $y = Y \cos\theta$, may be used.]

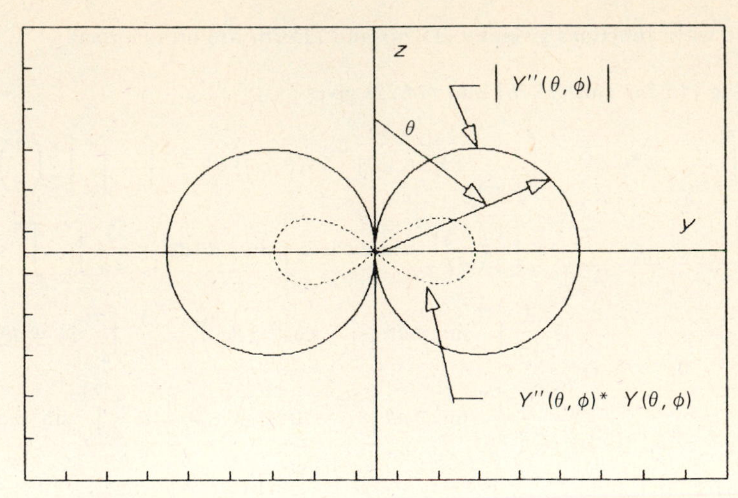

Fig. 15-1

15.10 Determine θ and ϕ for which the wave function given by *(15.7b)* is a maximum.

▌ Differentiating *(15.7b)* with respect to ϕ, setting the result equal to zero, and solving for the value of ϕ_{max} gives

$$\frac{\partial Y''(\theta,\phi)}{\partial\phi} = \frac{\partial}{\partial\phi}\left[\frac{1}{2}\left(\frac{3}{\pi}\right)^{1/2}\sin\theta\sin\phi\right] = \frac{1}{2}\left(\frac{3}{\pi}\right)^{1/2}\sin\theta\frac{\partial}{\partial\phi}\sin\phi = \frac{1}{2}\left(\frac{3}{\pi}\right)^{1/2}\sin\theta\cos\phi = 0$$

$$\cos\phi_{max} = 0 \qquad \phi_{max} = 90° \text{ or } 270°$$

The wave function is a maximum in the *yz* plane. Likewise, differentiating *(15.7b)* with respect to θ, setting the result equal to zero, and solving for the value of θ_{max} gives

$$\frac{\partial Y''(\theta,\phi)}{\partial\theta} = \frac{\partial}{\partial\theta}\left[\frac{1}{2}\left(\frac{3}{\pi}\right)^{1/2}\sin\theta\sin\phi\right] = \frac{1}{2}\left(\frac{3}{\pi}\right)^{1/2}\sin\phi\frac{\partial}{\partial\theta}\sin\theta = \frac{1}{2}\left(\frac{3}{\pi}\right)^{1/2}\sin\phi\cos\theta = 0$$

$$\cos\theta_{max} = 0 \qquad \theta_{max} = 90° \text{ or } 270°$$

The wave function is a maximum along the *y* axis.

15.11 Show that the angular distribution of a given p electron ($l = 1$) is spherical.

▌ A p electron is likely to be present in any of its three available states: either $Y(\theta,\phi)$ for $m = 0, \pm 1$ or $Y(\theta,\phi)$ for $m = 0$, $Y'(\theta,\phi)$, and $Y''(\theta,\phi)$. Choosing the latter set of orbitals, the total distribution will be given by

$$\phi = Y(\theta,\phi) + Y'(\theta,\phi) + Y''(\theta,\phi)$$

Because these angular wave functions are orthonormal,

$$\begin{aligned}\phi^*\phi &= [Y(\theta,\phi) + Y'(\theta,\phi) + Y''(\theta,\phi)]^*[Y(\theta,\phi) + Y'(\theta,\phi) + Y''(\theta,\phi)]\\ &= Y(\theta,\phi)^*Y(\theta,\phi) + Y'(\theta,\phi)^*Y'(\theta,\phi) + Y''(\theta,\phi)^*Y''(\theta,\phi)\end{aligned}$$

Substituting $Y(\theta,\phi)$ from Table 15-2 for $l = 1$ and $m = 0$, *(15.7a)* and *(15.7b)* give

$$\begin{aligned}\phi^*\phi &= \left[\frac{1}{2}\left(\frac{3}{\pi}\right)^{1/2}\cos\theta\right]^*\left[\frac{1}{2}\left(\frac{3}{\pi}\right)^{1/2}\cos\theta\right] + \left[\frac{1}{2}\left(\frac{3}{\pi}\right)^{1/2}\sin\theta\cos\phi\right]^*\left[\frac{1}{2}\left(\frac{3}{\pi}\right)^{1/2}\sin\theta\cos\phi\right]\\ &\quad + \left[\frac{1}{2}\left(\frac{3}{\pi}\right)^{1/2}\sin\theta\sin\phi\right]^*\left[\frac{1}{2}\left(\frac{3}{\pi}\right)^{1/2}\sin\theta\sin\phi\right]\\ &= \frac{3}{4\pi}(\cos^2\theta + \sin^2\theta\cos^2\phi + \sin^2\theta\sin^2\phi) = \frac{3}{4\pi}[\cos^2\theta + (\sin^2\theta)(\cos^2\phi + \sin^2\phi)]\\ &= \frac{3}{4\pi}(\cos^2\theta + \sin^2\theta) = \frac{3}{4\pi}\end{aligned}$$

Because there is no net angular dependence, the probability is radially symmetric (spherical).

15.12 The radial wave function describing an electron in a hydrogenlike atom is given by

$$R_{n,l}(r) = -\left(\frac{(2Z/na_0)^3(n-l-1)!}{2n[(n+l)!]^3}\right)^{1/2} e^{-\rho/2}\rho^l L_{n+l}^{2l+1}(\rho) \tag{15.8}$$

where $n = 1, 2, 3, \ldots,$

$$\rho = (2Z/na_0)r \tag{15.9}$$

$$a_0 = 4\pi\varepsilon_0\hbar^2/\mu e^2 \tag{15.10}$$

and the associated *Laguerre polynomials* are defined as

$$L_q^s(x) = \frac{d^s}{dx^s}\left(e^x \frac{d^q}{dx^q}(x^q e^{-x})\right) \tag{15.11}$$

What limits are placed on the azimuthal quantum number?

▌ Inspection of *(15.8)* shows that in order for $(n - l - 1)!$ to be defined, the permitted values of l are limited to $0, 1, 2, \ldots, (n - 1)$.

15.13 Confirm the entry in Table 15-3 for $n = 2$ and $l = 1$.

Table 15-3

n	l	L_{n+l}^{2l+1}
1	0	$L_1^1 = -1$
2	0	$L_2^1 = -2(2-\rho)$
2	1	$L_3^3 = -6$
3	0	$L_3^1 = -3(6 - 6\rho + \rho^2)$
3	1	$L_4^3 = -24(4-\rho)$
3	2	$L_5^5 = -120$

▌ For $n = 2$ and $l = 1$, substituting into *(15.11)* gives

$$L_{2+1}^{2(1)+1}(x) = L_3^3(x) = \frac{d^3}{dx^3}\left[e^x \frac{d^3}{dx^3}(x^3 e^{-x})\right] = \frac{d^3}{dx^3}\left[e^x \frac{d^2}{dx^2}(3x^2 e^{-x} - x^3 e^{-x})\right]$$

$$= \frac{d^3}{dx^3}\left[e^x \frac{d}{dx}(6x e^{-x} - 3x^2 e^{-x} - 3x^2 e^{-x} + x^3 e^{-x})\right]$$

$$= \frac{d^3}{dx^3} e^x(6e^{-x} - 6x e^{-x} - 12x e^{-x} + 6x^2 e^{-x} + 3x^2 e^{-x} - x^3 e^{-x})$$

$$= \frac{d^3}{dx^3}(6 - 18x + 9x^2 - x^3) = \frac{d^2}{dx^2}(0 - 18 + 18x - 3x^2) = \frac{d}{dx}(0 + 18 - 6x) = -6$$

15.14 An alternative form for determining the associated Laguerre polynomials is

$$L_{n+l}^{2l+1}(x) = \sum_{i=0}^{n-l-1} (-1)^{i+1} \frac{[(n+l)!]^2}{(n-l-1-i)!(2l+1+i)!\,i!} x^i \tag{15.12}$$

Use this relationship to confirm the entry in Table 15-3 for $n = 2$ and $l = 1$.

▌ Substituting $n = 2$ and $l = 1$ into *(15.12)* gives

$$L_{2+1}^{2(1)+1}(x) = L_3^3(x) = \sum_{i=0}^{2-1-1} (-1)^{i+1} \frac{[(2+1)!]^2}{(2-1-1-i)![(2)(1)+1+i)]!\,i!} x^i$$

$$= \sum_{i=0}^{0} (-1)^{i+1} \frac{(3!)^2}{(0-i)!(3+i)!\,i!} x^i = (-1)^1 \frac{(3!)^2}{0!3!0!} x^0 = -6$$

15.15 Confirm the entry in Table 15-1 for $R(r)$ for $n = 2$ and $l = 1$.

▌ Substituting $L_3^3(\rho) = -6$ (see Problem 15.14) into *(15.8)* gives

$$R_{2,l}(r) = -\left(\frac{[2Z/2a_0]^3(2-1-1)!}{2(2)[(2+1)!]^3}\right)^{1/2} e^{-\rho/2}\rho^1(-6) = \left(\frac{Z}{a_0}\right)^{3/2} \frac{1}{2(6)^{1/2}} \rho\, e^{-\rho/2}$$

15.16 Prepare plots of $R(r)$ and $R(r)^*R(r)r^2$ for a hydrogenlike atom for $n = 2$ and $Z = 1$.

▌ As a sample calculation, consider $r = 4a_0$. Substituting into *(15.9)* gives

$$\rho = \frac{(2)(1)}{(2)(a_0)} 4a_0 = 4$$

Using this value of ρ in the expression for $R_{2,0}(r)$ and $R_{2,1}(r)$ taken from Table 15-1 gives

$$R_{2,0}(r) = \frac{1}{2(2)^{1/2}}\left(\frac{1}{a_0}\right)^{3/2}(2-4)\,e^{-4/2} = -0.0957/a_0^{3/2}$$

$$R_{2,1}(r) = \frac{1}{2(6)^{1/2}}\left(\frac{1}{a_0}\right)^{3/2}(4)\,e^{-4/2} = 0.1105/a_0^{3/2}$$

$$R_{2,0}^*(r)R_{2,0}(r)r^2 = \frac{1}{8a_0^3}(2-4)^2\,e^{-(2)(4)/2}(4a_0)^2 = 0.1465/a_0$$

$$R_{2,1}^*(r)R_{2,1}(r)r^2 = \frac{1}{24a_0^3}(4)^2\,e^{-(2)(4)/2}(4a_0)^2 = 0.1954/a_0$$

The plots are shown in Fig. 15-2.

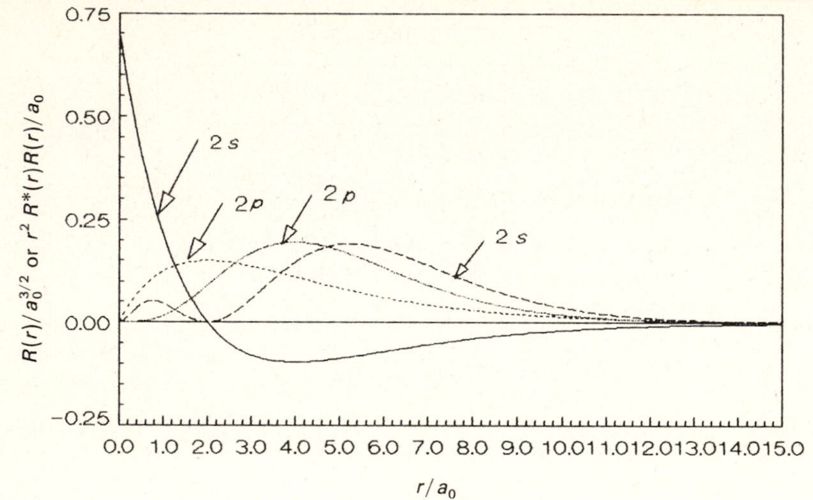

Fig. 15-2

15.17 Determine the number and locations of the nodes in the 2s wave function.

▌ The nodes will occur at the value(s) of ρ at which $R(r) = 0$. Using the wave function from Table 15-1 with $n = 2$ and $l = 1$ gives

$$R(r) = \frac{1}{2(2)^{1/2}}\left(\frac{Z}{a_0}\right)^{3/2}(2-\rho)\,e^{-\rho/2} = 0$$

$$\rho = 2$$

There is only one node occurring at $\rho = 2$ or at $r = 2a_0$. This result agrees with the plot shown in Fig. 15-1.

15.18 Determine the number and location of the maxima in the plot of $r^2R^*(r)R(r)$ for the 2s wave function.

▌ Substituting $n = 2$ and $Z = 1$ into *(15.9)* gives

$$r = \frac{2a_0}{2(1)}\rho = \rho a_0$$

Differentiating $r^2R^*(r)R(r)$ for the wave function with $n = 2$ and $l = 1$ from Table 15-1 with respect to ρ, setting the result equal to zero, and solving for ρ gives

$$\frac{\partial}{\partial\rho}\left[(\rho a_0)^2\left[\frac{1}{2(2)^{1/2}}\left(\frac{1}{a_0}\right)^{3/2}(2-\rho)\,e^{-\rho/2}\right]^*\left[\frac{1}{2(2)^{1/2}}\left(\frac{1}{a_0}\right)^{3/2}(2-\rho)\,e^{-\rho/2}\right]\right]$$

$$= \frac{1}{8a_0}\frac{\partial}{\partial\rho}[\rho^2(2-\rho)^2\,e^{-\rho}] = \frac{1}{8a_0}[2\rho(2-\rho)^2 - 2\rho^2(2-\rho) - \rho^2(2-\rho)^2]\,e^{-\rho}$$

$$= \frac{1}{8a_0}\rho(2-\rho)(\rho^2 - 6\rho + 4)\,e^{-\rho} = 0$$

The roots $\rho = 0$, $\rho = 2$, and $\rho = \infty$ correspond to the minima shown in Fig. 15-2, where $\rho = r/a_0$. Solving the quadratic expression gives the two maxima as

$$\rho = \frac{-(-6) \pm [(-6)^2 - 4(1)(4)]^{1/2}}{2(1)} = 3 \pm 5^{1/2} = 5.236 \text{ or } 0.764$$

which agree with the maxima shown in Fig. 15-2.

15.19 Determine $\langle r \rangle$ for 2p wave function.

▌ Substituting $r = \rho a_0$ (see Problem 15.18), $Z = 1$, and the wave function given by Table 15-1 for $n = 2$ and $l = 1$ into *(14.34)* with the denominator set equal to one gives

$$\langle r \rangle = \int_0^\infty R(r)^* r R(r) r^2\, dr = \int_0^\infty \left[\frac{1}{2(6)^{1/2}}\left(\frac{1}{a_0}\right)^{3/2} \rho\, e^{-\rho/2}\right](\rho a_0)\left[\frac{1}{2(6)^{1/2}}\left(\frac{1}{a_0}\right)^{3/2} \rho\, e^{-\rho/2}\right](\rho a_0)^2\, d(\rho a_0)$$

$$= \frac{a_0}{24}\int_0^\infty \rho^5\, e^{-\rho}\, d\rho = \frac{a_0}{24}(5!) = 5a_0$$

15.20 What is the probability of finding the 1s electron at $r > a_0$?

▌ For a 1s electron, *(15.9)* gives, for $Z = 1$,

$$r = \frac{(1)a_0}{2(1)}\rho = \frac{a_0}{2}\rho$$

Using the wave function given in Table 15-1 for $n = 1$ and $l = 0$, the probability will be given by

$$P = \int_{a_0}^\infty R(r)^* R(r) r^2\, dr = \int_2^\infty \left[2\left(\frac{1}{a_0}\right)^{3/2} e^{-\rho/2}\right]^*\left[2\left(\frac{1}{a_0}\right)^{3/2} e^{-\rho/2}\right]\left(\frac{a_0}{2}\rho\right)^2 d\left(\frac{a_0}{2}\rho\right)$$

$$= \frac{1}{2}\int_2^\infty \rho^2\, e^{-\rho}\, d\rho = \tfrac{1}{2}[e^{-\rho}(-\rho^2 - 2\rho - 2)]\Big|_2^\infty = \tfrac{1}{2}\{0 - e^2[-(2)^2 - 2(2) - 2]\} = 5\, e^{-2} = 0.6767$$

15.21 Determine the radius of the sphere that will contain a given probability of finding the 1s electron. What is the value for $P = 0.90$?

▌ As in Problem 15.20,

$$P = \int_0^r R(r)^* R(r) r^2\, dr = \frac{1}{2}\int_0^\rho \rho^2\, e^{-\rho}\, d\rho = \tfrac{1}{2}[e^{-\rho}(-\rho^2 - 2\rho - 2)]\Big|_0^\rho$$

$$= \tfrac{1}{2}[e^{-\rho}(-\rho^2 - 2\rho - 2) - (1)(-0 - 0 - 2)]$$

$$= \frac{-1}{2}[e^{-r}(\rho^2 + 2\rho + 2) + 2]$$

$$e^{-\rho}(\rho^2 + 2\rho + 2) = 2(1 - P)$$

Substituting $P = 0.90$ gives

$$e^{-\rho}(\rho^2 + 2\rho + 2) = 2(1 - 0.90) = 0.20$$

Solving gives $\rho = 5.3223$, or

$$r = \frac{a_0}{2}(5.3223) = 2.6612a_0$$

15.22 Place the orbitals in order of increasing energy for a hydrogenlike atom. What is the degeneracy of the level with $n = 3$?

▌ The energy of a hydrogenlike atom is given by *(14.21)*. Because E is a function of n only,

$$1s < 2s = 2p < 3s = 3p = 3d < 4s = 4p = 4d = 4f < \text{etc.}$$

For $n = 3$, the permitted values of l are 0, 1, and 2 (see Problem 15.12), giving three subshells. For $l = 0$, m will be 0; for $l = 1$, m will be $+1, 0, -1$; and for $l = 2$, m will be $+2, +1, 0, -1, -2$ (see Problem 14.75), giving a total of nine orbitals. For a hydrogenlike atom, the degeneracy of the third shell is 9.

15.2 QUANTUM THEORY OF POLYELECTRONIC ATOMS

15.23 The spin wave functions for a two-electron system are

$$\phi_1 = \alpha(1)\alpha(2) \tag{15.13a}$$

$$\phi_2 = \beta(1)\beta(2) \tag{15.13b}$$

$$\phi_3 = [(2)^{1/2}/2][\alpha(1)\beta(2) + \beta(1)\alpha(2)] \tag{15.13c}$$

$$\phi_4 = [(2)^{1/2}/2][\alpha(1)\beta(2) - \beta(1)\alpha(2)] \tag{15.13d}$$

Show that ϕ_1 and ϕ_4 are orthogonal.

▮ The spin wave functions (α and β) are defined such that

$$\int_{\text{spin space}} \alpha(i)\beta(i)\,dV = 0 \tag{15.14a}$$

$$\int_{\text{spin space}} \alpha(i)\alpha(i)\,dV = \int_{\text{spin space}} \beta(i)\beta(i)\,dV = 1 \tag{15.14b}$$

where, in this case, $dV = dV_1\,dV_2$. Substituting *(15.13a)* and *(15.13d)* into *(14.24)* gives

$$\int_{\text{spin space}} \alpha(1)\alpha(2)\frac{2^{1/2}}{2}[\alpha(1)\beta(2) - \beta(1)\alpha(2)]\,dV = \frac{2^{1/2}}{2}\int_{\text{spin space}} [\alpha(1)\alpha(1)\alpha(2)\beta(2) - \alpha(1)\alpha(2)\beta(1)\alpha(2)]\,dV$$

$$= \frac{2^{1/2}}{2}\left[\left(\int \alpha(1)\alpha(1)\,dV_1\right)\left(\int \alpha(2)\beta(2)\,dV_2\right) - \left(\int \alpha(1)\beta(1)\,dV_1\right)\left(\int \alpha(1)\alpha(2)\,dV_2\right)\right]$$

$$= \frac{2^{1/2}}{2}[(1)(0) - (0)(1)] = 0$$

15.24 Which of the spin wave functions given by *(15.13)* are symmetric with respect to the exchange of electrons?

▮ The effect of the *permutator operator* $[\hat{P}(1,2)]$ is the exchange of electrons. Performing this operation on *(15.13)* gives

$$\hat{P}(1,2)\phi_1 = \hat{P}(1,2)[\alpha(1)\alpha(2)] = \alpha(2)\alpha(1) = \phi_1$$

$$\hat{P}(1,2)\phi_2 = \hat{P}(1,2)[\beta(1)\beta(2)] = \beta(2)\beta(1) = \phi_2$$

$$\hat{P}(1,2)\phi_3 = \hat{P}(1,2)[\alpha(1)\beta(2) + \beta(1)\alpha(2)] = \alpha(2)\beta(1) + \beta(2)\alpha(1) = \phi_3$$

$$\hat{P}(1,2)\phi_4 = \hat{P}(1,2)[\alpha(1)\beta(2) - \beta(1)\alpha(2)] = \alpha(2)\beta(1) - \beta(2)\alpha(1) = -\phi_4$$

The wave functions ϕ_1, ϕ_2, and ϕ_3 are symmetric because the eigenvalue is $+1$, and the wave function ϕ_4 is antisymmetric because the eigenvalue is -1.

15.25 Show that the spin wave function given by *(15.13c)* is an eigenfunction of the total z component of spin angular momentum for a two-electron system. What is the eigenvalue?

▮ For a two-electron system, the total z component is given by

$$\hat{S}_z = \hat{S}_z(1) + \hat{S}_z(2) \tag{15.15}$$

where the spin operators and wave functions are related by

$$\hat{S}_z(i)\alpha(i) = \frac{\hbar}{2}\alpha(i) \tag{15.16a}$$

$$\hat{S}_z(i)\beta(i) = \frac{-\hbar}{2}\beta(i) \tag{15.16b}$$

Combining *(15.16)*, *(15.15)*, and *(15.13c)* gives

$$\hat{S}_z\phi_3 = [\hat{S}_z(1) + \hat{S}_z(2)]\left[\frac{2^{1/2}}{2}[\alpha(1)\beta(2) + \beta(1)\alpha(2)]\right]$$

$$= \frac{2^{1/2}}{2}[\beta(2)\hat{S}_z(1)\alpha(1) + \alpha(1)\hat{S}_z(2)\beta(2) + \alpha(2)\hat{S}_z(1)\beta(1) + \beta(1)\hat{S}_z(2)\alpha(2)]$$

$$= \frac{2^{1/2}}{2}\left[\beta(2)\frac{\hbar^2}{2}\alpha(1) + \alpha(1)\frac{-\hbar^2}{2}\beta(2) + \alpha(2)\frac{-\hbar^2}{2}\beta(1) + \beta(1)\frac{\hbar^2}{2}\alpha(2)\right]$$

$$= \frac{2^{1/2}}{2}\left[\alpha(1)\beta(2)\left(\frac{\hbar^2}{2} - \frac{\hbar^2}{2}\right) + \beta(1)\alpha(2)\left(\frac{\hbar^2}{2} - \frac{\hbar^2}{2}\right)\right]$$

$$= \frac{2^{1/2}}{2}[\alpha(1)\beta(2) + \beta(1)\alpha(2)]\left(\frac{\hbar^2}{2} - \frac{\hbar^2}{2}\right) = \phi_3\left(\frac{\hbar^2}{2} - \frac{\hbar^2}{2}\right)$$

Because *(14.32)* is satisfied, ϕ_3 is an eigenfunction. The eigenvalue is $\hbar^2/2 - \hbar^2/2 = 0$.

15.26 Write the Hamiltonian operator describing the electron motion for a two-electron system such as He.

▌ For electronic motion about a nucleus, *(14.36)* becomes

$$\mathscr{H} = \frac{-\hbar^2}{2m_e}\sum_{i=1}^{n}\nabla_i^2 - \sum_{i=1}^{n}\frac{Ze^2}{4\pi\varepsilon_0 r_i} + \sum_{j=2}^{n}\sum_{i<j}\frac{e^2}{4\pi\varepsilon_0 r_j} \tag{15.17}$$

where n is the number of electrons. For a two-electron system,

$$\mathscr{H} = \frac{-\hbar^2}{2m_e}(\nabla_1^2 + \nabla_2^2) - \frac{Ze^2}{4\pi\varepsilon_0}\left(\frac{1}{r_1} + \frac{1}{r_2}\right) + \frac{e^2}{4\pi\varepsilon_0}\left(\frac{1}{r_{12}}\right) \tag{15.18}$$

15.27 Use the Slater determinant to arrive at a wave function to describe the ground state of a two-electron system such as He.

▌ The *Slater determinant* is

$$\psi = \frac{1}{(n!)^{1/2}}\begin{vmatrix} \phi_{1s}(1)\alpha(1) & \phi_{1s}(2)\alpha(2) & \cdots & \phi_{1s}(n)\alpha(n) \\ \phi_{1s}(1)\beta(1) & \phi_{1s}(2)\beta(2) & \cdots & \phi_{1s}(n)\beta(n) \\ \phi_{2s}(1)\alpha(1) & \phi_{2s}(2)\alpha(2) & \cdots & \phi_{2s}(n)\alpha(n) \\ \vdots & \vdots & & \vdots \end{vmatrix}$$

where n is the number of electrons and ϕ_i represents atomic hydrogen wave functions. For the two-electron system,

$$\psi = \frac{1}{(2!)^{1/2}}\begin{vmatrix} \phi_{1s}(1)\alpha(1) & \phi_{1s}(2)\alpha(2) \\ \phi_{1s}(1)\beta(1) & \phi_{1s}(2)\beta(2) \end{vmatrix}$$

$$= \frac{1}{2^{1/2}}[\phi_{1s}(1)\alpha(1)\phi_{1s}(2)\beta(2) - \phi_{1s}(1)\beta(1)\phi_{1s}(2)\alpha(2)]$$

15.28 The *Pauli exclusion principle* states that an acceptable wave function must be antisymmetric with respect to the exchange of electrons. Show that the wave function determined in Problem 15.27 is antisymmetric.

▌ The effect of the permutator operator (see Problem 15.24) on ψ is

$$\hat{P}(1,2)\psi = \hat{P}(1,2)\left(\frac{1}{2^{1/2}}[\phi_{1s}(1)\alpha(1)\phi_{1s}(2)\beta(2) - \phi_{1s}(1)\beta(1)\phi_{1s}(2)\alpha(2)]\right)$$

$$= \frac{1}{2^{1/2}}[\phi_{1s}(2)\alpha(2)\phi_{1s}(1)\beta(1) - \phi_{1s}(2)\beta(2)\phi_{1s}(1)\alpha(1)] = -\psi$$

Because the eigenvalue is -1, the wave function is antisymmetric.

15.29 The variation method for He gives

$$E(\text{He}) = -[-2Z'^2 + \tfrac{5}{4}Z' + 4Z'(Z' - 2)]E(\text{H})$$

where Z' is an effective nuclear charge that is determined by minimization, and $E(\text{H})$ is the energy of the ground state of atomic H (-13.6 eV). Determine Z' and $E(\text{He})$ and compare $E(\text{He})$ to the experimental value of -79 eV.

▌ Differentiating the expression for $E(\text{He})$ with respect to Z', setting the result equal to zero, and solving for Z'

gives

$$\frac{\partial E(\text{He})}{\partial Z'} = -[-4Z' + \tfrac{5}{4} + 4(Z' - 2) + 4Z'(1)]E(\text{H}) = -(4Z' - \tfrac{27}{4})E(\text{H}) = 0$$

$$Z' = \tfrac{27}{16}$$

The corresponding value of $E(\text{He})$ is

$$E(\text{He}) = -[-2(\tfrac{27}{16})^2 + \tfrac{5}{4}(\tfrac{27}{16}) + 4(\tfrac{27}{16})(\tfrac{27}{16} - 2)](-13.6\text{ eV}) = -77.5\text{ eV}$$

which is 2% high compared to the experimental value.

15.30 The perturbation theory for He gives

$$E(\text{He}) = 2Z^2E(\text{H}) + H_{\text{nn}}^{(1)}$$

where $\mathscr{H}^{(1)} = e^2/r_{12}$, $H_{\text{nn}}^{(1)}$ is described in Problem 14.90, and $E(\text{H}) = -13.6\text{ eV}$. Evaluation of *(14.69)* using $\mathscr{H}^{(1)}$ gives $H_{\text{nn}}^{(1)} = -\tfrac{5}{4}ZE(\text{H})$. Evaluate $E(\text{He})$, and compare this value to the experimental value of -79 eV.

I The value of $E(\text{He})$ is

$$E(\text{He}) = 2(2)^2(-13.6\text{ eV}) + \tfrac{5}{4}(2)(-13.6\text{ eV}) = -74.8\text{ eV}$$

which is 5% high compared to the experimental value.

15.31 Write the electronic configuration for the ground state of atomic As.

I The order of filling atomic subshells in polyatomic atoms is given approximately by the sequence

$$1\text{s} < 2\text{s} < 2\text{p}(3) < 3\text{s} < 3\text{p}(3) < 4\text{s} < 3\text{d}(5) < 4\text{p}(3) < 5\text{s} < 4\text{d}(5)$$

$$< 5\text{p}(3) < 6\text{s} < 4\text{f}(7) < 5\text{d}(5) < 6\text{p}(3) < 7\text{s} < 5\text{f}(5) = 6\text{d}(5) < 7\text{p}(3) < \text{etc.}$$

where the number in parentheses represents the number of orbitals making up the subshell. Each orbital can hold two electrons. For 33 electrons the configuration is $(1\text{s})^2(2\text{s})^2(2\text{p})^6(3\text{s})^2(3\text{p})^6(4\text{s})^2(3\text{d})^{10}(4\text{p})^3$.

15.32 Determine the four quantum numbers for the 33rd electron in As. Is atomic As paramagnetic or diamagnetic?

I The last electron in the configuration given by Problem 15.31 is the third electron in the 4p subshell. For a 4p subshell, $n = 4$ and $l = 1$. (Note: s corresponds to $l = 1$, p to $l = 1$, d to $l = 2$, f to $l = 3$, etc.) *Hund's first rule* states that the most stable arrangement for a $(4\text{p})^3$ configuration is $m = +1$, $s = +\tfrac{1}{2}$; $m = 0$, $s = +\tfrac{1}{2}$; and $m = -1$, $s = +\tfrac{1}{2}$. The quantum numbers of the 33rd electron are $n = 4$, $l = 1$, $m = -1$, and $s = +\tfrac{1}{2}$. Because there are three unpaired electrons in the atom, the substance is paramagnetic.

15.33 Predict possible oxidation states for As based on its electron configuration.

I The order of removal of electrons is more like the order of filling of subshells on a hydrogenlike atom (see Problem 15.22). Rewriting the configuration determined in Problem 15.31 in this order gives $(1\text{s})^2(2\text{s})^2(2\text{p})^6(3\text{s})^2(3\text{p})^6(3\text{d})^{10}(4\text{s})^2(4\text{p})^3$. The oxidation state of 0 is assigned to the atom before considering the addition or removal of electrons. A total of three electrons could be added to the 4p subshell to give an oxidation state of -3. The removal of the electrons from the 4p subshell would give an oxidation state of $+3$, and the subsequent removal of the 4s electrons would give an oxidation state of $+5$.

15.34 Choose the species from each pair that has the larger radius: ***(a)*** Na, Mg; ***(b)*** Cl, Br; ***(c)*** Mo, W; ***(d)*** Ne, Na^+; ***(e)*** Cl^-, Ar; ***(f)*** Cu^+, Cu^{2+}.

I From *(14.19)*, the radius of a multielectron system is directly proportional to n^2 and inversely proportional to the effective nuclear charge. As electrons are removed, the overall Coulombic repulsion between the electrons decreases, and the resulting species is smaller. As electrons are added, the overall Coulombic repulsion between the electrons increases, and the resulting species is larger. Applying these rules gives

(a) $r(\text{Mg}) < r(\text{Na})$, because $n(\text{Mg}) = n(\text{Na})$ and $Z_{\text{eff}}(\text{Mg}) > Z_{\text{eff}}(\text{Na})$.
(b) $r(\text{Cl}) < r(\text{Br})$, because $n(\text{Cl}) < n(\text{Br})$.
(c) $r(\text{Mo}) \approx r(\text{W})$, because $n(\text{Mo}) < n(\text{W})$ and $Z_{\text{eff}}(\text{Mo}) \ll Z_{\text{eff}}(\text{W})$ as a result of the intervening lanthanide series elements.
(d) $r(\text{Ne}) < r(\text{Na}^+)$, even though these species are isoelectronic and the prediction that $r(\text{Na}^+) > r(\text{Ne})$ because $Z_{\text{eff}}(\text{Na}^+) > Z_{\text{eff}}(\text{Ne})$; this prediction is not correct.
(e) $r(\text{Ar}) < r(\text{Cl}^-)$, even though these species are isoelectronic; the additional Coulombic repulsion in Cl^- makes it larger.
(f) $r(\text{Cu}^{2+}) < r(\text{Cu}^+)$, because the successive removal of electrons decreases the Coulombic repulsion.

15.35 Choose the species from each pair that has the larger ionization energy:
(a) Br, Kr; **(b)** Al, Al^+; **(c)** Na, K.

▌ From *(14.21)*, the ionization energy for a multielectron system is directly proportional to the square of the effective nuclear charge and inversely proportional to n^2. The successive removal of electrons from an atom becomes increasingly difficult because of the increasing effective nuclear charge. Applying these rules gives

(a) $I(\text{Kr}) > I(\text{Br})$, because $n(\text{Kr}) = n(\text{Br})$ and $Z_{\text{eff}}(\text{Kr}) > Z_{\text{eff}}(\text{Br})$.
(b) $I(\text{Al}^+) > I(\text{Al})$, because of the increase in effective nuclear charge due to the successive removal of electrons.
(c) $I(\text{Na}) > I(\text{K})$, because $n(\text{K}) > n(\text{Na})$.

15.3 ATOMIC TERM SYMBOLS

15.36 Determine the values of L, the quantum number describing the *l-l coupling* of the angular momenta of two atomic orbitals, for two d electrons. What are the corresponding letter symbols?

▌ The permitted values of L are given by

$$L = l_1 + l_2, l_1 + l_2 - 1, \ldots, |l_1 - l_2| \tag{15.19}$$

The value of l for d electrons is 2. Substituting into *(15.19)* gives

$$L = \begin{cases} 2 + 2 = 4 \\ 2 + 2 - 1 = 3 \\ 2 + 2 - 2 = 2 \\ 2 + 2 - 3 = 1 \\ 2 + 2 - 4 = |2 - 2| = 0 \end{cases}$$

which correspond to the letters G, F, D, P, and S, respectively.

15.37 Determine the values of S, the quantum number describing the *s-s coupling* of the angular momenta of the spins of two electrons, for two d electrons.

▌ The permitted values of S are given by

$$S = 0 \text{ or } 1 \tag{15.20}$$

15.38 Determine the values of J, the quantum number describing the *Russell-Saunders coupling* between L and S, for two d electrons.

▌ The permitted values of J are

$$J = L + S, L + S - 1, \ldots, |L - S| \tag{15.21}$$

Using the results of Problems 15.36 and 15.37 gives

$$L = 4 \quad S = 0 \quad J = 4 + 0 = |4 - 0| = 4$$

$$S = 1 \quad J = \begin{cases} 4 + 1 = 5 \\ 4 + 1 - 1 = 4 \\ 4 + 1 - 2 = |4 - 1| = 3 \end{cases}$$

$$L = 3 \quad S = 0 \quad J = 3 + 0 = |3 - 0| = 3$$

$$S = 1 \quad J = \begin{cases} 3 + 1 = 4 \\ 3 + 1 - 1 = 3 \\ 3 + 1 - 2 = |3 - 1| = 2 \end{cases}$$

$$L = 2 \quad S = 0 \quad J = 2 + 0 = |2 - 0| = 2$$

$$S = 1 \quad J = \begin{cases} 2 + 1 = 3 \\ 2 + 1 - 1 = 2 \\ 2 + 1 - 2 = |2 - 1| = 1 \end{cases}$$

$$L = 1 \quad S = 0 \quad J = 1 + 0 = |1 - 0| = 1$$

$$S = 1 \quad J = \begin{cases} 1 + 1 = 2 \\ 1 + 1 - 1 = 1 \\ 1 + 1 - 2 = |1 - 1| = 0 \end{cases}$$

$$L = 0 \quad S = 0 \quad J = 0 + 0 = |0 - 0| = 0$$

$$S = 1 \quad J = 0 + 1 = |0 - 1| = 1$$

15.39 Write the complete atomic term symbols for each of the states determined in Problem 15.38.

▮ The format of an atomic term symbol is $^{2S+1}X_J$. For the states listed in Problem 15.38,

$L = 4$	$S = 0$	$J = 4$	$^1G_4$
$L = 4$	$S = 1$	$J = 5$	$^3G_5$
		$J = 4$	$^3G_4$
		$J = 3$	$^3G_3$
$L = 3$	$S = 0$	$J = 3$	$^1F_3$
$L = 3$	$S = 1$	$J = 4$	$^3F_4$
		$J = 3$	$^3F_3$
		$J = 2$	$^3F_2$
$L = 2$	$S = 0$	$J = 2$	$^1D_2$
$L = 2$	$S = 1$	$J = 3$	$^3D_3$
		$J = 2$	$^3D_2$
		$J = 1$	$^3D_1$
$L = 1$	$S = 0$	$J = 1$	$^1P_1$
$L = 1$	$S = 1$	$J = 2$	$^3P_2$
		$J = 1$	$^3P_1$
		$J = 0$	$^3P_0$
$L = 0$	$S = 0$	$J = 0$	$^1S_0$
$L = 0$	$S = 1$	$J = 1$	$^3S_1$

15.40 What are the L and S quantum numbers corresponding to ^{2}D?

▮ The symbol D corresponds to $L = 2$. The leading superscript representing the degeneracy gives

$$2J + 1 = 2 \qquad J = \tfrac{1}{2}$$

15.41 Prepare a list of the microstates that are permitted by the Pauli exclusion principle for an nd^2 configuration.

▮ The wave functions that are used for this configuration are $\phi_{2p,m}\gamma$, where $m = \pm 2, \pm 1$, and 0 and $\gamma = \alpha$ or β. Thus there are $5 \times 2 = 10$ choices of wave functions for each electron, or $10 \times 10 = 100$ choices for the pair. Of these, 10 give the same wave function to both electrons and so are ruled out by the *Pauli exclusion principle* (no two electrons can have identical values for all four quantum numbers in the same atom). Of the remaining 90, half must be eliminated because there is no way to distinguish between the "first" electron and the "second" electron. We are left with 45 microstates, which, using the notation $m\gamma$ instead of $\phi_{2p,m}\gamma$, can be indicated as follows:

$(+2\alpha, +2\beta)$	$(+2\alpha, +1\alpha)$	$(+2\alpha, +1\beta)$	$(+2\alpha, 0\alpha)$	$(+2\alpha, 0\beta)$
$(+2\alpha, -1\alpha)$	$(+2\alpha, -1\beta)$	$(+2\alpha, -2\alpha)$	$(+2\alpha, -2\beta)$	$(+1\alpha, +1\beta)$
$(+1\alpha, 0\alpha)$	$(+1\alpha, 0\beta)$	$(+1\alpha, -1\alpha)$	$(+1\alpha, -1\beta)$	$(+1\alpha, -2\alpha)$
$(+1\alpha, -2\beta)$	$(0\alpha, 0\beta)$	$(0\alpha, -1\alpha)$	$(0\alpha, -1\beta)$	$(0\alpha, -2\alpha)$
$(0\alpha, -2\beta)$	$(-1\alpha, -1\beta)$	$(-1\alpha, -2\alpha)$	$(-1\alpha, -2\beta)$	$(-2\alpha, -2\beta)$
$(+2\beta, +1\alpha)$	$(+2\beta, +1\beta)$	$(+2\beta, 0\alpha)$	$(+2\beta, 0\beta)$	$(+2\beta, -1\alpha)$
$(+2\beta, -1\beta)$	$(+2\beta, -2\alpha)$	$(+2\beta, -2\beta)$	$(+1\beta, 0\alpha)$	$(+1\beta, 0\beta)$
$(+1\beta, -1\alpha)$	$(+1\beta, -1\beta)$	$(+1\beta, -2\alpha)$	$(+1\beta, -2\beta)$	$(0\beta, -1\alpha)$
$(0\beta, -1\beta)$	$(0\beta, -2\alpha)$	$(0\beta, -2\beta)$	$(-1\beta, -2\alpha)$	$(-1\beta, -2\beta)$

15.42 Determine M_L and M_S for the microstates for an nd^2 configuration.

▮ For polyelectronic systems,

$$M_L = \sum m_i \quad \text{where } M_L = L, L-1, \ldots, -L \tag{15.22}$$

$$M_S = \sum s_i \quad \text{where } M_S = S, S-1, \ldots, -S \tag{15.23}$$

The microstates for the nd^2 system were determined in Problem 15.42. As a sample calculation, consider $(+2\alpha, +2\beta)$.

$$M_L = (+2) + (+2) = 4 \qquad M_S = (+\tfrac{1}{2}) + (-\tfrac{1}{2}) = 0$$

giving (4, 0). The respective values are

(4, 0) (3, 1) (3, 0) (2, 1) (2, 0)
(1, 1) (1, 0) (0, 1) (0, 0) (2, 0)
(1, 1) (1, 0) (0, 1) (0, 0) (−1, 1)
(−1, 0) (0, 0) (−1, 1) (−1, 0) (−2, 1)
(−2, 0) (−2, 0) (−3, 1) (−3, 0) (−4, 0)
(3, 0) (3, −1) (2, 0) (2, −1) (1, 0)
(1, −1) (0, 0) (0, −1) (1, 0) (1, −1)
(0, 0) (0, −1) (−1, 0) (−1, −1) (−1, 0)
(−1, −1) (−2, 0) (−2, −1) (−3, 0) (−3, −1)

15.43 Assign term symbols to the microstates for an nd^2 configuration.

I The values of M_L and M_S for the 45 microstates were determined in Problem 15.42. Starting with the microstate having the largest values of M_L and M_S, using $L = M_L$ and $S = M_S$, the corresponding term symbol for the (4, 0) microstate is $^{2(0)+1}G = {}^1G$. One group of microstates having M_L and M_S values related to the above values of L and S by *(15.22)* and *(15.23)* are

(3, 0) (2, 0) (1, 0) (0, 0) (−1, 0) (−2, 0) (−3, 0) (−4, 0)

These are assigned to this same term symbol, leaving 36 microstates. The value of J for ^1G is given by *(15.21)* as $J = 4 + 0 = |4 - 0| = 4$, giving the complete term symbol as $^1G_4$.

Repeating the procedure with the next highest M_L and M_S combination in the remaining microstates gives for the (3, 1) microstate

(3, 1) (3, 0) (3, −1) (2, 1) (2, 0) (2, −1)
(1, 1) (1, 0) (1, −1) (0, 1) (0, 0) (0, −1)
(−1, 1) (−1, 0) (−1, −1) (−2, 1) (−2, 0) (−2, −1)
(−3, 1) (−3, 0) (−3, −1)

The corresponding term symbol is $^{2(1)+1}F = {}^3F$, with $J = 4, 3,$ and 2 giving $^3F_4$, $^3F_3$, and $^3F_2$.

Repeating the procedure for the (2, 0) microstate gives

(2, 0) (1, 0) (0, 0) (−1, 0) (−2, 0)

The corresponding term symbol is $^{2(0)+1}D = {}^1D$, with $J = 2$ giving $^1D_2$.

Repeating the procedure for the (1, 1) microstate gives

(1, 1) (1, 0) (1, −1) (0, 1) (0, 0) (0, −1)
(−1, 1) (−1, 0) (−1, −1)

The corresponding term symbol is $^{2(1)+1}P = {}^3P$, with $J = 2, 1,$ and 0 giving $^3P_2$, $^3P_1$, and $^3P_0$.

Repeating the procedure for the remaining (0, 0) microstate gives $^{2(0)+1}S = {}^1S$, with $J = 0$ giving $^1S_0$.

Comparison of these term symbols with those listed in Problem 15.39 for two d electrons shows that several of the term symbols are not permitted if both of the d electrons have the same principal quantum number. The term symbols for two equivalent electrons are given in Table 15-4.

Table 15-4

configuration	term symbols
s^1	^2S
s^2	^1S
p^1 or p^5	^2P
p^2 or p^4	$^3P, {}^1D, {}^1S$
p^3	$^4S, {}^2D, {}^2P$
p^6	^1S
d^1 or d^9	^2D
d^2 or d^8	$^3F, {}^3P, {}^1G, {}^1D, {}^1S$
d^3 or d^7	$^4F, {}^4P, {}^2H, {}^2G, {}^2F, {}^2D(2), {}^2P$
d^4 or d^6	$^5D, {}^3H, {}^3G, {}^3F(2), {}^3D, {}^3P(2), {}^1I, {}^1G(2), {}^1F, {}^1D(2), {}^1S(2)$
d^5	$^6S, {}^4G, {}^4F, {}^4D, {}^4P, {}^2I, {}^2H, {}^2G(2), {}^2F(2), {}^2D(3), {}^2P, {}^2S$
d^{10}	^1S

15.44 Determine the term symbols for an nd^2 configuration.

▌ The electrons can either be paired giving $S = +\frac{1}{2} + (-\frac{1}{2}) = 0$ or unpaired giving $S = +\frac{1}{2} + (+\frac{1}{2}) = 1$. The mnemonic device shown in Fig. 15-3, which does not require the writing of all permitted microstates, can be used to find the term symbols. In Fig. 15-3a, all combinations of m are allowed for the paired electrons (except those excluded by the indistinguishability of the electrons) giving the values of M_L shown in the center of the figure. The three terms are ^{1}G, ^{1}D, and ^{1}S. In Fig. 15-3b, only those combinations for different values of m are allowed for the unpaired electrons (again excluding those by the indistinguishability), giving the values of M_L shown in the center of the figure. The two terms are ^{3}F and ^{3}P.

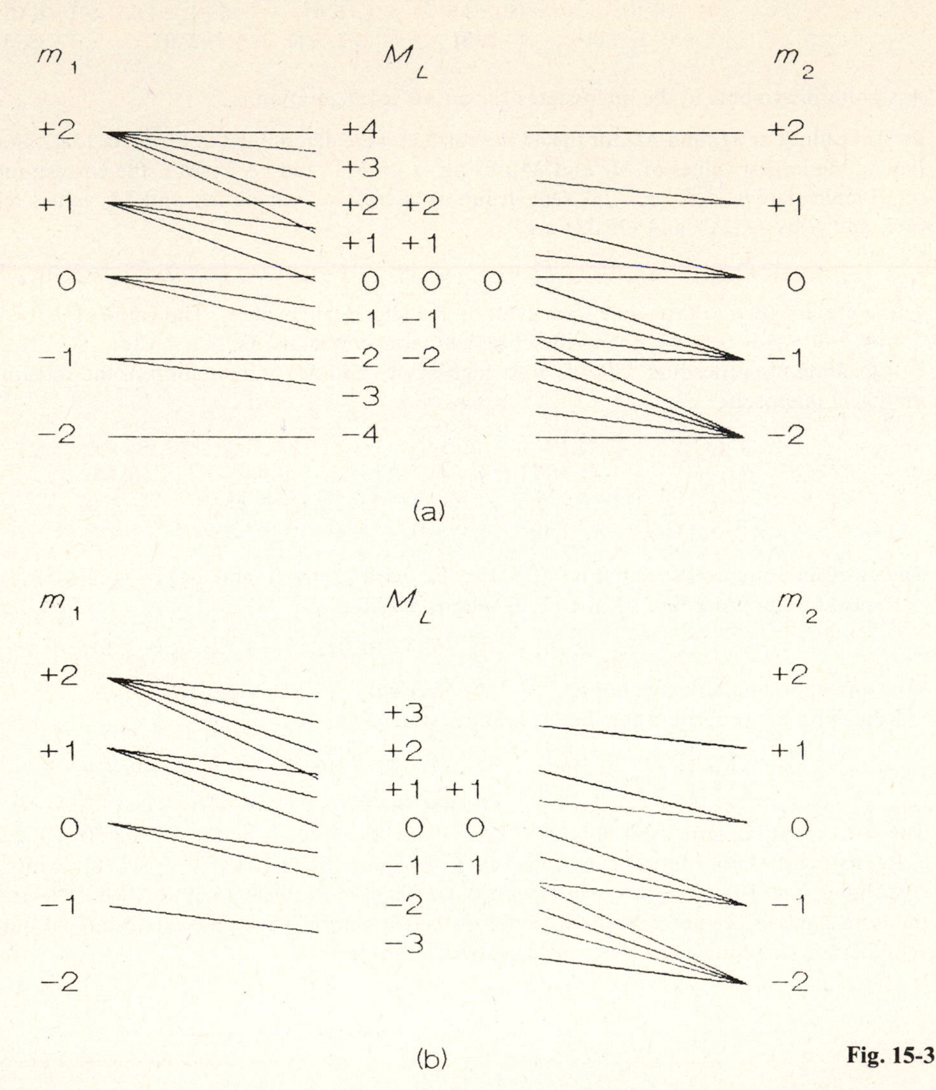

Fig. 15-3

15.45 What is the predicted term symbol for the ground state of an atom having an nd^2 configuration (e.g., Ti)?

▌ The ground state will correspond to the term symbol having the highest value of $2S + 1$, according to the rule of maximum multiplicity (*Hund's first rule*). From Problem 15.43, the terms with $2S + 1$ equal to 3 are ^{3}F and ^{3}P. For terms of equal multiplicity, the ground state will have the highest value of M_L (*Hund's second rule*). In this case, the ^{3}F term is chosen. The value of J for a given level is usually lowest for less-than-half-filled subshells and greatest for more-than-half-filled subshells. From Problem 15.43, the ^{3}F terms were $^3F_4$, $^3F_3$, and $^3F_2$. In this case, the $^3F_2$ term is chosen.

15.46 Determine the term symbol for the ground state of Ni, which has the configuration $\cdots (3d)^8$.

▌ The ground-state term symbol can be derived using a method that does not require writing all the permitted microstates. Substituting the values of m and s for the eight electrons ($m = +2$, $s = +\frac{1}{2}$; $m = +1$, $s = +\frac{1}{2}$; $m = 0$, $s = +\frac{1}{2}$; $m = -1$, $s = +\frac{1}{2}$; $m = -2$, $s = +\frac{1}{2}$; $m = +2$, $s = -\frac{1}{2}$; $m = +1$, $s = -\frac{1}{2}$; $m = 0$, $s = -\frac{1}{2}$) into *(15.22)* and

(15.23) gives

$$M_L = (+2) + (+1) + (0) + (-1) + (-2) + (+2) + (+1) + (0) = 3$$

$$M_S = (+\tfrac{1}{2}) + (+\tfrac{1}{2}) + (+\tfrac{1}{2}) + (+\tfrac{1}{2}) + (+\tfrac{1}{2}) + (-\tfrac{1}{2}) + (-\tfrac{1}{2}) + (-\tfrac{1}{2}) = 1$$

These values are assumed to be L and S, respectively, giving $^{2(1)+1}F = {}^3F$. The values of $L = 3$ and $S = 1$ correspond to $J = 4$, 3, and 2. Choosing the greatest value for J because the subshell is more than half-filled gives $^3F_4$.

15.47 The term symbol for the ground state of atomic N is $^4S_{3/2}$. Determine the electronic configuration for this element.

▮ The value of S is determined by

$$2(S) + 1 = 4 \qquad S = M_S = \tfrac{3}{2}$$

implying that there are three unpaired electrons. The S term gives $L = M_L = 0$, which agrees with one electron in each of the p orbitals.

$$M_L = (+1) + (0) + (-1) = 0$$

The configuration is $(1s)^2(2s)^2(2p)^3$ or $(1s)^2(2s)^2(2p_x)^1(2p_y)^1(2p_z)^1$. Note that this result agrees with the entry in Table 15-4.

15.48 The term symbol for the ground state of Nb is $^6D_{1/2}$. Does this agree with the predicted electronic configuration of $\cdots(5s)^2(4d)^3$?

▮ The d^3 configuration gives

$$M_L = L = (+2) + (+1) + (0) = +3 \qquad M_S = S = (+\tfrac{1}{2}) + (+\tfrac{1}{2}) + (+\tfrac{1}{2}) = \tfrac{3}{2}$$

which leads to the incorrect symbol ^4F. A configuration of $\cdots(5s)^1(4d)^4$ gives

$$M_L = L = (0) + (+2) + (+1) + (0) + (-1) = +2$$

$$M_S = S = (+\tfrac{1}{2}) + (+\tfrac{1}{2}) + (+\tfrac{1}{2}) + (+\tfrac{1}{2}) + (+\tfrac{1}{2}) = \tfrac{5}{2}$$

which leads to the correct symbol of $^{2(5/2)+1}D = {}^6D$.

15.4 SPECTRA OF POLYELECTRONIC ATOMS

15.49 Monochromatic $K_{\alpha 1 \alpha 2}$ radiation from Al at 1486.6 eV was absorbed by the photoelectronic emission of a 3d electron from Xe. Calculate the velocity of the emitted electron given that the binding energy of this electron is 678 eV.

▮ The energy of the incident radiation, the binding energy, and the kinetic energy of the emitted electron are related by

$$E_k = h\nu - E_b = \tfrac{1}{2}mv^2 \qquad (15.24)$$

Solving *(15.24)* for v and substituting the data gives

$$v = \left(\frac{2(1486.6\text{ eV} - 678\text{ eV})[(1.602 \times 10^{-19}\text{ J})/(1\text{ eV})][(1\text{ kg}\cdot\text{m}^2\cdot\text{s}^{-2})/(1\text{ J})]}{9.11 \times 10^{-31}\text{ kg}}\right)^{1/2} = 1.69 \times 10^7\text{ m}\cdot\text{s}^{-1}$$

15.50 During the internal conversion of ^{110m}Ag, K and L peaks in the X-ray spectrum were observed at 92 keV and 114 keV, respectively. The photon emitted by the internal conversion process is 118 keV. What are the approximate binding energies of the $n = 1$ and $n = 2$ electrons?

▮ The peaks in the spectrum correspond to the kinetic energies of the emitted electrons. Solving *(15.24)* for E_b and substituting the data gives

$$E_b(n = 1) = 118\text{ keV} - 92\text{ keV} = 26\text{ keV}$$

$$E_b(n = 2) = 118 - 114 = 4\text{ keV}$$

15.51 The three ionization energies of Li are 520.6 kJ · mol^{-1}, 7297.9 kJ · mol^{-1}, and 11 814.6 kJ · mol^{-1}. Determine the energy of the 2s electron with respect to infinity, and determine the total electronic energy of Li.

▮ The first ionization potential for Li is the removal of the 2s electron to infinity. Thus the energy with respect to infinity is −520.6 kJ · mol^{-1}. The total electronic energy is the negative of the sum of the three ionization energies, giving

$$E(\text{Li}) = -19\,633.1\text{ kJ}\cdot\text{mol}^{-1}$$

15.52 Choose the transitions that are allowed for a Li atom: **(*a*)** 4f ↔ 3d, **(*b*)** 3d ↔ 3s, **(*c*)** 3d ↔ 3p, **(*d*)** 3p ↔ 4s, **(*e*)** 3s ↔ 3p, **(*f*)** 3s ↔ 5s.

▮ The selection rules for allowed electronic transitions in an atom are

$$\Delta S = 0 \tag{15.25a}$$

$$\Delta L = 0, \pm 1, \quad \text{but } L = 0 \not\leftrightarrow L = 0 \tag{15.25b}$$

$$\Delta J = 0, \pm 1, \quad \text{but } J = 0 \not\leftrightarrow J = 0 \tag{15.25c}$$

$$\Delta l = \pm 1 \tag{15.25d}$$

(*a*) The term symbols for one electron in the 4f subshell are $^2F_{7/2}$ and $^2F_{5/2}$, and for the 3d subshell they are $^2D_{5/2}$ and $^2D_{3/2}$. For this change, $S = \frac{1}{2}$ in both cases, making *(15.25a)* valid; $L = 3$ and 2, making *(15.25b)* valid; $J = \frac{7}{2}$ and $\frac{5}{2}$ for the F term and $\frac{5}{2}$ and $\frac{3}{2}$ for the D term, making *(15.25c)* valid; and $l = 3$ and 2, making *(15.25d)* valid. This transition is one of the lines in the fundamental series.
(*b*) For this transition $\Delta l = 2$; *(15.25d)* is not valid.
(*c*), (*d*), and **(*e*)** All selection rules are obeyed for these transitions. The transitions each represent one of the lines in the diffuse, sharp, and principal series, respectively.
(*f*) For this transition, $\Delta l = 0$; *(15.25d)* is not valid.

15.53 The principal series in Li consists of the $^2P \leftrightarrow {}^2S$ transitions. Describe the expected spectrum.

▮ The term symbols for the states involved are $^2S_{1/2}$ for the s subshell and $^2P_{1/2}$ and $^2P_{3/2}$ for the p subshell. According to *(15.25c)*, the spectrum should appear as a closely spaced doublet (see Fig. 15-4a).

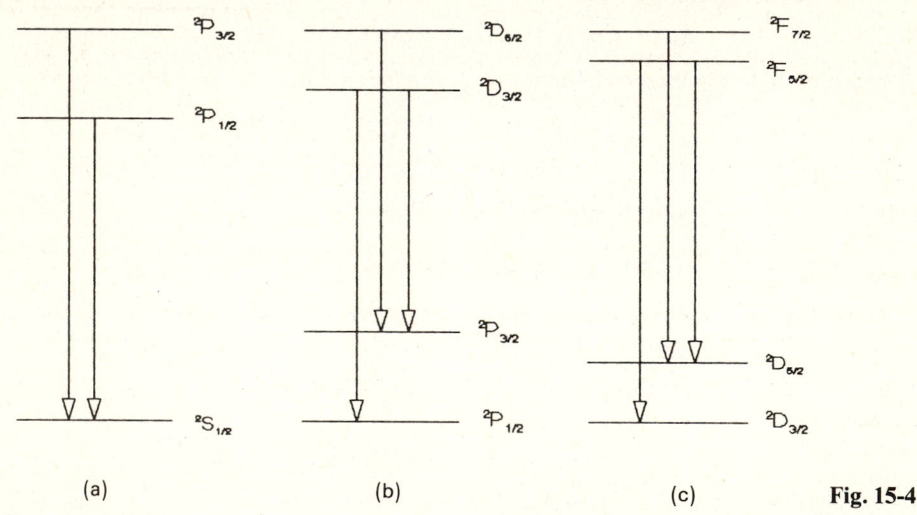

Fig. 15-4

15.54 The diffuse series in Li consists of the $^2D \leftrightarrow {}^2P$ transitions. Describe what the term *compound doublet* means.

▮ The term symbols for the states involved are $^2D_{5/2}$ and $^2D_{3/2}$ for the d subshell and $^2P_{1/2}$ and $^2P_{3/2}$ for the p subshell. According to *(15.25c)* the spectrum should appear as a three-component fine structure called a compound doublet (see Fig. 15-4b).

15.55 The fundamental series in Li consists of the $^2F \leftrightarrow {}^2D$ transitions. Describe the expected spectrum.

▮ The term symbols for the states involved are $^2F_{5/2} \leftrightarrow {}^2F_{7/2}$ for the f subshell and $^2D_{5/2}$ and $^2D_{3/2}$ for the d subshell. According to *(15.25c)* the spectrum should appear as a compound doublet (see Fig. 15-4c).

15.56 The first line in the principal series for Li is a doublet (6707.91 Å and 6707.76 Å) and the first line in the sharp series is a doublet (8126.23 Å and 8126.45 Å). Determine the energy spacings for the 2s, 2p, and 3s subshells. If the 2s subshell has an energy of $-43\,487.20\ \text{cm}^{-1}$ with respect to infinity, prepare an energy diagram for Li.

▮ Substituting the data for the principal series (2s ↔ 2p) into *(13.33)* gives

$$\bar{\nu} = \frac{1}{(6707.91\ \text{Å})[(1\ \text{cm})/(10^8\ \text{Å})]} = 14\,907.77\ \text{cm}^{-1}$$

$$\bar{\nu} = \frac{1}{(6707.76)(1/10^8)} = 14\,908.11\ \text{cm}^{-1}$$

The average distance between the 2p and 2s subshells is 14 907.94 cm^{-1}, and the 2p subshell will be located at $-43\,487.20\ cm^{-1} + 14\,907.94\ cm^{-1} = -28\,579.26\ cm^{-1}$. Likewise for the sharp series, $\bar{\nu} = 12\,305.83\ cm^{-1}$ and $12\,305.50\ cm^{-1}$, giving the average distance between the 3s and 2p subshells as 12 305.67 cm^{-1}. The total energy difference between the 3s and 2s subshells is $12\,305.67\ cm^{-1} + 14\,907.94\ cm^{-1} = 27\,213.61\ cm^{-1}$, and the 3s subshell will be located at $-43\,487.20\ cm^{-1} + 27\,213.61\ cm^{-1} = -16\,273.59\ cm^{-1}$. The energy diagram is shown in Fig. 15-5.

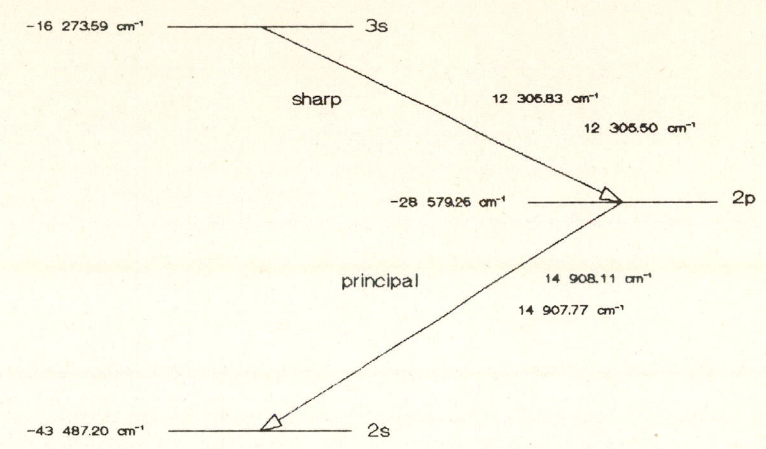

Fig. 15-5

15.57 Using the data given in Problem 15.56, calculate the spin-orbit coupling constant for the $^2P_{3/2}$ and $^2P_{1/2}$ states.

▮ The *spin-orbit coupling constant* (A) is related to the energy difference between two adjacent components $J - 1$ and J by

$$E_J - E_{J-1} = AJ \tag{15.26}$$

Solving *(15.26)* for A and substituting the data for the principal series gives

$$A = \frac{14\,908.11\ cm^{-1} - 14\,907.77\ cm^{-1}}{3/2} = 0.23\ cm^{-1}$$

15.58 Determine the Lande g factors for the following terms: **(*a*)** ^1S, **(*b*)** ^1P, **(c)** ^3S, **(*d*)** ^3P.

▮ The *Lande g factor* is defined as

$$g = 1 + \frac{J(J+1) + S(S+1) - L(L+1)}{2J(J+1)} \tag{15.27}$$

(Note that 0 divided by 0 in the second term is treated as equal to 0.)

(*a*) and **(*b*)** For the ^1S and ^1P terms, $J = L$ because $S = 0$. Substituting into *(15.27)* gives

$$g = 1 + \frac{L(L+1) + 0(0+1) - L(L+1)}{2L(L+1)} = 1$$

(*c*) For the ^3S term, $S = 1$ and $L = 0$, giving $J = 1$. Substituting into *(15.27)* gives

$$g = 1 + \frac{1(1+1) + 1(1+1) - 0(0+1)}{2(1)(1+1)} = 2$$

(*d*) For the ^3P term, $S = 1$ and $L = 1$, giving $J = 2, 1,$ or 0. Substituting into *(15.27)* gives

$$g(J=2) = 1 + \frac{2(2+1) + 1(1+1) - 1(1+1)}{2(2)(2+1)} = \tfrac{3}{2}$$

$$g(J=1) = 1 + \frac{1(1+1) + 1(1+1) - 1(1+1)}{2(1)(1+1)} = \tfrac{3}{2}$$

$$g(J=0) = 1 + \frac{0(0+1) + 1(1+1) - 1(1+1)}{2(0)(0+1)} = 1$$

15.59 Prepare a diagram showing the splitting of the $^1S_0 \leftrightarrow {}^1P_1$ transition in the presence of a magnetic field.

▮ The number of states that each term is split into is given by the M_J quantum number. There are $2J + 1$ values of M_J. For the $^1S_0$ term, $M_J = 0$ and no splitting occurs. For the ^1P, the $2(1) + 1 = 3$ values of M_J are +1,

0, and −1. These levels are shown in Fig. 15-6. The additional selection rule describing permitted transitions is

$$\Delta M_J = 0, \pm 1 \tag{15.28}$$

These transitions are also shown in Fig. 15-6. Note that the middle spectral line is at the original frequency and that the other two spectral lines differ by equal amounts from the original frequency (the *normal Zeeman effect*).

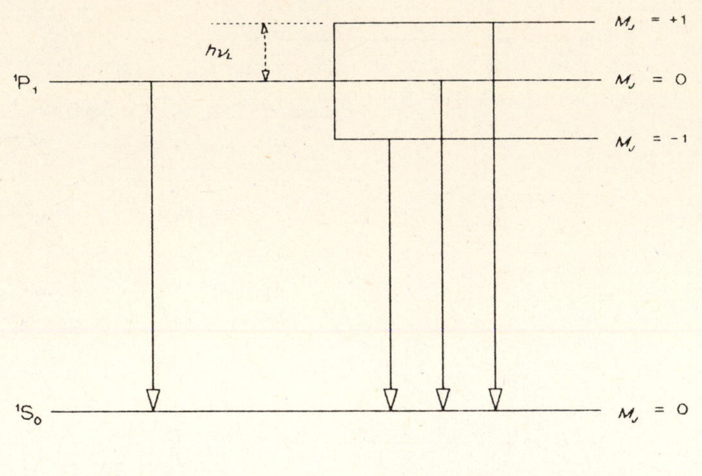

Fig. 15-6

15.60 Express the displacements of the $^1P_1$ lines in Problem 15.59 in terms of the Lamor frequency.

▮ The *Lamor frequency* is defined as

$$\nu_L = \mu_B B/h \tag{15.29}$$

where B is the magnetic flux density (often called the magnetic field strength) and μ_B is the Bohr magneton $[\mu_B = |\hbar(e/2m_e c)| = 9.274\,08 \times 10^{-24}\ \text{J}\cdot\text{T}^{-1}]$. The energy of each state is given by

$$E = E_0 + g\mu_B M_J B = E_0 + gM_J h\nu_L \tag{15.30}$$

where E_0 is the original frequency in the absence of the magnetic field. For the $^1P_1$ lines in Fig. 15-6, $M_J = +1$, 0, and −1, giving the three energies as

$$E = E_0 + (1)(+1)h\nu_L = E_0 + h\nu_L$$
$$E = E_0 + (1)(0)h\nu_L = E_0$$
$$E = E_0 + (1)(-1)h\nu_L = E_0 - h\nu_L$$

The energy spacings are $\pm h\nu_L$ from E_0.

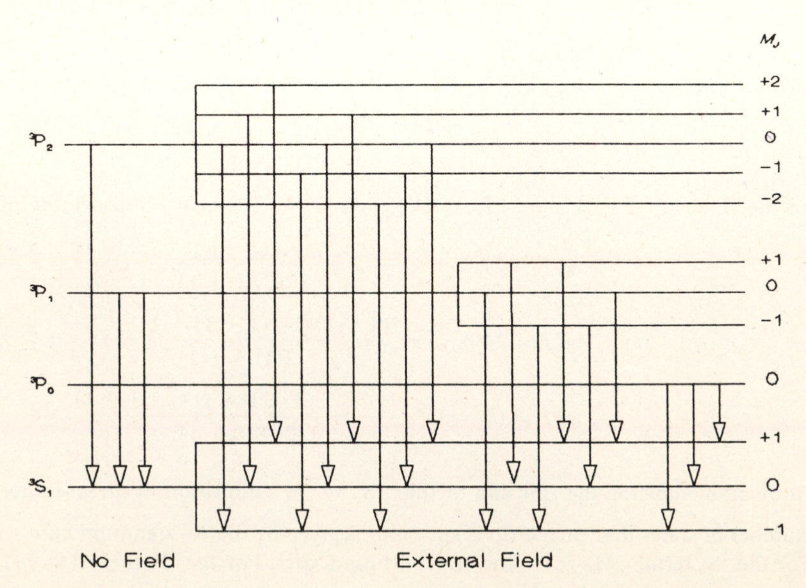

Fig. 15-7

15.61 Prepare a diagram showing the splitting of the $^3S_1 \leftrightarrow {}^3P$ transitions in the presence of a magnetic field.

▌ The number of states that each term is split into is given by the M_J quantum number. For the $^3S_1$ term, $2(1) + 1 = 3$ values of M_J are +1, 0, and −1. For the $^3P_0$ term, $M_J = 0$ and no splitting occurs. For the $^3P_1$ term, the $2(1) + 1 = 3$ values of M_J are +1, 0, and −1. For the $^3P_2$ term, the $2(2) + 1 = 5$ values of M_J are +2, +1, 0, −1, and −2. These levels are shown in Fig. 15-7. The transitions that are allowed by *(15.28)* are also shown in Fig. 15-7. Note that the $^3P_1(M_J = 0) \leftrightarrow {}^3S_1(M_J = 0)$ transition is not allowed because of *(15.25c)*. From the results of Problem 15.58, the displacements of the $^3P_2$, $^3P_1$, and ^3S lines are $\frac{3}{2}h\nu_L$, $\frac{3}{2}h\nu_L$, and $2h\nu_L$, respectively, from each other. Because g varies for the different terms, the normal triplet spectral lines that occurred in the normal Zeeman effect are not present (*anomalous Zeeman effect*).

15.62 Determine the magnetic flux density needed to separate the ^1P states in Problem 15.60 by 1.00 cm^{-1}.

▌ Solving *(15.30)* for B and substituting the data gives

$$B = \frac{(1.00\ \text{cm}^{-1})(6.626 \times 10^{-34}\ \text{J}\cdot\text{s})(2.9979 \times 10^{8}\ \text{m}\cdot\text{s}^{-1})[(10^{2}\ \text{cm})/(1\ \text{m})]}{(1)(9.274\,08 \times 10^{-24}\ \text{J}\cdot\text{T}^{-1})(1)} = 2.14\ \text{T}$$

CHAPTER 16

Electronic Structure of Diatomic Molecules

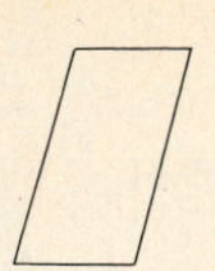

16.1 QUANTUM THEORY OF DIATOMIC MOLECULES

16.1 Write the Hamiltonian operator describing the He_2^+ molecule-ion.

▌ The complete Hamiltonian operator is given by

$$\mathscr{H} = \frac{-\hbar^2}{2}\sum_{\alpha}^{\text{all nuclei}}\frac{1}{m_\alpha}\nabla^2 + \frac{-\hbar^2}{2m_e}\sum_{i}^{\text{all electrons}}\nabla_i^2 + \sum_{\beta}\sum_{\alpha<\beta}\frac{Z_\alpha Z_\beta e^2}{4\pi\varepsilon_0 r_{\alpha\beta}} - \sum_{i}\sum_{\alpha}\frac{Z_\alpha e^2}{4\pi\varepsilon_0 r_{i\alpha}} + \sum_{j}\sum_{i<j}\frac{e^2}{4\pi\varepsilon_0 r_{ij}} \tag{16.1}$$

The *Born–Oppenheimer approximation* allows neglecting the ∇_α^2 terms and treating the $r_{\alpha\beta}^{-1}$ terms as constants at a given time. For the three electrons in He_2^+, *(16.1)* gives

$$\mathscr{H} = \frac{-\hbar^2}{2m_e}(\nabla_1^2 + \nabla_2^2 + \nabla_3^2) + \frac{e^2}{4\pi\varepsilon_0}\left(\frac{4}{r_{ab}} - \frac{2}{r_{1a}} - \frac{2}{r_{2a}} - \frac{2}{r_{3a}} - \frac{2}{r_{1b}} - \frac{2}{r_{2b}} - \frac{2}{r_{3b}} + \frac{1}{r_{12}} + \frac{1}{r_{13}} + \frac{1}{r_{23}}\right)$$

where $Z_a = Z_b = 2$.

16.2 Write the Hamiltonian operator describing the K_2 molecule.

▌ The electrons in this molecule can be classified into two types: the core electrons around each atom that do not contribute to the bonding and the two valence electrons that provide the bonding. Using this *valence-electron approximation, (16.1)* gives

$$\mathscr{H} = \frac{-\hbar^2}{2m_e}(\nabla_1^2 + \nabla_2^2) + \frac{e^2}{4\pi\varepsilon_0}\left(\frac{1}{r_{ab}} - \frac{1}{r_{1a}} - \frac{1}{r_{2a}} - \frac{1}{r_{1b}} - \frac{1}{r_{2b}} + \frac{1}{r_{12}}\right)$$

where the effective nuclear charge of 1 was used.

16.3 Write the expression for a trial wave function describing the He_2^+ system.

▌ The wave function ϕ can be written as the product of the wave functions for the individual electrons.

$$\phi = \phi_1\phi_2\phi_3\cdots \tag{16.2}$$

where the ϕ_i are produced by taking linear combinations of hydrogenlike atomic orbitals (LCAO):

$$\phi_i = a_{ia}\psi_a(i) + a_{ib}\psi_b(i) + \cdots \tag{16.3}$$

The $a_{i\alpha}$ terms are constants, and $\psi_\alpha(i)$ denotes a hydrogenlike atomic wave function describing electron i in terms of position with respect to nucleus α. In this case, using *(16.2)* and *(16.3)* gives

$$\phi = \phi_1\phi_2\phi_3$$
$$\phi_1 = a_1\psi_a(1) + a_2\psi_b(1)$$
$$\phi_2 = a_3\psi_a(2) + a_4\psi_b(2)$$
$$\phi_3 = a_5\psi_a(3) + a_6\psi_b(3)$$

where ψ_a and ψ_b would be the 1s wave function for H.

16.4 Write the expression for a trial wave function describing the K_2 molecule.

▌ Assuming the valence-electron approximation, using *(16.2)* and *(16.3)* gives

$$\phi = \phi_1\phi_2$$
$$\phi_1 = a_1\psi_a(1) + a_2\psi_b(1)$$
$$\phi_2 = a_3\psi_a(2) + a_4\psi_b(2)$$

where ψ_a and ψ_b would be the 4s wave function for H.

16.5 Write the expression for a trial wave function describing the HF molecule.

▮ Assuming the valence-electron approximation, using *(16.2)* and *(16.3)* gives

$$\phi = \phi_1\phi_2$$
$$\phi_1 = a_1\psi_a(1) + a_2\psi_b(1)$$
$$\phi_2 = a_3\psi_a(2) + a_4\psi_b(2)$$

where ψ_a would be the 1s wave function for H and ψ_b would be the 2p wave function for F.

16.6 Identify the ionic and covalent terms in the expression for ϕ determined in Problem 16.5.

▮ The complete trial wave function for HF is

$$\phi = \phi_1\phi_2 = [a_1\psi_a(1) + a_2\psi_b(1)][a_3\psi_a(2) + a_4\psi_b(2)]$$
$$= a_1a_3\psi_a(1)\psi_a(2) + a_2a_3\psi_b(1)\psi_a(2) + a_1a_4\psi_a(1)\psi_b(2) + a_2a_4\psi_b(1)\psi_b(2)$$

The first and fourth terms in the expression for ϕ each indicate that both electrons are around the same nucleus and are the ionic terms. The second and third terms each indicate that the electrons are around different nuclei and are the covalent terms.

16.7 Because He_2^+ is a homonuclear diatomic system, assume that the $a_{i\alpha}$ are all equal to a and that identical normalized hydrogenlike wave functions are chosen. Determine the value of a in terms of S where $S_{aa} = S_{bb} = 1$ and $S_{ab} = S_{ba} = S$ as defined in *(14.70)*.

▮ The integral of *(14.22)* is equal to

$$\int_{\text{all space}} \phi^*\phi\, dV = \int_{\text{all space}} (\phi_1\phi_2\phi_3)^*(\phi_1\phi_2\phi_3)\, dV_1\, dV_2\, dV_3 = \int_{\text{all space}} \phi_1^*\phi_2^*\phi_1\phi_2\, dV_1\, dV_2 \int_{\text{all space}} \phi_3^*\phi_3\, dV_3$$
$$= \int_{\text{all space}} \phi_1^*\phi_1\, dV_1 \int_{\text{all space}} \phi_2^*\phi_2\, dV_2 \int_{\text{all space}} \phi_3^*\phi_3\, dV_3 = \left(\int_{\text{all space}} \phi_1^*\phi_1\, dV_1\right)^3$$

Substituting $\psi_1 = a\psi_a(1) + a\psi_b(1)$ (see Problem 16.3) gives

$$\int_{\text{all space}} \phi_1^*\phi_1\, dV_1 = \int_{\text{all space}} [a\psi_a(1) + a\psi_b(1)]^*[a\psi_a(1) + a\psi_b(1)]\, dV_1$$
$$= a^2\left(\int_{\text{all space}} [\psi_a^*(1)\psi_a(1) + \psi_a^*(1)\psi_b(1) + \psi_b^*(1)\psi_a(1) + \psi_b^*(1)\psi_b(1)]\, dV\right)$$
$$= a^2(S_{aa} + S_{ab} + S_{ba} + S_{bb}) = a^2(1 + S + S + 1) = a^2(2 + 2S)$$

The integral of *(14.22)* becomes

$$\int_{\text{all space}} \phi^*\phi\, dV = [a^2(2 + 2S)]^3 = 1$$

Solving gives $a = (2 + 2S)^{-1/2}$.

16.2 APPLICATION OF THE VARIATION METHOD

16.8 Determine the energy for the H_2^+ molecule-ion in terms of H_{aa}, H_{ab}, and S using the variation method.

▮ The secular determinant for the system is given by *(14.71)* as

$$\begin{vmatrix} H_{aa} - ES_{aa} & H_{ab} - ES_{ab} \\ H_{ba} - ES_{ba} & H_{bb} - ES_{bb} \end{vmatrix} = 0$$

In this case $H_{aa} = H_{bb}$, $S_{ab} = S_{ba} = S$, and $S_{aa} = S_{bb} = 1$, giving

$$\begin{vmatrix} H_{aa} - E & H_{ab} - ES \\ H_{ab} - ES & H_{aa} - E \end{vmatrix} = 0$$

Expanding gives

$$(H_{aa} - E)^2 - (H_{ab} - ES)^2 = 0$$
$$[(H_{aa} - E) + (H_{ab} - ES)][(H_{aa} - E) - (H_{ab} - ES)] = 0$$

The two roots are

$$E_1 = \frac{H_{aa} + H_{ab}}{1 + S} \tag{16.4a}$$

and

$$E_2 = \frac{H_{aa} - H_{ab}}{1 - S} \tag{16.4b}$$

16.9 For the H_2^+ molecular ion described by the wave function

$$\phi = \phi_1 = a_1\psi_a(1) + a_2\psi_b(1) \tag{16.5}$$

it can be shown that

$$H_{aa} = E(\mathrm{H}) + J \tag{16.6a}$$

$$H_{ab} = E(\mathrm{H})S + K \tag{16.6b}$$

where $E(\mathrm{H})$ is the ground-state energy of the hydrogen atom (-13.60 eV $= -2.179 \times 10^{-18}$ J), the *Coulomb integral* is given by

$$J = \frac{-e^2}{4\pi\varepsilon_0} \int_{\text{all space}} \psi_a^*(1) \frac{1}{r_{ib}} \psi_a(1)\, dV \tag{16.7a}$$

the *exchange* (or *resonance*) *integral* is given by

$$K = \frac{-e^2}{4\pi\varepsilon_0} \int_{\text{all space}} \psi_b^*(1) \frac{1}{r_{ib}} \psi_a(1)\, dV \tag{16.8a}$$

and the *overlap integral* is given by

$$S = \int_{\text{all space}} \psi_a^*(1)\psi_b(1)\, dV \tag{16.9a}$$

Rewrite the expressions for energy given by *(16.4)* in terms of these integrals.

▮ Substituting *(16.5a)* and *(16.5b)* into *(16.4a)* and *(16.4b)* gives

$$E_1 = \frac{[E(\mathrm{H}) + J] + [E(\mathrm{H})S + K]}{1 + S} = E(\mathrm{H}) + \frac{J + K}{1 + S} \tag{16.10a}$$

$$E_2 = \frac{[E(\mathrm{H}) + J] - [E(\mathrm{H})S + K]}{1 - S} = E(\mathrm{H}) + \frac{J - K}{1 - S} \tag{16.10b}$$

16.10 The observed internuclear distance in H_2^+ is 0.1057 nm. Calculate the total energy of the system at this value of r_{ab} assuming that $\psi_\alpha(i)$ is given by 1s hydrogenlike orbitals.

▮ If the $\psi_\alpha(i)$ are hydrogenlike 1s atomic orbitals, the various integrals are given by

$$S = e^{-\rho}\left(1 + \rho + \frac{\rho^2}{3}\right) \tag{16.9b}$$

$$J = \frac{e^2}{4\pi\varepsilon_0 a_0}\left[\frac{-1}{\rho} + e^{-2\rho}\left(1 + \frac{1}{\rho}\right)\right] \tag{16.7b}$$

$$K = \frac{-e^2}{4\pi\varepsilon_0 a_0} e^{-\rho}(1 + \rho) \tag{16.8b}$$

where

$$\rho = r_{ab}/a_0 \tag{16.11}$$

and $a_0 = 0.052\,917\,706$ nm. Substituting $r_{ab} = 0.1057$ nm into *(16.11)* gives

$$\rho = \frac{0.1057\ \text{nm}}{0.052\,917\,706\ \text{nm}} = 1.997$$

Substitution of this value of ρ into *(16.9b)*, *(16.7b)*, and *(16.8b)* gives

$$S = e^{-1.997}\left(1 + 1.997 + \frac{(1.997)^2}{3}\right) = 0.5873$$

$$J = \left[\frac{(1.602 \times 10^{-19}\ \mathrm{C})^2[(1\ \mathrm{kg \cdot m \cdot s^{-2}})/(1\ \mathrm{N})][(1\ \mathrm{J})/(1\ \mathrm{kg \cdot m^2 \cdot s^{-2}})]}{(1.112 \times 10^{-10}\ \mathrm{C^2 \cdot N^{-1} \cdot m^{-2}})(5.292 \times 10^{-11}\ \mathrm{m})}\right]\left[\frac{-1}{1.997} + e^{-2(1.997)}\left(1 + \frac{1}{1.997}\right)\right]$$

$$= (4.360 \times 10^{-18}\ \mathrm{J})(-0.4731) = -2.063 \times 10^{-18}\ \mathrm{J}$$

$$K = (-4.360 \times 10^{-18}\ \mathrm{J})\, e^{-1.997}(1 + 1.997) = -1.774 \times 10^{-18}\ \mathrm{J}$$

Substitution of these values of S, J, and K into *(16.10)* gives

$$E_1 = -2.179 \times 10^{-18}\ \mathrm{J} + \frac{-2.063 \times 10^{-18}\ \mathrm{J} + (-1.774 \times 10^{-18}\ \mathrm{J})}{1 + 0.5873} = -4.596 \times 10^{-18}\ \mathrm{J}$$

$$E_2 = -2.179 \times 10^{-18}\ \mathrm{J} + \frac{-2.063 \times 10^{-18}\ \mathrm{J} - (-1.774 \times 10^{-18}\ \mathrm{J})}{1 - 0.5873} = -2.879 \times 10^{-18}\ \mathrm{J}$$

The total energy (E') is found by adding the nuclear repulsion term that was treated as a constant in *(16.1)* by the Born-Oppenheimer approximation:

$$E_1' = E_1 + \frac{e^2}{4\pi\varepsilon_0 r_{ab}} = -4.596 \times 10^{-18}\ \mathrm{J} + \frac{(1.602 \times 10^{-19})^2}{(1.112 \times 10^{-10})(1.057 \times 10^{-10})} = -2.413 \times 10^{-18}\ \mathrm{J}$$

$$E_2' = -2.879 \times 10^{-18} + 2.183 \times 10^{-18} = -0.696 \times 10^{-18}\ \mathrm{J}$$

The results of similar calculations are plotted in Fig. 16-1.

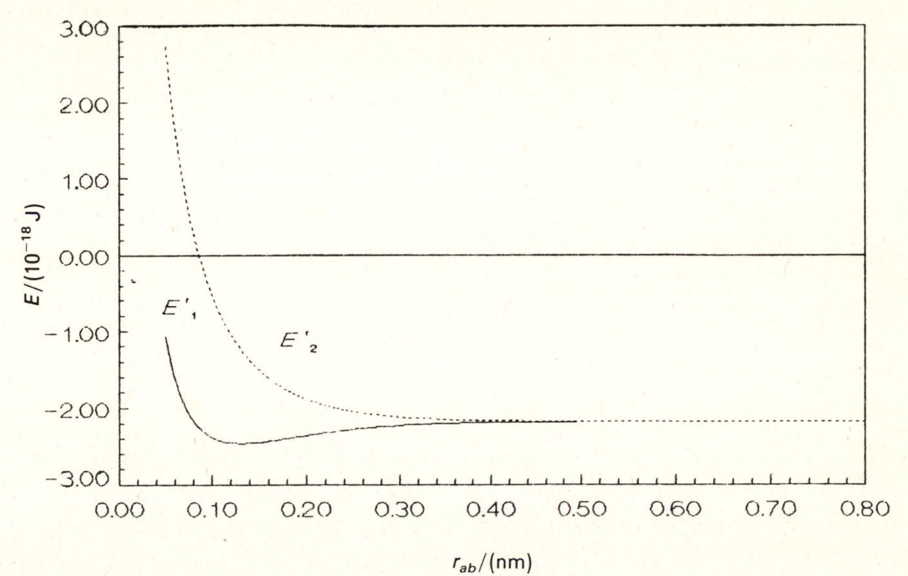

Fig. 16-1

16.11 What is the limit of the total energy for H_2^+ as $r_{ab} \to \infty$?

▌ Taking the limits of *(16.7b)*, *(16.8b)*, *(16.9b)*, and the nuclear repulsion term as r_{ab} or $\rho \to \infty$ gives

$$\lim_{\rho\to\infty}(S) = \lim_{\rho\to\infty}\left(\frac{1 + \frac{2}{3}\rho}{-e^{-\rho}}\right) = \lim_{\rho\to\infty}\left(\frac{\frac{2}{3}}{e^{-\rho}}\right) = 0$$

$$\lim_{\rho\to\infty}(J) = 0$$

$$\lim_{\rho\to\infty}(K) = \frac{-e^2}{4\pi\varepsilon_0 a_0}\lim_{\rho\to\infty}\left(\frac{1}{-e^{-\rho}}\right) = 0$$

$$\lim_{r_{ab}\to\infty}\left(\frac{e^2}{4\pi\varepsilon_0 r_{ab}}\right) = 0$$

The limiting values of the total energy are

$$E_1' = E(\mathrm{H}) + \frac{0+0}{1+0} + 0 = E(\mathrm{H}) = -2.179 \times 10^{-18}\ \mathrm{J}$$

$$E_2' = E(\mathrm{H}) + \frac{0-0}{1-0} + 0 = E(\mathrm{H})$$

16.12 Using Fig. 16-1, predict the equilibrium value of r_{ab} and the dissociation energy for the H_2^+ molecule-ion.

▌ The minimum value of E_1' in Fig. 16-1 appears at $r_{ab} = 0.132$ nm and -2.46×10^{-18} J. The predicted equilibrium value of r_{ab} is 0.132 nm, which is 1.24 times greater than the observed value of 0.1057 nm. In Problem 16.11, the limiting value of E_1' was determined to be -2.179×10^{-18} J as $r_{ab} \to \infty$. The difference gives the predicted value of the dissociation energy (D_e) as

$$D_e = (-2.179 \times 10^{-18}\ \text{J}) - (-2.46 \times 10^{-18}\ \text{J}) = 0.28 \times 10^{-18}\ \text{J}$$

which is considerably less than the observed value of 0.4472×10^{-18} J.

16.13 Assuming the H_2^+ molecule-ion to be described by *(16.5)*, determine the relation between a_1 and a_2.

▌ For this system *(14.68)* simplifies to

$$a_1(H_{aa} - E) + a_2(H_{ab} - ES) = 0$$

Substituting *(16.4a)* gives

$$a_1\left(H_{aa} - \frac{H_{aa} + H_{ab}}{1 + S}\right) + a_2\left(H_{ab} - \frac{H_{aa} + H_{ab}}{1 + S}\right) = 0$$

$$a_1[(1 + S)H_{aa} - H_{aa} - H_{ab}] + a_2[(1 + S)H_{ab} - SH_{aa} - SH_{ab}] = 0$$

$$a_1(SH_{aa} - H_{ab}) + a_2(H_{ab} - SH_{aa}) = 0$$

$$a_1 = a_2$$

for E_1. Likewise, substituting *(16.4b)* gives $a_1 = -a_2$ for E_2.

16.14 Evaluate a_1 and a_2 in the wave function given by *(16.5)*.

▌ For E_1, it was shown in Problem 16.13 that $a_1 = a_2$. The wave function becomes

$$\phi = a_1\psi_a(1) + a_1\psi_b(1) = a_1[\psi_a(1) + \psi_b(1)]$$

Substituting into *(14.22)* gives the normalization constant as

$$\int_{\text{all space}} \phi^*\phi\, dV = \int_{\text{all space}} a_1[\psi_a(1) + \psi_b(1)]^* a_1[\psi_a(1) + \psi_b(1)]\, dV$$

$$= a_1^2\left(\int_{\text{all space}} \psi_a^*(1)\psi_a(1)\, dV + \int_{\text{all space}} \psi_a^*(1)\psi_b(1)\, dV + \int_{\text{all space}} \psi_b^*(1)\psi_a(1)\, dV + \int_{\text{all space}} \psi_b^*(1)\psi_b(1)\, dV\right)$$

$$= a_1^2(S_{aa} + S_{ab} + S_{ba} + S_{bb}) = a_1^2(1 + S + S + 1) = a_1^2(2 + 2S)$$

$$a_1 = (2 + 2S)^{-1/2}$$

where $S_{aa} = S_{bb} = 1$ and $S_{ab} = S_{ba} = S$. The wave function corresponding to E_1 is

$$\phi_1 = (2 + 2S)^{-1/2}[\psi_a(1) + \psi_b(1)] \qquad (16.12a)$$

Likewise for E_2, $a_1 = (2 - 2S)^{-1/2}$, giving

$$\phi_2 = (2 - 2S)^{-1/2}[\psi_a(1) - \psi_b(1)] \qquad (16.12b)$$

16.15 Prepare a plot of $\phi^*\phi$ for *(16.12)* along the internuclear axis using $r_{ab} = 0.1057$ nm.

▌ From Tables 15-1 and 15-2,

$$\psi_{1s} = \left(\frac{Z}{a_0}\right)^{3/2} \frac{1}{\pi^{1/2}} e^{-Zr/a_0} = \frac{e^{-r/a_0}}{a_0^{3/2}\pi^{1/2}}$$

At $r_{ab} = 0.1057$ nm, $S = 0.5873$ (see Problem 16.10). Substituting *(16.12a)* gives

$$\phi_1 = \frac{e^{-r_a/a_0} + e^{-r_b/a_0}}{[2 + 2(0.5873)]^{1/2} a_0^{3/2}\pi^{1/2}} = \frac{e^{-r_a/a_0} + e^{-r_b/a_0}}{1.216 \times 10^{-15}\ \text{m}^{3/2}} \qquad (16.13a)$$

Likewise,

$$\phi_2 = \frac{e^{-r_a/a_0} - e^{-r_b/a_0}}{[2 - 2(0.5873)]^{1/2} a_0^{3/2}\pi^{1/2}} = \frac{e^{-r_a/a_0} - e^{-r_b/a_0}}{6.199 \times 10^{-16}\ \text{m}^{3/2}} \qquad (16.13b)$$

Because these wave functions consist of only real terms, $\phi^*\phi = \phi_i^2$. As a sample calculation, consider a point along the bonding axis that is 0.0500 nm from nucleus a. If nucleus a is placed at $r_{ab} = 0$ and nucleus b is placed at $r_{ab} = 0.1057$ nm, the values of r_a and r_b used in *(16.13)* will be

$$r_a = |r| = 0.0500\text{ nm}$$

$$r_b = |r - 0.1057\text{ nm}| = |0.0500\text{ nm} - 0.1057\text{ nm}| = 0.0557\text{ nm}$$

giving

$$\phi_1^*\phi_1 = \left[\frac{e^{-(0.0500\text{ nm})/(0.0529\text{ nm})} + e^{-(0.0557\text{ nm})/(0.0529\text{ nm})}}{1.216 \times 10^{-15}\text{ m}^{3/2}}\right]^2 = 3.68 \times 10^{29}\text{ m}^{-3}$$

$$\phi_2^*\phi_2 = \left[\frac{e^{-(0.0500)/(0.0529)} - e^{-(0.0557)/(0.0529)}}{6.199 \times 10^{-16}}\right]^2 = 4.24 \times 10^{27}\text{ m}^{-3}$$

This plot is shown in Fig. 16-2.

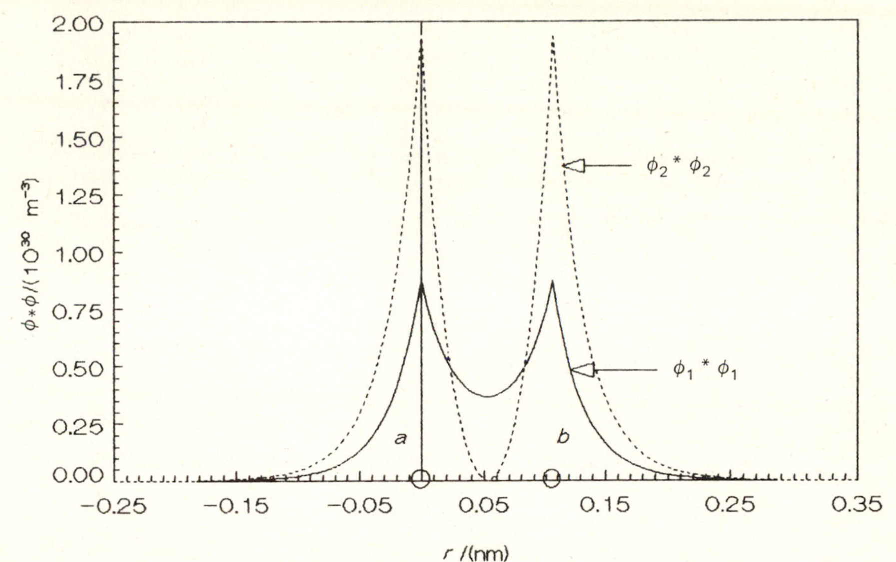

Fig. 16-2

16.16 Calculate the probability of finding the electron in a small volume element of 1 pm^3 located along the bonding axis in H_2^+ at $r = 0.0500$ nm. Assume that the nuclei are separated at a distance of 0.1057 nm.

▌ The probability will be given by $P = \phi^*\phi\, dV$, where the values of $\phi^*\phi$ were determined in Problem 16.15 for this system. The respective probabilities are

$$P_1 = (3.68 \times 10^{29}\text{ m}^{-3})(1 \times 10^{-12}\text{ m})^3 = 3.68 \times 10^{-7}$$

$$P_2 = (4.24 \times 10^{27})(1 \times 10^{-12})^3 = 4.24 \times 10^{-9}$$

16.17 Repeat the calculations of Problem 16.16 for the volume element located at $r = 0.0500$ nm from nucleus a and at a distance of 0.0200 nm off the bonding axis.

▌ The values of r_a and r_b to be used in *(16.13)* can be determined using the

$$r_a = [(0.0500\text{ nm})^2 + (0.0200\text{ nm})^2]^{1/2} = 0.0539\text{ nm}$$

$$r_b = [(0.1057\text{ nm} - 0.0500\text{ nm})^2 + (0.0200\text{ nm})^2]^{1/2} = 0.0592\text{ nm}$$

Substituting into *(16.13a)* gives

$$\phi_1 = \frac{e^{-(0.0539\text{ nm})/(0.0529\text{ nm})} + e^{-(0.0592\text{ nm})/(0.0529\text{ nm})}}{1.216 \times 10^{-15}\text{ m}^{3/2}} = 5.65 \times 10^{14}\text{ m}^{-3/2}$$

Likewise, $\phi_2 = 5.55 \times 10^{13}\text{ m}^{-3/2}$. The respective probabilities are

$$P_1 = (5.65 \times 10^{-14}\text{ m}^{-3/2})^2(1 \times 10^{-12}\text{ m})^3 = 3.19 \times 10^{-7}$$

$$P_2 = (5.55 \times 10^{-13})^2(1 \times 10^{-12})^3 = 3.08 \times 10^{-9}$$

16.18 Determine the energy of the H_2 molecule at infinite internuclear separation using the Heitler–London (valence bond) wave functions

$$\phi = \frac{1}{2^{1/2}}[\psi_a(1)\psi_b(2) \pm \psi_a(2)\psi_b(1)] \tag{16.14}$$

where ψ_α is the atomic 1s wave function.

▮ The Hamiltonian operator for this system is given by *(16.1)* as

$$\mathcal{H} = \frac{-\hbar^2}{2m_e}(\nabla_1^2 + \nabla_2^2) + \frac{e^2}{4\pi\varepsilon_0}\left(\frac{1}{r_{ab}} - \frac{1}{r_{1a}} - \frac{1}{r_{2a}} - \frac{1}{r_{1b}} - \frac{1}{r_{2b}} + \frac{1}{r_{12}}\right)$$

At infinite separation of the nuclei, the Hamiltonian reduces to

$$\mathcal{H} = \frac{-\hbar^2}{2m_e}(\nabla_1^2 + \nabla_2^2) + \frac{e^2}{4\pi\varepsilon_0}\left(-\frac{1}{r_{1a}} - \frac{1}{r_{2b}}\right)$$

Substituting into *(14.34)* with the denominator set equal to unity gives

$$\langle E\rangle = \int_{\text{all space}} \phi^*\left[\frac{-\hbar^2}{2m_e}(\nabla_1^2 + \nabla_2^2) - \frac{e^2}{4\pi\varepsilon_0}\left(\frac{1}{r_{1a}} + \frac{1}{r_{2b}}\right)\right]\phi\, dV = \int_{\text{all space}} \phi^*\left[\frac{-\hbar^2}{2m_e}\nabla_1^2 - \frac{e^2}{4\pi\varepsilon_0}\left(\frac{1}{r_{1a}}\right)\right]\phi\, dV$$

$$+ \int_{\text{all space}} \phi^*\left[\frac{-\hbar^2}{2m_e}\nabla_2^2 - \frac{e^2}{4\pi\varepsilon_0}\left(\frac{1}{r_{2b}}\right)\right]\phi\, dV = 2\int_{\text{all space}} \phi^*\mathcal{H}(\text{H})\phi\, dV$$

where $\mathcal{H}(\text{H})$ is the Hamiltonian operator for a hydrogen atom. Substituting *(16.14)* gives

$$\langle E\rangle = 2\int_{\text{all space}} \frac{1}{2^{1/2}}[\psi_a(1)\psi_b(2) \pm \psi_a(2)\psi_b(1)]^*\mathcal{H}(\text{H})\frac{1}{2^{1/2}}[\psi_a(1)\psi_b(2) \pm \psi_a(2)\psi_b(1)]\, dV$$

$$= \int_{\text{all space}} \psi_a^*(1)\psi_b^*(2)\mathcal{H}(\text{H})\psi_a(1)\psi_b(2)\, dV \pm \int_{\text{all space}} \psi_a^*(2)\psi_b^*(1)\mathcal{H}(\text{H})\psi_a(1)\psi_b(2)\, dV$$

$$\pm \int_{\text{all space}} \psi_a^*(1)\psi_b^*(2)\mathcal{H}(\text{H})\psi_a(2)\psi_b(1)\, dV + \int_{\text{all space}} \psi_a^*(2)\psi_b^*(1)\mathcal{H}(\text{H})\psi_a(2)\psi_b(1)\, dV$$

At infinite separation of the nuclei, the second and third integrals are equal to zero, giving

$$\langle E\rangle = \int_{\text{all space}} \psi_a^*(1)\mathcal{H}(\text{H})\psi_a(1)\, dV \int_{\text{all space}} \psi_b^*(2)\psi_b(2)\, dV$$

$$\pm 0 \pm 0 + \int_{\text{all space}} \psi_b^*(1)\mathcal{H}(\text{H})\psi_b(1)\, dV \int \psi_a^*(2)\psi_a(2)\, dV$$

$$= E(\text{H})S_{bb} + E(\text{H})S_{aa} = 2E(\text{H})$$

where $S_{aa} = S_{bb} = 1$ and $E(\text{H})$ is the energy of the hydrogen atom.

16.19 Write the complete wave functions for *(16.14)*, including the spin wave functions.

▮ The wave function with the positive sign is symmetric with respect to the exchange of electrons:

$$\hat{P}(1,2)\phi_+ = \hat{P}(1,2)\frac{1}{2^{1/2}}[\psi_a(1)\psi_b(2) + \psi_a(2)\psi_b(1)]$$

$$= \frac{1}{2^{1/2}}[\psi_a(2)\psi_b(1) + \psi_a(1)\psi_b(2)] = \phi_+$$

According to the Pauli exclusion principle (see Problem 15.28), an antisymmetric spin wave function must be used. Multiplying ϕ_+ by the antisymmetric spin wave function given by *(15.13d)* (see Problem 15.24) gives

$$\phi_1 = \frac{1}{2^{1/2}}[\psi_a(1)\psi_b(2) + \psi_a(2)\psi_b(1)]\frac{2^{1/2}}{2}[\alpha(1)\beta(2) - \beta(1)\alpha(2)]$$

$$= \tfrac{1}{2}[\psi_a(1)\psi_b(2) + \psi_a(2)\psi_b(1)][\alpha(1)\beta(2) - \beta(1)\alpha(2)]$$

Likewise, the wave function with the negative sign can be shown to be antisymmetric, and the symmetric spin wave functions given by *(15.13a)*, *(15.13b)*, and *(15.13c)* are used, giving

$$\phi_2 = \frac{1}{2^{1/2}}[\psi_a(1)\psi_b(2) - \psi_a(2)\psi_b(1)]\alpha(1)\beta(2)$$

$$\phi_3 = \frac{1}{2^{1/2}}[\psi_a(1)\psi_b(2) - \psi_a(2)\psi_b(1)]\beta(1)\beta(2)$$

$$\phi_4 = \tfrac{1}{2}[\psi_a(1)\psi_b(2) - \psi_a(2)\psi_b(1)][\alpha(1)\beta(2) + \beta(1)\alpha(2)]$$

16.20 What are the degeneracies of the levels described by the wave function given by *(16.14)*?

▌ There is only one spin state to the level with the wave function with the positive sign (see Problem 16.19), giving $g = 1$. For the wave function with the negative sign, there are three spin states, giving $g = 3$.

16.21 The bond length in the KF molecule is 0.217 155 nm. Determine the energy of the KF molecule with respect to the separated atom configuration. The ionization energy of K is 419 kJ · mol^{-1}, and the electron affinity of F is −322 kJ · mol^{-1}.

▌ The energy can be determined using the following steps in a Hess-law calculation,

$$\begin{array}{rl} \mathrm{KF} \rightarrow \mathrm{K^+} + \mathrm{F^-} & e^2/4\pi\varepsilon_0 r \\ \mathrm{K^+} + \mathrm{e^-} \rightarrow \mathrm{K} & -I(\mathrm{K}) \\ \mathrm{F^-} \rightarrow \mathrm{F} + \mathrm{e^-} & -E_a(\mathrm{F}) \\ \hline \mathrm{KF} \rightarrow \mathrm{K} + \mathrm{F} & E \end{array}$$

giving

$$E = \frac{(1.602 \times 10^{-19}\ \mathrm{C})^2 \dfrac{1\ \mathrm{kg \cdot m \cdot s^{-2}}}{1\ \mathrm{N}} \left(\dfrac{1\ \mathrm{J}}{1\ \mathrm{kg \cdot m^2 \cdot s^{-2}}}\right)(6.022 \times 10^{23}\ \mathrm{mol^{-1}})\left(\dfrac{10^{-3}\ \mathrm{kJ}}{1\ \mathrm{J}}\right)}{(1.113 \times 10^{-10}\ \mathrm{C^2 \cdot N^{-1} \cdot m^{-2}})(0.217\ 155 \times 10^{-9}\ \mathrm{m})}$$
$$- 419\ \mathrm{kJ \cdot mol^{-1}} - (-322\ \mathrm{kJ \cdot mol^{-1}})$$
$$= 543\ \mathrm{kJ \cdot mol^{-1}}$$

which agrees well with the observed value of 510 kJ · mol^{-1}.

16.22 Summarize the results of Problems 16.10 and 16.15.

▌ Figures 16-1 and 16-2 indicate that ϕ_1 corresponds to a bonding situation in that a potential energy well exists and there is a relatively high probability of finding the electron between the nuclei. The figures also indicate that ϕ_2 corresponds to an antibonding situation in that no potential energy well exists and there is a decreased probability of finding the electron between the nuclei ($P = 0$ at $r_{ab}/2$).

16.23 Using Fig. 16-3, write the molecular orbital designations for He_2 and He_2^+. Describe the bonding (if any) in each species.

▌ The configuration shown in Fig. 16-3a should be used if the number of electrons in the molecule is four or less, and the configuration shown in Fig. 16-3b should be used if the number of electrons is greater than four. Each of the molecular orbitals can hold a maximum of two electrons. For He_2, the four electrons would be assigned as

$$\sigma(1s)^2\sigma^*(1s)^2$$

and for He_2^+, the assignment for the three electrons is

$$\sigma(1s)^2\sigma^*(1s)$$

The net bond order in He_2 is 0 because there are exactly the same number of bonding and antibonding electrons. The net bond order in He_2^+ is $\frac{1}{2}$ because there is one more bonding electron than antibonding electrons present in the configuration. (Note: Traditionally a chemical bond is considered to consist of two electrons.) According to the molecular orbital theory, He_2 should not exist and He_2^+ should exist.

16.24 Using both configurations shown in Fig. 16-3, write the molecular orbital designation for B_2. Which configuration correctly predicts that B_2 should exist? Which configuration correctly predicts that B_2 is paramagnetic?

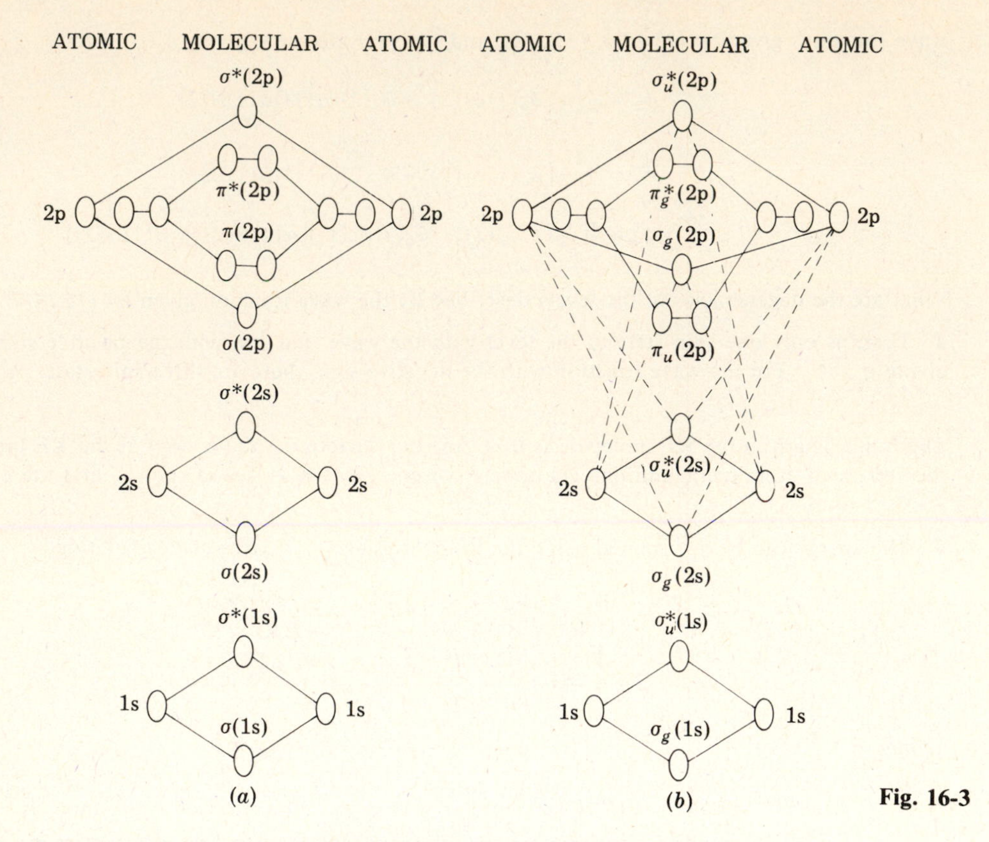

Fig. 16-3

▮ For the configuration given in Fig. 16-3a, the 10 electrons would be assigned as follows:

$$\sigma(1s)^2\sigma^*(1s)^2\sigma(2s)^2\sigma^*(2s)^2\sigma(2p)^2$$

Likewise, for the configuration given in Fig. 16-3b, the assignment is

$$\sigma_g(1s)^2\sigma_u^*(1s)^2\sigma_g(2s)^2\sigma_u^*(2s)^2\pi_u(2p)^2$$

Both configurations predict that B_2 should form a stable molecule. The first configuration predicts the bonding to result from two electrons occupying the $\sigma(2p)$ bonding orbital (a net bond order of one σ bond). The second configuration predicts the bonding to result from two electrons occupying the $\pi_u(2p)$ bonding orbitals with one electron in each (a net bond order of one π bond). The first configuration incorrectly predicts B_2 to be diamagnetic because all electrons have paired spin in the configuration. The second configuration correctly predicts the paramagnetic behavior of B_2 because the electrons in the equivalent $\pi_u(2p)$ orbitals would have parallel spins according to Hund's rule.

16.25 Write the molecular orbital designation for Li_2. Describe the bonding. Assuming that spin angular momentum is conserved, describe the electronic states of the Li atoms as Li_2 undergoes dissociation.

▮ The configuration for the six electrons is $\sigma_g(1s)^2\sigma_u^*(1s)^2\sigma_g(2s)^2$, resulting in the presence of a σ bond. As dissociation occurs, one Li atom will have a spin corresponding to $s = +\frac{1}{2}$ and the other atom will have a spin corresponding to $s = -\frac{1}{2}$.

16.26 Write the molecular orbital designations for N_2, N_2^+, and N_2^-. Which of these species is predicted to have the smallest bond length?

▮ The configurations are

$$N_2 \quad \sigma_g(1s)^2\sigma_u^*(1s)^2\sigma_g(2s)^2\sigma_u^*(2s)^2\pi_u(2p)^4\sigma_g(2p)^2$$
$$N_2^+ \quad \sigma_g(1s)^2\sigma_u^*(1s)^2\sigma_g(2s)^2\sigma_u^*(2s)^2\pi_u(2p)^4\sigma_g(2p)^1$$
$$N_2^- \quad \sigma_g(1s)^2\sigma_u^*(1s)^2\sigma_g(2s)^2\sigma_u^*(2s)^2\pi_u(2p)^4\sigma_g(2p)^2\pi_g^*(2p)^1$$

Bond length is inversely related to the net bond order. The net bond orders predicted by the configurations are one σ and two π bonds for N_2, $\frac{1}{2}$ σ and two π bonds for N_2^+, and one σ and $\frac{3}{2}$ π bonds for N_2^-. The bond length in N_2 should be the smallest. This prediction agrees with the observed values of 0.109 76 nm for N_2, 0.111 638 4 nm for N_2^+, and 0.1193 nm for N_2^-.

16.27 Assuming the configuration given by Fig. 16-3b to be valid, write the molecular orbital designation for NO and NO^+. Which is predicted to have the greater bond energy?

▌ The configurations are

$$\text{NO} \quad \sigma_g(1s)^2\sigma_u^*(1s)^2\sigma_g(2s)^2\sigma_u^*(2s)^2\pi_u(2p)^4\sigma_g(2p)^2\pi_g^*(2p)^1$$

$$\text{NO}^+ \quad \sigma_g(1s)^2\sigma_u^*(1s)^2\sigma_g(2s)^2\sigma_u^*(2s)^2\pi_u(2p)^4\sigma_g(2p)^2$$

Bond energy is directly related to the net bond order. The net bond order predicted by the configurations are one-half σ and two π bonds for NO and one σ and two π bonds for NO^+. The bond energy in NO^+ should be the greater.

16.28 Assuming that the 1s orbital for H is similar in energy with respect to the 2p orbitals for O, prepare a molecular orbital energy diagram to represent the bonding in OH.

▌ Assuming that the overlap of the hydrogen 1s orbital and one of the oxygen 2p orbitals forms a σ(sp) orbital and a σ^*(sp) orbital, the molecular orbital diagram shown in Fig. 16-4 is produced. The 1s and 2s orbitals of the oxygen are not involved in the bonding, and two of the 2p oxygen orbitals are nonbonding orbitals. The configuration of the nine electrons in OH is

$$(1s)^2(1s)^2\sigma(sp)^2(2p)^3$$

predicting that the atoms in the molecule are held together by one σ bond.

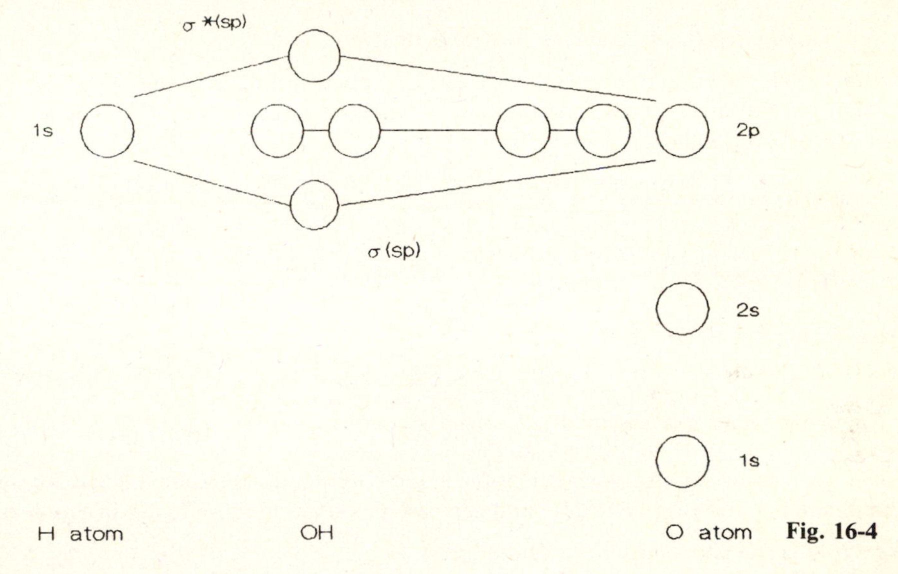

Fig. 16-4

16.29 Using Fig. 16-4, write the molecular orbital configurations for OH^+ and OH^-. Compare the relative bond lengths in OH^+, OH^-, and OH.

▌ The configurations are

$$\text{OH}^+ \quad (1s)^2(2s)^2\sigma(sp)^2(2p)^2$$

$$\text{OH}^- \quad (1s)^2(2s)^2\sigma(sp)^2(2p)^4$$

As with OH (see Problem 16.28), these species would be predicted to exist and each would contain one σ bond. Thus, there should not be much difference in the bond lengths of the three species. All four of the 2p electrons in OH^- are paired, OH contains one unpaired 2p electron, and OH^+ contains two unpaired 2p electrons. The presence of unpaired electrons causes a slight increase in the bond length, giving the prediction that OH^- should have the smallest bond length and OH^+ should have the greatest bond length. This prediction agrees with the observed values of 0.096 28 nm for OH^-, 0.097 06 nm for OH, and 0.102 89 nm for OH^+.

16.3 BOND DESCRIPTION

16.30 The ionization of Cl is 1251.1 kJ · mol^{-1}, and its electron affinity is −348.7 kJ · mol^{-1}. Using the Mulliken definition, calculate the electronegativity of Cl. Repeat the calculations for F given I = 1681.0 kJ · mol^{-1} and E_A = −322.2 kJ mol^{-1}.

▌ The *Mulliken electronegativity* of an atom is given by

$$EN_i = \tfrac{1}{2}(I_i + E_{A,i}) \qquad (16.15)$$

Substituting the data into *(16.15)* gives

$$EN(Cl) = (0.5)[1251.1\ kJ\cdot mol^{-1} + (-348.7\ kJ\cdot mol^{-1})] = 451.2\ kJ\cdot mol^{-1}$$

$$EN(F) = (0.5)[1681.0 + (-322.2)] = 679.4\ kJ\cdot mol^{-1}$$

16.31 The Cl—F, Cl—Cl, and F—F bond energies are 255 kJ · mol^{-1}, 243 kJ · mol^{-1}, and 159 kJ · mol^{-1}, respectively. Using the Pauling definition, calculate the difference in the electronegativities of Cl and F.

▌ The *Pauling electronegativity scale* assigns $EN(F) \equiv 4.0$ and defines differences in electronegativities between atoms by

$$EN_X - EN_Y = 0.102\Delta^{1/2} \qquad (16.16)$$

where the factor 0.102 is used to convert $(kJ\cdot mol^{-1})^{1/2}$ to $eV^{1/2}$, the usual unit for electronegativity, and

$$\Delta = BE(XY) - [BE(X_2)BE(Y_2)]^{1/2} \qquad (16.17)$$

Substituting the data into *(16.17)* gives

$$\Delta = 255\ kJ\cdot mol^{-1} - [(243\ kJ\cdot mol^{-1})(159\ kJ\cdot mol^{-1})]^{1/2} = 58\ kJ\cdot mol^{-1}$$

which gives

$$EN(F) - EN(Cl) = (0.102)(58)^{1/2} = 0.78\ eV^{1/2}$$

16.32 Compare the results of Problems 16.30 and 16.31.

▌ If EN(H) is assigned the value of 2.2 eV$^{1/2}$ on both scales, the Mulliken scale values can be converted to the Pauling scale values by dividing the former expressed in units of eV by 3.15 eV$^{1/2}$. Converting the values found in Problem 16.30 gives

$$EN(F) = \frac{(679.4\ kJ\cdot mol^{-1})[(1\ mol)/(6.022\times 10^{23})][(10^3\ J)/(1\ kJ)][(1\ eV)/(1.602\times 10^{-19}\ J)]}{3.15\ eV^{1/2}} = 2.24\ eV^{1/2}$$

$$EN(Cl) = \frac{(451.2)[1/(6.022\times 10^{23})](10^3)[1/(1.602\times 10^{-19})]}{3.15} = 1.48\ eV^{1/2}$$

and taking the difference gives

$$EN(F) - EN(Cl) = 2.24\ eV^{1/2} - 1.48\ eV^{1/2} = 0.76\ eV^{1/2}$$

The two methods give essentially the same results.

16.33 The usual values of electronegativity for F and Cl given in tables are 4.0 and 3.0, respectively. Use these values to determine $EN(F) - EN(Cl)$ and compare this result to those found in Problem 16.32.

▌ Taking the difference of the values gives

$$EN(F) - EN(Cl) = 4.0 - 3.0 = 1.0$$

This result is approximately 20% higher than the values determined in Problem 16.32.

16.34 Consider the ClF molecule to consist of two ions of opposite unit charge separated by the bond length of 0.162 81 nm. Calculate the dipole moment for this model.

▌ The *dipole moment* (μ) between charged particles separated by a distance r is given by

$$\mu = qr \qquad (16.18)$$

Substituting $q = 1.6022\times 10^{-19}\ C$ and $r = 1.6281\times 10^{-10}\ m$ into *(16.18)* gives

$$\mu = (1.6022\times 10^{-19}\ C)(1.6281\times 10^{-10}\ m) = 2.6085\times 10^{-29}\ C\cdot m$$

Usually the units of debyes are used to express dipole moments where $1\ D = 3.335\,641\times 10^{-30}\ C\cdot m$. Converting the above value gives

$$\mu = (2.6085\times 10^{-29}\ C\cdot m)\frac{1\ D}{3.335\,641\times 10^{-30}\ C\cdot m} = 7.8202$$

16.35 The permanent dipole moment of a H_2 molecule is zero. However, at a given instant the molecule can be polar. Calculate the instantaneous dipole moment for the system shown in Fig. 16-5, where $r_{ab} = 0.074\ 14$ nm.

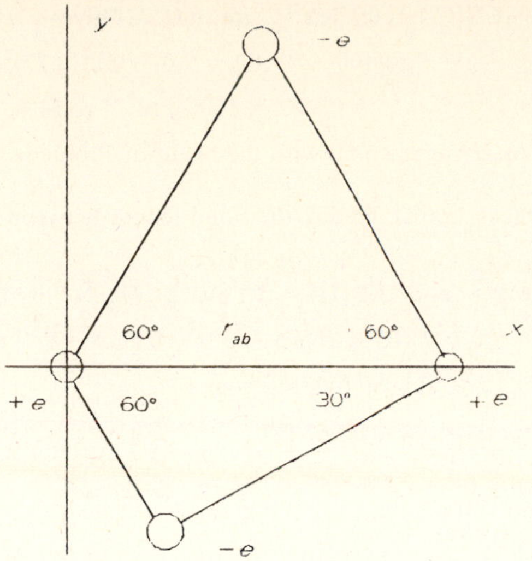

Fig. 16-5

I Dipole moments can be treated as vectors. Defining the component in one direction as

$$\mu_i = -\sum qr \tag{16.19}$$

the x and y components for this system are

$$\mu_x = -[(+e)(0) + (+e)(r_{ab}) + (-e)(r_{ab}/2) + (-e)(r_{ab}/4)] = -er_{ab}/4$$

$$\mu_y = -\left\{(+e)(0) + (+e)(0) + (-e)\left(\frac{r_{ab}}{2}\tan 60°\right) + (-e)\left[\frac{r_{ab}}{4}\tan(-60°)\right]\right\} = (-er_{ab}\tan 60°)/4$$

The net dipole moment is

$$\mu = (\mu_x^2 + \mu_y^2)^{1/2} = \left[\left(\frac{-er_{ab}}{4}\right)^2 + \left(\frac{er_{ab}\tan 60°}{4}\right)^2\right]^{1/2} = \frac{er_{ab}}{4}(1 + \tan^2 60°)^{1/2}$$

$$= \{[(1.602 \times 10^{-19}\ \text{C})(7.414 \times 10^{-11}\ \text{m})]/4\}(1 + \tan^2 60°)^{1/2} = 5.939 \times 10^{-30}\ \text{C}\cdot\text{m} = 1.781\ \text{D}$$

16.36 The observed dipole moment of ClF is 0.88 D. Calculate the percent ionic character of the bond.

I The percent ionic character is related to the actual dipole moment (μ_{actual}) by

$$\%\text{IC} = \frac{\mu_{\text{actual}}}{\mu_{\text{predicted}}} \times 100 \tag{16.20}$$

where $\mu_{\text{predicted}}$ is determined using *(16.18)* with $q = e$ and $r = r_{ab}$. Substituting the data and the results of Problem 16.34 into *(16.20)* gives

$$\%\text{IC} = \frac{0.88\ \text{D}}{7.8202\ \text{D}} \times 100 = 11\%$$

16.37 Using $\text{EN(F)} - \text{EN(Cl)} = 0.77\ \text{eV}^{1/2}$, predict the percent ionic character in the bond. Compare the prediction to the result of Problem 16.36.

I The relation between the electronegativity difference and the percent ionic character is

$$\%\text{IC} = 16|\text{EN}_i - \text{EN}_j| + 3.5|\text{EN}_i - \text{EN}_j|^2 \tag{16.21}$$

Substituting into *(16.21)* gives

$$\%\text{IC} = 16(0.77) + 3.5(0.77)^2 = 14\%$$

Although the values differ by nearly 30%, both methods predict that the Cl—F bonding will be mostly covalent (a polar covalent bond).

16.38 The following equation has been proposed to relate percent ionic character to electronegativity difference:

$$\log(100 - \%\text{IC}) = 2.0 - 0.11|\text{EN}_i - \text{EN}_j|^2 \tag{16.22}$$

Use this equation to predict the percent ionic character of the Cl—F bond, and compare the result to that of Problem 16.37.

▮ Substituting $EN(F) - EN(Cl) = 0.77\ eV^{1/2}$ into *(16.22)* gives

$$\log(100 - \%IC) = 2.0 - 0.11(0.77)^2 = 1.93$$

$$\%IC = 100 - 10^{1.93} = 14\%$$

The value predicted by *(16.22)* agrees well with the result of Problem 16.37.

16.39 The dipole moment of OH is 1.66 D. Predict the bond length between the hydrogen atom and the oxygen atom in this molecule.

▮ Using $EN(O) = 3.5\ eV^{1/2}$ and $EN(H) = 2.1\ eV^{1/2}$, *(16.21)* predicts that the percent ionic character is

$$\%IC = 16|3.5 - 2.1| + 3.5|3.5 - 2.1|^2 = 29\%$$

Solving *(16.20)* for $\mu_{predicted}$ and substituting the data gives

$$\mu_{predicted} = \frac{1.66\ D}{29} \times 100 = 5.7\ D = 1.9 \times 10^{-29}\ C \cdot m$$

Solving *(16.18)* for r_{ab} and substituting $q = e$ gives

$$r_{ab} = \frac{1.9 \times 10^{-29}\ C \cdot m}{1.602 \times 10^{-19}\ C} = 1.2 \times 10^{-10}\ m = 0.12\ nm$$

The predicted value of r_{ab} is approximately 25% greater than the observed value of 0.097 06 nm.

16.40 If an ionic bond is defined as one having at least 50% ionic character, what minimum difference in electronegativities will be defined by *(16.22)*?

▮ Solving *(16.22)* for $|EN_i - EN_j|$ and substituting $\%IC = 50\%$ gives

$$|EN_i - EN_j| = \left(\frac{2.0 - \log(100 - 50)}{0.11}\right)^{1/2} = 1.7$$

16.4 MOLECULAR TERM SYMBOLS

16.41 Identify the term symbols that correspond to $\Lambda = 1$ and $S = 1$.

▮ The molecular quantum number Λ is equal to $|M_L|$, and the designations Σ for $\Lambda = 0$, Π for $\Lambda = 1$, etc. are used. In this case, the symbol Π is chosen because $\Lambda = 1$. The multiplicity is $2S + 1 = 3$, giving $^3\Pi$. The permitted values of the $\mathcal{S}$ quantum number are

$$\mathcal{S} = S, S - 1, \ldots, -S \tag{16.23}$$

which in this case are

$$\mathcal{S} = \begin{cases} 1 \\ 1 - 1 = 0 \\ 1 - 2 = -S = -1 \end{cases}$$

A following subscript (Ω) on the molecular term symbol is determined by

$$\Omega = |\Lambda + \mathcal{S}| \tag{16.24}$$

which in this case will be

$$\Omega = \begin{cases} |1 + 1| = 2 & \text{for } \mathcal{S} = 1 \\ |1 + 0| = 1 & \text{for } \mathcal{S} = 0 \\ |1 - 1| = 0 & \text{for } \mathcal{S} = -1 \end{cases}$$

The complete term symbols are $^3\Pi_2$, $^3\Pi_1$, and $^3\Pi_0$. Note that the use of Ω is optional.

16.42 Which of the following electronic transitions are permitted?

$$(a)\ ^1\Sigma_u^+ \leftrightarrow {}^1\Sigma_g^+, \quad (b)\ ^1\Sigma_u^- \leftrightarrow {}^3\Sigma_g^-, \quad (c)\ ^1\Sigma_g^+ \leftrightarrow {}^1\Pi_u$$

▮ The selection rules for electronic transitions are

$$\Delta\Lambda = 0, \pm 1 \tag{16.25a}$$

$$\Delta S = 0 \tag{16.25b}$$

$$g \leftrightarrow u \tag{16.25c}$$

$$- \not\leftrightarrow + \tag{16.25d}$$

(a) For the $^1\Sigma_u^+ \leftrightarrow {}^1\Sigma_g^+$ transition, $\Lambda = 0$ for both levels and *(16.25a)* is valid; $S = 0$ for both levels and *(16.25b)* is valid; one level is g and the other is u and *(16.25c)* is valid; and both levels are + and *(16.25d)* is valid. This transition is permitted.

(b) For the $^1\Sigma_u^- \leftrightarrow {}^3\Sigma_g^-$ transition, $\Delta S = 1$ and the conditions expressed by *(16.25b)* are not met. This transition is not permitted by the usual selection rules.

(c) All conditions of *(16.25)* are met by the $^1\Sigma_g^+ \leftrightarrow {}^1\Pi_u$ transition, and it is permitted.

16.43 The ground-level term for a heteronuclear diatomic molecule is $^3\Sigma^-$. Write the term symbols of the electronic transitions permitted by *(16.25)* for this molecule.

▮ Only term symbols in which the following superscript is − are permitted by *(16.25d)*, and only those in which the leading superscript is 3 are permitted by *(16.25b)*. Transitions between levels with $\Lambda = 0$ and $\Lambda = 1$ are permitted by *(16.25a)*. The permitted transitions are $^3\Sigma^- \leftrightarrow {}^3\Sigma^-$ and $^3\Sigma^- \leftrightarrow {}^3\Pi$. Note that the notations g and u are not used for heteronuclear diatomic molecules.

16.44 Write the electronic configuration for Li_2, and predict the term symbol for the ground level.

▮ Using the configuration diagram shown in Fig. 16-3b, the molecular orbital designation is

$$\sigma_g(1s)^2\sigma_u^*(1s)^2\sigma_g(2s)^2$$

For this configuration, $M_L = 0$, giving $\Lambda = 0$ and $S = 0$. The basic term symbol is $^1\Sigma$. Because the wave function for a σ molecular orbital does not change sign upon reflection across an xz plane, the + following superscript is used. The parity can be found by multiplying the parities of the orbitals being used, according to the usual laws of odd (u for German *ungerade*, odd) and even (g for German *gerade*, even)

$$(g)^4(u)^2 = g$$

The complete term symbol is $^1\Sigma_g^+$.

16.45 Using Table 16-1, predict the term symbol for the ground level of N_2.

Table 16-1

configuration	term symbols
σ	$^2\Sigma^+$
σ^2	$^1\Sigma^+$
π	$^2\Pi$
π^2	$^1\Sigma^+, {}^1\Delta, {}^3\Sigma^-$
$\pi^2\sigma$	$^2\Sigma^+, {}^2\Sigma^-, {}^2\Delta, {}^4\Sigma$
$\pi^2\pi$	$^2\Pi(3), {}^2\Phi, {}^4\Pi$
$\pi^2\delta$	$^2\Sigma^+, {}^2\Sigma^-, {}^2\Delta(2), {}^2\Gamma, {}^4\Delta$
π^3	$^2\Pi$
$\pi^3\sigma$	$^1\Pi, {}^3\Pi$
$\pi^3\pi$	$^1\Sigma^+, {}^1\Sigma^-, {}^1\Delta, {}^3\Sigma^+, {}^3\Sigma^-, {}^3\Delta$
$\pi^3\delta$	$^1\Pi, {}^1\Phi, {}^3\Pi, {}^3\Phi$
π^4	$^1\Sigma^+$
δ	$^2\Delta$
δ^2	$^1\Sigma^+, {}^3\Sigma^-, {}^1\Gamma$
δ^3	$^2\Delta$
δ^4	$^1\Sigma^+$

▮ Using the configuration diagram shown in Fig. 16-3b, the molecular orbital designation is

$$\sigma_g(1s)^2\sigma_u^*(1s)^2\sigma_g(2s)^2\sigma_u^*(2s)^2\pi_u(2p)^4\sigma_g(2p)^2$$

The term symbol corresponding to σ^2 is $^1\Sigma^+$. The parity is $(g)^6(u)^8 = g$, giving the complete term symbol as $^1\Sigma_g^+$.

16.46 Write the molecular orbital designation for B_2 using both configuration patterns shown in Fig. 16-3. Which one correctly predicts the ground level to be $^3\Sigma_g^-$?

▮ Using the configuration given in Fig. 16-3a gives the molecular orbital designation as

$$\sigma(1s)^2\sigma^*(1s)^2\sigma(2s)^2\sigma^*(2s)^2\sigma(2p)^2$$

This configuration corresponds to $^1\Sigma^+$ (see Table 16-1), which is incorrect. Likewise, using the configuration given in Fig. 16-3b gives

$$\sigma_g(1s)^2\sigma_u^*(1s)^2\sigma_g(2s)^2\sigma_u^*(2s)^2\pi_u(2p)^2$$

which corresponds to the $^1\Sigma^+$, $^1\Delta$, and $^3\Sigma^-$ term symbols. Choosing the one with the greatest multiplicity and with the parity given by $(g)^4(u)^6 = g$ correctly gives the term symbol as $^3\Sigma_g^-$.

16.47 Using Table 16-1, deduce the ground-level electronic configurations for the molecules from the given term symbols:

$$(a)\ O_2,\ ^3\Sigma_g^-;\quad (b)\ O_2^+,\ ^2\Pi_g;\quad (c)\ O_2^-,\ ^2\Pi_g;\quad (d)\ O_2^{2-},\ ^1\Sigma_g^+.$$

▮ In each case, the table entry for

(*a*) $^3\Sigma_g^-$ corresponds to a π^2 system. This configuration can be attained by writing

$$\sigma_g(1s)^2\sigma_u^*(1s)^2\sigma_g(2s)^2\sigma_u^*(2s)^2\pi_u(2p)^4\sigma_g(2p)^2\pi_g(2p)^2$$

(*b*) $^2\Pi_g$ corresponds to a π^1 system, giving the configuration as

$$\cdots\ \pi_u(2p)^4\sigma_g(2p)^2\pi_g(2p)^3$$

(*c*) $^2\Pi_g$ corresponds to a π^3 system, giving the configuration as

$$\cdots\ \pi_u(2p)^4\sigma_g(2p)^2\pi_g(2p)^3$$

(*d*) $^1\Sigma_g^+$ corresponds to a π^4 system, giving the configuration as

$$\cdots\ \pi_u(2p)^4\sigma_g(2p)^2\pi_g(2p)^4$$

16.48 Write the molecular orbital configuration for the CN molecule, and predict the term symbol for the ground level.

▮ Using the configuration pattern given in Fig. 16-3b gives the molecular orbital designation as

$$\sigma(1s)^2\sigma^*(1s)^2\sigma(2s)^2\sigma^*(2s)^2\pi(2p)^4\sigma(2p)^1$$

For this configuration $M_L = 0$, giving $\Lambda = 0$, and $S = \frac{1}{2}$, giving $2S + 1 = 2$. Because the wave function for a σ orbital is symmetric with reflection across a plane, the + following superscript is used. The predicted term symbol is $^2\Sigma^+$.

16.49 The term for CN^- is $^1\Sigma^+$. Determine the ground level for this molecule.

▮ There are several entries in Table 16-1 for $^1\Sigma^+$, but the one of interest corresponds to a σ^2 system, giving the configuration as

$$\sigma(1s)^2\sigma^*(1s)^2\sigma(2s)^2\sigma^*(2s)^2\pi(2p)^4\sigma(2p)^2$$

16.50 Using the molecular orbital configuration written in Problem 16.28 for OH, predict the term symbol for the ground level.

▮ The $1s^2$ and $2s^2$ electrons are not important in determining the term symbol. The angular momentum and spin contributions from the $\sigma(sp)^2$ and $(2p)^3$ electrons are $M_L = 1$, giving $\Lambda = 1$, and $S = \frac{1}{2}$, giving $2S + 1 = 2$. The predicted term symbol is $^2\Pi$.

CHAPTER 17

Spectroscopy of Diatomic Molecules

17.1 ROTATIONAL AND VIBRATIONAL SPECTRA

17.1 Place the following rotational energy information in order of increasing energy:

$$B_e = 10\ \text{cm}^{-1}, \qquad B^* = 1 \times 10^{-23}\ \text{J}, \qquad \nu = 10\,000\ \text{MHz}, \qquad \lambda = 0.01\ \text{m}$$

▌ The comparison will be made by converting the given values of B_e, ν, and λ to corresponding values of B^*. The relation between B^* and B_e is

$$B_e/(\text{cm}^{-1}) = \frac{[B^*/(\text{J})](10^{-2})}{hc} \tag{17.1}$$

Solving *(17.1)* for B^* and substituting the data gives

$$B^* = (10\ \text{cm}^{-1})(6.626 \times 10^{-34}\ \text{J}\cdot\text{s})(2.9979 \times 10^8\ \text{m}\cdot\text{s}^{-1})\frac{1\ \text{cm}}{10^{-2}\ \text{m}} = 2 \times 10^{-22}\ \text{J}$$

Substituting into *(13.33)* with $E = B^*$ gives

$$B^* = (6.626 \times 10^{-34}\ \text{J}\cdot\text{s})(10\,000\ \text{MHz})\frac{10^6\ \text{Hz}}{1\ \text{MHz}}\left(\frac{1\ \text{s}^{-1}}{1\ \text{Hz}}\right) = 7 \times 10^{-24}\ \text{J}$$

$$B^* = \frac{(6.626 \times 10^{-34}\ \text{J}\cdot\text{s})(2.9979 \times 10^8\ \text{m}\cdot\text{s}^{-1})}{0.01\ \text{m}} = 2 \times 10^{-23}\ \text{J}$$

The order of increasing energy is $10\,000\ \text{MHz} < 1 \times 10^{-23}\ \text{J} < 0.01\ \text{m} < 10\ \text{cm}^{-1}$.

17.2 Identify which of the following diatomic molecules will have a pure rotational microwave spectrum: IF, O_2, KCl, Cl_2, K_2.

▌ A pure rotational spectrum will be observed only for those molecules that contain a permanent dipole moment. Because homonuclear diatomic molecules do not have permanent dipole moments and heteronuclear diatomic molecules do have permanent dipole moments, spectra will be observed only for IF and KCl.

17.3 The microwave $J = 1 \leftrightarrow 2$ transition is observed in the presence of an electric field. How many lines will appear in the spectrum?

▌ The selection rules describing rotational transitions in a $^1\Sigma$ electronic state are

$$\Delta J = \pm 1 \tag{17.2a}$$

$$\Delta M = 0, \pm 1 \tag{17.2b}$$

where there are $g_J = 2J + 1$ values of M ($M = 0, \pm 1, \ldots, \pm J$). The following transitions are permitted by *(17.2)*: $J = 1$ and $M = +1$ to $J = 2$ and $M = +2, +1$, and 0; $J = 1$ and $M = 0$ to $J = 2$ and $M = +1$, 0, and -1; and $J = 1$ and $M = -1$ to $J = 2$ and $M = 0, -1$, and -2. A total of nine lines should be observed.

17.4 Predict the location of the absorption peak(s) in the microwave rotational spectrum for $^{12}C^{16}O$ given $B_e = 1.9302\ \text{cm}^{-1}$.

▌ For a rigid rotator, *(14.56)* can be written as

$$E_J = B^*J(J + 1) \tag{17.3a}$$

where

$$B^* = \hbar^2/2I = \hbar^2/2\mu r_{ab}^2 \tag{17.4a}$$

or in terms of wave numbers as

$$F(J) = B_eJ(J + 1) \tag{17.3b}$$

where

$$B_e/(\text{cm}^{-1}) = \frac{\hbar^2 \times 10^{-2}}{4\pi c\mu r_{ab}^2} = \frac{2.7993 \times 10^{-46}}{[\mu/(\text{kg})][r_{ab}/(\text{m})]^2} \tag{17.4b}$$

Substituting the permitted values of ΔJ given by *(17.2a)*, the location of the various spectral peaks are

$$\bar{\nu} = F(J) - F(J-1) = B_e J(J+1) - B_e(J-1)(J-1+1) = 2B_e J \tag{17.5a}$$

Substituting the data into *(17.5a)* gives

$$\bar{\nu} = 2(1.9302\ \text{cm}^{-1})(1) = 3.8604\ \text{cm}^{-1}$$

$$\bar{\nu} = 2(1.9302\ \text{cm}^{-1})(2) = 7.7208\ \text{cm}^{-1}$$

and $\bar{\nu} = 11.5812\ \text{cm}^{-1}$, $15.4416\ \text{cm}^{-1}$, $19.3020\ \text{cm}^{-1}$, etc.

17.5 Assuming negligible centrifugal distortion, what should be the spacing between adjacent peaks in the microwave rotational spectrum for $^{12}C^{16}O$? See Problem 17.4 for additional data.

▮ Substituting the permitted values of ΔJ given by *(17.2a)* into *(17.5)*, the energy spacing will be

$$\Delta\bar{\nu} = 2B_e(J+1) - 2B_e J = 2B_e \tag{17.5b}$$

Substituting the data into *(17.5b)* gives

$$\Delta\bar{\nu} = 2(1.9302\ \text{cm}^{-1}) = 3.8604\ \text{cm}^{-1}$$

The peaks listed in Problem 17.4 are separated by $3.8604\ \text{cm}^{-1}$.

17.6 Calculate the bond length in $^{12}C^{16}O$ using $B_e = 1.9302\ \text{cm}^{-1}$. The molar masses are $15.994\,91\ \text{g}\cdot\text{mol}^{-1}$ for ^{16}O and $12.000\,00\ \text{g}\cdot\text{mol}^{-1}$ for ^{12}C.

▮ The reduced mass of the molecule is given by *(2.23)* as

$$\mu = \frac{(15.994\,91\ \text{g}\cdot\text{mol}^{-1})(12.000\,00\ \text{g}\cdot\text{mol}^{-1})[(10^{-3}\ \text{kg})/(1\ \text{g})]}{(15.994\,91\ \text{g}\cdot\text{mol}^{-1} + 12.000\,00\ \text{g}\cdot\text{mol}^{-1})(6.022\,045\times10^{23}\ \text{mol}^{-1})} = 1.138\,518\times10^{-26}\ \text{kg}$$

Solving *(17.4b)* for r_{ab} and substituting the data gives

$$r_{ab}/(\text{m}) = \left(\frac{2.7993\times10^{-46}}{(1.138\,518\times10^{-26})(1.9302)}\right)^{1/2} = 1.1286\times10^{-10}$$

The bond length is 0.112 86 nm.

17.7 Determine the center of mass of the $^{12}C^{16}O$ molecule. See Problem 17.6 for additional data.

▮ Because the O atom is heavier than the C atom, the O atom will be located slightly closer to the center of mass of the molecule than the C atom will. Letting r_O represent the distance between the center of mass and the O atom and r_C represent the distance between the center of mass and the C atom, the center of mass can be determined by solving the following simultaneous equations:

$$r_C + r_O = r_{ab} = 0.112\,86\ \text{nm}$$

$$(15.994\,91\ \text{g}\cdot\text{mol}^{-1})r_O = (12.000\,00\ \text{g}\cdot\text{mol}^{-1})r_C$$

giving $r_C = 0.064\,45\ \text{nm}$ and $r_O = 0.048\,36\ \text{nm}$.

17.8 What is the amount of time required for a $^{12}C^{16}O$ molecule to make one complete rotational cycle? Assume that the molecule can be treated classically and that $T = 298\ \text{K}$. See Problem 17.4 for additional information.

▮ Solving *(17.4b)* for I and substituting the data gives

$$I/(\text{kg}\cdot\text{m}^2) = \frac{2.7993\times10^{-46}}{1.9302} = 1.4503\times10^{-46}$$

The kinetic energy of a rigid rotator is given by

$$E = \tfrac{1}{2}I\omega^2 \tag{17.6}$$

Letting $E = kT/2$, solving *(17.6)* for ω, and substituting the data gives

$$\omega = \left[\frac{2(1.380\,66\times10^{-23}\ \text{J}\cdot\text{K}^{-1})(298.15\ \text{K})[(1\ \text{kg}\cdot\text{m}^2\cdot\text{s}^{-2})/(1\ \text{J})]}{2(1.4503\times10^{-46}\ \text{kg}\cdot\text{m}^2)}\right]^{1/2} = 5.3276\times10^{12}\ \text{s}^{-1}$$

The period of revolution is

$$\tau = \frac{2\pi}{5.3276\times10^{12}\ \text{s}^{-1}} = 1.1794\times10^{-12}\ \text{s}$$

17.9 What is the linear velocity of the O atom in a $^{12}C^{16}O$ molecule as a result of rotational motion with $J = 7$? See Problems 17.4, 17.7, and 17.8 for additional information.

▌ Solving *(17.1)* for B^* and substituting the data gives

$$B^* = (6.6262 \times 10^{-34}\ \text{J}\cdot\text{s})(2.9979 \times 10^{8}\ \text{m}\cdot\text{s}^{-1})(1.9302\ \text{cm}^{-1})[(10^{2}\ \text{cm})/(1\ \text{m})] = 3.8343 \times 10^{-23}\ \text{J}$$

Setting *(17.3a)* and *(17.6)* equal, solving for ω, and substituting the data gives

$$\omega = \left(\frac{2B^*J(J+1)}{I}\right)^{1/2} = \left(\frac{2(3.8343 \times 10^{-23}\ \text{J})(7)(7+1)[(1\ \text{kg}\cdot\text{m}^2\cdot\text{s}^{-2})/(1\ \text{J})]}{1.4503 \times 10^{-46}\ \text{kg}\cdot\text{m}^2}\right)^{1/2} = 5.4416 \times 10^{12}\ \text{s}^{-1}$$

The linear velocity is given by

$$v = r_0\omega = (0.048\,36\ \text{nm})[(10^{-9}\ \text{m})/(1\ \text{nm})](5.4416 \times 10^{12}\ \text{s}^{-1}) = 263.2\ \text{m}\cdot\text{s}^{-1}$$

17.10 The calibration of a spectrometer was such that the relative error in B^* was 1 ppt high. By what factor should the corresponding value of r_{ab} be changed?

▌ Solving *(17.4a)* for r_{ab} and taking the derivative with respect to B^* gives

$$\frac{dr_{ab}}{dB^*} = \frac{d}{dB^*}\left(\frac{\hbar^2}{2\mu B^*}\right)^{1/2} = \left(\frac{\hbar^2}{2\mu}\right)^{1/2}\frac{d(B^*)^{-1/2}}{dB^*} = -\frac{1}{2}\left(\frac{\hbar^2}{2\mu}\right)^{1/2}\left(\frac{1}{B^*}\right)^{3/2} = -\frac{1}{2}\left(\frac{\hbar^2}{2\mu B^*}\right)^{1/2}\left(\frac{1}{B^*}\right) = -\tfrac{1}{2}r_{ab}\frac{1}{B^*}$$

The relative error in r_{ab} is

$$\frac{dr_{ab}}{r_{ab}} = -\frac{dB^*}{2B^*} = -\frac{1\ \text{ppt}}{2} = -0.5\ \text{ppt}$$

The value of r_{ab} should be decreased by 0.5 ppt.

17.11 The 115 271.20-MHz absorption peak in the microwave spectrum af $^{12}C^{16}O$ corresponds to the $J = 0 \leftrightarrow 1$ transition. Determine r_{ab}. See Problem 17.6 for additional information.

▌ Solving *(13.32)* for $\bar{\nu}$ and substituting the data gives

$$\bar{\nu} = \frac{(115\,271.20\ \text{MHz})[(10^6\ \text{Hz})/(1\ \text{MHz})][(1\ \text{s}^{-1})/(1\ \text{Hz})][(10^{-2}\ \text{m})/(1\ \text{cm})]}{2.997\,925 \times 10^{8}\ \text{m}\cdot\text{s}^{-1}} = 3.845\,033\ \text{cm}^{-1}$$

Solving *(17.5a)* for B_e and substituting the data gives

$$B_e = \frac{3.845\,033\ \text{cm}^{-1}}{2(1)} = 1.922\,517\ \text{cm}^{-1}$$

Solving *(17.4b)* for r_{ab} and substituting $\mu = 1.138\,518 \times 10^{-26}\ \text{kg}$ (see Problem 17.6) gives

$$r_{ab} = \left(\frac{2.7993 \times 10^{-46}}{(1.138\,518 \times 10^{-26})(1.922\,517)}\right)^{1/2}\ \text{m} = 1.1309 \times 10^{-10}\ \text{m} = 0.113\,09\ \text{nm}$$

17.12 The 109 782.18-MHz absorption peak for $^{12}C^{18}O$ corresponds to the $J = 0 \leftrightarrow 1$ transition. Determine r_{ab}, and compare the value to the results of Problem 17.11. For ^{18}O, $M = 17.9916\ \text{g}\cdot\text{mol}^{-1}$.

▌ The reduced mass of the molecule is given by *(2.23)* as

$$\mu = \frac{(17.9916\ \text{g}\cdot\text{mol}^{-1})(12.000\,00\ \text{g}\cdot\text{mol}^{-1})[(10^{-3}\ \text{kg})/(1\ \text{g})]}{(17.9916\ \text{g}\cdot\text{mol}^{-1} + 12.000\,00\ \text{g}\cdot\text{mol}^{-1})(6.022\,045 \times 10^{23}\ \text{mol}^{-1})} = 1.195\,38 \times 10^{-26}\ \text{kg}$$

Following the procedure of Problem 17.11 gives

$$\bar{\nu} = \frac{(109\,782.18\ \text{MHz})[(10^6\ \text{Hz})/(1\ \text{MHz})][(1\ \text{s}^{-1})/(1\ \text{Hz})][(10^{-2}\ \text{m})/(1\ \text{cm})]}{2.997\,925 \times 10^{8}\ \text{m}\cdot\text{s}^{-1}} = 3.661\,939\ \text{cm}^{-1}$$

$$B_e = \frac{3.661\,939\ \text{cm}^{-1}}{2(1)} = 1.830\,970\ \text{cm}^{-1}$$

$$r_{ab} = \left(\frac{2.7993 \times 10^{-46}}{(1.195\,38 \times 10^{-26})(1.830\,970)}\right)^{1/2}\ \text{m} = 1.1309 \times 10^{-10}\ \text{m} = 0.113\,09\ \text{nm}$$

Isotopic substitution causes a negligible change in the equilibrium bond lengths.

17.13 A first absorption peak in the CO microwave spectrum was observed at $\Delta\nu = -5069.83$ MHz compared to the $J = 0 \leftrightarrow 1$ peak for $^{12}C^{16}O$. What is the source for this peak? See Problems 17.11 and 17.12 for additional information.

▮ The frequency of the observed peak is

$$\nu = 115\,271.20 \text{ MHz} - 5069.83 \text{ MHz} = 110\,201.37 \text{ MHz}$$

Solving *(13.32)* for $\bar{\nu}$ and substituting the data gives

$$\bar{\nu} = \frac{(110\,201.37 \text{ MHz})[(10^6 \text{ Hz})/(1 \text{ MHz})][(1 \text{ s}^{-1})/(1 \text{ Hz})][(10^{-2} \text{ m})/(1 \text{ cm})]}{2.997\,925 \times 10^8 \text{ m}\cdot\text{s}^{-1}} = 3.675\,922 \text{ cm}^{-1}$$

Solving *(17.5a)* for B_e and substituting the data gives

$$B_e = \frac{3.675\,922 \text{ cm}^{-1}}{2(1)} = 1.837\,961 \text{ cm}^{-1}$$

Solving *(17.4b)* for μ and substituting $r_{ab} = 1.1309 \times 10^{-10}$ (see Problem 17.12),

$$\mu = \frac{2.7993 \times 10^{-46}}{(1.837\,961)(1.1309 \times 10^{-10})^2} \text{ kg} = 1.1909 \times 10^{-26} \text{ kg}$$

The two naturally occurring isotopes of C are ^{12}C and ^{13}C, and those of O are ^{16}O, ^{17}O, and ^{18}O. The possible molecules are $^{12}C^{16}O$, $^{12}C^{17}O$, $^{12}C^{18}O$, $^{13}C^{16}O$, $^{13}C^{17}O$, and $^{13}C^{18}O$. Of these, $^{12}C^{16}O$ and $^{12}C^{18}O$ can be eliminated on the basis of the data given in Problems 17.11 and 17.12. The respective approximate values of reduced mass are $\mu/(10^{-26} \text{ kg}) = 1.17, 1.19, 1.22$, and 1.25. The source of the weak peak is $^{13}C^{16}O$.

17.14 Which rotational level, $J = 4$ or $J = 5$, is more highly occupied by CO molecules at $T = 298$ K? For $^{12}C^{16}O$, $B^* = 3.8343 \times 10^{-23}$ J.

▮ Substituting *(17.3a)* and $g_J = 2J + 1$ into *(2.22)* gives

$$\frac{N_i}{N_j} = \frac{2J_i + 1}{2J_j + 1} \exp\left(-\frac{[B^*J_i(J_i + 1) - B^*J_j(J_j + 1)]}{kT}\right) = \frac{2J_i + 1}{2J_j + 1} \exp\left(-\frac{[J_i(J_i + 1) - J_j(J_j + 1)]B^*}{kT}\right) \tag{17.7}$$

Substituting the data gives

$$\frac{N(J = 4)}{N(J = 5)} = \frac{2(4) + 1}{2(5) + 1} \exp\left(-\frac{[4(4 + 1) - 5(5 + 1)](3.8343 \times 10^{-23} \text{ J})}{(1.3807 \times 10^{-23} \text{ J}\cdot\text{K}^{-1})(298.15 \text{ K})}\right) = 0.8981$$

The $J = 5$ level is more highly occupied.

17.15 Calculate the ratio of the number of CO molecules with $J = 4$ to those with $J = 0$ at $T = 298$ K. For $^{12}C^{16}O$, $B^* = 3.8343 \times 10^{-23}$ J.

▮ Substituting $J_j = 0$ and $N_j = N_0$ into *(17.7)* gives

$$N_J = N_0(2J + 1)\, e^{-J(J+1)B^*/kT} \tag{17.8}$$

Solving *(17.8)* for N_J/N_0 and substituting the data into *(17.8)* gives

$$\frac{N(J = 4)}{N_0} = [2(4) + 1] \exp\left(-\frac{4(4 + 1)(3.8343 \times 10^{-23} \text{ J})}{(1.3807 \times 10^{-23} \text{ J}\cdot\text{K}^{-1})(298.15 \text{ K})}\right) = 7.4703$$

17.16 Predict the microwave spectrum of $^{12}C^{16}O$ that would be observed at $T = 298$ K.

▮ Neglecting centrifugal distortions, the absorption peaks will be located at the values of $\bar{\nu}$ listed in Problem 17.4. The intensities of each line will be proportional to the values of N_J/N_0 determined by *(17.8)*. For example, for the $J = 4 \leftrightarrow 5$ transition, the line will appear at $\bar{\nu} = 19.3020 \text{ cm}^{-1}$ (see Problem 17.4) and will have an intensity proportional to $N(J = 4)/N_0 = 7.4703$ (see Problem 17.15). The predicted spectrum is shown in Fig. 17-1.

17.17 Determine the value of J such that $N_J < 0.01 N_0$ for $^{12}C^{16}O$ at $T = 298$ K. See Problem 17.15 for additional information.

▮ Substituting the data into *(17.8)* gives

$$0.01 N = N_0(2J + 1) \exp\left(-\frac{J(J + 1)(3.8343 \times 10^{-23} \text{ J})}{(1.3807 \times 10^{-23} \text{ J}\cdot\text{K}^{-1})(298.15 \text{ K})}\right)$$

Solving gives $J = 30.1$ or $J \geq 31$. As can be seen from Fig. 17-1, for $J > 31$, $N_J/N_0 < 0.01$.

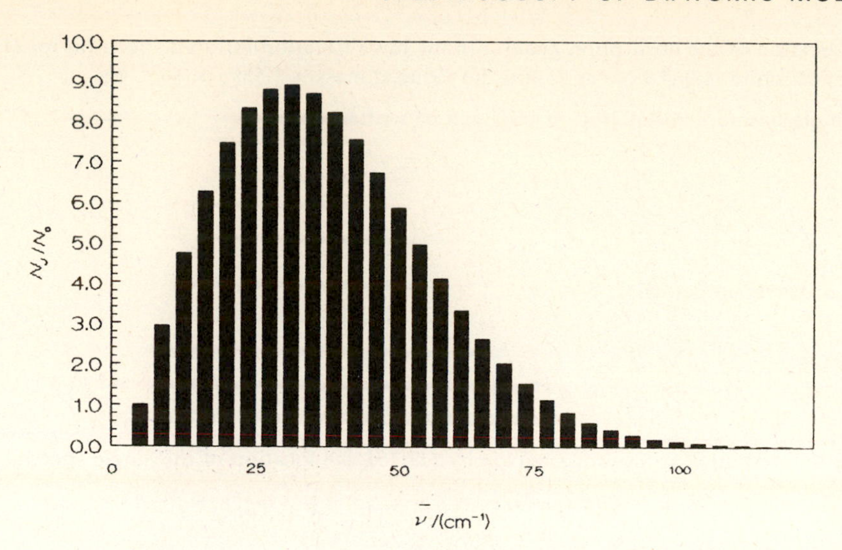

Fig. 17-1

17.18 Determine the value of J for which N_J/N_0 is a maximum for $^{12}C^{16}O$ at $T = 298$ K. See Problem 17.15 for additional information.

▌ Differentiating *(17.8)* with respect to J, setting the result equal to zero, and solving for $J_{\max}$ gives

$$J_{\max} = \frac{(2kT/B^*)^{1/2} - 1}{2} \tag{17.9}$$

Substituting the data into *(17.9)* gives

$$J_{\max} = \frac{[2(1.381 \times 10^{-23}\ \text{J} \cdot \text{K}^{-1})(298\ \text{K})/(3.834 \times 10^{-23}\ \text{J})]^{1/2} - 1}{2} = 6.827 \approx 7$$

The maximum value of J at 298 K is $J = 7$. This agrees with the plot shown in Fig. 17-1, where the eighth peak is the greatest (the $J = 7 \leftrightarrow 8$ transition will be most intense).

17.19 Using *(17.9)*, derive an expression for the rotational energy corresponding to $J_{\max}$. Assuming $B^* \ll kT$, simplify this expression and comment on the result.

▌ Substituting *(17.9)* into *(17.3a)* gives

$$E_J = B^*\left(\frac{(2kT/B^*)^{1/2} - 1}{2}\right)\left(\frac{(2kT/B^*)^{1/2} - 1}{2} + 1\right) = B^*\left(\frac{(2kT/B^*)^{1/2} - 1}{2}\right)\left(\frac{(2kT/B^*)^{1/2} + 1}{2}\right)$$

$$= B^*\frac{2kT/B^* - 1}{4}$$

If $B^* \ll kT$, then $2kT/B^* \gg 1$ and

$$E_J = B^*\frac{2kT/B^*}{4} = \frac{kT}{2}$$

This is the same as the classical result for the kinetic energy for one degree of freedom.

17.20 Which of the following vibrational transitions will be observed for a diatomic molecule:

$$v = 1 \leftrightarrow 3, \qquad v = 2 \leftrightarrow 3, \qquad v = 5 \leftrightarrow 4$$

▌ The selection rule for vibrational transitions is

$$\Delta v = \pm 1 \tag{17.10}$$

The transitions that obey *(17.10)* are $v = 2 \leftrightarrow 3$ and $v = 5 \leftrightarrow 4$.

17.21 Identify which of the following diatomic molecules will have a vibrational infrared spectrum: N_2, $^{35}Cl^{37}Cl$, CO, O_2^+.

▌ A vibrational infrared spectrum will be observed only for those molecules that contain a permanent dipole moment. Because homonuclear diatomic molecules do not have permanent dipole moments and heteronuclear diatomic molecules do, a spectrum will be observed only for CO.

17.22 Predict the location of the absorption peak(s) in the low-resolution infrared spectrum for HF. The force constant for the chemical bond is 964.9 $N \cdot m^{-1}$, and the reduced mass is 1.5895×10^{-27} kg.

▌ For a simple harmonic oscillator, *(14.63)* can be written as

$$E_v = \omega^*(v + \tfrac{1}{2}) \tag{17.11a}$$

where

$$\omega^* = h\nu_0 = \hbar\left(\frac{k}{\mu}\right)^{1/2} \tag{17.12a}$$

or in terms of wavenumbers as

$$G(v) = \omega_e(v + \tfrac{1}{2}) \tag{17.11b}$$

where

$$\omega_e/(\text{cm}^{-1}) = \frac{10^{-2}}{2\pi c}\left(\frac{k}{\mu}\right)^{1/2} = (5.3088 \times 10^{-12})\left(\frac{k/(\text{N}\cdot\text{m}^{-1})}{\mu/(\text{kg})}\right)^{1/2} \tag{17.12b}$$

Substituting the permitted values of Δv given by *(17.10)*, the location of the single spectral peak will be

$$\bar{\nu} = G(v) - G(v-1) = \omega_e(v + \tfrac{1}{2}) - \omega_e(v + \tfrac{1}{2} - 1) = \omega_e \tag{17.13a}$$

Substituting the data into *(17.12b)* and *(17.13a)* gives

$$\bar{\nu} = \omega_e = (5.3088 \times 10^{-12})\left(\frac{964.9}{1.5895 \times 10^{-27}}\right)^{1/2} \text{cm}^{-1} = 4136 \text{ cm}^{-1}$$

17.23 The low-resolution infrared spectrum of $^{12}C^{16}O$ has an absorption peak centered at 2142.61 cm^{-1}. What is the force constant for the CO molecule? The reduced mass is $1.138\,518 \times 10^{-26}$ kg.

▌ From *(17.13)*, $\omega_e = \bar{\nu}$. Solving *(17.12b)* for k and substituting the data gives

$$k = \left(\frac{2142.61}{5.3088 \times 10^{-12}}\right)^2 (1.138\,518 \times 10^{-26}) \text{ N}\cdot\text{m}^{-1} = 1854.5 \text{ N}\cdot\text{m}^{-1}$$

17.24 Predict the low-resolution infrared spectrum of $^{12}C^{18}O$ using the results of Problem 17.23. The reduced mass is $1.195\,38 \times 10^{-26}$ kg.

▌ The force constant is unchanged by isotopic substitution. Substituting $k = 1854.5 \text{ N}\cdot\text{m}^{-1}$ into *(17.12b)* gives

$$\omega_e = (5.3088 \times 10^{-12})\left(\frac{1854.5}{1.195\,38 \times 10^{-26}}\right)^{1/2} \text{cm}^{-1} = 2091.0 \text{ cm}^{-1}$$

17.25 The $v = 0 \leftrightarrow 1$ transition was observed for the $^3\Delta$ excited electronic level of $^{12}C^{16}O$. Given $\bar{\nu} = 1138.0 \text{ cm}^{-1}$ for this transition, determine k and compare the result to that determined in Problem 17.23 for the $^1\Sigma^+$ electronic level.

▌ Solving *(17.12b)* for k and substituting the data gives

$$k = \left(\frac{1138.0}{5.3088 \times 10^{-12}}\right)^2 (1.138\,518 \times 10^{-26}) \text{ N}\cdot\text{m}^{-1} = 523.2 \text{ N}\cdot\text{m}^{-1}$$

The decreased value of the force constant in the excited level indicates that the bonding in the excited molecule is not as strong as it is in the ground electronic level. One of the bonding electrons has been promoted to an antibonding π orbital to give a configuration of $\pi^3\pi$ corresponding to the $^3\Delta$ term symbol (see Table 16-1).

17.26 Determine the ratio of the populations of the $v = 0$ and $v = 1$ states for CO at $T = 298$ K given $\omega_e = 2169.52 \text{ cm}^{-1}$.

▌ Substituting *(17.11a)* and $g = 1$ into *(2.22)* gives

$$\frac{N_i}{N_j} = \frac{1}{1}\exp\left(-\frac{\omega^*(v_i + \frac{1}{2}) - \omega^*(v_j + \frac{1}{2})}{kT}\right) = e^{-(v_i - v_j)\omega^*/kT} \tag{17.14}$$

Solving *(17.12a)* and *(17.12b)* for $(k/\mu)^{1/2}$ and equating gives

$$\omega^* = 2\pi c\hbar\omega_e/10^{-2} \tag{17.15}$$

Substituting the data into *(17.15)* gives

$$\omega^* = \frac{2\pi(2.997\,925 \times 10^8 \text{ m}\cdot\text{s}^{-1})(1.0546 \times 10^{-34} \text{ J}\cdot\text{s})(2169.52 \text{ cm}^{-1})}{(10^{-2} \text{ m})/(1 \text{ cm})} = 4.3097 \times 10^{-20} \text{ J}$$

Substituting into *(17.14)* gives

$$\frac{N(v=1)}{N(v=0)} = \exp\left(-\frac{(1-0)(4.3097\times10^{-20}\ \text{J})}{(1.380\,66\times10^{-23}\ \text{J}\cdot\text{K}^{-1})(298.15\ \text{K})}\right) = 2.84\times10^{-5}$$

17.27 Consider a rotating molecule such that *(17.3b)* is written as

$$F(J) = B_eJ(J+1) - \bar{D}_eJ^2(J+1)^2 \qquad (17.3c)$$

where $\bar{D}_e$ is known as the centrifugal distortion. Derive a general equation for the location of the spectral lines. Using $B_e = 1.9302\ \text{cm}^{-1}$ and $\bar{D}_e = 6.14\times10^{-6}\ \text{cm}^{-1}$ for $^{12}C^{16}O$, determine $\bar{\nu}$ for $J=5$, and compare the answer to the value determined in Problem 17.4.

▌ Substituting *(17.3c)* into *(17.5a)* gives

$$\begin{aligned}\bar{\nu} &= F(J) - F(J-1) = [B_eJ(J+1) - \bar{D}_eJ^2(J+1)^2] - [B_e(J-1)(J-1+1) - \bar{D}_e(J-1)^2(J-1+1)^2]\\ &= B_e(J^2+J) - \bar{D}_eJ^2(J^2+2J+1) - B_e(J^2-J) + \bar{D}_eJ^2(J^2-2J+1)\\ &= 2B_eJ - 4\bar{D}_eJ^3 \qquad (17.5c)\end{aligned}$$

Substituting the data into *(17.5c)* gives

$$\bar{\nu} = 2(1.9032\ \text{cm}^{-1})(5) - 4(6.14\times10^{-6}\ \text{cm}^{-1})(5)^3 = 19.0289\ \text{cm}^{-1}$$

This value is 1.435% less than that for the rigid rotator.

17.28 Use the following data to confirm the values of B_e and $\bar{D}_e$ for $^{12}C^{16}O$ given in Problem 17.27: $J = 0 \leftrightarrow 1$, $\nu = 115\,271.20$ MHz; $J = 1 \leftrightarrow 2$, $\nu = 230\,537.97$ MHz; $J = 2 \leftrightarrow 3$, $\nu = 345\,795.9$ MHz; $J = 3 \leftrightarrow 4$, $\nu = 461\,040.7$ MHz; $J = 4 \leftrightarrow 5$, $\nu = 576\,267.8$ MHz.

▌ Dividing both sides of *(17.5c)* by J gives

$$\bar{\nu}/J = 2B_e - 4\bar{D}_eJ^2$$

Thus a plot of $\bar{\nu}/J$ against J^2 will be linear with an intercept equal to $2B_e$ and a slope equal to $-4D_e$. As a sample calculation, consider substituting $\nu = 115\,271.20$ MHz for $J = 0 \leftrightarrow 1$ into *(13.32)*:

$$\bar{\nu} = \frac{(115\,271.20\ \text{MHz})[(10^6\ \text{Hz})/(1\ \text{MHz})][(1\ \text{s}^{-1})/(1\ \text{Hz})]}{(2.997\,925\times10^8\ \text{m}\cdot\text{s}^{-1})[(10^2\ \text{cm})/(1\ \text{m})]} = 3.845\,033\ \text{cm}^{-1}$$

$$\bar{\nu}/J = (3.845\,033\ \text{cm}^{-1})/1 = 3.845\,033\ \text{cm}^{-1}$$

The plot is shown in Fig. 17-2. From the plot,

$$B_e = \frac{\text{intercept}}{2} = \frac{3.845\,057\ \text{cm}^{-1}}{2} = 1.922\,529\ \text{cm}^{-1}$$

$$\bar{D}_e = \frac{\text{slope}}{-4} = \frac{-2.45\times10^{-5}\ \text{cm}^{-1}}{-4} = 6.13\times10^{-6}\ \text{cm}^{-1}$$

Both of these values agree fairly well with the values given in Problem 17.27. It should be noted that the value of B_e determined here is actually the value of B_0 (see Problem 17.37).

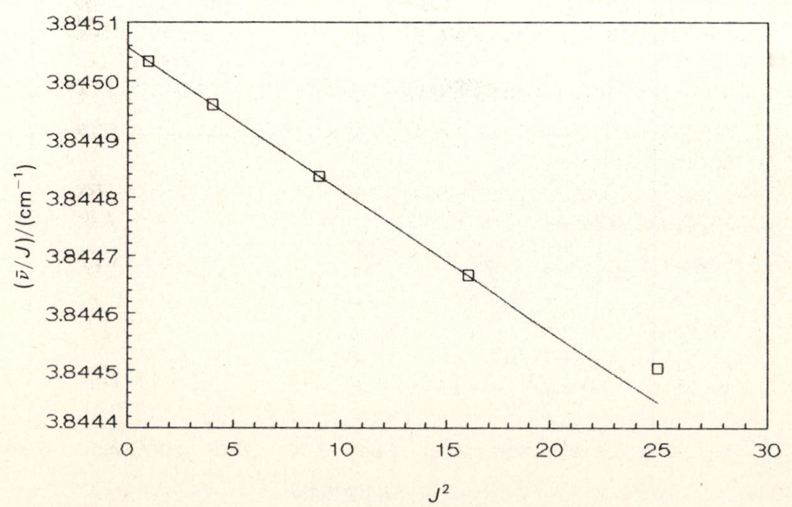

Fig. 17-2

17.29 The vibrational energy of a diatomic molecule can be written as

$$G(v) = \omega_e(v + \tfrac{1}{2}) - \omega_e x_e(v + \tfrac{1}{2})^2 + \omega_e y_e(v + \tfrac{1}{2})^3 + \cdots \qquad (17.11c)$$

where the $\omega_e x_e$, $\omega_e y_e$, etc. terms correct for the anharmonicity of the oscillator. Given $\bar{\nu}/(\text{cm}^{-1}) = 2142.71$ for $v = 0 \leftrightarrow 1$ and 4258.80 for $v = 0 \leftrightarrow 2$ (the first overtone) for $^{12}C^{16}O$, calculate the values of ω_e and $\omega_e x_e$.

▌ The location of the various spectral overtone peaks will be given by *(17.11c)* as

$$\bar{\nu}(v = 0 \leftrightarrow v) = G(v) - G(0)$$
$$= [\omega_e(v + \tfrac{1}{2}) - \omega_e x_e(v + \tfrac{1}{2})^2 + \omega_e y_e(v + \tfrac{1}{2})^3 + \cdots] - [\omega_e(\tfrac{1}{2}) - \omega_e x_e(\tfrac{1}{2})^2 + \omega_e y_e(\tfrac{1}{2})^3 + \cdots]$$
$$= \omega_e v - \omega_e x_e v + \tfrac{3}{4}\omega_e y_e v + \cdots - \omega_e x_e v^2 + \tfrac{3}{2}\omega_e y_e v^2 + \cdots + \omega_e y_e v^3 + \cdots$$
$$= \omega_0 v - \omega_0 x_0 v^2 + \omega_0 y_0 v^3 + \cdots \qquad (17.13b)$$

where

$$\omega_0 = \omega_e - \omega_e x_e + \tfrac{3}{4}\omega_e y_e + \cdots \qquad (17.16a)$$
$$\omega_0 x_0 = \omega_e x_e - \tfrac{3}{2}\omega_e y_e + \cdots \qquad (17.16b)$$
$$\omega_0 y_0 = \omega_e y_e + \cdots \qquad (17.16c)$$

Substituting the data into *(17.13b)* gives

$$2142.71 \text{ cm}^{-1} = \omega_0(1) - \omega_0 x_0(1)^2$$
$$4258.80 \text{ cm}^{-1} = \omega_0(2) - \omega_0 x_0(2)^2$$

Solving the simultaneous equations gives $\omega_0 = 2156.02 \text{ cm}^{-1}$ and $\omega_0 x_0 = 13.31 \text{ cm}^{-1}$. Solving *(17.16)* for $\omega_e x_e$ and ω_e and substituting the above results gives

$$\omega_e x_e \approx \omega_0 x_0 = 13.31 \text{ cm}^{-1}$$
$$\omega_e \approx \omega_0 + \omega_e x_e = 2156.02 \text{ cm}^{-1} + 13.31 \text{ cm}^{-1} = 2169.33 \text{ cm}^{-1}$$

The accepted values of the constants in *(17.11c)* are $\omega_e = 2169.52 \text{ cm}^{-1}$, $\omega_e x_e = 13.453 \text{ cm}^{-1}$, and $\omega_e y_e = 0.0308 \text{ cm}^{-1}$.

17.30 Prepare a plot of *(17.11c)* for $0 \le v \le 58$ using the data given in Problem 17.29 for $^{12}C^{16}O$.

▌ As a sample calculation, consider $v = 10$.

$$G(10) = (2169.52 \text{ cm}^{-1})(10 + \tfrac{1}{2}) - (13.453 \text{ cm}^{-1})(10 + \tfrac{1}{2})^2 - (0.0308 \text{ cm}^{-1})(10 + \tfrac{1}{2})^3 = 21\,332.4 \text{ cm}^{-1}$$

The lines are shown in Fig. 17-3 for $v = 0, 5, 10, 11, 15, 20, 25, 30, 35, 40, 45, 50, 55,$ and 58.

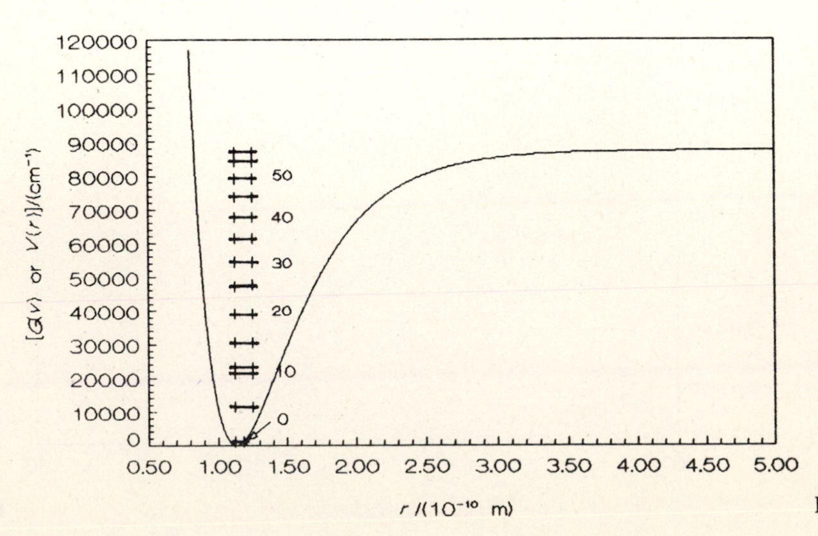

Fig. 17-3

17.31 What is the value of $\Delta G(v)$ for the $v = 10 \leftrightarrow 11$ transition? See Problem 17.29 for additional data.

▮ For the transition between v and $v + 1$, *(17.11c)* gives

$$\begin{aligned}\Delta G(v) &= [\omega_e(v+1+\tfrac{1}{2}) - \omega_e x_e(v+1+\tfrac{1}{2})^2 + \omega_e y_e(v+1+\tfrac{1}{2})^3 + \cdots] \\ &\quad - [\omega_e(v+\tfrac{1}{2}) - \omega_e x_e(v+\tfrac{1}{2})^2 + \omega_e y_e(v+\tfrac{1}{2})^3 + \cdots] \\ &= \omega_e - 2\omega_e x_e(v+1) + \omega_e y_e(3v^2 + 6v + \tfrac{13}{4}) + \cdots \end{aligned} \tag{17.17}$$

Substituting the data gives

$$\Delta G(v) = 2169.52\ \text{cm}^{-1} - 2(13.453\ \text{cm}^{-1})(10+1) + (0.0308\ \text{cm}^{-1})[3(10)^2 + 6(10) + \tfrac{13}{4}] = 1884.74\ \text{cm}^{-1}$$

This value agrees with the plots shown in Fig. 17-3 for $v = 10$ and $v = 11$.

17.32 Estimate the dissociation energy (D_e) for $^{12}C^{16}O$ from the spectroscopic constants given in Problem 17.29.

▮ Assuming that only the ω_e and $\omega_e x_e$ terms in *(17.17)* are important, the *Birge–Sponer equation* gives

$$D_e = \omega_e^2/4\omega_e x_e \tag{17.18}$$

Substituting the data gives

$$D_e = \frac{(2169.52\ \text{cm}^{-1})^2}{4(13.453\ \text{cm}^{-1})} = 87\,468\ \text{cm}^{-1} = 1046.3\ \text{kJ}\cdot\text{mol}^{-1}$$

This value is slightly lower than the accepted value of 1082 kJ · mol^{-1}.

17.33 Using the results of Problem 17.32, calculate the value of the dissociation energy (D_0) for $^{12}C^{16}O$.

▮ The difference between the two dissociation energies is that D_e is measured with respect to the bottom or minimum in the potential energy well and D_0 is measured with respect to the lowest vibrational energy state. These are related by

$$D_e = D_0 + \omega_e/2 \tag{17.19}$$

Solving *(17.19)* for D_0 and substituting the data gives

$$D_0 = 87\,468\ \text{cm}^{-1} - (2169.52\ \text{cm}^{-1})/2 = 86\,383\ \text{cm}^{-1} = 1033.3\ \text{kJ}\cdot\text{mol}^{-1}$$

17.34 Estimate the value of v for the last vibrational state of $^{12}C^{16}O$ before dissociation occurs. See Problem 17.32 for additional information.

▮ Substituting $D_e = 87\,468\ \text{cm}^{-1}$ into *(17.11c)* gives

$$87\,468\ \text{cm}^{-1} = (2169.52\ \text{cm}^{-1})(v+\tfrac{1}{2}) - (13.453\ \text{cm}^{-1})(v+\tfrac{1}{2})^2 + (0.0308\ \text{cm}^{-1})(v+\tfrac{1}{2})^3 + \cdots$$

Solving gives $v + \frac{1}{2} = 58.5$. The last vibrational level before dissociation occurs is $v = 58$, as seen in Fig. 17-3.

17.35 The *Morse potential*

$$V(r) = D_e(1 - e^{-a(r-r_e)})^2 \tag{17.20}$$

can be used to describe the potential energy well for an anharmonic oscillator where a is a constant given by

$$a = 10[\omega_e/(\text{cm}^{-1})](\pi c\mu/\hbar D_e)^{1/2} \tag{17.21}$$

Prepare a plot of $V(r)$ as a function of r for $^{12}C^{16}O$. See Problems 17.6 and 17.32 for additional information.

▮ Substituting the data into *(17.21)* gives

$$\begin{aligned} a &= \frac{10\ \text{cm}^{1/2}}{1\ \text{m}^{1/2}}(2169.52\ \text{cm}^{-1})\left(\frac{\pi(2.997\,925\times10^{8}\ \text{m}\cdot\text{s}^{-1})(1.138\,518\times10^{-26}\ \text{kg})[(1\ \text{J})/(1\ \text{kg}\cdot\text{m}^2\cdot\text{s}^{-2})]}{(1.0546\times10^{-34}\ \text{J}\cdot\text{s})(87\,468\ \text{cm}^{-1})}\right)^{1/2} \\ &= 2.3391\times10^{10}\ \text{m}^{-1}\end{aligned}$$

As a sample calculation, consider $r = 0.200$ nm.

$$V(r) = (87\,468\ \text{cm}^{-1})\{1 - \exp[-(2.3391\times10^{10}\ \text{m}^{-1})(2.00\times10^{-10}\ \text{m} - 1.1286\times10^{-10}\ \text{m})]\}^2 = 66\,166\ \text{cm}^{-1}$$

The plot is shown as part of Fig. 17-3.

17.36 Confirm *(17.21)* given that

$$k_e = 10^2\hbar c\left.\frac{d^2V(r)}{dr^2}\right|_{r_e}$$

▌ Taking the necessary derivatives of (17.20) and evaluating at $r = r_e$ gives

$$\left.\frac{d^2V(r)}{dr^2}\right|_{r_e} = \left[\frac{d^2}{dr^2} D_e(1 - e^{-a(r-r_e)})^2\right]\Bigg|_{r=r_e} = D_e\left[\frac{d}{dr} 2(1 - e^{-a(r-r_e)})a\, e^{-a(r-r_e)}\right]\Bigg|_{r=r_e}$$

$$= 2aD_e\left[\frac{d}{dr}(e^{-a(r-r_e)} - e^{-2a(r-r_e)})\right]\Bigg|_{r=r_e} = 2aD_e[(-a)\, e^{-a(r-r_e)} + 2a\, e^{-2a(r-r_e)}]|_{r=r_e} = 2a^2D_e$$

Substituting this result into (17.12b) gives

$$\omega_e = \frac{10^{-2}}{2\pi c}\left(\frac{(10^2hc)(2a^2D_e)}{\mu}\right)^{1/2} = 10^{-1}a\left(\frac{\hbar D_e}{\pi c\mu}\right)^{1/2}$$

which is the same as (17.21).

17.37 Because a real molecule is simultaneously undergoing both rotational and vibrational motion, the energy will be

$$T(v, J) = \omega_e(v + \tfrac{1}{2}) - \omega_e x_e(v + \tfrac{1}{2})^2 + \cdots + B_eJ(J + 1) - \bar{D}_eJ^2(J + 1)^2 + \cdots - \alpha_e(v + \tfrac{1}{2})J(J + 1) \quad (17.22)$$

Correct the value of B_0 determined in Problem 17.28 for this effect given $\alpha_e = 0.017\,46\ \text{cm}^{-1}$.

▌ The value of B_0 determined in Problem 17.28 included the last term of (17.22), the rotational-vibrational interaction. For $v = 0$,

$$B_0 = B_e - \tfrac{1}{2}\alpha_e$$

Solving for B_e and substituting the data gives

$$B_e = 1.922\,529\ \text{cm}^{-1} + \tfrac{1}{2}(0.017\,46\ \text{cm}^{-1}) = 1.931\,26\ \text{cm}^{-1}$$

17.38 The absorption for the $v = 0$ and $J = 0 \leftrightarrow 1$ transition is $3.845\,033\ \text{cm}^{-1}$, and for the $v = 1$ and $J = 0 \leftrightarrow 1$ transition it is $3.810\,009\ \text{cm}^{-1}$ for $^{12}C^{16}O$. Determine α_e for this molecule.

▌ Considering the third and last terms of (17.22) gives

$$3.845\,033\ \text{cm}^{-1} = B_e(1)(1 + 1) - \alpha_e(0 + \tfrac{1}{2})(1)(1 + 1)$$

$$3.810\,009\ \text{cm}^{-1} = B_e(1)(1 + 1) - \alpha_e(1 + \tfrac{1}{2})(1)(1 + 1)$$

Solving gives $\alpha_e = 0.017\,512\ \text{cm}^{-1}$. This agrees well with the accepted value of $0.017\,46\ \text{cm}^{-1}$.

17.39 For the $v = 0 \leftrightarrow 1$ infrared band of $^{12}C^{16}O$, the following data were obtained:

J	0	1	2	3	4
$\bar{\nu}_R/(\text{cm}^{-1})$	2147.084	2150.858	2154.599	2158.301	2161.971
$\bar{\nu}_P/(\text{cm}^{-1})$	—	2139.427	2135.548	2131.633	2127.684

J	5	6	7	8
$\bar{\nu}_R/(\text{cm}^{-1})$	2165.602	2169.200	2172.759	2176.287
$\bar{\nu}_P/(\text{cm}^{-1})$	2123.700	2119.681	2115.632	2111.545

where $\bar{\nu}_R$ refers to the rotational information in the R branch and $\bar{\nu}_P$ refers to the P branch. Determine α_e and B_e for CO assuming $\bar{D}_e = 0$.

▌ If $m = J + 1$ for the R branch and $m = -J$ for the P branch,

$$\bar{\nu} = \bar{\nu}_0 + (2B_e - 2\alpha_e)m - \alpha_e m^2 \quad (17.23)$$

where $\bar{\nu}_0$ is the wavenumber of the forbidden $v' = 1$, $J' = 0$, and $v'' = 0$, $J'' = 0$ transitions and $B'_e = B''_e$. (Note that the upper state is designated with a prime and the lower state with a double prime.) A plot of $\bar{\nu}$ against m is shown in Fig. 17-4. A least squares analysis of the data gives

$$\bar{\nu}/(\text{cm}^{-1}) = 2143.173 + 3.8264m - 0.017\,54m^2$$

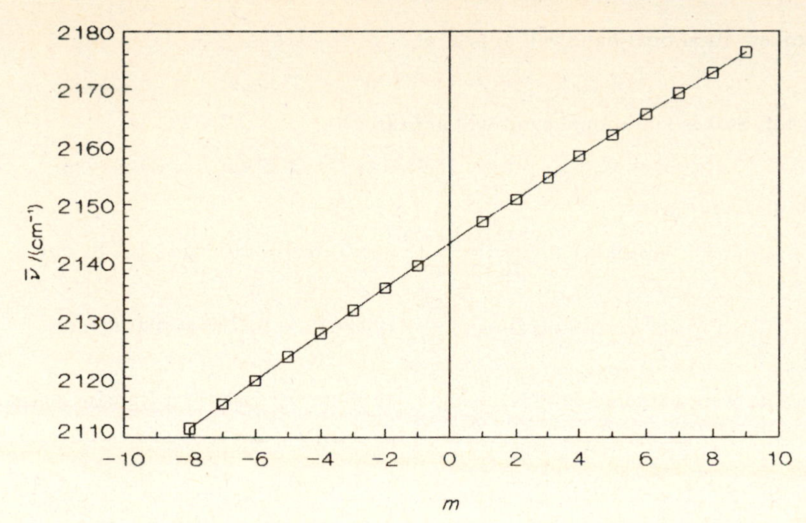

Fig. 17-4

The corresponding values of B_e and α_e are

$$\alpha_e = 0.017\,54\ \text{cm}^{-1} \qquad B_e = \frac{3.8264\ \text{cm}^{-1} + 2(0.017\,54\ \text{cm}^{-1})}{2} = 1.9307\ \text{cm}^{-1}$$

17.40 What will be the separation of the lines for $J = 10$ in the P and R branches of $^{12}C^{16}O$? For CO, $B_e = 1.9302\ \text{cm}^{-1}$ and $\alpha_e = 0.017\,46\ \text{cm}^{-1}$.

▮ Taking the difference of *(17.23)* for m and $m + 1$ gives

$$\Delta\bar{\nu} = [\bar{\nu}_0 + (2B_e - 2\alpha_e)(m+1) - \alpha_e(m+1)^2] - [\bar{\nu}_0 + (2B_e - 2\alpha_e)m - \alpha_e m^2]$$

$$= (2B_e - 3\alpha_e) - 2\alpha_e m \tag{17.24}$$

Substituting $m = -10$ for the P branch and $m = 11$ for the R branch into *(17.24)* gives

$$\Delta\bar{\nu}_P = [2(1.9302\ \text{cm}^{-1}) - 3(0.017\,46\ \text{cm}^{-1})] - 2(0.017\,46\ \text{cm}^{-1})(-10) = 4.1572\ \text{cm}^{-1}$$

$$\Delta\bar{\nu}_R = [2(1.9302) - 3(0.017\,46)] - 2(0.017\,46)(11) = 3.4239\ \text{cm}^{-1}$$

Note that the spacings in the P branch are greater than those in the R branch.

17.41 Let N_0 represent the number of molecules with $v = 0$ and $J = 0$. Calculate N for $v = 1$ and $J = 1$ for $^{12}C^{16}O$, at $T = 298$ K given $B^* = 3.8343 \times 10^{-23}$ J and $\omega^* = 4.3097 \times 10^{-20}$ J. Compare this result to that of Problem 17.26.

▮ Substituting *(17.7)*, *(17.11a)*, $g_v = 1$, $g_J = 2J + 1$ into *(2.22)* gives

$$\frac{N(v=1, J=1)}{N_0} = \left(\frac{1}{1}\right)\left(\frac{2(1)+1}{2(0)+1}\right)\exp\left[\frac{-[1(1+1)(3.8343\times10^{-23}\ \text{J}) + (1-0)(4.3097\times10^{-20}\ \text{J})]}{(1.380\,66\times10^{-23}\ \text{J}\cdot\text{K}^{-1})(298.15\ \text{K})}\right]$$

$$= 8.36 \times 10^{-5}$$

Even though this is a slightly higher energy level that that in Problem 17.26, this values is about three times as large because of the $2J + 1$ factor.

17.42 Compare the infrared spectrum for a $^1\Sigma$ molecule such as CO with that for a $^2\Pi$ molecule such as OH.

▮ For a $^1\Sigma$ molecule the selection rules are $\Delta v = \pm1$ and $\Delta J = \pm1$, giving the P and R branches. For $\Lambda \neq 0$, the selection rules are $\Delta v = \pm1$ and $\Delta J = 0, \pm1$, giving the P, Q, and R branches. The value of $\bar{\nu}_Q$ is a single peak if $B'_e \approx B''_e$.

17.43 A sample of $^{14}N_2$ was excited by a 540.80-nm Ar laser in a Raman spectrometer. At what wavelengths were the Stokes and anti-Stokes vibrational bands observed? For N_2, $\bar{\nu}_0 = 2330\ \text{cm}^{-1}$ for vibration.

▮ The wavenumber of the incident radiation is given by *(13.32)* as

$$\bar{\nu}_{\text{incident}} = \frac{(1\ \text{m})/(10^2\ \text{cm})}{540.88 \times 10^{-9}\ \text{m}} = 18\,488\ \text{cm}^{-1}$$

The intense Stokes vibrational band will appear at

$$\bar{\nu} = \bar{\nu}_{\text{incident}} - \bar{\nu}_0 = 18\,488\text{ cm}^{-1} - 2330\text{ cm}^{-1} = 16\,158\text{ cm}^{-1}$$

and the weak anti-Stokes vibrational band will appear at

$$\bar{\nu} = \bar{\nu}_{\text{incident}} + \bar{\nu}_0 = 18\,488\text{ cm}^{-1} + 2330\text{ cm}^{-1} = 20\,818\text{ cm}^{-1}$$

Substituting into *(13.32)* gives

$$\lambda(\text{Stokes}) = \frac{(1\text{ m})/(10^2\text{ cm})}{16\,158\text{ cm}^{-1}} = 6.1889 \times 10^{-7}\text{ m} = 618.89\text{ mm}$$

$$\lambda(\text{anti-Stokes}) = \frac{10^{-2}}{20\,818} = 4.8035 \times 10^{-7}\text{ m} = 480.35\text{ mm}$$

17.44 The following data were gathered for $^{14}N_2$ using a 540.80-nm Ar laser in a Raman spectrometer:

J	0	1	2	3
$\Delta\bar{\nu}_S/(\text{cm}^{-1})$	11.957	19.908	27.857	35.812

where $\Delta\bar{\nu}_S$ refers to the wavenumber shift for the excitation energy for the S branch. Determine r_{ab} for the N_2 molecule given $\mu = 1.162\,95 \times 10^{-26}$ kg.

▌ The wavenumber shift for the S branch is given by

$$\Delta\bar{\nu}_S = 6B_e + 4B_eJ \qquad (17.25a)$$

where $J = 0, 1, 2, \ldots$ and $B_e' \approx B_e''$. A plot of $\Delta\bar{\nu}_S$ against J will be linear, with an intercept of $6B_e$ and a slope of $4B_e$. The data are plotted in Fig. 17-5. From the plot,

$$B_e = \frac{\text{intercept}}{6} = \frac{11.956\text{ cm}^{-1}}{6} = 1.9927\text{ cm}^{-1}$$

$$B_e = \frac{\text{slope}}{4} = \frac{7.9514\text{ cm}^{-1}}{4} = 1.9879\text{ cm}^{-1}$$

The average value is 1.9903 cm^{-1}. Solving *(17.4b)* for r_{ab} and substituting the data gives

$$r_{ab} = \left(\frac{2.7993 \times 10^{-46}}{(1.9903)(1.162\,95 \times 10^{-26})}\right)^{1/2} = 1.0997 \times 10^{-10}\text{ m} = 0.109\,97\text{ mm}$$

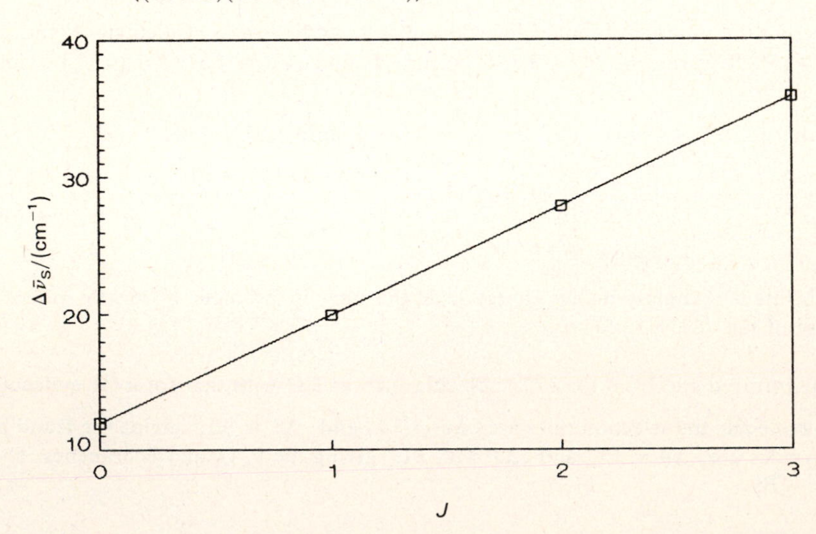

Fig. 17-5

17.45 Based on the results of Problem 17.44, what will be the location of the Q and O branches of the Raman spectrum for N_2?

▌ The wavenumber shift for the Q branch is given by

$$\Delta\bar{\nu}_Q = 0 \qquad (17.25b)$$

and the Q branch will consist of one peak at the center of the S and O branches. The wavenumber shift for the O branch is given by

$$\Delta\bar{\nu}_O = 2B_e - 4B_e J \qquad (17.25c)$$

where $J = 2, 3, \ldots$. Substituting into *(17.25c)* gives

$$\Delta\bar{\nu}_0 = 2(1.9903\ \text{cm}^{-1}) - 4(1.9903\ \text{cm}^{-1})(2) = -11.9418\ \text{cm}^{-1}$$

The first peak will be located at $11.9418\ \text{cm}^{-1}$ less than the Q branch, and each successive peak will be located at $4B_e = 7.9612\ \text{cm}^{-1}$ less.

17.2 ELECTRONIC SPECTRA

17.46 Several electronic levels for O_2 are shown in Fig. 17-6, where X represents the X $^3\Sigma_g^-$ level, a represents the $a^1\Delta_g$ level, b represents the $b^1\Sigma_g^+$ level, A represents the $A^3\Sigma_u^+$ level, and B represents the $B^3\Sigma_u^-$ level. What is the electron configuration corresponding to the X, a, and b levels?

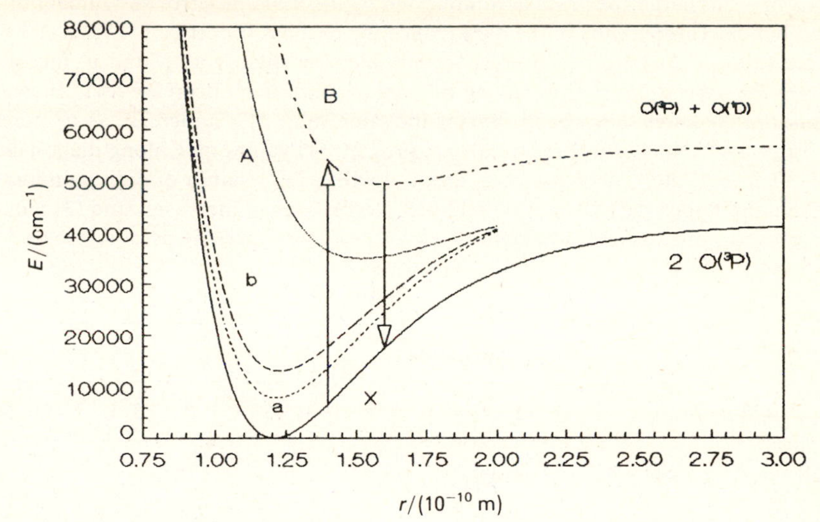

Fig. 17-6

▌ As in Problem 16.47, the configuration is

$$\sigma_g(1s)^2\sigma_u^*(1s)^2\sigma_g(2s)^2\sigma_u^*(2s)^2\pi_u(2p)^4\sigma_g(2p)^2\pi_g(2p)^2$$

From Table 16-1, the entry for the term symbols for a π^2 system is $^1\Sigma^+$, $^1\Delta$, and $^3\Sigma^-$. These three levels correspond to the ground level and excited levels of the configuration written above.

17.47 The spin-forbidden transitions $X \leftrightarrow a$ and $X \leftrightarrow b$ for O_2 shown in Fig. 17-6 occur in the atmosphere. What part of the spectrum is involved?

▌ The minima of the X and a potential energy wells are $\bar{\nu} \approx 8000\ \text{cm}^{-1}$ apart, giving $\lambda = 1.3 \times 10^6$ m for this transition. This transition occurs in the infrared region. The $X \leftrightarrow b$ transition occurs at $\bar{\nu} \approx 13\,000\ \text{cm}^{-1}$ or $\lambda = 8 \times 10^{-7}$ m and falls in the red region of the visible spectrum.

17.48 Consider the absorption/fluorescence shown by the vertical arrows in Fig. 17-6. What are the approximate wavelengths of the photons involved?

▌ The $X \leftrightarrow B$ transition absorbs at $\bar{\nu} \approx 47\,000\ \text{cm}^{-1}$, and the $B \to X$ transition emits at $\bar{\nu} \approx 28\,000\ \text{cm}^{-1}$. The corresponding wavelengths are given by *(13.32)* as

$$\lambda(\text{absorption}) = \frac{(10^{-2}\ \text{m})/(1\ \text{cm})}{47\,000\ \text{cm}^{-1}} = 2.1 \times 10^{-7}\ \text{m} = 210\ \text{nm}$$

$$\lambda(\text{emission}) = \frac{10^{-2}}{28\,000} = 3.6 \times 10^{-7}\ \text{m} = 360\ \text{nm}$$

17.49 A study of the convergence of the Schumann-Runge bands for the $X \to B$ transition in O_2 showed that the limit occurred at 1759 Å. Given that the energy difference between $O(^3P) + O(^1D)$ and $2O(^3P)$ is $15\,867.862\ \text{cm}^{-1}$, determine D_0 for the ground level of O_2.

▮ The energy corresponding to the convergence limit is given by *(13.32)* as

$$\tilde{\nu}_{\text{limit}} = \frac{1}{(1759\ \text{Å})[(10^{-10}\ \text{m})/(1\ \text{Å})][(10^{2}\ \text{cm})/(1\ \text{m})]} = 56\,850\ \text{cm}^{-1}$$

This energy represents the energy difference between the X and B levels plus the dissociation energy of the B level to form $O(^3P) + O(^1D)$ atoms. This energy also represents the dissociation energy of the X level to form $2O(^3P)$ atoms plus the difference in energy between the excited and ground-state atoms. Thus,

$$56\,850\ \text{cm}^{-1} = D_0 + 15\,867.862\ \text{cm}^{-1}$$

which gives $D_0 = 40\,980\ \text{cm}^{-1}$.

17.50 The following bands were observed as part of the $^3\Pi_u \leftrightarrow {}^3\Pi_g$ discharge spectrum of N_2: $\tilde{\nu}(\text{cm}^{-1}) = 31\,678.65$, 29 698.27, 28 600.03, 28 270.95, 27 991.60, 26 959.99, 26 638.25, 26 297.10, 25 700.99, 25 361.40, 25 028.16, and 24 642.07. Assign these bands and prepare a Deslandres table.

▮ The analysis of a discharge spectrum is complicated by the trial-and-error assignment of the values of $v' \rightarrow v''$ to the observed bands. Three trends in the data reduce the chances for error: (1) values of $\tilde{\nu}$ for *sequences* (bands having the same value of Δv) are nearly constant but decrease slightly as v' and v'' increase; (2) values of $\tilde{\nu}$ for *progressions from the upper state* (bands having the same values of v') decrease as v'' increases; and (3) values of $\tilde{\nu}$ for *progressions to the lower state* (bands having the same value of v'') increase as v' increases. The assignments are shown in Table 17-1. Note that three trends are present: (1) values of $\tilde{\nu}$ along diagonals, which correspond to sequences ($0 \rightarrow 0$, $1 \rightarrow 1$, etc.), decrease as v' and v'' increase; (2) values of $\tilde{\nu}$ in columns, which correspond to progressions from the upper state ($0 \rightarrow 0$, $0 \rightarrow 1$, etc.), decrease as v'' increases; and (3) values of $\tilde{\nu}$ in rows, which correspond to progressions to the lower state ($0 \rightarrow 1$, $1 \rightarrow 1$, etc.) increase as v' increases.

Table 17-1

lower state	upper state $v' = 0$		1		2	3	average separation in the lower state	$2(v'' + 1)$
$v'' = 0$	29 698.27	1980.38	31 678.65					
	1 706.67						1706.67	2
1	27 991.60							
	1 694.50						1694.50	4
2	26 297.10	1973.85	28 270.95					
	1 655.03		1 632.70				1643.87	6
3	24 642.07	1996.18	26 638.25	1961.78	28 600.03			
			1 610 09		1 640.04		1625.07	8
4			25 028.16	1931.83	26 959.99			
					1 598.59		1598.59	10
5					25 361.40			
6						25 700.99		
average separation in the upper state		1983.47		1946.81				
$2(v' + 1)$		2		4				

17.51 Using the Deslandres table prepared in Problem 17.50 for N_2, determine the average vibrational level separation in each of the states. Find ω_e and $\omega_e x_e$ for both states, and calculate $\tilde{\nu}_{00}$.

▮ The difference between the entries for the various rows and columns are given in Table 17-1. Equation *(17.17)* implies that a plot of these differences against $2(v + 1)$ will be linear with a slope of $-\omega_e x_e$ and an intercept of ω_e. From Fig. 17-7,

$$\omega_e' x_e' = -\text{slope} = -(-18\ \text{cm}^{-1}) = 18\ \text{cm}^{-1}$$

$$\omega_e' = \text{intercept} = 2020\ \text{cm}^{-1}$$

$$\omega_e'' x_e'' = -(-14\ \text{cm}^{-1}) = 14\ \text{cm}^{-1} \qquad \omega_e'' = 1739\ \text{cm}^{-1}$$

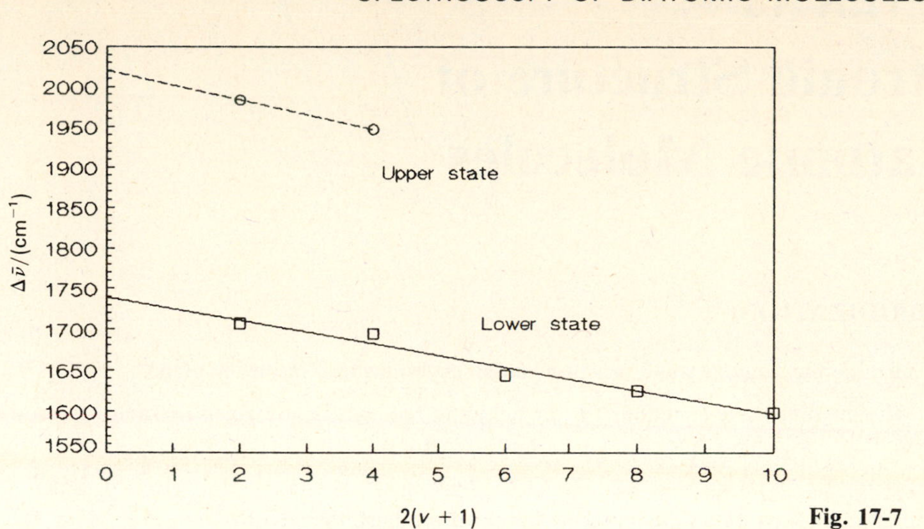

Fig. 17-7

The latter set of values agree well with the accepted values of $\omega_e x_e = 14.1221\ \text{cm}^{-1}$ and $\omega_e = 1733.391\ \text{cm}^{-1}$ for the $^3\Pi_g$ level.

CHAPTER 18
Electronic Structure of Polyatomic Molecules

18.1 HYBRIDIZATION

18.1 Determine the angular wave functions for sp^2 hybridization in the xy plane.

▌ The angular wave functions for the unhybridized orbitals are given in Table 15-2 and by *(15.7)* as

$$Y_s = \left(\frac{1}{4\pi}\right)^{1/2} \qquad Y_{p_x} = Y' = \left(\frac{3}{4\pi}\right)^{1/2} \sin\theta\cos\phi \qquad Y_{p_y} = Y'' = \left(\frac{3}{4\pi}\right)^{1/2} \sin\theta\sin\phi$$

The angular wave functions for the hybridized orbitals are given by

$$X_i = a_i Y_s + b_i Y_{p_x} + c_i Y_{p_y} \tag{18.1}$$

where, by requirements of normalization and orthogonality,

$$a_i^2 + b_i^2 + c_i^2 = 1 \tag{18.2a}$$

$$a_i a_j + b_i b_j + c_i c_j = 0 \qquad \text{for } i \neq j \tag{18.2b}$$

Assume that X_1 lies along the y axis, $X_1 = a_1 Y_s + c_1 Y_{p_y}$, where $b_1 = 0$. Two simultaneous equations relating a_1 and c_1 are

$$a_1^2 + c_1^2 = 1$$

$$c_1^2/a_1^2 = 2$$

where the first comes from *(18.2a)* and the second gives the ratio of the p to s "character" of the hybridization (2:1). Solving these equations gives $a_1 = \pm(\frac{1}{3})^{1/2}$ and $c_1 = \pm(\frac{2}{3})^{1/2}$. Choosing the positive sign, the first hybrid wave function is

$$X_1 = (\tfrac{1}{3})^{1/2} Y_s + (\tfrac{2}{3})^{1/2} Y_{p_y} \tag{18.3a}$$

For the second wave function, $X_2 = a_2 Y_s + b_2 Y_{p_x} + c_2 Y_{p_y}$. Two simultaneous equations relating a_2, b_2, and c_2 are

$$a_2^2 + b_2^2 + c_2^2 = 1$$

$$(b_2^2 + c_2^2)/a_2^2 = 2$$

Solving gives $a_2 = \pm(\frac{1}{3})^{1/2}$. Choosing the positive sign and substituting into *(18.2b)* for $i = 1$ and $j = 2$ gives

$$(\tfrac{1}{3})^{1/2}(\tfrac{1}{3})^{1/2} + (0)b_2 + (\tfrac{2}{3})^{1/2}c_2 = 0$$

Solving gives $c_2 = -(\frac{1}{6})^{1/2}$. Substituting into *(18.2a)* gives

$$[(\tfrac{1}{3})^{1/2}]^2 + b_2^2 + [-(\tfrac{1}{6})^{1/2}]^2 = 1$$

Solving gives $b_2 = \pm(\frac{1}{2})^{1/2}$. Choosing the positive sign, the second hybrid wave function is

$$X_2 = (\tfrac{1}{3})^{1/2} Y_s + (\tfrac{1}{2})^{1/2} Y_{p_x} - (\tfrac{1}{6})^{1/2} Y_{p_y} \tag{18.3b}$$

Likewise for the third wave function, $X_3 = a_3 Y_s + b_3 Y_{p_x} + c_3 Y_{p_y}$, the simultaneous equations are

$$a_3^2 + b_3^2 + c_3^2 = 1$$

$$(b_3^2 + c_3^2)/a_3^2 = 2$$

Solving gives $a_3 = \pm(\frac{1}{3})^{1/2}$. Choosing the positive sign and substituting into *(18.2b)* for $i = 1$ and $j = 3$ gives

$$(\tfrac{1}{3})^{1/2}(\tfrac{1}{3})^{1/2} + (0)(\tfrac{1}{2})^{1/2} + (\tfrac{2}{3})^{1/2}c_3 = 0$$

Solving gives $c_3 = -(\frac{1}{6})^{1/2}$. Substituting into *(18.2a)* gives

$$[(\tfrac{1}{3})^{1/2}]^2 + b_3^2 + [-(\tfrac{1}{6})^{1/2}]^2 = 1$$

Solving gives $b_3 = \pm(\frac{1}{2})^{1/2}$. If the positive sign is chosen, X_2 is obtained, giving no new information. Choosing the negative sign, the third hybrid wave function is

$$X_3 = (\tfrac{1}{3})^{1/2} Y_s - (\tfrac{1}{2})^{1/2} Y_{p_x} - (\tfrac{1}{6})^{1/2} Y_{p_y} \tag{18.3c}$$

18.2 Show that the hybrid angular wave function given by *(18.3a)* is normalized.

▌ Substituting *(18.3a)* into *(14.22)* gives

$$\int_{\text{all space}} X_1^* X_1 \, dV = \int_{\text{all space}} [(\tfrac{1}{3})^{1/2} Y_s + (\tfrac{2}{3})^{1/2} Y_{p_y}]^* [(\tfrac{1}{3})^{1/2} Y_s + (\tfrac{2}{3})^{1/2} Y_{p_y}] \, dV$$

$$= \tfrac{1}{3} \int_{\text{all space}} [Y_s^* Y_s + 2^{1/2} Y_{p_y}^* Y_s + 2^{1/2} Y_s^* Y_{p_y} + 2 Y_{p_y}^* Y_{p_y}] \, dV$$

$$= \tfrac{1}{3}[1 + 2^{1/2}(0) + 2^{1/2}(0) + 2(1)] = 1$$

18.3 Show that the hybrid angular wave functions given by *(18.3b)* and (18.3*c*) are orthogonal.

▌ Substituting *(18.3b)* and *(18.3c)* into *(14.24)* gives

$$\int_{\text{all space}} X_2^* X_3 \, dV = \int_{\text{all space}} [(\tfrac{1}{3})^{1/2} Y_s + (\tfrac{1}{2})^{1/2} Y_{p_x} - (\tfrac{1}{6})^{1/2} Y_{p_y}]^* [(\tfrac{1}{3})^{1/2} Y_s + (\tfrac{1}{2})^{1/2} Y_{p_x} - (\tfrac{1}{6})^{1/2} Y_{p_y}] \, dV$$

$$= \int_{\text{all space}} \left(\frac{Y_s^* Y_s}{3} - \frac{Y_s^* Y_{p_x}}{6^{1/2}} - \frac{Y_s^* Y_{p_y}}{18^{1/2}} + \frac{Y_{p_x}^* Y_s}{6^{1/2}} - \frac{Y_{p_x}^* Y_{p_x}}{2} \right.$$

$$\left. - \frac{Y_{p_x}^* Y_{p_y}}{2(3^{1/2})} - \frac{Y_{p_y}^* Y_s}{18^{1/2}} + \frac{Y_{p_y}^* Y_{p_x}}{2(3^{1/2})} + \frac{Y_{p_y}^* Y_{p_y}}{6} \right) dV$$

$$= \tfrac{1}{3} - 0 - 0 + 0 - \tfrac{1}{2} - 0 - 0 + 0 + \tfrac{1}{6} = 0$$

18.4 Prepare a plot of X_1 and $X_1^* X_1$ for the hybridized angular wave function given by *(18.3a)*.

▌ In the *xy* plane, $\theta = 90°$ and *(18.3a)* simplifies to

$$X_1 = (1/3)^{1/2}(1/4\pi)^{1/2} + (2/3)^{1/2}(3/4\pi)^{1/2}(1) \sin\phi = (1/12\pi)^{1/2}(1 + 6^{1/2} \sin\phi)$$

As a sample calculation, consider $\phi = 45°$.

$$X_1 = (1/12\pi)^{1/2}(1 + 6^{1/2} \sin 45°) = 0.445$$

The plot is shown in Fig. 18-1. For clarity, only absolute values of X_1 are shown. (The points can be located on Cartesian coordinate graph paper by $x = X_1 \cos\phi$ and $y = X_1 \sin\phi$.) Because *(18.3a)* contains no imaginary terms, $X_1^* X_1 = X_1^2$. The plot of $X_1^* X_1$ is also shown in Fig. 18-1.

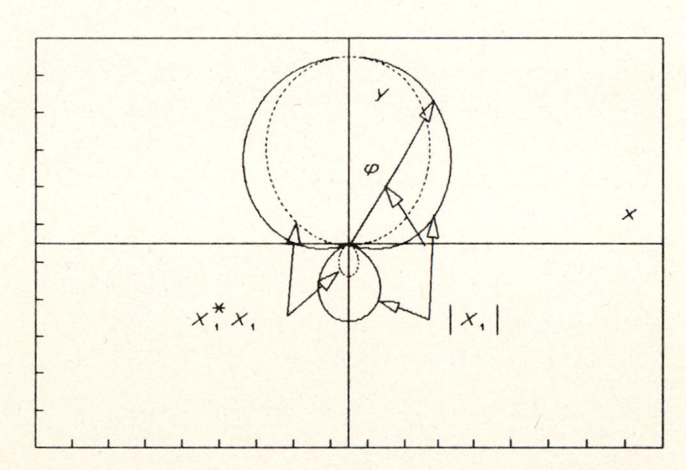

Fig. 18-1

18.5 Determine the value of ϕ for which X_2 given by *(18.3b)* is a maximum. Determine the angle between X_1 and X_2.

▌ In the *xy* plane, $\theta = 90°$ and *(18.3b)* simplifies to

$$X_2 = (1/3)^{1/2}(1/4\pi)^{1/2} + (1/2)^{1/2}(3/4\pi)^{1/2}(1) \cos\phi - (1/6)^{1/2}(3/4\pi)^{1/2}(1) \sin\phi$$

$$= (1/4\pi)^{1/2}[(1/3)^{1/2} + (3/2)^{1/2} \cos\phi - (1/2)^{1/2} \sin\phi]$$

Differentiating X_2 with respect to ϕ, setting the result equal to zero, and solving for ϕ_{max} gives

$$\frac{dX_2}{d\phi} = \left(\frac{1}{4\pi}\right)^{1/2}[0 + (\tfrac{3}{2})^{1/2}(-\sin\phi) - (\tfrac{1}{2})^{1/2}\cos\phi] = 0$$

$$3^{1/2}\sin\phi_{max} + \cos\phi_{max} = 0$$

$$\tan\phi_{max} = \frac{\sin\phi_{max}}{\cos\phi_{max}} = \frac{1}{3^{1/2}} \qquad \phi_{max} = -30°$$

The hybrid orbital represented by X_2 has a maximum at $\phi = -30°$, and the orbital represented by X_1 has a maximum at $\phi = 90°$, giving the angle between the two hybrid orbitals as 120°.

18.6 The angle between sp^n hybrid orbitals is given by

$$\cos\theta = -\lambda^2 \qquad (18.4)$$

where λ^2 is the ratio of the s and p "character" in the orbital (see Problem 18.1). Use this equation to confirm the calculations of Problem 18.5.

▌ For the sp^2 hybrid orbital, $\lambda^2 = \frac{1}{2}$, which upon substituting into *(18.4)* gives

$$\cos\theta = -\tfrac{1}{2} \qquad \theta = 120°$$

18.7 The bond angle in NH_3 is 106.67°. Determine the fraction of s and p "character" in the hybrid orbitals.

▌ Substituting the bond angle into *(18.4)* gives $\lambda^2 = -\cos 106.67° = 0.287$. The ratio of s to p "character" is 0.287. The fractions are found by solving the simultaneous equations

$$s/p = 0.287$$

$$s + p = 1$$

giving 0.223 s "character" and 0.777 p "character."

18.8 Prepare a molecular orbital diagram for CH_3 assuming sp^2 hybridization of the C atom. Assume similar energies for the 1s orbitals of the H atoms and the sp^2 hybrid orbitals.

▌ The diagram is shown in Fig. 18-2. The orbitals of the unhybridized C atom at the far left are shown to form the sp^2 hybridized orbitals, which overlap with the three 1s orbitals of the H atoms to form the three $\sigma(1s, sp^2)$ bonds. One electron is in the nonbonding $n(2p)$ orbital.

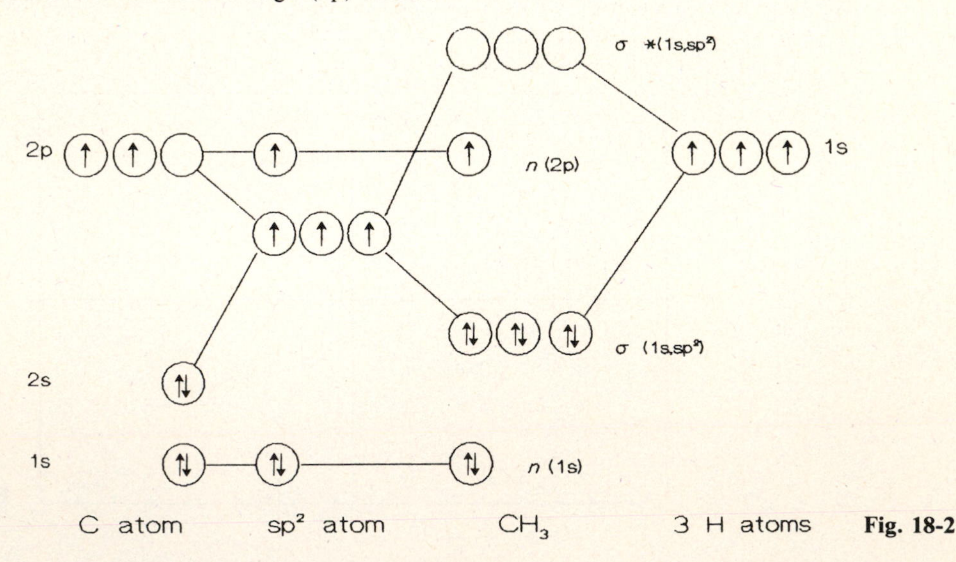

Fig. 18-2

18.9 Determine the relative bond strength of the X_1 hybrid orbital given by *(18.3a)*.

▌ The *relative bond strength* (rbs) is a ratio of the overlap along the bonding axis of the hybrid orbital compared to an unhybridized 1s orbital.

$$\text{rbs} = \frac{\text{maximum of } X_i}{Y_s} \qquad (18.5)$$

The maximum of X_1 will be the expression for X_1 given in Problem 18.4 evaluated at $\phi = 0°$. Substituting into *(18.5)* gives

$$\text{rbs} = \frac{(1/12\pi)^{1/2}(1 + 6^{1/2}\sin 0°)}{(1/4\pi)^{1/2}} = 1.992$$

18.2 MULTIPLE BONDS

18.10 Prepare a molecular orbital diagram representing the bonding in C_2H_4. Predict the types of electronic transitions that could occur. Assume that the C atoms are sp^2 hybridized.

▌ The diagram is shown in Fig. 18-3. For clarity, only the H atoms and the two valence electrons on the sp^2 hybridized C atoms are shown. The double bond between the C atoms consists of a σ bond between the sp^2 orbitals that are oriented along the bonding axis and a π bond resulting from the parallel overlap of the unhybridized p orbitals. The most probable electronic transitions are $\pi \rightarrow \pi^*$ and $\pi \rightarrow \sigma^*$ within the C=C double bond and $\sigma \rightarrow \sigma^*$ within the C—H bonds.

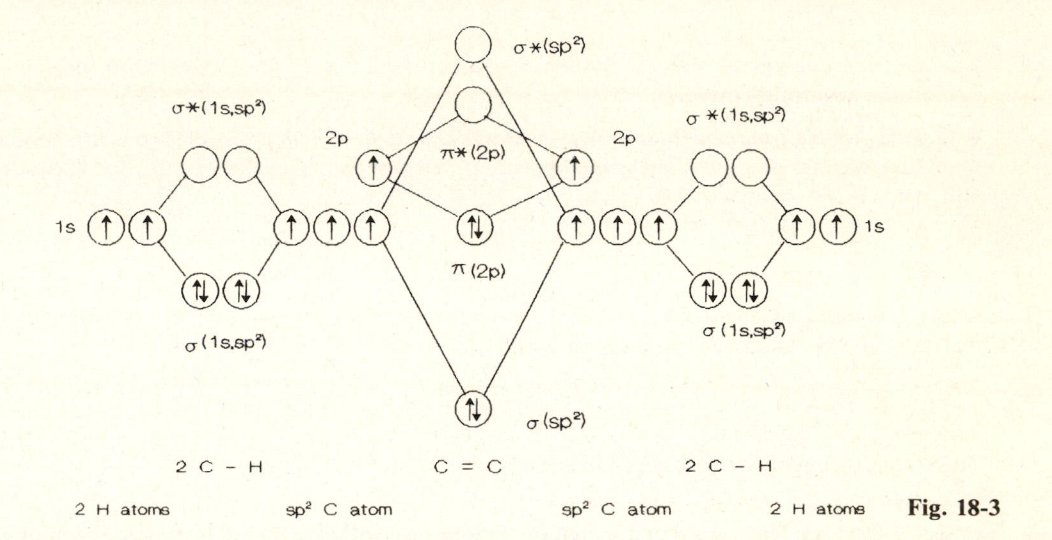

Fig. 18-3

18.11 Write the wave functions describing the π bonding in C_2H_4.

▌ This is a two atom–two electron problem. Using *(14.67)* to write the wave function and *(14.71)* to write the secular determinant gives

$$\phi_\pi = c_1\chi_1 + c_2\chi_2 \tag{18.6}$$

$$\begin{vmatrix} H_{11} - ES_{11} & H_{12} - ES_{12} \\ H_{21} - ES_{21} & H_{22} - ES_{22} \end{vmatrix} = 0 \tag{18.7}$$

where χ_i represents the ψ_{2p_z} wave function for the carbon atoms. The *simplified Hückel molecular orbital* approach assumes $H_{ij} = \beta$ if the atoms are bonded, $H_{ij} = 0$ if the atoms are not bonded, $H_{ii} = \alpha$, $S_{ij} = 0$, and $S_{ii} = 1$. Substituting these approximations into *(18.7)* gives

$$\begin{vmatrix} \alpha - E & \beta \\ \beta & \alpha - E \end{vmatrix} = 0$$

Solving gives

$$E_i = \alpha \pm \beta \tag{18.8}$$

The secular equations are given by *(14.68)* as

$$c_1(H_{11} - ES_{11}) + c_2(H_{12} - ES_{12}) = 0 \tag{18.9a}$$

$$c_1(H_{21} - ES_{21}) + c_2(H_{22} - ES_{22}) = 0 \tag{18.9b}$$

Substituting $E_1 = \alpha + \beta$ into *(18.9a)* gives

$$c_1[\alpha - (\alpha + \beta)(1)] + c_2[\beta - (\alpha + \beta)(0)] = 0$$

$$c_1 = c_2$$

and likewise substituting $E_2 = \alpha - \beta$ gives $c_1 = -c_2$. Normalization requires

$$c_1^2 + c_2^2 = c_1^2 + c_1^2 = 2c_1^2 = 1$$

giving $c_1 = (\frac{1}{2})^{1/2}$ in both cases. The two wave functions are given by *(18.6)* as

$$\phi_1 = (\tfrac{1}{2})^{1/2}(\chi_1 + \chi_2) \qquad \phi_2 = (\tfrac{1}{2})^{1/2}(\chi_1 - \chi_2)$$

Recalling the results of Problem 16.15, ϕ_1 is the bonding orbital and ϕ_2 is the antibonding orbital.

18.12 Empirically, $\beta = -75\ \text{kJ} \cdot \text{mol}^{-1}$ for C_2H_4. Use *(18.8)* to determine the π-electron energy.

▌ Both electrons are placed in the orbital defined by ϕ_1 (see Fig. 18-3) with an energy of

$$E_\pi = 2E_1 = 2(\alpha + \beta) = 2\alpha + 2\beta$$

Because α is essentially the energy of an electron in an isolated C atom $2p_z$ orbital, it is used to set a zero reference for the bonding and antibonding orbitals, giving

$$E_\pi = 2(0) + 2(-75\ \text{kJ} \cdot \text{mol}^{-1}) = -150\ \text{kJ} \cdot \text{mol}^{-1}$$

18.13 Consider the conjugated polyene system 1,3,5-hexatriene. Using the free-electron molecular orbital approach, predict the absorption bond.

▌ In the *free-electron molecular orbital* (FEMO) *approach*, the $n_\pi = 6$ electrons are considered to occupy the three lowest states of a one-dimensional potential well with infinite walls. For the first excitation between the $n_\pi/2$ and $(n_\pi/2) + 1$ states, *(14.44)* gives

$$\bar{\nu} = \frac{h(n_\pi + 1)}{8m_e ca^2} \approx \frac{153\,000\ \text{cm}^{-1}}{n_C + 1} \qquad (18.10a)$$

where a is replaced by a factor combining the single- and double-bond lengths between carbon atoms and n_C is the number of C atoms. Substituting into *(18.10a)* gives

$$\bar{\nu} = \frac{153\,000\ \text{cm}^{-1}}{6 + 1} = 21\,900\ \text{cm}^{-1}$$

This value is roughly 41% lower than the observed value of 37 300 cm^{-1}.

18.14 If the π electrons are considered to have a greater probability of being in the double-bond regions than in the single-bond regions, an improved FEMO model gives

$$\bar{\nu} = \frac{153\,000\ \text{cm}^{-1}}{n_C + 1} + (16\,000\ \text{cm}^{-1})\left(1 - \frac{1}{n_C}\right) \qquad (18.10b)$$

Use this equation to calculate $\bar{\nu}$ for hexatriene.

▌ Substituting into *(18.10b)* gives

$$\bar{\nu} = \frac{153\,000\ \text{cm}^{-1}}{6 + 1} + (16\,000\ \text{cm}^{-1})\left(1 - \frac{1}{6}\right) = 35\,200\ \text{cm}^{-1}$$

This value is only 6% lower than the observed value.

18.15 For polymethine ions, $R\overset{+}{N}{=}CH({-}CH{=}CH)_k{-}\ddot{N}R$, in which the carbon-carbon bond lengths are identical, the simple FEMO approach gives

$$\bar{\nu} = \frac{(2k + 5)(155\,000\ \text{cm}^{-1})}{(2k + 4)^2} \qquad (18.11)$$

Calculate $\bar{\nu}$ for $k = 4$, and compare your answer to the experimental value of 16 000 cm^{-1}.

▌ Substituting into *(18.11)* gives

$$\bar{\nu} = \frac{[2(4) + 5](155\,000\ \text{cm}^{-1})}{[2(4) + 4]^2} = 14\,000\ \text{cm}^{-1}$$

which is 13% low compared to the experimental value.

18.16 Write the secular determinant for hexatriene, and determine the energy states.

▮ For six electrons, *(14.71)* becomes

$$\begin{vmatrix} \alpha - E & \beta & 0 & 0 & 0 & 0 \\ \beta & \alpha - E & \beta & 0 & 0 & 0 \\ 0 & \beta & \alpha - E & \beta & 0 & 0 \\ 0 & 0 & \beta & \alpha - E & \beta & 0 \\ 0 & 0 & 0 & \beta & \alpha - E & \beta \\ 0 & 0 & 0 & 0 & \beta & \alpha - E \end{vmatrix} = 0$$

where the notation introduced in Problem 18.11 is used. Letting

$$x = (\alpha - E)/\beta \tag{18.12}$$

the determinant becomes

$$\begin{vmatrix} x & 1 & 0 & 0 & 0 & 0 \\ 1 & x & 1 & 0 & 0 & 0 \\ 0 & 1 & x & 1 & 0 & 0 \\ 0 & 0 & 1 & x & 1 & 0 \\ 0 & 0 & 0 & 1 & x & 1 \\ 0 & 0 & 0 & 0 & 1 & x \end{vmatrix} = 0$$

Expanding the determinant gives $x^6 - 5x^4 + 6x^2 = 1$. The general solutions for x for this type of determinant are given by

$$x_j = -2\cos\frac{j\pi}{n_C + 1} \tag{18.13}$$

where $j = 1, 2, \ldots, n_C$. Solving *(18.12)* for E and substituting *(18.13)* gives

$$E_j = \alpha + 2\beta\cos\frac{j\pi}{n_C + 1} \tag{18.14}$$

For hexatriene,

$$E_1 = \alpha + 2\beta\cos\frac{(1)\pi}{6 + 1} = \alpha + 1.802\beta$$

$$E_2 = \alpha + 2\beta\cos\frac{(2)\pi}{6 + 1} = \alpha + 1.247\beta$$

$E_3 = \alpha + 0.445\beta$, $E_4 = \alpha - 0.445\beta$, $E_5 = \alpha - 1.247\beta$, and $E_6 = \alpha - 1.802\beta$.

18.17 Write the Hückel molecular orbital wave functions for hexatriene.

▮ Assuming the wave functions to be in the form given by *(14.67)*,

$$\phi_\pi = c_1\chi_1 + c_2\chi_2 + c_3\chi_3 + c_4\chi_4 + c_5\chi_5 + c_6\chi_6$$

where χ_i represents the ψ_{2p_z} wave function for the carbon atoms, the coefficients can be determined using the general equation

$$c_{jr} = \left(\frac{2}{n_C + 1}\right)^{1/2}\sin\frac{jr\pi}{n_C + 1} \tag{18.15}$$

For ϕ_1, using *(18.15)* gives

$$c_1 = \left(\frac{2}{6 + 1}\right)^{1/2}\sin\frac{(1)(1)\pi}{6 + 1} = 0.232$$

$$c_2 = \left(\frac{2}{6 + 1}\right)^{1/2}\sin\frac{(2)(1)\pi}{6 + 1} = 0.418$$

$c_3 = 0.521$, $c_4 = 0.521$, $c_5 = 0.418$, and $c_6 = 0.232$, giving

$$\phi_1 = 0.232\chi_1 + 0.418\chi_2 + 0.521\chi_3 + 0.521\chi_4 + 0.418\chi_5 + 0.232\chi_6$$

Likewise,

$$\phi_2 = 0.418\chi_1 + 0.521\chi_2 + 0.232\chi_3 - 0.232\chi_4 - 0.521\chi_5 - 0.418\chi_6$$
$$\phi_3 = 0.521\chi_1 + 0.232\chi_2 - 0.418\chi_3 - 0.418\chi_4 + 0.232\chi_5 + 0.521\chi_6$$
$$\phi_4 = 0.521\chi_1 - 0.232\chi_2 - 0.418\chi_3 + 0.418\chi_4 + 0.232\chi_5 - 0.521\chi_6$$
$$\phi_5 = 0.418\chi_1 - 0.512\chi_2 + 0.232\chi_3 + 0.232\chi_4 - 0.521\chi_5 + 0.418\chi_6$$
$$\phi_6 = 0.232\chi_1 - 0.418\chi_2 + 0.521\chi_3 - 0.521\chi_4 + 0.418\chi_5 - 0.232\chi_6$$

As in Problem 18.11, ϕ_1, ϕ_2, and ϕ_3 are bonding orbitals and ϕ_4, ϕ_5, and ϕ_6 are antibonding orbitals.

18.18 Determine the total π-electron energy and the delocalization energy for hexatriene.

▌ The six π electrons are assigned to ϕ_1, ϕ_2, and ϕ_3 (see Problem 18.17). Substituting the corresponding expressions determined in Problem 18.16 gives the total π electron energy as

$$E_\pi = 2E_1 + 2E_2 + 2E_3 = 2(\alpha + 1.802\beta) + 2(\alpha + 1.247\beta) + 2(\alpha + 0.445\beta) = 6\alpha + 6.988\beta$$

The *delocalization energy* (DE) is the energy by which the molecule is stabilized relative to isolated double bonds. For hexatriene,

$$\text{DE} = E_\pi(\text{hexatriene}) - 3E_\pi(\text{ethylene}) = (6\alpha + 6.988\beta) - 3(2\alpha + 2\beta) = 0.988\beta$$

where E_π(ethylene) was determined in Problem 18.12. If β is assumed to be $-75\ \text{kJ}\cdot\text{mol}^{-1}$, then $\text{DE} = -74\ \text{kJ}\cdot\text{mol}^{-1}$ for hexatriene.

18.19 Determine $\bar{\nu}$ for the first excitation of an electron between the $n_\pi/2$ and $n_\pi/2 + 1$ states for hexatriene using *(18.14)*.

▌ The energy change is

$$\Delta E = E_{n_\pi/2+1} - E_{n_\pi/2} = \alpha + 2\beta\cos\frac{[(n_\pi/2) + 1]\pi}{n_\text{C} + 1} - \left(\alpha + 2\beta\cos\frac{(n_\pi/2) + \pi}{n_\text{C} + 1}\right)$$
$$= 2\beta\left(\cos\frac{(n_\text{C} + 2)\pi}{2(n_\text{C} + 1)} - \cos\frac{n_\text{C}\pi}{2(n_\text{C} + 1)}\right)$$
$$= 2\beta(-2)\sin\left(\frac{1}{2}\left(\frac{(n_\text{C} + 2)\pi}{2(n_\text{C} + 1)} + \frac{n_\text{C}\pi}{2(n_\text{C} + 1)}\right)\right)\sin\left(\frac{1}{2}\left(\frac{(n_\text{C} + 2)\pi}{2(n_\text{C} + 1)} - \frac{n_\text{C}\pi}{2(n_\text{C} + 1)}\right)\right)$$
$$= -4\beta\sin\left(\frac{1}{2}\left(\frac{(2n_\text{C} + 2)\pi}{2(n_\text{C} + 1)}\right)\right)\sin\left(\frac{1}{2}\left(\frac{2\pi}{2(n_\text{C} + 1)}\right)\right) = -4\beta(1)\sin\frac{\pi}{2n_\text{C} + 2}$$

where $n_\text{C} = n_\pi$. From *(13.32)*,

$$\bar{\nu} = \frac{-4\beta}{hc}\sin\frac{\pi}{2n_\text{C} + 2} = (149\,000\ \text{cm}^{-1})\sin\frac{\pi}{2n_\text{C} + 2} \qquad (18.16)$$

where β is determined empirically. Substituting into *(18.16)* gives

$$\bar{\nu} = (149\,000\ \text{cm}^{-1})\sin\frac{\pi}{2(6) + 2} = 33\,200\ \text{cm}^{-1}$$

This value is 11% lower than the observed value. (An improved Hückel theory can reduce the difference to 3%.)

18.20 According to the aromaticity rules, how is cyclobutadiene classified?

▌ To have the extra stability associated with aromatic compounds, the number of π electrons must be $n_\pi = 4m + 2$, where $m = 0, 1, 2, \ldots$. If $n_\pi = 4m \pm 1$, the compound is a free radical with a singlet ground state; and if $n_\pi = 4m$, the compound is a diradical with a triplet ground state. Cyclobutadiene has four π electrons and would be classified as a diradical.

18.21 Write the secular determinant for cyclobutadiene, and determine the energy states.

▌ For four electrons, *(14.71)* becomes

$$\begin{vmatrix} \alpha - E & \beta & 0 & \beta \\ \beta & \alpha - E & \beta & 0 \\ 0 & \beta & \alpha - E & \beta \\ \beta & 0 & \beta & \alpha - E \end{vmatrix} = 0$$

where the notation introduced in Problem 18.11 is used. Note the two extra β terms in this secular determinant compared to those written for noncyclic polyenes (see Problem 18.16). Substituting *(18.12)* gives

$$\begin{vmatrix} x & 1 & 0 & 1 \\ 1 & x & 1 & 0 \\ 0 & 1 & x & 1 \\ 1 & 0 & 1 & x \end{vmatrix} = 0$$

Expanding the determinant gives $x^4 - 4x^2 = 0$. The general solutions for x for this type of determinant are given by

$$x_k = -2\cos\frac{2\pi k}{n_C} \tag{18.17}$$

where $k = 0, 1, \ldots, n_C - 1$. Note that the index k does not correspond to the actual order of the molecular orbitals in terms of increasing order—the order usually follows the pattern $k = 0 < k = 1 = n_C - 1 < k = 2 = n_C - 2$, etc. Substituting into *(18.17)* gives

$$x_0 = -2\cos\frac{2\pi(0)}{4} = -2$$

$x_1 = 0$, $x_2 = 2$, and $x_3 = 0$. From *(18.12)*, $E_0 = \alpha + 2\beta$, $E_1 = \alpha$, $E_2 = \alpha - 2\beta$, and $E_3 = \alpha$.

18.22 Write the Hückel molecular orbital wave functions for cyclobutadiene.

▮ Assuming the wave functions to be in the form given by *(14.67)*,

$$\phi_\pi = c_1\chi_1 + c_2\chi_2 + c_3\chi_3 + c_4\chi_4$$

where χ_i represents the ψ_{2p_z} wave function for the carbon atoms, the coefficients are given by

$$c_{kr} = \frac{1}{n_C^{1/2}} e^{2\pi irk/n_C} \tag{18.18}$$

giving

$$\phi_k = \frac{1}{n_C^{1/2}} \sum_{r=1}^{n_C} e^{2\pi i(r-1)k/n_C}\chi_r \tag{18.19}$$

Substituting into *(18.19)* gives

$$\phi_0 = \frac{1}{4^{1/2}}[e^{2\pi i(1-1)(0)/4}\chi_1 + e^{2\pi i(2-1)(0)/4}\chi_2 + e^{2\pi i(3-1)(0)/4}\chi_3 + e^{2\pi i(4-1)(0)/4}\chi_4]$$

$$= \tfrac{1}{2}(\chi_1 + \chi_2 + \chi_3 + \chi_4) \tag{18.20a}$$

$$\phi_1 = \frac{1}{4^{1/2}}[e^{2\pi i(1-1)(1)/4}\chi_1 + e^{2\pi i(2-1)(1)/4}\chi_2 + e^{2\pi i(3-1)(1)/4}\chi_3 + e^{2\pi i(4-1)(1)/4}\chi_4]$$

$$= \tfrac{1}{2}(\chi_1 + e^{\pi i/2}\chi_2 + e^{\pi i}\chi_3 + e^{3\pi i/2}\chi_4) = \tfrac{1}{2}(\chi_1 + e^{\pi i/2}\chi_2 - \chi_3 + e^{3\pi i/2}\chi_4) \tag{18.20b}$$

where the equalities $e^{4\pi i} = e^{2\pi i} = 1$, $e^{3\pi i} = e^{\pi i} = -1$, etc., were used. Likewise,

$$\phi_2 = \tfrac{1}{2}(\chi_1 - \chi_2 + \chi_3 - \chi_4) \tag{18.20c}$$

$$\phi_3 = \tfrac{1}{2}(\chi_1 + e^{3\pi i/2}\chi_2 - \chi_3 + e^{\pi i/2}\chi_4) \tag{18.20d}$$

18.23 Often real forms of *(18.20)* are used. Find these wave functions.

▮ Because *(18.20a)* and *(18.20c)* are real, nothing needs to be done to generate real wave functions from them. Taking linear combinations of *(18.20b)* and *(18.20d)* as in Problem 15.7, *(15.6)* gives

$$\phi'_{\text{real}} = 2^{1/2}\frac{\phi_1 + \phi_3}{2} = 2^{1/2}\frac{1}{2}\left(\frac{(\chi_1 + e^{\pi i/2}\chi_2 - \chi_3 + e^{3\pi i/2}\chi_4) + (\chi_1 + e^{3\pi i/2}\chi_2 - \chi_3 + e^{\pi i/2}\chi_4)}{2}\right)$$

$$= \frac{1}{2^{1/2}}(\chi_1 - \chi_3)$$

$$\phi''_{\text{real}} = 2^{1/2}\frac{\phi_1 - \phi_3}{2i} = 2^{1/2}\frac{1}{2}\left(\frac{(\chi_1 + e^{\pi i/2}\chi_2 - \chi_3 + e^{3\pi i/2}\chi_4) - (\chi_1 + e^{3\pi i/2}\chi_2 - \chi_3 + e^{\pi i/2}\chi_4)}{2i}\right)$$

$$= \frac{1}{2^{1/2}}(\chi_2 + \chi_4)$$

18.24 Calculate the π-electron energy and the resonance energy for cyclobutadiene.

▌ There are four π electrons to be placed in the molecular orbitals determined in Problem 18.23. The ground state is ϕ_0, and two electrons will have the energy given by E_0 (see Problem 18.21). The remaining two electrons will be assigned to the doubly degenerate level given by ϕ_1 and ϕ_3 and will have the energy given by E_1 and E_3. The π-electron energy is

$$E_\pi = 2E_0 + E_1 + E_3 = 2(\alpha + 2\beta) + \alpha + \alpha = 4\alpha + 4\beta$$

Substituting the energy equivalent to two ethylene molecules gives the delocalization energy as

$$\text{DE} = E_\pi(\text{cyclobutadiene}) - 2E_\pi(\text{ethylene}) = (4\alpha + 4\beta) - 2(2\alpha + 2\beta) = 0$$

18.25 Prepare a molecular orbital sketch of cyclobutadiene showing the π-electron energy and the assignment of the electrons.

▌ A mnemonic device is available to determine the energies of the Hückel molecular orbitals for C_nH_n. A vertex of the regular polygon of $n = n_C$ sides inscribed in a circle of radius 2β is placed at the bottom of the sketch, and an energy state is drawn beside it. Additional energy states are drawn corresponding to the other corners of the polygon, creating n energy states. The value of α in the diagram corresponds to the center of the polygon, and the value of $\alpha + 2\beta$ corresponds to the lowest energy level. The square for cyclobutadiene is shown in Fig. 18-4. Note that the electronic configuration shown in the figure agrees with the result predicted in Problem 18.20.

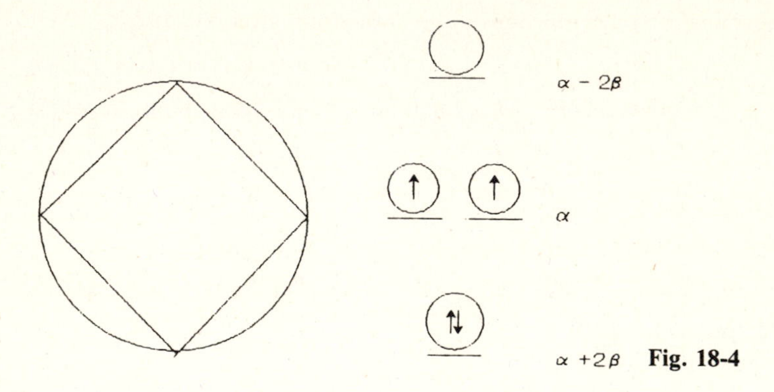

Fig. 18-4

18.26 Consider the π electrons in cyclobutadiene to be particles in a two-dimensional square potential energy well with infinite walls. Predict the absorption band, assuming the length of a side to be 0.135 nm.

▌ The expression for the energy states of a square potential energy well is

$$E = \frac{h^2}{8ma^2}(n_x^2 + n_y^2) \tag{18.21}$$

The assignment of the four π electrons is two electrons in $n_x = 1,\ n_y = 1,$ and one π electron in each of $n_x = 2,\ n_y = 1$ and $n_x = 1,\ n_y = 2.$ The lowest unoccupied state is $n_x = 2,\ n_y = 2.$ Taking the energy difference given by *(18.21)* and writing this difference in terms of $\bar{\nu}$ gives

$$\bar{\nu} = \frac{h}{8ma^2c}[(n_{x,2}^2 + n_{y,2}^2) - (n_{x,1}^2 + n_{y,1}^2)]$$

$$= \frac{(6.626 \times 10^{-34}\ \text{J}\cdot\text{s})[(1\ \text{kg}\cdot\text{m}^2\cdot\text{s}^{-2})/(1\ \text{J})][(10^{-2}\ \text{m})/(1\ \text{cm})]}{8(9.11 \times 10^{-31}\ \text{kg})(0.135 \times 10^{-9}\ \text{m})^2(2.9979 \times 10^8\ \text{m}\cdot\text{s}^{-1})}[(2^2 + 2^2) - (2^2 + 1^2)] = 499\ 000\ \text{cm}^{-1}$$

18.27 Consider the π electrons in cyclobutadiene to be particles on a circle. Predict the absorption band, assuming the radius of the circle to be 0.095 nm.

▌ The assignment of the four π electrons is two electrons in the $n = 0$ state with the remaining two electrons assigned to the doubly degenerate $n = 1$ level. The lowest unoccupied state is $n = 2.$ Taking the energy difference given by *(14.53)* and writing this difference in terms of $\bar{\nu}$ gives

$$\bar{\nu} = \frac{h}{8\pi^2mr^2c}(n_2^2 - n_1^2) = \frac{(6.626 \times 10^{-34}\ \text{J}\cdot\text{s})[(1\ \text{kg}\cdot\text{m}^2\cdot\text{s}^{-2})/(1\ \text{J})][(10^{-2}\ \text{m})/(1\ \text{cm})]}{8\pi^2(9.11 \times 10^{-31}\ \text{kg})(0.095 \times 10^{-9}\ \text{m})^2(2.9979 \times 10^8\ \text{m}\cdot\text{s}^{-1})}(2^2 - 1^2) = 102\ 000\ \text{cm}^{-1}$$

18.28 Determine the π-electron bond order, the total bond order, and the bond length for the 1-2 bonds in 1,3,5-hexatriene.

▮ The *total bond order* (P_{rs}) between two atoms, r and s, including a bond order of 1 for the σ bond, is given by

$$P_{rs} = 1 + p_{rs} \tag{18.22}$$

where the *π-electron bond order* (p_{rs}) is given by

$$p_{rs} = \sum_j \frac{n_j}{2}(c_{jr}^* c_{js} + c_{js}^* c_{jr}) \tag{18.23}$$

The sum is over all the π molecular orbitals, and n_j represents the number of electrons in the jth orbital. The values of c_{jr} and c_{js} are given by *(18.15)* for chain molecules and by *(18.18)* for cyclic molecules. Substituting the values for c_{j_1} and c_{j_2} determined in Problem 18.17 into *(18.23)* gives

$$\begin{aligned} p_{12} &= \frac{2}{2}[(0.232)(0.418) + (0.418)(0.232)] + \frac{2}{2}[(0.418)(0.521) + (0.521)(0.418)] \\ &\quad + \frac{2}{2}[(0.521)(0.232) + (0.232)(0.521)] + \frac{0}{2}[(0.521)(-0.232) + (-0.232)(0.521)] \\ &\quad + \frac{0}{2}[(0.418)(-0.521) + (-0.521)(0.418)] + \frac{0}{2}[(0.232)(-0.418) + (-0.418)(0.232)] \\ &= 0.871 \end{aligned}$$

Substituting into *(18.22)* gives $P_{12} = 1 + 0.871 = 1.871$. The bond length for conjugated π bonds is given by

$$r_{rs}/(\text{nm}) = 0.1707 - 0.0186 P_{rs} \tag{18.24}$$

Substituting into *(18.24)* gives

$$r_{rs}/(\text{nm}) = 0.1707 - (0.0186)(1.871) = 0.1359$$

18.3 COORDINATION COMPOUNDS

18.29 Prepare a diagram showing the configuration of $Mn(H_2O)_6^{2+}$, and predict the number of unpaired electrons.

▮ The electronic configuration of Mn is $(1s)^2(2s)^2(2p)^6(3s)^2(3p)^6(4s)^2(3d)^5$. The Mn^{2+} ion is formed by the removal of the two 4s electrons. The six pairs of electrons from the water molecules, a weak ligand, simply fill the 4s, 4p(3), and 4d(2) orbitals, producing an outer-orbit complex with five unpaired electrons. The configurations are summarized in Fig. 18-5.

	3d					4s	4p			4d				
Mn	↑	↑	↑	↑	↑	↑↓								
Mn^{2+}	↑	↑	↑	↑	↑									
$Mn(H_2O)_6^{2+}$	↑	↑	↑	↑	↑	xx	xx	xx	xx	xx	xx			
$Mn(SCN)_6^{4-}$	↑↓	↑↓	↑	xx	xx	xx	xx	xx	xx					

Fig. 18-5

18.30 Repeat Problem 18.29 for $Mn(SCN)_6^{4-}$.

▮ The six pairs of electrons from the thiocyanate ions, a strong ligand, pair the 3d electrons and fill the 3d(2), 4s, and 4p(3) orbitals, producing an inner-orbit complex with one unpaired electron. The configurations are summarized in Fig. 18-5.

18.31 Using the crystal field splitting diagram shown in Fig. 18-6, write the electronic configuration for $Mn(H_2O)_6^{2+}$ and determine the crystal field stabilization energy.

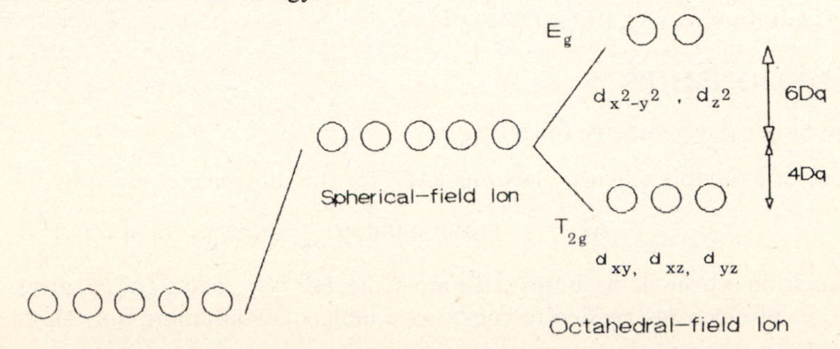

Fig. 18-6

▮ Because H_2O is a weak ligand, the value of $10Dq$ (the magnitude of the splitting in terms of the field strength Dq) is not large, and the two sets of orbitals are close to each other, resulting in a high-spin complex (see Problem 18.29). The configuration is $(T_{2g})^3(E_g)^2$, giving the crystal field stabilization energy (CFSE) as

$$\text{CFSE} = 3(-4Dq) + 2(6Dq) = 0$$

18.32 Repeat Problem 18.31 for $Mn(SCN)_6^{4-}$.

▮ Because the SCN^- ion is a strong ligand, the value of $10Dq$ is large and the two sets of orbitals are separated sufficiently that a low-spin complex results (see Problem 18.30). The configuration is $(T_{2g})^5$, giving

$$\text{CFSE} = 5(-4Dq) + 2P = -20Dq + 2P$$

where P is the energy required to form a pair of electrons.

18.33 The value of P for $Mn(H_2O)_6^{2+}$ is 25 500 cm^{-1}. What must be the minimum value of $10Dq$ in order for a low-spin complex to form? If $10Dq = 7800\ cm^{-1}$ for $Mn(H_2O)_6^{2+}$, will $Mn(H_2O)_6^{2+}$ be a low-spin or a high-spin complex?

▮ Setting the result of Problem 18.32 equal to zero, substituting the data, and solving for the minimum value of $10Dq$ needed to form the low-spin complex gives

$$0 = -20Dq + 2P = -2(10Dq) + 2P = -2(10Dq) + 2(25\,500\ cm^{-1})$$

$$10Dq = 25\,500\ cm^{-1}$$

Because the actual value of $10Dq$ is less than 25 500 cm^{-1}, a high-spin complex will form.

18.34 Using the molecular orbital diagram shown in Fig. 18-7, write the configuration for $Mn(H_2O)_6^{2+}$.

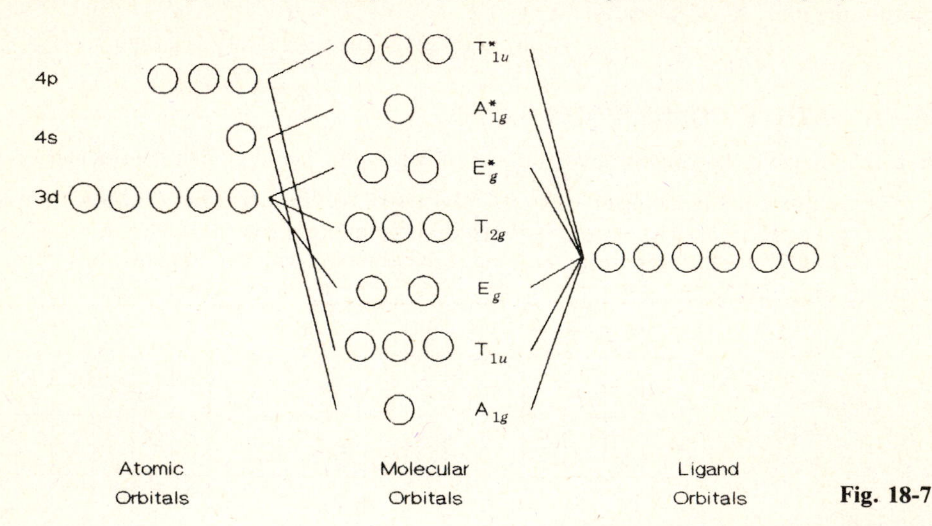

Fig. 18-7

▮ The six pairs of electrons from the ligands fill the A_{1g}, T_{1u}, and E_g orbitals. Because $Mn(H_2O)_6^{2+}$ is a high-spin complex (see Problem 18.29), the five electrons from the Mn^{2+} ion occupy each of the T_{2g} and E_g^* orbitals, giving the configuration as $(A_{1g})^2(T_{1u})^6(E_g)^4(T_{2g})^3(E_g^*)^2$.

18.35 Repeat Problem 18.34 for $Mn(SCN)_6^{4-}$.

▮ Because $Mn(SCN)_6^{4-}$ is a low-spin complex, the five electrons from the Mn^{2+} ion occupy the T_{2g} orbitals, giving the configuration as $(A_{1g})^2(T_{1u})^6(E_g)^4(T_{2g})^5$.

18.4 SPATIAL RELATIONSHIPS

18.36 Determine the molecular geometry for $NF_3(g)$.

▮ The number of available valence electrons (AE) for the molecule is given by

$$\text{AE} = \sum_x (\text{group number})_x - (\text{charge on species}) \tag{18.25}$$

where the summation is over all the atoms in the molecule. For NF_3, using *(18.25)* gives $\text{AE} = 5 + 7 + 7 + 7 - 0 = 26$. The number of electrons needed to construct a molecule containing only simple bonds (NE) is given by

$$\text{NE} = 2(\text{number of H atoms}) + 8(\text{number of non-H atoms}) - 2(\text{number of atoms} - 1) \tag{18.26}$$

where the factors of 2 and 8 represent the number of electrons needed to complete the "octets" of H and non-H atoms, respectively. For NF_3, using *(18.25)* gives

$$\mathrm{NE} = 2(0) + 8(4) - 2(4 - 1) = 26$$

Comparing the values of AE and NE shows that this is a "saturated" molecule (see Fig. 18-8), and the following Lewis structure can be drawn:

:F̤̈—N̈—F̤̈:
|
:F̤̈:

in which all octets are satisfied and all bonds are single bonds. The structure number of each atom in the molecule is next calculated using

$$\mathrm{SN}_x = (\text{number of bonds})_x + \mathrm{LP}_x \qquad (18.27)$$

where LP represents the number of lone pairs or nonbonding pairs of electrons. Because the arrangement of electrons around all three fluorine atoms in the Lewis structure is the same, only one value of SN(F) and the value of SN(N) need to be calculated. Using *(18.27)* gives

$$\mathrm{SN(F)} = 1 + 3 = 4 \qquad \mathrm{SN(N)} = 3 + 1 = 4$$

From the values of the structure numbers, the respective geometric shape for each atom in the molecule is found from Table 18-1 and a three-dimensional sketch for the molecule can be prepared (see Fig. 18-9a). As seen in Table 18-1, even though the nitrogen atom is considered to be sp^3 hybridized and to have a tetrahedral electronic

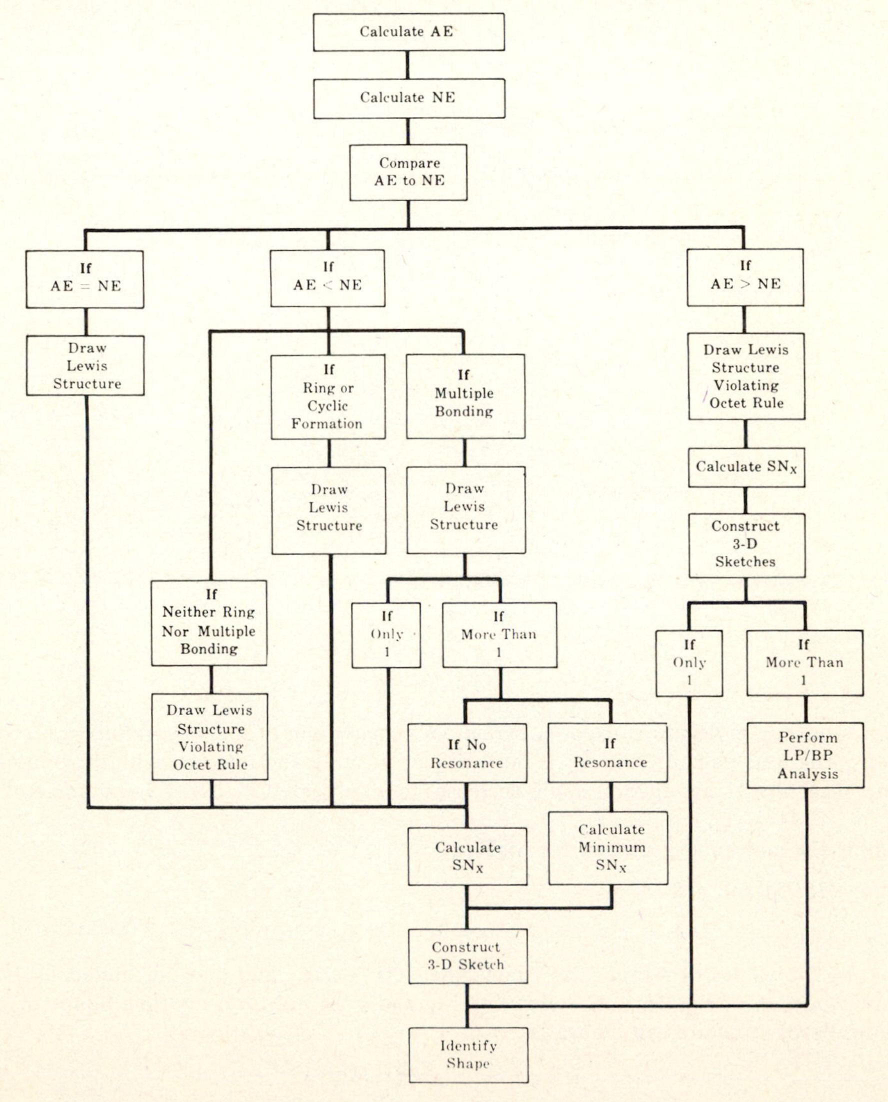

Fig. 18-8

Table 18-1

Structure Number SN_X	Geometric Shape for X	Hybridization	Sketch of Shape	Molecular Shape Possibilities	
1	Spherical	—			
2	Linear	sp		LP = 0, BP = 2 LP = 1, BP = 1	linear linear
3	Triangular	sp^2		LP = 0, BP = 3 LP = 1, BP = 2 LP = 2, BP = 1	triangular bent (angle ~ 120°) linear
4	Tetrahedral	sp^3		LP = 0, BP = 4 LP = 1, BP = 3 LP = 2, BP = 2 LP = 3, BP = 1	tetrahedral trigonal pyramidal bent (angle ~ 109°) linear
4	Square planar	dsp^2 or sp^2d		LP = 0, BP = 4	square planar
5	Trigonal bipyramidal	dsp^3 or sp^3d		LP = 0, BP = 5 LP = 1, BP = 4 LP = 2, BP = 3 LP = 3, BP = 2 LP = 4, BP = 1	trigonal bipyramidal seesaw T-shaped linear linear
6	Octahedral	d^2sp^3 or sp^3d^2		LP = 0, BP = 6 LP = 1, BP = 5 LP = 2, BP = 4	octahedral square pyramidal square planar
7	Pentagonal bipyramidal	...			
8	Cubic	...			
8	Square antiprismal	...			

Source: After C. R. Metz, "Molecular Geometry and Bonding," STRC-352, *Modular Laboratory Program*, Chemical Education Resources, Inc., Palmyra, PA, 1988.

configuration, the molecule is trigonal pyramidal because one of the sp^3 orbitals contains a nonbonded pair of electrons. The lone pair of electrons is more diffuse in space and interacts with the bonding pairs of electrons so that the usual 109.47° tetrahedral angle decreases to an observed F—N—F bond angle of 102.9°.

18.37 Determine the molecular geometry for $MgF_2(g)$.

▌ Using *(18.25)* and *(18.26)* gives

$$AE = 2 + 7 + 7 - 0 = 16 \qquad NE = 2(0) + 8(3) - 2(3 - 1) = 20$$

Comparing the values of AE and NE shows that $AE < NE$ and that this molecule should be analyzed using the center portion of Fig. 18-8. The elements Mg and F do not form multiple bonds or cyclic structures, so the following Lewis structure can be drawn:

$$:\ddot{\underset{..}{F}}—Mg—\ddot{\underset{..}{F}}:$$

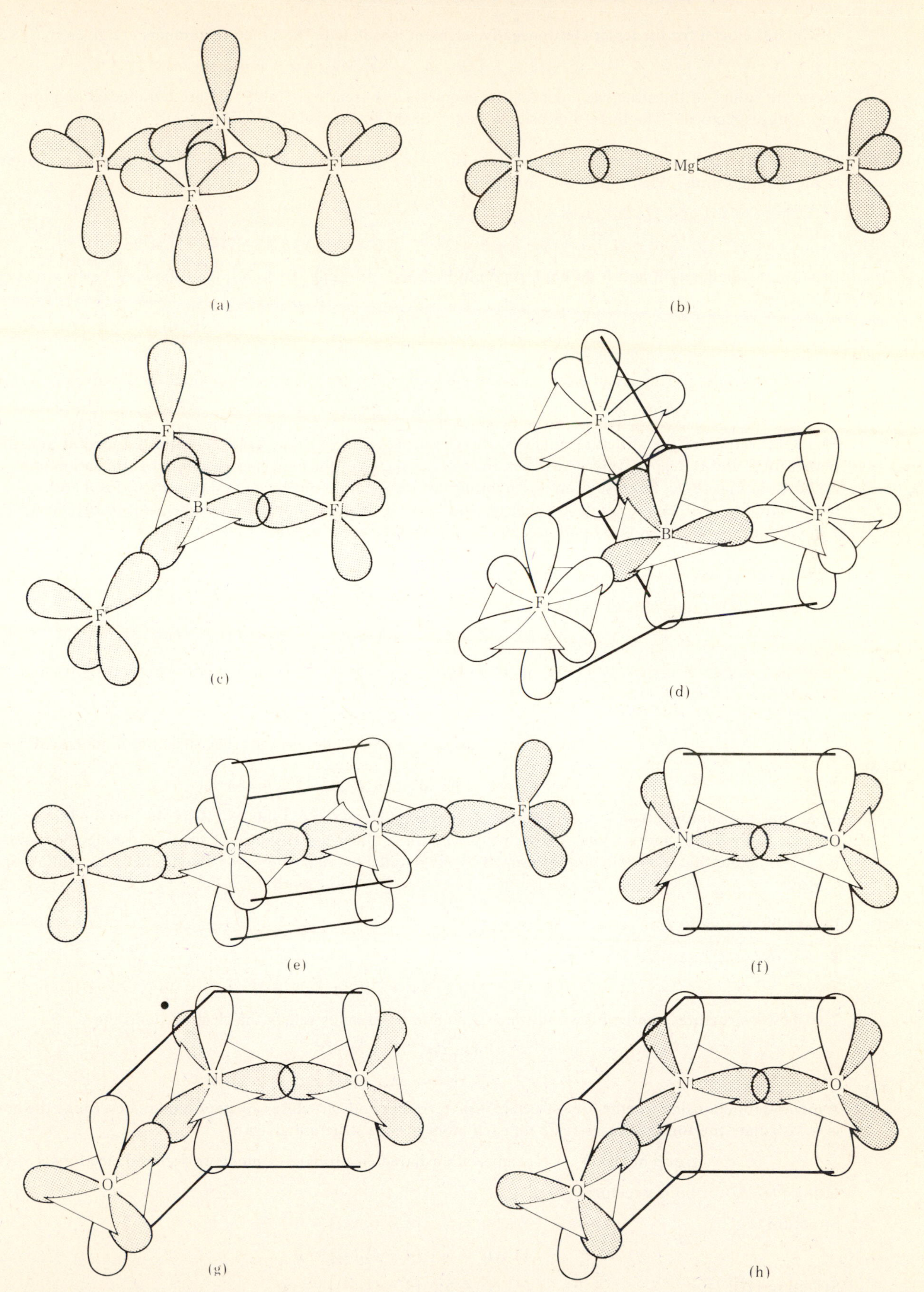

Fig. 18-9

in which the octet for the lesser electronegative element is violated. The structure numbers are given by *(18.27)* as

$$SN(F) = 1 + 3 = 4 \qquad SN(Mg) = 2 + 0 = 2$$

From the values of the structure numbers, the geometries chosen from Table 18-1 are tetrahedral and linear for F and Mg, respectively. The three-dimensional sketch of the linear molecule is shown in Fig. 18-9b.

18.38 Determine the molecular geometry for $BF_3(g)$. The observed B—F bond length in this compound is shorter than in most compounds. What might this observation imply?

▌ Using *(18.25)* and *(18.26)* gives

$$AE = 3 + 7 + 7 + 7 - 0 = 24 \qquad NE = 2(0) + 8(4) - 2(4 - 1) = 26$$

Because the elements B and F do not form multiple bonds or cyclic structures, the following Lewis structure can be drawn:

```
:F̤̈—B—F̤̈:
     |
    :F̤:
```

in which the octet for the lesser electronegative element is violated. The structure numbers are given by *(18.27)* as

$$SN(F) = 1 + 3 = 4 \qquad SN(B) = 3 + 0 = 3$$

From the values of the structure numbers, the geometries chosen from Table 18-1 are tetrahedral and trigonal planar for F and B, respectively. The three-dimensional sketch of the molecule assuming hybridization of all atoms is shown in Fig. 18-9c. Because of the sp^2 hybridization of B atom, the $2p_z$ orbital is unhybridized and is available for π bonding if other p orbitals are aligned parallel to it. If the F atoms are assumed not to be hybridized, the π bonding shown in Fig. 18-9d will account for the smaller bond length.

18.39 Determine the molecular geometry for $C_2F_2(g)$.

▌ Using *(18.25)* and *(18.26)* gives

$$AE = 7 + 4 + 4 + 7 - 0 = 22 \qquad NE = 2(0) + 8(4) - 2(4 - 1) = 26$$

Because C is one of the elements that can form multiple bonds (C, N, O, S), the following Lewis structure can be drawn:

:F̤̈—C≡C—F̤̈:

in which the six electrons in the triple bond satisfy the octets of the C atoms. The structure numbers are given by *(18.27)* as

$$SN(F) = 1 + 3 = 4 \qquad SN(C) = 2 + 0 = 2$$

Note that the multiple bond is considered as one bond in *(18.27)*. Table 18-1 lists the tetrahedral and linear geometries for the F and C atoms, respectively. The three-dimensional sketch of the linear molecule is shown in Fig. 18-9e. Note that the triple bond is formed by the parallel overlap of the two unhybridized 2p orbitals on each C atom.

18.40 Determine the molecular geometry for NO(g).

▌ Using *(18.25)* and *(18.26)* gives

$$AE = 5 + 6 - 0 = 11 \qquad NE = 2(0) + 8(2) - 2(2 - 1) = 14$$

Two Lewis structures can be drawn in which a double bond helps satisfy the electron deficiency:

·N̈=Ö:	:N̈=Ö·
(I)	(II)

A method for choosing whether one structure is preferred or both structures are important in describing the bonding is to calculate the formal charge (FC) for each atom is each structure given by

$$FC_x = (\text{group number})_x - [(\text{number of unshared electrons}) + \tfrac{1}{2}(\text{number of shared electrons})]_x \qquad (18.28)$$

Using *(18.28)* for both structures gives

Structure (I): $FC(N) = 5 - [3 + (\frac{1}{2})(4)] = 0$

$FC(O) = 6 - [4 + (\frac{1}{2})(4)] = 0$

Structure (II): $FC(N) = 5 - [4 + (\frac{1}{2})(4)] = -1$

$FC(O) = 6 - [3 + (\frac{1}{2})(4)] = +1$

Structures for which the sum of the absolute values of FC_x is small are usually the important structures to consider, and those with higher formal charge are discarded. In this case, structure I is chosen (as could have been predicted on the basis of electronegativity), and using *(18.27)* gives

$$\mathrm{SN(N)} = 1 + 2 = 3 \qquad \mathrm{SN(O)} = 1 + 2 = 3$$

Note that the single electron is considered a lone pair in *(18.27)*. Table 18-1 lists linear geometries for both atoms, and the three-dimensional sketch of the linear molecule is shown in Fig. 18-9f. Note that the double bond is formed by the parallel overlap of the unhybridized $2p_z$ orbital on each atom.

18.41 Determine the molecular geometry for $NO_2(g)$. How does the nitrogen–oxygen bond length compare to the lengths for a single bond and a double bond?

▌ Using *(18.25)* and *(18.26)* gives

$$\mathrm{AE} = 5 + 6 + 6 - 0 = 17 \qquad \mathrm{NE} = 2(0) + 8(3) - 2(3 - 1) = 20$$

Two Lewis structures can be drawn in which a double bond helps satisfy the electron deficiency:

:Ö=Ṅ—Ö̤:   :Ö̤—Ṅ=Ö:
(I)   (II)

A formal charge analysis indicates that both structures should be considered in describing the bonding. This is indicated by writing the contributing "resonance hybrid" structures separated by a double-headed arrow:

:Ö=Ṅ—Ö̤: ↔ :Ö̤—Ṅ=Ö:
(I)   (II)

Using *(18.27)* gives

Structure I:	SN(O) = 1 + 2 = 3	Structure II:	SN(O) = 1 + 3 = 4
	SN(N) = 2 + 1 = 3		SN(N) = 2 + 1 = 3
	SN(O) = 1 + 3 = 4		SN(O) = 1 + 2 = 3

In the case of "resonance," the minimum structure number is chosen for each atom, giving SN(O) = 3, SN(N) = 3, and SN(O) = 3. Table 18-1 lists sp^2 hybridization for each of these atoms, and the three-dimensional sketch for the "bent" molecule is shown in Fig. 18-9g. Note that the sketch shows delocalized π bonding accounting for the "resonance." Each bond between the N and O atoms should be shorter than a single N—O bond but longer than a double N=O bond. This prediction is confirmed by an observed bond length of 0.1197 nm compared to 0.146 nm for N—O and 0.115 nm for N=O. Because a single electron occupies one of the orbitals on N, the O—N—O bond angle is predicted to be greater than 120°. The observed bond angle is 134.25°.

18.42 Determine the molecular geometry for $NO_2^-(g)$. How does the bond angle compare to that for NO_2?

▌ The analysis for NO_2^- (AE = 18, NE = 20) is identical to that for NO_2 given in Problem 18.41 except that there are a pair of unshared electrons associated with the N atom instead of a single unpaired electron. The three-dimensional sketch for the "bent" molecule is shown in Fig. 18-9h. Because a pair of unshared electrons now occupy the orbital on the N atom, the O—N—O bond angle is predicted to be less than 120°. The observed bond angle is 115°.

18.43 Determine the molecular geometry for IF_3.

▌ Using *(18.25)* and *(18.26)* gives

$$\mathrm{AE} = 7 + 7 + 7 + 7 - 0 = 28 \qquad \mathrm{NE} = 2(0) + 8(4) - 2(4 - 1) = 26$$

Comparing the values of AE and NE shows that AE > NE and that this molecule should be analyzed using the right-hand portion of Fig. 18-8. The following Lewis structure can be drawn:

:F̤̈—Ï—F̤̈:
  |
 :F̤̈:

in which the octet for the central atom is violated. The structure numbers are given by *(18.27)* as

$$\mathrm{SN(F)} = 1 + 3 = 4 \qquad \mathrm{SN(I)} = 3 + 2 = 5$$

From the values of the structure numbers, the geometries chosen from Table 18-1 are tetrahedral and trigonal bipyramidal for F and I, respectively. As can be seen in Figs. 18-10a, b, and c, there are three possible three-dimensional sketches for this molecule. Structure (b) is eliminated because of the highly unfavorable lone pair–lone

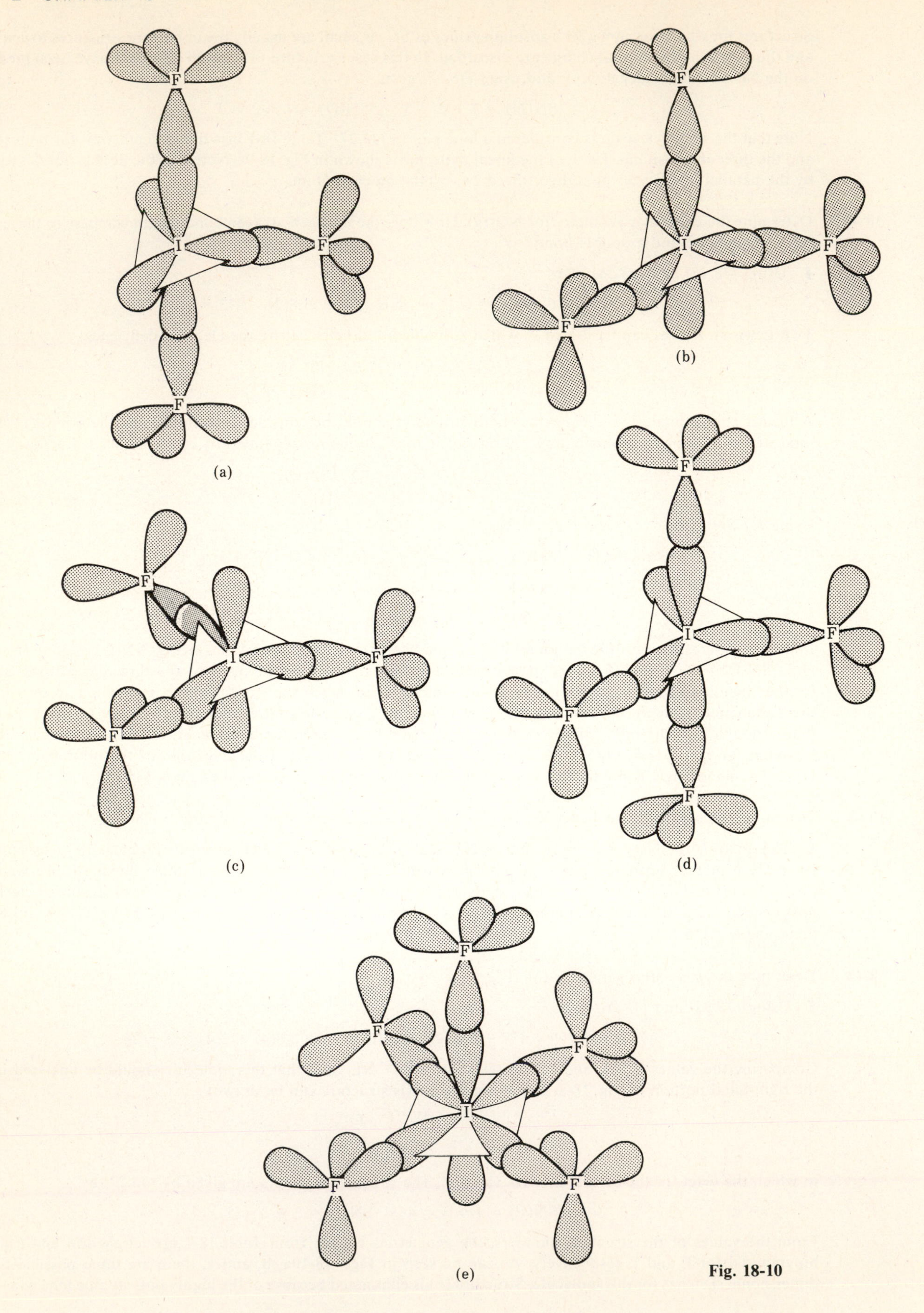

Fig. 18-10

pair electron interactions at 90° in the molecule that the other two structures do not contain. Structure (c) is eliminated on the basis of the six unfavorable lone pair–bonded pair electron-pair interactions at 90° in the molecule compared to only four similar interactions in structure (a). The predicted shape for the molecule is "T-shaped," which agrees with the entry in Table 18-1 for LP = 2, BP = 3. The lone pair–bond pair interactions are strong enough in this molecule to make the F—I—F bond angles slightly less than 90°.

18.44 Determine the molecular geometry for $IF_4^+(g)$.

▮ Using *(18.25)* and *(18.26)* gives

$$AE = 7 + 7 + 7 + 7 + 7 - 1 = 34$$

$$NE = 2(0) + 8(5) - 2(5 - 1) = 32$$

Violating the octet rule for the central atom, the following Lewis structure can be written:

```
:F̤̈—Ï—F̤̈:
    / \
 :F̤̈:  :F̤̈:
```

The structure numbers are given by *(18.27)* as

$$SN(F) = 1 + 3 = 4 \qquad SN(I) = 4 + 1 = 5$$

Choosing the tetrahedral geometry for the F atoms and the "seesaw" geometry (corresponding to LP = 1, BP = 4 as determined by a lone pair–bond pair analysis) for the I atom, the three-dimensional sketch shown in Fig. 18-10d can be drawn.

18.45 Determine the molecular geometry for $IF_5(g)$.

▮ Using *(18.25)* and *(18.26)* gives

$$AE = 7 + 7 + 7 + 7 + 7 + 7 - 0 = 42$$

$$NE = 2(0) + 8(6) - 2(6 - 1) = 38$$

Violating the octet rule for the central atom, the following Lewis structure can be written:

```
:F̤̈—Ï—F̤̈:
    /|\
 :F̤̈ | F̤̈:
    :F̤̈:
```

The structure numbers are given by *(18.27)* as

$$SN(F) = 1 + 3 = 4 \qquad SN(I) = 5 + 1 = 6$$

Choosing the sp^3 hybridization for the F atom and the sp^3d^2 hybridization for the I atom, the three-dimensional sketch shown in Fig. 18-10e can be drawn. No lone pair–bond pair analysis need be performed for this square pyramidal molecule.

CHAPTER 19

Spectroscopy of Polyatomic Molecules

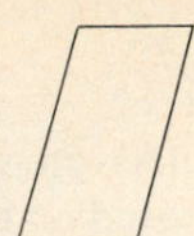

19.1 ROTATIONAL SPECTRA

19.1 Determine the coordinates of the atoms in $N_2O(g)$ with respect to the center of the mass of the molecule. The masses of the atoms are $m/(10^{-26}\text{ kg}) = 2.3253$ for ^{14}N and 2.6561 for ^{16}O, and the bond lengths are 0.112 82 nm for N—N and 0.118 42 nm for N—O.

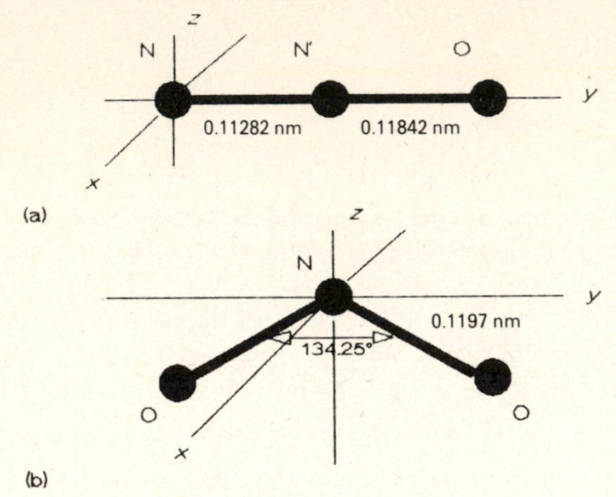

Fig. 19-1

The coordinates (in nanometers) of the atoms with respect to the coordinate system shown in Fig. 19-1a are N (0, 0, 0), N′ (0, 0.112 82, 0), and O (0, 0.231 24, 0). The coordinates of the center of mass with respect to an arbitrary Cartesian coordinate system are given by

$$x_{\text{cm}} = \frac{\sum_{i=1}^{n} m_i x_i}{\sum_{i=1}^{n} m_i} \tag{19.1a}$$

$$y_{\text{cm}} = \frac{\sum_{i=1}^{n} m_i y_i}{\sum_{i=1}^{n} m_i} \tag{19.1b}$$

$$z_{\text{cm}} = \frac{\sum_{i=1}^{n} m_i z_i}{\sum_{i=1}^{n} m_i} \tag{19.1c}$$

where the summations are done over all the atoms in the molecule. Substituting the data into *(19.1)* gives

$$x_{\text{cm}} = \frac{(2.3253)(0) + (2.3253)(0) + (2.6561)(0)}{2(2.3253) + 2.6561} = 0$$

$$y_{\text{cm}} = \frac{(2.3253)(0) + (2.3253)(0.112\,82) + (2.6561)(0.231\,24)}{2(2.3253) + 2.6561} = 0.119\,96$$

$$z_{\text{cm}} = \frac{(2.3253)(0) + (2.3253)(0) + (2.6561)(0)}{2(2.3253) + 2.6561} = 0$$

where all coordinates are expressed in nanometers and all masses in units of 10^{-26} kg. The center of mass lies slightly to the right of the central N atom. The coordinates of the atoms with respect to the center of mass are

$$\text{N} \quad [(0 + 0), (0 - 0.119\,96), (0 + 0)] = (0, -0.119\,96, 0)$$

$$\text{N}' \quad [(0 + 0), (0.112\,82 - 0.119\,96), (0 + 0)] = (0, -0.007\,14, 0)$$

$$\text{O} \quad [(0 + 0), (0.231\,24 - 0.119\,96), (0 + 0)] = (0, -0.111\,28, 0)$$

19.2 Determine the coordinates of the atoms in $NO_2(g)$ with respect to the center of mass of the molecule. The masses of the atoms are $m/(10^{-26}\text{ kg}) = 2.3253$ for ^{14}N and 2.6561 for ^{16}O, the N—O bond length is 0.1197 nm, and the O—N—O bond angle is 134.25°.

▌ The coordinates (in nm) of the atoms with respect to the coordinate system shown in Fig. 19-1b are N (0, 0, 0), O (0, $-b$, $-c$), and O′ (0, b, $-c$), where

$$b = 0.1197 \sin (134.25°/2) = 0.1103$$

$$c = 0.1197 \cos (134.25°/2) = 0.0465$$

Substituting the data into *(19.1)* gives

$$x_{cm} = \frac{(2.6561)(0) + (2.6561)(0) + (2.3253)(0)}{2(2.6561) + 2.3253} = 0$$

$$y_{cm} = \frac{(2.6561)(-0.1103) + (2.6561)(0.1103) + (2.3253)(0)}{2(2.6561) + 2.3253} = 0$$

$$z_{cm} = \frac{(2.6561)(-0.0465) + (2.6561)(-0.0465) + (2.3253)(0)}{2(2.6561) + 2.3253} = 0$$

where all coordinates are expressed in nanometers and all masses in units of 10^{-26} kg. The center of mass lies below the N atom and slightly above the O atoms. The coordinates of the atoms with respect to the center of mass are

$$\text{N} \quad [(0+0), (0+0), (0-0.0323)] = (0, 0, -0.0323)$$

$$\text{O} \quad [(0+0), (-0.1103+0), (-0.0465+0.0323)] = (0, -0.1103, -0.0142)$$

$$\text{O}' \quad [(0+0), (0.1103+0), (-0.0465+0.0323)] = (0, 0.1103, -0.0142)$$

19.3 Determine the coordinates of the atoms in $CClF_3(g)$ with respect to the center of mass of the molecule. The masses of the atoms are $m/(10^{-26}\text{ kg}) = 3.1548$ for ^{19}F, 5.8068 for ^{35}Cl, and 1.9927 for ^{12}C, the C—Cl bond length is 0.1751 nm, the C—F bond lengths are 0.1328 nm, the Cl—C—F bond angle is 110.33°, and the F—C—F bond angles are 108.6°.

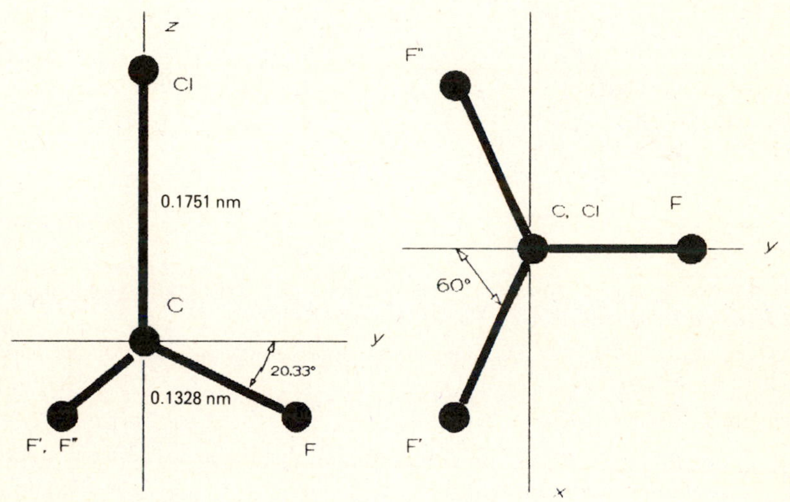

Fig. 19-2

▌ The coordinates (in nm) of the atoms with respect to the coordinate system shown in Fig. 19-2 are C (0, 0, 0), Cl (0, 0, 0.1751), F (0, b, $-c$), F′ (a, $-b/2$, $-c$), and F″ ($-a$, $-b/2$, $-c$), where

$$b = 0.1328 \cos 20.33° = 0.1245$$

$$c = 0.1328 \sin 20.33° = 0.0461$$

$$a = 0.1245 \sin 60° = 0.1078$$

Substituting the data into *(19.1)* gives

$$x_{cm} = \frac{(5.8068)(0) + (1.9927)(0) + (3.1548)(-0.1078) + (3.1548)(0.1078) + (3.1548)(0)}{5.8068 + 1.9927 + 3(3.1548)} = 0$$

$$y_{cm} = \frac{(5.8068)(0) + (1.9927)(0) + (3.1548)(-0.0623) + (3.1548)(-0.0623) + (3.1548)(0.1245)}{5.8068 + 1.9927 + 3(3.1548)} = 0$$

$$z_{cm} = \frac{(5.8068)(0.1751) + (1.9927)(0) + (3.1548)(-0.0461) + (3.1548)(-0.0461) + (3.1548)(-0.0461)}{5.8068 + 1.9927 + 3(3.1548)} = 0.0336$$

where all coordinates are expressed in nanometers and all masses in units of 10^{-26} kg. The center of mass lies slightly above the C atom along the C—Cl bond. The coordinates of the atoms with respect to the center of mass are

$$\begin{array}{lrl} \text{C} & [(0+0),(0+0),(0-0.0336)] & = (0,0,-0.0336) \\ \text{Cl} & [(0+0),(0+0),(0.1751-0.0336)] & = (0,0,0.1415) \\ \text{F} & [(0+0),(0.1245+0),(-0.0461-0.0336)] & = (0,0.1245,-0.0797) \\ \text{F}' & [(0.1078+0),(-0.0623+0),(-0.0461-0.0336)] & = (0.1078,-0.0623,-0.0797) \\ \text{F}'' & [(-0.1078+0),(-0.0623+0),(-0.0461+0.0336)] & = (-0.1078,-0.0623,-0.0797) \end{array}$$

19.4 Determine the coordinates of the atoms in $NF_3(g)$ with respect to the center of mass of the molecule. The masses of the atoms are $m/(10^{-26}\text{ kg}) = 3.1548$ for ^{19}F and 2.3253 for ^{14}N, and the F—N—F bond length is 102.15°.

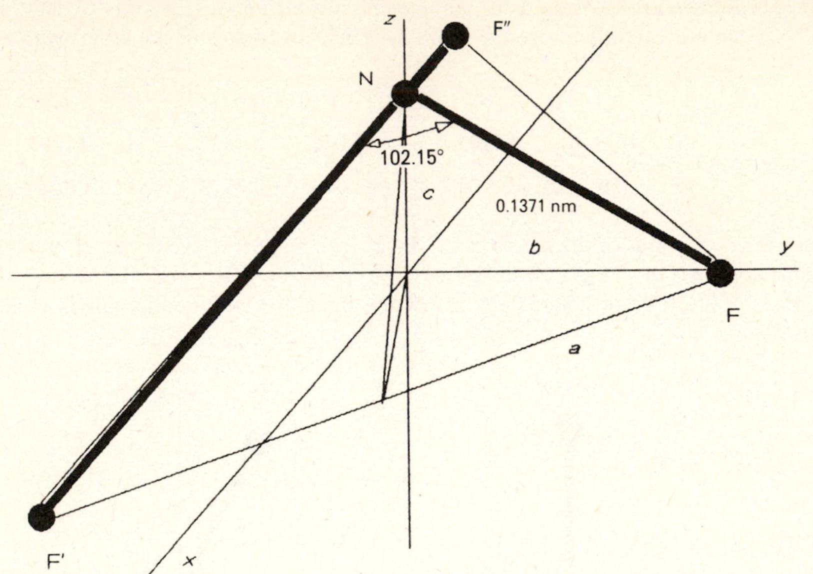

Fig. 19-3

▮ The coordinates (in nm) of the atoms with respect to the coordinate system shown in Fig. 19-3 are N $(0, 0, c)$, F $(0, b, 0)$, F'$(a, -b/2, 0)$, and F'' $(-a, -b/2, 0)$, where

$$a = 0.1371 \sin(102.15°/2) = 0.1067$$

$$b = 0.1067/(\cos 30°) = 0.1232$$

$$c = [(0.1371)^2 - (0.1232)^2]^{1/2} = 0.0602$$

Substituting the data into *(19.1)* gives

$$x_{\text{cm}} = \frac{(3.1548)(-0.1067) + (3.1548)(0.1067) + (3.1548)(0) + (2.3253)(0)}{3(3.1548) + 2.3253} = 0$$

$$y_{\text{cm}} = \frac{(3.1548)(-0.0616) + (3.1548)(-0.0616) + (3.1548)(0.1232) + (2.3253)(0)}{3(3.1548) + 2.3253} = 0$$

$$z_{\text{cm}} = \frac{(3.1548)(0) + (3.1548)(0) + (3.1548)(0) + (2.3253)(0.0602)}{3(3.1548) + 2.3253} = 0.0119$$

where all coordinates are expressed in nanometers and all masses in units of 10^{-26} kg. The center of mass lies slightly above the plane of F atoms. The coordinates of the atoms with respect to the center of mass are

$$\begin{array}{lrl} \text{N} & [(0+0),(0+0),(0.0602-0.0119)] & = (0,0,0.0483) \\ \text{F} & [(0+0),(0.1232+0),(0-0.0119)] & = (0,0.1232,-0.0119) \\ \text{F}' & [(0.1067+0),(-0.0616+0),(0-0.0119)] & = (0.1067,-0.0616,-0.0119) \\ \text{F}'' & [(-0.1067+0),(-0.0616+0),(0-0.0119)] & = (-0.1067,-0.0616,-0.0119) \end{array}$$

19.5 Determine the location of the center of mass for $SF_6(g)$, $CH_4(g)$, $CO_2(g)$.

▮ Because of the symmetry of these molecules, no calculations are needed to find the centers of mass. By inspection, the centers are located at the position of the S atoms in SF_6, the C atom in CH_4, and the C atom in CO_2.

19.6 The moment of inertia for $CH_4(g)$ was listed in a table as $I = 0.5313 \times 10^{-39}\ \mathrm{g \cdot cm^2}$. Express this value in the usual units of $\mathrm{kg \cdot m^2}$, Hz, and $\mathrm{cm^{-1}}$.

▮ Converting from the cgs units to the SI units gives

$$I = (0.5313 \times 10^{-39}\ \mathrm{g \cdot cm^2})\left(\frac{10^{-3}\ \mathrm{kg}}{1\ \mathrm{g}}\right)\left(\frac{10^{-2}\ \mathrm{m}}{1\ \mathrm{cm}}\right)^2 = 5.313 \times 10^{-47}\ \mathrm{kg \cdot m^2}$$

Letting B' represent the moment of inertia in spectroscopic units,

$$B'/(\mathrm{s^{-1}}) = h/8\pi^2 B \tag{19.2a}$$

$$B'/(\mathrm{cm^{-1}}) = (h \times 10^{-2})/8\pi^2 Bc \tag{19.2b}$$

where h, B, and c are expressed in the normal SI units. Substituting the data gives

$$B' = \frac{(6.626 \times 10^{-34}\ \mathrm{J \cdot s})[(1\ \mathrm{kg \cdot m^2 \cdot s^{-2}})/(1\ \mathrm{J})]}{8\pi^2(5.313 \times 10^{-47}\ \mathrm{kg \cdot m^2})} = 1.580 \times 10^{11}\ \mathrm{s^{-1}} = 1.580 \times 10^{11}\ \mathrm{Hz}$$

$$B' = \frac{(6.626 \times 10^{-34})(10^{-2})}{8\pi^2(5.313 \times 10^{-47})(2.9979 \times 10^8)}\ \mathrm{cm^{-1}} = 5.269\ \mathrm{cm^{-1}}$$

19.7 Calculate the principal moments of inertia for $N_2O(g)$, and classify this molecule.

▮ The *principal moments of inertia* (I_{xx}, I_{yy}, I_{zz}) are calculated with reference to the center of mass of the molecule and are given by

$$I_{xx} = \sum_{i=1}^{n} m_i(y_i^2 + z_i^2) \tag{19.3a}$$

$$I_{yy} = \sum_{i=1}^{n} m_i(x_i^2 + z_i^2) \tag{19.3b}$$

$$I_{zz} = \sum_{i=1}^{n} m_i(x_i^2 + y_i^2) \tag{19.3c}$$

Substituting the coordinates of the atoms determined in Problem 19.1 into *(19.3)* gives

$$I_{xx}/(10^{-26}\ \mathrm{kg \cdot nm^2}) = (2.3253)[(-0.1196)^2 + (0)^2] + (2.3253)[(-0.007\,14)^2 + (0)^2] + (2.6561)[(-0.1128)^2 + (0)^2] = 0.066\,472$$

$$I_{yy}/(10^{-26}\ \mathrm{kg \cdot nm^2}) = (2.3253)[0^2 + 0^2] + (2.3253)[0^2 + 0^2] + (2.6561)[0^2 + 0^2] = 0$$

$$I_{zz}/(10^{-26}\ \mathrm{kg \cdot nm^2}) = (2.3253)[0^2 + (-0.1196)^2] + (2.3253)[0^2 + (-0.007\,14)^2] + (2.6561)[0^2 + (0.111\,28)^2] = 0.066\,472$$

In normal SI units,

$$I_{xx} = I_{zz} = (0.066\,472 \times 10^{-26}\ \mathrm{kg \cdot nm^2})\left(\frac{10^{-9}\ \mathrm{m}}{1\ \mathrm{nm}}\right)^2 = 6.6472 \times 10^{-46}\ \mathrm{kg \cdot m^2}$$

and $I_{yy} = 0$. Using the spectroscopic notation, $C = B = I_{xx} = I_{zz}$ and $A = I_{yy} = 0$. Because $C = B > A$, the molecule is classified as a prolate symmetrical top.

19.8 For a linear triatomic molecule ABC, the moment of inertia is given by

$$I = \frac{m_A m_B r_{AB}^2 + m_B m_C r_{BC}^2 + m_A m_C (r_{AB} + r_{BC})^2}{m_A + m_B + m_C} \tag{19.4}$$

Using the data given in Problem 19.1, confirm the results of Problem 19.7.

▮ Substituting the data into *(19.4)* gives

$$I = \frac{\left[\begin{array}{l}(2.3253 \times 10^{-26}\ \mathrm{kg})(2.3253 \times 10^{-26}\ \mathrm{kg})(0.112\,82 \times 10^{-9}\ \mathrm{m})^2 \\ \quad + (2.3253 \times 10^{-26}\ \mathrm{kg})(2.6561 \times 10^{-26}\ \mathrm{kg})(0.118\,42 \times 10^{-9}\ \mathrm{m})^2 \\ \quad\quad + (2.3253 \times 10^{-26}\ \mathrm{kg})(2.6561 \times 10^{-26}\ \mathrm{kg})(0.112\,82 \times 10^{-9}\ \mathrm{m} + 0.118\,42 \times 10^{-9}\ \mathrm{m})^2\end{array}\right]}{2(2.3253 \times 10^{-26}\ \mathrm{kg}) + 2.652\,61 \times 10^{-26}\ \mathrm{kg}}$$

$$= 6.6503 \times 10^{-46}\ \mathrm{kg \cdot m^2}$$

The values differ by 0.05%.

19.9 Calculate the principal moments of inertia for $NO_2(g)$, and classify this molecule.

▌ Substituting the results of Problem 19.2 into *(19.3)* gives

$$I_{xx}/(10^{-26}\ \text{kg}\cdot\text{nm}^2) = (2.6561)[(-0.1103)^2 + (-0.0142)^2] + (2.6561)[(0.1103)^2 + (-0.0142)^2] + (2.3252)[(0)^2 + (0.0323)^2] = 0.068\ 12$$

$$I_{yy}/(10^{-26}\ \text{kg}\cdot\text{nm}^2) = (2.6561)[0^2 + (-0.1422)^2] + (2.6561)[0^2 + (-0.1422)^2] + (2.3253)[0^2 + (0.0323)^2] = 0.003\ 50$$

$$I_{zz}/(10^{-26}\ \text{kg}\cdot\text{nm}^2) = (2.6561)[0^2 + (-0.1103)^2] + (2.656)[0^2 + (0.1103)^2] + (2.3252)(0^2 + 0^2) = 0.064\ 63$$

Using the spectroscopic notation, $C = I_{xx} = 6.812 \times 10^{-46}\ \text{kg}\cdot\text{m}^2$, $B = I_{zz} = 6.463 \times 10^{-46}\ \text{kg}\cdot\text{m}^2$, and $A = I_{yy} = 3.50 \times 10^{-47}\ \text{kg}\cdot\text{m}^2$. Because $C > B > A$, this molecule is classified as an asymmetrical top.

19.10 Calculate the principal moments of inertia for $CClF_3(g)$, and classify this molecule.

▌ Substituting the results of Problem 19.3 into *(19.3)* gives

$$I_{xx}/(10^{-26}\ \text{kg}\cdot\text{nm}^2) = (5.8068)[0^2 + (0.1415)^2] + (1.9927)[0^2 + (-0.0336)^2] + (3.1548)[(-0.0623)^2 + (-0.0797)^2] + (3.1548)[(-0.0623)^2 + (-0.0797)^2] + (3.1548)[(0.1245)^2 + (-0.0797)^2] = 0.2520$$

$$I_{yy}/(10^{-26}\ \text{kg}\cdot\text{nm}^2) = (5.8068)[0^2 + (0.1415)^2] + (1.9927)[0^2 + (-0.0336)^2] + (3.1548)[(-0.1078)^2 + (-0.0797)^2] + (3.1548)[(0.1078)^2 + (-0.0797)^2] + (3.1548)[0^2 + (-0.0797)^2] = 0.2520$$

$$I_{zz}/(10^{-26}\ \text{kg}\cdot\text{nm}^2) = (5.8068)[0^2 + 0^2] + (1.9927)[0^2 + 0^2] + (3.1548)[(-0.1078)^2 + (-0.0623)^2] + (3.1548)[(0.1078)^2 + (-0.0623)^2] + (3.1548)[0^2 + (0.1245)^2] = 0.1467$$

Using the spectroscopic notation, $C = I_{xx} = 2.520 \times 10^{-45}\ \text{kg}\cdot\text{m}^2$, $B = I_{yy} = 2.520 \times 10^{-45}\ \text{kg}\cdot\text{m}^2$, and $A = I_{zz} = 1.467 \times 10^{-45}\ \text{kg}\cdot\text{m}^2$. Because $C = B > A$, this molecule is classified as a prolate symmetrical top.

19.11 Calculate the principal moments of inertia for $NF_3(g)$, and classify this molecule.

▌ Substituting the results of Problem 19.4 into *(19.3)* gives

$$I_{xx}/(10^{-26}\ \text{kg}\cdot\text{nm}^2) = (3.1548)[(-0.0616)^2 + (-0.0119)^2] + (3.1548)[(-0.0616)^2 + (-0.0119)^2] + (3.1548)[(0.1232)^2 + (-0.0119)^2] + (2.3253)[0^2 + (0.0483)^2] = 0.078\ 59$$

$$I_{yy}/(10^{-26}\ \text{kg}\cdot\text{nm}^2) = (3.1548)[(-0.1067)^2 + (-0.0119)^2] + (3.1548)[(0.1067)^2 + (-0.0119)^2] + (3.1548)[0^2 + (-0.0119)^2] + (2.3253)[0^2 + (0.0483)^2] = 0.078\ 60$$

$$I_{zz}/(10^{-26}\ \text{kg}\cdot\text{nm}^2) = (3.1548)[(-0.1067)^2 + (-0.0616)^2] + (3.1548)[(0.1067)^2 + (-0.0616)^2] + (3.1548)[0^2 + (0.1232)^2] + (2.3253)[0^2 + 0^2] = 0.1437$$

Using the spectroscopic notation, $C = I_{zz} = 1.437 \times 10^{-45}\ \text{kg}\cdot\text{m}^2$, $B = I_{xx} = 7.859 \times 10^{-46}\ \text{kg}\cdot\text{m}^2$, and $A = I_{yy} = 7.860 \times 10^{-46}\ \text{kg}\cdot\text{m}^2$. Because $C > B = A$, this molecule is classified as an oblate symmetrical top.

19.12 The moments of inertia in $CH_4(g)$ and $SF_6(g)$ are such that $C = B = A$. Classify these molecules.

▌ Molecules for which $C = B = A$ are classified as spherical tops.

19.13 The observed moment of inertia for $CO_2(g)$ is $B' = 0.391\ 514\ \text{cm}^{-1}$. Given the atomic masses as $m/(10^{-26}\ \text{kg}) = 2.656\ 06$ for ^{16}O and $1.992\ 68$ for ^{12}C, calculate the C=O bond length.

▌ Solving *(19.2b)* for B and substituting the data gives

$$B/(\text{kg}\cdot\text{m}^2) = \frac{(6.626\ 18 \times 10^{-34})(10^{-2})}{8\pi^2(0.391\ 514)(2.997\ 925 \times 10^8)} = 7.149\ 99 \times 10^{-46}$$

The coordinates of the two O atoms with respect to the center of mass are $(0, r, 0)$ and $(0, -r, 0)$. Substituting into *(19.3a)* gives

$$I_{xx} = (2.656\ 06 \times 10^{-26}\ \text{kg})[r^2 + 0^2] + (2.656\ 06 \times 10^{-26}\ \text{kg})[(-r)^2 + 0^2] + (1.992\ 68 \times 10^{-26}\ \text{kg})(0^2 + 0^2)$$
$$= (5.312\ 12 \times 10^{-26}\ \text{kg})r^2$$

Equating these values for the moment of inertia and solving for r gives

$$r = \left(\frac{7.149\,99 \times 10^{-46}\ \text{kg}\cdot\text{m}^2}{5.312\,12 \times 10^{-26}\ \text{kg}}\right)^{1/2} = 1.160\,16 \times 10^{-10}\ \text{m} = 0.116\,016\ \text{nm}$$

19.14 The observed moment of inertia for $CH_4(g)$ is $I_{xx} = I_{yy} = I_{zz} = 5.313 \times 10^{-47}\ \text{kg}\cdot\text{m}^2$. Given the atomic masses as $m/(10^{-26}\ \text{kg}) = 0.167\,36$ for ^1H and $1.992\,68$ for ^{12}C, calculate the C—H bond length.

▮ For a molecule AB_4 in the shape of a regular tetrahedron, the moment of inertia is given by

$$I = \tfrac{8}{3} m_B r^2 \tag{19.5}$$

Solving *(19.5)* for r and substituting the data gives

$$r = \left(\frac{5.313 \times 10^{-47}\ \text{kg}\cdot\text{m}^2}{(8/3)(0.167\,36 \times 10^{-26}\ \text{kg})}\right)^{1/2} = 1.0911 \times 10^{-10}\ \text{m} = 0.109\,11\ \text{nm}$$

19.15 A table lists $I_{xx}I_{yy}I_{zz} = 2.941\,168 \times 10^{-113}\ \text{g}^3\cdot\text{cm}^6$ for $SF_6(g)$. Given the atomic masses as $m/(10^{-26}\ \text{kg}) = 3.154\,81$ for ^{19}F and $5.309\,17$ for ^{32}S, calculate the S—F bond length.

▮ Because SF_6 is a spherical top molecule, $I_{xx} = I_{yy} = I_{zz}$ and

$$I = \{(2.941\,168 \times 10^{-113}\ \text{g}^3\cdot\text{cm}^6)[(10^{-3}\ \text{kg})/(1\ \text{g})]^3[(10^{-2}\ \text{m})/(1\ \text{cm})]^6\}^{1/3} = 3.086\,787 \times 10^{-45}\ \text{kg}\cdot\text{m}^2$$

The coordinates of the six F atoms with respect to the center of mass are $(r, 0, 0)$, $(-r, 0, 0)$, $(0, r, 0)$, $(0, -r, 0)$, $(0, 0, r)$, and $(0, 0, -r)$. Substituting into *(19.3a)* gives

$$\begin{aligned} I_{xx} &= (3.154\,81 \times 10^{-26}\ \text{kg})(r^2 + 0^2) + (3.154\,81 \times 10^{-26}\ \text{kg})(0^2 + r^2) \\ &\quad + (3.154\,81 \times 10^{-26}\ \text{kg})[(-r)^2 + 0^2] + (3.154\,81 \times 10^{-26}\ \text{kg})[0^2 + (-r)^2] \\ &\quad + (3.154\,81 \times 10^{-26}\ \text{kg})(0^2 + 0^2) + (3.154\,81 \times 10^{-26}\ \text{kg})(0^2 + 0^2) + (5.309\,17 \times 10^{-26}\ \text{kg})(0^2 + 0^2) \\ &= (12.619\,24 \times 10^{-26}\ \text{kg})r^2 \end{aligned}$$

Equating these values for the moment of inertia and solving for r gives

$$r = \left(\frac{3.086\,787 \times 10^{-45}\ \text{kg}\cdot\text{m}^2}{12.619\,24 \times 10^{-26}\ \text{kg}}\right)^{1/2} = 1.564\,000 \times 10^{-10}\ \text{m} = 0.156\,400\,0\ \text{nm}$$

19.16 The moments of inertia of $^{16}O^{12}C^{32}S$ and $^{18}O^{12}C^{34}S$ were determined to be $1.379\,14 \times 10^{-45}\ \text{kg}\cdot\text{m}^2$ and $1.508\,44 \times 10^{-45}\ \text{kg}\cdot\text{m}^2$, respectively. Given the atomic masses as $m/(10^{-26}\ \text{kg}) = 2.656\,06$ for ^{16}O, $2.987\,62$ for ^{18}O, $1.992\,68$ for ^{12}C, $5.309\,17$ for ^{32}S, and $5.640\,59$ for ^{34}S, calculate the O=C and C=S bond lengths.

▮ To simplify the calculations, let the atomic masses be expressed in units of 10^{-26} kg and the bond lengths in units of nanometers. The moments of inertia in this system of reduced units become 0.137 914 and 0.150 844, respectively. Substituting into *(19.4)* gives

$$0.137\,914 = \frac{[(2.656\,06)(1.992\,68)r(\text{OC})^2 + (1.992\,68)(5.309\,17)r(\text{CS})^2(2.656\,06)(5.309\,17)[r(\text{OC}) + r(\text{CS})]^2]}{2.656\,06 + 1.992\,68 + 5.309\,17}$$

$$0.150\,844 = \frac{[(2.987\,62)(1.992\,68)r(\text{OC})^2 + (1.992\,68)(5.640\,59)r(\text{CS})^2(2.987\,62)(5.640\,59)[r(\text{OC}) + r(\text{CS})]^2]}{2.987\,62 + 1.992\,68 + 5.640\,59}$$

After multiplication and collection of common terms, roots must be found for the following simultaneous equations:

$$19.394\,14r(\text{OC})^2 + 28.202\,94r(\text{OC})r(\text{CS}) + 24.680\,95r(\text{CS})^2 - 1.373\,34 = 0$$

$$22.805\,31r(\text{OC})^2 + 33.703\,88r(\text{OC})r(\text{CS}) + 28.091\,83r(\text{CS})^2 - 1.602\,10 = 0$$

Solving these equations is not trivial. The most straightforward approach is to solve the first equation for $r(\text{OC})$ in terms of $r(\text{CS})$ using the quadratic formula

$$\begin{aligned} r(\text{OC}) &= \frac{[-28.202\,94r(\text{CS}) \pm \{(28.202\,94)^2 - 4(19.394\,14)[24.680\,95r(\text{CS})^2 - 1.373\,34]\}^{1/2}]}{2(19.394\,14)} \\ &= \frac{-28.202\,94r(\text{CS}) + [106.539 - 1119.257r(\text{CS})^2]^{1/2}}{38.788\,28} \end{aligned}$$

Choosing the positive sign so that $r(\text{OC}) > 0$, substituting this expression for $r(\text{OC})$ into the second equation, and simplifying gives

$$(0.015\,157\,76)\{-28.202\,94r(\text{CS}) + [106.539 - 1119.257r(\text{CS})^2]^{1/2}\}^2$$
$$+ 0.868\,919\,2\{-28.202\,94r(\text{CS}) + [106.539 - 1119.257r(\text{CS})^2]^{1/2}\}r(\text{CS}) + 28.091\,83r(\text{CS})^2 - 1.602\,10 = 0$$

Solving this equation for $r(\text{CS})$ by trial and error gives $r(\text{CS}) = 0.155\,844$ nm. Substituting this value into the equation for $r(\text{OC})$ gives $r(\text{OC}) = 0.116\,347$ nm.

19.17 Choose the molecules that will exhibit a pure rotational absorption microwave spectrum: N_2O, NO_2, $CClF_3$, NF_3, SF_6, CH_4, CO_2.

▮ Those molecules that have a permanent dipole moment will exhibit a rotational spectrum: N_2O, NO_2, $CClF_3$, and NF_3.

19.18 What information about the molecular geometry for N_2O can be determined from the knowledge that a pure rotational absorption spectrum is observed for this substance?

▮ There are four possible arrangements for the atoms in N_2O: linear N—N—O, linear N—O—N, bent N—N—O, and bent N—O—N. The linear N—O—N molecule can be eliminated as a possibility because this molecule would not have a permanent dipole moment and would not exhibit a pure rotational spectrum.

19.19 What is the degeneracy of a rotational level for a spherical top molecule?

▮ The rotational energy of a spherical top molecule is given by

$$E_J = \frac{J(J+1)\hbar^2}{2B} \tag{19.6}$$

where $J = 0, 1, 2, \ldots$. For a given value of J there are $2J + 1$ values of the M_J quantum number and $2J + 1$ values of the K quantum number. The degeneracy is $g_J = (2J + 1)^2$.

19.20 Determine the value of J for which N_J/N_0 is a maximum for CH_4 at $T = 298$ K. See Problems 19.6 and 19.19 for additional information.

▮ Substituting *(19.6)* and $g_J = (2J + 1)^2$ into *(2.22)* gives

$$\frac{N_i}{N_j} = \frac{(2J_i + 1)^2}{(2J_j + 1)^2} \exp\left\{-\frac{[J_i(J_i + 1) - J_j(J_j + 1)]\hbar^2}{2BkT}\right\} \tag{19.7a}$$

Substituting $J_j = 0$ and $N_j = 0$ gives

$$N_j = N_0(2J + 1)^2\, e^{-J(J+1)\hbar^2/2BkT} \tag{19.7b}$$

Differentiating *(19.7b)* with respect to J, setting the result equal to zero, and solving for the maximum value of J gives

$$\frac{d(N_j/N_0)}{dJ} = 2(2J + 1)(2)\, e^{-J(J+1)\hbar^2/2BkT} + (2J + 1)^2[-(2J + 1)(\hbar^2/2IkT)\, e^{-J(J+1)\hbar^2/2BkT}] = 0$$

$$J_{\max} = \frac{(8BkT/\hbar^2)^{1/2} - 1}{2} \tag{19.8}$$

Substituting the data gives

$$J_{\max} = \frac{\left[\dfrac{8(5.313 \times 10^{-47}\ \text{kg}\cdot\text{m}^2)(1.381 \times 10^{-23}\ \text{J}\cdot\text{K}^{-1})(298.15\ \text{K})}{(1.0546 \times 10^{-34}\ \text{J}\cdot\text{s})^2[(1\ \text{kg}\cdot\text{m}^2\cdot\text{s}^{-2})/(1\ \text{J})]}\right] - 1}{2} = 5.77 \approx 6$$

19.21 Sketch the rotational energy levels for CF_4(g), a spherical top molecule with $B' = 0.1910\ \text{cm}^{-1}$.

▮ The levels are given by *(19.6)* as

$$F(J, K) = B'J(J + 1) \tag{19.9}$$

As a sample calculation, consider $J = 4$:

$$F(4, K) = (0.1910\ \text{cm}^{-1})(4)(4 + 1) = 3.8200\ \text{cm}^{-1}$$

Values of $F(J, K)$ for $J = 0$ to $J = 5$ are shown in Fig. 19-4a.

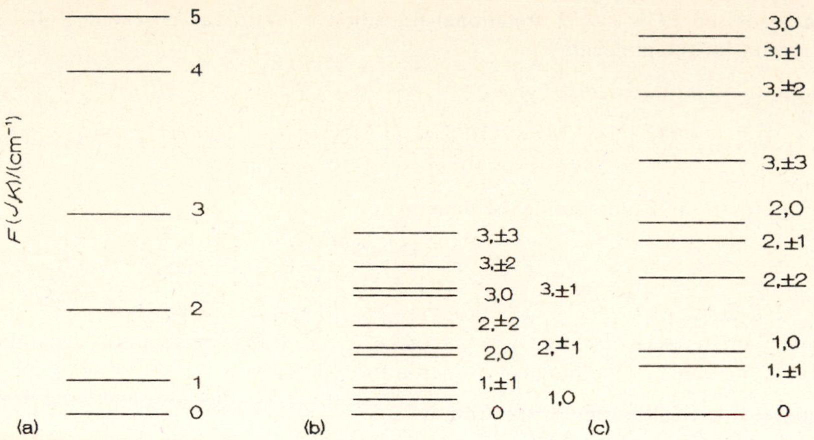

Fig. 19-4

19.22 Sketch the rotational energy levels for $CClF_3(g)$, a prolate symmetrical top molecule with $B' = 0.1111\ \text{cm}^{-1}$ and $A' = 0.1908\ \text{cm}^{-1}$.

▮ The rotational energy levels for a prolate symmetrical top are given by

$$E_{J,K} = \frac{J(J+1)\hbar^2}{B} + K^2\hbar^2\left(\frac{1}{2A} - \frac{1}{2B}\right) \tag{19.10a}$$

which can be written as

$$F(J,K) = B'J(J+1) + (A' - B')K^2 \tag{19.10b}$$

where $K = 0, \pm1, \pm2, \ldots, \pm J$. As a sample calculation, consider $J = 2$, $K = \pm1$.

$$F(4,\pm1) = (0.1111\ \text{cm}^{-1})(2)(2+1) + (0.1908\ \text{cm}^{-1} - 0.1111\ \text{cm}^{-1})(\pm1)^2 = 0.7463\ \text{cm}^{-1}$$

Values of $F(J, K)$ for $J = 0$ to $J = 3$ are shown in Fig. 19-4b.

19.23 Sketch the rotational energy levels for $NF_3(g)$, an oblate symmetrical top molecule with $B' = 0.3561\ \text{cm}^{-1}$ and $C' = 0.1948\ \text{cm}^{-1}$.

▮ The rotational energy levels for an oblate symmetrical top are given by

$$E_{J,K} = \frac{J(J+1)\hbar^2}{2B} - K^2\hbar^2\left(\frac{1}{2B} - \frac{1}{2C}\right) \tag{19.11a}$$

which can be written as

$$F(J,K) = B'J(J+1) - (B' - C')K^2 \tag{19.11b}$$

where $K = 0, \pm1, \pm2, \ldots, \pm J$. As a sample calculation, consider $J = 2$, $K = \pm1$.

$$F(2,\pm1) = (0.3561\ \text{cm}^{-1})(2)(2+1) - (0.3561\ \text{cm}^{-1} - 0.1948\ \text{cm}^{-1})(\pm1)^2 = 1.9753\ \text{cm}^{-1}$$

Values of $F(J, K)$ for $J = 0$ to $J = 3$ are shown in Fig. 19-4c.

19.24 What will be the values of the first few lines in the rotational spectrum of $CClF_3(g)$? See Problem 19.22 for additional information.

▮ The selection rule for a symmetrical top are

$$\Delta J = \pm1 \tag{19.12a}$$

$$\Delta K = 0 \tag{19.12b}$$

Substituting *(19.12)* into *(19.10b)* gives

$$\bar{\nu} = 2B'(J+1) \tag{19.13}$$

Substituting the data gives

$$\bar{\nu} = 2(0.1111\ \text{cm}^{-1})(0+1) = 0.2222\ \text{cm}^{-1}$$

$$\bar{\nu} = 2(0.1111)(1+1) = 0.4444\ \text{cm}^{-1}$$

$$\bar{\nu} = 2(0.1111)(2+1) = 0.6666\ \text{cm}^{-1}$$

19.25 The frequency for the $J = 1 \leftrightarrow 2$ rotational transition of NF_3 was observed at 42 723.84 MHz. What is the moment of inertia for the molecule?

▌ Substituting the data into *(13.22)* gives

$$\bar{\nu} = \frac{(42\,723.84 \text{ MHz})[(10^6 \text{ Hz})/(1 \text{ MHz})][(1 \text{ s}^{-1})/(1 \text{ Hz})]}{(2.997\,925 \times 10^8 \text{ m} \cdot \text{s}^{-1})[(10^2 \text{ cm})/(1 \text{ m})]} = 1.425\,114 \text{ cm}^{-1}$$

Solving *(19.13)* for B' and substituting the data gives

$$B' = \frac{1.425\,114 \text{ cm}^{-1}}{2(1+1)} = 0.356\,279 \text{ cm}^{-1}$$

19.26 The rotational Raman spectrum of $CO_2(g)$ showed a series of absorption peaks separated by 3.16 cm^{-1} in the S branch. What is the value of the moment of inertia for CO_2?

▌ The selection rule for the Raman spectrum is

$$\Delta J = 0, \pm 2 \tag{19.14}$$

which upon substituting into *(19.10b)* gives for the S branch

$$\bar{\nu}_S = \{B'(J+2)[(J+2)+1] + (A' - B')K^2\} - [B'J(J+1) + (A' - B')K^2] = 2B'(2J+3)$$

Because of the zero spin of the ^{16}O nuclei, only even values of J are observed. Thus for the $J = J$ to $J = J + 2$ transitions, the peaks will be separated by

$$\Delta\bar{\nu}_S = 2B'[2(J+2)+3] - 2B'(2J+3) = 8B'$$

Solving for B' and substituting the data gives

$$B' = (3.16 \text{ cm}^{-1})/8 = 0.395 \text{ cm}^{-1}$$

19.2 VIBRATIONAL SPECTRA

19.27 How many vibrational degrees of freedom will be observed for $H_2O(g)$, and which of these will be infrared-active?

▌ The number of vibrational degrees of freedom is

$$3\Lambda - 6 = 3(3) - 6 = 3$$

(see Problem 3.11). They are sketched in Fig. 19-5a. Because H_2O has a permanent dipole moment, all three vibrational modes are observed in the infrared. All three modes are Raman-active, also.

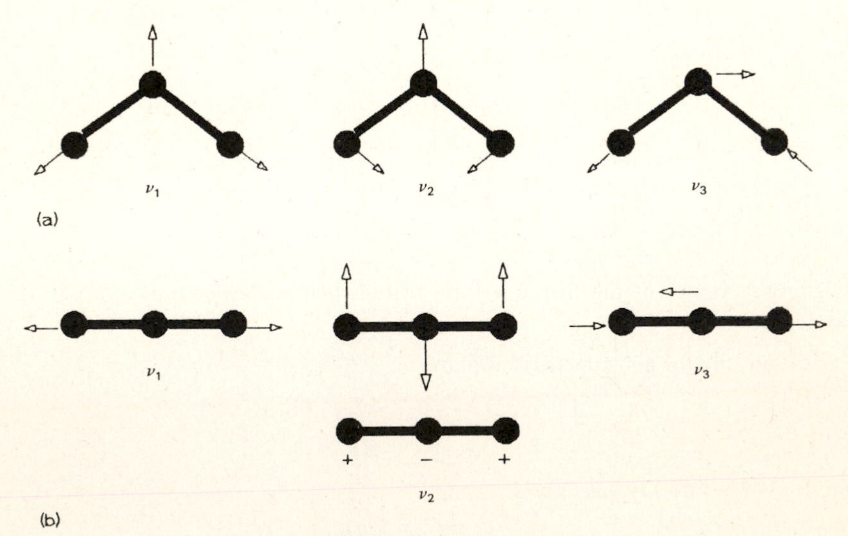

Fig. 19-5

19.28 How many vibrational degrees of freedom will be observed for $CO_2(g)$, and which of these will be infrared-active?

▌ The number of vibrational degrees of freedom is

$$3\Lambda - 5 = 3(3) - 5 = 4$$

(see Problem 3.10). They are sketched in Fig. 19-5b. Because CO_2 does not contain a permanent dipole moment, only those vibrations that induce a dipole moment will be infrared-active. The ν_2 and ν_3 vibrations will be infrared-active, and the ν_1 vibration will be Raman-active.

19.29 How many vibrational degrees of freedom will be observed for $C_2H_2(g)$, and which of these will be infrared-active?

▮ The number of vibrational degrees of freedom is

$$3\Lambda - 5 = 3(4) - 5 = 7$$

(see Problem 3.10). They are sketched in Fig. 19-6. Because C_2H_2 does not contain a permanent dipole moment, only those vibrations that induce a dipole moment will be infrared-active. The ν_3 and ν_5 vibrations will be infrared-active, and the ν_1, ν_2, and ν_4 vibrations will be Raman-active.

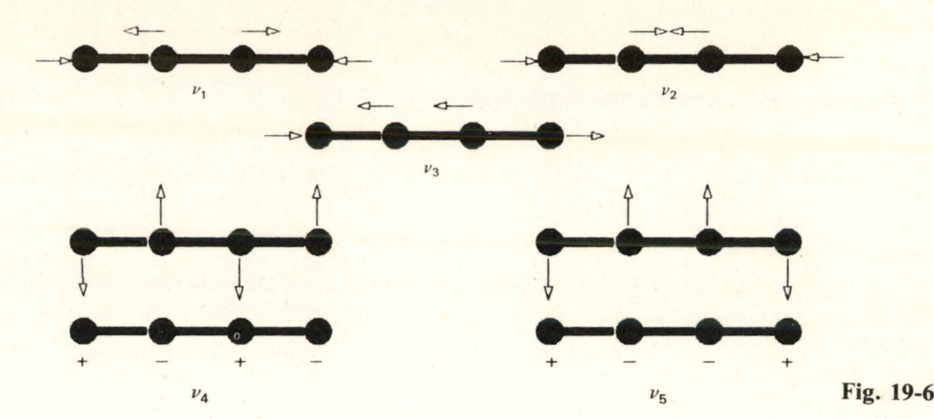

Fig. 19-6

19.30 How many vibrational degrees of freedom will be observed for $BF_3(g)$, and which of these will be infrared-active?

▮ The number of vibrational degrees of freedom is

$$3\Lambda - 6 = 3(4) - 6 = 6$$

(see Problem 3.11). They are sketched in Fig. 19-7. Because BF_3 does not contain a permanent dipole moment, only those vibrations that induce a dipole moment will be infrared-active. The ν_2, ν_3, and ν_4 vibrations will be infrared-active, and the ν_1 vibration will be Raman-active.

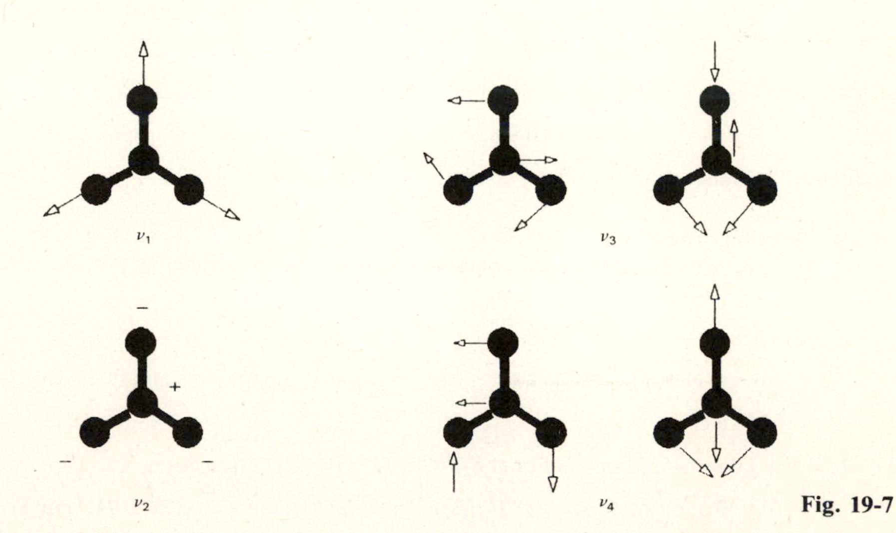

Fig. 19-7

19.31 In Problem 19.18, a linear N—O—N geometry for $N_2O(g)$ was eliminated on the basis of the observation of a pure rotational spectrum in the microwave region. Using the additional information that three vibrations are infrared-active (two of those are also Raman-active), determine the geometry of the molecule.

▮ If the molecule were bent, either N—N—O or N—O—N, there would be $3(3) - 6 = 3$ vibrational degrees of freedom that would be both infrared- and Raman-active, which is not observed. Assuming a linear N—N—O geometry, there are $3(3) - 5 = 4$ vibrational degrees of freedom (one is doubly degenerate, similar to ν_2 for CO_2 in Fig. 19-5b), giving three infrared-active vibrations, which is observed. The molecule is linear N—N—O.

19.32 The fundamental frequencies for $SO_2(g)$ are $\bar{\nu}_1 = 1151.38\ \text{cm}^{-1}$, $\bar{\nu}_2 = 517.69\ \text{cm}^{-1}$, and $\bar{\nu} = 1361.76\ \text{cm}^{-1}$. Determine the ratios of the population of molecules with (1, 0, 0), (0, 1, 0), (0, 2, 0), and (0, 0, 1) to that for (0, 0, 0) at 298 K.

▮ Substituting the values of the wavenumber into *(13.32)* gives

$$E_1 = (6.626\,18 \times 10^{-34}\ \text{J}\cdot\text{s})(2.997\,925 \times 10^8\ \text{m}\cdot\text{s}^{-1})(1151.38\ \text{cm}^{-1})[(10^2\ \text{cm})/(1\ \text{m})] = 2.287\,19 \times 10^{-20}\ \text{J}$$

$E_2 = 1.028\,38 \times 10^{-20}\ \text{J}$, and $E_3 = 2.705\,11 \times 10^{-20}\ \text{J}$. Substituting into *(17.14)* gives

$$\frac{N(1,0,0)}{N(0,0,0)} = \exp\left[-\frac{(1-0)(2.287\,19 \times 10^{-20}\ \text{J})}{(1.380\,66 \times 10^{-23}\ \text{J}\cdot\text{K}^{-1})(298.15\ \text{K})}\right] = 3.863 \times 10^{-3}$$

$$\frac{N(0,1,0)}{N(0,0,0)} = 8.223 \times 10^{-2} \qquad \frac{N(0,2,0)}{N(0,0,0)} = 6.762 \times 10^{-3} \qquad \frac{N(0,0,1)}{N(0,0,0)} = 1.400 \times 10^{-3}$$

19.33 The approximate position of the infrared absorption based on the N—H stretch is 3300-3370 cm^{-1}. Predict the approximate position for the N—D stretch.

▮ The force constant for the vibration is independent of the hydrogen isotope present. Taking a ratio of *(17.12)* gives

$$\bar{\nu}(\text{N—D})/\bar{\nu}(\text{N—H}) = (\mu(\text{N—H})/\mu(\text{N—D}))^{1/2}$$

The reduced mass of each system is essentially $m(\text{H})$ and $m(\text{D})$. Solving the above equation for $\bar{\nu}(\text{N—D})$ and substituting the data gives

$$\bar{\nu}(\text{N—D}) = (3300\text{-}3370\ \text{cm}^{-1})(\tfrac{1}{2})^{1/2} = 2330\text{-}2380\ \text{cm}^{-1}$$

19.34 One of the fundamental frequencies of $SO_2(g)$ is $\bar{\nu}_3 = 1361.76\ \text{cm}^{-1}$. When predicting the valence of various overtone frequencies such as $2\bar{\nu}_3$ and $3\bar{\nu}_3$, the observed values are not exactly 2 and 3 times the value of $\bar{\nu}_3$ but are slightly lower because of the anharmonicity factor. Given $2\bar{\nu}_3 = 2715.46\ \text{cm}^{-1}$ and $3\bar{\nu}_3 = 4054.26\ \text{cm}^{-1}$, determine ω_e and $\omega_e x_e$ for this mode of vibration for SO_2.

▮ For the $v = 0$ to v transition, *(17.11c)* gives

$$\bar{\nu} = [(v + \tfrac{1}{2})\omega_e - (v + \tfrac{1}{2})^2\omega_e x_e] - [(0 + \tfrac{1}{2})\omega_e - (0 + \tfrac{1}{2})^2\omega_e x_e] = v\omega_e - (v^2 + v)\omega_e x_e$$

Substituting the data into the equation for $\bar{\nu}$ gives

$$1361.76\ \text{cm}^{-1} = (1)\omega_e - (1^2 + 1)\omega_e x_e = \omega_e - 2\omega_e x_e$$

$$2715.46\ \text{cm}^{-1} = (2)\omega_e - (2^2 + 2)\omega_e x_e = 2\omega_e - 6\omega_e x_e$$

and solving gives $\omega_e = 1369.82\ \text{cm}^{-1}$ and $\omega_e x_e = 4.03\ \text{cm}^{-1}$. As a check, for $3\bar{\nu}_3$,

$$\bar{\nu} = 3(1369.82\ \text{cm}^{-1}) - (3^2 + 3)(4.03\ \text{cm}^{-1}) = 4061.10\ \text{cm}^{-1}$$

which is 0.2% higher than the observed value.

19.35 The fundamental frequencies of $H_2O(g)$ are $\bar{\nu}_1 = 3651.1\ \text{cm}^{-1}$, $\bar{\nu}_2 = 1594.7\ \text{cm}^{-1}$, and $\bar{\nu}_3 = 3755.9\ \text{cm}^{-1}$. Predict the values of the following combination frequencies neglecting the anharmonicity factor: (0, 0, 0) → (0, 1, 1), (0, 2, 1), (1, 0, 1), (1, 1, 1), (2, 0, 1), and (2, 1, 1).

▮ For the (0, 0, 0) → (0, 1, 1) transition,

$$\bar{\nu} = (0-0)\bar{\nu}_1 + (1-0)\bar{\nu}_2 + (1-0)\bar{\nu}_3 = (0)(3651.1\ \text{cm}^{-1}) + (1)(1594.7\ \text{cm}^{-1}) + (1)(3755.9\ \text{cm}^{-1})$$
$$= 5350.6\ \text{cm}^{-1}$$

The observed value is 5331 cm^{-1}. For the (0, 0, 0) → (0, 2, 1) transition,

$$\bar{\nu} = (0-0)\bar{\nu}_1 + (2-0)\bar{\nu}_2 + (1-0)\bar{\nu}_3 = (0)(3651.1\ \text{cm}^{-1}) + (2)(1594.7\ \text{cm}^{-1}) + (1)(3755.9\ \text{cm}^{-1})$$
$$= 6945.3\ \text{cm}^{-1}$$

The observed value is 6872 cm^{-1}. Likewise for the other transitions, the wavenumbers are 7407.0 (7252) cm^{-1}, 9001.7 (8807) cm^{-1}, 11 058.1 (10 613) cm^{-1}, and 12 652.8 (12 151) cm^{-1}, respectively. The numbers in parentheses are the observed values.

19.36 The following peaks appear in the infrared spectrum of BF_3: $\bar{\nu}/(\text{cm}^{-1})$ = 482, 718, 1370, 1505, 1838, 1985, 2243, 2385, 3008, and 3263. Using $\bar{\nu}_1 = 888\ \text{cm}^{-1}$, $\bar{\nu}_2 = 696.7\ \text{cm}^{-1}$, $\bar{\nu}_3 = 1463.3\ \text{cm}^{-1}$, and $\bar{\nu}_4 = 480.7\ \text{cm}^{-1}$, determine the assignment of the observed peaks.

▌ This is a strictly trial-and-error process. Several of the peaks can be assigned by looking for the fundamentals and the overtones of the four given wavenumbers:

$\bar{\nu}/(\text{cm}^{-1})$	482	718	1505	3008
	$\bar{\nu}_4$	$\bar{\nu}_2$	$\bar{\nu}_3$	$2\bar{\nu}_3$
	(480.7)	(696.7)	(1463.3)	(2926.6)

where the numbers in parentheses represent the predicted values (neglecting the anharmonicity factor). Note that $\bar{\nu}_1$ is only Raman-active and is not observed as a fundamental or overtone. Simple combinations of two given wavenumbers accounts for several more of the peaks:

$\bar{\nu}/(\text{cm}^{-1})$	1370	1985	2243	2385
	$\bar{\nu}_1 + \bar{\nu}_4$	$\bar{\nu}_3 + \bar{\nu}_4$	$\bar{\nu}_2 + \bar{\nu}_3$	$\bar{\nu}_1 + \bar{\nu}_3$
	(1369)	(1944.0)	(2160.0)	(2351)

Note that combination bonds involving $\bar{\nu}_1$ are infrared-active. The final assignments are

$\bar{\nu}/(\text{cm}^{-1})$	1838	3263
	$\bar{\nu}_1 + 2\bar{\nu}_4$	$2\bar{\nu}_1 + \bar{\nu}_3$
	(1849)	(3239)

19.37 The Raman spectrum of $CF_4(g)$ was observed using the 435.83-nm Hg excitation energy. Four bands, corresponding to the four fundamental vibrations, were observed at 444.25, 448.16, 453.80, and 461.64 nm. Determine the wavenumbers of these fundamental vibrations.

▌ The energy difference in wavenumbers between the incident radiation and the scattered radiation is given by

$$\bar{\nu} = \frac{1}{\lambda(\text{incident})} - \frac{1}{\lambda(\text{scattered})} \tag{19.15}$$

Substituting the data into *(19.15)* gives

$$\bar{\nu} = \left(\frac{1}{435.83\text{ nm}} - \frac{1}{444.25\text{ nm}}\right)\left(\frac{10^9\text{ nm}}{1\text{ m}}\right)\left(\frac{10^{-2}\text{ m}}{1\text{ cm}}\right) = 434.88\text{ cm}^{-1}$$

Likewise, $\bar{\nu}/(\text{cm}^{-1}) = 631.27$, 908.59, and 1282.83.

19.38 The Raman spectrum of $C_2H_2(g)$ was observed using 435.83-nm Hg incident radiation. A band at 510.96 nm corresponding to $\bar{\nu}_1$ in Fig. 19-6 was observed at 510.96 nm. Determine the wavenumber of this vibration.

▌ Substituting the data into *(19.15)* gives

$$\bar{\nu} = \left(\frac{1}{435.83\text{ nm}} - \frac{1}{510.96\text{ nm}}\right)\left(\frac{10^9\text{ nm}}{1\text{ m}}\right)\left(\frac{10^{-2}\text{ m}}{1\text{ cm}}\right) = 3373.7\text{ cm}^{-1}$$

19.3 ELECTRON MAGNETIC PROPERTIES

19.39 The magnetic susceptibility of Pt is listed in a handbook as 2.019×10^{-4} "cgs units" $\cdot\ \text{mol}^{-1}$. Calculate the SI value of the magnetic susceptibility and the non-SI parameters known as the mass magnetic susceptibility and molar magnetic susceptibility. The density of Pt is $21.45\text{ g}\cdot\text{cm}^{-3}$.

▌ The SI value of the magnetic susceptibility (χ_v), or volume susceptibility, is a dimensionless quantity and can be evaluated from the cgs system by

$$\chi_v = \frac{4\pi[\chi/(\text{cgs units}\cdot\text{mol}^{-1})][\rho/(\text{g}\cdot\text{cm}^{-3})]}{[M/(\text{g}\cdot\text{mol}^{-1})]} \tag{19.16}$$

where M is the molar mass and ρ is the density. Substituting the data into *(19.16)* gives

$$\chi_v = \frac{4\pi(2.019\times10^{-4})(21.45)}{195.08} = 2.790\times10^{-4}$$

The mass magnetic susceptibility (χ_g) is related to χ_v by

$$\chi_g/(\text{m}^3\cdot\text{g}^{-1}) = \frac{\chi_v(10^{-3})}{[\rho/(\text{kg}\cdot\text{m}^{-3})]} \tag{19.17}$$

Substituting the data into *(19.17)* gives

$$\chi_g = \frac{(2.970 \times 10^{-4})(10^{-3})}{21.45 \times 10^3}\ \mathrm{m^3 \cdot g^{-1}} = 1.385 \times 10^{-11}\ \mathrm{m^3 \cdot g^{-1}}$$

The molar magnetic susceptibility (χ_m) is related to χ_g by

$$\chi_m = \chi_g M \tag{19.18}$$

Substituting the data into *(19.18)* gives

$$\chi_m = (1.385 \times 10^{-11}\ \mathrm{m^3 \cdot g^{-1}})(195.08\ \mathrm{g \cdot mol^{-1}}) = 2.702 \times 10^{-9}\ \mathrm{m^3 \cdot mol^{-1}}$$

19.40 Paramagnetic susceptibility data may be presented in terms of *effective Bohr magnetons* (μ_{eff}) defined as

$$\mu_{eff} = (3kT\chi_m / L\mu_B^2)^{1/2} \tag{19.19}$$

where μ_B is the Bohr magneton ($\mu_B = 9.274\,08 \times 10^{-24}\ \mathrm{J \cdot T^{-1}}$). The ground state of the Ce^{3+} ion is $\cdots 5s^2 5p^6 4f^1$ corresponding to the $^2F_{5/2}$ term symbol. Predict the value of μ_{eff}.

▮ For the rare earth ions,

$$\mu_{eff} = g[J(J+1)]^{1/2} \tag{19.20}$$

where g is given by *(15.27)* and J is the quantum number corresponding to the subscript on the atomic term symbol. Substituting $J = \frac{5}{2}$, $L = 3$, and $S = \frac{1}{2}$ into *(15.27)* gives

$$g = 1 + \frac{\frac{5}{2}(\frac{5}{2}+1) + \frac{1}{2}(\frac{1}{2}+1) - 3(3+1)}{2(\frac{5}{2})(\frac{5}{2}+1)} = 0.857$$

Substituting this value of g into *(19.20)* gives

$$\mu_{eff} = 0.857[\tfrac{5}{2}(\tfrac{5}{2}+1)]^{1/2} = 2.54$$

The observed value, as calculated using the experimental value of χ_m in *(19.19)*, is 2.4.

19.41 Calculate μ_{eff} and χ_m at 298 K for the V^{3+} ion. The ground-state electron configuration is $\cdots 3d^2$ corresponding to the $^3F_2$ term symbol.

▮ For ions of transition elements, only the spin degeneracy is important in contributing to the magnetic susceptibility and *(19.20)* becomes

$$\mu_{eff} = 2[S(S+1)]^{1/2} \tag{19.21}$$

where g has been set equal to 2. Substituting $S = 1$ into *(19.21)* gives

$$\mu_{eff} = 2[1(1+1)]^{1/2} = 2.83$$

Solving *(19.19)* for χ_m and substituting the data gives

$$\chi_m = \frac{(6.022 \times 10^{23}\ \mathrm{mol^{-1}})(9.274 \times 10^{-24}\ \mathrm{J \cdot T^{-1}})^2(2.83)^2}{3(1.381 \times 10^{-23}\ \mathrm{J \cdot K^{-1}})(298\ \mathrm{K})} = 3.36 \times 10^{-2}\ \mathrm{J \cdot T^{-2} \cdot mol^{-1}}$$

Substituting $1\ \mathrm{J} = 1\ \mathrm{kg \cdot m^2 \cdot s^{-2}}$ and $1\ \mathrm{T} = 1\ \mathrm{kg \cdot s^{-2} \cdot A^{-1}}$ and multiplying the above result by $\mu = 4\pi \times 10^{-7}\ \mathrm{H \cdot m^{-1}}$, where $1\ \mathrm{H} = 1\ \mathrm{m^2 \cdot kg \cdot s^{-2} \cdot A^{-2}}$, gives

$$\chi_m = (3.36 \times 10^{-2}\ \mathrm{J \cdot T^{-2} \cdot mol^{-1}})(4\pi \times 10^{-7}\ \mathrm{H \cdot m^{-1}})\left(\frac{1\ \mathrm{kg \cdot m^2 \cdot s^{-2}}}{1\ \mathrm{J}}\right)\left(\frac{1\ \mathrm{T}}{1\ \mathrm{kg \cdot s^{-2} \cdot A^{-1}}}\right)^2\left(\frac{1\ \mathrm{m^2 \cdot kg \cdot s^{-2} \cdot A^{-2}}}{1\ \mathrm{H}}\right)$$

$$= 4.22 \times 10^{-8}\ \mathrm{m^3 \cdot mol^{-1}}$$

19.42 For $O_2(g)$,

$$\chi_m = \frac{1.22 \times 10^{-5}}{T}\ \mathrm{m^3 \cdot mol^{-1}}$$

Determine the number of unpaired electrons in the O_2 molecule.

▮ The relation between χ_m and the spin quantum number is

$$\chi_m = (4L\mu_0\mu_B^2/3kT)S(S+1) \tag{19.22}$$

Solving *(19.22)* for $S(S+1)$ and substituting the data gives

$$S(S+1) = \frac{\left[\begin{array}{l}[(1.22\times10^{-5}\ \mathrm{m^3\cdot mol^{-1}\cdot K})/T](3)(1.381\times10^{-23}\ \mathrm{J\cdot K^{-1}})(T)\\ \qquad\times[(1\ \mathrm{kg\cdot s^{-2}\cdot A^{-1}})/(1\ \mathrm{T})]^2[(1\ \mathrm{H})/(1\ \mathrm{m^2\cdot kg\cdot s^{-2}\cdot A^{-2}})]\end{array}\right]}{4(6.022\times10^{23}\ \mathrm{mol^{-1}})(4\pi\times10^{-7}\ \mathrm{H\cdot m^{-1}})(9.274\times10^{-24}\ \mathrm{J\cdot T^{-1}})^2\dfrac{1\ \mathrm{kg\cdot m^2\cdot s^{-2}}}{1\ \mathrm{J}}} = 1.94$$

Solving for S gives $S = 0.98 \approx 1$. There are two unpaired electrons.

19.43 The mass magnetic susceptibility of $CuSO_4{\cdot}5H_2O$ is $7.7\times10^{-8}\ \mathrm{m^3\cdot kg^{-1}}$ at 25 °C. Determine the number of unpaired electrons in the substance.

▮ Substituting into *(19.18)* gives

$$\chi_m = (7.7\times10^{-8}\ \mathrm{m^3\cdot kg^{-1}})[(10^{-3}\ \mathrm{kg})/(1\ \mathrm{g})](249.68\ \mathrm{g\cdot mol^{-1}}) = 1.92\times10^{-8}\ \mathrm{m^3\cdot mol^{-1}}$$

Solving *(19.22)* for $S(S+1)$ and substituting the data gives

$$S(S+1) = \frac{\left[\begin{array}{l}(1.92\times10^{-8}\ \mathrm{m^3\cdot mol^{-1}})(3)(1.381\times10^{-23}\ \mathrm{J\cdot K^{-1}})(298\ \mathrm{K})\\ \qquad\times[(1\ \mathrm{kg\cdot s^{-2}\cdot A^{-1}})/(1\ \mathrm{T})]^2[(1\ \mathrm{H})/(1\ \mathrm{m^2\cdot kg\cdot s^{-2}\cdot A^{-2}})]\end{array}\right]}{4(6.022\times10^{23}\ \mathrm{mol^{-1}})(4\pi\times10^{-7}\ \mathrm{H\cdot m^{-1}})(9.274\times10^{-24}\ \mathrm{J\cdot T^{-1}})^2\dfrac{1\ \mathrm{kg\cdot m^2\cdot s^{-2}}}{1\ \mathrm{J}}} = 0.91$$

Solving for S gives $S = 0.58 \approx 0.5$. There is one unpaired electron.

19.44 In a Gouy balance, the apparent weight of a sample in the presence and absence of a magnetic field is

$$w(\text{field}) - w(\text{no field}) = \chi_v\mu_0H^2A/2 \qquad (19.23)$$

where H is the magnetic field strength and A is the cross-sectional area of the sample. Using the following equation for the mass magnetic susceptibility of a $NiCl_2$ solution with a mass fraction of p and a density of ρ,

$$\chi_g/(\mathrm{m^3\cdot kg^{-1}}) = \left[\frac{10\,030p}{T/(\mathrm{K})} - 0.720(1-p)\right](4\pi\times10^{-9}) \qquad (19.24)$$

determine the calibration constant $\mu_0H^2A/2$ given that $\Delta w = 0.1526\ \mathrm{g}$ for a 25.3% aqueous solution of $NiCl_2$ at 25 °C. The density of the solution is $1.282\ \mathrm{g\cdot cm^{-3}}$.

▮ The mass susceptibility is given by *(19.24)* as

$$\chi_g = \left[\frac{(10\,030)(0.253)}{298} - (0.720)(1-0.253)\right](4\pi\times10^{-9})\ \mathrm{m^3\cdot kg^{-1}}$$
$$= 1.002\times10^{-7}\ \mathrm{m^3\cdot kg^{-1}} = 1.002\times10^{-10}\ \mathrm{m^3\cdot g^{-1}}$$

Solving *(19.17)* for χ_v and substituting the data gives

$$\chi_v = \frac{(1.002\times10^{-10})(1.282\times10^3)}{10^{-3}} = 1.285\times10^{-4}$$

Solving *(19.23)* for the calibration constant and substituting the data gives

$$\frac{\mu_0H^2A}{2} = \frac{2(0.1526\ \mathrm{g})}{1.285\times10^{-4}} = 2375\ \mathrm{g}$$

19.45 A sample of 0.495 M $MnSO_4$ was analyzed using the Gouy balance described in Problem 19.44. The observed change in apparent weight was 0.1939 g. Determine the number of unpaired electrons in the substance. The density of the solution is $1.069\ \mathrm{g\cdot cm^{-3}}$, and the temperature is 25 °C.

▮ Solving *(19.23)* for χ_v and substituting the data gives

$$\chi_v = \frac{0.1939\ \mathrm{g}}{2375\ \mathrm{g}} = 8.16\times10^{-5}$$

For aqueous solutions with molarity C,

$$\chi_v = 10^3[C/(\mathrm{mol\cdot L^{-1}})][\chi_m/(\mathrm{m^3\cdot mol^{-1}})]$$
$$- (0.720)(4\pi\times10^{-9})\{[\rho/(\mathrm{kg\cdot m^{-3}})] - 10^3[C/(\mathrm{mol\cdot L^{-1}})][M/(\mathrm{kg\cdot mol^{-1}})]\} \qquad (19.25)$$

Solving for χ_m and substituting the data gives

$$\chi_m = \frac{8.16 \times 10^{-5} + (0.720)(4\pi \times 10^{-9})[1.069 \times 10^3 - 10^3(0.495)(0.151\,00)]}{10^3(0.495)} \text{ m}^3 \cdot \text{mol}^{-1} = 1.83 \times 10^{-7} \text{ m}^3 \cdot \text{mol}^{-1}$$

Solving *(19.22)* for $S(S+1)$ and substituting the data gives

$$S(S+1) = \frac{\begin{bmatrix} (1.83 \times 10^{-7} \text{ m}^3 \cdot \text{mol}^{-1})(3)(1.381 \times 10^{-23} \text{ J} \cdot \text{K}^{-1})(298 \text{ K}) \\ \times [(1 \text{ kg} \cdot \text{s}^{-2} \cdot \text{A}^{-1})/(1 \text{ T})]^2[(1 \text{ H})/(1 \text{ m}^2 \cdot \text{kg} \cdot \text{s}^{-2} \cdot \text{A}^{-2})] \end{bmatrix}}{4(6.022 \times 10^{23} \text{ mol}^{-1})(4\pi \times 10^{-7} \text{ H} \cdot \text{m}^{-1})(9.274 \times 10^{-24} \text{ J} \cdot \text{T}^{-1})^2 \dfrac{1 \text{ kg} \cdot \text{m}^2 \cdot \text{s}^{-2}}{1 \text{ J}}} = 8.7$$

Solving for S gives $S = 2.5$. There are five unpaired electrons.

19.46 A sample of $CCl_4(l)$ was placed in the Gouy balance described in Problem 19.44. Given $\chi_v = -8.9 \times 10^{-6}$ at 25 °C, describe what will happen.

▌ Substituting into *(19.23)* gives

$$w(\text{field}) - w(\text{no field}) = (-8.9 \times 10^{-6})(2375 \text{ g}) = -0.021 \text{ g}$$

The sample will be repelled by the magnetic field with a force equivalent to

$$f = (0.021 \text{ g})(9.81 \text{ m} \cdot \text{s}^{-2})\left(\frac{10^{-3} \text{ kg}}{1 \text{ g}}\right)\left(\frac{1 \text{ N}}{1 \text{ kg} \cdot \text{m} \cdot \text{s}^{-2}}\right) = 2.1 \times 10^{-4} \text{ N}$$

19.47 Calculate the resonance frequency for a free electron in a magnetic field with $B = 1.00$ T.

▌ The energy related to the electronic spin magnetic moment is

$$E = g_e \mu_B s B \qquad (19.26)$$

where $s = \pm\frac{1}{2}$ and g_e is the g factor for an electron ($g_e = 2.002\,319$). The energy separation between the states is

$$\Delta E = g_e \mu_B B(\tfrac{1}{2}) - g_e \mu_B B(-\tfrac{1}{2}) = g_e \mu_B B \qquad (19.27a)$$

which upon substitution into *(13.32)* gives

$$\nu = g_e \mu_B B / h \qquad (19.27b)$$

Substituting the data into *(19.27b)* gives

$$\nu = \frac{(2.002\,319)(9.274\,08 \times 10^{-24} \text{ J} \cdot \text{T}^{-1})(1.00 \text{ T})}{6.626 \times 10^{-34} \text{ J} \cdot \text{s}} = 2.80 \times 10^{10} \text{ s}^{-1}$$

This is equivalent to 28.0 GHz.

19.48 The resonance of a free electron was to be observed in the microwave region at 5.00 mm. What is the value of B needed?

▌ Combining *(19.27a)* and *(13.32)* gives

$$\lambda = hc/g_e \mu_B B \qquad (19.27c)$$

Solving *(19.27c)* for B and substituting the data gives

$$B = \frac{(6.626 \times 10^{-34} \text{ J} \cdot \text{s})(2.9979 \times 10^8 \text{ m} \cdot \text{s}^{-1})}{(2.002)(9.274 \times 10^{-24} \text{ J} \cdot \text{T}^{-1})(5.00 \times 10^{-3} \text{ m})} = 2.14 \text{ T}$$

or 2400 G where $1 \text{ G} = 10^{-4}$ T.

19.49 What is the ratio of the number of free electrons in each of the spin states in a magnetic field with $B = 1.00$ T at 298 K? What is the ratio at 4 K?

▌ Substituting *(19.27a)* into *(2.22)* gives

$$N_\alpha / N_\beta = e^{-g_e \mu_B B / kT} \qquad (19.28)$$

Substituting the data into *(19.28)* gives

$$\frac{N_\alpha}{N_\beta} = \exp\left[-\frac{(2.002)(9.274 \times 10^{-24} \text{ J} \cdot \text{T}^{-1})(1.00 \text{ T})}{(1.381 \times 10^{-23} \text{ J} \cdot \text{s})(298 \text{ K})}\right] = 0.9955$$

at 298 K and $N_\alpha / N_\beta = 0.71$ at 4 K.

19.50 Assume that the detection limit of the ESR spectrometer described in Problem 19.49 is 1.000×10^{-12} M. What would be the number of radicals in each of the spin states?

▮ The total number of radicals is

$$N_\alpha + N_\beta = (1.000 \times 10^{-12}\ \text{mol} \cdot \text{L}^{-1})(6.022 \times 10^{23}\ \text{mol}^{-1}) = 6.022 \times 10^{11}\ \text{L}^{-1}$$

where

$$N_\alpha + N_\beta = 0.9955 N_\beta + N_\beta = 1.9955 N_\beta$$

Solving gives $N_\beta = 3.018 \times 10^{11}\ \text{L}^{-1}$ and $N_\alpha = 3.004 \times 10^{11}\ \text{L}^{-1}$.

19.51 The center of the ESR spectrum of the hydrogen atom occurs at 1.000 26 T, where $\nu = 28.031$ GHz. What is the effective value for g?

▮ Solving *(19.27b)* for g and substituting the data gives

$$g_{\text{eff}} = \frac{(6.6262 \times 10^{-34}\ \text{J} \cdot \text{s})(28.031\ \text{GHz})[(10^9\ \text{Hz})/(1\ \text{GHz})][(1\ \text{s}^{-1})/(1\ \text{Hz})]}{(9.2741 \times 10^{-24}\ \text{J} \cdot \text{T}^{-1})(1.000\ 26\ \text{T})} = 2.0022$$

19.52 Predict the characteristics of the ESR spectrum for atomic H.

▮ In the absence of a magnetic field, the spin states are degenerate. In the presence of an external field, the degeneracy is destroyed with the $s = -\frac{1}{2}$ state taken as lower in energy (by convention) and the $s = +\frac{1}{2}$ state taken as higher in energy. In the presence of n equivalent protons, each of these states is split into $n + 1$ components. By convention, the nuclear spin states with positive spin are assumed to be lower than those having negative spin in the $s = -\frac{1}{2}$ state, and vice versa in the $s = +\frac{1}{2}$ state. The diagram is shown in Fig. 19-8a.

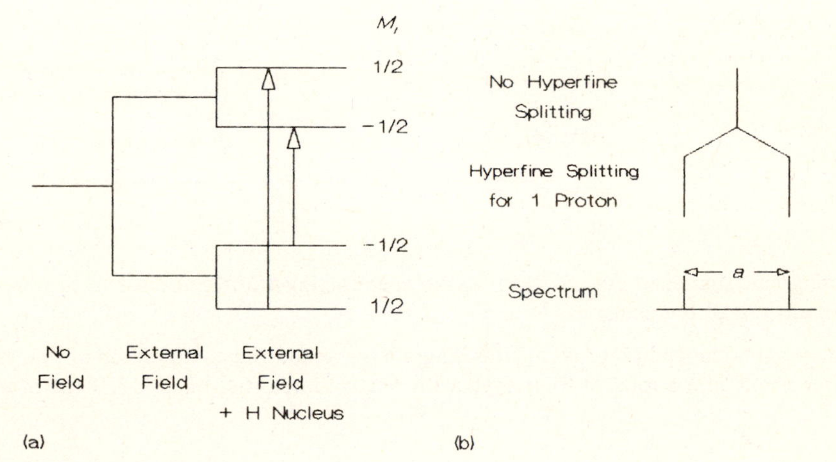

Fig. 19-8

The two allowed transitions follow the selection rule $\Delta M_I = 0$, where $M_I = I, I - 1, \ldots, -I$, and where $I = n(\frac{1}{2})$. The intensities of the resonance peaks can be shown to be proportional to the number of combinations of nuclear spins that will create the value of M_I and can be found using Pascal's triangle of binomial coefficients given in Table 19-1.

Table 19-1

n	relative intensities
0	1
1	1 1
2	1 2 1
3	1 3 3 1
4	1 4 6 4 1
5	1 5 10 10 5 1
6	1 6 15 20 15 6 1

A second method used to illustrate the hyperfine splitting is shown in Fig. 19-8b. The advantages of this second method are that the number of lines, their relative appearance in the spectrum, and their relative intensities can be illustrated. The center of the spectrum for atomic hydrogen occurs midway between the two hyperfine peaks separated by a distance a.

19.53 The ESR spectrum of atomic hydrogen described in Problem 19.51 consists of two lines. One is at 1.0256 T, and the other is located at 0.9749 T. What is the hyperfine coupling constant for the atom?

▮ Taking the difference between the two lines (see Fig. 19-8b) gives

$$a = 1.0256\text{ T} - 0.9749\text{ T} = 0.0507\text{ T} = 50.7\text{ mT}$$

or 1.421 GHz.

19.54 Prepare diagrams similar to Fig. 19-8b for a system consisting of two equivalent and two nonequivalent protons.

▮ The diagrams are shown in Fig. 19-9. In both cases, the splitting by the first proton produces two levels, which are separated by a. For the equivalent proton case shown in Fig. 19-9a, each of these levels is split into two levels separated by a. Because two of these levels are at the same energy, only three peaks will be observed, with the intensity of the center peak being twice as great as the others. For the nonequivalent proton case shown in Fig. 19-9b, each of the two levels is split into two levels separated by a' by the second proton. Because all of these levels are at different energies, four peaks will be observed, with the intensities being identical.

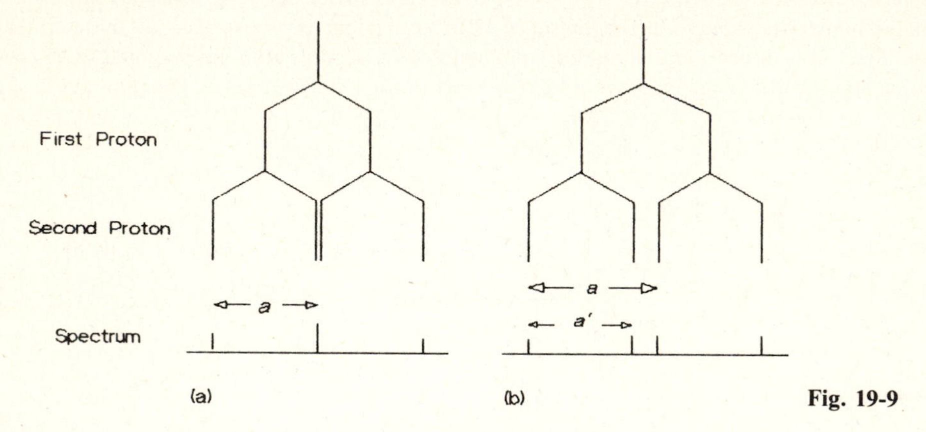

Fig. 19-9

19.55 Prepare diagrams similar to Fig. 19-8b for a system consisting of three equivalent protons. Predict the number and relative intensities of the peaks.

▮ The diagram is shown in Fig. 19-10. This diagram is essentially a continuation of Fig. 19-9a showing the splitting of the three levels into a total of four levels with relative intensities of 1, 3, 3, 1 (as predicted by Table 19-1).

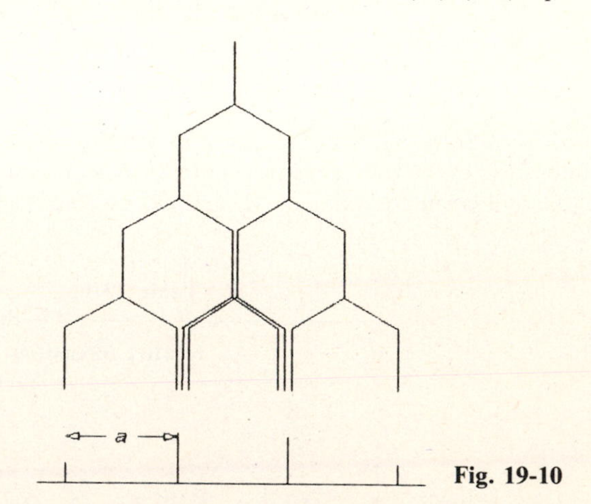

Fig. 19-10

19.56 The ESR spectrum of the methyl radical, $CH_3\cdot$, consisted of four peaks that appeared from 327.0 mT to 333.9 mT. What is the hyperfine coupling constant for the proton in this radical?

▮ From Fig. 19-10, $3a = (333.9\text{ mT} - 327.0\text{ mT}) = 6.9\text{ mT}$, or $a = 2.3\text{ mT}$.

19.57 The coupling constant for the protons in the benzene anion, $C_6H_6^-$, is 3.75 G. Describe the ESR spectrum for this substance.

▌ The six equivalent protons will generate seven peaks separated by 3.75 G or 0.375 mT. The relative intensities will be given by Table 19-1 as 1, 6, 15, 20, 15, 6, 1.

19.58 The coupling constants for the protons in an ethyl radical, $CH_3CH_2\cdot$, are 26.9 G and 22.4 G, respectively. Describe the ESR spectrum for this substance.

▌ As seen in Fig. 19-10, the three equivalent protons will generate four peaks separated by $a_1 = 26.9$ G ($=2.69$ mT) with relative intensities of 1, 3, 3, 1. As seen in Fig. 19-9a, each of these peaks is split into three peaks separated by $a_2 = 22.4$ G ($=2.24$ mT) with relative intensities of 1, 2, 1. The predicted spectrum is shown in Fig. 19-11.

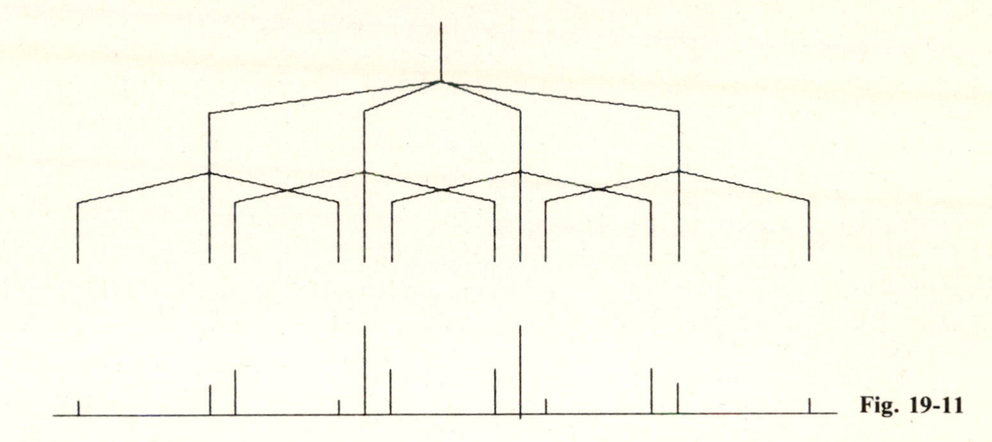

Fig. 19-11

19.59 The ESR spectrum for the isopropyl radical, CH_3CHCH_3, is shown in Fig. 19-12. Determine the coupling constants.

▌ There are two sets of equivalent protons, which generate the 14 lines of the spectrum (see Fig. 19-12). The value of a_2 is simply 2.21 mT. The spacing between the doublets is

$$\frac{2.73\text{ mT} - 2.21\text{ mT}}{2} = 0.26\text{ mT}$$

giving $a_1 = 2.21\text{ mT} + 0.26\text{ mT} = 2.47$ mT.

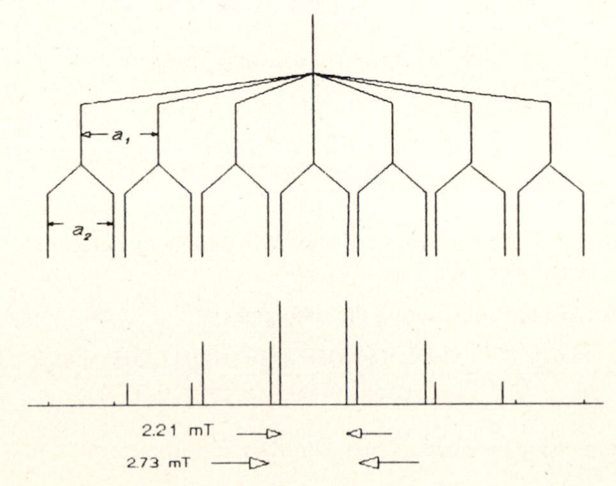

Fig. 19-12

19.60 The nuclear g factor is 5.5857 for H and 0.857 45 for D. If the hyperfine coupling constant is $CH_3\cdot$ is 0.230 mT, calculate the hyperfine coupling constant for $CD_3\cdot$.

▌ The splitting is directly proportional to the value of g. For $CD_3\cdot$,

$$a(CD_3\cdot) = a(CH_3\cdot)\frac{g(D)}{g(H)} = (0.230\text{ mT})\frac{0.857\,45}{5.5857} = 0.0353\text{ mT}$$

19.61 How does the diagram similar to Fig. 19-9a differ for two equivalent deuterons?

▌ The nuclear spin of H is $\frac{1}{2}$ and that of D is 1. Because I can be +1, 0, and −1, each splitting involves three levels instead of two. The diagram is shown in Fig. 19-13.

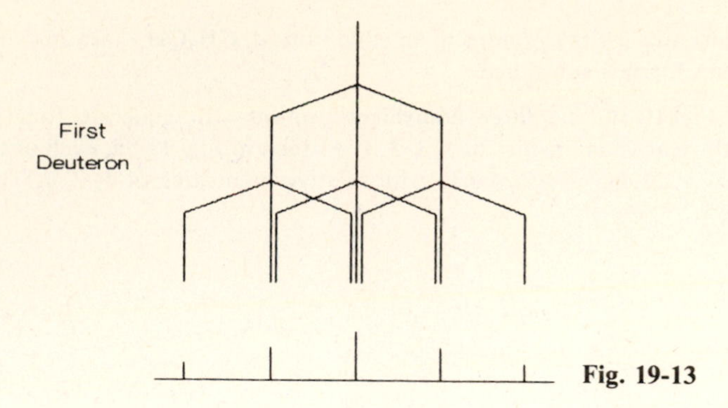

Fig. 19-13

19.4 NUCLEAR MAGNETIC RESONANCE

19.62 Calculate the energies of a ^{35}Cl nucleus in a magnetic field of $B_0 = 1.00$ T. The ^{35}Cl nucleus has $I = \frac{3}{2}$ and $g_N = 0.5479$.

▌ The energy levels of an isolated nuclear magnetic moment in an applied magnetic field of flux density B_0 are given by

$$E = -g_N \mu_N B_0 M_I \tag{19.29}$$

where g_N is the nuclear g factor, μ_N is the nuclear magneton ($\mu_N = 5.050\,824 \times 10^{-27}\ \mathrm{J \cdot T^{-1}}$), and $M_I = I, I-1, \ldots, -I$, where I is the nuclear spin. Substituting $M_I = \frac{3}{2}, \frac{1}{2}, -\frac{1}{2}$, and $-\frac{3}{2}$ for ^{35}Cl into *(19.21)* gives

$$E = -(0.5479)(5.051 \times 10^{-27}\ \mathrm{J \cdot T^{-1}})(1.00\ \mathrm{T})(\tfrac{3}{2}) = -4.151 \times 10^{-27}\ \mathrm{J}$$

$E = -1.384 \times 10^{-27}$ J, 1.384×10^{-27} J, and 4.151×10^{-27} J, respectively.

19.63 At what frequency would the NMR spectrum of ^{35}Cl be observed in the system described in Problem 19.62?

▌ The permitted transitions between the energy levels follow the selection rule $\Delta M_I = \pm 1$. Combining *(19.29)* and *(13.32)* gives

$$\nu = g_N \mu_N B_0 / h \tag{19.30}$$

Substituting the data into *(19.30)* gives

$$\nu = \frac{(0.5479)(5.051 \times 10^{-27}\ \mathrm{J \cdot T^{-1}})(1.00\ \mathrm{T})}{6.626 \times 10^{-34}\ \mathrm{J \cdot s}} = 4.177 \times 10^{6}\ \mathrm{s^{-1}} = 4.177\ \mathrm{MHz}$$

19.64 Assume that a 60-MHz NMR spectrometer was being used to detect the ^{35}Cl spectrum. What must be the corresponding value of B_0? For ^{35}Cl, $g_N = 0.5479$.

▌ Solving *(19.30)* for B_0 and substituting the data gives

$$B_0 = \frac{(6.626 \times 10^{-34}\ \mathrm{J \cdot s})(60\ \mathrm{MHz})[(10^6\ \mathrm{Hz})/(1\ \mathrm{MHz})][(1\ \mathrm{s^{-1}})/(1\ \mathrm{Hz})]}{(0.5479)(5.051 \times 10^{-27}\ \mathrm{J \cdot T^{-1}})} = 14.37\ \mathrm{T}$$

19.65 For an isolated proton, resonance occurs in a 220-MHz spectrometer at 5.1672 T. Determine the nuclear g factor for ^{1}H.

▌ Solving *(19.30)* for g_N and substituting the data gives

$$g_N = \frac{(6.626\,176 \times 10^{-34}\ \mathrm{J \cdot s})(220\ \mathrm{MHz})[(10^6\ \mathrm{Hz})/(1\ \mathrm{MHz})][(1\ \mathrm{s^{-1}})/(1\ \mathrm{Hz})]}{(5.050\,824 \times 10^{-27}\ \mathrm{J \cdot T^{-1}})(5.1672)} = 5.5856$$

19.66 What is the ratio of the numbers of protons in each spin level at 298 K? What is the ratio at 4 K? Assume that a 60-MHz spectrometer is being used. For ^{1}H, $g_N = 5.5856$.

▮ Substituting *(13.22)* into *(2.22)* gives

$$\frac{N_-}{N_+} = \exp\left[-\frac{(6.626\times10^{-23}\ \mathrm{J\cdot s})(60\ \mathrm{MHz})[(10^6\ \mathrm{Hz})/(1\ \mathrm{MHz})][(1\ \mathrm{s^{-1}})/(1\ \mathrm{Hz})]}{(1.381\times10^{-23}\ \mathrm{J\cdot K^{-1}})(298\ \mathrm{K})}\right] = 0.999\,990\,34$$

Likewise, at $T = 4$ K, $N_-/N_+ = 0.999\,280$.

19.67 The low-resolution NMR spectrum of cyclohexane, C_6H_{12}, appeared 96 Hz downfield from the TMS standard in a 60-MHz spectrometer. What is the value of the chemical shift?

▮ The chemical shift is often represented by δ_i and is expressed in parts per million.

$$\delta_i/(\mathrm{ppm}) = \frac{\Delta\nu}{\nu}\times10^6 \qquad (19.31)$$

Substituting the data into *(19.31)* gives

$$\delta_i = \frac{96\ \mathrm{Hz}}{(60\ \mathrm{MHz})[(10^6\ \mathrm{Hz})/(1\ \mathrm{MHz})]}(10^6\ \mathrm{ppm}) = 1.6\ \mathrm{ppm}$$

19.68 Describe the low-resolution NMR spectrum of an equimolar mixture of benzene, C_6H_6, and cyclohexane. The chemical shift for C_6H_6 is 6.9 ppm. See Problem 19.67 for additional information.

▮ The C_6H_{12} peak will appear 96 Hz downfield from the TMS, and the location of C_6H_6 peak will be given by *(19.31)* as

$$\Delta\nu = \frac{(6.9)(60\ \mathrm{MHz})[(10^6\ \mathrm{Hz})/(1\ \mathrm{MHz})]}{10^6} = 414\ \mathrm{Hz}$$

The ratio of the intensities of the peaks will be 2:1, the same as the ratio of the number of H atoms in each compound (12/6).

19.69 The two peaks in the low-resolution NMR spectrum of ethanal, CH_3CHO, were separated by 7.60 ppm. What is the frequency difference corresponding to this difference in chemical shifts in a 60-MHz spectrometer? What is the frequency difference in a 100-MHz instrument? How do the relative intensities of the peaks compare?

▮ Solving *(19.31)* for $\Delta\nu$ and substituting the data gives

$$\Delta\nu = \frac{(7.60)(60\ \mathrm{MHz})[(10^6\ \mathrm{Hz})/(1\ \mathrm{MHz})]}{10^6} = 456\ \mathrm{Hz}$$

for the 60-MHz instrument and 760 Hz for the 100-MHz spectrometer. The peak corresponding to the CH_3 absorption will be three times as great as the peak corresponding to the CHO absorption, the same ratio as the number of H atoms in each group.

19.70 Describe the spin-spin splitting in the NMR spectrum of ethanal. See Problem 19.69 for additional information.

▮ For a molecule having m protons of type A and n protons of type X on adjacent nuclei, a set of $n+1$ lines centered about the frequency for proton type A will appear with relative intensities as given in Table 19-1, and a set of $m+1$ lines centered about the frequency for the proton type X will appear with similar relative intensities. For ethanal, the CH_3 peak will be split into a doublet as a result of the single CHO proton, and the CHO peak will be split into a quartet as a result of the three CH_3 protons.

19.71 A substance with the empirical formula $C_4H_{10}O$ had an NMR spectrum consisting of a triplet and a quartet with relative intensities of 3 and 2, respectively. Identify the compound.

▮ The predicted spectrum of *tert*-butyl alcohol, $(CH_3)_3COH$, would consist of two peaks with no spin-spin splitting. The predicted spectrum of 2-butanol, $CH_3CH_2CHOHCH_3$, would be very complex with two doublets, a triplet, an octet, and one peak split into 24 parts (if all spin-spin splittings actually occurred). Neither of these predicted spectra agree with the observed spectrum. The predicted spectrum for diethyl ether would consist of two peaks (one for the CH_2 groups and one for the CH_3 groups), and the predicted splittings would agree with the observed spectrum. The substance is diethyl ether.

19.72 A substance with the empirical formula CH_xF_y had an NMR spectrum consisting of two peaks—one a quartet and one a doublet. Identify the substance.

▮ The substances CF_4 and CH_4 are both eliminated from consideration because both of these will have a spectrum consisting of a single peak. The substance CH_2F_2 is also eliminated because it will have a spectrum consisting of two peaks, each a triplet. Both CHF_3 and CH_3F will have spectra consisting of two peaks, and the splitting will consist of a quartet and a doublet. Unless information is known about which peak is the doublet and which is the quartet, the final identification cannot be made.

19.5 PHOTOELECTRON SPECTROSCOPY

19.73 A sample of $H_2(g)$ was excited by 21.22-eV radiation from a helium source. The observed band consisted of several closely spaced peaks and had a minimum value of 5.77 eV. What is the value of the adiabatic ionization energy for H_2? Write the chemical equation describing this process. What is the source for the peaks in the band?

▮ The ionization energy is related to the kinetic energy of the emitted electron by *(14.2)* as

$$I = 21.22\text{ eV} - 5.77\text{ eV} = 15.45\text{ eV}$$

This process corresponds to the chemical equation $H_2(g, v = 0) \rightarrow H_2^+(g, v) + e^-$, where the various peaks in the band will be separated by the vibrational energy of $H_2^+(g)$.

19.74 A second peak in the photoelectron spectrum of H_2 was observed at $I = 13.60$ eV. What does this value of energy represent?

▮ This is the ionization energy of the atomic H formed from the dissociation process of the molecular H_2 described in Problem 19.73.

19.75 The molecular orbital diagram for $H_2O(g)$ as determined from self-consistent-field calculations is shown in Fig. 19-14. Predict the location of the photoelectron peaks using 21.22-eV He radiation.

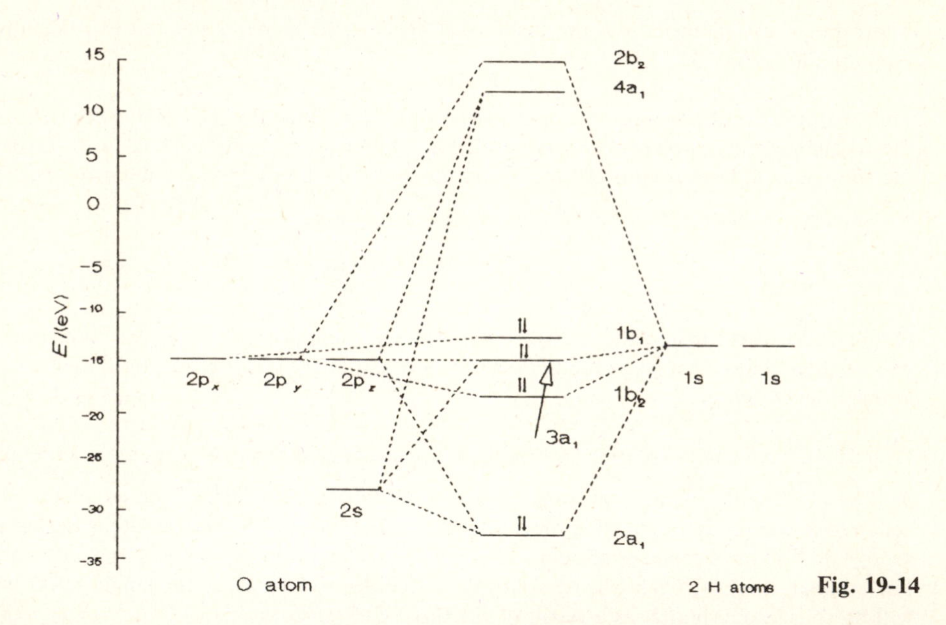

Fig. 19-14

▮ Only those ionizations with $I < 21.22$ eV will be observed. Using *(14.2)* gives

$$E(1b_2) = 21.22\text{ eV} - 18.2\text{ eV} = 3.0\text{ eV} \qquad E(3a_1) = 21.22 - 14.8 = 6.4\text{ eV} \qquad E(1b_1) = 21.22 - 12.6 = 8.6\text{ eV}$$

19.76 The X-ray photoelectron spectrum of NaN_3 showed two peaks, one at 405 eV and the second at 398 eV, resulting from the ionizations of the nitrogen atoms. The area under the second peak was twice that of the first peak. Explain.

▮ The two peaks are the ionizations of the 1s electrons from the three N atoms in the azide ion. The two end N atoms are chemically equivalent and account for the larger 398-eV peak, and the central N atom, which has a slightly different chemical environment, accounts for the smaller 405-eV peak.

19.77 Predict the X-ray photoelectron spectrum for acetone, CH_3COCH_3.

▮ There would be one peak for the O atom and two peaks for the C atoms. The areas under the C peaks would be in a ratio of 2:1.

19.78 The X-ray photoelectron spectrum for $ClCOOCH_2CH_3$ consisted of four peaks assigned to the 1s ionization of the carbonyl C at 289 eV, the methylene C atom at 285 eV, the methyl C atom at 283 eV, and the C atom at 270 eV. The spectrum for $F_3CCOOCH_2CH_3$ consisted of four peaks at 292 eV, 289 eV, 286 eV, and 285 eV. Assign these peaks.

▮ The 285-eV peak corresponds to the methyl C atom, the 286-eV peak corresponds to the methylene C atom, and the 289-eV peak corresponds to the carbonyl C atom. Because 292 eV falls into the range of the other C atoms, it probably corresponds to the 1s ionization of the C atom in the CF_3 group.

19.6 OPTICAL ACTIVITY

19.79 The optical rotation of light through a 1.00-cm sample of liquid nicotine was −16.4°. What is the value of the specific rotation? The density of nicotine is $1.0097\ g \cdot cm^{-3}$, the molar mass is $162.24\ g \cdot mol^{-1}$, the temperature is 20 °C, and the D line from a sodium lamp was used ($\lambda = 589.3$ nm). What does the negative sign in the value of the optical rotation mean?

▮ The relation between the specific rotation ($[\alpha]$), the optical rotation (α), the path length usually expressed in decimeters (l), and the density (ρ) is given by

$$[\alpha]_\lambda^T = \alpha / l\rho \tag{19.32}$$

Substituting the data into *(19.32)* gives

$$[\alpha]_D^{20\,°C} = \frac{-16.4°}{(1.00\ cm)[(10^{-1}\ dm)/(1\ cm)](1.0097\ g \cdot cm^{-3})} = -162° \cdot cm^3 \cdot dm^{-1} \cdot g^{-1}$$

Negative values of α refer to levorotatory substances—those that rotate plane-polarized light counterclockwise (left) as viewed toward the light source.

19.80 Using the results of Problem 19.79, calculate the molar rotation of nicotine.

▮ The molar rotation ($[\Phi]$) is given by

$$[\Phi]_\lambda^T = [\alpha]_\lambda^T M/100 \tag{19.33}$$

Substituting the data into *(19.33)* gives

$$[\Phi]_D^{20\,°C} = \frac{(-162° \cdot cm^3 \cdot dm^{-1} \cdot g^{-1})(162.24\ g \cdot mol^{-1})}{100} = -263° \cdot cm^3 \cdot dm^{-1} \cdot mol^{-1}$$

19.81 A handbook gives the following equation for the specific rotation of aqueous solutions of sucrose:

$$[\alpha]_D^{20\,°C}/(° \cdot dm^{-1}) = 66.412 + 0.012\,67d - 0.000\,376d^2$$

where d is the concentration of the sucrose in grams per 100 g of solution. What is the value of $[\alpha]_D^{20\,°C}$ for a solution containing 5.00 g of sucrose in 95.00 g of H_2O? What does the positive sign in the value mean?

▮ Substituting the data into the expression for $[\alpha]$ gives

$$[\alpha]_D^{20\,°C} = 66.412 + (0.1267)(5.00) - (0.000\,376)(5.00)^2 = 66.466° \cdot dm^{-1}$$

The units of degrees per decimeter are commonly used and are the same as $° \cdot cm^3 \cdot dm^{-1} \cdot g^{-1}$. The positive value refers to a dextrotatory substance—one that rotates plane-polarized light clockwise (right) as viewed toward the light source.

19.82 The solution described in Problem 19.81 was placed in a 2.00-dm polarimeter cell. At what angle was the light rotated? The density of the solution is $1.018\ g \cdot cm^{-3}$.

▮ For solutions, the relation between α and $[\alpha]$ is given by

$$[\alpha]_\lambda^T = \alpha / lc \tag{19.34}$$

where c is the concentration of the solute in grams per 100 cm^3. The concentration of the solution is

$$c = \left(\frac{5.00\ g\ sucrose}{100.00\ g\ solution}\right)\left(\frac{1.018\ g\ solution}{1\ cm^3\ solution}\right) = 5.09\ g\ sucrose\ per\ 100\ cm^3$$

Solving *(19.34)* for α and substituting the data gives

$$\alpha = (66.466° \cdot cm^3 \cdot dm^{-1} \cdot g^{-1})(2.00\ cm)[(5.09\ g)/(100\ cm^3)] = 6.77°$$

19.83 Calculate the difference in the indices of refraction for left- and right-polarized light for the solution described in Problem 19.82.

▮ The optical rotation is related to the difference in the indices of refraction $(n_L - n_R)$ by

$$\alpha = \frac{180°}{\lambda}(n_L - n_R)l \tag{19.35}$$

Solving *(19.35)* for $n_L - n_R$ and substituting the data from Problem 19.82 gives

$$n_L - n_R = \frac{(6.77°)(589.3\ \text{nm})[(10^{-9}\ \text{m})/(1\ \text{nm})]}{(180°)(2.00\ \text{dm})[(10^{-1}\ \text{m})/(1\ \text{dm})]} = 1.11 \times 10^{-7}$$

19.84 The optical rotation of a sample of α-D-glucose is 112.0°, and that of β-D-glucose is 18.7°. A mixture of the two sugars had an optical rotation of 52.7°. What is the composition of the mixture?

▮ The total optical activity of a mixture is the sum of the optical activities of the individual components multiplied by the respective mole fractions. Recognizing that the sum of the mole fractions for a binary mixture is unity,

$$52.7° = x(\alpha)(112.0°) + x(\beta)(18.7°) = x(\alpha)(112.0°) + [1 + x(\alpha)](18.7°)$$

which gives $x(\alpha) = 0.364$.

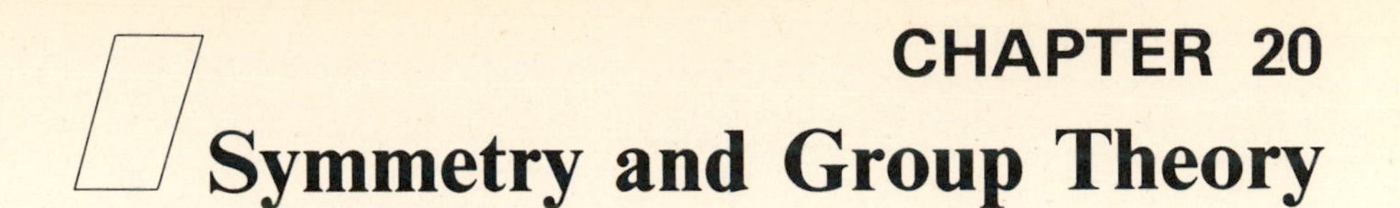

CHAPTER 20
Symmetry and Group Theory

20.1 SYMMETRY OPERATIONS AND ELEMENTS

20.1 Describe the symmetry elements found in the letter D.

▮ There are four operations that generate a configuration that is indistinguishable from the original letter (see Fig. 20-1a): (1) If the letter is simply left alone, an equivalent letter is certainly obtained—thus the identity element is present; (2) if the letter is rotated by 180° out of and into the plane of the paper around the axis shown, an equivalent letter is formed—thus an axis of proper rotation is present; (3) if a plane is constructed perpendicular to the plane of the letter as shown and the various parts of the letter are reflected, an equivalent letter is formed—thus a mirror plane is present; (4) the plane of the letter is, by the same reasoning, a mirror plane.

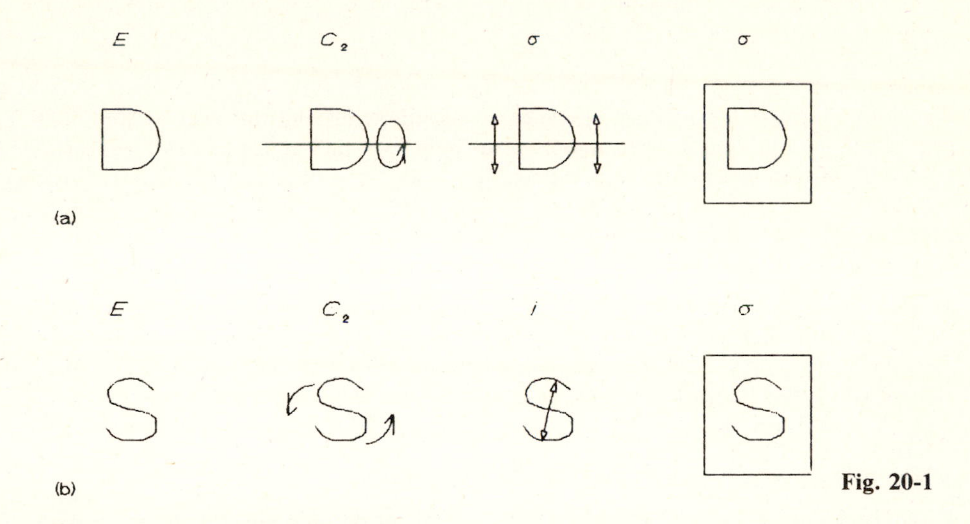

Fig. 20-1

20.2 Describe the symmetry elements found in the letter S.

▮ There are four operations that generate a configuration that is indistinguishable from the original letter (see Fig. 20-1b): (1) If the letter is simply left alone, an equivalent letter is certainly obtained—thus the identity element is present; (2) if the letter is rotated 180° around an axis perpendicular to the letter, an equivalent letter is formed—thus an axis of proper rotation is present; (3) if the various points defining the letter are projected equivalently through a point located at the middle of the letter, an equivalent letter is formed—thus a center of inversion is present; (4) the plane of the letter is a mirror plane.

20.3 Describe the symmetry elements found in *cis*-dichloroethylene and in *trans*-dichloroethylene.

▮ These two molecules

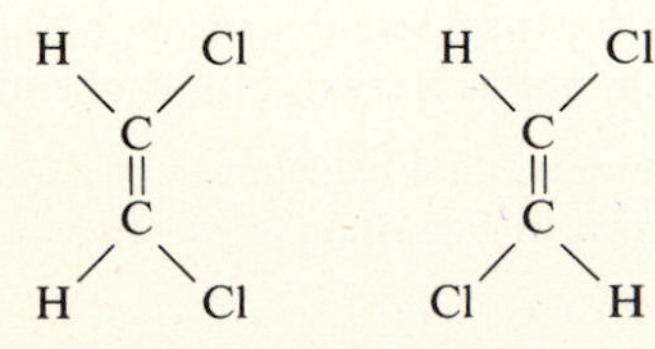

are exactly like the letters D and S, respectively, and contain the same symmetry elements as specified in Problems 20.1 and 20.2. The only major difference in the structures is that the positions of the atoms are the important factors when discussing molecules.

20.4 Prepare an orthographic projection for the identity symmetry operation (E). What happens to the coordinates (x, y, z) of the motif as a result of this operation?

▮ An orthographic projection is essentially a top view of a three-dimensional sketch with a plane represented by a dashed circle, an axis perpendicular to this plane by a dot, and a generally placed motif by a solid dot above

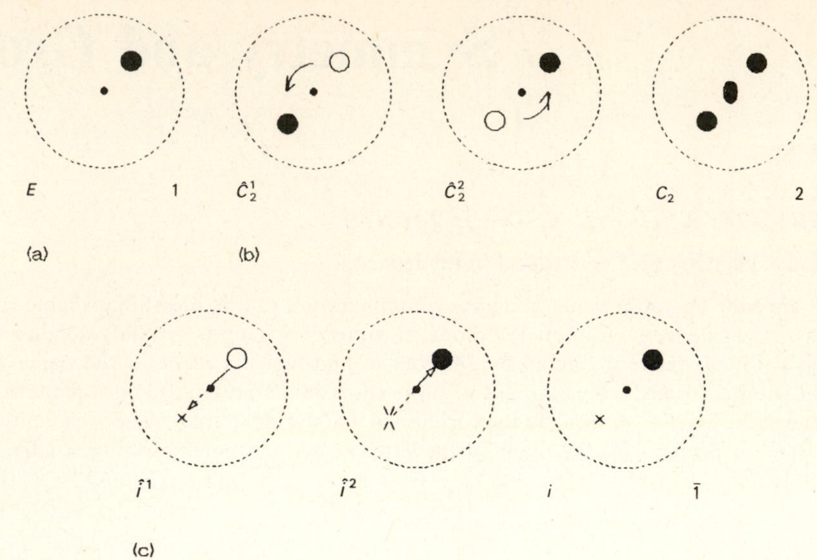

Fig. 20-2

the plane. Because the identity operation is essentially leaving the system alone, only one equivalent point is generated (see Fig. 20-2a). The symbols at the bottom of the projection represent the Schönflies notation and the Hermann–Mauguin notation, respectively. The (x, y, z) coordinates are unchanged by this operation.

20.5 Prepare orthographic projections for the rotation about a twofold axis ($\hat{C}_2^k$). What happens to the coordinates (x, y, z) of the motif as a result of this operation?

▌ The basic operation for a C_n axis of symmetry rotates the motif around the axis by an angle of $2\pi/n$ from the original point. This process is performed k times until the original point is recreated (the identity operation). The operations and the set of equivalent points for the C_2 element are shown in Fig. 20-2b. The first operation changes the coordinates from (x, y, z) to $(-x, -y, z)$, and the second operation changes the coordinates from $(-x, -y, z)$ to (x, y, z).

20.6 Which of the operations shown in Fig. 20-2b are distinct?

▌ An operation is said to be *distinct* if the equivalent position generated for the system cannot be generated in another way that is less complicated. The C_2^1 operation is distinct, but because the C_2^2 operation is essentially the same as E, it is not distinct.

20.7 Prepare orthographic projections for the inversion operation ($\hat{i}^k$). What happens to the coordinates (x, y, z) of the motif as a result of this operation? Are any of the operations distinct?

▌ The basic operation corresponding to the element i or $\bar{1}$ consists of projecting a motif through a center of symmetry located at the point of intersection between the plane and the axis (see Fig. 20-2c). Because the point now lies below the plane, it is represented by an ×. As a result of the $\hat{i}^1$ operation, the coordinates change from (x, y, z) to $(-x, -y, -z)$, and as a result of the $\hat{i}^2$ operation, the coordinates change from $(-x, -y, -z)$ to (x, y, z). Only $\hat{i}^1$ is distinct.

20.8 Prepare orthographic projections for the operation corresponding to a vertical mirror plane ($\hat{\sigma}_v^k$). What happens to the coordinates (x, y, z) of the motif as a result of this operation?

▌ The basic reflection operation creates a mirror image of a motif equivalent from and perpendicular to a plane. As a result of the operations shown in Fig. 20-3a, σ_v^1 changes (x, y, z) to $(x, -y, z)$ and σ_v^2 changes $(x, -y, z)$ to (x, y, z).

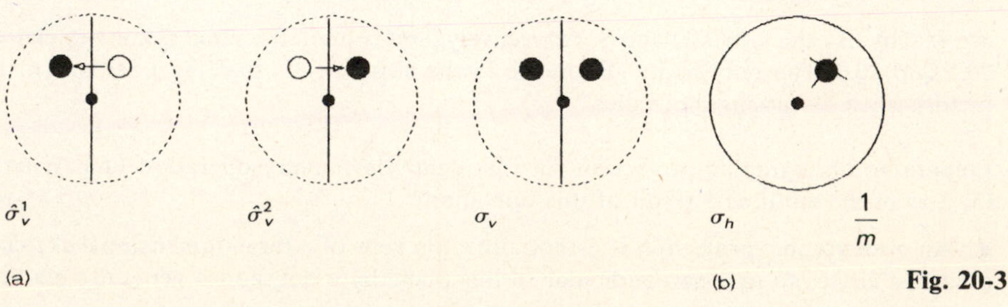

Fig. 20-3

20.9 Prepare an orthographic projection for the equivalent points corresponding to a horizontal mirror plane (σ_h).

▌ The reflection is done directly into the plane of the paper in this case to give two equivalent points at (x, y, z) and $(x, y, -z)$. The sketch is shown in Fig. 20-3b.

20.10 Prepare orthographic projections for the equivalent points for the rotoreflection operations $\hat{S}_1^k$ and $\hat{S}_2^k$.

▌ The basic operation corresponding to the element S_n or $\tilde{n}$ consists of a rotation about an axis of $2\pi/n$ and a reflection through a plane perpendicular to the axis. (For some values of n, the operation $\hat{S}_n^k$ will not produce the same result as $\hat{E}$ until $k = 2n$.) The projections are shown in Fig. 20-4a.

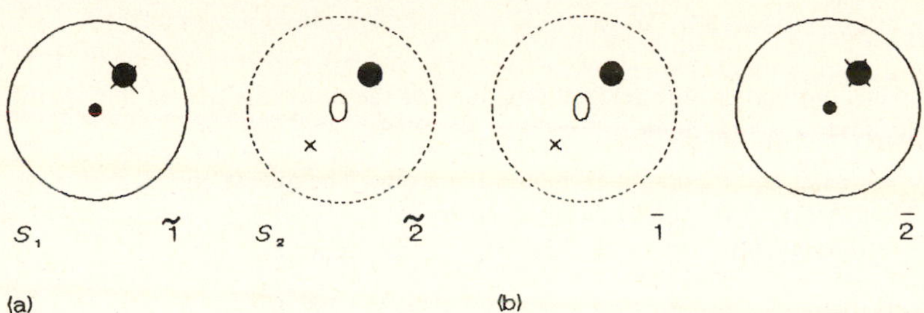

Fig. 20-4

20.11 Prepare orthographic projections for the equivalent points for the rotoinversion operations $\bar{1}$ and $\bar{2}$. Compare the results to those obtained in Problem 20.10.

▌ The basic operation corresponding to the element $\bar{n}$ consists of a rotation of $2\pi/n$ about an axis and an inversion through the point of interaction between the plane and the axis. (There will be either n or $2n$ points alternating above and below the orthographic projection plane.) The projections are shown in Fig. 20-4b. Note that $\bar{1} = \tilde{2}$ and $\bar{2} = \tilde{1}$. Rotoreflection is used primarily for describing molecular geometry, and rotoinversion is used primarily in crystallography.

20.12 Determine what happens to the relative positions of the atoms in *cis*-dichloroethylene as the operations $\hat{E}$, $\hat{C}_2^1$, $\hat{\sigma}_v^1$, and $\hat{\sigma}_h^1$ are performed. See Fig. 20-5a.

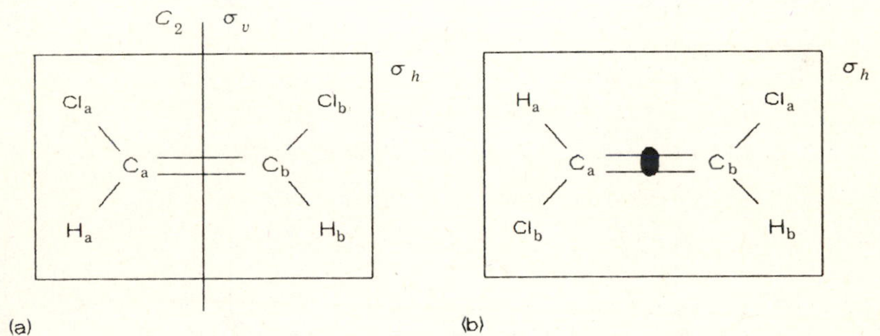

Fig. 20-5

▌ The identity operation leaves all atoms unchanged. Both the $\hat{C}_2^1$ operation and the σ_v^1 operation exchange all like atoms: $Cl_a \to Cl_b$, $Cl_b \to Cl_a$, $C_a \to C_b$, $C_b \to C_a$, $H_a \to H_b$, and $H_b \to H_a$. The σ_h operation leaves all atoms unchanged.

20.13 Determine what happens to the relative positions of the atoms in *trans*-dichloroethylene as the operations $\hat{E}$, $\hat{C}_2^1$, $\hat{\sigma}_h^1$, and $\hat{i}^1$ are performed. See Fig. 20-5b.

▌ The identity operation and the σ_h operation leave all atoms unchanged. Both the C_2^1 (the axis is represented by the dark oval) operation and the i^1 operation exchange all like atoms: $Cl_a \to Cl_b$, $Cl_b \to Cl_a$, $H_a \to H_b$, $H_b \to H_a$, $C_a \to C_b$, and $C_b \to C_a$.

20.14 Determine what happens to the relative positions of the atoms in CO_2 as the operations $\hat{E}$, $\hat{i}^1$, σ_h^1, σ_v^1, C_∞^k, and C_2^1 are performed. See Fig. 20-6a.

▌ The operations $\hat{E}$, σ_v^1, and C_∞^h do not exchange any atoms. The operations $\hat{i}^1$, σ_h^1, and C_2^1 exchange the O atoms: $O_a \to O_b$ and $O_b \to O_a$.

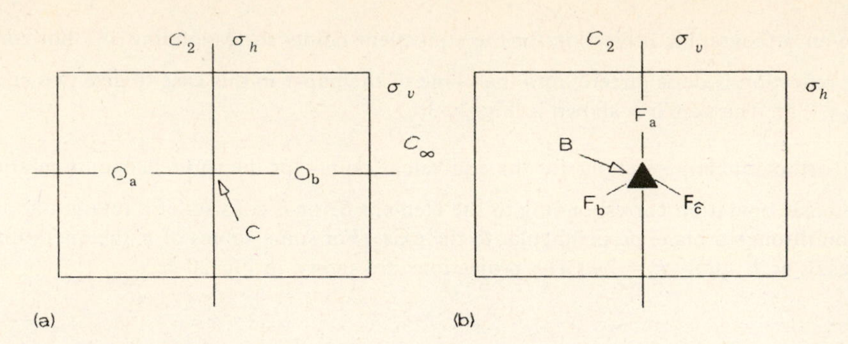

Fig. 20-6

20.15 Determine what happens to the relative positions of the atoms in BF_3 as the operations $\hat{E}$, $\hat{\sigma}_h^1$, $\hat{\sigma}_v^1$, $\hat{C}_3^1$, $\hat{C}_3^2$, and $\hat{C}_2^1$ are performed. See Fig. 20-6b.

▮ The operations $\hat{E}$ and σ_h^1 do not exchange any atoms. The operations C_2^1 and σ_v^1 exchange F_b and F_c: $F_b \rightarrow F_c$, $F_c \rightarrow F_b$. The operation C_3^1 rotates all F atoms by 120°: $F_a \rightarrow F_b$, $F_b \rightarrow F_c$, and $F_c \rightarrow F_a$. The operation C_3^2 rotates all F atoms by 240°: $F_a \rightarrow F_c$, $F_b \rightarrow F_a$, and $F_c \rightarrow F_b$.

20.2 POINT GROUPS

20.16 The elements in the $\mathscr{D}_2$ point group are E, $C_2(x)$, $C_2(y)$, and $C_2(z)$. Show that $\hat{A} \times \hat{B}$ (where the product means that the operation $\hat{B}$ is performed and then operation $\hat{A}$ is performed on the result) generates $\hat{F}$, where $\hat{F}$ is also a symmetry operation in the group.

▮ As an example, consider $\hat{C}_2(z) \times \hat{C}_2(z) = \hat{F}$. As shown in Fig. 20-7a, the two rotations carried out in succession generates the original motif, giving $\hat{F} = \hat{E}$ as the result. Note that all the rest of the combinations of operations simply generate another element in the group. The projections for $\hat{C}_2(y) \times \hat{C}_2(y) = \hat{C}_2(x) \times \hat{C}_2(x) = \hat{E}$ and for $\hat{E} \times \hat{B} = \hat{B} \times \hat{E} = \hat{B}$ are not shown.

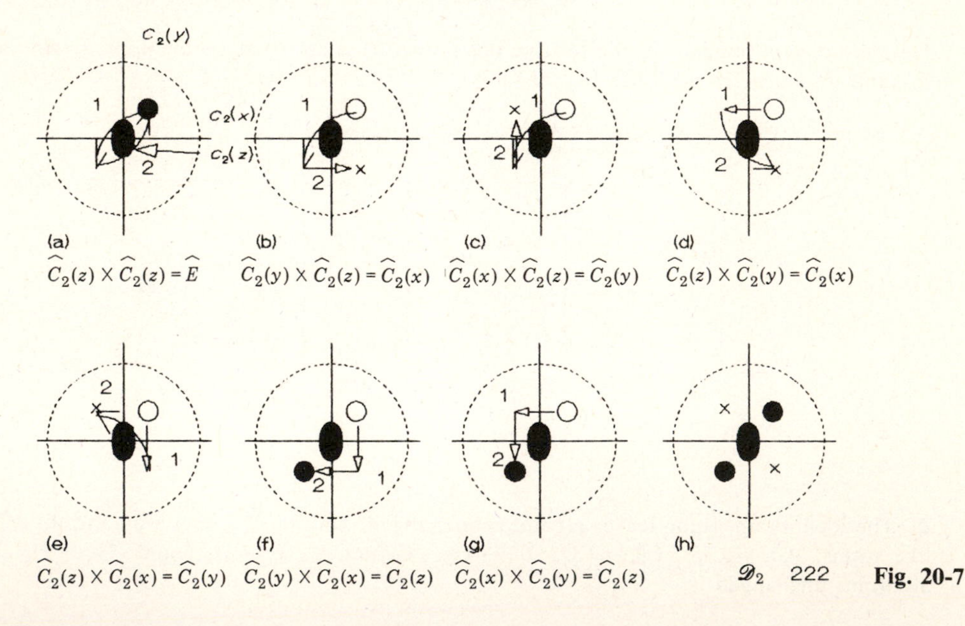

Fig. 20-7

20.17 Show that the point group described in Problem 20.16 contains an identity operation.

▮ The concept of the identity operation is such that $\hat{E} \times \hat{B} = \hat{B} \times \hat{E} = \hat{B}$. Because the operation $\hat{E}$ means to leave the motif alone, performing $\hat{B}$ and then leaving the motif alone must equal the result of performing $\hat{B}$ alone. Likewise, performing $\hat{E}$ first does not change the system, and then performing the operation $\hat{B}$ must equal the result of performing $\hat{B}$ alone. (Note that drawings involving the identity operation were not included in Fig. 20-7.)

20.18 Does the *commutative law*, $\hat{A} \times \hat{B} = \hat{B} \times \hat{A}$, hold for the point group described in Problem 20.16?

▮ A comparison of the various pairs of operations done in opposite order shows that the commutative law is valid for this point group—this is *not* generally true for many point groups. For example, the sketches in Fig. 20-7b and Fig. 20-7d show that $\hat{C}_2(y) \times \hat{C}_2(z) = \hat{C}_2(z) \times \hat{C}_2(y) = \hat{C}_2(x)$.

20.19 Determine the inverses of the operations for the point group described in Problem 20.16.

▌ The *inverse operation* ($\hat{A}^{-1}$) is defined such that $\hat{A}^{-1} \times \hat{A} = \hat{A} \times \hat{A}^{-1} = \hat{E}$. Inspection of Fig. 20-7a shows that the inverse of $\hat{C}_2(z)$ is $\hat{C}_2(z)^{-1} = \hat{C}_2(z)$; no other combination of $\hat{A} \times \hat{C}_2(z)$ will generate the result of the operation $\hat{E}$. In this point group, all of the inverses are the operations themselves. This is *not* generally true for many point groups.

20.20 Does the *associative law*, $\hat{A} \times (\hat{B} \times \hat{C}) = (\hat{A} \times \hat{B}) \times \hat{C}$, hold for the point group described in Problem 20.16?

▌ This relation is generally true for all point groups. As an example, consider the combination of operations given by $\hat{C}_2(z) \times [\hat{C}_2(x) \times \hat{C}_2(y)]$. Inspection of Fig. 20-7g shows that $\hat{C}_2(x) \times \hat{C}_2(y) = \hat{C}_2(z)$, so that $\hat{C}_2(z) \times [\hat{C}_2(x) \times \hat{C}_2(y)] = \hat{C}_2(z) \times \hat{C}_2(z)$, and from Fig. 20.7a, $\hat{C}_2(z) \times \hat{C}_2(z) = \hat{E}$. For the combination of $[\hat{C}_2(z) \times \hat{C}_2(x)] \times \hat{C}_2(y)$, inspection of Fig. 20-7e shows that $\hat{C}_2(z) \times \hat{C}_2(x) = \hat{C}_2(y)$, and it is easily shown that $\hat{C}_2(y) \times \hat{C}_2(y) = \hat{E}$.

20.21 The *conjugate* of operation $\hat{A}$ is $\hat{B}$ if $\hat{X}^{-1} \times \hat{A} \times \hat{X} = \hat{B}$. Determine the conjugate of $\hat{C}_2(z)$ in the point group described in Problem 20.16.

▌ Performing the operations in succession (from right to left), with the help of Fig. 20-7, gives

$$\hat{E}^{-1} \times \hat{C}_2(z) \times \hat{E} = \hat{E}^{-1} \times \hat{C}_2(z) = \hat{C}_2(z)$$

$$\hat{C}_2(z)^{-1} \times \hat{C}_2(z) \times \hat{C}_2(z) = \hat{C}_2(z) \times \hat{E} = \hat{C}_2(z)$$

$$\hat{C}_2(y)^{-1} \times \hat{C}_2(z) \times \hat{C}_2(y) = \hat{C}_2(y) \times \hat{C}_2(x) = \hat{C}_2(z)$$

$$\hat{C}_2(x)^{-1} \times \hat{C}_2(z) \times \hat{C}_2(x) = \hat{C}_2(x) \times \hat{C}_2(y) = \hat{C}_2(z)$$

In this point group, all of the conjugates of the operations are the operations themselves. This is *not* generally true for many point groups.

20.22 Prepare the multiplication table for the group of distinct symmetry operations in the point group described in Problem 20.16.

▌ The first two steps in preparing the table are to make projections of the known distinct operations and for the various multiplications (see Fig. 20-7). The third step is to prepare a table with the operations $\hat{A}$ listed at the left, the operations $\hat{B}$ listed at the top, and the products $\hat{F}$ listed within the table (see Table 20-1).

Table 20-1

$\mathscr{D}_2$	$\hat{E}$	$\hat{C}_2(z)$	$\hat{C}_2(y)$	$\hat{C}_2(x)$ $= \hat{B}$
$\hat{A} = \hat{E}$	$\hat{E}$	$\hat{C}_2(z)$	$\hat{C}_2(y)$	$\hat{C}_2(x)$
$\hat{C}_2(z)$	$\hat{C}_2(z)$	$\hat{E}$	$\hat{C}_2(x)$	$\hat{C}_2(y)$
$\hat{C}_2(y)$	$\hat{C}_2(y)$	$\hat{C}_2(x)$	$\hat{E}$	$\hat{C}_2(z)$
$\hat{C}_2(x)$	$\hat{C}_2(x)$	$\hat{C}_2(y)$	$\hat{C}_2(z)$	$\hat{E}$

20.23 Prepare a projection of the equivalent points of the $\mathscr{D}_2$ point group.

▌ All of the various operations shown in Fig. 20-7a–g deal with the same four motifs over and over. These are located at (x, y, z), $(-x, y, -z)$, $(-x, -y, z)$, and $(x, -y, -z)$. The projection is shown in Fig. 20-7h. Usually one begins with the knowledge of the equivalent points—atoms in a molecule or the faces on a crystal—and works backward to identify the point group.

20.24 The symmetry elements in the $\mathscr{C}_{2h}$ point group include E; C_2, which also is an S_2; i; and σ_h. Prepare the multiplication table for the group of distinct symmetry operations and a projection of the equivalent points.

▌ The orthographic projections of the operations are shown in Figs. 20-8a–g, and the multiplication table is given in Table 20-2. The projection of the equivalent points is shown in Fig. 20-8h.

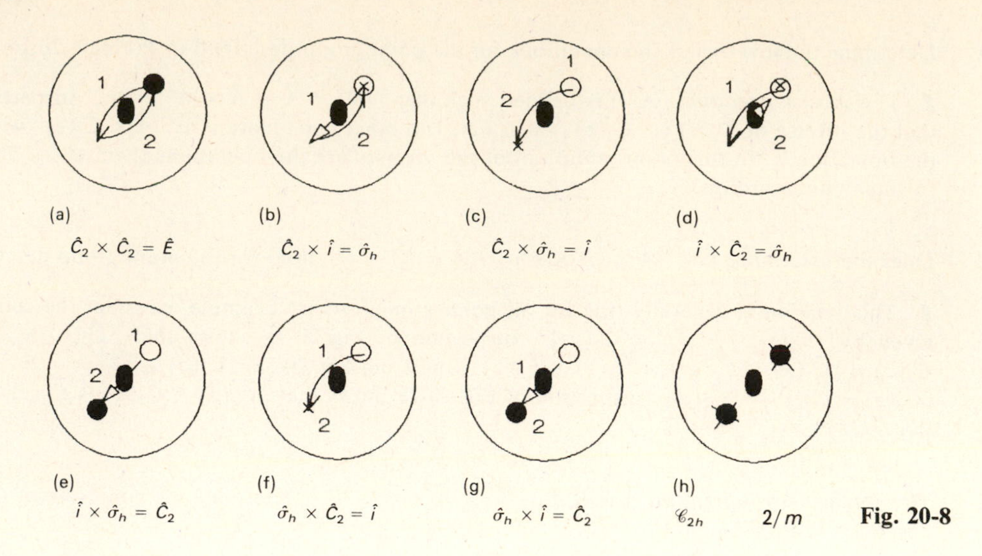

Fig. 20-8

Table 20-2

$\mathscr{C}_{2h}$	$\hat{E}$	$\hat{C}_2$	$\hat{i}$	$\hat{\sigma}_h = \hat{B}$
$\hat{A} = \hat{E}$	$\hat{E}$	$\hat{C}_2$	$\hat{i}$	$\hat{\sigma}_h$
$\hat{C}_2$	$\hat{C}_2$	$\hat{E}$	$\hat{\sigma}_h$	$\hat{i}$
$\hat{i}$	$\hat{i}$	$\hat{\sigma}_h$	$\hat{E}$	$\hat{C}_2$
$\hat{\sigma}_h$	$\hat{\sigma}_h$	$\hat{i}$	$\hat{C}_2$	$\hat{E}$

20.25 Determine the point group for NF_3(g) (see Fig. 18-9a).

▌ The flowchart in Fig. 20-9 can be used to determine the point group to which a molecule belongs. Using the flowchart, the following analysis can be made: (1) Are there ∞ C_∞ axes present? no; (2) is there a pentagonal dodecahedron or icosahedron present? no; (3) are there four C_3 axes at 50°44′? no; (4) is there at least one C_n where $n \geq 2$? yes, C_3; (5) is there an S_n present? no; (6) are there n C_2 axes perpendicular to C_n? no; (7) are there any σ_h planes present? no; (8) are there n σ_v planes present? yes, 3; therefore $\mathscr{C}_{3v}$.

20.26 Determine the point group for MgF_2(g) (see Fig. 18-9b).

▌ Using Fig. 20-9, the following analysis can be made: (1) Are there ∞ C_∞ axes present? no; (2) is there a pentagonal dodecahedron or icosahedron present? no; (3) are there four C_3 axes at 50°44′? no; (4) is there at least one C_n where $n \geq 2$? yes, C_∞; (5) is there an S_{2n} present? yes; (6) are there other elements excluding i? yes; (7) are there n C_2 axes perpendicular to C_n? yes; (8) is there a σ_h perpendicular to C_n? yes; therefore $\mathscr{D}_{\infty h}$.

20.27 Determine the point group for BF_3(g) (see Fig. 18-9c).

▌ Using Fig. 20-9, the following analysis can be made: (1) Are there ∞ C_∞ axes present? no; (2) is there a pentagonal dodecahedron or icosahedron present? no; (3) are there four C_3 axes at 50°44′? no; (4) is there at least one C_n where $n \geq 2$? yes, C_3; (5) is there an S_{2n} present? no; (6) are there n C_2 axes perpendicular to C_n? yes; (7) is there a σ_h perpendicular to C_n? yes; therefore $\mathscr{D}_{3h}$.

20.28 Determine the point group for C_2F_2(g) (see Fig. 18-9e).

▌ Using Fig. 20-9, the following analysis can be made: (1) Are there ∞ C_∞ axes present? no; (2) is there a pentagonal dodecahedron or icosahedron present? no; (3) are there four C_3 axes at 50°44′? no; (4) is there at least one C_n where $n \geq 2$? yes, C_∞; (5) is there an S_{2n} present? yes; (6) are there other elements excluding i? yes; (7) are there n C_2 axes perpendicular to C_n? yes; (8) is there a σ_h perpendicular to C_n? yes; therefore $\mathscr{D}_{\infty h}$.

20.29 Determine the point group for NO(g) (see Fig. 18-9f).

▌ Using Fig. 20-9, the following analysis can be made: (1) Are there ∞ C_2 axes present? no; (2) is there a pentagonal dodecahedron or icosahedron present? no; (3) are there four C_3 axes at 50°44′? no; (4) is there at least

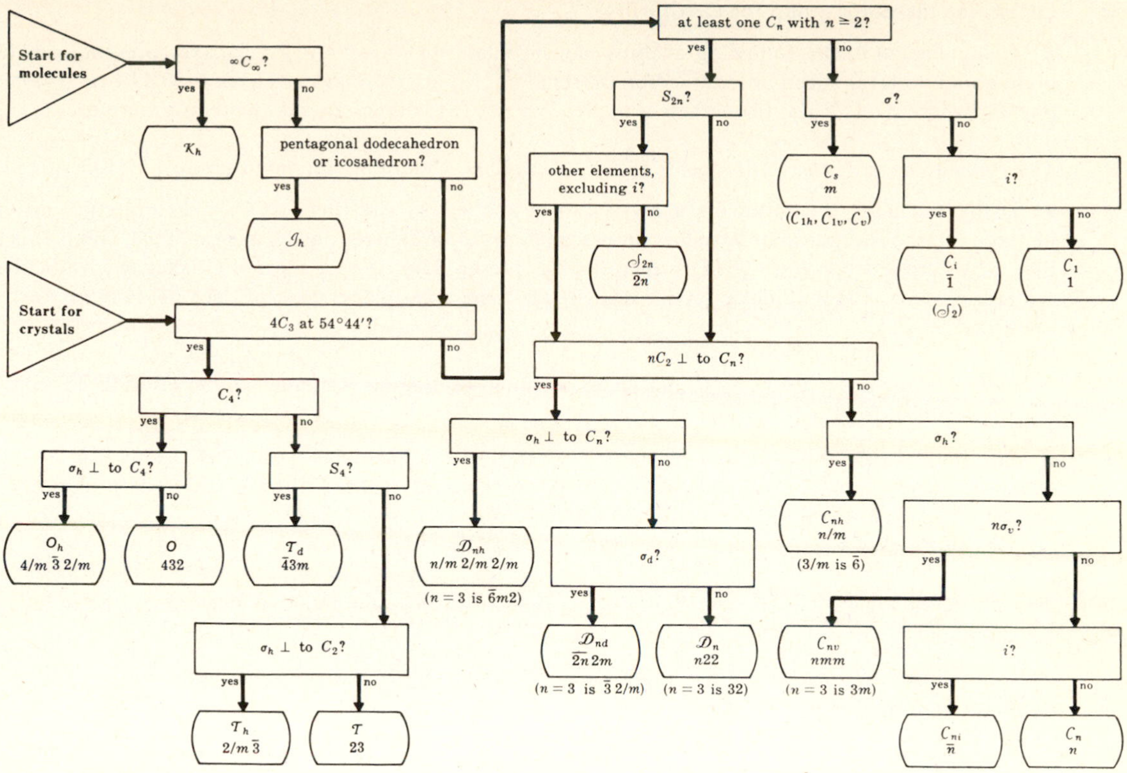

Fig. 20-9

one C_n where $n \geq 2$? yes; C_∞; (5) is there an S_{2n} present? no; (6) are there n C_2 axes perpendicular to C_n? no; (7) is there a σ_h perpendicular to C_n? no; (8) are there n σ_v planes present? yes; therefore $\mathscr{C}_{\infty v}$.

20.30 Determine the point group for $NO_2(g)$ (see Fig. 18-9g).

▮ Using Fig. 20-9, the following analysis can be made: (1) Are there ∞ C_∞ axes present? no; (2) is there a pentagonal dodecahedron or icosahedron present? no; (3) are there four C_3 axes at 50°44′? no; (4) is there at least one C_n where $n \geq 2$? yes, C_2; (5) is there an S_{2n} present? no; (6) are then n C_2 axes perpendicular to C_n? no; (7) is there a σ_h perpendicular to C_n? no; (8) are there n σ_v planes present? yes; therefore $\mathscr{C}_{2v}$.

20.31 Determine the point group for $NO_2^-(g)$ (see Fig. 18-9h).

▮ The analysis using Fig. 20-9 is exactly the same as for $NO_2(g)$ given in Problem 20.30. The point group is $\mathscr{C}_{2v}$.

20.32 Determine the point group for $IF_5(g)$ (see Fig. 18-10e).

▮ Using Fig. 20-9, the following analysis can be made: (1) Are there ∞ C_∞ axes present? no; (2) is there a pentagonal dodecahedron or icosahedron present? no; (3) are there four C_3 axes at 50°44′? no; (4) is there at least one C_n where $n \geq 2$? yes, C_4; (5) is there an S_{2n} present? no; (6) are there n C_2 axes perpendicular to C_n? no; (7) is there a σ_h perpendicular to C_n? no; (8) are there n σ_v planes present? yes; therefore $\mathscr{C}_{4v}$.

20.33 Determine the point group for *cis*-dichloroethylene(g) (see Fig. 20-5a).

▮ Using Fig. 20-9, the following analysis can be made: (1) Are there ∞ C_∞ axes present? no; (2) is there a pentagonal dodecahedron or icosahedron present? no; (3) are there four C_3 axes at 50°44′? no; (4) is there at least one C_n where $n \geq 2$? yes, C_2; (5) is there an S_{2n} present? no; (6) are there n C_2 axes perpendicular to C_n? no; (7) is there a σ_h perpendicular to C_n? no; (8) are there n σ_v planes present? yes; therefore $\mathscr{C}_{2v}$.

20.34 Determine the point group for *trans*-dichloroethylene(g) (see Fig. 20-5b).

▮ Using Fig. 20-9, the following analysis can be made: (1) Are there ∞ C_∞ axes present? no; (2) is there a pentagonal dodecahedron or icosahedron present? no; (3) are there four C_3 axes at 50°44′? no; (4) is there at least one C_n where $n \geq 2$? yes, C_2; (5) is there an S_{2n} present? no; (6) are there n C_2 axes perpendicular to C_n? no; (7) is there a σ_h perpendicular to C_n? yes; therefore $\mathscr{C}_{2h}$.

20.35 Determine the point group for CHClBrF(g).

▮ Using Fig. 20-9, the following analysis can be made: (1) Are there ∞ C_∞ axes present? no; (2) is there a pentagonal dodecahedron or icosahedron present? no; (3) are there four C_3 axes at 50°44′? no; (4) is there at least one C_n where $n \geq 2$? no; (5) is there a σ present? no; (6) is there an i present? no; therefore $\mathscr{C}_1$.

20.36 The four atoms in H_2O_2(g) do not lie in the same plane. Determine the point group for H_2O_2(g).

▮ Using Fig. 20-9, the following analysis can be made: (1) Are there ∞ C_∞ axes present? no; (2) is there a pentagonal dodecahedron or icosahedron present? no; (3) are there four C_3 axes at 50°44′? no; (4) is there at least one C_n where $n \geq 2$? yes, C_2; (5) is there an S_{2n} present? no; (6) are there n C_2 axes perpendicular to C_n? no; (7) is there a σ_h perpendicular to C_n? no; (8) are there n σ_v planes present? no; (9) is there an i present? no; therefore $\mathscr{C}_2$.

20.37 Elemental phosphorus exists as P_4 molecules in which each atom is bonded to the other three atoms. Determine the point group for P_4(g).

▮ The shape of the molecule is a regular tetrahedron with a P atom at each vertex. Using Fig. 20-9, the following analysis can be made: (1) Are there ∞ C_∞ axes present? no; (2) is there a pentagonal dodecahedron or icosahedron present? no; (3) are there four C_3 axes at 50°44′? yes; (4) is there a C_4 axis present? no; (5) is there an S_4 present? yes; therefore $\mathscr{T}_d$.

20.38 The six ligands in $Fe(CN)_6^{4-}$ are aligned along each respective bonding axis. Determine the point group for this complex ion.

▮ Using Fig. 20-9, the following analysis can be made: (1) Are there ∞ C_∞ axes present? no; (2) is there a pentagonal dodecahedron or icosahedron present? no; (3) are there four C_3 axes at 50°44′? yes; (4) is there a C_4 axis present? yes; (5) is there a σ_h perpendicular to the C_4 axis? yes; therefore $\mathscr{O}_h$.

20.39 Both IF_3(g) and IF_4(g) belong to the $\mathscr{C}_{2v}$ point group. Identify the symmetry elements present.

▮ The entries in Table 20-3 show that the elements present are E, C_2, and two types of σ_v planes.

20.40 Both CF_4(g) and CH_4(g) belong to the $\mathscr{T}_d$ point group, CH_3F(g) and CHF_3(g) belong to the $\mathscr{C}_{3v}$ point group, and CH_2F_2(g) belongs to the $\mathscr{C}_{2v}$ point group. Identify the symmetry elements present in each set of molecules.

▮ For CF_4(g) and CH_4(g), Table 20-3 lists E, $4C_3$, $3C_2(S_4)$, and $6\sigma_d$. For CH_3F(g) and CHF_3(g), Table 20-3 lists E, C_3, and $3\sigma_v$. For CH_2F_2(g), Table 20-3 lists E, C_2, and $2\sigma_v$.

20.3 REPRESENTATIONS OF GROUPS

20.41 Consider the symmetry elements for the $\mathscr{D}_2$ point group: E, $C_2(x)$, $C_2(y)$, and $C_2(z)$. Show that the matrix expression for $\hat{E}$ is

$$\begin{bmatrix} 1 & 0 & 0 \\ 0 & 1 & 0 \\ 0 & 0 & 1 \end{bmatrix}$$

▮ A point given by (x_1, y_1, z_1) is taken by $\hat{E}$ into the point (x_2, y_2, z_2), where

$$x_2 = \hat{E} \times x_1 = x_1 \qquad y_2 = \hat{E} \times y_1 = y_1 \qquad z_2 = \hat{E} \times z_1 = z_1$$

which can be stated as

$$x_2 = 1x_1 + 0y_1 + 0z_1 \qquad y_2 = 0x_1 + 1y_1 + 0z_1 \qquad z_2 = 0x_1 + 0y_1 + 1z_1$$

or in matrix form as

$$\begin{bmatrix} x_2 \\ y_2 \\ z_2 \end{bmatrix} = \begin{bmatrix} 1 & 0 & 0 \\ 0 & 1 & 0 \\ 0 & 0 & 1 \end{bmatrix} \begin{bmatrix} x_1 \\ y_1 \\ z_1 \end{bmatrix}$$

20.42 Determine the matrix expression for $\hat{C}_2(x)$ in the $\mathscr{D}_2$ point group.

▮ A point given by (x_1, y_1, z_1) is taken by $\hat{C}_2(x)$ into the point (x_2, y_2, z_2) (see Fig. 20-7d), where

$$x_2 = \hat{C}_2(x) \times x_1 = x_1 \qquad y_2 = \hat{C}_2(x) \times y_1 = -y_1 \qquad z_2 = \hat{C}_2(x) \times z_1 = -z_1$$

Table 20-3

Point Group	Special Comments	E	Axes of Rotation	σ_h	σ_v	i
$\mathscr{K}_h$	Sphere	√	∞C_∞	√	$\infty\sigma_d$	√
$\mathscr{I}_h$	Regular pentagonal dodecahedron (12 pentagons) or icosahedron (20 triangles)	√	$6C_5\,(S_{10})$ $15C_2\,(S_4)$		$15\sigma_d$	√
$\mathscr{O}_h$ $4/m\;\bar{3}\;2/m$	Octahedron or cube	√	$4C_3\,(S_6)$ $3C_4\,(S_4)$ $6C_2$	√	$6\sigma_d$	√
$\mathscr{O}$ 432		√	$4C_3$ $3C_4$ $6C_2$			
$\mathscr{T}_d$ $\bar{4}3m$	Tetrahedron	√	$4C_3$ $3C_2\,(S_4)$		$6\sigma_d$	
$\mathscr{T}_h$ $2/m\;\bar{3}$		√	$4C_3\,(S_6)$ $3C_2$	√		√
$\mathscr{T}$ 23		√	$4C_3$ $3C_2$			
$\mathscr{S}_{2n}$ $\overline{2n}$	$n = 2, 3, 4, \ldots$ $\mathscr{S}_1$ is $\mathscr{C}_s$, $\mathscr{S}_2$ is $\mathscr{C}_i$ $\mathscr{S}_n$ is $\mathscr{C}_{nh}$ if n odd	√	$S_{2n}\,(C_n)$			If n odd
$\mathscr{D}_{nh}$ $n/m\;2/m\;2/m$	$n = 2, 3, 4, \ldots$ $\mathscr{D}_{1h}$ is $\mathscr{C}_{2v}$	√	$C_n\,(S_n)$ nC_2	√	$n\sigma_v$	If n even
$\mathscr{D}_{nd}$ $\overline{2n}\;2m$	$n = 2, 3, 4, \ldots$ D_{1d} is $\mathscr{C}_{2h}$	√	$C_n\,(S_{2n})$ nC_2		$n\sigma_d$	If n odd
$\mathscr{D}_n$ $n22$	$n = 2, 3, 4, \ldots$ $\mathscr{D}_1$ is $\mathscr{C}_2$	√	C_n nC_2			
$\mathscr{C}_{nh}$ n/m	$n = 2, 3, 4, \ldots$ $\mathscr{C}_{1h}$ is $\mathscr{C}_s$	√	$C_n\,(S_n)$	√		If n even
$\mathscr{C}_{nv}$ nmm	$n = 2, 3, 4, \ldots$ $\mathscr{C}_{1v}$ is $\mathscr{C}_s$	√	C_n		$n\sigma_v$	
$\mathscr{C}_{ni}$ $\bar{n}$	$n = 3, 5, 7, \ldots$ $\mathscr{C}_{1i}$ is $\mathscr{C}_i$ For even values of n, $\mathscr{C}_{ni}$ is $\mathscr{C}_{nh}$ if $n/2$ is odd and $\mathscr{S}_{2n}$ if $n/2$ is even.	√	$C_n\,(S_{2n})$			√
$\mathscr{C}_n$ n	$n = 2, 3, 4, \ldots$	√	C_n			
$\mathscr{C}_s$ m		√		√		
$\mathscr{C}_i$ $\bar{1}$		√				√
$\mathscr{C}_1$ 1		√				

which can be stated as

$$x_2 = 1x_1 + 0y_1 + 0z_1 \qquad y_2 = 0x_1 + (-1)y_1 + 0z_1 \qquad z_2 = 0x_1 + 0y_1 + (-1)z_1$$

or in matrix form as

$$\begin{bmatrix} x_2 \\ y_2 \\ z_2 \end{bmatrix} = \begin{bmatrix} 1 & 0 & 0 \\ 0 & -1 & 0 \\ 0 & 0 & -1 \end{bmatrix} \begin{bmatrix} x_1 \\ y_1 \\ z_1 \end{bmatrix}$$

20.43 Using Table 20-4, write the matrix expression for $\hat{C}_2(z)$ in the $\mathscr{D}_2$ point group.

▮ Letting $m = 1$ and $n = 2$, the table entry becomes

$$\begin{bmatrix} \cos[2\pi(1)/2] & -\sin[2\pi(1)/2] & 0 \\ \sin[2\pi(1)/2] & \cos[2\pi(1)/2] & 0 \\ 0 & 0 & 1 \end{bmatrix} = \begin{bmatrix} -1 & 0 & 0 \\ 0 & -1 & 0 \\ 0 & 0 & 1 \end{bmatrix}$$

Table 20-4

$\hat{E} \longrightarrow \begin{bmatrix} 1 & 0 & 0 \\ 0 & 1 & 0 \\ 0 & 0 & 1 \end{bmatrix}$	$\hat{C}_n(z)^m \longrightarrow \begin{bmatrix} \cos(2\pi m/n) & -\sin(2\pi m/n) & 0 \\ \sin(2\pi m/n) & \cos(2\pi m/n) & 0 \\ 0 & 0 & 1 \end{bmatrix}$	
$\hat{i} \longrightarrow \begin{bmatrix} -1 & 0 & 0 \\ 0 & -1 & 0 \\ 0 & 0 & -1 \end{bmatrix}$	$\hat{S}_n(z)^m \longrightarrow \begin{bmatrix} \cos(2\pi m/n) & -\sin(2\pi m/n) & 0 \\ \sin(2\pi m/n) & \cos(2\pi m/n) & 0 \\ 0 & 0 & -1 \end{bmatrix}$	for odd m
$\hat{\sigma}_h \longrightarrow \begin{bmatrix} 1 & 0 & 0 \\ 0 & 1 & 0 \\ 0 & 0 & -1 \end{bmatrix}$	$\hat{\sigma}_v \longrightarrow \begin{bmatrix} \cos 2\beta & \sin 2\beta & 0 \\ \sin 2\beta & -\cos 2\beta & 0 \\ 0 & 0 & 1 \end{bmatrix}$	β = angle between σ_v and the x axis

20.44 Using Table 20-4, determine the matrix representations for the elements in the $\mathscr{C}_{2h}$ point group: E, C_2, σ_h, and i.

▮ For $\hat{E}$, σ_h, and i, the table gives

$$\hat{E} \rightarrow \begin{bmatrix} 1 & 0 & 0 \\ 0 & 1 & 0 \\ 0 & 0 & 1 \end{bmatrix} \qquad \hat{i} \rightarrow \begin{bmatrix} -1 & 0 & 0 \\ 0 & -1 & 0 \\ 0 & 0 & -1 \end{bmatrix} \qquad \hat{\sigma}_h \rightarrow \begin{bmatrix} 1 & 0 & 0 \\ 0 & 1 & 0 \\ 0 & 0 & -1 \end{bmatrix}$$

and letting $m = 1$ and $n = 2$, the table gives

$$\hat{C}_2 \rightarrow \begin{bmatrix} \cos[2\pi(1)/2] & -\sin[2\pi(1)/2] & 0 \\ \sin[2\pi(1)/2] & \cos[2\pi(1)/2] & 0 \\ 0 & 0 & 1 \end{bmatrix} = \begin{bmatrix} -1 & 0 & 0 \\ 0 & -1 & 0 \\ 0 & 0 & 1 \end{bmatrix}$$

20.45 Using the matrix representations for $\hat{C}_2$ and $\hat{i}$ determined in Problem 20.44 for the $\mathscr{C}_{2h}$ point group, show that $\hat{C}_2 \times \hat{i} = \hat{\sigma}_h$ as given in Table 20-2.

▮ Performing the multiplication of the respective matrices gives

$$\hat{C}_2 \times \hat{i} = \begin{bmatrix} -1 & 0 & 0 \\ 0 & -1 & 0 \\ 0 & 0 & 1 \end{bmatrix} \begin{bmatrix} -1 & 0 & 0 \\ 0 & -1 & 0 \\ 0 & 0 & -1 \end{bmatrix} = \begin{bmatrix} 1 & 0 & 0 \\ 0 & 1 & 0 \\ 0 & 0 & -1 \end{bmatrix}$$

which is the matrix for $\hat{\sigma}_h$.

20.46 The matrices determined in Problems 20.41–20.43 as well as the matrix expression for $\hat{C}_2(y)$ given by

$$\begin{bmatrix} -1 & 0 & 0 \\ 0 & 1 & 0 \\ 0 & 0 & -1 \end{bmatrix}$$

for the $\mathscr{D}_2$ point group are known as the *representation* of the group (V). Show that this set of square matrices is reducible.

▮ The *order* of the group (h) is equal to the number of distinct operations in the group. In this case, $h = 4$. If l_i represents the dimension of the ith irreducible representation (the number of rows or columns in the square matrices), then

$$\sum_{V_i} l_i^2 = h \tag{20.1}$$

where the sum is done over the irreducible representations V_i. In this case, $l_i = 3$ for each of the four matrices, and $3^2 = 9$. Because *(20.1)* is not satisfied, this representation is reducible.

20.47 Is the representation for the $\mathscr{C}_{2h}$ group given in Problem 20.44 reducible?

▮ The dimension of each matrix is $l_i = 3$, giving $3^2 = 9$. The order of the group is $h = 4$. Because *(20.1)* is not satisfied, this representation is reducible.

20.48 Find the irreducible representations of the $\mathscr{D}_2$ space group. See Problem 20.46 for additional information.

▌ Three of the irreducible representations of this group can be found by converting all of the matrices to the block-diagonal form shown below:

$$V:\quad \hat{E} \to \begin{bmatrix} 1 & 0 & 0 \\ 0 & 1 & 0 \\ 0 & 0 & 1 \end{bmatrix} \qquad \hat{C}_2(z) \to \begin{bmatrix} -1 & 0 & 0 \\ 0 & -1 & 0 \\ 0 & 0 & 1 \end{bmatrix}$$

$$\hat{C}_2(y) \to \begin{bmatrix} -1 & 0 & 0 \\ 0 & 1 & 0 \\ 0 & 0 & -1 \end{bmatrix} \qquad \hat{C}_2(x) \to \begin{bmatrix} 1 & 0 & 0 \\ 0 & -1 & 0 \\ 0 & 0 & -1 \end{bmatrix}$$

Each row of blocks provides an irreducible representation. In this case the blocks are 1×1 matrices. Labeling these representations using the subscripts corresponding to the entries in the character table for this group (see Table 20-5),

$$V_4:\quad \hat{E} \to [1] \qquad \hat{C}_2(z) \to [-1] \qquad \hat{C}_2(y) \to [-1] \qquad \hat{C}_2(x) \to [1]$$
$$V_3:\quad \hat{E} \to [1] \qquad \hat{C}_2(z) \to [-1] \qquad \hat{C}_2(y) \to [1] \qquad \hat{C}_2(x) \to [-1]$$
$$V_2:\quad \hat{E} \to [1] \qquad \hat{C}_2(z) \to [1] \qquad \hat{C}_2(y) \to [-1] \qquad \hat{C}_2(x) \to [-1]$$

Table 20-5

$\mathscr{D}_2$ **representation**	$\hat{E}$	$\hat{C}_2(z)$	$\hat{C}_2(y)$	$\hat{C}_2(x)$	
$V_1 = A_1$	1	1	1	1	x^2, y^2, z^2
$V_2 = B_1$	1	1	−1	−1	z, R_z, xy
$V_3 = B_2$	1	−1	1	−1	y, R_x, xz
$V_4 = B_3$	1	−1	−1	1	x, R_y, yz

The reducible representation V is said to be the *direct sum* of V_2, V_3, and V_4,

$$V = V_2 \oplus V_3 \oplus V_4$$

For the dimension of the fourth irreducible representation, V_1, *(20.1)* gives $l_1^2 + 1^2 + 1^2 + 1^2 = 4$, or $l_1 = 1$. Using methods not discussed here, it can be shown that

$$V_1:\quad \hat{E} \to [1] \qquad \hat{C}_2(z) \to [1] \qquad \hat{C}_2(y) \to [1] \qquad \hat{C}_2(x) \to [1]$$

20.49 Find the irreducible representation of the $\mathscr{C}_{2h}$ point group. See Problem 20.44 for additional information.

▌ Two of the irreducible representations of this group can be formed by converting all of the matrices to the block-diagonal form:

$$V:\quad \hat{E} \to \begin{bmatrix} 1 & 0 & 0 \\ 0 & 1 & 0 \\ 0 & 0 & 1 \end{bmatrix} \qquad \hat{C}_2 \to \begin{bmatrix} -1 & 0 & 0 \\ 0 & -1 & 0 \\ 0 & 0 & 1 \end{bmatrix}$$

$$\hat{i} \to \begin{bmatrix} -1 & 0 & 0 \\ 0 & -1 & 0 \\ 0 & 0 & -1 \end{bmatrix} \qquad \hat{\sigma}_h \to \begin{bmatrix} 1 & 0 & 0 \\ 0 & 1 & 0 \\ 0 & 0 & -1 \end{bmatrix}$$

Assigning each row of blocks to an irreducible representation and using the subscripts corresponding to the entries in the character table for this group (see Table 20-6) gives

$$V_4:\quad \hat{E} \to [1] \qquad \hat{C}_2 \to [-1] \qquad \hat{i} \to [-1] \qquad \hat{\sigma}_h \to [1]$$
$$V_4:\quad \hat{E} \to [1] \qquad \hat{C}_2 \to [-1] \qquad \hat{i} \to [-1] \qquad \hat{\sigma}_h \to [1]$$
$$V_3:\quad \hat{E} \to [1] \qquad \hat{C}_2 \to [1] \qquad \hat{i} \to [-1] \qquad \hat{\sigma}_h \to [-1]$$

Table 20-6

$\mathscr{C}_{2h}$ representation	$\hat{E}$	$\hat{C}_2$	$\hat{i}$	$\hat{\sigma}_h$	
$V_1 = A_g$	1	1	1	1	R_z, x^2, y^2, z^2, xy
$V_2 = B_g$	1	−1	1	−1	R_x, R_y, xz, yz
$V_3 = A_u$	1	1	−1	−1	z
$V_4 = B_u$	1	−1	−1	1	x, y

The reducible representation V is $V = V_3 \oplus 2V_4$. The dimensions of the remaining irreducible representations, V_1 and V_2, are given by *(20.1)* as

$$l_1^2 + l_2^2 + 1^2 + 1^2 = 4$$

or $l_1 = l_2 = 1$. Using methods not discussed here, it can be shown that

$$V_1:\quad \hat{E} \rightarrow [1] \qquad \hat{C}_2 \rightarrow [1] \qquad \hat{i} \rightarrow [1] \qquad \hat{\sigma}_h \rightarrow [1]$$

$$V_2:\quad \hat{E} \rightarrow [1] \qquad \hat{C}_2 \rightarrow [-1] \qquad \hat{i} \rightarrow [1] \qquad \hat{\sigma}_h \rightarrow [-1]$$

20.50 Demonstrate that the elements of the irreducible representation V_4 for the $\mathscr{D}_2$ point group multiply as the symmetry operations of the group.

▮ The elements found in Table 20-5 are

$$\hat{E} \rightarrow [1] \qquad \hat{C}_2(z) \rightarrow [-1] \qquad \hat{C}_2(y) \rightarrow [-1] \qquad \hat{C}_2(x) \rightarrow [1]$$

Examples of the 16 multiplications are shown below and are confirmed by the entries in Table 20-1.

$$\begin{array}{ccc} \hat{E} \times \hat{C}_2(z) = \hat{C}_2(z) & \hat{C}_2(y) \times \hat{C}_2(z) = \hat{C}_2(x) & \hat{C}_2(x) \times \hat{C}_2(z) = \hat{C}_2(y) \\ \downarrow \quad \downarrow \quad \downarrow & \downarrow \quad \downarrow \quad \downarrow & \downarrow \quad \downarrow \quad \downarrow \\ [1] \times [-1] = [-1] & [-1] \times [-1] = [1] & [1] \times [-1] = [-1] \end{array}$$

20.51 Determine the characters (or traces) of the reducible matrices given in Problem 20.48 for the $\mathscr{D}_2$ point group.

▮ The *character* for the operation $\hat{R}$, $\chi(V, \hat{R})$, is defined as

$$\chi(V, \hat{R}) = \sum r_{ij} \tag{20.2}$$

where the r_{ij} are the diagonal elements of the matrix corresponding to $\hat{R}$ in the representation V. The characters of the reducible representation are

$$\chi(V, \hat{E}) = [1] + [1] + [1] = 3 \qquad \chi(V, \hat{C}_2(z)) = [-1] + [-1] + [1] = -1$$

$$\chi(V, \hat{C}_2(y)) = [-1] + [1] + [-1] = -1 \qquad \chi(V, \hat{C}_2(x)) = [1] + [-1] + [-1] = -1$$

20.52 Determine the characters of the irreducible representations given in Problem 20.48 for the $\mathscr{D}_2$ point group.

▮ Applying *(20.2)* gives

$$\begin{array}{llll} \chi(V_1, \hat{E}) = 1 & \chi(V_1, C_2(z)) = 1 & \chi(V_1, C_2(y)) = 1 & \chi(V_1, C_2(x)) = 1 \\ \chi(V_2, \hat{E}) = 1 & \chi(V_2, C_2(z)) = 1 & \chi(V_2, C_2(y)) = -1 & \chi(V_2, C_2(x)) = -1 \\ \chi(V_3, \hat{E}) = 1 & \chi(V_3, C_2(z)) = -1 & \chi(V_3, C_2(y)) = 1 & \chi(V_3, C_2(x)) = -1 \\ \chi(V_4, \hat{E}) = 1 & \chi(V_4, C_2(z)) = -1 & \chi(V_4, C_2(y)) = -1 & \chi(V_4, C_2(x)) = 1 \end{array}$$

20.53 Determine the characters of the matrices of the reducible representation and of the irreducible representations for the $\mathscr{C}_{2h}$ point group given in Problem 20.49.

▮ Using *(20.2)* for the reducible representation gives

$$\chi(V, \hat{E}) = [1] + [1] + [1] = 3 \qquad \chi(V, \hat{C}_2) = [-1] + [-1] + [1] = -1$$

$$\chi(V, \hat{i}) = [-1] + [-1] + [-1] = -3 \qquad \chi(V, \hat{\sigma}_h) = [1] + [1] + [-1] = 1$$

and for the irreducible representations,

$$\begin{array}{llll}
\chi(V_1, \hat{E}) = 1 & \chi(V_1, \hat{C}_2) = 1 & \chi(V_1, \hat{i}) = 1 & \chi(V_1, \hat{\sigma}_h) = 1 \\
\chi(V_2, \hat{E}) = 1 & \chi(V_2, \hat{C}_2) = -1 & \chi(V_2, \hat{i}) = 1 & \chi(V_2, \hat{\sigma}_h) = -1 \\
\chi(V_3, \hat{E}) = 1 & \chi(V_3, \hat{C}_2) = 1 & \chi(V_3, \hat{i}) = -1 & \chi(V_3, \hat{\sigma}_h) = -1 \\
\chi(V_4, \hat{E}) = 1 & \chi(V_4, \hat{C}_2) = -1 & \chi(V_4, \hat{i}) = -1 & \chi(V_4, \hat{\sigma}_h) = 1
\end{array}$$

20.54 Show that the statement $V = V_2 \oplus V_3 \oplus V_4$ given in Problem 20.48 for the $\mathscr{D}_2$ point group is valid. See Problems 20.51 and 20.52 for additional information.

▌ The number of times that an irreducible representation occurs in a reducible representation (v_i) is given by

$$v_i = \frac{1}{h}\sum_{\hat{R}} \chi(V, \hat{R})\chi(V_i, \hat{R}) \tag{20.3}$$

Substituting the results of Problems 20.51 and 20.52 into *(20.3)* for V_1 gives

$$\begin{aligned} v_1 &= \tfrac{1}{4}[\chi(V, \hat{E})\chi(V_1, \hat{E}) + \chi(V, \hat{C}_2(z))\chi(V_1, \hat{C}_2(z)) + \chi(V, \hat{C}_2(y))\chi(V_1, \hat{C}_2(y)) + \chi(V, \hat{C}_2(x))\chi(V_1, \hat{C}_2(x))] \\ &= \tfrac{1}{4}[(3)(1) + (-1)(1) + (-1)(1) + (-1)(1)] = 0 \end{aligned}$$

Likewise, for V_2, V_3, and V_4,

$$v_2 = \tfrac{1}{4}[(3)(1) + (-1)(1) + (-1)(-1) + (-1)(-1)] = 1$$

$$v_3 = \tfrac{1}{4}[(3)(1) + (-1)(-1) + (-1)(1) + (-1)(-1)] = 1$$

$$v_4 = \tfrac{1}{4}[(3)(1) + (-1)(-1) + (-1)(-1) + (-1)(1)] = 1$$

which means that V can be reduced to $V = 0V_1 \oplus 1V_2 \oplus 1V_3 \oplus 1V_4 = V_2 \oplus V_3 \oplus V_4$.

20.55 Show that the statement $V = V_3 \oplus 2V_4$ given in Problem 20.49 for the $\mathscr{C}_{2h}$ point group is valid. See Problem 20.53 for additional information.

▌ Substituting the results of Problem 20.53 into *(20.3)* for V_1, V_2, V_3, and V_4 gives

$$v_1 = \tfrac{1}{4}[(3)(1) + (-1)(1) + (-3)(1) + (1)(1)] = 0$$

$$v_2 = \tfrac{1}{4}[(3)(1) + (-1)(-1) + (-3)(1) + (1)(-1)] = 0$$

$$v_3 = \tfrac{1}{4}[(3)(1) + (-1)(1) + (-3)(-1) + (1)(-1)] = 1$$

$$v_4 = \tfrac{1}{4}[(3)(1) + (-1)(-1) + (-3)(-1) + (1)(1)] = 2$$

which means that V can be reduced to $V = 0V_1 \oplus 0V_2 \oplus 1V_3 \oplus 2V_4 = V_3 \oplus 2V_4$.

20.56 One of the properties of the characters of irreducible representations is

$$\sum_{\hat{R}} \chi(V_i, \hat{R})\chi(V_j, \hat{R}) = h\delta_{ij} \tag{20.4}$$

where $\delta_{ij} = 0$ if $i \neq j$ and $\delta_{ij} = 1$ if $i = j$. Use this property to determine the characters of V_1 for the $\mathscr{D}_2$ point group using the characters of V_2, V_3, and V_4 (see Problem 20.48).

▌ The order of V_1 was determined in Problem 20.48 as $l_1 = 1$, which means that $\chi(V_1, \hat{E}) = 1$. Substituting the results of Problem 20.52 into *(20.4)* for V_1 and V_2 gives

$$\begin{aligned} &\chi(V_1, \hat{E})\chi(V_2, \hat{E}) + \chi(V_1, \hat{C}_2(z))\chi(V_2, \hat{C}_2(z)) + \chi(V_1, \hat{C}_2(y))\chi(V_2, \hat{C}_2(y)) + \chi(V_1, \hat{C}_2(x))\chi(V_2, \hat{C}_2(x)) \\ &\quad = (1)(1) + \chi(V_1, \hat{C}_2(z))(1) + \chi(V_1, \hat{C}_2(y))(-1) + \chi(V_1, \hat{C}_2(x))(-1) = 0 \end{aligned}$$

Likewise for V_1 and V_3 and for V_1 and V_4,

$$(1)(1) + \chi(V_1, \hat{C}_2(z))(-1) + \chi(V_1, \hat{C}_2(y))(1) + \chi(V_1, \hat{C}_2(x))(-1) = 0$$

$$(1)(1) + \chi(V_1, \hat{C}_2(z))(-1) + \chi(V_1, \hat{C}_2(y))(-1) + \chi(V_1, \hat{C}_2(x))(1) = 0$$

Solving simultaneously gives $\chi(V_1, \hat{C}_2(z)) = 1$, $\chi(V_1, \hat{C}_2(y)) = 1$, and $\chi(V_1, \hat{C}_2(x)) = 1$, which agree with the entries in Table 20-5 for V_1.

20.57 Confirm the Mulliken designations for the irreducible representations of the $\mathscr{D}_2$ point group given in Table 20-5.

▌ Upon comparing the entries in Table 20-5 with the code given in Table 20-7, the letter A should be chosen for V_1 because $\chi(V_1, \hat{E}) = 1$ (one-dimensional) and $\chi(V_1, \hat{C}_2(z)) = 1$ [symmetric about the $C_2(z)$ axis], and the

Table 20-7 Mulliken Designation for Irreducible Representations

Designation	Interpretation
Capital letter	
A	One-dimensional and symmetric to rotation of $2\pi/n$ about the principal C_n axis as indicated by $\chi(V_i, \hat{C}_n) = 1$
B	One-dimensional and antisymmetric to rotation of $2\pi/n$ about the principal C_n axis as indicated by $\chi(V_i, \hat{C}_n) = -1$
E	Two-dimensional
T or F	Three-dimensional
G	Four-dimensional
H	Five-dimensional
Numeral subscript	
1	Symmetric to rotation about a secondary C_2 axis as indicated by $\chi(V_i, \hat{C}_2) = 1$
2	Antisymmetric to rotation about a secondary C_2 axis as indicated by $\chi(V_i, \hat{C}_2) = -1$ (if a secondary C_2 axis does not exist, used for σ_v if present)
Letter subscript	
g	Symmetric to inversion as indicated by $\chi(V_i, \hat{i}) = 1$
u	Antisymmetric to inversion as indicated by $\chi(V_i, \hat{i}) = -1$
Superscript	
′	Symmetric to σ_h as indicated by $\chi(V_i, \hat{\sigma}_h) = 1$
″	Antisymmetric to σ_h as indicated by $\chi(V_i, \hat{\sigma}_h) = -1$

subscript 1 should be chosen because $\chi(V_1, \hat{C}_2(y)) = \chi(V_1, \hat{C}_2(x)) = 1$ (symmetric about a secondary C_2 axis). (The use of the subscript for V_1 is optional because there is only one A designation.) Likewise, V_2 is designated B_1 because $\chi(V_2, \hat{E}) = 1$, $\chi(V_2, \hat{C}_2(y)) = -1$, and $\chi(V_2, \hat{C}_2(z)) = 1$. The other irreducible representations essentially involve the same characters for the other $\hat{C}_2$ operations as does V_2, but in a different order. To distinguish between these, the numerical subscripts of 2 and 3 are used. (These can be justified by selecting an alternate C_2 axis as the principal axis and applying the rules in Table 20-7.)

20.58 Confirm the Mulliken designations for the irreducible representations of the $\mathscr{C}_{2h}$ point group given in Table 20-6.

▌ Upon comparing the entries in Table 20-6 with the code given in Table 20-7, the letter A should be chosen for V_1 because $\chi(V_1, \hat{E}) = 1$ and $\chi(V_1, \hat{C}_2) = 1$, and the subscript g should be chosen because $\chi(V_1, \hat{i}) = 1$. Likewise, V_2 is designated B_g because $\chi(V_2, \hat{E}) = 1$, $\chi(V_2, \hat{C}_2) = -1$, and $\chi(V_2, \hat{i}) = 1$; V_3 is designated A_u because $\chi(V_2, \hat{E}) = 1$, $\chi(V_2, \hat{C}_2) = 1$, and $\chi(V_2, \hat{i}) = -1$; and V_4 is designated B_u because $\chi(V_2, \hat{E}) = 1$, $\chi(V_2, \hat{C}_2) = -1$, and $\chi(V_2, \hat{i}) = -1$. The prime and double prime superscripts could be added, but it is not necessary to distinguish between the various designations.

20.4 APPLICATIONS OF GROUP THEORY TO MOLECULAR PROPERTIES

20.59 The point group for *cis*-dichloroethylene is $\mathscr{C}_{2v}$, and the point group for *trans*-dichloroethylene is $\mathscr{C}_{2h}$. Is either of these molecules optically active?

▌ Any molecule containing the elements corresponding to $\hat{S}_n$, $\hat{\sigma}$ $(=\hat{S}_1)$, or $\hat{i}$ $(=\hat{S}_2)$ will not be optically active. The $\mathscr{C}_{2v}$ point group contains two σ elements, and so *cis*-dichloroethylene will not be optically active. In the $\mathscr{C}_{2h}$ point group the C_2 axis is the same as an S_n axis, so *trans*-dichloroethylene will not be optically active.

20.60 Will *cis*-dichloroethylene or *trans*-dichloroethylene contain an electric dipole moment? See Problem 20.59 for additional information.

▌ Only molecules belonging to the $\mathscr{C}_n$, $\mathscr{C}_{nv}$, and $\mathscr{C}_s$ point groups may have an electric dipole moment. Thus *cis*-dichloroethylene ($\mathscr{C}_{2v}$ point group) will have an electric dipole moment.

20.61 What will be the location of the dipole moment in *cis*-dichloroethylene?

▌ The point group of the molecule is $\mathscr{C}_{2v}$, and the symmetry elements include a C_2 axis in the plane of the molecule, two σ_v mirror planes (one in the plane of the molecule and one perpendicular to the plane of its molecule), and the identity element (see Fig. 29.5a). The dipole moment must lie along the principal axes of rotation and, since present, must lie in both of the planes. This occurs at the intersection of the planes, which is the C_2 axis.

Table 20-8

$\mathscr{C}_{2v}$ representation	$\hat{E}$	$\hat{C}_2$	$\hat{\sigma}(xz)$	$\hat{\sigma}(yz)$	
A_1	1	1	1	1	z, x^2, y^2, z^2
A_2	1	1	−1	−1	R_z, xy
B_1	1	−1	1	−1	x, R_x, xy
B_2	1	−1	−1	1	y, R_y, yz

20.62 The character table for the $\mathscr{C}_{2v}$ point group is given in Table 20-8. Which electric-dipole transitions from B_2 are permitted?

▌ A transition will be permitted if the *direct product* of the representations

$$(a_1, b_1, c_1, \ldots) \otimes (a_2, b_2, c_2, \ldots) \equiv (a_1a_2, b_1b_2, c_1c_2, \ldots) \tag{20.5}$$

of the two states has the same symmetry as listed for the entries for x, y, or z in the character table. For the $B_2 \to A_1$ transition,

$$(1, -1, -1, 1) \otimes (1, 1, 1, 1) = (1, -1, -1, 1)$$

which corresponds to B_2, which will be active in the y direction. Likewise for the $B_2 \to A_2$ transition,

$$(1, -1, -1, 1) \otimes (1, 1, -1, -1) = (1, -1, 1, -1)$$

which corresponds to B_1, which will be active in the x direction. The $B_2 \to B_1$ transition,

$$(1, -1, -1, 1) \otimes (1, -1, 1, -1) = (1, 1, -1, -1)$$

corresponds to A_2 and will not be permitted. The $B_2 \to B_2$ transition,

$$(1, -1, -1, 1) \otimes (1, -1, -1, 1) = (1, 1, 1, 1)$$

corresponds to A_1 and will be active in the z direction.

20.63 Confirm the assignments of the irreducible representations given in Table 20-6 for the various translations for a molecule in the $\mathscr{C}_{2h}$ point group.

▌ In Problem 20.55 it was shown that $V_{\text{trans}} = V_3 \oplus 2V_4$. To find which of these V_i corresponds to the x component of the translational motion, the four operations of the group are performed in x, giving

$$\hat{E} \times x = (1)x \qquad \hat{C}_2 \times x = (-1)x \qquad \hat{i} \times x = (-1)x \qquad \hat{\sigma}_h \times x = (1)x$$

which are the same as the transformations contained in V_4. Likewise for the y component of the translational motion,

$$\hat{E} \times y = (1)y \qquad \hat{C}_2 \times y = (-1)y \qquad \hat{i} \times y = (-1)y \qquad \hat{\sigma}_h \times y = (1)y$$

which are the same as the transformations contained in V_4. For the z component,

$$\hat{E} \times z = (1)z \qquad \hat{C}_2 \times z = (1)z \qquad \hat{i} \times z = (-1)z \qquad \hat{\sigma}_h \times z = (-1)z$$

which are the same as the transformations contained in V_3.

20.64 Confirm the assignments of the irreducible representations given in Table 20-6 for the various rotations for a molecule in the $\mathscr{C}_{2h}$ point group.

▌ The simplest technique for finding these assignments is based on the values of the characters in the V_i for the corresponding translational motion: $\chi(V_R, \hat{R})$ will be given by $\chi(V_i, \hat{R})$ for $\hat{R} = \hat{E}$ and $\hat{C}_n$ and by $-\chi(V_i, \hat{R})$ for $\hat{R} = \hat{S}_n$, σ, and $\hat{i}$. Applying these rules to determine R_x using V_4, to which the x component of translation is assigned, gives

$$\chi(V_{R_x}, \hat{E}) = \chi(V_4, \hat{E}) = 1 \qquad \chi(V_{R_x}, \hat{C}_2) = \chi(V_4, \hat{C}_2) = -1$$

$$\chi(V_{R_x}, \hat{i}) = -\chi(V_4, \hat{i}) = -(-1) = 1 \qquad \chi(V_{R_x}, \hat{\sigma}_h) = -\chi(V_4, \hat{\sigma}_h) = -(1) = -1$$

which corresponds to V_2 in Table 20-6. For R_y, using V_4 gives identical results. Likewise for R_z, using V_3 gives

$$\chi(V_{R_z}, \hat{E}) = \chi(V_3, \hat{E}) = 1 \qquad \chi(V_{R_z}, \hat{C}_2) = \chi(V_3, \hat{C}_2) = 1$$

$$\chi(V_{R_z}, \hat{i}) = -\chi(V_3, \hat{i}) = -(-1) = 1 \qquad \chi(V_{R_z}, \hat{\sigma}_h) = -\chi(V_3, \hat{\sigma}_h) = -(-1) = 1$$

which corresponds to V_1 in Table 20-6.

20.65 Describe the vibration motion of $NO_2(g)$.

▮ The point group for this molecule was determined in Problem 20.30 as $\mathscr{C}_{2v}$. To determine which irreducible representations will describe the motion of the molecule, the following characters must be determined for Λ atoms in the molecule:

$$\chi(V_{3\Lambda}, \hat{E}) = 3\Lambda_{\hat{E}} \tag{20.6a}$$

$$\chi(V_{3\Lambda}, \hat{C}_n^m) = \Lambda_{\hat{C}_n^m}[1 + 2\cos(2\pi m/n)] \tag{20.6b}$$

$$\chi(V_{3\Lambda}, \hat{S}_n^m) = \Lambda_{\hat{S}_n^m}[-1 + 2\cos(2\pi m/n)] \tag{20.6c}$$

$$\chi(V_{3\Lambda}, \hat{i}) = -3\Lambda_{\hat{i}} \tag{20.6d}$$

$$\chi(V_{3\Lambda}, \hat{\sigma}) = \Lambda_{\hat{\sigma}} \tag{20.6e}$$

where $\Lambda_{\hat{R}}$ represents the number of unmoved atoms for the operation $\hat{R}$. For NO_2, *(20.6)* gives

$$\chi(V_{3\Lambda}, \hat{E}) = 3(3) = 9 \qquad \chi(V_{3\Lambda}, \hat{C}_2) = 1[1 + 2\cos(2\pi(1)/2)] = -1$$

$$\chi(V_{3\Lambda}, \sigma_v(xz)) = 1 \qquad \chi(V_{3\Lambda}, \sigma_v(yz)) = 3$$

where the $\sigma_v(yz)$ plane contains the molecule. Substituting the above results along with the data from Table 20-8 into *(20.1)* gives

$$v_1 = \tfrac{1}{4}[(9)(1) + (-1)(1) + (1)(1) + (3)(1)] = 3$$

$$v_2 = \tfrac{1}{4}[(9)(1) + (-1)(1) + (1)(-1) + (3)(-1)] = 1$$

$$v_3 = \tfrac{1}{4}[(9)(1) + (-1)(-1) + (1)(1) + (3)(-1)] = 2$$

$$v_4 = \tfrac{1}{4}[(9)(1) + (-1)(-1) + (1)(-1) + (3)(1)] = 3$$

or $V_{3\Lambda} = 3V_1 \oplus V_2 \oplus 2V_3 \oplus 3V_4$. Subtracting V_1 for z, V_3 for x, V_4 for y, V_2 for R_z, V_3 for R_x, and V_4 for R_y (see Table 20-8) leaves

$$V_{vib} = 3V_1 \oplus V_2 \oplus 2V_3 \oplus 3V_4 \ominus V_1 \ominus V_2 \ominus 2V_3 \ominus 2V_4 = 2V_1 \oplus V_4 = 2A_1 \oplus B_2$$

The three modes of vibration are singly degenerate, with two of them being completely symmetrical and one being symmetric with respect to $\hat{E}$ and $\hat{\sigma}(yz)$ but antisymmetric with respect to $\hat{C}_2$ and $\hat{\sigma}(xz)$. These modes are shown in Fig. 19-5a and are labeled ν_1, ν_2, and ν_3, respectively.

20.66 The Mulliken convention for numbering the vibrational frequencies is to assign them according to the appearance of the corresponding irreducible representations in the character table. For modes of the same symmetry, the assignment is made according to decreasing frequency. An exception to this convention is made for linear triatomic molecules, where ν_2 has traditionally been assigned to the doubly degenerate irreducible representation. Using this convention, assign the molecular vibrational frequencies for $NO_2(g)$: 589.2 cm^{-1} and 1276.5 cm^{-1} for the completely symmetric modes and 2223.7 cm^{-1} for the other mode.

▮ For the symmetric modes, ν_1 will be assigned to 1276.5 cm^{-1} corresponding to the symmetric stretch and ν_2 will be assigned to 589.2 cm^{-1} corresponding to the bending motion (see Fig. 19-5a). The value of 2223.7 cm^{-1} is assigned to ν_3, the asymmetric stretch.

20.67 Gaseous C_2F_2 belongs to the $\mathscr{D}_{\infty h}$ point group. Using Fig. 19-6 and Table 20-9, assign each given mode to an irreducible representation.

Table 20-9

$\mathscr{D}_{\infty h}$ representation	$\hat{E}$	$2\hat{C}_\infty^\phi$	$\hat{\sigma}_v$	$\hat{i}$	$\hat{\sigma}_h$	$2\hat{S}_\infty^\phi$	$\infty\hat{C}_2$	
A_{1g}	1	1	1	1	1	1	1	$x^2 + y^2, z^2$
A_{1u}	1	1	1	1	−1	−1	−1	
A_{2g}	1	1	−1	−1	1	1	−1	R_z
A_{2u}	1	1	−1	−1	−1	−1	1	z
E_{1g}	2	$2\cos\phi$	0	0	−2	$-2\cos\phi$	0	$(R_x, R_y), (xz, yz)$
E_{1u}	2	$2\cos\phi$	0	0	2	$2\cos\phi$	0	(x, y)
E_{2g}	2	$2\cos\phi$	0	2	2	$2\cos\phi$	0	$(x^2 - y^2, xy)$
⋮								

The modes identified as ν_1 and ν_2 in Fig. 19-6 are completely symmetric and would be assigned to A_{1g}. The mode identified as ν_3 is antisymmetric with respect to $\hat{\sigma}_h$ and the C_2 axes but is symmetric with respect to $\hat{E}$, $\hat{C}_\infty$, $\hat{\sigma}_v$, and $\hat{i}$ and would be assigned to A_{1u}. The doubly degenerate mode identified as ν_4 is antisymmetric with respect to $\hat{\sigma}_h$ and is assigned to E_{1g}, whereas ν_5 is symmetric with respect to $\hat{\sigma}_h$ and is assigned to E_{1u}.

20.68 There is a mode of vibration for *trans*-dichloroethylene in which the C atoms do not move, the Cl atoms move upward from the original plane of the molecule, and the H atoms move downward from the original plane of the molecule. Using Table 20-6, assign this mode to one of the irreducible representations.

The mode is symmetric with respect to $\hat{E}$ and $\hat{C}_2$ and antisymmetric with respect to $\hat{i}$ and $\hat{\sigma}_h$. The entries for A_u agree with this information.

20.69 Using Table 20-10 for $BF_3(g)$, determine the symmetry of the vibrational modes. Assign the modes shown in Fig. 19-7 to the proper irreducible representations.

Table 20-10

$\mathscr{D}_{3h}$ representation	$\hat{E}$	$2\hat{C}_3$	$3\hat{C}'$	$\hat{\sigma}_h$	$2\hat{S}_3$	$3\hat{\sigma}_v$	
$V_1 = A_1'$	1	1	1	1	1	1	$x^2 + y^2, z^2$
$V_2 = A_2'$	1	1	−1	1	1	−1	R_z
$V_3 = E'$	2	−1	0	2	−1	0	$(x, y), (x^2 - y^2, xy)$
$V_4 = A_1''$	1	1	1	−1	−1	−1	
$V_5 = A_2''$	1	1	−1	−1	−1	1	z
$V_6 = E''$	2	−1	0	−2	1	0	$(R_x, R_y), (xz, yz)$

For BF_3, *(20.6)* gives

$$\chi(V_{3A}, \hat{E}) = 3(4) = 12 \qquad \chi(V_{3A}, \hat{C}_3) = 1[1 + 2\cos(2\pi(1)/3)] = 0$$
$$\chi(V_{3A}, \hat{C}_2) = 2[1 + 2\cos(2\pi(1)/2)] = -2 \qquad \chi(V_{3A}, \hat{\sigma}_h) = 4$$
$$\chi(V_{3A}, \hat{S}_3) = 1[-1 + 2\cos(2\pi(1)/3)] = -2 \qquad \chi(V_{3A}, \hat{\sigma}_v) = 2$$

Note that Table 20-10 does not list each operation separately but uses a coefficient to represent the number of operations in each "class." Substituting the above results along with the data from Table 20-10 into *(20.3)* gives

$$v_1 = \tfrac{1}{12}[(12)(1) + (2)(0)(1) + (3)(-2)(1) + (4)(1) + (2)(-2)(1) + (3)(2)(1)] = 1$$
$$v_2 = \tfrac{1}{12}[(12)(1) + (2)(0)(1) + (3)(-2)(-1) + (4)(1) + (2)(-2)(1) + (3)(2)(-1)] = 1$$
$$v_3 = \tfrac{1}{12}[(12)(2) + (2)(0)(-1) + (3)(-2)(0) + (4)(2) + (2)(-2)(-1) + (3)(2)(0)] = 3$$
$$v_4 = \tfrac{1}{12}[(12)(1) + (2)(0)(1) + (3)(-2)(1) + (4)(-1) + (2)(-2)(-1) + (3)(2)(-1)] = 0$$
$$v_5 = \tfrac{1}{12}[(12)(1) + (2)(0)(1) + (3)(-2)(-1) + (4)(-1) + (2)(-2)(-1) + (3)(2)(1)] = 2$$
$$v_6 = \tfrac{1}{12}[(12)(2) + (2)(0)(-1) + (3)(-2)(0) + (4)(-2) + (2)(-2)(1) + (3)(2)(0)] = 1$$

or $V_{3A} = V_1 \oplus V_2 \oplus 3V_3 \oplus 2V_5 \oplus V_6$. Subtracting V_5 and V_3 for translation and V_2 and V_6 for rotation (see Table 20-10) leaves

$$V_{vib} = V_1 \oplus 2V_3 \oplus V_5$$

The completely symmetric mode identified as ν_1 in Fig. 19-7 is assigned to A_1', and the mode identified as ν_2, which is antisymmetric with respect to $\hat{C}_2$, σ_h, and $\hat{S}_3$, is assigned to A_2''. The other two doubly degenerate modes are assigned to E'.

20.70 Which of the modes described in Problem 20.69 are infrared-active? Which of the infrared-active vibrations will be the parallel type, and which the perpendicular type?

The vibrations that have symmetry corresponding to entries of x, y, or z in the character table will be infrared-active. Because $\nu_2(A_2'')$ has the entry of z in Table 20-10, it will be infrared-active, and because the vibrations generate an oscillating dipole along the C_3 axis, it is of the parallel type. For ν_3 and ν_4, both E', the combined entry is (x, y), and these modes will be infrared-active. These vibrations generate an oscillating dipole perpendicular to the principal axis in the xy plane.

20.71 Which of the modes described in Problem 20-69 are Raman-active?

▮ The vibrations have symmetry corresponding to entries of x^2, y^2, z^2, xy, xz, yz, or combinations of these will be Raman-active. Because $\nu_1(A_1')$ has the entry of $x^2 + y^2, z^2$ in Table 20-10, it will be Raman-active. For ν_3 and ν_4, both E', the combined entry is $(x^2 - y^2, xy)$, and these modes will be Raman-active.

20.72 What is the symmetry of the $(0, 1, 1, 0)$ vibrational state of BF_3? See Problem 20.69 for additional information.

▮ The symmetry of the state is determined by taking the direct product $A_2'' \otimes E'$. Substituting the character from Table 20-10 into *(20.5)* gives

$$(1, 1, -1, -1, -1, 1) \otimes (2, -1, 0, 2, -1, 0) = (2, -1, 0, -2, 1, 0)$$

which corresponds to E''.

20.73 Will the transition from $(0, 0, 0, 0)$ to $(0, 1, 1, 0)$ vibrational states of BF_3 be allowed? See Problem 20.72 for additional information.

▮ The symmetry of the ground state is A_1', and from Problem 20.72 the symmetry of the $(0, 1, 1, 0)$ state is E''. Substituting the characters from Table 20-10 into *(20.5)* for $A_1' \otimes E''$ gives

$$(1, 1, 1, 1, 1, 1) \otimes (2, -1, 0, -2, 1, 0) = (2, -1, 0, -2, 1, 0)$$

which corresponds to E''. Because the entries in Table 20-10 for E'' do not include x, y, or z, this transition is not permitted.

20.74 Determine the symmetry of the two energy levels into which the d orbitals are split in the presence of an octahedral field.

▮ From Table 20-11 for the $\mathcal{O}_h$ point group, the entries for (xz, yz, xy), corresponding to the d_{xy}, d_{yz}, and d_{xy} orbitals, are assigned T_{2g}, and the entry for $(2z^2 - x^2 - y^2, x^2 - y^2)$, corresponding to the d_{z^2} and $d_{x^2-y^2}$ orbitals, is assigned E_g. These results confirm the assignments made in Fig. 18-6.

Table 20-11

$\mathcal{O}_h$ representation	$\hat{E}$	$8\hat{C}_3$	$6\hat{C}_4$	$3\hat{C}''$ $(=\hat{C}_4^2)$	$6\hat{C}'$	$\hat{i}$	$8\hat{S}_6$	$6\hat{S}_4$	$3\hat{\sigma}_h$	$6\hat{\sigma}_d$	
A_{1g}	1	1	1	1	1	1	1	1	1	1	$(x^2 + y^2 + z^2)$
A_{2g}	1	1	−1	1	−1	1	1	−1	1	−1	
E_g	2	−1	0	2	0	2	−1	0	2	0	$(2z^2 - x^2 - y^2, x^2 - y^2)$
T_{1g}	3	0	1	−1	−1	3	0	1	−1	−1	(R_x, R_y, R_z)
T_{2g}	3	0	−1	−1	1	3	0	−1	−1	1	(xz, yz, xy)
A_{1u}	1	1	1	1	1	−1	−1	−1	−1	−1	
A_{2u}	1	1	−1	1	−1	−1	−1	1	−1	1	
E_u	2	−1	0	2	0	−2	1	0	−2	0	
T_{1u}	3	0	1	−1	−1	−3	0	−1	1	1	(x, y, z)
T_{2u}	3	0	−1	−1	1	−3	0	1	1	−1	

20.75 Which orbitals would be used to form six equivalent hybridized orbitals in an octahedral field?

▮ The reducible representation V_{orb} is determined by considering only unmoved bonding orbitals for the various operations.

$$\chi(V_{\text{orb}}, \hat{E}) = 6 \qquad \chi(V_{\text{orb}}, \hat{C}_3) = 0 \qquad \chi(V_{\text{orb}}, \hat{C}_4) = 2$$

$$\chi(V_{\text{orb}}, \hat{C}_4^2) = 2 \qquad \chi(V_{\text{orb}}, \hat{C}_2^1) = 0 \qquad \chi(V_{\text{orb}}, \hat{i}) = 0$$

$$\chi(V_{\text{orb}}, \hat{S}_6) = 0 \qquad \chi(V_{\text{orb}}, \hat{S}_4) = 0 \qquad \chi(V_{\text{orb}}, \hat{\sigma}_h) = 4 \qquad \chi(V_{\text{orb}}, \hat{\sigma}_d) = 2$$

Substituting the above results and the entries from Table 20-11 into *(20.3)* gives for the 48 operations

$$v_1 = \tfrac{1}{48}[(6)(1) + (8)(0)(1) + (6)(2)(1) + (3)(2)(1) + (6)(0)(1) + (0)(1)$$
$$+ (8)(0)(1) + (6)(0)(1) + (3)(4)(1) + (6)(2)(1)] = 1$$

Likewise, $v_2 = v_4 = v_5 = v_6 = v_7 = v_8 = v_{10} = 0$ and $v_3 = v_9 = 1$, giving

$$V_{\text{orb}} = A_{1g} \oplus E_g \oplus T_{1u}$$

Inspection of Table 20-11 shows that these irreducible representations correspond to the $(x^2 + y^2 + z^2)$, $(2z^2 - x^2 - y^2, x^2 - y^2)$, and (x, y, z) entries. Thus the s, d_{z^2}, $d_{x^2-y^2}$, p_x, p_y, and p_z orbitals are used.

20.76 Which orbitals would be used to form the three equivalent hybridized orbitals in $BF_3(g)$?

▮ The point group for BF_3 is $\mathscr{D}_{3h}$ (see Problem 20.27). The reducible representation V_{orb} is determined as in Problem 20.75 as

$$\chi(V_{orb}, \hat{E}) = 3 \qquad \chi(V_{orb}, \hat{C}_3) = 0 \qquad \chi(V_{orb}, \hat{C}_2^1) = 1$$

$$\chi(V_{orb}, \hat{\sigma}_h) = 3 \qquad \chi(V_{orb}, \hat{S}_3) = 0 \qquad \chi(V_{orb}, \hat{\sigma}_v) = 1$$

Substituting these results and the entries from Table 20-10 into *(20.3)* gives for the 12 operations

$$v_1 = \tfrac{1}{12}[(3)(1) + (2)(0)(1) + (3)(1)(1) + (3)(1) + (2)(0)(1) + (3)(1)(1)] = 1$$

Likewise, $v_2 = v_4 = v_5 = v_6 = 0$ and $v_3 = 1$, giving $V_{orb} = A_1' \oplus E'$. Inspection of Table 20-10 shows that these irreducible representations correspond to the $x^2 + y^2, z^2$ and $(x, y), (x^2 - y^2, xy)$ entries. Thus the s, p_x, and p_y orbitals are used. The entries for z^2 and $(x^2 - y^2, xy)$ are symmetrically equivalent, but the ground state of B does not involve the d_{z^2}, $d_{x^2-y^2}$, and d_{xy} orbitals.

20.77 How do the s and p valence-shell orbitals of the O atom combine with the s orbitals of the attached H atoms to give molecular orbitals in the $H_2O(g)$ molecule?

▮ The valence-shell 2s atomic orbital is completely symmetric, and this corresponds to the entry of A_1 in Table 20-8. The symmetries of p_x, p_y, and p_z correspond to B_1, B_2, and A_1, respectively. The symmetries of the two H atoms are obtained by

$$\chi(V, \hat{E}) = 2 \qquad \chi(V, \hat{C}_2) = 0 \qquad \chi(V, \hat{\sigma}(xy)) = 2 \qquad \chi(V, \hat{\sigma}(yz)) = 0$$

$$v_1 = \tfrac{1}{4}[(2)(1) + (0)(1) + (2)(1) + (0)(1)] = 1$$

$$v_2 = \tfrac{1}{4}[(2)(1) + (0)(1) + (2)(-1) + (0)(-1)] = 0$$

$$v_3 = \tfrac{1}{4}[(2)(1) + (0)(-1) + (2)(1) + (0)(-1)] = 1$$

$$v_4 = \tfrac{1}{4}[(2)(1) + (0)(-1) + (2)(-1) + (0)(1)] = 0$$

as $V = A_1 \oplus B_1$. The orbitals on the central atom and attached atoms with the same symmetry can be combined to give bonding and antibonding molecular orbitals with the same symmetry. Thus the A_1 orbital from the H atoms combines with the A_1 orbitals from the 2s and $2p_z$ to form three a_1 molecular orbitals, the B_1 orbitals from the H atoms combines with the B_1 orbital from the $2p_x$ to form two b_1 molecular orbitals, and the B_2 orbital from the p_y does not combine with the H atoms. This prediction agrees with the molecular orbital diagram shown in Fig. 19-14.

20.78 Use the symmetry of the π molecular orbitals in *trans*-dichloroethylene to determine the wave functions.

▮ The reducible representation for the π electrons in the $\mathscr{C}_{2h}$ point group is obtained for the four distinct operations (see Table 20-6) by assigning the value of +1 for an unchanged orbital, a 0 for an orbital that is moved, and a −1 for an orbital that has been inverted into itself. The unhybridized p orbitals lie parallel to the C_2 axis and are perpendicular to σ_h, giving

$$\chi(V_\pi, \hat{E}) = 2 \qquad \chi(V_\pi, \hat{C}_2) = 0 \qquad \chi(V_\pi, \hat{i}) = 0 \qquad \chi(V_\pi, \hat{\sigma}_h) = -2$$

Using *(20.3)* and the entries from Table 20-6 gives $V_\pi = A_u \oplus B_g$. One of the orbitals will have A_u symmetry, represented by an arrangement of $\pm\ \pm$ for the p orbitals, which is symmetric with respect to $\hat{E}$ and $\hat{C}_2$ and antisymmetric with respect to $\hat{i}$ and $\hat{\sigma}_h$. The other orbital will have B_g symmetry, represented by an arrangement of $\pm\ \mp$ for the p orbitals, which is symmetric with respect to $\hat{E}$ and $\hat{i}$ and antisymmetric with respect to $\hat{C}_2$ and $\hat{\sigma}_h$. The respective functions for the molecular orbitals can be written as

$$\phi_a = N(\phi_1 + \phi_2) \qquad \phi_b = N(\phi_1 - \phi_2)$$

where ϕ_i is the $2p_z$ wave function and N is a normalization constant. Note that ϕ_a is lower in energy and represents the bonding orbital and ϕ_b is higher in energy and is the antibonding orbital.

20.79 Use the symmetry of the π molecular orbitals in *trans*-butadiene ($\mathscr{C}_{2h}$ point group) to determine the wave functions.

▮ The reducible representation describing the four unhybridized p orbitals can be determined as in Problem 20.78 as

$$\chi(V_\pi, \hat{E}) = 4 \qquad \chi(V_\pi, \hat{C}_2) = 0 \qquad \chi(V_\pi, \hat{i}) = 0 \qquad \chi(V_\pi, \hat{\sigma}_h) = -4$$

Using *(20.3)* and the entries from Table 20-6 gives

$$V_\pi = 2A_u \oplus 2B_g$$

Two of the orbitals will have A_u symmetry, represented by $\pm\ \pm\ \pm\ \pm$ and $\pm\ \mp\ \mp\ \pm$ for the p orbitals, which are symmetric with respect to $\hat{E}$ and $\hat{C}_2$ and antisymmetric with respect to $\hat{i}$ and $\hat{\sigma}_h$. The other orbitals will have B_g symmetry, represented by $\pm\ \pm\ \mp\ \mp$ and $\pm\ \mp\ \pm\ \mp$, which are symmetric with respect to $\hat{E}$ and $\hat{i}$ and antisymmetric with respect to $\hat{C}_2$ and $\hat{\sigma}_h$. The respective wave functions for the molecular orbitals can be written as

$$\phi_a = N(\phi_1 + \phi_2 + \phi_3 + \phi_4) \qquad \phi_b = N(\phi_1 - \phi_2 - \phi_3 + \phi_4)$$

$$\phi_c = N(\phi_1 + \phi_2 - \phi_3 - \phi_4) \qquad \phi_d = N(\phi_1 - \phi_2 + \phi_3 - \phi_4)$$

where ϕ_i is the $2p_z$ wave function and N is a normalization constant. The order of increasing energy is directly related to the number of nodes in ϕ_i. Thus ϕ_a is the lowest in energy, ϕ_c is next lowest, ϕ_b is next to the highest, and ϕ_d is the highest in energy. Note that other wave functions can be obtained from these by taking linear combinations and are commonly used to describe the molecular orbitals.

20.80 Use the symmetry of the π molecular orbitals in benzene ($\mathscr{D}_{6h}$ point group) to determine the wave functions.

▮ The reducible representation describing the six unhybridized p orbitals can be determined as in Problem 20.78 as

$$\chi(V_\pi, \hat{E}) = 6 \qquad \chi(V_\pi, \hat{C}_6) = 0 \qquad \chi(V_\pi, \hat{C}_3) = 0 \qquad \chi(V_\pi, \hat{C}_2'') = 0$$

$$\chi(V_\pi, \hat{C}_2) = -2 \qquad \chi(V_\pi, \hat{C}_2^1) = 0 \qquad \chi(V_\pi, \hat{\sigma}_h) = -6 \qquad \chi(V_\pi, \hat{\sigma}_v) = 2$$

$$\chi(V_\pi, \hat{\sigma}_d) = 0 \qquad \chi(V_\pi, \hat{S}_6) = 0 \qquad \chi(V_\pi, \hat{S}_3) = 0 \qquad \chi(V_\pi, \hat{i}) = 0$$

Using *(20.3)* and the entires from Table 20-12 gives

$$V_\pi = A_{2u} \oplus B_{2g} \oplus E_{1g} \oplus E_{2u}$$

Table 20-12

$\mathscr{D}_{6h}$ representation	$\hat{E}$	$2\hat{C}_6$	$2\hat{C}_3$	$\hat{C}_2''$	$3\hat{C}_2$	$3\hat{C}_2'$	$\hat{\sigma}_h$	$3\hat{\sigma}_v$	$3\hat{\sigma}_d$	$2\hat{S}_6$	$2\hat{S}_3$	$\hat{i}$	
A_{1g}	1	1	1	1	1	1	1	1	1	1	1	1	$x^2 + y^2, z^2$
A_{1u}	1	1	1	1	1	1	−1	−1	−1	−1	−1	−1	
A_{2g}	1	1	1	1	−1	−1	1	−1	−1	1	1	1	R_z
A_{2u}	1	1	1	1	−1	−1	−1	1	1	−1	−1	−1	z
B_{1g}	1	−1	1	−1	1	−1	−1	−1	1	1	−1	1	
B_{1u}	1	−1	1	−1	1	−1	1	1	−1	−1	1	−1	
B_{2g}	1	−1	1	−1	−1	1	−1	1	−1	1	−1	1	
B_{2u}	1	−1	1	−1	−1	1	1	−1	1	−1	1	−1	
E_{1g}	2	1	−1	−2	0	0	−2	0	0	−1	1	2	$(R_x, R_y), (xz, yz)$
E_{1u}	2	1	−1	−2	0	0	2	0	0	1	−1	−2	(x, y)
E_{2g}	2	−1	−1	2	0	0	2	0	0	−1	−1	2	$(x^2 - y^2, xy)$
E_{2u}	2	−1	−1	2	0	0	−2	0	0	1	1	−2	

The lowest in energy is the A_{2u} orbital, represented by $\pm\ \pm\ \pm\ \pm\ \pm\ \pm$ for the six p orbitals in the ring, and the highest in energy is the B_{2g} orbital, represented by $\pm\ \mp\ \pm\ \mp\ \pm\ \mp$. It can be shown that the next lowest in energy are the two E_{1g} orbitals, represented by $2\pm,\ \pm,\ \mp,\ 2\mp,\ \mp,\ \pm$ and by $\pm,\ 2\pm,\ \pm,\ \mp,\ 2\mp,\ \mp$, and the next to the highest in energy are the two E_{2u} orbitals, represented by $2\pm,\ \mp,\ \mp,\ 2\pm,\ \mp, \mp$ and by $\mp,\ 2\pm,\ \mp,\ \mp,\ 2\pm,\ \mp$. The respective wave functions are

$$\phi_a = N(\phi_1 + \phi_2 + \phi_3 + \phi_4 + \phi_5 + \phi_6) \qquad \phi_b = N(\phi_1 - \phi_2 + \phi_3 - \phi_4 + \phi_5 - \phi_6)$$

$$\phi_c = N(2\phi_1 + \phi_2 - \phi_3 - 2\phi_4 - \phi_5 + \phi_6) \qquad \phi_d = N(\phi_1 + \phi_2 + \phi_3 - \phi_4 - 2\phi_5 - \phi_6)$$

$$\phi_e = N(2\phi_1 - \phi_2 - \phi_3 + 2\phi_4 - \phi_5 - \phi_6) \qquad \phi_f = N(-\phi_1 + 2\phi_2 - \phi_3 - \phi_4 + 2\phi_5 - \phi_6)$$

where ϕ_i is the $2p_z$ wave function and N is a normalization constant. Note that other wave functions can be obtained from these by taking linear combinations and are commonly used to describe the molecular orbitals.

CHAPTER 21
Intermolecular Bonding

21.1 EXTENDED COVALENT BONDING

21.1 Each silicon atom in Si(s) is held by four covalent bonds to four other Si atoms in a tetrahedral arrangement as in the diamond structure for carbon. Given $\Delta_f H° = 455\ \text{kJ}\cdot\text{mol}^{-1}$ for Si(g), calculate the average bond enthalpy in crystalline Si.

▮ For the equation $Si(s) \rightarrow Si(g)$, using *(4.5)* gives

$$\Delta H° = (1\ \text{mol})(455\ \text{kJ}\cdot\text{mol}^{-1}) - (1\ \text{mol})(0) = 455\ \text{kJ}$$

where $\Delta_f H° = 0$ for Si(s). On the average, two bonds must be broken for each atom that is vaporized, giving

$$\overline{\text{BE}} = \frac{455\ \text{kJ}}{2\ \text{mol}} = 228\ \text{kJ}\cdot\text{mol}^{-1}$$

21.2 Predict whether amorphous SiO_2 or α-cristobalite should have the greater density.

▮ The crystal structure of α-cristobalite is similar to that for elemental Si except that O atoms are midway between each pair of Si atoms. Amorphous SiO_2 is a "glass" in which there is no long-range order, but rather a disordered array of polymeric chains, sheets, or tetrahedra of atoms. The density of the glass should be less than that of the solid. The handbook values are $2.32\ \text{g}\cdot\text{cm}^{-3}$ and $2.2\ \text{g}\cdot\text{cm}^{-3}$, respectively.

21.3 Predict whether amorphous SiO_2 or α-cristobalite will have the greater enthalpy of sublimation. Given $\Delta_f H°/(\text{kJ}\cdot\text{mol}^{-1}) = -903.49$ for amorphous SiO_2, -322 for Si(g), and -909.48 for α-cristobalite, calculate the enthalpies of sublimation and confirm your prediction.

▮ The amorphous form described in Problem 21.2 should have the lower heat of vaporization because there are not as many bonds to break. This prediction is confirmed using *(4.5)* for the equation $SiO_2(s) \rightarrow SiO_2(g)$.

$$\Delta_s H°(\text{amorphous } SiO_2) = (1\ \text{mol})(-322\ \text{kJ}\cdot\text{mol}^{-1}) - (1\ \text{mol})(-903.49\ \text{kJ}\cdot\text{mol}^{-1}) = 581\ \text{kJ}$$

$$\Delta_s H°(\alpha\text{-cristobalite}) = (1)(-322) - (1)(-909.48) = 587\ \text{kJ}$$

21.4 Why is acetic acid, a relatively polar molecule, soluble in a nonpolar solvent like benzene?

▮ Acetic acid forms dimers in which two hydrogen bonds are formed between the carboxyl groups:

```
          O···H—O—C—CH3
          ‖       ‖
    CH3—C—O—H···O
```

The resulting dimer acts as a nonpolar molecule, which is soluble in the nonpolar solvent.

21.5 Which substance is predicted to have the greater heat of vaporization—ethanol (CH_3CH_2OH) or dimethyl ether (CH_3OCH_3)?

▮ Even though ethanol has a slightly greater dipole moment than dimethyl ether, it is expected to have a significantly greater heat of vaporization because of the hydrogen bonding between molecules:

```
CH3—CH2—O—H
        ⋮
        H—O—CH2—CH3
```

This prediction is correct, the enthalpies of vaporization are $40.5\ \text{kJ}\cdot\text{mol}^{-1}$ for CH_3CH_2OH and $22.6\ \text{kJ}\cdot\text{mol}^{-1}$ for CH_3OCH_3.

21.6 The enthalpies of vaporization of the noble gases are

Substance	Ne	Ar	Kr	Xe
$\Delta_{vap}H°/(\text{kJ}\cdot\text{mol}^{-1})$	1.77	6.518	9.029	12.64

Prepare a plot of these data against molar mass. Assuming that the noble gases represent a general trend, prepare similar plots for the hydrogen halides and the hydrogen chalcogens

Substance	HI	HBr	HCl	H_2Te	H_2Se	H_2S
$\Delta_{vap}H°/(kJ \cdot mol^{-1})$	19.77	17.61	16.15	23.2	19.3	18.67

and predict $\Delta_{vap}H°$ for HF and H_2O. The observed values are 30.1 kJ · mol^{-1} and 40.656 kJ · mol^{-1}, respectively. Why is there such a difference between the predicted and observed values?

▮ The plots are shown in Fig. 21-1. The predicted values of $\Delta_{vap}H°$ are roughly 11.0 kJ · mol^{-1} and 13.3 kJ · mol^{-1} for HF and H_2O, respectively. The predicted values are significantly lower than the observed values because of the strong hydrogen bonding in these molecules. Note that HCl and H_2S also demonstrate some hydrogen bonding in these molecules.

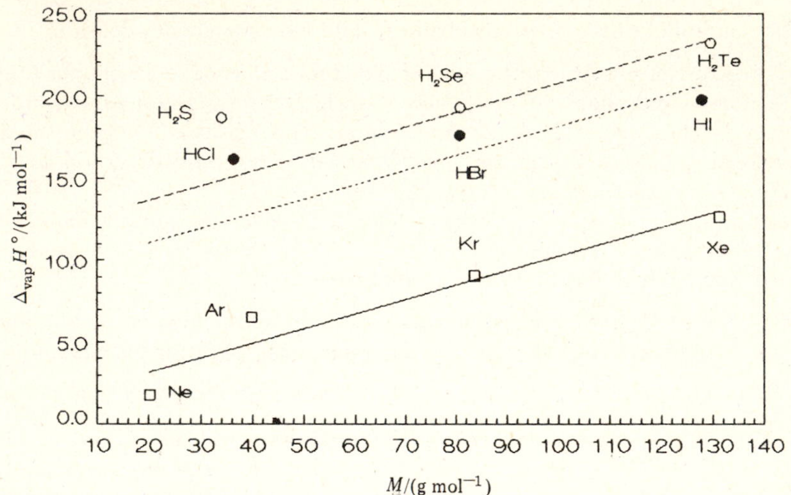

Fig. 21-1

21.7 Match the following values of first and second ionization constants to the respective acid: $K_{a1} = 1.2 \times 10^{-2}$, $K_{a2} = 6.0 \times 10^{-7}$, and $K_{a1} = 9.6 \times 10^{-4}$, $K_{a2} = 4.1 \times 10^{-5}$.

```
       O                    O
       ||                   ||
  H—C—C—OH            H—C—C—OH
     ||                  ||
  H—C—C—OH          HO—C—C—H
       ||                 ||
       O                  O
   Maleic acid         Fumaric acid
```

▮ The carboxylic acid groups on fumaric acid are rather independent of each other, and the two ionization constants should be rather similar, with the second somewhat smaller than the first, so $K_{a1} = 9.6 \times 10^{-4}$ and $K_{a2} = 4.1 \times 10^{-5}$. The first ionization for maleic acid is highly favored, and the second ionization is highly restricted because of the formation of intramolecular hydrogen bonding,

```
        O
        ||
  H—C—C—O
    ||      ·.
    ||        H
    ||       /
  H—C—C—O
        ||
        O
```

so $K_{a1} = 1.2 \times 10^{-2}$ and $K_{a2} = 6.0 \times 10^{-7}$.

21.2 METALLIC BONDING

21.8 Calculate the value of the Fermi energy for Mg(s). The density of Mg is $1.74\ \text{g}\cdot\text{cm}^{-3}$.

▌ The Fermi energy (E_F) is given by

$$E_F = \frac{h^2}{8m_e}\left(\frac{3N}{\pi V_m}\right)^{2/3} \tag{21.1}$$

where N is the number of conducting or valence electrons per mole of metal. Substituting the data into *(21.1)* gives

$$E_F = \frac{(6.626\times10^{-34}\ \text{J}\cdot\text{s})^2}{8(9.11\times10^{-31}\ \text{kg})}\left(\frac{3(2)(6.022\times10^{23}\ \text{mol}^{-1})[10^2\ \text{cm}/1\ \text{m}]^3}{\pi[(24.305\ \text{g}\cdot\text{mol}^{-1})/(1.74\ \text{g}\cdot\text{cm}^{-3})]}\right)^{2/3}\left(\frac{1\ \text{kg}\cdot\text{m}^2\cdot\text{s}^{-2}}{1\ \text{J}}\right) = 1.14\times10^{-18}\ \text{J}$$

21.9 Prepare a plot of the density of levels for Mg(s) against energy. What is the temperature dependence? See Problem 21.8 for additional data information.

▌ The number of levels per unit energy range is given by

$$\frac{dN}{dE} = \frac{4\pi V_m(2m_e)^{3/2}}{h^3}E^{1/2} \tag{21.2}$$

As a sample calculation, consider $E = E_F = 1.14\times10^{-18}$ J:

$$\frac{dN}{dE} = \frac{4\pi\left(\frac{24.305\ \text{g}\cdot\text{mol}^{-1}}{1.74\ \text{g}\cdot\text{cm}^{-3}}\right)\left(\frac{10^{-2}\ \text{m}}{1\ \text{cm}}\right)^3[2(9.11\times10^{-31}\ \text{kg})]^{3/2}(1.14\times10^{-18}\ \text{J})^{1/2}}{(6.626\times10^{-34}\ \text{J}\cdot\text{s})^3[(1\ \text{kg}\cdot\text{m}^2\cdot\text{s}^{-2})/(1\ \text{J})]^{3/2}} = 1.58\times10^{42}\ \text{J}^{-1}\cdot\text{mol}^{-1}$$

The plot is shown in Fig. 21-2. There is no temperature dependence indicated by *(21.2)*, but because V_m is a function of T, dN/dE will have a small temperature dependence.

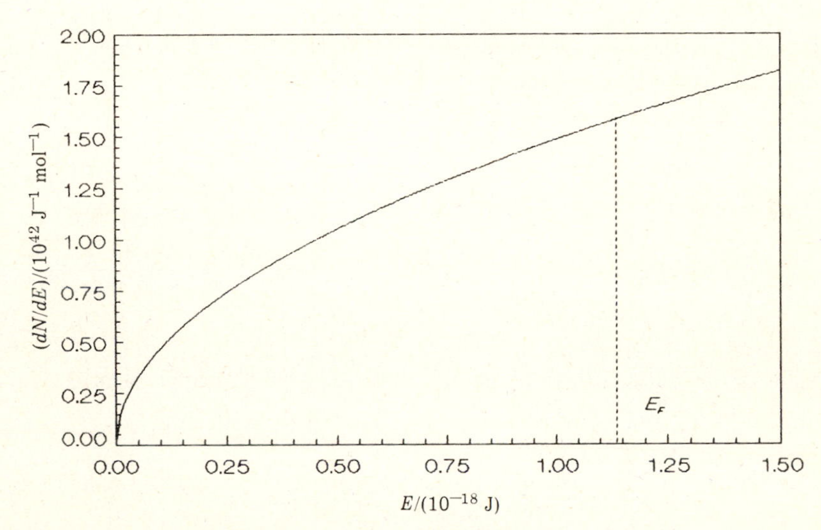

Fig. 21-2

21.10 How does E_F compare to kT for Mg(s) at room temperature? What is the value of the Fermi temperature? See Problem 21.8 for additional information.

▌ The ratio of E_F ($= 1.14\times10^{-18}$ J, see Problem 21.8) to kT is

$$\frac{E_F}{kT} = \frac{1.14\times10^{-18}\ \text{J}}{(1.381\times10^{-23}\ \text{J}\cdot\text{K}^{-1})(298\ \text{K})} = 277$$

The *Fermi temperature* is the temperature at which $E_F = kT_F$. Solving for T and substituting the data gives

$$T_F = \frac{1.14\times10^{-18}\ \text{J}}{1.381\times10^{-23}\ \text{J}\cdot\text{K}^{-1}} = 82\,600\ \text{K}$$

21.11 Prepare a plot of the Fermi-Dirac distribution function for Mg(s) at $T = 298$ K and $T = 1000$ K. See Problem 21.8 for additional information.

▮ The *Fermi–Dirac distribution function*

$$P(E) = (1 + e^{(E-E_F)/kT})^{-1} \tag{21.3}$$

gives the probability that a state of energy E is occupied by an electron. As a sample calculation, consider $E = 1.10 \times 10^{-18}$ J and $E_F = 1.14 \times 10^{-18}$ J (see Problem 21.8) at $T = 1000$ K.

$$P(E) = \left[1 + \exp\left(\frac{1.10 \times 10^{-18}\ \text{J} - 1.14 \times 10^{-18}\ \text{J}}{(1.381 \times 10^{-23}\ \text{J}\cdot\text{K}^{-1})(1000\ \text{K})}\right)\right]^{-1} = 0.9477$$

The plot is shown in Fig. 21-3.

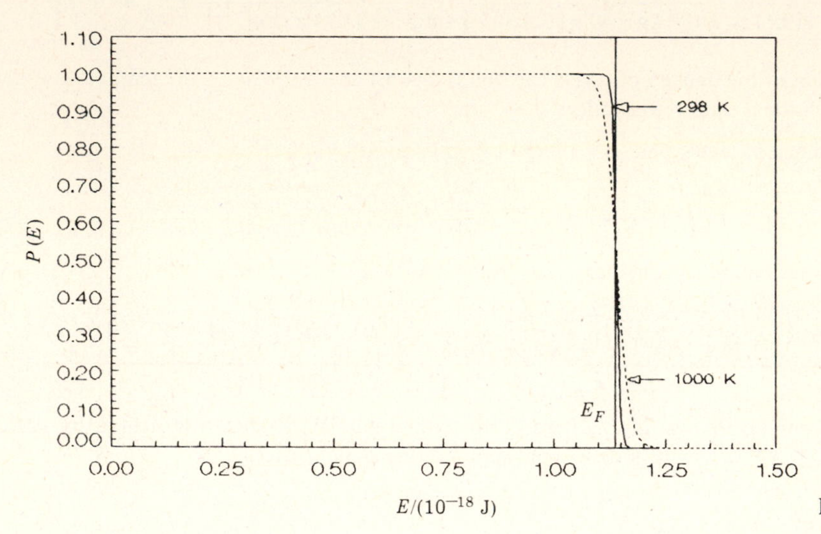

Fig. 21-3

21.12 What is the value of the electronic kinetic energy of Mg(s) at 298 K? See Problem 21.8 for additional information.

▮ The electronic kinetic energy of a metal is given by

$$E = \tfrac{3}{5}NE_F = \left[1 + \frac{5\pi^2}{12}\left(\frac{kT}{E_F}\right)^2 + \cdots\right] \tag{21.4}$$

Substituting $E_F = 1.14 \times 10^{-18}$ J (see Problem 21.8) and the data into *(21.4)* gives

$$E = \tfrac{3}{5}(2)(6.022 \times 10^{23}\ \text{mol}^{-1})(1.14 \times 10^{-18}\ \text{J})\left[1 + \frac{5\pi^2}{12}\left(\frac{(1.381 \times 10^{-23}\ \text{J}\cdot\text{K}^{-1})(298\ \text{K})}{1.14 \times 10^{-18}\ \text{J}}\right)^2\right]$$
$$= 8.24 \times 10^5\ \text{J}\cdot\text{mol}^{-1}$$

21.13 What is the value of the thermal electronic energy of Mg(s) at 298 K? See Problem 21.8 for additional information.

▮ Substituting $T = T$ and $T = 0$ into *(21.4)* gives the thermal energy as

$$E(\text{thermal}) = E - E_0 = \tfrac{3}{5}NE_F\left[1 + \frac{5\pi^2}{12}\left(\frac{kT}{E_F}\right)^2 + \cdots\right] - \tfrac{3}{5}NE_F(1 + 0 + \cdots) = \frac{\pi^2 Nk^2T^2}{4E_F} \tag{21.5}$$

Substituting $E_F = 1.14 \times 10^{-18}$ J (see Problem 21.8) into *(21.15)* gives

$$E(\text{thermal}) = \frac{\pi^2(2)(6.022 \times 10^{23}\ \text{mol}^{-1})(1.381 \times 10^{-23}\ \text{J}\cdot\text{K}^{-1})^2(298\ \text{K})^2}{4(1.14 \times 10^{-18}\ \text{J})} = 44.2\ \text{J}\cdot\text{mol}^{-1}$$

21.14 Determine the average electronic kinetic energy for Mg(s). See Problem 21.8 for additional information.

▮ Substituting *(21.1)* into *(2.2)* and integrating gives

$$\bar{E} = \frac{\displaystyle\int_0^{E_F} E\frac{4\pi V_m(2m_e)^{3/2}}{h^3}E^{1/2}\,dE}{\displaystyle\int_0^{E_F}\frac{4\pi V_m(2m_e)^{3/2}}{h^3}E^{1/2}\,dE} = \frac{\displaystyle\int_0^{E_F}E^{3/2}\,dE}{\displaystyle\int_0^{E_F}E^{1/2}\,dE} = \frac{\frac{2}{5}E^{5/2}\big|_0^{E_F}}{\frac{2}{3}E^{3/2}\big|_0^{E_F}} = \frac{\frac{2}{5}E_F^{5/2}}{\frac{2}{3}E_F^{3/2}} = \tfrac{3}{5}E_F \tag{21.6}$$

Substituting $E_F = 1.14 \times 10^{-18}$ J (see Problem 21.8) into *(21.6)* gives

$$\bar{E} = \tfrac{3}{5}(1.14 \times 10^{-18}\ \text{J}) = 6.84 \times 10^{-19}\ \text{J}$$

21.15 Estimate γ in *(3.40)* for Mg(s). See Problem 21.8 for additional information.

▌ Differentiating *(21.5)* with respect to T gives

$$C_{V,m}(\text{elec}) = \frac{2\pi^2 Nk^2 T}{4E_F} = \frac{\pi^2 Nk^2}{2E_F} T$$

which upon comparison to *(3.40)* gives

$$\gamma = \pi^2 Nk^2/2E_F \tag{21.7}$$

Substituting $E_F = 1.14 \times 10^{-18}$ J (see Problem 21.8) into *(21.7)* gives

$$\gamma = \frac{\pi^2(2)(6.022 \times 10^{23}\ \text{mol}^{-1})(1.381 \times 10^{-23}\ \text{J}\cdot\text{K}^{-1})^2}{2(1.14 \times 10^{-18}\ \text{J})} = 9.94 \times 10^{-4}\ \text{J}\cdot\text{K}^{-2}\cdot\text{mol}^{-1}$$

21.16 A handbook lists the value of the resistivity of Mg(s) as $4.45 \times 10^{-6}\ \Omega\cdot\text{cm}$. Calculate the "relaxation time" for electrons in Mg. See Problem 21.8 for additional information.

▌ The resistivity of a metallic conductor (ρ) is related to the time between collisions of the electrons with atoms and with defects in the crystal by

$$\rho = m_e V_m / Ne^2\tau \tag{21.8}$$

where τ is the relaxation time. Solving *(21.8)* for τ and substituting the data gives

$$\tau = \frac{(9.11 \times 10^{-31}\ \text{kg})\left(\dfrac{24.305\ \text{g}\cdot\text{mol}^{-1}}{1.74\ \text{g}\cdot\text{cm}^{-3}}\right)\left(\dfrac{1\ \text{C}}{1\ \text{A}\cdot\text{s}}\right)^2\left(\dfrac{1\ \Omega}{1\ \text{kg}\cdot\text{m}^2\cdot\text{s}^{-3}\cdot\text{A}^{-2}}\right)}{2(6.022 \times 10^{23}\ \text{mol}^{-1})(1.602 \times 10^{-19}\ \text{C})^2(4.45 \times 10^{-6}\ \Omega\cdot\text{cm})\left(\dfrac{10^2\ \text{cm}}{1\ \text{m}}\right)^2} = 9.25 \times 10^{-15}\ \text{s}$$

21.17 Prepare an energy diagram representing an insulator.

▌ An insulator has no electrons that occupy a partially filled metal band. This is represented in Fig. 21-4a by a solid filled box representing a completely filled band separated by an energy gap from an empty box representing a completely empty band.

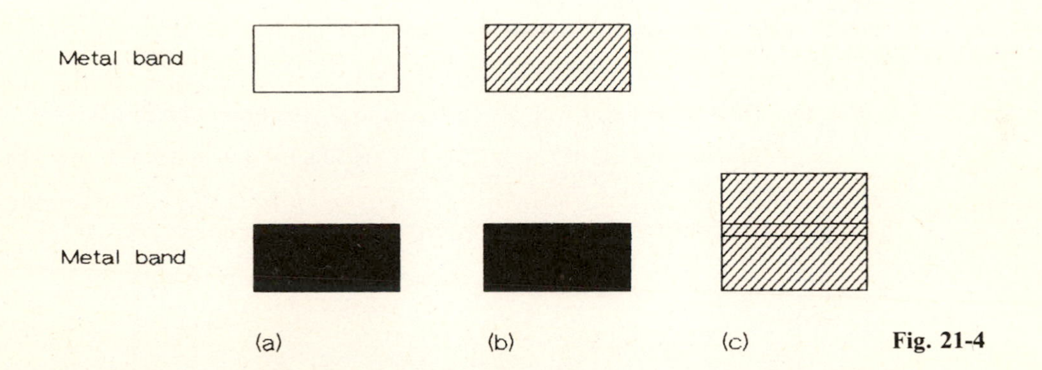

Fig. 21-4

21.18 Prepare an energy diagram representing a conducting metal such as Na(s).

▌ A conductor has electrons that occupy a partially filled metal band. This is represented in Fig. 21-4b by a cross-hatched box representing a partially filled band that is a result of the $3s^1$ electrons.

21.19 Prepare an energy diagram representing a conducting metal such as Mg(s).

▌ Because of the $3s^2$ electrons, the metal band would be expected to be completely filled and the diagram shown in Fig. 21-4a for an insulator would be predicted. However, because of the overlap of the empty metal band with the filled band to form a partially filled band (see Fig. 21-4c), Mg is a conductor.

21.20 Prepare an energy diagram representing an n-type semiconductor.

▌ An n-type semiconductor operates by thermally promoting an electron from an impurity level to an empty metal band so that conduction can occur (see Fig. 21-5a).

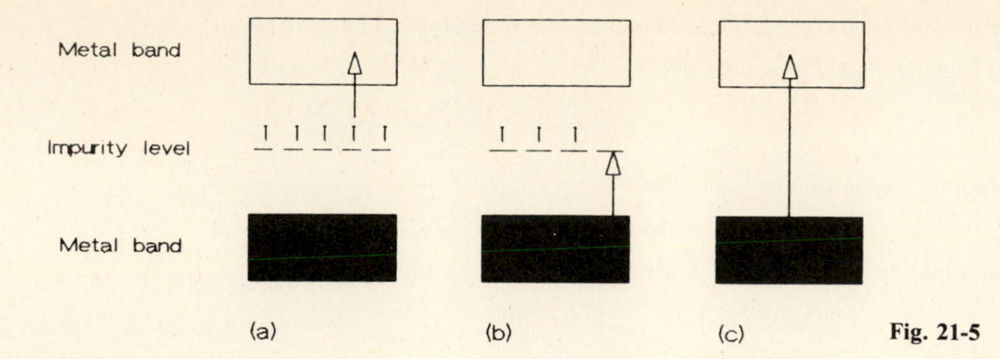

Fig. 21-5

21.21 Prepare an energy diagram representing a p-type semiconductor.

▌ A p-type semiconductor operates by thermally promoting an electron from a completely filled band to a partially empty impurity level so that conduction can occur in the originally filled band (see Fig. 21-5b). The concept of a "hole" is often used to describe conduction in this type of semiconductor.

21.22 Prepare an energy diagram representing an intrinsic semiconductor. What might cause the promotion of the electron?

▌ If sufficient energy is available to promote an electron from the completely filled band to the completely empty band, conduction can occur (see Fig. 21-5c). The source of the energy might be thermal or light energy.

21.23 Use the following resistivity data to determine the energy for the energy gap between the bands for Si(s):

$T/(\mathrm{K})$	500	600	800	1200
$\rho/(\Omega\cdot\mathrm{cm})$	33	5.0	0.13	0.013

▌ The resistivity of an intrinsic semiconductor is given by

$$\rho = A\,e^{E_g/2kT} \qquad (21.9a)$$

where E_g is the energy difference between the bands. Taking logarithms gives

$$\ln\rho = \ln A + (E_g/2k)T^{-1} \qquad (21.9b)$$

Thus a plot of $\ln\rho$ against T^{-1} will be linear with a slope equal to $E_g/2k$. From Fig. 21-6,

$$E_g = 2k(\text{slope}) = 2(1.381\times10^{-23}\ \mathrm{J\cdot K^{-1}})(6950\ \mathrm{K}) = 1.92\times10^{-19}\ \mathrm{J}$$

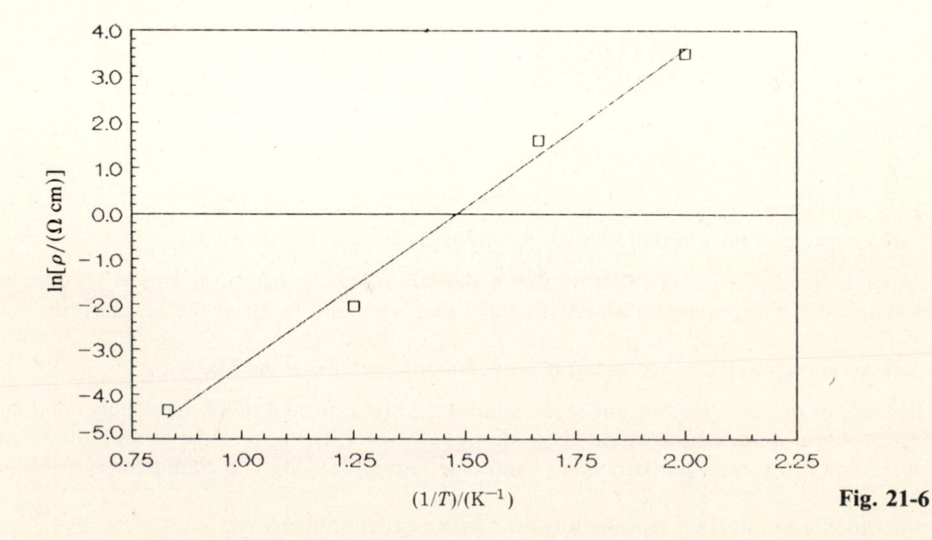

Fig. 21-6

21.24 What is the wavelength of light needed to promote an electron from the filled band to the empty band in Si(s)? The energy difference between these bands is 1.107 eV.

▮ Converting E_g from electronvolts to joules gives

$$E_g = (1.107\ \text{eV})\left(\frac{1.602 \times 10^{-19}\ \text{J}}{1\ \text{eV}}\right) = 1.774 \times 10^{-19}\ \text{J}$$

Solving *(13.32)* for λ and substituting the data gives

$$\lambda = \frac{(6.626 \times 10^{-34}\ \text{J}\cdot\text{s})(2.9979 \times 10^{8}\ \text{m}\cdot\text{s}^{-1})}{1.774 \times 10^{-19}\ \text{J}} = 1.120 \times 10^{-6}\ \text{m} = 1120\ \text{nm}$$

21.25 Calculate the number of electrons that have been promoted to the conduction band in Si(s) at $T = 298\ \text{K}$. What is the number of holes present? For Si(s), $E_g = 1.774 \times 10^{-19}\ \text{J}$.

▮ The number of electrons that have been promoted is given by

$$N = \frac{2(2\pi m_e kT)^{3/2}}{h^3} e^{-E_g/2kT} \tag{21.10}$$

Substituting the data into *(21.10)* gives

$$N = \frac{2[2\pi(9.11 \times 10^{-31}\ \text{kg})(1.381 \times 10^{-23}\ \text{J}\cdot\text{K}^{-1})(298\ \text{K})]^{3/2}}{(6.626 \times 10^{-34}\ \text{J}\cdot\text{s})^3[(1\ \text{kg}\cdot\text{m}^2\cdot\text{s}^{-2})/(1\ \text{J})]^{3/2}} \exp\left(\frac{-1.774 \times 10^{-19}\ \text{J}}{(2)(1.381 \times 10^{-23}\ \text{J}\cdot\text{K}^{-1})(298\ \text{K})}\right)$$

$$= 1.09 \times 10^{16}\ \text{m}^{-3}$$

For a pure intrinsic semiconductor, the number of holes is equal to the number of electrons that have been promoted.

21.26 Calculate the ratio of the number of holes at $T = 1000\ \text{K}$ to that at $T = 298\ \text{K}$. See Problem 21.25 for additional information.

▮ Taking a ratio of *(21.10)* gives

$$\frac{N(1000\ \text{K})}{N(298\ \text{K})} = \frac{(1000\ \text{K})^{3/2} \exp[-(1.774 \times 10^{-19}\ \text{J})/(2)(1.381 \times 10^{-23}\ \text{J}\cdot\text{K}^{-1})(1000\ \text{K})]}{(298\ \text{K})^{3/2} \exp[-(1.774 \times 10^{-19}\ \text{J})/(2)(1.381 \times 10^{-23}\ \text{J}\cdot\text{K}^{-1})(298\ \text{K})]} = 2.27 \times 10^{7}$$

21.3 IONIC BONDING

21.27 What is the energy needed to separate the ions in a gaseous K^+Cl^- molecule? The bond length is 0.266 67 nm.

▮ The *Coulombic energy* between two ions is given by

$$U = q_1 q_2/(4\pi\varepsilon_0)r \tag{21.11}$$

where $4\pi\varepsilon_0$ is the permittivity constant ($4\pi\varepsilon_0 = 1.112\,650\,056 \times 10^{-10}\ \text{C}^2\cdot\text{N}^{-1}\cdot\text{m}^{-2}$), q_i is the ionic charge, and r is the distance between the ions. Substituting the data into *(21.11)* gives

$$U = \frac{(1.602 \times 10^{-19}\ \text{C})(-1.602 \times 10^{-19}\ \text{C})[(1\ \text{J})/(1\ \text{N}\cdot\text{m})]}{(1.113 \times 10^{-10}\ \text{C}^2\cdot\text{N}^{-1}\cdot\text{m}^{-2})(0.266\,67\ \text{nm})[(10^{-9}\ \text{m})/(1\ \text{nm})]} = -8.652 \times 10^{-19}\ \text{J}$$

21.28 At what interatomic distance will the bonding in KCl(g) change from covalent to ionic? The ionization energy for K(g) is $419\ \text{kJ}\cdot\text{mol}^{-1}$, and the electron affinity of Cl(g) is $-349\ \text{kJ}\cdot\text{mol}^{-1}$.

▮ The bonding in KCl(g) will be essentially ionic for values of r such that

$$I_M + E_A(X) + U < 0 \tag{21.12}$$

where U is given by *(21.11)*. Substituting *(21.11)* into *(21.12)*, solving for r, and substituting the data gives

$$r < \frac{-(1.602 \times 10^{-19}\ \text{C})(-1.602 \times 10^{-19}\ \text{C})[(1\ \text{J})/(1\ \text{N}\cdot\text{m})][(10^{-3}\ \text{kJ})/(1\ \text{J})]}{(1.113 \times 10^{-10}\ \text{C}^2\cdot\text{N}^{-1}\cdot\text{m}^{-2})\left[\dfrac{419\ \text{kJ}\cdot\text{mol}^{-1} + (-349\ \text{kJ}\cdot\text{mol}^{-1})}{6.022 \times 10^{23}\ \text{mol}^{-1}}\right]} = 1.98 \times 10^{-9}\ \text{m}$$

For values of $r < 1.98\ \text{nm}$, an ionic band is energetically favorable.

21.29 Using the following data, calculate the lattice energy for KCl(s): $E_A(\text{Cl}) = -349\ \text{kJ}\cdot\text{mol}^{-1}$, $I(\text{K}) = 419\ \text{kJ}\cdot\text{mol}^{-1}$, $\text{BE}(\text{Cl}_2) = 242.604\ \text{kJ}\cdot\text{mol}^{-1}$, $\Delta_f H°(\text{KCl(s)}) = -436.684\ \text{kJ}\cdot\text{mol}^{-1}$, and $\Delta_{\text{sub}}H°(\text{K}) = 89.000\ \text{kJ}\cdot\text{mol}^{-1}$.

▮ Figure 21-7 represents the *Born–Haber–Fajans cycle* for an ionic substance with the empirical formula MX. The lattice energy is given by

$$\Delta_{\text{lat}}H(\text{MX}) = -\Delta_f H(\text{MX(s)}) + \Delta_{\text{sub}}H(\text{M}) + (\text{BE})/2 + I_M + E_A(X) \tag{21.13}$$

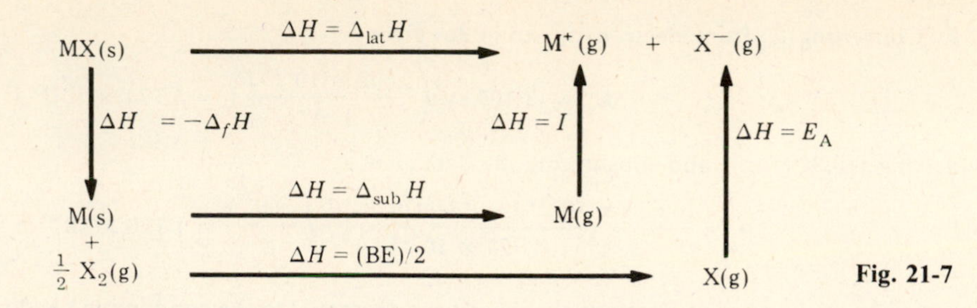

Fig. 21-7

If the dissociation energy is used instead of the bond energy, an additional $\frac{1}{2}RT$ must be placed on the right-hand side of *(21.13)*. Substituting the data into *(21.13)* gives

$$\Delta_{\text{lat}}H(\text{KCl}) = -(-436.684\ \text{kJ}\cdot\text{mol}^{-1}) + (89.000\ \text{kJ}\cdot\text{mol}^{-1}) + \frac{242.604\ \text{kJ}\cdot\text{mol}^{-1}}{2} + 419\ \text{kJ}\cdot\text{mol}^{-1} + (-349\ \text{kJ}\cdot\text{mol}^{-1}) = 717\ \text{kJ}\cdot\text{mol}^{-1}$$

21.30 The lattice energy for MgO(s) is $3866\ \text{kJ}\cdot\text{mol}^{-1}$. Given $\Delta_f H°(\text{MgO(s)}) = -601.70\ \text{kJ}\cdot\text{mol}^{-1}$, $\Delta_{\text{sub}}H°(\text{Mg}) = 147.70\ \text{kJ}\cdot\text{mol}^{-1}$, $\text{BE}(\text{O}_2) = 498.340\ \text{kJ}\cdot\text{mol}^{-1}$, and $I(\text{Mg}) = 2200.80\ \text{kJ}\cdot\text{mol}^{-1}$ for two electrons, calculate $E_A(\text{O})$ for two electrons.

▮ The cycle shown in Fig. 21-7 is valid for MgO. Solving *(21.13)* for $E_A(\text{O})$ and substituting the data gives

$$E_A(\text{O}) = 3866\ \text{kJ}\cdot\text{mol}^{-1} + (-601.70\ \text{kJ}\cdot\text{mol}^{-1}) - 147.70\ \text{kJ}\cdot\text{mol}^{-1} - \frac{498.340\ \text{kJ}\cdot\text{mol}^{-1}}{2} - 2200.80\ \text{kJ}\cdot\text{mol}^{-1}$$
$$= 677\ \text{kJ}\cdot\text{mol}^{-1}$$

21.31 Prepare a plot of the attractive potential energy between ions, the repulsive energy of the ions, and the total potential energy for solid KCl. The unit cell length of KCl(s) is 0.629 31 nm.

▮ The Coulombic attraction is given by

$$U_{\text{att}} = \frac{L\mathcal{M}q_1q_2}{(4\pi\varepsilon_0)r} \tag{21.14}$$

where $\mathcal{M}$ is the Madelung constant ($\mathcal{M}$ = 1.747 56 for KCl) and the repulsion between the ions is given by

$$U_{\text{rep}} = \frac{-4L\mathcal{M}q_1q_2\rho}{(4\pi\varepsilon_0)a^2}e^{(a/2-r)/\rho} \tag{21.15}$$

where $\rho = 0.034$ nm and a is the unit cell length. The total energy is the sum of $U_{\text{att}} + U_{\text{rep}}$. As a sample calculation, consider $r = 0.650$ nm.

$$U_{\text{att}} = \frac{(6.022\times10^{23}\ \text{mol}^{-1})(1.747\ 56)(1.602\times10^{-19}\ \text{C})(-1.602\times10^{-19}\ \text{C})\dfrac{1\ \text{J}}{1\ \text{N}\cdot\text{m}}}{(1.113\times10^{-10}\ \text{C}^2\cdot\text{N}^{-1}\cdot\text{m}^{-2})(0.650\times10^{-9}\ \text{m})[(10^3\ \text{J})/(1\ \text{kJ})]} = -373\ \text{kJ}\cdot\text{mol}^{-1}$$

$$U_{\text{rep}} = \frac{[-4(6.022\times10^{23}\ \text{mol}^{-1})(1.747\ 56)(1.602\times10^{-19}\ \text{C})(-1.602\times10^{-19}\ \text{C})(0.034\times10^{-9}\ \text{m})]}{(1.113\times10^{-10}\ \text{C}^2\cdot\text{N}^{-1}\cdot\text{m}^{-2})(0.629\ 31\times10^{-9}\ \text{m})^2[(1\ \text{N}\cdot\text{m})/(1\ \text{J})]}$$
$$\times \exp\left(\frac{0.629\ 31\ \text{nm}/2 - 0.650\ \text{nm}}{0.034\ \text{nm}}\right)$$
$$= 4.34\ \text{J}\cdot\text{mol}^{-1}$$

$$U = -373\ \text{kJ}\cdot\text{mol}^{-1} + 4.43\ \text{J}\cdot\text{mol}^{-1} = -373\ \text{kJ}\cdot\text{mol}^{-1}$$

The plots are shown in Fig. 21-8.

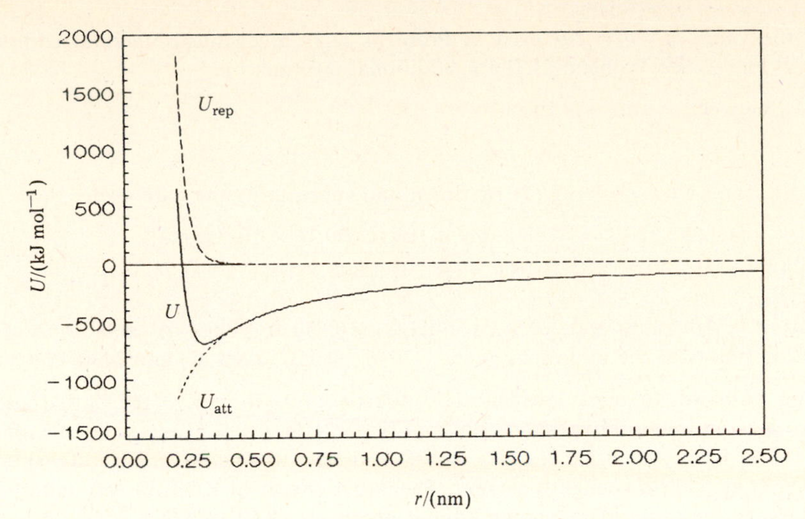

Fig. 21-8

21.32 Using the information given in Problem 21.31, calculate the value of $\Delta_{\rm lat}H(\mathrm{KCl})$ at 298 K.

▌ The lattice energy is given by *(21.14)* and *(21.15)* as

$$\Delta_{\rm lat}H = \frac{2L\mathscr{M}q_1q_2}{(4\pi\varepsilon_0)a}\left(\frac{2\rho}{a} - 1\right) + 2RT \qquad (21.16)$$

Substituting the data into *(21.16)* gives

$$\Delta_{\rm lat}H = \frac{2(6.022\times10^{23}\ \mathrm{mol^{-1}})(1.747\,56)(1.602\times10^{-19}\ \mathrm{C})(-1.602\times10^{-19}\ \mathrm{C})[(10^{-3}\ \mathrm{kJ})/(1\ \mathrm{J})]}{(1.113\times10^{-10}\ \mathrm{C^2\cdot N^{-1}\cdot m^{-2}})(0.629\,31\times10^{-9}\ \mathrm{m})[(1\ \mathrm{N\cdot m})/(1\ \mathrm{J})]}$$

$$\times\left(\frac{2(0.034\ \mathrm{nm})}{0.629\,31\ \mathrm{nm}} - 1\right) + 2(8.314\ \mathrm{J\cdot K^{-1}\cdot mol^{-1}})(298\ \mathrm{K})\frac{10^{-3}\ \mathrm{kJ}}{1\ \mathrm{mol}}$$

$$= 693\ \mathrm{kJ\cdot mol^{-1}}$$

which is 3% lower than the value determined in Problem 21.29.

21.33 An alternative form for *(21.15)* is

$$U_{\rm rep} = \frac{-L\mathscr{M}q_1q_2(a/2)^{n-1}}{(4\pi\varepsilon_0)nr^n} \qquad (21.17)$$

where n is an integer, usually between 6 and 12. Using $n = 10$ for KCl(s), calculate $U_{\rm rep}$ for $r = 0.650$ nm and compare the answer to that obtained using *(21.15)*. See Problem 21.31 for additional information.

▌ Substituting the data into *(21.17)* gives

$$U_{\rm rep} = \frac{[-(6.022\times10^{23}\ \mathrm{mol^{-1}})(1.747\,56)(1.602\times10^{-19}\ \mathrm{C})(-1.602\times10^{-19}\ \mathrm{C})[(0.629\,31\times10^{-9}\ \mathrm{m})/2]^9]}{(1.113\times10^{-10}\ \mathrm{C^2\cdot N^{-1}\cdot m^{-2}})(6)(0.650\times10^{-9}\ \mathrm{m})^{10}[(1\ \mathrm{N\cdot m})/(1\ \mathrm{J})]}$$

$$= 90.9\ \mathrm{J}$$

21.34 By what factor does $U_{\rm rep}$ change if the value of n is changed to 12 for KCl(s)? What is the corresponding change in U for the crystal?

▌ Substituting $n = 12$ into *(12.17)* gives

$$U_{\rm rep} = \frac{[-(6.022\times10^{23}\ \mathrm{mol^{-1}})(1.747\,56)(1.602\times10^{-19}\ \mathrm{C})(-1.602\times10^{-19}\ \mathrm{C})[(0.629\,31\times10^{-9}\ \mathrm{m})/2]^{11}]}{(1.113\times10^{-10}\ \mathrm{C^2\cdot N^{-1}\cdot m^{-2}})(12)(0.650\times10^{-9}\ \mathrm{m})^{12}[(1\ \mathrm{N\cdot m})/(1\ \mathrm{J})]}$$

$$= 10.6\ \mathrm{J}$$

The value of $U_{\rm rep}$ changes by a factor of

$$\frac{U_{\rm rep}(n=6)}{U_{\rm rep}(n=12)} = \frac{90.9\ \mathrm{J}}{10.6\ \mathrm{J}} = 8.6$$

Even though the two values of n give values of $U_{\rm rep}$ that differ considerably, this 80.3-J difference corresponds to 0.02% of the value of U (see Problem 21.31).

21.35 Confirm that the value of $n = 10$ used in Problem 21.33 given the isothermal compressibility of KCl(s) is $\kappa = 4.8 \times 10^{-11}\ \text{Pa}^{-1}$. See Problem 21.31 for additional information.

▮ The relation between κ and n is given by

$$\kappa = (4\pi\varepsilon_0)18r^4/(n-1)\,e^2\mathcal{M} \qquad (21.18)$$

where $r = a/2$ for KCl(s). Solving *(21.18)* for n and substituting the data gives

$$n = \frac{(1.113 \times 10^{-10}\ \text{C}^2 \cdot \text{N}^{-1} \cdot \text{m}^{-2})(18)[(0.629\,31 \times 10^{-9}\ \text{m})/2]^4}{(4.8 \times 10^{-11}\ \text{Pa}^{-1})[(1\ \text{Pa})/(1\ \text{N} \cdot \text{m}^{-2})](1.602 \times 10^{-19}\ \text{C})^2(1.747\,56)} + 1 = 9.1 + 1 = 10.1$$

21.36 Using $a/(\text{nm}) = 0.510$ for SrO(s), 0.533 for KF(s), 0.550 for BaO(s), and 0.628 for KCl(s), assign the following melting points to the ionic substances: 776 °C, 880 °C, 1923 °C, and 2430 °C.

▮ The melting point of an ionic compound is directly proportional to the lattice energy, which is directly proportional to q_1q_2 and inversely proportional to a [see *(21.16)*]. Because q_1q_2 is four times as large for SrO and BaO as for KF and KCl, the lattice energies of SrO and BaO will be the greatest. Of these, the lattice energy of SrO will be greater because a is slightly smaller. The lattice energy of KCl will be smaller than that of KF because a for KCl is larger. The order of increasing lattice energies is $\text{KCl} < \text{KF} < \text{BaO} < \text{SrO}$, which is the same as the order of increasing melting point. Thus the melting points are 776 °C for KCl, 880 °C for KF, 1923 °C for BaO, and 2430 °C for SrO.

21.4 VAN DER WAALS FORCES

21.37 Describe the important intermolecular forces for NF_3.

▮ The forces for the trigonal pyramidal molecule (see Fig. 18-9a) are London forces and dipole-dipole interactions.

21.38 Describe the important intermolecular forces for MgF_2.

▮ The forces for the linear species (see Fig. 18-9b) are London forces.

21.39 Describe the important intermolecular forces for BF_3.

▮ The forces for the trigonal planar molecule (see Fig. 18-9c) are London forces.

21.40 Describe the important intermolecular forces for C_2F_2.

▮ The forces for the linear molecule (see Fig. 18-9e) are London forces.

21.41 Describe the important intermolecular forces for NO.

▮ The forces for the linear molecule (see Fig. 18-9f) are London forces and dipole-dipole interactions.

21.42 Describe the important intermolecular forces for NO_2.

▮ The forces for the bent molecule (see Fig. 18-9g) are London forces and dipole-dipole interactions.

21.43 Describe the important intermolecular forces for IF_3.

▮ The forces for the T-shaped molecule (see Fig. 18-10a) are London forces and dipole-dipole interactions.

21.44 Describe the important intermolecular forces for IF_5.

▮ The forces for the square pyramidal molecule (see Fig. 18-10e) are London forces and dipole-dipole interactions.

21.45 The dipole moment of toluene is 0.36 D. Predict the dipole moment of *o*-xylene and *p*-xylene.

▮ Dipole moments can be treated vectorially. For *p*-xylene, where the angle between the vectors is 180°, assume that the two vectors lie along the y axis and along the $-y$ axis. The predicted dipole moment is

$$\mu_x = (0.36\ \text{D}) \cos 90° + (0.36\ \text{D}) \cos(-90°) = 0$$

$$\mu_y = (0.36\ \text{D}) \sin 90° + (0.36\ \text{D}) \sin(-90°) = 0$$

$$\mu = (\mu_x^2 + \mu_y^2)^{1/2} = (0^2 + 0^2)^{1/2} = 0$$

which agrees with the observed value. For *o*-xylene, where the angle between the vectors is 60°, assume that the two vectors lie along the y axis and in the positive xy quadrant. The predicted dipole moment is

$$\mu_x = (0.36\ \text{D}) \cos 90° + (0.36\ \text{D}) \cos 30° = 0.31\ \text{D}$$

$$\mu_y = (0.36\ \text{D}) \sin 90° + (0.36\ \text{D}) \sin 30° = 0.54\ \text{D}$$

$$\mu = [(0.31\ \text{D})^2 + (0.54\ \text{D})^2]^{1/2} = 0.62\ \text{D}$$

which agrees with the observed value 0.62 D.

21.46 The dipole moment of H_2O is 1.85 D. Given the bond angle as 104.5°, calculate the O—H bond moment.

▮ If the molecule is oriented as in Fig. 19-5a, the angles with respect to the x axis will be $-37.75°$ and $-142.25°$. The dipole moment in terms of the bond moment $\mu(\text{OH})$ is

$$\mu_x = \mu(\text{OH}) \cos(-37.75°) + \mu(\text{OH}) \cos(-142.25°) = 0$$

$$\mu_y = \mu(\text{OH}) \sin(-37.75°) + \mu(\text{OH}) \sin(-142.25°) = -1.224\mu(\text{OH})$$

$$\mu = \{0^2 + [-1.224\mu(\text{OH})]^2\}^{1/2} = 1.85\ \text{D}$$

Solving gives $\mu(\text{OH}) = 1.51\ \text{D}$.

21.47 A 15-V electric potential was placed across the plates of a capacitor. Given that the plates are separated by 0.010 m and the surface area of each plate is 0.010 m^2, what are the electric field strength and the capacitance? Calculate the force on an electron between the plates and the charge density.

▮ The electric field strength (E) is given by

$$E = V/l \tag{21.19}$$

where V is the potential difference and l is the distance between the plates. Substituting the data into *(21.19)* gives

$$E_0 = \frac{15\ \text{V}}{0.010\ \text{m}} = 1500\ \text{V} \cdot \text{m}^{-1}$$

or, upon substituting the definitions of the various desired units,

$$E_0 = (1500\ \text{V} \cdot \text{m}^{-1})\left(\frac{1\ \text{kg} \cdot \text{m}^2 \cdot \text{A}^{-1} \cdot \text{s}^{-3}}{1\ \text{V}}\right)\left(\frac{1\ \text{A} \cdot \text{s}}{1\ \text{C}}\right)\left(\frac{1\ \text{N}}{1\ \text{kg} \cdot \text{m} \cdot \text{s}^{-2}}\right) = 1500\ \text{N} \cdot \text{C}^{-1}$$

The capacitance (C) is given by

$$C = \varepsilon_0 A/l \tag{21.20}$$

where ε_0 is the permittivity of a vacuum ($8.854\ 187\ 18 \times 10^{-12}\ \text{C}^2 \cdot \text{N}^{-1} \cdot \text{m}^{-2}$). Substituting the data into *(21.20)* gives

$$C_0 = \frac{(8.854 \times 10^{-12}\ \text{C}^2 \cdot \text{N}^{-1} \cdot \text{m}^{-2})(0.010\ \text{m}^2)\left(\frac{1\ \text{N} \cdot \text{C}^{-1}}{1\ \text{V} \cdot \text{m}^{-1}}\right)\left(\frac{1\ \text{F}}{1\ \text{C} \cdot \text{V}^{-1}}\right)}{0.010\ \text{m}} = 8.9 \times 10^{-12}\ \text{F}$$

where the conversion factor $1\ \text{N} \cdot \text{C}^{-1} = 1\ \text{V} \cdot \text{m}^{-1}$ was derived above for E_0 and F is the SI unit farad. The force exerted on an electron by the field is

$$f = E_0 q \tag{21.21}$$

Substituting the data into *(21.21)* gives

$$f = (1500\ \text{N} \cdot \text{C}^{-1})(1.602 \times 10^{-19}\ \text{C}) = 2.4 \times 10^{-16}\ \text{N}$$

The surface charge density (σ) is given by

$$\sigma = E_0 \varepsilon_0 \tag{21.22}$$

Substituting the data into *(21.22)* gives

$$\sigma = (1500\ \text{N} \cdot \text{C}^{-1})(8.854 \times 10^{-12}\ \text{C}^2 \cdot \text{N}^{-1} \cdot \text{m}^{-2}) = 1.3 \times 10^{-8}\ \text{C} \cdot \text{m}^{-2}$$

21.48 Suppose the capacitor described in Problem 21.47 contained liquid $SiCl_4$ between the plates with $\varepsilon/\varepsilon_0 = 2.4$. Calculate E, C, the charge on the surface of the dielectric, and the electric field in a cavity in the $SiCl_4$.

▮ The electric field strength and the capacitance will be given by

$$E = E_0(\varepsilon_0/\varepsilon) = (1500\ \text{V} \cdot \text{m}^{-1})/2.4 = 630\ \text{V} \cdot \text{m}^{-1}$$

$$C = C_0(\varepsilon/\varepsilon_0) = (8.9 \times 10^{-12}\ \text{F})(2.4) = 2.1 \times 10^{-11}\ \text{F}$$

The charge on the surface of the $SiCl_4$ is given by

$$p = (E_0 - E)\varepsilon_0 \tag{21.23}$$

Substituting the data into *(21.23)* gives

$$p = (1500\ \text{V}\cdot\text{m}^{-1} - 600\ \text{V}\cdot\text{m}^{-1})(8.854 \times 10^{-12}\ \text{C}^2\cdot\text{N}^{-1}\cdot\text{m}^{-2})\left(\frac{1\ \text{N}\cdot\text{C}^{-1}}{1\ \text{V}\cdot\text{m}^{-1}}\right) = 8.0 \times 10^{-9}\ \text{C}\cdot\text{m}^{-2}$$

The electric field in a cavity is given by

$$E(\text{cavity}) = p/3\varepsilon_0 \tag{21.24}$$

Substituting the data into *(21.24)* gives

$$E(\text{cavity}) = \frac{8.0 \times 10^{-9}\ \text{C}\cdot\text{m}^{-2}}{3(8.854 \times 10^{-12}\ \text{C}^2\cdot\text{N}^{-1}\cdot\text{m}^{-2})} = 3.0 \times 10^2\ \text{N}\cdot\text{C}^{-1} = 3.0 \times 10^2\ \text{V}\cdot\text{m}^{-1}$$

21.49 The dielectric constant for benzene is 2.284 at 20 °C. Calculate the molar polarization and the polarizability. The density of $C_6H_6(l)$ is $0.879\ \text{g}\cdot\text{cm}^{-3}$.

I The molar polarization ($\mathcal{P}$) is related to the dielectric constant by

$$\mathcal{P} = \frac{\varepsilon/\varepsilon_0 - 1}{\varepsilon/\varepsilon_0 + 2}\left(\frac{M}{\rho}\right) \tag{21.25}$$

Substituting the data into *(21.25)* gives

$$\mathcal{P} = \frac{2.284 - 1}{2.284 + 2}\left(\frac{78.11\ \text{g}\cdot\text{mol}^{-1}}{0.879\ \text{g}\cdot\text{cm}^{-3}}\right)\left(\frac{10^{-2}\ \text{m}}{1\ \text{cm}}\right)^3 = 2.663 \times 10^{-5}\ \text{m}^3\cdot\text{mol}^{-1}$$

The molar polarization is related to the polarizability (α) and the dipole moment by

$$\mathcal{P} = \frac{L}{3\varepsilon_0}\left(\alpha + \frac{\mu^2}{3kT}\right) \tag{21.26}$$

For benzene, $\mu = 0$. Solving *(21.26)* for α and substituting the data gives

$$\alpha = \frac{3(8.854 \times 10^{-12}\ \text{C}^2\cdot\text{N}^{-1}\cdot\text{m}^{-2})(2.663 \times 10^{-5}\ \text{m}^3\cdot\text{mol}^{-1})}{6.022 \times 10^{23}\ \text{mol}^{-1}} = 1.175 \times 10^{-39}\ \text{C}^2\cdot\text{N}^{-1}\cdot\text{m}$$

21.50 The dielectric constant of He is 1.000 065 0 and that of H_2 is 1.000 253 8 at 20.00 °C and 1.00 atm. Calculate the polarizability for these gases and briefly discuss the results.

I Substituting *(1.20)* into *(21.25)* for M/ρ and substituting the data gives

$$\mathcal{P}(\text{He}) = \frac{1.000\,065\,0 - 1}{1.000\,065\,0 + 2}\left(\frac{(8.314\ \text{J}\cdot\text{K}^{-1}\cdot\text{mol}^{-1})(293.15\ \text{K})}{(101\,325\ \text{Pa})[(1\ \text{J})/(1\ \text{Pa}\cdot\text{m}^3)]}\right) = 5.21 \times 10^{-7}\ \text{m}^3\cdot\text{mol}^{-1}$$

$$\mathcal{P}(\text{H}_2) = \frac{1.000\,253\,8 - 1}{1.000\,253\,8 + 2}\left(\frac{(8.314)(293.15)}{101\,325}\right) = 2.035 \times 10^{-6}\ \text{m}^3\cdot\text{mol}^{-1}$$

Solving *(21.26)* for α and substituting $\mu = 0$ and the above results gives

$$\alpha(\text{He}) = \frac{3(8.854 \times 10^{-12}\ \text{C}^2\cdot\text{N}^{-1}\cdot\text{m}^{-2})(5.21 \times 10^{-7}\ \text{m}^3\cdot\text{mol}^{-1})}{6.022 \times 10^{23}\ \text{mol}^{-1}} = 2.30 \times 10^{-41}\ \text{C}^2\cdot\text{N}^{-1}\cdot\text{m}$$

$$\alpha(\text{H}_2) = \frac{3(8.854 \times 10^{-12})(2.035 \times 10^{-6})}{6.022 \times 10^{23}} = 8.976 \times 10^{-41}\ \text{C}^2\cdot\text{N}^{-1}\cdot\text{m}$$

The greater polarizability of H_2 is the result of the two electrons being more diffuse in the linear molecule than in the spherical atom.

21.51 The dielectric constant of $H_2O(g)$ at 1.00 atm was determined as a function of temperature.

$T/(°C)$	111.2	147.0	171.6	211.0	248.9
$\varepsilon/\varepsilon_0$	1.005 47	1.004 66	1.004 12	1.003 54	1.003 02

Determine α and μ for H_2O.

▌ The n/ρ term in *(21.25)* can be replaced by RT/P (see Problem 21.50). As a sample calculation, consider $T = 111.2\ °\mathrm{C}$.

$$\mathcal{P} = \frac{1.005\,47 - 1}{1.005\,47 + 2}\left[\frac{(8.314\ \mathrm{J\cdot K^{-1}\cdot mol^{-1}})(384.4\ \mathrm{K})}{(101\,325\ \mathrm{Pa})[(1\ \mathrm{J})/(1\ \mathrm{Pa\cdot m^3})]}\right] = 5.74 \times 10^{-5}\ \mathrm{m^3\cdot mol^{-1}}$$

Likewise, $\mathcal{P}/(10^{-5}\ \mathrm{m^3\cdot mol^{-1}}) = 5.35$ at 147.0 °C, 5.01 at 171.6 °C, 4.68 at 211.0 °C, and 4.31 at 248.9 °C. From *(21.26)*, a plot of $\mathcal{P}$ against T^{-1} will be linear with an intercept of $\alpha L/3\varepsilon_0$ and a slope of $L\mu^2/9\varepsilon_0 k$. From Fig. 21-9,

$$\alpha = \frac{3\varepsilon_0(\text{intercept})}{L} = \frac{3(8.854 \times 10^{-12}\ \mathrm{C^2\cdot N^{-1}\cdot m^{-2}})(3.5 \times 10^{-6}\ \mathrm{m^3\cdot mol^{-1}})}{6.022 \times 10^{23}\ \mathrm{mol^{-1}}} = 1.54 \times 10^{-40}\ \mathrm{C^2\cdot N^{-1}\cdot m}$$

$$\mu^2 = 9\varepsilon_0 k(\text{slope})/L = \frac{9(8.854 \times 10^{-12}\ \mathrm{C^2\cdot N^{-1}\cdot m^{-2}})(1.381 \times 10^{-23}\ \mathrm{J\cdot K^{-1}})(0.0208\ \mathrm{m^3\cdot mol^{-1}\cdot K})}{(6.022 \times 10^{23}\ \mathrm{mol^{-1}})[(1\ \mathrm{J})/(1\ \mathrm{N\cdot m})]}$$

$$= 3.80 \times 10^{-59}\ \mathrm{C^2\cdot m^2}$$

The dipole moment is $\mu = 6.16 \times 10^{-30}\ \mathrm{C\cdot m} = 1.85\ \mathrm{D}$ (see Problem 16.34).

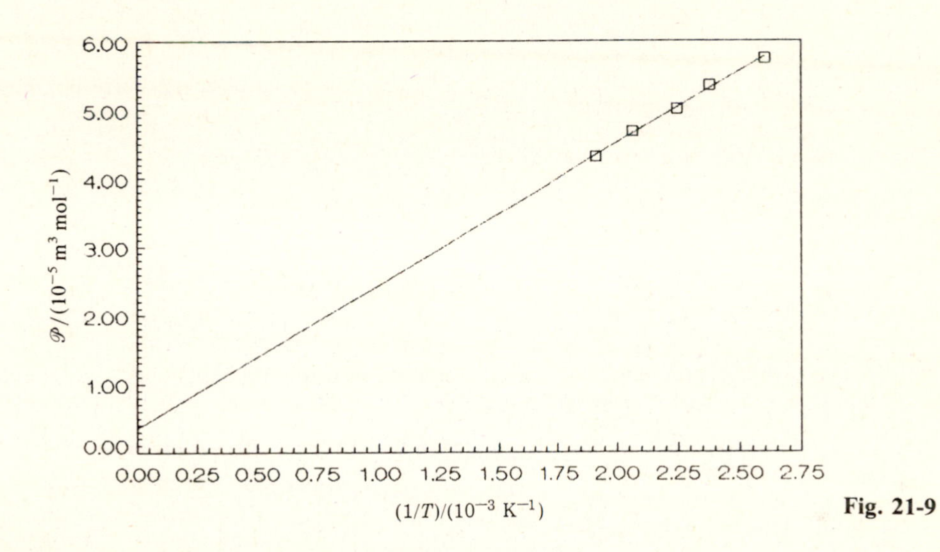

Fig. 21-9

21.52 The molar refraction of water is $3.76\ \mathrm{cm^3\cdot mol^{-1}}$. Calculate the polarizability.

▌ The molar refraction (R_m) is related to the polarizability by

$$R_m = L\alpha/3\varepsilon_0 \qquad (21.27)$$

Solving *(21.27)* for α and substituting the data gives

$$\alpha = \frac{3(8.854 \times 10^{-12}\ \mathrm{C^2\cdot N^{-1}\cdot m^{-2}})(3.76\ \mathrm{cm^3\cdot mol^{-1}})[(10^{-2}\ \mathrm{m})/(1\ \mathrm{cm})]^3}{6.022 \times 10^{23}\ \mathrm{mol^{-1}}} = 1.65 \times 10^{-40}\ \mathrm{C^2\cdot N^{-1}\cdot m}$$

21.53 The index of refraction for acetone, CH_3COCH_3, is 1.3588. Calculate the polarizability. The density is $0.7899\ \mathrm{g\cdot cm^{-3}}$.

▌ The molar refraction is related to the index of refraction (n) by

$$R_m = \frac{n^2 - 1}{n^2 + 2}\left(\frac{M}{\rho}\right) \qquad (21.28)$$

Substituting the data into *(21.28)* gives

$$R_m = \frac{(1.3588)^2 - 1}{(1.3588)^2 + 2}\left(\frac{58.08\ \mathrm{g\cdot mol^{-1}}}{0.7899\ \mathrm{g\cdot cm^{-3}}}\right)\left(\frac{10^{-2}\ \mathrm{m}}{1\ \mathrm{cm}}\right)^3 = 1.618 \times 10^{-5}\ \mathrm{m^3\cdot mol^{-1}}$$

Solving *(21.27)* for α and substituting the data gives

$$\alpha = \frac{3(8.854 \times 10^{-12}\ \mathrm{C^2\cdot N^{-1}\cdot m^{-2}})(1.618 \times 10^{-5}\ \mathrm{m^3\cdot mol^{-1}})}{6.022 \times 10^{23}\ \mathrm{mol^{-1}}} = 7.14 \times 10^{-40}\ \mathrm{C^2\cdot N^{-1}\cdot m}$$

21.54 Using the following molar refractivities

$$\text{C—H},\ 1.67\ \text{cm}^3\ \text{mol}^{-1} \qquad \text{C—C},\ 1.21\ \text{cm}^3\cdot\text{mol}^{-1} \qquad \text{C=O},\ 3.38\ \text{cm}^3\cdot\text{mol}^{-1}$$

predict the molar refraction for acetone.

▌ The molecule consists of six C—H bonds, two C—C bonds, and one C=O bond. Adding the contributions gives

$$R_m = 6(1.67\ \text{cm}^3\cdot\text{mol}^{-1}) + 2(1.21\ \text{cm}^3\cdot\text{mol}^{-1}) + 1(3.38\ \text{cm}^3\cdot\text{mol}^{-1}) = 15.82\ \text{cm}^3\cdot\text{mol}^{-1}$$

which agrees to within 2% with the value determined in Problem 21.53.

21.55 The polarizability of CO(g) is $2.20\times10^{-40}\ \text{C}^2\cdot\text{N}^{-1}\cdot\text{m}$, and the dipole moment is $3.90\times10^{-31}\ \text{C}\cdot\text{m}$. Calculate the value of the potential energy corresponding to the London dispersion forces at $r = 0.4224$ nm.

▌ The potential energy resulting from the induced dipole-induced dipole effect is given by

$$U_L = (-1.8\times10^{-18}\ \text{J})\alpha^2 L/(4\pi\varepsilon_0)^2 r^6 \tag{21.29}$$

Substituting the data into *(21.29)* gives

$$U_L = \frac{(-1.8\times10^{-18}\ \text{J})(2.20\times10^{-40}\ \text{C}^2\cdot\text{N}^{-1}\cdot\text{m})^2(6.022\times10^{23}\ \text{mol}^{-1})}{(1.113\times10^{-10}\ \text{C}^2\cdot\text{N}^{-1}\cdot\text{m}^{-2})^2(0.4224\times10^{-9}\ \text{m})^6} = -750\ \text{J}\cdot\text{mol}^{-1}$$

21.56 Using the data given in Problem 21.55, calculate the value of the potential energy corresponding to the dipole-dipole interactions at $r = 0.4224$ nm and $T = 298$ K.

▌ The potential energy resulting from the dipole-dipole effect is given by

$$U_{\text{d-d}} = -(2\mu^4/3kT)L/(4\pi\varepsilon_0)^2 r^6 \tag{21.30}$$

Substituting the data into *(21.30)* gives

$$U_{\text{d-d}} = \frac{-2(3.90\times10^{-31}\ \text{C}\cdot\text{m})^4(6.022\times10^{23}\ \text{mol}^{-1})[(1\ \text{J})/(1\ \text{N}\cdot\text{m})]^2}{3(1.381\times10^{-23}\ \text{J}\cdot\text{K}^{-1})(298\ \text{K})(1.113\times10^{-10}\ \text{C}^2\cdot\text{N}^{-1}\cdot\text{m}^{-2})^2(0.4224\times10^{-9}\ \text{m})^6} = -0.0322\ \text{J}\cdot\text{mol}^{-1}$$

21.57 Using the data given in Problem 21.55, calculate the value of the potential energy corresponding to the dipole-induced dipole interactions at $r = 0.4224$ nm.

▌ The potential energy resulting from the dipole-induced dipole effect is given by

$$U_{\text{d-id}} = -2\alpha\mu^2 L/(4\pi\varepsilon_0)^2 r^6 \tag{21.31}$$

Substituting the data into *(21.31)* gives

$$U_{\text{d-id}} = \frac{-2(2.20\times10^{-40}\ \text{C}^2\cdot\text{N}^{-1}\cdot\text{m})(3.90\times10^{-31}\ \text{C}\cdot\text{m})^2(6.022\times10^{23}\ \text{mol}^{-1})[(1\ \text{J})/(1\ \text{N}\cdot\text{m})]}{(1.113\times10^{-10}\ \text{C}^2\cdot\text{N}^{-1}\cdot\text{m}^{-2})^2(0.4224\times10^{-9}\ \text{m})^6}$$
$$= -0.573\ \text{J}\cdot\text{mol}^{-1}$$

21.58 Determine the total intermolecular potential energy for CO(g) at $T = 298$ K. Compare the relative importance of the various contributions.

▌ From Problems 21.55, 21.56, and 21.57,

$$U_t = U_L + U_{\text{d-d}} + U_{\text{d-id}} = (-750\ \text{J}\cdot\text{mol}^{-1}) + (-0.0322\ \text{J}\cdot\text{mol}^{-1}) + (-0.573\ \text{J}\cdot\text{mol}^{-1}) = -750\ \text{J}\cdot\text{mol}^{-1}$$

The contributions are 99.92%, 0.004%, and 0.076%, respectively.

21.59 Using the Lennard-Jones potential, calculate the total intermolecular potential energy for CO(g) at $r = 0.4224$ nm. For CO, $\varepsilon = 1.383\times10^{-21}$ J and $\sigma = 0.3763$ nm.

▌ Substituting the data into *(1.40)* gives

$$U = 4(1.383\times10^{-21}\ \text{J})(6.022\times10^{23}\ \text{mol}^{-1})\left[\left(\frac{0.3763\ \text{nm}}{0.4224\ \text{nm}}\right)^{12} - \left(\frac{0.3763\ \text{nm}}{0.4224\ \text{nm}}\right)^{6}\right] = -833\ \text{J}\cdot\text{mol}^{-1}$$

which differs by 11% from the result of Problem 21.58.

21.60 Many spherical, nonpolar molecules obey the following empirical relations:

$$\varepsilon \approx 1.3kT_b \tag{21.32}$$

$$T_c \approx 1.6T_b \tag{21.33}$$

where T_b is the normal boiling point and T_c is the critical temperature. Using $\varepsilon = 1.96 \times 10^{-21}$ J for $CH_4(g)$, predict T_b and T_c.

▮ Solving *(21.32)* for T_b and substituting the data gives

$$T_b = \frac{1.96 \times 10^{-21}\ \text{J}}{1.3(1.381 \times 10^{-23}\ \text{J} \cdot \text{K}^{-1})} = 109\ \text{K}$$

which is in good agreement with the accepted value of 109.2 K. Substituting this result into *(21.33)* gives

$$T_c = 1.6(109\ \text{K}) = 174\ \text{K}$$

which is 8.9% lower than the observed value of 191.1 K.

CHAPTER 22
Crystals

22.1 UNIT CELL

22.1 Consider the parallelograms shown in Fig. 22-1a representing two-dimensional unit cells. Which of these are primitive, and which are multiple unit cells? Are any orthogonal?

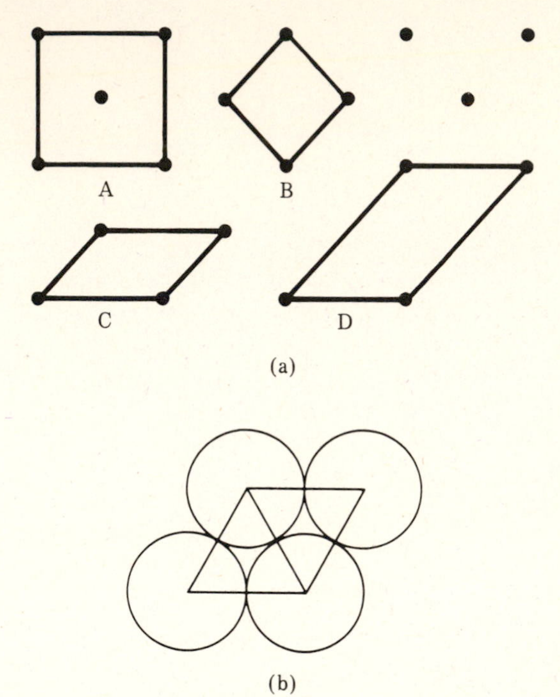

Fig. 22-1

▮ Unit cells B and C are *primitive unit cells* because the only points contained in the unit cells are at the corners of the parallelograms. Unit cells A and D are *multiple unit cells* because there are points in the unit cells in addition to the corners of the parallelogram. Unit cell A is an *orthogonal unit cell* because it contains angles of 90°.

22.2 What is the content of the unit cell A shown in Fig. 22-1a?

▮ The *unit cell content* (Z) is the total number of atoms contained within the unit cell. As can be seen in Fig. 22-1b, a space-filling drawing, only part of each circle at each corner is contained within the parallelogram unit cell. Adding up these fractions of circles gives $Z = 1$.

22.3 What is the relation between the radius of the circle and the length of the parallelogram representing the unit cell for the unit cell shown in Fig. 22-1b?

▮ The radius of the circle (R) and the length of the side of the parallelogram (a) are related by $a = 2R$.

22.4 How many nearest neighbor circles does a given circle have in Fig. 22-1b?

▮ The *crystal coordination number* (CN) is the number of nearest neighbors around a given atom, ion, or molecule in the crystal. Looking at any point in Fig. 22-1a shows that each atom (except the "surface" atoms) is touching six other atoms, giving CN = 6.

22.5 What is the radius of the triangular hole shown in Fig. 21-1b?

▮ The length of a line (l) drawn from the vertex through the centroid to the base of the triangle shown in Fig. 22-1b is $l = a(3)^{1/2}/2$. The centroid is located at a distance of $2l/3$. The radius of a circle that can be drawn at the centroid that is tangent to the three larger circles is

$$r = \frac{2l}{3} - R = \frac{2[a(3)^{1/2}/2]}{3} - R = \frac{2[2R(3)^{1/2}/2]}{3} - R = \left(\frac{2}{(3)^{1/2}} - 1\right)R = 0.1547R$$

22.6 What is the packing fraction of the unit cell shown in Fig. 22-1b?

▮ The *packing fraction* (or *packing efficiency*) is determined by dividing the area actually occupied by the circles by the area of the unit cell. Because $Z = 1$ (see Problem 22.2), the area occupied is $A_{occ} = \pi R^2$ and the area of the parallelogram (see Problem 22.5) is

$$A_{cell} = (2R)[(3)^{1/2}R] = 2(3)^{1/2}R^2$$

giving

$$\text{Packing fraction} = \frac{\pi R^2}{2(3)^{1/2}R^2} = \frac{\pi}{2(3)^{1/2}} = 0.907$$

22.7 What fraction of the surface of a crystal of Cd at $T = 298$ K consists of vacancies? Assume that the energy needed to form a vacancy is approximately $0.5\,\Delta_{sub}H°$. For Cd(s), $\Delta_{sub}H° = 112.01\ \text{kJ}\cdot\text{mol}^{-1}$.

▮ Substituting into *(22.2)* gives the fraction (n/N) as

$$\frac{n}{N} = \exp\left(\frac{-(0.5)(112.01 \times 10^3\ \text{J}\cdot\text{mol}^{-1})}{(8.314\ \text{J}\cdot\text{K}^{-1}\cdot\text{mol}^{-1})(298\ \text{K})}\right) = 1.5 \times 10^{-10}$$

22.8 Calcium crystallizes in a face-centered cubic unit cell with $a = 0.556$ nm. What is the unit cell content for Ca?

▮ One-eighth of each corner atom and one-half of each face-centered atom are contained within the unit cell under consideration (see Fig. 22-2), giving

$$Z = 8(1/8) + 6(1/2) = 4$$

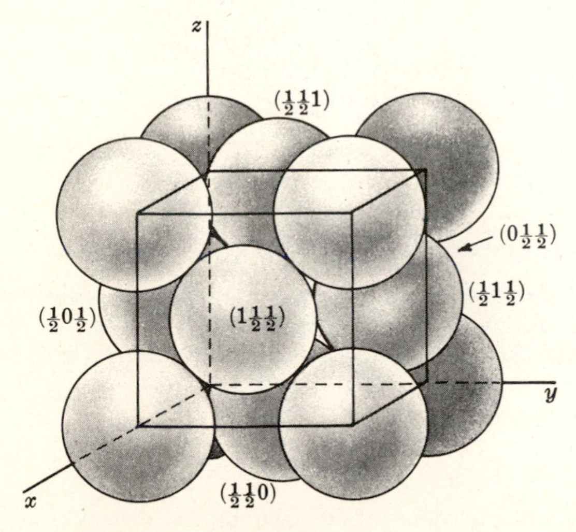

Fig. 22-2

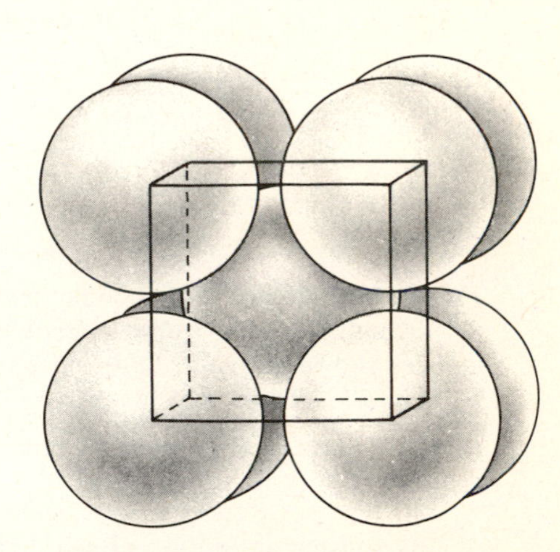

Fig. 22-3

22.9 Potassium crystallizes in a body-centered cubic unit cell with $a = 0.5333$ nm. What is the unit cell content for K?

▮ One-eighth of each corner atom and the entire body-centered atom are contained within the unit cell under consideration (see Fig. 22-3) giving

$$Z = 8(1/8) + 1(1) = 2$$

22.10 Silicon crystallizes in a unit cell similar to that of diamond (a face-centered cubic unit cell containing a tetrahedron of atoms) with $a = 0.541\,73$ nm. What is the unit cell content for Si?

▮ One-eighth of each corner atom, one-half of each face-centered atom, and the entire body-centered atom are contained within the unit cell under consideration, giving

$$Z = 8(1/8) + 6(1/2) + 4(1) = 8$$

22.11 Lanthanum crystallizes in a hexagonal closest-packed unit cell with $a = 0.372$ nm and $c = 0.606$ nm. What is the unit cell content for La?

▌ There are two types of corner atoms (see Fig. 22-4), one-sixth of some and one-twelfth of some contained within the unit cell. The body-centered atom is also contained with the unit cell. Adding these contributions gives

$$Z = 4(1/12) + 4(1/6) + 1(1) = 2$$

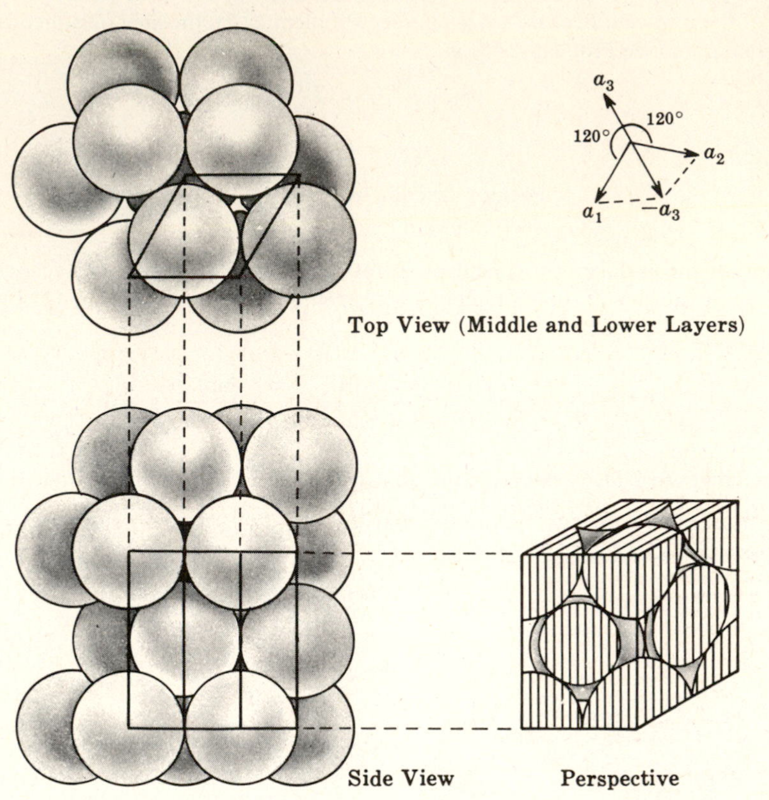

Fig. 22-4

22.12 Write the coordinates of the atoms in the Ca unit cell. See Problem 22.8 for additional information.

▌ The location of each atom in the unit cell can be specified by using the x, y, z coordinates of the atom given as multiples of a, b, and c, respectively. For example, for the atom in the rear lower left-hand corner of the unit cell shown in Fig. 22-2, the coordinates would be (000). Some of the coordinates are shown next to the atoms in Fig. 22-2. The complete set is

$$(000) \quad (100) \quad (010) \quad (001) \quad (110) \quad (101) \quad (011) \quad (111)$$

$$(0\tfrac{1}{2}\tfrac{1}{2}) \quad (\tfrac{1}{2}0\tfrac{1}{2}) \quad (\tfrac{1}{2}\tfrac{1}{2}0) \quad (1\tfrac{1}{2}\tfrac{1}{2}) \quad (\tfrac{1}{2}1\tfrac{1}{2}) \quad (\tfrac{1}{2}\tfrac{1}{2}1)$$

By convention, the set of coordinates (000) stands for the locations of all eight corners; the set of coordinates $(0\tfrac{1}{2}\tfrac{1}{2})$ stands for $(0\tfrac{1}{2}\tfrac{1}{2})$ and $(1\tfrac{1}{2}\tfrac{1}{2})$, etc. The unique coordinates are (000), $(0\tfrac{1}{2}\tfrac{1}{2})$, $(\tfrac{1}{2}0\tfrac{1}{2})$, $(\tfrac{1}{2}\tfrac{1}{2}0)$. Note that the minimum number of coordinate sets necessary to express the location of all atoms in the unit cell is equal to Z.

22.13 Write the coordinates of the atoms in the K unit cell. See Problem 22.9 for additional information.

▌ The locations of the corner atoms are given by (000) and those of the body-centered atom by $(\tfrac{1}{2}\tfrac{1}{2}\tfrac{1}{2})$; see Fig. 22-3.

22.14 Write the coordinates of the atoms in the Si unit cell. See Problem 22.10 for additional information.

▌ The locations of the corner atoms are given by (000), and those of the face-centered atoms by $(\tfrac{1}{2}\tfrac{1}{2}0)$, $(\tfrac{1}{2}0\tfrac{1}{2})$, $(0\tfrac{1}{2}\tfrac{1}{2})$. A set of coordinates that generates a tetrahedron of body-centered atoms is $(\tfrac{1}{4}\tfrac{1}{4}\tfrac{1}{4})$, $(\tfrac{3}{4}\tfrac{3}{4}\tfrac{1}{4})$, $(\tfrac{1}{4}\tfrac{3}{4}\tfrac{3}{4})$, and $(\tfrac{3}{4}\tfrac{1}{4}\tfrac{3}{4})$.

22.15 Write the coordinates of the atoms in the La unit cell. See Problem 22.11 for additional information.

▌ Even though this is a nonorthogonal unit cell, the same system of determining the coordinates is applicable, giving (000) and $(\tfrac{2}{3}\tfrac{1}{3}\tfrac{1}{2})$.

22.16 Prepare a crystallographic projection for the Ca unit cell. See Problem 22.8 for additional information.

▌ A crystallographic projection shows the shape of the unit cell in two dimensions, with three-dimensional information being given by the coordinate of each constituent along the axis perpendicular to the projections. If the projection is made along the z axis, the combined symbol ① represents an atom in the xy plane (the plane of the paper) and an atom above the xy plane by one unit length c. For cubic systems, only one projection is required (see Fig. 22-5a).

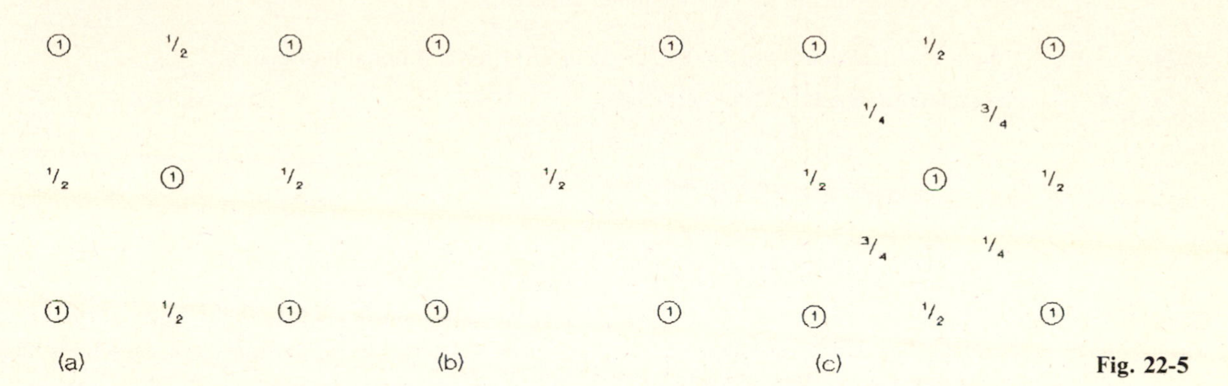

Fig. 22-5

22.17 Prepare a crystallographic projection for the K unit cell. See Problem 22.9 for additional information.

▌ The projection is shown in Fig. 22-5b.

22.18 Prepare a crystallographic projection for the Si unit cell. See Problem 22.10 for additional information.

▌ The projection is shown in Fig. 22-5c.

22.19 Determine the crystal coordination number for an atom in the Ca unit cell. See Problem 22.8 for additional information.

▌ Consider the atom in the center of the bottom face shown in Fig. 22-5a. The four corner atoms and the face atoms with $z = \frac{1}{2}$ are touching the atom. In addition, the four atoms with $z = -\frac{1}{2}$ are also touching the atom, giving the total number of atoms as 12. For the Ca unit cell, CN = 12.

22.20 Determine the crystal coordination number for an atom in the K unit cell. See Problem 22.9 for additional information.

▌ Consider the body-centered atom shown in Fig. 22-5b. The eight corner atoms are touching the atom. For the K unit cell, CN = 8.

22.21 Determine the crystal coordination number for an atom in the Si unit cell. See Problem 22.10 for additional information.

▌ Consider the atom having the coordinates ($\frac{1}{4}\frac{1}{4}\frac{1}{4}$). From Fig. 22-5c, there are four atoms touching it—those with coordinates (000), ($\frac{1}{2}\frac{1}{2}0$), ($\frac{1}{2}0\frac{1}{2}$), and ($0\frac{1}{2}\frac{1}{2}$). For the Si unit cell, CN = 4.

22.22 Calculate the theoretical density of Ca. See Problem 22.8 for additional information.

▌ The theoretical density of a solid substance can be calculated from the mass and volume of the unit cell using

$$\rho = ZM/LV \tag{22.1}$$

where

$$V = abc(1 - \cos^2\alpha - \cos^2\beta - \cos^2\gamma + 2\cos\alpha\cos\beta\cos\gamma)^{1/2} \tag{22.2a}$$

For a cubic unit cell, *(22.2a)* becomes

$$V = a^3 \tag{22.2b}$$

Substituting $Z = 4$, *(22.2b)*, and the data into *(22.1)* gives

$$\rho = \frac{(4)(40.08\ \text{g}\cdot\text{mol}^{-1})}{(6.022\times10^{23}\ \text{mol}^{-1})(0.556\ \text{nm})^3[(10^{-7}\ \text{cm})/(1\ \text{nm})]^3} = 1.549\ \text{g}\cdot\text{cm}^{-3}$$

which is 0.6% greater than the experimental value.

22.23 The experimental value of the density of K is $0.86\ \mathrm{g\cdot cm^{-3}}$. If K crystallizes in a cubic unit cell with $a = 0.5333$ nm, as determined by X-ray analysis, show that this is a body-centered cubic unit cell.

▮ Solving *(22.1)* for Z and substituting *(22.2b)* and the data gives

$$Z = \frac{(0.86\ \mathrm{g\cdot cm^{-3}})(6.022\times 10^{23}\ \mathrm{mol^{-1}})(0.5333\ \mathrm{nm})^3[(10^{-7}\ \mathrm{cm})/(1\ \mathrm{nm})]^3}{39.0983\ \mathrm{g\cdot mol^{-1}}} = 2.01$$

The value of Z corresponds to the body-centered unit cell.

22.24 Calculate the theoretical density of La. See Problem 22.11 for additional information.

▮ For a hexagonal unit cell, *(22.2a)* becomes

$$V = a^2 c \sin\gamma \tag{22.2c}$$

Substituting $Z = 2$, *(22.2c)*, and the data into *(22.1)* gives

$$\rho = \frac{(2)(138.906\ \mathrm{g\cdot mol^{-1}})}{(6.022\times 10^{23}\ \mathrm{mol^{-1}})(0.372\ \mathrm{nm})^2(0.606\ \mathrm{nm})\sin 120°[(10^{-7}\ \mathrm{cm})/(1\ \mathrm{nm})]^3} = 6.35\ \mathrm{g\cdot cm^{-3}}$$

22.25 Methane crystallizes in a cubic unit cell with $a = 0.589$ nm. Calculate the theoretical density of CH_4 assuming $Z = 1$, 2, and 4. The density of liquid CH_4 is $0.466\ \mathrm{g\cdot cm^{-3}}$. Assuming that the density of the solid is nearly that of the liquid, predict which type of cubic unit cell will result.

▮ Substituting *(22.2b)* and the data into *(22.1)* gives

$$\rho(\mathrm{PC}) = \frac{(1)(16.04\ \mathrm{g\cdot mol^{-1}})}{(6.022\times 10^{23}\ \mathrm{mol^{-1}})(0.589\ \mathrm{nm})^3[(10^{-7}\ \mathrm{cm})/(1\ \mathrm{nm})]^3} = 0.130\ \mathrm{g\cdot cm^{-3}}$$

$$\rho(\mathrm{BCC}) = \frac{(2)(16.04)}{(6.022\times 10^{23})(0.589\times 10^{-7})^3} = 0.261\ \mathrm{g\cdot cm^{-3}}$$

$$\rho(\mathrm{FCC}) = \frac{(4)(16.04)}{(6.022\times 10^{23})(0.589\times 10^{-7})^3} = 0.521\ \mathrm{g\cdot cm^{-3}}$$

The theoretical density predicted for the face-centered cubic unit cell agrees best with the experimental density data.

22.26 The following crystallographic data were obtained for a protein: $V = 1.50\times 10^{-19}\ \mathrm{cm^3}$, $\rho = 1.35\ \mathrm{g\cdot cm^{-3}}$, $Z = 4$, and protein fraction $= 0.75$. Calculate the molar mass of the protein.

▮ Solving *(22.1)* for M and substituting the data gives

$$M = \frac{(1.35\ \mathrm{g\cdot cm^{-3}})(6.022\times 10^{23}\ \mathrm{mol^{-1}})(1.50\times 10^{-19}\ \mathrm{cm^3})(0.75)}{4} = 2.3\times 10^4\ \mathrm{g\cdot mol^{-1}}$$

22.27 Suppose the sample of Ca described in Problem 22.22 contained 0.1% Frenkel defects in the crystal. What is the density of this crystal?

▮ Frenkel defects do not change the value of the theoretical density because the atom is only occupying an interstitial position instead of a lattice position.

22.28 Repeat Problem 22.27 for a sample of Ca that contained 0.1% Schottky defects.

▮ The value of Z in *(22.1)* in this case becomes $Z = (0.999)(4) = 3.996$ because of the absence of atoms in the crystal. Substituting into *(22.1)* gives

$$\rho = \frac{(3.996)(40.08\ \mathrm{g\cdot mol^{-1}})}{(6.022\times 10^{23}\ \mathrm{mol^{-1}})(0.556\ \mathrm{nm})^3[(10^{-7}\ \mathrm{cm})/(1\ \mathrm{nm})]^3} = 1.547\ \mathrm{g\cdot cm^{-3}}$$

22.29 Using the information given in Problem 22.8, calculate the crystal radius of a Ca atom.

▮ The face diagonal (see Fig. 22-2) is equal to $4R$ and to $a(2)^{1/2}$. Equating these expressions, solving for R, and substituting the data gives

$$R = \frac{a(2)^{1/2}}{4} = \frac{(0.556\ \mathrm{nm})(2)^{1/2}}{4} = 0.197\ \mathrm{nm}$$

22.30 Using the information given in Problem 22.9, calculate the crystal radius of a K atom.

▮ The body diagonal (see Fig. 22-3) is equal to $4R$ and to $a(3)^{1/2}$. Equating these expressions, solving for R, and substituting the data gives

$$R = \frac{a(3)^{1/2}}{4} = \frac{(0.5333\text{ nm})(3)^{1/2}}{4} = 0.2309\text{ nm}$$

22.31 Using the information given in Problem 22.10, calculate the crystal radius of a Si atom.

▮ Atoms that are touching lie along the body diagonal, for example, (000) and $(\frac{1}{4}\frac{1}{4}\frac{1}{4})$. The distance $2R$ is equivalent to one-fourth of the body diagonal or $a(3)^{1/2}/4$. Equating these expressions, solving for R, and substituting the data gives

$$R = \frac{a(3)^{1/2}/4}{2} = \frac{(0.541\,73\text{ nm})(3)^{1/2}/4}{2} = 0.117\,29\text{ nm}$$

22.32 Using the information given in Problem 22.11, calculate the crystal radius of a La atom.

▮ As seen in Fig. 22-1b, $2R$ is equal to a. Solving for R and substituting the data gives

$$R = \frac{a}{2} = \frac{0.372\text{ nm}}{2} = 0.186\text{ nm}$$

22.33 For an ideal hexagonal closest-packed unit cell,

$$c = 2a(2/3)^{1/2} \tag{22.3}$$

How ideal is the unit cell of La? See Problem 22.11 for additional information.

▮ Substituting the value of a into *(22.3)* gives

$$c = 2(0.372\text{ nm})(2/3)^{1/2} = 0.607\text{ nm}$$

which is in excellent agreement with the experimental value of 0.606 nm.

22.34 Calculate the packing fraction for the Ca unit cell. See Problems 22.8 and 22.29 for additional information.

▮ The three-dimensional equivalent of Problem 22.6 is to divide the volume actually occupied by the atoms by the volume of the unit cell. Using $Z = 4$ and the relationship between a and R determined in Problem 22.29 gives

$$\text{Packing fraction} = \frac{4V}{a^3} = \frac{4(4/3)\pi R^3}{[4R/(2)^{1/2}]^3} = 0.740$$

22.35 Calculate the packing fraction for the K unit cell. See Problems 22.9 and 22.30 for additional information.

▮ Using $Z = 2$ and the relationship between a and R determined in Problem 22.30 gives

$$\text{Packing fraction} = \frac{2V}{a^3} = \frac{2(4/3)\pi R^3}{[4R/(3)^{1/2}]^3} = 0.680$$

22.36 In Problem 22.10, Z was determined to be 8 for the unit cell of Si. Would the Si unit cell be predicted to be more efficiently packed than the face-centered unit cell? See Problem 22.31 for additional information.

▮ Using $Z = 8$ and the relationship between a and R determined in Problem 22.31 gives

$$\text{Packing fraction} = \frac{8V}{a^3} = \frac{8(4/3)\pi R^3}{[8R/(3)^{1/2}]^3} = 0.340$$

This unit cell is very inefficiently packed.

22.37 The face-centered cubic unit cell is often referred to as the cubic closest-packed unit cell. Is the packing fraction for the cubic closest-packed unit cell the same as for the hexagonal closest-packed unit cell? See Problems 22.8, 22.32, 22.33, and 22.34 for additional information.

▮ Using $Z = 2$, *(22.2c)*, the relationship between a and R determined in Problem 22.32, and *(22.3)* gives

$$\text{Packing fraction} = \frac{2V}{a^2c\sin\gamma} = \frac{2(4/3)\pi R^3}{a^2(2a)(2/3)^{1/2}\sin\gamma} = \frac{2(4/3)\pi R^3}{(2R)^3(2)(2/3)^{1/2}\sin 120^\circ} = 0.740$$

The two unit cells have identical packing fractions.

22.38 What fraction of Ca atoms lie on the surface of a cubic crystal that is 1.00 cm in length? See Problem 22.8 for additional information.

▮ The number of unit cells present in the crystal is

$$\text{Number of unit cells} = \frac{V(\text{crystal})}{V(\text{unit cell})} = \frac{(1.00\ \text{cm})^3}{\{(0.556\ \text{nm})[(10^{-7}\ \text{cm})/(1\ \text{nm})]\}^3} = 5.82 \times 10^{21}$$

and the number of atoms present is

$$N(\text{crystal}) = (4)(5.82 \times 10^{21}) = 2.33 \times 10^{22}$$

The number of unit cells on the surface of the crystal is

$$\text{Number of surface cells} = \frac{A(\text{crystal})}{A(\text{unit cell})} = \frac{6(1.00\ \text{cm})^2}{\{(0.556\ \text{nm})[(10^{-7}\ \text{cm})/(1\ \text{nm})]\}^2} = 1.94 \times 10^{15}$$

On the surface of a unit cell, there are two atoms—the face-centered atom and one-fourth of each of the four corner atoms—giving the number of atoms present on the surface as

$$N(\text{surface}) = 2(1.94 \times 10^{15}) = 3.88 \times 10^{15}$$

The fraction of the atoms on the surface is

$$\frac{N(\text{surface})}{N(\text{crystal})} = \frac{3.88 \times 10^{15}}{2.33 \times 10^{22}} = 1.67 \times 10^{-7}$$

22.39 The empty center of the face-centered cubic unit cell is known as an octahedral hole. In terms of the radius of the atoms making up the unit cell, what is the radius of the largest atom that could just fit into the octahedral hole? See Problem 22.29 for additional information.

▮ If the octahedral hole were occupied by a sphere of radius r, the distance between two faces of the unit cell ($=a$) would be equal to $R + 2r + R$. Equating these expressions, solving for r, and substituting the relation between R and a determined in Problem 22.29 gives

$$r = \frac{a - 2R}{2} = \frac{4R/(2)^{1/2} - 2R}{2} = 0.414R$$

22.40 Determine the distance between the (000) atom and the (111) atom in La. See Problem 22.11 for additional information.

▮ The distance (l) separating two atoms in a unit cell is given by

$$l = [a^2(x_2 - x_1)^2 + b^2(y_2 - y_1)^2 + c^2(z_2 - z_1)^2 - 2ab(x_2 - x_1)(y_2 - y_1)\cos\gamma$$
$$- 2ac(x_2 - x_1)(z_2 - z_1)\cos\beta - 2bc(y_2 - y_1)(z_2 - z_1)\cos\alpha]^{1/2} \qquad (22.4a)$$

where x_i, y_i, and z_i are the respective coordinates of the two atoms. Substituting the data into *(22.4a)* gives

$$l = [(0.372\ \text{nm})^2(1 - 0)^2 + (0.372\ \text{nm})^2(1 - 0)^2 + (0.606\ \text{nm})^2(1 - 0)^2$$
$$- 2(0.372\ \text{nm})(0.372\ \text{nm})(1 - 0)(1 - 0)\cos 120°$$
$$- 2(0.372\ \text{nm})(0.606\ \text{nm})(1 - 0)(1 - 0)\cos 90°$$
$$- 2(0.372\ \text{nm})(0.606\ \text{nm})(1 - 0)(1 - 0)\cos 90°]^{1/2} = 0.885\ \text{nm}$$

22.41 Confirm the statement made in Problem 22.39 that the distance between two faces of the face-centered cubic unit cell is equal to a.

▮ For a pair of atoms on opposite faces, the coordinates are $(\frac{1}{2}0\frac{1}{2})$ and $(\frac{1}{2}1\frac{1}{2})$. For a cube, $\alpha = \beta = \gamma = 90°$, and *(22.4a)* simplifies to

$$l = a[(x_2 - x_1)^2 + (y_2 - y_1)^2 + (z_2 - z_1)^2]^{1/2} \qquad (22.4b)$$

Substituting the coordinates into *(22.4b)* gives

$$l = a[(\tfrac{1}{2} - \tfrac{1}{2})^2 + (1 - 0)^2 + (\tfrac{1}{2} - \tfrac{1}{2})^2]^{1/2} = a$$

22.42 Calculate the Si—Si bond length using the information given in Problem 22.10.

▮ For a pair of atoms bonded together, the coordinates are (000) and $(\frac{1}{4}\frac{1}{4}\frac{1}{4})$. Substituting the data into *(22.4b)* gives

$$l = (0.541\,73\ \text{nm})[(\tfrac{1}{4} - 0)^2 + (\tfrac{1}{4} - 0)^2 + (\tfrac{1}{4} - 0)^2]^{1/2} = 0.234\,58\ \text{nm}$$

22.43 Using *(22.4b)*, determine the tetrahedral angle.

▮ The tetrahedral angle will be present between the three atoms in the Si unit cell having the coordinates of (000), $(\frac{1}{4}\frac{1}{4}\frac{1}{4})$, and $(\frac{1}{2}\frac{1}{2}0)$. The respective distances between the (000) and $(\frac{1}{4}\frac{1}{4}\frac{1}{4})$, the (000) and $(\frac{1}{2}\frac{1}{2}0)$ atoms, and the $(\frac{1}{4}\frac{1}{4}\frac{1}{4})$ and $(\frac{1}{2}\frac{1}{2}0)$ atoms are given by *(22.4b)* as

$$l = a[(\tfrac{1}{4} - 0)^2 + (\tfrac{1}{4} - 0)^2 + (\tfrac{1}{4} - 0)^2]^{1/2} = a(3)^{1/2}/4$$

$$l' = a[(\tfrac{1}{2} - 0)^2 + (\tfrac{1}{2} - 0)^2 + (0 - 0)^2]^{1/2} = a/(2)^{1/2}$$

$$l'' = a[(\tfrac{1}{2} - \tfrac{1}{4})^2 + (\tfrac{1}{2} - \tfrac{1}{4})^2 + (0 - \tfrac{1}{4})^2]^{1/2} = a(3)^{1/2}/4$$

Substituting into the law of cosines gives

$$\cos A = \frac{[a(3)^{1/2}/4]^2 + [a(3)^{1/2}/4]^2 - [a/(2)^{1/2}]^2}{2[a(3)^{1/2}/4][a(3)^{1/2}/4]} = -\frac{1}{3}$$

Solving for the angle gives $A = 109.47°$.

22.2 CRYSTAL FORMS

22.44 Silver crystallizes in a face-centered cubic unit cell with $a = 0.408\,62$ nm. Given that the density is $10.501\ \mathrm{g \cdot cm^{-3}}$, calculate the value of L.

▮ Solving *(22.1)* for L and substituting the data gives

$$L = \frac{4(107.868\ \mathrm{g \cdot mol^{-1}})}{(10.501\ \mathrm{g \cdot cm^{-3}})[(0.408\,62\ \mathrm{nm})[(10^{-7}\ \mathrm{cm})/(1\ \mathrm{nm})]]^3} = 6.0223 \times 10^{23}\ \mathrm{mol^{-1}}$$

22.45 The data given in Problem 22.44 are for $T = 25$ °C. What is the unit cell length at $T = 100$ °C? The linear coefficient of thermal expansion is $\alpha_l = 19 \times 10^{-6}\ \mathrm{K^{-1}}$.

▮ The linear coefficient of thermal expansion is given by

$$\alpha_l = \frac{1}{l}\frac{dl}{dT} \tag{22.5}$$

Solving *(22.5)* for dl, integrating, and substituting the data gives

$$\ln\left[\frac{l_2}{l_1}\right] = \alpha_l(T_c - T_1)$$

$$\ln\left[\frac{a}{0.408\,62\ \mathrm{nm}}\right] = (19 \times 10^{-6}\ \mathrm{K^{-1}})(373\ \mathrm{K} - 298\ \mathrm{K}) = 1.4 \times 10^{-3}$$

$$\frac{a}{0.408\,62\ \mathrm{nm}} = 1.0014 \qquad a = 0.409\,20\ \mathrm{nm}$$

22.46 Boron nitride crystallizes in a hexagonal unit cell with $a = 0.251$ nm, $b = 0.669$ nm, and $Z = 2$. Calculate the theoretical density of BN.

▮ Substituting *(22.2c)* and the data into *(22.1)* gives

$$\rho = \frac{(2)(24.82\ \mathrm{g \cdot mol^{-1}})}{(6.022 \times 10^{23}\ \mathrm{mol^{-1}})[(0.251\ \mathrm{nm})^2(0.669\ \mathrm{nm})(\sin 120°)[(10^{-7}\ \mathrm{cm})/(1\ \mathrm{nm})]]^3} = 2.26\ \mathrm{g \cdot cm^{-3}}$$

22.47 The unit cell for BN is similar to that of graphite except that each layer consists of alternating B and N atoms in hexagonal rings. Given $a = 0.2455$ nm and that the distance between C atoms is 0.1415 nm in graphite, predict the distance between a B atom and an N atom in BN. See Problem 22.46 for additional information.

▮ For similar substances,

$$l(\mathrm{B{-}N}) = l(\mathrm{C{-}C})\frac{a(\mathrm{B{-}N})}{a(\mathrm{C{-}C})} = (0.1415\ \mathrm{nm})\frac{0.251\ \mathrm{nm}}{0.2455\ \mathrm{nm}} = 0.145\ \mathrm{nm}$$

22.48 The unit cell length along the z axis for BN is approximately equal to 4 times the average van der Waals radii of B and N. Using the data given in Problem 22.46, estimate the average van der Waals radius.

▮ The average radius is $(0.669\ \mathrm{nm})/4 = 0.167$ nm.

22.49 The coordinates of the ions in a compound consisting of 29.48% Ca by mass, 35.22% Ti, and 35.30% O are Ca at (000), Ti at ($\frac{1}{2}\frac{1}{2}\frac{1}{2}$), and O at ($\frac{1}{2}\frac{1}{2}0$), ($\frac{1}{2}0\frac{1}{2}$), and ($0\frac{1}{2}\frac{1}{2}$). Determine the empirical formula of the compound.

▮ The number of sets of unique coordinates for each ion is equal to Z for that ion. Thus $Z(\text{Ca}) = 1$, $Z(\text{Ti}) = 1$, and $Z(\text{O}) = 3$, giving the empirical formula as $CaTiO_3$. This formula agrees with that determined using the percentage composition data

element	Ca	Ti	O
m in 100 g compound	29.48 g	35.22 g	35.50 g
n of atoms	$\frac{29.48}{40.08} = 0.7355$	$\frac{35.22}{47.88} = 0.7356$	$\frac{35.30}{16.00} = 2.206$
$n(\text{Ca}):n(\text{Ti}):n(\text{O})$	$\frac{0.7355}{0.7355} = 1.000$	$\frac{0.7356}{0.7355} = 1.000$	$\frac{2.206}{0.7355} = 2.999$

22.50 The olivine series of minerals consists of crystals in which Fe and Mg ions may substitute for each other. The density of forsterite (Mg_2SiO_4) is 3.21 g · cm^{-3}, and that of fayalite (Fe_2SiO_4) is 4.34 g · cm^{-3}. What is the percentage of fayalite in an olivine with a density of 3.88 g · cm^{-3}?

▮ For minerals in which simple substitution can occur, the density is a linear function of composition.

$$\rho = \rho_A X_A + \rho_B X_B \qquad (22.6)$$

Substituting $x(\text{forsterite}) = 1 - x(\text{fayalite})$ into *(22.6)*, solving *(22.6)* for x(forsterite), and substituting the data gives

$$x(\text{fayalite}) = \frac{\rho - \rho(\text{forsterite})}{\rho(\text{fayalite}) - \rho(\text{forsterite})} = \frac{3.88\ \text{g}\cdot\text{cm}^{-3} - 3.21\ \text{g}\cdot\text{cm}^{-3}}{4.34\ \text{g}\cdot\text{cm}^{-3} - 3.21\ \text{g}\cdot\text{cm}^{-3}} = 0.59$$

22.51 The cubic unit cell of NaCl contains Na^+ ions at (000), ($\frac{1}{2}\frac{1}{2}0$), ($\frac{1}{2}0\frac{1}{2}$), and ($0\frac{1}{2}\frac{1}{2}$) and Cl^- ions at ($\frac{1}{2}\frac{1}{2}\frac{1}{2}$), ($\frac{1}{2}00$), ($0\frac{1}{2}0$), and ($00\frac{1}{2}$). Describe the unit cell in words. What is the unit cell content?

▮ The coordinates listed for Na^+ correspond to a face-centered cubic unit cell with its origin at (000); see Problem 22.12. The coordinates listed for Cl^- also correspond to a face-centered cubic unit cell with its origin at ($\frac{1}{2}00$). For the unit cell, $Z_+ = Z_- = 4$ or $Z(\text{NaCl}) = 4$.

22.52 Using $a = 0.564\,02$ nm for NaCl, estimate the ionic radii. See Problem 22.51 for additional information.

▮ Assume that the larger Cl^- ions touch along a face diagonal. From Problem 22.29,

$$R(\text{Cl}^-) = \frac{(0.564\,02\ \text{nm})(2)^{1/2}}{4} = 0.199\ \text{nm}$$

The unit cell length is equal to $a = 2R(\text{Na}^+) + 2R(\text{Cl}^-)$. Solving for $R(\text{Na}^+)$ and substituting the data gives

$$R(\text{Na}^+) = \frac{0.564\,02\ \text{nm} - 2(0.199\ \text{nm})}{2} = 0.083\ \text{nm}$$

This calculation gives a value of $R(\text{Cl}^-)$ that is 9.9% high and a value of $R(\text{Na}^+)$ that is 14% low. The Cl^- ions are not actually touching along the face diagonal.

22.53 The cubic unit cell of CsCl contains Cs^+ ions at (000) and Cl^- ions at ($\frac{1}{2}\frac{1}{2}\frac{1}{2}$). Describe the unit cell in words. What is the unit cell content?

▮ The coordinates for Cs^+ correspond to a primitive cubic unit cell with its origin at (000). The coordinates listed for Cl^- also correspond to a primitive cubic unit cell with its origin at ($\frac{1}{2}\frac{1}{2}\frac{1}{2}$). For the unit cell, $Z_+ = Z_- = 1$ or $Z(\text{CsCl}) = 1$.

22.54 Calculate the theoretical density of CsCl given $a = 0.4110$ nm. See Problem 22.53 for additional information.

▮ Substituting $Z = 1$ into *(22.1)* gives

$$\rho = \frac{(1)(168.358\ \text{g}\cdot\text{mol}^{-1})}{(6.022\times10^{23}\ \text{mol}^{-1})\{(0.4110\ \text{nm})^3[(10^{-7}\ \text{cm})/(1\ \text{nm})]\}^3} = 4.027\ \text{g}\cdot\text{cm}^{-3}$$

22.55 The density of CaF_2 is $3.180\ g \cdot cm^{-3}$ and $a = 0.5451$ nm for the cubic unit cell. Determine the unit cell content for the fluorite structure.

▌ Solving *(22.1)* for Z and substituting the data gives

$$Z = \frac{(3.180\ g \cdot cm^{-3})(6.022 \times 10^{23}\ mol^{-1})(0.5451\ nm)^3[(10^{-7}\ cm)/(1\ nm)]^3}{78.08\ g \cdot mol^{-1}} = 3.972 \approx 4$$

The unit cell contains four Ca^{2+} ions and eight F^- ions.

22.56 The Ca^{2+} ions in the fluorite structure (see Problem 22.55) define a face-centered cubic unit cell at (000), and the F^- ions occupy all of the tetrahedral holes in the Ca^{2+} structure. What are the coordinates of the ions?

▌ The Ca^{2+} ion coordinates are (000), $(\frac{1}{2}\frac{1}{2}0)$, $(\frac{1}{2}0\frac{1}{2})$, and $(0\frac{1}{2}\frac{1}{2})$. The eight F^- ions in all of the tetrahedral holes define a primitive cube with atoms located at $(\frac{1}{4}\frac{1}{4}\frac{1}{4})$, $(\frac{1}{4}\frac{3}{4}\frac{1}{4})$, $(\frac{3}{4}\frac{1}{4}\frac{1}{4})$, $(\frac{3}{4}\frac{3}{4}\frac{1}{4})$, $(\frac{3}{4}\frac{1}{4}\frac{3}{4})$, $(\frac{1}{4}\frac{3}{4}\frac{3}{4})$, $(\frac{1}{4}\frac{1}{4}\frac{3}{4})$, and $(\frac{3}{4}\frac{3}{4}\frac{3}{4})$.

22.57 The ionic radius of the Mn^{2+} ion is 0.080 nm and that of the S^{2-} ion is 0.184 nm. Predict the structure of the cubic unit cell of MnS.

▌ The radius ratio of the ions is

$$\frac{R_+}{R_-} = \frac{0.080\ nm}{0.184\ nm} = 0.435$$

which falls into the $0.414 < R_+/R_- < 0.732$ range of values given in Table 22-1. The location of the ions in the MnS unit cell is predicted to be identical to the location of the ions in the NaCl unit cell.

Table 22-1

empirical formula	radius ratio	CN	cubic structure
MX	$0.225 < R_+/R_- < 0.414$	4	ZnS (sphalerite)
	$0.414 < R_+/R_- < 0.732$	6	NaCl (halite)
	$0.732 < R_+/R_- < 1.000$	8	CsCl

22.58 The unit cell of ammonium chloride is similar to the unit cell of CsCl. Given $R(Cl^-) = 0.181$ nm, estimate the ionic radius of the NH_4^+ ion.

▌ The range of R_+/R_- values for the CsCl unit cell is 0.732–1.000 (see Table 22-1). The corresponding range of R_+ values is

$$R(NH_4^+) > (0.181\ nm)(0.732) = 0.132\ nm$$

$$R(NH_4^+) < (0.181\ nm)(1.000) = 0.181\ nm$$

The observed value of 0.148 nm falls within the predicted range of $0.132\ nm < R(NH_4^+) < 0.181\ nm$.

22.59 Determine the value of the Madelung constant for a one-dimensional "crystal" consisting of alternating positive and negative unit charges separated by a distance R.

▌ The potential energy of a given ion with respect to all of the other ions is given by Coulomb's law as

$$U(R) = -\frac{2e^2}{R} + \frac{2e^2}{2R} - \frac{2e^2}{3R} + \frac{2e^2}{4R} - \frac{2e^2}{5R} + \cdots$$

$$= -\frac{e^2}{R}2[1 - \tfrac{1}{2} + \tfrac{1}{3} - \tfrac{1}{4} + \tfrac{1}{5} - \cdots]$$

where e is the electronic charge, negative signs are used for attractions, and positive signs are used for repulsions. In each case, the given ion interacts with two equidistant ions. Often $U(R)$ is expressed in terms of the Madelung constant $U(R) = -e^2\mathcal{M}/R$. Comparison of the two expressions gives

$$\mathcal{M} = 2[1 - \tfrac{1}{2} + \tfrac{1}{3} - \tfrac{1}{4} + \tfrac{1}{5} - \cdots]$$

Comparison to the series expansion for $\ln(1 + x)$,

$$\ln(1 + x) = x - \frac{x^2}{2} + \frac{x^3}{3} - \frac{x^4}{4} + \cdots$$

gives

$$\mathcal{M} = 2\ln(1 + 1) = 2\ln(2) = 1.386\,29\ldots$$

22.60 Determine the first five terms for the Madelung constant for the NaCl unit cell. How does this value agree with $\mathcal{M} = 1.762\,67$?

▮ The nearest neighbor ions around a given ion are six ions of the opposite charge at a distance R. The next nearest neighbor ions are 12 ions of the same charge at a distance of $(2)^{1/2}R$. The first terms are

$$\mathcal{M} = 6 - \frac{12}{(2)^{1/2}} + \frac{8}{(3)^{1/2}} - \frac{6}{(4)^{1/2}} + \frac{24}{(5)^{1/2}} - \frac{24}{(6)^{1/2}} + \cdots$$
$$= 0.069$$

The value of 1.762 67 is obtained only after including many terms.

22.61 Molecular Cl_2 crystallizes in a face-centered orthorhombic unit cell with $a = 0.629$ nm, $b = 0.450$ nm, and $c = 0.821$ nm. Calculate the theoretical density of Cl_2.

▮ For an orthorhombic unit cell, *(22.2a)* becomes

$$V = abc \qquad (22.2d)$$

Substituting *(22.2d)* and the data into *(22.1)* gives

$$\rho = \frac{4(70.906\ \text{g}\cdot\text{mol}^{-1})}{(6.022\times10^{23}\ \text{mol}^{-1})(0.629\ \text{nm})(0.450\ \text{nm})(0.821\ \text{nm})[(10^{-7}\ \text{cm})/(1\ \text{nm})]^3} = 2.03\ \text{g}\cdot\text{cm}^{-3}$$

22.62 In the unit cell described in Problem 22.61, the center of one Cl_2 molecule is located at (000) and one of the atoms in the molecule is located at $(0yz)$, where $y = 0.1173$ and $z = 0.1016$. Calculate the Cl—Cl bond length.

▮ The distance between the center of the molecule and the atom is given by *(22.4a)* as

$$l = [a^2(x_2 - x_1)^2 + b^2(y_2 - y_1)^2 + c^2(z_2 - z_1)^2]^{1/2} \qquad (22.4c)$$

$$l = [(0.629\ \text{nm})^2(0-0)^2 + (0.450\ \text{nm})^2(0.1173-0)^2 + (0.821\ \text{nm})^2(0.1016-0)^2]^{1/2} = 0.0987\ \text{nm}$$

The bond length is $2(0.0987\ \text{nm}) = 0.197$ nm.

22.63 In the cell described in Problem 22.61, atoms in adjacent molecules are located at $(0\ y\ z)$ and $(\frac{1}{2}\ y\ \frac{1}{2} - z)$. Calculate the van der Waals radius for a Cl atom.

▮ The distance between the atoms is given by *(22.4c)* as

$$l = \{(0.629\ \text{nm})^2(0-\tfrac{1}{2})^2 + (0.450\ \text{nm})^2(0.1173-0.1173)^2 + (0.821\ \text{nm})^2[(\tfrac{1}{2}-0.1016)-0.1016]^2\}^{1/2} = 0.398\ \text{nm}$$

The van der Waals radius is $(0.398\ \text{nm})/2 = 0.199$ nm.

22.64 Water crystallizes in a hexagonal unit cell with $a = 0.451\,35$ nm and $c = 0.735\,21$ nm at 0 °C and $a = 0.450\,85$ nm and $c = 0.7338$ nm at −66 °C. Calculate the linear coefficient of thermal expansion along each axis.

▮ Letting $(\partial l/\partial T)_p \approx \Delta l/\Delta T$, *(1.11)* gives

$$\alpha_a = \frac{1}{[(0.451\,35\ \text{nm} + 0.450\,85\ \text{nm})/2]}\ \frac{0.451\,35\ \text{nm} - 0.450\,85\ \text{nm}}{273\ \text{K} - 207\ \text{K}} = 1.68\times10^{-5}\ \text{K}^{-1}$$

$$\alpha_c = \frac{1}{[(0.735\,21 + 0.7338)/2]}\ \frac{0.735\,21 - 0.7338}{66\ \text{K}} = 2.91\times10^{-5}\ \text{K}^{-1}$$

22.65 The density of ice at 0 °C is 0.917 g · cm^{-3}. Determine the unit cell content of the unit cell described in Problem 22.64.

▮ Solving *(22.1)* for Z and substituting *(22.2c)* and the data gives

$$Z = \frac{[(0.917\ \text{g}\cdot\text{cm}^{-3})(6.022\times10^{23}\ \text{mol}^{-1})(0.451\,35\ \text{nm})^2(0.735\,21\ \text{nm})(\sin 120°)[(10^{-7}\ \text{cm})/(1\ \text{nm})]^3]}{18.015\ \text{g}\cdot\text{mol}^{-1}} = 3.976 \approx 4$$

22.66 The unit cell of $TlAl(SO_4)_2 \cdot xH_2O$ is face-centered cubic with $a = 1.221$ nm. If $\rho = 2.32\ \text{g}\cdot\text{cm}^{-3}$, determine x.

▮ Solving *(22.1)* for M and substituting the data gives

$$M = \frac{(2.32\ \text{g}\cdot\text{cm}^{-3})(6.022\times10^{23}\ \text{mol}^{-1})(1.221\ \text{nm})^3[(10^{-7}\ \text{cm})/(1\ \text{nm})]^3}{4} = 636\ \text{g}\cdot\text{mol}^{-1}$$

Subtracting the molar mass of the anhydrous salt and dividing by the molar mass of water gives

$$x = \frac{636\ \text{g}\cdot\text{mol}^{-1} - 423.48\ \text{g}\cdot\text{mol}^{-1}}{18.015\ \text{g}\cdot\text{mol}^{-1}} = 11.8 \approx 12$$

22.3 CRYSTALLOGRAPHY

22.67 Consider the two-dimensional "unit cell" shown in Fig. 22-6a. What are the Miller indices of the line shown?

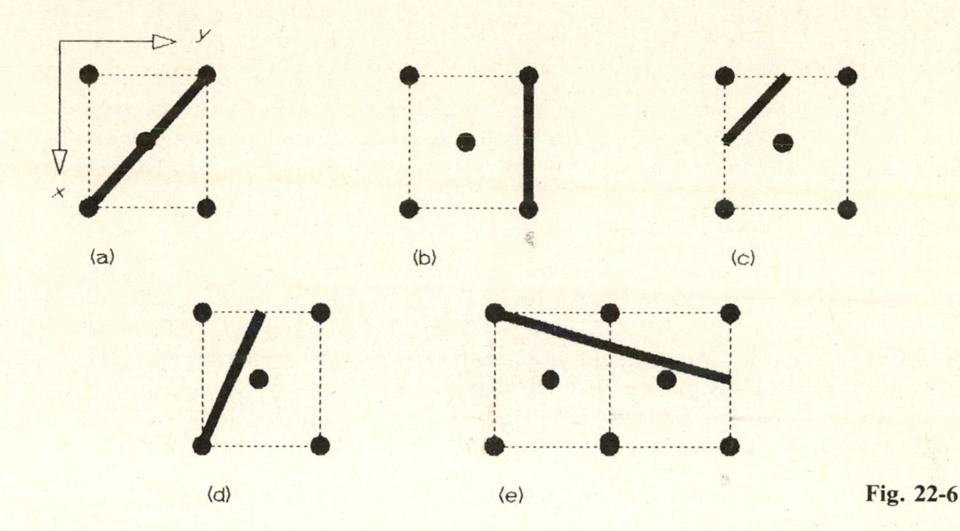

Fig. 22-6

▮ The *Miller indices* are a set of integers *hkl* that are used to describe a given plane in a crystal. For this two-dimensional case, only *hk* will be used. The procedure for determining the Miller indices for a plane is: (1) Prepare a table with the unit cell axes at the top of columns; (2) enter in each column the intercept (expressed as a multiple of *a*, *b*, or *c*) of the plane with that axis; (3) invert all numbers; and (4) clear fractions to obtain *h*, *k*, and *l*. Preparing the table as described above,

a	*b*	
1	1	intercepts
1	1	reciprocals
1	1	clear fractions

gives the indices as 11.

22.68 What are the Miller indices of the line shown in Fig. 22-6b?

▮ Preparing the table as described in Problem 22.67,

a	*b*	
∞	1	intercepts
0	1	reciprocals
0	1	clear fractions

gives the indices as 01.

22.69 What are the Miller indices of the line shown in Fig. 22-6c?

▌ Preparing the table as described in Problem 22.67,

a	b	
$\frac{1}{2}$	$\frac{1}{2}$	intercepts
2	2	reciprocals
2	2	clear fractions

gives the indices as 22.

22.70 Prepare a sketch showing the 12 line in a two-dimensional "unit cell."

▌ The Miller indices represent the number of equal segments into which each respective axis has been divided by the plane. For the 12 line, the a axis has been intersected at unity and the b axis has been divided into two equal segments. The line from $x = a/1$ to $y = b/2$ shown in Fig. 22-6d is the 12 line.

22.71 Prepare a sketch showing the $2\bar{1}$ line in a two-dimensional "unit cell."

▌ The bar over the 1 in the set of indices represents -1. The $2\bar{1}$ line divides the a axis into two equal segments and intersects the b axis at $-b$. The line from $x = a/2$ to $y = -b/1$ is shown in Fig. 22-6e.

22.72 The coordinates of the three corners of an octahedral face on a cubic unit cell are $(\frac{1}{2}\frac{1}{2}1)$, $(01\frac{1}{2})$, and $(11\frac{1}{2})$. Determine the Miller indices of this plane.

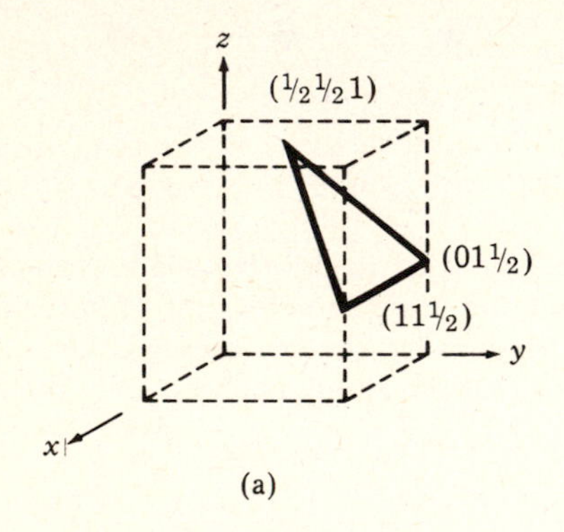

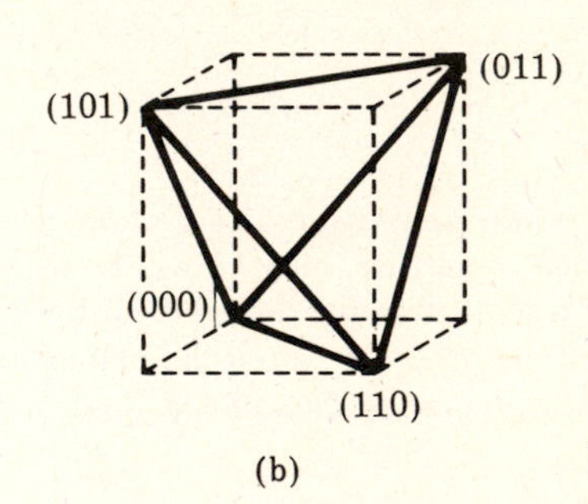

Fig. 22-7

▌ The plane is shown in Fig. 22-7a. The intercepts with the axes are ∞, $2b$, and $2c$. Preparing the table as described in Problem 22.67,

a	b	c	
∞	2	2	intercepts
0	$\frac{1}{2}$	$\frac{1}{2}$	reciprocals
0	1	1	clear fractions

giving the indices as 011.

22.73 The planes that define a tetrahedron are shown in Fig. 22-7b. Determine the Miller indices of the four planes.

▌ Extension of the plane between (101), (011), and (110) shows that it will intersect the axes at $x = 2a$, $y = 2b$, and $z = 2c$. The table is

a	b	c	
2	2	2	intercepts
$\frac{1}{2}$	$\frac{1}{2}$	$\frac{1}{2}$	reciprocals
1	1	1	clear fractions

giving the indices as 111. A plane parallel to the plane between (000), (110), and (101) will intersect the axes at $x = a$, $y = -b$, and $z = -c$. The table is

a	b	c	
1	−1	−1	intercepts
1	−1	−1	reciprocals
1	−1	−1	clear fractions

giving the indices as $1\bar{1}\bar{1}$. Likewise, a plane parallel to the plane between (000), (110), and (011) will intersect the axes at $x = a$, $y = -b$, and $z = c$, giving the indices as $1\bar{1}1$ and a plane parallel to the plane between (000), (011), and (101) will intersect the axes at $x = a$, $y = b$, and $z = -c$, giving the indices as $11\bar{1}$. (Note that the choices of the parallel planes are not unique; others could have been chosen to show that $1\bar{1}\bar{1} = \bar{1}11$, $1\bar{1}1 = \bar{1}1\bar{1}$, and $11\bar{1} = \bar{1}\bar{1}1$.)

22.74 Calculate the distance between the 111 planes in a crystal of Ca. See Problem 22.8 for additional information. Repeat the calculation for the 222 planes. Which planes are closer?

I For a cubic unit cell, the distance (d_{hkl}) between planes with Miller indices hkl is given by

$$\frac{1}{d_{hkl}^2} = \frac{h^2 + k^2 + l^2}{a^2} \tag{22.7}$$

Substituting the data into *(22.7)* for the 111 plane gives

$$\frac{1}{d_{111}^2} = \frac{1^2 + 1^2 + 1^2}{(0.556\ \text{nm})^2} = 9.70\ \text{nm}^{-2} \qquad d_{111} = 0.321\ \text{nm}$$

Likewise for the 222 plane,

$$\frac{1}{d_{222}^2} = \frac{2^2 + 2^2 + 2^2}{(0.556\ \text{nm})^2} = 38.8\ \text{nm}^{-2} \qquad d_{222} = 0.161\ \text{nm}$$

The separation of the 111 planes is twice as great as that of the 222 planes.

22.75 The d_{111} spacing for crystalline K is 0.3079 nm. Determine the length of the cubic unit cell.

I Solving *(22.7)* for a and substituting the data gives

$$a = (1^2 + 1^2 + 1^2)^{1/2}(0.3079\ \text{nm}) = 0.5333\ \text{nm}$$

22.76 Determine hkl for planes in Ca with $d_{hkl} = 0.185$ nm. See Problem 22.8 for additional information.

I Solving *(22.7)* for $h^2 + k^2 + l^2$ and substituting the data gives

$$h^2 + k^2 + l^2 = \frac{(0.556\ \text{nm})^2}{(0.185\ \text{nm})^2} = 9.03 \approx 9$$

There are several planes in the cubic system for which the sum of the squares of the Miller indices equals 9: 300, 030, 003, 221, 212, and 122.

22.77 Calculate the d spacings for the 101, 110, and 011 planes in crystalline Cl_2. Which are closer? See Problem 22.61 for additional information.

I For an orthorhombic unit cell,

$$\frac{1}{d_{hkl}^2} = \frac{h^2}{a^2} + \frac{k^2}{b^2} + \frac{l^2}{c^2} \tag{22.8}$$

Substituting the data into *(22.8)* for the 101 planes gives

$$\frac{1}{d_{101}^2} = \frac{1^2}{(0.629\ \text{nm})^2} + \frac{0^2}{(0.450\ \text{nm})^2} + \frac{1^2}{(0.821\ \text{nm})^2} = 4.01\ \text{nm}^{-2} \qquad d_{101} = 0.499\ \text{nm}$$

Likewise for the 110 and 011 planes,

$$\frac{1}{d_{110}^2} = \frac{1^2}{(0.629)^2} + \frac{1^2}{(0.450)^2} + \frac{0^2}{(0.821)^2} = 7.47 \text{ nm}^{-2} \qquad d_{110} = 0.366 \text{ nm}$$

$$\frac{1}{d_{011}^2} = \frac{0^2}{(0.629)^2} + \frac{1^2}{(0.450)^2} + \frac{1^2}{(0.821)^2} = 6.42 \text{ nm}^{-2} \qquad d_{011} = 0.395 \text{ nm}$$

The 110 planes are closer.

22.78 What are the d spacings for the 100, 110, and 111 planes in a cubic unit cell with unit cell length a?

▌ Solving *(22.7)* for d_{hkl} and substituting the data gives

$$d_{100} = \frac{a}{(1^2 + 0^2 + 0^2)^{1/2}} = a$$

$$d_{110} = \frac{a}{(1^2 + 1^2 + 0^2)^{1/2}} = \frac{a}{2^{1/2}}$$

$$d_{111} = \frac{a}{(1^2 + 1^2 + 1^2)^{1/2}} = \frac{a}{3^{1/2}}$$

22.79 What symmetry elements are present in crystalline Ca? The unit cell of Ca is cubic and belongs to the $4/m\,\bar{3}\,2/m$ or $\mathcal{O}_h$ point group.

▌ Table 20-3 lists the elements as E, $4C_3(S_6)$, $3C_4(S_4)$, $6C_2$, σ_h, $6\sigma_d$, and i.

22.80 Determine the point group for pyrite (FeS_2) using the orthographic projections given in Fig. 22-8. The lines shown on the large faces are striations formed during the crystallization process.

▌ The following analyses can be made using Fig. 20-9: (1) Are there four C_3 axes at 54°44′? yes; (2) is there a C_4 axis? no; (3) is there an S_4 axis? no; (4) is there a σ_h perpendicular to C_2? yes; therefore $\mathcal{T}_h$ or $2/m\,\bar{3}$.

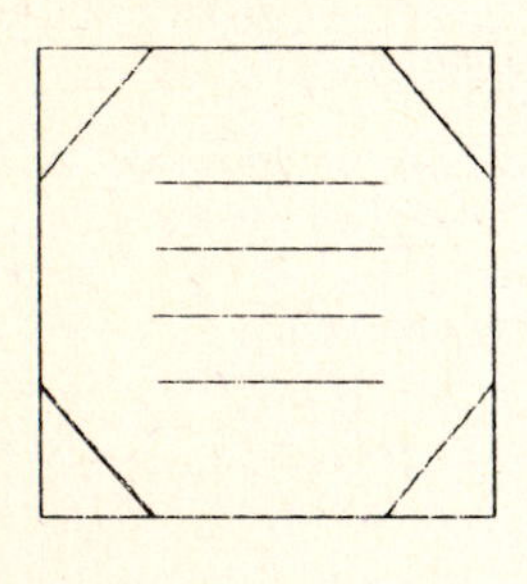

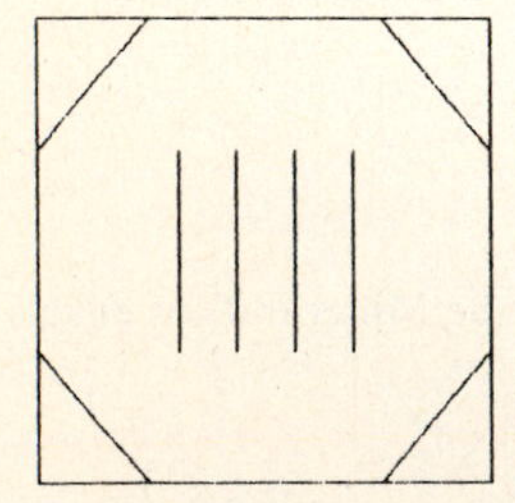

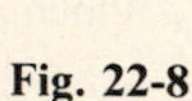

Fig. 22-8

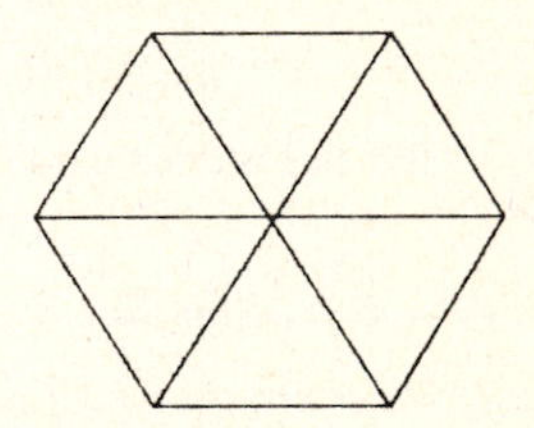

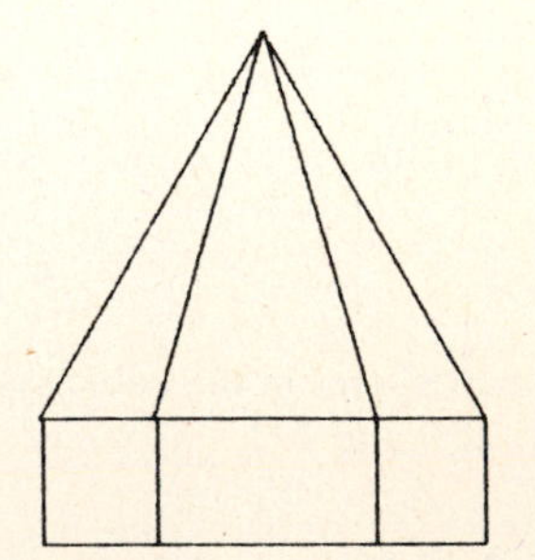

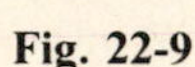

Fig. 22-9

22.81 Determine the point group for zincite (ZnO) using the orthographic projections given in Fig. 22-9.

▌ The following analyses can be made using Fig. 20-9: (1) Are there four C_3 axes at 54°44′? no; (2) is there at least one C_n axis with $n \geq 2$? yes, C_6; (3) is there an S_{2n}? no; (4) are there six C_2 axes perpendicular to C_6? no; (5) is there a σ_v? no; (6) are there six σ_v? yes; therefore $\mathcal{C}_{nv}$ or $6\,mm$.

22.82 Determine the point group for sulfur using the orthographic projections given in Fig. 22-10.

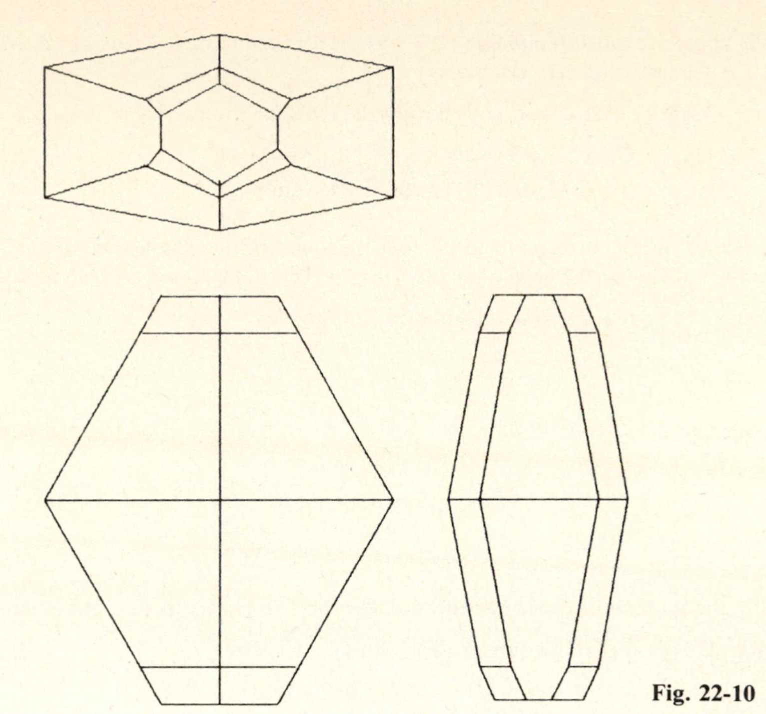

Fig. 22-10

▮ The following analysis can be made using Fig. 20-9: (1) Are there four C_3 axes at 54°44′? no; (2) is there at least one C_n axis with $n \geq 2$? yes, C_2; (3) is there an S_{2n}? no; (4) are there two C_2 axes perpendicular to C_2? yes; (5) is there a σ_h perpendicular to C_2? yes; therefore $\mathscr{D}_{2h}$ or $2/m\, 2/m\, 2/m$.

22.4 X-RAY SPECTRA

22.83 At what Bragg angle will "reflection" occur for the 111 planes in crystalline Ca using filtered Cu K_α radiation? The wavelength of the radiation is $\lambda = 0.154\,18$ nm. See Problem 22.74 for additional information.

▮ The relation among the order of reflection (n), the wavelength (λ), the d spacing, and the angle of reflection (θ) is given by the *Bragg equation*

$$n\lambda = 2d_{hkl} \sin\theta \tag{22.9}$$

Solving *(22.9)* for $\sin\theta$ and substituting the data gives

$$\sin\theta = \frac{1(0.154\,18\text{ nm})}{2(0.321\text{ nm})} = 0.240 \qquad \theta = 13.9°$$

22.84 Repeat the calculation of Problem 22.83 for the second-order reflection of the 111 planes and for the first-order reflection of 222 planes. What conclusion can be made?

▮ Substituting the data into *(22.9)* for $n = 2$ gives for the 111 planes

$$\sin\theta = \frac{2(0.154\,18\text{ nm})}{2(0.321\text{ nm})} = 0.480 \qquad \theta = 28.7°$$

Likewise for the 222 planes,

$$\sin\theta = \frac{1(0.154\,18\text{ nm})}{2(0.161\text{ nm})} = 0.479 \qquad \theta = 28.6°$$

The two sets of planes are identical. The notation 222 is simply another way of representing the second-order reflection of the 111 planes.

22.85 The Bragg angle for the d_{200} reflection of CaO is 21.81° for $\lambda(\text{Co}) = 0.178\,69$ nm. Determine d_{200}.

▮ Solving *(22.9)* for d_{hkl} and substituting the data gives

$$d_{200} = \frac{1(0.178\,69\text{ nm})}{2\sin 21.81°} = 0.2405\text{ nm}$$

22.86 The Bragg angle for the 110 plane in BeO was determined to be 8.72° using $\lambda(\text{Mo}) = 0.070\,93$ nm radiation and 24.44° using Fe radiation. Determine $\lambda(\text{Fe})$.

▮ Taking a ratio of *(22.9)* and substituting the data for the two experiments gives

$$\frac{(1)\lambda(\text{Fe})}{(1)(0.070\,93\text{ nm})} = \frac{2d_{110}\sin 24.44°}{2d_{110}\sin 8.72°} \qquad \lambda(\text{Fe}) = 0.1936\text{ nm}$$

22.87 The 111 planes in KI correspond to the first reflection in the X-ray spectrum. Calculate the minimum value of θ for KI using $\lambda(\text{Cu}) = 0.154\,05$ nm and $a = 0.706\,55$ nm for the cubic unit cell.

▮ Solving *(22.9)* for $\sin\theta$ and substituting *(22.7)* gives

$$\sin\theta = \frac{n\lambda}{2a}(h^2 + k^2 + l^2)^{1/2} \qquad (22.10)$$

Substituting the data gives

$$\sin\theta = \frac{1(0.154\,05\text{ nm})}{2(0.706\,55\text{ nm})}(1^2 + 1^2 + 1^2)^{1/2} = 0.188\,82$$

$$\theta = 10.884°$$

22.88 Determine the maximum order of reflection for the 111 planes in the KI system described in Problem 22.87.

▮ Solving *(22.10)* for n and substituting $\sin\theta = 1$ gives

$$n = \frac{2(0.706\,55\text{ nm})}{(0.154\,05\text{ nm})(1^2 + 1^2 + 1^2)^{1/2}} = 5.2960 \approx 5$$

which corresponds to the 555 plane.

22.89 The distances between the lines on a powder X-ray diffraction pattern for Al were $S/(\text{mm}) = 96.8$, 112.3, 163.6, 196.8, and 207.8. The circumference of the film was determined to be $z = 452.25$ mm, and $\lambda(\text{Cu}) = 0.154\,18$ nm radiation was used. Calculate the d spacings corresponding to these lines.

▮ The values of θ are determined using

$$\theta = (S/2)(180°/z) \qquad (22.11)$$

Substituting the data for the first set of lines into *(22.11)* gives

$$\theta = \left(\frac{96.8\text{ mm}}{2}\right)\left(\frac{180°}{452.25\text{ mm}}\right) = 19.26°$$

Likewise, $\theta = 22.35°, 32.56°, 39.16°$, and $41.35°$, respectively. Solving *(22.9)* for d_{hkl} and substituting the data for the first set of lines gives

$$d_{hkl} = \frac{1(0.154\,18\text{ nm})}{2\sin 19.26°} = 0.2337\text{ nm}$$

Likewise, $d_{hkl}/(\text{nm}) = 0.2027, 0.1432, 0.1221$, and 0.1167, respectively.

22.90 The distances between the lines on a powder X-ray diffraction pattern for NH_4Cl were $S/(\text{mm}) = 58.3$, 82.7, 101.6, 118.2, 133.4, 147.2, 172.4, 184.6, 196.4, 208.2, and 219.8. The circumference of the film was determined to be $z = 452.42$ mm, and $\lambda(\text{Cu}) = 0.154\,18$ nm radiation was used. Calculate the d spacings corresponding to these lines.

▮ Substituting the data for the first set of lines into *(22.11)* gives

$$\theta = \left(\frac{58.3\text{ mm}}{2}\right)\left(\frac{180°}{452.42\text{ mm}}\right) = 11.60°$$

Likewise, $\theta = 16.45°, 20.21°, 23.51°, 26.54°, 29.28°, 34.30°, 36.72°, 39.07°, 41.42°$, and $43.73°$, respectively. Solving *(22.9)* for d_{hkl} and substituting the data for the first set of lines gives

$$d_{hkl} = \frac{1(0.154\,18\text{ nm})}{2\sin 11.60°} = 0.3834\text{ nm}$$

Likewise, $d_{hkl}/(\text{nm}) = 0.2722, 0.2231, 0.1932, 0.1725, 0.1576, 0.1368, 0.1289, 0.1223, 0.1165$, and 0.1115, respectively.

22.91 Reflections for the following planes were observed in the X-ray spectrum of cubic W: 110, 200, 211, 220, 310, 222, 321, 400, 411, 420, 332, and 431. What type of cubic unit cell do these data imply?

▮ If a nonprimitive unit cell is used to describe a substance, certain reflections are not allowed. For example, if a body-centered unit cell is chosen, the values of hkl that are permitted on these that satisfy $h + k + l =$ an even number; and if an end-centered cell designated as C is chosen, $h + k =$ an even number; and if a face-centered unit cell is chosen, all indices in a set must be even or all must be odd. In this case, $h + k + l =$ an even number for all observed reflections (e.g., $1 + 1 + 0 = 2$, $2 + 0 + 0 = 2$, $2 + 1 + 1 = 4$), which implies that a body-centered cubic unit cell is present.

22.92 Prepare a list of the indices of the planes that are permitted in a face-centered cubic unit cell.

▮ For a face-centered cubic unit cell, sets of indices in which all indices are even or all indices are odd are permitted. The first few planes are 111, 200, 220, 222, 311, 331, and 400.

22.93 Place the planes listed in Problem 22.92 in order of their appearance on an X-ray powder spectrum.

▮ As implied by *(22.10)*, the appearance of the planes will be in order of increasing values of $h^2 + k^2 + l^2$. For example, for the 111 plane,

$$h^2 + k^2 + l^2 = 1^2 + 1^2 + 1^2 = 3$$

Likewise, $h^2 + k^2 + l^2 = 4, 8, 12, 11, 19,$ and 16, respectively. The order of appearance is 111, 200, 220, 311, 222, 400, and 331.

22.94 Given that Al is known to crystallize in the cubic system, determine a and index the lines given in Problem 22.89.

▮ The *method of Ito* is a trial-and-error approach for indexing the X-ray powder pattern of a substance and inferring the dimensions of the unit cell. If

$$Q_{hkl} = 1/d_{hkl}^2 \tag{22.12}$$

then for a cubic crystal

$$Q_{hkl} = a^{*2}(h^2 + k^2 + l^2) \tag{22.13}$$

where

$$a^* = a^{-1} \tag{22.14}$$

Substituting the d spacings determined in Problem 22.89 into *(22.12)* gives

$$Q_1 = \frac{1}{(0.2337\ \text{mm})^2} = 19.98\ \text{nm}^{-2}$$

$Q_2 = 24.34\ \text{nm}^{-2}$, $Q_3 = 48.77\ \text{nm}^{-2}$, $Q_4 = 67.08\ \text{nm}^{-2}$, and $Q_5 = 73.43\ \text{nm}^{-2}$, respectively. If Q_1 is assumed to equal Q_{100}, then *(22.13)* gives

$$a^{*2} = \frac{19.98\ \text{nm}^{-2}}{1^2 + 0^2 + 0^2} = 19.98\ \text{nm}^{-2}$$

and

$$Q_{200} = (19.98\ \text{nm}^{-2})(2^2 + 0^2 + 0^2) = 79.92\ \text{nm}^{-2}$$
$$Q_{110} = 19.98(1^2 + 1^2 + 0^2) = 39.96\ \text{nm}^{-2}$$
$$Q_{111} = 19.98(1^2 + 1^2 + 1^2) = 59.94\ \text{nm}^{-2}$$

None of these predicted d spacings are observed, so the assumption that $Q_1 = Q_{100}$ is not valid. Trying $Q_1 = Q_{110}$ for a body-centered cubic unit cell,

$$a^{*2} = \frac{19.98\ \text{nm}^{-2}}{1^2 + 1^2 + 0^2} = 9.99\ \text{nm}^{-2}$$

$$Q_{200} = (9.99\ \text{nm}^{-2})(2^2 + 0^2 + 0^2) = 39.96\ \text{nm}^{-2}$$
$$Q_{211} = 9.99(2^2 + 1^2 + 1^2) = 59.94\ \text{nm}^{-2}$$
$$Q_{220} = 9.99(2^2 + 2^2 + 0^2) = 79.92\ \text{nm}^{-2}$$

Again, none of these d spacings are observed, so the assumption that $Q_1 = Q_{110}$ is not valid. Assuming that

$Q_1 = Q_{111}$ for a face-centered cubic unit cell,

$$a^{*2} = \frac{19.98\ \text{nm}^{-2}}{1^2 + 1^2 + 1^2} = 6.66\ \text{nm}^{-2}$$

$$Q_{200} = (6.66\ \text{nm}^{-2})(2^2 + 0^2 + 0^2) = 26.64\ \text{nm}^{-2} \approx Q_2$$

$$Q_{220} = 6.66(2^2 + 2^2 + 0^2) = 53.28\ \text{nm}^{-2} \approx Q_3$$

$$Q_{311} = 6.66(3^2 + 1^2 + 1^2) = 73.26\ \text{nm}^{-2} \approx Q_4$$

$$Q_{222} = 6.66(2^2 + 2^2 + 2^2) = 79.92\ \text{nm}^{-2} \approx Q_5$$

These predicted values are in fair agreement with the observed peaks. Using these assignments to determine additional values of a^{*2} gives

$$a^{*2} = \frac{24.34\ \text{nm}^{-2}}{2^2 + 0^2 + 0^2} = 6.09\ \text{nm}^{-2}$$

and $a^{*2} = 6.10\ \text{nm}^{-2}$, $6.10\ \text{nm}^{-2}$, and $6.12\ \text{nm}^{-2}$, respectively. The average value of a^{*2} is

$$\overline{a^{*2}} = \frac{6.66\ \text{nm}^{-2} + 6.09\ \text{nm}^{-2} + 6.10\ \text{nm}^{-2} + 6.10\ \text{nm}^{-2} + 6.12\ \text{nm}^{-2}}{5} = 6.21\ \text{nm}^{-2}$$

which upon substitution into *(22.14)* gives

$$a = \frac{1}{(6.21\ \text{nm}^{-2})^{1/2}} = 0.401\ \text{nm}$$

22.95 Given that NH_4Cl is known to crystallize in the cubic system, determine a and index the lines given in Problem 22.90.

▌ Substituting the d spacings determined in Problem 22.90 into *(22.12)* gives

$$Q_1 = \frac{1}{(0.3834\ \text{mm})^2} = 6.803\ \text{nm}^{-2}$$

$Q_2 = 13.50\ \text{nm}^{-2}$, $Q_3 = 20.09\ \text{nm}^{-2}$, $Q_4 = 26.79\ \text{nm}^{-2}$, and $Q_5 = 33.61\ \text{nm}^{-2}$, $Q_6 = 40.26\ \text{nm}^{-2}$, $Q_7 = 53.44\ \text{nm}^{-2}$, $Q_8 = 60.19\ \text{nm}^{-2}$, $Q_9 = 66.86\ \text{nm}^{-2}$, $Q_{10} = 73.68\ \text{nm}^{-2}$, and $Q_{11} = 80.44\ \text{nm}^{-2}$. If Q_1 is assumed to be Q_{100} for a CsCl-type unit cell, then *(22.13)* gives

$$a^{*2} = \frac{6.803\ \text{nm}^{-2}}{1^2 + 0^2 + 0^2} = 6.803\ \text{nm}^{-2}$$

and

$$Q_{110} = (6.803\ \text{nm}^{-2})(1^2 + 1^2 + 0^2) = 13.61\ \text{nm}^{-2} \approx Q_2$$

$$Q_{111} = 6.803(1^2 + 1^2 + 1^2) = 20.41\ \text{nm}^{-2} \approx Q_3$$

$$Q_{200} = 6.803(2^2 + 0^2 + 0^2) = 27.21\ \text{nm}^{-2} \approx Q_4$$

These predicted values are in good agreement with the observed peaks. Using the additional assignments $Q_5 = Q_{210}$, $Q_6 = Q_{211}$, $Q_7 = Q_{220}$, $Q_8 = Q_{300}$, $Q_9 = Q_{310}$, $Q_{10} = Q_{311}$, and $Q_{11} = Q_{222}$, additional values of a^{*2} can be determined as

$$a^{*2} = \frac{13.50\ \text{nm}^{-2}}{1^2 + 1^2 + 0^2} = 6.750\ \text{nm}^{-2}$$

and $a^{*2} = 6.697\ \text{nm}^{-2}$, $6.698\ \text{nm}^{-2}$, $6.722\ \text{nm}^{-2}$, $6.710\ \text{nm}^{-2}$, $6.680\ \text{nm}^{-2}$, $6.688\ \text{nm}^{-2}$, $6.686\ \text{nm}^{-2}$, $6.698\ \text{nm}^{-2}$, and $6.703\ \text{nm}^{-2}$, respectively. The average value of a^{*2} is $6.712\ \text{nm}^{-2}$, which upon substitution into *(22.14)* gives

$$a = \frac{1}{(6.712\ \text{nm}^{-2})^{1/2}} = 0.3860\ \text{nm}$$

22.96 Which peak, 110 or 200, would be predicted to be most intense in the X-ray spectrum of W? Tungsten forms a body-centered cubic unit cell with $a = 0.316\,48\ \text{nm}$.

▌ The 200 plane in a unit cell contains only the body-centered atom, whereas the 110 plane contains the body-centered atom and portions of the four corner atoms. Because the intensity of the peak is proportional to the number of atoms, the 110 plane is predicted to have the greater relative intensity. The observed values are $I/I_0 = 100$ for 110 and 15 for 200.

22.97 Determine the structure factor for the 110 and 200 planes in W. For the experimental values of θ and λ, the atomic scattering factor (f) is 58 for the 110 plane and 51 for the 200 plane. See Problem 22.96 for additional information.

▌ The *structure factor* for a plane *hkl* is defined as

$$F(hkl) = \sum_j f_j e^{2\pi i(hx_j + ky_j + lz_j)} \qquad (22.15)$$

where the summation is performed over all the atoms in the unit cell. The coordinates of atoms in the body-centered cubic unit cell are (000) and $(\frac{1}{2}\frac{1}{2}\frac{1}{2})$. For the 110 plane,

$$F(110) = 58(e^{2\pi i[(1)(0)+(1)(0)+(0)(0)]} + e^{2\pi i[(1)(1/2)+(1)(1/2)+(0)(1/2)]}) = 58(1 + e^{2\pi i}) = 58(1 + 1) = 116$$

where $e^{2\pi i}$ was evaluated using *(14.25)*. Likewise, for the 200 plane,

$$F(200) = 51(e^{2\pi i[(2)(0)+(0)(0)+(0)(0)]} + e^{2\pi i[(2)(1/2)+(0)(1/2)+(0)(1/2)]}) = 51(1 + e^{2\pi i}) = 51(1 + 1) = 102$$

22.98 Using the results of Problem 22.97, show that the intensity of the 110 peak is greater than that of the 200 peak.

▌ The intensity of a scattered X-ray beam is proportional to $F(hkl)^*F(hkl)$. Taking a ratio gives

$$\frac{I(110)}{I(200)} = \frac{F(110)^*F(110)}{F(200)^*F(200)} = \frac{116^2}{102^2} = 1.29$$

22.99 Determine $F(111)$ for W. See Problem 22.96 for additional information.

▌ Substituting into *(22.15)* gives

$$F(111) = f(e^{2\pi i[(1)(0)+(1)(0)+(1)(0)]} + e^{2\pi i[(1)(1/2)+(1)(1/2)+(1)(1/2)]})$$
$$= f[1 + e^{2\pi i(3/2)}] = f(1 + e^{3\pi i}) = f(1 - 1) = 0$$

Note that this peak is predicted not to exist in the X-ray spectrum, which agrees with the extinction rule given in Problem 22.91 that $h + k + l$ must be an even number for a body-centered unit cell.

22.100 The relative intensities observed for the planes containing all even values of *hkl* (e.g., 200, 220, 222, and 400) are considerably greater than those observed for the planes containing all odd values of *hkl* (e.g., 111, 311, and 331) in NaCl. Why is this true?

▌ The coordinates of the Na^+ ions are (000), $(\frac{1}{2}\frac{1}{2}0)$, $(\frac{1}{2}0\frac{1}{2})$, $(0\frac{1}{2}\frac{1}{2})$, and those of the Cl^- ions are $(\frac{1}{2}00)$, $(0\frac{1}{2}0)$, $(00\frac{1}{2})$, $(\frac{1}{2}\frac{1}{2}\frac{1}{2})$. Substituting even values of *hkl* into *(22.15)* gives

$$\begin{aligned}F(\text{even } hkl) &= f(Na^+)(e^{2\pi i[(\text{even})(0)+(\text{even})(0)+(\text{even})(0)]} + e^{2\pi i[(\text{even})(1/2)+(\text{even})(1/2)+(\text{even})(0)]} \\ &\quad + e^{2\pi i[(\text{even})(1/2)+(\text{even})(0)+(\text{even})(1/2)]} + e^{2\pi i[(\text{even})(0)+(\text{even})(1/2)+(\text{even})(1/2)]}) \\ &\quad + f(Cl^-)(e^{2\pi i[(\text{even})(1/2)+(\text{even})(0)+(\text{even})(0)]} + e^{2\pi i[(\text{even})(0)+(\text{even})(1/2)+(\text{even})(0)]} \\ &\quad + e^{2\pi i[(\text{even})(0)+(\text{even})(0)+(\text{even})(1/2)]} + e^{2\pi i[(\text{even})(1/2)+(\text{even})(1/2)+(\text{even})(1/2)]}) \\ &= f(Na^+)[1 + 3\,e^{\pi i(\text{even})}] + f(Cl^-)[4\,e^{\pi i(\text{even})}] = f(Na^+)[1 + (3)(1)] + f(Cl^-)[(4)(1)] \\ &= 4f(Na^+) + 4f(Cl^-) = 4[f(Na^+) + f(Cl^-)]\end{aligned}$$

Likewise for odd values of *hkl*,

$$\begin{aligned}F(\text{odd } hkl) &= f(Na^+)(e^{2\pi i[(\text{odd})(0)+(\text{odd})(0)+(\text{odd})(0)]} + e^{2\pi i[(\text{odd})(1/2)+(\text{odd})(1/2)+(\text{odd})(0)]} \\ &\quad + e^{2\pi i[(\text{odd})(1/2)+(\text{odd})(0)+(\text{odd})(1/2)]} + e^{2\pi i[(\text{odd})(0)+(\text{odd})(1/2)+(\text{odd})(1/2)]}) \\ &\quad + f(Cl^-)(e^{2\pi i[(\text{odd})(1/2)+(\text{odd})(0)+(\text{odd})(0)]} + e^{2\pi i[(\text{odd})(0)+(\text{odd})(1/2)+(\text{odd})(0)]} \\ &\quad + e^{2\pi i[(\text{odd})(0)+(\text{odd})(0)+(\text{odd})(1/2)]} + e^{2\pi i[(\text{odd})(1/2)+(\text{odd})(1/2)+(\text{odd})(1/2)]}) \\ &= f(Na^+)[1 + 3\,e^{\pi i(\text{even})}] + f(Cl^-)[4\,e^{\pi i(\text{odd})}] = f(Na^+)[1 + (3)(1)] + f(Cl^-)[(4)(-1)] \\ &= 4f(Na^+) - 4f(Cl^-) = 4[f(Na^+) - f(Cl^-)]\end{aligned}$$

As can be seen from the values of the structure factors, the scattering from all of the ions for the even values of *hkl* is constructive and the scattering for the odd values of *hkl* is destructive.

22.5 DIFFRACTION SPECTROSCOPY

22.101 The wavelengths for the $K_{\alpha 1}$ line for several elements are tabulated below.

element	$_{21}Sc$	$_{22}Ti$	$_{23}V$	$_{24}Cu$	$_{25}Mn$	$_{26}Fe$	$_{27}Co$	$_{28}Ni$	$_{29}Cu$
λ/(nm)	0.303 114	0.274 841	0.250 348	0.228 962	0.210 175	0.193 597	0.178 892	0.165 784	0.154 051

Prepare a plot of $\nu^{1/2}$ against Z and determine the constants for *Moseley's law*:

$$\nu^{1/2} = a + bZ \tag{22.16}$$

I As a sample calculation, consider the data for $_{21}$Sc, where using *(14.1)* gives

$$\nu^{1/2} = \left(\frac{c}{\lambda}\right)^{1/2} = \left(\frac{(2.997\,924\,58 \times 10^8\ \mathrm{m\cdot s^{-1}})[(1\ \mathrm{Hz})/(1\ \mathrm{s^{-1}})]}{(0.303\,114\ \mathrm{nm})[(10^{-9}\ \mathrm{m})/(1\ \mathrm{nm})]}\right)^{1/2} = 9.945\,06 \times 10^8\ \mathrm{Hz^{1/2}}$$

The plot is shown in Fig. 22-11. From the graph, a = intercept = $-5.70 \times 10^{-7}\ \mathrm{Hz^{1/2}}$ and b = slope = $5.006 \times 10^7\ \mathrm{Hz^{1/2}}$.

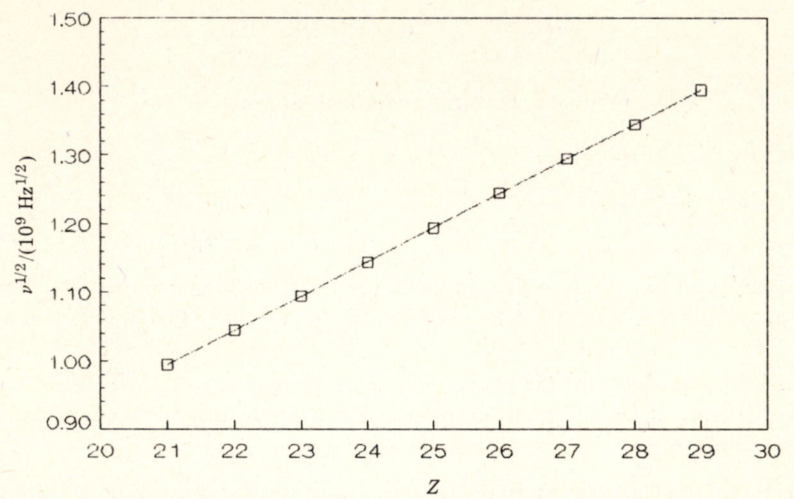

Fig. 22-11

22.102 For atomic iodine, $\lambda = 0.043\,329$ nm for the $K_{\alpha 1}$ line. Determine the atomic number of I.

I Substituting the data into *(14.1)* gives

$$\nu^{1/2} = \left(\frac{(2.997\,924\,58 \times 10^8\ \mathrm{m\cdot s^{-1}})[(1\ \mathrm{Hz})/(1\ \mathrm{s^{-1}})]}{(0.043\,329\ \mathrm{nm})[(10^{-9}\ \mathrm{m})/(1\ \mathrm{nm})]}\right)^{1/2} = 2.6304 \times 10^9\ \mathrm{Hz^{1/2}}$$

Solving *(22.16)* for Z and substituting the data gives

$$Z = \frac{2.6304 \times 10^9\ \mathrm{Hz^{1/2}} - (-5.70 \times 10^7\ \mathrm{Hz^{1/2}})}{5.006 \times 10^7\ \mathrm{Hz^{1/2}}} = 53.68 \approx 53$$

Early periodic tables based on atomic masses placed I in the chalcogen family and Te in the halogen family. Values of Z determined from X-ray spectra confirmed the proposed reassignment of these elements to their current locations based on chemical properties.

22.103 What is the minimum voltage needed to generate the Cu $K_{\alpha 1}$ line ($\lambda = 0.154$ nm) from a copper target?

I Substituting the data into *(13.32)* gives

$$E = \frac{(6.626 \times 10^{-34}\ \mathrm{J\cdot s})(2.9979 \times 10^8\ \mathrm{m\cdot s^{-1}})}{(0.154\ \mathrm{nm})[(10^{-9}\ \mathrm{m})/(1\ \mathrm{nm})][(1.602 \times 10^{-19}\ \mathrm{J})/(1\ \mathrm{eV})]} = 8050\ \mathrm{eV}$$

The voltage requirement to promote the electron is 8050 V.

22.104 The mass absorption coefficient for Cu K_α radiation is $\mu/\rho = 313\ \mathrm{cm^2\cdot g^{-1}}$ for Co. Determine the thickness of Co needed to reduce the intensity of radiation to below 1% of the incident radiation. The density of Co is $8.9\ \mathrm{g\cdot cm^{-3}}$.

▮ The absorption of X-rays is described by

$$I/I_0 = e^{-\mu x} \tag{22.17}$$

where I is the transmitted intensity, I_0 is the incident intensity, x is the thickness of the absorber, and μ is the absorption coefficient. Solving *(22.17)* for x and substituting $I/I_0 = 0.01$ gives

$$x = \frac{-\ln(0.01)}{(313\ \text{cm}^2 \cdot \text{g}^{-1})(8.9\ \text{g} \cdot \text{cm}^{-3})} = 1.65 \times 10^{-3}\ \text{cm}$$

22.105 What is the energy associated with "thermal" neutrons ($T = 298$ K) with $\lambda = 0.14$ nm? The mass of a neutron is 1.675×10^{-27} kg.

▮ The kinetic energy is given by

$$E = h^2/2\lambda^2 m \tag{22.18}$$

Substituting the data into *(22.18)* gives

$$E = \frac{(6.626 \times 10^{-34}\ \text{J} \cdot \text{s})^2[(1\ \text{kg} \cdot \text{m}^2 \cdot \text{s}^{-2})/(1\ \text{J})]}{2(0.14 \times 10^{-9}\ \text{m})^2(1.675 \times 10^{-27}\ \text{kg})} = 6.7 \times 10^{-21}\ \text{J}$$

or

$$E = (6.7 \times 10^{-21}\ \text{J})\frac{1\ \text{eV}}{1.602 \times 10^{-19}\ \text{J}} = 0.042\ \text{eV}$$

22.106 Calculate the wavelength of 40-kV electrons. The mass of an electron is 9.11×10^{-31} kg.

▮ Solving *(22.18)* for λ and substituting the data gives

$$\lambda = \left(\frac{(6.626 \times 10^{-34}\ \text{J} \cdot \text{s})^2[(1\ \text{kg} \cdot \text{m}^2 \cdot \text{s}^{-2})/(1\ \text{J})]}{2(40 \times 10^3\ \text{eV})[(1.602 \times 10^{-19}\ \text{J})/(1\ \text{eV})](9.11 \times 10^{-31}\ \text{kg})}\right)^{1/2} = 6 \times 10^{-12}\ \text{m} = 6\ \text{pm}$$

22.107 Prepare an electron diffraction scattering curve for the chlorine molecule.

▮ The intensity of the scattered beam $[I(\theta)]$ at an angle θ is given by the *Wierl equation*

$$I(\theta) = I_0\left(\frac{8\pi m_e e^2}{h}\right)^2\left(\frac{1}{s^4}\right)\sum_i\sum_j (Z_i - f_j)\frac{\sin sr_{ij}}{sr_{ij}} \tag{22.19a}$$

where Z_i is the atomic number, f_i is the atomic scattering factor for X-rays, and

$$s = \frac{4\pi}{\lambda}\sin\frac{\theta}{2} \tag{22.20}$$

For structural analysis *(22.19)* can be approximated by

$$I(\theta) \approx \sum_i\sum_j Z_i Z_j \frac{\sin sr_{ij}}{sr_{ij}} \tag{22.19b}$$

where $i \neq j$. As a sample calculation, consider $sr_{ij} = 5.00$,

$$I(\theta) = (17)(17)\frac{\sin 5.00}{5.00} = -55.43$$

The plot is shown in Fig. 22-12.

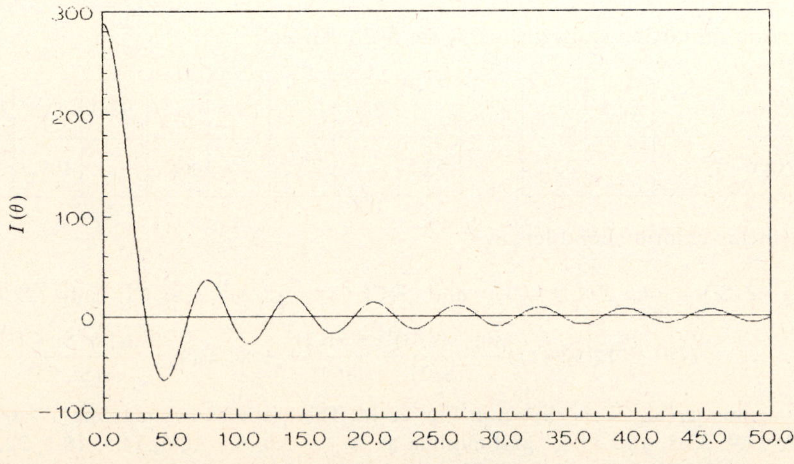

Fig. 22-12

22.108 Prepare an electron diffraction scattering curve for CO_2. Given

s(maxima)/(nm⁻¹)		67			122			178			230
s(weak maxima)/(nm⁻¹)			85			140			190		
s(minima)/(nm⁻¹)	44			100			154			210	

determine the carbon-oxygen bond length.

▮ Substituting $Z(\mathrm{C}) = 6$, $Z(\mathrm{O}) = 8$, and $r(\mathrm{O,O}) = 2r(\mathrm{C{=}O})$ into *(22.19b)* gives

$$I(\theta) = 2(6)(8)\frac{\sin[sr(\mathrm{C{=}O})]}{sr(\mathrm{C{=}O})} + (8)(8)\frac{\sin[2sr(\mathrm{C{=}O})]}{2sr(\mathrm{C{=}O})}$$

A plot of $I(\theta)$ against $sr(\mathrm{C{=}O})$ is shown in Fig. 22-13. The graph shows maxima at $sr_{ij} = 7.3$, 13.7, 20.0, and 26.2, minima at $sr_{ij} = 5.0$, 11.4, 17.7, and 24.0, and weak maxima as "shoulders" on the curve between each maximum and minimum. The value of $r(\mathrm{C{=}O})$ is found by dividing each value of sr_{ij} from the plot by the respective observed value of s to give

$$r(\mathrm{C{=}O}) = \frac{7.3}{67\ \mathrm{nm}^{-1}} = 0.109\ \mathrm{nm}$$

and 0.122 nm, 0.112 nm, 0.114 nm, 0.114 nm, 0.114 nm, 0.115 nm, 0.114 nm. The average value is 0.114 nm, which is in good agreement with the accepted value of 0.116 nm.

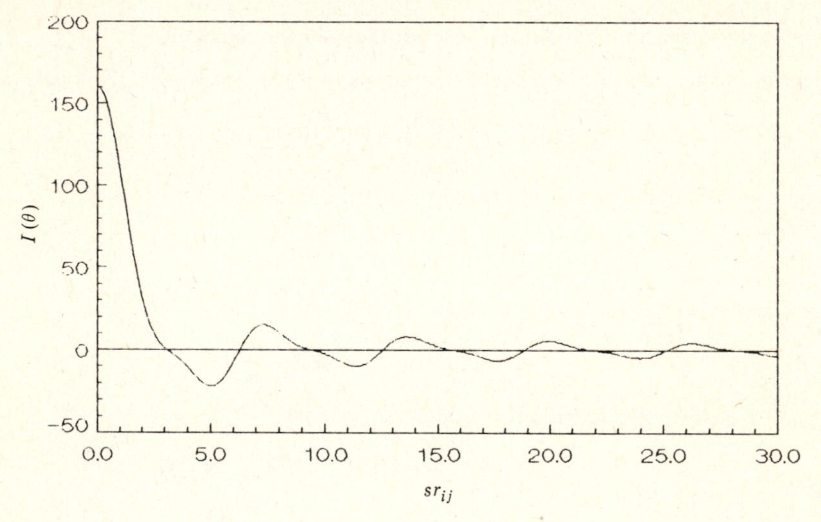

Fig. 22-13

22.109 Prepare an electron diffraction scattering curve for $SiCl_4$. Given

s(maximum)/(nm⁻¹)	23.8		42.5		62.9		80.6		98.3		118.5	137.7		157.7
s(minimum)/(nm⁻¹)		33.2		53.5		72.2		89.9		109.3			147.7	

determine the silicon-chlorine bond length.

▮ Substituting $Z(\mathrm{Si}) = 14$, $Z(\mathrm{Cl}) = 17$, and $r(\mathrm{Cl, Cl}) = (8/3)^{1/2}r(\mathrm{Si, Cl})$ into *(22.19b)* gives

$$I(\theta) = 12(17)(17)\frac{\sin[s(8/3)^{1/2}r(\mathrm{SiCl})]}{s(8/3)^{1/2}r(\mathrm{SiCl})} + 8(14)(17)\frac{\sin[sr(\mathrm{Si, Cl})]}{sr(\mathrm{Si, Cl})}$$

A plot of $I(\theta)$ against $sr(\mathrm{Si, Cl})$ is shown in Fig. 22-14. The graph shows maxima at $sr_{ij} = 4.8$, 8.5, 12.7, 16.1, 20.2, 24.2, 27.7, 31.9, 35.5, and 39.4, and minima at $sr_{ij} = 3.0$, 6.5, 10.6, 14.5, 18.1, 22.3, 25.8, 29.8, 33.8, and 37.3. The value of $r(\mathrm{Si, Cl})$ is found by dividing each value of sr_{ij} from the plot by the respective observed value

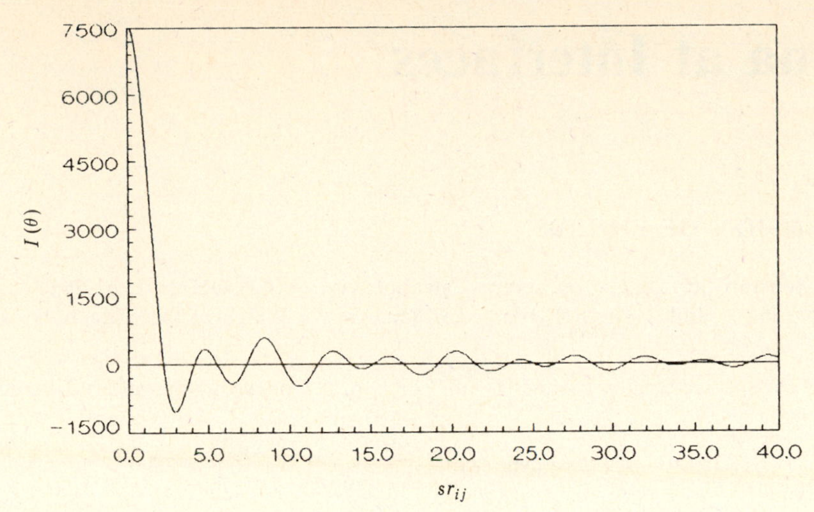

Fig. 22-14

of s to give

$$r(\mathrm{Si, Cl}) = \frac{4.8}{23.8\ \mathrm{nm}^{-1}} = 0.202\ \mathrm{nm}$$

and $r(\mathrm{Si, Cl})$ = 0.200 nm, 0.202 nm, 0.200 nm, 0.205 nm, 0.204 nm, 0.201 nm, 0.203 nm, 0.196 nm, 0.198 nm, 0.201 nm, 0.201 nm, 0.204 nm, and 0.202 nm. The average value is 0.201 nm, which is in excellent agreement with the accepted value of 0.2017 nm.

CHAPTER 23
Phenomena at Interfaces

23.1 SURFACE TENSION OF LIQUIDS

23.1 The surface tension at 20 °C for several alcohols is $\gamma(CH_3OH) = 22.61\ \text{dyne}\cdot\text{cm}^{-1}$, $\gamma(C_2H_5OH) = 2.275 \times 10^{-2}\ \text{N}\cdot\text{m}^{-1}$, and $\gamma(n\text{-}C_3H_7OH) = 23.78\ \text{mJ}\cdot\text{m}^{-2}$. Which of these alcohols has the greatest surface tension?

▌ The SI unit for surface tension is either $\text{N}\cdot\text{m}^{-1}$ or $\text{J}\cdot\text{m}^{-2}$. Converting the values for methanol and n-propanol gives

$$\gamma(CH_3OH) = (22.61\ \text{dyne}\cdot\text{cm}^{-1})\left(\frac{10^{-5}\ \text{N}}{1\ \text{dyne}}\right)\left(\frac{10^{2}\ \text{cm}}{1\ \text{m}}\right) = 2.261 \times 10^{-2}\ \text{N}\cdot\text{m}^{-1}$$

$$\gamma(n\text{-}C_3H_7OH) = (23.78\ \text{mJ}\cdot\text{m}^{-2})\left(\frac{10^{-3}\ \text{J}}{1\ \text{mJ}}\right)\left(\frac{1\ \text{N}\cdot\text{m}}{1\ \text{J}}\right) = 2.378 \times 10^{-2}\ \text{N}\cdot\text{m}^{-1}$$

The alcohol having the greatest value is n-propanol.

23.2 A 0.2-mm-diameter tube was placed in a sample of water at 20 °C. As a result of capillary action, the water rose inside the tube to a height of 11.40 cm. Determine the exact diameter of the tube given $\gamma = 72.75 \times 10^{-3}\ \text{N}\cdot\text{m}^{-1}$ and $\rho = 0.998\ \text{g}\cdot\text{cm}^{-3}$ for H_2O.

▌ The height (h) to which a liquid will rise or be depressed in a capillary tube is related to the radius of the tube (r) by

$$h = 2\gamma/\rho r g \tag{23.1}$$

where ρ is the density of the liquid [assuming $\rho \gg \rho$ (air or vapor)] and g is the gravitational constant. Solving *(23.1)* for r and substituting the data gives

$$r = \frac{2(72.75 \times 10^{-3}\ \text{N}\cdot\text{m}^{-1})[(1\ \text{kg}\cdot\text{m}\cdot\text{s}^{-2})/(1\ \text{N})]}{(0.998\ \text{g}\cdot\text{cm}^{-3})[(10^{-3}\ \text{kg})/(1\ \text{g})][(10^{2}\ \text{cm})/(1\ \text{m})]^3(9.81\ \text{m}\cdot\text{s}^{-2})(11.40 \times 10^{-2}\ \text{m})} = 1.304 \times 10^{-4}\ \text{m}$$

The diameter of the tube is

$$d = 2(1.304 \times 10^{-4}\ \text{m})[(10^{3}\ \text{mm})/(1\ \text{m})] = 0.2608\ \text{mm}$$

23.3 The capillary tube described in Problem 23.2 was placed in a sample of chloroform at 20 °C. The $CHCl_3$ rose to a height of 2.86 cm in the tube. What is the value of γ given $\rho = 1.4832\ \text{g}\cdot\text{cm}^{-3}$ for $CHCl_3$?

▌ For a comparison method, *(23.1)* can be written as

$$\frac{\gamma}{\gamma_0} = \frac{\rho h}{\rho_0 h_0} \tag{23.2}$$

where the subscript refers to the reference substance. Solving *(23.2)* for γ and substituting the data from Problem 23.2 and those given above gives

$$\gamma(CHCl_3) = (72.75 \times 10^{-3}\ \text{N}\cdot\text{m}^{-1})\frac{(1.4832\ \text{g}\cdot\text{cm}^{-3})(2.86\ \text{cm})}{(0.998\ \text{g}\cdot\text{cm}^{-3})(11.40\ \text{cm})} = 27.1 \times 10^{-3}\ \text{N}\cdot\text{m}^{-1}$$

23.4 A 0.20-mm capillary tube and a 0.10-mm capillary tube were placed into a sample of liquid hydrogen peroxide. The difference between the heights of the liquid in the tubes is 5.50 cm. Given that $\rho = 1.41\ \text{g}\cdot\text{cm}^{-3}$ for H_2O_2 at 25 °C, determine γ.

▌ Let $\Delta h = h_2 - h_1$, where h_i is given by *(23.1)*. Solving for γ and substituting the data gives

$$\gamma = \frac{\rho g\,\Delta h}{2(r_1^{-1} - r_2^{-1})} \tag{23.3}$$

$$\gamma = \frac{(1.41\ \text{g}\cdot\text{cm}^{-3})\left(\frac{10^{-3}\ \text{kg}}{1\ \text{g}}\right)\left(\frac{10^{2}\ \text{cm}}{1\ \text{m}}\right)^3(9.81\ \text{m}\cdot\text{s}^{-2})(5.50 \times 10^{-2}\ \text{m})\left(\frac{1\ \text{N}}{1\ \text{kg}\cdot\text{m}\cdot\text{s}^{-2}}\right)}{2[(0.10 \times 10^{-3}\ \text{m})^{-1} - (0.20 \times 10^{-3}\ \text{m})^{-1}]} = 7.61 \times 10^{-2}\ \text{N}\cdot\text{m}^{-1}$$

23.5 A sample of water at 70 °C was placed in a U-tube consisting of one arm with a radius of 0.10 mm and the other with a radius of 0.15 mm. What pressure must be applied to the water in the smaller arm to make the water levels equal? See Problem 23.2 for additional information.

▮ Substituting *(1.4)* into *(23.3)*, solving for *P*, and substituting the data gives

$$P = 2(72.75 \times 10^{-3}\ \mathrm{N \cdot m^{-1}})[(0.10 \times 10^{-3}\ \mathrm{m})^{-1} - (0.15 \times 10^{-3}\ \mathrm{m})^{-1}]\left(\frac{1\ \mathrm{Pa}}{1\ \mathrm{N \cdot m^{-2}}}\right) = 490\ \mathrm{Pa}$$

23.6 A 1.00-cm-radius glass rod was sealed into the center of a piece of 1.01-cm inside radius glass tubing. If the assembly is placed into a sample of water at 20 °C, to what height will the liquid rise? See Problem 23.2 for additional information.

▮ Correcting *(23.1)* for the different shape of the miniscus by eliminating the factor of 2 and substituting the data gives

$$h = \frac{(72.75 \times 10^{-3}\ \mathrm{N \cdot m^{-1}})[(1\ \mathrm{kg \cdot m \cdot s^{-2}})/(1\ \mathrm{N})]}{(0.998\ \mathrm{g \cdot cm^{-3}})\left(\frac{10^{-3}\ \mathrm{kg}}{1\ \mathrm{g}}\right)\left(\frac{10^{2}\ \mathrm{cm}}{1\ \mathrm{m}}\right)^{3}(9.81\ \mathrm{m \cdot s^{-2}})(1.01\ \mathrm{cm} - 1.00\ \mathrm{cm})\left(\frac{10^{-2}\ \mathrm{m}}{1\ \mathrm{cm}}\right)} = 7 \times 10^{-2}\ \mathrm{m} = 7\ \mathrm{cm}$$

23.7 The surface tension of water was determined using the "drop-weight" method in which 100 drops of the liquid from a capillary tube are collected and weighed. Given that the mass of the drops is 2.53 g and the radius of the tube is 8.78×10^{-2} cm, determine $\gamma(H_2O)$. The correction factor for the system is $F' = 0.2657$.

▮ The relation of γ to the average mass of a drop (m) and the radius of the tip is

$$\gamma = mgF'/r \tag{23.4}$$

Substituting $m = (2.53\ \mathrm{g})/100 = 2.53 \times 10^{-2}\ \mathrm{g}$ into *(23.4)* gives

$$\gamma = \frac{(2.53 \times 10^{-2}\ \mathrm{g})[(10^{-3}\ \mathrm{kg})/(1\ \mathrm{g})](9.81\ \mathrm{m \cdot s^{-2}})(0.2657)[(1\ \mathrm{N})/(1\ \mathrm{kg \cdot m \cdot s^{-2}})]}{(8.78 \times 10^{-2}\ \mathrm{cm})[(10^{-2}\ \mathrm{m})/(1\ \mathrm{cm})]} = 7.51 \times 10^{-2}\ \mathrm{N \cdot m^{-1}}$$

23.8 An eyedropper delivered 20 drops to 1 mL of various dilute aqueous solutions at 20 °C. What is the radius of the tip of the dropper? See Problem 23.2 for additional information. Assume $F' = 0.25$ for the dropper.

▮ The mass of one drop is approximately

$$m = (0.998\ \mathrm{g \cdot cm^{-3}})\left(\frac{1\ \mathrm{mL}}{20}\right)\left(\frac{1\ \mathrm{cm^3}}{1\ \mathrm{mL}}\right) = 5 \times 10^{-2}\ \mathrm{g}$$

Solving *(23.4)* for *r* and substituting the data gives

$$r = \frac{(5 \times 10^{-2}\ \mathrm{g})[(10^{-3}\ \mathrm{kg})/(1\ \mathrm{g})](9.81\ \mathrm{m \cdot s^{-2}})(0.25)}{(72.75 \times 10^{-3}\ \mathrm{N \cdot m^{-1}})[(1\ \mathrm{kg \cdot m \cdot s^{-2}})/(1\ \mathrm{N})]} = 2 \times 10^{-3}\ \mathrm{m} = 2\ \mathrm{mm}$$

23.9 The surface tension of glacial acetic acid was determined using the "bubble-pressure" method in which the pressure needed to dislodge bubbles of air from the end of a capillary tube immersed in the liquid is measured. Given that the radius of the tube (r) is 1.1 mm, the depth of the tube in the liquid (h) is 3.56 cm, the pressure is 420 Pa, and $\rho = 1.0492\ \mathrm{g \cdot cm^{-3}}$ for CH_3COOH, determine γ.

▮ The surface tension is given by the *Laplace equation*,

$$\gamma = \frac{r}{2}(P - h\rho g) \tag{23.5}$$

Substituting the data into *(23.5)* gives

$$\gamma = \frac{1.1 \times 10^{-3}\ \mathrm{m}}{2}\left[(420\ \mathrm{Pa})\left(\frac{1\ \mathrm{N \cdot m^{-2}}}{1\ \mathrm{Pa}}\right) - (3.56 \times 10^{-2}\ \mathrm{m})(1.0492\ \mathrm{g \cdot cm^{-3}})\right.$$

$$\left. \times \left(\frac{10^{-3}\ \mathrm{kg}}{1\ \mathrm{g}}\right)\left(\frac{10^{2}\ \mathrm{cm}}{1\ \mathrm{m}}\right)^{3}(9.81\ \mathrm{m \cdot s^{-2}})\left(\frac{1\ \mathrm{N}}{1\ \mathrm{kg \cdot m \cdot s^{-2}}}\right)\right]$$

$$= 2.9 \times 10^{-2}\ \mathrm{N \cdot m^{-1}}$$

23.10 The surface tension of carbon tetrachloride was determined using a du Noüy tensiometer in which the force (F) needed to break the surface of the liquid with a ring of mean radius R is measured. Given $R = 5.03$ cm and $F = 17.7 \times 10^{-3}$ N, determine $\gamma(CCl_4)$. The correction factor for the system is $F' = 0.953$.

▮ The surface tension is given by

$$\gamma = FF'/4\pi R \tag{23.6}$$

Substituting the data into *(23.6)* gives

$$\gamma = \frac{(17.7 \times 10^{-3}\ \mathrm{N})(0.953)}{4\pi(5.03 \times 10^{-2}\ \mathrm{m})} = 2.67 \times 10^{-2}\ \mathrm{N \cdot m^{-1}}$$

23.11 A film of pyridine filled a rectangular wire loop in which one side could be moved. Given that the wire loop is 8.53 cm wide and that a force of 6.48×10^{-3} N is needed to move the side, determine the value of the surface tension.

▮ The force (F) is related to γ and the width (l) by

$$F = 2l\gamma \tag{23.7}$$

Solving *(23.7)* for γ and substituting the data gives

$$\gamma = \frac{6.48 \times 10^{-3}\ \mathrm{N}}{2(8.53 \times 10^{-2}\ \mathrm{m})} = 3.80 \times 10^{-2}\ \mathrm{N \cdot m^{-1}}$$

23.12 What is the work necessary to stretch the film described in Problem 23.11 a distance of 0.10 cm?

▮ From Table 3-5 for surface work,

$$w = \gamma\,\Delta A = 2\gamma l\,\Delta x \tag{23.8}$$

Substituting the data into *(23.1)* gives

$$w = 2(3.80 \times 10^{-2}\ \mathrm{N \cdot m^{-1}})(8.53 \times 10^{-2}\ \mathrm{m})(0.10 \times 10^{-2}\ \mathrm{m})[(1\ \mathrm{J})/(1\ \mathrm{N \cdot m})] = 6.5 \times 10^{-6}\ \mathrm{J}$$

23.13 How much work is required to break up a mole of water at 20 °C into spherical droplets of radius 1 cm? See Problem 23.2 for additional information.

▮ The volume of a spherical molar sample of water is

$$V_{\mathrm{m}} = \frac{18.015\ \mathrm{g \cdot mol^{-1}}}{(0.998\,23\ \mathrm{g \cdot cm^{-3}})[(10^2\ \mathrm{cm})/(1\ \mathrm{m})]^3} = 1.8047 \times 10^{-5}\ \mathrm{m^3}$$

which corresponds to

$$r = \left(\frac{3(1.8047 \times 10^{-5}\ \mathrm{m^3})}{4\pi}\right)^{1/3} = 1.6272 \times 10^{-2}\ \mathrm{m}$$

$$A = 4\pi(1.6272 \times 10^{-2}\ \mathrm{m})^2 = 3.3273 \times 10^{-3}\ \mathrm{m^2}$$

The volume of each droplet is

$$V = \tfrac{4}{3}\pi(1 \times 10^{-2}\ \mathrm{m})^3 = 4 \times 10^{-6}\ \mathrm{m^3}$$

corresponding to

$$N = \frac{1.8047 \times 10^{-5}\ \mathrm{m^3}}{4 \times 10^{-6}\ \mathrm{m^3}} = 5\ \text{droplets}$$

The total surface area of the droplets is

$$A = 5(4\pi)(1 \times 10^{-2}\ \mathrm{m})^2 = 6 \times 10^{-3}\ \mathrm{m^2}$$

Substituting the data into *(23.8)* gives

$$w = (72.75 \times 10^{-3}\ \mathrm{N \cdot m^{-1}})(6 \times 10^{-3}\ \mathrm{m^2} - 3.3273 \times 10^{-3}\ \mathrm{m^2})[(1\ \mathrm{J})/(1\ \mathrm{N \cdot m})] = 2 \times 10^{-4}\ \mathrm{J}$$

23.14 Repeat the calculations of Problem 23.13 assuming the radius of the droplets is 1×10^{-7} m.

▮ The volume of each droplet is

$$V = \tfrac{4}{3}\pi(1 \times 10^{-7}\ \mathrm{m})^3 = 4 \times 10^{-21}\ \mathrm{m^3}$$

corresponding to

$$N = \frac{1.8047 \times 10^{-5}\ \mathrm{m^3}}{4 \times 10^{-21}\ \mathrm{m^3}} = 5 \times 10^{15}\ \text{droplets}$$

$$A = (5 \times 10^{15})(4\pi)(1 \times 10^{-6}\ \mathrm{m})^2 = 6 \times 10^4\ \mathrm{m^2}$$

Substituting into *(23.8)* gives

$$w = (72.75 \times 10^{-3}\ \mathrm{N\cdot m^{-1}})(6 \times 10^{4}\ \mathrm{m^2} - 3.3273 \times 10^{-3}\ \mathrm{m^2})[(1\ \mathrm{J})/(1\ \mathrm{N\cdot m})] = 4 \times 10^{3}\ \mathrm{J}$$

23.15 Assume that the droplets in the system described in Problem 23.14 coalesce to form the 1-cm spheres described in Problem 23.13. If the process is reversible and adiabatic, calculate the temperature change. Assume $C_V = 4\ \mathrm{J\cdot K^{-1}\cdot g^{-1}}$ for $H_2O(l)$.

▮ Setting $q = 0$ in *(3.47)* and substituting *(23.8)* into *(3.18)* gives

$$\Delta U = 0 + w = \gamma\,\Delta A = nC_V\,\Delta T$$

Solving for ΔT and substituting the data gives

$$\Delta T = \frac{(72.75 \times 10^{-3}\ \mathrm{N\cdot m^{-1}})(6 \times 10^{4}\ \mathrm{m^2} - 6 \times 10^{-3}\ \mathrm{m^2})[(1\ \mathrm{J})/(1\ \mathrm{N\cdot m})]}{(1.00\ \mathrm{mol})(4\ \mathrm{J\cdot K^{-1}\cdot g^{-1}})(18\ \mathrm{g\cdot mol^{-1}})} = 60\ \mathrm{K}$$

23.16 The temperature dependence of the surface tension of a liquid is given by the *Sugden equation*

$$\gamma = \gamma_0(1 - T/T_c)^n \qquad (23.9)$$

where γ_0 is a constant for a given liquid, T/T_c is the reduced temperature, and n is a constant. Use the following data for benzene to determine n and γ_0:

T/(°C)	0.0	10.0	20.0	30.0	40.0	50.0	60.0	70.0	80.0
$\gamma/(10^{-3}\ \mathrm{N\cdot m^{-1}})$	31.58	30.22	28.88	27.56	26.26	24.98	23.72	22.48	21.26

The critical temperature for C_6H_6 is 288.9 °C.

▮ Equation *(23.9)* can be transformed into linear form by taking logarithms of both sides, giving

$$\log\gamma = \log\gamma_0 + n\log(1 - T/T_c)$$

Thus a plot of $\log\gamma$ against $\log(1 - T/T_c)$ will be linear with a slope equal to n and an intercept equal to $\log\gamma_0$. The plot is shown in Fig. 23-1. From the graph,

$$n = \text{slope} = 1.221$$

$$\log\gamma_0 = \text{intercept} = -1.1484$$

$$\gamma_0 = 71.06 \times 10^{-3}\ \mathrm{N\cdot m^{-1}}$$

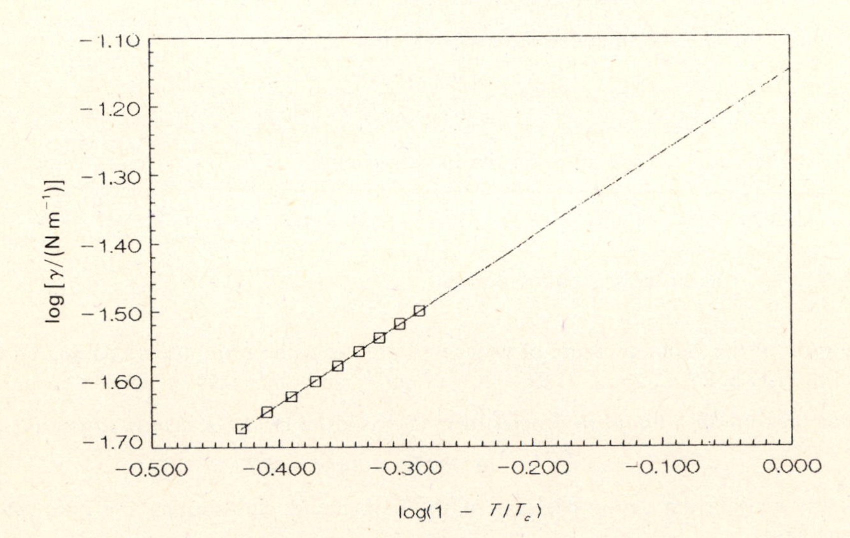

Fig. 23-1

23.17 Predict the value of γ for C_6H_6 at 25.0 °C. See Problem 23.16 for additional information.

▮ Substituting $T = 298.15\ \mathrm{K}$ into *(23.9)* gives

$$\gamma = (71.06 \times 10^{-3}\ \mathrm{N\cdot m^{-1}})\left(1 - \frac{298.15\ \mathrm{K}}{(288.9 + 273.15)\ \mathrm{K}}\right)^{1.221} = 28.23 \times 10^{-3}\ \mathrm{N\cdot m^{-1}}$$

23.18 Determine $d\gamma/dT$ for C_6H_6 at 25 °C. See Problem 23.16 for additional information.

▮ Taking the derivative of *(23.9)* with respect to T and substituting the data gives

$$\frac{d\gamma}{dT} = -\frac{\gamma_0}{T_c}\left(1 - \frac{T}{T_c}\right)^{n-1} = \frac{-(71.06 \times 10^{-3}\ \mathrm{N \cdot m^{-1}})}{(288.9 + 273.15)\ \mathrm{K}}\left(1 - \frac{298.15\ \mathrm{K}}{(288.9 + 273.15)\ \mathrm{K}}\right)^{1.221-1} \tag{23.10}$$

$$= -1.070 \times 10^{-4}\ \mathrm{N \cdot m^{-1} \cdot K^{-1}}$$

23.19 Determine the value of the Gibbs energy per unit area for C_6H_6 at 25.0 °C. What is ΔG for an increase of 0.10 m^2 in surface area? See Problem 23.17 for additional information.

▮ The Gibbs energy per unit area is given by

$$G_\gamma/A = \gamma \tag{23.11}$$

Substituting the data into *(23.11)* gives

$$\frac{G_\gamma}{A} = (28.23 \times 10^{-3}\ \mathrm{N \cdot m^{-1}})\left(\frac{1\ \mathrm{J}}{1\ \mathrm{N \cdot m}}\right) = 28.23 \times 10^{-3}\ \mathrm{J \cdot m^{-2}}$$

For the increase in area,

$$\Delta G_\gamma = (28.23 \times 10^{-3}\ \mathrm{J \cdot m^{-2}})(0.10\ \mathrm{m^2}) = 2.8 \times 10^{-3}\ \mathrm{J}$$

23.20 Determine the value of the entropy per unit area for C_6H_6 at 25.0 °C. Calculate ΔS for the process described in Problem 23.19. See Problem 23.18 for additional data.

▮ The entropy per unit area is given by

$$\frac{S_\gamma}{A} = -\frac{d\gamma}{dT} \tag{23.12}$$

Substituting the data into *(23.12)* gives

$$\frac{S_\gamma}{A} = -(-1.070 \times 10^{-4}\ \mathrm{N \cdot m^{-1} \cdot K^{-1}})\left(\frac{1\ \mathrm{J}}{1\ \mathrm{N \cdot m}}\right) = 1.070 \times 10^{-4}\ \mathrm{J \cdot K^{-1} \cdot m^{-2}}$$

For the increase in area,

$$\Delta S_\gamma = (1.070 \times 10^{-4}\ \mathrm{J \cdot K^{-1} \cdot m^{-2}})(0.10\ \mathrm{m^2}) = 1.1 \times 10^{-5}\ \mathrm{J \cdot K^{-1}}$$

23.21 Is the process described in Problem 23.19 exothermic or is it endothermic?

▮ Solving *(6.1a)* for H_γ/A and substituting *(23.11)* and *(23.12)* gives

$$\frac{U_\gamma}{A} \approx \frac{H_\gamma}{A} = \gamma - T\frac{d\gamma}{dT} \tag{23.13}$$

Substituting the data into *(23.13)* gives for the increase in area

$$\Delta H_\gamma = [(28.23 \times 10^{-3}\ \mathrm{N \cdot m^{-1}}) - (298.15\ \mathrm{K})(-1.070 \times 10^{-4}\ \mathrm{N \cdot m^{-1} \cdot K^{-1}})]\left(\frac{1\ \mathrm{J}}{1\ \mathrm{N \cdot m}}\right) = 6.0 \times 10^{-3}\ \mathrm{J}$$

Because $\Delta H_\gamma > 0$, the process is endothermic.

23.22 What is the ratio of the vapor pressure of water in a droplet with $r = 1.0\ \mu\mathrm{m}$ compared to that for bulk water at 25.0 °C? For H_2O at 25 °C, $\gamma = 71.97 \times 10^{-3}\ \mathrm{N \cdot m^{-1}}$ and $\rho = 0.997\ \mathrm{g \cdot cm^{-3}}$.

▮ The vapor pressure of a liquid in droplet form (P_s) is given by the *Kelvin equation* as

$$\ln(P_s/P) = 2\gamma M/r\rho RT \tag{23.14}$$

where P is the equilibrium vapor pressure of the bulk liquid. Substituting the data into *(23.14)* and taking antilogarithms gives

$$\ln\frac{P_s}{P} = \frac{2(71.97 \times 10^{-3}\ \mathrm{N \cdot m^{-1}})(18.015\ \mathrm{g \cdot mol^{-1}})[(1\ \mathrm{J})/(1\ \mathrm{N \cdot m})]}{(1.0 \times 10^{-6}\ \mathrm{m})(0.997\ \mathrm{g \cdot cm^{-3}})\left(\frac{10^2\ \mathrm{cm}}{1\ \mathrm{m}}\right)^3(8.314\ \mathrm{J \cdot K^{-1} \cdot mol^{-1}})(298.15\ \mathrm{K})} = 1.05 \times 10^{-3}$$

$$P_s/P = 1.001$$

23.23 What must the radius of a water droplet be if the vapor pressure is 2.7 times greater than that of the bulk liquid at 25.0 °C? See Problem 23.22 for additional information.

▮ Solving *(23.14)* for r and substituting the data gives

$$r = \frac{2(71.97 \times 10^{-3}\ \mathrm{N \cdot m^{-1}})(18.015\ \mathrm{g \cdot mol^{-1}})[(1\ \mathrm{J})/(1\ \mathrm{N \cdot m})]}{(\ln 2.7)(0.997\ \mathrm{g \cdot cm^{-3}})[(10^2\ \mathrm{cm})/(1\ \mathrm{m})]^3(8.314\ \mathrm{J \cdot K^{-1} \cdot mol^{-1}})(298.15\ \mathrm{K})} = 1.1 \times 10^{-9}\ \mathrm{m}$$

23.24 Pure water vapor must be compressed so that $P_s/P = 2.7$ before condensation will occur. Using the results of Problem 23.23, determine the number of water molecules in the droplet being formed.

▮ The approximate number of molecules will be given by

$$N = \tfrac{4}{3}\pi r^3 \rho L/M = \frac{\frac{4}{3}\pi(1.1 \times 10^{-9}\ \mathrm{m})^3(0.997\ \mathrm{g \cdot cm^{-3}})[(10^2\ \mathrm{cm})/(1\ \mathrm{m})]^3(6.022 \times 10^{23}\ \mathrm{mol^{-1}})}{18.015\ \mathrm{g \cdot mol^{-1}}} = 190\ \text{molecules}$$

23.25 Equation *(23.14)*, with an appropriate sign change, can be used to describe bubble formation within a boiling liquid. Calculate the pressure within a bubble of water vapor at 25.0 °C that contains about 100 molecules ($r \approx 9 \times 10^{-10}$ m). The vapor pressure of H_2O at 25.0 °C is 23.756 torr. See Problem 23.22 for additional information.

▮ The pressure in the bubble (P_b) will be given by *(23.14)* as

$$\ln \frac{P_b}{23.756\ \mathrm{torr}} = \frac{-2(71.97 \times 10^{-3}\ \mathrm{N \cdot m^{-1}})(18.015\ \mathrm{g \cdot mol^{-1}})[(1\ \mathrm{J})/(1\ \mathrm{N \cdot m})]}{(9 \times 10^{-10}\ \mathrm{m})(0.997\ \mathrm{g \cdot cm^{-3}})\left(\frac{10^2\ \mathrm{cm}}{1\ \mathrm{m}}\right)^3(8.314\ \mathrm{J \cdot K^{-1} \cdot mol^{-1}})(298.15\ \mathrm{K})} = -1.17$$

$$P_b = (23.756\ \mathrm{torr})(0.31) = 7.4\ \mathrm{torr}$$

23.26 Determine the ratio of the number of water molecules on the surface of the droplet to the total number of water molecules present in the droplet described in Problem 23.25. Assume that the radius of the water molecule is 1×10^{-10} m.

▮ The surface area of the droplet is $A = 4\pi r^2$, and the volume of the droplet is $V = \frac{4}{3}\pi r^3$, where r is the radius of the droplet. For one molecule, the area covered is $A' = \pi R^2$ and the volume is $\frac{4}{3}\pi R^3$, where R is the radius of the molecule. The fraction of surface molecules is

$$\frac{N(\text{surface})}{N} = \frac{4\pi r^2/\pi R^2}{\frac{4}{3}\pi r^3/\frac{4}{3}\pi R^3} = \frac{4R}{r} = \frac{4(1 \times 10^{-10}\ \mathrm{m})}{9 \times 10^{-10}\ \mathrm{m}} = 0.4 \qquad (23.15)$$

23.27 Repeat the calculations of Problem 23.26 for the droplets described in Problems 23.22 and 23.13.

▮ Substituting the data into *(23.15)* gives

$$\frac{N(\text{surface})}{N} = \frac{4(1 \times 10^{-10}\ \mathrm{m})}{1 \times 10^{-6}\ \mathrm{m}} = 4 \times 10^{-4}$$

$$\frac{N(\text{surface})}{N} = \frac{4(1 \times 10^{-10}\ \mathrm{m})}{1 \times 10^{-2}\ \mathrm{m}} = 4 \times 10^{-8}$$

$$\frac{N(\text{surface})}{N} = \frac{4(1 \times 10^{-10}\ \mathrm{m})}{1.6272 \times 10^{-2}\ \mathrm{m}} = 2 \times 10^{-8}$$

for the 1.0-μm, 1-cm, and molar droplets, respectively.

23.28 Calculate the pressure inside a soap bubble having a radius of 1.0 cm. Assume $\gamma = 40 \times 10^{-3}\ \mathrm{N \cdot m^{-1}}$ for the soap solution.

▮ Because the bubble has an outer and an inner surface, the factor of 2 in *(23.5)* is replaced by a factor of 4. Assuming that the $h\rho g$ term is zero, substituting into *(23.5)* gives

$$P = \frac{4(40 \times 10^{-3}\ \mathrm{N \cdot m^{-1}})[(1\ \mathrm{Pa})/(1\ \mathrm{N \cdot m^{-2}})]}{1.0 \times 10^{-2}\ \mathrm{m}} = 16\ \mathrm{Pa}$$

23.29 Repeat the calculations of Problem 23.28 for a soap bubble having a radius of 2.0 cm. If the two bubbles are connected by a drinking straw, describe what will happen.

▮ The pressure in the 2.0-cm bubble is given by the revised form of *(23.5)* as

$$P = \frac{4(40 \times 10^{-3}\ \mathrm{N \cdot m^{-1}})[(1\ \mathrm{Pa})/(1\ \mathrm{N \cdot m^{-2}})]}{2.0 \times 10^{-2}\ \mathrm{m}} = 8\ \mathrm{Pa}$$

Because the pressure inside the smaller bubble is greater, the smaller bubble will decrease in size as the larger bubble increases in size.

23.30 The *parachor*, $\{P\}$, of a substance is defined as

$$\{P\} = \frac{[M/(\mathrm{g \cdot mol^{-1}})][\gamma/(10^{-3}\ \mathrm{N \cdot m^{-1}})]^{1/4}}{\rho/(\mathrm{g \cdot cm^{-3}})} \qquad (23.16)$$

Using $\gamma = 31 \times 10^{-3}\ \mathrm{N \cdot m^{-1}}$, $\rho = 1.2351\ \mathrm{g \cdot cm^{-3}}$, and $M = 98.96\ \mathrm{g \cdot mol^{-1}}$, calculate $\{P\}$ for 1,2-dichloroethane.

▮ Substituting the data into *(23.16)* gives

$$\{P\} = (98.96)(31)^{1/4}/1.2351 = 189$$

23.31 The parachor of a substance is an additive property based on the molecular constituents and geometry of the substance. Using the values given in Table 23-1, predict $\{P\}$ for 1,2-dichloroethane.

Table 23-1

Contribution	C atom	H atom	O atom	Cl atom	N atom	Double bond	Triple bond	Six-membered ring
Parachor equivalent	4.8	17.1	20.0	53.8	12.5	23.2	46.6	6.1

▮ Using the values for the parachor equivalents given in the table, the predicted value is

$$\{P\} = (2\ \text{carbon atoms @ } 4.8) + (4\ \text{hydrogen atoms @ } 17.1) + (2\ \text{chlorine atoms @ } 53.8) = 185.6$$

which is slightly less than 2% lower than the observed value determined in Problem 23.30.

23.32 Predict the value of $\{P\}$ for 1,1-dichloroethane. Given $\gamma = 24.7 \times 10^{-3}\ \mathrm{N \cdot m^{-1}}$ and $\rho = 1.1757\ \mathrm{g \cdot cm^{-3}}$ for this substance, calculate $\{P\}$ and compare the results.

▮ The equivalents listed in Table 23-1 do not include contributions for various isomers; thus the predicted value is $\{P\} = 185.6$ (see Problem 23.31). Substituting the data into *(23.16)* gives the observed value as

$$\{P\} = \frac{(98.96)(24.7)^{1/4}}{1.1757} = 188$$

which is in excellent agreement with the predicted value.

23.33 Predict the value of $\{P\}$ for methanol. Given $\gamma = 22.61 \times 10^{-3}\ \mathrm{N \cdot m^{-1}}$ and $\rho = 0.7914\ \mathrm{g \cdot cm^{-3}}$ for this substance, calculate $\{P\}$ and compare the results.

▮ The sum of the equivalents from Table 23-1 is

$$\{P\} = (2\ \text{carbon atoms @ } 4.8) + (4\ \text{hydrogen atoms @ } 17.1) + (1\ \text{oxygen atom @ } 20.0) = 93.2$$

Substituting the data into *(23.16)* gives the observed value as

$$\{P\} = \frac{(32.04)(22.61)^{1/4}}{0.7914} = 88.3$$

The 5% discrepancy may be attributed to the presence of hydrogen bonding within the alcohol.

23.2 SURFACE TENSIONS IN BINARY SYSTEMS

23.34 The interfacial tension (γ_{AB}) for bromoform on water is $40.85 \times 10^{-3}\ \mathrm{N \cdot m^{-1}}$ and for chloroform on water is $32.80 \times 10^{-3}\ \mathrm{N \cdot m^{-1}}$ at 20 °C. Given $\gamma/(10^{-3}\ \mathrm{N \cdot m^{-1}}) = 41.53$, 27.13, and 72.75 for $CHBr_3$, $CHCl_3$, and H_2O, respectively, describe what will happen as a drop of each organic liquid is placed on the surface of water.

▌ The *speading coefficient* (S_{BA}) of B on A is defined as

$$S_{BA} = \gamma_A - \gamma_B - \gamma_{AB} \tag{23.17}$$

Substituting the data into *(23.17)* gives

$$S(CHBr_3, H_2O) = 72.75 \times 10^{-3}\ N\cdot m^{-1} - 41.53 \times 10^{-3}\ N\cdot m^{-1} - 40.85 \times 10^{-3}\ N\cdot m^{-1} = -9.63 \times 10^{-3}\ N\cdot m^{-1}$$

$$S(CHCl_3, H_2O) = 72.75 \times 10^{-3} - 27.13 \times 10^{-3} - 32.80 \times 10^{-3} = 12.82 \times 10^{-3}\ N\cdot m^{-1}$$

Because $S(CHBr_3, H_2O) < 0$, a droplet of $CHBr_3$ will form on the surface of the H_2O (a "nonwetting" situation), and because $S(CHCl_3, H_2O) > 0$, the $CHCl_3$ will spread out over the surface of the H_2O ("wetting" occurs).

23.35 The interfacial tension for diethyl ether on water is $10.70 \times 10^{-3}\ N\cdot m^{-1}$ at 20 °C. Given $\gamma/(10^{-3}\ N\cdot m^{-1}) = 17.10$ for diethyl ether and 72.75 for H_2O, describe what will happen as a drop of ether is placed on a water surface and as a drop of water is placed on an ether surface.

▌ Substituting the data into *(23.17)* gives

$$S(\text{ether}, H_2O) = 72.75 \times 10^{-3}\ N\cdot m^{-1} - 17.10 \times 10^{-3}\ N\cdot m^{-1} - 10.70 \times 10^{-3}\ N\cdot m^{-1} = 44.95 \times 10^{-3}\ N\cdot m^{-1}$$

$$S(H_2O, \text{ether}) = 17.10 \times 10^{-3} - 72.75 \times 10^{-3} - 10.70 \times 10^{-3} = -66.35 \times 10^{-3}\ N\cdot m^{-1}$$

Because $S(\text{ether}, H_2O) > 0$, the drop of ether will wet the surface of the H_2O, and because $S(H_2O, \text{ether}) < 0$, the water drop will not wet the surface of the ether.

23.36 Calculate the work of adhesion between diethyl ether and water. See Problem 23.35 for additional information.

▌ The work of adhesion is given by the *Dupré equation*

$$w_{AB} = \gamma_A + \gamma_B - \gamma_{AB} \tag{23.18}$$

Substituting the data into *(23.18)* gives

$$w(\text{ether}, H_2O) = (72.75 \times 10^{-3}\ N\cdot m^{-1} + 17.10 \times 10^{-3}\ N\cdot m^{-1} - 10.70 \times 10^{-3}\ N\cdot m^{-1})\left(\frac{1\ J}{1\ N\cdot m}\right)$$

$$= 79.15 \times 10^{-3}\ J\cdot m^{-2}$$

23.37 Calculate the work of cohesion for diethyl ether on a water surface. Compare the work of cohesion to the work of adhesion, and determine whether or not wetting will occur. See Problems 23.35 and 23.36 for additional information.

▌ The work of cohesion is given by

$$w_i = 2\gamma_i \tag{23.19}$$

Substituting the data into *(23.19)* gives

$$w(\text{ether}) = 2(17.10 \times 10^{-3}\ N\cdot m^{-1})[(1\ J)/(1\ N\cdot m)] = 34.20 \times 10^{-3}\ J\cdot m^{-2}$$

Because $w(\text{ether}, H_2O) > w(\text{ether})$, a drop of ether will wet the surface of the water.

23.38 Repeat Problem 23.37 for water on an ether surface.

▌ Substituting the data into *(23.19)* for H_2O gives

$$w(H_2O) = 2(72.75 \times 10^{-3}\ N\cdot m^{-1})[(1\ J)/(1\ N\cdot m)] = 145.50 \times 10^{-3}\ J\cdot m^{-2}$$

Because $w(\text{ether}, H_2O) < w(H_2O)$, a drop of water will not wet the surface of ether.

23.39 Calculate the surface excess concentration for a 1.00 M aqueous solution of NH_4NO_3 using the following data at 20.0 °C:

$C/(M)$	0.50	1.00	2.00	3.00	4.00
$\gamma/(10^{-3}\ N\cdot m^{-1})$	73.25	73.75	74.65	75.52	76.33

▌ The *surface excess concentration* (Γ_2) is defined as

$$\Gamma_2 = \frac{-C}{RT}\frac{d\gamma}{dC} \tag{23.20}$$

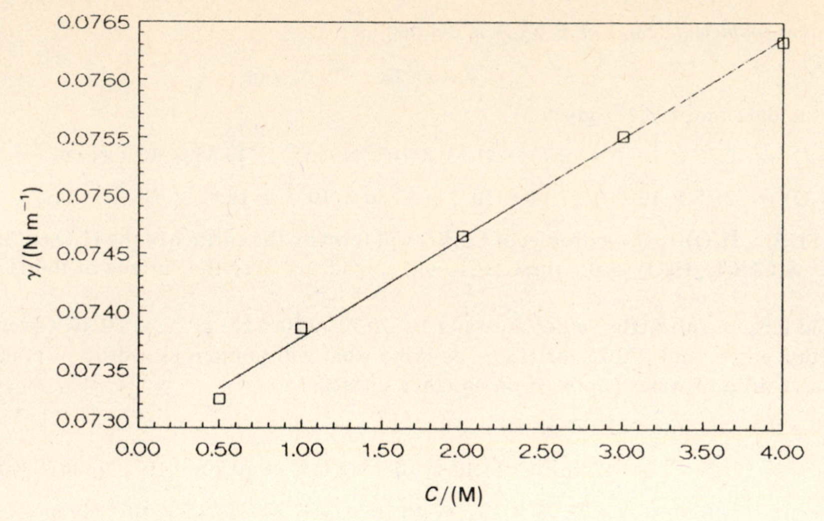

Fig. 23-2

From the plot of γ against C shown in Fig. 23-2, $d\gamma/dC = 8.7 \times 10^{-4}\ \mathrm{N \cdot m^{-1} \cdot M^{-1}}$ and is independent of C. Substituting into *(23.30)* gives

$$\Gamma(NH_4NO_3) = \frac{-(1.00\ \mathrm{M})(8.7 \times 10^{-4}\ \mathrm{N \cdot m^{-1} \cdot M^{-1}})}{(8.314\ \mathrm{J \cdot K^{-1} \cdot mol^{-1}})(293.2\ \mathrm{K})[(1\ \mathrm{N \cdot m})/(1\ \mathrm{J})]} = -3.6 \times 10^{-7}\ \mathrm{mol \cdot m^{-2}}$$

The negative sign indicates that electrostatic attraction between the ions tends to draw the ions together and away from the surface.

23.40 Determine the effective empty-layer thickness for the system described in Problem 23.39.

I The *effective thickness* is given by

$$x = |\Gamma_2/C| \tag{23.21}$$

and represents the distance from the surface over which the actual concentration is significantly different from the bulk concentration. Substituting the data into *(23.21)* gives

$$x = \left|\frac{-3.6 \times 10^{-7}\ \mathrm{mol \cdot m^{-2}}}{1.00\ \mathrm{mol \cdot L^{-1}}}\right|\left(\frac{10^{-3}\ \mathrm{m^3}}{1\ \mathrm{L}}\right) = 3.6 \times 10^{-10}\ \mathrm{m}$$

23.41 Calculate the surface excess concentration for a 0.250 M aqueous solution of 1-butanol using the following data at 20.0 °C:

C/(M)	0.105	0.211	0.433	0.854
$f(C_4H_9OH)$	0.930	0.916	0.887	0.832
$\gamma/(10^{-3}\ \mathrm{N \cdot m^{-1}})$	56.03	48.08	40.38	28.57

I For real solutions, *(23.20)* is often written as

$$\Gamma_2 = \frac{-1}{RT}\frac{d\gamma}{d(\ln a)} \tag{23.22}$$

where a is given by *(11.15)*. From the plot of γ against $\ln a$ shown in Fig. 23-3, $d\gamma/d(\ln a) = -0.014\ \mathrm{N \cdot m^{-1}}$ and is independent of C. Substituting into *(23.22)* gives

$$\Gamma_2 = \frac{-(-0.014\ \mathrm{N \cdot m^{-1}})[(1\ \mathrm{J})/(1\ \mathrm{N \cdot m})]}{(8.314\ \mathrm{J \cdot K^{-1} \cdot mol^{-1}})(293.2\ \mathrm{K})} = 5.7 \times 10^{-6}\ \mathrm{mol \cdot m^{-2}}$$

The positive sign indicates that the solute molecules tend to accumulate at the surface, where, in many cases, they form a unimolecular layer of adsorbed molecules.

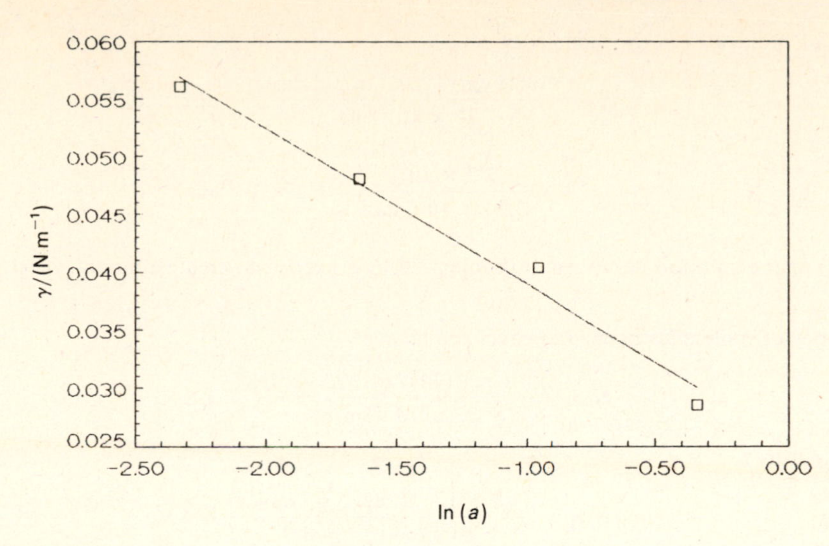

Fig. 23-3

23.42 Calculate the effective thickness of the 1-butanol layer in the system described in Problem 23.41.

▮ Substituting into *(23.21)* gives

$$x = \left|\frac{5.7 \times 10^{-6}\ \mathrm{mol \cdot m^{-2}}}{0.250\ \mathrm{mol \cdot L^{-1}}}\right|\left(\frac{10^{-3}\ \mathrm{m^3}}{1\ \mathrm{L}}\right) = 2.3 \times 10^{-8}\ \mathrm{m}$$

23.43 Determine the average surface area of a molecule of 1-butanol. See Problem 23.41 for additional information.

▮ Assuming closest packing of the molecules,

$$A = \frac{1}{(5.7 \times 10^{-6}\ \mathrm{mol \cdot m^{-2}})(6.022 \times 10^{23}\ \mathrm{mol^{-1}})} = 2.9 \times 10^{-19}\ \mathrm{m^2}$$

23.44 Calculate the approximate length of a 1-butanol molecule. The density of the alcohol is $0.8098\ \mathrm{g \cdot cm^{-3}}$. See Problem 23.43 for additional information.

▮ The volume occupied by 1 molecule is

$$V = \frac{(74.12\ \mathrm{g \cdot mol^{-1}})[(10^{-2}\ \mathrm{m})/(1\ \mathrm{cm})]^3}{(0.8098\ \mathrm{g \cdot cm^{-3}})(6.022 \times 10^{23}\ \mathrm{mol^{-1}})} = 1.520 \times 10^{-28}\ \mathrm{m^3}$$

which corresponds to

$$l = \frac{1.520 \times 10^{-28}\ \mathrm{m^3}}{2.9 \times 10^{-19}\ \mathrm{m^2}} = 5.2 \times 10^{-10}\ \mathrm{m}$$

23.45 Compare the answer to Problem 23.44 to that predicted using average bond lengths of 0.14 nm for C—O, 0.15 nm for C—C, 0.10 nm for C—H, and 0.10 nm for O—H.

▮ Assuming all of the bond angles to be approximately equal to 109°, the length of the molecule is predicted to be

$$\begin{aligned} l &= [l(\mathrm{H{-}O}) + l(\mathrm{O{-}C}) + 3l(\mathrm{C{-}C}) + l(\mathrm{C{-}H})] \sin(109°/2) \\ &= [0.10\ \mathrm{nm} + 0.14\ \mathrm{nm} + 3(0.15\ \mathrm{nm}) + 0.10\ \mathrm{nm}] \sin(109°/2) \\ &= 0.64\ \mathrm{nm} \end{aligned}$$

which is in good agreement with the answer to Problem 23.44.

23.46 A 1.00-mL sample of a dilute solution of oleic acid dissolved in benzene was placed on the surface of water, and a monolayer of the acid formed after the benzene evaporated. Given that the radius of the monolayer is 38.6 cm, the concentration of the solution is $1.00\ \mathrm{g \cdot L^{-1}}$, and the cross-sectional area of a molecule is $22 \times 10^{-20}\ \mathrm{m^2}$, determine the value of Avogadro's constant.

▮ The amount of oleic acid in the monolayer is

$$n = \frac{(1.00\ \mathrm{g \cdot L^{-1}})(1.00\ \mathrm{mL})[(10^{-3}\ \mathrm{L})/(1\ \mathrm{mL})]}{282.47\ \mathrm{g \cdot mol^{-1}}} = 3.54 \times 10^{-6}\ \mathrm{mol}$$

The number of molecules in the monolayer is

$$N = \frac{\pi(38.6\ \text{cm})^2[(10^{-2}\ \text{m})/(1\ \text{cm})]^2}{22 \times 10^{-20}\ \text{m}^2} = 2.2 \times 10^{18}$$

giving

$$L = \frac{2.2 \times 10^{18}}{3.54 \times 10^{-6}\ \text{mol}} = 6.2 \times 10^{23}\ \text{mol}^{-1}$$

23.47 What volume of the solution described in Problem 23.46 is needed to create a monolayer over a lake with an area of 1.0 acre?

▌ The number of molecules needed to cover the lake is

$$N = \frac{(1.0\ \text{acre})[(4047\ \text{m}^2)/(1\ \text{acre})]}{22 \times 10^{-20}\ \text{m}^2} = 1.8 \times 10^{22}$$

which corresponds to

$$m = \frac{(1.8 \times 10^{22})(282.47\ \text{g} \cdot \text{mol}^{-1})}{6.022 \times 10^{23}\ \text{mol}^{-1}} = 8.4\ \text{g}$$

The volume of the solution needed is

$$V = (8.4\ \text{g})/(1.00\ \text{g} \cdot \text{L}^{-1}) = 8.4\ \text{L}$$

23.3 ADSORPTION

23.48 The amount of acetic acid adsorbed by 1 g of activated charcoal at various concentrations of acetic acid in water are

$C/(\text{M})$	0.015 36	0.015 24	0.014 52	0.011 80	0.008 40	0.005 72
$n/(10^{-3}\ \text{mol} \cdot \text{g}^{-1})$	1.551	1.504	1.447	1.180	0.834	0.565

Determine the constants for the Freundlich isotherm.

▌ For this system, the *Freundlich isotherm* is

$$n = kC^a \qquad (23.23a)$$

where k and a are constants. Taking logarithms of both sides gives $\ln n = \ln k + a \ln C$. Thus a plot of $\ln n$ against $\ln C$ will be linear, with the intercept equal to $\ln k$ and the slope equal to a. The plot is shown in Fig. 23-4. From the graph,

$$a = \text{slope} = 1.010$$

$$k = e^{\text{intercept}} = e^{-2.264} = 0.1039\ \text{mol} \cdot \text{g}^{-1} \cdot \text{M}^{-1.010}$$

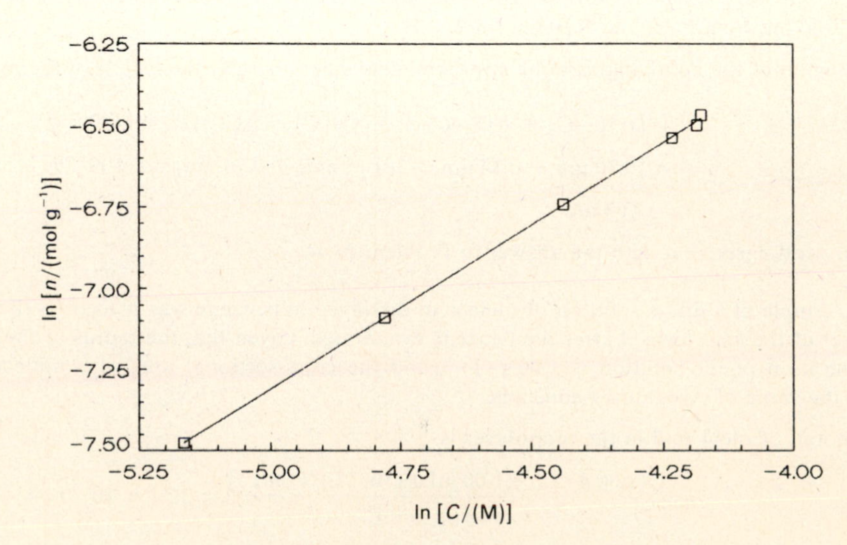

Fig. 23-4

23.49 The volume of CH_4 (corrected to STP) adsorbed per gram of charcoal at 240 K at various pressures of CH_4 is

$P/(\text{torr})$	36	53	76	102	132	171	215	270
$V/(\text{cm}^3 \cdot \text{g}^{-1})$	12.64	16.02	19.87	23.21	26.48	29.87	33.07	36.22

Which isotherm, the Freundlich or Langmuir isotherm, describes the system better?

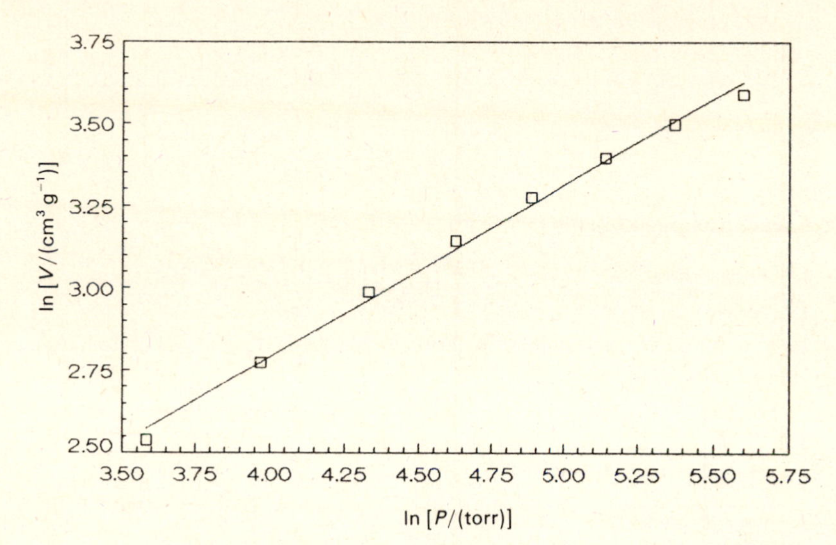

Fig. 23-5

▮ A plot of ln V against ln P is shown in Fig. 23-5. The linearity of the plot indicates that the Freundlich isotherm given by

$$V = kP^a \tag{23.23b}$$

is valid for this system, where

$$a = \text{slope} = 0.52$$

$$k = e^{\text{intercept}} = e^{0.70} = 2.01\ \text{cm}^3 \cdot \text{g}^{-1} \cdot \text{torr}^{-0.52}$$

The *Langmuir isotherm* is

$$\theta = bP/(1 + bP) \tag{23.24}$$

where θ is the fraction of the solid surface covered by adsorbed molecules and b is a constant. Assuming

$$\theta = V/V_m \tag{23.25}$$

where V_m is the volume of gas per gram of solid required to form a monolayer, *(23.24)* can be written in linear form as

$$P/V = P/V_m + 1/bV_m \tag{23.26a}$$

Thus a plot of P/V against P will be linear, with the slope equal to $1/V_m$ and the intercept equal to $1/bV_m$. The linearity of the plot shown in Fig. 23-6 indicates that the Langmuir isotherm is valid for this system. From the graph,

$$V_m = \frac{1}{\text{slope}} = \frac{1}{0.0195\ \text{cm}^{-3} \cdot \text{g}} = 51.3\ \text{cm}^3 \cdot \text{g}^{-1}$$

$$b = \frac{1}{V_m(\text{intercept})} = \frac{1}{(51.3\ \text{cm}^3 \cdot \text{g}^{-1})(2.3\ \text{torr} \cdot \text{cm}^{-3} \cdot \text{g})} = 8.5 \times 10^{-3}\ \text{torr}^{-1}$$

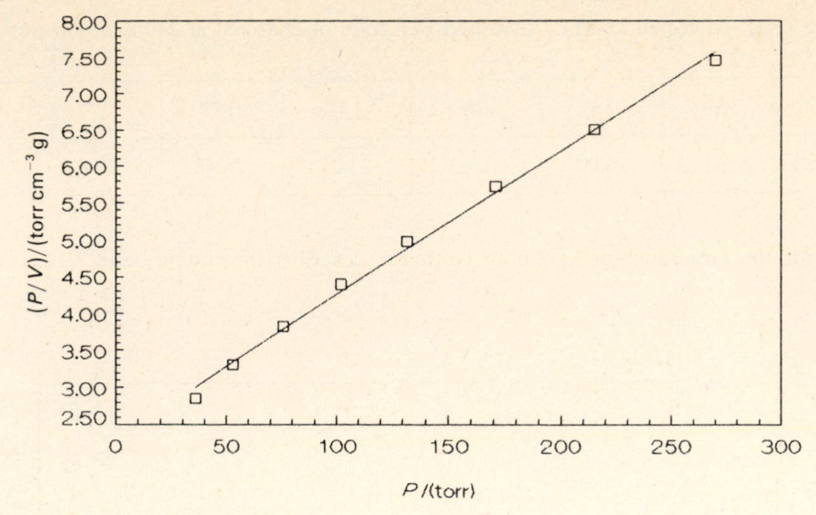

Fig. 23-6

23.50 An alternative linear form of *(23.24)* is

$$\frac{1}{V} = \frac{1}{V_m bP} + \frac{1}{V_m} \tag{23.26b}$$

Confirm the values of b and V_m determined in Problem 23.49 for the CH_4-charcoal system.

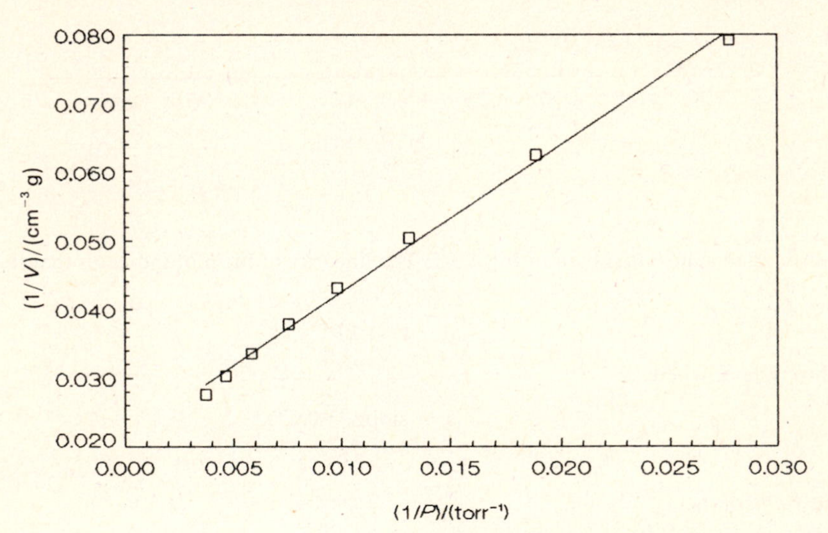

Fig. 23-7

A plot of $1/V$ against $1/P$ is shown in Fig. 23-7. From the graph,

$$V_m = \frac{1}{\text{intercept}} = \frac{1}{0.021\ \text{cm}^{-3}\cdot\text{g}} = 48\ \text{cm}^3\cdot\text{g}^{-1}$$

$$b = \frac{1}{V_m(\text{slope})} = \frac{1}{(48\ \text{cm}^3\cdot\text{g}^{-1})(2.14\ \text{cm}^{-3}\cdot\text{g}\cdot\text{torr})} = 9.7\times 10^{-3}\ \text{torr}^{-1}$$

The values of V_m differ by 7%, and the values of b differ by 13%.

23.51 What fraction of the charcoal surface described in Problem 23.49 is covered by CH_4 molecules at $P = 150$ torr?

Substituting the data into *(23.24)* gives

$$\theta = \frac{(8.5\times 10^{-3}\ \text{torr}^{-1})(150\ \text{torr})}{1 + (8.5\times 10^{-3}\ \text{torr}^{-1})(150\ \text{torr})} = 0.56$$

23.52 The density of liquid CH_4 is $0.466\times 10^3\ \text{kg}\cdot\text{m}^{-3}$. Determine the approximate cross-sectional area of a CH_4 molecule.

▌ The approximate area for a molecule is given by

$$A = (M/\rho L)^{2/3} \tag{23.27}$$

Substituting the data into *(23.37)* gives

$$A = \left(\frac{16.0 \times 10^{-3}\ \mathrm{kg \cdot mol^{-1}}}{(0.466 \times 10^{3}\ \mathrm{kg \cdot m^{-3}})(6.022 \times 10^{23}\ \mathrm{mol^{-1}})}\right)^{2/3} = 14.8 \times 10^{-20}\ \mathrm{m^2}$$

23.53 Use the results of Problem 23.52 to determine the surface area of the charcoal in the system described in Problem 23.49.

▌ The amount of CH_4 on the surface of the charcoal is given by *(1.18)* as

$$n = \frac{(1.013\,25\ \mathrm{bar})(51.3\ \mathrm{cm^3 \cdot g^{-1}})[(10^{-2}\ \mathrm{m})/(1\ \mathrm{cm})]^3}{(8.314 \times 10^{-5}\ \mathrm{bar \cdot m^3 \cdot K^{-1} \cdot mol^{-1}})(293\ \mathrm{K})} = 2.13 \times 10^{-3}\ \mathrm{mol \cdot g^{-1}}$$

which corresponds to

$$N = (2.13 \times 10^{-3}\ \mathrm{mol \cdot g^{-1}})(6.022 \times 10^{23}\ \mathrm{mol^{-1}}) = 1.28 \times 10^{21}\ \mathrm{g^{-1}}$$

The area of adsorption is

$$A = (1.28 \times 10^{21}\ \mathrm{g^{-1}})(14.8 \times 10^{-20}\ \mathrm{m^2}) = 189\ \mathrm{m^2 \cdot g^{-1}}$$

23.54 The *Tempkin isotherm* for the adsorption of gases at low pressures on solids is

$$V = r \ln sP = r \ln s + r \ln P \tag{23.28}$$

Using the data given in Problem 23.49, determine r and s for the CH_4-charcoal system.

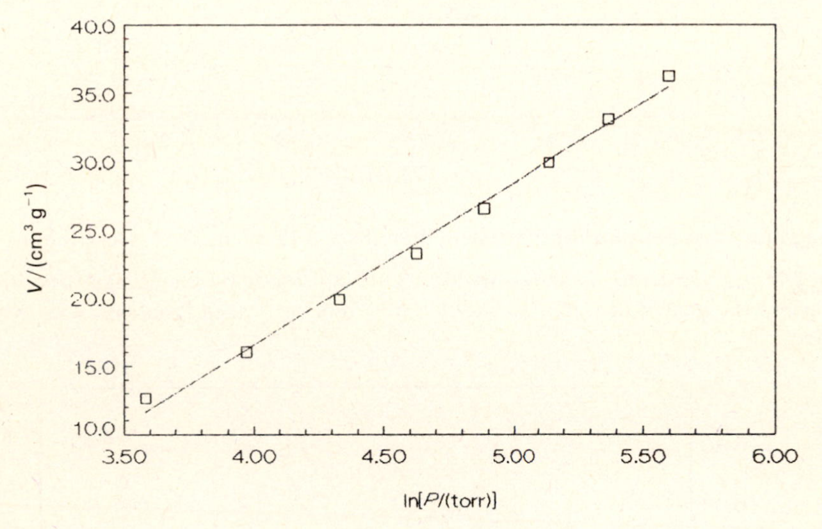

Fig. 23-8

▌ A plot of V against $\ln P$ is shown in Fig. 23-8. From the plot,

$$r = \text{slope} = 11.8\ \mathrm{cm^3 \cdot g^{-1}}$$

$$s = e^{\text{intercept}/r} = e^{-30.8/11.8} = 7.35 \times 10^{-2}\ \mathrm{torr^{-1}}$$

23.55 The following data were obtained for the adsorption of N_2 on Al_2O_3 at 77.3 K:

***P*/(torr)**	31.7	40.1	56.6	64.5	82.7	96.7	112.4	128.8	148.6	169.3
***n*/(10^{-4} mol · g^{-1})**	8.31	8.53	8.90	9.03	9.53	9.85	10.15	10.45	10.81	11.18

If the vapor pressure ($P°$) of liquid N_2 is 759.0 torr at this temperature, show that the data can be described by the *Brunauer-Emmett-Teller* (BET) *isotherm*:

$$V = \frac{V_m cP}{(P° - P)[1 + (c - 1)(P/P°)]} \tag{23.29}$$

where c is a constant.

▌ Equation *(23.29)* can be transformed into the linear form

$$\frac{P}{n(P^\circ - P)} = \frac{(c-1)}{cn_m}\left(\frac{P}{P^\circ}\right) + \frac{1}{cn_m}$$

where n is the amount of N_2 adsorbed. A plot of $P/n(P^\circ - P)$ against P/P° is shown in Fig. 23-9. From the graph,

$$cn_m = \frac{1}{\text{intercept}} = \frac{1}{6.4\ \text{mol}^{-1}\cdot\text{g}} = 0.16\ \text{mol}\cdot\text{g}^{-1}$$

$$c = 1 + (\text{slope})cn_m = 1 + \frac{\text{slope}}{\text{intercept}} = 1 + \frac{1118\ \text{mol}^{-1}\cdot\text{g}}{614\ \text{mol}^{-1}\cdot\text{g}} = 176$$

$$n_m = \frac{cn_m}{c} = \frac{0.16\ \text{mol}\cdot\text{g}^{-1}}{176} = 9.1\times10^{-4}\ \text{mol}\cdot\text{g}^{-1}$$

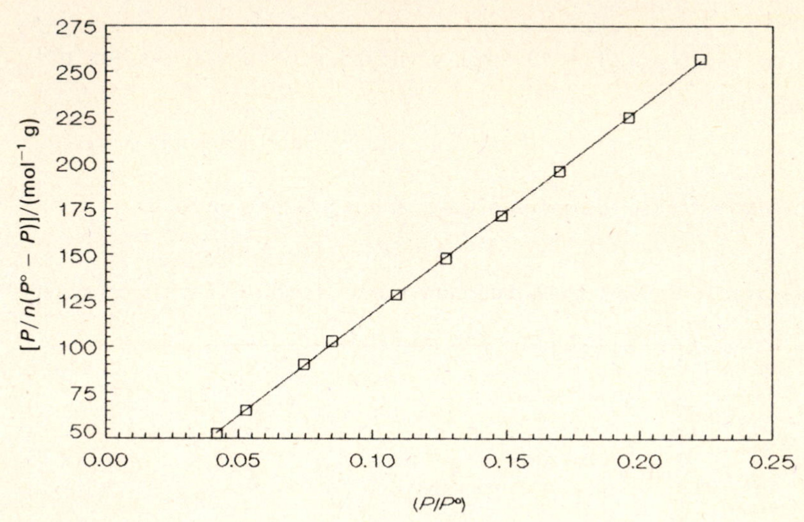

Fig. 23-9

23.56 Prepare a Langmuir isotherm plot for the system described in Problem 23.55. Describe the plot.

▌ The plot of $\ln n$ against $\ln P$ is shown in Fig. 23-10. As shown by the straight line, the Langmuir isotherm correctly describes the system for $n < n_m = 9.1\times10^{-4}\ \text{mol}\cdot\text{g}^{-1}$ (see Problem 23.55) where only a monolayer is being formed.

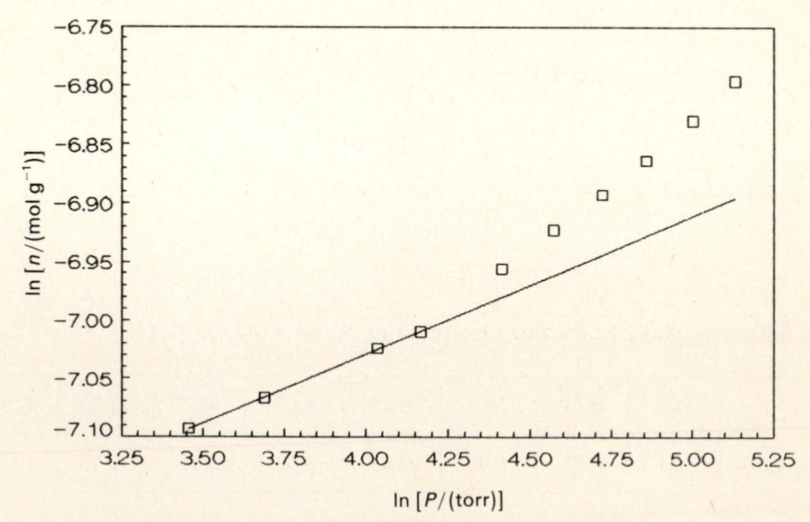

Fig. 23-10

23.57 Using the results of Problem 23.55, determine the enthalpy of adsorption given $\Delta_{vap}H = 5.6\ \text{kJ}\cdot\text{mol}^{-1}$ for N_2. Does chemisorption or physisorption occur in this system?

▌ The constant c in *(23.29)* is related to $\Delta_{ads}H(\text{mono})$ by

$$c = \exp\{[\Delta_{ads}H(\text{mono}) - \Delta_{vap}H]/RT\} \qquad (23.30)$$

Solving *(23.30)* for $\Delta_{ads}H°$(mono) and substituting the data gives

$$\Delta_{ads}H(\text{mono}) = (8.314\ \text{J}\cdot\text{K}^{-1}\cdot\text{mol}^{-1})\left(\frac{10^{-3}\ \text{kJ}}{1\ \text{J}}\right)(77.3\ \text{K})\ln 176 + 5.6\ \text{kJ}\cdot\text{mol}^{-1} = 8.9\ \text{kJ}\cdot\text{mol}^{-1}$$

Values of $\Delta_{ads}H(\text{mono}) < 20\ \text{kJ}\cdot\text{mol}^{-1}$ usually indicate physisorption of the gas.

23.58 The pressures of CH_4 required to adsorb 12.5 cm^3 of CH_4 on charcoal at different temperatures are given below:

***P*/(torr)**	33	66	145	294	582
***T*/(K)**	240	255	273	293	319

Determine the value of $\Delta_{ads}H$.

▌ A plot of ln P against $1/T$ as suggested by *(4.19a)* is shown in Fig. 23-11. From the graph,

$$\Delta_{ads}H = -R(\text{slope}) = -(8.314\ \text{J}\cdot\text{K}^{-1}\cdot\text{mol}^{-1})(-2.81\times 10^3\ \text{K})[(10^{-3}\ \text{kJ})/(1\ \text{J})] = 23.4\ \text{kJ}\cdot\text{mol}^{-1}$$

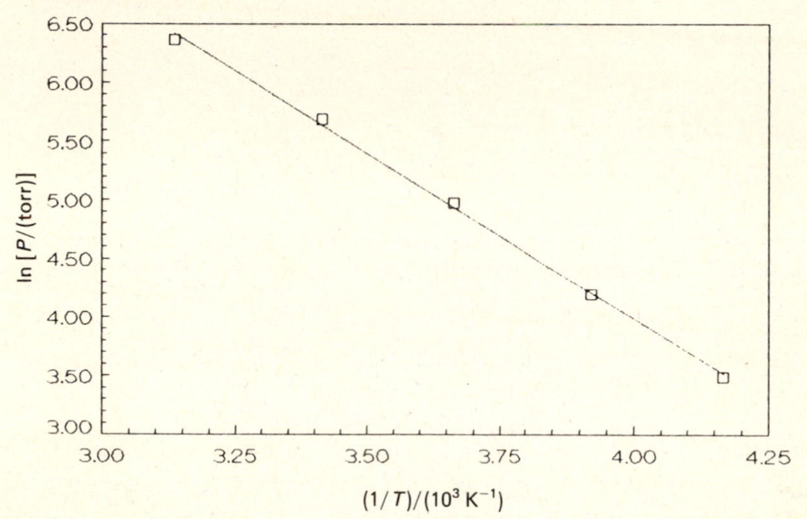

Fig. 23-11

23.59 What is the approximate mean lifetime of a CH_4 molecule on the surface of the system described in Problem 23.58 at $T = 298$ K?

▌ The mean lifetime (τ) is given approximately by

$$\tau \approx \tau_0\, e^{\Delta_{ads}H/RT} \tag{23.31}$$

where $\tau_0 \approx 10^{-12}$ s (the period of a typical vibrational frequency of a weak molecule–surface bond). Substituting the data into *(23.31)* gives

$$\tau \approx (10^{-12}\ \text{s})\exp\left[\frac{23.4\times 10^3\ \text{J}\cdot\text{mol}^{-1}}{(8.314\ \text{J}\cdot\text{K}^{-1}\cdot\text{mol}^{-1})(298\ \text{K})}\right] = 1\times 10^{-8}\ \text{s}$$

23.60 Write the general rate equation describing the decomposition reaction $A(g) \rightarrow$ products in the presence of a heterogeneous catalyst. Derive the integrated rate equations for the special cases in which A is moderately adsorbed, A is slightly adsorbed, and A is strongly adsorbed. Assume that the product gases are not adsorbed.

▌ The general form of *(23.24)* in the presence of a mixture of gases is

$$\theta_i = \frac{b_i P_i}{1 + b_A P_A + b_B P_B + \cdots} \tag{23.32}$$

where, in this problem, $b_B P_B = b_C P_C = \cdots = 0$. The general rate law describing the decomposition reaction is

$$\frac{d\xi}{dt} = k'\theta_A \tag{23.33}$$

where k' is a constant.

For the system in which A is moderately adsorbed, substituting *(23.32)* into *(23.33)* gives

$$-\frac{dP_A}{dt} = k'\frac{b_A P_A}{1 + b_A P_A} = \frac{kP_A}{1 + b_A P_A} \tag{23.34}$$

The integrated form of *(23.34)* is

$$\ln\frac{P_{A,0}}{P_A} + b_A(P_{A,0} - P_A) = kt \tag{23.35}$$

For the system in which A is slightly adsorbed, $b_A P_A \ll 1$ and *(23.34)* becomes a pseudo-first-order reaction described by

$$\frac{-dP_A}{dt} = \frac{kP_A}{1 + b_A P_A} = kP_A$$

Integration gives *(12.5)*, which can be written as

$$\ln P_A = \ln P_{A,0} - kt \tag{23.36}$$

For the system in which A is strongly adsorbed, $b_A P_A \gg 1$, and *(23.34)* becomes a pseudo-zero-order reaction described by

$$\frac{-dP_A}{dt} = \frac{kP_A}{b_A P_A} = k'$$

Integration gives *(12.4)*, which can be written as

$$P_A = P_{A,0} - k't \tag{23.37}$$

23.61 Stibine (SbH_3) is moderately adsorbed on the surface of Sb during decomposition at $T = 298$ K. Assuming that the adsorption of H_2 on Sb is negligible during the initial stages of the reaction, use the following data to determine k and b in *(23.35)*:

t/(s)	0	100	200	300	400	500	600
P(SbH₃)/(bar)	1.013	0.919	0.832	0.741	0.660	0.584	0.509

▌ The format of *(23.35)* does not lend itself to graphical analysis. However, by substituting each set of data into *(23.35)*, a series of six equations can be written from which the two unknowns can be determined. The average values are

$$k = 2.67 \times 10^{-3}\ \text{s}^{-1} \qquad \text{and} \qquad b = 1.824\ \text{bar}^{-1}$$

23.62 Because of the difficulties encountered using *(23.35)*, *(23.34)* is often written in terms of *(23.23)* as

$$\frac{-dP_A}{dt} = kP_A^n \tag{23.38}$$

Determine k and n for the reaction $C_3H_6(g) \rightarrow CH_3CHCH_2(g)$ using the following initial rate data:

$P(C_3H_6)_0$/(torr)	6.8	18.6	24.1	59.7	65.8	80.7
$-[dP(C_3H_6)/dt]_0$	1.00	1.04	1.20	2.09	2.21	2.26

▌ Taking logarithms of both sides of *(23.38)* gives

$$\log\{-[dP(C_3H_6)/dt]_0\} = \log k + n\log[P(C_3H_6)_0]$$

A plot of $\log\{-[dP(C_3H_6)/dt]_0\}$ against $\log[P(C_3H_6)_0]$ is given in Fig. 23-12. From the graph,

$$n = \text{slope} = 0.39 \qquad \text{and} \qquad k = 10^{\text{intercept}} = 10^{-0.39} = 0.41$$

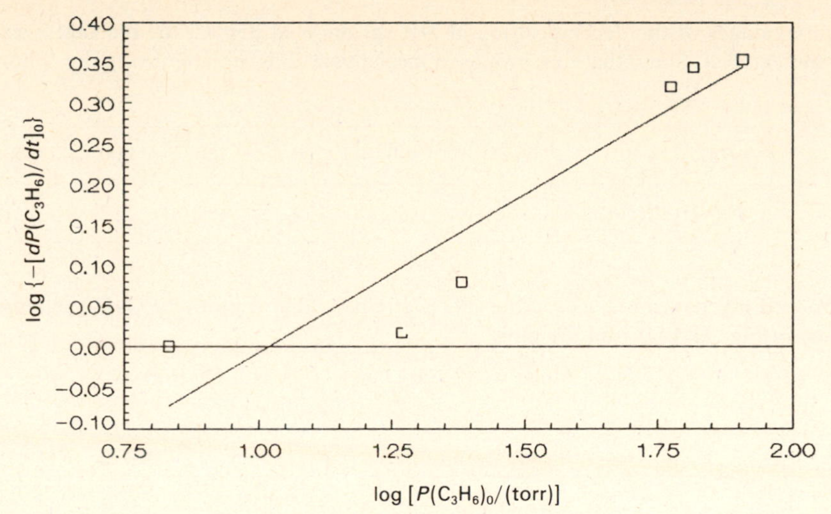

Fig. 23-12

23.63 Assuming that *(23.38)* is used to describe the decomposition of stibine on Sb, determine k and n using the data given in Problem 23.61.

▌ A plot of $\alpha = P(\mathrm{SbH_3})/P(\mathrm{SbH_3})_0$ against $\log t$ was prepared as described in Problem 12.40, and from Fig. 12-14, $n \approx 0.5$. Solving *(12.4)* for k and substituting one set of data gives

$$k = \frac{2(1.013^{0.5} - 0.919^{0.5})}{100} = 9.57 \times 10^{-4}\,\mathrm{bar}^{1/2}\cdot\mathrm{s}^{-1}$$

The average value of k is $9.65 \times 10^{-4}\,\mathrm{bar}^{1/2}\cdot\mathrm{s}^{-1}$.

23.64 The decomposition of $N_2O(g)$ on Au at 900 °C is expected to occur only with slight adsorption of the gas. Show that this proposed mechanism is acceptable using the following data:

$t/(\mathrm{s})$	0	1 000	2 000	3 000	4 000
$P(N_2O)/(\mathrm{Pa})$	26 700	21 500	17 500	14 100	11 300

▌ If the proposed mechanism is acceptable, *(23.36)* implies that a plot of $\ln[P(N_2O)]$ against t will be linear. The graph is shown in Fig. 23-13. From the plot,

$$k = -\text{slope} = -(-2.14 \times 10^{-4}\,\mathrm{s}^{-1}) = 2.14 \times 10^{-4}\,\mathrm{s}^{-1}$$

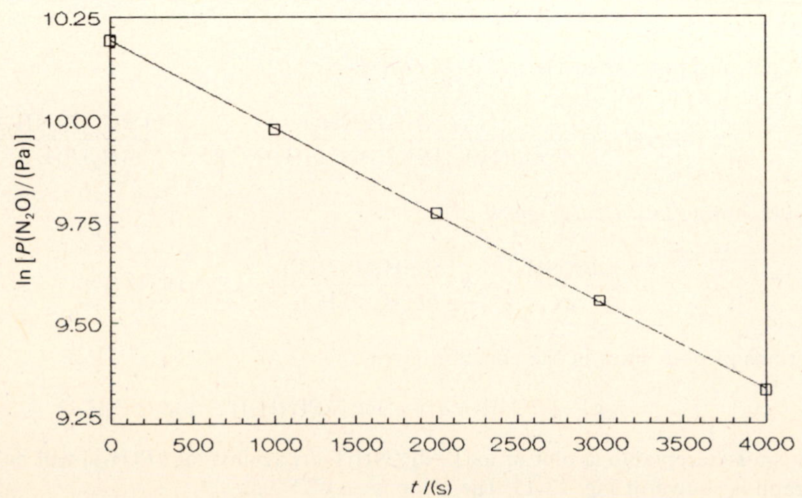

Fig. 23-13

23.65 During the initial stages of the decomposition of $NH_3(g)$ on W at 856 °C, the reactant is expected to be strongly adsorbed on the catalyst. Show that this proposed mechanism is acceptable using the following data:

$t/(s)$	0	100	200	300	400	500
$P(NH_3)/(torr)$	200	186	173	162	152	141

▮ If the proposed mechanism is acceptable, *(23.37)* implies that a plot of $P(NH_3)$ against t will be linear. The graph is shown in Fig. 23-14. From the plot,

$$k = -\text{slope} = -(-0.117 \text{ torr} \cdot \text{s}^{-1}) = 0.117 \text{ torr} \cdot \text{s}^{-1}$$

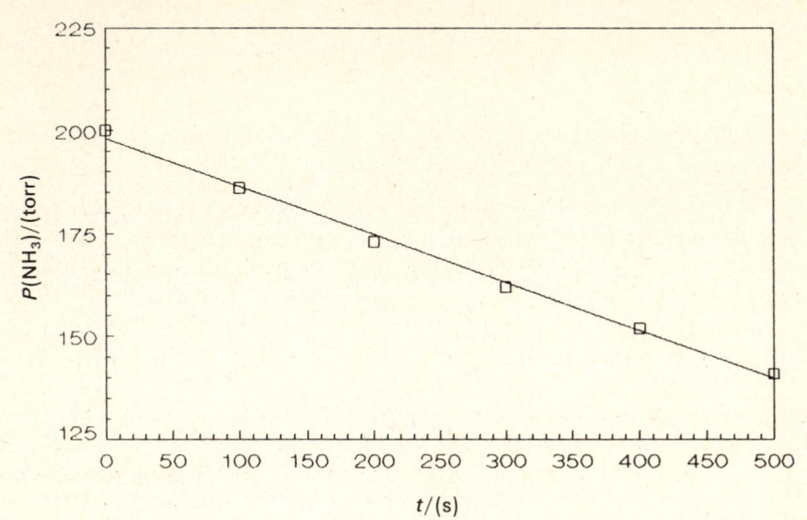

Fig. 23-14

23.66 During the final stages of the decomposition of $NH_3(g)$ on Pt at 1138 °C, the $H_2(g)$ formed is expected to be very strongly adsorbed on the catalyst. Show that this proposed mechanism is acceptable using the following data:

$-[dP(NH_3)/dt]/(\text{torr} \cdot \text{s}^{-1})$	0.28	0.23	0.13	0.083
$P(H_2)/(torr)$	50	75	100	150

▮ For the very strongly adsorbed H_2, *(23.32)* becomes

$$\theta(NH_3) = \frac{b(NH_3)P(NH_3)}{1 + b(NH_3)P(NH_3) + b(H_2)P(H_2)} \approx \frac{b(NH_3)P(NH_3)}{b(H_2)P(H_2)}$$

which upon substituting into *(23.33)* gives

$$\frac{-dP(NH_3)}{dt} = k' \frac{b(NH_3)P(NH_3)}{b(H_2)P(H_2)} = kP(NH_3)P(H_2)^{-1}$$

Taking logarithms of both sides of the equation gives

$$\log[-dP(NH_3)/dt] = \log[kP(NH_3)] - 1 \log[P(H_2)]$$

If the mechanism is acceptable, a plot of $\log[-dP(NH_3)/dt]$ against $\log[P(H_2)]$ will be linear and have a slope of -1. The graph is shown in Fig. 23-15. The slope is -1.2 ± 0.2.

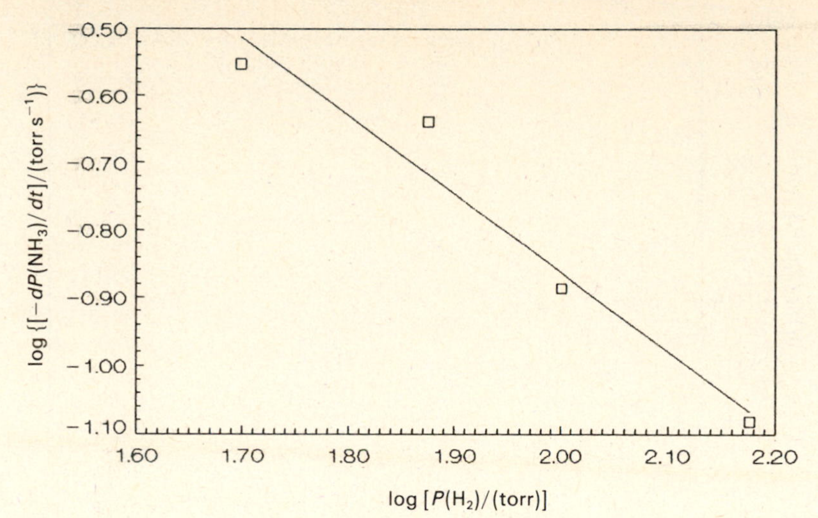

Fig. 23-15

23.67 During the decomposition of $N_2O(g)$ on Pt at 741 °C, the $O_2(g)$ formed is expected to be moderately adsorbed on the catalyst. Assuming only a weak adsorption for the $N_2O(g)$, write the rate equation for the decomposition. Show that the following data support the proposed mechanism:

$t/(s)$	0	315	750	1400	2250	3450
$P(N_2O)/(torr)$	95	85	75	65	55	45

▮ For the proposed mechanism, assuming $b(N_2O)P(N_2O) \approx 0$, *(23.32)* becomes

$$\theta(O_2) = \frac{b(O_2)P(O_2)}{1 + b(O_2)P(O_2)}$$

Equation *(23.33)* becomes

$$\frac{-dP(N_2O)}{dt} = k'[1 - \theta(O_2)]P(N_2O) = k'\left(1 - \frac{b(O_2)P(O_2)}{1 + b(O_2)P(O_2)}\right)P(N_2O) = \frac{k'P(N_2O)}{1 + b(O_2)P(O_2)}$$

Letting x represent the decrease in the pressure of N_2O,

$$P(N_2O) = P(N_2O)_0 - x \qquad P(O_2) = x/2$$

and the rate equation becomes

$$\frac{-dP(N_2O)}{dt} = \frac{-d[P(N_2O)_0 - x]}{dt} = \frac{dx}{dt} = \frac{k'[P(N_2O)_0 - x]}{1 + b(O_2)[x/2]} = \frac{k'[P(N_2O)_0 - x]}{1 + b'x}$$

The integrated rate equation is

$$[1 + b'P(N_2O)_0] \ln\left(\frac{P(N_2O)_0}{P(N_2O)}\right) = k't + b'[P(N_2O)_0 - P(N_2O)]$$

which can be written in linear form as

$$\frac{P(N_2O)_0 - P(N_2O)}{t} = \frac{1 + b'P(N_2O)_0}{b'}\left(\frac{\ln [P(N_2O)_0/P(N_2O)]}{t}\right) - \frac{k'}{b'}$$

If this mechanism is acceptable, a plot of $[P(N_2O)_0 - P(N_2O)]/t$ against $(1/t)\ln[P(N_2O)_0/P(N_2O)]$ will be linear. The graph is shown in Fig. 23-16. From the plot,

$$b' = \frac{1}{\text{slope} - P(N_2O)_0} = \frac{1}{126\text{ torr} - 95\text{ torr}} = 0.032\text{ torr}^{-1}$$

$$k' = -b'(\text{intercept}) = -(0.032\text{ torr}^{-1})(-0.0127\text{ torr}\cdot\text{s}^{-1}) = 4.1\times 10^{-4}\text{ s}^{-1}$$

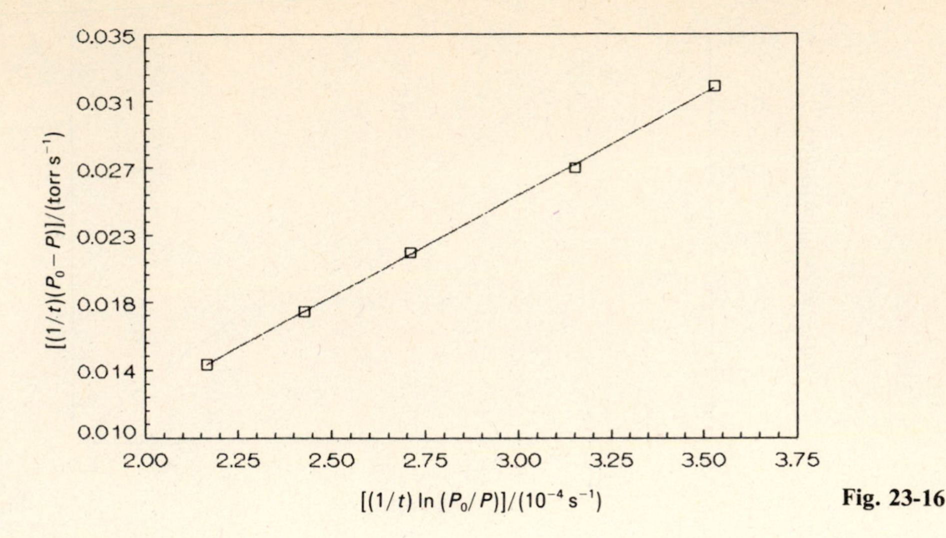

Fig. 23-16

23.68 For the reaction $C_3H_6(g) + H_2(g) \rightarrow C_3H_8(g)$ on a nickel catalyst at 122 °C, the following initial rate data were collected:

$P(C_3H_6)$ = 200 torr

$P(H_2)$/(torr)	100	200	400
$-[dP(C_3H_6)/dt]_0$	1.1	1.1	1.25

$P(H_2)$ = 200 torr

$P(C_3H_6)$/(torr)	100	200	400
$-[dP(C_3H_6)/dt]_0$	1	1.95	3.80

Predict an acceptable mechanism for this reaction.

▌ Using the procedure outlined in Problem 12.29, the data indicate that the reaction is first-order with respect to $P(C_3H_6)$ and zero-order with respect to $P(H_2)$.

For a proposed mechanism in which both reactants must undergo adsorption, *(23.33)* becomes $-dP(C_3H_6)/dt = k'\theta(C_3H_6)\theta(H_2)$, which upon substitution of *(23.32)* gives

$$\frac{-dP(C_3H_6)}{dt} = \frac{k'b(C_3H_6)b(H_2)P(C_3H_6)P(H_2)}{[1 + b(C_3H_6)P(C_3H_6) + b(H_2)P(H_2)]^2}$$

This mechanism incorrectly predicts a first-order reaction with respect to C_3H_6 and with respect to H_2.

If a mechanism is assumed in which the H_2 molecules are strongly adsorbed, *(23.33)* becomes

$$\frac{-dP(C_3H_6)}{dt} = k'[1 - \theta(H_2)]P(C_3H_6) = \frac{k'P(C_3H_6)}{1 + b(H_2)P(H_2)}$$

upon substitution of *(23.32)*, as in Problem 23.67. This rate equation agrees with the experimental data.

23.69 The rate equation describing the formation of $CH_4(g)$ from $CO(g)$ and $H_2(g)$ on a nickel catalyst is

$$\frac{d\xi}{dt} = \frac{kP(CO)[P(H_2)]^{1/2}}{1 + bP(H_2)}$$

Describe what is occurring on the surface of the catalyst.

▌ As in Problem 23.68, the H_2 molecules are being strongly adsorbed on the catalyst and inhibit the reaction.

23.70 The rate equation describing the oxidation of $NO(g)$ on an activated carbon catalyst is

$$\frac{d\xi}{dt} = \frac{k[P(NO)]^2P(O_2)}{1 + b(NO)[P(NO)]^2 + b(NO_2)P(NO_2)}$$

where $b(NO) \gg b(NO_2)$. Describe what is happening on the surface of the catalyst.

▌ The NO molecules are being strongly adsorbed, and, once formed, the NO_2 molecules are being weakly adsorbed on the surface of the catalyst.

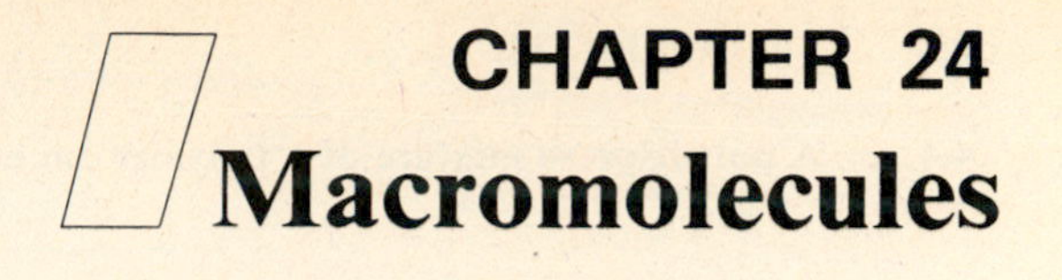

CHAPTER 24
Macromolecules

24.1 MOLAR MASS AND SIZE

24.1 To a good approximation, a polystyrene molecule can be considered to be a sphere. A molecule was observed to have a 10.0-mm diameter using a 5.0×10^4 power microscope. Given $\rho = 1.06\ \mathrm{g \cdot cm^{-3}}$ for polystyrene, calculate the molecular mass.

▮ The radius of the molecule is

$$r = \frac{(10.0\ \mathrm{mm})[(0.1\ \mathrm{cm})/(1\ \mathrm{mm})]}{2(5.0 \times 10^4)} = 1.00 \times 10^{-5}\ \mathrm{cm}$$

The mass of the molecule is

$$m = (1.06\ \mathrm{g \cdot cm^{-3}})\tfrac{4}{3}\pi(1.00 \times 10^{-5}\ \mathrm{cm})^3 = 4.44 \times 10^{-15}\ \mathrm{g}$$

which corresponds to $M = 2.67 \times 10^9\ \mathrm{g \cdot mol^{-1}}$.

24.2 Hemoglobin contains 0.35 mass % Fe. What is the minimum molar mass of hemoglobin? If $M = 64\,450\ \mathrm{g \cdot mol^{-1}}$ for hemoglobin, how many Fe atoms are present in the molecule?

▮ Assuming that one Fe atom is present, the minimum molar mass is

$$M = \frac{55.847\ \mathrm{g \cdot mol^{-1}}}{0.0035} = 1.6 \times 10^4\ \mathrm{g \cdot mol^{-1}}$$

Dividing the actual molar mass by the minimum value gives

$$N(\mathrm{Fe}) = \frac{64\,450\ \mathrm{g \cdot mol^{-1}}}{1.6 \times 10^4\ \mathrm{g \cdot mol^{-1}}} = 4$$

24.3 A polydisperse mixture of a polymer can be described by the following composition–molar mass data:

$n_i/(\mathrm{mol})$	0.10	0.20	0.40	0.20	0.10
$M_i/(\mathrm{kg \cdot mol^{-1}})$	1.00	1.20	1.40	1.60	1.80

Calculate the number-average, mass-average, and z-average molar masses.

▮ The *number-average molar mass* ($\bar{M}_n$) is defined as

$$\bar{M}_n = \sum n_i M_i / \sum n_i \tag{24.1}$$

the *mass-average molar mass* ($\bar{M}_m$) is defined as

$$\bar{M}_m = \sum n_i M_i^2 / \sum n_i M_i \tag{24.2}$$

and the *z-average molar mass* ($\bar{M}_z$) is defined as

$$\bar{M}_z = \sum n_i M_i^3 / \sum n_i M_i^2 \tag{24.3}$$

Substituting the data into *(24.1)–(24.3)* gives

$$\bar{M}_n = \frac{\begin{bmatrix}(0.10\ \mathrm{mol})(1.00\ \mathrm{kg \cdot mol^{-1}}) + (0.20\ \mathrm{mol})(1.20\ \mathrm{kg \cdot mol^{-1}}) \\ +(0.40\ \mathrm{mol})(1.40\ \mathrm{kg \cdot mol^{-1}}) + (0.20\ \mathrm{mol})(1.60\ \mathrm{kg \cdot mol^{-1}}) \\ +(0.10\ \mathrm{mol})(1.80\ \mathrm{kg \cdot mol^{-1}})\end{bmatrix}}{0.10\ \mathrm{mol} + 0.20\ \mathrm{mol} + 0.40\ \mathrm{mol} + 0.20\ \mathrm{mol} + 0.10\ \mathrm{mol}} = 1.40\ \mathrm{kg \cdot mol^{-1}}$$

$$\bar{M}_m = \frac{[(0.10)(1.00)^2 + (0.20)(1.20)^2 + (0.40)(1.40)^2 + (0.20)(1.60)^2 + (0.10)(1.80)^2]\ \mathrm{kg^2 \cdot mol^{-1}}}{[(0.10)(1.00) + (0.20)(1.20) + (0.40)(1.40) + (0.20)(1.60) + (0.10)(1.80)]\ \mathrm{kg}} = 1.43\ \mathrm{kg \cdot mol^{-1}}$$

$$\bar{M}_z = \frac{[(0.10)(1.00)^3 + (0.20)(1.20)^3 + (0.40)(1.40)^3 + (0.20)(1.60)^3 + (0.10)(1.80)^3]\ \mathrm{kg^3 \cdot mol^{-2}}}{[(0.10)(1.00)^2 + (0.20)(1.20)^2 + (0.40)(1.40)^2 + (0.20)(1.60)^2 + (0.10)(1.80)^2]\ \mathrm{kg^2 \cdot mol^{-1}}} = 1.47\ \mathrm{kg \cdot mol^{-1}}$$

24.4 A polydisperse mixture of a polymer can be described by the following composition–molar mass data:

$(\text{mass }\%)_i$	25.0	50.0	25.0
$M_i/(\text{kg}\cdot\text{mol}^{-1})$	1.00	1.20	1.40

Calculate the number-average and the mass-average molar masses.

▮ Each of the mass % values must be converted to values of n_i. Assuming 100.0 g of the mixture, the amounts of each polymer are

$$n_i = \frac{(100.0\ \text{g})(0.250)}{1.00 \times 10^3\ \text{g}\cdot\text{mol}^{-1}} = 2.50 \times 10^{-2}\ \text{mol}$$

4.17×10^{-2} mol, and 1.79×10^{-2} mol, respectively. Substituting into *(24.1)* and *(24.2)* gives

$$\bar{M}_n = \frac{\left[\begin{array}{r}(2.50 \times 10^{-2}\ \text{mol})(1.00\ \text{kg}\cdot\text{mol}^{-1}) + (4.17 \times 10^{-2}\ \text{mol})(1.20\ \text{kg}\cdot\text{mol}^{-1}) \\ +(1.79 \times 10^{-2}\ \text{mol})(1.40\ \text{kg}\cdot\text{mol}^{-1})\end{array}\right]}{2.50 \times 10^{-2}\ \text{mol} + 4.17 \times 10^{-2}\ \text{mol} + 1.79 \times 10^{-2}\ \text{mol}} = 1.18\ \text{kg}\cdot\text{mol}^{-1}$$

$$\bar{M}_m = \frac{[(2.50 \times 10^{-2})(1.00)^2 + (4.17 \times 10^{-2})(1.20)^2 + (1.79 \times 10^{-2})(1.40)^2]\ \text{kg}^2\cdot\text{mol}^{-1}}{[(2.50 \times 10^{-2})(1.00) + (4.17 \times 10^{-2})(1.20) + (1.79 \times 10^{-2})(1.40)]\ \text{kg}} = 1.20\ \text{kg}\cdot\text{mol}^{-1}$$

24.5 Calculate the polydispersity index for the mixture described in Problem 24.4.

▮ The *polydispersity index* is the ratio of $\bar{M}_m$ to $\bar{M}_n$:

$$\text{Polydispersity index} = \bar{M}_m/\bar{M}_n \qquad (24.4)$$

Substituting the results of Problem 24.4 into *(24.4)* gives

$$\text{Polydispersity index} = \frac{1.20\ \text{kg}\cdot\text{mol}^{-1}}{1.18\ \text{kg}\cdot\text{mol}^{-1}} = 1.02$$

24.6 An equimolar mixture of two polymers has a number-average molar mass of $1.00\ \text{kg}\cdot\text{mol}^{-1}$ as determined from osmotic pressure measurements and a mass-average molar mass of $1.20\ \text{kg}\cdot\text{mol}^{-1}$ as determined by ultracentrifugation measurements. Determine the molar masses of the two polymers.

▮ Substituting $n_1 = n_2 = n$ into *(24.1)* and *(24.2)* gives

$$\bar{M}_n = \frac{nM_1 + nM_2}{n + n} = \frac{M_1 + M_2}{2}$$

$$\bar{M}_m = \frac{nM_1^2 + nM_2^2}{nM_1 + nM_1} = \frac{M_1^2 + M_2^2}{M_1 + M_2}$$

Solving the first equation for M_1 and substituting into the second equation gives

$$\bar{M}_m = \frac{(2\bar{M}_n - M_2)^2 + M_2^2}{(2\bar{M}_n - M_2) + M_2}$$

which can be rewritten as

$$M_2^2 - 2\bar{M}_n M_2 + \bar{M}_n(2\bar{M}_n - \bar{M}_m) = 0$$

Solving for M_2 using the quadratic formula and substituting the data gives

$$M_2 = \tfrac{1}{2}[-(-2)(1.00\ \text{kg}\cdot\text{mol}^{-1}) \pm \{[(-2)(1.00\ \text{kg}\cdot\text{mol}^{-1})]^2 - 4(1)(1.00\ \text{kg}\cdot\text{mol}^{-1})[2(1.00\ \text{kg}\cdot\text{mol}^{-1}) - 1.20\ \text{kg}\cdot\text{mol}^{-1}]\}^{1/2}]$$

$$= (2.00 \pm 0.89)/2\ \text{kg}\cdot\text{mol}^{-1} = (1.45\ \text{or}\ 0.56)\ \text{kg}\cdot\text{mol}^{-1}$$

The corresponding values of M_1 are

$$M_1 = 2(1.00\ \text{kg}\cdot\text{mol}^{-1}) - (1.45\ \text{or}\ 0.56)\ \text{kg}\cdot\text{mol}^{-1} = (0.55\ \text{or}\ 1.44)\ \text{kg}\cdot\text{mol}^{-1}$$

The molar masses are 0.55 and $1.45\ \text{kg}\cdot\text{mol}^{-1}$.

24.7 A polymer of molar mass $1.00\ \text{kg}\cdot\text{mol}^{-1}$ is known to dimerize to a limited extent. If $\bar{M}_n = 1.25\ \text{kg}\cdot\text{mol}^{-1}$ as determined from osmotic pressure measurements for the mixture of the monomer and dimer, determine the equilibrium constant for the reaction $2\text{P} \rightleftarrows \text{P}_2$.

▮ Assume that the amount of monomer at equilibrium is $n = 1.00\ \text{mol}$ and the amount of dimer is n_2. Substituting $M_2 = 2M$, where M is the molar mass of the monomer into *(24.1)* gives

$$\bar{M}_n = \frac{(1.00\ \text{mol})M + m_2(2M)}{1.00\ \text{mol} + n_2}$$

Solving for n_2 and substituting the data gives

$$n_2 = \frac{(1.00\ \text{mol})(1.25\ \text{kg}\cdot\text{mol}^{-1} - 1.00\ \text{kg}\cdot\text{mol}^{-1})}{2(1.00\ \text{kg}\cdot\text{mol}^{-1}) - 1.25\ \text{kg}\cdot\text{mol}^{-1}} = 0.33\ \text{mol}$$

Writing the expression for the equilibrium constant in terms of amounts of substance and substituting the data gives

$$K = \frac{n_2}{n_1^2} = \frac{0.33}{(1.00)^2} = 0.33$$

24.8 The distribution of molar masses for a synthetic polymer is given by

$$\frac{dN/N}{dM} = \left(\frac{0.612\ 35}{a}\right) e^{-[(M-a)/a]^2} \tag{24.5}$$

where $a = 1.00\ \text{kg}\cdot\text{mol}^{-1}$. Prepare a plot of this distribution function as a function of *n*.

▮ As a sample calculation, consider $M = 1.50\ \text{kg}\cdot\text{mol}^{-1}$.

$$\frac{dN/N}{dM} = \left(\frac{0.612\ 35}{1.00\ \text{kg}\cdot\text{mol}^{-1}}\right) \exp\left\{-\left[\frac{1.50\ \text{kg}\cdot\text{mol}^{-1} - 1.00\ \text{kg}\cdot\text{mol}^{-1}}{1.00\ \text{kg}\cdot\text{mol}^{-1}}\right]^2\right\} = 0.477\ \text{kg}^{-1}\cdot\text{mol}$$

The plot is shown in Fig. 24-1.

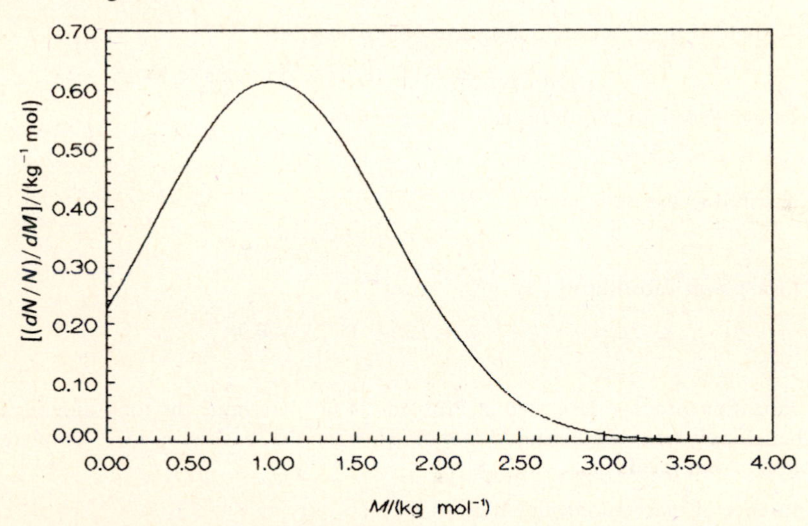

Fig. 24-1

24.9 Using the distribution function given in Problem 24.8, determine the most probable molar mass.

▮ Taking the derivative of *(24.5)* with respect to M, setting the result equal to zero, and solving for M gives

$$\frac{d[(dN/N)/dM]}{dM} = \frac{0.612\ 35}{a} e^{-[(M-a)/a]^2}(-2)\left(\frac{M-a}{a}\right) = 0$$

$$M = a = 1.00\ \text{kg}\cdot\text{mol}^{-1}$$

This value agrees with the maximum of the curve shown in Fig. 24-1.

24.10 Calculate the number-average molar mass of the polymer system described in Problem 24.8.

▮ Substituting *(24.5)* into *(2.2)* gives

$$\bar{M}_m = \int_0^\infty M\frac{dN}{N} = \int_0^\infty M\frac{0.612\ 35}{a} e^{-[(M-a)/a]^2}\, dM$$

Letting $x = (M - a)/a$, the expression for $\bar{M}_n$ becomes

$$\bar{M}_n = \frac{0.612\,35}{a}\int_{-1}^{\infty}(ax + a)\,e^{-x^2}a\,dx = \frac{0.612\,35}{a}\int_{-1}^{\infty}(x+1)\,e^{-x^2}\,dx$$

$$= \frac{0.612\,35}{a}\left[\int_{-1}^{\infty}x\,e^{-x^2}\,dx + \int_{0}^{1}e^{-x^2}\,dx + \int_{0}^{\infty}e^{-x^2}\,dx\right]$$

$$= \frac{0.612\,35}{a}\left[-\tfrac{1}{2}e^{-x^2}\Big|_{-1}^{\infty} + \frac{\pi^{1/2}}{2}\operatorname{erf}(1) + \frac{\pi^{1/2}}{2}\right]$$

$$= \frac{0.612\,35}{a}\left[-\tfrac{1}{2}(e^{-\infty^2} - e^{-(-1)^2}) + \frac{\pi^{1/2}}{2}(0.842\,70) + \frac{\pi^{1/2}}{2}\right]$$

$$= \frac{0.612\,35}{a}\left(\frac{1}{2e} + \frac{\pi^{1/2}}{2}(1.842\,70)\right)$$

where the integrals were evaluated using Table 2-1. Substituting the data gives

$$\bar{M}_n = \frac{0.612\,35}{1.00\ \text{kg}\cdot\text{mol}^{-1}}\left(\frac{1}{2e} + \frac{\pi^{1/2}}{2}(1.842\,70)\right) = 1.11\ \text{kg}\cdot\text{mol}^{-1}$$

24.11 Calculate the probability that a given molecule of a polymer formed by condensation of monomers consists of exactly 25 monomers when the reaction is 75% complete. What is the value of the number-average degree of polymerization under these conditions? What must the extent of reaction be in order for the number-average degree of polymerization to exceed 25?

▌ The probability that a given molecule will consist of x monomers is given by the *mole fraction* or *number distribution* $[X(x)]$

$$X(x) = p^{x-1}(1 - p) \tag{24.6}$$

where p is the extent of the reaction. Substituting the data gives

$$X(25) = (0.75)^{25-1}(1 - 0.75) = 2.5 \times 10^{-4}$$

The *number-average degree of polymerization* $\bar{x}_n$ is given by

$$\bar{x}_n = 1/(1 - p) \tag{24.7}$$

which upon substituting the data gives

$$\bar{x}_n = 1/(1 - 0.75) = 4.0$$

Solving *(24.7)* for p and substituting $\bar{x}_n = 25$ gives

$$p = (25 - 1)/25 = 0.96$$

24.12 For the polymerization process described in Problem 24.11, determine the total number of macromolecules and the total number of polymers containing 25 monomers when the reaction is 75% complete. Assume that a 1-mole sample of monomer was used.

▌ The total number of macromolecules is given by

$$N = N_0(1 - p) \tag{24.8}$$

which upon substituting the data gives

$$N = (6.022 \times 10^{23})(1 - 0.75) = 1.5 \times 10^{23}$$

The total number of polymers containing x monomers is given by

$$N(x) = N(1 - p)p^{x-1} = N_0(1 - p)^2p^{x-1} \tag{24.9}$$

which upon substituting the data gives

$$N(25) = (6.022 \times 10^{23})(1 - 0.75)^2(0.75)^{25-1} = 3.8 \times 10^{19}$$

24.13 Calculate the value of the mass distribution for the polymerization process described in Problem 24.11 when the reaction is 75% complete. Prepare plots of the number distribution and the mass distribution against the number of monomers contained in the polymer for the reaction at $p = 0.75$.

▮ The *mass distribution* $[W(x)]$ is given by

$$W(x) = xp^{x-1}(1-p)^2 \tag{24.10}$$

Substituting the data gives

$$W(25) = (25)(0.75)^{25-1}(1-0.75)^2 = 1.6 \times 10^{-3}$$

Plots of $X(x)$ and $W(x)$ against x using values calculated from *(24.6)* and *(24.10)* are shown in Fig. 24-2. Note that the maximum in the $W(x)$ curve falls very near $\bar{x}_n = 4$ (see Problem 24.11).

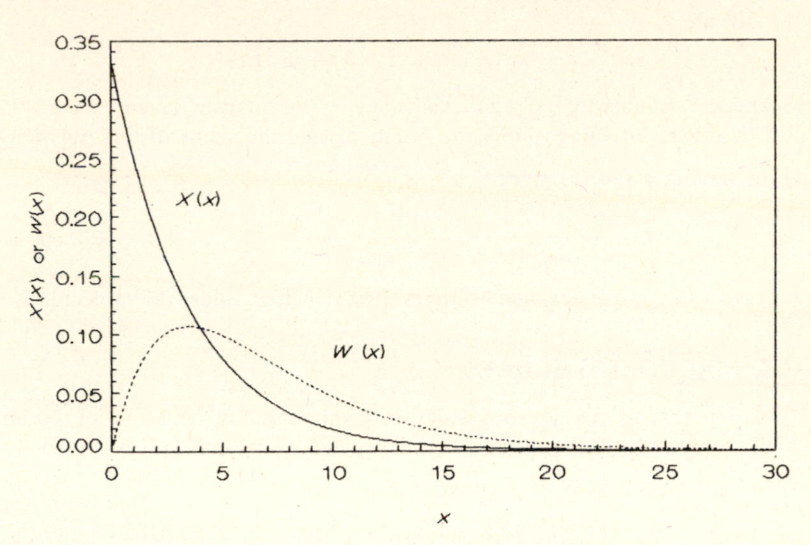

Fig. 24-2

24.14 Calculate the values of $\bar{M}_n$ and $\bar{M}_m$ for the polymerization process described in Problem 24.11 when the reaction is 75% complete. Assume that the molar mass of the monomer is $100\ \text{g} \cdot \text{mol}^{-1}$.

▮ The molar masses are related to p and the molar mass of the monomer (M_0) by

$$\bar{M}_n = M_0/(1-p) \tag{24.11}$$

$$\bar{M}_m = M_0(1+p)/(1-p) \tag{24.12}$$

Substituting the data gives

$$\bar{M}_n = \frac{100\ \text{g} \cdot \text{mol}^{-1}}{1-0.75} = 400\ \text{g} \cdot \text{mol}^{-1}$$

$$\bar{M}_m = (100\ \text{g} \cdot \text{mol}^{-1})\frac{1+0.75}{1-0.75} = 700\ \text{g} \cdot \text{mol}^{-1}$$

24.15 What is the probability that a given monomer molecule is located at the end of a polymer molecule in the polymerization process described in Problem 24.11 when the reaction is 75% complete?

▮ The probability of the monomer being at a specific location in the polymer is given by $1-p$. Because there are two ends, the probability of the monomer being on an end is $2(1-0.75) = 0.50$. This value is in good agreement with that predicted for $\bar{x}_n = 4$ (see Problem 24.11) where the probability is $2/4 = 0.50$.

24.16 Calculate the root-mean-square end-to-end distance in a polymer with $\bar{M}_n = 1.00\ \text{kg} \cdot \text{mol}^{-1}$ given that the length of the monomer $(M = 14.0\ \text{g} \cdot \text{mol}^{-1})$ is 0.15 nm.

▮ The *root-mean-square end-to-end distance* (r_{rms}) for a polymer containing tetrahedral bonding is given by

$$r_{\text{nm}} = (2N)^{1/2} l \tag{24.13}$$

where l is the length of the monomer. Substituting the data into *(24.13)* gives

$$r_{\text{rms}} = \left(\frac{2(1.00 \times 10^3\ \text{g} \cdot \text{mol}^{-1})}{14\ \text{g} \cdot \text{mol}^{-1}}\right)^{1/2}(0.15\ \text{nm}) = 1.8\ \text{nm}$$

24.17 Calculate the radius of gyration for the polymer described in Problem 24.16.

The average root-mean-square distance of the atoms from the center of mass of the molecule is known as the *radius of gyration* and is given by

$$r_g = (N/3)^{1/2} l \tag{24.14}$$

Substituting the data into *(24.14)* gives

$$r_g = \left(\frac{1.00 \times 10^3\ \text{g} \cdot \text{mol}^{-1}}{3(14\ \text{g} \cdot \text{mol}^{-1})}\right)^{1/2} (0.15\ \text{nm}) = 0.73\ \text{nm}$$

24.18 For a spherical molecule,

$$r_g = (3/5)^{1/2} (3Mv/4\pi L)^{1/3} \tag{24.15}$$

where v is the specific volume of polymer. Calculate r_g for myosin given $M = 493\ \text{kg} \cdot \text{mol}^{-1}$ and $v = 0.73\ \text{cm}^3 \cdot \text{g}^{-1}$. If the observed value is 46.8 nm, briefly discuss the shape of this polymer.

Substituting the data into *(24.15)* gives

$$r_g = \left(\frac{3}{5}\right)^{1/2} \left(\frac{3(493 \times 10^3\ \text{g} \cdot \text{mol}^{-1})(0.73\ \text{cm}^3 \cdot \text{g}^{-1})}{4\pi(6.022 \times 10^{23}\ \text{mol}^{-1})}\right)^{1/3} = 4.0 \times 10^{-7}\ \text{cm} = 4.0\ \text{nm}$$

Because the observed value is greater than 10 times the calculated value, the molecule is not spherical.

24.2 SOLUTIONS OF MACROMOLECULES

24.19 Calculate $\Delta_{\text{soln}} S$ for preparing an aqueous solution containing 1.00 mass % of inulin, $(C_6H_{10}O_5)_x$. Assume $M = 5200\ \text{g} \cdot \text{mol}^{-1}$ and $\rho = 1.35\ \text{g} \cdot \text{cm}^{-3}$ for inulin, and $\rho = 0.9982\ \text{g} \cdot \text{cm}^{-3}$ for H_2O.

For solutions of polymers, *(5.13)* is often written as

$$\Delta_{\text{soln}} S = -R\left(\frac{\rho(1-\phi_2)}{\rho + \phi_2(1-\rho)} \ln(1-\phi_2) + \frac{\phi_2}{\rho + \phi_2(1-\rho)} \ln \phi_2\right) \tag{24.16}$$

where $\rho = V^\circ_{m,2}/V^\circ_{m,1}$ and ϕ_2 is the volume fraction of polymer. The respective molar volumes are

$$V^\circ_{m,2} = \frac{5200\ \text{g} \cdot \text{mol}^{-1}}{1.35\ \text{g} \cdot \text{cm}^{-3}} = 3900\ \text{cm}^3 \cdot \text{mol}^{-1}$$

$$V^\circ_{m,1} = \frac{18.015\ \text{g} \cdot \text{mol}^{-1}}{0.9982\ \text{g} \cdot \text{cm}^{-3}} = 18.047\ \text{cm}^3 \cdot \text{mol}^{-1}$$

which corresponds to

$$\rho = \frac{3900\ \text{cm}^3 \cdot \text{mol}^{-1}}{18.047\ \text{cm}^3 \cdot \text{mol}^{-1}} = 220$$

Assuming 100.00 g of solution,

$$\phi_2 = \frac{n_2 V^\circ_{m,2}}{n_1 V^\circ_{m,1} + n_2 V^\circ_{m,2}} = \frac{[(1.00\ \text{g} \cdot \text{inulin})/(5200\ \text{g} \cdot \text{mol}^{-1})](3900\ \text{cm}^3 \cdot \text{mol}^{-1})}{\left(\frac{99.00\ \text{g} \cdot H_2O}{18.015\ \text{g} \cdot \text{mol}^{-1}}\right)(18.047\ \text{cm}^3 \cdot \text{mol}^{-1}) + \left(\frac{1.00\ \text{g} \cdot \text{inulin}}{5200\ \text{g} \cdot \text{mol}^{-1}}\right)(3900\ \text{cm}^3 \cdot \text{mol}^{-1})}$$

$$= 7.5 \times 10^{-3}$$

Substituting into *(24.16)* gives

$$\Delta_{\text{soln}} S = -(8.314\ \text{J} \cdot \text{K}^{-1} \cdot \text{mol}^{-1})\left(\frac{220(1 - 7.5 \times 10^{-3})}{220 + (7.5 \times 10^{-3})(1 - 220)} \ln(1 - 7.5 \times 10^{-3}) + \frac{7.5 \times 10^{-3}}{220 + (7.5 \times 10^{-3})(1 - 220)} \ln(7.5 \times 10^{-3})\right)$$

$$= 0.064\ \text{J} \cdot \text{K}^{-1} \cdot \text{mol}^{-1}$$

24.20 Assuming $\Delta_{\text{soln}} H = 0$ for the solution described in Problem 24.19, calculate $\Delta_{\text{soln}} G$ at 25 °C.

Substituting the results of Problem 24.19 and *(24.16)* into *(6.2a)* gives

$$\Delta_{\text{soln}} G = 0 - (298\ \text{K})(0.064\ \text{J} \cdot \text{K}^{-1} \cdot \text{mol}^{-1}) = -19\ \text{J} \cdot \text{mol}^{-1}$$

24.21 What is the vapor pressure of the aqueous solution described in Problem 24.19 at 25 °C? For H_2O, $P_1^\circ = 23.756$ torr.

■ Assuming 100.00 g of solution, the mole fraction of polymer is

$$x_2 = \frac{(1.00\ \text{g})/(5200\ \text{g}\cdot\text{mol}^{-1})}{(99.00\ \text{g})/(18.015\ \text{g}\cdot\text{mol}^{-1}) + (1.00\ \text{g})/(5200\ \text{g}\cdot\text{mol}^{-1})} = 3.5 \times 10^{-5}$$

which upon substituting into *(11.22)* gives

$$\Delta P = -(23.756\ \text{torr})(3.5 \times 10^{-5}) = -8.3 \times 10^{-4}\ \text{torr}$$

The vapor pressure of the solution is

$$P_{\text{soln}} = P_1^\circ + \Delta P = 23.756\ \text{torr} + (-8.3 \times 10^{-4}\ \text{torr}) = 23.755\ \text{torr}$$

a negligible change.

24.22 What is the freezing point of the aqueous solution described in Problem 24.19? For H_2O, $K_{x,\text{fp}} = 103$ K.

■ Substituting $x_2 = 3.5 \times 10^{-5}$ (see Problem 24.21) into *(11.30)* gives

$$T_{\text{fp,soln}} = 0.0000\ ^\circ\text{C} - (103\ \text{K})(3.5 \times 10^{-5}) = -0.0036\ ^\circ\text{C}$$

a negligible change.

24.23 What is the osmotic pressure at 25 °C of the aqueous solution described in Problem 24.19?

■ Substituting $x_1 = (1 - 3.5 \times 10^{-5})$ and $V_{m,1}^\circ = 18.047\ \text{cm}^3\cdot\text{mol}^{-1}$ into *(11.35)* gives

$$\Pi = \frac{-(0.083\,14\ \text{bar}\cdot\text{L}\cdot\text{K}^{-1}\cdot\text{mol}^{-1})(298\ \text{K})}{(18.047\ \text{cm}^3\cdot\text{mol}^{-1})[(10^{-3}\ \text{L})/(1\ \text{cm}^3)]} \ln(1 - 3.5 \times 10^{-5}) = 0.048\ \text{bar}$$

a small but measurable value.

24.24 The following osmotic pressures were observed for solutions of polyisobutylene in cyclohexane at 25 °C:

$C'/(\text{g}\cdot\text{L}^{-1})$	20.0	15.0	10.0	5.0
$\Pi/(10^{-3}\ \text{bar})$	16.09	9.92	5.29	2.03

Determine $\bar{M}_n$ and the second osmotic virial coefficient for this polymer.

■ As implied by *(11.37)*, a plot of Π/C' against C' is shown in Fig. 24-3. From the graph,

$$\bar{M}_n = \frac{RT}{\text{intercept}} = \frac{(0.083\,14\ \text{bar}\cdot\text{L}\cdot\text{K}^{-1}\cdot\text{mol}^{-1})(298\ \text{K})}{0.268 \times 10^{-3}\ \text{bar}\cdot\text{g}^{-1}\cdot\text{L}} = 9.25 \times 10^4\ \text{g}\cdot\text{mol}^{-1}$$

$$b = \text{slope} = 2.66 \times 10^{-5}\ \text{bar}\cdot\text{g}^{-2}\cdot\text{L}^2$$

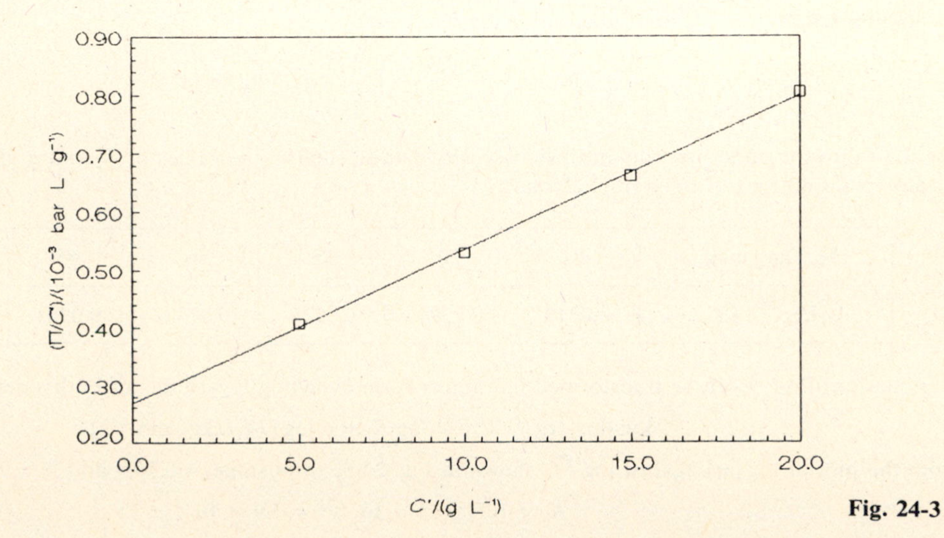

Fig. 24-3

24.25 The following relative viscosity data were collected for solutions of poly(vinyl alcohol) in water:

η_r	1.409	1.198	1.098
$C'/(g \cdot L^{-1})$	10.00	5.00	2.50

Determine the intrinsic viscosity of this polymer.

▌ The *intrinsic viscosity* ($[\eta]$) is given by

$$[\eta] = \lim_{C' \to 0} \left(\frac{\eta_{sp}}{C'} \right) \tag{24.17}$$

where the *specific viscosity* (η_{sp}) is given by

$$\eta_{sp} = \eta_r - 1 \tag{24.18}$$

A plot of η_{sp}/C' against C' is given in Fig. 24-4, and from the intercept, $[\eta] = 0.0386\ g^{-1} \cdot L$.

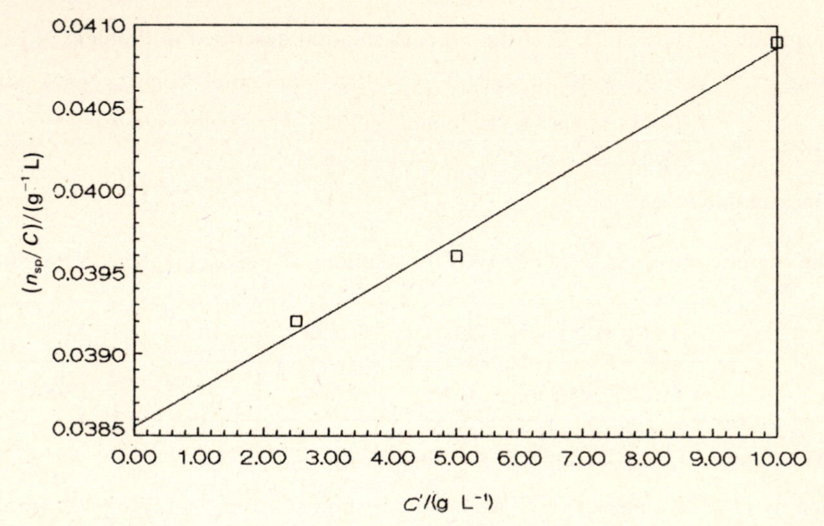

Fig. 24-4

24.26 For solutions of polymers, the molar mass and the specific viscosity are related by the *Mark–Houwink equation*

$$\{[\eta]/(g^{-1} \cdot L)\} = K[\bar{M}_v/(kg \cdot mol^{-1})]^a \tag{24.19}$$

where $\bar{M}_v$ is the *viscosity molar mass*. Given $K = 3.8 \times 10^{-3}$ and $a = 0.76$ for aqueous solutions of poly(vinyl alcohol), calculate the molar mass of the polymer described in Problem 24.25.

▌ Solving *(24.19)* for $\bar{M}_v$ and substituting the data gives

$$\bar{M}_v = \left(\frac{0.0386}{3.8 \times 10^{-3}} \right)^{1/0.76} = 21\ kg \cdot mol^{-1}$$

24.27 Use the following concentration–intrinsic viscosity data to confirm the values of K and a given in Problem 24.26 for aqueous solutions of poly(vinyl alcohol):

$\bar{M}_v/(kg \cdot mol^{-1})$	10	20	35	50	65	80
$[\eta]/(g^{-1} \cdot L)$	0.0213	0.0394	0.0575	0.0740	0.0901	0.1059

▌ Equation *(24.19)* can be transformed into linear form by taking logarithms of both sides, giving

$$\log \{[\eta]/(g^{-1} \cdot L)\} = \log K + a \log [\bar{M}_v/(kg \cdot mol^{-1})]$$

From the plot of $\log [\eta]$ against $\log \bar{M}_v$ shown in Fig. 24-5, $a = \text{slope} = 0.76$, and

$$K = 10^{\text{intercept}} = 10^{-2.41} = 3.9 \times 10^{-3}$$

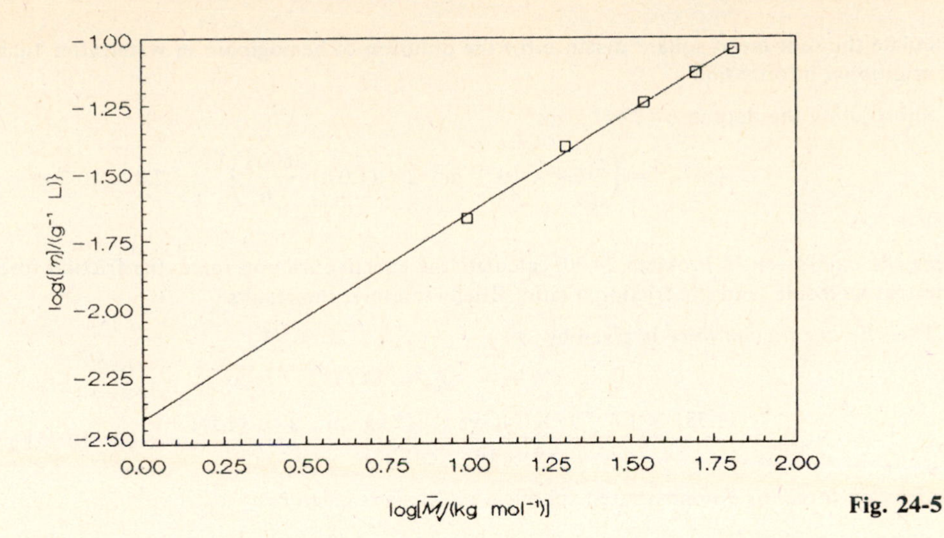

Fig. 24-5

24.28 Using the data given in Problem 24.25, determine the shape of the poly(vinyl alcohol) molecule.

▌ Typical values of a in *(24.19)* are 0 for spherical molecules, 0.5 for random coil molecules, and 1.8 for rigid rod molecules. Because $a = 0.76$ for this system, the molecule must be a random coil.

24.29 Assuming a spherical molecule, predict the value of the diffusion coefficient of myoglobin ($M = 16.9\ \text{kg}\cdot\text{mol}^{-1}$) in water at 25 °C. The specific volume of the protein is $0.74\ \text{cm}^3\cdot\text{g}^{-1}$, and the viscosity of water is $\eta = 0.8904$ cP.

▌ Equation *(2.59)* can be rewritten in terms of the specific volume (v) as

$$D_{ij}^{\infty} = \frac{RT}{6\pi\eta_j L}\left(\frac{4\pi L}{3\bar{M}_v v}\right)^{1/3} \tag{24.20}$$

Substituting the data gives

$$D_{ij}^{\infty} = \frac{(8.314\ \text{J}\cdot\text{K}^{-1}\cdot\text{mol}^{-1})(298\ \text{K})[(1\ \text{Pa}\cdot\text{m}^3)/(1\ \text{J})]}{6\pi(0.8904\ \text{cP})[(10^{-2}\ \text{P})/(1\ \text{cP})][(0.1\ \text{Pa}\cdot\text{s})/(1\ \text{P})](6.022\times10^{23}\ \text{mol}^{-1})}$$
$$\times\left(\frac{4\pi(6.022\times10^{23}\ \text{mol}^{-1})^{1/3}}{3(16.9\times10^3\ \text{g}\cdot\text{mol}^{-1})(0.74\ \text{cm}^3\cdot\text{g}^{-1})[(10^{-2}\ \text{m})/(1\ \text{cm})]^3}\right)^{1/3}$$
$$= 1.4\times10^{-10}\ \text{m}^2\cdot\text{s}^{-1}$$

which agrees fairly well with the actual value of $1.13\times10^{-10}\ \text{m}^2\cdot\text{s}^{-1}$.

24.30 For aqueous solutions of hemoglobin, $D_{ij}^{\infty} = 6.9\times10^{-11}\ \text{m}^2\cdot\text{s}^{-1}$ and $v = 0.75\ \text{cm}^3\cdot\text{g}^{-1}$ at 25 °C. Assuming a spherical molecule, estimate the molar mass given $\eta = 0.8904$ cP for H_2O.

▌ Solving *(24.20)* for M and substituting the data gives

$$\bar{M}_v = \left[\frac{(8.314\ \text{J}\cdot\text{K}^{-1}\cdot\text{mol}^{-1})(298\ \text{K})[(1\ \text{Pa}\cdot\text{m}^3)/(1\ \text{J})]}{6\pi(0.8904\ \text{cP})\left(\frac{10^{-2}\ \text{P}}{1\ \text{cP}}\right)\left(\frac{0.1\ \text{Pa}\cdot\text{s}}{1\ \text{P}}\right)(6.022\times10^{23}\ \text{mol}^{-1})(6.9\times10^{-11}\ \text{m}^2\cdot\text{s}^{-1})}\right]^3$$
$$\times\frac{4\pi(6.022\times10^{23}\ \text{mol}^{-1})}{3(0.75\ \text{cm}^3\cdot\text{g}^{-1})[(10^{-2}\ \text{m})/(1\ \text{cm})]^3}$$
$$= 1.5\times10^5\ \text{g}\cdot\text{mol}^{-1}$$

This value is roughly twice the actual value of $64.5\ \text{kg}\cdot\text{mol}^{-1}$.

24.31 Using the data given in Problem 24.30, estimate the approximate diameter of a hemoglobin molecule.

▌ Upon comparison of *(2.59)* and *(24.20)*,

$$\sigma_i = 2(3\bar{M}_v/4\pi L)^{1/3} = 2\left[\frac{(3)(64.5\times10^3\ \text{g}\cdot\text{mol}^{-1})(0.75\ \text{cm}^3\cdot\text{g}^{-1})[(10^{-2}\ \text{m})/(1\ \text{cm})]^3}{4\pi(6.022\times10^{23})}\right]^{1/3} \tag{24.21}$$
$$= 5.4\times10^{-9}\ \text{m} = 5.4\ \text{nm}$$

24.32 Calculate the root-mean square distance for the diffusion of hemoglobin in water after 1.0 h. See Problem 24.30 for additional information.

▌ Substituting the data into *(2.61)* gives

$$(\overline{z^2})^{1/2} = \left(2(6.9 \times 10^{-11}\ \text{m}^2 \cdot \text{s}^{-1})(1.0\ \text{h})\frac{3600\ \text{s}}{1\ \text{h}}\right)^{1/2} = 7.0 \times 10^{-4}\ \text{m}$$

24.33 Using the data given in Problem 24.30, calculate the effective friction force, the friction force for a nonhydrated spherical molecule, and the frictional ratio. Briefly interpret the results.

▌ The *effective friction force* is given by

$$f_{\text{eff}} = kT/D_{ij}^{\infty} \tag{24.22}$$

$$f_{\text{eff}} = \frac{(1.381 \times 10^{-23}\ \text{J} \cdot \text{K}^{-1})(298\ \text{K})[(1\ \text{kg} \cdot \text{m}^2 \cdot \text{s}^{-2})/(1\ \text{J})]}{6.9 \times 10^{-11}\ \text{m}^2 \cdot \text{s}^{-1}} = 6.0 \times 10^{-11}\ \text{kg} \cdot \text{s}^{-1}$$

The friction force for a nonhydrated spherical molecule is given by

$$f_0 = 6\pi(\sigma_i/2)\eta_j \tag{24.23}$$

which upon substitution of $\sigma_i = 5.4\ \text{nm}$ (see Problem 24.31) gives

$$f_0 = 6\pi\frac{5.4 \times 10^{-9}\ \text{m}}{2}(0.8904\ \text{cP})\left(\frac{10^{-2}\ \text{P}}{1\ \text{cP}}\right)\left(\frac{0.1\ \text{Pa} \cdot \text{s}}{1\ \text{P}}\right)\left(\frac{1\ \text{kg} \cdot \text{m}^{-1} \cdot \text{s}^{-2}}{1\ \text{Pa}}\right) = 4.5 \times 10^{-11}\ \text{kg} \cdot \text{s}^{-1}$$

The frictional ratio is

$$\frac{f_{\text{eff}}}{f_0} = \frac{6.0 \times 10^{-11}\ \text{kg} \cdot \text{s}^{-1}}{4.5 \times 10^{-11}\ \text{kg} \cdot \text{s}^{-1}} = 1.3$$

which is related to the ratio of the major axis to the minor axis of an ellipsoid. It can be concluded that hemoglobin is not a spherical molecule.

24.34 What is the rate of sedimentation as a result of gravity in an aqueous solution of fumarase ($M = 218\ \text{kg} \cdot \text{mol}^{-1}$) at 25 °C? For the solution, $D_{ij}^{\infty} = 4.05 \times 10^{-11}\ \text{m}^2 \cdot \text{s}^{-1}$, $v = 0.75\ \text{cm}^3 \cdot \text{g}^{-1}$, and $\rho = 0.998\ \text{g} \cdot \text{cm}^{-3}$.

▌ The *steady-state sedimentation velocity* as a result of the force of gravity is given by

$$v_{\text{sed}} = \bar{M}_z Dg(1 - v\rho)/RT \tag{24.24}$$

where g is the gravitational constant. Substituting the data gives

$$v_{\text{sed}} = \frac{(218\ \text{kg} \cdot \text{mol}^{-1})(4.05 \times 10^{-11}\ \text{m}^2 \cdot \text{s}^{-1})[1 - (0.75\ \text{cm}^3 \cdot \text{g}^{-1})(0.998\ \text{g} \cdot \text{cm}^{-3})]}{(8.314\ \text{J} \cdot \text{K}^{-1} \cdot \text{mol}^{-1})[(1\ \text{kg} \cdot \text{m}^2 \cdot \text{s}^{-2})/(1\ \text{J})](298\ \text{K})} = 8.96 \times 10^{-13}\ \text{m} \cdot \text{s}^{-1}$$

24.35 Calculate the concentration ratio in the polymer solution described in Problem 24.34 between the top and bottom of a 1.00-cm tube.

▌ The concentration ratio at two points (x_i) in a gravitational equilibrium system

$$\ln\frac{C_2}{C_1} = \frac{\bar{M}_z g(x_2 - x_1)(1 - \rho v)}{RT} \tag{24.25}$$

Substituting the data into *(24.25)* and taking antilogarithms gives

$$\ln\frac{C_2}{C_1} = \frac{(218\ \text{kg} \cdot \text{mol}^{-1})(9.81\ \text{m} \cdot \text{s}^{-2})(1.00 \times 10^{-2}\ \text{m})[1 - (0.75\ \text{cm}^3 \cdot \text{g}^{-1})(0.998\ \text{g} \cdot \text{cm}^{-3})]}{(8.314\ \text{J} \cdot \text{K}^{-1} \cdot \text{mol}^{-1})[(1\ \text{kg} \cdot \text{m}^2 \cdot \text{s}^{-2})/(1\ \text{J})](298\ \text{K})} = 2.2 \times 10^{-3}$$

$$C_2/C_1 = 1.002$$

24.36 Repeat the calculations of Problem 24.35 except assume that the tube is placed in a centrifuge operating at $1.0 \times 10^4\ \text{min}^{-1}$.

▌ In a centrifuge, *(24.25)* becomes

$$\ln\frac{C_2}{C_1} = \frac{\bar{M}_z\omega^2(x_2^2 - x_1^2)(1 - \rho v)}{2RT} \tag{24.26}$$

where ω is the angular velocity. Substituting the data into *(24.26)* and taking antilogarithms gives

$$\ln \frac{C_2}{C_1} = \frac{\left[\begin{array}{r}(218\ \text{kg}\cdot\text{mol}^{-1})[(1.0\times 10^4\ \text{min}^{-1})[(1\ \text{min})/(60\ \text{s})](2\pi)]^2[(1.00\times 10^{-2}\ \text{m})^2 \\ -(0.00\ \text{m})^2][1-(0.75\ \text{cm}^3\cdot\text{g}^{-1})(0.998\ \text{g}\cdot\text{cm}^{-3})]\end{array}\right]}{2(8.314\ \text{J}\cdot\text{K}^{-1}\cdot\text{mol}^{-1})[(1\ \text{kg}\cdot\text{m}^2\cdot\text{s}^{-2})/(1\ \text{J})](298\ \text{K})} = 1.21$$

$$C_2/C_1 = 3.4$$

24.37 Calculate the rate of sedimentation in the centrifuge described in Problem 24.36. Assume that the initial distance of the sample boundary from the center of rotation is 5.0 cm. For fumarase, the sedimentation coefficient (S) is 9.06×10^{-13} s. Compare the result to that obtained from gravity sedimentation in Problem 24.34.

▌ In the centrifuge, the sedimentation velocity is given by

$$v_{\text{sed}} = \frac{dr}{dt} = S\omega^2 r \tag{24.27}$$

Substituting the data gives

$$v_{\text{sed}} = (9.06\times 10^{-13}\ \text{s})\left[(1.0\times 10^4\ \text{min})\left(\frac{1\ \text{min}}{60\ \text{s}}\right)(2\pi)\right]^2(5.0\times 10^{-2}\ \text{m}) = 5.0\times 10^{-8}\ \text{m s}^{-1}$$

The use of the centrifuge increases the rate of sedimentation by

$$\frac{5.0\times 10^{-8}\ \text{m}\cdot\text{s}^{-1}}{8.96\times 10^{-13}\ \text{m}\cdot\text{s}^{-1}} = 5.6\times 10^4$$

24.38 Approximately how long would it take for the boundary described in Problem 24.37 to move 1.00 mm?

▌ The time required is

$$t = \frac{(1.00\ \text{mm})[(10^{-3}\ \text{m})/(1\ \text{mm})]}{5.0\times 10^{-8}\ \text{m}\cdot\text{s}^{-1}} = 2.0\times 10^4\ \text{s}$$

24.39 Use the following sedimentation velocity data at 20 °C to determine the molar mass of fumarase:

$C'/(\text{g}\cdot\text{L}^{-1})$	0.70	1.60	3.40
$S/(10^{-13}\ \text{s})$	9.03	8.96	8.87

Additional measurements on these solutions gave $D_{ij}^{\infty} = 4.05\times 10^{-11}\ \text{m}^2\cdot\text{s}^{-1}$, $v = 0.75\ \text{cm}^3\cdot\text{g}^{-1}$, and $\rho = 0.993\ \text{g}\cdot\text{cm}^{-3}$.

▌ A plot of S against C' is shown in Fig. 24-6. The extrapolated value of S is 9.06×10^{-13} s. The molar mass is related to S by

$$\bar{M}_{\text{m}} = RTS/D_{ij}^{\infty}(1-v\rho) \tag{24.28}$$

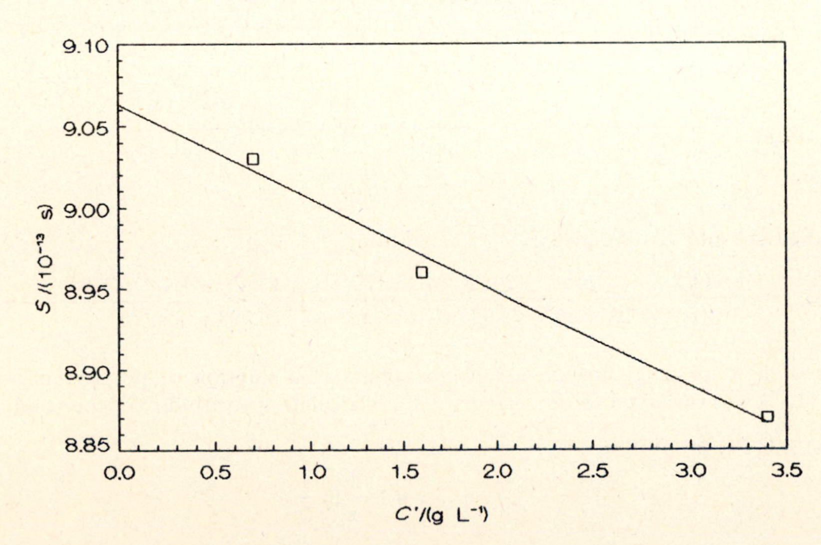

Fig. 24-6

Substitution of the data into *(24.28)* gives

$$\bar{M}_m = \frac{(8.314\ \text{J}\cdot\text{K}^{-1}\cdot\text{mol}^{-1})(293\ \text{K})(9.06\times10^{-13}\ \text{s})[(1\ \text{kg}\cdot\text{m}^2\cdot\text{s}^{-2})/(1\ \text{J})]}{(4.05\times10^{-11}\ \text{m}^2\cdot\text{s}^{-1})[1-(0.75\ \text{cm}^3\cdot\text{g}^{-1})(0.998\ \text{g}\cdot\text{cm}^{-3})]} = 213\ \text{kg}\cdot\text{mol}^{-1}$$

24.40 How fast must a centrifuge spin in order for a fumarase surface originally 5.00 cm from the center of the centrifuge to move at a rate of $1.0\times10^{-4}\ \text{cm}\cdot\text{s}^{-1}$? See Problem 24.39 for additional information.

▮ Solving *(24.27)* for ω and substituting the data gives

$$\omega = \left(\frac{1.0\times10^{-4}\ \text{cm}\cdot\text{s}^{-1}}{(9.06\times10^{-13}\ \text{s})(5.00\ \text{cm})}\right)^{1/2} = 4700\ \text{s}^{-1}$$

which corresponds to an operating speed of

$$\frac{(4700\ \text{s}^{-1})[(60\ \text{s})/(1\ \text{min})]}{2\pi} = 45\times10^3\ \text{min}^{-1}$$

24.41 The sedimentation of hemoglobin was monitored at 25 °C. The initial radius of the solute surface was 5.00 cm, and during the centrifugation process at $\omega = 5.00\times10^3\ \text{s}^{-1}$ the following data were collected:

$t/(\text{s})$	0	1000	2000	3000	4000	5000
$r/(\text{cm})$	5.00	5.06	5.12	5.15	5.23	5.27

Determine the sedimentation constant. Given $D_{ij}^{\infty} = 6.9\times10^{-11}\ \text{m}^2\cdot\text{s}^{-1}$, $\rho = 0.998\ \text{g}\cdot\text{cm}^{-3}$, and $v = 0.75\ \text{cm}^3\cdot\text{g}^{-1}$, calculate the molar mass of hemoglobin.

▮ The integral form of *(24.27)* is

$$\ln r = \ln r_0 + \omega^2 St \qquad (24.29)$$

A plot of $\ln r$ against t is given in Fig. 24-7. From the plot,

$$S = \frac{\text{slope}}{\omega^2} = \frac{1.05\times10^{-5}\ \text{s}^{-1}}{(5.0\times10^3\ \text{s}^{-1})^2} = 4.2\times10^{-13}\ \text{s}$$

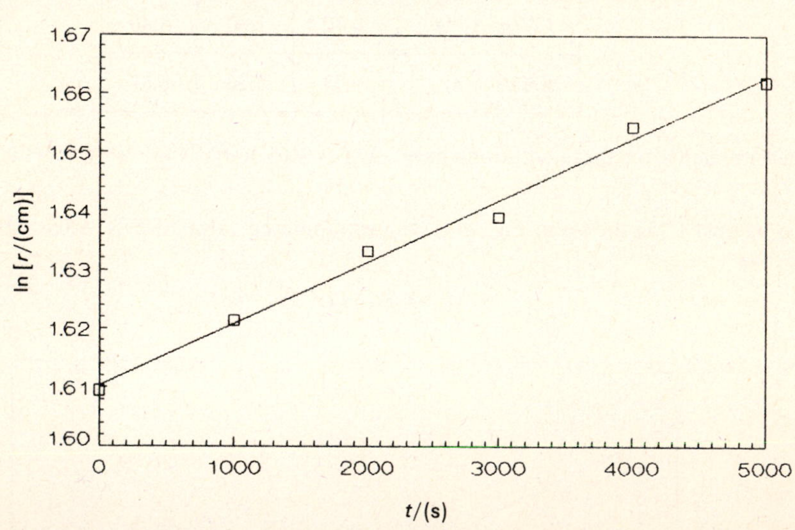

Fig. 24-7

Substituting the data into *(24.28)* gives

$$\bar{M}_m = \frac{(8.314\ \text{J}\cdot\text{K}^{-1}\cdot\text{mol}^{-1})(298\ \text{K})(4.2\times10^{-13}\ \text{s})[(1\ \text{kg}\cdot\text{m}^2\cdot\text{s}^{-2})/(1\ \text{J})]}{(6.9\times10^{-11}\ \text{m}^2\cdot\text{s}^{-1})[1-(0.75\ \text{cm}^3\cdot\text{g}^{-1})(0.998\ \text{g}\cdot\text{cm}^{-3})]} = 60\ \text{kg}\cdot\text{mol}^{-1}$$

24.42 The intensity of light passing through a 1.00-cm sample of a solution of polystyrene in methyl ethyl ketone ($C' = 12.3\ \text{g}\cdot\text{L}^{-1}$) was observed to decrease by 2.3%. Calculate the turbidity of the solution.

▮ The *turbidity* (τ) is given by

$$\tau = -\frac{1}{x}\ln\frac{I}{I_0} \qquad (24.30)$$

where x is the path length. Substituting the data into *(24.30)* gives

$$\tau = -\frac{1}{(1.00\ \text{cm})[(10^{-2}\ \text{m})/(1\ \text{cm})]} \ln \frac{0.997}{1.000} = 2.33\ \text{m}^{-1}$$

24.43 The scattering factor for the solution described in Problem 24.42 is $K = 3.38 \times 10^{-5}\ \text{m}^2 \cdot \text{mol} \cdot \text{kg}^{-2}$. Calculate the molar mass of the polystyrene.

▮ The turbidity is related to the molar mass of the polymer by

$$\tau = H\bar{M}_\text{m}C' \tag{24.31}$$

where

$$H = \frac{32\pi^3 n_0^2 (dn/dC')^2}{3L\lambda^4} = \frac{16\pi K}{3} \tag{24.32}$$

and n_0 is the index of refraction for the solvent. Substituting the data into *(24.32)* gives

$$H = \frac{16\pi(3.38 \times 10^{-5}\ \text{m}^2 \cdot \text{mol} \cdot \text{kg}^{-2})}{3} = 5.66 \times 10^{-4}\ \text{m}^2 \cdot \text{mol} \cdot \text{kg}^{-2}$$

Solving *(24.31)* for $\bar{M}_\text{m}$ and substituting the data gives

$$\bar{M}_\text{m} = \frac{2.33\ \text{m}^{-1}}{(5.66 \times 10^{-4}\ \text{m}^2 \cdot \text{mol} \cdot \text{kg}^{-2})(12.3\ \text{g} \cdot \text{L}^{-1})[(10^{-3}\ \text{kg})/(1\ \text{g})][(10^3\ \text{L})/(1\ \text{m}^3)]} = 335\ \text{kg} \cdot \text{mol}^{-1}$$

24.44 Calculate the value of H for the system described in Problem 24.43 given $\lambda = 546.1\ \text{nm}$, $n_0 = 1.377$, and $dn/dC' = 2.20 \times 10^{-4}\ \text{m}^3 \cdot \text{kg}^{-1}$.

▮ Substituting the data into *(24.32)* gives

$$H = \frac{32\pi^3(1.377)^2(2.20 \times 10^{-4}\ \text{m}^3 \cdot \text{kg}^{-1})^2}{3(6.022 \times 10^{23}\ \text{mol}^{-1})(546.1 \times 10^{-9}\ \text{m})^4} = 5.67 \times 10^{-4}\ \text{m}^2 \cdot \text{mol} \cdot \text{kg}^{-2}$$

24.45 The following turbidity data were collected for solutions of polystyrene in methyl ethyl butane

$C'/(\text{g} \cdot \text{L}^{-1})$	1.00	0.75	0.50	0.25
$\tau/(\text{m}^{-1})$	0.125	0.096	0.063	0.032

Given $H = 5.67 \times 10^{-4}\ \text{m}^2 \cdot \text{mol} \cdot \text{kg}^{-2}$ for these solutions, calculate the molar mass of this sample of polystyrene.

▮ Equation *(24.31)* implies that a plot of τ against C' will be linear with a slope equal to $H\bar{M}_\text{m}$. From the graph shown in Fig. 24-8,

$$\bar{M}_\text{m} = \frac{\text{slope}}{H} = \frac{(0.125\ \text{m}^{-1} \cdot \text{L} \cdot \text{g}^{-1})[(10^{-3}\ \text{m}^3)/(1\ \text{L})][(10^3\ \text{g})/(1\ \text{kg})]}{5.67 \times 10^{-4}\ \text{m}^2 \cdot \text{mol} \cdot \text{kg}^{-2}} = 220\ \text{kg} \cdot \text{mol}^{-1}$$

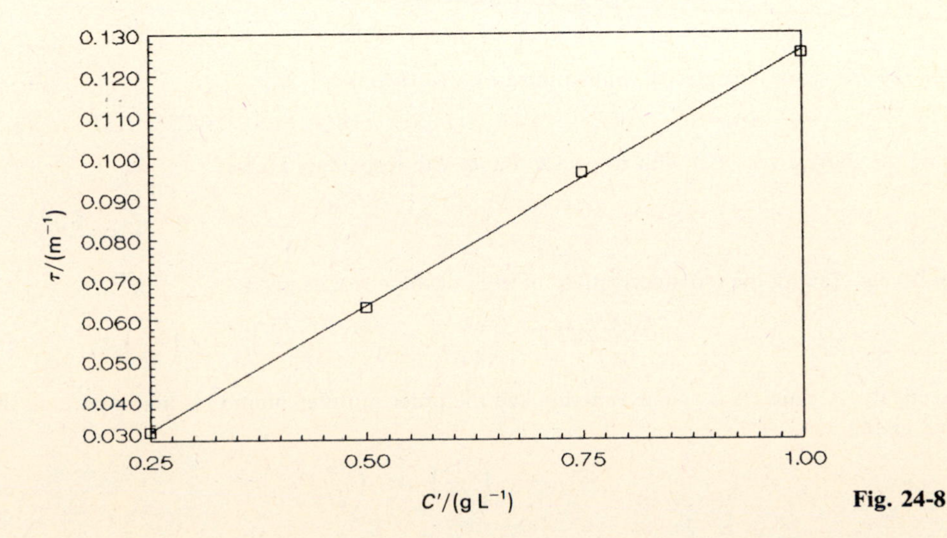

Fig. 24-8

24.3 THERMODYNAMIC PROPERTIES AND KINETICS

24.46 The force needed to stretch a sample of rubber at 30 °C from l_0 to $1.50l_0$ is $3.7 \times 10^5\ \text{N}\cdot\text{m}^{-2}$. Calculate the number of chain segments per unit volume in the polymer.

▮ The force exerted by a sample of rubber for elongations less than ~350% is given by

$$F = 2KkT(l/l_0 - l_0^2/l^2) \tag{24.33}$$

where K is the number of chain segments per unit volume. Solving *(24.33)* for K and substituting the data gives

$$K = \frac{3.7 \times 10^5\ \text{N}\cdot\text{m}^{-2}}{2(1.381 \times 10^{-23}\ \text{J}\cdot\text{K}^{-1})(303\ \text{K})[(1\ \text{N}\cdot\text{m})/(1\ \text{J})][1.50 - (1.50)^{-2}]} = 4.2 \times 10^{25}\ \text{m}^{-3}$$

24.47 Predict the value of the force for the rubber sample described in Problem 24.46 at $l = 2.50l_0$.

▮ Substituting the data into *(24.33)* gives

$$F = 2(4.2 \times 10^{25}\ \text{m}^{-3})(1.381 \times 10^{-23}\ \text{J}\cdot\text{K}^{-1})\left(\frac{1\ \text{N}\cdot\text{m}}{1\ \text{J}}\right)(303\ \text{K})[2.50 - (2.50)^{-2}] = 8.2 \times 10^5\ \text{N}\cdot\text{m}^{-2}$$

24.48 Describe what will happen to an elongated sample of rubber as the temperature increases.

▮ Because the elongation and T are inversely related [see *(24.33)*], as the temperature increases the rubber sample will undergo contraction.

24.49 Assuming an average force of $6.0 \times 10^5\ \text{N}\cdot\text{m}^{-2}$ for the rubber sample described in Problem 24.46, calculate the work needed to stretch the sample from 1.50 cm to 2.50 cm.

▮ Neglecting any PV work, the expression for work is given by Table 3-5 as

$$w = \int F\,dl \tag{24.34}$$

which upon integration and substitution of data gives

$$w = (6.0 \times 10^5\ \text{N}\cdot\text{m}^{-2})(2.50\ \text{cm} - 1.50\ \text{cm})\left(\frac{10^{-2}\ \text{m}}{1\ \text{cm}}\right)\left(\frac{1\ \text{J}}{1\ \text{N}\cdot\text{m}}\right) = 6.0 \times 10^3\ \text{J}\cdot\text{m}^{-2}$$

24.50 Assume that the process described in Problem 24.49 is performed adiabatically and reversibly. If $C_p = 1.8 \times 10^4\ \text{J}\cdot\text{K}^{-1}\cdot\text{m}^{-2}$ for the 1-cm segment, calculate the temperature change in the rubber.

▮ Dividing the work by the heat capacity gives

$$\Delta T = \frac{6.0 \times 10^3\ \text{J}\cdot\text{m}^{-2}}{1.8 \times 10^4\ \text{J}\cdot\text{K}^{-1}\cdot\text{m}^{-2}} = 0.33\ \text{K}$$

24.51 Derive the relationship for $-(\partial S/\partial l)_T$ in terms of F for a reversible stretching process.

▮ For a reversible process, *(3.47)* and *(5.6)* give

$$dU = đq + đw = T\,dS - P\,dV + F\,dl \approx T\,dS + F\,dl$$

where the PV work is neglected. Substituting into *(6.1b)* gives

$$dA = dU - T\,dS - S\,dT = T\,dS + F\,dl - T\,dS - S\,dT = F\,dl - S\,dT$$

Taking the derivatives of A with respect to l and with respect to T gives

$$\left(\frac{\partial A}{\partial l}\right)_T = F \qquad \left(\frac{\partial A}{\partial T}\right)_l = -S$$

respectively. Taking the cross-derivatives of each of these results gives

$$\left(\frac{\partial(\partial A/\partial l)_T}{\partial T}\right)_l = \left(\frac{\partial F}{\partial T}\right)_l \qquad \left(\frac{\partial(\partial A/\partial T)_l}{\partial l}\right)_T = -\left(\frac{\partial S}{\partial l}\right)_T$$

respectively. Because A is a state function and the order of differentiation is not important, these results must be equal, giving

$$\left(\frac{\partial F}{\partial T}\right)_l = -\left(\frac{\partial S}{\partial l}\right)_T \tag{24.35}$$

24.52 For the rubber sample described in Problem 24.49, $(\partial F/\partial T)_l = 3.7 \times 10^3\ \mathrm{N \cdot K^{-1} \cdot m^{-2}}$. Calculate ΔS for the process.

▌ Solving *(24.35)* for dS and integrating gives

$$\Delta S = -\left(\frac{\partial F}{\partial T}\right)_l \Delta l = -(3.7 \times 10^3\ \mathrm{N \cdot K^{-1} \cdot m^{-2}})(1.00\ \mathrm{cm})\left(\frac{10^{-2}\ \mathrm{m}}{1\ \mathrm{cm}}\right)\left(\frac{1\ \mathrm{J}}{1\ \mathrm{N \cdot m}}\right) = -37\ \mathrm{J \cdot K^{-1} \cdot m^{-2}}$$

Note that the entropy decreases as the rubber sample is stretched.

24.53 Derive the expression for $(\partial U/\partial l)_T$ for a reversible stretching process. Calculate ΔU for the process described in Problem 24.49. See Problem 24.52 for additional information.

▌ From Problem 24.51, $dA = F\,dl - S\,dT$. Solving for F and imposing constant T gives

$$F = \left(\frac{\partial A}{\partial l}\right)_T = \left(\frac{\partial(U - T\,dS)}{dl}\right)_T = \left(\frac{\partial U}{\partial l}\right)_T - T\left(\frac{\partial S}{\partial l}\right)_T$$

Upon rearrangement and substitution of *(24.35)*,

$$\left(\frac{\partial U}{\partial l}\right)_T = F - T\left(\frac{\partial F}{\partial T}\right)_l \qquad (24.36)$$

Solving *(24.36)* for dU and integrating gives

$$\Delta U = \left[F - T\left(\frac{\partial F}{\partial T}\right)_l\right]\Delta l$$

$$= [(6.0 \times 10^5\ \mathrm{N \cdot m^{-2}}) - (303\ \mathrm{K})(3.7 \times 10^3\ \mathrm{N \cdot K^{-1} \cdot m^{-2}})](1.00\ \mathrm{cm})\left(\frac{10^{-2}\ \mathrm{m}}{1\ \mathrm{cm}}\right)\left(\frac{1\ \mathrm{J}}{1\ \mathrm{N \cdot m}}\right)$$

$$= -5200\ \mathrm{J \cdot m^{-2}}$$

24.54 The general rate equation describing the condensation polymerization

$$\mathrm{A + B \rightarrow A{-}B \xrightarrow{+A} A{-}B{-}A \xrightarrow{+B} A{-}B{-}A{-}B \rightarrow etc.}$$

catalyzed by C is

$$\frac{d\xi}{dt} = kC(\mathrm{A})C(\mathrm{B})C(\mathrm{C}) \qquad (24.37)$$

Derive the integrated rate equations for the special cases in which the reaction is autocatalytic in A and in which $C(\mathrm{C})$ is essentially a constant. Assume $C(\mathrm{A})_0 = C(\mathrm{B})_0$. Describe the graphical analysis of pt data.

▌ For the autocatalytic case, $C(\mathrm{A}) = C(\mathrm{C})$ and *(24.37)* becomes

$$\frac{-dC(\mathrm{A})}{dt} = k[C(\mathrm{A})]^2 C(\mathrm{B}) = k[C(\mathrm{A})]^3$$

The integrated rate law is given by *(12.4)* as

$$1/[C(\mathrm{A})]^2 = 1/[C(\mathrm{A})]_0^2 + 2kt$$

Substituting $C(\mathrm{A}) + C(\mathrm{A})_0(1 - p)$ gives

$$1/(1 - p)^2 = 1 + 2[C(\mathrm{A})]_0^2 kt$$

A plot of $(1 - p)^{-2}$ against t will be linear with a slope of $2[C(\mathrm{A})]_0^2 k$ and an intercept of 1.

For the second case, $-dC(\mathrm{A})/dt = k'[C(\mathrm{A})]^2$, where $k' = kC(\mathrm{C})$. The integrated rate law is given by *(12.4)* as

$$1/C(\mathrm{A}) = 1/C(\mathrm{A})_0 + k't$$

which can be written as

$$\frac{1}{1 - p} = 1 + k'C(\mathrm{A})_0 t$$

A plot of $(1 - p)$ against t will be linear, with a slope of $k'C(\mathrm{A})_0$ and an intercept of 1.

24.55 The general rate equation describing the free-radical polymerization of A is

$$\frac{-dC(\mathrm{A})}{dt} = k_p\left(\frac{\nu_i}{k_t}\right)^{1/2} = C(\mathrm{A}) \qquad (24.38)$$

where k_p and k_t are the rate constants for the propagation and termination steps, respectively, and ν_i is the rate of the initiation step. What is the order of reaction with respect to the monomer for the following special cases: thermal initiation of A, a bimolecular initiation reaction between A and a catalyst C, and photochemical initiation?

▌ For the thermal initiation reaction, $\nu_i = k_i[C(\mathrm{A})]^2$, which upon substitution into *(24.38)* gives the second-order reaction

$$\frac{-dC(\mathrm{A})}{dt} = k_p\left(\frac{k_i C(\mathrm{A})}{k_t}\right)^{1/2} C(\mathrm{A}) = k_p\left(\frac{k_i}{k_t}\right)^{1/2}[C(\mathrm{A})]^2$$

For the catalyzed reaction, $\nu_i = k_i C(\mathrm{A})C(\mathrm{C})$, which upon substitution into *(24.38)* gives the 3/2-order reaction

$$\frac{-dC(\mathrm{A})}{dt} = k_p\left(\frac{k_i C(\mathrm{A})C(\mathrm{C})}{k_t}\right)^{1/2} C(\mathrm{A}) = k_p\left(\frac{k_i}{k_t}\right)^{1/2}[C(\mathrm{A})]^{3/2}[C(\mathrm{C})]^{1/2}$$

For the photochemical reaction, $\nu_i = I$, which upon substitution into *(24.38)* gives the first-order reaction

$$\frac{-dC(\mathrm{A})}{dt} = k_p\left(\frac{I}{k_t}\right)^{1/2} C(\mathrm{A})$$

24.56 For a free-radical polymerization,

$$\bar{x}_n = \frac{k_p C(\mathrm{A})}{(2k_t \nu_i)^{1/2}} \qquad (24.39)$$

Calculate $\bar{x}_n$ for a catalyzed polymerization with $k_p = 1 \times 10^3\ \mathrm{L \cdot mol^{-1} \cdot s^{-1}}$, $k_t = 2 \times 10^2\ \mathrm{L \cdot mol^{-1} \cdot s^{-1}}$, $k_i = 3 \times 10^{-5}\ \mathrm{L \cdot mol^{-1} \cdot s^{-1}}$, $C(\mathrm{A}) = 1\ \mathrm{mol \cdot L^{-1}}$, and $C(\mathrm{C}) = 1 \times 10^{-3}\ \mathrm{mol \cdot L^{-1}}$. What concentration changes can be made on the system to increase the value of $\bar{x}_n$?

▌ For the catalyzed polymerization, $\nu_i = k_i C(\mathrm{A})C(\mathrm{C})$ (see Problem 24.55). Substituting this expression for ν_i and the data into *(24.39)* gives

$$\bar{x}_n = \frac{(1 \times 10^3\ \mathrm{mol \cdot L^{-1} \cdot s^{-1}})(1\ \mathrm{mol \cdot L^{-1}})}{[2(2 \times 10^2\ \mathrm{mol \cdot L^{-1} \cdot s^{-1}})(3 \times 10^{-5}\ \mathrm{s^{-1}})(1\ \mathrm{mol \cdot L^{-1}})(1 \times 10^{-3}\ \mathrm{mol \cdot L^{-1}})]^{1/2}} = 3 \times 10^5$$

The value of the number-average degree of polymerization can be increased by decreasing the concentration of the catalyst relative to the concentration of the monomer.

Index